# Internal Combustion Engines, Air Pollution and E-Vehicles

## — An Innovative Approach

(SI Units)

(For B.E./B.Tech., I.E.S., I.A.S. and I.F.S. Examinations)

(Sixth Revised Edition)

# Internal Combustion Engines, Air Pollution and E-Vehicles – An Innovative Approach

**(SI Units)**

**(For B.E./B.Tech., I.E.S., I.A.S. and I.F.S. Examinations)**

**(Sixth Revised Edition)**

**Prof. (Dr.) R. Yadav**
*Ph.D, F.E.I., M.I.S.T.E*
Emeritus Professor, M.N.N.I.T., Allahabad (UP)
Former Professor & Head
M.N.R. Engg. College (now M.N.N.I.T.), Allahabad
Former Principal
Regional Institute of Technology (now N.I.T.), Jamshedpur
Former Dean, CET, AAI-DU, Allahabad
Former Director General, MIT, Moradabad

**Prof. (Dr.) Sanjay**
*B.E., M.E., Ph.D.*
Professor of Mech. Engg.
N.I.T. Jamshedpur

**Er. Rajay**
*B.Tech, IT, BHU, (now IIT BHU)*
Manager, B.H.E.L, Hardwar

**I.K. International Pvt. Ltd.**
NEW DELHI

*Published by*
I.K. International Pvt. Ltd.
4435-36/7, Ansari Road, Daryaganj
New Delhi-110 002 (India)
E-mail: info@ikinternational.com
Website: www.ikbooks.com

ISBN 978-93-90620-73-9



Published by Krishan Makhijani for I.K. International Pvt. Ltd., 4435-36/7, Ansari Road, Daryaganj, New Delhi–110 002

Printed by Shree Maitrey Printech Pvt. Ltd., Noida

*Dedicated to*
*My Parents*

# Preface to the Sixth Revised Edition

The present form of the book is the outgrowth of long teaching and research experience of the main author. The purpose of this book is to make it broad based keeping the following objectives in view:

- To present a thorough treatment of IC engines to cover the syllabus of all universities.
- To prepare students for design concepts.
- To develop creativity and engineering judgement.

The present edition consists of 22 chapters in about 900 pages. One new topic "E-Vehicles" has been included to cover the possible revised syllabus of all of the universities.

The special features of the book include the following:

- A large number of typical and real-life solved examples in each chapter with good diagrammatic approach have been included.
- The language is simple and lucid.
- For an idea of model solutions, some have been given as solved examples in some chapters which include: **Schematic, Data given, to Find, Assumptions, Analysis and Comments on results**
- Emphasis on physical explanation of equations involved has been given.
- Emphasis has been given for fundamentals involved in the subject.
- Unsolved problem include **Things to Think about, Short and Long Theoretical Questions, Numerical Questions. Software Based Questions, Design and Open Ended Question and Multiple-Choice Questions.**

The questions given in **"Things to Think About"** force the students to think about the physics involved and this may enhance the reasoning capability. **"Short and Long Theoretical Questions"** help in developing the core knowledge of the subject while **Numerical Questions** enhance the analytical capability in the subject. **Software Based Questions** enhance the capability of students to develop software for engineering problems by developing models, going through parametric study and finally designing. These questions may take the form of **Projects and Research Topics**. The **Design and Open Ended Questions** provide brief design experience that offer students opportunities to develop their creativity and engineering judgement, formulate design task statements, apply realistic constraints, and consider alternatives. These questions are such that some of them may lead the students up to **high research level. Multiple-Choice Questions** put some check on the understanding of the subject.

The depth, richness, emphasis on fundamentals, creativity, innovative approach and judgement enhancement capabilities are the beauty and strength of the book.

While writing of the book, the authors have consulted a large number of books, research papers and journals which are listed in the references. Sincere acknowledgements are made to all the authors listed in the references and elsewhere in the book. Though, printing mistakes have been avoided, but the authors will appreciate, if any mistakes and errors found in the text are brought to their notice.

Finally, the authors thank Meera Yadav, wife of the main author and mother of last two co-authors who has been a consistent source of encouragement and pillar of strength for the authors and endured certain hardships due to preoccupation in this venture.

**R Yadav**
**Sanjay**
**Rajay**

# Contents

# Nomenclature

| | |
|---|---|
| $A$ | – area |
| $a$ | – sound velocity |
| $a$ | – constant |
| $B$ | – constant |
| bp | – brake power |
| bmep | – brake mean effective pressure |
| bsfc | – brake specific fuel consumption |
| BDC | – bottom dead centre (position) |
| $C$ | – velocity |
| $c$ | – constant |
| $c_d$ | – coefficient of discharge |
| $c_p$ | – specific heat at constant pressure |
| CV | – calorific value of fuel |
| $C_m$ | – mean piston speed |
| $D$ | – cylinder bore or diameter |
| $d$ | – diameter of orifice |
| $E$ | – energy |
| $e$ | – specific energy |
| EVC | – exhaust valve closing |
| EVO | – exhaust valve opening |
| F | – force |
| f | – coefficient of friction |
| fmep | – frictional mean effective pressure |
| fp | – frictional power |
| F/A | – fuel-air ratio |
| $g$ | – acceleration due to gravity |
| $H$ | – enthalpy |
| $h$ | – specific enthalpy |
| $h$ | – pressure differential |
| $h$ | – convective heat transfer coefficient |
| ip | – indicated power |
| imep | – indicated mean effective pressure |
| isfc | – indicated specific fuel consumption |
| IDC | – inner dead centre |
| IVC | – inlet valve closing |
| IVO | – inlet valve opening |
| $k$ | – thermal conductivity |
| $k$ | – constant |
| $L$ | – stroke |
| $l$ | – distance, characteristics length |
| $M$ | – molecular weight, Mach. no. |
| $m$ | – mass |
| $\dot{m}$ | – mass flow rate |
| mep | – mean effective pressure |
| $N$ | – revolution per minute (speed) |

$n$ – polytropic index
$n$ – number of power strokes
ODC – outer dead centre
$P$ – power
$p$ – pressure
$p_e$ – exhaust pressure
$PN$ – performance number
$Q$ – heat transfer
$Q_A$ – heat added
$Q_R$ – heat rejected
$r$ – compression ratio
$r_p$ – pressure ratio
$R$ – characteristics of gas constant
$R_0$ – universal gas constant
$sfc$ – specific fuel consumption
$T$ – absolute temperature
$T$ – torque
TDC – top dead centre
$U$ – internal energy
$u$ – specific internal energy
$V$ – volume
$V_c$ – clearance volume
$V_s$ – stroke volume
$W$ – weight, load, work done
$w$ – specific work transfer
$x$ – number of cylinders
$y$ – distance
$z$ – datum, height of nozzle exit

GREEK LETTERS

$\alpha$ – coefficient ratio
$\beta$ – coefficient ratio
$\gamma$ – ratio of specific heat
$\varepsilon$ – effectiveness
$\eta$ – efficiency
$\theta$ – crank angle
$\rho$ – density
$\phi$ – equivalence ratio
$\omega$ – angular velocity
$\mu$ – kinematic viscosity
$\nu$ – dynamic viscosity

CHAPTER

# 1 Introduction and Thermochemistry

## OVERVIEW

As per the Second Law of Thermodynamics, a heat engine working in a complete cycle can only produce continuous work when it takes heat from a single thermal reservoir at uniform high temperature and rejects heat to a low uniform temperature reservoir. The combustion chamber of IC engine acts as a heat source and the ambient temperature acts as a heat sink.

*Energy is the wheel of civilization and the growth of civilization is closely linked with improvements in transportation.* With the advent of more rapid forms of transportation, greater exchange of goods has resulted and more people have profited. The world has come closer and there has been a rapid rise in tourism. Cultural and scientific gains among the people of the world have been particularly noteworthy.

*The present transportation system consisting of railroads, ships, automobiles and airplanes depends on mechanical power which is derived from combustion engines. A combustion (heat) engine is one which converts the chemical energy of fuel into mechanical energy.* It is worth to mention that transport sector consumes about 25% of energy in India thus, efficient transportation is a must. IC engines cause pollution, so E-vehicles are coming up in a big way. Now, electric locomotives are common.

Electric cars and buses based on battery and fuel cells are plying on roads. U.K. has decided to run only E-cars by 2030.

**Applications.** Internal combustion engines are extensively used in road transport (cars, buses trucks, tractors, motorcycles, scooters, etc.), rail transport (diesel locomotives), air transport (planes), space vehicles, water transport (ships), power plants, agriculture pump drives, lawn mover, industrial engines, military tanks, earth moving equipment, etc.

This chapters deals with IC engine and thermochemistry of fuels.

## 1.1 SUBJECTS INVOLVED

The subject of combustion engine is very complex. Figure 1.1 shows the subjects involved in the design of internal combustion engine comprising of science and many subjects of mechanical engineering. This book will deal only with such materials which are necessary for the understanding of theory and design of internal combustion (IC) engines.

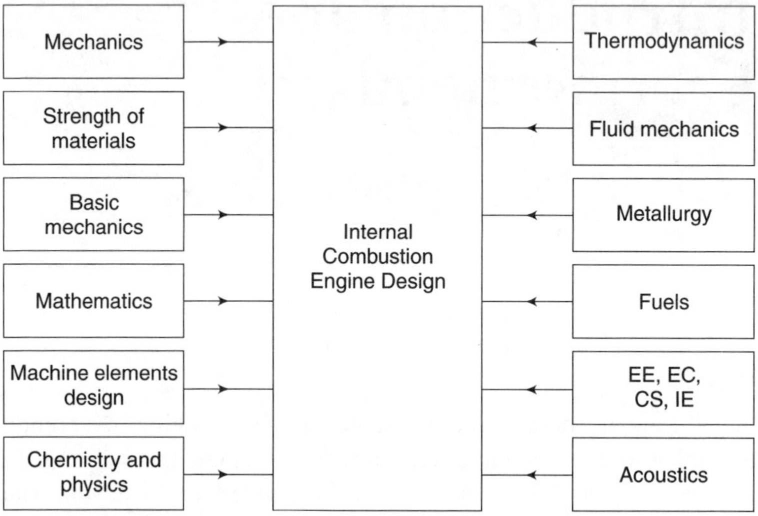


**Fig. 1.1** Subjects involved in the design of IC engines

## 1.2 CLASSIFICATION OF COMBUSTION (HEAT) ENGINES

Combustion engines may be broadly classified in two categories, namely:

(a) Internal combustion engines (IC engines)

(b) External combustion engines (EC engines)

The classification is given below in chart form.

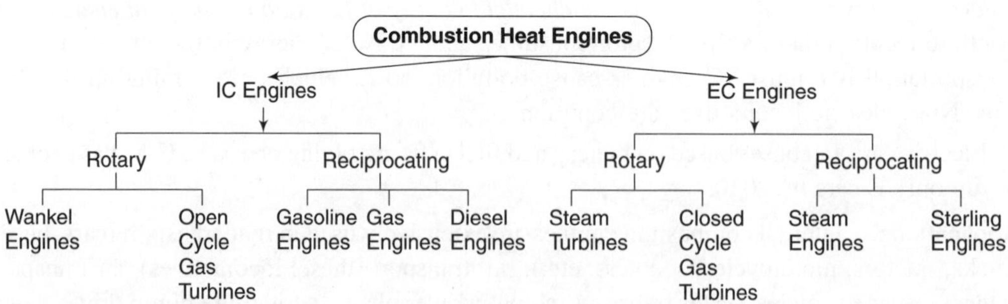

### 1.2.1 Internal Combustion Engines

As the name implies, the combustion of fuel takes place inside the engine. The engine inducts air from the atmosphere and the combustion of fuel and air occurs in the combustion chamber of engine which converts heat energy into mechanical work by expanding through the engine and leaves the engine for the rejection of heat into atmosphere. The examples of such engines are petrol, diesel and gas engines used in scooter, car, bus, truck, airplane, rail, marine, etc. Another examples are open cycle gas turbines, jets engines and rockets.

### 1.2.2 External Combustion Engines

As the name implies, the combustion of fuel takes place outside the engine and product of combustion of fuel is not used as the working fluid. A secondary fluid called the working fluid is utilised to transfer some of the heat energy released due to combustion of fuel in the presence of air to the portion of engine (reciprocating or rotary) wherein this heat is transformed into mechanical work. A common example of external combustion engines is steam power plant employing boiler in which steam is generated as a result of combustion and a turbine or reciprocating engine in which steam expands and does work. Steam ***locomotive*** (now obsolete) used in rail transport is one of the example. Another example of this engine is ***closed cycle gas turbine plant*** in which the heat released in an external combustion is utilised to heat the air/gas which expands in gas turbine and performs work.

### 1.2.3 Advantages and Disadvantages

The main advantages of internal combustion engines over external combustion engines are lower ratio of weight to output power due to absence of boiler and condenser, greater mechanical simplicity, high thermal efficiency, lesser requirement of water for cooling system, etc. This gives rise to the wide use of internal combustion engines for mobile power plants such as automobiles, locomotives, aircraft, etc.

The external combustion engines have also some advantages over internal combustion engines. They use cheaper fuels, need higher starting torque and develop higher power output. Steam power plant is most suited for electricity generation.

### 1.2.4 Detailed Classifications with Size, Efficiency and Applications

Table 1.1 shows the detailed classification of combustion engines with the maximum possible size, thermal efficiency and prime applications.

## 1.3 HISTORICAL BACKGROUND

The historical background of IC engines is given as follows in chronological order.

### 1.3.1 Huygens Gunpowder Engine (1680)

The credit for discovery of ancient internal combustion engines goes to famous Dutch physicist Christian Huygens in 1680, employing energy of *gunpowder*. It consisted of a vertical cylinder having a sliding fit type piston. Explosion of a charge of gunpowder in the cylinder drove the piston on its upward stroke and the return stroke was caused by atmospheric pressure acting

**Table 1.1** Classification of Combustion Engines

| *Classification* | *Name of Engines* | *Reciprocating or Ratory* | *Maximum size in kW* | *Principle Application* |
|---|---|---|---|---|
| (A) Internal Combustion Engines | (1) Gasoline or petrol (SI) | Reciprocating | 5000 | Road vehicles, small aircraft |
| | (2) Gas Engines (SI) | Reciprocating | 5000 | Industrial, electric power |
| | (3) Diesel Engines (CI) | Reciprocating | 1,00,000 | Road vehicles, industrial, locomotives, electric power, marine |
| | (4) Wankel Engines (SI/CI) | Rotary | 500 | Road vehicles, small aircraft |
| | (5) Open Cycle Gas Turbines | Rotary | 2,00,000 | Electric power, aircraft |
| | (6) Jet Engines | Rotary | 10,000 | Aircraft |
| | (7) Rockets | No mechanism | Very big | Missiles, space vehicles |
| (B) External Combustion Engines | (1) Steam Engines | Reciprocating | 5000 | Locomotives, ships (**obsolete**) |
| | (2) Steam Turbines | Rotary | 10,00,000 | Electric power, large marine |
| | (3) Stirling Engines | Reciprocating | 1000 | Experimental, power in space, vehicles |
| | (4) Closed Cycle gas Turbines | Rotary | 2,00,000 | Electric power, marine |

on the piston. The useful power was produced on the downward stroke of the piston and by a pulley and rope attached with load and piston.

### 1.3.2 Lenoir Gas Engine (1860)

In 1860, Frenchman Lenoir developed a *non-compression combustion gas engine* which later on was called Lenoir engine. It was essentially similar to a double acting steam engine in which steam was replaced by the gas formed by the combustion of the charge of air–gas mixture.

### 1.3.3 Free Piston Otto-Langen Engine (1866)

Otto-Langen of Germany developed a *free piston engine* in 1866. It consisted of a piston which was not connected to any crankshaft but was free to move vertically to outwards during the explosion and expansion strokes. After expansion, the products get cooled down and the piston falls under gravity causing downward stroke. This is transmitted to flywheel shaft during downward stoke through piston rod connected by ratchet and a rocker and pinion device to the flywheel. The details have been discussed in the chapter "IC Engine Under Development".

### 1.3.4 Otto Cycle (1876)

Nikolas Augusti Otto, a German scientist built in 1876 a practical engine based on the principle proposed by Bean de Rochas in 1862. This engine consisted of piston-cylinder engine arrangement having four strokes. The four strokes are suction, compression, expansion and exhaust. Modern petrol engine is nothing but the refinement and improvement of Otto engine which has been discussed in detail in subsequent chapters.

### 1.3.5 Brayton Engine (1873)

Brayton, an American engineer developed an engine in 1873 based on constant pressure combustion and complete expansion in a separate cylinder piston arrangement. The present gas turbine cycle is basically a refinement and development of the Brayton engine cycle which consists of compressor, combustion chamber and gas turbine. This is discussed in detail in a subsequent chapter.

### 1.3.6 Atkinson Engine (1883)

Atkinson, British scientist built an engine in 1883 which used a short stroke for induction and compression and a larger stroke for expansion and exhaust. It consists of single piston-cylinder arrangement with complicated linkage mechanism.

### 1.3.7 Diesel Engine (1892)

Rudolf Diesel, a German engineer developed compression ignition engine in 1892 which later on was named as diesel engine. It was a piston-cylinder arrangement. Diesel first tested the engine with coal dust but later tested with oil. The present diesel engine is the refinement and modification of the earlier engine. This has been discussed in detail in a subsequent chapter.

### 1.3.8 Clerk's Engine (1881)

Dugald Clerk, a Scottish engineer patented a two-stroke engine in 1881. One revolution of the crankshaft was achieved in two strokes. The modern two-stroke engine is the refinement and development of this engine which is discussed in detail in a subsequent chapter.

### 1.3.9 Wankel Engine (1957)

The credit for invention of Wankel engine in 1957 goes to Dr. Felix Wankel, a German engineer. It is a rotary engine. It has a three lob rotors driven eccentrically in a casing in such a way that three separate volumes of working fluid get trapped between the rotor and the casing performing four processes, namely, induction, compression, combustion, expansion and exhaust process in sequence. The working principle of this engine has been discussed in detail in a separate chapter (IC Engines Under Development).

### 1.3.10 Stirling Engine (1816/1938/1965)

Robert Stirling, a Scottish clergyman developed a pumping engines for mines in 1816 using air as working fluid. The engine was inefficient and bulky. In 1938, Philips revived the engine using modern technology. The practical engine could take its modern shape in 1965. It is an external combustion engine. The expansion and compression processes occur in opposite ends of the same cylinder. The details have been discussed in the chapter "IC Engine Under Development".

## 1.4 COMPARISON BETWEEN IC ENGINES AND STEAM ENGINES

The following points bring out the difference between IC engines and steam engines (Table 1.2).

**Table 1.2** Comparison between IC Engines and Steam Engines

| *Steam Engines* | *IC Engines* |
|---|---|
| (i) In steam engines, the combustion of fuel is external and the working fluid (steam) is generated outside the power unit. | (i) In IC engine, the combustion is internal. The correct amount of fuel air ratio is introduced into the cylinder and combustion takes place in it. |
| (ii) Steam engines operate at a lesser temperature (about 600°C). | (ii) IC engine operates at a temperature of (about 2300°C). |
| (iii) Steam engine does not require cooling of the cylinder, but its cylinder is steam jacketed to prevent condensation of steam. | (iii) Due to high operating temperature, IC engine cylinder requires cooling for efficient operation and to prevent burning of overheated components. |
| (iv) Steam engine cannot be put into operation instantaneously. A sufficient amount of fuel is consumed in generation of steam. | (iv) IC engine can be put into operation instantaneously. In addition to this, starting is very easy and quick. |
| (v) The exhaust of steam engine may be condensed in a condenser and it may be used as feed water or ejected to atmosphere. | (v) The exhaust gas of IC engine is led to the atmosphere. |
| (vi) Steam engine is bulky. The ratio of weight to power is very high. | (vi) IC engine is compact in design. The ratio of weight to power is low. |
| (vii) The efficiency of steam engine is very low (about 15 to 20%). **Now it is obsolete**. | (vii) The efficiency of IC engine is higher (about 30 to 35%). |

## 1.5 CLASSIFICATION OF IC ENGINES

The internal combustion engines are generally classified on the basis of the following.

**(A) Number of strokes per cycle**

(i) *Four-stroke engines*. An engine in which the cycle is completed in four strokes of the piston.

(ii) *Two-stroke engines*. An engine in which the cycle is completed in two strokes of the piston.

**(B) Nature of thermodynamic cycle**

(i) *Constant-volume combustion* (Otto cycle)

(ii) *Constant-pressure combustion* (Diesel cycle)

(iii) *Partly constant volume and pressure combustion* (Dual or mixed cycles).

**(C) Ignition system**

(i) *Spark ignition (SI) engines*. These are the engines in which a spark plug is used for combustion (petrol engine). The mixture of A/F is achieved in a carburettor previously. Now, in most advanced engines, the fuel is injected in the induction manifold through electronic injection system called multipoint fuel injection **(MPFI) system**.

(ii) *Compression ignition (CI) engines.* In these engines, combustion starts automatically due to compression of air and injection of fuel at the end of compression process (diesel engine).

**(D) Fuel used**

(i) *Gas engines.* These engines use gaseous fuel, such as, methane.

(ii) *Petrol engines.* These engines use highly volatile liquid fuel, such as petrol.

(iii) *Oil engines.* These engines use less volatile liquid fuel, such as, diesel oil, kerosene, heavier residual fuel.

(iv) *Bi-fuel engines.* In these engines, gas is used as basic fuel while liquid fuel is used for starting purposes.

**(E) Cylinder arrangement**

(i) *In-line engines.* These engines contain one bank of cylinders with their axes parallel. They transmit power to a single crankshaft.

(ii) *V-engines.* In these engines, the cylinders are arranged in the form of the letter V. In other words, these engines contain two banks of cylinders set at an angle and attached to the same crankcase with the same crankshaft.

(iii) *Opposed cylinder.* These engines consist of two or more cylinders on opposite sides of a common crankshaft.

(iv) *X engines.* These engines contain four banks of cylinders with a single crankcase and a single crankshaft.

(v) *Opposed piston.* When a single cylinder houses two pistons each of which drives a separate crankshaft, it is called an opposed piston.

(vi) *Radial engines.* In these engines, the cylinders are arranged radially like the spokes of a wheel and are connected to a single crankshaft. These engines are used in aircraft. Figure 1.2 shows the classification of engines by cylinder arrangement.

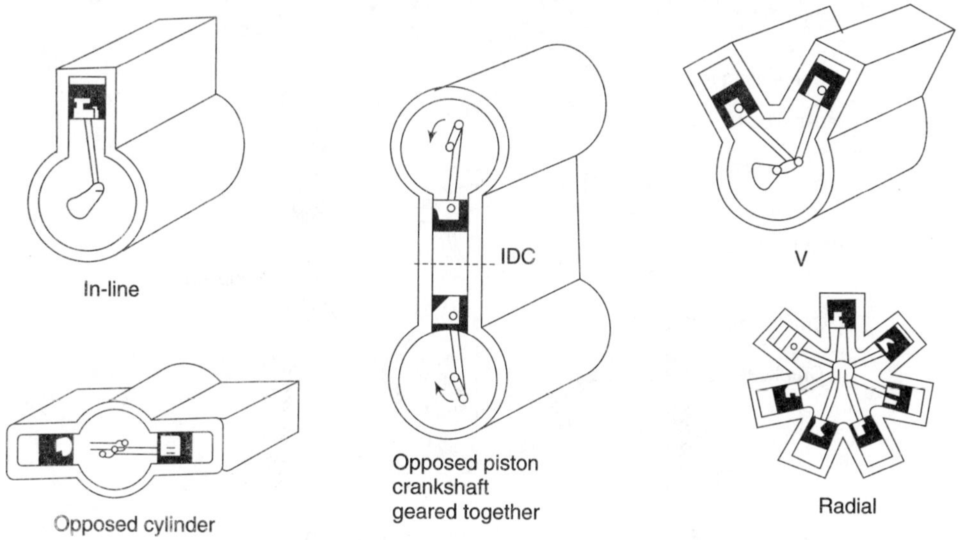


**Fig. 1.2** Classification of engines by cylinder arrangement

**(F) Cooling system**

(i) *Water or coolant cooled engines.* In these engines, the cylinder wall is cooled by water or coolant.

(ii) *Air cooled engines.* In air cooled engines, the cylinder wall is cooled by air.

**(G) Field of application**

(i) *Stationary engines.* These engines are used as small and medium capacity electric power plants and to drive pumping units in agriculture.

(ii) *Mobile engines.* These engines are used in motor vehicles, motorcycles, aeroplanes, ships, locomotive, etc.

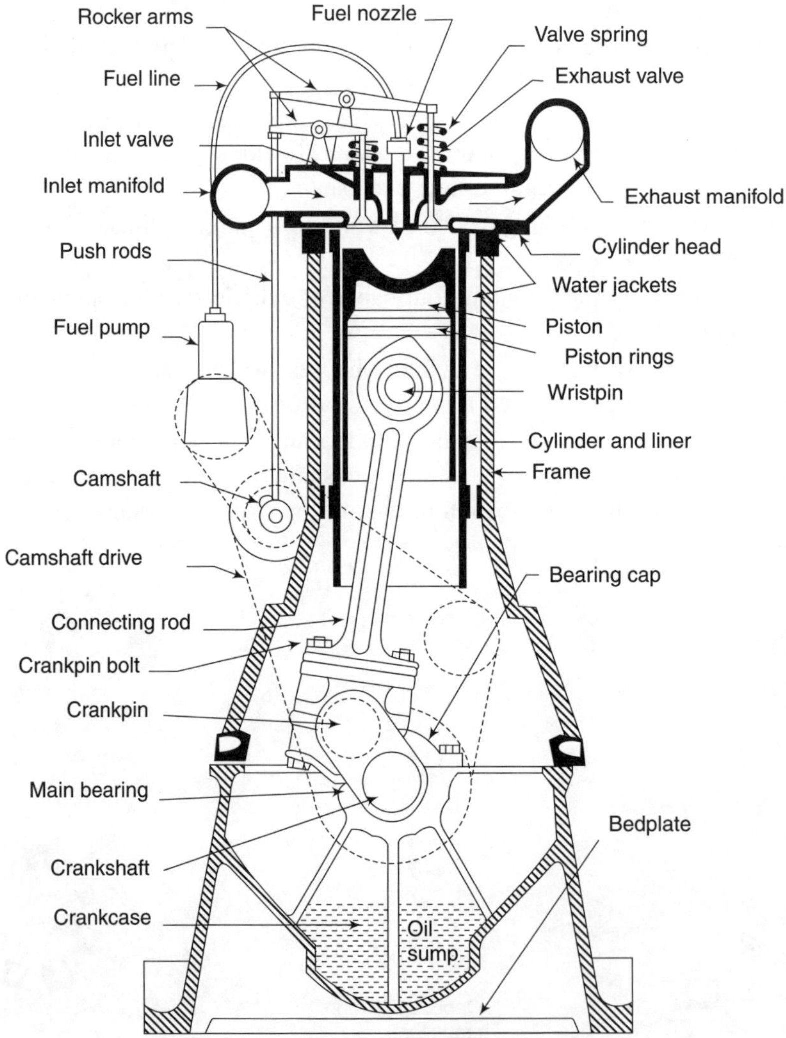


**Fig. 1.3** Details of four-stroke CI engine

**(H) Fuel supply system**

(i) *Carburettor engines.* In these engines (petrol engines), the air and fuel is properly mixed in a carburettor or before induction to the cylinder **(now major obsolete)**.

(ii) *Air injection engines.* In some cases of diesel engines, the fuel is supplied under pressure to the engine cylinder—by using compressed air.

(iii) *Airless or solid or mechanical injection system.* In these diesel engines, the fuel is injected with mechanical or pump injection.

(iv) *Multi-point fuel injection* (MPFI) SI engine. Electronic multi-point fuel injection SI engine, equipped with 3 way catalytic converter, is now the standard passenger car engine.

**(I) Method of control**

(i) *Quality control engines.* In these engines, the composition of the mixture is changed by admitting more or less fuel according to the load.

(ii) *Quantity control engines.* In these engines, the composition of mixture remains constant but the quantity of the mixture is changed.

(iii) *Quality-quantity control engines.* In quality-quantity control engines, both the quality and quantity of the mixture is changed to affect control.

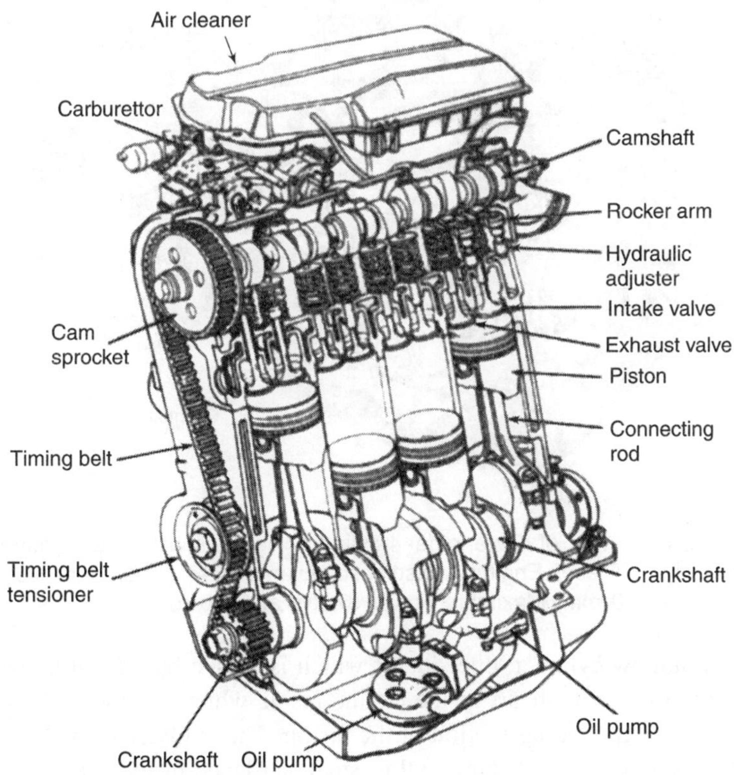


**Fig. 1.4(a)** A cut-way view of SI engine

**(J) Lubrication system**

(i) *Wet sump lubrication engines*

(ii) *Dry sump lubrication engines*

(iii) *Pressurised lubrication engines.*

## 1.6 ENGINE COMPONENTS

Figure 1.3 shows the details of a four stroke compression ignition engine. A cut-way view of a spark ignition engine and compression ignition engines are shown in Fig. 1.4.

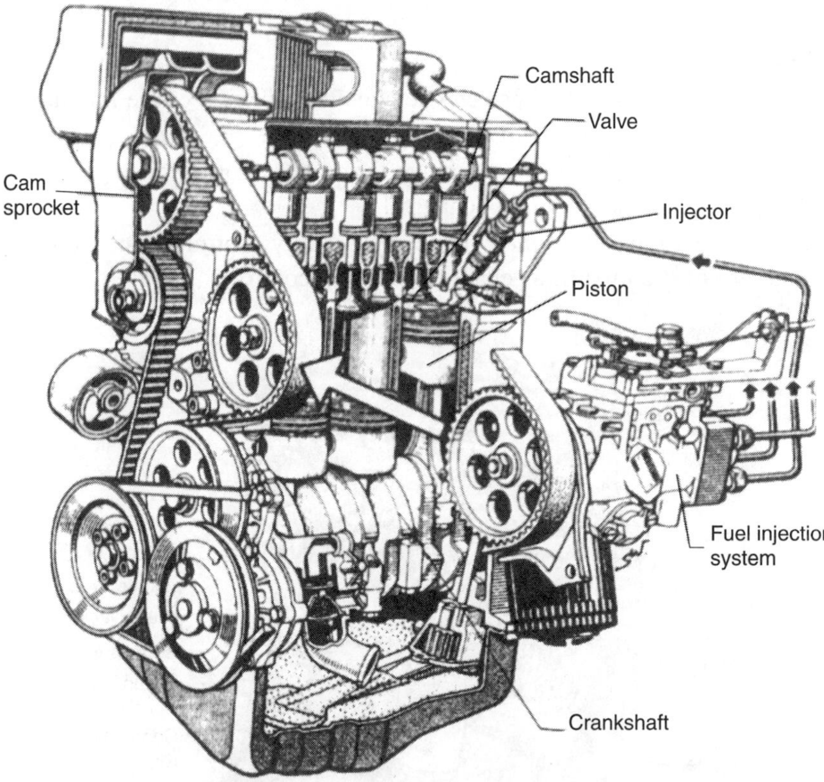


**Fig. 1.4(b)** A cut-way view of CI engine 4-cylinder Naturally Aspirated Indirected-injection Automobile Volkswagen Diesel Engines. Displaced volume 1.47 litres, Bore 76.5 mm, Stroke 80 mm, Maximum Power 37 kW at 5000 rev/min.

**Cylinder.** It is a hollow cylinder, one side of which is closed by a head known as cylinder head. It is supposed to be the heart of the engine inside which the fuel is burnt, and power is developed due to reciprocating motion of the piston. The cylinder is made of cast iron for automobile engines. For aircraft engines, alloy, steel, hardened inside against wear resistance is used. The inside diameter of the cylinder is known as **bore**. Figure 1.5 shows air and water cooled cylinder.

**Cylinder head.** It covers one side of the cylinder and usually contains both the valves, valve ports and fuel injector or spark plug. It is made of cast iron and aluminium alloy.

**Cylinder liner.** Generally, a cylinder is equipped with cylinder liner, so that, if the wear and tear that takes place only the liner is damaged. This is easily replaceable as compared to the cylinder. It is also made of cast iron and alloy steel.

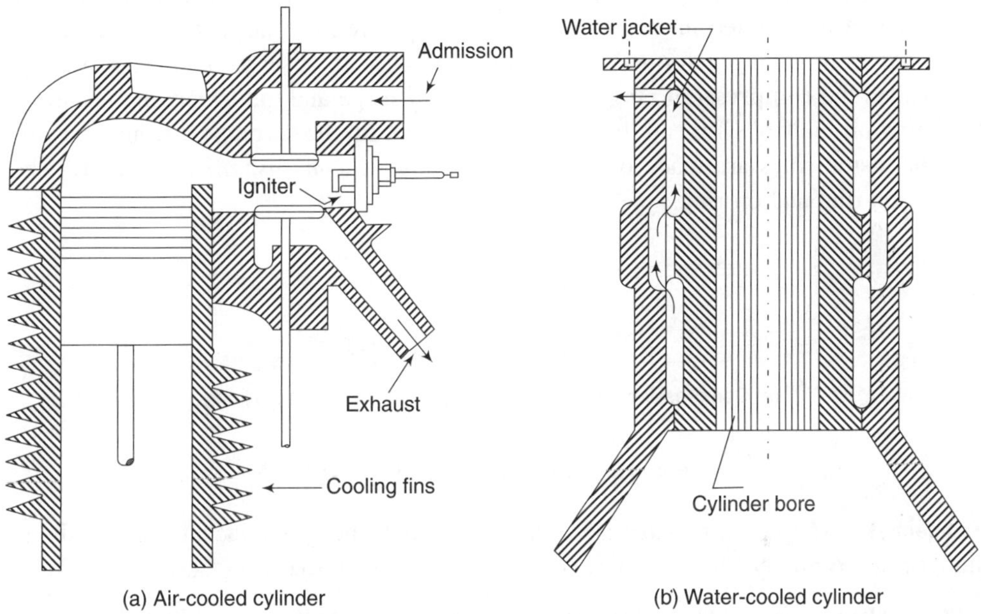


**Fig. 1.5** Air and water cooled cylinder

**Piston.** The piston is a cylindrical casting of iron, aluminium alloy, or steel, with one end closed by a **crown**, which forms the lower surface of the combustion chamber (Fig. 1.6). The outer periphery of each piston is accurately machined to a running fit in the cylinder bore and is provided with several grooves into which piston rings are fitted. The pistons are attached to the connecting rods by **hardened steel piston pins** fitted into bosses cast in opposite sides of the piston. Piston pins are said to be floating when they are free to turn in bearings both in the upper ends of the connecting rods in the piston bosses but are restrained by suitable retainers from contacting the cylinder wall. In some cases, the piston pin is secured in either the connecting rod or the piston bosses and is not free to turn in both. The portion

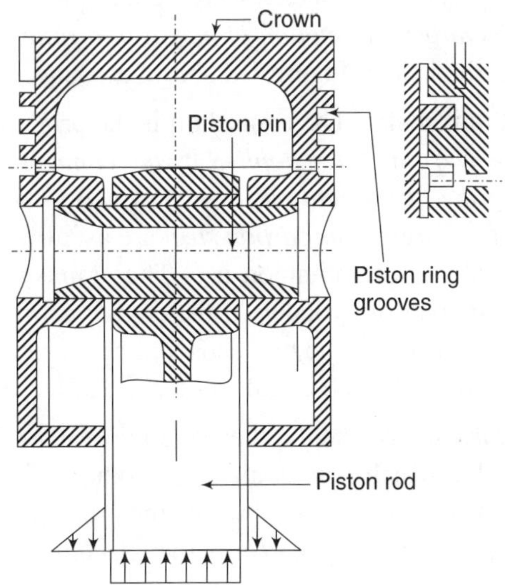


**Fig. 1.6** Piston assembly

of the piston below the pin bosses is called the **skirt** and serves as a bearing surface of the piston. *The main function of the piston is to transmit the force created by combustion process to the connecting rod.*

**Piston rings.** *The function of the piston ring is to prevent any leakage of gas past the piston and to prevent the wear of the piston.* Whatever the wear that takes place, it takes place only in piston rings which is less costly to replace than the piston. They have a slot cut so that they can be easily inserted in the piston grooves. Moreover, the slot cut prevents the metal piston rings from getting deformed. The piston is fitted with at least three rings. The upper ring is called the **compression ring** which prevents leakage and the lower ring is called the **lubricating oil-control ring** and the third which **scraps up** the surplus oil from the cylinder wall. The piston rings are made of *grey cast iron and silicon cast iron.* Figure 1.7 shows compression ring and oil ring.

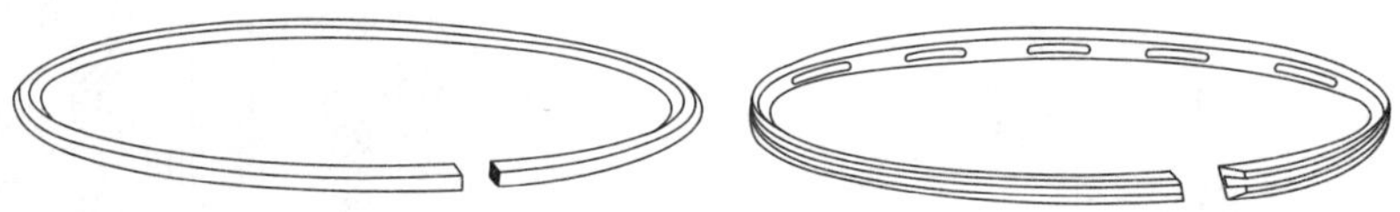

(a) Compression ring (b) Oil ring

**Fig. 1.7** Compression and oil rings

**Water jacket.** The cooling water is circulated through the water jackets surrounding the cylinder head. Actually, the water jacket is cast integral with the cylinder.

**Connecting rod.** The connecting rod connects, the piston and the crank pin. One end, called the small end, is attached to the wrist pin located in the piston; the other end, called the big end is connected to the crank pin. *The function of the connecting rod is to convert the reciprocating motion of piston into rotary motion of crankshaft.* It is made of forged steel or carbon steel. Figure 1.8 shows a connecting rod.

**Crankshaft.** The crankshaft is the principal rotating part of the engine (Fig. 1.9). *It serves to convert the force applied by the connecting rods into a rotating force. It controls the motion of the piston and must be designed to cause them to reciprocate in proper sequence. Alloy steel forgings or carbon steel* are usually employed in the crankshaft, although cast-alloy-iron shafts are in common use. The crankshaft is according to the throws for each cylinder or each pair of cylinders if the engine is of the V-type, and a suitable number of main bearings are located at the two ends and between adjacent throws. The crankshaft is usually drilled to permit oil to be circulated from entrances in the main bearing areas to the crankpins.

**Valves.** Normally, there are two valves for every cylinder. (Fig. 1.10 (a,b). One is an inlet valve which admits air or a mixture of air and fuel in the suction stroke. The other is the exhaust valve through which the product of combustion after doing work on the piston escapes to the atmosphere. *The valve mechanism consists of cams, cam follower, push rod, rocker*

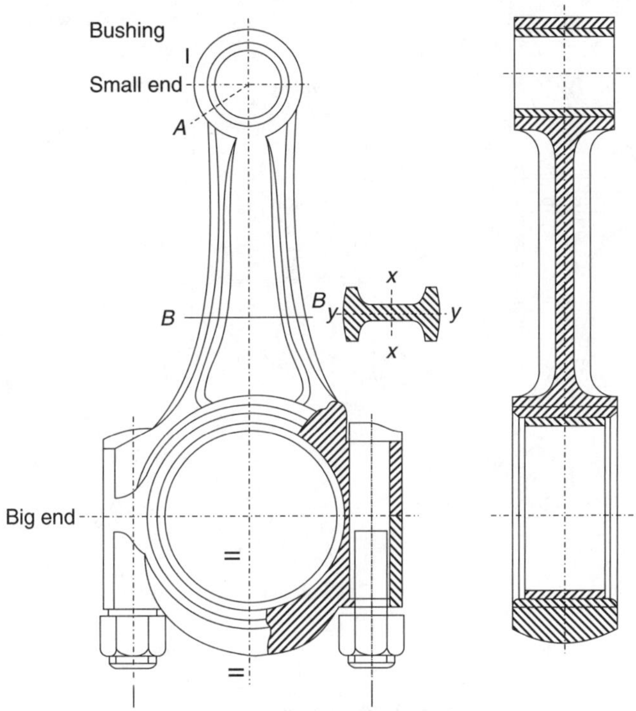


**Fig. 1.8** Connecting rod

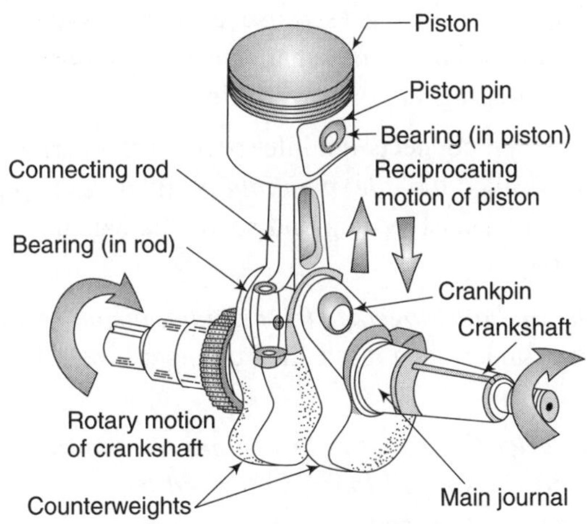


**Fig. 1.9** Crankshaft and other parts

*arms and spring*. The *valves are made of special alloy steel.* Nowadays, 2 valves for suction and one or two valves for exhaust are becoming common for better volumetric efficiency and scavenging. *Thus, in a 4 cylinders engine, 16 valves are common.*

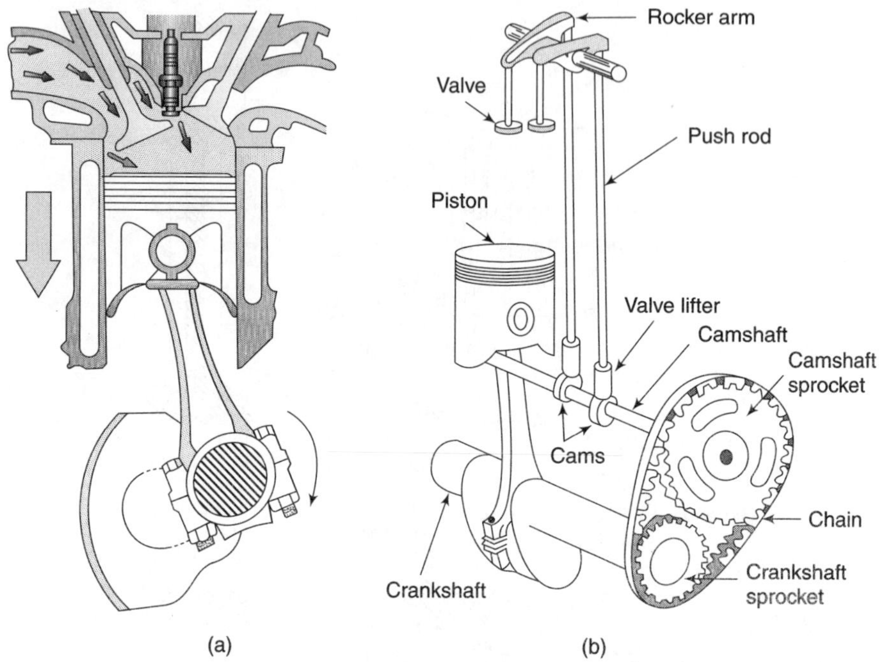


**Fig. 1.10** Engine valves

**Cams and Camshaft.** Each valve requires a cam to open and close it at the proper point in the engine cycle. The shape of the cam is established by the valve timing and the lift imparted to the valves. The cams are mounted on a shaft known as camshaft which is driven by the crankshaft at half speed through gears or by a **timing chain**. It is made of alloy steel.

**Manifolds.** The piping which connects the inlet ports of the various cylinders to a common air intake for the engine is called the *inlet manifold*. Similarly, the piping which connects the exhaust ports to a common exhaust system and sends the exhaust products to the muffler is called the exhaust manifold.

**Crankcase.** *The crankcase holds together the cylinder, piston and crankshaft.* If a cylinder liner is inserted into the cylinder, then it is called cylinder block. It is made of cast iron steel and aluminium alloy.

**Carburettor.** *This is a device employed in SI engines to atomize, vaporize and mix the fuel with air in required proportions at all loads and speed before entering the engine cylinder. It is more or less obsolete except racing cars and scooty.*

**Gasoline injection, Multipoint Fuel injection (MPFI) System.** Instead of carburettor, now all the passenger car engines working on gasoline (petrol) fuel use MPFI system. With the help of pump, injector and electronic devices, the required quantity of fuel is pumped in the suction manifold before suction valve according to load and speed.

**Fuel pump and injector unit.** This unit is used in CI engines. The fuel pump supplies the fuel under pressure to the injector or atomizer, which consists of one or more orifices through which the fuel is sprayed into the cylinder. Now, multi-point fuel injection is used in SI engines to meet the EURO-III or IV emission standard.

**Spark plug.** In SI engine, a spark plug is located near the top of the cylinder, and initiates combustion. It is one of the elements of ignition system.

**Flywheel.** *It is a heavy wheel mounted on the crankshaft to minimize the cyclic variations in speed. It absorbs energy during the working stroke and supplies it during non-working strokes.* By employing a flywheel, *the turning moment becomes uniform at the crankshaft. It is made of cast iron, steel and aluminium alloy.*

**Ignition system.** This is provided in SI engine. It controls the firing order in the SI engine cylinder through spark plug.

**Bed plate.** The lower portion of the crankcase which is bolted to the foundation is called the bed plate. A summary of list of engine parts and materials is tabulated in Table 1.3.

**Table 1.3** Summary of Engine Parts and Materials

| *S. No.* | *Name of parts* | *Materials used* | *Method of manufacturing* |
|---|---|---|---|
| 1. | Cylinder | Cast iron, alloy steel | Casting |
| 2. | Cylinder liner | Cast iron, alloy steel | Casting |
| 3. | Cylinder head | Cast iron, aluminium alloy | Casting, forging |
| 4. | Piston | Cast iron, aluminium alloy | Casting, forging |
| 5. | Piston rings | Grey cast iron, silicon CI | Casting |
| 6. | Piston pins | Hardened steel | Forging |
| 7. | Valves | Special alloy steel | Forging |
| 8. | Connecting rod | Carbon steel, alloy steel | Forging |
| 9. | Crankshaft | Alloy steel, carbon steel, cast alloy of iron | Forging |
| 10. | Cams and camshaft | Alloy steel, carbon steel | Forging |
| 11. | Crankcase | Cast iron, steel, aluminium alloy | Casting |
| 12. | Flywheel | Cast iron, steel, aluminium alloy | Casting |
| 13. | Bearing | White metal, leaded bronze | Casting |

## 1.7 BASIC ENGINES NOMENCLATURE

Internal combustion engines make use of piston-cylinder arrangement in which the piston is observed to undergo cycles or revolutions. Figure 1.11 shows the various nomenclature used in the piston cylinder arrangement.

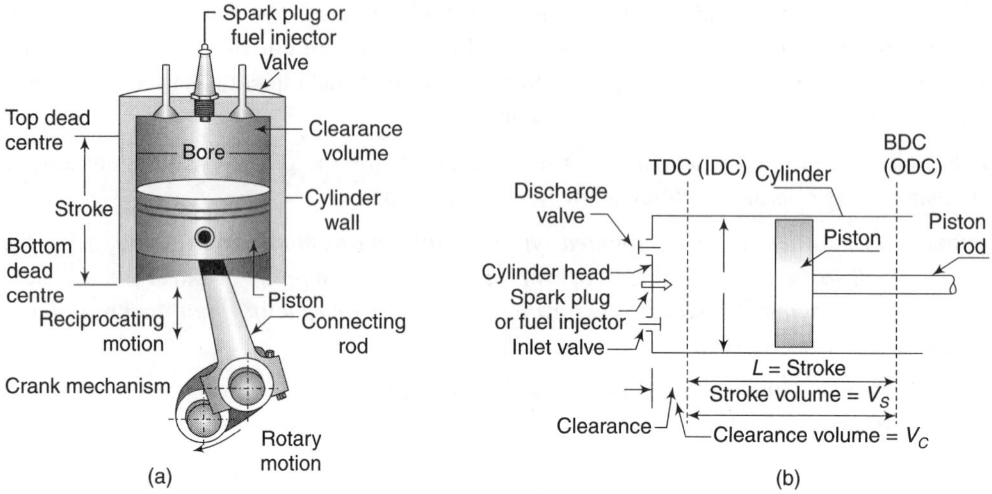


**Fig. 1.11** Piston-cylinder arrangements

**(i) Cylinder bore (D).** The cylinder or piston nominal diameter (D) is known as bore.

**(ii) Piston area (A).** The area of circle having diameter equal to cylinder bore is called piston area.

**(iii) Dead centre.** The extreme positions of the piston at which its motion is reversed is called dead centre.

- **(a) Top Dead Centre (TDC).** The extreme position of the piston farthest from the crankshaft is called Top Dead Centre position (TDC) or Inner Dead Centre position (IDC).
- **(b) Bottom Dead Centre (BDC).** The extreme position of the piston nearest to the crankshaft is called Bottom Dead Centre position (BDC) or Outer Dead Centre position (ODC).

**(iv) Stroke (L).** The nominal distance that the piston moves in the direction between two extreme positions is called the stroke.

**(v) Displacement volume or piston swept volume ($V_s$).** The volume swept by the piston in moving from TDC to BDC or vice versa is called the swept volume or displacement volume.

**(vi) Clearance volume ($V_c$).** Theoretically, the piston may touch the cylinder head but in actual practice, there is a clearance left between the cylinder head and the piston at TDC position. Thus, when the piston is at the TDC position the minimum volume, i.e., the volume of the clearance is called the clearance volume.

**(vii) Cylinder volume (V).** The cylinder volume is the sum of clearance volume and swept volume. Thus,

$$V = V_c + V_s \qquad ...(1.1)$$

**(viii) Compression ratio (r).** This is defined as the cylinder volume to the clearance volume. Thus,

$$\text{Compression ratio, } r = \frac{\text{Full cylinder volume}}{\text{Clearance volume}} = \frac{V}{V_c}$$

or

$$r = \frac{V_c + V_s}{V_c} = 1 + \frac{V_s}{V_c} \qquad ...(1.2)$$

**(ix) Clearance ratio (c).** This is defined as the ratio of clearance volume to swept volume. Thus,

$$\text{Clearance ratio, } c = \frac{V_c}{V_s} \qquad ...(1.3a)$$

or

$$r = 1 + \frac{1}{c} = \frac{1 + c}{c} \qquad ...(1.3b)$$

## 1.8 WORKING PRINCIPLE OF FOUR-STROKE CYCLE ENGINES

A large number of internal combustion engines, both petrol and diesel, operate on a four-stroke cycle. The sequence of the four strokes is as follows:

(i) Intake or suction stroke, (ii) compression stroke, (iii) working or expansion or power stroke and (iv) exhaust stroke.

(i) *Intake or suction stroke.* Suppose that the piston is very near to the top dead centre position (TDC) as shown in Fig. 1.12(a). During suction stroke, the inlet valve is opened and the discharge valve is closed and the piston moves down (i.e., outward) due to rotation of the crankshaft either getting energy from the flywheel or a motor starter. As the piston moves, vacuum is created between the piston and cylinder and the pressure in the cylinder drops below atmospheric.

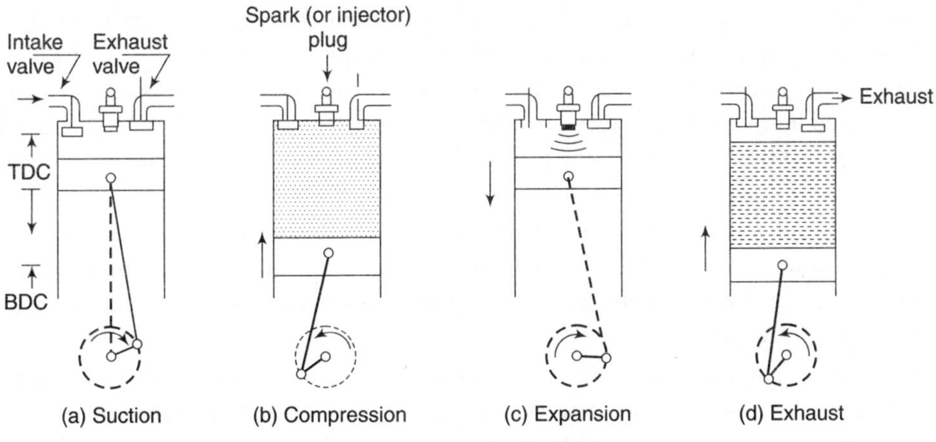


**Fig. 1.12** Operation of four-stroke engines (SI and CI)

In the case of SI engine, a mixture of air and fuel in proper proportions rushes inside the cylinder through the inlet manifold and valve which is cam operated. The suction process continues till the piston reaches the bottom dead centre position (BDC. This is represented by process 0-1 on the indicator diagram shown in Fig. 1.13(a).

In the case of CI engines, only air enters the engine cylinder through inlet valve during suction process as represented by process 0-1 in Fig. 1.13(b).

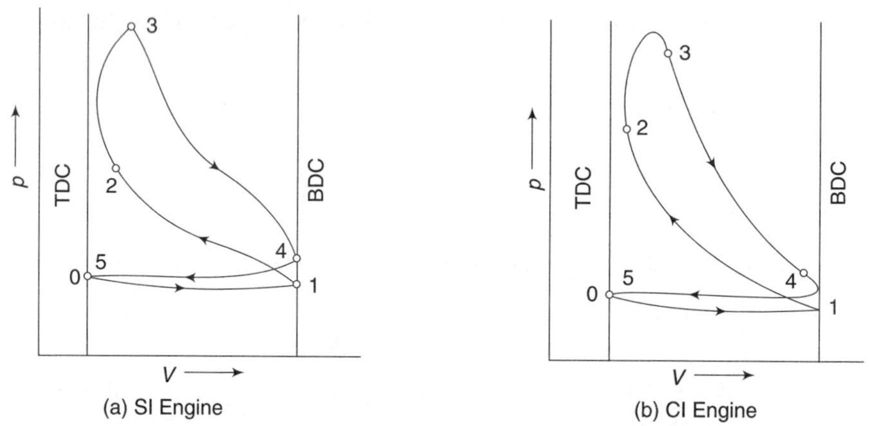


**Fig. 1.13** Indicator diagram for 4-stroke engines (SI and CI)

(ii) *Compression stroke.* During a compression stroke, both the valves are closed and the piston moves from bottom to top dead centre position, i.e., inwards [Fig. 1.12(b)]. The charge or air is compressed up to a compression ratio which depends upon the type and need of the engine. Since both the valves are closed and the piston moves inwards, there is a reduction in volume of the charge or air which results in an increase of pressure and temperature of the cylinder contents. In petrol engines, the compression ratio varies from 5 to 10.5, and pressure and temperature at the end of compression are about 7 to 14 bar and 250°C to 300°C respectively. For diesel engines the compression ratio is about 12 to 20 and the pressure and temperature at the end of compression are about 28 bar to 59 bar and 600°C to 700°C respectively. The compression process is represented by 1-2 on the indicator diagram (Fig. 1.13).

(iii) *Working on Expansion or Power Stroke.* During power stroke, both the valves are closed. The power stroke includes combustion of fuel and expansion of the products of combustion. The combustion starts at the end of the compression stroke when the piston is approaching the TDC position.

In SI engines, a spark plug initiates the combustion. During combustion, the chemical energy of the fuel is released and there is a rise in temperature and pressure of the gas. The temperature of the gases is increased between 1800–2300°C and the pressure to 30–40 bar. The volume of the gas however remains constant during combustion and this is represented by process 2-3 as shown in Fig. 1.13(a).

In CI engine, (mechanical airless injection), one or more jets of fuel, compressed to a pressure of 105–210 bar by an injector pump are injected into the combustion chamber by a fuel nozzle at the end of the compression stroke. The injected fuel is vaporized in the combustion chamber. When the fuel vapour is raised to self ignition temperature, the combustion then starts automatically and there is a sudden rise of pressure at approximately constant volume. However, the latter part of combustion occurs at almost constant pressure. The combustion pressure is about 45–90 bar. In air injection system (not used nowadays) the combustion process is more or less at constant pressure.

The high pressure and the high temperature of the products of combustion, thus obtained, pushes the piston outward from TDC to BDC position (Fig. 1.12(c)) for expansion stroke, This reciprocating motion of the piston is converted into rotary motion by the crankshaft, connecting rod and crank mechanism. The expansion process is represented by process 3-4 on the indicator diagram, during which, the pressure drops due to increase of volume of gas and absorption of heat by the cylinder walls.

(iv) *Exhaust stroke.* Refer to Fig. 1.12 (d) for exhaust stroke. When 7/8 of the power stroke is completed, the exhaust valve begins to open and the inlet valve remains closed. This is represented by the point 4 on the indicator diagram. At this instant, the pressure of the gas is about 4-5 bar which is higher than the exhaust manifold pressure. Thus, about 60% of the products of combustion are discharged into the exhaust manifold through the discharge valve. The process 4-5 represents the drop in the cylinder when the exhaust gases are discharged to atmosphere. The remaining products of combustion are exhausted into atmosphere when the piston moves from BDC to TDC position. During ths process, the pressure is slightly more than atmospheric pressure and is represented by process 5-0. When the piston is approaching TDC, the exhaust valve closes and some of the burnt gases are left in the clearance space.

At the end of the exhaust stroke, the inlet valve opens, and the cylinder is ready to receive the fresh charge to begin a new cycle.

It is obvious from the above discussion, that the crankshaft makes two revolutions during the four strokes, and in these four strokes, there is only one power stroke. This means that *for every two revolutions of the crankshaft, there is only one power stroke.*

## 1.9 VALVE TIMINGS DIAGRAM FOR FOUR-STROKE ENGINES (SI AND CI)

While discussing the operation of the four-stroke cycle engine, theoretically, it was assumed, that the opening and closing of the valves and the sparking or injection occurs at the engine dead centre positions. But, in actual operation, the time necessary for opening and closing the valves and the effects of high speeds upon the flow of the gases require that these events be displaced somewhat from the dead extreme points due to mechanical and dynamic factors.

The timing of these events referred in terms of crank angles from dead centre positions is represented on a *valve timing diagram.* For the efficient designing of an IC engine, correct valve timings are absolutely necessary. Figures 1.14 and 1.15 show the valve timing diagrams of a petrol and diesel engine respectively. This diagram shows all the four valve events, i.e., inlet opening, inlet closure, exhaust opening and exhaust closure, to be displaced appreciably from the top and bottom dead centre points, where they would otherwise theoretically occur. *The opening events are earlier and closing events later with respect to dead centre to allow for the time required to open and close the valves without excessive inertia loading or slamming.*

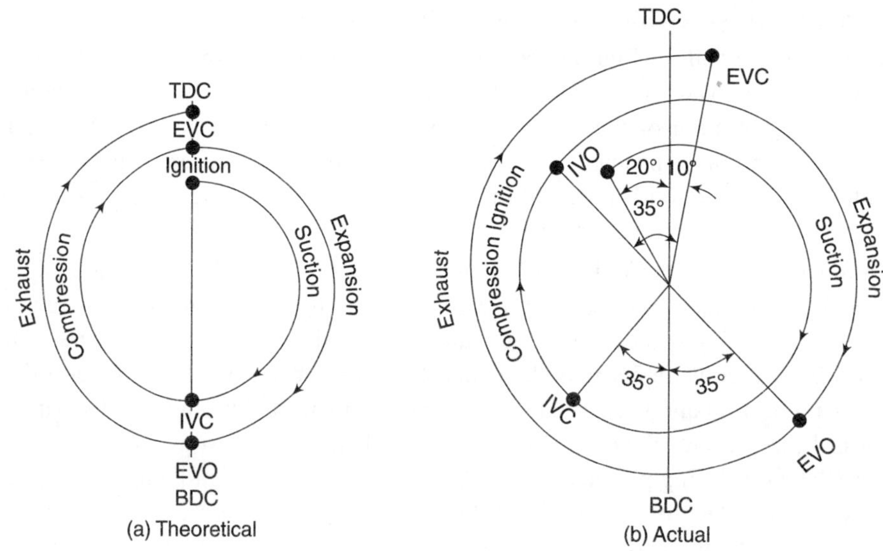


**Fig. 1.14** Valve timing diagram for 4-stroke SI engines

**Inlet Valve Timings.** Owing to the effect of inertia and time required to attain the full opening, the inlet valve opens somewhat earlier than at TDC, so that, by the time the piston reaches TDC, the inlet valve is fully open. The valves are made to open and close slowly to provide quiet operation under high speed conditions. The rapid decrease in pressure in the cylinder due to outward stroke of the piston, causes the gases to rush in and fill up the space above the piston. At high speed at which most IC engines run, the piston will reach the end of the suction stroke, i.e., at BDC before the time it takes in a complete charge to enter the cylinder through the valve opening; and the gases will be moving in the direction of the motion of the piston with high velocity. Moreover, there will be a considerable resistance to the flow of the charge through the air cleaner, carburettor, inlet manifold and the port.

This clearly shows, that if the inlet valve is closed at BDC, the cylinder will receive a charge less than its capacity and the pressure inside the cylinder would be somewhat less than atmospheric pressure. It is for this reason, the inlet valve is kept open until the piston reaches a point in its next incoming stroke (i.e., after BDC) at which the pressure in the cylinder is equal to the atmospheric pressure. It may appear, that since the inlet valve is open even during compression, some of the charge may be sent back to the inlet manifold, but, instead of this,

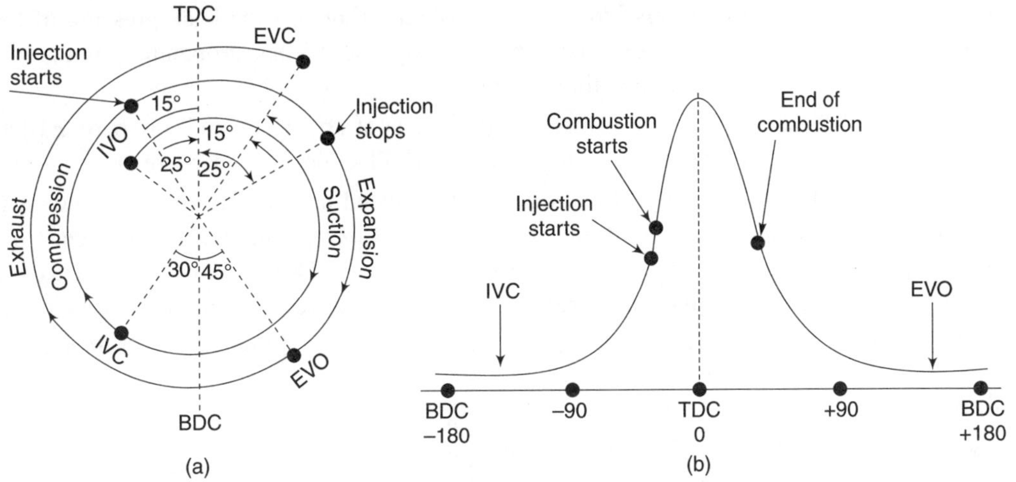


**Fig. 1.15** Valve timing diagram for 4-stroke CI engines

the kinetic energy of the air-fuel mixture (or air) produces a ramming effect which enables more charge to enter the cylinder. Thus, theoretically, it is possible to induce a charge more than the capacity volume of the combustion space.

In a particular petrol engine, the inlet valve opens at about 20°before TDC and closes at about 35° after BDC, while in CI engine, the inlet valve opens at about 25° before TDC and closes at about 30° after BDC This means that in CI engines the crankshaft has to rotate 235° for a suction stroke.

**Ignition (or injection).** Theoretically, it is assumed, that ignition in SI engine, or injection of fuel in CI engine, occurs at TDC position. But in actual practice, it is not so. There is a time lag between the spark and actual ignition. In order to have better result, the ignition of charge is timed to occur as early as the engine permits. In the case of an engine running at higher speed, the ignition is further advanced, ahead of TDC, and this shifting of the ignition timing is called the *ignition advance*. In petrol engine, the ignition is further advanced, ahead of TDC In petrol engine, the ignition occurs at 35° before TDC

In diesel engine, the fuel is injected ahead of TDC at about 15°, so that, by the time the piston reaches the TDC the actual combustion of fuel starts. The injection of fuel ends at about 25° after TDC

**Exhaust valve.** The exhaust valve opens in advance, i.e., before piston reaches the BDC position and closes later, i.e., after TDC position. The earlier opening of the exhaust valve makes it possible for the exhaust gases to leave by virtue of their pressure being higher than atmospheric. Due to late opening of the exhaust valve, the kinetic energy of the fresh charge pushes the remaining combustion products to discharge from the cylinder; hence by this way the maximum exhausting of cylinder is possible. This results in some loss of work done on the piston by the expanding gases, but at the same time it affects a greater reduction in the negative work that the piston must do in being pushed up against the back pressure of the gases during the early part of the exhaust stroke. The optimum exhaust-opening point is that at which the net work of the cycle is the maximum.

In petrol and diesel engines, the exhaust valve opens at about 35° and 45° before BDC position and exhaust valve closes 10° and 15° after the TDC position respectively.

It is to be noted that the values of angular positions listed in the valve-timing diagram, is a **mean value**, and a considerable difference exists with different engines. It is evident from the valve timing diagram, that at some part of the cycle near TDC, both the valves are open and this period is known as **overlap period**. These timings affect the volumetric efficiency.

Table 1.4 gives typical valve timing for 4-stroke low speed and high speed SI engines.

**Table 1.4** Typical Valve Timing of 4-Stroke SI Engines

| *Position of valves timing* | *Theoretical* | *Actual timings* | |
|---|---|---|---|
| | | *Low speed engines* | *High speed engines* |
| Inlet valves opens (IVO) | TDC | 10° b TDC | 10° b TDC |
| Inlet valves closes (IVC) | BDC | 10° a BDC | 60° a BDC |
| Inlet valve is open for | 180° | 200° | 250° |
| Exhaust valve opens | BDC | 25° b BDC | 55° b BDC |
| Exhaust valve closes | TDC | 5° a TDC | 20° a TDC |
| Exhaust valve is open for | 180° | 210° | 255° |
| Valve overlap | Nil | 15° | 30° |
| Spark | 15° b TDC | 30° b TDC | 30° b TDC |

Table 1.5 gives the typical valve timings of a 4-stroke diesel engines.

**Table 1.5** Typical Valve Timings of 4-Stroke Diesel Engines

| *Position of valves* | *Theoretical timings* | *Actual timings* |
|---|---|---|
| Inlet valves opens (IVO) | TDC | 30° b TDC |
| Inlet valves closes (IVC) | BDC | 50° a BDC |
| Inlet valves is open for | 180° | 260° |
| Exhaust valve opens (EVO) | BDC | 45° b BDC |
| Exhaust valve closes (EVC) | TDC | 30° a TDC |
| Exhaust valve is open for | 180° | 255° |
| Valve overlap | Nil | 15° |
| Injection | Nil | 15° b TDC |

## 1.10 COMPARISON BETWEEN 4-STROKE SI AND CI ENGINES

While SI and CI engines have much in common, there are certain fundamental differences that cause their operation to vary considerably. On the basis of the following points, their difference may be discussed.

*(i) Basic cycle.* The SI engine, in general, is based on the Otto cycle while the CI engine, in general, is based on the Diesel cycle.

*(ii) Introduction of fuel.* In most SI engines, a mixture of air fuel in proper proportion is introduced to the combustion chamber. A carburettor serves this purpose of mixing air and fuel in proper proportions. A throttle controls the quantity of the mixture introduced. In new technology of SI engines, now electronic fuel injection system is used at inlet manifold. It is called multi-point fuel injection (MPFI). In cars and motorcycles, MPFI is common. In a CI engine, the air alone is introduced in the cylinder during suction stroke and it is compressed during compression stroke, the mixing takes place in the combustion chamber itself. The fuel control valve controls the quantity of fuel introduced. Thus, the CI engine does not need a carburettor, but instead, it needs fuel pump and injector nozzle.

*(iii) Ignition system.* The SI engine is a spark ignition engine. It needs a spark plug along with other accessories of ignition system to initiate combustion at the end of a compression stroke.

The CI engine is a compression ignition engine which utilizes the condition of high temperature produced by compression of air in the cylinder to initiate combustion, when the fuel is injected.

*(iv) Compression ratio range.* Present day, SI engines operate in a range of compression ratios from about 5 to 10.5; the higher value being limited by detonation. On the other hand, the detonation in the CI engines does not restrict the compression ratio. In fact, increasing the compression ratio tends to reduce the possibility of detonation. Moreover, ignition is by compression ignition, hence it needs a higher compression ratio. At present, the range of compression ratio in CI engines varies from 12 to 20. The higher the compression ratio employed in CI engines the higher are thermal efficiencies. This results in a greater power output for a certain quantity of fuel consumed.

*(v) Operating compression ratio, temperature and pressure.* Since the compression ratio is less in SI engines hence the compression pressure varies from 7 to 14 bar, and temperature varies from 250 to 300°C.

On the other hand, higher pressures and temperatures are found in the CI engine as a result of higher compression ratio utilised. In fact, a higher compression pressure must exist to heat the air in the cylinder to temperature greater than the ignition temperature of the fuel. Therefore, in CI engines, the compression pressure varies from 28 to 49 bar, and compression temperature is about 500°C. These high pressures require a stronger engine structure, and therefore, a CI engine is heavier and more sturdy.

*(vi) Operating speed.* The SI engines are normally called high speed engines with operating speed between the range of 3000 to 5000 rpm. The CI engine on the other hand is a low speed

engine with operating speeds ranging between 400 to 1200 rpm. In recent years, however, the higher speed diesel engine (1200 to 3000 rpm) has seen rapid development and is becoming increasingly popular.

*(vii) Power control.* The governing (i.e., speed and load control) of SI engine is performed by regulating a throttle which controls the quantity of air-fuel mixture going to cylinder (quantity governing), while in CI engine, the governing is performed by regulating the quantity of fuel injected at the end of the compression; the air quantity remaining constant (quantity governing).

*(viii) Distribution of fuel to cylinders.* In the multi-cylinder SI engines, considerable difficulty is encountered in distributing the mixture of air and fuel to the various cylinders in the case of carburettor engines but in MPFI system, the distribution is very good. On the other hand, in CI engine, the injection system provides a generally excellent distribution of fuel to the cylinders.

*(ix) Supercharging.* To increase the power output of the engine, both the SI and the CI engines may be supercharged. *The amount of supercharging in SI engine is limited by detonation.* In CI engines, supercharging acts as an antiknock. A limit is imposed by the amount of the power required to drive the supercharge.

*(x) Exhaust gas temperature.* Compared to SI engines, the exhaust gases in the CI engine are returned to the atmosphere at low temperature. Since a greater portion of the available energy goes into useful work and the losses to the exhaust are lower, the excess air in CI engine lowers the exhaust temperature.

*(xi) Starting.* The starting of CI engine is inherently more difficult than SI engine due to greater cranking effort required to overcome the higher compression ratios used. The starting problem of CI engine is mainly in cold weather.

*(xii) Noise.* Due to higher operating pressure and vibration, noise problem in a CI engine are more pronounced than a SI engine.

*(xiii) Weight per unit power output.* In SI engines, the compression ratio and pressure are less, hence the weight per unit power is smaller and it amounts to about 0.6 to 4.3 kg/kW. In CI engines, the compression ratio and pressure are higher, hence, it requires for the same output a larger stronger and heavier engine and parts. The CI engines, generally weigh about 3.2 to 12.2 kg/kW, or even more.

*(xiv) Power output per unit piston displacement.* Most high speed CI engines deliver approximately 0.026 kW for each cubic cm of piston displacement. Whereas this ratio for SI automobiles engines is about 0.035, and some SI aircraft engines reach as much as 0.04. This means that the SI engines occupy less space than CI engines for the same power output.

*(xv) Fuel economy.* The CI engines give better fuel economy at both full and part load than the SI engines. This is the greatest advantage of the CI engine. Another, and very important advantage is reduced fuel fire hazard. In SI engines with MPFI system also give better fuel economy.

*(xvi) Initial and maintenance cost.* Due to the fact that the CI engine is of a more sturdier and heavier construction than the SI engine and because the CI engine has an expensive fuel supply system, the initial cost of the CI engine is higher. Moreover, the maintenance and repair of the CI engine are also more than that of the SI engine.

*(xvii) Acceleration.* Better acceleration is possible from CI engine than SI engine. This is because of the fact that the injection system offers direct control over the quantity of fuel injected. However, SI engine with MPFI offers better acceleration.

*(xviii) Application.* The SI engines are extensively used in automobiles and aeroplanes because of low initial costs, low weight power output, smaller port sizes for power output, low starting or cranking effort and less objectionable exhaust gas odour. The CI engines are extensively used for heavy duty equipment such as large trucks, buses, bulldozers, tractors, marine, rail-road engines and earth moving machineries, because of their low specific fuel consumption, reduced fire hazard, less expensive fuel, larger power output. They are better suited for two-stroke operations and have lower operating life.

## 1.11 WORKING PRINCIPLE OF TWO-STROKE CYCLE ENGINES

*In a two-stroke engine, the working cycle is completed on two strokes of the piston or one revolution of the crankshaft, whereas in a 4-strokes engine, it is completed in 4-strokes or two revolutions of the crankshaft. In the 2-stroke engine, the intake and compression processes are completed during the inward stroke and expansion and exhaust processes during the outward stroke.* The two-stroke cycle engines can be either a spark ignition or a compression ignition engine. For spark ignition, a carburettor and a spark plug are used while for a CI engine, a pump and an injector are used.

In a 4-stroke engine, the admission of fresh charge and the removal of exhaust gases are controlled by intake and exhaust valves, generally located in the cylinder head. But in 2-stroke engine, these operations are performed by ports located in the cylinder walls, which are opened or closed by the piston.

Generally speaking, two-stroke cycle engines are classified on the method of **scavenging**. The working principle of crankcase scavenged engine is given below.

Figure 1.16 shows a crankcase scavenged engine. *The term scavenging refers to the process of removal of the burnt gases from the cylinder at the end of the exhaust stroke, by deflecting the fresh charge across the cylinder. The purpose of scavenging is to reduce the dilution of the fresh charge with the residual gases.* The engine consists of ports cut into the cylinder walls instead of valves. The air (in the case of CI) or air-fuel mixture (in the case of SI engine) is led to the crank case through the inlet port. The compressed charge from the crankcase, enters the cylinder through the transfer port. The spark plug is provided to initiate the combustion in S.I engine. The products of combustion after expansion are forced out of the cylinder through the exhaust port. The piston controls the closing and opening of the port. The projection on the piston is called the ***deflector***. *The function of the deflector is to deflect the fresh charge to the top before flowing to the exhaust port.* In every stroke, two processes are performed.

The various processes performed are explained as follows.

**Ignition and induction.** In Fig. 1.16(a), the piston is almost at the TDC position when the air or mixture of air and fuel is in compressed form. In the case of SI engine, spark occurs and ignition takes place. In the case of CI engine, the fuel is injected and combustion takes place automatically. At this same position of the piston, the lower edge of the piston uncovers the inlet port and fresh charge enters the crankcase through the inlet port.

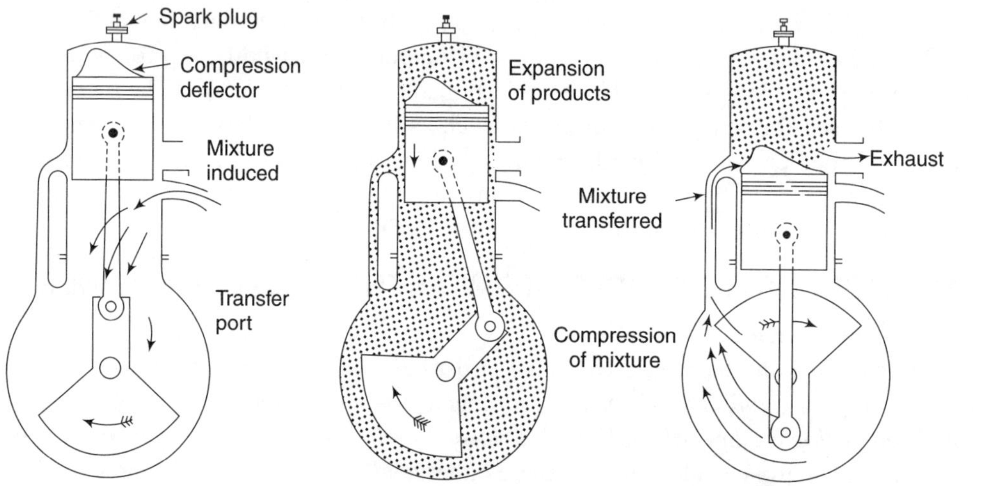


(a) Ignition and induction (b) Expansion and compression (c) Exhaust and transference

**Fig. 1.16** Crankcase scavenged two-stroke SI engines

**Expansion and compression.** Due to combustion of the charge, the pressure increases which pushes the piston down, i.e., the expansion of the gas takes place and work is done. During this downward stroke of the piston, the charge contained in the crankcase is compressed to a pressure of about 1.35 bar.

**Exhaust and transference.** After completion of about 4/5$^{th}$ of the power stroke, the exhaust port is uncovered by the piston and the products of combustion start leaving the cylinder through the exhaust port. When the piston is approaching the BDC position, the exhaust port is still open and the transfer port is uncovered by the piston. The charge compressed in the crankcase enters the cylinder through the transfer port.

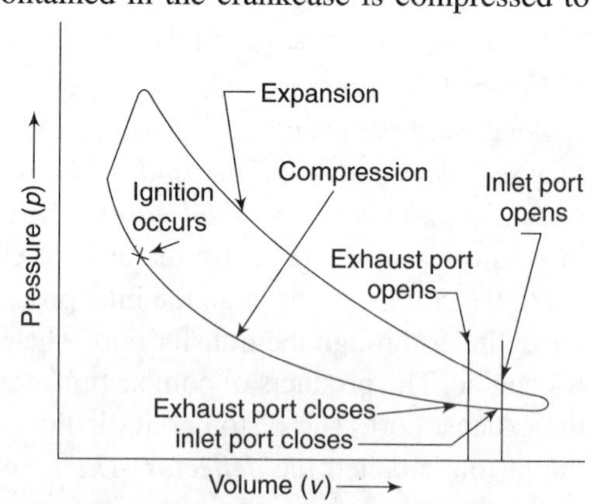


**Fig. 1.17** Ideal and actual indicator diagram of two-stroke engines

As soon as the piston reaches the BDC position, it moves upward. During its upward motion, it first closes the transfer port and then exhaust port. After this, the compression of the charge begins in the cylinder till the piston reaches the TDC position. Thus, during the two strokes of the piston, there is one power stroke.

Recently **five-port engines** are gaining popularity. **However, as per emissions standard, recent laws say all two wheelers are four-stroke engines from 2004 in India**.

The various processes of the two-stroke engines are shown on *p-V* diagram in Fig. 1.17 and the valve timing diagram is shown in Fig. 1.18.

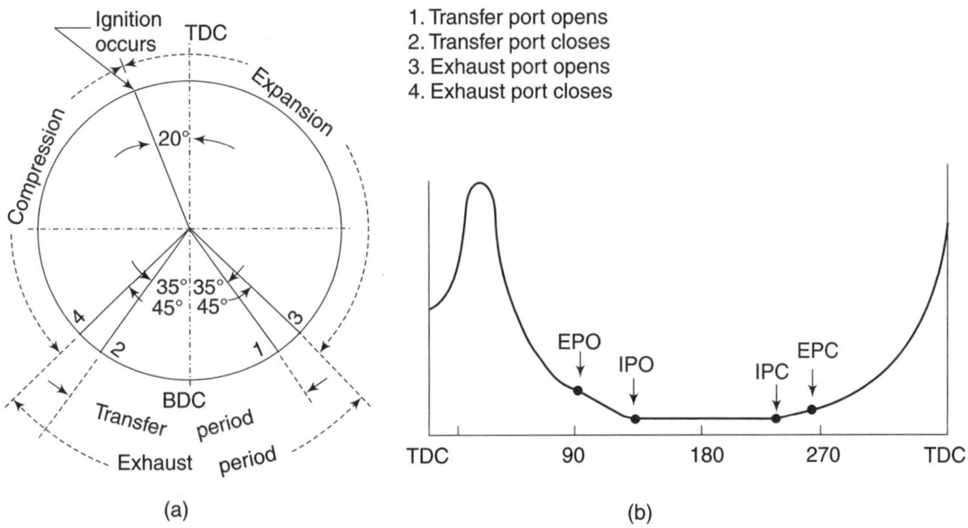


**Fig. 1.18** Valve timing diagram for two-stroke engines

## 1.12 COMPARISON BETWEEN 4-STROKE AND 2-STROKE ENGINES

Table 1.6 gives the comparison between 4-stroke and 2-stroke engines.

**Table 1.6** Comparison between 4-Stroke and 2-Stroke Engines

| *Basis* | *4-stroke engines* | *2-stroke engines* |
|---|---|---|
| 1. Basic stroke | Cycle is completed in 4 strokes of the piston, i.e., in two revolutions of the crankshaft. This means one power stroke is obtained in every two revolutions. | Cycle is completed in 2 strokes of the piston, i.e., in one revolution of crankshaft. This means one power stroke is obtained in every revolution of the crankshaft. |
| 2. Power per cycle | Power produced per cycle for the same size of the engine is small because one power stroke is achieved in every two revolutions of crankshaft. | Power produced per cycle for the same size of the engine is large (theoretically twice of 4-stroke) as power stroke is achieved in every revolution. |

*Contd..*

*Table 1.6 Contd.*

| | | |
|---|---|---|
| 3. Turning moment | Due to 4 strokes, the turning moment is not uniform, so a heavier flywheel is needed. | Due to 2 strokes, the turning moment is more uniform, so light flywheel is needed. |
| 4. Cooling and lubrication | Due to one power stroke in two revolutions lesser cooling and lubrication of engine are required. | Due to one power stroke in every revolution, more cooling and lubrication of engine are required. |
| 5. Rate of wear and tear | Less rate of wear and tear due to one power stroke in two revolutions. | Greater rate of wear and tear due to one power stroke in every revolution. |
| 6. Mechanical efficiency | The valves, valve mechanism, camshaft offer more friction losses results in less mechanical efficiency. | Absence of valves, valve mechanism and camshaft offers less friction loss resulting in higher mechanical efficiency. |
| 7. Initial cost | The heavy weight of the engine and complicated design result in higher initial cost. | The light weight of the engine and simple design results in lower initial cost. |
| 8. Volumetric efficiency | Greater time of induction of charge results in better volumetric efficiency. | Lesser time of induction of charge results in poor volumetric efficiency. |
| 9. Thermal efficiency | Higher thermal efficiency and better part-load efficiency are possible. | Poor thermal efficiency and lesser part-load efficiency are possible. |
| 10. Foundation requirement | It occupies more space, more parts and strong foundation. | It occupies less space, less parts and lighter foundation. |
| 11. Application | Used in cars, buses, trucks, tractors, rail, aeroplanes, power generation, etc. | Used in scooters, motor cycles, mopeds, lawn movers, ships, etc. |

## 1.13 APPLICATIONS OF IC ENGINES

The applications of IC engines are very wide. These cover the following:

(a) Transport (land, sea and air) such as passenger cars, buses, trucks, diesel locomotive marine, ships, motorcycles, scooters, mopeds aeroplanes, etc., and agriculture farming

(b) Industries

(c) Prime movers for electric generators and source of drives for earth moving equipment

All these above **applications** are met by either four-stroke or two-stroke SI/CI engines. Table 1.7 gives the details of application and Table 1.8 gives typical design and performance data of IC engines. Table 1.9 gives **specification of engines** used in some passenger cars.

## 1.14 ENERGY FLOW THROUGH IC ENGINE: ENERGY BALANCE

Energy flow through IC engine helps in locating the energy input, output and various losses resulting in calculating the energy balance. Figure 1.19 shows a schematic diagram of this flow through a reciprocating type IC engine.

**Table 1.7** Applications of IC Engines

| *S. No.* | *Application* | *Approx. Engine Power kW, cc.* | *Type SI or CI* | *Number of Strokes 2, 4* | *Method of cooling* |
|---|---|---|---|---|---|
| 1. | **Road Vehicles** | | | | |
| | (i) Scooters, motorcycles, mopeds | 1–9 kW, 50–250 cc | SI | 2, 4 | Air |
| | —Bajaj Chetak scooter 5300 rpm | 5 kW, 150 cc | SI | 2 | Air |
| | —Yamaha motorcycle | 4 kW, 100 cc | SI | 2 | Air |
| | —Luna moped | 1.5 kW, 50 cc | SI | 2 | Air |
| | (ii) Small passenger car | 15–75 kW 800–2000 cc | SI | 4 | Water/special coolant |
| | —Fiat car 4 cylinders, 5000 rpm | 32 kW 1089 cc | SI | 4 | Special coolant |
| | —Maruti car 3 to 4 cylinders | 800 to 1500 cc | SI | 4 | Special coolant |
| | (iii) Large passenger car | 75–200 kW | SI CI | 4 | Special coolant |
| | Six or eight cylinders Mercedes, BMW, Ceilo, Peugeot Opel Astra, Rover, Toyata, etc. | 2500–4000 cc | CI | 4 | |
| | (iv) Light commercial, | 30–150 kW | SI/CI | 4 | Water |
| | Jeeps | 55 kW | CI | 4 | Water |
| | (v) Heavy commerical trucks, buses | 120–400 kW | CI | 2, 4 | Water |
| 2. | **Railway Locomotives** | 500–6000 kW | CI | 2, 4 | Water |
| 3. | **Small aircraft** | | | | |
| | (i) Helicopters | 50–1500 kW | SI | 4 | Air |
| | (ii) Airplanes | 50–2500 kW | SI | 4 | Air |
| 4. | **Marine** | | | | |
| | (i) Motor boats | 1–80 kW | SI, CI | 4 | Water |
| | (ii) Ships and submarines | 3000–40000 kW | CI | 2, 4 | Water |
| | —Nordberg 2-stroke, 12 cylinder, 80 cm, bore 155 cm stroke, 120 rpm engine | 20,000 kW | CI | 2 | Water |

*Contd..*

*Table 1.7 Contd..*

| 5. | **Off-road vehicles** | | | | |
|---|---|---|---|---|---|
| | (i) Light vehicles factory, airport | 2–15 kW | SI | 2, 4 | Air/water |
| | (ii) Agriculture | 3–150 kW | SI, CI | 2, 4 | Air/water |
| | —Tractors | 15–40 kW | CI | 4 | Air |
| | (iii) Earth moving equipment (Excavators, Backhoe, Loaders, Bulldozers, Skid-Steer loaders, Trenchers) | 40–800 kW<br>50–2000 kW | CI<br>CI | 2, 4<br>2, 4 | Air/water<br>Air/water |
| | (iv) Military (Tanks, etc.) | 1400–1500 kW | | | |
| | (v) Others (Construction m/c, compressors drilling rigs) | 5–100 kW | CI | 2, 4 | Air/water |
| 6. | **Industry** | | | | |
| | (i) Electric power | 40–25,000 kW | CI | 2, 4 | Water |
| | (ii) Gas pipeline | 800–5000 kW | CI | 2, 4 | Water |
| | (iii) Others | 5–400 kW | CI | 2, 4 | Water |
| 7. | **Household** | | | | |
| | (i) Lawn mowers | 0.5–3 kW | SI | 4 | Air |
| | (ii) Small generating set | 2–10 kW | SI | 4 | Air |

**Table 1.8** Typical Design and Performance Data of IC Engines

| *Engine type* | *Operating cycle stroke* | *Compressor ratio* | *Bore (cm)* | *Stroke/bore ratio* | *Rated maximum speed (atm)* | *Rated maximum bmep (bar)* | *Weight/ power ratio (kg/kW)* | *Best bsfc (kg/kWh)* |
|---|---|---|---|---|---|---|---|---|
| **A. SI Engines** | | | | | | | | |
| (i) Small (scooters & motorcycles) | 2/4 | 6–10 | 5–8.5 | 0.9–1.2 | 4000–8500 | 4–10 | 2.5–5.5 | 0.35 |
| (ii) Passenger cars | 4 | 8–10 | 7–10 | 0.9–1.1 | 4000–6500 | 7–10 | 2–4 | 0.27 |
| (iii) Large gas engines | 2/4 | 8–12 | 22–45 | 1.1–1.4 | 300–900 | 7–12 | 23–35 | 0.2 |
| (iv) Wankel engine | 4 | 9 | – | – | 6000–8000 | 0.9–10.5 | 0.9–1.6 | 0.3 |
| **B. CI Engines** | | | | | | | | |
| (i) Passenger cars | 4 | 16–20 | 7.5–10 | 0.9–1.2 | 4000–5000 | 5–7.5 | 2.5–5 | 0.25 |
| (ii) Trucks/Buses | 4 | 16–20 | 10–15 | 0.8–1.3 | 2000–4000 | 6–9 | 4–7 | 0.21 |
| (iii) Locomotives | 4/2 | 16–18 | 15–40 | 1.1–1.3 | 400–1800 | 7–23 | 6–18 | 0.19 |
| (iv) Large engines (marine) | 2 | 10–12 | 40–100 | 1.2–3 | 100–400 | 9–17 | 12–50 | 0.18 |

**Table 1.9** Engine Specification of Various Car Models (Subject to Updating)

| *S. No.* | *Specification* | *Car Models* | | | |
|---|---|---|---|---|---|
| | | *Maruti Alto (Petrol)* | *Santro Xing (Petrol)* | *Accent (Diesel)* | *Getz (Petrol)* |
| 1 | Swept volume (CC) | 796 CC | 1086 CC | 1493 CC | 1341 CC |
| 2 | Engine type | FC engines | Hyundai Epsilon engine | CRDi | SOHC engine |
| 3 | No. of cylinders | 3 | 4 | 3 | 4 |
| 4 | Valves per cylinder | 4 | 3 | 4, SOHC | 4, SOHC |
| 5 | Ignition/fuel system | MPFI per cylinder | MPFI, distributor less | Direct injection common rail | MPFI |
| 6 | Engine control | 32 bit computer | 16 bit computer | Electronically controlled injection timing | 16 bit |
| 7 | Maximum power | 47 PS ≤ (a) 6200 rpm | 63 PS/5500 rpm | 82 PS/4000 rpm | 82.9 PS/5500 rpm |
| 8 | Maximum torque | 62 Nm (a) 3000 rpm | 96 Nm/3000 rpm | 187 Nm/2000-2700 rpm | 115 Nm/3200 rpm |
| 9 | Transmission | 5 speeds, all synchromesh/ manual | 5 speeds, manual/ automatic | 5 speeds, manual OD | 5 speeds, manual |
| 10 | Kerls weight | 720 kg | 854 (M/T), 868 (A/T) | – | – |
| 11 | Gross vehicle weight | 1165 kg | – | – | – |
| 12 | Capacity | 5 persons | 5 persons | 5 persons | 5 persons |
| 13 | Overall length | 3495 mm | 3565 mm | 4250 mm | – |
| 14 | Overall width | 1475 mm | 1525 mm | 1670 mm | – |
| 15 | Power to CC ratio | 59.05 | 58.01 | – | – |
| 16 | Emission norm | BS III | EURO. III | EURO. III | – |
| 17 | Power steering | Electronic | Hydraulic | Rack & pinion (powered) | – |
| 18 | Ground clearance | 160 mm | – | 172 mm | – |
| 19 | Comfort | AC & heater | AC & heater | AC & heater | AC & heater |
| 20 | Fuel tank capacity | 28 litre | 35 litre | 45 litre | 45 litre |

The fuel is supplied to the combustion chamber ($A$), where it is burnt and chemical energy of fuel is converted into heat energy. It would be desirable that all the liberated energy should drive the piston but there are various heat losses, namely:

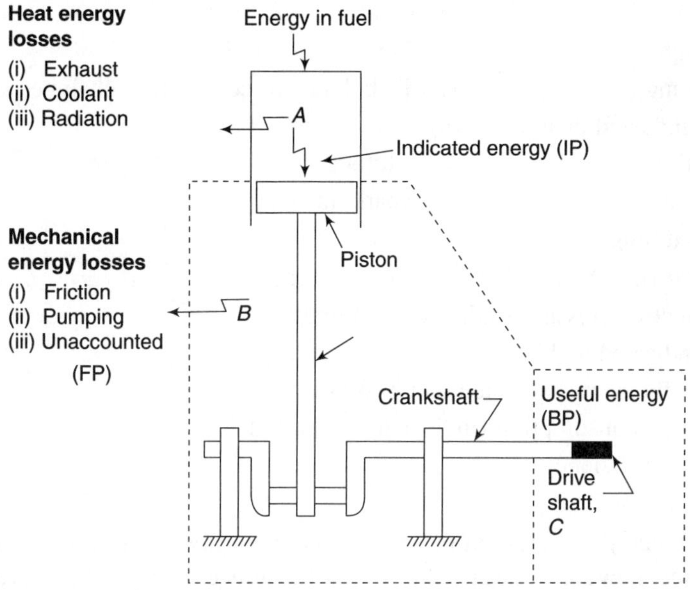


**Fig. 1.19** Energy flow through reciprocating internal combustion engine

(i) Losses to the exhaust (ii) Losses to the coolant (iii) Losses to radiation

which reduce the power developed inside the engine cylinder. The remaining energy known as *indicated power* (*ip*) is utilised to drive the piston. The ratio of indicated power to the energy input by fuel is known as *indicated thermal efficiency* ($\eta_{ith}$).

The energy applied to the piston passes through the connecting rod and crankshaft (area *B*) and finally to the drive shaft (area *C*). Again, there are various mechanical losses due to friction, pumping and other causes, sum of these, converted to power is termed *friction power* (*fp*). The remaining energy (delivered to area *C*) available to the drive shaft is the mechanical energy which can be utilised to drive the vehicles/pumps compressors/generators, etc. This energy is called *brake power* (bp). The ratio of brake power to indicate power is known as *mechanical efficiency,* $\eta_m$.

A general visualization of this overall picture of energy flow helps in understanding the subject matter. There has been a constant endeavour by engineers to increase the efficiency by reducing losses.

## 1.15 BASIC PERFORMANCE PARAMETERS

Basic performance parameters of IC engines are classified in two categories, namely:

(i) Dependent variables

(ii) Independent variables

The dependent variables are functions of independent variables. The ***independent variables*** include the following:

(a) Speed
(b) Load
(c) Crank angle
(d) Compression ratio
(e) Valve timing
(f) Turbulence in combustion chamber
(g) Mass of inducted charge
(h) Heat losses
(i) Dissociation
(j) Altitude
(k) Air-fuel ratio
(l) Spark advance
(m) Injection timing

The ***dependent variables*** includes the following:

(a) Mean effective pressure (indicated and brake)
(b) Power (indicated and brake)
(c) Thermal efficiency (indicated and brake)
(d) Specific fuel consumption (indicated and brake)
(e) Mechanical efficiency
(f) Volumetric efficiency

Though, a detailed discussion on these parameters will be dealt in the chapter of **Testing and Performance**, however a brief definition of the above parameters is given below.

**(i) Mean effective pressure ($p_m$):** The mean effectiveness pressure is a useful parameter in the study of the reciprocating devices that are used for power production. *It is defined as average pressure, if acted, on the piston during the entire power or outward stroke, would produce the same work output as the net work output for the actual cyclic process.* Thus,

$$\text{Work done per cycle} = p_m \times \text{Piston area} \times \text{Stroke}$$

or
$$W_{\text{cycle}} = p_m.A.L = p_m.V_s$$

or
$$p_m = \frac{\text{Work done per cycle}}{\text{Stroke volume}} = \frac{W_{\text{cycle}}}{V_s} \qquad ...(1.4)$$

The mean effective pressure may be indicated ($p_{mi}$) or brake ($p_{mb}$).

**(ii) Indicated power (*ip*):**

$$ip = \frac{p_{mi}\, A\, L\, n\, x}{60} \times 10^2 \text{ ; kW} \qquad ...(1.5)$$

where $p_{mi}$ in bar, $A$ in m$^2$, $L$ is in metre and $n$ is cycles per minute which is

$n$ = N/2 for four strokes, $x$ = number of cylinders

$n$ = N for two strokes

*It is to be noted that for reciprocating engines of comparable size, a large $p_{mi}$ is an indication of better performance in terms of power produced at the same rated speed.*

**(iii) Brake power (*bp*):**

$$bp = \frac{p_{mb}\, A\, L\, n}{60} \times 10^2 \; ; \quad \text{kW} \qquad ...(1.6)$$

**(iv) Air-fuel ratio (A/F):** *It is the ratio of mass of air to the mass of fuel consumed per unit time by engine.* Thus,

$$A/F = \frac{\text{Mass of air per unit time}}{\text{Mass of fuel per unit time}} = \frac{\dot{m}_a}{\dot{m}_f} \qquad ...(1.7)$$

A mixture that contains just sufficient air to support complete combustion of all the fuel is called a chemically correct or *stoichiometric air-fuel ratio.*

**(v) Indicated thermal efficiency ($\eta_{ith}$):** *It is the ratio of energy in the indicated power to the energy supplied by fuel.* Thus,

$$\eta_{ith} = \frac{\text{Energy equivalent } ip}{\text{Energy supplied by fuel}} = \frac{i.p}{\dot{m}_f \times q_{LHV}} \qquad ...(1.8)$$

where $\dot{m}_f$ = mass of fuel (kg/s), $q_{LHV}$ = lower calorific value of fuel (kJ/kg of fuel)

**(vi) Brake thermal efficiency ($\eta_{ith}$):** *It is the ratio of energy in the brake power to the energy supplied by the fuel.* Thus,

$$\eta_{bth} = \frac{\text{Energy equivalent } bp}{\text{Energy supplied by fuel}} = \frac{bp}{\dot{m}_f \times q_{LHV}} \qquad ...(1.9)$$

**(vii) Relative efficiency or efficiency ratio ($\eta_{rel}$):** *It is the ratio of actual thermal efficiency to the air standard efficiency.* Thus,

$$\eta_{rel} = \frac{\text{Actual thermal efficiency}}{\text{Air standard efficiency}} = \frac{\eta_{tha}}{\eta_{\text{air stand}}} \qquad ...(1.10)$$

**(viii) Mechanical efficiency ($\eta_m$):** It is the ratio of brake power to indicated power. Thus,

$$\eta_m = \frac{bp}{ip} \qquad ...(1.11)$$

Also $\quad fp = ip - bp \qquad ...(1.12)$

**(ix) Volumetric efficiency ($\eta_{vol}$):** It is related to breathing of engine. *It is the ratio of air actually induced at ambient conditions to the swept volume of the engine.* Thus,

$$\eta_{vol} = \frac{\text{Air actually induced at ambient condition}}{\text{Swept volume of the engine}} = \frac{V}{V_s} \qquad ...(1.13)$$

**(x) Specific fuel consumption (*sfc*).** *It is the ratio of fuel consumption expressed in kg/kWh.* Thus, brake and indicated *sfc* are given by

$$bsfc = \frac{\dot{m}_f \text{ in kg/h}}{bp \text{ in kW}} \quad ...(1.14)$$

$$isfc = \frac{\dot{m}_f \text{ in kg/h}}{ip \text{ in kW}} \quad ...(1.15)$$

## 1.16 VARIOUS AUXILIARY SYSTEMS FOR IC ENGINES

Apart from the piston-cylinder, connecting rod, crankshaft, valve mechanism, flywheel arrangement, there are various auxiliary systems which are used for optimum, safe and reliable operation of IC engines. These are as follows:

(i) Carburetion system/gasoline injection system (MPFI system for SI only)
(ii) CI injection system
(iii) Ignition system (for SI only)
(iv) Cooling system
(v) Lubricating system
(vi) Governing system
(vii) Starting system
(viii) Intake and exhaust system
(ix) Electrical system
(x) Air pollution control system

## 1.17 ENVIRONMENT FRIENDLY (GREEN TECHNOLOGY) ENGINES

Since internal combustion engines pollute the environment due to emission of CO, $CO_2$, $NO_x$, etc., now environmental friendly car engines are coming up. These are:

(a) Battery operated electric car (see chapter 22)
(b) Solar cells operated electric car
(c) Fuel cells operated electric car (see chapter 22)
(d) Blended fuel vehicles
(e) Dual fuel vehicles
(f) Micro and mild hybrid vehicles
(g) Bio-fuel vehicles
(h) Hydrogen fuel based vehicles
(i) Hybrid electric and fuel cell vehicles
(j) Hybrid electric and gasoline vehicles

These car engines have many limitations but researches are going on to overcome these. All these cars, although very less in number, are running in many parts of the world.

Figure 1.20 shows a hybrid electric-gasoline vehicles arrangement.

**Tata motors** of India and other companies of the world have announced to manufacture **green technology vehicles** (passenger cars, pick-ups, buses and trucks). Many such cars are plying on roads.

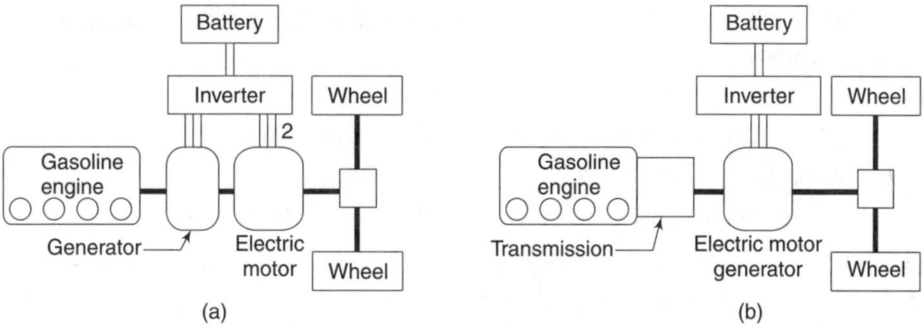


**Fig. 1.20** Hybrid electric-gasoline vehicle

## 1.18 THERMOCHEMISTRY

The power output, efficiency and the exhaust emission of an I.C engine depend upon the quality of combustion of a fuel-air mixture inside the engine cylinder. In this chapter, we shall review the thermodynamics of reactive mixture which may also termed **thermochemistry**.

A *conventional (fossil) fuel may be defined as a substance which once raised to ignition temperature, it will continue to burn without any external support provided sufficient quantity of oxygen is available for the combustion. Most the fuels belong to hydrocarbon family of which the major constituents are carbon and hydrocarbon.* The fuels may either be (a) solid, (b) liquid or (c) gas. These may be natural or artificial as discussed in the chapter "Fuels".

Before we define the term combustion, let us first understand what is chemical reaction. *A chemical reaction may be defined as a process in which the inter-atomic bonds of the reactant molecules are broken and a re-arrangement of atoms in a new molecular combination takes place. A chemical reaction which releases heat is called **exothermic** and the other which absorbs heat is known as **endothermic**.*

*Combustion means burning* and it involves a chemical reaction which releases heat. *Hence, combustion of fuel is an exothermic reaction.* In order to understand the combustion process fully, we have to turn to chemistry.

### 1.18.1 Complete Combustion Equation (or Stoichiometric Equation)

In a combustion process, when the fuel is completely oxidized, the combustion is said to be complete. For example, if a fuel contains carbon and hydrogen, complete combustion means, all carbon is burnt to carbon dioxide and all hydrogen, is reduced to water.

The combustion reaction is written in the form of a chemical reaction expressing the conservation of mass in terms of the conservation of atoms. If we consider the complete combustion of carbon fuel with oxygen, it is written as

$$C + O_2 \rightarrow CO_2 \qquad ...(1.16)$$

This balanced equation for the complete combustion of fuel such as carbon is called ***stoichiometric equation***. It is obvious from the Eq. (1.16) that the number of carbon and oxygen atoms are the same before and after the reaction. The following meanings may be drawn from the above equation.

(a) *On molal basis*

$$(1 \text{ mole C}) + (1 \text{ mole } O_2) \rightarrow (1 \text{ mole } CO_2)$$

(b) *On volume basis*

$$(1 \text{ volume C}) + (1 \text{ volume } O_2) \rightarrow (1 \text{ volume } CO_2)$$

The volume (or mole) of the carbon solid fuel is negligible with the volume of air. Hence, 1 volume of $O_2$ with carbon gives 1 volume of $CO_2$. Similarly, 1 mole of $O_2$ with 1 mole of carbon gives 1 mole of $CO_2$. Thus, we see that there is no molecular expansion of contraction.

(c) *On mass basis*

$$(12 \text{ kg C}) + (32 \text{ kg } O_2) \rightarrow (44 \text{ kg } CO_2)$$

or

$$1 \text{ kg C} + 8/3 \text{ kg } O_2 \rightarrow 11/3 \text{ kg } CO_2$$

This shows that the total mass of the reactants is equal to the total mass of the products.

Though, we have seen that the number of moles (or volumes) of the reactants is the same as products but this is not always true. For example, consider the combustion of hydrogen with oxygen

$$2H_2 + O_2 \rightarrow 2H_2O$$

In this case, 3 moles of reactants give only 2 moles of the products.

In general, the stoichiometric equation may be written as

$$n_A.A + n_B.B + \ldots \rightarrow n_L.L + n_M.M \qquad \text{...(1.17)}$$

where the upper case letters *A*, *B* (not the suffixes), represent the chemical species of the reactants and *L*, *M*, (not suffixes), etc., represent the chemical species of the products. The $n_i$ terms are known as the stoichiometric coefficients of the various species.

In the combustion of carbon, $n_A = 1$, $n_B = 1$ and $n_L = 1$, $M = 0$

In an incomplete combustion of carbon in a fuel, the carbon reacts according to the reaction

$$C + \frac{1}{2}O_2 = CO$$

In actual practice, there is usually CO in the products, even through an excess of oxygen was supplied. This may be attributed either to incomplete mixing during the process or to insufficient time for complete combustion.

In industrial practice, the combustion of fuels is normally accomplished by applying air is the oxidizer rather than pure oxygen. The molecular weight of air is 28.967. For computation work it is sufficiently accurate to take $M_{air} = 29$.

The composition of air may be taken as follows:

*Composition on mass basis:* $O_2 = 23.2\%$ and $N_2 = 76.85\%$

∴ For each kg of $O_2$ supplied by air, we have

$$76.85/23.2 = 3.31 \text{ kg of } N_2 \text{ and } 100/23.2 = 4.31 \text{ kg of air}$$

*Composition on volume basis:* $O_2 = 21\%$, $N_2 = 79\%$

∴ For each mole of $O_2$, we have

$$79/21 = 3.76 \text{ moles of } N_2 \text{ and } 100/21 = 4.76 \text{ moles of air.}$$

We can write this as

$$1 \text{ kg of } O_2 + 3.31 \text{ kg of } N_2 = 4.31 \text{ kg of air}$$

and

$$1 \text{ mole of } O_2 + 4.76 \text{ mole of } N_2 = 5.76 \text{ moles of air}$$

Although, nitrogen of air does not contribute to the oxidation reactions, it leaves the reacting system at the same temperature from its original energy entrance. This factor is taken into consideration in the energy balance of the combustion process.

*The minimum amount of air required for the complete oxidation of a fuel is called the* ***theoretical*** *or* ***stoichiometric air***. For example, the chemical equation for the stoichiometric reaction of propane ($C_3H_8$) with air is

$$C_3H_8 + 5O_2 + 18.8\,N_2 = 3CO_2 + 4H_2O + 18.8\,N_2 \qquad ...(1.18)$$

This equation indicates that a minimum of 5 moles of $O_2$ accompanied by 18.8 moles of $N_2$ are necessary for complete oxidization of 1 mole of $C_3H_8$ to form 3 moles of $CO_2$ and 4 moles of $H_2O$.

The amount of air actually supplied may exceed or may be less than the theoretical amount required. If excess of air is used, oxygen will appear in the products of combustion, but if insufficient air is used, the combustion will be incomplete and the products such as CO, CH, $H_2$ and unburnt fuel can also appear. In the previous example, when 150 per cent theoretical air or 50% excess air was supplied, the theoretical combustion reaction was

$$C_3H_8 + 7.5O_2 + 28.2\,N_2 \rightarrow 3CO_2 + 4H_2O + 2.5O_2 + 28.2\,N_2$$

*In actual practice, some excess air is generally supplied in order to ensure that no unburnt fuel is contained in the products.* But at the same time any large amount of excess air is also avoided because the additional oxygen and nitrogen absorb energy from the system, and so lower the temperature of the combustion products resulting in the lowering of the rate of heat transfer from the combustion products.

### 1.18.2 Air-Fuel Ratio and Equivalence Ratio

*Air-fuel ratio is defined as the ratio of the mass of the air to the mass of fuel.* The reciprocal to the above definition is the fuel-air ratio. Hence,

$$\text{A/F} = \text{Air-fuel ratio} = \frac{\text{Mass of air}}{\text{Mass of fuel}} = \frac{\dot{m}_a}{\dot{m}_f} \qquad ...(1.19)$$

$$F/A = \text{Fuel-air ratio} = \frac{\text{Mass of fuel}}{\text{Mass of air}} = \frac{\dot{m}_f}{\dot{m}_a} \qquad ...(1.20)$$

For example, in a 100% theoretical combustion of propane, the chemical Eq. (1.18) shows that 23.8 moles of air (5 moles of $O_2$ + 18.8 moles of $N_2$) are required per mole of fuel. Therefore, the air-fuel for theoretical combustion of this fuel is

$$A/F = \frac{\text{Mass of } (5 + 18.8) \text{ moles of air}}{\text{Mass of 1 mole of } C_3\,H_8} = \frac{(5 + 18.8) \times 29}{44} = 15.686$$

Also,

$$A/F = \frac{\text{Mass of 7.5 moles of } O_2 + \text{Mass of 28.2 mole of } N_2}{\text{Mass of 1 mole of } C_3H_8}$$

$$= (5 \times 32 + 18.8 \times 28)/44 = 15.6$$

The fuel-air ratio for the same combustion process will be 0.0641 kg fuel/kg of air. *It is interesting to know that many hydrocarbon fuels obtained from petroleum or natural gas require an air-fuel ratio in the neighbourhood of 15 to 16 for stoichiomeric combustion.*

***Excess air*** *means the amount of air supplied in excess to that theoretically needed air to burn the fuel completely.* **Lean mixture** means less fuel than stoichiometric and *rich mixture* means more fuel than stoichiometric. Sometimes, a term called **equivalence ratio** is used instead of A/F ratio. It is the ratio of actual fuel ratio to the stoichiometric (i.e., theoretical) A/F ratio.

### 1.18.3 Enthalpy of Formation

In applying the conservation of energy principle to reacting systems, it is necessary to evaluate their enthalpy, internal energy and entropy.

An enthalpy datum for the study of reaching systems can be established by assigning arbitrarily a value of zero to the enthalpy of the stable elements at a state called the *standard reference state and defined by* $T_{ref}$ = 298.15 K (25°C) and $p_{ref}$ = 1 atm. *Enthalpy of formation is defined as the energy released or absorbed when chemical compound is formed from its elements if the temperature and pressure of the reactants and products are at the reference state. In other words, the enthalpy of a compound at the standard state is the enthalpy of formation, symbolized as* $\bar{h}_f^{\circ}$.

In order to consider the enthalpy of formation concept, refer to a simple combustion chamber as shown in Fig. 1.21 in which carbon and oxygen each enter at $T_{ref}$ (25°C) and $p_{ref}$ (1 atm) and react completely at steady state to form carbon dioxide at the same temperature and pressure. The reaction is as follows

$$C + O_2 \rightarrow CO_2$$

This reaction would be exothermic and there would be heat transfer ($\dot{Q}_{cv}$) from the combustion chamber to its surroundings.

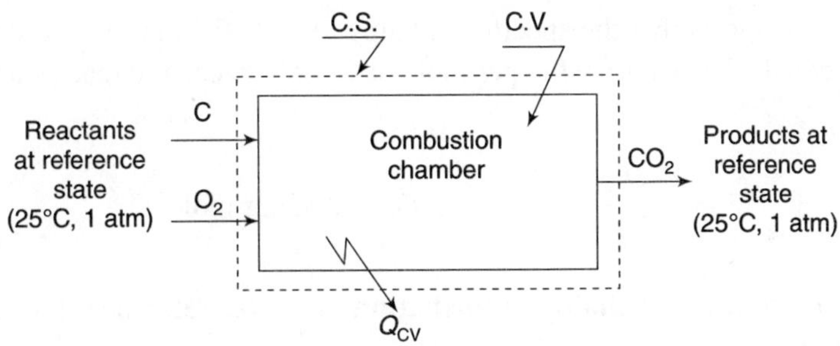


**Fig. 1.21** Enthalpy of formation

The measured value of heat transfer is –393,522 kJ/kg mole of $CO_2$ formed. If $H_r$ and $H_p$ refer to the total enthalpy of reactants and products respectively, then applying the first law to the reaction gives

$$H_r + Q_{cv} = H_p \qquad ...(1.21)$$

For all the reactants and products in a reaction, the above equation may be written in *rate form* based on molar flow rate as

$$\sum_r \dot{n}_i \bar{h}_i + \dot{Q}_{cv} = \sum_p \dot{n}_e \bar{h}_e \qquad ...(1.22)$$

where $\dot{n}$ and $\bar{h}$ represent the molar flow rate and enthalpy per mole, respectively. The suffixes $r$ and $p$ represent reactants and products, respectively. The suffixes $i$ and $e$ denote inlet and exit, respectively.

For enthalpies on mass rate basis, Eq. (1.21) is written as

$$\sum_r \dot{m}_i h_i + \dot{Q}_{cv} = \sum_p \dot{m}_e h_e \qquad ...(1.23)$$

where $\dot{m}$ and $h$ denote respectively mass flow rate and specific enthalpy.

For the above chemical Eq. (1.22) takes the form as

$$\dot{n}_C \bar{h}_C + \dot{n}_{O_2} \bar{h}_{O_2} + \dot{Q}_{cv} = \dot{n}_{CO_2} \bar{h}_{CO_2}$$

or

$$\bar{h}_{(CO_2)} = \frac{\dot{Q}_{cv}}{\dot{n}_{CO_2}} + \frac{\dot{n}_C}{\dot{n}_{CO_2}} \bar{h}_C + \frac{\dot{n}_{O_2}}{\dot{n}_{CO_2}} \bar{h}_{O_2} \qquad ...(1.24)$$

Since carbon and oxygen are stable elements at the reference state

$$\bar{h}_C = \bar{h}_{O_2} = h_r = 0,$$

So,

$$\bar{h}_{CO_2} = \bar{h}_p = \frac{\dot{Q}_{cv}}{\dot{n}_{CO_2}} = \frac{\dot{Q}_{cv}}{\dot{n}_p} \qquad ...(1.25)$$

Equation 1.25 suggests that the specific enthalpy of $CO_2$ (i.e., products) at the standard or reference state equals the heat transfer, per mole of $CO_2$, between the reactor and surrounding. Thus,

$$\frac{\dot{Q}_{cv}}{\dot{n}_{CO_2}} = \frac{\dot{Q}_{cv}}{\dot{n}_p} = -\ 393,\ 522 \text{ kJ/kg mole}$$

This is what is known as **enthalpy of formation** of $CO_2$ at reference state 25°C and 1 atm and designated by the symbol $\bar{h}_f^0$. Thus,

$$\left(\bar{h}_f^0\right)_{CO_2} = \frac{\dot{Q}_{cv}}{\dot{n}_{CO_2}} = \frac{\dot{Q}_{cv}}{\dot{n}_p} = -\ \mathbf{393,\ 522\ kJ/kg\ mole}$$

The sign convention is related to the sign convention of heat transfer. If there is heat transfer from the combustion chamber (reactor) in which compound is formed from its element (an exothermic reaction, $\bar{h}_f^0$ has a negative sign. If the heat transfer to the reactor is required (an endothermic reaction), $\bar{h}_f^0$ is positive.

In real situations, the reactants and products are not at standard state. The *specific enthalpy of a compound at a state other than the standard state is found by adding specific enthalpy change* ($\overline{\Delta h}$) between the standard state and the state of interest to the enthalpy formation. Thus,

$$\bar{h}(T, p) = \bar{h}_f^0 + [\bar{h}(T, p) - \bar{h}(T_{ref}, p_{ref})] = \bar{h}_f^0 + \overline{\Delta h} \qquad ...(1.26)$$

An arbitrary choice of datum can be used to determine $\overline{\Delta h}$, since it is a difference at constant composition.

Table $D_1$ of Appendices gives the value of enthalpy of formation ($\bar{h}_f^0$) for a number of substances. The values of $\overline{\Delta h}$ is given in Table $D_2$.

### 1.18.4 First Law Applied to Reactive System

#### *1.18.4.1 Open System (Steady Flow Process)*

Consider a reaction system for steady flow process as shown in Fig. 1.22. From the First Law of Thermodynamics, we have

$$H_r + Q_{cv} = H_p + W_{cv} \qquad ...(1.27)$$

Equation (1.27) may be written in rate form on molar basis.

$$\sum_r \dot{n}_i \bar{h}_i + \dot{Q}_{cv} = \sum_p \dot{m}_e \bar{h}_e + \dot{W}_{cv} \qquad ...(1.28)$$

On mass rate basis, Eq. (1.27) may be written as

$$\sum_r \dot{m}_1 h_i + \dot{Q}_{cv} = \sum_p \dot{m}_e h_e + \dot{W}_{cv} \qquad ...(1.29)$$

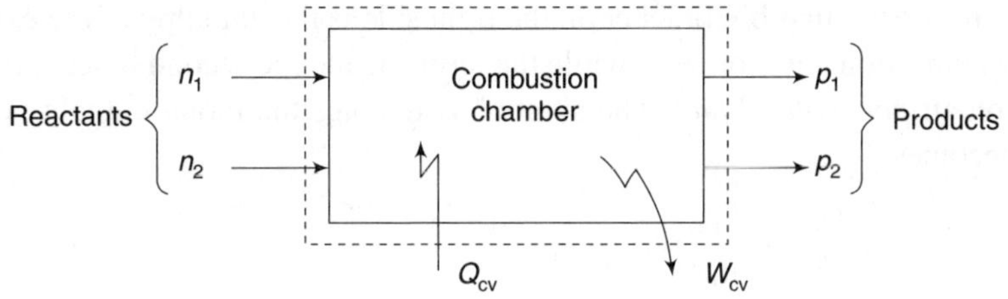


**Fig. 1.22** First law for reactive system

### *1.18.4.2 Combustion of Hydrocarbon under Steady State*

Consider a steady state reactor (combustion chamber) shown in Fig. 1.23 in which a hydrocarbon fuel $C_aH_b$ burns completely with the theoretical amount of air. The corresponding reaction is expressed as

$$C_aH_b + \left(a + \frac{b}{4}\right)(O_2 + 3.76\ N_2) \rightarrow a\ CO_2 + \frac{b}{2}\ H_2O + \left(a + \frac{b}{4}\right) 3.76\ N_2 \qquad ...(1.30)$$

Assume that reactants and products modelled as ideal gas and changes in kinetic and potential energies are ignored. The energy equation per mole of fuel basis is given by

$$\frac{\dot{Q}_{cv}}{\dot{n}_f} - \frac{\dot{W}_{cv}}{\dot{n}_F} = \left[a\ \bar{h}_{CO_2} + \frac{b}{2}\bar{h}_{O_2} + \left(a + \frac{b}{4}\right) 3.76\ \bar{h}_{N_2}\right]$$

$$\dot{n}_F \cdot \bar{h}_F - \left[\left(a + \frac{b}{4}\right)\bar{h}_{O_2} + \left(a + \frac{b}{4}\right) 3.76\ \bar{h}_{N_2}\right] \qquad ...(1.31)$$

where $\dot{n}_F$ represents the molar flow rate of the *fuel*.

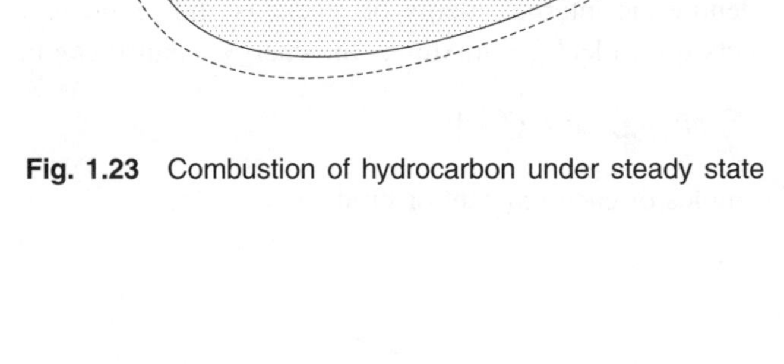


**Fig. 1.23** Combustion of hydrocarbon under steady state

The terms inside the first big bracket on the right side is the enthalpy of the exiting gaseous products of combustion per mole fuel while the terms inside the second bracket is the enthalpy of combustion air *per mole of fuel*. The symbol $\overline{h}_F$ denotes the molar enthalpy of fuel. Thus, Eq. (1.31) becomes

$$\frac{\dot{Q}_{cv}}{\dot{n}_F} - \frac{\dot{W}_{cv}}{\dot{n}_F} = \overline{h_p} - \overline{h_r} \qquad ...(1.32)$$

where $\overline{h}_p$ and $\overline{h}_r$ represent, respectively, the enthalpies of products and reactants.

Since each components of combustion products assumed to behave as an ideal gas, its contribution to enthalpy of the products depends solely on the temperature of the products $T_p$. Thus,

$$\overline{h} = \overline{h}^0_f + [(\overline{h}(T_p) - \overline{h}(T_{ref})] \qquad ...(1.33)$$

In Eq. (1.33), $\overline{h}^0_f$ is the enthalpy of formation. The second terms in bracket account for the change in enthalpy from the temperature $T_{ref}$ to the temperature $T_p$. It is to be noted that $\overline{h}^0_f = 0$ for $N_2$ and $O_2$. Using Eq. (1.33), the energy Eq. (1.31) for $C_aH_b$ takes the form as

$$\frac{\dot{Q}_{cv}}{\dot{n}_F} - \frac{\dot{W}_{cv}}{\dot{n}_F} = a(\overline{h}^0_f + \overline{\Delta h})_{CO_2} + \frac{b}{2}(\overline{h}^0_f + \overline{\Delta h})_{H_2O} + \left(a + \frac{b}{4}\right) 3.76\, (\overline{\Delta h})_{N_2}$$

$$- \left(\overline{h}^0_f + \overline{\Delta h}\right)_F - \left(a + \frac{b}{4}\right)(\overline{\Delta h})_{O_2} - \left(a + \frac{b}{4}\right) 3.76\, (\overline{\Delta h})_{N_2} \qquad ...(1.34)$$

Equation (1.34) can be written more concisely as

$$\frac{\dot{Q}_{cv}}{\dot{n}_F} - \frac{\dot{W}_{cv}}{\dot{n}_F} = \sum_p n_e \left(\overline{h}^0_f + \overline{\Delta h}\right)_e - \sum_r n_i \left(\overline{h}^0_f + \overline{\Delta h}\right)_i \qquad ...(1.35)$$

where $i$ represents incoming fuel and air streams and $e$ the exiting combustion products. The coefficients $n_i$ and $n_e$ denote the respective coefficients of the reaction equation giving the moles of reactants and products *per mole of fuel*, respectively.

### 1.18.4.3 Closed System

Consider a closed system involving a combustion process. In the absence of kinetic and potential energy effects, we have,

$$U_p - U_r = Q - W \qquad ...(1.36)$$

where $U_r$ and $U_p$ denote the internal energy of reactants and products, respectively. If the reactants and products form ideal gas mixtures, the energy balance can be expressed as

$$\sum_p n\overline{u} - \sum_r n\overline{u} = Q - W \qquad ...(1.37)$$

where $n$ stands for moles of each reactant or product.

Since $u = h - pv = h - RT$, so Eq. (1.37) becomes,

$$Q - W = \sum_p n\,(\bar{h} - \bar{R}\,T_p) - \sum_r n\,(\bar{h} - \bar{R}\,T_r) \qquad ...(1.38)$$

where $T_p$ and $T_r$ represent the temperature of product and reactants, respectively. Equation (1.38) may be expressed involving $\bar{h}_f^0$ as follows

$$Q - W = \sum_p n(\bar{h}_f^0 + \overline{\Delta h} - \bar{R}\,T_p) - \sum_r n\,(\bar{h}_f^0 + \overline{\Delta h} - \bar{R}\,T_r)$$

or

$$Q - W = \sum_p n(\bar{h}_f^0 + \overline{\Delta h}) - \sum_r n(\bar{h}_f^0 + \overline{\Delta h}) - \bar{R}\,T_p \sum_p n + \bar{R}\,T_r \sum_r n \qquad ...(1.39)$$

The above concepts are applicable where gaseous mixtures burn in a closed, rigid container.

### 1.18.5 Enthalpy and Internal Energy of Reaction (Combustion)

The concept of enthalpy of formation underlies the formulations of the energy balances for reactive systems but the enthalpy of formation ($\bar{h}_f^0$) of fuel is not always known owing to the wide variation in composition. In many cases of practical interest, however, the enthalpy of reaction (or combustion), which is accessible experimentally, can be used to conduct an energy analysis when $\bar{h}_f^0$ data are lacking.

*The enthalpy of reaction (combustion) is defined as the difference between enthalpy of products and enthalpy of reactants when complete combustion occurs in a steady flow process at a given temperature and pressure. It is denoted by* $\overline{\Delta h}_R$ or $\bar{h}_{rp}$. Thus,

$$\bar{h}_{rp} = \overline{\Delta h}_R = H_p - H_r = \sum_p n_e\,\bar{h}_e - \sum_r n_i\,\bar{h}_i \qquad ...(1.40)$$

The unit of $\bar{h}_{rp}$ (or $\overline{\Delta h}_R$) is expressed in kJ/kg mole or kJ/kg of fuel. Here the suffix *rp* stands for reactants products while suffix *R* stands for reaction. The term *n* corresponds to the respective coefficients of the reaction giving the moles of reactants and products *per mole of fuel*. The suffixes, *i* and *e*, denote inlet and exit conditions. When the enthalpy of combustion (or reaction) is expressed on a unit mass of fuel basis, it is designated as $h_{rp}$ or $\Delta h_R$. The values of enthalpy of reaction (or combustion) of different hydrocarbon fuels at standard state, 25°C and 1 atm designated as $\bar{h}^0_{rp}$ or $\overline{\Delta h}_R$ are given in Table $D_3$. It is represented on *h–T* diagram as shown in Fig. 1.24.

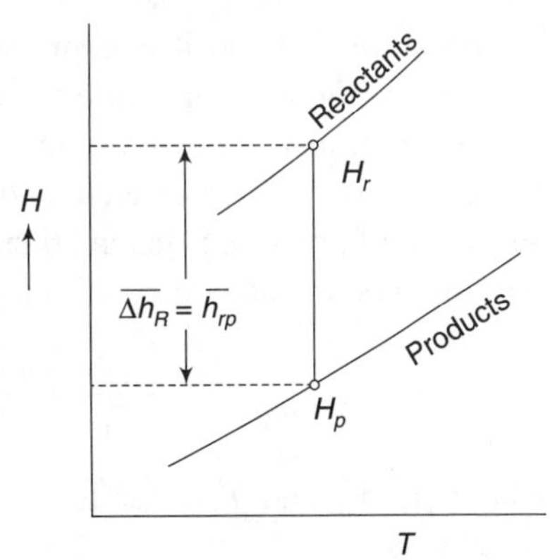


**Fig. 1.24** Representation of $\overline{\Delta H}_R$

Equation (1.35) may be remanaged to read

$$\frac{\dot{Q}_{cv}}{\dot{n}_F} - \frac{\dot{W}_{cv}}{\dot{n}_F} = \left[\sum_p n_e\,(\bar{h}_f^0)_e - \sum_r n_i(\bar{h}_f^0)_i\right] + \sum_p n_e\,(\overline{\Delta h})_e - \sum_r n_i\,(\overline{\Delta h})_i$$

The first two terms of right side inside big bracket is just the enthalpy of combustion at $T_{ref}$ and $p_{ref}$. Thus, the above equation becomes

$$\frac{\dot{Q}_{cv}}{\dot{n}_f} - \frac{\dot{W}_{cv}}{\dot{n}_F} = \bar{h}^0_{rp} + \sum_p n_e(\overline{\Delta h})_e - \sum_r n_i(\overline{\Delta h})_i \quad ...(1.41)$$

The right side of Eq. (1.41) can be evaluated with an experimentally determined value of $\bar{h}^0_{rp}$ and $\overline{\Delta h}$.

*The internal energy of reaction $\bar{u}_{rp}$ is defined as the difference between internal energy of products and reactants in a closed system at a given temperature and pressure.* Thus,

$$\bar{u}_{rp} = \Delta\bar{u}_R = u_p - u_r \quad ...(1.42)$$

$$= \sum_p n_e[\bar{h}_f^0 + \overline{\Delta h} - p\bar{v}]_e - \sum_r n_i[\bar{h}_f^0 + \overline{\Delta h} - p\bar{v}]_i \quad ...(1.43)$$

If all the gaseous constituents are considered ideal gases, and the volume of liquid and solid considered is assumed to be negligible compared to gaseous volume, then

$$\bar{u}_{rp} = \bar{h}_{rp} - \bar{R}T\,(n_{\text{gaseous products}} - n_{\text{gaseous reactants}}) \quad ...(1.44)$$

$\bar{u}_{rp}$ is also generally evaluated at standard state (25°C and 1 atm)

In the case of constant pressure or steady flow process, the *negative of enthalpy of combustion* is frequently called the ***heating value at constant pressure***. Similarly, the *negative value of internal energy of combustion* is frequently called ***heating value at constant volume*** in the case of combustion.

### 1.18.6 Calorific (Heating) Value of Fuel

*The calorific value of fuel is defined as the amount of heat liberated in kJ due to burning of one kg of fuel* and it is expressed in kJ/kg of fuel. It is also called *heat of combustion. In general, the calorific value of gaseous fuel is based on 1 m³ of fuel instead of 1 kg of fuel under standard condition of temperature and pressure.* It is expressed in kJ/m³ of fuel, however, it is also expressed in kJ/kg. The standard temperature is taken as 15°C (288K) and the standard pressure is taken 76 cm of mercury. We therefore, calculate the value of gaseous fuel first under standard condition using the gas equation

$$\frac{p_1V_1}{T_1} = \frac{p_0V_0}{T_0}$$

where $p_1$, $V_1$ and $T_1$ refer to the given conditions and $p_0$, $V_0$ and $T_0$ refer to standard conditions.

Combustion of a fuel may be at constant volume or constant pressure.

**Calorific Value at Constant Pressure.** When the fuel is burnt at constant pressure, the heat liberated per kg of fuel is called the *calorific value at constant pressure.* Thus, calorific value at constant pressure is equal to the enthalpy of reaction $\Delta h_R$ with negative sign showing the liberation of heat

$$(\text{C.V.})_p = -\overline{\Delta h}_R = -\bar{h}_{rp} \quad ...(1.45)$$

**Calorific Value at Constant Volume.** When a fuel is allowed to burn in a closed vessel, the combustion takes place at constant volume and the heat liberated per kg of fuel which is stored as internal energy is *called the calorific value at constant volume.* It is equal to the internal energy of reaction with negative sign showing the liberation of heat.

$$(\text{C.V.})_v = -\Delta\bar{u}_R = -\bar{u}_{rp} \quad ...(1.46)$$

The relationship between calorific value at constant pressure and constant volume is given as follows.

We have seen that the enthalpy of reaction is related to internal energy of reaction as under

$$\Delta h_R = \Delta u_R + \Delta n\, R_0\, T_0 \quad ...(1.47)$$

where $\Delta n$ may be zero, –ve and +ve

But $(\text{C.V.})_p = -\Delta h_R$ and $(\text{C.V.})_v = -\Delta u_R$

or $(\text{C.V.})_p = (\text{C.V.})_v - \Delta n\, R_0 T_0 = (\text{C.V.})_v - \Delta n \times 8314\, T_0$

or $(\text{C.V.})_p = (\text{C.V.})_v - 8.314\, \Delta n\, R_0 T_0$ ...(1.48)

## 1.18.7 Higher and Lower Calorific Values of a Fuel

Most of the fuels contain hydrogen. During the combustion process, the hydrogen combines with oxygen and forms steam (water vapour). *If this vapour is condensed at constant temperature a large amount of additional heat will be released than if it exits complete in the vapour phase.* On account of this, two types of calorific values are defined; the **higher or gross calorific value** *and the* **lower or net calorific value**.

***(i) Higher Calorific Value (H.C.V.):*** *The higher calorific value is defined as the total heat liberated when the water vapour formed by combustion is completely condensed at constant temperature releasing a large amount of additional heat.* The H.C.V. of fuel may be determined accurately by experimental method. The calorific values of elementary fuels at N.T.P. are given below.

| *Fuel* | *Higher calorific value (H.C.V.) kJ/kG* | *Lower calorific value (L.C.V.) kJ/kG* |
|---|---|---|
| C | 33923 | – |
| CO | 10174 | – |
| S | 92947 | – |
| $H_2$ | 144026 | 121417 |

If a solid fuel consists of C kg carbon, H kg of hydrogen, S kg of sulphur, O kg of oxygen in 1 kg of fuel, then higher calorific value will be written as

$$\text{HCV} = \left[33913\ \text{C} + 144026\left(\text{H}_2 - \frac{\text{O}_2}{8}\right) + 92947\ \text{S}\right] \text{kJ/kg} \quad ...(1.49)$$

The quantity $\left(H_2 - \frac{O_2}{8}\right)$ in the Eq. (1.49) represents the quantity of hydrogen available for combustion and that oxygen in the fuel is already wholly in combination with hydrogen. Also, 8 parts of oxygen by weight combine with 1 part of hydrogen by weight to form water. Therefore, hydrogen in combination with oxygen will amount to only $O_2/8$ leaving $(H_2 - O_2/8)$ available for combustion.

**Dulong** suggested the following empirical formula for calculating the higher calorific value of fuel.

$$\text{H.C.V.} = \left[33829\text{ C} + 144026\left(H_2 - \frac{O_2}{8}\right) + 92947\text{ S}\right] \text{kJ/kg} \qquad ...(1.50)$$

***(ii) Lower Calorific Value (L.C.V.):*** *The lower calorific value of fuel may be defined as the net heat liberated when the water formed by combustion exists completely in the vapour phase.* Thus, the lower calorific value is the difference between higher calorific value (H.C.V.) and heat absorbed by hydrogen for formation of steam. In actual practice, this is the net value of heat produced and available for heating purpose.

Now, if we consider the formation of steam by the combination of $H_2$ and $O_2$, we find that 5 kg of hydrogen combines with 32 kg of oxygen ($2H_2 + O_2 = 2H_2O$) to form 36 kg of steam. In other words, 1 kg of hydrogen combines with 8 kg of oxygen to form 9 kg value of hydrogen in fuel weight. Thus, the lower calorific value may be expressed as

$$\text{L.C.V.} = \text{H.C.V.} - m_{H20} \times h_{fg}$$

or

$$\text{L.C.V.} = \text{H.C.V.} - h_{fg}\text{ (9 times the mass of hydrogen in 1 kg of fuel)}$$

where $h_{fg}$ = Latent heat of steam at saturation temperature of 15°C (N.T.P.)

= 2466.1 kJ/kg

$$\therefore \quad \text{L.C.V.} = 33913\text{C} + 144026\left(H_2 - \frac{O_2}{8}\right) + 92947\text{ S} - 9 \times 2466.1\text{ H}_2 \qquad ...(1.51)$$

The H.C.V. and L.C.V. at constant pressure and constant volume are governed by the following formula

$$(\text{L.C.V.})_p = (\text{H.C.V.})_p - m_{H_2O} \times h_{fg} \qquad ...(1.52)$$

$$(\text{L.C.V})_y = (\text{H.C.V.})_v - m_{H_2O}(u_g - u_f) \qquad ...(1.53)$$

### 1.18.8 Adiabatic Combustion (Flame) Temperature

In some engineering applications of reacting system, it is necessary to evaluate the highest possible temperature the products of a reaction can attain. In some case, we sometimes need high temperature products as in the case of rocket propulsion and at other instances, there may be a metallurgical limitation on the temperature that a solid bounding surface can withstand in which case lower temperature has to be managed.

The *maximum temperature is possible only when the reaction is an adiabatic process. Under adiabatic conditions, the temperature attained in a reaction is called the adiabatic combustion (or reaction or flame) temperature.* In actual practice, the combustion temperature is less than the adiabatic combustion temperature because of (i) heat loss, (ii) incomplete combustion and (iii) dissociation of the products into other chemical species.

The adiabatic reaction temperature may be determined from the first law which reduces to

$$\sum_{p} n_e(\bar{h}_f^\circ + \Delta\bar{h}^\circ)_e = \sum_{r} n_i\,(\bar{h}_f^\circ + \Delta\bar{h})_i$$

An iteration procedure is to be adopted to find flame temperature.

## 1.18.9 Thermodynamics Charts

Thermodynamic properties of unburnt and burnt gas mixtures for IC engines have been represented on charts. These charts are very useful for illustrations and provide easy method to solve problems.

### *1.18.9.1 Unburnt Mixture Charts*

Heywood of MIT [73] has prepared charts on the basis of the following assumptions:

(i) The compression process is reversible and adiabatic.

(ii) The fuel exits in vapour phase.

(iii) The mixture composition in homogeneous and no reaction takes place between air and fuel.

(iv) Each constituent behaves as a perfect gas.

Such charts are shown in Fig. 1.25 and Fig. 1.26. These are drawn for $\psi$ and $\phi$, respectively. In these charts $\psi$ and $\phi$ are a function of $T$.

Hence,
$$\psi(t) = \int_{T0}^{T} c_v \cdot \frac{dT}{T}$$

$$\phi(T) = \int_{T0}^{T} c_p \cdot \frac{dT}{T}$$

### *1.18.9.2 Burnt Mixture Charts*

The charts for the products of combustion are designed for the analysis of working fluid during the expansion process. The charts have been prepared under the following assumptions:

(i) Each constituent behaves like an ideal gas

(ii) The mixture is thermodynamic equilibrium at temperature above 1700K

(iii) No reaction takes place below 1700K

(iv) The chemical elements such on $N_2$, $O_2$, $H_2$ are treated as diatomic gases and C as a solid graphite.

Figures 1.27 to 1.29 depict the charts for burnt mixture for $\phi = 0.8$, 1.0 and 1.2 in the form of internal energy vs. entropy for the iso-octane fuel.

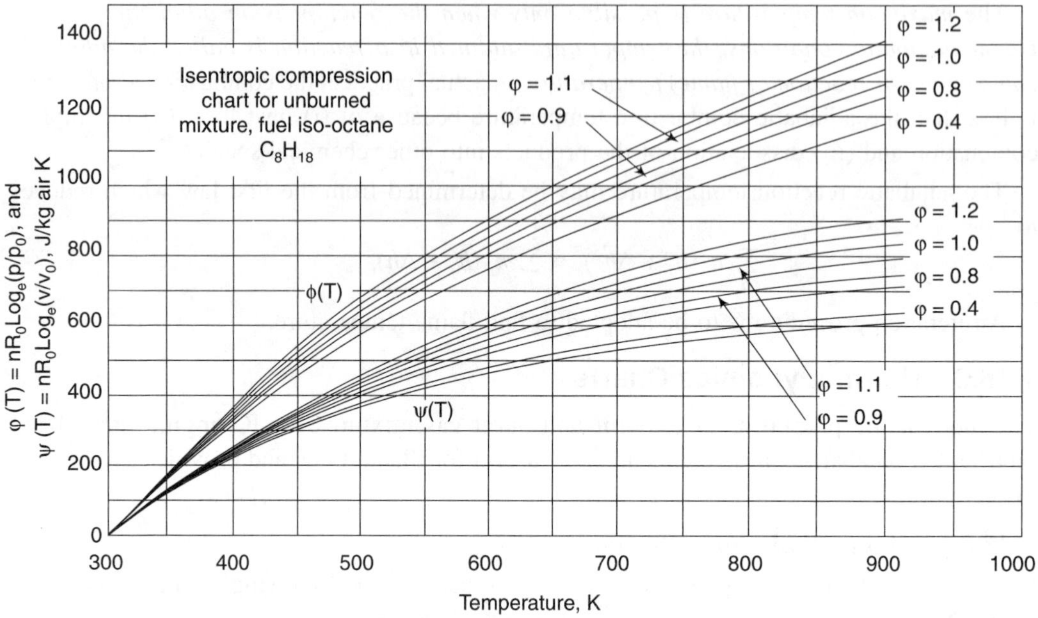


**Fig. 1.25** Isentropic compression function $\phi$ and $\psi$ [73]

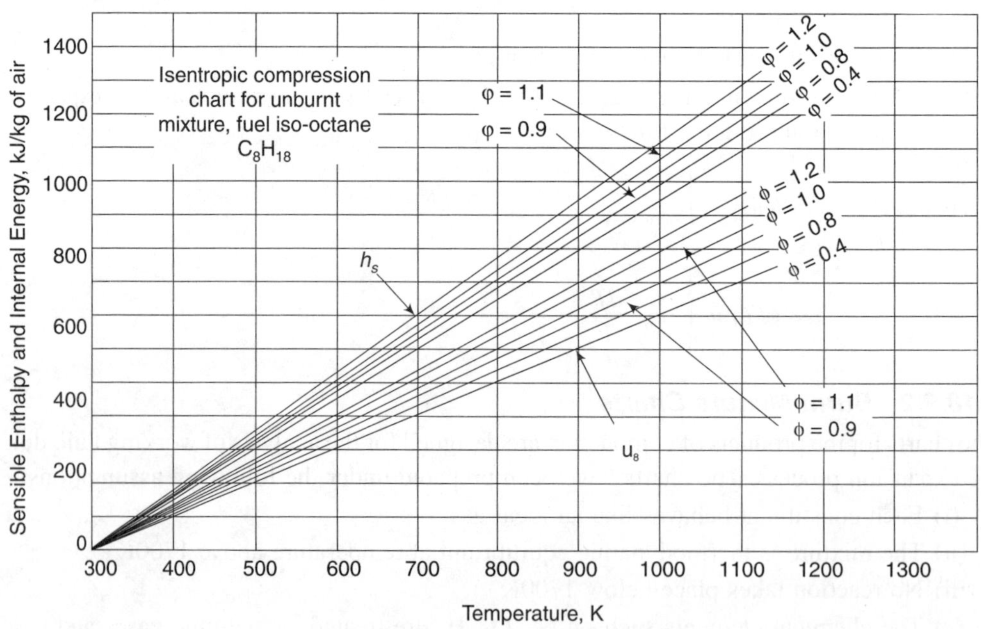


**Fig. 1.26** Sensible enthalpy ($h_s$) and internal energy ($u_s$) of unburnt iso-octane air and mixtures as a function of temperature [73]

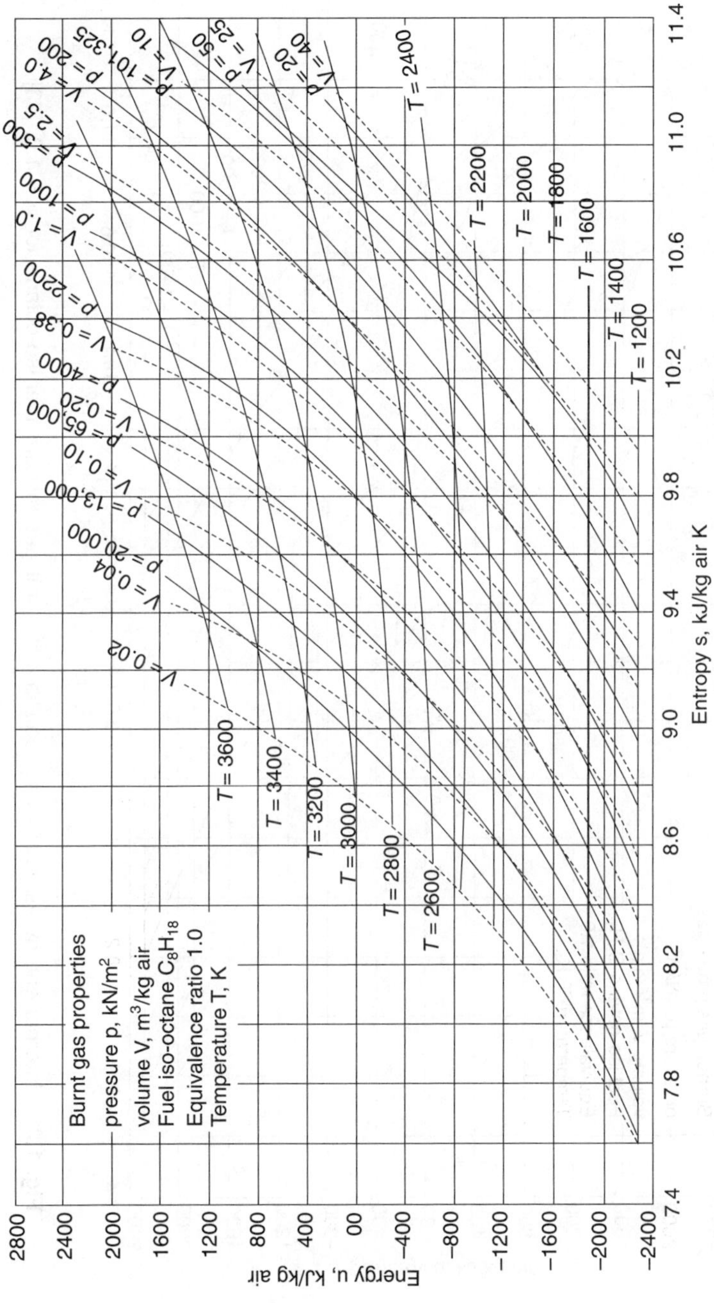


**Fig. 1.27** Internal energy vs. entropy for equilibrium burnt gas mixture, fuel iso-octane equivalence ratio (ϕ) = 0.8. [73]

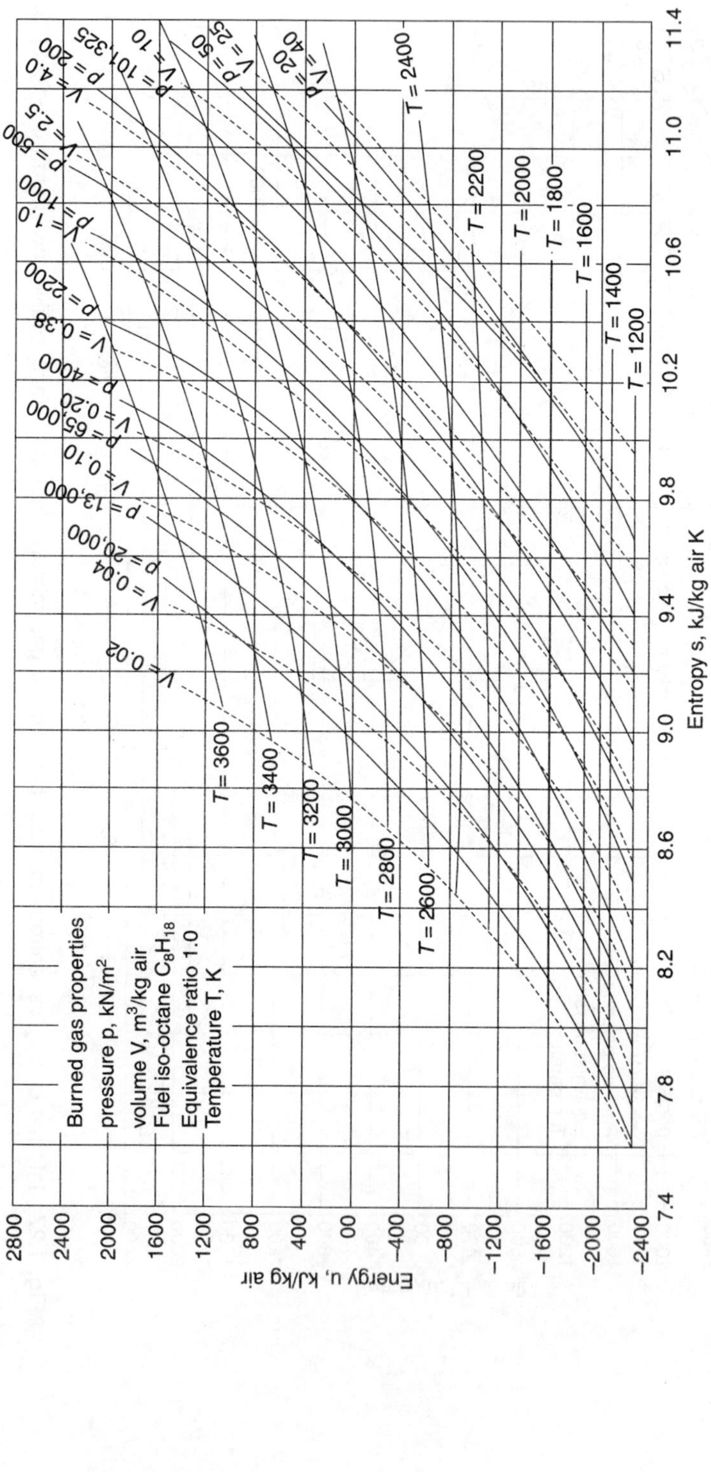


**Fig. 1.28** Internal energy vs. entropy for equilibrium burnt gas mixture fuel iso-octane for $\phi = 1.0$. [73]

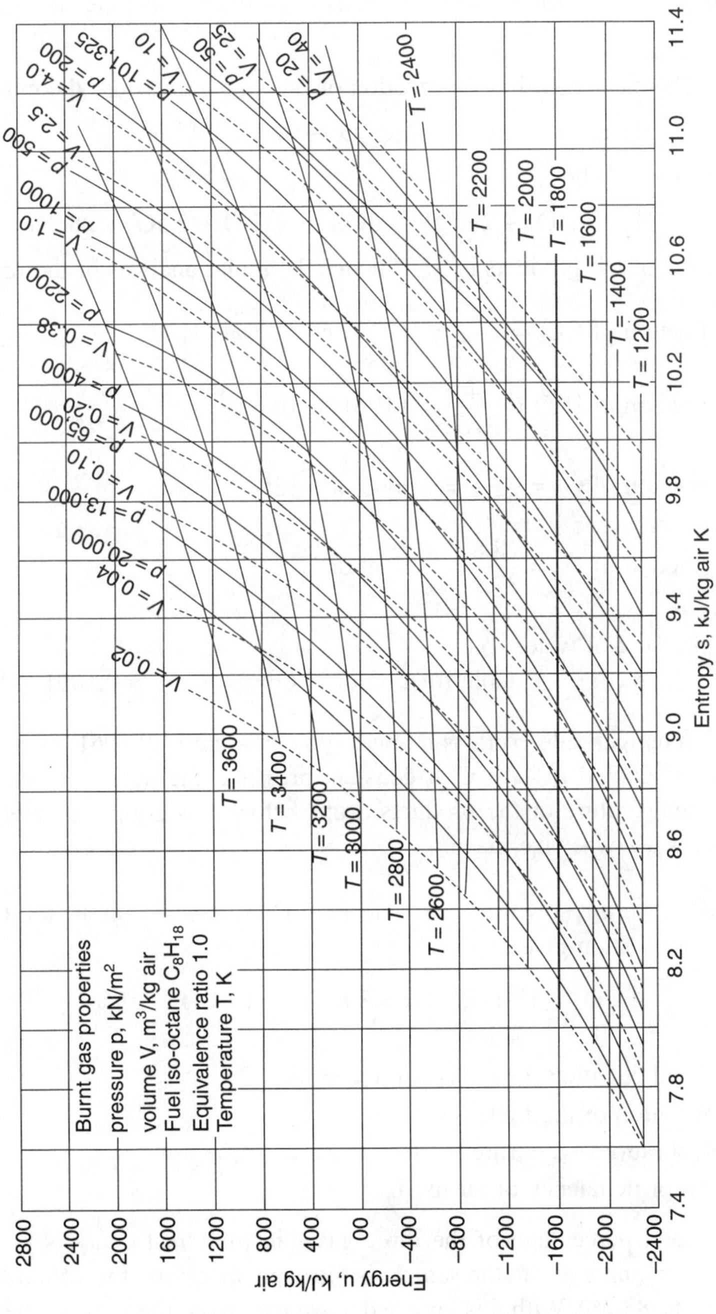


**Fig. 1.29** Internal energy vs. entropy for equilibrium burnt gas mixture, fuel Iso-octane for $\phi$ = 1.2. [73]

**EXAMPLE 1.1** Propane gas is allowed to burn with 150% theoretical air at a pressure of 1 bar. If the air is dry, determine the mole analysis of the product gas and the dew point of the gas mixture.

***Solution*** The chemical equation for reaction of propane with 100% theoretical air is

$$C_3H_8 + 5O_2 + 18.8N_2 \rightarrow 3CO_2 + 4H_2O + 18.8N_2$$

With 150% air it will be

$$C_3H_8 + 7.5O_2 + 28.2N \rightarrow 3CO_2 + 4H_2O + 2.5O_2 + 28.2N_2$$

The total moles of the products is 37.7, hence the molar analysis of the product is

$$\left.\begin{aligned} \text{Mole fraction of } CO_2 &= \frac{3}{37.7} = 0.0795 \text{ or } 7.95\% \\ \text{Mole fraction of } H_2O &= \frac{4}{37.7} = 0.1061 \text{ or } 10.61\% \\ \text{Mole fraction of } O_2 &= \frac{2.5}{37.7} = 0.0663 \text{ or } 6.63\% \\ \text{Mole fraction of } N_2 &= \frac{28.8}{37.7} = 0.7639 \text{ or } 76.39\% \end{aligned}\right\} \quad \textit{Ans.}$$

The partial pressure of water

$$= \text{mole fraction of } H_2O \times \text{pressure} = 0.1061 \times 1 = 0.1061 \text{ bar.}$$

The saturation temperature corresponding to this pressure of 0.1061 bar is 47.3°C. When the gas mixture is cooled to 47.3°C at constant pressure, the dew point of the mixture is reached. If the temperature of the products drops below dew point temperature of 47.3°C, condensation of water will take place.

**EXAMPLE 1.2** An unknown hydrocarbon fuel, $C_xH_y$ reacts with air the Orsate analysis of products of combustion gives:

$$CO_2 = 12.1\%; \quad O_2 = 3.8\% \quad \text{and} \quad CO = 0.9\%$$

Calculate

(a) the chemical equation for the actual reaction,
(b) the composition of the fuel,
(c) the air-fuel ratio used during the test, and
(d) the excess or deficiency of air used.

***Solution*** The total percentage of the three gases in the Orsat analysis is 16.8%. Let us assume that the remaining gas in the sample is nitrogen, therefore, the volumetric percentage of nitrogen must be 83.2%. With this knowledge we may write the general chemical equation for the reaction.

$$C_xH_y + a\ O_2 + 3.76a\ N_2 \rightarrow 12.1\ CO_2 + 3.8\ O_2 + 0.9\ CO + 83.2\ N_2 + bH_2O$$

From the equation we find that 83.2 moles of $N_2$ enter the reaction chamber. However, the composition of air is known to be 1 mole of $O_2$ per 3.76 moles nitrogen.

Hence, $a = \dfrac{83.2}{3.76} = 22.1$ moles of $O_2$

$O_2$ balance gives: $2 \times 22.1 = 2 \times 12.1 + 2 \times 3.8 + 0.9 + b$

$\therefore$ $b = 11.5$

To find the value of $x$ and $y$, carbon and hydrogen balance should be used.

C balance gives : $x = 12.1 + 0.9 = 13$

$H_2$ balance gives : $y = 2 \times 11.5 = 23$

Putting the value of $x$, $y$, $a$ and $b$, we get the correct chemical equation as

$$C_{13}H_{23} + 22.1O_2 + 83.2\ N_2 \rightarrow 12.1CO_2 + 3.8O_2 + 0.9CO + 83.2\ N_2 + 11.5\ H_2O$$ ***Ans.***

The air-fuel ratio can be computed from the above balance equation. As 105.3 moles of air are used per mole of fuel, so

$$A/F = \frac{\text{Mass of (22.1 + 83.2) moles of air}}{\text{Mass of one mole of } C_{13}\,H_{23}}$$

$$= \frac{105.3 \times 29}{179} = 17.059 \text{ kg of air/kg of fuel}$$

The excess or deficiency of air can be found computing the theoretical A/F value of the fuel first. The chemical equation for the combustion of $C_{13}\,H_{23}$ with the theoretical air requires

$$C_{13}\,H_{23} + 18.75O_2 + 3.76 \times 18.75\ N_2 > 13CO_2 + 11.5\ H_2O + 18.75 \times 3.76\ N_2$$

Since the total requirement of $O_2$ was 18.75 moles but 22.1 moles of $O_2$ was actually used, the percentage excess air used during the test is

$$\text{Percentage excess} = \frac{22.1 - 18.75}{18.75} = 17.86\%$$ ***Ans.***

**EXAMPLE 1.3** Propane gas is allowed to react with air in such a ratio that the Orsat analysis of the products of combustion gives $CO_2$, 11.5% $O_2$, 2.7% and CO, 0.7%. Calculate the percentage of theoretical air used during the test.

***Solution*** The right way to begin the analysis is to write down the chemical equation for 100 moles of the dry product gases. The total of the product is 14.9%, hence 85.1% is the nitrogen in the product. With this data, the chemical equation may be written as

$$x\ C_3H_8 + a\ O_2 + 3.76a\ N_2 \rightarrow 11.5\ CO_2 + 0.7\ CO + 2.7\ O_2 + 85.1\ N_2 + b\ H_2O$$

Since $N_2$ is inert, it is written down as a side product. On the basis of the composition of air

$$a = \frac{85.1}{3.76} = 22.63 \text{ moles of air}$$

Carbon balance gives : $3x = 11.5 + 0.7 = 12.2$

or $$x = 4.066$$

$O_2$ balance gives : $2 \times 22.63 = 2 \times 11.5 + 0.7 + 2 \times 2.7 + b$

or $$b = 16.16$$

Now, employing the hydrogen balance, we check the computation

$$8x = 2b$$

or $$8 \times 4.066 = 2 \times 16.16$$

or $$32.28 = 32.32$$

This check is fairly accurate and the actual reaction may be written as

$$4.066\ C_3H_8 + 22.63O_2 + 85.1N_2 \rightarrow 11.5CO_2 + 0.7CO + 85.1N_2 + 16.16H_2O$$

Dividing throughout by 4.066, the equation could be reduced to the basis of fuel. The stoichiometric equation for propane is

$$C_3H_8 + 5O_2 + 18.8\ N_2 \rightarrow CO_2 + 4H_2O + 18.8\ N_2$$

The moles of oxygen used theoretically per mole of fuel is 5. For the actual combustion, the ratio is 22.63/4.066 = 5.565. Hence, the percentage of theoretical air used in the actual combustion process to percentage of theoretical air

$$= \frac{5.565 \times 100}{5} = 111.3\%$$

The percentage of excess air is 111.3%.

**EXAMPLE 1.4** Calculate the stiochiometric air required for the complete combustion of 1 kg of normal heptane $C_7H_{16}$. Also, calculate the percentage analysis of the products on a mass and a molal basis.

***Solution*** The chemical equation for the combustion of heptane is

$$C_7H_{16} + 11O_2 + 11(3.76)\ N_2 \rightarrow 7CO_2 + 8H_2O + 11(3.76)\ N_2$$

$$\text{The air-fuel ratio on mass basis} = \frac{11 \times 32 + 11 \times 3.76 \times 28}{7 \times 12 + 16 \times 1}$$

$$= 15.10 \text{ kg of air/kg of fuel}$$

The analysis of the product on mass and molal basis may be computed as follows.

| Gas | | By mole | By mass | |
|---|---|---|---|---|
| $CO_2$ | 7 | 12.42% | 7 × 44 = 308 | 19.13% |
| $H_2O$ | 8 | 14.20% | 8 × 18 = 144 | 8.94% |
| $N_2$ | 41.36 | 73.38% | 41.36 × 28 = 1158.08 | 71.93% |
| Total | 56.37 | 100% | 1610.08 | 100% |

**EXAMPLE 1.5** Five moles of propane $C_3H_8$ are completely burnt at atmospheric pressure in the theoretical amount of air. Calculate:

(a) the volume of air used in the combustion process measured at 1 bar and 25°C,
(b) the partial pressure of each constituent of the product, and
(c) the volumetric analysis of the dry products.

***Solution*** The chemical equation of the complete combustion of 5 moles of propane is

$$5\ C_3H_8 + 25\ O_2 + 25(3.76)\ N_2 \rightarrow 15\ CO_2 + 20\ H_2O + 25 \times (3.76)\ N_2$$

or

$$5\ C_3H_8 + 25\ O_2 + 94\ N_2 \rightarrow 15\ CO_2 + 20\ H_2O + 94\ N_2$$

From the above equation, we see that 25 moles of $O_2$ is present in (94 + 25) = 119 moles of air. If air is assumed as perfect gas, then

(a)
$$V = \frac{nR_0T}{p} = \frac{119 \times 8.314\ (273 + 25)}{1 \times 10^2} = 2948.31\ \text{m}^3 \qquad \textit{Ans.}$$

(b) Sine the partial pressure of the product (constituents) is proportional to the mole fraction, hence

$$p_{CO_2} = x_{CO_2} \cdot p = \frac{15}{15 + 20 + 94} \times p = \frac{15}{129} \times 1 = 0.1162\ \text{bar}$$

$$p_{H_2O} = x_{H_2O} \cdot p = \frac{20}{120} \times p = \frac{20}{129} \times 1 = 0.1550\ \text{bar}$$

$$p_{N_2} = x_{N_2} \cdot p = \frac{94}{129} \times 1 = 0.7286\ \text{bar}$$

Total pressure = 1 bar

(c) The volumetric analyses of the dry product are:

$$\frac{V_{CO_2}}{V_{CO_2} + V_{N_2}} = \frac{15}{15 + 94} = 0.1376 \text{ or } 13.76\% \qquad \textit{Ans.}$$

$$\frac{V_{N_2}}{V_{CO_2} + V_{N_2}} = 1 - 0.1376 = 0.8624 \text{ or } 86.24\%$$

**EXAMPLE 1.6** Calculate the air-fuel ratio on both a molar and mass basis for the complete combustion of octane, $C_8H_{18}$, with (a) the theoretical amount of air, (b) 160 of theoretical air.

***Solution*** (a) For the complete combustion of octane, $C_8H_{18}$ with the theoretical (stoichiometric) amount of air, the products of combustion will be carbon dioxide, water and nitrogen only and the chemical equation is given as follows:

$$C_8H_{18} + a(O_2 + 3.76\ N_2) \rightarrow bCO_2 + cH_2O + dN_2$$

Applying the principle of conservation of mass to the carbon, hydrogen, oxygen and nitrogen, respectively, we get

C: $b = 8$ H: $2c = 18$ O: $2b + c = 2a$ N: $d = 3.76a$

Solving the above equations, we get $a = 12.5$, $b = 8$, $c = 9$, $d = 47$

Thus, the balanced chemical equation is expressed as

$$C_8H_{18} + 12.5(O_2 + 3.76\ N_2) \rightarrow 8\ CO_2 + 9\ H_2O + 47\ N_2$$

The air-fuel ratio on a molar basis is given by

$$A/F = \frac{12.5 + 12.5\ (3.76)}{1} = 59.5 \text{ kgmole (air)/kg mole (fuel)}$$ **Ans.**

The air-fuel ratio expressed on mass basis is given by

$$A/F = \left[59.5\ \frac{\text{kgmole (air)}}{\text{kg mole (fuel)}}\right]\left[\frac{28.97\ \dfrac{\text{kg (air)}}{\text{kg mole (air)}}}{114.22\ \dfrac{\text{kg (fuel)}}{\text{kg mole (fuel)}}}\right] = 15.1\ \frac{\text{kg of air}}{\text{kg of fuel}}$$ **Ans.**

(b) For 160% theoretical air (60% excess air), the chemical equation for complete combustion may be written as follows:

$$C_8H_{18} + 1.6(12.5)(O_2 + 3.76\ N_2) \rightarrow bCO_2 + cH_2O + dN_2 + eO_2$$

Applying conservation of mass to C, H, O and N, we have

C : $b = 8$ H : $2c = 18$

O : $2b + c + 2e = 1.6\ (12.5)\ (2)$ N : $d = 1.6\ (12.5)\ (3.76)$

Solution of the above equation yields: $b = 8$, $c = 9$, $d = 75.2$, $e = 7.5$

Thus, the balanced chemical equation is expressed as

$$C_8H_{18} + 1.6(12.5)\ (O_2 + 3.76\ N_2) \rightarrow 8CO_2 + 9\ H_2O + 75.2\ N_2 + 7.5\ O_2$$

The air-fuel ratio on molar basis is given by

$$(A/F)_{molar} = \frac{1.6 \times 12.5\ [1 + 3.76]}{1} = 95.2\ \frac{\text{kgmole (air)}}{\text{kg mole (fuel)}}$$ **Ans.**

On the mass basis, air-fuel ratio is given by

$$(A/F)_{mass} = \left[95.2\ \frac{\text{kgmole (air)}}{\text{kg mole (fuel)}}\right]\left[\frac{28.97\ \dfrac{\text{kg (air)}}{\text{kgmole (air)}}}{114.22\ \dfrac{\text{kg (fuel)}}{\text{kgmole (fuel)}}}\right] = 24.14\ \frac{\text{kg (air)}}{\text{kg mole (fuel)}}$$ **Ans.**

**EXAMPLE 1.7** In a combustion chamber methane, $CH_4$ is burnt with dry air. The molar analysis of the products on a dry basis is $CO_2$, 9.6%, CO, 0.6%, $O_2$, 2.95, and $N_2$, 86.85%. Calculate

(a) the air-fuel ratio on both a molar and a mass basis,
(b) the percentage theoretical air, and
(c) the dew point temperature of product if the pressure is 1 atm.

***Solution*** It is assumed that each mole of oxygen in the combustion air is accompanied by 3.76 moles of nitrogen which is inert.

(a) The chemical equation may be written on the basis of 100 kg mole of dry products. In addition to the assumed 100 kg mole of dry products, water must be included as a product,

$$aCH_4 + b(O_2 + 3.76\ N_2) \rightarrow 9.6\ CO + 2.95\ O_2 + 86.85\ N_2 + cH_2O$$

Applying conservation of mass to carbon, hydrogen and oxygen, respectively, we have

$$C : 9.6 + 0.6 = a \qquad\qquad H : 2c = 4a$$

$$O : (9.6)(2) + 0.6 + 2\,(2.95) + c = 26$$

Solution of above equation yields $a = 10.2$, $b = 23.05$, $c = 20.4$

The balanced chemical equation is written as

$$10.2CH_4 + 23.05(O_2 + 3.76N_2) \rightarrow 9.6CO_2 + 0.6CO + 2.95\ O_2 + 86.85\ N_2 + 20.4H_2O$$

On molar basis, the air-fuel ratio is given by

$$(A/F)_{molar} = \frac{23.05\ (1 + 3.76)}{10.2} = 10.75\ \frac{\text{kgmole (air)}}{\text{kg mole (fuel)}} \qquad \textbf{\textit{Ans.}}$$

On mass basis,

$$(A/F)\text{mass} = \left[10.75\ \frac{\text{kgmole (air)}}{\text{kgmole (fuel)}}\right]\left[\frac{28.97\ \frac{\text{kg(air)}}{\text{kgmole air}}}{16.04\ \frac{\text{kg (fuel)}}{\text{kgmole fuel)}}}\right] = 19.41\ \frac{\text{kg (air)}}{\text{kg (fuel)}} \qquad \textbf{\textit{Ans.}}$$

(b) The balanced chemical equation for the complete combustion of methane with theoretical amount of air is expressed as

$$CH_4 + 2\ (O_2 + 3.76\ N_2) \rightarrow CO_2 + 2H_2O + 7.5\ N_2$$

The theoretical air-fuel ratio on molar basis is given by

$$(A/F)_{theo} = \frac{1 \times 4.76}{1} = 9.57\ \frac{\text{kgmole (air)}}{\text{kg mole (fuel)}}$$

The percentage of theoretical air is given by

$$\text{Percentage of theoretical air} = \frac{\text{A/F}}{(\text{A/F})_{\text{theo}}} \times 100 = \frac{10.75 \times 100}{9.52} = 112.9\%$$

(c) The dew point temperature may be evaluated when partial pressure of water vapour is known. The partial pressure of water vapour is given by

$$p_v = y_v\, p \text{ where } y \text{ is the mole fraction of water vapour}$$

But $$y_v = \frac{20.4}{100 + 20.4} = 0.169$$

Thus, $$p_v = 0.169 \times 1 = 0.169 \text{ atm}$$

At 0.169 atm, the saturation temperature from steam table is 56.6°C. ***Ans.***

**EXAMPLE 1.8** In a well-insulated combustion chamber (reactor), liquid octane enters at 25°C and 1 atm. The air also enters at the same pressure and temperature. The combustion takes place and the product comes out of the combustion chamber. For steady-state operation and neglecting the effect of changes in kinetic and potential energies, calculate the temperature of combustion products (or adiabatic temperature) for complete combustion with (a) theoretical amount of air and (b) 200% theoretical air.

***Solution*** Since the combustion chamber is insulated and does not exchange any work, so $\dot{Q}_{cv} = 0$ and $\dot{W}_{cv} = 0$. It is assumed that combustion is accompanied by 3.76 moles of nitrogen, which is inert.

At steady state, the control volume energy rate balance yields

$$\sum_p n_e \bar{h}_e = \sum_r n_i \bar{h}_i$$

or $$\sum_p n_e(\bar{h}_f^0 + \overline{\Delta h}_e) = \sum_r n_i(\bar{h}_f^0 + \overline{\Delta h}_i)$$

After re-arranging the above equation, we have

$$\sum_p n_e(\overline{\Delta h})_e = \sum_r n_i(\overline{\Delta h})_i + \sum_r n_i \bar{h}_{fi}^0 - \sum_p n_e \bar{h}_{fe}^0$$

Since the reactants enters at 25°C and 1 bar, so $(\overline{\Delta h})_i = 0$, thus

$$\sum_p n_e(\overline{\Delta h})_e = \sum_r n_i \bar{h}_{fi}^0 - \sum_p n_e \bar{h}_{fe}^0 \quad \text{...(a)}$$

The chemical equation for the combustion of liquid $C_8H_{18}$ with the theoretical amount of air is given by

$$C_8H_{18}(l) + 12.5\, O_2 + 47\, N_2 = 8\, CO_2 + 9\, H_2O\,(g) + 47\, N_2$$

Using the chemical equation in Eq (a), we have

$$8(\overline{\Delta h})_{CO_2} + 9(\overline{\Delta h})_{H_2O(g)} + 47\ (\overline{\Delta h})_{N_2}$$

$$= \left[(\overline{h}^0_f)_{C_8H_{18}(l)} + 12.5(\overline{h}^0_f)_{02} + 47(\overline{h}^0_f)_{N_2}\right] - \left[8(\overline{h}^0_f)_{CO_2} + 9\ (\overline{h}^0_f)_{H_2O(g)} + 47(\overline{h}^0_f)_{N_2}\right]$$

Since $(\overline{h}°_f)_{O_2} = 0$, $(\overline{h}^0_f)_{N_2} = 0$, so the above equation becomes

$$8(\overline{\Delta h})_{CO_2} + 9\ (\overline{\Delta h})_{H_2O(g)} + 47(\overline{\Delta h})_{N_2} = \left[(\overline{h}^0_f)_{C_8H_{18}(l)}\right] - \left[8(\overline{h}^0_f)_{CO_2} + 9(\overline{h}^0_f)_{H_2O(g)}\right]$$

$$= -\ 249{,}910 - [8(-393,\ 520) + 9\ (-241,\ 820)] = 5{,}074,\ 630 \text{ kJ/kg mole of fuel}$$

Now each term of L.H.S. of equation depends on the temperature of the product, $T_p$ and this value can be obtained only by trial and error through balancing L.H.S. = R.H.S. Let at 2400 K.

| | |
|---|---|
| $8\,(\overline{\Delta h})_{CO_2}$ | 926,304 |
| $9\,(\overline{\Delta h})_{H_2O(g)}$ | 842,436 |
| $47\,(\overline{\Delta h})_{N_2}$ | 3,320,597 |
| $\sum_p n_e\ (\overline{\Delta h})_e$ | 5,089,337 kJ/kg mole of fuel |

Thus, at $T_p$ = 2400 K, L.H.S. is slightly (about 14707) more than R.H.S., so it is estimated that $T_p$ = 2395 K will satisfy the equation. Thus, the adiabatic flame temperature or temperature of combustion product will be 2395 K. ***Ans.***

**EXAMPLE 1.9** An internal combustion engine uses liquid octane as fuel and operating at steady state consumes 0.002 kg/s. It is mixed with theoretical amount of air for combustion in engine cylinder. The fuel and air enters the engine at 25°C and 1 atm. The mixture burns completely and combustion products leave the engine at 500°C. The engine develops a power output of 45 kW. Calculate the rate of heat transfer from the engine neglecting the changes in kinetic and potential energies.

***Solution*** Assume each mole of oxygen in the combustion air is accompanied by 3.76 mole of nitrogen.

The chemical equation for complete combustion with theoretical air is given by

$$C_8\,H_{18} + a(O_2 + 3.76\ N_2) \rightarrow bCO_2 + cH_2O + dN_2$$

Applying *conservation of mass* to C, H, O and N, we get balanced chemical equation as

$$C_8H_{18} + 12.5\ O_2 + 47\ N_2 \rightarrow 8CO_2 + 9H_2O + 47\ N_2$$

The energy rate balance in the absence of KE and PE effect is given by

$$\frac{\dot{Q}_{cv}}{n_F} = \frac{\dot{W}_{cv}}{n_F} + \overline{h}_p - \overline{h}_r$$

or $$\frac{\dot{Q}_{cv}}{n_F} = \frac{\dot{W}_{cv}}{n_F} + \left\{8\,[\bar{h}^0_f + \Delta\bar{h}]_{CO_2} + 9\,[\bar{h}^0_f + \Delta\bar{h}]_{H_2O(g)} + 47\,[\bar{h}^0_f + \Delta\bar{h}\}_{N_2}\right\}$$

$$- \left\{[\bar{h}^0_f + \Delta\bar{h}]_{C_8H_{18}(l)} + 12.5\,[\bar{h}^0_f + \Delta\bar{h}]_{O_2} + 47\,[\bar{h}^0_f + \Delta\bar{h}]_{N_2}\right\}$$

In the product $[\bar{h}^0_f]_{N_2} = 0$ and in the reactants

$$[\Delta\bar{h}]_{C_8H_{18}(i)},\ [\bar{h}^0_f + \Delta\bar{h}]_{O_2}$$

and $$[\bar{h}^0_f + \Delta\bar{h}]_{N_2} = 0$$

From Table D1 in reference, we have

$$H_r = [\bar{h}^0_f]_{C_8H_{18}(l)} = -\ 249{,}910 \text{ kJ/kgmole}$$

Substituting the values of $\bar{h}^0_f$ and $\Delta\bar{h}$ for various products from Tables A8 and A9 of gas tables by using

$$\Delta\bar{h} = h(T, p) - h\,(T_{ref}, p_{ref}),$$

we have $$\bar{h}_p = 8\,[-393{,}520 + (36{,}811 - 9{,}366)]$$

$$+9\,[-241{,}820 + (31{,}381 - 9{,}902)] + 47\,[170801]$$

$$= -\ 40{,}75002 \text{ kJ/kg mole (fuel)}$$

The molecular weight of fuel is 114.23, thus the molar flow rate of fuel is

$$\dot{n}_F = \frac{0.002}{114.22} = 1.75 \times 10^{-5} \text{ kgmole fuel/s}$$

Substituting the value of $\bar{h}_p$, $\bar{h}_r$, $\dot{W}_{cv}$ and $\dot{n}_F$, we have

$$\dot{Q}_{cv} = \dot{W}_{cv} + \dot{n}_F\,(\bar{h}_p - \bar{h}_r)$$

$$= 45 + 1.75 \times 10^{-5}\,[-\ 40{,}75002 - (-249{,}910)] = -21.939 \text{ kW.}$$ ***Ans.***

## 1.19 SUMMARY OF FORMULAE

(1) $c = \frac{V_c}{V_s}$

(2) $r = \frac{V_c + V_s}{V_c} = \frac{1 + c}{c}$

(3) $p_m = \frac{W_{cycle}}{V_s}$

(4) $ip = \frac{p_{mi}\,ALn}{60}$

(5) $bp = \frac{p_{mb}\,ALn}{60}$

(6) $A/F = \frac{\dot{m}_a}{\dot{m}_f}$

(7) $\eta_{ith} = \frac{ip}{\dot{m}_f \times q_{LHV}}$

(8) $\eta_{bth} = \frac{bp}{\dot{m}_f \times q_{LHV}}$

(9) $\eta_{rel} = \dfrac{\eta_{act}}{\eta_{as}}$ (10) $fp = ip - bp$

(11) $\eta_v = \dot{m}_a / \dot{m}_{th}$ (12) $bsfc = \dfrac{\dot{m}_f(\text{kg/h})}{bp(\text{kW})}$

## PROBLEMS

### Things to Think About

1.1 What is the future scope of IC engines in the light of global warming?

1.2 List the various types of eco-friendly engines.

1.3 What will happen if the inlet and exhaust valves open before and after of the optimum timing?

1.4 What is the reason that the size of inlet valve is kept larger than exhaust valve?

1.5 What parameters govern the output and efficiency of IC engines?

### Short and Long Theoretical Questions

1.6 Define an internal combustion engines.

1.7 Differentiate between a steam engine and IC engine.

1.8 What are the two basic types of internal combustion engines? What are the fundamental differences between them?

1.9 How are the IC engines classified?

1.10 What do you mean by MPFI system?

1.11 List the name of important parts of an IC engine and give simple sketches. Also mention the materials used and method of manufacturing.

1.12 What are the functions of piston, piston rings, crankshaft, connecting rod, flywheel and crankshaft?

1.13 Discuss the principle of working of four-stroke CI and SI engines.

1.14 Explain the valve timing diagram of SI and CI engines for 4-stroke engines.

1.15 Differentiate between SI and CI engines.

1.16 Discuss the working principle of 2-stroke engines.

1.17 Discuss the valve timing diagram of 2-stroke engines.

1.18 Give a comparison between 4-stroke and 2-stroke engines.

1.19 List the applications of IC engine in detail.

1.20 List and define the performance parameters of IC engines.

1.21 By suitable sketches, show the various possible cylinder arrangement of IC engines.

1.22 List the various auxiliary system of IC engines

1.23 List the environment-friendly car engines

1.24 Define enthalpy of reaction and enthalpy of formation.

1.25 Define enthalpy and internal energy of reaction.

1.26 What is the difference between higher and lower heating values of fuel?

1.27 What is the reason that 2-stroke engines are not used in passenger cars?

### Numerical Questions

1.28 In a petrol engine, the dry exhaust analysis gave 1.5% CO, and the $O_2$ was negligible. If the fuel has an ultimate analysis by weight of C = 0.84 and $H_2$ = 0.16, calculate the air-fuel ratio supplied. Assume air is 23.1% $O_2$ by weight and all the carbon in the fuel may be assumed to have been burnt either completely or incompletely. [*Ans.* A/F = 14.75 kg/kg of fuel]

1.29 The dry exhaust gas analysis from an engine using hydrocarbon fuel is $CO_2$ = 8.7%, $O_2$ = 0.3%, CO = 8.9%, $CH_4$ = 0.3%, $H_2$ = 3.7%, $N_2$ = 78.1%. Calculate the air-fuel ratio (a) given that fuel is $C_8 H_{17}$ and (b) fuel composition not known. Solve the question by using

(i) carbon balance,

(ii) hydrogen balance, and

(iii) carbon-hydrogen balance. [*Ans.* (a) 11.31. 11.31, 11.31 (b) 11.31]

1.30 From a test carried on a four-cylinder petrol engine running on a rich mixture the following volumetric analyses obtained are $CO_2$ = 9.8%, $O_2$ = 0.2%, CO = 7.6% and $N_2$ = 82%. Calculate the air-fuel ratio by weight and the probable composition of the fuel on the assumption that this consists solely of hydrogen and carbon. [*Ans.* C = 86.9%; $H_2$ = 13.1%; A/F = 12.5%]

1.31 The following data show the enthalpy of formation at 25°C:

$$C_2H_{2(g)} = 226899.43 \text{ kJ/kg mole}$$

$$C_6H_{6(g)} = 82982 \text{ kJ/kg mole}$$

$$C_6H_{14(g)} = -\ 467304.42 \text{ kJ/kg mole}$$

$$NH_{3(g)} = -\ 46222.27 \text{ kJ/kg mole}$$

$$CO_{2(n)} = -\ 393776.9 \text{ kJ/kg mole}$$

$$H_2O_{(g)} = -241988.58 \text{ kJ/kg mole}$$

Calculate the constant pressure lower heating values of $C_6\,H_{14}$, and the enthalt of reaction of

$$4NH_3 + 3O_2 \rightarrow 6H_2O + 2N_2$$

1.32 Calculate the H.C.V. and L.C.V. of a fuel having the following compositions:

C = 80%, $H_2$ = 10% $O_2$ = 3%, S = 2%, $N_2$ = 20% and rest incombustible matter.

[*Ans.* H.C.V. = 41132.84 kJ/kg, L.C.V. = 38924.9 kJ/kg]

1.33 Calculate the enthalpy of reaction at 25°C when 1 mole of propane $C_3\,H_8$ is burnt according to the following equation

$$C_3H_8(g) \rightarrow CO_2 + 4H_2O(l)$$

Assume steady flow combustion process.

### Software Based Questions

1.34 Develop a software to calculate the theoretical air-fuel ratio required for the combustion of any hydrocarbon fuel.

1.35 Develop a software to calculate the adiabatic flame temperature due to combustion in petrol engines.

1.36 Develop a software for a design battery operated electric car engine.

### Design and Open Ended Question

1.37 Prepare a comprehensive report on solar cell operated car engines. Include the design procedure for such an engine.

### Multiple-Choice Questions

1.38 A bi-fuel engines uses

(a) Liquid fuel for starting and gaseous fuel for running

(b) Gaseous fuel for starting and liquid fuel for running

(c) Two different fuels in different chambers

(d) Liquid fuel for starting and solid fuel for running

1.39 The brake thermal efficiency for SI engines varies from

(a) 15–25% (b) 30–35% (c) 40–50% (d) 45–60%

1.40 In SI engine, the compression ratio ranges from

(a) 2 to 7 (b) 5 to 10 (c) 10 to 16 (d) 15 to 20

1.41 In CI engine, the compression ratio ranges from

(a) 4 to 12 (b) 8 to 16 (c) 12 to 20 (d) 20 to 30

1.42 In 4-stroke SI engine

(a) Inlet vapour opens about 5° bTDC and closes 5° aBDC

(b) Inlet valve opens about 10° bTDC and closes 20° aBDC

(c) Inlet valve opens about 15° bTDC and closes 25° aBDC

(d) Inlet valve opens about 25° bTDC and closes about 35° aBDC

## Key to Multiple-Choice Questions

1.38 (a) 1.39 (b) 1.40 (b) 1.41 (c) 1.42 (d)

CHAPTER

# Air Standard Cycles

# 2

## OVERVIEW

In the modern world, there is an ever increasing demand of power, and in order to meet it, a large number of power producing devices have been developed. Broadly, they are classified as*

(a) *Indirect energy converters*

(b) *Direct energy converters.*

Power plants may also be classified as

(a) *Conventional power plants*

(b) *Non-conventional power plants.*

In the ***conventional system (indirect energy conversion)***, the chemical energy of the fuel is first converted into heat through combustion process, and then heat energy is converted into work as shown in Fig. 2.1. The ***indirect energy conversion system*** includes the following.

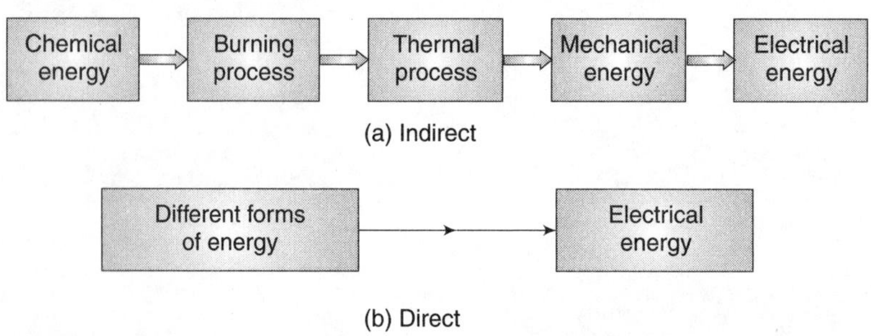


**Fig. 2.1** Indirect and direct energy converters

---

* *See author's book* **"Steam and Gas Turbines, Turbomachines and Power Plant Engineering"** for various power plants.

(i) Internal combustion engines
(ii) Gas turbines
(iii) Steam power plants (engine, turbine)
(iv) Nuclear power plants
(v) Hydel power plants (no fuel requirement)
(vi) Geothermal power plants, utilizing earth's internal energy (no fuel requirement)
(vii) Tidal and wave power plants (no fuel requirement)
(viii) Ocean thermal energy power plants (no fuel requirement)
(ix) Wind mill (no fuel requirement)
(x) Solar thermal power plants (no fuel requirement)

In the ***direct energy conversion system***, different types of energy are converted directly into electrical energy. This includes:

(i) Fuel cells
(ii) Thermoelectric generators
(iii) Thermo-ionic generators
(iv) Magneto hydro dynamic power plant (M.H.D.)
(v) Solar cells.

Amongst the above devices, the working fluid of some of them *executes a cyclic process* and as we know, the essential elements of a thermodynamic power cycles are:

(a) ***Working fluid*** which undergoes a change of state while receiving or rejecting heat and thus produces work.
(b) ***Heat sources and sink.*** These are essential for working fluid to receive heat from a heat source and reject it into a heat sink.
(c) ***Arrangement.*** An arrangement is necessary for getting the work out of the working fluid. This may be done by a *piston-cylinder* or a *turbine* or any other arrangement. Piston-cylinder nomenclatures have already been discussed.

In this chapter, we will restrict ourselves to an analysis of the power plants which employ gas as the working fluid and execute a cyclic process, for example, *internal combustion engines and gas turbines.*

## 2.1 AIR STANDARD CYCLES

*In gas power plants such as internal combustion engines and gas turbines, the working fluid remains gas throughout the cycle.* The plants take in either a mixture of air and fuel or air and fuel separately, compress it to a high pressure, and cause it to burn in a combustion chamber liberating heat energy. The products of combustion at high pressure, temperature expand and do work and the exhaust leaves the engine or turbine.

In the most gas power cycles, the working fluid consists of mainly of air. For example, in an internal combustion engine, the air fuel ratio is generally 15 : 1 and in the gas turbines 50 : 1 to 250 : 1. *The properties of the working fluid closely follow the properties of air. Therefore, it is convenient to analyse the gas power cycles by devising an idealized cycle known as* ***air standard cycle*** *by modelling the actual engines.*

**Modelling:** The air standard cycles are based on the following assumptions:

(i) The working fluid is air throughout the cycle and follows the perfect gas law $pV = mRT$.

(ii) The working fluid is homogeneous throughout at all time and no chemical reaction takes place.

(iii) The specific heats of air do not vary with temperature.

(iv) The mass of air in the cycle remains fixed, i.e., the engine operates as a closed cycle so that working fluid is restored to its initial state at the end of each cycle. Thus, there are no intake and exhaust processes.

(v) Combustion process which may appear in actual process is replaced by an equivalent heat addition process. This heat addition process is due to heat transfer alone.

(vi) Since there is no exhaust process, it is replaced by an equivalent heat rejection process. Further, this heat rejection process is due to heat transfer alone.

(vii) All processes are internally reversible, and compression and expansion processes are reversible adiabatic processes.

The following parameters are used for the analysis of air standard cycles.

**(i) *Thermal or Air Standard Efficiency (ASE)***

The thermal efficiency of air standard cycle is given by:

$$\eta_{th} = \text{ASE} = \frac{\text{Work done}}{\text{Heat supplied}} = \frac{\text{Heat added–Heat rejected}}{\text{Heat supplied}}$$

or

$$\eta_{th} = \text{ASE} = \frac{Q_A - Q_R}{Q_A} = \frac{q_A - q_R}{q_A} \quad \text{...(2.1)}$$

The thermal efficiency of the ideal air standard cycles is also known as the air standard efficiency.

**(ii) *Relative Efficiency.*** It is defined as

$$\text{Relative efficiency} = \eta_{rel} = \frac{\text{Actual thermal efficiency}}{\text{Air standard efficiency}} \quad \text{...(2.2)}$$

**(iii) *Specific Work Output (w).*** *It is the work output per unit mass of working substance (air).* This indicates the size of the plant needed. It is expressed in kW/kg of air.

**(iv) *Specific Air Consumption (SAC).*** *It is the amount of working substance needed per unit work output express in kg/kWh*

$$\text{SAC} = \frac{3600}{w_{net}} : \text{kg/kWh} \quad \text{...(2.3)}$$

**(v) *Work Ratio (WR).*** *It is defined as the ratio of network transfer in a cycle to positive work transfer in a cycle.*

$$\text{WR} = \frac{\text{Net work}}{\text{Positive work}} \qquad \text{...(2.4)}$$

The concept of air standard cycle helps us

(i) to establish a standard of perfection by which the performance of actual engines can be estimated; and

(ii) to study qualitatively the influence of a number of variables by performance on actual power plant.

The air standard cycles includes:

(i) Carnot cycle, (ii) Otto cycle, (iii) Diesel cycle, (iv) Dual cycle, (v) Stirling cycle, (vi) Ericsson cycle, (vii) Atkinson cycle, and (viii) Joule or Brayton cycle.

Usually, the devices which work on gas power cycles are called heat engines.

## 2.2 CARNOT CYCLE

*The Carnot cycle is an ideal cycle which is not possible in practice and has the maximum efficiency compared to other cycles.* Though, the analysis of Carnot cycle does not give the performance of actual cycle, however, it permits the designer to arrive at the maximum possible efficiency that can be obtained in any heat engine under the given conditions. The cycle is represented on $p$–$V$ and $T$–$s$ diagrams as shown in Fig. 2.2. The various processes are:

1-2: Isentropic compression of air when the piston moves from BDC to TDC position.

2-3: Addition of heat at constant temperature, $T_H$, i.e., isothermal addition of heat at temperature $T_H$, the highest temperatures.

3-4: Isentropic expansion of air from highest temperature $T_H$ to lowest temperature, $T_L$.

4-1: Rejection of heat at constant temperature $T_L$, i.e., isothermal heat rejection at sink temperature, $T_L$.

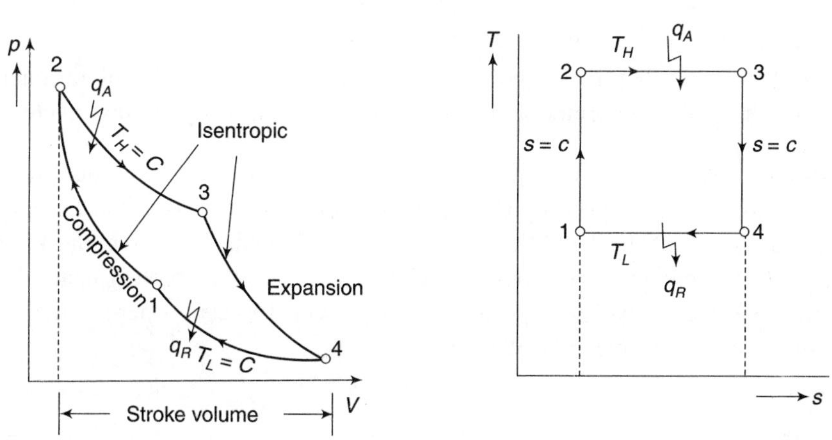


**Fig. 2.2** Carnot cycle on $p$-$V$ and $T$-$s$ diagrams

Consider 1 kg of working fluid.

From first law of thermodynamics for isothermal process

$$q = w \quad \text{since} \quad \Delta u = 0 \text{ at constant temperature}$$

$$\text{Heat added} = q_A = q_{2-3} = RT_2 \ln \frac{V_3}{V_2} = RT_H \ln \frac{v_3}{v_2}$$

$$\text{Heat rejected} = q_R = q_{4-1} = RT_4 \ln \frac{V_4}{V_1} = RT_L \ln \frac{v_4}{v_1}$$

$$\text{Work done per cycle} = w = \text{Heat added} - \text{Heat rejected}$$

or
$$w = RT_H \ln \frac{v_3}{v_2} - RT_L \ln \frac{v_4}{v_1}$$

*Note:* If the heat rejected is written as $q_R = q_{4-1} = RT_L \frac{v_1}{v_4}$ which is a negative value, then the work done will be the summation of heat supplied and heat rejected resulting in the same equation for work done as above.

But
$$\frac{v_3}{v_4} = \left(\frac{T_4}{T_3}\right)^{1/(\gamma-1)} = \left(\frac{T_L}{T_H}\right)^{1/(\gamma-1)} = \left(\frac{T_2}{T_1}\right)^{1/(\gamma-1)} = \frac{v_2}{v_1}$$

∴ Hence,
$$\frac{v_3}{v_2} = \frac{v_4}{v_1}$$

∴
$$w = \left[R \ln \frac{v_3}{v_2}\right] [T_H - T_L] \text{ kJ/kg}$$

$$\text{Thermal efficiency} = \frac{\text{Work done}}{\text{Heat added}} = \frac{\left(R \ln \frac{v_3}{v_2}\right)[T_H - T_L]}{RT_H \ln \frac{v_3}{v_2}}$$

or
$$\eta_{th} = \frac{T_H - T_L}{T_H} = 1 - \frac{T_L}{T_H} \quad \text{...(2.5)}$$

Equation (2.5) shows that the thermal efficiency of a Carnot cycle is only a funcion of the maximum and the minimum temperature in the cycle. **It is not a practical cycle.**

**EXAMPLE 2.1** A Carnot cycle has lowest pressure and temperature equal to 1 bar and 20°C. Pressure after isothermal compression is 4 bar. Pressure after isentropic compression is 12 bar and after isothermal heat addition process is 6 bar. Calculate (a) the highest temperature in the cycle, (b) the change in entropy during isothermal expansion, (c) heat added to the cycle, (d) heat rejected by cycle, (e) power developed with 150 cycles per minute, (f) thermal efficiency of the cycle, (g) mean effective pressure.

***Solution*** Refer to Fig. 2.2.

Given, $p_4 = 1$ bar, $T_L = T_4 = T_1 = 293\text{K}$, $p_2 = 12$ bar, $p_1 = 4$ bar and $p_3 = 6$ bar

Applying characteristics gas equation at point 4, we have

$$v_4 = \frac{RT_L}{p_4} = \frac{RT_4}{p_4} = \frac{0.287 \times 293}{1 \times 10^2} = 0.8409 \text{ m}^3/\text{kg}.$$

From isothermal heat rejection process 4–1, we have

$$v_1 = \frac{p_4 v_4}{p_1} = \frac{1 \times 10^2 \times 0.8409}{4 \times 10^2} = 0.2102 \text{ m}^3/\text{kg}$$

From isentropic compression process 1–2, we have

$$v_2 = v_1 \left(\frac{p_1}{p_2}\right)^{1/\gamma} = 0.2102\left(\frac{4}{12}\right)^{1/1.4} = 0.0959 \text{ m}^3/\text{kg}$$

$\therefore$ Stroke volume $= v_s = v_4 - v_2 = 0.8409 - 0.0959 = 0.745 \text{ m}^3/\text{kg}$

Again,
$$T_2 = T_H = T_1\left(\frac{v_1}{v_2}\right)^{\gamma-1} = T_L\left(\frac{v_1}{v_2}\right)^{\gamma-1} = 293\left(\frac{0.2102}{0.0959}\right)^{0.4} = 401.15 \text{ K}$$

(a) $T_H = 401.15$ K

(b) From isothermal heat addition process 2–3, we have

$$v_3 = v_2 \frac{p_2}{p_3} = 0.0959 \times \frac{12}{6} = 0.1918 \text{ m}^3/\text{kg}.$$

Change in entropy during isothermal expansion (heat addition) process is given by

$$s_3 - s_2 = \frac{p_2 v_2}{T_H} \ln \frac{v_3}{v_2} = \frac{12 \times 10^2 \times 0.0959}{401.15} \ln \frac{0.1918}{0.0959} = 0.1984 \text{ kJ/kg K}$$ ***Ans.***

(c) Heat added $= q_A = p_2 v_2 \ln \frac{v_3}{v_2} = T_H (s_3 - s_2) = 401.15 \times 0.1984 = 79.56$ kJ/kg ***Ans.***

(d) Heat rejected $= q_R = p_4 v_4 \ln \frac{v_1}{v_4} = T_L (s_1 - s_4) = - T_L (s_3 - s_2)$

$= -293 \times 0.1984 = -58.13$ kJ/kg ***Ans.***

–ve sign is used to show the heat rejection.

(e) Work done per cycle = Head added – Heat rejected = 79.56 –58.13 = 21.6 kJ/kg

Power developed $= \frac{21.6 \times 150}{60} = 54$ kW/kg ***Ans.***

(f) Thermal efficiency $= 1 - \frac{T_L}{T_H} = 1 - \frac{293}{401.15} = 0.2695$ or 26.95% ***Ans.***

(g) *m.e.p.* $= \frac{\text{Work done/kg}}{\text{Stroke volume}} = \frac{21.6}{v_4 - v_2} = \frac{21.6}{0.849 - 0.0959} = 28.99 \text{ kN/m}^2 = 0.2899$ bar

***Ans.***

**EXAMPLE 2.2** It is claimed that an engine works on Carnot cycle with air as the working fluid. The ratio of greatest to least volume is fixed and say *x*. The lower temperature of the cycle is fixed but the isentropic compressor ratio ($r$) is variable. The ratio of specific heat is $y$. Prove that for maximum work done, the following equation holds good.

$$(\gamma - 1) \ln \left(\frac{x}{r}\right) + \frac{1}{r^{\gamma-1}} - 1 = 0.$$

***Solution*** Refer to Fig. 2.2. Assume 1 kg of working fluid.

Work done = Heat supplied – Heat rejected

$$w = p_2 v_2 \ln \frac{v_3}{v_2} - p_4 v_4 \ln \frac{v_4}{v_1} = p_2 v_2 \ln \frac{v_3}{v_2} + p_4 v_4 \ln \frac{v_1}{v_4}$$

$$= RT_2 \ln \frac{v_3}{v_2} + RT_4 \ln \frac{v_1}{v_4} = RT_H \ln \frac{v_3}{v_2} \ln \frac{v_3}{v_2} + RT_L \ln \frac{v_1}{v_4}$$

But $$T_2 = T_H = T_1 \left(\frac{v_1}{v_2}\right)^{\gamma-1} = T_L \left(\frac{v_1}{v_2}\right)^{\gamma-1} = T_L \, r^{\gamma-1} \quad \text{...(i)}$$

$$T_3 = T_H = T_4 \left(\frac{v_4}{v_3}\right)^{\gamma-1} = T_L \left(\frac{v_4}{v_3}\right)^{\gamma-1} \quad \text{...(ii)}$$

Equations (i) and (ii) result in

$$\frac{v_1}{v_2} = \frac{v_4}{v_3} \quad \text{or} \quad \frac{v_3}{v_2} = \frac{v_4}{v_1}$$

We have $$\frac{v_4}{v_2} = x \quad \text{and} \quad \frac{v_1}{v_2} = r = \frac{v_4}{v_3}$$

∴ $$\frac{v_3}{v_2} = \frac{v_3}{v_4} \times \frac{v_4}{v_2} = \frac{x}{r} = \frac{v_4}{v_1}$$

∴ $$w = RT_H \ln \frac{x}{r} + RT_L \ln \frac{r}{x} = RT_L \, r^{\gamma-1} \ln \frac{x}{r} + RT_L \ln \frac{r}{x}$$

$$= RT_L \left(r^{\gamma-1} - 1\right) \ln \frac{x}{r}$$

In this equation, the only variable is $r$, hence, for the maximum work

$$\frac{dw}{dr} = 0 = RT_L \left\{\left(\ln \frac{x}{r}\right) (\gamma - 1)\, r^{\gamma-2} + \left(r^{\gamma-1} - 1\right) \left(\frac{r}{x}\right) \frac{-x}{r^2}\right\}$$

or $$\left(\ln \frac{x}{r}\right) (\gamma - 1)\, r^{\gamma-2} - r^{\gamma-2} + \frac{1}{r} = 0$$

or $$\left(\ln \frac{x}{r}\right) (\gamma - 1)\, r^{\gamma-1} - r^{\gamma-1} + 1 = 0$$

or $$\left(\ln \frac{x}{r}\right) (\gamma - 1) - 1 + \frac{1}{r^{\gamma-1}} = 0$$

or $$(\gamma - 1) \ln \frac{x}{r} + \frac{1}{r^{\gamma-1}} - 1 = 0$$ **Proved**

## 2.3 OTTO CYCLE (CONSTANT VOLUME CYCLE)

The Otto cycle is named after its originator Dr. A.N. Otto (1876), a German scientist who devised this idealized cycle for the spark ignition internal combustion engines. Figure 2.3

shows the representation of this cycle on *p-v* or *p-V* and *T-s* or *T-S* diagrams. It is composed of four internally reversible processes, two adiabatic and two constant volume processes (hence the name constant volume cycle). The various process are:

Process 1–2: Isentropic compression of air raising its pressure and temperature.

Process 2–3: Addition of heat at constant volume causing rise in pressure and temperature of air.

Process 3–4 : Isentropic expansion of air from high pressure and temperature to lower pressure and temperature to do work.

Process 4–1: Rejection of heat at constant volume causing drop in pressure and temperature of air restoring initial state.

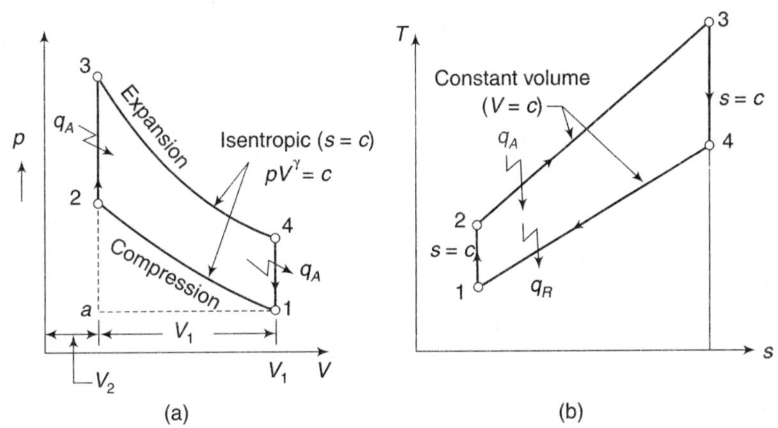


**Fig. 2.3** Otto cycle on *p-V* and *T-s* diagrams

The proximity of air standard Otto cycle to the actual spark ignition cycle can be better understood by assuming a piston cylinder as shown in Fig. 2.4. The actual engine consists of four strokes namely (i) suction, (ii) compression, (iii) expansion or power and (iv) exhaust stroke. *The actual engine operates as an open cycle. But as discussed earlier, the idealized cycle is free from suction and exhaust process, being assumed a closed cycle.* We will now see that the closed assumption is not a wrong one.

Consider that the piston is situated at BDC position and the cylinder is full of air. This is represented as point 1 on the diagram. As the piston moves from BDC position to TDC position, compression of the air occurs isentropically ending at 2 and thus raising the pressure and temperature of air. In actual spark ignition engine, the mixture of air and fuel which has entered during suction stroke (i.e., movement of piston from TDC to BDC) is compressed (Fig. 2.3).

The piston is momentarily at rest at TDC position and heat is then added to air from an external source which is brought into contact with the cylinder head instantaneously so that both pressure and temperature rise to the highest during a constant volume process 2–3. In an actual spark ignition engine, heat is added by igniting the fuel-air mixture.

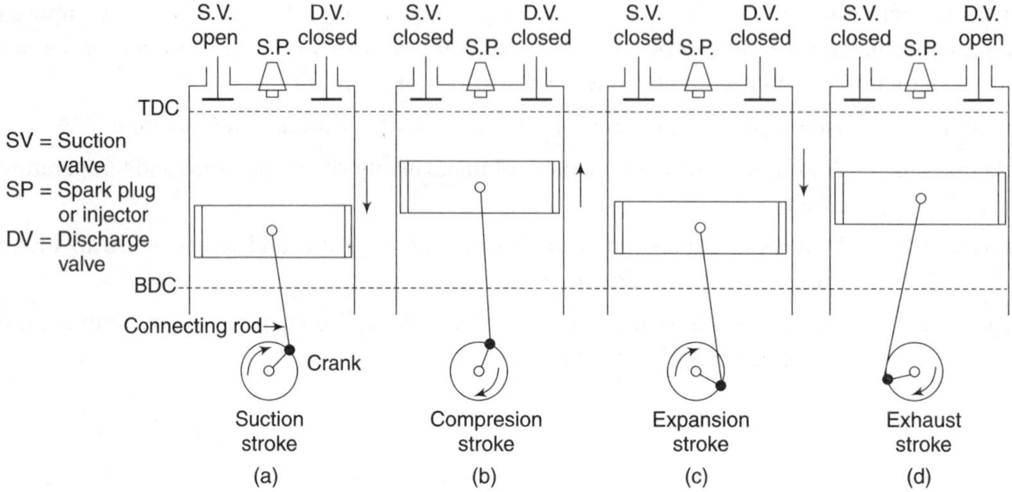


**Fig. 2.4** Engine working on Otto cycle

On removal of the heat source, the air at high temperature and pressure expands isentropically causing the piston to move from TDC to BDC position. This process is represented as 3–4. It is also called *power stroke* because it is the only process in which the work is done. In actual engine, the products of combustion at high pressure and temperature forces the piston to move from TDC to BDC

The piston is momentarily at BDC position when heat is rejected to external heat sink by bringing it into contact with the cylinder head at constant volume until the initial state is reached. This process is represented by 4–1. In actual engine, the gases contain the products of combustion so that the exhaust stroke is essential. Consequently, the exhaust valve opens at the end of the expansion stroke and the exhaust gases leave the cylinder while the pressure drop from $p_4$ to $p_1$. Thus, heat is rejected to outside atmosphere at constant volume during the process 4–1. The remaining exhaust gases are now expelled from the cylinder as the piston moves again from BDC to TDC This process is called the exhaust process. In the case of actual engine, once the piston reaches the TDC position, the exhuast valve closes and the intake valve opens while the piston returns to BDC position. During suction stroke the cylinder is filled with fresh mixture of air and fuel as shown by dotted line *a*–1.

In the air standard cycles, recharging of cylinder is not necessary since the same fluid continuously undergoes cyclic variation. It is to be noted that the work required to push the charge out of cylinder is the same in magnitude but opposite in sign to that required to suck the new charge. Hence, these two functions, suction and exhaust, do not affect the net work done by the cycle and it is the only cycle 1–2–3–4–1 which is important in the theoretical cycle.

**Thermal Efficiency or Air Standard Efficiency.** The thermal efficiency or air standard efficiency of the Otto cycle can be calculated as follows.

Consider 1 kg of air flowing through the cycle. Since the air in the cylinder acts as a closed system, hence from first law for isentropic compression and expansion process.

$$-w = \Delta u$$

For constant volume heat addition and rejection processes, since $w = 0$

$$\therefore \quad q = \Delta u = c_v \, \Delta T$$

$$\therefore \quad \text{Heat added} = q_A = q_{2\text{–}3} = c_v \, (T_3 - T_2) \text{ kJ/kg}$$

$$\text{Heat rejected} = q_R = q_{4\text{–}1} = c_v \, (T_4 - T_1) \text{ kJ/kg}$$

$$\therefore \text{ Work done per cycle} = w = \text{Heat added} - \text{Heat rejected}$$

or

$$w = c_v \, (T_3 - T_2) - c_v \, (T_4 - T_1) \text{ kJ/kg}$$

***Note:*** If the heat rejected is written as $q_R = q_{1\text{–}4} = c_v \, (T_1 - T_4)$ which is a negative value, then the work done will be a summation of heat supplied and heat rejected resulting in the same equation as for work done given above.

$$\therefore \quad \text{Thermal efficiency} = \frac{\text{Work done}}{\text{Heat added}}$$

or

$$\eta_{th} = \frac{c_v(T_3 - T_2) - c_v(T_4 - T_1)}{c_v \, (T_3 - T_2)} = 1 - \frac{T_4 - T_1}{T_3 - T_2} \qquad ...(2.6)$$

But

$$T_2 = T_1 \left(\frac{V_1}{V_2}\right)^{\gamma-1} = T_1 \left(\frac{v_1}{v_2}\right)^{\gamma-1} = T_1 \, r^{\gamma-1} \qquad ...(2.7)$$

where

$$r = \frac{v_1}{v_2} = \frac{V_1}{V_2} = \text{compression ratio}$$

$$T_3 = T_2 \frac{p_3}{p_2} = \alpha T_2 = \alpha T_1 \, r^{\gamma-1} \qquad ...(2.8)$$

where

$$\alpha = \frac{p_3}{p_2} = \frac{T_3}{T_2}$$

and

$$T_4 = T_3 \left(\frac{1}{r}\right)^{\gamma-1} = \alpha T_1 \qquad ...(2.9)$$

$\therefore$ Putting the above values of $T_2$, $T_3$ and $T_4$ in Eq. (2.6), we get

$$\eta_{th} = 1 - \frac{\alpha T_1 - T_1}{\alpha T_1 \, r^{\gamma-1} - T_1 \, r^{\gamma-1}}$$

$$\eta_{th} = \text{ASE} = 1 - \frac{1}{r^{\gamma-1}} \qquad ...(2.10)$$

From Eq. (2.10), it is obvious that the *air standard efficiency of the Otto cycle depends only on the compression ratio (r) and thus efficiency increases as the compression ratio increases as shown in Fig. 2.5.* In the beginning, the efficiency increases rapidly and after that it is slow.

*In actual engines, working on Otto cycle, the compression ratio varies from 5 to 10.5 and depends upon the quality of the fuel.*

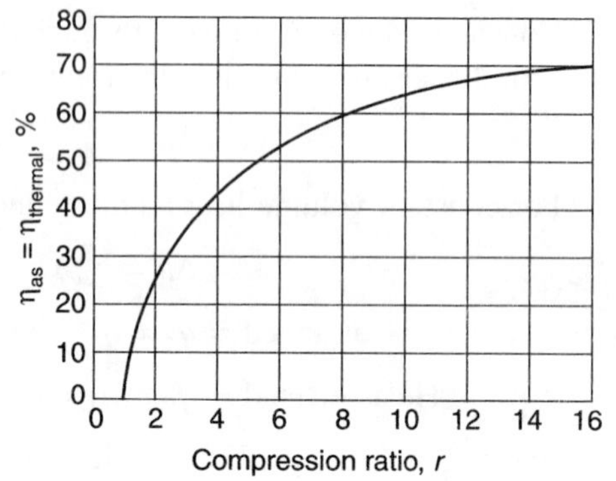


**Fig. 2.5** Thermal efficiency of Otto cycle vs. compression ratio

**Mean Effective Pressure.** *The mean effective pressure may be defined as the ratio of work done to the displacement volume of the piston.* Consider 1 kg of working fluid.

$$\text{Work done} = c_v\,[(T_3 - T_2) - (T_4 - T_1)]$$

$$\text{Displacement volume} = v_s = v_1 - v_2 = V_s = V_1 - V_2$$

$$\text{m.e.p.} = \frac{\text{Work done per cycle}}{\text{Displacement volume}} = \frac{c_v[(T_3 - T_2) - (T_4 - T_1)]}{v_1 - v_2}$$

But $$v_1 - v_2 = v_1\left(1 - \frac{1}{r}\right) = \frac{RT_1}{p_1}\left(\frac{r-1}{r}\right) = \frac{(\gamma - 1)T_1}{p_1}\left(\frac{r-1}{r}\right)c_v$$

Putting the values or $T_2$, $T_3$ and $T_1$, we get

$$\therefore \quad \text{m.e.p.} = \frac{p_1 r[\alpha r^{\gamma-1} - r^{\gamma-1}) - (\alpha - 1)}{(\gamma - 1)\,(r - 1)}$$

$$= p_1\, r\left[\frac{(r)^{\gamma-1}\,(\alpha - 1) - (\alpha - 1)}{(r - 1)\,(\gamma - 1)}\right]$$

or $$\text{m.e.p.} = p_1 . r\frac{(\alpha - 1)\,(r^{\gamma-1} - 1)}{(r - 1)\,(\gamma - 1)} \text{ bar} \quad (\text{if } p_1 \text{ is in bar}) \qquad ...(2.11)$$

This is the required expression for m.e.p. for Otto cycle. The m.e.p. may also be expressed as follows.

The net work done in Otto cycle is expressed as

$$w = \frac{(p_3 v_3 - p_4 v_4)}{\gamma - 1} - \frac{(p_2 v_2 - p_1 v_1)}{\gamma - 1}$$

So, $$\text{m.e.p.} = \frac{\left[\dfrac{(p_3 v_3 - p_4 v_4)}{\gamma - 1} - \dfrac{(p_2 v_2 - p_1 v_1)}{\gamma - 1}\right]}{(v_1 - v_2)} \qquad ...(2.12)$$

**Work Ratio (WR).** The work ratio of Otto cycle is given by

$$WR = \frac{\text{Net work}}{\text{Positive work}}$$

$$= \frac{c_v(T_3 - T_4) - c_v(T_2 - T_1)}{c_v(T_3 - T_4)} = 1 - \frac{T_2 - T_1}{T_3 - T_4} = 1 - \frac{T_1\,(r^{\gamma-1} - 1)}{\alpha\, T_1\,(r^{\gamma-1} - 1)}$$

$$= 1 - \frac{1}{\alpha} \qquad ...(2.13)$$

*For the high value of work ratio, α should be as high as possible but this is limited by the metallurgical condition of the cylinder and piston.*

**EXAMPLE 2.3** Compare the change in efficiency of an Otto engine when the compression ratio changes from 4 to 5. Take $\gamma = 1.4$ for air.

***Solution*** For Otto cycle, we know that

$$\eta_{th} = 1 - \frac{1}{r^{\gamma-1}}$$

where $r$ = Compression ratio and $\gamma$ = Index for isentropic process

$$\therefore \qquad \eta_{th1} = 1 - \frac{1}{r_1^{\gamma-1}} = 1 - \frac{1}{4^{0.4}} = 1 - 0.575 = 0.425 \quad \text{or} \quad 42.5\%$$

$$\eta_{th2} = 1 - \frac{1}{r_2^{\gamma-1}} = 1 - \frac{1}{5^{0.4}} = 1 - 0.5253 = 0.4747 \quad \text{or} \quad 47.47\%$$

Hence, percentage increase in efficiency

$$= \frac{\eta_{th2} - \eta_{th1}}{\eta_{th1}} \times 100 = \frac{0.4747 - 0.425}{0.425} \times 100 = 11.69\% \qquad \textit{Ans.}$$

**EXAMPLE 2.4** The pressure limits in an Otto air cycle air 100 kN/m$^2$ and 2000 kN/m$^2$ respectively. The compression ratio is 4. Calculate (a) thermal efficiency, (b) mean effective pressure, and (c) worke ratio. Take $\gamma = 1.4$ for air.

***Solution*** Refer to Fig. 2.3.

(a) Thermal efficiency for Otto cycle is

$$\eta_{th} = 1 - \frac{1}{r^{\gamma-1}} = 1 - \frac{1}{4^{0.4}} = 1 - 0.575 = 0.425 \quad \text{or} \quad 42.5\% \qquad \textit{Ans.}$$

(b) We have, $p_1 = 100$ kN/m$^2$ and $p_3 = 2 \times 10^3$ kN/m$^2$

$$\frac{p_2}{p_1} = \left(\frac{v_1}{v_2}\right)^{\gamma} = (4)^{1.4} = 6.964$$

$$\therefore \qquad p_2 = p_1 \times 6.964 = 100 \times 6.964 = 6.964 \times 10^2 \text{ kN/m}^2$$

$$p_3 = 2 \times 10^3 \text{ kN/m}^2$$

$$\therefore \quad \frac{p_3}{p_2} = \alpha = \frac{2 \times 10^3}{6.964 \times 10^2} = 2.871$$

The mean effective pressure is given by

$$\text{m.e.p.} = p_1 r \frac{(\alpha - 1)(r^{\gamma-1} - 1)}{(r-1)(\gamma - 1)}$$

$$= 100 \times 4 \frac{(2.871 - 1)(4^{0.4} - 1)}{(4-1)(1.4-1)} = 4.62 \times 10^2 \text{ kN/m}^2 = 4.62 \text{ bar} \qquad \textit{Ans.}$$

**Alternative Method.** Mean effective pressure may also be calculated as follows

$$\frac{p_4}{p_3} = \left(\frac{v_3}{v_4}\right)^{\gamma} = -\left(\frac{1}{4}\right)^{1.4} = \frac{1}{6.964}$$

or $$p_4 = 20 \times 100 \times \frac{1}{6.964} = 2.871 \times 10^2 \text{ kN/m}^2$$

Let $v_3 = v_2 = v$ and $r = \frac{v_1}{v_2} = 4$, hence $v_1 = 4v$

$$\text{Work done} = w = \frac{(p_3 v_3 - p_4 v_4)}{\gamma - 1} - \frac{(p_2 v_2 - p_1 v_1)}{\gamma - 1}$$

$$= 10^2 \left[\frac{(20 \times v - 2.871 \times 4v)}{0.4} - \frac{(6.964 \times v - 1 \times 4v)}{0.4}\right] = 13.88 \times v \times 10^2 \text{ kJ/kg}$$

$$\therefore \quad \text{m.e.p} = \frac{w}{v_1 - v_2} = \frac{w}{3v} = \frac{13.88 \times v \times 10^2}{3v} = 4.62 \times 10^2 \text{ kN/m}^2 = 4.62 \text{ bar} \qquad \textit{Ans.}$$

$$\text{Work ratio, } w_R = 1 - \frac{1}{\alpha} = 1 - \frac{1}{2.871}$$

$$= 0.652 \qquad \textit{Ans.}$$

**EXAMPLE 2.5** An air standard Otto cycle has a compression ratio of 8. At the start of the compression process, the temperature is 26°C and the pressure is 1 bar. If the maximum temperature of the cycle is 1080°C, calculate (a) the heat supplied per kg of air, (b) the net work done per kg of air and (c) the thermal efficiency of the cycle.

***Solution*** Refer to Fig. 2.3. From isentropic compression process 1–2,

$$\frac{T_2}{T_1} = \left(\frac{v_1}{v_2}\right)^{\gamma-1} = \left(\frac{V_1}{V_2}\right)^{\gamma-1} = (8)^{0.4} = 2.2973$$

or $$T_2 = (273 + 26) \times 2.2973 = 686.89 \text{ K}$$

From isentropic expansion process 3–4,

$$\frac{T_4}{T_3} = \left(\frac{v_3}{v_4}\right)^{\gamma-1} = \left(\frac{V_2}{V_1}\right)^{\gamma-1} = \left(\frac{1}{8}\right)^{0.4} = \frac{1}{2.2973}$$

or $$T_4 = (1080 + 273)/2.2973 = 588.95 \text{ K}$$

(a) Assuming $c_v = 0.718$ for air, the heat added is

$$q_A = q_{2-3} = c_v(T_3 - T_2) = 0.718\ (1353 - 686.89) = 478.266 \text{ kJ/kg} \qquad \textbf{Ans.}$$

(b) Heat rejected is $q_R = q_{4-1} = c_v\ (T_4 - T_1) = 0.718\ (588.95 - 299) = 208.18$ kJ/kg

Work done = $w$ = Heat supplied – Heat rejected

$$= 478.266 - 208.18 = 270.08 \text{ kJ/kg} \qquad \textbf{Ans.}$$

(c) The thermal efficiency = $\eta_{th} = \dfrac{\text{Work done}}{\text{Heat supplied}} = \dfrac{270.08}{478.266} = 56.47\%$ **Ans.**

**EXAMPLE 2.6** In the constant volume cycle, the temperatures at the beginning and end of the compression are 43°C and 323°C respectively. Calculate the (a) air standard efficiency and (b) the compression ratio. Assume $\gamma = 1.4$ for air.

A petrol engine, with the above compression ratio develops 22 kW indicated power and consumes 7.5 litres of fuel per hour. The specific gravity of fuel is 0.78 and its calorific value is 44000 kJ/kg. Calculate the indicated thermal efficiency and the relative efficiency.

***Solution*** Refer to Fig. 2.3. $T_1 = 273 + 43 = 316$ K, $T_3 = 273 + 323 = 596$ K

From isentropic compression process 1–2

$$\frac{T_2}{T_1} = \left(\frac{v_1}{v_2}\right)^{\gamma-1} \quad \text{or} \quad \frac{v_1}{v_2} = r = \left(\frac{T_2}{T_1}\right)^{1/(\gamma-1)} = \left(\frac{596}{316}\right)^{1/0.4}$$

(a) or $$r = 4.884 \qquad \textbf{Ans.}$$

(b) ∴ Air standard efficiency $= 1 - \dfrac{1}{r^{\gamma-1}} = 1 - \dfrac{1}{4.884^{0.4}} = 46.98\%$ **Ans.**

Fuel consumption = 7.5 litres/h

We know that 1 liter of water = 1 kg of water

∴ Fuel consumption = 7.5 × 0.78 × 1 = 5.85 kg/h

$$\text{Indicated thermal efficiency} = \frac{\text{Indicated work}}{\text{Heat added}} = \frac{22 \times 3600}{44000 \times 5.85} = 30.76\%$$

$$\text{Relative efficiency} = \frac{\text{Thermal efficiency}}{\text{Air standard efficiency}} = \frac{30.76}{46.98} \times 100 = 65.74\% \qquad \textbf{Ans.}$$

**EXAMPLE 2.7** Show that, for the maximum work to be done per kg of air in Otto cycle between given upper and lower limits of absolute temperatures $T_3$ and $T_1$ respectively, the ratio of compression should have the value.

$$r = \left(\frac{T_3}{T_1}\right)^{1.25} \quad \text{when} \quad \gamma = 1.4$$

***Solution*** Refer to Fig. 2.3.

Work done per cycle = Heat added – Heat rejected

$$w = c_v(T_3 - T_2) - c_v(T_4 - T_1) \text{ kJ/kg}$$

From isentropic compression process 1–2,

$$T_2 = T_1 r^{\gamma-1} \quad \text{where} \quad r = \frac{v_1}{v_2} = \frac{v_4}{v_2}$$

From isentropic expansion process 3–4, $T_4 = T_3\left(\frac{1}{r}\right)^{\gamma-1}$

$$\therefore \quad w = c_v\left[T_3 - T_1\, r^{\gamma-1} - T_3\left(\frac{1}{r}\right)^{\gamma-1} - T_1\right]$$

To obtain the maximum work, differentiate work done $w$ with respect to $r$ and equal to zero. Thus,

$$\frac{dw}{dr} = 0 = c_v\left[-T_1\,(\gamma-1)\,r^{\gamma-2} - T_3\,(1-\gamma)r^{-\gamma}\right]$$

or $$T_1(\gamma-1)\,r^{\gamma-2} = T_3(\gamma-1)r^{-\gamma}$$

or $$r^{2-2\gamma} = \frac{T_1}{T_3}$$

or $$\frac{T_3}{T_1} = r^{2\gamma-2} = r^{2.8-2} = r^{0.8}$$

or $$r = \left(\frac{T_3}{T_1}\right)^{1/0.8} = \left(\frac{T_3}{T_1}\right)^{1.25}$$ **Proved**

**EXAMPLE 2.8** A Maruti car working on Otto cycle has initial pressure 1 bar and pressure after compression 10 bar.

Calculate

(a) the compression ratio,

(b) the percentage clearance, and

(c) the thermal efficiency.

If the initial temperature is 26°C find

(d) the temperature at the end of compression, and

(e) the mean effective pressure if the maximum pressure in the cycle is 20 bar.

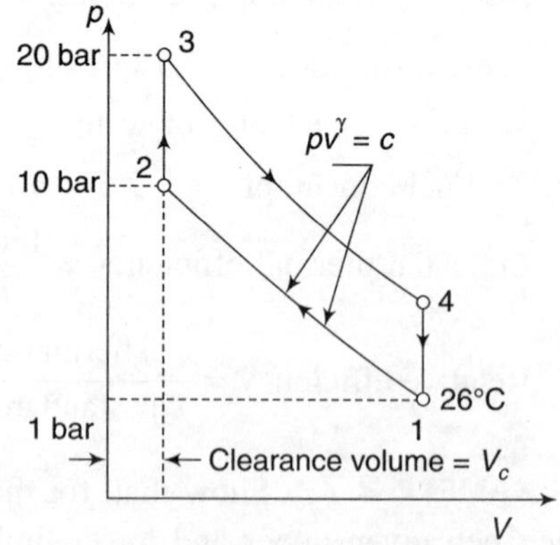


**Fig. 2.6** Example 2.8

***Solution*** Refer to Fig. 2.6. From isentropic compression process 1–2, we have

(a) Compression ratio $= r = \frac{v_1}{v_2} = \frac{V_1}{V_2} = \left(\frac{p_2}{p_1}\right)^{1/\gamma} = \left(\frac{10}{1}\right)^{1/1.4} = 5.178$ ***Ans.***

(b) Clearance volume $= v_2 = v_3$

$\therefore$ Percentage clearance $= \frac{v_2}{v_1} \times 100 = \frac{100}{5.178} = 19.31$ of cylinder volume ***Ans.***

(c) Thermal efficiency $= \eta_{th} = 1 - \left(\frac{1}{r}\right)^{\gamma-1} = 1 - \left(\frac{1}{5.178}\right)^{0.4} = 0.428 = 48.22\%$ ***Ans.***

(d) $T_2 = T_1 (r)^{\gamma-1} = (273 + 26)(5.178)^{0.4} = 577.21$ K ***Ans.***

(e) Mean effective pressure is given by

$$\text{m.e.p.} = p_1 r \frac{(\alpha - 1)(r^{\gamma-1} - 1)}{(r-1)(\gamma - 1)} \text{ bar}$$

where $\alpha = \frac{p_3}{p_2} = \frac{20}{10} = 2$

$$\therefore \quad \text{m.e.p.} = 1 \times 5.178 \frac{(2-1)(5.178^{0.4} - 1)}{(5.178 - 1)(1.4 - 1)} = 2.882 \text{ bar}$$ ***Ans.***

**EXAMPLE 2.9** A two-cylinder, two-stroke cycle engine working on the constant volume cycle has cylinder diameter of 35 cm and stroke 50 cm. The clearance volume 1/15 of the swept volume, and the exhaust ports are closed by the piston when it has travelled 8 cm from the dead centre on the compression stroke. The indicated mean effective pressure is 4.5 bar and the specific fuel consumption is 0.27 kg/kWh, the calorific value of the fuel is 44000 kJ/kg. Calculate (a) the indicated power at 180 rpm, and (b) the relative efficiency based on the effective compression ratio of the engine.

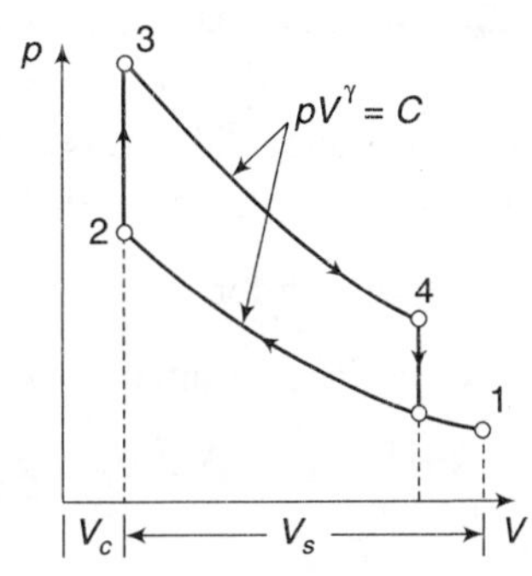


**Fig. 2.7** Example 2.9

***Solution*** Refer to Fig. 2.7.

Swept volume = Stroke volume $= V_s = V_1 - V_2$

Clearance volume $= V_2 = V_3 = \frac{V_s}{15} = V_c$

Stroke volume $= V_s = V_1 - V_2 = \frac{\pi}{4} d^2 \times L = \frac{\pi}{4} \times (0.35)^2 \times 0.5 = 0.0481 \text{ m}^3$

$\therefore$ Clearance volume $= V_2 = \frac{V_s}{15} = \frac{0.0481}{15} = 0.0032 \text{ m}^3$

$\therefore$ Cylinder volume $= V_1 = V_s + V_2 = 0.0481 + 0.0032 = 0.0513 \text{ m}^2$

Cylinder area $= A = \frac{\pi}{4} d^2 = \frac{\pi}{4} (0.35)^2 = 0.0962 \text{ m}^2$

$\therefore V_1 - V_4 = 0.0962 \times 0.08 = 0.007696 \text{ m}^3$

$\therefore V_4 = 0.0513 - 0.007696 = 0.0436 \text{ m}^3$

$\therefore$ Effective compression ratio $= r = \frac{V_4}{V_2} = \frac{0.0436}{0.0032} = 13.62$

(a) i.p. $= \frac{p_m LAnx}{60} = \frac{4.5 \times 100 \times 0.5 \times 0.0962 \times 180}{60} \times 2 = 129.87 \text{ kW}$ ***Ans.***

Here, $n = N$ for two-stroke engine

(b) Thermal efficiency $= \frac{\text{Work done}}{\text{Heat added}} = \frac{3600}{0.27 \times 44000} = 30.3\%$ ***Ans.***

Air standard efficiency $= 1 - \frac{1}{r^{\gamma-1}} = 1 - \left(\frac{1}{13.62}\right)^{0.4} = 64.82\%$

$\therefore$ Relative efficiency $= \frac{\text{Thermal efficiency}}{\text{Air standard efficiency}} = \frac{30.3}{64.82} = 46.75\%$ ***Ans.***

**EXAMPLE 2.10** The bore and stroke of the cylinder of an engine working on Otto cycle are 18 cm and 32 cm respectively. The clearance volume is 0.002 m³. Find compression ratio and the thermal efficiency of the engine. Take $\gamma = 1.4$ for air.

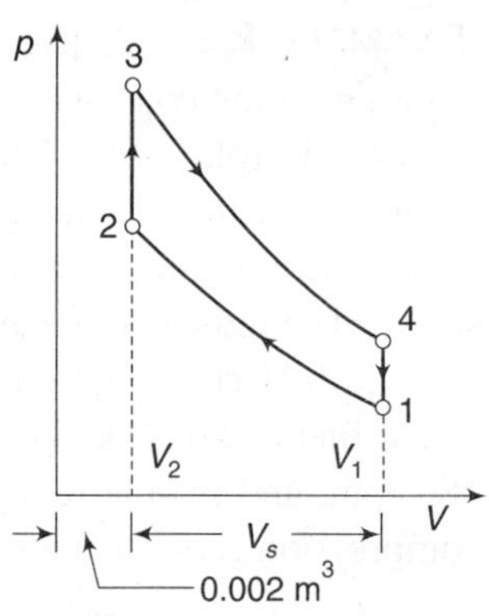


**Fig. 2.8** Example 2.10

***Solution*** Refer to Fig. 2.8.

Clearance volume $= V_2 = V_3 = V_c = 0.002 \text{ m}^3$

Stroke volume $= V_s = V_1 - V_2$

$= \frac{\pi}{4} d^2 \times L = \frac{\pi}{4} \times (0.18)^2 \times 0.32$

$= 0.00814 \text{ m}^3$

Cylinder volume $= V_1 = V_c + V_s = 0.002 + 0.00814 = 0.01014 \text{ m}^3$

$\therefore$ Compression ratio $= r = \frac{V_1}{V_2} = \frac{0.01014}{0.002} = 5.07$ ***Ans.***

Thermal efficiency $= \eta_{th} = 1 - \frac{1}{r^{\gamma-1}} = 1 - \frac{1}{5.07^{0.4}} = 47.77\%$ ***Ans.***

## 2.4 DIESEL CYCLE (CONSTANT PRESSURE HEAT ADDITION CYCLE)

This cycle is named after its inventor Rudolph Diesel (1893) who devised the idealized cycle for the compression ignition thermal combustion engines or the engines working on diesel oil.

The cycle like the Otto cycle consists of four internally reversible processes, two adiabatic, one constant pressure and other constant volume. Figure 2.9 shows the representation Diesel cycle on *p–V* and *T–s* diagrams. The various processes are

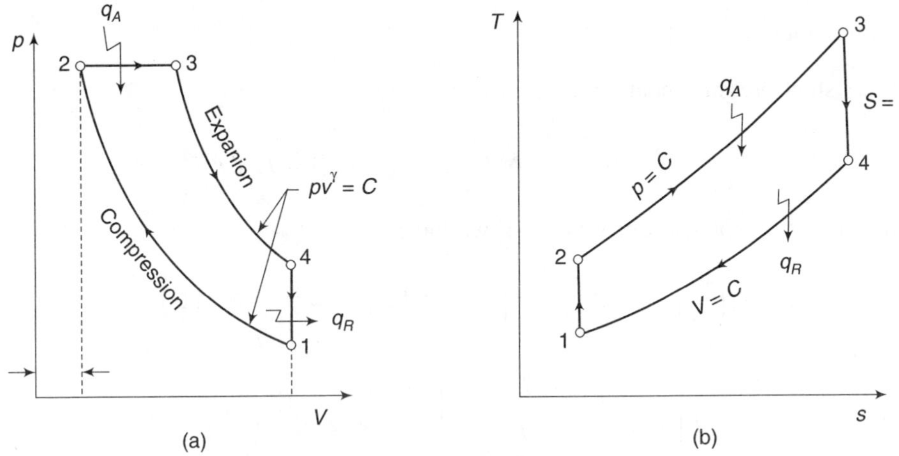


**Fig. 2.9** Diesel cycle on *p–V* and *T–s* diagrams

1–2: Isentropic compression of air during which the piston moves from BDC to TDC position, thus, raising pressure and temperature of air.

2–3: Addition of heat at constant pressure from an external heat source which is brought into contact with cylinder head. During this process, pressure remains constant but volume and temperature increase. In actual diesel engine, the heat is added by injecting and burning of fuel (due to compression ignition) in the cylinder.

At point 3, the heat addition is stopped and volume ratio $v_3/v_2$ is called ***cut-off ratio***.

3–4: Isentropic expansion of air at high pressure and temperature to lower pressure and temperature till the piston reaches the BDC position. Due to expansion, work is done, and this stroke is called the *power stroke*. In actual engine, the products of combustion expand.

4–1: Rejection of heat at constant volume to an external heat sink and the air restores its initial state and thus the cycle is completed.

Heat added = $q_A = q_{2-3} = c_p\,(T_3 - T_2)$ and

Heat rejected = $q_R = q_{4-1} = c_v\,(T_4 - T_1)$

Work done per cycle = $w = c_p\,(T_3 - T_2) - c_v\,(T_4 - T_1)$

Hence, the thermal efficiency is given by

$$\eta_{th} = \frac{\text{Work done}}{\text{Heat supplied}} = \frac{c_p(T_3 - T_2) - c_v(T_4 - T_1)}{c_p(T_3 - T_2)} = 1 - \frac{(T_4 - T_1)}{\gamma\,(T_3 - T_2)} \qquad ...(2.14)$$

Let $$r = \frac{v_1}{v_2} = \frac{v_4}{v_2} = \text{Compression ratio}$$

From isentropic compression process 1–2, we have, $T_2 = T_1 \left(\frac{v_1}{v_2}\right)^{\gamma-1} = T_1 r^{\gamma-1}$

Let cut-off ratio $= \frac{v_3}{v_2} = \beta$

From constant pressure heat addition process 3–2, we have

$$\frac{T_3}{T_2} = \frac{v_3}{v_2} = \beta \qquad \text{or} \qquad T_3 = T_2\,\beta = T_1\,\beta r^{\gamma-1}$$

From isentropic expansion process 3–4, we have

$$T_4 = T_3\left(\frac{v_3}{v_4}\right)^{\gamma-1} = T_3\left(\frac{v_3}{v_2}\frac{v_2}{v_4}\right)^{\gamma-1} = T_3\left(\frac{\beta}{r}\right)^{\gamma-1} = T_1\,\beta r^{\gamma-1}\left(\frac{\beta}{r}\right)^{\gamma-1} = T_1\,\beta^{\gamma}$$

$$\therefore \qquad \eta_{th} = 1 - \frac{1}{\gamma}\left[\frac{T_1\,\beta^{\gamma} - T_1}{\beta\,T_1\,r^{\gamma-1} - T_2\,r^{\gamma-1}}\right]$$

or $$\text{ASE} = \eta_{th} = 1 - \frac{1}{r^{\gamma-1}}\left[\frac{\beta^{\gamma} - 1}{\gamma(\beta - 1)}\right] \qquad ...(2.15)$$

Equation (2.15) gives the required expression for thermal efficiency in terms of compression ratio, cut-off ratio ($\beta$) and adiabatic exponent, $\gamma$. Figure 2.10 shows the variation of Diesel cycle efficiency with compression ratio ($r$) for various cut-off ratios, $\beta$. It is obvious that the thermal efficiency of Diesel cycle increases with increase of compression ratio. It is also evident that as the cut-off ratio ($\beta$) increases, the value of terms in the bracket increases hence the thermal efficiency decreases. *When $\beta = 1$, the efficiency of the Diesel cycle is equal to the efficiency of Otto cycle. For a given compression ratio r, under all circumstances when $\beta$ is greater than unity, the efficiency of the Diesel cycle is less than that for Otto cycle. This is true because the term in the bracket in Eq. (2.15) is always positive. This comparative statement is somewhat misleading for a modern Diesel engine which has a higher efficiency than a spark ignition engine found in modern automobiles. This is due to the fact that Diesel or compression ignition engines normally operate at considerably higher compression ratio than spark ignition engines.* In practice, the compression ratio is nearly twice as large. Theoretically, the lower cut-off ratios lead to better efficiency but high value of $\beta$ leads to more power output. Large values of cut-off ratio are not possible, since in practice, smoking will occur due to improper combustion.

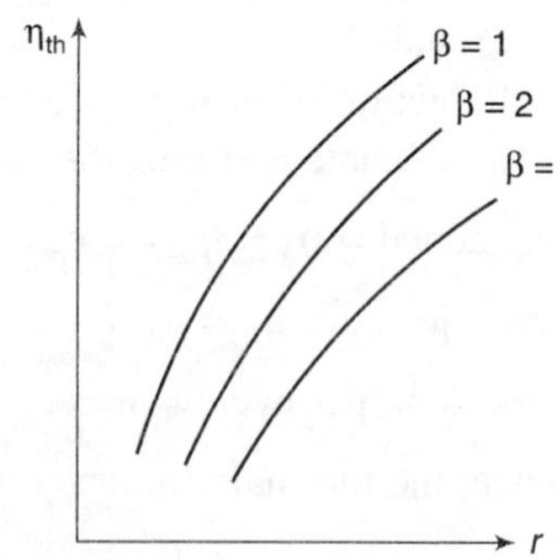


**Fig. 2.10** Diesel cycle efficiencies

$$\text{Mean effective pressure} = \frac{\text{Work done}}{\text{Displacement volume}}$$

or $$\text{m.e.p.} = \frac{c_p(T_3 - T_2) - c_v(T_4 - T_1)}{v_1 - v_2}$$

But $$v_1 - v_2 = v_1\left(1 - \frac{v_2}{v_1}\right) = v_1\left(1 - \frac{1}{r}\right) = \frac{RT_1}{p_1}\left(\frac{r-1}{r}\right) = \frac{c_v(\gamma - 1)T_1}{p_1}\left(\frac{r-1}{r}\right)$$

$\therefore$ $$\text{m.e.p.} = \frac{c_p(T_3 - T_2) - c_v(T_4 - T_1)}{c_v\, T_1 \frac{(\gamma - 1)}{p_1}\left(\frac{r-1}{r}\right)}$$

$$= p_1\left(\frac{r}{r-1}\right)\left(\frac{1}{\gamma - 1}\right)\left[\gamma\left(\frac{T_3 - T_2}{T_1} - \frac{T_4 - T_1}{T_1}\right)\right]$$

or $$\text{m.e.p.} = p_1\, r\left[\frac{\gamma . r^{\gamma-1}(\beta - 1) - (\beta^\gamma - 1)}{(r-1)\,(\gamma - 1)}\right] \quad ...(2.16)$$

**EXAMPLE 2.11** An air standard Diesel cycle has a compression ratio of 16. The pressure at the beginning of the compression stroke is 1 for and the temperature is 20°C. The maximum temperature is 1430°C. Determine the thermal efficiency and the mean effective pressure for this cycle.

***Solution*** Refer to Fig. 2.9.

Given, $r = \frac{v_1}{v_2} = 16$, $p_1 = 1$ bar, $T_2 = 273 + 20 = 293$ K, $T_3 = 273 + 1430 = 1703$ K

The thermal efficiency of Diesel cycle is given by

$$\eta_{th} = 1 - \frac{1}{r^{\gamma-1}}\left[\frac{(\beta^\gamma - 1)}{\gamma\,(\beta - 1)}\right] \quad \text{where} \quad \beta = \frac{V_3}{V_2} = \frac{v_3}{v_2} = \text{cut-off ratio}$$

Also, $$\frac{T_2}{T_1} = \left(\frac{v_1}{v_2}\right)^{\gamma-1} = r^{\gamma-1} = 16^{0.4} = 3.03 \quad \text{or} \quad T_2 = 293 \times 3.03 = 887.79 \text{ K}$$

From constant pressure heat addition process 2–3,

$$\frac{v_2}{T_2} = \frac{v_3}{T_2} \text{ or } \beta = \frac{v_3}{v_2} = \frac{T_3}{T_2} = \frac{1703}{887.79} = 1.918$$

$\therefore$ $$\eta_{th} = 1 - \frac{1}{16^{0.4}}\left[\frac{(1.918^{1.4} - 1)}{1.4(1.918 - 1)}\right] = 61.8\%$$ ***Ans.***

Mean effective pressure for Diesel cycle is

$$\text{m.e.p.} = p_1 r \left[\frac{\gamma . r^{\gamma-1}(\beta - 1) - (\beta^{\gamma} - 1)}{(r - 1)(\gamma - 1)}\right]$$

$$= 1 \times 16 \left[\frac{1.4 \times 16^{0.4}(1.918 - 1) - (1.918^{1.4} - 1)}{(16 - 1)(1.4 - 1)}\right] = 6.414 \text{ bar} \qquad \textit{Ans.}$$

**EXAMPLE 2.12** A Diesel engine operates on the air-standard Diesel cycle. The engine has six cylinder of 11 cm bore and 13 cm stroke. The engine runs at 2000 rpm. At the beginning of compression, the air is at 1 bar and 26°C. If the clearance volume is 12.5 per cent of the stroke volume, find (a) compression ratio, (b) pressure and temperature of the air after compression, and (c) thermal efficiency and power output if the air is heated to 1370°C.

***Solution*** Refer to Fig. 2.9.

Stroke volume = $v_s = v_1 - v_2$, Clearance volume = $v_c = v_2$

$$\text{Hence, } 0.125 = \frac{v_2}{v_s} = \frac{v_2}{v_1 - v_2} = \frac{1}{(v_1/v_2) - 1} = \frac{1}{r - 1}$$

(a) ∴ Compression ratio = $r = 9$ ***Ans.***

$$\text{(b)} \qquad v_1 = \frac{RT_1}{p_1} = \frac{0.287 \times 299}{1 \times 10^2} = 0.858 \text{ m}^3\text{/kg} \quad \therefore v_2 = \frac{v_1}{r} = \frac{0.858}{9} = 0.0953 \text{ m}^3\text{/kg}$$

$$\frac{T_2}{T_1} = \left(\frac{v_1}{v_2}\right)^{\gamma-1} = (9)^{0.4} = 2.408 \quad \text{or} \quad T_2 = 299 \times 2.408 = 719.992 \text{ K}$$

$$\text{Also,} \quad \frac{p_2}{p_1} = \left(\frac{v_1}{v_2}\right)^{\gamma} = (9)^{1.4} = 21.674$$

$$\text{or} \quad p_2 = 1 \times 21.674 = 21.674 \text{ bar} \qquad \textit{Ans.}$$

(c) From constant pressure heat addition 2–3

$$\frac{v_2}{T_2} = \frac{v_3}{T_3}$$

$$\text{or} \quad \beta = \frac{v_3}{v_2} = \frac{T_3}{T_2} = \frac{1370 + 273}{719.992} = 2.281$$

Thermal efficiency is given by

$$\eta_{th} = 1 - \frac{1}{r^{\gamma-1}}\left[\frac{(\beta^{\gamma} - 1)}{\gamma(\beta - 1)}\right] = 1 - \frac{1}{9^{0.4}}\left[\frac{(2.281^{1.4} - 1)}{1.4(2.2.81 - 1)}\right] = 49.7\% \qquad \textit{Ans.}$$

$$\text{Stroke volume} = V_s = \frac{\pi}{4} \times (0.11)^2 \times 0.13 = 0.001235 \text{ m}^3$$

$$\text{and} \quad V_s = V_1 - V_c = V_1 - V_2$$

$$\text{or} \quad V_1 = V_s + V_2 = V_s + 0.125\, V_s = 0.001235 \times 1.125 = 0.001389 \text{ m}^3$$

The mass of air for the six cylinders = $m = \frac{6V_1}{v_1} = \frac{6 \times 0.001389}{0.858} = 0.00971$ kg.

Head added = $Q_A = mc_p (T_3 - T_2) = 0.00971 \times 1.005 (1643 - 719.992) = 9.007$ kJ

Now $\frac{T_4}{T_3} = \left(\frac{v_3}{v_4}\right)^{\gamma-1} = \left(\frac{v_3}{v_1}\right)^{\gamma-1}$

But $v_3 = 2.281\ v_2 = \frac{2.281 v_1}{r}$

or $\frac{v_3}{v_1} = \frac{2.281}{9} = 0.2534$

$\therefore$ $T_4 = 1643\ (0.2534)^{0.4} = 948.76$ K

Heat rejected = $mc_v (T_4 - T_1) = 0.00971 \times 0.718\ (948.76 - 299) = 4.5299$ kJ

Work done/cycle = $W$ = Heat supplied – Heat rejected = 9.007 – 4.5299 = 4.4771 kJ

Power developed = $\frac{W \times N}{2 \times 60} = \frac{4.4771 \times 2000}{2 \times 60} = 74.618$ kW **_Ans._**

**EXAMPLE 2.13** A Diesel engine working on air-standard Diesel cycle takes air at 1 bar at 15°C. The specific volume of air at the inlet is 0.8 m$^3$/kg. The compression ratio is 14 and 840 kJ/kg of heat is added at constant pressure.

Determine (a) pressure, sp. volume and temperature at each point of the cycle, (b) thermal efficiency, (c) mean effective pressure, and (d) power developed by the engine if it consumes 0.2 kg/s of air.

***Solution*** Refer to Fig. 2.9.

(a) ***Point 1:*** $p_1 = 1$ bar, $T_1 = 273 + 15 = 288$ K and $v_1 = 0.8$ m$^3$/kg **_Ans._**

***Point 2:*** $v_2 = \frac{v_1}{r} = \frac{0.8}{14} = 0.05714$ m$^3$/kg

From isentropic compression 1–2, $p_2 = p_1 \left(\frac{v_1}{v_2}\right)^{\gamma} = 1 \times (14)^{1.4} = 40.23$ bar **_Ans._**

Temperature $T_2$ can be calculated by, $T_2 = T_1 \left(\frac{v_1}{v_2}\right)^{\gamma-1} = 288\ (14)^{0.4} = 827.64$ K **_Ans._**

***Point 3:*** Heat added = $q_A = 840 = c_p (T_3 - T_2) = 1.005\ (T_3 - 827.64)$

or $T_3 = 1663.46$ K **_Ans._**

From constant pressure heat addition process 2–3

$\frac{v_3}{v_2} = \frac{T_3}{T_2} = \frac{1663.46}{827.64} = 2.0098$ or $v_3 = 2.0098 \times 0.05714 = 0.1148$ m$^3$ **_Ans._**

$p_3 = p_2 = 40$ bar **_Ans._**

***Point 4:*** From isentropic expansion process 3–4

$$T_4 = T_3\left(\frac{v_3}{v_4}\right)^{\gamma-1} = \frac{T_3}{(v_4/v_3)^{\gamma-1}} = \frac{1663.46}{(0.8/0.1148)^{0.4}} = 756.16 \text{ K} \qquad \textbf{\textit{Ans.}}$$

From constant volume heat rejection process 4–1

$$= p_4 = \frac{p_1 \times T_4}{T_1} = \frac{1 \times 765.16}{288} = 2.656 \text{ bar} \qquad \textbf{\textit{Ans.}}$$

(b) Heat rejected $= q_R = c_v\,(T_4 - T_1) = 0.718(765.16 - 288) = 342.6$ kJ/kg

$\therefore$ Work done $= w = q_A - q_R = 840 - 342.6 = 497.4$ kJ/kg

$$\therefore \text{ Thermal efficiency} = \frac{w}{q_A} = \frac{497.4}{840} = 59.2\% \qquad \textbf{\textit{Ans.}}$$

(c)
$$\text{m.e.p.} = \frac{w}{v_1 - v_2} = \frac{497.4}{(0.8 - 0.05714)} = 669.57 \text{ kN/m}^2 = 6.6957 \text{ bar} \qquad \textbf{\textit{Ans.}}$$

**EXAMPLE 2.14** The German Mercedes '1900' car has a four-stroke cylinder in line Diesel engine with compression ratio 20 : 1 and expansion ratio 10 : 1. Calculate, the (a) cut-off ratio and (b) air standard efficiency.

***Solution*** Refer to Fig. 2.9.

$$\text{Cut-off ratio} = \beta = \frac{v_3}{v_2} = \frac{v_3}{v_4} \times \frac{v_4}{v_2} = \frac{v_3}{v_2} \times \frac{v_1}{v_2}$$

But compression ratio $= r = \dfrac{v_1}{v_2} = 20$ and expansion ratio $\dfrac{v_4}{v_3} = 10$

(a) $\therefore \quad \beta = \dfrac{20}{10} = 2$ ***Ans.***

(b) Air standard efficiency of Diesel cycle is given by

$$\eta_{Air\ stand.} = 1 - \frac{1}{r^{\gamma-1}}\left[\frac{(\beta^{\gamma} - 1)}{\gamma(\beta - 1)}\right] = 1 - \frac{1}{20^{0.4}}\left[\frac{(2^{1.4} - 1)}{1.4(2 - 1)}\right] = 64.67\% \qquad \textbf{\textit{Ans.}}$$

**EXAMPLE 2.15** A Diesel engine operating on a standard Diesel cycle has 20 cm bore and 30 cm stroke. The clearance volume is 420 cm$^3$. The fuel is injected at constant pressure for 5% the stroke. Calculate the air-standard efficiency. If the cut-off is delayed from 5 to 8% what will be the percentage loss in efficiency. In both cases, the compression ratio is the same.

***Solution*** Refer to Fig. 2.9.

Swept volume $= V_s = V_1 - V_2 = \dfrac{\pi}{4} d^2 \times L = \dfrac{\pi}{4} \times 20^2 \times 30 = 9424.778$ cm$^3$

Clearance volume $= V_c = V_2 = 420$ cm$^3$

Cylinder volume $= V_1 = V_s + V_2 = 9424.778 + 420 = 9844.778$ cm$^3$

But $\quad V_3 = V_2 + 0.05 \times V_2 = 420 + 0.05 \times 9424.778 = 891.238$ cm$^3$

$$\therefore \text{ Cut-off ratio} = \beta = \frac{V_3}{V_2} = \frac{891.238}{420} = 2.121$$

$$\text{Compression ratio} = r = \frac{V_1}{V_2} = \frac{9844.778}{420} = 23.439$$

Efficiency of Diesel cycle is given by

$$\eta_{th1} = 1 - \frac{1}{r^{\gamma-1}}\left[\frac{(\beta^{\gamma} - 1)}{\gamma(\beta - 1)}\right] = 1 - \frac{1}{23.439^{0.4}}\left[\frac{(2.121^{1.4} - 1}{1.4(2.121 - 1)}\right] = 69.98\% \qquad \textbf{\textit{Ans.}}$$

If the fuel is cut-off at 8% of the stroke, $V_3 = 420 + 0.08 \times 9424.778 = 1173.982 \text{ m}^3$

$$\text{Cut-off ratio} = \beta = \frac{V_3}{V_2} = \frac{1173.982}{420} = 2.795$$

Hence, the new thermal efficiency is

$$\eta_{th2} = 1 - \frac{1}{r^{\gamma-1}}\left[\frac{(\beta^{\gamma} - 1)}{\gamma(\beta - 1)}\right] = 1 - \frac{1}{23.439^{0.4}}\left[\frac{(2.795^{1.4} - 1)}{1.4 \times 1.795}\right] = 63.759\% \qquad \textbf{\textit{Ans.}}$$

$$\text{Percentage loss} = \frac{\eta_{th1} - \eta_{th2}}{\eta_{th1}} \times 100 = \frac{0.6998 - 0.63759}{0.6998} \times 100 = 8.88\% \qquad \textbf{\textit{Ans.}}$$

**EXAMPLE 2.16** A Diesel engine has a compression ratio of 16. The fuel is cut-off at 0.08 of stroke. The relative efficiency is 52%. Find the mass of the fuel of calorific value 41800 kJ/kg which would be required per kW per hour. Also find actual thermal efficiency.

***Solution*** Refer to Fig. 2.11.

$$\text{Compression ratio} = r = \frac{V_1}{V_2} = \frac{v_1}{v_2} = 16$$

Let $\quad V_s = \text{Stroke volume} = V_1 - V_2$

and $\quad V_3 - V_2 = 0.08V_s = 0.08\,(V_1 - V_2)$

or $$\frac{V_3}{V_2} - 1 = 0.08\left(\frac{V_1}{V_2} - 1\right)$$

$$= 0.08\,(16 - 1) = 1.2$$

or $$\beta = \frac{V_3}{V_2} = 2.2$$

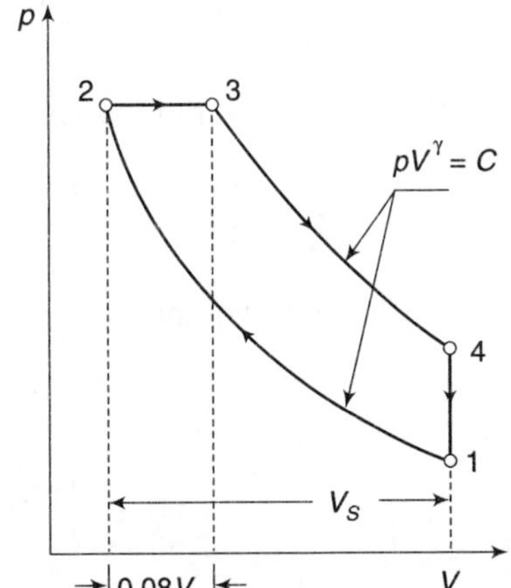

**Fig. 2.11** Example 2.16

The air standard efficiency of Diesel cycle is given by

$$\text{Air standard efficiency} = 1 - \frac{1}{r^{\gamma-1}}\left[\frac{(\beta^{\gamma} - 1)}{\gamma(\beta - 1)}\right]$$

$$= 1 - \frac{1}{16^{0.4}}\left[\frac{(2.2^{1.4} - 1)}{1.4(2.2 - 1)}\right] = 0.6042 \text{ or } 60.42\%$$

$$\text{Relative efficiency} = \frac{\text{Actual thermal efficiency}}{\text{Air standard efficiency}} \text{ or } 0.52 = \frac{\text{Actual thermal efficiency}}{0.6042}$$

or Actual thermal efficiency = 0.52 × 0.6042 = 0.3141 = 31.41% ***Ans.***

$$\text{or Actual thermal efficiency} = \frac{\text{Actual work done, kJ}}{\text{Heat added, kJ}}$$

or $$0.3141 = \frac{\text{Work done}}{\text{C.V.} \times \dot{m}_f} = \frac{\text{Work done}}{41800 \times \dot{m}_f}$$

or $$\frac{\text{Work done}}{\dot{m}_f} = 0.3141 \times 41800 \text{ kJ/kg of fuel}$$

$$= \frac{0.3141 \times 41800}{3600} \text{ kWh/kg of fuel}$$

∴ $$\text{Mass of fuel} = \dot{m}_f = \frac{3600}{0.3141 \times 41800} = 0.2741 \text{ kg/kWh.}$$ ***Ans.***

**EXAMPLE 2.17** A Diesel engine working on Diesel cycle takes air at pressure 1 bar and temperature 30°C. The pressure at the end of compression is 30 bar and the cut-off is 6% of the stroke. Calculate (a) the compression ratio, (b) the percentage clearance, (c) the heat supplied, (d) the heat rejected, (e) the thermal efficiency and (f) the mean effective pressure.

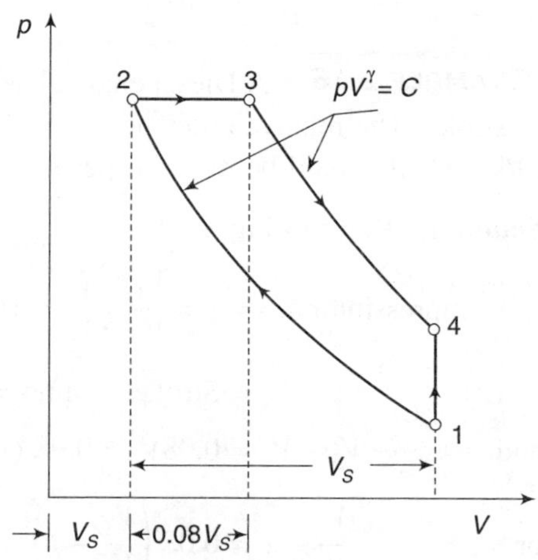


**Fig. 2.12** Example 2.17

***Solution*** Refer to Fig. 2.12. Assume 1 kg of working fluid.

(a) Compression ratio is given by

$$r = \frac{v_1}{v_2} = \left(\frac{p_2}{p_1}\right)^{1/\gamma} = \left(\frac{30}{1}\right)^{1/1.4} = 11.352 \text{ ***Ans.***}$$

(b) Clearance volume = $v_c = v_2$

$$\therefore \text{Percentage clearance} = \frac{v_2}{v_s} \times 100 = \frac{v_2}{v_1 - v_2} = \frac{100}{r-1} = \frac{100}{10.352} = 9.659\%$$ ***Ans.***

(c) $$\frac{T_2}{T_1} = \left(\frac{v_1}{v_2}\right)^{\gamma-1} = (11.352)^{0.4} = 2.6425$$

or $$T_2 = (273 + 30) \times 2.6425 = 800.67 \text{ K}$$

$$v_3 - v_2 = 0.06\ v_s$$

But $$v_2 = \frac{v_s}{r-1} = \frac{v_s}{10.352} = 0.0965\ v_s$$

$\therefore$ $$v_3 = 0.0965\ v_s + 0.06\ v_s = 0.1565\ v_s$$

And $$\frac{v_3}{v_2} = \beta = \frac{0.1565}{0.0965} = 1.6217$$

From constant pressure heat addition process 2–3.

$$T_3 = T_2 \frac{v_3}{v_2} = 800.67 \times 1.6217 = 1298.44 \text{ K}$$

(c) Heat added = $q_A = c_p\ (T_3 - T_2)$

$$= 1.005\ (1298.44 - 800.67) = 500.25 \text{ kJ/kg}$$ ***Ans.***

From isentropic process 3–4,

$$\frac{T_4}{T_3} = \left(\frac{v_3}{v_4}\right)^{\gamma-1} = \left(\frac{v_3}{v_1}\right)^{0.4} = \left(\frac{0.1565\ v_s}{11.352 \times 0.0965\ v_s}\right)^{0.4} \quad \text{as } v_4 = v_1$$

or $$T_4 = 1298.44\ (0.1428)^{0.4} = 596.09 \text{ K}$$

(d) Heat rejected = $q_R = c_v(T_4 - T_1) = 0.718\ (596.09 - 303) = 210.43$ kJ/kg ***Ans.***

(e) $$\eta_{th} = 1 - \frac{1}{r^{\gamma-1}}\left[\frac{(\beta^\gamma - 1)}{\gamma(\beta - 1)}\right] = 1 - \frac{1}{11.352^{0.4}}\left[\frac{(1.6217^{1.4} - 1)}{1.4(0.6217)}\right] = 57.93\%$$ ***Ans.***

(f) Mean effective pressure is given by

$$\text{m.e.p.} = p_1 r\left[\frac{\gamma r^{\gamma-1}\ (\beta - 1) - (\beta^\gamma - 1)}{(r-1)\ (\gamma - 1)}\right]$$

$$= 1 \times 11.352\left[\frac{1.4 \times 11.352^{0.4}(1.6217 - 1) - (1.6217^{1.4} - 1)}{(11.352 - 1)(1.4 - 1)}\right] = 3.65 \text{ bar}$$ ***Ans.***

Also, m.e.p. can also be calculated by

$$\text{m.e.p.} = \frac{\text{Work done}}{v_1 - v_2} = \frac{q_A - q_R}{v_1 - v_2} = 3.65 \text{ bar}$$

## 2.5 DUAL CYCLE (LIMITED PRESSURE CYCLE)

Since it is difficult to achieve either constant volume or constant pressure heat addition in actual spark ignition or compression ignition engines due to appreciable amount of time required for the completion of combustion process, therefore, a dual cycle has been proposed. Thus, special cycle combines the features of the theoretical Otto and Diesel cycles and the end result is a theoretical cycle which probably comes a little closer to matching the actual performance of modern compression ignition engine. In other words, in dual cycle, part of heat addition takes place at constant volume and rest at constant pressure. The dual cycle is also called ***mixed or limited pressure cycle***. Figure 2.13 shows the representation of dual cycle on *p–V* and *T–s* diagrams. The various processes are following.

1–2: Isentropic compression of air resulting in increase in pressure and temperature.

2–3: Addition of heat at constant volume resulting in increase in pressure and temperature.

3–4: Addition of heat at constant pressure resulting in increase in volume and temperature.

4–5: Isentropic expansion of air from high temperature and pressure to lower pressure and temperature to do work.

5–1: Rejection of heat at constant volume till the air is restored to its initial state.

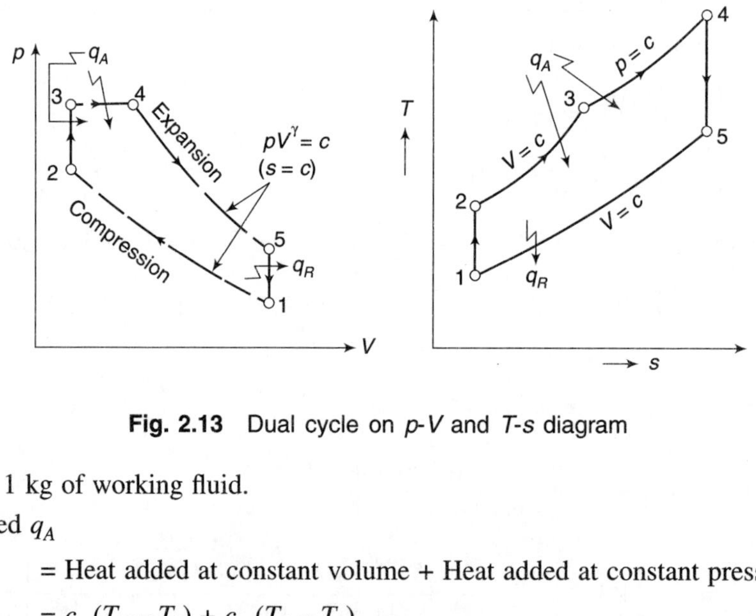


**Fig. 2.13** Dual cycle on *p-V* and *T-s* diagram

Consider 1 kg of working fluid.

Heat added $q_A$

$$= \text{Heat added at constant volume} + \text{Heat added at constant pressure}$$

$$= c_v\,(T_3 - T_2) + c_p\,(T_4 - T_3)$$

Heat rejected $= q_R = c_v\,(T_5 - T_1)$

Work done per cycle = Heat added – Heat rejected

$$= c_v\,(T_3 - T_2) + c_p\,(T_4 - T_3) - c_p\,(T_5 - T_1)$$

$$\text{Thermal efficiency} = \frac{\text{Work done}}{\text{Heat supplied}}$$

or $$\eta_{th} = \frac{c_v(T_3 - T_2) + c_p(T_4 - T_3) - c_p(T_5 - T_1)}{c_v(T_3 - T_2) + c_p(T_4 - T_3)}$$

or $$\eta_{th} = 1 - \frac{(T_5 - T_1)}{(T_3 - T_2) + c_p\,(T_4 - T_3)} \qquad ...(2.17)$$

Let $\alpha = \frac{p_1}{p_2}$ = Pressure ratio for constant volume heating

$\beta = \frac{v_4}{v_3} = \frac{v_4}{v_2}$ = Cut-off ratio and $r = \frac{v_1}{v_2}$ = Compression ratio

From isentropic compression process 1–2, $T_2 = T_1\, r^{\gamma-1}$

From constant volume process 2–3, $\frac{T_3}{T_2} = \frac{p_3}{p_2} = \alpha$

or $$T_3 = T_2 . \alpha = T_1\, \alpha\, r^{\gamma-1}$$

From constant pressure process 3–4, $\frac{T_4}{T_3} = \frac{v_4}{v_3} = \frac{v_4}{v_2} = \beta$

or $$T_4 = T_3\, \beta = T_1\, \alpha\, \beta\, r^{\gamma-1}$$

From isentropic expansion process 4–5.

$$\frac{T_5}{T_4} = \left(\frac{v_4}{v_5}\right)^{\gamma-1} = \left(\frac{v_4}{v_2}\frac{v_2}{v_5}\right)^{\gamma-1} = \left(\frac{v_4}{v_2}\frac{v_2}{v_1}\right)^{\gamma-1} = \left(\frac{\beta}{r}\right)^{\gamma-1}$$

or $$T_5 = T_4 \left(\frac{\beta}{r}\right)^{\gamma-1} = T_1\, \alpha\, \beta^{\gamma}$$

$\therefore$ $$\eta_{th} = 1 - \frac{\alpha\beta^{\gamma} - 1}{(\alpha r^{\gamma-1} - r^{\gamma-1}) + (\alpha\beta r^{\gamma-1} - \alpha r^{\gamma-1})}$$

or $$\eta_{th} = 1 - \frac{1}{r^{\gamma-1}}\left[\frac{(\alpha\beta^{\gamma} - 1)}{(\alpha - 1) + \gamma\,\alpha\,(\beta - 1)}\right] \qquad ...(2.18)$$

Figure 2.14 shows the variation of dual cycle efficiency with compression ratio. At cut-off ratio $\beta = 1$, the efficiency of Dual cycle is equal so that of the Otto and at pressure ratio $\alpha = 1$ it is equal to Diesel cycle efficiency.

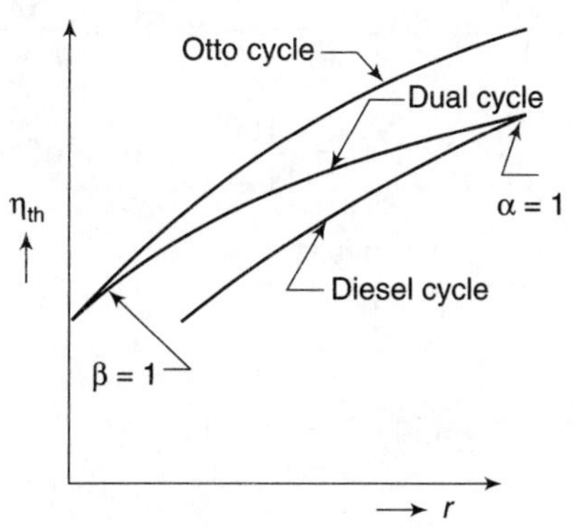


**Fig. 2.14** Efficiency of Dual cycle

Mean effective pressure is calculated as follows:

$$\text{m.e.p.} = \frac{\text{Work done}}{\text{Displacement volume}} = \frac{w}{v_1 - v_2}$$

$$= \frac{c_v(T_2 - T_2) + c_p(T_4 - T_3) - c_v(T_5 - T_1)}{v_1 - v_2}$$

But $$v_1 - v_2 = v_1\left(1 - \frac{1}{r}\right)$$

$$= v_1\left(\frac{r-1}{r}\right) = \frac{RT_1}{p_1}\left(\frac{r-1}{r}\right) = \frac{c_v(r-1)T_1}{p_1}\left(\frac{r-1}{r}\right)$$

∴ $$\text{m.e.p.} = \frac{p_1 . r}{(\gamma - 1)(r - 1)}\left[\frac{T_3 - T_2}{T_1} + \frac{\gamma(T_4 - T_3)}{T_1} - \frac{T_5 - T_1}{T_1}\right]$$

$$= \frac{p_1 . r}{(\gamma - 1)(r - 1)}[r^{\gamma-1}(\alpha - 1) + \alpha(\beta - 1)\gamma r^{\gamma-1} - (\alpha\beta^{\gamma} - 1)]$$

or $$\text{m.e.p.} = \frac{p_1 . r}{(\gamma - 1)(r - 1)}\left[r^{\gamma-1}\{(\alpha - 1) + \gamma\alpha(\beta - 1)\} - (\alpha . \beta^{\gamma} - 1)\right] \quad ...(2.19)$$

This is required expression for m.e.p. of dual cycle.

**EXAMPLE 2.18** An air standard dual cycle has pressure and temperature at the beginning of compression as 1 bar and 35°C respectively. The compression ratio is 11, the pressure ratio during heat addition 1.6 and cut-off ratio 1.7. Calculate (a) the percentage clearance, (b) the pressure, temperature and volume at each point of the cycle, (c) heat added, (d) heat rejected, (e) work done, (f) the thermal efficiency and (g) the mean effective pressure. Assume 0.5 kg of air flowing in the cycle.

***Solution*** Refer to Fig. 2.13.

(a) $$\text{Percentage clearance} = \frac{V_2}{V_s} \times 100 = \frac{V_2}{V_1 - V_2} \times 100 = \frac{100}{r-1}$$

$$= \frac{1}{r-1} \times 100 = \frac{100}{11-1} = 10\% \text{ of the stroke volume}$$

(b) **Point 1:** $$V_1 = \frac{mRT_1}{p_1} = \frac{0.5 \times 0.287 \times 308}{1 \times 10^2} = 0.4419 \text{ m}^3$$

**Point 2:** $$p_2 = p_1\left(\frac{V_1}{V_2}\right)^{\gamma} = 1 \times (11)^{1.4} = 28.7 \text{ bar}$$

$$T_2 = T_1\left(\frac{V_1}{V_2}\right)^{\gamma-1} = 308\,(11)^{0.4} = 803.72\,\text{K and } V_3 = \frac{V_1}{r} = \frac{0.4419}{11} = 0.0401 \text{ m}^3$$

**Point 3:** $p_3 = p_2 \times \alpha = 28.7 \times 1.6 = 45.92$ bar

$$T_3 = T_2 \left(\frac{p_3}{p_2}\right) = 803.72 \times 1.6 = 1285.95 \text{ K}$$

**Point 4:** $V_4 = V_2 \times \beta = 0.0401 \times 1.7 = 0.0681 \text{ m}^3$

$$T_4 = T_3 \left(\frac{V_4}{V_3}\right) = 1285.95 \left(\frac{0.0681}{0.0401}\right) = 2183.87 \text{ K}$$

**Point 5:** $$T_5 = T_4 \left(\frac{V_4}{V_5}\right)^{\gamma-1} = 2183.87 \left(\frac{0.0681}{0.4419}\right)^{0.4} = 1033.6 \text{ K}$$

$$p_5 = p_4 \left(\frac{V_4}{V_5}\right)^{\gamma} = 45.92 \left(\frac{0.0681}{0.4419}\right)^{1.4} = 3.349 \text{ bar}$$

(c) Heat added $= Q_A = mc_v (T_3 - T_2) + m\, c_p (T_4 - T_3)$

or $$Q_A = 0.5 \times 0.718 (1285.95 - 803.72)$$

$$+ 0.5 \times 1.005 (2183.87 - 1285.95) = 624.24 \text{ kJ}$$ ***Ans.***

(d) Heat rejected $= Q_R = m\, c_v (T_5 - T_1)$

$$= 0.5 \times 0.718 (1033.6 - 308) = 260.49 \text{ kJ}$$ ***Ans.***

(e) Work done $= Q_A - Q_R = 624.24 - 260.49 = 363.75$ kJ ***Ans.***

(f) Thermal efficiency $= \dfrac{\text{Work done}}{\text{Heat added}} = \dfrac{363.75}{624.24} = 58.27\%$ ***Ans.***

(g) m.e.p. $= \dfrac{\text{Work done}}{\text{Stroke volume}} = \dfrac{363.75}{0.4419 - 0.0401} = 905.3 \text{ kN/m}^2 = 9.053$ bar ***Ans.***

**EXAMPLE 2.19** During a test on a four-stroke six-cylinder engine having 11 cm bore and 16 cm stroke and working on the dual combustion cycle, the fuel consumption had been found to 0.18 kg/min. The engine runs at 1600 r.p.m. The compression ratio is 10 and calorific value of fuel used is 44000 kJ/kg. At the beginning of compression, the pressure is 0.95 bar and temperature is 45°C. The law of compression and expansion is polytropic $pV^{1.35}$ = constant. Assume that 2/3rds of heat released by the combustion of fuel is at constant volume and remainder 1/3rds at constant pressure. Calculate

(a) Heat supplied per cylinder charge, (b) pressure and temperature at each corner of the cycle, (c) work output, (d) cycle efficiency and (e) mean effective pressure.

Assume $c_p = 1.005$ and $c_v = 0.718$ kJ/kg K of air.

***Solution*** Refer to Fig. 2.13.

$$\text{Stroke volume} = V_s = V_1 - V_2 = \frac{\pi}{4} d^2 \times L = \frac{\pi}{4} \times (0.11)^2 \times 0.16$$

$$= 0.0015205 = 1.5205 \times 10^{-3} \text{ m}^3$$

Fuel used per cycle per cylinder $= \dfrac{\dot{m}_f}{x.N/2} = \dfrac{0.18}{6 \times 1600/2} = \dfrac{0.18}{6 \times 800}$ kg

Since there are $\dfrac{N}{2}$ cycles in N r.p.m.

Heat supplied per cylinder per cycle is $Q_A = 44000 \times \dfrac{0.18}{6 \times 800} = 1.65$ kJ

(a) Heat supplied per cycle in all six cylinders = 1.65 × 6 = 9.9 kJ **_Ans._**

(b) $V_s = V_1 - v_2 = 1.5205 \times 10^{-3}$ m$^3$

$$V_2 = \frac{V_s}{r-1}. \qquad \text{But} \quad r = \frac{V_1}{V_2} = 10$$

$$\therefore \qquad V_2 = \frac{1.5205 \times 10^{-3}}{9} = 0.16894 \times 10^{-3} \text{ m}^3$$

$$\therefore \qquad V_1 = V_5 = 10 \times V_2 = 10 \times 0.16894 \times 10^{-3} = 1.6894 \times 10^{-3} \text{ m}^3$$

The mass of air per cylinder is given by

$$m = \frac{p_1 V_1}{RT_1} = \frac{0.95 \times 10^2 \times 1.6894 \times 10^{-3}}{0.287 \times 318} = 0.00175 \text{ kg}$$

Heat supplied per cycle in each cylinder per kg of air

$$Q_A = \frac{1.65}{0.00175} = 942.85 \text{ kJ/kg}$$ **_Ans._**

From polytropic compression process 1–2, we have

$$T_2 = T_1\left(\frac{V_1}{V_2}\right)^{n-1} = 318 \times 10^{0.35} = 711.91 \text{ K}$$ **_Ans._**

$$p_2 = p_1\left(\frac{V_1}{V_2}\right)^{n} = 0.95 \times (10)^{1.35} = 21.26 \text{ bar}$$ **_Ans._**

Heat added per cycle per kg of air in each cylinder volume process 2–3, is

$$q_{2\text{-}3} = c_v(T_3 - T_2) = 942.85 \times 2/3 = 682.566 \text{ kJ/kg}$$

or $\qquad 0.718(T_3 - 719.91) = 682.566$

or $\qquad T_3 = 1595.35$ K **_Ans._**

$$p_3 = \frac{T_3}{T_2}p_2 = \frac{1595.35}{711.91} \times 21.26 = 47.64 \text{ bar}$$ **_Ans._**

Heat added per cycle at constant volume per kg of air in each cylinder is

$$q_{3\text{-}4} = c_p\,(T_4 - T_3) = 1/3 \times 942.85 = 314.28 \text{ kJ/kg}$$

or $\qquad 1.005(T_4 - 1595.35) = 314.28$

or $\qquad T_4 = 1908.06$ K **_Ans._**

Also, $V_4 = V_3 \frac{T_4}{T_3} = 0.16894 \times 10^{-3} \times \frac{1908.6}{1595.35} = 0.202 \times 10^{-3}\ \text{m}^3$

From expansion process 4–5, we have

$$T_5 = T_4 \left(\frac{V_4}{V_5}\right)^{\gamma-1} = 1908.06 \left(\frac{0.202 \times 10^{-3}}{1.6894 \times 10^{-3}}\right)^{0.35} = 907.31\ \text{K} \qquad \textbf{\textit{Ans.}}$$

$$p_5 = p_4 \left(\frac{V_4}{V_5}\right)^{\gamma} = 47.6 \times \left(\frac{0.202 \times 10^{-3}}{1.6894 \times 10^{-3}}\right)^{1.35} = 2.706\ \text{bar} \qquad \textbf{\textit{Ans.}}$$

It is to be noted that the work done cannot be calculated from the relation $(Q_A - Q_R)$ during polytropic process because the process is not adiabatic

Work done by each cylinder/cycle = Area enclosed by the cycle

$$= p_4 (V_4 - V_3) + \frac{p_4V_4 - p_5V_5}{n-1} - \frac{p_2V_2 - p_1V_1}{n-1}$$

$$= mR\left[(T_4 - T_3) + \frac{(T_4 - T_5)}{n-1} - \frac{(T_2 - T_1)}{n-1}\right]$$

$$= 0.00175 \times 0.287 \left[(1908.06 - 1595.35) + \frac{(1908.06 - 907.31)}{0.35} - \frac{(711.91 - 318)}{0.35}\right]$$

$$= 1.0278\ \text{kJ}$$

Total work done = 1.0278 × 6 = 6.1668 kJ ***Ans.***

(d) Thermal efficiency $= \frac{\text{Work done}}{\text{Heat added}} = \frac{1.0278}{1.65} = 62.29\%$ ***Ans.***

(e) m.e.p. $= \frac{\text{Work done}}{\text{Stroke volume}} = \frac{1.0278}{1.5205 \times 10^{-3}} = 6.7596\ \text{bar}$ ***Ans.***

**EXAMPLE 2.20** An engine works on the dual combustion cycle, the compression ratio being 11. The pressure and the commencement of compression is 1 bar and the temperature is 90°C. The maximum pressure in the cycle is 50 bar and the constant pressure heat reception continues for 1/20 of the stroke. Calculate the work done per kg of air and the ideal thermal efficiency. Assume $c_p = 1.005$, $c_v = 0.718$ kJ/kg K.

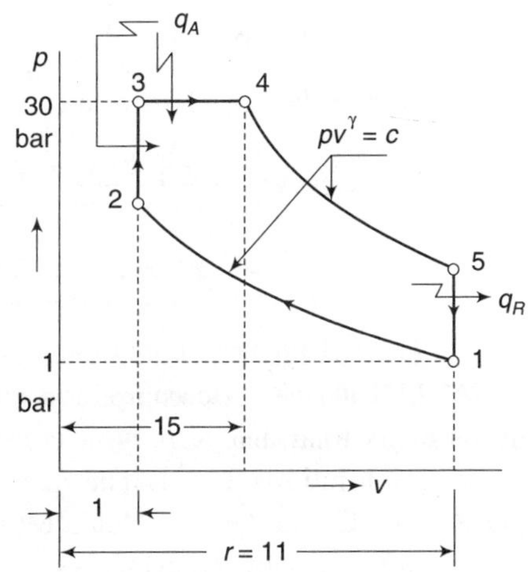


**Fig. 2.15** Example 2.20

***Solution*** Refer to Fig. 2.15.

$$\gamma = \frac{c_p}{c_v} = \frac{1.005}{0.718} = 1.4$$

Let $v_2 = v_3 = 1$ unit

$\therefore$ $v_1 = r = 11 \text{ unit} = v_5$

Stroke volume $= v_s = v_1 - v_2 = 10$ units

Hence, $v_4 = v_3 + (v_4 - v_3) = 1 + \dfrac{11-1}{20} = 1.5$

$\therefore$ $\dfrac{v_4}{v_3} = 1.5$

From isentropic compression process 1–2, we have

$$T_2 = T_1\, r^{0.4} = 363 \times 11^{0.4} = 947.24 \text{ K}$$

$$p_2 = p_1\, r^{\gamma} = 1 \times 11^{1.4} = 28.704 \text{ bar}$$

From constant heat addition process 2–3, we have

$$T_3 = T_2 \frac{p_3}{p_2} = 947.24 \times \frac{50}{28.704} = 1650 \text{ K}$$

From constant pressure heat addition process 3–4, we have

$$T_4 = T_3 \frac{v_4}{v_3} = 1650 \times 1.5 = 2475 \text{ K}$$

From isentropic expansion process 4–5, we have

$$T_5 = T_4 \left(\frac{v_4}{v_5}\right)^{\gamma-1} = 2475\left(\frac{1.5}{11}\right)^{0.4} = 1115.46 \text{ K}$$

Work done = Heat added – Heat rejected

$= c_v\,(T_3 - T_2) + c_p\,(T_4 - T_3) - c_v\,(T_5 - T_1)$

$= 0.718(1650 - 947.24) + 1.005(2475 - 1650) - 0.718(1115.46 - 363)$ ***Ans.***

$= 793.44$ kJ/kg of air ***Ans.***

$$\text{Thermal efficiency} = \frac{\text{Work done}}{\text{Heat added}}$$

$$= \frac{793.44}{0.718\,(1650 - 947.24) + 1.005\,(2475 - 1650)} = 59.94\%.$$ ***Ans.***

**EXAMPLE 2.21** A compression ignition engine working on the dual cycle takes in two-thirds of its total heat supply at constant volume and one-third at constant pressure. The compression ratio is 13 : 1. The maximum pressure in the cycle is 45 bar. Air intake takes place at 15°C and 1 bar. Calculate (a) the temperature at the each point of cycle and (b) thermal efficiency of the cycle. Assume $c_p = 1.005$ and $c_v = 0.718$ kJ/kg K.

***Solution*** Refer to Fig. 2.16

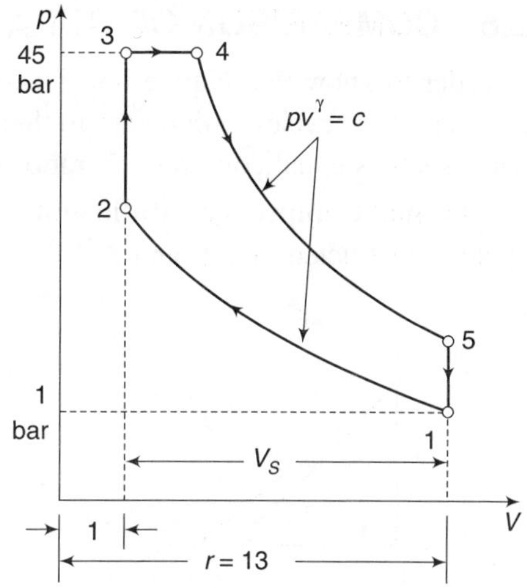


**Fig. 2.16** Example 2.21

(a) The clearance volume

$$v_c = v_2 = v_3 = 1 \text{ unit}$$

$$\therefore \quad v_1 = 13 \text{ unit}, \quad r = \frac{v_1}{v_2} = 13$$

and $$\gamma = \frac{c_p}{c_v} = \frac{1.005}{0.718} = 1.4$$

From isentropic compression 1-2,

$$T_2 = T_1 \left(\frac{v_1}{v_1}\right)^{\gamma-1} = 288(13)^{0.4}$$

$$= 824.47 \text{ K} \qquad \textbf{\textit{Ans.}}$$

$$\therefore \quad p_2 = p_1 \left(\frac{v_1}{v_2}\right)^{\gamma} = (13)^{1.4} = 36.26 \text{ bar}$$

***Ans.***

From constant volume heat addition process 2-3,

$$T_3 = T_2 \frac{p_3}{p_2} = 824.47 \times \frac{45}{36.26} = 1023.19 \text{ K} \qquad \textbf{\textit{Ans.}}$$

Heat added at constant volume,

$$q_{2-3} = c_v(T_3 - T_2) = 0.718\ (1023.19) - 824.47) = 142.68 \text{ kJ/kg of air.}$$

Heat supplied at constant pressure during 3-4, is

$$q_{3-4} = \frac{1}{3} \times 142.68 = 71.34 \text{ kJ/kg, because } q_{2-3} \text{ is 2/3 of the total heat supplied.}$$

$$\therefore \quad q_{3-4} = 71.34 = c_p(T_4 - T_3) = 1.005\ (T_4 - 1023.19)$$

or $$T_4 = 1094.17 \text{ K}$$

$$v_4 = v_3 \frac{T_4}{T_3} = 1 \times \frac{1094.17}{1023.14} = 1.069 \text{ m}^3\text{/kg}$$

From isentropic expansion process 3-4,

$$T_5 = T_4 \left(\frac{v_4}{v_5}\right)^{\gamma-1} = 1094.17 \left(\frac{1.069}{13}\right)^{0.4} = 402.8 \text{ K} \qquad \textbf{\textit{Ans.}}$$

(b) Thermal efficiency is given by

$$\eta_{th} = 1 - \frac{\text{Heat rejected}}{\text{Heat added}} = 1 - \frac{c_v(T_5 - T_1)}{c_v(T_3 - T_2) + c_p(T_4 - T_2)}$$

$$= 1 - \frac{0.718(402.8 - 288)}{142.68 + 71.34} = 61.5\% \qquad \textbf{\textit{Ans.}}$$

## 2.6 COMPARISON OF OTTO, DIESEL AND DUAL CYCLES

In order to know the relative importance of Otto, Diesel and Dual cycles, it is necessary to compare these cycles with regard to thermal efficiency. The comparison can be made in many ways such as equal compression ratio or equal final temperature or equal final pressures.

**(i) Equal Compression Ratio and Heat Input.** Figure 2.17 shows these three theoretical cycles constructed on *p-v* and *T-s* diagrams for the same compression ratio and heat input.

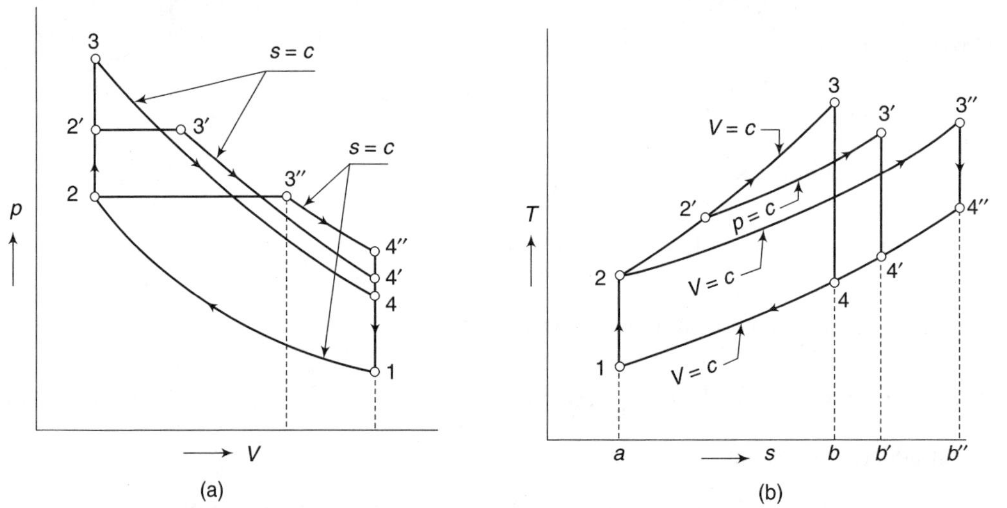


**Fig. 2.17** Comparison of Otto, Diesel and dual cycle for equal compression ratio and heat input

The Otto cycle is shown as 1–2–3–4, the dual cycles as 1–2–2′–3′–4′ and the Diesel cycles as 1–2–3″–4″. All the cycles start at the same initial temperature, pressure and volume at point 1 with isentropic compression to 2. Heat is added under varying conditions for the different cycles. In order to fulfil the condition of equal heat input, the *T-s* areas *a*–2–3*b*, *a*–2–2′–3′–*b*′ and *a*–2–3″–*b*″ must be equal. Since constant volume lines are steeper on the *T-s* diagram than constant pressure lines, construction of equal areas for the heat supplied shows point 3″ will lie at the point of greatest entropy. With isentropic expansion to the constant volume line 1–4″. As the same amount of heat is supplied to each cycle, the cycle will be more efficient which rejects the least amount of heat after expansion, as obvious from the expression of thermal efficiency.

$$\eta_{th} = 1 - \frac{Q_R}{Q_A} = 1 - \frac{Q_R}{\text{Constant}}$$

We see that it is the Otto cycle which rejects least amount of heat shown by area *a*–1–4–*b*. Thus, the Otto cycle has the maximum efficiency. The order of the efficiencies for the three cycles is given as

$$(\eta_{th})_{Otto} > (\eta_{th})_{Dual} > (\eta_{th})_{Diesel}$$

It is to be noted that the condition laid down here for same compression ratio is purely theoretical. The actual compression ignition engine is operated at a high compression ratio than spark ignition which is limited by detonation

**(ii) Constant Maximum Pressure and Heat Input.** Sometimes, pressure is the limiting factor in the engine design. Thus, it necessary to compare the cycles on the basis of same maximum pressure and heat input. First consider only the Otto and Diesel cycles constructed as shown in Fig. 2.18.

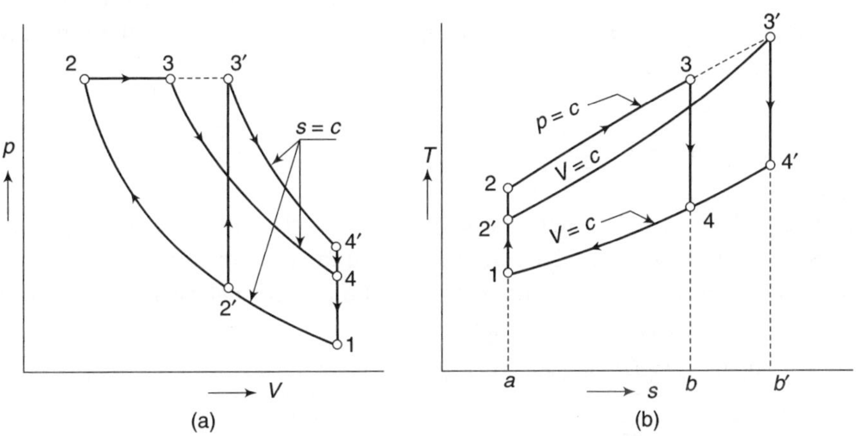


**Fig. 2.18** Comparison of Otto and Diesel cycles with same maximum pressure and heat input

The Diesel cycle is shown as 1–2–3–4 whereas the Otto cycle as 1–2′–3′–4′. In order to fulfil the given conditions, the point 3 and 3′ must lie on a constant pressure line while the *T-s* areas *a*–2–3–*b* and *a*–2′–3′–*b* must be equal for the same pressure and the same input respectively. The thermal efficiency is given as

$$\eta_{th} = 1 - \frac{Q_R}{Q_A} = 1 - \frac{Q_R}{\text{Constant}}$$

This means that, that cycle will be more efficient which will reject the least amount of heat. It is the Diesel cycle which rejects the least amount of heat as shown by the area *a*–1–4–*b*. A similar argument would show that the performance of the Dual cycle will fall between the two others. Hence, the order of efficiencies are

$$(\eta_{th})_{Diesel} > (\eta_{th})_{Dual} > (\eta_{th})_{Otto}$$

**(iii) Constant Maximum Pressure and Output.** In order to compare the thermal efficiencies of the cycles for constant maximum pressure and output, refer to Fig. 2.18. From the *T-s* diagram, it will be seen that work area 1–2–3–4 can be only equal to 1–2′–3′–4′ when point 3′ lies on the greater entropy. In order to fulfil the constant maximum pressure condition, point 3 must lie on the pressure line passing through 2–3. Hence,

$$\eta_{th} = \frac{W}{Q_A} = \frac{W}{W + Q_R} = \frac{\text{Constant}}{\text{Constant} + Q_R}$$

This equation shows that the cycle will be most efficient which rejects least amount of heat. Hence, the Diesel cycle is more efficient than Otto.

Hence, $(\eta_{th})_{Diesel} > (\eta_{th})_{Dual} > (\eta_{th})_{Otto}$

**(iv) Constant Maximum Pressure and Temperature.** In order to compare the cycles with constant maximum pressure and temperature, refer to Fig. 2.19. First Otto and Diesel cycles are considered. To fulfil the condition of (c) the same pressure, the Otto cycle must be limited to low compression ratio so that point is at a common state for both the cycles. It is clear from construction of Fig. 2.19 that both the cycles will reject the same amount of heat. The thermal efficiency is given as

$$\eta_{th} = 1 - \frac{Q_R}{Q_A} = 1 - \frac{\text{Constant}}{Q_A}$$

This equation shows with greater heat addition the cycle will be more efficient. We also see that it is the Diesel cycle in which the heat addition is the maximum. Hence, the Diesel cycle is more efficient than Otto. Similarly, it can be proved that Dual cycles will be more efficient than Otto cycle. Hence, the order is

$$(\eta_{th})_{Diesel} > (\eta_{th})_{Dual} > (\eta_{th})_{Otto}$$

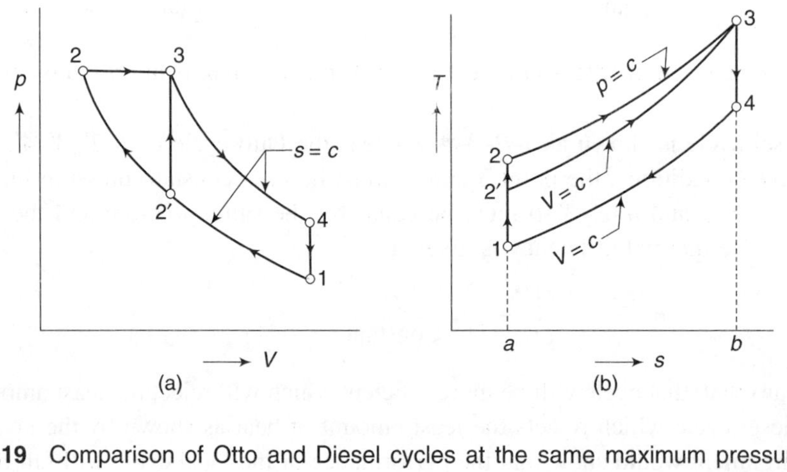


**Fig. 2.19** Comparison of Otto and Diesel cycles at the same maximum pressure and temperature

**EXAMPLE 2.22** A six-cylinder SI engine working on Otto cycle has a swept volume of 300 cm$^3$ per cylinder. The compression ratio is 10 and the engine running at 3500 rpm develops 75 kW. Determine the cycle efficiency, the heat addition, m.e.p. and the maximum temperature in the cycle. Assume inlet pressure and temperature as 1 bar and 150°C.

If the above engine is CI engine operating on Diesel cycle and receiving heat at the same rate, determine cycle efficiency, m.e.p., the maximum temperature in the cycle and the power output.

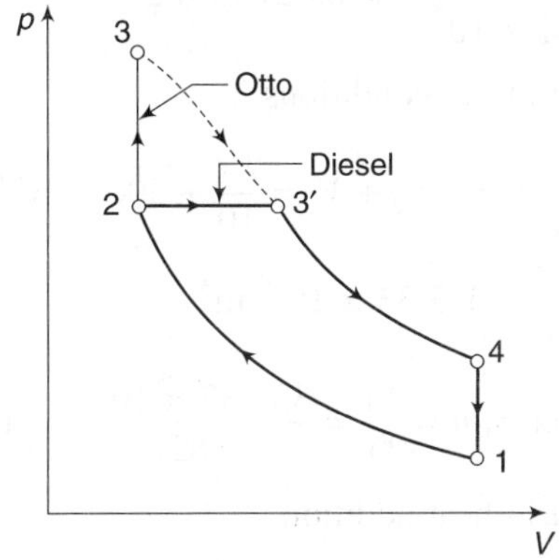


**Fig. 2.20** Example 2.22

***Solution*** **(A) Otto Cycle (1–2–3–4–1)**. Refer to Fig. 2.20.

The air standard efficiency $= \eta_{th} = 1 - \dfrac{1}{r^{\gamma-1}} = 1 - \dfrac{1}{10^{0.4}} = 60.2\%$ ***Ans.***

Heat added $= \dfrac{\text{Power output}}{\eta_{th}} = \dfrac{75}{0.602} = 124.55$ kW ***Ans.***

The number of cycles/ssc $= n = \dfrac{N}{2 \times 60} = \dfrac{3500}{2 \times 60} = 29.17$

Net power output per cycle per cylinder.

$$= \frac{\text{Power}}{\text{No. of cylinders}} \times \frac{1}{n} = \frac{75}{6 \times 29.17} = 0.428 \text{ kJ}$$

$$\text{m.e.p.} = \frac{\text{Work done}}{\text{Swept volume}} = \frac{0.428 \times 10^3}{300 \times 10^{-6} \times 10^5} = 14.27 \text{ bar}$$

The temperature of the end of compression is

$$T_2 = T_1 \left(\frac{V_1}{V_2}\right)^{\gamma-1} = T_1 (r)^{\gamma-1} = 288 \times 10^{0.4} = 723 \text{ K}$$

$$Q_A = Q_{2-3} = \frac{\text{Total heat added in the engine}}{\text{No. of cylinders} \times \text{No. of cycles/sec}}$$

$$= \frac{124.6}{6 \times (3500/2 \times 60)} = 0.712 \text{ kJ}$$

Heat added per cylinder per cycle = $Q_{2-3} = mc_v (T_3 - T_2)$

Here, $T_3$ cannot be calculated unless and until $m$ is known from gas law,

$$v_1 = \frac{RT_1}{p_1} = \frac{0.287 \times 288}{1 \times 10^2} = 0.827 \text{ m}^3/\text{kg}$$

The cylinder volume at the initial conditions is

$$V_1 = V_2 + (V_1 - V_2) = V_2 + V_s = \frac{V_1}{10} + 300 \times 10^{-6}$$

or
$$V_1 = \frac{300 \times 10^{-6}}{9} = 333.33 \times 10^{-6} \text{ m}^3$$

The mass of air per cylinder = $m = \dfrac{V_1}{v_1} = \dfrac{333.33 \times 10^{-6}}{0.827} = 403 \times 10^{-6}$ kg

Substituting the value of $m$ in heat addition

$$0.712 = 403 \times 0.718\,(T_3 - 723) \times 10^{-6}$$

$$T_3 = 3183.65 \text{ K}$$ ***Ans.***

**B. Diesel cycle (1–2–3′–4–1)**

Temperature at the end of compression, $T_2$ will be the same, thus,

$$T_2 = 723 \text{ K.}$$

As per question, the heat added in the cycle is the same, hence,

$$Q_{2-3'} = Q_{2-3} = 0.712 \text{ kJ}$$

In the Diesel cycle, the heat is added at constant pressure, so

$$Q_{2-3'} = m\,c_p(T_3' - T_2) \quad \text{or} \quad 0.712 = 403 \times 10^{-6} \times 1.005\,(T_3' - 723)$$

or
$$T_3' = 2481.1 \text{ K}$$

The cut-off ratio is expressed as, $\beta = \dfrac{V_3'}{V_2} = \dfrac{T_3'}{T_2} = \dfrac{2481.1}{723} = 3.432$

The air standard efficiency is given by

$$\text{ASE} = \eta_{th} = 1 - \frac{1}{r^{\gamma-1}}\left[\frac{\beta^{\gamma} - 1}{\gamma(\beta - 1)}\right] = 1 - \frac{1}{10^{0.4}}\left[\frac{3.432^{1.4} - 1}{1.4(3.432 - 1)}\right] = 46.1\%$$ ***Ans.***

The total power output = $\eta_{th} \times Q_{total} = 0.461 \times 124.55 = 57.4$ kW

Power output per cylinder = $\dfrac{57.4}{6} = 9.56$ kW ***Ans.***

Work done per cylinder per cycle = $\dfrac{9.56 \times 10^3 \times 60}{(3500/2)} = 327.77$ Nm

Mean effective pressure in the cylinder (m.e.p) is given by

$$\text{m.e.p.} = \frac{\text{Work done}}{\text{Swept volume}} = \frac{327.77}{300 \times 10^{-6} \times 10^5} = 10.92 \text{ bar}$$

The above result is tabulated below to give a glimpse for comparison.

| | *SI engine* | *CI engine* |
|---|---|---|
| Cycle efficiency (%) | 60.2% | 46.1% |
| m.e.p. (bar) | 14.27 | 10.92 |
| $T_{max}$ (K) | 3183.65 | 2481.1 |

**EXAMPLE 2.23** An SI engine working on Otto cycle has compression ratio 10 receives 2700 kJ of heat. The pressure and temperature at the beginning of compression is 1 bar and 40°C. Calculate the mean effective pressure and thermal efficiency of the cycle.

If the same engine works as dual cycle and the maximum pressure is limited to 70 bar and heat supplied remains the same, recalculate the above.

In the case of Otto cycle, calculate the additional work developed if it were possible to expand isentropically the exhaust gases to their original pressure of 1 bar.

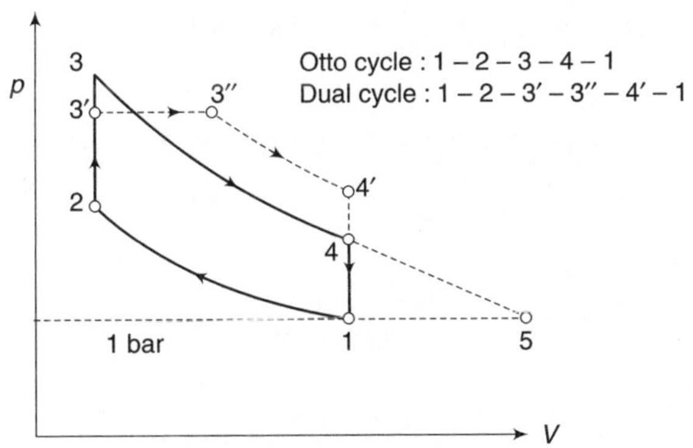


**Fig. 2.21** Example 2.23

***Solution*** Refer to Fig. 2.21.

**Otto cycle:** 1–2–3–4–1 **Dual cycle:** 1–2–3′–3″–4′–1

From the gas laws assuming $m$ = 1 kg

$$V_1 = \frac{mRT_1}{p_1} = \frac{1 \times 0.287 \times 313}{1 \times 10^2} = 0.898 \text{ m}^3/\text{kg}$$

$$\text{Swept volume} = V_s = V_1 - V_2 = V_1 \left(1 - \frac{V_2}{V_1}\right) = 0.9 \times 0.898 = 0.8082 \text{ m}^3/\text{kg}$$

and $V_2 = 0.0898 \text{ m}^3/\text{kg}$

At the end of compression process, the temperature and pressure are given by

$$T_2 = T_1 \left(\frac{V_1}{V_2}\right)^{\gamma - 1} = 313(10)^{0.4} = 786.2 \text{ K}$$

$$p_2 = p_1 \left(\frac{V_1}{V_2}\right)^{\gamma} = 1 \times (10)^{1.4} = 25.1 \text{ bar}$$

**A. Otto cycle**

$$Q_A = Q_{2-3} = m\,c_v\,(T_3 - T_2) \quad \text{or} \quad 2700 = 1 \times 0.718\,(T_3 - 786.2)$$

or $$T_3 = 4546.6 \text{ K}$$

For process 2–3, $\dfrac{p_2 V_2}{T_2} = \dfrac{p_3 V_3}{T_3}$ or $\dfrac{p_2}{T_2} = \dfrac{p_3}{T_3}$

or $$p_3 = T_3\left(\frac{p_2}{T_2}\right) = 4545.6 \times \frac{25.1}{786.2} = 145.4 \text{ bar}$$

For process 3–4, we have, $\dfrac{T_3}{T_4} = \left(\dfrac{p_3}{p_4}\right)^{\gamma - 1}$

or $$T_4 = T_3\left(\frac{1}{r}\right)^{\gamma - 1} = 4546.6\left(\frac{1}{10}\right)^{0.4} = 1809.8 \text{ K}$$

For process 4–1, we have

$$Q_R = Q_{4-1} = m\,c_v(T_4 - T_1) = 1 \times 0.718\,(1809.8 - 313) = 1047.7 \text{ kJ/kg}$$

$$\text{Work done} = Q_A - Q_R = 2700 - 1074.7 = 1625.3 \text{ kJ/Kg}$$

Thus, air standard efficiency

$$\text{ASE} = \eta_{th} = \frac{\text{Work done}}{Q_A} = \frac{1625.3}{2700} = 60\%$$ ***Ans.***

$$\text{m.e.p.} = \frac{\text{Work done}}{\text{Stroke volume}} = \frac{1625.3 \times 10^3}{0.8082 \times 10^5} = 20.1 \text{ bar}$$ ***Ans.***

**B. Dual cycle:** 1–2–3′–3″–4′–1

For process 2–3′, we have

$$\frac{p_2}{T_2} = \frac{p_3'}{T_3'} \quad \text{or} \quad T_3' = 786.2 \times \frac{70}{25.1} = 2192.58 \text{ K}$$

$$Q_A = Q_{2-3'} + Q_{3'-3''} = m\,c_v\,(T_3 - T_2) + m\,c_p\,(T_3'' - T_3')$$

or $$2700 = 1 \times 0.718\,(2192.58 - 786.2) + 1 \times 1.005\,(T_3'' - 2192.58)$$

or $$T_3'' = 3874.39 \text{ K}$$

$$V_3'' = V_3' \frac{T_3''}{T_3'} = 0.0898 \times \frac{3874.39}{2192.58} = 0.1586 \text{ m}^3\text{/kg}$$

For process 3″–4, we have,

$$\frac{T_3''}{T_4'} = \left(\frac{V_4'}{V_3''}\right)^{\gamma-1}$$

or $$T_4' = 3874.39 \times \left(\frac{0.1586}{0.898}\right)^{0.4} = 1936.48 \text{ K}$$

Heat rejected, $q_R = Q_{4'-1} = c_v(T_4' - T_1) = 0.718\ (1936.48 - 313) = 1165.65$ kJ/kg

Work done $= Q_A - Q_R = 2700 - 1165.65 = 1534.35$ kJ/kg

Air standard efficiency is given by

$$\eta = \frac{\text{Work done}}{\text{Heat supplied}} = \frac{1534.35}{2700} = 56.8\%$$ ***Ans.***

$$\text{m.e.p.} = \frac{\text{Work done}}{\text{Swept volume}} = \frac{1534.35 \times 10^3}{0.8082 \times 10^5} = 18.98 \text{ bar}$$ ***Ans.***

In the case of Otto cycle, if the gases were expanded isentropically to 1 bar, then

$$\frac{T_3}{T_5} = \left(\frac{p_3}{p_5}\right)^{\frac{\gamma-1}{\gamma}} \quad \text{or} \quad T_5 = T_3\left(\frac{p_5}{p_3}\right)^{\frac{\gamma-1}{\gamma}} = 4546.6\left(\frac{1}{145.1}\right)^{\frac{0.4}{1.4}} = 1096.6 \text{ K}$$

Heat rejecting during 5–1 $= c_p\ (T_5 - T_1) = 1.005\ (1096.6 - 313) = 787.5$ kJ/kg

Reduction in heat rejected = 1047.7 – 787.5 = 287.2 kJ/kg ***Ans.***

## 2.7 STIRLING CYCLE (REGENERATIVE CYCLES)

This cycle is named after Dr. Robert Stirling and his brother James Stirling (1816) who devised this cycle to work on the principle of regeneration by using a regenerator within the engine itself which would store the rejected heat energy during heat rejection process and supply the same, during heat addition process. Figure 2.22 shows the representation of Stirling cycle on *p-V* and *T-s* diagrams. It consists of two isothermal and two constant volume processes as given below.

1–2 : Reversible isothermal compression.

2–3 : Heat addition at constant volume.

3–4: Reversible isothermal expansion.

4–1: Heat rejection at constant volume = Heat addition at constant volume.

The regenerator used in the Stirling cycle consists of a wire mesh stacked in a duct as shown in Fig. 2.23. The air in the space at the lower end is kept at a high temperature ($T_H$) by heating with a furnace below it. In the upper space, a low-temperature ($T_L$) of air is maintained by a supply of cold water passing through the tubes in contact with it. Hence, two heat reservoirs, one at high temperature ($T_H$) and other low temperature ($T_L$) are obtained.

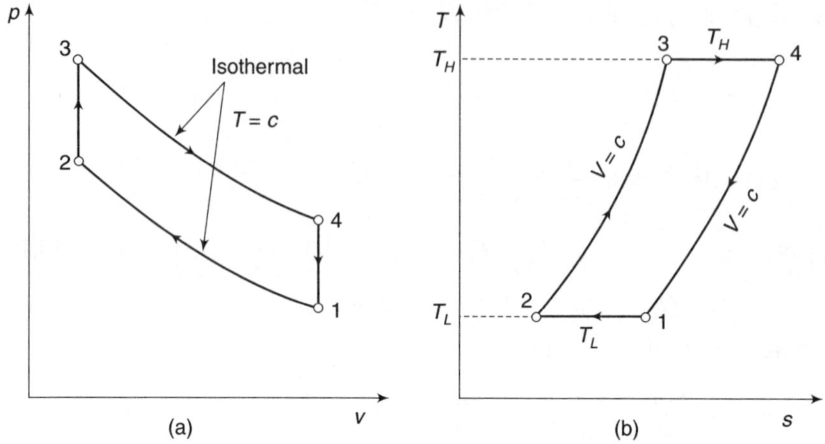


**Fig. 2.22** Stirling cycle on *p-V* and *T-s* diagrams

The wire mesh interposed ensures uniform temperature gradient from $T_H$ to $T_L$. This provides large heat transfer and a high heat capacity.

The operation of the cycle is as follows:

Assume the piston at BDC and its position is marked by point 1 on *p-v* and *T-s* diagrams. The air is isothermally compressed from state 1 to state 2 by keeping the cylinder head in contact with heat sink at temperature, $T_L$. At 2, the air at temperature $T_1 = T_2 = T_L$ enters the regenerators from top and while flowing through the wire mesh gets heated to temperature $T_3 = T_H$. At constant volume. Now the hot air is again admitted to the engine cylinder from the bottom portion of the regenerator and undergoes an isothermal expansion in the cylinder

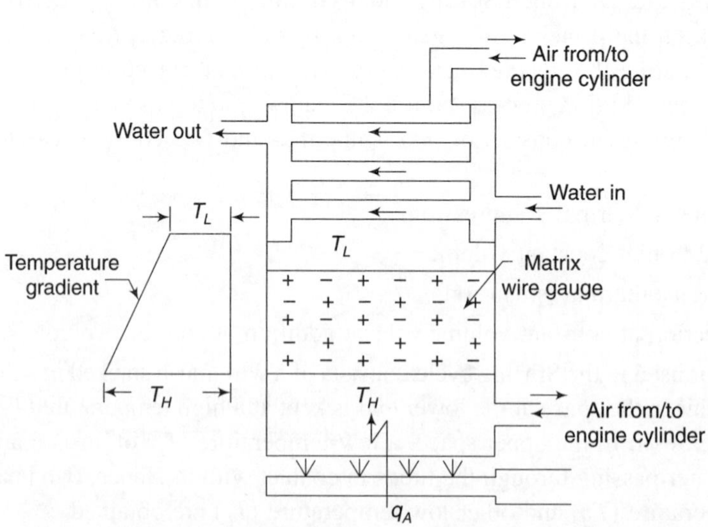


**Fig. 2.23** Regenerator

by keeping the cylinder head in contact with the heat source at $T_H$. The expansion of air forces the piston to the position BDC thus producing work. The expansion process is represented as 3–4. At point 4, the air at temperature $T_4 = T_H$ enters the regenerators from the lower end and while passing upwards transfer its heat to the wire mesh and thus gets cooled at constant volume. Thus, we see that heat rejected is stored in the regenerator and the same is supplied during heat addition.

It is to be noted that the temperature difference between the working fluid and the wire mesh all along the flow path is infinitesimally small and thus the transfer process may be considered as reversible. In addition, if the regenerator does not lose any heat to the surroundings the cycle heat exchanges take place isothermally at $T_L$ and $T_H$.

Consider 1 kg of working fluid.

In determining the heat added, we note that heat addition process from 2 to 3 and heat rejected from 4 to 1, cancel each other since the energy taken from the regenerator is returned in the latter. An alternate interpretation is to consider the regenerator as a part of the system; in this way heat exchange with the surrounding involves only the source at temperature, $T_H$ and the sink at $T_L$. The heat supplied per cycle is therefore during the process from 3 to 4 and heat rejection is during the process 1–2. Note that during the isothermal process, $\Delta u = 0$ since $T = c$. From first law

$$\therefore \quad \text{Head added} = q_A = w_{3-4} = RT_H \ln \frac{V_4}{V_3} = RT_H \ln \frac{v_4}{v_3}$$

$$\text{Heat rejected} = q_R = w_{1-2} = RT_L \ln \frac{V_1}{V_2} = RT_L \ln \frac{v_1}{v_2}$$

$$\therefore \quad \eta_{thermal} = \frac{RT_H \ln v_4/v_3 - RT_L \ln v_1/v_2}{RT_H \ln v_4/v_3}$$

But $\quad v_4 = v_1, \quad v_3 = v_2, \quad \text{and} \quad \frac{v_1}{v_2} = r$, hence

$$\text{or} \quad \eta_{thermal} = \frac{T_H - T_L}{T_H} = 1 - \frac{T_L}{T_H} \qquad \text{...(2.20)}$$

Thus, we see that *efficiency of Stirling cycle is equal to Carnot cycle between the same temperature limits, $T_H$ and $T_L$*. This is because of the fact that all processes are thermodynamically reversible.

It is to be noted that in the above derivation, the efficiency of regenerator is assumed as 100%. But in actual practice, there is a loss of heat varying from 10 to 20% due to radiation and poor conductivity of air resulting in lower value of the regenerator efficiency. If $\eta_{reg}$ be the efficiency of regenerative, then

$$\text{Heat supplied} = RT_H \ln r + (1 - \eta_{reg})\, c_v(T_H - T_L)$$

$$\text{Heat rejected} = RT_L \ln r + (1 - \eta_{reg})\, c_v(T_H - T_L)$$

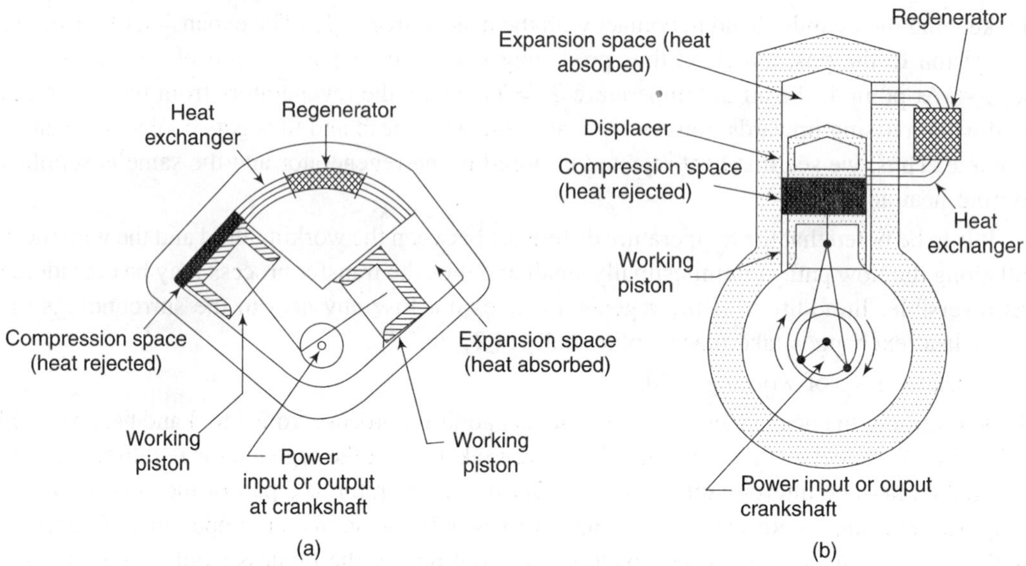


**Fig. 2.24** Double-cylinder and single-cylinder Stirling engines

$$\therefore \quad \eta_{\text{thermal}} = \frac{R(T_H - T_L)\ln r}{RT_H \ln r + (1 - \eta_{reg})\, c_v(T_H - T_L)} \quad ...(2.21)$$

If $\eta_{reg} = 100\%$, we get same expression at Eq. (2.20).

Figure 2.24 shows the two working models of Stirling engines. Referring to the double-cylinder engine, the main charge of working fluid is in the right-hand cylinder at the state represented by point 1 in *p-v* or *T-s* diagram.The linkage to the piston is in such a way that the left hand piston does not move when the right hand piston compresses the gas from 1 to 2. Only a small pressure rise is obtained due to heat removal. The left-hand piston now begins its downward movement while the right-hand piston further compresses the gas and forces it through the regenerator. The net result is a constant volume process in which the gas is transferred to the left hand cylinder and changed to state 3. Heat is next added to the cycle at a rate which allows the regenerator into the cooling cylinder.

In single cylinder, the expansion from 3 to 4 to 1 and compression from 1 to 2 to 3 occurs in opposite ends of the same cylinder. A displacer which acts as a second piston is used to alter the proportion of the gas in the expansion and compression spaces. Due to complex drive mechanism, this engine has not been put for commercial use.

**EXAMPLE 2.24** The following particulars relate to an engine working on Stirling cycle.

Compression ratio = 3 — Speed = 50 r.p.m.

Regenerator efficiency = 90% — Heat added = 2300 kJ/min

Lower temperature = 30°C Higher temperature = 550°C

Initial pressure = 1 bar

Calculate (a) work done/kg, (b) thermal efficiency, (c) mean effective pressure and (d) indicated power.

***Solution*** Refer to Fig. 2.22.

(a) Work done = $R(T_H - T_L) \ln r = 0.287(823 - 303) \ln 3 = 163.95$ kJ/kg ***Ans.***

Heat added = $RT_4 \ln r + (1 - \eta_{reg})\, c_v(T_H - T_L)$ kJ/kg

$$= 0.287 \times 823 \times \ln 3 + (1 - 0.9)\, 0.718\, (823 - 303) = 296.829 \text{ kJ/kg}$$

(b) Thermal efficiency = $\dfrac{\text{Work done}}{\text{Heat added}} = \dfrac{164.9}{296.829} = 55.55\%$ ***Ans.***

(c) Mass of air = $\dfrac{2300}{296.896} = 7.7485$ kg/min

Net working cycles/min = $\dfrac{N}{2} = \dfrac{50}{2} = 25$

$\therefore$ Mass of air per working cycle = $m = \dfrac{7.7485}{25} = 0.3099$ kg

Volume at state 1 can be calculated by characteristic gas equation.

$$V_1 = \frac{mRT_1}{p_1} = \frac{0.3099 \times 0.287 \times 303}{1 \times 10^2} = 0.2694 \text{ m}^3$$

$$V_2 = \frac{V_1}{r} = \frac{0.2694}{3} = 0.0898 \text{ m}^3$$

$\therefore$ Stroke volume = $V_1 - V_2 = 0.2694 - 0.0898 = 0.1796$ m$^3$

$$\text{m.e.p.} = \frac{\text{Work done}}{\text{Stroke volume}} = \frac{164.9 \times 0.3099}{0.1796 \times 10^2} = 2.845 \text{ bar}$$ ***Ans.***

(d) $$\text{i.p.} = \frac{100\, p_m\, LAN}{60} = \frac{2.845 \times 0.1796 \times 50 \times 10^2}{60} = 42.58 \text{ kW}$$ ***Ans.***

**EXAMPLE 2.25** An engine working on regenerative Stirling cycle takes in air at 1 bar and 20°C. The compression takes place isothermally and then receives heat at constant volume till the pressure rises to 12 bar and temperature 320°C. The cycle is completed by an isothermal expansion and constant volume heat rejection. Calculate the efficiency and mean effective pressure.

***Solution*** Refer to Fig. 2.25. Let clearance volume = $v_c = v_2 = v_3$

$$\therefore \quad v_1 = v_4 \quad \text{and} \quad \frac{v_1}{v_2} = \frac{v_4}{v_3} = \frac{v_4}{v_2} = \frac{v_1}{v_3} = r$$

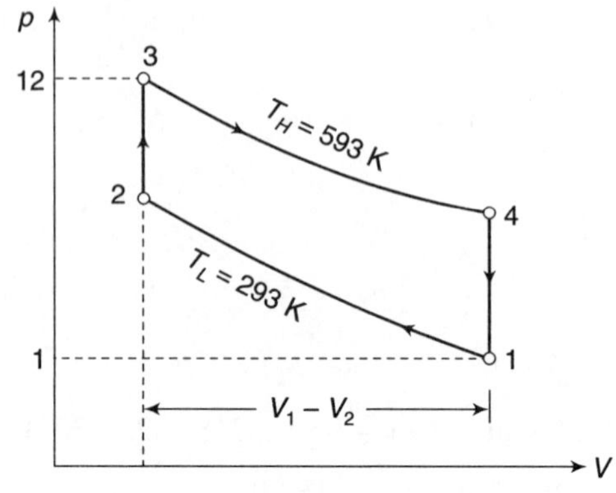


**Fig. 2.25** Example 2.25

$$T_1 = T_2 = T_L = 293 \text{ K} \quad \text{and} \quad T_3 = T_4 = T_L = 593 \text{ K}$$

$$v_1 = \frac{RT_1}{p_1} = \frac{RT_L}{p_1} = \frac{0.287 \times 293}{1 \times 10^2} = 0.8409 \text{ m}^3\text{/kg}$$

$$v_3 = \frac{RT_3}{p_3} = \frac{RT_H}{p_3} = \frac{0.287 \times 593}{12 \times 10^2} = 0.1418 \text{ m}^3\text{/kg}$$

$$r = \frac{v_1}{v_2} = \frac{v_1}{v_3} = \frac{0.8409}{0.1418} = 5.9301$$

$$\text{m.e.p.} = \frac{RT_H \ln v_4/v_3 - RT_L \ln v_1/v_2}{v_1 - v_2} = \frac{[R \ln r]\,[T_H - T_L]}{v_1 - v_2}$$

$$= \frac{[0.287 \times \ln 5.9301]\,[593 - 293]}{(0.8409 - 0.1418)10^2} = 2.192 \text{ bar.}$$ ***Ans.***

Thermal efficiency is given by

$$\eta_{th} = 1 - \frac{T_L}{T_H} = 1 - 293/593 = 0.5059 \text{ or } 50.59\%$$ ***Ans.***

## 2.8 ERICSSON REGENERATIVE CYCLE

This cycle also works on the principle of regeneration. Figure 2.26 shows the representation of the cycle on *p-v* and *T-s* diagrams. It consists of two isothermal and two constant pressure lines as follows

1–2 : Reversible isothermal compression of air.

2–3 : Addition of heat at constant pressure.

3–4 : Reversible isothermal expansion of air.

4–1 : Rejection of heat at constant pressure.

The heat rejected during the heat rejection process 4–1 at constant pressure is stored in the regenerator and the same is supplied during heat addition process 2–3. Thus this cycle is thermally reversible by the use of regenerator. The efficiency of regenerator is assumed as 100%.

Consider 1 kg of working fluid

$$q_{3-4} = w_{3-4} = R_{TH} \ln\frac{p_3}{p_4} = R_{TH} \ln\frac{p_2}{p_1}$$

$$q_{4-1} = c_p(T_1 - T_4) = c_p(T_L - T_H) = -\, c_p(T_H - T_L)$$

$$w_{4-1} = p_4(v_1 - v_4) = R(T_1 - T_4) = R(T_H - T_L)$$

$$q_{1-2} = w_{1-2} = R_{T_L} \ln\frac{p_1}{p_2} = -\, R_{T_L} \ln\frac{p_2}{p_1}$$

$$q_{2-3} = c_p(T_3 - T_2) = c_p(T_H - T_L)$$

$$w_{2-3} = p_2(v_3 - v_2) = R(T_3 - T_2) = R(T_H - T_L)$$

In the case of ideal regenerator, $q_{4-1} = q_{2-3}$, so that all the heat added from the external source at $T_H$ and all the heat rejection to an external sink at $T_L$ yields efficiency equal to Carnot cycle as given below.

$$\eta_{th} = 1 - \frac{Q_R}{Q_A} = 1 - \frac{q_{1-2}}{q_{3-4}} = 1 - \frac{RT_L \ln p_2/p_1}{RT_H \ln p_2/p_1}$$

or $$\eta_{th} = 1 - \frac{T_L}{T_H} \qquad ...(2.22)$$

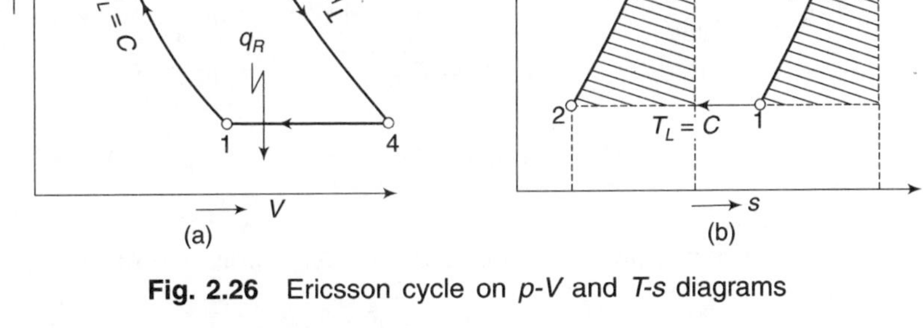


**Fig. 2.26** Ericsson cycle on *p-V* and *T-s* diagrams

## 2.9 ATKINSON CYCLE

This cycle is named after its inventor Dr. J. Atkinson, a British engineer. Figure 2.27 shows the representation of the Atkinson cycle on the *p-v* and *T-s* diagrams. It consists of the following four processes.

1–2 : Isentropic compression of air.

2–3 : Heat is supplied at constant volume from an external source. The pressure ratio $p_3/p_2$ is called the **explosion ratio** and denoted by $\alpha$

3–4 : Isentropic expansion of air from high temperature and pressure to lowest pressure to obtain work.

4–1 : Rejection of heat at constant pressure with some preliminary compression so that the initial state of air is restored.

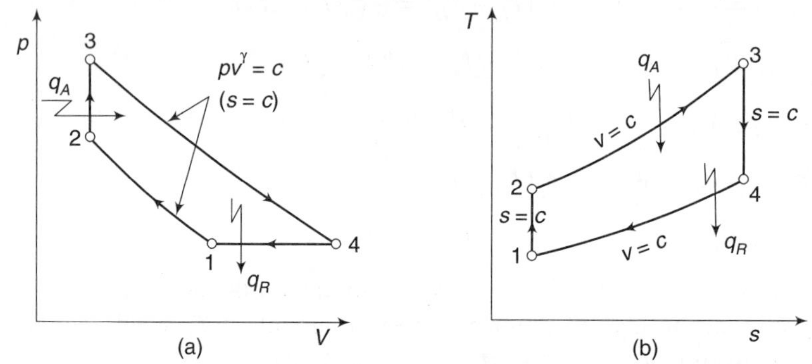


**Fig. 2.27** Atkinson cycle on *p-V* and *T-s* diagrams

The Atkinson cycle differs from the Otto cycle in so far as process 3–4 and 4–1 are concerned. This cycle is an ideal cycle for an Otto engine exhausting to a gas turbine. Expansion of exhaust gases from Otto engine occurs in the turbine beyond volume $v_1$. For example,

Consider 1 kg of working fluid

Heat supplied = $q_A = c_v\,(T_3 - T_2)$ and Heat rejected = $q_R = c_p\,(T_4 - T_1)$

Work done per cycle = $q_A - q_R = c_v(T_3 - T_2) - c_p(T_4 - T_1)$

$$\therefore \quad \eta_{\text{thermal}} = \frac{\text{Work done}}{\text{Heat added}} = \frac{c_v(T_3 - T_2) - (T_4 - T_1)}{c_v(T_3 - T_2)} = 1 - \frac{\gamma(T_4 - T_1)}{(T_3 - T_2)} \qquad ...(2.23)$$

where $\gamma$ = Adiabatic exponent = $c_p/c_v$

Let $\dfrac{V_1}{V_2} = \dfrac{v_1}{v_2} = r$ = compression ratio and $\dfrac{p_3}{p_2} = \alpha$ = explosion ratio

From compression process 1–2, $T_2 = T_1\left(\dfrac{v_1}{v_2}\right)^{\gamma-1} = T_1\, r^{\gamma-1}$

From constant volume heat supplied process 2–3, $\dfrac{p_2}{T_2} = \dfrac{p_3}{T_3}$

$$\therefore \qquad T_3 = T_2 \frac{p_3}{p_2} = T_2.\alpha = \alpha.T_1.r^{\gamma-1}$$

From expansion process 3–4

$$\frac{T_3}{T_4} = \left(\frac{p_3}{p_4}\right)^{(\gamma-1/\gamma} = \left(\frac{p_3}{p_2}\frac{p_2}{p_4}\right)^{(\gamma-1)/\gamma} = \left(\alpha.\frac{p_2}{p_4}\right)^{(\gamma-1)/\gamma} = \alpha^{(\gamma-1)\gamma}\left(\frac{p_2}{p_1}\right)^{(\gamma-1)/\gamma}$$

But $\quad p_1v_1^\gamma = p_2v_2^\gamma \quad \therefore \quad$ or $\quad \left(\frac{p_2}{p_1}\right) = \left(\frac{v_1}{v_2}\right)^\gamma = r^\gamma$

$$\therefore \qquad \frac{T_3}{T_4} = \alpha^{(\gamma-1)/\gamma}.r^{(\gamma-1)}$$

or

$$T_4 = \frac{T_3}{\alpha^{(\gamma-1)/\gamma}.r^{\gamma-1}} = \frac{\alpha\, T_1\, r^{\gamma-1}}{\alpha^{(\gamma-1)/\gamma}.r^{(\gamma-1)}} \quad \text{or} \quad T_4 = T_1\,\alpha^{1/\gamma}$$

Putting the values of $T_2$, $T_3$ and $T_4$ in Eq. (2.23), we get

$$\text{Thermal efficiency} = 1 - \frac{\gamma\left(T_1\,\alpha^{1/\gamma} - T_1\right)}{\left(\alpha\, T_1 r^{\gamma-1} - T_1\, r^{\gamma-1}\right)}$$

$$\eta_{th} = 1 - \frac{1}{r^{\gamma-1}}\left[\frac{\gamma\,(\alpha^{1/\gamma} - 1)}{(\alpha - 1)}\right] \qquad ...(2.24)$$

This is the required equation for the thermal efficiency of Atkinson cycle.

The operation of Atkinson cycle is achieved by using a special linkage mechanism alternatively providing short and long strokes. The isentropic compression is achieved on a short stroke while the expansion is on long stroke. The main purpose of devising this cycle is to get more work than the Otto cycles shown by area 1-4-5. Note that heat rejection also occurs here at lower to average temperature which gives more efficiency than the Otto cycle. But due to the complex drive mechanism, the engines have not been commercialized yet.

**EXAMPLE 2.26** An engine working on Atkinson cycle takes air at 1 bar and 25°C. The air is compressed isentropically by a compression ratio 6 : 1. The heat is added at constant volume resulting the final pressure to be 20 bar. Now, the air is expanded isentropically to pressure 1 bar. The heat is rejected at constant pressure. Calculate (a) pressure and temperature at various points, (b) work done per kg of air and (c) cycle efficiency. Compare the efficiency with that of Carnot working with same temperature limits.

***Solution*** Refer to Fig. 2.27.

(a) $\qquad p_1 = 1 \text{ bar}, \quad T_1 = 298 \text{ K}$

From isentropic compression process 1–2, we have

$$T_2 = T_1\,(r)^{\gamma-1} = (273 + 25)\,(6)^{0.4} = 610.206 \text{ K}$$ ***Ans.***

$$p_2 = p_1 (r)^\gamma = 1 \times 6^{1.4} = 12.286 \text{ bar} \qquad \textbf{\textit{Ans.}}$$

$$p_3 = 20 \text{ bar} \qquad \textbf{\textit{Ans.}}$$

From constant volume heat addition 2–3, we have

$$T_3 = \frac{p_3}{p_2} T_2 = \frac{20}{12.286} \times 610.206 = 993.335 \text{ K} \qquad \textbf{\textit{Ans.}}$$

$$p_4 = p_1 = 1 \text{ bar} \qquad \textbf{\textit{Ans.}}$$

From isentropic expansion process 3–4, we have

$$T_4 = \frac{T_3}{(p_3/p_4)^{(\gamma-1)/\gamma}} = \frac{993.335}{(20)^{0.2857}} = 422.076 \text{ K} \qquad \textbf{\textit{Ans.}}$$

Heat added $= q_A = c_v(T_3 - T_2) = 0.718\ (993.335 - 610.206) = 275.08$ kJ/kg

Heat rejected $= q_R = c_p(T_4 - T_1) = 1.005\ (422.076 - 298) = 124.696$ kJ/kg

(b) Work done $= w = 275.08 - 124.696 = 150.38$ kJ **_Ans._**

(c) Thermal efficiency $= \dfrac{w}{q_A} = \dfrac{150.384}{275.08} = 54.66\%$

Efficiency of Carnot cycle $= \dfrac{T_H - T_L}{T_H} = \dfrac{993.335 - 298}{993.335} = 70\%$ **_Ans._**

## 2.10 BRAYTON OR JOULE CYCLE

The air standard Brayton cycle or Joule cycle or constant pressure cycle is an idealised cycle for simple gas turbine power plant shown in Fig. 2.28. Basically, it consists of a compressor, a combustion chamber and a turbine. *This will be discussed in detail in a separate chapter,* **"Gas Turbine and Jet Propulsion"**.

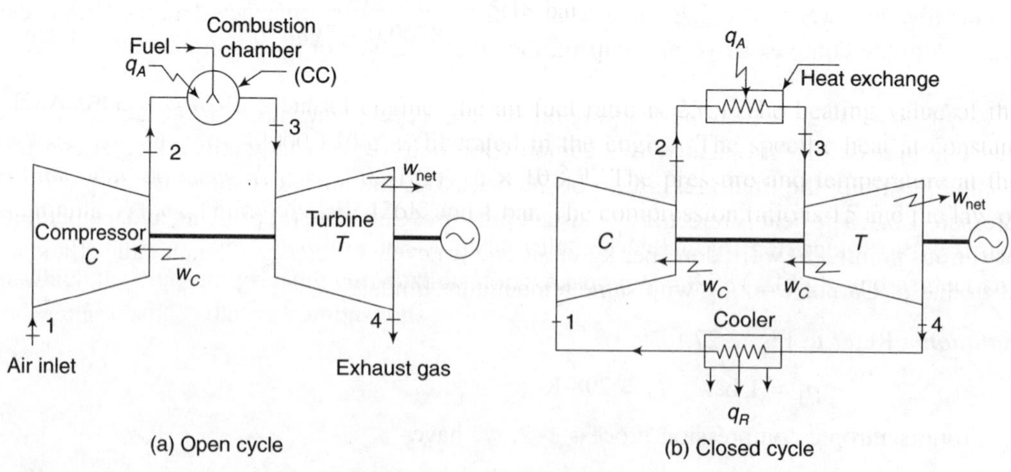


**Fig. 2.28** Simple gas turbine

Atmospheric air is compressed from pressure $p_1$ to a high pressure $p_2$ in the compressor and delivered to the combustion chamber where fuel is injected and burnt. The combustion process occurs nearly at constant pressure. Due to combustion, heat is added to the working fluid raising its temperature from $T_2$ to $T_3$. The products of combustion from the combustion chamber are expanded in the turbine from pressure $p_3$ to atmospheric pressure and then discharged into the atmosphere. The turbine and compressor and mechanically coupled, hence the net work is the difference between turbine work and compressor work. *Actual gas turbine cycles are open cycle, since new air must be continuously introduced into the compressor.* In a closed cycle, heat is added by an indirect method called the heat exchanger, so that, there is no mixing of fuel and air. Similarly, the heat is rejected in another heat exchanger called the precooler.

In the ideal Brayton or Joule cycle, the cycle is assumed to be a closed one and heat addition or rejection is only through heat transfer process. Figure 2.29 shows the representation of ideal Brayton cycle on the *p-v* and *T-s* diagrams.

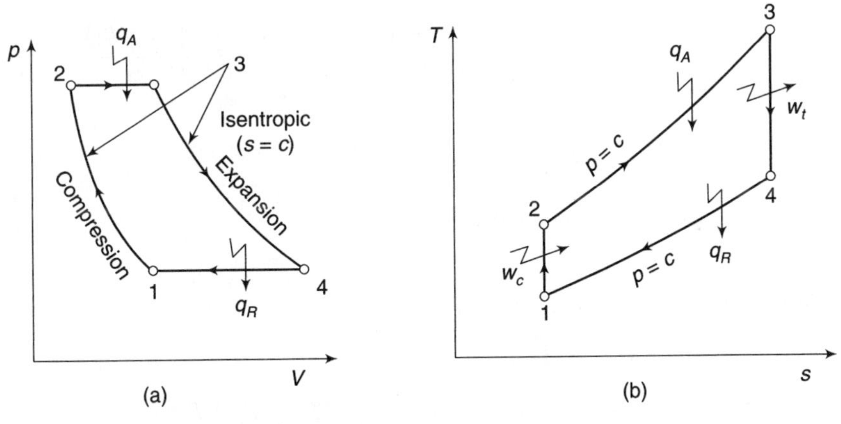


**Fig. 2.29** Brayton cycle on *p-V* and *T-s* diagrams

The various process are:

1–2 : Isentropic compression in the compressor thus raising pressure and temperature from $p_1$, $T_1$ to $p_2$ $T_2$.

2–3 : Addition of heat at constant pressure raising temperature from $T_2$ to $T_3$.

3–4 : Isentropic expansion of air from high pressure and temperature to low pressure and temperature and thus doing work.

4–1 : Rejection of heat at constant pressure to restore the original state of air.

For example, consider 1 kg of working fluid.

From first law for steady flow, neglecting ΔKE and ΔPE

$$\delta q - \delta w = dh$$

During heat addition or heat rejection no work is done, hence

$$q = \Delta h$$

$\therefore$ Heat added $= q_A = h_3 - h_2 = c_p\,(T_3 - T_2)$

Heat rejected $= q_R = h_4 - h_1 = c_p\,(T_4 - T_1)$

Net work = Heat supplied – Heat rejected $= c_p\,\{(T_3 - T_2) - (T_4 - T_1)\}$

The net work is also found from the concept of turbine and compressor work.

From first law, for isentropic process

$$w = -\,\Delta h$$

$\therefore$ Work done by turbine $= w_t = h_3 - h_4 = c_p\,(T_3 - T_4)$

Work consumed by compressor $= w_c = h_2 - h_1 = c_p\,(T_2 - T_1)$

Net work $= w_t - w_c = c_p\,\{(T_3 - T_4) - (T_2 - T_1)\} = c_p\,[(T_3 - T_2) - (T_4 - T_1)]$

$$\therefore \quad \eta_{\text{net}} = \frac{\text{Net work}}{\text{Heat added}} = \frac{w_t - w_c}{q_A} = \frac{c_p(T_3 - T_2) - (T_4 - T_1)}{c_p(T_3 - T_2)}$$

$$\text{or} \quad \eta_{\text{thermal}} = 1 - \frac{T_4 - T_1}{T_3 - T_2} \qquad \text{...(2.25)}$$

From isentropic compression process 1–2, $\dfrac{T_2}{T_1} = \left(\dfrac{p_2}{p_1}\right)^{(\gamma-1)/\gamma}$

And from isentropic expansion process 3–4, $\dfrac{T_3}{T_4} = \left(\dfrac{p_2}{p_4}\right)^{(\gamma-1)/\gamma}$

But $\quad p_2 = p_3 \quad$ and $\quad p_1 = p_4$

$$\therefore \quad \frac{T_2}{T_1} = \frac{T_3}{T_4} = \left(\frac{p_2}{p_1}\right)^{(\gamma-1)/\gamma} \quad \text{or} \quad \frac{T_1}{T_2} = \frac{T_4}{T_3} = \frac{T_4 - T_1}{T_3 - T_2}$$

Putting the value of $(T_4 - T_1)/(T_3 - T_2)$ in Eq. (2.25), we get

$$\eta_{th} = 1 - \frac{T_1}{T_2} = 1 - \frac{T_4}{T_3} \qquad \text{...(2.26)}$$

Let $\quad r_p = \text{Pressure ratio} = \dfrac{p_2}{p_1}$

$$\frac{T_2}{T_1} = \left(\frac{p_2}{p_1}\right)^{(\gamma-1)/\gamma} = (r_p)^{(\gamma-1)/\gamma} \quad \text{or} \quad \frac{T_1}{T_2} = \frac{1}{(r_p)^{(\gamma-1)/\gamma}}$$

$$\therefore \quad \eta_{th} = 1 - \frac{1}{(r_p)^{(\gamma-1)/\gamma}} \qquad \text{...(2.27)}$$

From the above equation, we find that the thermal efficiency of Brayton cycle is the same as that of the Otto cycle. Here, $r$ is replaced by $r_p$. The variation of thermal efficiency, $\eta_{th}$ as a function of pressure ratio, $r_p$ as shown in Fig. 2.30. It is obvious from Fig. 2.30 that the thermal efficiency progressively increases with increasing value of pressure ratio.

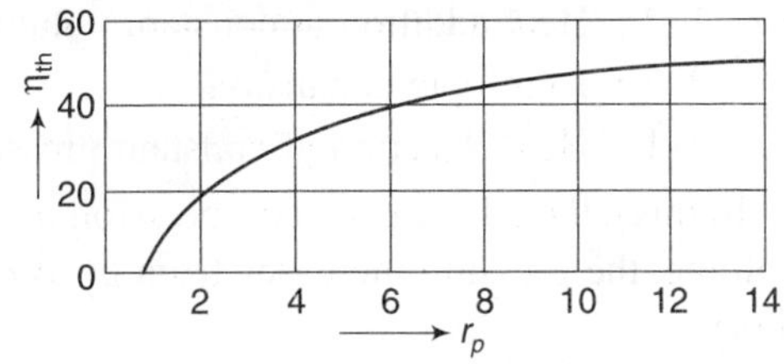


**Fig. 2.30** Brayton cycle vs pressure ratio

**EXAMPLE 2.27** A simple open cycle gas turbine power plant takes in air at 1 bar and 15°C and compresses the air to 5 bar. This is then heated in a combustion chamber to a temperature of 800°C. The hot air expands in the turbine to 1 bar.

Find (a) the power developed.

(b) thermal efficiency of the plant.

Take $\gamma = 1.4$ and $c_p = 1.005$ kJ/kg K for air.

***Solution*** Refer to Fig. 2.29. The flow is a steady flow process

$$T_2 = T_1\,(r_p)^{(\gamma-1)/\gamma} = (273 + 15)\,(5)^{0.4/1.4} = 456.14 \text{ K}$$

$$\text{Compressor work} = w_c = h_2 - h_1 = c_p(T_2 - T_1)$$

$$= c_p(456.14 - 288) = c_p \times 168.14 \text{ kJ/kg}$$

$$T_4 = \frac{T_3}{r^{(\gamma-1)/\gamma}} = \frac{800 + 273}{5^{0.4/1.4}} = 677.47 \text{ K}$$

$$\text{Turbine work} = w_t = h_3 - h_4 = c_p\,(T_3 - T_4)$$

$$= c_p(1073 - 677.74) = c_p \times 395.53 \text{ kJ/kg}$$

$$\text{Heat added} = q_A = q_{3-2} = h_3 - h_2 = c_p(T_3 - T_2)$$

$$= c_p\,(1073 - 456.14) = c_p \times 616.86 \text{ kJ/kg}$$

$$\text{Net work} = w_T - w_c = c_p\,(395.53 - 168.14) = c_p \times 277.39 \text{ kJ/kg}$$

$$\therefore \quad \text{Thermal efficiency} = \eta_{th} = \frac{w_{net}}{q_A} = \frac{c_p \times 277.39}{c_p \times 1616.86} = 36.86\% \qquad \textit{Ans.}$$

$$\text{Output} = 1.005 \times 227.39 = 228.52 \text{ kW/kg} \qquad \textit{Ans.}$$

## 2.11 LENOIR CYCLE

The Lenoir cycle is the idealized cycle for the pulse jet engine. The representation of the Lenoir cycle on $p$-$v$ and $T$-$s$ diagrams is shown in Fig. 2.31. It consists of three processes as given below:

1–2 : Heat addition at constant volume.

2–3 : Isentropic expansion.

3–1 : Heat rejection at constant pressure.

In this cycle, there is no compression of the working fluid. Due to heat addition at constant volume, the pressure increases from $p_1$ to $p_2$ while expanding from pressure $p_2$ to $p_3$ work is done.

For example, consider 1 kg of working fluid

$$\text{Heat added} = c_v\,(T_2 - T_1)$$

and $$\text{Heat rejected} = c_p\,(T_3 - T_1)$$

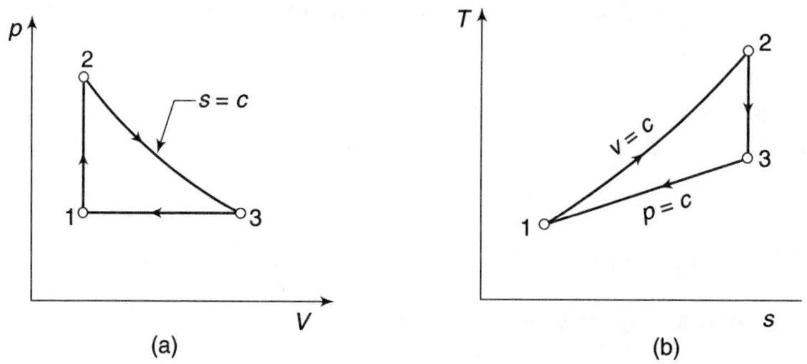


**Fig. 2.31** Lenoir cycle on *p-V* and *T-s* diagrams

$$\therefore \text{Thermal efficiency} = \frac{c_v(T_2 - T_1) - c_p(T_3 - T_1)}{c_v(T_2 - T_1)}$$

Let $$\frac{p_2}{p_1} = \alpha$$

From constant volume heat addition process,

$$\frac{T_2}{T_1} = \frac{p_2}{p_1} = \alpha$$

$$\therefore \qquad T_2 = \alpha\, T_1$$

From isentropic expansion process 2–3,

$$\frac{T_3}{T_2} = \left(\frac{p_3}{p_2}\right)^{(\gamma-1)/\gamma} = \left(\frac{p_1}{p_2}\right)^{(\gamma-1)/\gamma} = \left(\frac{1}{\alpha}\right)^{(\gamma-1)/\gamma} = \alpha^{(1/\gamma)-1}$$

$$\therefore \qquad T_3 = T_2\,\alpha^{(1/\gamma)-1} = T_1\,\alpha^{1/\gamma}$$

$$\therefore \qquad \eta_{th} = 1 - \frac{\gamma(\alpha^{1/\gamma} - 1)}{(\alpha - 1)} \qquad ...(2.28)$$

This is the required expression for thermal efficiency.

**EXAMPLE 2.28** The specific internal energy of a certain simple fluid substance is given by

$$u = u_0 \frac{p \cdot v}{\gamma - 1}$$

where $p$ = pressure,
$v$ = specific volume,
$u_0$ = a constant, and
$\gamma$ = adiabatic index.

It follows the law $pv^\gamma$ = constant along an isentropic path. It is proposed to study the efficiency of a reversible cycle using this substance. The cycle consists of three parts as follows in sequence:

(i) Constant volume pressure rise.
(ii) Isentropic expansion to initial pressure.
(iii) Decrease of volume to initial pressure at constant pressure.

Plot the process on $p$-$v$ and $h$-$s$ charts and prove the formula for efficiency.

$$\eta_{th} = 1 - \frac{\gamma(r - 1)}{\alpha - 1}$$

If $\gamma = 1.25$ and $r = 16$, and $\alpha = 40$ what is the efficiency?

***Solution*** Refer to Fig. 2.31 for $p$-$v$ and $h$-$s$ charts

$$\text{Heat added} = q_{1-2} = c_v(T_2 - T_1) = u_2 - u_1$$

$$\text{Heat rejected} = q_{3-1} = c_p(T_3 - T_1) = h_3 - h_1$$

$$\therefore \qquad \eta_{th} = \frac{\text{Heat supplied} - \text{Heat rejected}}{\text{Heat supplied}} = 1 - \frac{q_R}{q_A}$$

$$= 1 - \frac{h_3 - h_1}{u_2 - u_1} = 1 - \frac{(u_3 + p_3v_3 - u_1 - p_1v_1)}{(u_2 - u_1)}$$

Since

$$u = u_0 + \frac{p \cdot v}{\gamma - 1}$$

$$\therefore \qquad \eta_{th} = 1 - \frac{u_0 + \dfrac{p_3v_3}{\gamma - 1} + p_3v_3 - u_0 - \dfrac{p_1v_1}{\gamma - 1} - p_1v_1}{u_0 + \dfrac{p_2v_2}{\gamma - 1} - u_0 - \dfrac{p_1v_1}{\gamma - 1}}$$

$$= 1 - \frac{(p_3v_3 - p_1v_1)(\gamma - 1) + (p_3v_3 - p_1v_1)}{(p_2v_2 - p_1v_1)} = 1 - \frac{\gamma(p_3v_3 - p_1v_1)}{(p_2v_2 - p_1v_1)}$$

But $\quad p_3 = p_1, \quad v_3 = r\,v_1, \quad p_2 = p_1\,\alpha, \quad v_2 = v_1$ and $p_2/p_1 = \alpha$

$$\eta_{th} = 1 - \frac{\gamma(p_1.rv_1 - p_1v_1)}{p_1\alpha v_1 - p_1v_1} = 1 - \frac{\gamma(r-1)}{\alpha - 1}$$

**Proved**

If $\quad r = 16, \gamma = 1.25$, and $\alpha = 40$

then $\quad \eta_{th} = 1 - \dfrac{1.25(16-1)}{40-1} = 0.52$ or 52% ***Ans.***

## 2.12 SUMMARY OF FORMULAE

(1) $\eta_{as} = \eta_{th} = 1 - \dfrac{1}{r^{\gamma-1}}$ for Otto cycle

(2) $\text{m.e.p.} = p_1.r\,\dfrac{(\alpha-1)(r^{\gamma-1}-1)}{(r-1)(\gamma-1)}$ for Otto cycle

(3) $\eta_{as} = \eta_{th} = 1 - \dfrac{1}{r^{\gamma-1}}\left[\dfrac{\beta^{\gamma}-1}{\gamma(\beta-1}\right]$ for Diesel cycle

(4) $\text{m.e.p.} = p_1.r\left[\dfrac{\gamma\,r^{\gamma-1}(\beta-1)-(\beta^{\gamma}-1)}{(r-1)(\gamma-1)}\right]$ for Diesel cycle

(5) $\eta_{as} = \eta_{th} = 1 - \dfrac{1}{r^{\gamma-1}}\left[\dfrac{\alpha\beta^{\gamma}-1}{(\alpha-1)+\gamma\times(\beta-1)}\right]$ for Dual cycle

(6) $\text{m.e.p.} = \dfrac{p_1r}{(\gamma-1)(r-1)}\,[r^{\gamma-1}\{(\alpha-1)+\gamma\,\alpha\,(\beta-1)-(\alpha\beta^{\gamma}-1)]$ for Dual cycle

(7) $\text{m.e.p.} = \dfrac{\text{Work done}}{\text{Displacement volume}}$, a general expression

(8) $\eta_{as} = \eta_{th} = \dfrac{T_H - T_L}{T_H}$ for Carnot, Stirling or regenerative cycle

(9) $\eta_{as} = \eta_{th} = 1 - \dfrac{1}{r^{\gamma-1}}\left[\dfrac{\gamma(\alpha^{1/y}-1)}{(\alpha-1)}\right]$ for Atkinson cycle

(10) $\eta_{as} = \eta_{th} = 1 - \dfrac{1}{(r_p)^{\gamma - 1}}$ for Brayton cycle

(11) $\eta_{th} = 1 - \dfrac{(\alpha^{1}/\gamma - 1)}{(\alpha - 1)}$

## PROBLEMS

### Things to Think About

2.1 For the analysis of IC engine, the concept of air standard cycle has been modelled. Do you suggest any other better model?

2.2 Out of Otto, Diesel and Dual cycles which cycle is more close to the real engine?

2.3 What parameters limit the efficiency of IC engines?

2.4 Which type of direct energy conversion system may be applicable to IC engines?

### Short and Long Theoretical Questions

2.5 List the names of various thermal power plants in use.

2.6 What do you mean by air standard cycles? What are the assumptions for air standard cycles?

2.7 Prove that the air standard efficiency for an engine working on the Otto cycles is $1 - \dfrac{1}{r^{\gamma - 1}}$. where $r$ is the ratio of compression and $\gamma$ is the ratio of specific heats of air.

2.8 An engine working on the air standard Otto cycle in which the salient points are 1, 2, 3 and 4 has upper and lower temperature limits $T_3$ and $T_1$. If the maximum work per kg of air is to be done show that the intermediate temperature $T_2 = \sqrt{T_1 T_3} = T_4$.

2.9 Derive an expression for the efficiency of the ideal cycle for the Diesel engine in terms of the compression ratio $r$, the cut-off ratio $\beta$ and $\gamma$.

2.10 Show that the mean effective pressure for an engine working on an ideal Diesel cycle is given by the following expression.

$$\text{m.e.p.} = p_1\, r \left[\frac{\gamma . r^{\gamma - 1}(\beta - 1) - r\,(\beta^{\gamma} - 1)}{(r - 1)\,(\gamma - 1)}\right]$$

where $p_1$ is the pressure at the beginning of compression, $r$ compression ratio, $\beta$ cut-off ratio and $\gamma$ is the adiabatic index.

2.11 Derive the expression for mean effective pressure of Otto cycle.

2.12 Derive an expression for the efficiency of the dual combustion cycle in terms of compression ratio, cut-off ratio and other relevant parameters. Show that this expression reduces to that for Diesel cycle if there is no constant volume rise of pressure and that of the Otto cycle if there is no constant pressure combustion. Also derive the expression for mean effective pressure.

2.13 Compare the efficiency of the Otto, the Diesel and the dual cycle under the conditions of

(a) equal compression ratio and heat input,

(b) constant maximum pressure and heat input,

(c) constant maximum pressure and output, and

(d) constant maximum pressure and temperature.

2.14 Discuss the causes of deviations of the actual Otto cycles from that of ideal.

2.15 Show that the efficiency of Stirling and Erricson regenerative cycles are equal to that of the Carnot.

2.16 Derive the air-standard efficiency of the Joule or Brayton cycle and explain why it is the true constant pressure cycle. Explain its use in modern engines.

2.17 In an ideal cycle, the charge is compressed adiabatically, explosion occurs at the end of the stroke and is followed by adiabatic expansion to such a condition that subsequent isothermal compression restores the charge to its original pressure and volume. Show that the air-standard efficiency of the cycle is

$$\text{ASE} = 1 - \frac{\ln x^{\gamma-1} - \ln r^{\gamma-1}}{x^{\gamma-1} - r^{\gamma-1}}$$

where $r$ = ratio of compression and $x$ = ratio of expansion.

### Numerical Questions

2.18 In a Carnot cycle, the maximum pressure and temperature of the cycle are 17.5 bar and 175°C. The isothermal expansion ratio is 3 while the isentropic expansion ratio is 2. Calculate the thermal efficiency, work done and mean effective pressure.

[***Ans.*** 24.25%, 34.3 kJ and 4.642 bar]

2.19 Find the source and sink temperature of an ideal engine working on Carnot cycle in which 1/5 of the heat input is converted into work and the efficiency becomes doubled when the temperature of the sink is reduced by 60°C.

2.20 In an air-standard Otto cycle the compression ratio 10, the compression begins at 37.8°C, 1 bar and the maximum temperature of the cycle is 1060°C. Determine:

(a) the heat supplied per kg of air,

(b) the work done per kg of air,

(c) the maximum pressure of the cycle, and

(d) the thermal efficiency. [***Ans.*** (a) 795.8 kJ/kg, (b) 478.62 kJ/kg, (d) 60.15% (c) 60.2 bar]

2.21 At the beginning of compression in an air standard Otto cycle engine cylinder the temperature is 37.8°C, the pressure is 0.99 bar and volume is 0.000707 $m^3$. At the end of compression the pressure is 10.55 bar. The heat supplied to the cycle is 1.5 kJ. Calculate:

(a) the compression ratio,

(b) the net work per cycle, and

(c) the mean effective pressure. [***Ans.*** (a) 5.42, (b) 0.735 kJ, (c) 12.2 bar]

2.22 The compression ratio of an Otto cycle is 8 : 1. Before the compression stroke of the cycle begins, the pressure is 1 bar and the temperature is 26.7°C. The heat added to the air per cycle is 2079 kJ/kg air. Determine:

(a) the pressure and temperature at the end of each process of the cycle, and

(b) the thermal efficiency and the mean affective pressure.

[***Ans.*** (a) 688.5 K, 18.37 bar, 3543 K, 94.52 bar, 1542 K, 5.14 bar, (b) 56.47%, 15.26 bar]

2.23 An air standard Otto cycle operates with a compression ratio of 8.5, At the beginning of the compression the air is at 1 bar and 32°C, and during the heat addition process the pressure is tripled. Calculate:

(a) the thermal efficiency of the cycle, and

(b) the efficiency of the Carnot engine operating between the same overall temperature limits. [*Ans.* (a) 57.5%, (b) 85.4%]

2.24 In an air standard Otto cycle, the compression ratio is 9 and a heat addition of 2312.6 kJ/kg of air. If the pressure and temperature at the beginning of the compression process are 1 bar and 4.45°C. Calculate:

(a) the maximum pressure and temperature for the cycle,

(b) the thermal efficiency, and

(c) the mean effective pressure.

2.25 Calculate the ratio of actual efficiency to the air standard for an engine working on Carnot cycle in which the compression ratio ($r$) is 6 and which uses 0.161 kg of oil of calorific value 41820 kJ/kg.

2.26 In gas engine working on Otto cycle, the compression ratio is 10 and the suction conditions are 1 bar and 50°C. If the heat rejection equals 3517 kJ/kg, calculate the air standard efficiency and work ratio. How these quantities would compare with those of a Diesel cycle working under similar conditions? [*Ans.* 60.18%, 0.785; 47.9% 0.688]

2.27 The intake conditions for a theoretical Otto cycle operating with a compression ratio of 8 : 1 are 1 bar and 15°C. At the beginning of the compression stroke, the cylinder volume is 0.00444 m$^3$ and 9.45 kJ of heat is added to the gas during the constant volume heating process. Determine the pressure and temperature at the end of each process of the cycle and determine the thermal efficiency and mean effective pressure. [*Ans.* $T_2$ = 661.65 K, $p_2$ = 18.37 bar, $\eta_{th}$ = 56.47%]

2.28 An air standard Diesel cycle has a compression ratio of 16 and a cut-off ratio of 2. At the beginning of compression the cylinder volume is 0.001415 m$^3$, the air pressure 1 bar and the temperature is 4.45°C. Calculate the thermal efficiency. [*Ans.* 61.4%]

2.29 A Diesel engine with cut-off ratio 1.75 and compression ratio 13.5 has a polytropic exponent $n$ for both compression and expression of 1.35. The pressure at the beginning of compression is 1 bar. Calculate the mean effective pressure and the maximum pressure of the cycle.

[*Ans.* 4.23 bar, 33 bar]

2.30 A Diesel engine with compression ratio of 15 is operating with a fuel having a heating value of 44505 kJ/kg. At half load, this engine requires a 70 to 1 ratio of air to fuel.

(a) What value of the cut-off ratio should be used in an ideal cycle and the temperature at the beginning of compression is 93.3°C?

(b) What is the ideal cycle efficiency? [*Ans.* (a) 1.592, (b) 60.6%]

2.31 In an air standard Otto cycle, the maximum and minimum temperatures are 1400°C and 15°C. The amount of heat supplied is 800 kJ/kg of air. Find the compression ratio and the thermal efficiency. Find also the ratio of the maximum to minimise pressure in the cycle.

[*Ans.* 5.26, 48.6% 30.5:1]

2.32 The swept volume of a four-cylinder petrol engine is 2000 cm$^3$ the clearance volume of each cylinder is 60 cm$^3$. Determine the air standard thermal efficiency. If the inlet conditions are

1 bar and 24°C, and the maximum cycle temperature is 1400°C, determine the mean effective pressure. [*Ans.* 59%, 5.27 bar]

2.33 Calculate the thermal efficiency and the mean effective pressure of a dual combustion cycle when the maximum temperature is 2000°C and the maximum pressure is 70 bar. The pressure and temperature at the start of compression are 1 bar and 17°C respectively.

*Ans.* 63.6%, 10.5 bar

2.34 An engine operates on the theoretical Diesel cycle with a compression ratio of 12 : 1 and fuel is injected for 10% of the stroke. The pressure of the air entering the cylinder is 0.98 bar and its temperatures is 15°C. Calculate:

(a) cut-off ratio, and

(b) the temperatures at the end of compression process, and (c) the heat input.

[*Ans.* (a) 2.1, (b) 753 K,′ (c) 953.5 kJ/kg]

2.35 A Diesel cycle operating on air standard cycle has a compression ratio of 15. The pressure and temperature at the beginning of the compression are 1.04 bar and 15°C. If the maximum temperature of the cycle is 2330 K, determine

(a) the thermal efficiency, and

(b) the mean effective pressure. [*Ans.* (a) 56.9%, (b) 16.7 bar]

2.36 In an air standard Diesel cycle, the compression ratio is 15 and the pressure and temperatures of the air at the beginning of compression is 1 bar and 288 K. The peak temperatures in the cycle is 2700 K. Calculate:

(a) heat supplied,

(b) work done,

(c) cycle efficiency,

(d) peak pressure of the cycle,

(e) cut-off ratio, and

(f) m.e.p.

[*Ans.* (a) 1858.62 kJ/kg, (b) 1022.3 kJ/kg, (c) 55.1%
(d) 44.3 bar, (e) 3.17 and (f) 13.25 bar]

2.37 An air standard Diesel cycles operates with a compression ratio of 14.8 and a cut-off ratio 2. At the beginning of the compression, the air temperatures are 37.8°C and 1 bar respectively. Calculate (a) the maximum temperature in the cycle and the heat input per cycle.

2.38 In a dual combustion air standard cycle for a Diesel engine, the maximum pressure 56 bar. At the beginning of compression, the pressure is 0.98 bar and the temperature is 28.6°C. the compression ratio is 14 and the heat supplied is 1265 kJ/kg of air. Find the cut-off ratio and the ratio of expansion. [*Ans.* 1.805, 7.8] (A.M.ICE. 1955)

2.39 The compression ratio is 8 for an engine operating on the air standard dual cycle. The diameter of the cylinder is 25.4 cm and the stroke is 30 cm. At the start of compression, the air temperature is 21.6°C and the pressure is 1.03 bar. At the end of constant volume process the pressure is 56 bar. If heat is added at constant pressure during 3% of the stroke, Calculate:

(a) the net work of the cycle, and

(b) the thermal efficiency. [*Ans.* (a) 779 kJ/kg, (b) 55.8%]

2.40 An engine operating on the ideal air standard dual cycle has a compression ratio of 11 to 1 and the piston displacement is 0.00328 $m^3$. The initial pressure and temperature at the start of the compression are 1.04 bar and 26.7°C. The volume at the end of constant pressure heat addition process is 1.5 times the clearance volume. The pressure at the end of the adiabatic expansion is 4.2 bar. Calculate the heat added during the constant volume process, (b) the heat added during the constant pressure process and the net work per cycle. [***Ans.*** (a) 6.8 kJ, (b) 8.65 kJ]

2.41 The following data refer to an air standard cycle:

Pressure and temperature at the commencement of compression = 1 bar and 37°C

Compression ratio, $r = 10$

Heat added = 2802 kJ/kg of air

The compression and expansion processes are both isentropic and heat is rejected at constant volume. Find the mean effective pressure and thermal efficiency of the cycle for the following cases:

(a) The maximum pressure is 70 bar and heat is added at constant volume and constant pressure.

(b) Heat is added at constant volume only. For this condition also calculate the extra work available if the expansion takes place isentropically to pressure 1 bar.

[***Ans.*** (a) $\eta_{th}$ = 56.82%, m.e.p. = 19.91 bar (b) 60.18%, m.e.p. = 21.17 bar, 302 kJ]

2.42 An air standard dual cycle has a compression ratio of 6 and the pressure and temperature at the beginning of compression are 1 bar and 29°C. After constant pressure heat addition, the temperature becomes 1177°C and constant volume heat addition raises the temperature to 1627°C. Determine

(a) cut-off ratio,

(b) constant volume pressure ratio,

(c) heat added and rejected per kg of air, and

(d) thermal efficiency.

[***Ans.*** (a) 1.31, (b) 2.344, (c) 1059.29 kJ/kg, (d) 526.8 kJ/kg, (e) 49.93%]

2.43 A four-stroke semi-Diesel engine operates on dual cycles. At the commencement of compression, the pressure and temperature are 1 bar and 37.8°C. The pressure after the end of isentropic compression is 16.4 bar. Heat is added first at constant volume and then at constant pressure of 29.3 bar. The heat of combustion is shared equally between the two phases. If the relative efficiency is 55% estimate

(a) the actual specific fuel consumption in kg/kWh, and

(b) the theoretical power developed per $m^3$ displacement at an engine speed of 180 r.p.m. the calorific value of fuel is 43493 kJ/kg of fuel.

[***Ans.*** (a) 0.293 kJ/kWh, (b) 779 kW/$m^3$]

2.44 Calculate the index for compression and the mass of the mixture in the cylinder of a gas engine where bore is 18 cm and stroke is 32.5 cm. The ratio of compression is 5. Pressure and temperature at the start of compression is 1 bar and 100°C and the pressure at the end of compression is 6 bar. Determine the work of compression and heat interchange through cylinder wall during compression.

[***Ans.*** $m$ = 0.00945 kg, $n$ = 1.13, Compression work = – 40.5 kJ, $Q_R$ = 27 kJ]

2.45 An internal combustion engine has a bore of 56 cm and a stroke of 76 cm. The compression ratio is 13.5 Pressure and temperature at the end of suction stroke are 1 bar and 43°C respectively. The compression index is 1.37. Calculate

(a) pressure and temperature at the end of compression,

(b) mass of air in the cylinder,

(c) work done, and

(d) heat rejected during compression.

[***Ans.*** (a) 35.2 bar, 828 K, (b) 0.212 kg, (c) 82.6 kJ (d) 7.56 kJ]

$$\left[\textbf{Hint: } \text{Heat rejected during polytropic process} = \left(\frac{\gamma - n}{\gamma - 1}\right) \times \text{work done}\right]$$

2.46 A Stirling regenerative air engine works between temperature of 400°C and 15°C. The ratio of isothermal expansion is 3.

Calculate the ideal efficiency when

(a) the engine is filled with a perfect regenerator, and

(b) when the efficiency of the regenerator is 0.8. [***Ans.*** (a) 57.3%, (b) 45.4%]

2.47 Find the air standard efficiency of Atkinson cycle if the compression ratio is 5 and pressure at the end of heat reception is 2.5 times that at the beginning. Take $\gamma = 1.41$. [***Ans.*** 57.5%]

2.48 In an air standard Brayton cycle, the air enters the compressor at 1 bar and 25°C. The pressure after compression is 3 bar. The temperature at turbine inlet is 650°C. Calculate per kg of air–

(a) heat supplied,

(b) heat rejected,

(c) work available at the shaft,

(d) temperature of air leaving the turbine, and

(e) cycle efficiency.

[***Ans.*** (a) 510 kJ/kg, (b) 378 kJ/kg, (c) 138.5 kJ/kg, (d) 875 K, (e) 26.9%]

2.49 A gas turbine power plant operates under the following conditions: Air enters the compressor at 1 bar and 15°C and leaves the compressor at 15 bar. Air leaves the combustion chamber at 1600 K. Calculate the thermal efficiency.

2.50 For an ideal Otto engine operating on air, the temperature at the end of isentropic compression is 452°C and at the end of expansion is 1347°C. The compression ratio is 7.5. Calculate the cycle work and efficiency. Use gas tables. [***Ans.*** 1229 kJ/kg, 47.6%]

2.51 The fuel-air ratio of an ideal Otto engine with 12% clearance is 0.06 kg fuel/kg of air. The calorific value of the fuel is 41840 kJ/kg. The engine operates on 25 kg/min of air initially at 1 bar and 37°C. Treating the working substance as air, calculate:

(a) displacement volume/min,

(b) pressure and temperature of all important points,

(c) work obtained,

(d) thermal efficiency of the cycle, and

(e) the m.e.p.

[***Ans.*** (a) 20.89 m$^3$/min, (b) $T_2 = 755$ K, $p_2 = 22$ bar, $p_3 = 99$ bar, $T_3 = 3377$ K, $p_4 = 5.64$ bar, $T_4 = 1755$ k, (c) 1276.6 kJ/kg, (b) 50.85% (e) 15.558 bar]

2.52 In air IC engine operating on the dual cycle, the temperature of the working fluid (air) at the beginning of compression is 27°C. The ratio of the maximum and minimum pressures of the cycle is 70 and the compression ratio is 15. The amounts of heat added at constant volume and at constant pressure are equal. Compute the air standard thermal efficiency of the cycle. State the main reason why the actual thermal efficiency is different from theoretical value?

[*ES 1993*]

2.53 A four-stroke limited pressure cycle (diesel) engine draws 1.2 kg/s of air at 1.03 bar and 27°C. Compression ratio of cycle is 16. Pressure ratio during constant volume heat addition is 2.0. The total heat added is equal to 2300 kJ/kg of air in the cylinder. Determine:

(i) pressure, volume and temperature at all salient points,

(ii) percentage of heat added during constant pressure process,

(iii) cut-off ratio,

(iv) thermal efficiency, and

(v) m.e.p. [*ES 1988*]

2.54 A six-cylinder four-stroke petrol engine has a swept volume of 300 $cm^3$ per cylinder, a compression ratio 10 and operates at a speed of 3500 rpm. If the engine is required to develop an output of 75 kW at this speed, calculate:

(a) The cycle efficiency, the necessary rate of heat addition, the mean effective pressure and the maximum temperature of the cycle. Assume that the engine operates on the Otto cycle and that the pressure and temperature before isentropic compression are 1 bare and 15°C respectively take $c_v = 0.718$, $\gamma = 1.4$.

(b) If the above engine is a compression ignition engine operating on the diesel cylinder and receiving heat at the same rate, calculate the efficiency, power output and mean effective pressure. [*ES 1993*]

### Software Based Questions

2.55 Develop a computer software for the analysis of all types of air standard cycles.

2.56 Develop a software for the sizing of SI and CI engines.

### Design and Open Ended Questions

2.57 Device a cycle for an internal combustion which you have not heard of. Prepare a comprehensive report on this new cycle with design feature of the engine.

2.58 Design on eco-friendly electric car engine.

### Multiple-Choice Questions

2.59 The air standard efficiency of a Diesel cycle increases when

(a) the cut-off ratio is increased (b) the cut-off ratio is decreased

(c) the speed increases (d) the more heat energy is supplied

2.60 When the pressure ratio is increased keeping the compression ratio and heat input same the thermal efficiency of dual cycle will

(a) increase (b) decrease

(c) remain the same (d) approach the Diesel cycle.

2.61 The performance of an IC engine cycle should be judged by its

(a) ideal cycle efficiency (b) work ratio
(c) ratio of mep/peak pressure (d) size per unit power output

2.62 An Otto cycle operates with volume of 30 cc and 300 cc at TOC and BDC respectively. If the power output is 100 kW, the input in kW will be

(a) 93 (b) 110 (c) 135 (d) 165

2.63 For the maximum specific output of on Otto cycle

(a) the working fluid should be air
(b) the speed should be high
(c) suction temperature should be high
(d) the temperature of working fluid at the end of compression and expression should be equal.

## Key to Multiple-Choice Questions

2.59 (b) 2.60 (a) 2.61 (a) 2.62 (c) 2.63 (d)

CHAPTER

# Fuel-Air and Actual Cycles

# 3

## OVERVIEW

The actual cycle of internal combustion engine is very complex. In the previous chapter, for the approximate and quick analysis, the actual cycle has been modelled as air standard cycle and the resulting governing equations developed give much higher indicated thermal efficiency as compared to actual engine. For example, the air standard Otto cycle's efficiency for compression ratio of 7 results in 54% while actual cycle efficiency is around 32%. *The lower value of thermal efficiency of actual cycle as compared to air standard cycle is attributed to many factors such as non-consideration of properties of real working fluid, i.e., mixture of air fuel and residual gases, non-instantaneous burning of fuel and valve operation, incomplete combustion, variation of specific heat with temperature, etc.* These factors contribute to losses which are given below.

(i) Real working fluid.
(ii) Losses due to variation of specific heat of charge with temperature.
(iii) Losses due to dissociation of gases.
(iv) Time losses due to mixing of fuel and air and combustion.
(v) Losses due to incomplete combustion.
(vi) Heat losses.
(vii) Exhaust gas effect.
(viii) Exhaust blow-down losses.
(ix) Pumping losses.

*These losses can be estimated on the basis of extensive experimental results.* If the losses due to variation of specific heat of charge with temperature and dissociation of gases (**i to iii**) are subtracted from the air standard cycle, the cycle is called ***fuel-air cycle***, the analysis of which will be dealt in this chapter. If the other remaining losses (iv to ix) are subtracted from the fuel cycle, the resulting cycle will closely approximate the ***actual cycle*** which will also be discussed in this chapter. A study of various losses will enable to model the fuel-air and actual cycles. A block diagram shown in Fig. 3.1 depicts the fuel-air and actual cycles.

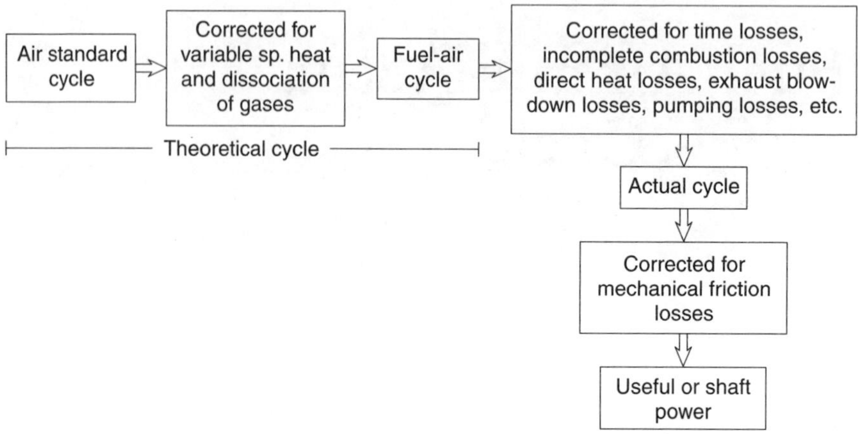


**Fig. 3.1** Air, fuel-air and actual cycle relationship

## (A) FUEL-AIR CYCLE

## 3.1 MODELLING OF FUEL-AIR CYCLE

The following assumptions are made for the fuel-air cycle.

(i) The charge consists of mixture of fuel, air, water vapour and residual gases.

(ii) The specific heat of the gas varies with temperature.

(iii) Dissociation of gases takes place at high temperature.

(iv) The number of molecules of the gas present in the cylinder is a function of pressure and temperature.

(v) There is a negligible chemical change in either fuel or air prior to combustion.

(vi) After the combustion, the charge is assumed to be in chemical equilibrium.

(vii) The compression and expansion processes are adiabatic with negligible friction.

(viii) The fuel is assumed completely vaporized and perfectly mixed with air.

(ix) The combustion takes place instantaneously maintaining constant volume heat addition, in the case of petrol engine.

The above assumptions in modelling fuel-air cycle are very near (about 80%) to actual cycle as compared to air standard cycle.

## 3.2 EFFECT OF VARIABLE SPECIFIC HEAT ON OTTO CYCLE

From the knowledge of basic thermodynamics, we know that specific heat, in general, is not constant. There is significantly large variation of specific heats with temperature but no significant change with pressure except at very high value. This variation may be expressed as a function of temperature in the form of polynomials given below.

$$c_p = a + bT + cT^2 + dT^3 + ....$$

The increase in the specific heat with temperature is attributed to the fact that as the temperature is raised, large and large fractions of heat input are consumed in producing the motion of the atoms within the molecules, hence more heat is required to raise the temperature of gas resulting in higher value of specific heat.

For an example, $c_p$ of air at 0°C = 1.005 kJ/kgK while at 2000°C. $c_p$ = 1.264 kJ/kgK.

The effect of variable specific heat on an ideal Otto cycle is shown in Fig. 3.2. Process 1–2–3–4 is the ideal cycle when the specific heat of working fluid is assumed to have constant value. We know that specific heats of all gases increase with temperature and the ratio of specific heats ($\gamma$) decreases with temperature. If variable specific heats are taken into account for the compression stroke of the engine, the final temperature will be lower and consequently the pressure will also be lower (2′) than if constant values of heat capacity had been used. The temperature at the end of combustion will reach only 3′ lower than 3 because the temperature at the start of the combustion is lower and specific heat increases with temperature. Process 3′–4′ shows the isentropic expansion for constant specific heat, the isentropic expansion process for variable specific heat will be 3′–4″ which lies above 3′–4′, because of the fact that the temperature falls during expansion and specific heat decreases. We see that due to the variable specific heat consideration, more work is delivered but this is no gain, since had the specific heat been constant, expansion would have started from 3 and not from 3′.

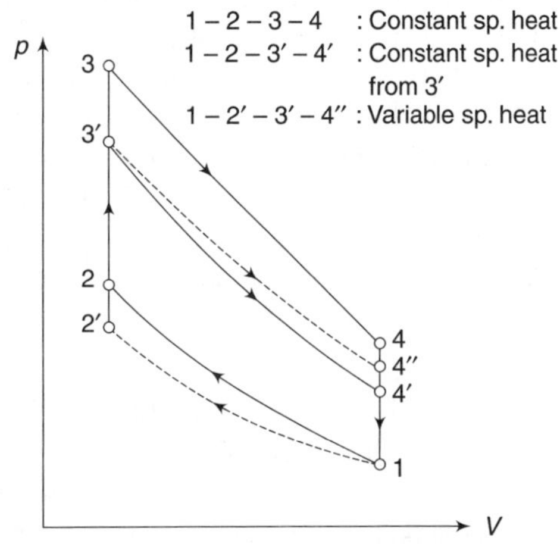

**Fig. 3.2** Effect of variable specific heat

It is clear from the above analysis that *the consideration of the variable specific heat lowers most of the pressure and temperature in the cycle and delivers less work than for the corresponding cycle with constant specific heat.*

## 3.3 EFFECT OF DISSOCIATION OF CHEMICAL EQUILIBRIUM ON OTTO CYCLE

*Dissociation means disintegration of burnt gases into simpler components at higher temperature and it increases with temperature. A considerable amount of heat is absorbed during dissociation and this heat is liberated when the elements recombine as the temperature falls.* This means that dissociation is a reversible process. The effect of dissociation is identical with the effects produced by variable specific heat. However, the effect of dissociation is much smaller as compared to variable specific heat. The dissociation is mainly of $CO_2$ into $CO$ and $O_2$ and commences at 1500°C as given below:

$$2CO + O_2 \longleftrightarrow 2CO_2 + \text{Heat}$$

At 1500°C, the dissociation amounts to about 1%.

There is a very little dissociation of $H_2O$ and $H_2$ and $O_2$ as given below

$$2H_2 + O_2 \leftrightarrow 2H_2O + \text{Heat}$$

It is to be noted that the arrow in the above equation is in both the directions. It signifies that a limiting temperature is attained when the reaction has the same rate for either direction. The reaction is then said to be in equilibrium in either direction. In other words, a definite proportion of CO, $O_2$ and $CO_2$ is present in the equilibrium mixture at any temperature, at low temperature the portion of $CO_2$ is high while at high temperature the proportion of CO is high. *It is this reason that combustion temperature does not go still higher because dissociation limits the release of chemical energy.* Figure 3.3 shows the equilibrium composition of the products of combustion of octane ($C_8H_{18}$) with air over the range from 1500 K to 3000 K. The products are $CO_2$, $H_2O$, $H_2$, $N_2$, CO, $O_2$, $OH_2$, H and O.

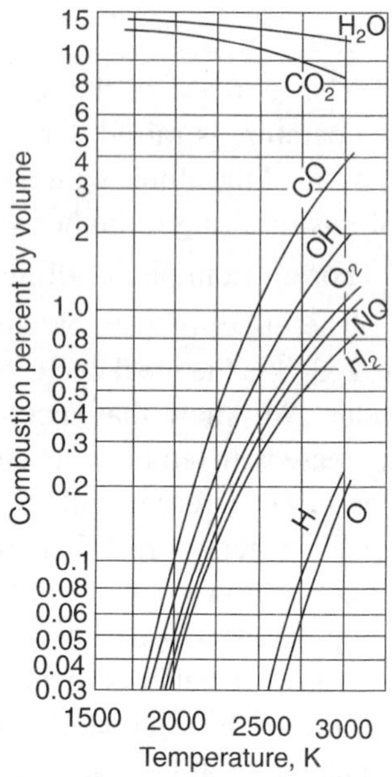


**Fig. 3.3** Equilibrium composition of combustion of octane with air

Figure 3.4 shows the effect of dissociation of gas on *p-V* diagram of Otto cycle. The point 3 corresponds to theoretical temperature of combustion while the lower point 3′ corresponds to the temperature for chemical equilibrium or dissociation as the pressure and temperature get reduced due to dissociation. Processes 3–4 and 3′–4′ show isentropic expansion of gases without reassociation due to fall in temperature. But isentropic expansion with continued equilibrium (i.e., dissociation) would start at 3′ and end at 4″ because of the liberation of energy throughout the expansion stroke as the temperature falls and reaction proceeds towards completion. It is to be noted that expansion line 3′–4″ lies above 3′–4′ which means that more work is done during expansion stroke with continuing dissociation than with no dissociation. But if dissociation had not been present the combustion process would have finished at 3 and the work of expansion would be that from 3 to 4 which is greater than from 3′–4″. It is reasoned that chemical energy released in process 2–3 is available for doing work throughout the entire stroke 3–4, on the other hand the same amount of chemical energy would have liberated, partially in process 2–3′ and the remainder forms 3′–4″. Thus, only a portion of the chemical energy is

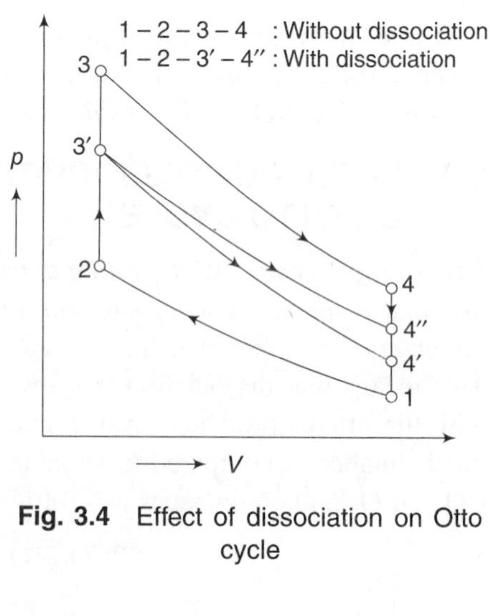


**Fig. 3.4** Effect of dissociation on Otto cycle

released at TDC, the position for full expansion, while a part of energy is released late in the expansion stroke and therefore with less opportunity to do expansion work. Since the work done is less with the consideration of chemical equilibrium the energy left in the gas must be higher and therefore, the temperature and pressure are higher at 4″ than 4′. *Thus, the effect of dissociation is to reduce the maximum temperature, maximum pressure, power output and thermal efficiency of the cycle.*

It is to be *noted that the effect of dissociation on Diesel cycle is very less as compared to Otto cycle.*

## 3.4 COMPARISON OF AIR STANDARD AND FUEL-AIR CYCLES FOR SI ENGINES

A comparative study of air standard cycle, air standard cycle with variable specific heat and fuel air cycles may be made by plotting these on $p$-$V$ diagram as shown in Fig. 3.5. The fuel-air cycle is the combined effect of variable specific heat and dissociation of fuel-air or combustion products. In this comparison, compression ratio and heat input have been kept the same.

*For a compression ratio of 7 and heat input of 3180 kJ/kg, the cycle efficiencies of air standard cycle, air standard cycle with variable specific heat and fuel-air cycle are 57%, 49.4% and 35.5% respectively.*

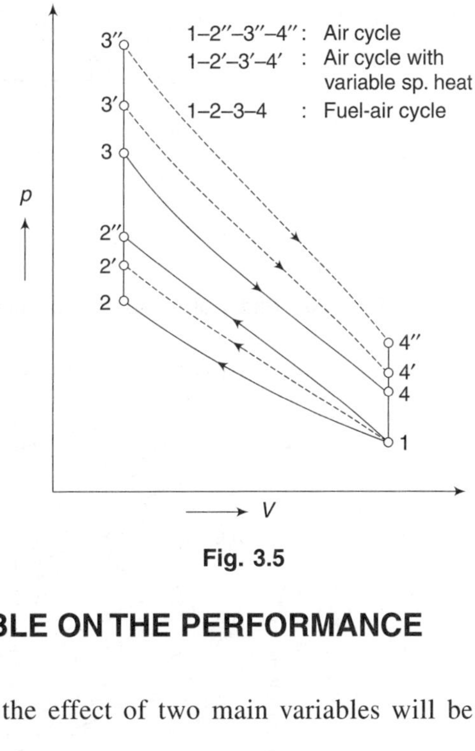


**Fig. 3.5**

## 3.5 EFFECT OF OPERATING VARIABLE ON THE PERFORMANCE OF AIR-FUEL OTTO CYCLE

Though, the operating variables are many, but the effect of two main variables will be considered here. These are:

(a) *Compression ratio;* (b) *Air-fuel ratio.*

### 3.5.1 Effect of Compression Ratio on Thermal Efficiency

Figure 3.6 shows the effect of compression ratio on thermal efficiency of air standard and air-fuel Otto cycles for various values of relative air-fuel ratio, $\phi_R$, which is defined as the actual air-fuel ratio in consideration to the chemically correct air-fuel ratio on mass basis. Figure 3.7 shows the above results depicted as thermal efficiency vs. $\phi_R$ for various values of compression ratio. From the above results, it is clear that *thermal efficiency increases with the increase of compression ratio. Further, the rich fuel-air mixture results in lower thermal efficiency as compared to lean mixture. Air standard cycle results in maximum efficiency.*

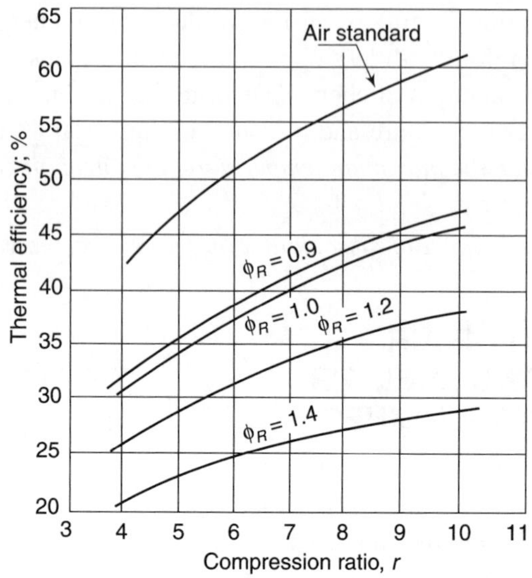


**Fig. 3.6** Effect of compression ratio and mixture strength on thermal efficiency

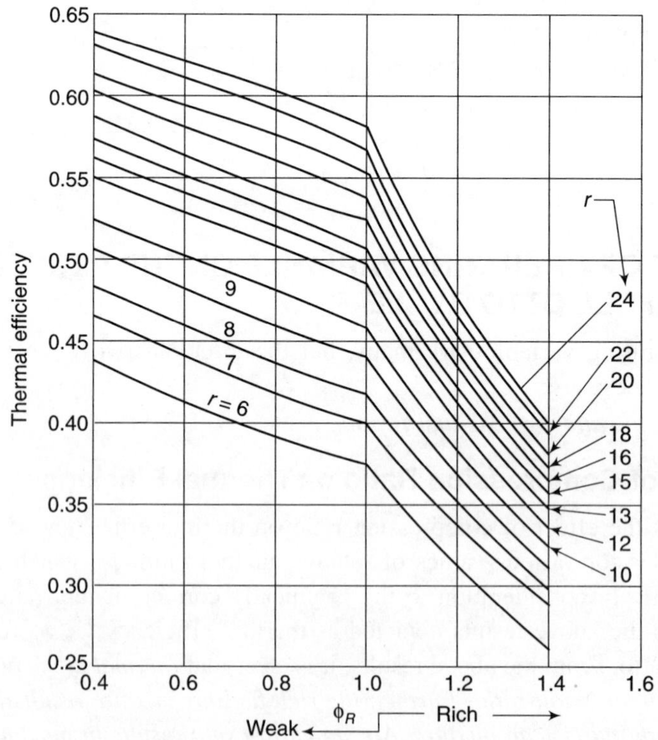


**Fig. 3.7** Thermal efficiency *vs.* relative fuel-air ratio for various compression ratio

## 3.5.2 Effect of Fuel-Air Ratio

### 3.5.2.1 On Thermal Efficiency

Figure 3.8 shows the effect of mixture strength on thermal efficiency for a given compression ratio. The thermal efficiency of air standard cycle is invariant of mixture strength. In the case of fuel-air cycle, as the fuel-air ratio reduces (i.e., lean mixture), the thermal efficiency improves. This is because of the fact that with very lean mixture temperature rise at the end of combustion is less. The low temperature of gas results in low specific heat and no dissociation loss. As a result, the thermal efficiency approaches to air standard cycle efficiency.

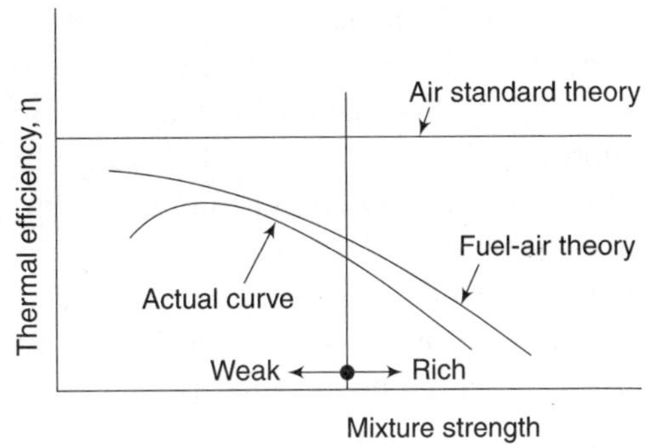


**Fig. 3.8** Effect of mixture strength on thermal efficiency

### 3.5.2.2 On the Maximum Power

The effect of fuel-air ratio on cycle power is depicted in Fig. 3.9. *Air standard cycle results in the maximum power by chemically correct mixture but air-fuel cycle yields the maximum power by a mixture having about 10% richness.*

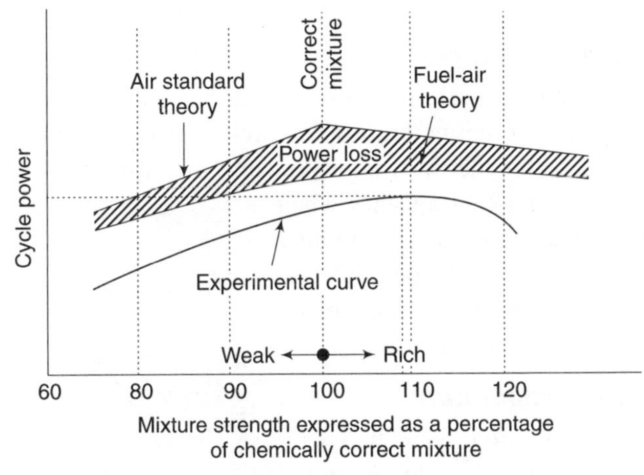


**Fig. 3.9** Effect of mixture strength on power

With the increase of richness, the efficiency of air-fuel cycle falls rapidly. This is attributed to the combined effect of higher specific heat, loss due to dissociation of gases and incomplete combustion on account of insufficient air. The power loss due to dissociation as compared to air standard cycle is shown by hatched area. Actual power measured is lesser than that predicted by air-fuel cycle.

### 3.5.2.3 On Maximum Temperature

Figure 3.10 shows no effect of mixture strength on temperature. There is no dissociation in the burnt gases of a weak fuel-air mixture because the temperature rise is too low to initiate dissociation. Rich fuel mixture suppresses dissociation of $CO_2$ because of the presence of CO and $O_2$ due to incomplete combustion which again results in less temperature rise. The maximum temperature rise takes place with chemically correct mixture when no dissociation is taken into account. In the case of fuel-air cycle, the maximum temperature rise occurs for all compression ratios when the mixture is about 6% rich. This is because of the fact that at chemically correct mixture, some oxygen is left which combines with extra 6% fuel and raises the temperature further. Dissociation reduces the maximum temperature of gas by about 300°C at correct mixture strength and compression ratio of 7.

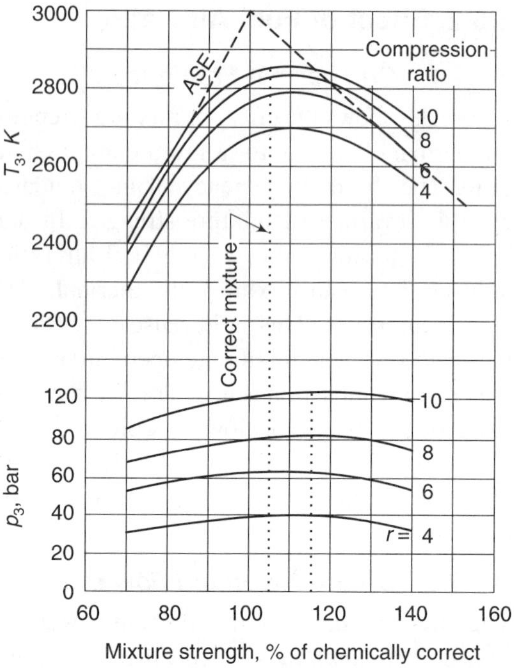


**Fig. 3.10** Effect of mixture strength on the maximum temperature and pressure

### 3.5.2.4 On Maximum Pressure

In the case of Otto cycle, the volume remains constant during combustion and so the maximum pressure achieved depends on temperature and number of molecules as governed by gas law, $pv = nR_oT$. The curve for the maximum pressure, therefore, follows the nature of the maximum temperature and its peak reaches when mixture is 10% richer in strength because number of molecules start decreasing only beyond this condition (Fig. 3.10). *As expected, the maximum pressure increases with compression ratio.*

### 3.5.2.5 On Exhaust Temperature

The exhaust temperature is directly related with the maximum temperature in the cycle. Since the maximum temperature is less either in lean or rich mixture, so is the nature of exhaust temperature. The highest exhaust temperature occurs when mixture is chemically correct as the exhaust temperature is not high which may initiate dissociation. *The effect of compression ratio on temperature is just different from the maximum temperature as shown in Fig.*

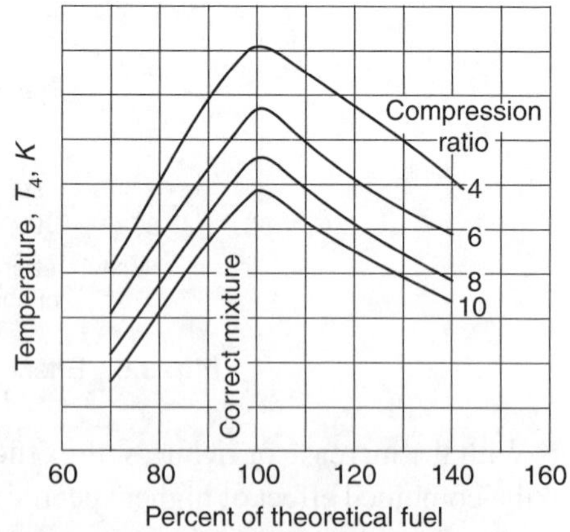


**Fig. 3.11** Effect of fuel-ratio on exhaust temperature

*3.11. In this case, this exhaust temperature is higher at lower compression ratio. This is because of the fact that larger expansion ratio results in more work done and less heat to exhaust which yields lower temperature.*

### 3.5.2.6 On Mean Effective Pressure (m.e.p.) and SFC

Figure 3.12 shows the effect of mixture of strength on mean effective pressure for various compression ratios. *For all compression ratios, the mean effective pressure increases with fuel-ratio, reaches the maximum value when the mixture is slightly rich and then decreases.* The nature of variation is similar to that of the maximum temperature and pressure. The mean effective pressure increases with compression ratio.

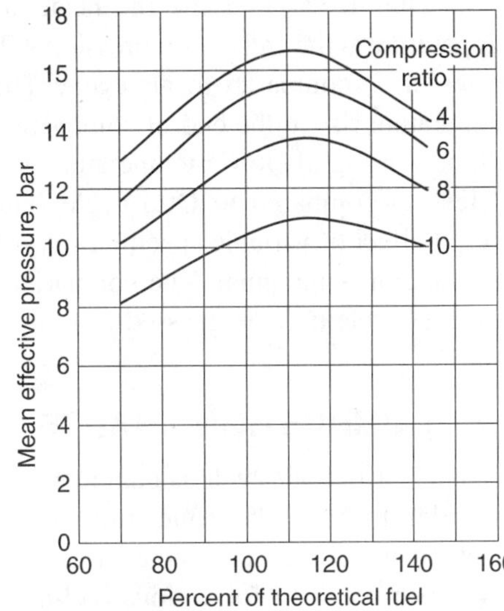


**Fig. 3.12** Effect of mixture strength on mean effective pressure

The variation of mean effective pressure with specific fuel consumption (sfc) for various mixture strength and cycles are shown in Fig. 3.13. There is no variation of thermal efficiency with mixture strength in the case of air standard cycle. *But in the case of fuel-air cycle, the specific fuel consumption first increases with increase of m.e.p. and then after optimum value of m.e.p., it increases with the decrease in m.e.p. This indicates that the thermal efficiency will deteriorate as the mixture supplied to an engine is enriched.* This is attributed due to the fact that with the enrichment of mixture strength losses caused by variable specific heat and dissociation though decreases as compared to chemically correct mixture but the effect of unused excess fuel is more pronounced resulting in lower thermal efficiency. *From this loop, it is also possible that the maximum efficiency is possible in the neighbourhood of chemically correct mixture.*

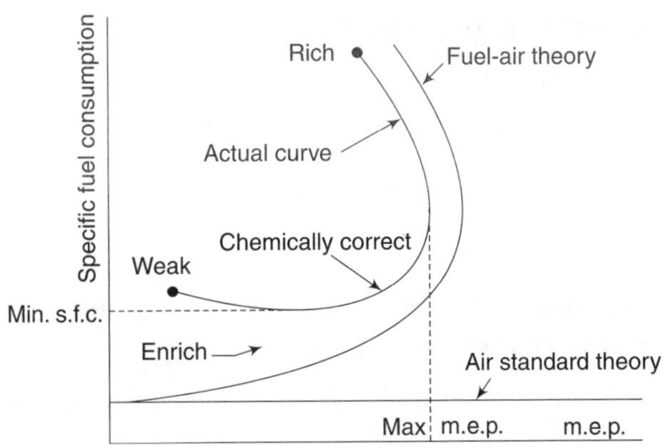


**Fig. 3.13** Effect of mixture strength on mean effect pressure and specific fuel combustion

## 3.6 COMPARISON OF DUAL PRESSURE AIR-FUEL CYCLE WITH AIR STANDARD CYCLE

Figure 3.14 shows the comparison of air-fuel dual pressure cycle with air standard cycle. Due to variable specific heat, the temperature at the end of compression will be lesser as compared to air cycle. The temperature rise at the end of combustion will be at a slightly lower temperature, not so less as compared to Otto cycle. The overall effect of variable specific heat and dissociation is marginal. The comparison of diesel cycle may be guessed from dual cycle.

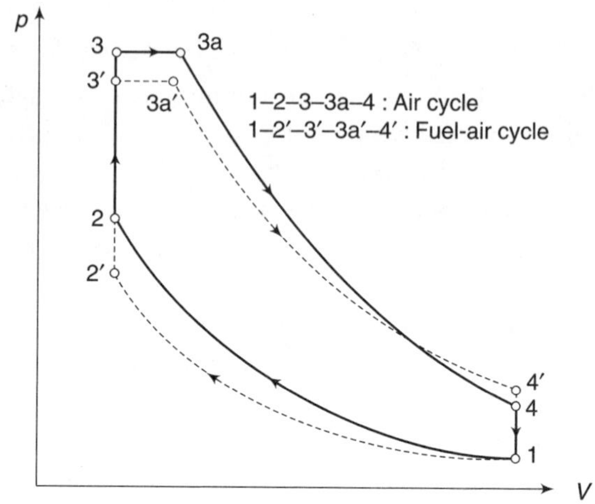


**Fig. 3.14** Comparison of fuel-air dual cycle and corresponding air cycle

## 3.7 COMBUSTION CHARTS

Analysis of fuel-air cycle is cumbersome. In order to save the time and labour involved in the analysis, combustion charts have been prepared for various fuel-air ratios to calculate the thermodynamic properties of the cycle which are given in chapter 1.

## 3.8 GAS TABLES

It is possible to express the thermodynamic properties of gases that obey the law of ideal gas in terms of temperature alone. These properties when tabulated are called the gas tables. Hence, gas tables may be conveniently used to calculate property relations between different states of a system. The tabulated values take into consideration the variation of specific heat with temperature. The most widely used gas tables are of Keenan and Kaye. The gas tables are constructed in the following manner.

**(a) Enthalpy and Internal Energy (h and u)**

We know from the previous article

$$dh = c_p . dT$$

or $$h = \int c_p \, dT + \text{constant}$$

where $$c_p = f(T)$$

If the variation of $c_p$ with temperature is known then enthalpy may be calculated. The constant of integration may be chosen arbitrarily.

If $$h = h_0 \quad \text{at} \quad T = T_0 \quad \text{then constant} = h_0$$

∴ $$h = \int c_p \, dT + h_o \qquad ...(3.1)$$

Here enthalpy is a function of temperature. The internal energy can be written as

$$u = h - pv = h - RT \qquad ...(3.2)$$

Therefore, with the help of Eqs. (3.1) and (3.2), $h$ and $u$ can be tabulated as a function of temperature only.

**(b) Pressure ratio ($p_r$).** We can express the pressure ratio along an isentropic path as a function of temperature only. For an ideal gas, we have

$$ds = c_p \frac{dT}{T} - R \frac{dp}{p}$$

For an isentropic process, $ds = 0$, hence

$$c_p \frac{dT}{T} = R \frac{dp}{p} \qquad ...(3.3)$$

Integrating this equation between temperature $T_0$ to $T$, we have

$$\ln \frac{p}{p_0} = \frac{1}{R} \int c_p \frac{dT}{T} \qquad ...(3.4)$$

The ratio $(p/p_0)$ is called the **relative pressure, $p_r$**. In this equation, $p_0$ and $p$ are pressures corresponding to $T_0$ and $T$ respectively. If we select a certain value of $T_0$, the ratio $(p/p_0)$ for an isentropic path is a function of $T$ alone because $c_p = f(T)$ only. Hence,

$$p_r = \frac{p}{p_0} = \exp\left[\frac{1}{R} \int c_p \frac{dT}{T}\right]$$

Integrating Eq. (3.3), we get

$$\ln \frac{p_2}{p_1} = \frac{1}{R} \int_{T_1}^{T_2} c_p \frac{dT}{T} \qquad ...(3.5)$$

or $$\ln \frac{p_2}{p_0} - \ln \frac{p_1}{p_0} = \frac{1}{R} \int_{T_1}^{T_2} c_p \frac{dT}{T}$$

or $$\ln \frac{p_{r2}}{p_{r1}} = \frac{1}{R} \int_{T_1}^{T_2} c_p \frac{dT}{T} \qquad ...(3.6)$$

Comparing Eqs. (3.5) and (3.6), we get for isentropic process

$$\frac{p_2}{p_1} = \frac{p_{r2}}{p_{r1}} \qquad ...(3.7)$$

**(c) Volume Ratio ($v_r$)** Following the same procedure as that of pressure ratio, the **volume ratio** ($v_r$) along an isentropic path is

$$ds = 0 = c_v \frac{dT}{T} + R \frac{dv}{v} \qquad ...(3.8)$$

or $$\frac{dv}{v} = \frac{1}{R}\left(c_v \frac{dT}{T}\right)$$

or $$\ln \frac{v}{v_0} = \frac{1}{R}\int_{T_1}^{T_2} c_p \frac{dT}{T} = \ln v_r$$

Here $v_r = v/v_0$ = relative volume. From the above equation, $v_r$ is a function of temperature alone. Similar to pressure ratio, we may get for volume ratio as

$$\frac{v_2}{v_1} = \frac{v_{r2}}{v_{r1}} \text{ for isentropic path} \qquad ...(3.9)$$

**(d) Change of Entropy.** We have

$$Tds = dh - vdp$$

or $$ds = c_p \frac{dT}{T} - R\frac{dp}{p}$$

Integrating this, we get

$$s = \int c_p \frac{dT}{T} - R \ln p + c \qquad ...(3.10)$$

Equation (3.10) reveals that entropy is a function of both $T$ and $p$, hence in general, entropy cannot have a single value for any given value of $T$. Let us define a new function

$$\psi = \int_{T_0}^{T} c_p \frac{dT}{T} \qquad ...(3.11)$$

Here temperature ($T_0$) is an arbitrary reference temperature. For a constant pressure process we have

$$s_2 - s_1 = \int_{T_1}^{T_2} c_p \frac{dT}{T} = \int_{T_0}^{T_2} c_p \frac{dT}{T} - \int_{T_0}^{T_1} c_p \frac{dT}{T} = \psi_2 - \psi_1$$

In the general case, we may write

$$s_2 - s_1 = \int_{T_1}^{T_2} c_p \frac{dT}{T} - R \ln\left(\frac{p_2}{p_1}\right) = (\psi_2 - \psi_1) - R \ln \frac{p_2}{p_1} \qquad ...(3.12)$$

With the help of Eq. (3.12), we can calculate $s_2 - s_1$.

**EXAMPLE 3.1** Calculate the percentage change in the efficiency of air standard Otto cycle having a compression ratio of 7 for the following cases:

(a) The specific heat at constant volume increases by 2%.

(b) The specific heat at constant pressure increases by 2% assuming $\gamma$ to be invariant.

***Solution*** The air standard efficiency of Otto cycle is given by

$$\eta = 1 - \frac{1}{r^{\gamma-1}} = 1 - \frac{1}{7^{0.4}} = 54\%$$

We know that, $\gamma - 1 = \dfrac{R}{c_v}$

$$\therefore \qquad \eta = 1 - \frac{1}{r^{R/c_v}} = 1 - r^{-R/c_v}$$

Taking log on both sides, we get

$$\ln(1-\eta) = \frac{-R}{c_v}\ln r$$

Differentiating the above equation, we get

$$-\frac{1}{1-\eta}\,d\eta = \frac{R}{c_v^2}\ln r\,dc_v = \frac{(\gamma-1)c_v \ln r}{c_v^2}\,dc_v$$

$$\text{or} \qquad \frac{d\eta}{\eta} = \frac{-(1-\eta)(\gamma-1)\ln r}{\eta}\frac{dc_v}{c_v}$$

(a) $$\text{or} \qquad \frac{d\eta}{\eta} = \frac{-(1-0.54)(1.4-1)\ln 7}{0.54} \times \frac{2}{100} = -1.326\%$$ ***Ans.***

$$\gamma - 1 = \frac{r\gamma}{c_p}$$

The thermal efficiency is expressed as

$$\eta = 1 - \frac{1}{r^{\gamma-1}} = 1 - \frac{1}{r^{R\gamma/c_p}}$$

$$1 - \eta = r^{-R\gamma/c_p}$$

Taking log by both sides, we get

$$\ln(1-\eta) = \frac{-R\gamma}{c_p}\ln r$$

Differentiating, we get assuming $\gamma$ to be constant

$$-\frac{1}{1-\eta}\,d\eta = \frac{R\gamma}{c_p^2}\ln r\,dc_p$$

(b) $$\text{or} \qquad \frac{d\eta}{\eta} = -\frac{(1-\gamma)(\gamma-1)}{\eta}\ln r\frac{dc_p}{c_p} = -1.326\%$$ ***Ans.***

**EXAMPLE 3.2** In an engine, working on the Otto cycle, the compression ratio is 5 and the ratio of air to fuel at the beginning of the compression is 15:1. The calorific value of fuel is 44000 kJ/kg. At the beginning of the compression, the pressure and temperature of the charge is 60°C and 1 bar. Calculate the maximum pressure in the cylinder, if the compression index $n$ = 1.355 and the specific heat of constant volume is given as

$$c_v = 0.718 + 20 \times 10^{-5}\, T$$

where $T$ is the temperature in K.

Compare the value of the maximum pressure if the value of specific heat at constant volume remained constant at 0.718

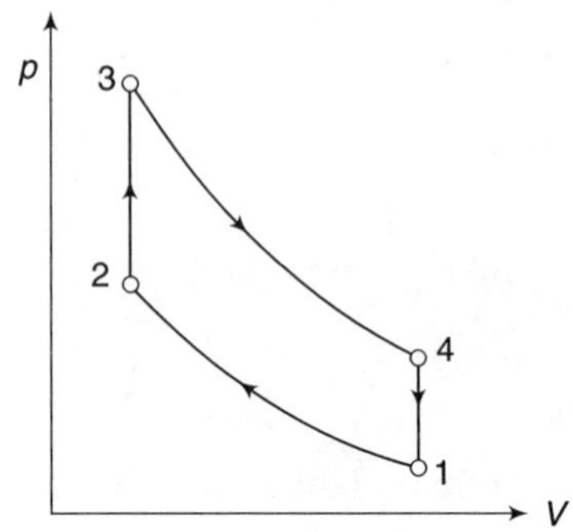


**Fig. 3.15** Example 3.2

***Solution*** Refer to Fig. 3.15.

The maximum pressure in the cycle is attained at point 3, i.e., after the heat addition.

We have, $$\frac{T_2}{T_1} = \left(\frac{v_1}{v_2}\right)^{n-1} = (5)^{0.355} = 1.7706$$

$$\therefore \quad T_2 = (60 + 273) \times 1.7706 = 589.609 \text{ K}$$

Since heat is added in the Otto engine at constant volume, hence

$$\int_2^3 \delta q = \int_2^3 du = \int_2^3 c_v\, dT$$

The air-fuel ratio is 15:1. Hence, 1 kg of mixture will have 1/16 kg of fuel. Taking 1 kg of mixture

$$\int_2^3 c_v\, dT = \int_2^3 \delta q$$

or $$\int_{589.609}^{3} (0.718 + 20 \times 10^{-5}\, T)\, dT = \frac{44000}{16}$$

or $$0.718\,(T_3 - 589.609) + 10 \times 10^{-5}\,(T_3^2 - 589.609^2) = \frac{44000}{16}$$

or $$T_3^2 + 7180\, T_3 - 320.809 \times 10^5 = 0$$

or $$T_3 = \frac{-\,7180 + \sqrt{(7180)^2 + 4(320.809 \times 10^5)]}}{2}$$

Taking +ve sign, $T_3 = 3115.89$ K

Now $$\frac{p_3 v_3}{T_3} = \frac{p_1 v_1}{T_1}$$

or $$p_3 = \frac{T_3}{T_1}\frac{v_1}{v_3} \times p_1 = \frac{3115.89}{333} \times 5 \times 1 = 46.78 \text{ bar}$$ ***Ans.***

For constant value of $c_v = 0.718$, we have

$$\int_{589.609}^{T_3} 0.718\, dT = \frac{44000}{16}$$

or $$0.718\,(T_3 - 589.609) = \frac{44000}{16}$$

or $$T_3 = 4419.69 \text{ K}$$

Using $$\frac{p_3 v_3}{T_3} = \frac{p_1 v_1}{T_1}$$

or $$p_3 = \frac{T_3}{T_1} \times \frac{v_1}{v_3} \times p_1 = \frac{4419.69}{333} \times 5 \times 1 = 66.36 \text{ bar}$$ ***Ans.***

**EXAMPLE 3.3** In an air standard Otto cycle, the air is at 37°C and 1 bar. The pressure at the end of compression is 30 bar and at the end of combustion the pressure is 80 bar. Using the gas tables, calculate (a) compression ratio, (b) percentage clearance, (c) highest temperature of the cycles and (d) thermal efficiency.

***Solution*** Refer to Fig. 3.15. From the gas tables, at $T_1 = 273 + 37 = 310$ K,

$$p_{r1} = 1.5546, \quad u_1 = 221.27 \text{ kJ/kg}, \quad v_{r1} = 0.5837 \quad \text{and} \quad h_1 = 310.24 \text{ kJ/kg}$$

Now $$p_{r2} = p_{r1}\left(\frac{p_2}{p_1}\right) = 1.5546 \times 30 = 46.638$$

From gas tables at $p_{r2} = 46.638$, we have

$$T_2 = 795 \text{ K}, \quad h_2 = 373.211 \text{ kJ/kg}, \quad u_2 = 588.27 \text{ kJ/kg} \quad \text{and} \quad v_{r2} = 0.04992$$

(a) ∴ Compression ratio $= r = \dfrac{v_1}{v_2} = \dfrac{v_{r1}}{v_{r2}} = \dfrac{0.5837}{0.04992} = 11.7$ ***Ans.***

(b) Clearance ratio is defined as

$$c = \frac{1}{r-1} = \frac{1}{11.7 - 1} = 0.0935 \quad \text{or} \quad 9.35\%$$ ***Ans.***

(c) $T_3 = T_2 \dfrac{p_3}{p_2} = 795 \times \dfrac{80}{30} = 2120$ K ***Ans.***

(d) From gas tables at $T_3 = 2120$ K.

$$h_3 = 2401.92 \text{ kJ/kg}, \ p_{r3} = 2649.9, \ u_3 = 1794.04 \text{ kJ/kg and } v_{r3} = 0.002331$$

∴ $$v_{r4} = v_{r3}\left(\frac{v_4}{v_3}\right) = 0.002331 \times 11.7 = 0.0272$$

From gas tables, $v_{r4} = 0.0272$, we have

$$T_4 = 979.5 \text{ K}, \ p_{r4} = 105.2, \ h_4 = 1023.04 \text{ kJ/kg and } u_4 = 741.98 \text{ kJ/kg}$$

We know that in Otto cycle, heat is added and rejected at constant volume. Hence,

Heat added $= q_A = u_3 - u_2 = 1794.04 - 583.27 = 1210.77$ kJ/kg

Heat rejected $= q_R = u_4 - u_1 = 741.98 - 221.27 = 520.71$ kJ/kg

$\therefore$ Efficiency $= \eta = 1 - \dfrac{q_R}{q_A} = 1 - \dfrac{520.71}{1210.77} = 56.99\%$. ***Ans.***

**EXAMPLE 3.4** An Otto cycle operates under the suction condition of 1 bar and 26°C. The compression ratio is 8. Calculate the pressure and temperature at the end of compression, (a) if cold air ($\gamma = 1.4$) is the working substance, (b) if hot air ($\gamma = 1.32$) is the working substance, (c) using the gas tables.

Also find (d) the ideal cycle efficiency for the case (a) and (b).

***Solution*** Refer to Fig. 3.15.

(a) $$p_2 = p_1\left(\frac{v_1}{v_2}\right)^{\gamma} = 1 \times (8)^{1.4} = 18.379 \text{ bar}$$ ***Ans.***

$$T_2 = T_1\left(\frac{v_1}{v_2}\right)^{\gamma-1} = 299\ (8)^{0.4} = 686.92 \text{ K}$$ ***Ans.***

(b) When $\gamma = 1.32$

$$p_2 = p_1\left(\frac{v_1}{v_2}\right)^{\gamma} = 1 \times (8)^{1.32} = 15.56 \text{ bar}$$ ***Ans.***

$$T_2 = T_1\left(\frac{v_1}{v_2}\right)^{0.32} = 299\ (8)^{0.32} = 581.64 \text{ K}$$ ***Ans.***

(c) At $T_1 = 299$ K from gas tables, $v_{r1} = 0.6380$, $p_{r1} = 1.3228$

Since the compression ratio $= \dfrac{v_1}{v_2} = 8$

$$\therefore \quad \frac{v_{r1}}{v_{r2}} = \frac{v_1}{v_2} = 8 \quad \text{and} \quad v_{r2} = \frac{v_{r1}}{8} = \frac{0.6380}{8} = 0.07985$$

From the gas tables at $v_{r2} = 0.07985$, $T_2 = 669$ K and $p_{r2} = 24.2$ ***Ans.***

Again, $$\frac{p_2}{p_1} = \frac{p_{r2}}{p_{r1}} = \frac{24.2}{1.3228}$$

$$\therefore \quad p_2 = 1 \times \frac{24.2}{1.3228} = 18.25 \text{ bar}$$ ***Ans.***

(d) We have, $\eta = 1 - \left(\dfrac{1}{r}\right)^{\gamma-1}$

For cold air, $\gamma = 1.4$

$$\therefore \quad \eta = 1 - \left(\frac{1}{8}\right)^{0.4} = 0.5647 \quad \text{or} \quad 56.47\%$$ ***Ans.***

For hot air, $\gamma = 1.32$,

$$\therefore \quad \eta = 1 - \left(\frac{1}{8}\right)^{0.32} = 0.4859 \quad \text{or} \quad 48.59\%$$ ***Ans.***

**EXAMPLE 3.5** A Diesel cycle operates under the suction of 1 bar and 37°C. At the end of the compression, the pressure is 35 bar. The cut-off takes place at 6% of the strokes from the inner dead centre position. Using the gas tables, calculate (a) compression ratio, (b) the percentage clearance, (c) the heat supplied, (d) the heat rejected, (c) thermal efficiency and mean effective pressure.

***Solution*** Refer to Fig. 3.16. From gas tables, $T_1 = 273 + 27 = 310$ K, we have

$$h_1 = 310.24 \text{ kJ/kg},\ u_1 = 221 \text{ kJ/kg},\ p_{r1} = 1.5546 \text{ and } v_{r1} = 0.5837$$

(a) ∴ $$p_{r2} = p_{r1}\left(\frac{p_2}{p_1}\right) = 1.5546 \times \frac{35}{1} = 54.5$$

From gas tables, for $p_{r2} = 54.5$, we have

$$T_2 = 828.03 \text{ K},\ u_2 = 450.985 \text{ kJ/kg},\ v_{r2} = 0.0445 \text{ and } h_2 = 857.959 \text{ kJ/kg}$$

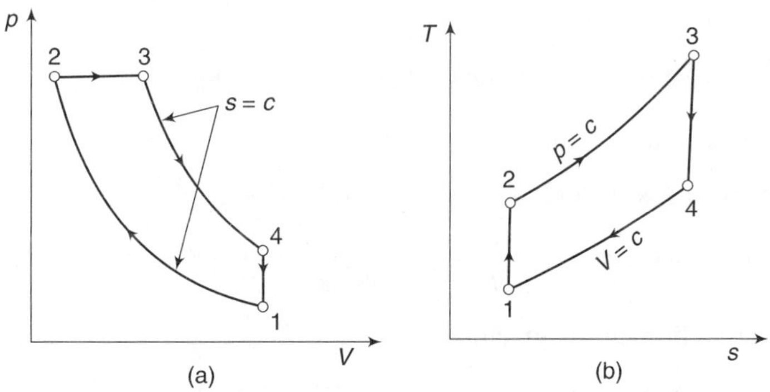


**Fig. 3.16** Example 3.5

The compression ratio $= r = \dfrac{v_1}{v_2} = \dfrac{v_{r1}}{v_{r2}} = \dfrac{0.5837}{0.0445} = 13.1$ ***Ans.***

(b) The percentage clearance is given by $\left(v_1 = \dfrac{RT_1}{p_1} = 0.8897 \text{ m}^3\text{/kg}\right)$

$$\frac{1 + c}{c} = r \quad \text{where } c \text{ is clearance ratio}$$

or $$c = \frac{1}{r - 1} = \frac{1}{13.1 - 1} = 0.08264 \quad \text{or} \quad 8.264\% \text{ of stroke volume}$$ ***Ans.***

(c) We have, $v_2 = \dfrac{RT_2}{p_2} = \dfrac{0.287 \times 828.03}{35 \times 10^2} = 0.0678 \text{ m}^3\text{/kg}$

and $$v_3 = \text{fraction cut-off} \times v_s + v_2 = \text{fraction cut-off} \times (v_1 - v_2) + v_2$$

$$= 0.06\times (0.8897 - 0.0678) + 0.0678 = 0.11711 \text{ m}^3/\text{kg}$$

Since process 2–3 is a constant pressure process, hence

$$T_3 = T_2 \times \frac{v_3}{v_2} = 828.03 \times \frac{0.11717}{0.0678} = 1430 \text{ K}$$

From gas tables, $T_3$ = 1425 K, we have

$$h_3 = 1545.44 \text{ kJ/kg}, \; u_3 = 1136.48 \text{ kJ/kg}, \; p_{r3} = 485.1, \text{ and } v_{r3} = 0.008598.$$

Since the heat addition in the case of Diesel cycle is at constant pressure, hence heat addition is given by the change in enthalpy during the process. Hence,

$$q_A = h_3 - h_2 = 1545.44 - 857.959 = 687.48 \text{ kJ/kg}$$ ***Ans.***

(d) We have $v_{r4} = v_{r3}\left(\frac{v_4}{v_3}\right) = v_{r3}\left(\frac{v_1}{v_3}\right) = 0.008595 \times \frac{0.8897}{0.11711} = 0.00653$

From the gas tables, for $v_{r1}$ = 0.078, we have

$$T_4 = 676 \text{ K}, \; h_4 = 690 \text{ kJ/kg}, \; u_4 = 492.84 \text{ kJ/kg}$$

Heat rejected at constant volume is given by

$$q_R = u_4 - u_1 = 492.84 - 221 = 271.84 \text{ kJ/kg}$$

(e) Thermal efficiency $= \eta = 1 - \frac{q_R}{q_A} = 1 - \frac{271.84}{687.84} = 60.4\%$ ***Ans.***

(f) Mean effective pressure is given by

$$p_m = \frac{\text{Work done}}{\text{Swept volume}} = \frac{w}{(v_1 - v_2)} = \frac{(q_A - q_R)}{(v_1 - v_2)}$$

$$= \frac{687.48 - 271.84}{10^2(0.8897 - 0.0678)} = 5.18 \text{ bar}$$

**EXAMPLE 3.6** In a Diesel engine, the air-fuel ratio is 25 : 1 The heating value of the fuel may be taken as 42000 kJ/kg as liberated in the engine. The specific heat at constant volume may be taken as $c_v$ = 0 0.718 + 16 × $10^{-5}$ T. The pressure and temperature at the beginning of the compression are 426K and 1 bar. The compression ratio is 15 and the law of compression is $pv^n = c$ where $n$ = 1.34 a mean value. Calculate the percentage of the stroke at which the constant pressure combustion stops. Assume $R$ = 0.287 kJ/kgK and neglect any molecular changes during combustion.

***Solution*** Refer to Fig. 3.16.

$$T_2 = T_1\left(\frac{v_1}{v_2}\right)^{n-1} = 426\,(15)^{0.34} = 1069.74 \text{ K}$$

For the combustion process 2–3, we have, from first law

$$q_{2-3} = (u_2 - u_2) + w_{2-3}$$

But $$q_{2-3} = \frac{C.V.}{\text{Mass of charge}} = \frac{42000}{25+1} = \frac{42000}{26} \text{ kJ/kg charge}$$

$$w_{2-3} = p_2 (v_3 - v_2) = R (T_3 - T_2) = 0.287 (T_3 - 1069.74) \text{ kJ/kg charge}$$

$$u_3 - u_2 = \int_2^3 c_v \, dT \text{ kJ/kg charge}$$

$$= \int_2^3 (0.718 + 16 \times 10^{-5} T) dT = 0.718 (T_3 - T_2) + 8 \times 10^{-5} (T_3^2 - T_2^2)$$

or $$\frac{42000}{26} = 0.718 (T_3 - 1069.74) + 8 \times 10^{-5} [T_3^2 - (1069.74)^2] + 0.287 (T_3 - 1069.74)$$

or $$T_3^2 + 0.1256 \times 10^5 T_3 - 337.93 \times 10^5 = 0$$

or $$T_3 = \frac{-0.1256 \times 10^5 + \sqrt{[(4 \times 0.1256 \times 10^5)^2 + 4 \times 337.93 \times 10^5)]}}{2} = 2277.5 \text{ K}$$

Also $$\frac{v_3}{v_2} = \frac{T_3}{T_2} = \frac{2277.5}{1069.74} = 2.129$$

The percentage of cut-off at which constant pressure combustion stops is given by

$$= \frac{v_3 - v_2}{v_1 - v_2} \times 100 = \frac{(v_3/v_2) - 1}{(v_1/v_2) - 1} \times 100 = \frac{2.129 - 1}{15 - 1} \times 100 = 8.064\% \quad \textit{Ans.}$$

**EXAMPLE 3.7** The following data refer to a Diesel engine for cold air standard: The compression ratio is 16. The cut-off occurs at 5% of the stroke. The mean specific heat at constant volume for the cycle is 0.718 kJ/kgK.

Assume $R = 0.287$ kJ/kgK

Knowing that the specific heat increases with temperature, if the mean specific heat $c_v$ for hot air cycle increases by 2%, calculate the percentage variation in air standard efficiency.

***Solution*** According to example,

$$\beta = \text{Cut-off ratio} = \frac{\%\text{cut-off}}{100} (r - 1) + 1 = \frac{5}{100} (16 - 1) + 1 = 1.75$$

$$\gamma = 1 + \frac{R}{c_v} = 1 + \frac{0.287}{0.718} = 1.4$$

∴ The air standard efficiency of the engine is given by

$$\eta = 1 - \frac{1}{r^{\gamma-1}}\left[\frac{\beta^{\gamma}-1}{\gamma(\beta-1)}\right] = 1 - \frac{1}{16^{0.4}}\left[\frac{(1.75)^{1.4}-1}{1.4(1.75-1)}\right] = 62.64\%$$

If the value of $c_v$ increases by 2%, then the new value of $\gamma$ will be

$$\gamma = 1 + \frac{R}{c_v} = 1 + \frac{0.287}{\left[\left(\frac{102}{100}\right)\times 0.718\right]} = 1.392$$

The value of $\beta$ would not change. Now the air standard efficiency is

$$\eta = 1 - \frac{1}{r^{\gamma-1}}\left[\frac{\beta^{\gamma}-1}{\gamma(\beta-1)}\right] = 1 - \frac{1}{16^{0.392}}\left[\frac{(1.75)^{1.392}-1}{1.392\,(1.75-1)}\right] = 61.902\%$$

The above equation gives the percentage variation in air standard efficiency due to the percentage variation in $c_v$.

∴ Percentage decrease in efficiency $= \frac{62.64 - 61.902}{62.64} \times 100 = 1.178\%$ ***Ans.***

***Altier:*** We have $\frac{c_p}{c_v} = \gamma$ and $c_p = c_v + R$

$$\therefore \quad R/c_v = \gamma - 1 \quad \text{or} \quad -\frac{R}{c_v^2}\,dc_v = d\gamma$$

$$\text{or} \quad -(\gamma-1)\frac{dc_v}{c_v} = d\gamma$$

We see that increase in $c_v$ decreases the value of $\gamma$.

The air standard efficiency $= \eta = 1 - \frac{1}{r^{\gamma-1}}\left[\frac{\beta^{\gamma}-1}{\gamma(\beta-1)}\right]$

$$\text{or} \quad 1-\eta = \left(\frac{1}{r}\right)^{\gamma-1}\left[\frac{\beta^{\gamma}-1}{\gamma(\beta-1)}\right]$$

Taking log from both sides

$$\ln(1-\eta) = -(\gamma-1)\ln r + \ln(\beta^{\gamma}-1) - \ln\gamma - \ln(\beta-1)$$

Differentiating both sides w.r.t. $\gamma$, we get

$$\frac{d\eta/d\gamma}{1-\eta} = -\ln r + \frac{\beta^{\gamma}\ln\beta}{\beta^{\gamma}-1} + \frac{1}{\gamma}$$

Multiplying both sides the above equation by $\frac{d\gamma}{\eta}$, we get

$$\frac{d\eta}{\eta} = \frac{(1-\eta)d\gamma}{\eta}\left[\ln r - \frac{\beta^\gamma \ln \beta}{\beta^\gamma - 1} + \frac{1}{\gamma}\right]$$

Putting the value of $d\gamma$, we get

$$\frac{d\eta}{\eta} = -\frac{(1-\eta)}{\eta}(\gamma - 1)\frac{dc_v}{c_v}\left[\ln r - \frac{\beta^\gamma \ln \beta}{\beta^\gamma - 1} + \frac{1}{\gamma}\right]$$

$$\therefore \quad \frac{d\eta}{\eta} = -\frac{(1-0.6264)}{0.6264}(1.4-1)\times 0.02\left[\ln 16 - \frac{(1.75)^{1.4}\ln 1.75}{(1.75)^{1.4}-1} + \frac{1}{1.4}\right]$$

$$= 1.178\%$$

***Ans.***

**EXAMPLE 3.8** A dual cycle has compression ratio 15:1. The cut-off takes place at 6% of the stroke. The maximum pressure obtained is 52 bar. The temperature and pressure at the beginning of the compression is 90°C and 1 bar. Calculate the pressure and temperature at all points of the cycle. Calculate also the heat supplied per kg of the working fluid. The specific heat at constant volume is $c_v = 0.718 + 12 \times 10^{-5}\ T$.

Assume $R = 0.287$ kJ/kg K.

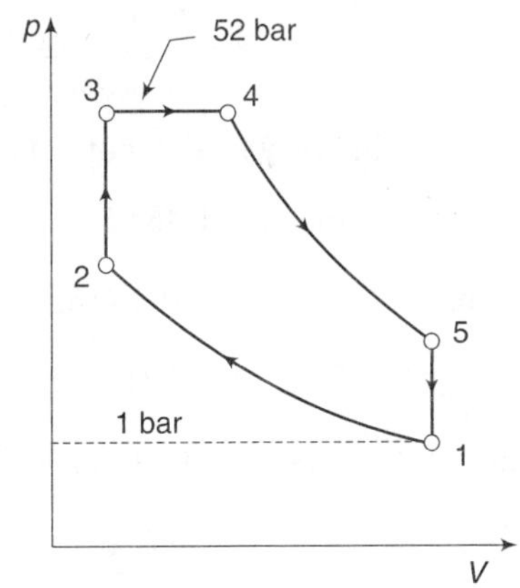


**Fig. 3.17** Example 3.8

***Solution*** Refer to Fig. 3.17. Since $c_v$ and $c_p$ are variable and function of temperature, hence $\gamma = \frac{c_p}{c_v}$ is not known either for compression or expansion. Under the circumstance, we assume the mean value of $c_p$ and $c_v$ and then calculate mean value of $\gamma$.

Process 1–2. Let us assume $\gamma = 1.4$, then

$$T_2 = T_1\left(\frac{v_1}{v_2}\right)^{\gamma-1} = (273 + 90)\,(15)^{0.4} = 1072.36 \text{ K}$$

The mean value of $c_v$ for the temperature range 363 K to 1072.36 K is given by

$$(c_v) = \frac{\int_{T_1}^{T_2} c_v\, dT}{(T_2 - T_1)} = \frac{1}{T_2 - T_1}\int_{T_1}^{T_2}(0.718 + 12\times 10^{-5}\,T)\,dT$$

$$= \frac{1}{T_2 - T_1}\left[0.718\,(T_2 - T_1) + 12\times 10^{-5}\left(\frac{T_2^2 - T_1^2}{2}\right)\right]$$

$$= 0.718 + 6\times 10^{-5}\,(T_2 + T_1) = 0.718 + 6\times 10^{-5}\,(1072.36 + 363)$$

$$= 0.80412$$

and $$(c_p)_{mean} = (c_v)_{mean} + R = 0.80412 + 0.287 = 1.09112$$

∴ $$\gamma = \frac{(c_p)_{mean}}{(c_v)_{mean}} = \frac{1.09112}{0.80412} = 1.3569$$

Thus, we see that the assumed value of $\gamma = 1.4$ is not correct.

Now taking $\gamma = 1.3569$, we have to recalculate the temperature $T_2$ and then the mean value of $\gamma$ which will be reasonably correct

∴ $$T_2 = 363(15)^{0.3569} = 954.23 \text{ K}$$

$$(c_v)_{mean} = 0.718 + 6 \times 10^{-5}(T_2 + T_1) = 0.718 + 6 \times 10^{-5}(954.23 + 363) = 0.797$$

and $$(c_p)_{mean} = c_v + R = 0.797 + 0.287 = 1.084$$

∴ $$\gamma = \frac{(c_p)_{mean}}{(c_v)_{mean}} = \frac{1.084}{0.797} = 1.36$$

This value of $\gamma$ is also not very correct.

Now assume, $\gamma = 1.36$

Then $$T_2 = 363\,(15)^{0.36} = 962.27$$

∴ $$(c_v)_{mean} = 0.718 + 6 \times 10^{-5}\,(962.27 + 363) = 0.7975$$

and $$(c_p)_{mean} = c_v + R = 0.7975 + 0.287 = 1.0845$$

∴ $$\gamma = \frac{(c_p)_{mean}}{(c_v)_{mean}} = \frac{1.0845}{0.7975} = 1.3598$$

Hence, the mean value of $\gamma$ for the compression stroke is 1.3598 and $T_2 = 962.27$ K

We have, $$p_2 = \frac{T_2}{T_1} \times \frac{v_1}{v_2} \times p_1 = \frac{962.27 \times 15 \times 1}{363} = 39.763$$ ***Ans.***

For process 2–3 : We have $p_3 = 52$ bar ***Ans.***

∴ $$T_3 = \frac{p_3}{p_2} \times T_2 = \frac{52}{39.763} \times 955 = 1248.899 \text{ K}$$ ***Ans.***

For process 3–4 : We have $p_3 = 52$ bar

$$\frac{v_4 - v_2}{v_1 - v_2} = 0.06 \quad \text{or} \quad \frac{\left(\frac{v_4}{v_2}\right) - 1}{\left(\frac{v_1}{v_2}\right) - 1} = 0.06$$

or $$\frac{v_4}{v_3} - 1 = (15 - 1) \times 0.06 = 0.84 \quad \text{or} \quad \frac{v_4}{v_2} = 1.84$$

Also $\dfrac{p_4 v_4}{T_4} = \dfrac{p_3 v_3}{T_3}$, but $p_4 = p_3$

$\therefore$ $T_4 = \dfrac{v_v}{v_3} \times T_3 = 1.84 \times 48.889 = 2297.97$ K

For process 4–5 : Value of $\gamma$ will not be same for compression process. We know that at higher temperature, the value of $\gamma$ decrease. Let $\gamma = 1.3$ and check this. We have

$$\frac{v_5}{v_4} = \frac{v_5 - v_3}{v_3 - v_4} = \frac{15}{1.84} = 8.152$$

Also $\dfrac{T_4}{T_b} = \left(\dfrac{v_5}{v_1}\right)^{\gamma - 1} = (8.152)^{0.3} = 1.8766$

or $T_5 = \dfrac{T_4}{1.878} = \dfrac{2297.97}{1.8766} = 1224.539$

$\therefore$ $(c_v)_{\text{mean}} = 0.718 + 6 \times 10^{-5}\,(T_4 + T_5)$

$$= 0.718 + 6 \times 10^{-5}\,(2297.97 + 1224.539) = 0.\,9293$$

and $(c_p)_{\text{mean}} = c_v + R = 0.9293 + 0.287 = 1.2163$

$\therefore$ $\gamma = \dfrac{(c_p)_{\text{mean}}}{(c_v)_{\text{mean}}} = \dfrac{1.2163}{0.9293} = 1.3088$

We see that the value of $\gamma = 1.3088$ checks closely with the assumed value of $\gamma = 1.3$ and further trial is not necessary. Taking $\gamma = 1.3088$ as the correct value, we have

$$T_4 = T_5 \left(\frac{v_4}{v_5}\right)^{0.3088} = T_5\,(8.152)^{0.3088} = 1.9116 \times T_5$$

$\therefore$ $T_5 = \dfrac{T_4}{1.9116} = \dfrac{2297.97}{1.9116} = 1202.118$ K

Also $\dfrac{p_1 v_1}{T_1} = \dfrac{p_5 v_5}{T_5} = \dfrac{v_1 p_5}{T_5}$

$\therefore$ $p_5 = \dfrac{T_5}{T_1} \times p_1 = \dfrac{1202.118}{363} \times 1 = 3.311$ bar ***Ans.***

## (B) ACTUAL CYCLE

## 3.9 VARIATION BETWEEN AIR STANDARD OTTO CYCLE AND ACTUAL CYCLE

There is great variation between the air standard Otto cycles and the actual cycles. Figure 3.18 shows the *p-V* diagram for an actual Otto engine (spark ignition engine). The air standard Otto cycle has been also incorporated in the diagram by dotted line to mark the deviation.

The main differences between actual and theoretical cycles are as given below:

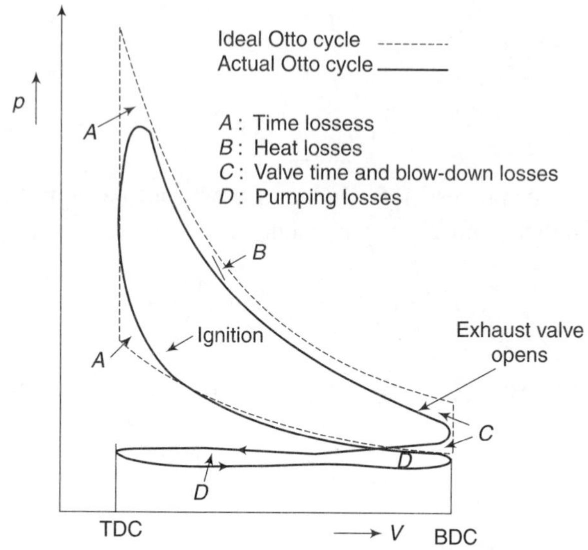


**Fig. 3.18** Actual Otto cycle

(i) *Working fluid.* The working fluid is not air but fuel air-mixture and the products of combustion. Thus the properties of the actual working fluid are different from that of air.

(ii) *Variable specific heats.* The specific heats are not constant but in actual engine it is a function of temperature and it increases as temperature increases.

(iii) *Dissociation.* At the high temperatures involved, the products of combustion tend to break up or dissociate, into other compounds such as CO, $H_2$ and $O_2$ in an endothermic reaction. This reduces the available energy and tends to reduce the thermal efficiency and the net work of the cycle.

(iv) *Time loss or non-instantaneous burning.* In actual engine, the addition of heat is not instantaneous, i.e., the combustion does not occur at constant volume or constant pressure. Combustion takes finite time. This creates a reduction in the net area of the *p-v* diagram (area *A* on top).

(v) *Heat losses.* In ideal cycle, there is no heat loss and compression and expansion are isentropic processes. But friction is present in the actual engines and there is a heat transfer to the cylinder wall. This reduces the net work (area *B*).

(vi) *Exhaust gas effect.* In clearance volume, there exists the products of combustion. The succeeding fresh charge is, consequently, heated and diluted by the remaining exhaust gas, thereby tending to lower the net work due to smaller mass admitted.

(vi) *Valve timing and blow down losses.* In an ideal cycle, instantaneous rejection of heat is assumed. But in actual engines it is not possible due to inertia of the exhaust valves and exhaust gases. The exhaust valve opens prior to time at which the piston reaches BDC This reduces the net work of the actual cycle (area *C*.)

(vii) *Pumping losses.* In the actual cycle, the pressure in the cylinder is different during the intake stroke from that during the exhaust stroke (area *D*).

## 3.10 DETAILED ANALYSIS OF DIFFERENCES BETWEEN FUEL-AIR OTTO CYCLES AND REAL CYCLE

As discussed above, non-consideration of the following losses in fuel-air cycle is responsible for the difference between real cycle and fuel-air cycle. These are:

(a) Burning time losses
(b) Heat losses
(c) Exhaust blow-down losses
(d) Pumping losses
(e) Rubbing friction losses

### 3.10.1 Burning Time Losses

This loss arises due to time required for mixing of fuel-air and for combustion in real engine.

In the fuel-air cycle, the combustion is assumed to be instantaneous whereas in actual cycle, combustion process takes some finite time (about 40° of crankshaft rotation) and during this time change in volume of the combustion products takes place. The pressure rises up to point which lies after TDC where in fuel-air cycle, pressure rises up to point 3 which lies at TDC. The loss of work by actual cycle is shown by the hatched area which reduces the power output and efficiency and is called burning time loss (Fig. 3.19). This loss depends upon the velocity of flame front and the distance covered by the flame travel during combustion.

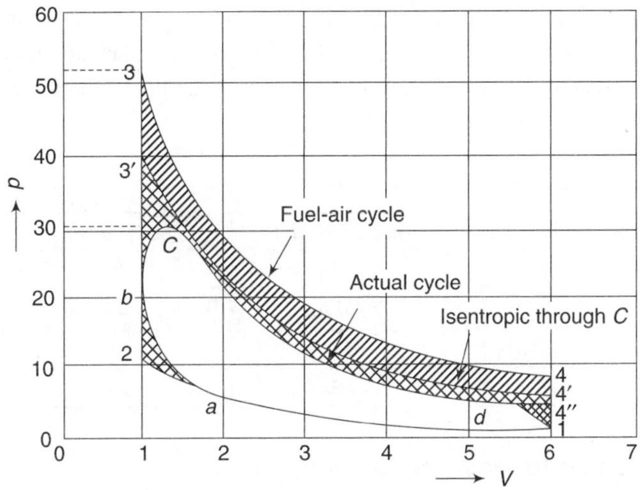


**Fig. 3.19** The effect of time loss on Otto cycle

The above discussion suggests that spark timing affects the performance of the engine which is discussed as follows.

#### *3.10.1.1 Effect of Spark Timing*

The effect of spark timing on the performance of Otto cycle may be studied by varying the spark timing say 0° advance, 40° advance and optimum (13° to 26°) advance as shown in

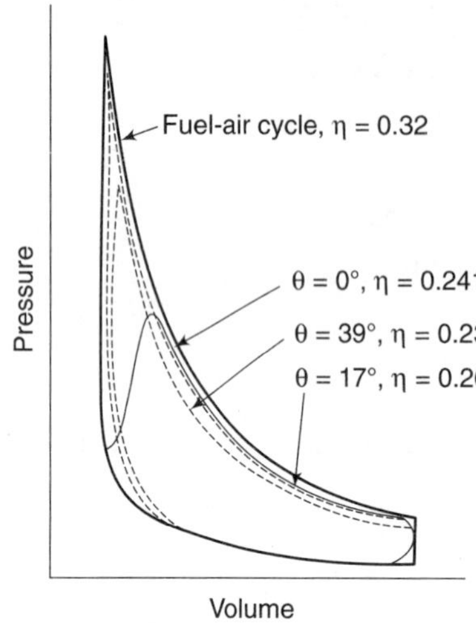


**Fig. 3.20** Power loss due to spark advance 0°, 39° and 17°

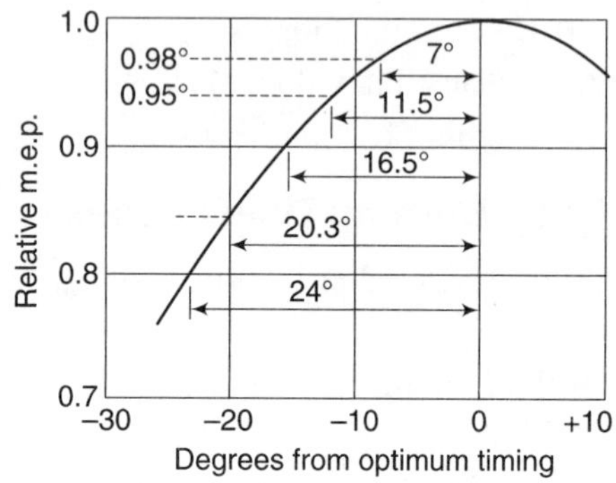


**Fig. 3.21** m.e.p. vs spark advance

Figs. 3.20 and 3.21 for the same condition of fuel-air cycle. From the results, it is obvious that the minimum loss of work is from optimum advance (13°–26°) whereas the maximum loss is at 0° advance. This is because of the fact that with spark at TDC, the pressure rise is less due to the expansion of gases whereas in high degree of spark advance (39°), additional work is required to compress the gases. Table 3.1 shows the performance of Otto cycle under various spark advance for compression ratio of 6.

**Table 3.1** Effect of Spark Advance for $r = 6$

| *Cycle advance* | *Spark advance* | *Max. cycle (bar)* | *m.e.p. bar* | *Efficiency* |
|---|---|---|---|---|
| Fuel-air | 0° | 46 | 10 | 32% |
| Actual | 0° | 23.5 | 7.5 | 24% |
| Actual | 39° | 42 | 7.6 | 23.8% |
| Actual | 17° | 34 | 8.1 | 26% |

### *3.10.1.2 Losses due to Incomplete Combustion*

*Loss due to incomplete combustion is accounted under time loss.* In real situation, perfect homogeneity in the mixture is an impossibility and as a result, some fuel burns completely while some fuel does not burn or burns partially resulting in presence of CO and $O_2$ in exhaust (Fig. 3.22), even for stoichiometric fuel-air ratio only 95% energy is released due to combustion. It is worthwhile to note that to eliminate the fuel waste, a lean mixture is used. *A slightly leaner mixture results in the maximum efficiency* but a too lean mixture will burn slowly resulting in increased time losses or no burning of at all causing total waste of fuel. *A rich mixture does not burn completely due to lack of oxygen resulting in wastage of fuel.* Further, the travel of the flame speed in the mixture is low which increases the time losses and lowers the efficiency of the engine.

There is a possibility that the unburnt fuel may combine with oxygen during exhaust stroke and may burn but the energy released cannot force the piston and this is also a wastage of energy.

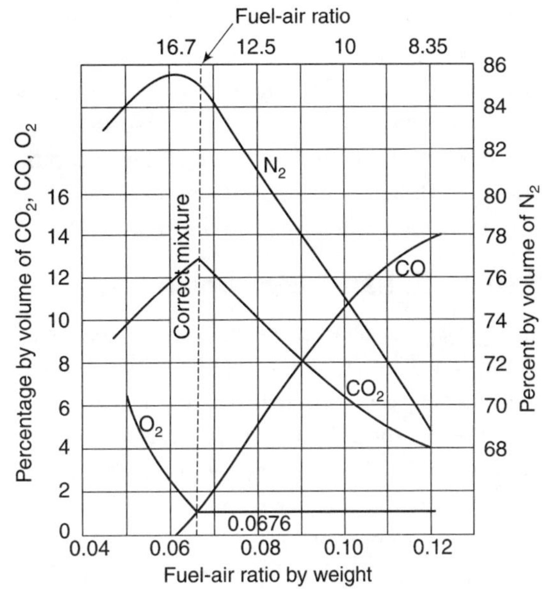


**Fig. 3.22** Composition of exhaust gases

### 3.10.2 Heat Losses

As a result of combustion of air-fuel mixture, a large amount of heat is released which increases the temperature of combustion products. *To prevent the failure of materials due to excessive temperature, the engines are cooled either by flowing water through water jacket in the cylinder wall or by air providing fins over the cylinder heads.* During the combustion process and the subsequent expansion processes, the heat flows from hot gases through cylinder walls and cylinder head into the water jacket or cooling fins. A portion of the heat flows through the piston head and then to cylinder through piston rings and is carried away by lubricating engine oil which is splashed on the underside of the piston. This heat loss is shown on the *p-V* diagram in Fig. 3.23.

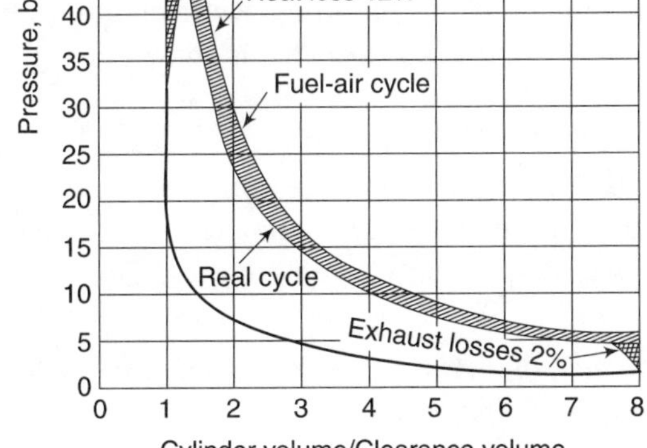


**Fig. 3.23** Representation of time loss, heat loss and exhaust loss in SI engines

*The maximum effect of heat loss is observed during the combustion and in the beginning of expansion stroke as the chance of doing useful work is very high. About 12% of the work is lost during combustion and expansion.* The effect of heat loss reduces the maximum temperature in the cycle which results in lowering down the specific heats. This means that this heat loss may ensure complete combustion to certain extent which results in a slight improvement in cycle efficiency to offset some of the heat loss. *The ultimate effect of the heat loss is the reduction in engine power output and cycle efficiency.*

### 3.10.3 Exhaust Blowdown Losses

In ideal cycle, there is no exhaust stroke but in actual cycle, exhaust stroke is a must for admission of fresh charge. If the exhaust valve is opened at the bottom dead centre (BDC) piston has to do work against a high cylinder pressure of about 7 bar costing part of the exhaust stroke. In the case, when the exhaust valve is opened too early there is a chance that the entire part of the expansion stroke is lost. In order to reduce the blow-down loss, the best compromise is to open the exhaust valve about 35° to 75° before BDC thereby reducing the cylinder pressure from about 7 bar to 3 bar before the exhaust stroke begins. The blow-down loss is shown in *p-V* diagram in Fig. 3.24 by the roundness of the end of the diagram. *About 2% of cycle work is lost due to exhaust blow down.*

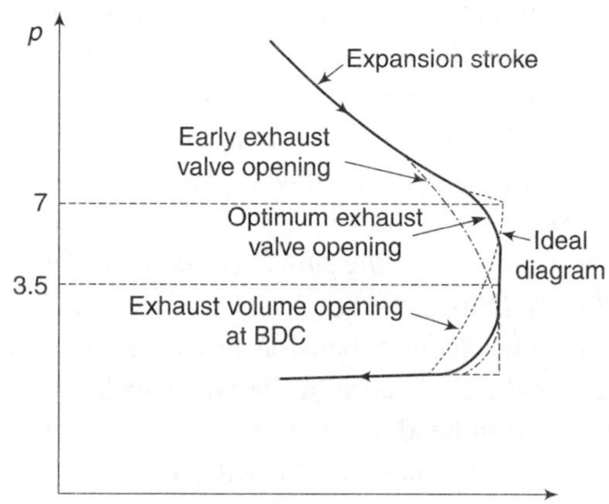


**Fig. 3.24** Exhaust blowdown losses

### 3.10.4 Pumping Losses

As mentioned earlier, the ideal cycle does not include suction and exhaust valves but in real cycles these are a must to allow the suction of the charge and exhaust of the gas. For the suction process, the pressure inside the cylinder has to be more than atmospheric. *The difference in work done in expelling the exhaust gases and the work done by the fresh charge during suction process is termed as the* **pumping work** or **pumping losses**. *This pumping loss increases with speed and at part throttle.* This is shown in Fig. 3.18.

### 3.10.5 Rubbing Friction Losses

In real engine, losses arise due to friction between piston and cylinder walls, friction in various bearings and friction in the auxiliary equipment such as pumps and fans. The friction losses reduce the power output and hence efficiency. *The friction losses increase with engine speed.*

*The part load performance of engine is poor as compared to full load. This is because of the fact that direct heat loss, pumping loss and rubbing friction losses increase at part loads.*

## Summary of Losses

Table 3.2 depicts the approximate losses as a percentage of fuel energy input for a compression ratio of 8 of SI engine.

**Table 3.2** Approximate Losses as a Percentage of Fuel Energy Input in SI Engine

| *Items in %* | *% Load* | |
|---|---|---|
| | *Full load* | *Half load* |
| A. Air standard cycle efficiency (%) | 56.5 | 56.5 |
| 1. Losses due to variation of specific heat and dissociation | 12.0 | 12.0 |
| 2. Time losses | 4.0 | 4.0 |
| 3. Incomplete combustion losses | 2.5 | 2.5 |
| 4. Direct heat losses | 3.5 | 5.0 |
| 5. Exhaust blow-down losses | 0.5 | 0.5 |
| 6. Pumping losses | 0.5 | 0.2 |
| 7. Rubbing friction losses | 3.0 | 6.0 |
| B. Fuel-air cycle efficiency $\eta_{fuel} = \eta_{air\ standard}$ – item no. 1 | 44.5 | 44.5 |
| C. Indicated thermal efficiency $\eta_{ith} = \eta_{fuel}$ – (Item 2 + 3 + 4 + 5 + 6) | 33.5 | 30.5 |
| D. Actual brake thermal efficiency $\eta_{bth} = \eta_{ith}$ – (Item 7) | 30.5 | 24.5 |

## 3.11 VARIATION BETWEEN THE AIR CYCLES AND THE ACTUAL CYCLES OF DIESEL ENGINES

There is much variation in the air cycles and the actual cycle of Diesel engines due to the under mentioned factors not considered in the air cycle.

(i) Working fluid, (ii) variable specific heat, (iii) exhaust gas effect, (iv) heat losses, (v) dissociation, (vi) non-constant pressure combustion of the fuel, (vii) blow-down losses and (viii) pumping losses.

The effects of *first four factors are* similar to the actual Otto cycle.

***Non-constant pressure combustion.*** In actual cycle, combustion does not take place at constant pressure. In fact, in the case of high speed engines, the combustion approaches a constant volume. In some low speed engines, the heat addition is at constant pressure as shown in Fig. 3.25.

***Dissociation.*** Dissociation apparently exists in Diesel engines but it is less pronounced than in petrol engines. This is due to the heterogeneous mixture existing in the combustion chamber, as well as the excess air present. Both these factors apparently, tend to reduce dissociation by lowering the overall temperature of the combustion.

***Blow-down losses.*** This is similar to spark ignition engines. The two-stroke Diesel engine shows greater losses than the four-stroke one.

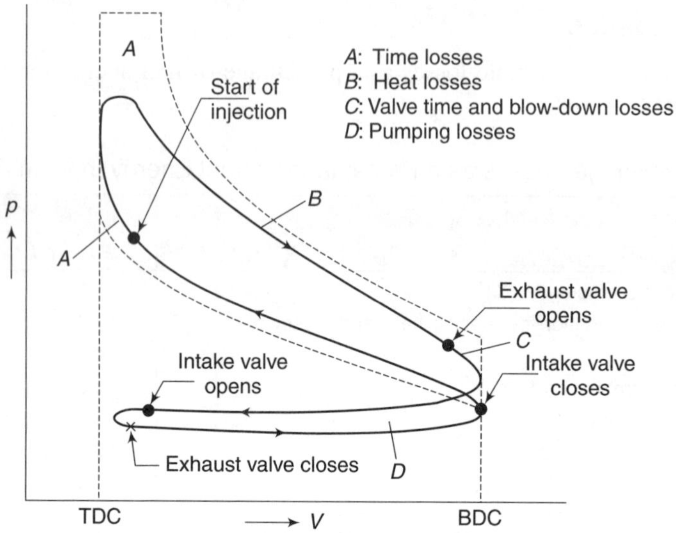


**Fig. 3.25** Actual diesel cycle

***Pumping losses.*** Pumping loss also exists in the Diesel engine but it is rather small compared to petrol engine.

**Note:** Valve timing diagrams have been discussed in Chapter 1.

## PROBLEMS

### Things to Think About

3.1 List the factors which contribute to losses in the IC engines cycle.

3.2 In what way exhaust blow-down losses affect the cycle?

3.3 What is the significance of fuel-air cycle?

3.4 Give the physical significance of losses to dissociations and incomplete combustion.

### Short and Long Theoretical Questions

3.5 List the factors which differentiate between air-standard and actual cycles.

3.6 List the assumptions for modelling of fuel-air cycles.

3.7 Discuss the effect of variables specific heat and dissociation of gases on Otto cycle.

3.8 Give a comparison of air-standard and fuel-air cycle for SI engines.

3.9 Discuss the effect of operating variables on the performance of air-fuel Otto cycle.

3.10 Give a comparison of dual pressure air-fuel cycle with standard cycle.

3.11 With the help of *p-V* diagram discuss the variation between air-standard Otto cycle and actual cycle.

3.12 Discuss in detail the effect of the following on fuel-air Otto cycle.

(a) Burning time losses (b) Heat losses

(c) Exhaust blow-down losses (d) Pumping losses

(e) Rubbing friction losses

3.13 Discuss the variation between air-standard Diesel cycle and actual cycle.

3.14 What is the reason that directs heat losses increase with load?

3.15 Give a relationship between air, fuel-air and actual cycle.

### Numerical Question

3.16 A Otto, Diesel and Dual cycle relate to the following data : The compression ratio is 5, 20 and 15 respectively. The air-fuel cycle ratio is 15:1. The calorific value of 44 MJ/kg. The beginning of compression, the temperature and pressure of the charge is 60°C and 1 bar. Take $n = 1.335$. and $c_v = 0.718 + 20 \times 10^{-5}\ T$ where $T$ is in K.

Calculate the ASE and specific fuel consumption for each of the three cases.

### Software Based Question

3.17 Develop a software for the analysis of real Otto, Diesel and Dual cycles

### Design and Open Ended Question

3.18 Make a comprehensive report for the design parameters which affect the performance of SI and CI four-stroke engines. Based on these, develop a design procedure to design efficient SI and CI engines.

### Multiple-Choice Questions

3.19 Blow-down in real cycle occurs because
- (a) exhaust valve is opened suddenly
- (b) exhaust valve opens a few degree before piston reaches BDC
- (c) combustion is never complete
- (d) period of overlap is substantial

3.20 For a constant volume fuel-air cycle, the thermal efficiency increases when
- (a) equivalence ratio ($\phi$) increase
- (b) compression ratio increase when $\phi > 1$
- (c) compression ratio increase when $\phi$ is kept constant
- (d) compression ratio decreases and $\phi$ increases

3.21 Choose the correct answers

The efficiency of constant volume fuel-air cycle
- (a) increases with both $r$ and $\phi$ are increasing
- (b) increases with $r$ but decrease with increasing $\phi$
- (c) decreases when both $r$ and $\phi$ are increasing
- (d) increases with increasing $\phi$ and decreases with increasing $r$.

3.22 For a given compression ratio, the efficiency of an Otto cycle with a decrease in $c_v$
- (a) decreases
- (b) remains the same
- (c) increases
- (d) will depend upon the value of $c_v$

## Key to Multiple-Choice Questions

3.19 (c) 3.20 (b) 3.21 (b) 3.22 (c)

CHAPTER

# Combustion in SI Engines

# 4

## OVERVIEW

Heat energy which drives the internal combustion engine is obtained through the combustion of fuel in the combustion chamber of IC engines. Thus, it is very essential to understand the combustion phenomena in SI engines in detail for better designing of combustion chamber resulting in the minimum losses. The combustion may be *normal and abnormal. The abnormal combustion is undesirable*. The subject of combustion of mixture of air and fuel in a SI engine will be discussed in this chapter in a relatively detailed manner.

## 4.1 GENERAL COMBUSTION THEORY

*Combustion phenomenon may be defined as a relatively rapid chemical combination of the hydrogen and carbon in the fuel with the oxygen in the air, resulting in liberation of energy in the form of heat.* The heat energy so released is utilised to drive the internal combustion engine.

Combustion is a highly complex phenomenon. An extensive research is still going on to unfold the mystery behind it, so a highly detailed theory is beyond the scope of this book.

There are three conditions necessary for combustion. These are:

(a) the presence of a combustible mixture,

(b) some means of initiation of combustion, and

(c) stabilization and propagation of flame.

In an SI engine, the combustible mixture is generally supplied by the carburettor and a spark plug is used to initiate the combustion. Nowadays, instead of carburettor, *multi-point fuel injection* **(MPFI)** is gaining the popularity to meet the emission norms (EURO-IV and VI). A chemical reaction for combustion of any hydrocarbon (iso-octane) is given as

$$C_8H_{18} + 12.5\ O_2 = 8CO_2 + 9H_2O$$

Though, such an equation indicates that the combustion process is a simple and direct combination of the atoms involved but this is not sufficient to explain the combustion phenomenon.

There are two general phases in the combustion process, namely:

(a) Preparation phase

(b) Actual burning phase.

As the name implies, during the preparation phase, the spark sets of a local reaction resulting in the formation of some intermediate reactors, which in turn, apparently both prepare the mixture for, and initiate, the actual burning. The experimental results show that the ignition of charge is only possible within certain limits of fuels-air ratio (Fig. 4.1).

These ignition limits correspond approximately to those mixture ratios, at lean and reach ends of the scale where the heat released by spark is no longer sufficient to initiate combustion in the neighbourhood of unburnt mixture. The experimental evidences further suggest that the flame will propagate only if the temperature of the burnt gases exceeds approximately 1500 K in the case of hydrocarbons – air mixture. To achieve this required temperature, for a mixture at room temperature, the relative fuel-air ratio should lie between 0.5 and 2.1 which corresponds to fuel-air ratio 1 : 7 and 1 : 30 as stoichiometric fuel-air ratio is 1 : 15 as shown in Fig. 4.1.

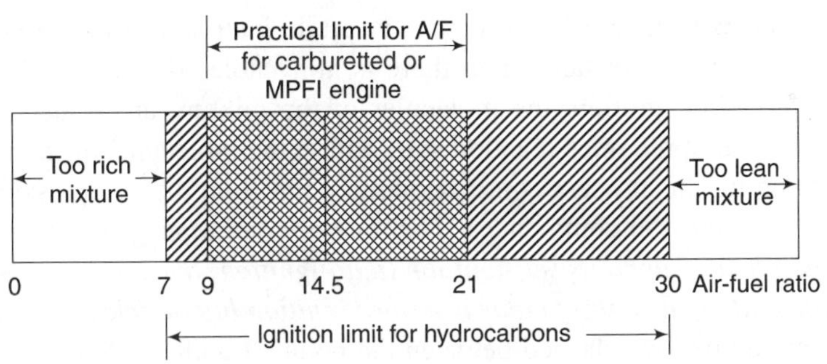


**Fig. 4.1** Range of ignition limits of hydrocarbon

It is to be noted that these ignition limits depend upon mixture ratio and temperature. The range of ignition limits increases with temperature of the mixture because the rate of reaction and thermal diffusivity also increase.

## 4.2 COMBUSTION IN SI ENGINES

To start combustion in spark ignition engines, a single high temperature spark is provided to the combustible mixture. The spark is provided at some crank degree before the piston reaches the TDC position during the compression stroke as shown by the point "*A*" on the pressure—time curve shown in Fig. 4.2 (a). Due to spark a sufficient amount of heat is provided to the combustible mixture, which raises its temperature sufficiently to bring about a combustion reaction. Combustion in the SI engine depends upon the flame created by spark plug being

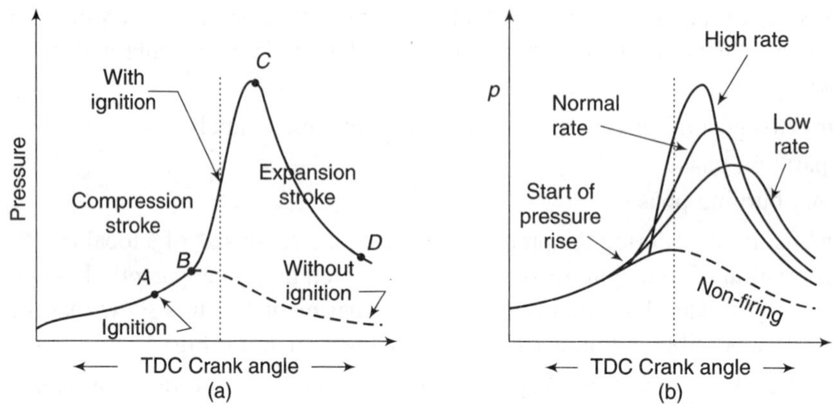


**Fig. 4.2** Combustion stages in SI engines

able to advance into the unburnt mixture. Propagation of a flame through the unburnt mixture, will only follow, provided, that the release of heat from the spark is greater enough, to heat the adjoining portion of the mixture, to a temperature at which the reaction rate overcomes the heat losses. The combustion in the SI engines is divided in two parts **(i) ignition delay** and **(ii) flame propagation**.

It has been seen experimentally, that there is a certain interval of time between the instant the spark is passed, and the instant when there is noticeable rise in cylinder gas pressure, due to combustion. After ignition, the molecules of the mixture are heated to self ignition temperature, and certain preflames reaction start within the molecules of the mixture, before they are ready to burn. This process continues up to point *B* on the compression curve where the actual fuel of combustion commences. *The time interval during which the mixture near the spark plug gets heated up to its self-ignition (auto-ignition) temperature, ignites and gives rise to a self-propagating nucleus of flame is termed* ***ignition lag*** *or* ***delay period.*** This ignition delay is shown by the periods elapsed between the points *A* and *B*. After the ignition period, the combustion starts causing a rapid increase in pressure and temperature of the gas.

The number of degrees before TDC position, at which ignition occurs is called the ***angle of advance***. If the angle of advance is increased, the engine is said to have been ***over advanced*** and if the *angle of advance* is decreased, it is called ***retarded***. The instant of the spark plays an important role on the peak pressure of combustion shown in Fig. 4.2 (b). *The instant of the spark has to be so adjusted that the heat released on combustion produces a high peak pressure at a point closer to the TDC which is a desirable feature.* Higher peak pressure close to TDC produces a greater force acting through a large portion of the power stroke, and hence, increases the power output. In practice, the engine is so designed that about one-half of the pressure rise takes place by the time the piston reaches TDC. **The delay period is very short, about 0.002 sec**. For a high speed engines running at 3000 rpm, *this delay corresponds to* ***36 degrees*** *of* ***crank rotation***. *It is for this reason, the point of ignition is always well in advance of TDC*

Once the combustion starts and the reaction is well underway, a spherical flame front (boundary of luminous zone) will advance from the spark plug. The edges of the sphere are somewhat rugged (Fig. 4.3) because of the convective current in the highly turbulent mixture. The heat from the flame front is transferred to the surrounding of the unburnt mixture. When the unburnt mixture attains self-ignition temperature, it starts burning and a new flame front is developed. By this way, the flame spreads throughout the mixture and heat is released which increases the temperature and pressure of the gas. The combustion stage is said to commence at *B* when some pressure rise above the compression curve can be seen as the indicator card (Fig. 4.2a), and ends at *C*, after the attainment of the peak pressure near TDC Thus, the flame propagation is shown by *B–C*. In some cases, the combustion is not complete at the end of the second stage i.e., at *C*. And some fuel may burn after point *C* during the expansion stroke, this phenomenon is called ***after burning***.

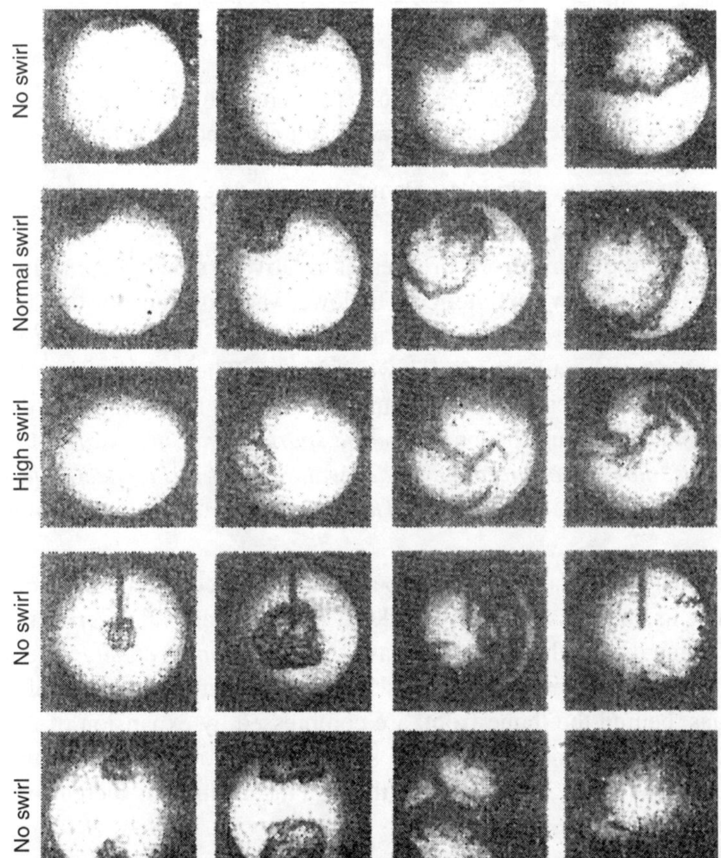


**Fig. 4.3** Combustion process in SI engines

An analysis of the actual photographic flame front traces show that the flame velocity across the chamber follows a pattern similar to that shown in Fig. 4.4. The flame travel appears to pass through **three rather distinct stages**.

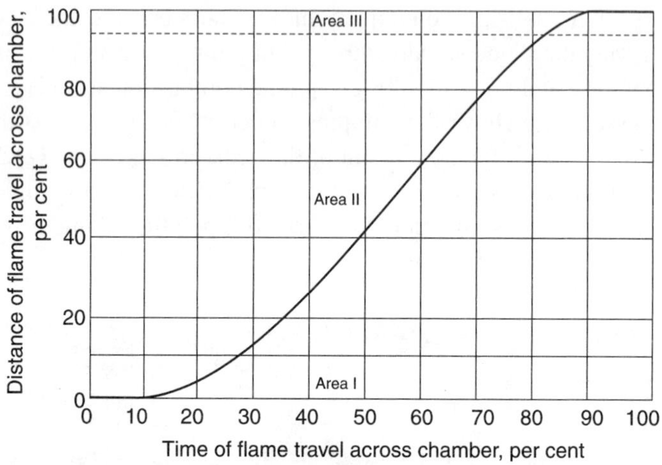


**Fig. 4.4** Flame travel vs. time

Initially **(area I)**, the flame front progresses relatively slowly, primarily due to a low **"transposition rate"** and **low turbulence**. The lower value of transposition rate is due to the relative small mass of charge burnt at the start. This means, that flame is propagated almost entirely by reaction rate, resulting in a slower advance. The turbulence in the combustion chamber rises from the velocity of air-fuel mixture entering the cylinder at the intake stroke and from the cylinder head and piston. *Since the spark plug is necessarily located in a quiescent layer of gas, that is close to the cylinder wall, lack of turbulence reduces the reaction rate, which further lowers the flame speed. This is one of the reasons that spark is advanced before TDC*

As the flame front proceeds into more turbulent areas and commences to consume a greater mass of mixture, its progress is more rapid but at a constant rate **(Area II)**. The average velocity of the flame during this period of travel is called ***flame velocity***. The flame is fastest when the mixture is the richest. Note at this stage, the unburnt gas ahead of the flame front and the burnt gas, behind the flame front, are compressed by expansion of the burning mixture. Here as before, the combustion of any small element of mixture, occurs at essentially constant pressure, even through the pressure through the chamber is increasing for the overall process.

At the end of flame travel **(area III)**, the volume of the unburnt charge is reduced appreciably, and the transposition rate again reduces to a negligible value, which in turn reduces the flame speed. Moreover, the flame enters a zone of relatively low turbulence which, in turn, reduces the reaction rate.

## 4.3 EFFECT OF ENGINE VARIABLES ON IGNITION DELAY IN SI ENGINES

We have seen that ignition delay plays an important part in the combustion of SI engines. Thus, it is necessary to study the effects of engine variables on ignition delay (or lag). Ignition delay depends upon the following engine variables.

**(i)** ***Temperature and pressure of the gas at the time of spark.*** The ignition lag decreases if the pressure and temperature of the gas are increased. This is so because of the fact, that the rate of chemical reaction increases markedly with increase in temperature and to a lesser extent with increase in pressure. This means, that if we can produce high temperature and pressure at the time of sparking by *increasing the compression ratio, intake temperature and pressure and retarding the spark timing,* we can reduce the ignition lag.

**(ii)** ***Speed.*** The ignition lag increases almost linearly with the engine speed. So, the spark timing is advanced at higher speeds.

**(iii)** ***Chemical nature of the fuel and air-fuel ratio.*** The higher the self-ignition temperature of the fuel, the longer is the ignition delay. A rich mixture gives a minimum ignition lag as shown in Fig. 4.5.

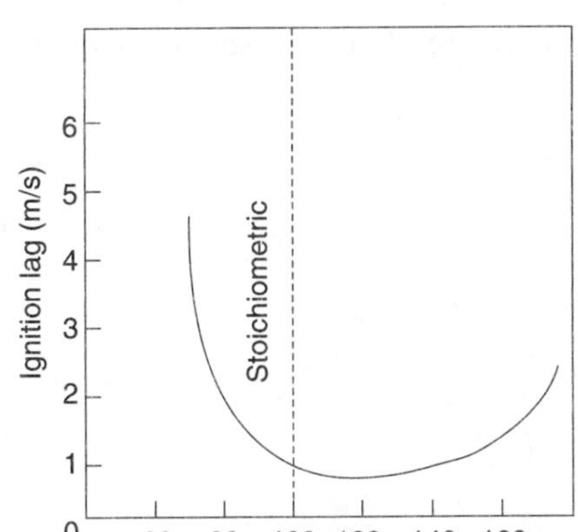


**Fig. 4.5** Effect of mixture ratio on ignition lag

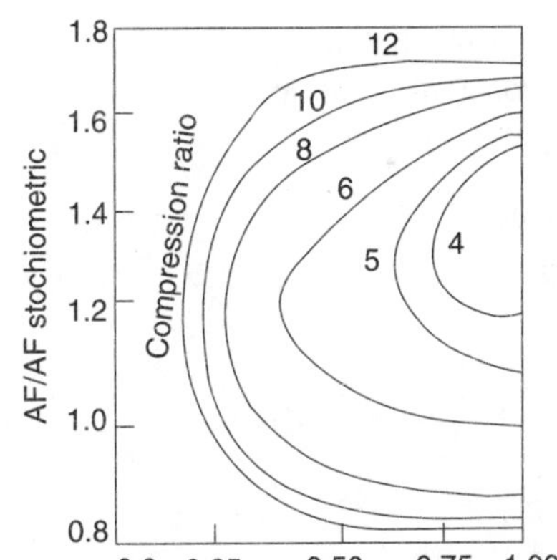


**Fig 4.6** Effect of electrode gap on A/F ratio

**(iv)** ***Electrode gap.*** If the electrode gap is too small, quenching of the flame nucleus may occur and the range of the fuel-air ratio for the development of a flame nucleus reduces. A gap of about 0.625 mm is generally found satisfactory for a compression ratio of 6.0. At very low loads (idling), a larger gap is essential for satisfactory ignition as the compression ratio is increased. The effect of the electrode gap for various compression ratios is shown in Fig. 4.6.

## 4.4 EFFECT OF ENGINE VARIABLES ON FLAME SPEED

After the ignition delay, it is the flame propagation which is responsible for the combustion of the combustible mixture in SI engines. There are many operating variables which affect the flame speed. They are:

(i) **Compression ratio.** With increase in compression ratio, the clearance volume decreases and, a greater portion of the spent up gases are exhausted. This means, that there is less dilution of the charge and the charge density is greater during the burning process. In addition to this, the temperature of the charge is high at higher compression ratio. Since the heat transfer increases with greater charge density and increased temperature, the flame speed increases with the increase of compression ratio.

(ii) **Intake pressure.** Inlet pressure influences the density of the charge. During throttling and supercharging, there always occurs a change in intake pressure. With high intake pressures, the charge densities are high, resulting in high rates of heat transfer and consequently, high flame speed.

(iii) **Engine speed.** Figure 4.7 shows the effect of engine speed on flame speed. The higher the engine speed, the greater is the inlet velocity of the charge. Because of this fact, the flame speed increases with the engine speed. Thus, the time required for the flame to travel the combustion space would be halved, if the engine speed is doubled. This means, that the number of the crank degrees required for flame propagation will remain constant at all speeds.

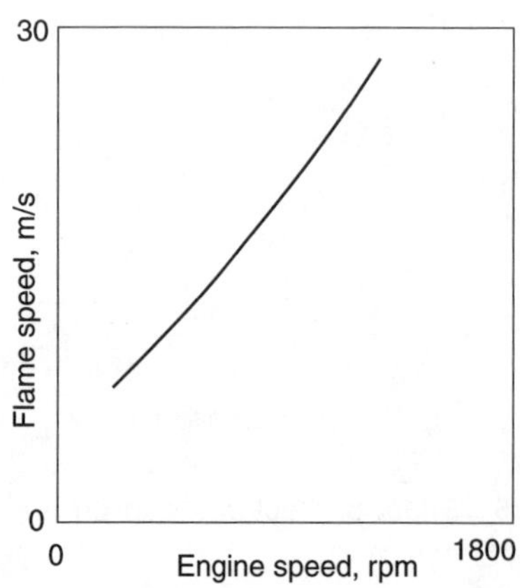


**Fig. 4.7** Effect of speed on flame travel

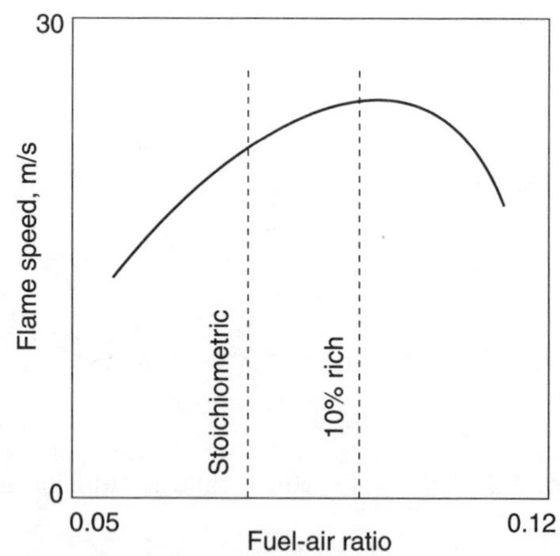


**Fig 4.8** Effect of A/F on flame speed

(iv) **Air-fuel ratio.** Figure 4.8 shows the effect of air-fuel ratio on flame speed. The maximum flame speed is found in a mixture when it is 10% richer than stoichiometric. With lean mixture, less thermal energy is released, so that, the flame temperature and flame speed are both low. A very rich mixture burns to CO instead of $CO_2$ and so produces less sensible heat or low flame temperature which results in low speed. Thus, a lean or very rich mixtures produces low flame speed.

(v) **Turbulence.** It has been found that in quiescent mixture, the flame speed is very low, about 10 m/s. With turbulent motion, there is a rapid intermingling of the burning and unburnt portion of the charge. This results in a high rate of heat transfer, and thus a high flame speed. The turbulence of the mixture is caused mainly due to the intake velocity of the charge, geometry of the cylinder head and piston.

(vi) **Engine load.** As the engine load increases, the cycle pressure also increases. Hence, the flame speed increases.

(vii) **Engine size.** The speed of the engine depends upon the size of the engine; The large is the engine, the smaller is the rpm and vice versa. Thus, engines of the similar design will run at the same piston speed regardless of the size of the engine. This means, that inlet air velocity or turbulence will be nearly the same and so will be the flame speed. However, in large sized engines, the flame has to travel a longer distance. Thus, if the engine size is double, the time required for flame propagation will also be doubled. But the doubled size engine will run at half the speed while the number of crank degrees required for flame travel will be the same.

## 4.5 ABNORMAL COMBUSTION

Whatever discussed in the previous articles, it is concerning the **normal combustion** process in SI engines. In normal combustion, the flame initiated by the spark travels across the combustion chamber in a fairly uniform manner. But under certain engine operating condition **"abnormal combustion"** may take place which is detrimental to life and performance of the engine. *This abnormal combustion includes detonation or knock, run on wild ping, rumble, etc. Among the various outcome of abnormal combustion,* ***detonation*** *or* ***knock*** *is the most important because it puts a limit on the selection of compression ratio at which engine can be operated, which in turn, governs power output and efficiency.* ***The abnormal combustion is undesirable.*** All the parameters affecting the abnormal combustion may be selected or controlled in a such way that it does not initiate.

## 4.6 DETONATION OR KNOCK THEORY

*Knock or detonation are terms used to signify an unusual sharp metallic sound which is clearly audible under abnormal conditions of engine operation. This is a very harmful phenomenon of the combustion process, which gives a direct blow to the engine structure and, may lead to failure of the engine.* The exact mechanism of knocking phenomenon is as yet incomplete; however, analysis of high speed photographs of combustion process and other experimental data led to two general theories of knock. They are:

(a) The auto-ignition theory. (b) The detonation wave theory.

### 4.6.1 The Auto-ignition Theory of Knock

The auto-ignition theory of knock, is the well established and widely accepted theory of knock. In order to understand the knocking phenomenon, consider the combustion process of the engine.

After ignition, once a stable flame nucleus is formed, the flame will propagate from molecule to molecule. When the flame reaches the far side of the combustion chamber and passes through the last part of the unburnt charge, the peak pressure is normally attained.

Now, let us consider the burning of the ***last part of the charge*** so called ***end gas***. The end gas experiences rapid compression first by piston during the compression stroke and later by the advancing flame front. The pressure of the end gas when the flame front is about to reach it will be almost equal to the peak pressure of the cycle, as by this time, most of the charge (about 90–95%) has been burnt. Also the temperature and density of the end gas will increase both by compression and by radiation from the advancing flame. Thus, the end gas exists at a very high pressure and temperature.

It has been found experimentally that if the fuel-air mixture is rapidly compressed, so that, its temperature exceeds the self-ignition temperature of the mixture, ***auto-ignition*** or ***self-ignition*** takes place without the assistance of any initiating flame or spark plug.

*It has also been found that auto-ignition of the mixture does not occur instantaneously as soon as its temperature exceeds the self-ignition temperature, but an ignition delay or induction period occurs before the reaction becomes explosive.* Evidently, preflame reaction must occur during the delay period to condition the mixture for self-ignition. Accordingly, the recent chemical theory, during the preflame reaction, extensive decomposition of fuel takes place producing aldehydes, peroxide, formaldehyde, ketones, hydrogen peroxide and free-radicals. The energy released by this reaction, raises the mixture temperature and thus auto-ignition occurs. *The delay period depends upon the air-fuel ratio, temperature and pressure. As the temperature increases,* the delay period reduces, this is obvious from Fig. 4.9. Similarly, as the pressure increases, the delay period decreases.

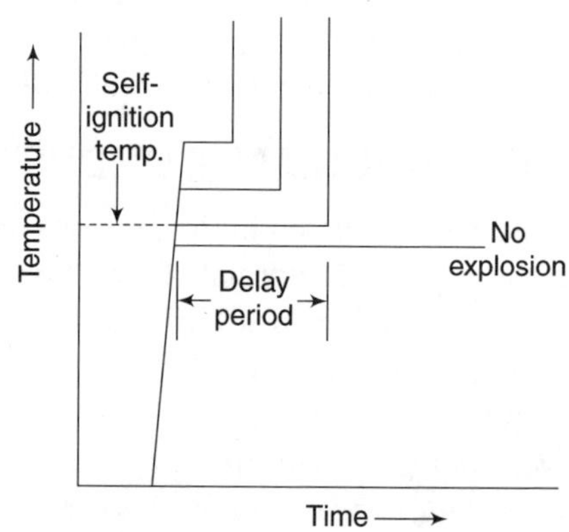


**Fig. 4.9** Effect of temperature on ignition lag

From the above analysis, *it follows that the last part of the charge may burn in two ways,* one **normal and other abnormal** (Fig. 4.10). *If the delay period of the end charge is sufficiently long, the flame front will reach it before the end of the delay period and the combustion of the end charge (area BB′D in Fig. 4.10(a) will be in normal manner.*

*If the ignition delay of the end charge is consumed before the flame (CC′) has arrived, auto-ignition of the end gas (area CC′D) takes place. With this auto-ignition, the orderly combustion* is **abnormal** (Fig. 4.10b). Energy may be liberated at a rate faster than the anticipated by the designer with the result that knocking occurs. *Thus, knock in the SI engine is characterised by sudden auto-ignition of the charge near the end of the combustion process.*

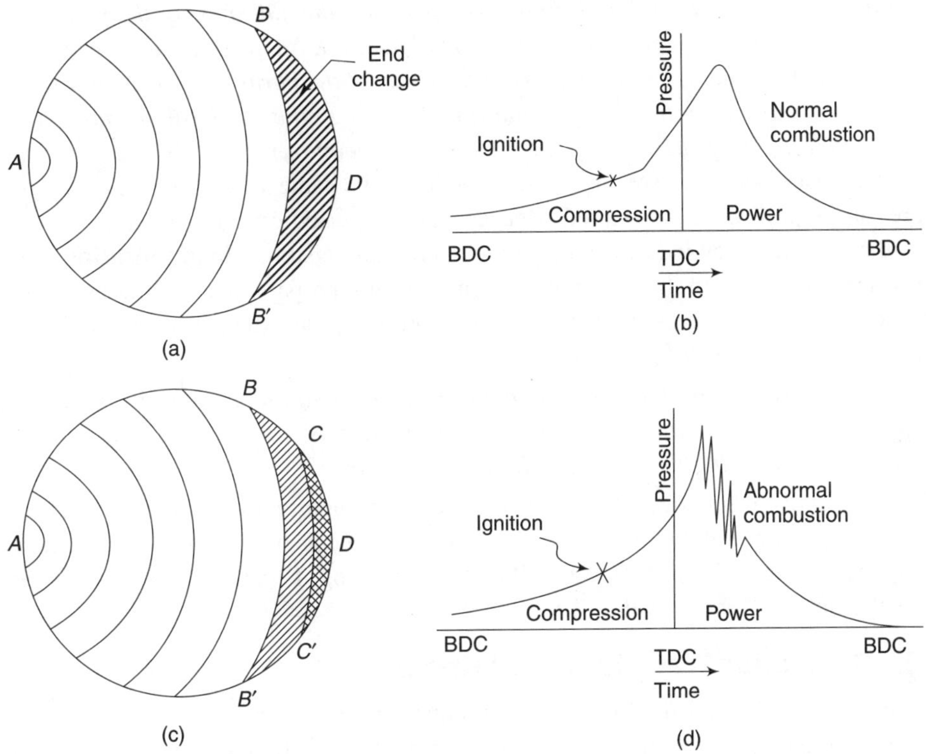


**Fig. 4.10** Normal and abnormal combustion

Figure 4.11 shows the pressure-volume diagram for a normal cycle (*A B C D*) and a cycle with auto-ignition (*A B E F G H*). The peak pressure of the normal cycle $p_C$ – is (50–60 atm.). The end charge is compressed from $p_B$ to $p_E$ by the advancing flame front; $p_E$ being almost equal to $p_C$. The auto-ignition of the end gas raises the pressure very rapidly from $p_E$ to $p_F$ (120 – 200 atm.). The rapid pressure rise caused by auto-ignition may result in *two different types of vibrations of the engine. In one case, a large amount of mixture may auto-ignite and so give rise to a rapid increase in pressure in throughout the combustion chamber (i.e., the pressure equalization keeps pace with the rate of high pressure rise), the engine parts will experience a direct blow of the high pressure gas and the ear will detect a thudding sound from the impact and consequent, free vibration to the engine parts. In the second*

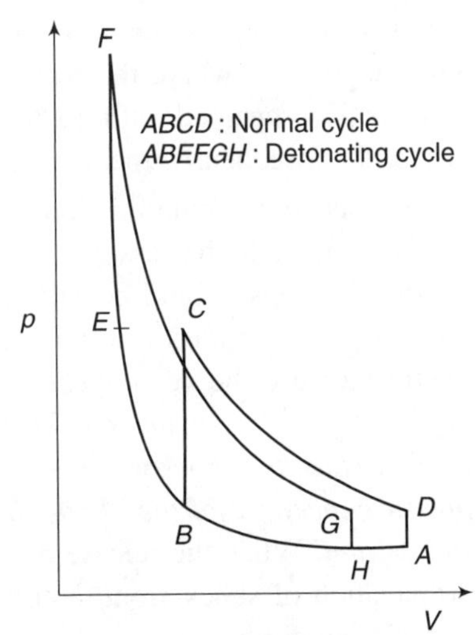


**Fig. 4.11** *p-v* diagram with detonation

*case, if the pressure equalization cannot keep pace with the quickness of the pressure rise of the end gas (due to auto-ignition at several locations), a large pressure difference ($p_F - p_E$). comes into existence between the end gas and the rest of the burnt mixture. As a result, a high pressure wave starts from the end gas and travels through the combustion chamber towards the plug with a tremendous velocity (i.e., acoustic velocity* or more). The pressure wave will be reflected back and forth by the chamber walls in the same way as an echo in a room till the pressure equilibrium is restored. Because of the periodic impact of high speed waves across the wall of the cylinder, combustion chamber and piston, **a loud pulsating noise, like a hammer striking a metal**, is produced and the engine parts start vibrating at the same frequency as that of the pressure wave. In automotive engines, the gas vibration frequency is of the order of 5000 cycles per second.

*The factors that control the auto-ignition are; (i) temperature, (ii) density, (iii) time (ignition delay), (iv) fuel-air ratio, (v) supplementary materials (inert gases, catalysts, etc,) and (vi) turbulence (affecting the homogeneity of the mixture).*

From the above discussion, it follows that the most important factors for the auto-ignition to take place are; the *delay period of the end gas and the time of flame travel across the combustion chamber.* There is less likelihood of knocking if there is longer delay period or shorter the time of flame travel.

### 4.6.2 Detonation Wave Theory of Knock

The objectors' to the auto-ignition theory, point to the fact, that the mixture is not entirely homogeneous and therefore, the probability is remote that a sufficient number of neighbouring elements will simultaneously react and cause perceptible knock. They propose a **detonation wave theory**. According to this theory, due to **shock wave** or **some other disturbance** within the combustion chamber, a ***true detonation*** *wave* is established which propagates at a very high velocity (supersonic) through the unburnt charge and compresses the charge to a pressure and temperature where the reaction should be practically instantaneous. Such a detonation wave is accompanied with a ***sharp-pressure-discontinuity*** which results in vibration of the gas, and consequently, the vibration of engine structure.

In order to examine this hypothesis, suppose, that a small element explodes and creates, momentarily, a high pressure. The pressure discontinuity between the exploding element and its surroundings, would constitute a **shock wave** that would advance into the surrounding mixture and compress it to a higher pressure. Now if reaction occurs, the release of energy would increase the velocity of the shock front as well as increase the pressure difference. *This stronger shock causes a considerable rise in temperature of the mixture through which it passes. Thus, the shock wave is accelerated to an extremely high velocity, with attending high pressure, because of the almost instantaneous reaction that occurs in the high-temperature shock front.* When the release of chemical energy keeps pace with the energy demanded for propagation of shock front, a constant and a supersonic velocity will be attained. If such a a stable condition prevails, the disturbance is called a true detonation wave which will give rise to gas vibration, and consequently, to engine structure vibration. *This vibration is known as* **knock** or **detonation**.

Some authors believed that neither auto-ignition nor detonation wave is necessary for the occurrence of detonation. They argue that when a fuel undergoes extensive preflame reactions, the flame velocity increases and occupies a velocity of the order of 100 to 400 m/s., i.e., 10 to 20 times the normal rate. This increases the rate of energy release tremendously which results in knock in the engine.

## 4.7 CHEMISTRY OF KNOCK OR DETONATION

It is worth to know the chemistry of knock as it will help in controlling it. While discussing the combustion process in SI engines, we have come to know that *preflame reactions precede the auto-ignition. As expected, during the pre-flame reactions many intermediate products appear which help in the initiation of auto-ignition.* While conducting the spectrum analysis of burning gases, Recardo and Thornycraft detected the presence of *aldehydes* in the cylinder contents prior to the main combustion of gases. Many researchers support this findings and they concluded that when *aldehydes and peroxides* were present in cylinder gases, knocking occurred but no knocking occurred when they were absent.

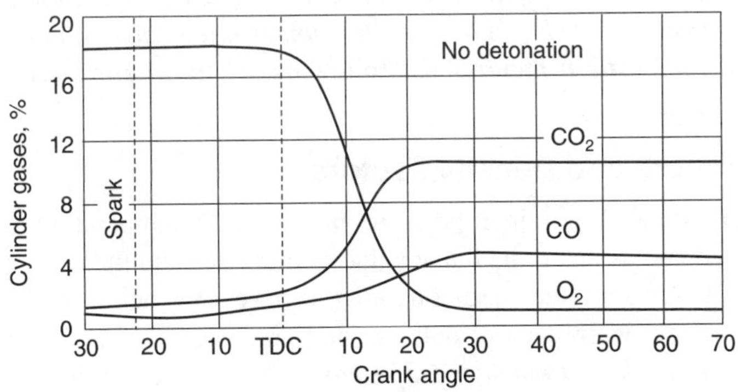


**Fig. 4.12** Composition of cylinder gases during normal combustion (without detonation)

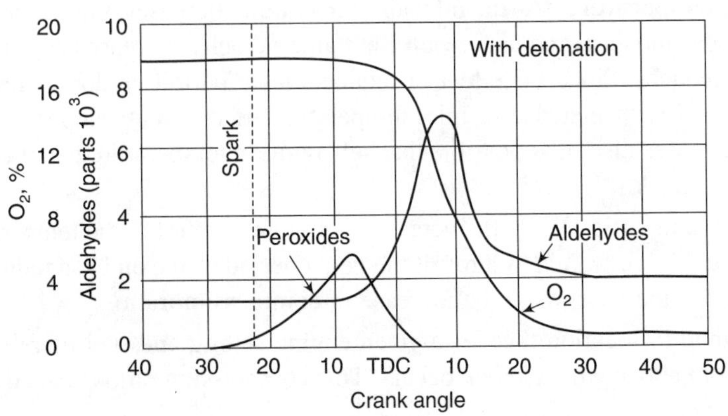


**Fig. 4.13** Composition of cylinder gases during abnormal combustion (with detonation)

Figure 4.12 and Fig. 4.13 show the composition of the cylinder gases without and with detonation respectively. During normal combustion (i.e., no detonation), as obvious from the figure the amounts of $CO_2$ and CO increase while the amount of $O_2$ falls with the proceeding of combustion.

During abnormal combustion (with detonation), Fig. 4.13 shows the presence of aldehydes and peroxides and their concentration abruptly varies as the combustion proceeds. Further, there is rapid decline of concentration of $O_2$.

*This suggests that to avoid detonation the presence of aldehydes and peroxides must be avoided either by selecting proper fuel or adding additives in fuels or controlling auto-ignition or any other factor.*

## 4.8 VARIABLES AFFECTING KNOCK IN SI ENGINES

On the basis of the fundamental factors that control auto-ignition, it follows, that to prevent knock in the SI engines, the gas, which is essentially a homogeneous mixture, should have—*(a) a low temperature (b) a low density (c) a long ignition delay and (d) either a lean or rich mixture. These conditions are met within many engine variables. Thus, the various engine variables affecting knock can be classified under a few factors, namely,* ***temperature, density, time and composition factors***. The following are the engine variables which affect the knock in SI engines.

### 4.8.1 Temperature and Density Factors

**(i) Inlet air-pressure.** As the inlet pressure increases, the temperature and density of the end gas get affected and as result the flame velocity increases and the delay period of the end gas reduces. But these two effects are contradictory to each other. A higher value of flame velocity reduces knock, while shorter delay period promotes knock. However, *the effect of shorter delay period is more pronounced. Therefore, the increase in inlet pressure increases the tendency of knock.* For example, supercharging of engines increases the inlet air pressure that promotes knock while throttling the intake reduces it.

**(ii) Inlet air temperature.** As the inlet air temperature increases, the temperature and density of the end gas increase and as a result the flame velocity increases and the delay period of the end gas reduces. Since the effect of reduced delay period predominates, the tendency of knock increases with increase in inlet temperature. If the inlet temperature reduces, the delay period increases also, it is possible that self-ignition temperature is not achieved, hence there will be no knock.

**(iii) Compression ratio.** With the increase in compression ratio, the temperature and pressure of the end gas increase and as a result the delay period of the end gas reduces. Therefore, knocking tendency increases with the increase in compression ratio.

It is worthwhile to mention that for a given engine setting and fuel there exists a ***critical compression ratio*** above which knock occurs. This compression ratio is called ***highest useful compression ratio***.

**(iv) Spark timings.** If the spark timing is advanced, the peak pressure of the cycle increases which increases the temperature and density of the end gas, which in turn, reduces the delay period of the end gas. Thus, the knocking tendency increases with the advanced spark timing.

**(v) Air-fuel ratio.** The most important effect of fuel-air ratio is on the *ignition delay*. When the mixture is about 10% richer (A/F 12.5), the delay period of the end gas is the minimum and the velocity of the flame propagation is the maximum. Since the effect of shorter delay period predominates, the knocking tendency is the maximum. This means that by lean mixture or rich mixture varying more than 10% the tendency of the knock is *decreased.*

It is worthwhile to note that air-fuel ratio over and above affecting the temperature of end gas it also affects time.

**(vi) Supercharging.** As a result of supercharging, the inlet pressure and temperature of air increase, which in turn, increase both temperature and density of end gas consequently the delay period decreases which promotes knock.

**(vii) Coolant temperature.** As the coolant temperature increases, the end gas temperature increases. This reduces the delay period and promotes knock.

**(viii) Combustion chamber and cylinder walls temperature.** The combustion chamber spark and exhaust valves are the hottest parts. As the temperature of cylinder wall and combustion chamber increases, delay period decreases, which in turn, promotes knock.

**(ix)** ***Engine Load.*** With an increase in engine load, the temperature of the cylinder wall and combustion chamber increases thereby, raising the temperature of the end gas. This reduces the delay period and thus promotes the tendency of knock.

### 4.8.2 Time Factors

**(i) Engine speed.** With the increase in engine speed, the inlet velocity or turbulence increases. The higher the turbulence, the greater is the heat transfer rate and hence, higher will be the flame speed. The delay period is not much influenced by high engine speed and hence knock is relieved at higher speed.

**(ii) Engine size.** A cylinder of smaller size needs less time of the flame front to travel the combustion chamber and reach the end gas. Delay period of end gas is the same irrespective of the cylinder size. Hence, *tendency of knock reduces with smaller engine size.*

**(iii) Combustion chamber design.** A compact combustion chamber design is one which reduces the distance the flame front has to travel. This means, that in a compact combustion chamber, the time required for the flame front to reach the end gas is decreased which has the effect to reduce the knocking tendency. Further, if the combustion chamber is properly designed, there will be an adequate turbulence of the mixture and as a result the speed of the flame front will be high and thus, the tendency of the knock will be still further reduced.

**(iv) Location of spark plug.** A centrally located spark plug in the combustion chamber has the minimum tendency to knock as the *flame travel is the minimum.*

**(v) Location of exhaust valve.** The exhaust valve should be located close to the spark plug ensuring that is not the end gas region, otherwise there will be a tendency to knock.

### 4.8.3 Composition Factor

**(i) Fuel-type.** The type of the fuel used greatly influences the knocking tendency. A fuel, with a higher self-ignition temperature or lower preflame reactivity, has less tendency of knock. *Octane number of fuel is the measure of resistance to knock.*

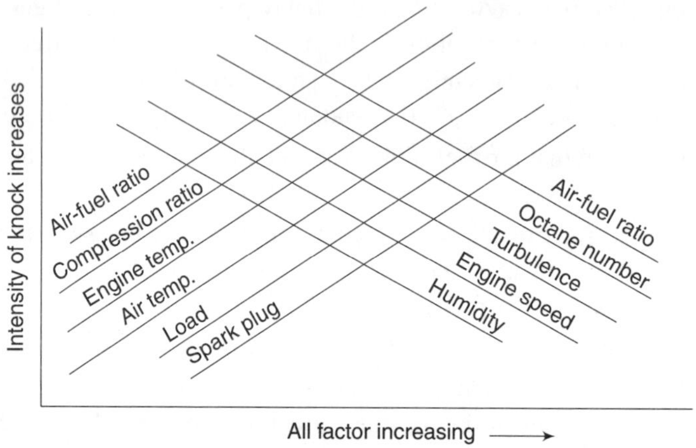


**Fig. 4.14** Influence of engine variables on knocking tendency

**Table 4.1** Summary of the Variables Affecting Detonation

| *S.No.* | *Variables* | *Major effects on unburnt charge* | *Action to be taken to reduce detonation* | *Can operater usually control* |
|---|---|---|---|---|
| 1. | Inlet air pressure | Temperature and density | Reduce | In some cases |
| 2. | Inlet air temperature | Temperature and density | Reduce | In some cases |
| 3. | Compression ratio | Temperature and density | Reduce | No |
| 4. | Spark timing | Temperature and density | Retard | In some cases |
| 5. | Air-fuel ratio | Temperature and time | Make very rich or weak | In some cases |
| 6. | Supercharging | Temperatures time and density | Reduce | In some cases |
| 7. | Coolant temperature | Temperature and density | Reduce | In some cases |
| 8. | Cylinder wall temperature | Temperature and density | Reduce | Not ordinarily |
| 9. | Engine load | Temperature and density | Reduce | Yes |
| 10. | Engine speed | Time | Increase | Yes |
| 11. | Engine size | Time | Select compact | No |
| 12. | Combustion chamber design | Time | Select compact | No |
| 13. | Location of spark plug | Time | Select optimum | No |
| 14. | Location of exhaust valve | Time | Select optimum | No |

Higher the octane number, lesser is the tendency to knock. The fuels may be arranged as follows in order of decreasing tendency of knock—paraffins, naphthenes, aromatics. By adding a small amount of certain additives called **dopes**, the knocking tendency decreases.

**(ii) Humidity of air.** *As the humidity of air increases, the reaction time decreases which reduces the tendency to knock.*

Influences of above engine variables on knocking tendency is depicted in Fig. 4.14.

The effect of variables affecting detonation has been summarised in Table 4.1 with the comments whether it is possible by operator to control these or not.

## 4.9 EFFECT OF KNOCK IN SI ENGINES

Knocking phenomenon in SI engines gives rise to the following effects:

**(i) *Load pulsating noise and mechanical failure.*** Due to the violent pressure fluctuations, there is a loud pulsating noise. As a consequence, the engine parts are set in vibration and if, the detonation continues for a longer period, the vibrating engine parts may be damaged due to fatigue and high stresses.

**(ii) *Local melting of piston and rings.*** Rapid heat transfer from the high pressure and high temperature gases, tend to force their way past the piston rings which can cause local melting of piston and piston rings.

**(iii) *Overheating of spark plug.*** As a result of violent gas vibrations, the hot gases time and again travel in an out of the spark plug cavity greatly increasing its temperature. This high temperature adversely affects the electrical and thermal properties of the plug and render it useless.

**(iv) Pre-ignition.** *Pre-ignition is the ignition of a fresh combustible charge before the normal spark is passed. The presence of overheated spark plug, exhaust valve or carbon deposits on the cylinder wall acts as an ignition source similar to the regular spark, and thus, ignite the adjacent fresh combustible charges.* An entirely distinct flame front is, thus produced, and if *this ignition occurs earlier than the time of normal spark, the process is termed **pre-ignition***. Such a condition is undesirable as the combustion becomes both erratic and uncontrollable. Moreover, pre-ignition tends to raise the temperature and pressure in the chamber which causes the temperature of the hot spot to rise further and produce still earlier pre-ignition on succeeding cycle. The cumulative effect not only tends to raise peak pressure and increases the possibility of detonation, but also causes the peak pressure to occur progressively earlier in the cycle. Thus, *detonation can lead to pre-ignition and pre-ignition encourages detonation (i.e., auto-ignition), which in turn, leads to more severe detonation causing considerable damages of the engine parts.*

*The phenomena of pre-ignition and detonation are totally different from each other. The detonation follows the spark while pre-ignition precedes it.* Again pre-ignition can occur with any type of fuel is due to the presence of hot sports in the combustion chamber. These hot spots may be an overheated spark plug, an exhaust valve or carbon with fuels which are volatile in nature and which are affected by the composition and molecular structure of the fuel.

*Pre-ignition may be detected by shutting of the ignition. If the engine continues firing, this means, that pre-ignition is taking place. If there is a sudden loss in power without any mechanical malfunctioning of the engine, then this is another indication of pre-ignition.*

## 4.10 METHODS TO REDUCE DETONATION

The following methods are generally used to reduce detonation.

(i) **Proper location of spark plug.** A compact combustion chamber with proper central location of the spark reduces the path of the flame front travel from the spark plug to the remotest part of the combustion chamber and thereby, reduces the tendency of knocking. The same results could be achieved by using two or more spark plugs at each side of the wall.

(ii) **Proper selection of material for piston and cylinder head.** The end charge cooling is much better when piston and cylinder heads are made of *aluminium alloys as they have a good thermal conductivity. A better cooling of end charges reduces the tendency of knock.*

(iii) **Injecting water into the intake manifold.** Injection of water in the cylinder reduces the temperature of the end gas which increases the delay period, thereby decreasing the tendency of knock.

(iv) **Retarding the spark timing.** By retarding the spark timing, the peak pressures are reached only during the power stroke and are of power magnitude, this reduces knock.

(v) **Extremely rich or lean mixture.** By using extremely rich or lean mixture, the flame temperature can be kept low thus eliminating considerable tendency of knock.

(vi) **Squish recesses in combustion chamber.** The provision of squish recesses in combustion chamber cools the last portion of the charge and reduces the tendency of knock.

## 4.11 ABNORMAL COMBUSTION KNOCK-SURFACE IGNITION

The knock or detonation phenomena discussed above may be termed as *abnormal combustion knock due to surface ignition or surface ignition knock.*

*Surface ignition is defined as the initiation of a flame front of charge by a hot surface other than spark.* The hot surface may be spark plug, electrode exhaust valve heat, any hot carbon particle either deposited on the combustion chamber surface or floating in the charge.

*It is the carbon deposits which cause mostly surfaces ignition.* It is obvious that these deposits occupy some space in the combustion chamber and as a result there is a decrease in clearance volume and increase in compression ratio. They are very poor heat conductors which prevent proper cooling and as result surface temperature increases. The lead compounds contained in the fuel as additives catalyze the glowing oxidation of the carbon like material in deposit which raise temperature still more and provide material for a variety of unwanted sources. Further, some of the glowing deposits may take off and float about the combustion chamber enhancing sporadic ignition.

It is worthwhile to note that the surface ignition may occur before (pre-ignition) or after (post-ignition) normal combustion and the surface ignition caused by spark plug electrodes can only be pre-ignition and not post-ignition.

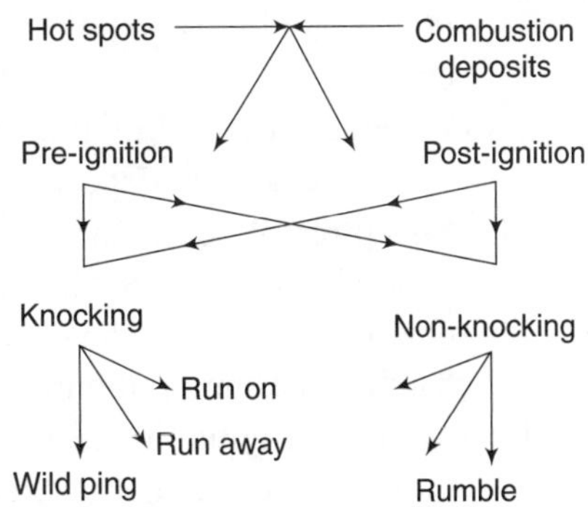


**Fig. 4.15** Effects of surface ignition

The effects of surface ignition are shown in Fig. 4.15 and Fig. 4.16. Both pre- and post-surface ignition may or may not cause knocking. Also, both knocking and non-knocking surface ignition may give rise to phenomenon known as *"run or surface ignition"* and "run-away surface ignition." Over and above, knocking surface may give rise "wild ping" and non-knocking surface ignition to *rumble*. Following are the definitions of various effects of surface ignition.

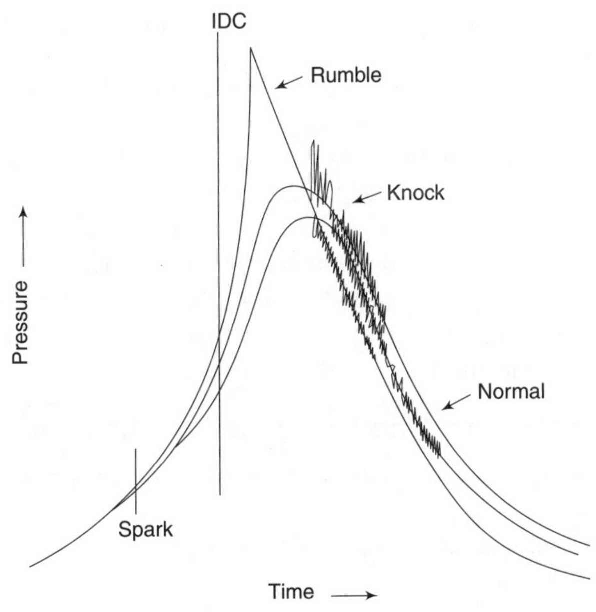


**Fig. 4.16** *p-t* diagram for normal combustion, rumble and knock

**(i) Run-on surface ignition.** If the engine continues to fire even when ignition system is switched off, it is called *run-on surface ignition*. The *continuity* of the firing is only possible when there is a presence of surface ignition even after the ignition system is switched off.

**(ii) Run-away surface ignition.** If the surface ignition in one combustion cycle heats the source of ignition to a still higher temperature and a series of earlier and earlier surface ignition is setup, it is called *run-away surface ignition.* It is serious case of surface ignition and may lead to disastrous results such as seizure or melting of piston or engine catching fire. It is due to overheated or defective spark plug or exhaust valve or an incandescent deposit of high melting ash.

(iii) **Wild ping.** When a knock occurs in a particularly erratic and unpredictable way, it is called *wild ping*. Such knock is very sharp in nature. This is due to the fact that some fragments of a deposit break free and move erratically through the combustion chamber causing ignition now and then.

(iv) **Rumble.** If ignition occurs at a number of points in the combustion chamber either before or after the normal spark, the fuel burns leading almost a heavy explosion of the mixture. The rise in pressure is very rapid and the resulting vibration and noise are known as *engine rumble*. Rumble noise is low-pitched noise distinctly different from normal combustion and from spark knock or detonation.

## 4.12 COMBUSTION CHAMBER FOR SI ENGINES

As the name implies, combustion chamber is a chamber where combustion of air-fuel mixture takes place resulting in release of heat energy which will be converted into mechanical work through a mechanism of piston, cylinder, connecting rod, crankshaft, etc., of engine. This means that the design of combustion chamber has an important influence on the engine performance and its knocking characteristics. *The design includes the shape of combustion chamber, the location of spark plug and the location, shape and rise of inlet and exhaust valves.* Because of this importance, the combustion chamber design has been a subject of considerable amount of research and development in the last fifty years.

These research and development work have resulted in the raising of the compression ratio of the engine from 4 before the first world war period to 10.5 with special combustion chamber design and suitable anti-knock fuels.

### 4.12.1 Requirements of a Good Combustion Chamber

The main requirements of a good combustion chamber for SI engines are to provide the following:

(a) High power output with minimum octane requirement.
(b) High thermal efficiency
(c) Smooth engine operation.

These requirements are achieved by considering the following factors.

**A. High Power Output.** It requires the following:

(i) *High compression ratio.* High compression ratio results in high power output and efficiency. But the compression ratio is limited by the phenomenon of detonation. Detonation is governed by the design of combustion chamber and fuel quality. Any design that improves the anti-knock characteristics of a combustion chamber permits the use of higher compression ratio.

(ii) *A high degree of turbulence.* This is needed to achieve a high flame front velocity. Turbulence is induced by inlet flow configuration or squish. Squish is the rapid injection of trapped gas which can be induced by having a bowl in piston or with dome shaped cylinder head.

(iii) *High volumetric efficiency.* This can be achieved by providing ample clearance round the valve heads, large diameter valves and straight passages with minimum pressure drop. At present, the passenger car engines employ two suction and two exhaust valves. **The modern passenger car cylinder engines are using 16 valves to achieve high volumetric efficiency**.

(iv) *Small or no excess air.* It will be ensured by proper design of carburetion system or electronic injection system.

(v) *Complete utilisation of air.* This ensures no dead pockets of air.

**B. High Thermal Efficiency.** This is governed by the following factors:

(i) *High compression ratio.* This is already discussed above.

(ii) *Minimum heat loss.* This is achieved by a compact combustion chamber which provides small surface-volume ratio. The compact combustion also reduces flame travel. A hemispherical shape provides the minimum surface-volume ratio and ensures less pollution.

(iii) *Good scavenging of exhaust gas.* This is achieved by the proper design and location of exhaust valve.

**C. Smooth Engine Operation.** This is governed by the following factors:

(i) *Gradual and moderate pressure rise.* The normal combustion ensures gradual and moderate pressure rise of combustion products.

(ii) *No detonation.* This is achieved by compact combustion chamber, proper locating of spark plug and exhaust valve and satisfactory cooling of spark plug points and exhaust valve head. There should not be hot surface in the end gas region. The exhaust valve should be located near the spark plug and should be kept small in size as compared to intake valve.

## 4.12.2 Type of Combustion Chambers

The following types of (combustion chambers) valves are commonly used in SI engines, (a) T-head type, (b) L-head, (c) F-head type, (d) I-head type, (e) overhead valve (f) overhead inclined, (g) Recardo turbulent type and (h) Rover head type.

Figure 4.17 shows the above-mentioned valves.

***T-head.*** In the T-head type, both the inlet and exhaust valves are located in the engine block on opposite sides. The valves are operated by two camshafts, one for each valve. *The characteristics of this type of valve are (i) good turbulence, (ii) short cylinder block, (iii) long flame travel and so greater tendency to knock and (iv) unsatisfactory fuel and air utilization.*

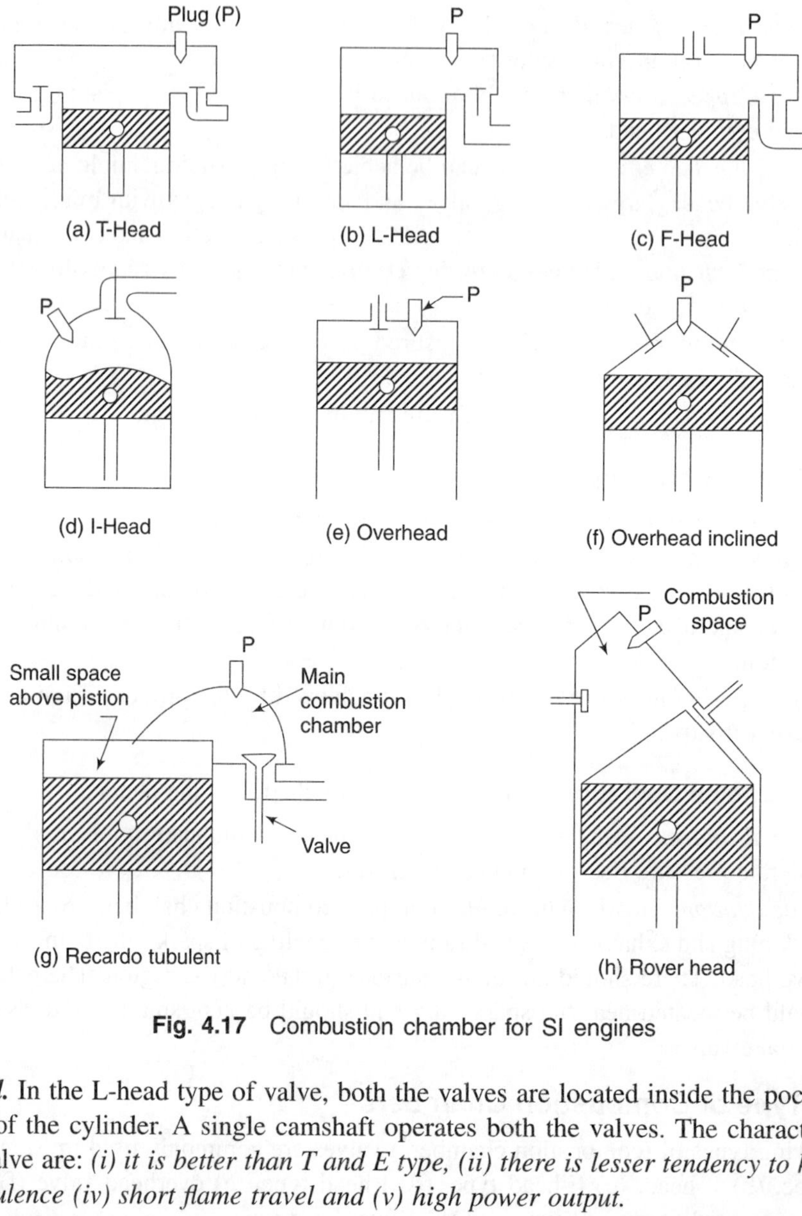


**Fig. 4.17** Combustion chamber for SI engines

***L-head.*** In the L-head type of valve, both the valves are located inside the pockets along one side of the cylinder. A single camshaft operates both the valves. The characteristics of L-head valve are: *(i) it is better than T and E type, (ii) there is lesser tendency to knock, (iii) high turbulence (iv) short flame travel and (v) high power output.*

***F-head.*** In the F-head valve, one of the valves is located in the side pockets and the other is in the cylinder head. A single camshaft operates the valve. The F-head type valve is better than T-head type but valve operating mechanism is complicated.

***Overhead and overhead inclined.*** *The overhead valves are mostly used in aircraft, automobile and racing car engines.* Both the valves are placed in the cylinder head, and in overhead inclined types, the valves are inclined at a certain angle to the cylinder axis. Obviously, overhead inclined type is complex and costly. *The characteristics of overhead valves are:*

*(i) very high turbulence, (ii) smooth operation, (iii) high volumetric efficiency and (iv) high power output.*

***Side valves.*** The advantages of the side valve engines are:

(a) easy lubrication, and (b) ready removal of cylinder head

But the disadvantages are:

(1) unsatisfactory fuel and air utility,

(2) longer flame travel,

(3) higher production cost, and

(4) greater heat loss.

***Recardo turbulent combustion chamber.*** It is an improvement over the side valve arrangement. By using hemispherical combustion space, the heat losses are reduced to a minimum and a minimum surface to volume ratio is obtained. The location of the spark plug is central and near the exhaust valve. The most favourable position for the spark plug is near the exhaust valve from the detonation point of view, the reason being that for any other location of the spark plug, the charge near the exhaust valve will pick up heat from the hot valve, and the delay period of this portion of the charge will be shortened considerably. There is always a possibility of detonation of the charge near the exhaust valve. The charge located farthest from the spark plug is the thin layer between the piston and cylinder head which is kept cool by the cooler head. This portion of the charge due to its lower temperature has a relatively longer delay period which permits the flame to reach it before detonation occurs. Thus, when the charge is forced into the main combustion chamber by piston, a certain amount of turbulence is produced which increases the flame speed and thus eliminates the tendency of knock.

***The Rover head combustion chamber.*** It is shown in Fig. 4.17 (h) and has the following characteristics:

(i) It has a compact combustion space and shorter length of flame travel.

(ii) The combustion space is approximately hemispherical which gives a low surface volume ratio.

(iii) The spark plug is centrally located.

(iv) Good turbulence is achieved and so a high flame speed is obtained.

(v) The thin layers of the charge between the piston and the cylinder head is kept cool by the relatively cool piston and cool inlet valve. This layer of charge running cool, has a long delay period which permits normal combustion avoiding the possibility of detonation.

The main shortcomings of this valve are:

(i) Complicated valve mechanism.

(ii) Uncommon shape of the piston.

(iii) Cylinder head cannot be removed without disturbing the valve gear mechanism.

## PROBLEMS

### Things to Think About

4.1 What do you mean by normal and abnormal combustion in SI engine?

4.2 What parameters govern the normal combustion in SI engines?

4.3 What should be the optimum shape of combustion chamber which will yield higher efficiency?

4.4 What are the duration of ignition delay and combustion in CI engine?

4.5 During abnormal combustion which undesirable gases get evolved in the cycle.

### Short and Long Theoretical Questions

4.6 Define combustion and mention the general conditions necessary for combustion.

4.7 Mention the two phases of combustion.

4.8 Discuss the range of ignition limits of hydrocarbon.

4.9 Describe the combustion phenomena in SI engines.

4.10 Discuss the effect of engine variables on ignition delay in the case of SI engines.

4.11 Discuss the effect of engine variables on flame speed in the case of SI engines

4.12 What do you mean by normal and abnormal combustion in SI engines?

4.13 Discuss detonation or knock theory of SI engines

4.14 What is the chemistry of knock in SI engines?

4.15 Discuss the variables affecting knock in SI engines.

4.16 Discus the effect of knock in SI engines.

4.17 Describe the methods to reduce detonation.

4.18 Discuss surface ignition knock in SI engines.

4.19 What are the requirements of good combustion chambers?

4.20 Describe the various types of combustion chamber in SI engines with sketches.

### Software Based Question

4.21 Develop a software for the design of various types of combustion chamber of SI engines.

### Design and Open Ended Question

4.22 Give a conceptual design of new type of combustion chamber which ensures efficient normal combustion in SI engine without initiating knock.

### Multiple-Choice Questions

4.23 Knocking tendency in SI engine increases with any one or more.

(a) increase in compression ratio (b) decrease in compression ratio
(c) increase in inlet temperature (d) retarding of spark timing

4.24 A SI engine is likely to knock when the

(a) ignition timing is advanced (b) engine is air-cooled
(c) carburettor or MPFI supplies a rich mixture (d) spark plug is centrally located

4.25 Cylinders equipped with sleeve valve gives
   (a) high knocking tendency
   (b) slow knocking tendency
   (c) opportunity for pre-ignition
   (d) none of the above

4.26 Knock in SI engine can be reduced by
   (a) supercharging
   (b) retarding spark timing
   (c) low self ignition temperature fuel
   (d) decreasing engine space

4.27 Flame development and subsequent propagation varies cycle to cycle when
   (a) engine cooling is improved
   (b) period of overlap is increased
   (c) the local mixture motion and compression varies
   (d) engine has more than one cylinder

4.28 The best possible location for the spark plug is
   (a) near the inlet valve
   (b) near the exhaust valve
   (c) at the centre of cylinder head
   (d) any place of the combustion chamber

## Key to Multiple-Choice Questions

4.23 (a) and (c) 4.24 (a) 4.25 (b) 4.26 (b) 4.27 (c) 4.28 (b)

CHAPTER

# Combustion in CI Engines

# 5

## OVERVIEW

As the name implies, CI engine is a compression ignition engine. There is a fundamental difference between the SI and CI engines. One of the major differences is the method by which the fuel is ignited. In the SI engine, a spark plug is used to initiate combustion of air-fuel (petrol) mixture. In the CI engine, the high temperature of air caused by compression process due to high compression ratio causes the injected diesel fuel to ignite. Because of this different types of ignition and fuels used, the combustion process and the operating variables in the CI engines are different than that of the SI engines.

This chapter deals with the combustion process in CI engines, the parameters influencing the combustion process and type of combustion chambers.

## 5.1 COMBUSTION PHENOMENA IN CI ENGINES

### 5.1.1 General Theory of Combustion in CI Engines

The combustion process in the CI engines is entirely different from that of the SI engines. In the SI engine, a homogeneous carburetted mixture of gasoline vapour and air or by multipoint fuel injection in certain proportion required by load and speed and controlled electronically is compressed in the compression stroke through a small compression ratio (6 to 10.5) and the mixture is ignited at or before the end of compression stroke (about 30° bTDC) by means of an electric spark. A single flame front advances through the air-fuel mixture after ignition and by this way combustion is completed.

In the CI engine, only air is compressed through a high compression ratio (12 to 22) during compression stroke raising its pressure and temperature to a high value (above ignition point of diesel fuel). Fuel is first pumped to a delivery pressure of 110 to 200 bar and then injected by injector through one or more jets into this highly compressed air in the combustion chamber. Here, the fuel jet disintegrates into a core of fuel surrounded by a spray envelope

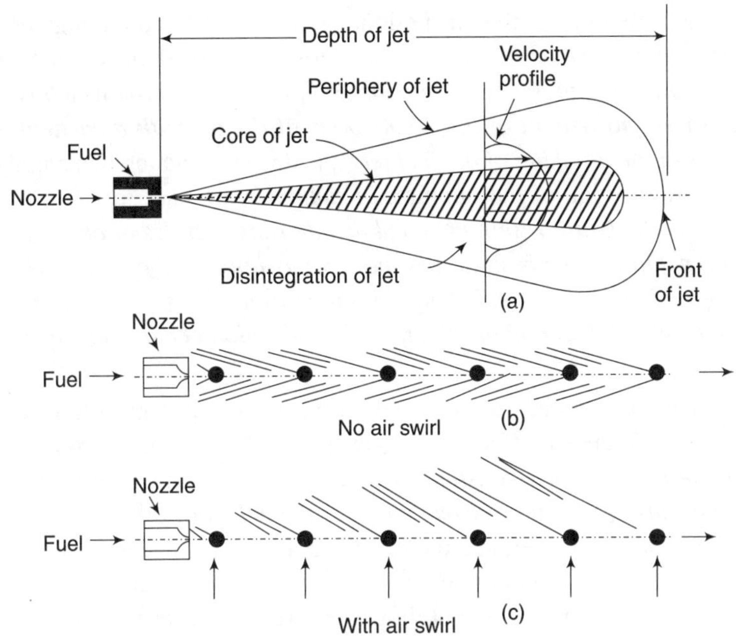


**Fig. 5.1** Disintegrates of fuel jet

of air and fuel particles as shown in Fig. 5.1(a). It is the atomization and vaporization of fuel that creates spray envelope. The turbulence of air present in the combustion chamber passing across the jet separates the fuel particles from the core resulting in better mixing of fuel and air. By this ways, a proper mixture of air and fuel forms at some location in the spray envelope which provides a scope to start oxidation of mixture.

Evaporation of the liquid fuel droplets starts by absorbing the latent heat of vaporization from the surrounding air which reduces the temperature of this layer of air surrounding the droplet and some get elapsed before this temperature can be raised again by absorbing heat from the bulk of air. As soon as this vapour and the air reach the level of auto-ignition temperature and if the local air-fuel ratio is within the combustible range, initiation of ignition starts. Thus, it is concluded that at *first there is a certain delay period before the ignition starts.*

The fuel-air mixture is *heterogeneous*. This is because of the fact that the fuel droplets cannot be injected and distributed uniformly throughout the combustion space. In a heterogeneous mixture, if the air within the cylinder were motionless, there will not be enough oxygen in the burning zone and the burning of the fuel would be either slow or totally fail as it would be surrounded by its own products of combustion [Fig. 5.1(b)]. This suggests that an orderly and controlled movement must be imparted to the air and the fuel so that a continuous flow of fresh air is given to each burning droplet and the products of combustion are swept away. The air motion is known as ***air swirl*** *which is produced by the shape of combustion chamber.* Figure 5.1(c) shows the effect of swirl on the fuel jet.

There is a basic difference in the air motions required in SI and CI engines. *In SI engine, it is called* **turbulence** *whereas in CI it is called* **swirl***.* The turbulence of SI engines has a *disorderly air motion with no general direction of flow. But the swirl which is required in CI engines, is an* ***orderly movement of the whole body of the air with a particular direction of flow*** *and it assists in the breaking of the fuel jet.* It is this swirl which helps in the intermixing of burnt and unburnt portion of the mixture. *In summary, we may say that in a SI engine, a homogeneous air-fuel mixture within the combustible range ignites at one point and contains the progress of a definite flame front across the combustion chamber with a slow rise in pressure whereas in a CI engine, the A/F ratios in the various parts of the combustion chamber vary widely, no definite flame front in evident and combustion occurs in many locations within the chamber.*

It is worth to mention that air-fuel ratio (A/F) plays an important role in CI as well as in SI engine operation, because it affects both power output and engine economy. *The air supply in a CI engine is essentially constant for a given speed and changing the quality of the fuel injected by means of injection system changes the A/F ratio.* With decreasing amount of fuel, the A/F ratio becomes leaner, and the indicated thermal efficiency approaches that of air cycle but the resulting smaller amount of available energy lowers m.e.p. and power output of the engine. The m.e.p. and power output, however, are increased with decreasing A/F ratios up to a point in the neighbourhood of stoichiometric value.

As the amount of fuel is increased and the stoichiometric value is approached, the CI engine begins to produce a noticeably **black smoke**. This phenomenon is attributed to the fact that in certain areas within the combustion chamber, the A/F ratio becomes so rich that some of the carbon and oxygen particles is unable to combine in the time allotted for combustion and these uncombined carbon particles are the cause of the engine smoke. Though, higher power may be obtained by operating in the vicinity of stoichiometric value of A/F but this objectionable smoke necessitates operation at a slightly leaner A/F ratios with consequent reduction in power output. *Thus, the overall A/F ratio of CI engine varies from* ***18:1*** *at full load to about* ***80:1*** *at no load.* A CI engine is always designed to operate with an *excess air of 15 to 40% depending upon the application but the emphasis should be to utilise trapped oxygen in the combustion chamber as far as possible, thereby reducing the excess air to be a minimum.*

The power output curve for a typical CI engine operating at constant speed is depicted in Fig. 5.2. The approximate region of A/F ratios in which visible black smoke will occur is indicated by the shaded area and the operation in this region should be avoided.

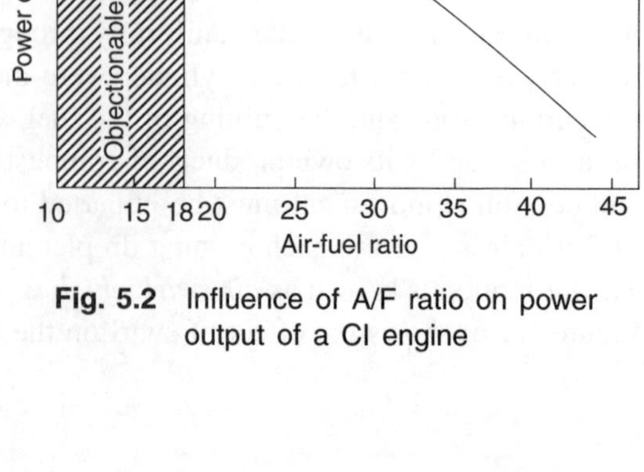


**Fig. 5.2** Influence of A/F ratio on power output of a CI engine

### 5.1.2 Stages of Combustion in CI Engines

The combustion process in CI engine is divided into **four stages**. These are as follows (Fig. 5.3).

(i) Ignition delay, (ii) rapid pressure rise or period of uncontrolled combustion, (iii) controlled or constant pressure rise (i.e., period of controlled combustion), and (iv) burning on the expansion stroke.

**(i) *Ignition delay.*** In the CI engine, air alone is compressed and raised to a high temperature on the compression stroke. Then, one or more jets of fuel, compressed to a pressure of 110–200 bar, are introduced into the combustion chamber towards the end of compression stroke where the temperature and pressure of air are high. In Fig. 5.3, the point *A represents the time at which injection starts. Here, the jet disintegrates into a core of fuel surrounded by a spray envelope of air and fuel particles* (Fig. 5.1). This latter zone is created both by the atomization and vaporization of the fuel and the turbulence of the air in the combustion chamber passing across the jet and stripping the fuel particles from the core. The fuel sprayed into the combustion space just before the end of compression stroke does not ignite immediately even though the air temperature inside the cylinder is above self-ignition temperature of the fuel.

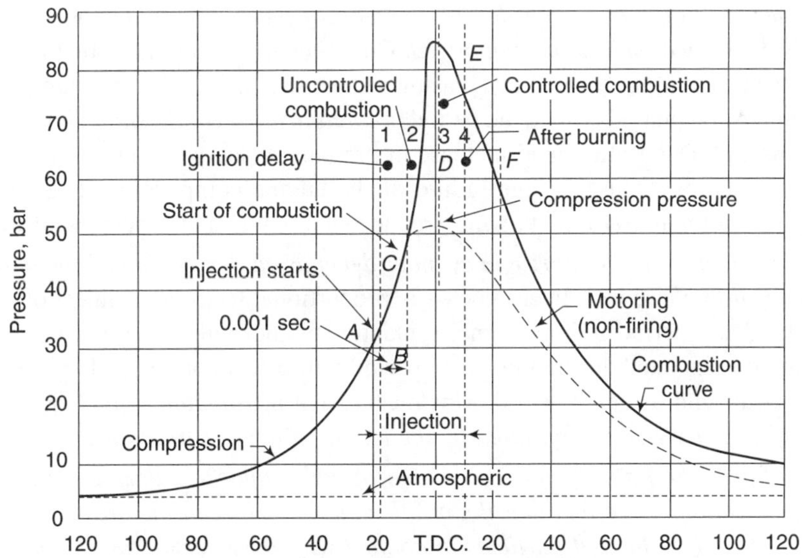


**Fig. 5.3** Stages of combustion in CI engine

*The fuel sprayed first attains self-ignition temperature and then undergoes preliminary chemical reaction before the actual ignition takes place. The time interval between the start of fuel injection (A) and the commencement of actual combustion (C) is called the* ***delay period*** *or* ***ignition delay***. Thus, ignition delay is divided into two parts, the ***physical delay*** or

**the mixing period and the *chemical delay*** or the interaction period. In the physical delay period, the fuel is atomized, vaporized, mixed with air and raised to self ignition temperature. In the chemical delay, reaction starts slowly and then accelerates until inflammation or ignition takes place (about 5° bTDC), Initially, these reactions are slow and rate of energy release is no longer able to keep pace with the rate of energy absorption by the fuel. This results in a drop in cylinder pressure to a value which is lower than what it would have been without injection of fuel (process *AB*). Once the reaction proceeds at a rapid rate, the released energy becomes proportionately more and the gas pressure begins to build. At point *C* the gas pressure attains a value equal to that without fuel injection, *i.e.*, ***motoring***.

**(ii) Rapid pressure rise or period of uncontrolled combustion.** *At the end of delay period, ignition takes place in some region or regions where the concentration of the fuel is the optimum.* This is generally in the spray envelope. The fuel-air mixture at the time of inflammation is very *heterogeneous* and if we consider a small region, the concentration of fuel in this region may range from 0 to 100%. Once the ignition starts the flame propagates itself through continuous regions in which the fuel vapour and air are already mixed. The adjoining regions will ignite either by contact with the advancing flame front or by accelerated auto-ignition caused by rapid increase in temperature due to radiant heat transferred from the flame.

*During the delay period, a considerable amount of fuel accumulates in the combustion space and is properly atomized and vaporized.* Consequently, when the actual ignition starts, the fuel burns at an extremely rapid rate, casing a steep rise in cylinder pressure and temperature. The combustion phase is called the period of ***uncontrolled combustion*** and is represented by the process *CD*. Due to the rapid increase in pressure, a sudden load is applied to engine structure and, this may cause fatigue failure of the engine components. The violent noise produced is called the ***diesel knock***, which reduces the power and efficiency of the engine.

*The magnitude of the pressure rise in cylinder depends on the amount of fuel burnt and the volume occupied by the gases.* Fuel injected is proportional to (time of injection) × (Engine speed). If the delay period is longer, more quantity of fuel will be injected and, a greater portion of the fuel injected during this delay period will be evaporated and, will be ready for combustion at the end of delay period. This means, that *the amount of fuel representing the magnitude of pressure rise is controlled by the length of delay period and rate of injection.*

*The rate of pressure rise depends on amount of fuel evaporated during the delay period, degree of atomization of the fuel and the relative motion between the air and the fuel, as it is responsible to bring the fuel particles in contact with oxygen. So, for smooth running of the engine, the delay period must be the minimum.* Moreover, engine speed also influences the relative motion and the rate of fuel injection. Therefore, it is necessary to keep the engine speed low, so that, less fuel is injected and the pressure rise is less.

*It is to be noted that in the SI engines, ignition occurs at one point with consequent slow rise in pressure, whereas, in CI engine ignition occurs at many points with the consequent rapid rise in pressure.*

**(iii) Period of controlled combustion.** At the end of combustion of main bulk of the fuel, the flame spreads throughout the whole combustion space and the cylinder contents are at very high temperatures. Any further quantity of fuel injected during this period burns as soon as it leaves the injector nozzle, provided it finds sufficient quantity of air. Hence, it is essential to bring the fuel immediately in contact with oxygen to minimize the time required for completion of combustion and thus the time losses.

Note that, this requirement is contrary to the requirement of the second stage combustion. The rate of fuel injection and the intensity of air motion controls the rate of burning and pressure rise. *The process DE shows the period of controlled combustion.*

**(iv) After burning.** It has been found that combustion is not complete when the piston reaches T.D.C position. *There is always an after burning, i.e., the burning at the end of injection causing smoky exhaust.* It is believed that the principal cause of diesel smoke is the incomplete combustion of the products. The after burning process is represented by *EF* in Fig. 5.3. *After burning reduces the power output and therefore, it is avoided by using high turbulence (swirl).*

The whole combustion process in CI engine may be summarized as shown in Fig. 5.4. Figure 5.5 shows the photographic study of CI combustion.

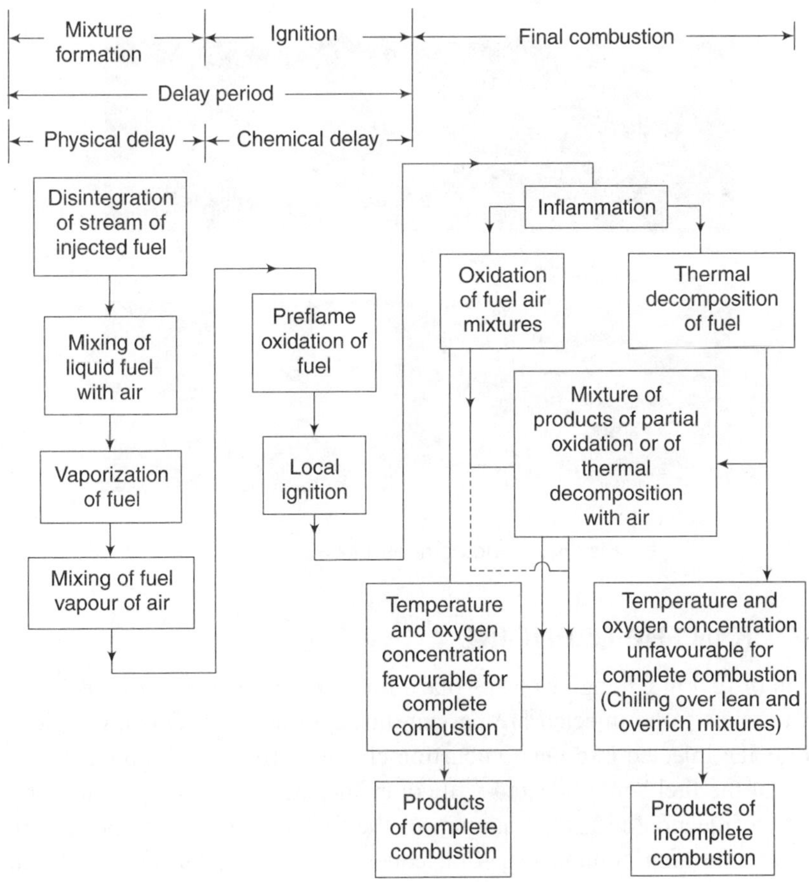


**Fig. 5.4** Outline of combustion process in CI engine

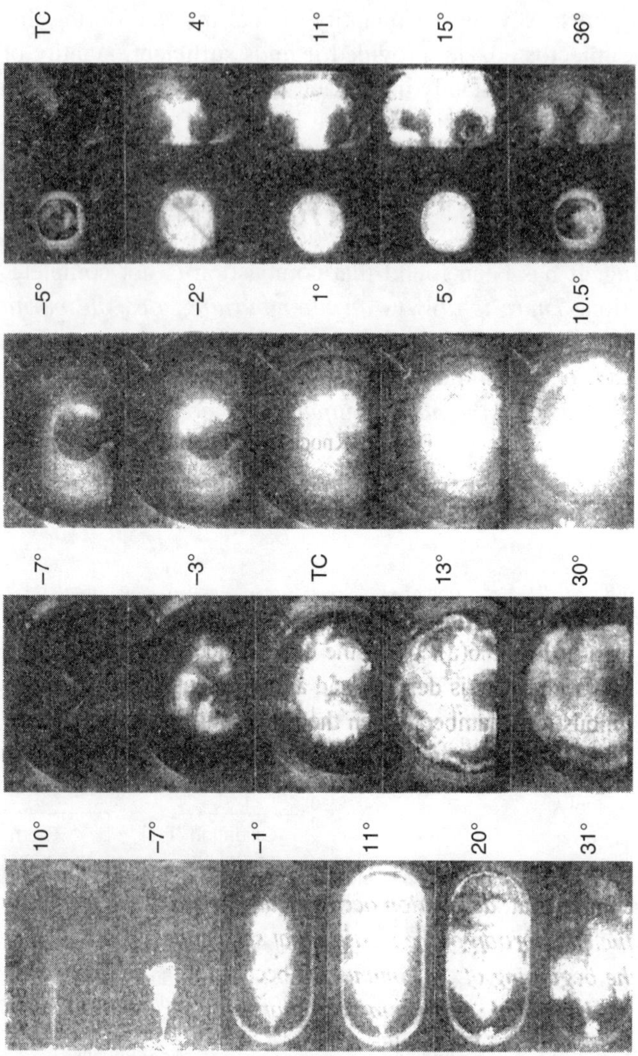


**Fig. 5.5** Photographs of CI combustion

## 5.2 KNOCK IN THE CI ENGINE

The injection of fuel in CI engines takes place for a certain interval of time. This means that as the first droplets to be injected are passing through the ignition delay period, additional droplets are being injected into the combustion chamber. In case of shorter delay period, the first droplets of the fuel being injected will commence actual "burning" phase in a relatively short time after injection and small amount of the fuel will be accumulated in the chamber when the actual burning commences. Consequently, the mass burning of the mixture will

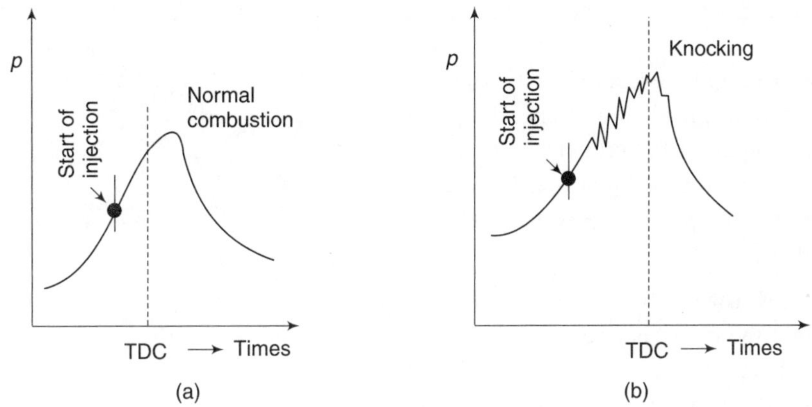


**Fig. 5.6** Knock in CI engine

produce a smooth pressure rise and the combustion will be normal as shown in Fig. 5.6(a). If, on the other hand, ignition delay is longer, the actual burning of the first droplets is delayed and a greater quantity of fuel droplets is accumulated in the combustion chamber. When the actual burning starts, the ignition of the large amount of accumulated fuel causes a violent and instantaneous rise of pressure as shown in Fig. 5.6(b). Under such condition, extreme pressure differentials are produced and violent gas vibrations known as ***detonation or knock*** occurs. This phenomenon is similar to that of SI engine discussed earlier.

It is to be noted that *detonation occurs near the end of the combustion in SI engine, because it is the end portion of the mixture that self-ignites. But in CI engine, detonation occurs near the beginning of the combustion, because, it is the first portion of the mixture that self-ignites. In the CI engines since ignition is by self-ignition, a knocking process is more inevitable than in the SI engine.*

Thus, *to decrease the tendency of knock, it is essential to decrease the ignition delay.* Ignition delay of the CI engine can be decreased by addition of small amounts of certain compounds to the diesel fuel such as *ethyl nitrate, amyl thionitrite, etc.* The function of these additives is to affect the combustion process by speeding the molecular interaction period, thereby decreasing the ignition delay.

## 5.3 EFFECT OF OPERATING VARIABLES ON DELAY PERIOD AND DIESEL KNOCK

*The combustion process in CI engines is mainly controlled by the length of the delay period. If the delay period is short, the engine runs smoothly. A longer delay period makes the engine run in a rougher and noiser fashion giving rise to knock.* This leads to the conclusion that the *delay period should be as short as possible.* But there is a practical limit beyond which the delay period cannot be lowered. Let us consider what would happen if there is no delay period at all and, the droplets of fuel ignite as soon as they leave the injector nozzle. In this case, the concentration of the burning droplets will be so closely packed, that it would be impossible to distribute among them the air needed for their complete combustion. Therefore, a certain minimum delay period is essential, so that, the fuel can undergo certain physical and chemical transformations to prepare it for proper combustion. It has already been mentioned that the *delay period may by divided in two parts (i)* **physical delay** and *(ii)* **chemical delay**.

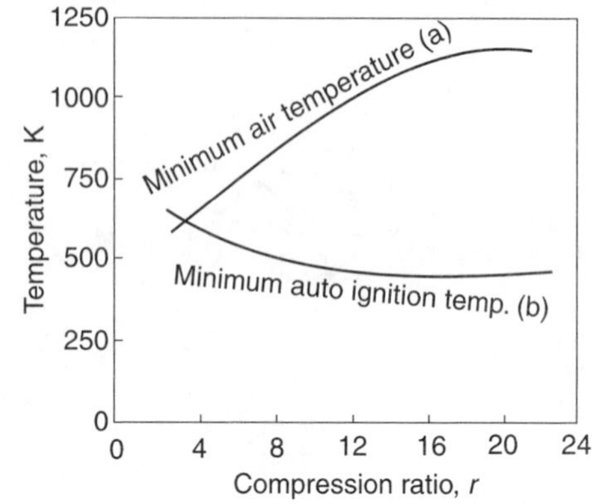


**Fig. 5.7** Effect of compression ratio

**Operating variables:** The following operating variables affect the delay period and cause diesel knock. We will see that the factors that affect the delay period and one other factor speed *(i.e., njection rate),* after the knock.

**(i) Inlet temperature.** A higher inlet air temperature increases the cylinder air temperature at the moment of injection. This decreases the delay period, which in turn, decreases the tendency of knock.

**(ii) Inlet pressure.** High inlet pressure results in high compression pressure. The delay period is reduced and the pressure rise is normal. Normal pressure rise eliminates knock. However, the peak pressures will be high, and so, the engine has to be made strong. *Supercharging suppresses knock while throttling hastens it.*

**(iii) Compression ratio.** A high compression ratio brings about high pressure and temperature in the air at the moment of injection. As a result, the delay period is reduced and the rate of pressure rise is normal. In this condition, the knocking tendency is less. *One of the practical disadvantages of using higher compression ratio is that the mechanical efficiency tends to decrease due to increase in weight so a compromise is made for compression ratio.*

**(iv) Injection timing.** If the injection of fuel is too early (i.e., much before TDC) the delay period will be longer due to the existence of low pressure and temperature of air. During this long delay period most of the fuel is injected and the mixing is relatively thorough. This results in an abnormal pressure rise. *This abnormal pressure rise due to early injection is favourable for knocking.* In the case of late injection, it will also result in a long delay period and as such, the rate of pressure rise should be high enough, but because the piston is already moving down in its expansion stroke, the increase in cylinder volume will reduce the pressure rise. Figure 5.8 shows the effect of three injection advances timings of

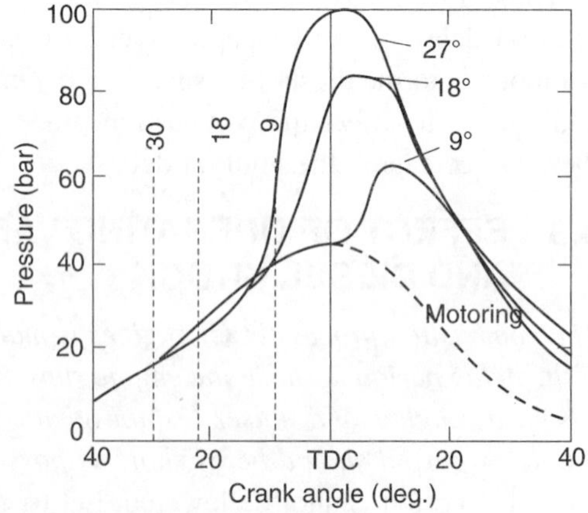


**Fig. 5.8** Effect of injection timings on pressure

9°, 18° and 27° before TDC on pressure variation. Generally, injection timings of fuel varying from about 11.5° to 20° degrees before TDC gives the minimum delay period and the tendency of knocking is less.

**(v) Engine speed.** *The amount of fuel injected in the engine depends upon the engine's speed because the fuel pump is driven by the engine's crankshaft.* At high speed, more fuel will be injected during the delay period, and the mixing of air and fuel will also be better, both of which cause a high rate of pressure rise giving rise to knocking (Fig. 5.9).

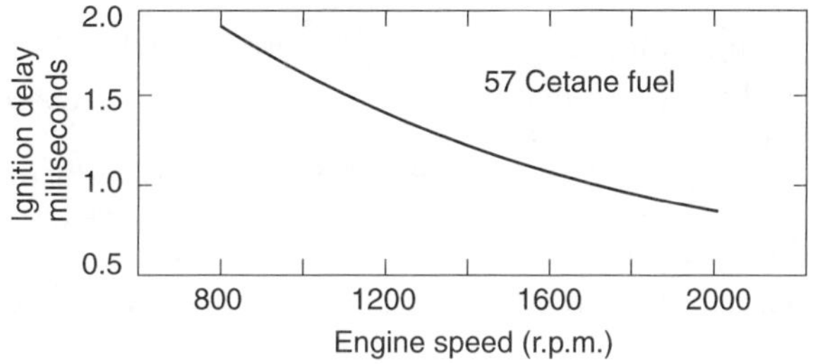


**Fig. 5.9** Effect of speed on ignition delay

At low engine speed, the fuel injected during the delay period will be much less, and also its mixing with air will be poor resulting in a normal rate of pressure rise which decreases the tendency of knock. *At low speed, the engine efficiency is high.* Figure 5.9 shows the effect of speed on ignition delay.

**(vi) Engine size.** A large engine runs at low speed and hence the quantity of fuel injecting during delay period is less. Moreover, in large engines, the compression temperature is higher. Both these effects have the tendency to reduce knocking.

**(vii) Jacket water temperature (cooling).** Increase in jacket water temperature decreases the delay period. Decrease in delay period reduces the tendency of knock.

**(viii) Injection pressures.** Increase in injection pressure generally decreases the delay period due to higher degree of atomization and higher surface volume ratio. The decrease in delay period reduces the tendency of knock.

**(ix) Type of combustion chamber.** In general, precombustion gives a shorter delay period (i.e., less tendency to knock) compared to an open type combustion chamber due to its compactness.

**(x) Turbulence.** The effect of turbulence is to strip the fuel from the injected spray and, therefore, promotes a homogeneous mixture. A high turbulence thus decreases the delay period and the tendency of knock.

**(xi) Fuel volatility.** Highly volatile fuels with low viscosity are usually desirable in order to promote a homogeneous mixture quickly and so reduce physical delay. On the other hand, a highly volatile fuel may knock badly because a longer supply of combustible mixture is furnished during the delay period. With these contradictions, it is difficult to forecast diesel fuel behaviour. In general, however, a low volatile fuel increases the probability of knock.

**(xii) Chemical composition of fuel (Cetane Number).** Fuels having high self-ignition temperature give longer delay periods. This longer delay period increases knocking. *A higher cetane number of fuel reduces delay period due to lower value of self ignition temperature* (Fig. 5.10).

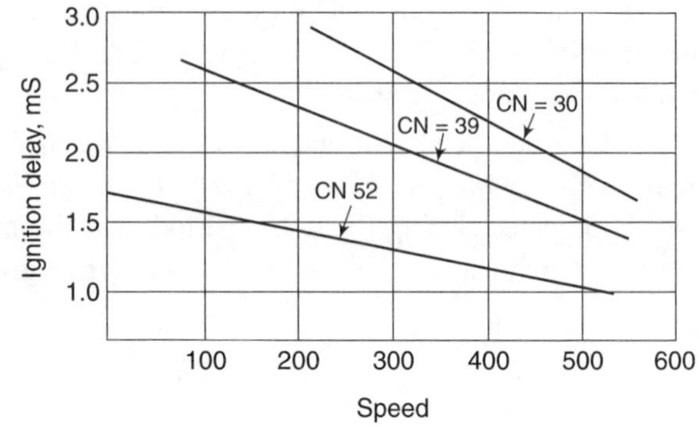


**Fig. 5.10** Effect of cetane number on ignition delay

**(xiii) Additives.** Some additives such as amylnitrate, reduce the ignition delay and hence the tendency of knock.

**(xiv) F/A (Load).** With increased in load (F/A), operating temperature increases and hence delay period decreases. (Fig. 5.11).

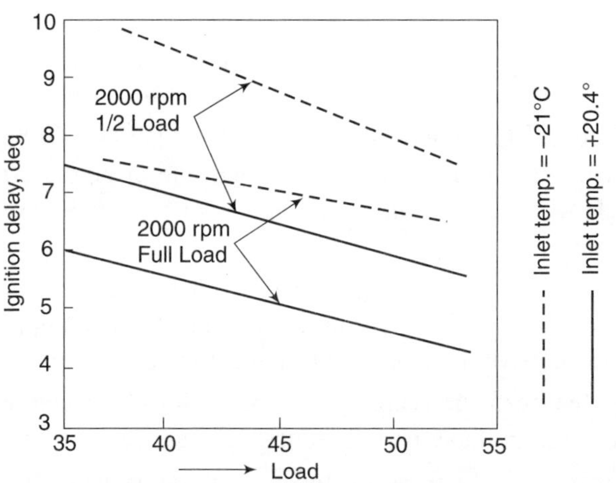


**Fig. 5.11** Ignition delay vs load

**(xv) Swirl rate.** Change in swirl rate change the fuel evaporation and fuel-air mixing processes. They also affect wall heat transfer during compression and hence, change the temperature at injection. The good swirl rate reduces ignition delay.

**(xvi) Oxygen concentration.** The oxygen concentration in the charge into which the fuel is injected would be expected to influence the delay period. The ignition delay decreases with oxygen concentration which may be varied by recycling exhaust gases.

*From the above discussion, it is obvious that to reduce the possibility of knock in the CI engine, the first element of fuel and air should have (a) high temperature, (b) high density, (c) short delay period and (d) reactive mixture.*

Table 5.1 depicts the summary of factors which influence the delay period in CI engines.

**Table 5.1** Effect of Variables on Delay Period

| *S.No.* | *Increase in variables* | *Effect on delay period* | *Reason* |
|---|---|---|---|
| 1. | Inlet temperature | Reduces | Increases air temperature at the moment of injection. |
| 2. | Inlet pressure (Supercharging) | Reduces | Increases density and also reduces auto-ignition temperature. |
| 3. | Compression ratio | Reduces | Increases air temperature and pressure and reduces auto-ignition temperature. |
| 4. | Injecting timing (advance angle) | Increases | Pressure and temperature of air are lower when injection begins. |
| 5. | Injection pressure | Reduces | Reduces physical delay due to greater surface volume ratio. |
| 6. | Engine speed | Reduces in terms of milli-sec. | Reduces loss of heat. |
| 7. | Engine size | Decrease in terms of crank angle. Little effect in terms of millisec. | Larger engines operate normally at low speed. |
| 8. | Jacket water temperature | Reduces | Increases density and also reduces auto-ignition temperature. |
| 9. | Combustion chamber | Lower for engines with pre-combustion chamber | Due to compactness of chamber. |
| 10. | Turbulence | Reduces | Produces homogenous mixture. |
| 11. | Fuel cetane number | Reduces | Reduces self ignition temperature. |
| 12. | Fuel volatility | Reduces | Due to production of homogenous mixture. |
| 13. | Fuel additives | Reduces | Reduces self ignition temperature. |
| 14. | F/A (Load) | Reduces | Operating temperature increases. |

## 5.4 A COMPARATIVE STUDY OF KNOCK IN SI AND CI ENGINES

It is worthwhile to mention that the knocking phenomenon in SI and CI engines is fundamentally due to the auto-ignition subject to the ignition time-lag characteristics of the fuel-air mixture. However, there is a difference in the knocking process of SI and CI engines. Figure 5.12 shows the difference in knocking process of these two engines.

The basic difference may be made on the basis of the following.

**(i) Auto-ignition location and timings.** In SI engines, it is the auto-ignition of the end gas away from the spark plug, most likely near the end of combustion causes knocking. But in CI engines, knocking occurs near the beginning of the process because it is the first portion of the mixture that self-ignites as shown in Fig. 5.2(b). It is obvious from the figure that the explosive auto-ignition is more or less over before the peak pressure in the case of CI engines. But for SI engines, the condition for explosive auto-ignition of the end charge is more favourable after the peak pressure. From the above facts, it is concluded that in order to avoid knocking in SI engine, it is necessary to prevent auto-ignition of the end gas to take place while in CI engine earliest possible auto-ignition prevents the knock.

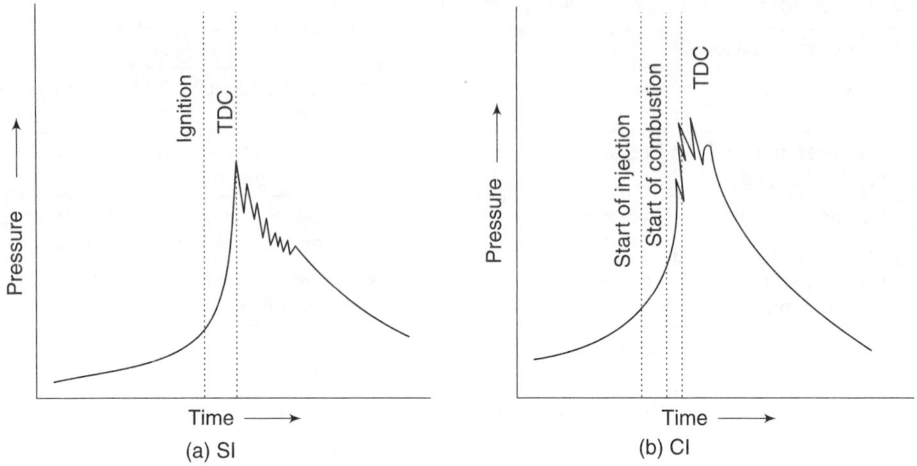


**Fig. 5.12** Comparison of knocking process in SI and CI engines

**(ii) Intensity of knocking.** The intensity of knocking depends upon whether the mixture is homogeneous or heterogeneous in nature. In the SI engines, the mixture is homogeneous, so the auto-ignition of such charge increases the intensity of knocking, i.e., the rate of pressure much more than the CI engine where the mixture is heterogeneous. The maximum rate of pressure rise in the CI engine may reach as high as 10 bar per crank degree angle.

**(iii) Noise level.** In the SI engine, it is relatively easy to distinguish between knocking and non-knocking operation by human ear as the noise level in knocking is very high as compared to non-knocking condition. However, in the case of the CI engine, the normal ignition is itself by auto-ignition and hence most CI engines have a sufficiently high rate of pressure rise per degree of crank angle to cause audible noise. The engine is said to be knocking if this noise is more, it means that, only the personal judgement is involved to judge knocking. Thus, in the CI engines, there is no definite distinction between normal and knocking combustion.

**(iv) Main factors tending to reduce knocking.** It is interesting to note that the factors that tend to reduce knocking in the SI engine increase knocking in the CI engines and vice versa because of opposite nature of auto-ignition location and timings. As mentioned earlier, the knocking in the SI engine is due to simultaneous *auto-ignition of the last part of the charge. In order to eliminate knocking, it is necessary to prevent altogether the auto-ignition of the last part of the charge which is possible from a fuel having a long delay and high self ignition temperature of the fuel. To eliminate knocking in the CI engines it is necessary to achieve auto-ignition as early as possible which may be possible from a fuel having a short delay period and low, self ignition temperature.*

Table 5.2 depicts the main factors which reduce knocking in SI and CI engines.

**Table 5.2** Main Factors to Reduce Knocking in SI and CI Engines

| S.No. | Factors | SI Engine | CI Engine |
|---|---|---|---|
| 1. | Auto-ignition temperature of fuel | High | Low |
| 2. | Delay period for fuel | Long | Short |
| 3. | Inlet temperature | Low | High |
| 4. | Inlet pressure | Low | High |
| 5. | Compression ratio | Low | High |
| 6. | Combustion wall temperature | Low | High |
| 7. | Speed | High | Low |
| 8. | Cylinder size | Small | Large |
| 9. | A/F | Very low or Very high | Very low |

## 5.5 COMBUSTION CHAMBER FOR CI ENGINES

*The design of the combustion chamber should be able to meet the requirement of greater power output, economy and smooth engine operation.* To achieve high engine output, the fuel-air charge should be efficiently prepared in the combustion chamber itself because air alone is compressed in the cylinder and the fuel is injected during a period of some 20 to 35 degrees of crank angle. In other words, the fuel on injection, should be uniformly distributed throughout the combustion chamber and, there should be good mixing between fuel and air and the combustion chamber should meet these conditions. To achieve smooth engine operation without diesel knock, the rate of pressure rise during the second stage of combustion should be a moderate one. This is possible by reducing the ignition delay. Moreover, during the delay period, there should be a limited contact only between the fuel and air. After the completion of second stage combustion, the mixing of the charge should be extremely rapid so that the entire combustion is complete during the third combustion stage and after burning is avoided.

The combustion chambers of CI engines are broadly classified as:

(A) Direct Injection (DI) combustion type.

(i) Open combustion chambers. (Bowl-in piston).

(B) Indirect Injection (IDI) combustion chambers.

(i) Pre-combustion chambers.

(ii) Turbulence combustion chambers.

(iii) Air-cell combustion chambers.

### 5.5.1 Open Direct Injection (DI) Combsution Chambers

In open direction combustion chambers, the fuel is injected directly into the upper portion of the cylinder which acts as the combustion chamber. There are various shapes of combustion space such as (i) hemispherical, (ii) toroidal and (iii) chamber with side injector as shown in Fig. 5.13. The combustion space is usually recessed into the crown of the piston, called *bowl-in-piston chamber. This type depends little on turbulence to perform the mixing. In order*

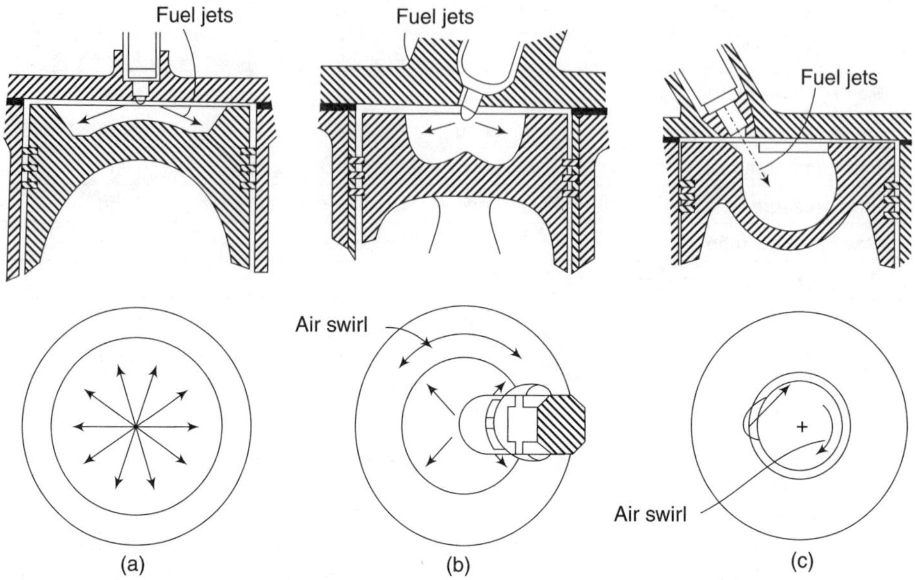


**Fig. 5.13** Direct injection (DI) Combustion chambers. (a) Hemispherical, (b) Torodial, (c) Chamber with side injector

*to obtain proper penetration and dispersal of fuel necessary for mixing, higher injection pressures (200 atm.) and multi-orifice nozzles are used.* Attempts are being made to improve the air-motion in open combustion chambers. The two methods most commonly employed are:

(i) By shrouding the inlet valve. (ii) By providing squish.

For shrouding the inlet valve, it is masked or shrouded, to give swirling motion to the air entering the cylinder which is believed to persist during compression stroke and at the time of injection. *This method is suitable for low speed but at high speed is objectionable.* For providing ***squish*** (i.e., a rapid) movement of air from the walls of the cylinder towards the centre of combustion chamber) the air at the end of compression stroke is pushed into the main combustion space whose diameter is smaller than the cylinder bore.

*The open combustion chamber is ordinarily used on low speed engine where injection is spread through a greater period of time and, ignition delay is a relatively less, important factor.* If attempts are made to use this type of chamber on high speed engines, the possibility of knock increases, as the allowable fuel injection time must be shortened.

The **advantages** *of open combustion chambers* are: (i) Small direct heat loss because the mixing of air and fuel depends little on turbulence, (ii) high thermal efficiency due to use of excess air, (iii) ease of cold starting, and (iv) low fuel consumption.

The **disadvantages** of an open combustion chamber are: (i) a longer delay period which results in a higher rate of pressure rise, and (ii) the multi-orifice nozzle is more liable to frequent clogging or diversion of the fuel spray by accumulated carbon particles, with consequent higher maintenance costs.

There is a *special type of open combustion chamber known* as **MAN** combustion, shown in Fig. 5.14, which was developed in West Germany. The principles employed in the design of this chamber are: (i) to prevent the initial formation of heterogeneous fuel-air mixture so that the reaction rate is normal, (ii) to prevent access of oxygen to the fuel to eliminate formation of peroxide, and (iii) to assist early formation of slowly reacting hydrocarbon compounds.

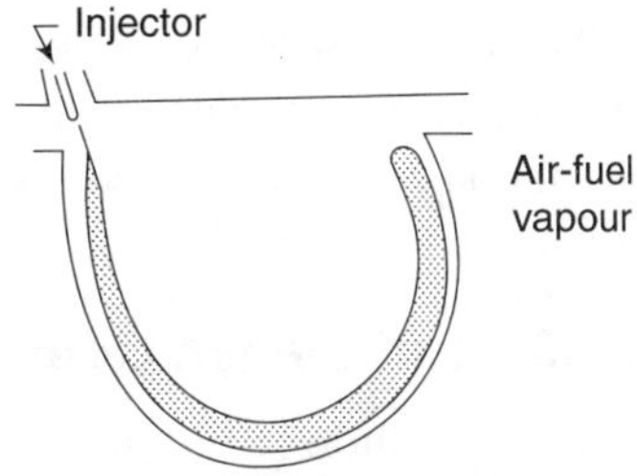


**Fig. 5.14** MAN combustion chamber

The **MAN** *combustion chamber is of spherical shape situated in the piston crown.* Due to the presence of strong swirling action of air, a thin liquid film of fuel is formed over the combustion space, *this eliminates the formation of fine mist.* For this type of chamber, it is claimed that combustion takes place in a homogeneous gas phase; thermal decomposition is prevented and there is no soot formation. The pressure rises is at low rate. There is an ease of starting.

### 5.5.2 Pre-combustion Indirect Injection (IDI) Chamber

This consists of two chambers: *pre-combustion chamber and main combustion chamber* as shown in Fig. 5.15. These two chambers communicate with each other through a number of small passages. The pre-combustion chamber (ante-chamber) occupies about 25 to 35% of the total volume of the chamber. During the compression stroke nearly one-third of the cylinder air is forced into the pre-chamber and the fuel is injected into it. The pre- combustion chamber is designed to run hot which reduces the delay period. The combustion first starts in the pre-combustion chamber and the resulting pressure rise forces the products of combustion, air and flaming fuel droplets into the main combustion chamber through small passages and creates a violent air motion. This violent air motion helps in rapid mixing and burning in the main combustion chamber. Since the fuel reaching the main combustion chamber has practically no delay period, the combustion in the main chamber in the third stage is rapid and complete.

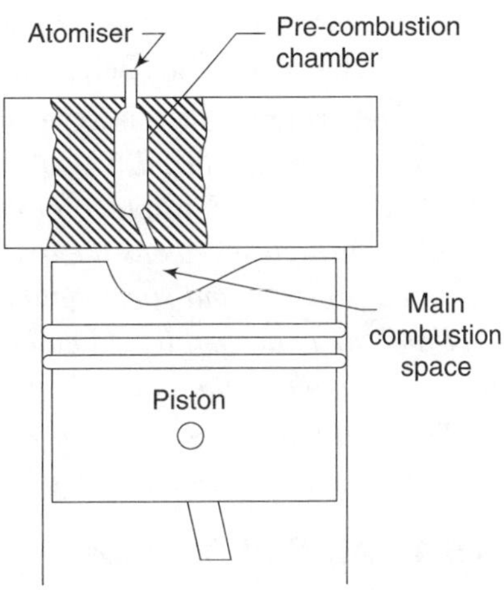


**Fig. 5.15** Pre-combustion chamber

The **advantages** *of this type of combustion chamber are (i) tendency of knock is the minimum because of short delay period, (ii) rapid combustion in third stage, (iii) there is no need of high injection pressure, (iv) the spray pattern is not critical and (v) it is suitable for high speed engines.*

The **disadvantages** include (i) heat loss is very high because the velocity of the burning mixture is greater during its flow from the pre-combustion chamber, (ii) there is a more likelihood of thermal decomposition of fuel, (iii) it is difficult to cold start and (iv) thermal efficiency is lower.

### 5.5.3 Turbulence Combustion Chamber

Figure 5.16 shows a turbulence combustion chamber. *Turbulence or swirl is created by pushing almost all the air or 70 to 80% of it at a great velocity into the combustion chamber during the compression stroke. The combustion chamber is separated from the engine cylinder.* The turbulence chamber is insulated so that it may run hot. The air enters the chamber in a tangential direction, and so the air is given a rotary motion. The fuel which is injected into the rotating air is rapidly mixed with the air and combustion starts. Since the chamber is insulated, a higher temperature is maintained which reduces the delay period. This results in a normal rate of pressure rise and smooth engine operation. The pressure rise in the combustion chamber forces the products of combustion into cylinder through the communicating passage. The communicating passage offers resistance to the flow, which in turn, imparts high turbulence and thus helps further combustion.

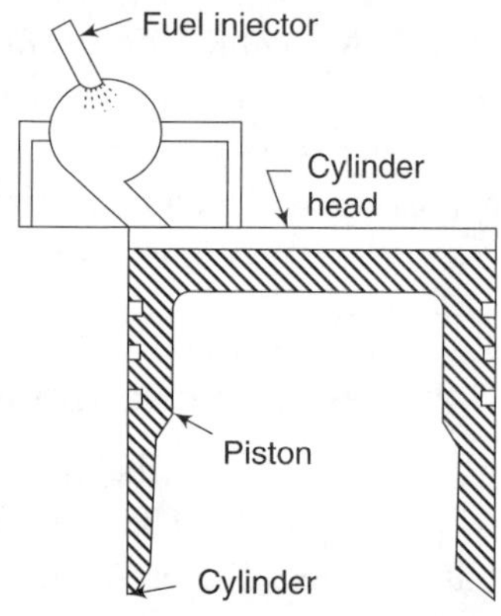

**Fig. 5.16** Turbulence chamber

The **advantages** of this type of combustion chamber are: *(i) the engine runs smoothly because the delay period is short, (ii) high turbulence results in rapid mixing and burning of fuel during the third stage of combustion and hence, there is no after burning (iii) it is suitable for high speeds as the amount of turbulence is proportional to the engine speed, (iv) injection pressure and spray pattern are not so important in this chamber because mixing of fuel and air is accomplished mainly by the air swirl, (v) better utilization of air is possible, and (vi) higher m.e.p. and volumetric efficiency are possible.*

The **disadvantages** are: *(i) it results in a* ***greater*** *direct heat loss and hence lower thermal efficiency, and (ii) cold starting is difficult.*

### 5.5.4 Air Cell Chamber

Figure 5.17 shows an air cell combustion chamber. *It consists of two chamber's namely*, a small chamber known as **air cell** and a large chamber known as the **main combustion chamber**.

*Both the chambers communicate with each other by a narrow restricted neck.* The fuel injector (nozzle) is placed in the main combustion chamber and just opposite to the communicating neck. At the end of compression stroke, some of the air is forced into the major and minor chambers. When fuel is injected, part of the fuel passes across the main combustion chamber and enters the cell where it is mixed with the entering air. Combustion first commences in the main combustion chamber where the temperature is higher but the rate of

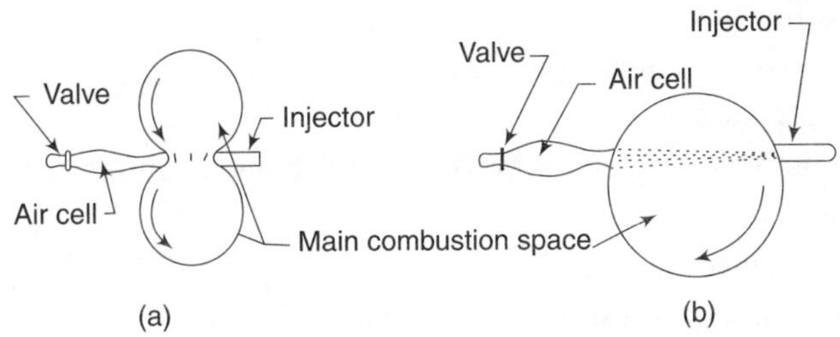


**Fig. 5.17** Air cell (a) Lonva type, (b) Aerotype

burning is slow due to insufficient mixing of fuel and air on account of its location. Combustion in the air cell occurs immediately at a rapid rate due to better mixing of air and fuel. The pressure rise in the minor cells forces the burning gases out into the main combustion chamber, thereby creating added turbulence and producing better combustion in this chamber. All the air is not able to come out of air cell by the general pressure rise in the main combustion chamber due to simultaneous combustion. This entrapped air continues burning even when the piston is on the expansion stroke. The power and efficiency of engine are reduced due to late burning.

The **advantages** of this combustion chamber are: (i) engine runs smoothly due to short ignition delay in the air cell, and (ii) starting is easier. The **disadvantages** are: (i) both power and efficiency decline due to late burning in the air cell.

## 5.6 COLD STARTING OF CI ENGINES

*One of the most important requirement of a CI engine is the ease for cold starting.*

To ensure easy cold starting, frequently compression ratios, higher than necessary, are used but even then, cold starting may become difficult in the following situations.

(a) Extreme cold climate
(b) When the cylinder liner is heavily worn
(c) When the valves are leaky.

Basically, *three types of* ***starting aids*** are used on modern high speed diesel engines.

(i) **Electric glow plugs** in the combustion chambers.
(ii) **Manifold heaters** which ignite a small feed of fuel.
(iii) The injection into the intake of controlled amount of low ignition temperature liquids usually *ethyl-ether* with addition of other fuels.

## 5.7 FACTORS TO BE CONSIDERED FOR DESIGN OF COMBUSTION CHAMBER

A particular combustion chamber design must be chosen to perform a given job. No combustion chamber can fulfil all the requirements. As in most engineering works, the design of the combustion chamber must be based on a compromise, after full consideration of the following factors.

(i) High thermal efficiency
(ii) Minimum heat loss to combustion chamber walls
(iii) High injection pressure
(iv) Nozzle design number, size and arrangement of the holes in the nozzle
(v) Ease of starting
(vi) Minimum maintenance
(vii) Fuel requirement ability to use less expensive fuel
(viii) Weight relation of engine to power output
(ix) Capable to cope with variables speed operation
(x) Smoothness with which forces created by expanding gases are transmitted to the piston
(xi) Low exhaust emission
(xii) High volumetric efficiency

## 5.8 CHARACTERISTICS OF COMMON DIESEL ENGINE COMBUSTION SYSTEMS

Table 5.3 shows the characteristics of common diesel engine combustion systems. The number for dimensions and operating characteristics are typical ranges for which different types of diesel engines and combustion systems have been considered.

**Table 5.3** Characteristics of Common Diesel Engine Combustion Systems

| *System* | *Direct injection (DI)* | | | | *Indirect injection (IDI)* | |
|---|---|---|---|---|---|---|
| | *Quiscent* | *Medium swirl* | *High swirl MAN* | *High swirl multi-spray* | *Swirl chamber* | *Prechamber* |
| • Size | Largest | Medium | Medium-smaller | Medium small | Smallest | Smallest |
| • Cycle | 2/4 Stroke | 4 stroke | 4 stroke | 4 stroke | 4 stroke | 4 stroke |
| • Turbocharged (TC) Supercharged (S) Naturally aspirated (NA) | TC/S | TC/NA | TC/NA | NA/TC | NA/TC | – |
| • Maximum speed (rpm) | 120–2100 | 1800–3500 | 2500–5000 | 3500–4300 | 3600–4800 | 4500 |
| • Bore (mm) | 900–150 | 180–350 | 130–80 | 100–80 | 95–75 | 95–70 |
| • Stroke/Bore | 3.5–1.2 | 1.3–1.0 | 1.2–09 | 1.1–0.9 | 1.1–0.9 | 1.1–0.9 |
| • Compression ratio (*r*) | 12–15 | 15–16 | 16–18 | 16–22 | 20–24 | 22–24 |
| • Chamber | Open or shallow dish | Bowl-in piston | Deep bowl-in-piston | Deep bowl-in piston | Swirl pre-chamber | Single/multi-orifice pre-chamber |

*Contd.*

*Table 5.3 Contd.*

| | | | | | | |
|---|---|---|---|---|---|---|
| • Air-flow pattern | Qiescent | Medium swirl | High swirl | Highest swirl | Very high swirl in pre-chamber | Very turbulent in pre-chamber |
| • Number of nozzle holes | Multi | Multi | Single | Multi | Single | Single |
| • Injection pressure | Very high | High | Medium | High | Lowest | Lowest |

*For the largest, slowest speed, engines for power generation or marine applications (usually two-stroke engines) open quiescent chambers are used which are essentially disc shaped; the motion of the fuel jets is responsible for distributing and mixing the fuel. Large truck and locomotive engines use a quiescent chamber consisting of shallow dish or bowl in the piston crown often used. It is worth to note that air utilisation in these engines is low, but are invariably supercharged to obtain high power density.*

Further, in the DI category, with the decrease in engine size, the maximum speed rises hence, swirl is used increasingly to obtain high enough fuel-air mixing ratios. The swirl is generated by suitably shaped inlet ports, and is amplified during compression by forcing most of the air towards the cylinder axis into the deep bowl-in-piston combustion chamber (Fig. 5.13 (b) and c). In this size range engine, MAN combustion chamber may also be used. *In the smallest engine sizes, IDI engines has traditionally been used increasingly to obtain high enough fuel-air mixing rates. These include pre-combustion chambers.* They either generate substantial swirl in the auxiliary chamber during the latter part of compression stroke using a nozzle or connecting passage that enters the auxiliary combustion chambers tangential or they generate intense turbulence in the prechambers.

## PROBLEMS

### Things to Think About

5.1 What should be the pressure of a jet in the combustion chamber of CI engine?

5.2 What is the effect of turbulence or swirl on the formation of mixture in the case of CI engine?

5.3 What is the importance of injection timing on the Diesel knock?

5.4 Is it possible to have a CI engine free from knock? How you will meet this?

### Short and Long Theoretical Questions

5.5 Define homogeneous and heterogeneous mixtures.

5.6 Discuss general theory of combustion in the CI engines.

5.7 Discuss in detail the various stages of combustion in the CI engines.

5.8 What is the range of overall A/F in a CI combustion chambers?

5.9 What do you mean by delay period?

5.10 Discuss the parameters that affect delay period.

5.11 Explain the phenomenon of diesel knock.

5.12 Compare the knocking process in the SI and CI engines.

5.13 "Factors tending to increase knocking in the SI engines tend to reduce knock in the CI engines". Discuss this statement.

5.14 Discuss the various types of combustion chambers used in CI engines.

5.15 Compare air swirl of CI engines with air swirl of SI engines.

5.16 Mention the three methods to produce general swirl and to solve cold starting problems for each in the CI engines.

5.17 Mention the factors to be considered for the design of combustion chambers of the CI engines.

### Software Based Question

5.18 Develop a software to design the various types of combustion chambers for CI engines.

### Design and Open Ended Question

5.19 Give a conceptual design of direct and indirect injection type combustion chambers which may give better fuel economy, high efficiency and less undesirable emission.

### Multiple-Choice Questions

5.20 In diesel engine, the duration between the start of injection and ignition is known as

(a) physical delay (b) chemical delay (c) ignition delay (d) period of ignition

5.21 In a CI engine, squish is created

(a) at the beginning of expansion process (b) near the end of the expansion process

(c) at the start of compression process (d) towards the end of compression process

5.22 Assertion : (A) By decreasing the load on a CI engine, the possibility of knock increases.

Reasoning: (R) Because the temperature of the cylinder and combustion wall increases.

**Answer Code:** (a) Both A and R true, (b) A is true but R is false, (c) Both A and R false, (d) R is true but A is false.

5.23 In a CI engine, the ignition delay

(1) increases with increasing air inlet temperature

(2) decreases with increasing air inlet temperature

(3) decreases with increasing speed

(4) decreases with increasing Cetane number of fuel

(a) 1, 2 and 3 are correct (b) 2, 3 and 4 are correct

(c) 1, 2 and 4 are correct (d) All are correct

5.24 Which of the following factors encourage knocking tendency in CI engine?

(1) High self-ignition temperature.

(2) Long ignition delay period.

(3) Low volatility of fuel.

(a) 1, 2 and 3, (b) 2 and 1 (c) 2 and 3 (d) 1 and 3

## Key to Multiple-Choice Questions

5.20 (c) 5.21 (d) 5.22 (b) 5.23 (b) 5.24 (b)

CHAPTER

# Fuels for IC Engines

6

## OVERVIEW

*Internal combustion engine is a prime mover which converts heat energy into mechanical work.* Since the heat supplied to the engine is obtained from the fuel used, it is desirable to have a fundamental knowledge of fuel types and their characteristics. It is worth to mention here that the use of particular fuel in internal combustion engines is determined not by its physical and chemical characteristics but also by its availability and cost of production. The fuel used in an internal combustion engine must have the following **properties**:

(i) Good combustion qualities
(ii) High energy density
(iii) High thermal stability
(iv) Low deposit forming tendencies
(v) Compatibility with the engine hardware
(vi) Good fire safety
(vii) Low toxicity
(viii) Easy transferability and unload vehicle storage
(ix) Low pollution
(x) Cheap and easily available.

Almost of all the fuels for the internal combustion engines of today are derived from *petroleum*, which is a complex mixture of hydrocarbon compounds. This chapter deals with the types, structure, refining process, properties and testing of fuels of internal combustion engines.

## 6.1 THE NATURAL FUELS

It is interesting to note that the origin and subsequent evolution of petroleum in liquid and gaseous forms found inside the earth are not definitely known. However, *petroleum is gener-*

*ally found in certain rock formations that, thousands of years ago, were the floors of oceans. It is believed that marine organic material on the sea bottom was enfolded by rock layers and subjected to high pressure. Here, the organic matter subjected to high pressure and under a temperature of 70°C to 150°C was cracked over a period of 1–2 million years to a lower molecular weight bituminous material. During the evolution process, the* ***heavy oils or tar sands first formed certain few light hydrocarbons but with the passage of time underwent*** *progressive change. Thus, crude oils found in the oldest rock formatives contain the* ***light hydrocarbons;*** *the younger rock formations contain* **heavier oils** *with little or no light portions, and with appreciable amounts of nitrogen and oxygen containing compounds. It is believed that oxygen was eliminated due to continuous exposure through the ages to slight radioactivity from neighbouring rock formation or bacterial action might have been responsible.*

Today crude petroleum is found accumulated in porous rocks or sands or limestones and covered with an impervious rock cap. Over and above of crude petroleum, these **underground rock** traps also contain large amount of **natural gas** and **salt water**.

***Geological evidence suggests that about 6 million square kilometers of the earth's surface, which were once under water, probably contain oil.*** Most of it originated in the shallow waters of the landrocked seas because such water would naturally contain abundant organic material which would be buried through the years by the sediment deposits from the surrounding land. Further, small amount of petroleum is also formed without the necessity of a burial ground due to chemical action on the remains of aquatic organisms in both fresh and salt water.

It is worth to note that there are **four great oil regions**, each near a landrocked sea. These are:

**(a) Eastern Mediterranean basin** **(b) Caribbean basin**

**(c) Far-East basin** **(d) North pole basin**

***The richest source lies at the eastern end of the Mediterranean sea—the cradle of the mankind.*** *This includes the lands which surround the Caspian Sea, the Red Sea, the Black Sea and the Persian Gulf, Iran, Iraq, Southwest Russia, Arabia and Romania fall under this. The Caribbean basin includes the Gulf state of US, Mexico, Venezuela, Columbia and Central America. Far-east basin lies between Asia and Australia.*

It has been estimated that the petroleum will last for about ***next 50 years*** with the known reservoirs.

## 6.2 TYPES OF FUELS

Internal combustion engines operate on liquid and gaseous fuels. However, solid fuels may also be used but it is not common due to many problems. Based on the type of the fuels to be used, the engines are designed accordingly.

### 6.2.1 Solid Fuels

There is a little practical application of solid fuels in IC engines because of problem in handling the fuel and disposing of the ash after combustion. As a result, efforts are made to generate gaseous or liquid fuels from charcoal for use in the IC engines.

### 6.2.2 Liquid Fuels

*All the modern internal combustion engines use liquid fuels which are the derivatives of liquid petroleum. There are three types of commercial liquid fuels.* These are *benzyl alcohol and petroleum products.* But, *the petroleum products such as gasoline (petrol), diesel, kerosene form the main fuels for IC engines.* The important factors which favour the use of liquid fuels are their relative abundance, ease of storing, ease of transportation, low cost, high heating value and excellent performance.

### 6.2.3 Gaseous Fuels

*Gaseous fuels being more cleaner are considered ideal fuels but pose a few problems when used in internal combustion engines.* Even though, the use of gaseous fuels in IC engines are restricted due to storage and handling problems, however, they are commonly used in stationary power plant engines where its availability is possible through pipe line. Some of the gaseous fuels may be liquefied under pressure but this is very expensive and risky. IC engines using gaseous fuels are generally termed ***gas engines***.

## 6.3 CHEMICAL STRUCTURE OF PETROLEUM

*Petroleum is basically a mixture of infinite number of hydrocarbon compounds which are made up exclusively of carbon and hydrogen atoms.* It also contains small amount of sulphur, oxygen, nitrogen and impurities such as water and sand. However, in some petroleum, metallic compounds such as derivatives of vanadium, iron, nickel, arsenic, etc., are also found.

The carbon and hydrogen combine in different proportions and molecular structures to form a variety of hydrocarbons.

*The carbon to hydrogen ratio is one of the important parameters and their nature of bounding determine the energy characteristics of the hydrocarbon fuels.* Based on the number of carbon and hydrogen atoms, the petroleum products are classified into different groups ranging from light gases to heavy tar. *Hydrocarbons with 6 to 15 carbon atoms can be classified as* ***light oils which are volatile****, while hydrocarbons with carbon atoms varying* ***from 16 to 26 are termed heavy oils****. The viscosity of oil is proportional to the number of carbon atoms in the hydrocarbon molecule.* These hydrocarbons belong to the ***paraffin, naphthene, aromatic*** **and *olefin* family. The general chemical formula for all hydrocarbons is $C_mH_n$.**

Table 6.1 shows the basic family of hydrocarbon indicating their structure, saturated, unsaturated and stability.

**Table 6.1** Basic Families of Hydrocarbons

| *Family of hydrocarbons* | *General formula* | *Molecular structure* | *Saturated/ unsaturated* | *Stability* |
|---|---|---|---|---|
| Paraffin | $C_n H_{2n+2}$ | Chain | Saturated | Stable |
| Olefin | $C_n H_{2n}$ | Chain | Unsaturated | Unstable |
| Naphthene | $C_n H_{2n}$ | Ring | Saturated | Stable |
| Aromatic | $C_n H_{2n-6}$ | Ring | Highly unsaturated | Most unstable |

### 6.3.1 Paraffin ($C_n H_{2n+2}$) Series

*Paraffins have a straight or open (chain type) structure, with one bond between each atom, and are stable.* The molecules' structure of some members of paraffin hydrocarbon are given below.

From the molecular structure, we conclude that the valency of all carbon atoms is fully utilised by single bonds with hydrogen atoms, and that is why the paraffin hydrocarbons are saturated compounds and are very stable by characteristics. The hydrocarbons which have the same chemical formulae but different structure formula are known as **isomers**. Thus, isobutene is isomer of butane.

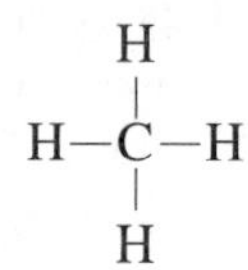

**Methane** ($CH_4$)

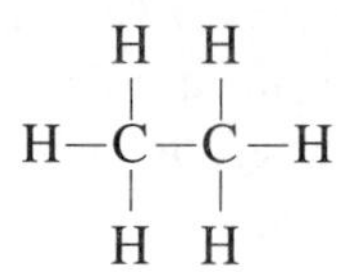

**Ethane** ($C_2H_6$)

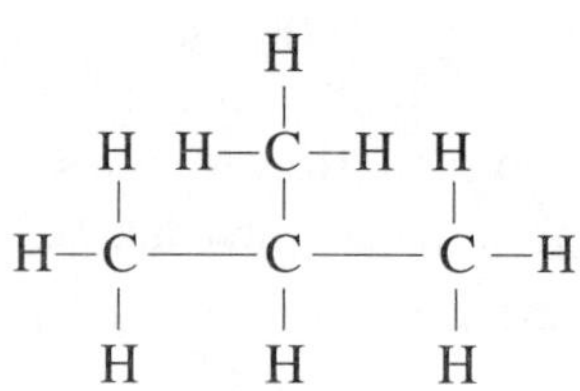

**Iso-butane** ($C_4H_{10}$)

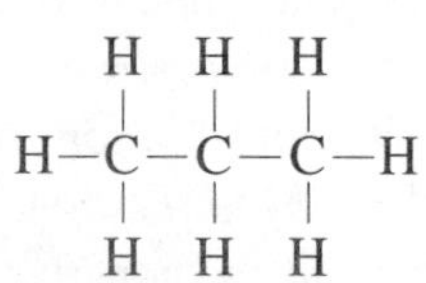

**Propane** ($C_3H_8$)

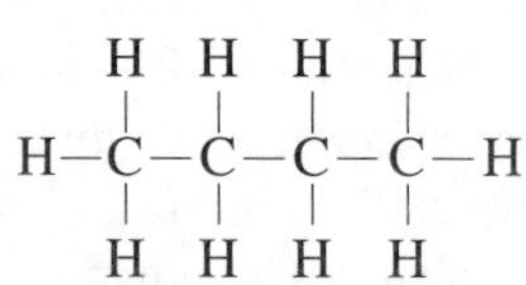

**Butane** ($C_4H_{10}$)

### 6.3.2 Olefin ($C_n H_{2n}$) Series

*Olefins are also straight chain compounds similar to paraffins but are unsaturated because of the fact that they contain one or more double bonds between carbon atoms which are capable of uniting with other materials without displacing hydrogen.* The names of the hydrocarbons having one double bond end in *'lene'* as in ethylene, hexane called mono-olefins and in *'adi-ene'* for two double bonds as in butadiene called diolefin. The chemical structure of hexene and butadiene are given below.

Diolefins are more unstable than mono-olefins. Being unstable in nature, olefins are readily oxidized in storage to form gummy deposits. That is why, olefins content in certain petroleum products is kept low by specification. *For example, in turbine fuel, it is kept less than 3%.*

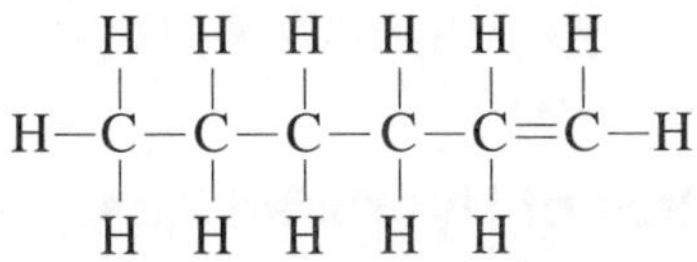

**Hexane** (Mono-olefin) ($C_6H_{12}$)

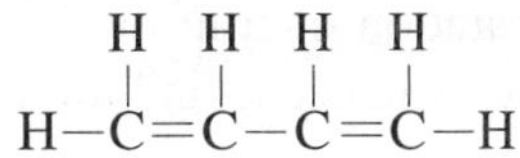

**Butadiene** (Di-olefin) ($C_4H_6$)

### 6.3.3 Naphthene ($C_n H_{2n}$) Series

*The naphthene series has the same chemical formula as the olefin series but have a ring structure.* They are sometimes also called cyclo-paraffins. The naphthene are saturated compound whereas olefins are unsaturated. Some of the hydrocarbon of naphthene series are cyclopropene ($C_3H_6$), cyclobutene ($C_4H_8$), cyclopentane, etc.

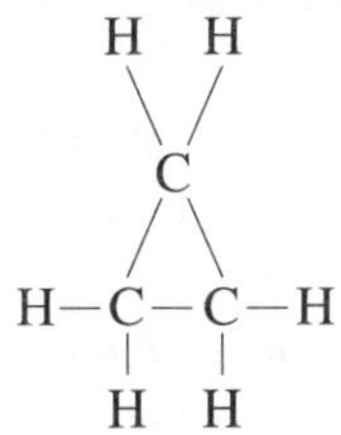

**Cyclo-propane** ($C_3H_6$)

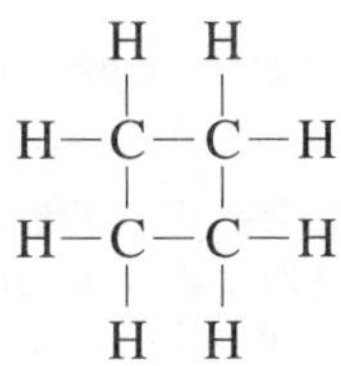

**Cyclo-butane** ($C_4H_8$)

### 6.3.4 Aromatic ($C_n H_{2n-6}$) Series

*Aromatic series is a ring structure compound having the benzene molecule ($C_6H_6$) as their central structure which have three double bonds.* Though, the presence of three double bonds indicates that they are unsaturated but there exists a peculiar nature of these double bonds which cause them to be highly stable than the other unsaturated compounds. It is possible to form to large number of aromatic compounds by replacing one or more of the hydrogen atoms of the benzene molecule with or organic radical such as paraffins, naphthenes and olefins. By addition of methyl group ($CH_3$), benzene is converted to toluene ($C_6H_5CH_3$) which is the

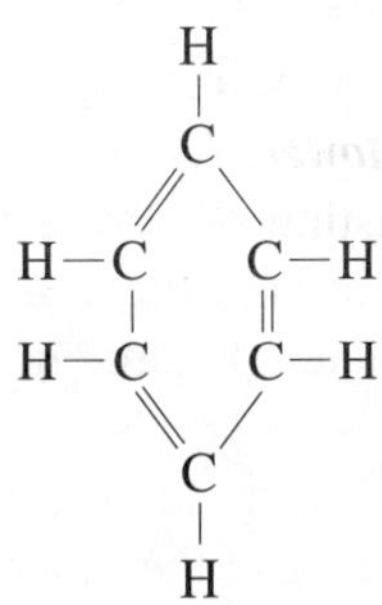

Benzene ($C_6H_6$)

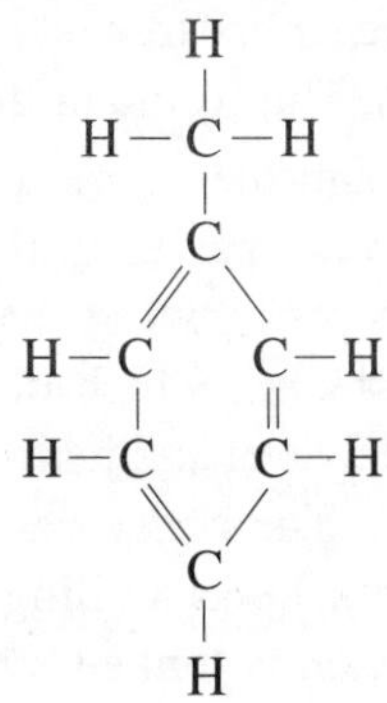

Toluene ($C_6H_5CH_3$) or ($C_7H_8$)

base for the preparation of trinitrotoluene **(TNT)**, a well-known explosive compound. The chemical structure of benzene and toluene are given above.

### 6.3.5 General Characteristics of Families of Hydrocarbons

The following are the general characteristics of families of hydrocarbons due to their molecular structure.

(a) ***Anti-knock quality for SI engines.*** The poorest anti-knock quality is exhibited by normal paraffins when used in SI engines. But the anti-knock quality improves with the increasing number of carbon atoms and the compactness of molecular structure. *The family of aromatic offers the best resistance to anti-knock when used in SI engines.*

(b) ***Anti-knock quality for CI engines.*** In the CI engines, the story is just reverse. The *normal paraffins prove to be fuels which exhibit the highest anti-knock quality while the aromatics are the least desirable fuels as they exhibit the poorest anti-knock quality.*

(c) ***Volatility.*** With the increase in number of atoms in the molecular structure the boiling temperature of fuel increases. This means that the fuels with fewer atoms in the molecule tend to be more volatile.

(d) ***Heating value.*** The *heating value of the fuel generally increases with the increase of the proportion of hydrogen atoms to carbon atoms in the molecule.* This is because of the fact that the heating value of hydrogen is more than that of carbon. Therefore, *paraffins exhibit the highest heating value while the aromatics the lowest.*

Table 6.2 gives the properties of various families of hydrocarbon fuels. (see p. 213)

## 6.4 PETROLEUM REFINING PROCESS

The crude oil is obtained from the oil wells. It contains impurities such as water, solid and gases consisting mainly of ethane and methane. The crude oil is separated into gasoline, kerosene, fuel oil, etc., by the process of ***fractional distillation*** which *is based on* the fact that the *boiling point of various hydrocarbons increases with the increase in molecular weight.*

The refining process of crude petroleum is shown in Fig. 6.1. As the petroleum is taken from the ground, the crude contains a certain amount of wet gas. In the first step, the crude is passed through a separator in which a product, known as natural gasoline, is absorbed from the wet gas removed. The resultant dry gas is known as natural gas.

The liquid petroleum then goes to **still** where it is vaporized by raising the temperature about 600°C, and vapour passes to the ***fractionating tower***. As the vapour rises through the tower, it is forced to pass through zig-zag path consisting of labyrinth like arrangement of plates or similar obstacles, which direct the vapour through trays of liquid fuel located at definite predetermined level and maintained of different temperature in decreasing order from lower to higher level. The compounds with higher boiling points are separated out a lower level while those with lower boiling points are removed at higher levels continuously. In general, the top fraction is termed *straight run gasoline* and the other fractions in descending order through the tower are *naptha, kerosene, gas oils, lubricating stocks and residue.* It is to be noted that each of these fractions covers a certain boiling point range and each may

**Table 6.2** Properties of Hydrocarbons Fuels (Adopted Form)

| S.No. | Fuel family formula | Fuel & weight | Molecular weight | Specific gravity | Boiling temp. at 1 atm °C | Ignition temp. at 1 atm °C | Constant pressure heating value kJ/kg | | Critical compression ratio | A/F ratio | Octane rating research/motor | | | |
|---|---|---|---|---|---|---|---|---|---|---|---|---|---|---|
| | | | | | | | Higher | Lower | | | 0 | 3 | 0 | 3 |
| 1 | 2 | 3 | 4 | 5 | 6 | 7 | 8 | 9 | 10 | 11 | 12 | 13 | 14 | 15 |
| 1 | Paraffins $C_n H_{2n+2}$ | Methane, $CH_4$ | 16 | 0.3 | –162 | 732 | 55010 | 49538 | 12.6 | 17.2 | 120 | | 120 | |
| | | Ethane, $C_2H_6$ | 30 | 0.37 | –89 | 504 | 51565 | 47148 | 12.4 | 16.1 | 115 | | 99 | |
| | | Propane, $C_3H_8$ | 44 | 0.5 | –42 | – | 49974 | 45979 | 12.2 | 15.7 | 112 | | 97 | |
| | | Butane, $C_4H_8$ | 58 | 0.579 | –0.6 | 431 | 49128 | 45343 | 5.5 | 15.5 | 94 | 104 | 90 | 104 |
| | | Heptane, $C_7H_{10}$ | 100 | 0.684 | 98 | 247 | 60633 | 44560 | 30 | 15.2 | 25 | 65 | 26 | 65 |
| | | Iso-Octane, $C_8H_{18}$ | 114 | 0.692 | 99.4 | 447 | 47813 | 44347 | 7.3 | 15.1 | 100 | 116 | 100 | 116 |
| 2 | Naphthene $C_n H_{2n}$ | Cyclo-heptane, $C_7H_{14}$ | 98 | 0.816 | 119 | – | 46708 | 43524 | 3.4 | 14.8 | 39 | 60 | 41 | 65 |
| | | Cyclo-octane $C_8H_{16}$ | 112 | 0.841 | 151 | – | 46725 | 43551 | – | 14.8 | 71 | – | 58 | – |
| 3 | Aromatics $C_n H_{2n-6}$ | Benzene, $C_6H_6$ | 78 | 0.879 | 80 | 592 | 41835 | 40143 | – | 13.3 | – | – | 115 | – |
| | | Toluene $C_7H_8$ | 92 | 0.867 | 111 | 568 | 42437 | 40327 | 15 | 13.5 | 120 | 120 | 109 | 113 |
| 4 | Alcohol | Methanol $CH_3$ OH | 32 | 0.792 | – | – | 22726 | 20105 | – | 6.4 | 107 | – | – | – |

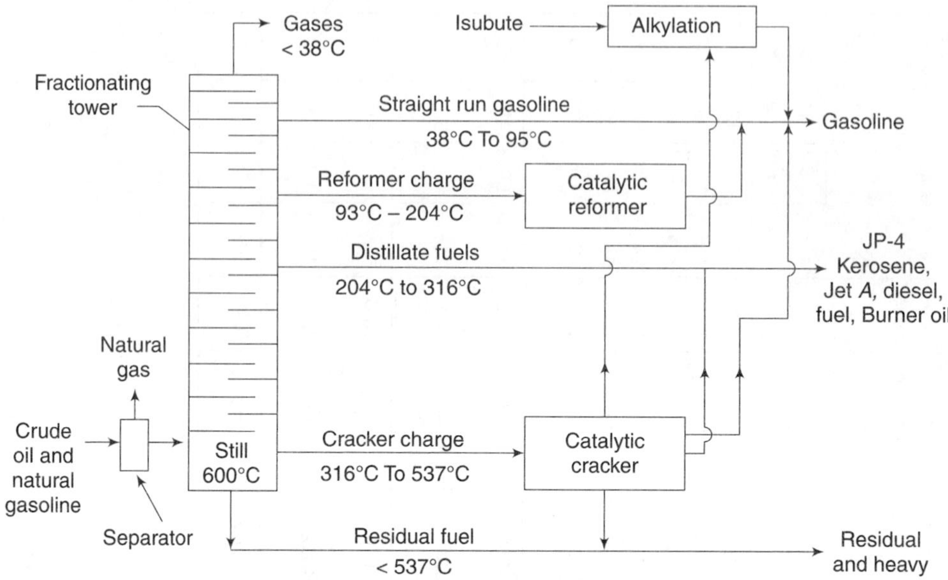


**Fig. 6.1** Petroleum refining process

be further refined by again fractionating within a narrow range of boiling points. *Table 6.3 shows the important products of refining process.*

These are some commercial products in great demand which are not readily available by fractional distillation process. To meet this demand, the following **processes** can be used.

**Table 6.3** Products Obtained from Refining Process

| *S.No.* | *Fraction* | *Approximate boiling range, °C* | *Remarks for applications* |
|---|---|---|---|
| 1. | Fuel gas | −160 to 44 | Methane, ethane and some propane used as refinery fuel |
| 2. | Propane | −40 | Liquefied petroleum gas (LPG) mostly used as cooking gas |
| 3. | Butane | −12 to 30 | Blended with motor gasoline to increase its volatility |
| 4. | Light Naphtha | 0 to 150 | Motor gasoline for catalytic reforming |
| 5. | Heavy Naphtha | 150 to 200 | Catalytic reforming fuel, blended with light gas oil to form jet fuels |
| 6. | Kerosene-Middle distillate | 200 to 300 | Domestic and aviation fuels |
| 7. | Light gas oil Middle distillate | 200 to 315 | Furnace fuel oil and diesel fuels |
| 8. | Heavy gas oil | 315 to 425 | Feed for catalytic cracking |
| 9. | Vacuum gas oil | 426 to 600 | Feed for catalytic cracking |
| 10. | Pitch | >600 | Heavy fuel oil and asphalts |

(i) **Cracking.** It consists of breaking down large and complex molecules into lighter and simpler compounds with lower boiling points. *Thermal cracking* subjects the heavy hydrocarbons to rather high temperatures and pressure. *Catalytic cracking* is performed at somewhat lower temperature and pressure but due to presence of catalytic agent, the fuels produced exhibit higher anti-knock qualities.

(ii) **Hydrogenation.** In this process, hydrogen atoms are added to certain hydrocarbons, under high pressure and temperature, to produce more desirable compound. Conversion of unstable to stable compound is often done by this process.

(iii) **Polymerization.** It brings together light, unsaturated gases of one family, in the presence of catalyst, to produce liquid.

(iv) **Alkylation.** It combines light gases of different families in the presence of a catalyst. In general, an olefin is combined with a paraffin to give a branch chain paraffin.

(v) **Isomerization.** This process changes the relative position of the atoms within the molecule of a hydrocarbons without changing its molecular formula. It produces isomers of the original hydrocarbons.

(vi) **Cyclization.** This process essentially joins together the ends of a straight chain molecule to form a ring compound of the naphthene family.

(vii) **Aromatization.** This process is similar to cyclization with the exception that the product is an aromatic compound.

(viii) **Reforming.** It is a type of cracking process in which naptha or straight run gasoline is converted into gasoline of higher octane rating.

(ix) **Blending.** This is a process of mixing certain refinery products to obtain a commercial product of desired quality.

## 6.5 IMPORTANT QUALITIES OF SI FUELS

Present day, SI engines use gasoline fuels. ***In fact, gasoline is a blend of several refinery products containing paraffins, naphthenes and aromatics in varying proportion.*** A particular blend depends upon the characteristics desired by the fuel. The following are the most important qualities in the selection of gasoline fuels for SI engines: Fig. 6.2 shows the ASTM distillation curve.

**1. Volatility.** Volatility is one of the most important properties of a gasoline as it produces important effects on several phases of the operation and maintenance of the engine. Since gasoline is a mixture of many hydrocarbons with different boiling points, hence, its constituents will boil at a wide range of temperature. Volatility is usually measured by distillation method.

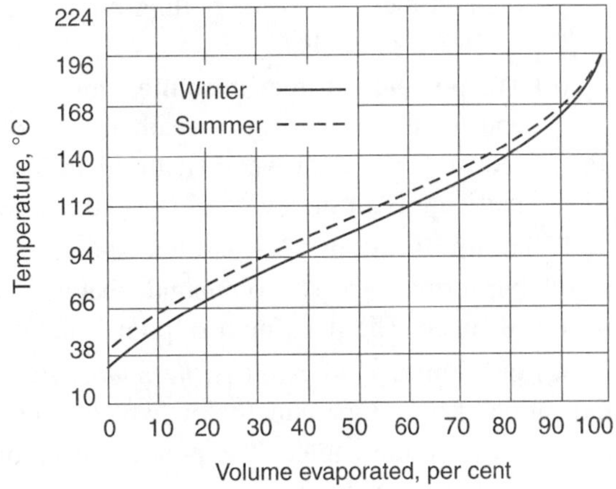


**Fig. 6.2** ASTM distillation curve

The following are the important aspects of volatility:

(a) **Starting and warm up.** For ease in starting, it is necessary that some of the gasoline should vaporize at the starting temperature. This means that the portion of the distillation curve between about 9 and 10 per cent boiled off, should have a relatively low boiling temperature. As the engine warms up, the temperature gradually increases to the operating temperature. For best warm up, low distillation temperatures are desirable throughout the range of the distillation curve. The maximum boiling point is 215°C.

(b) **Operating range, performance, acceleration and distribution.** To obtain good vaporization for gasoline, it is desirable to have low distillation temperature well within the engine's operating range. If the vaporization is better, there is a more uniform distribution of fuel to the cylinders and better acceleration characteristics are obtained by reducing the quantity of liquid droplets in the intake manifold.

(c) **Crankcase dilution.** The presence of liquid gasoline in the cylinder produces undesirable effect such as : it washes away the lubrication oil from cylinder wall thereby impairing the lubrication and causing damage to the engine through increased friction between the piston rings and the cylinders. Further, it may also dilute the lubricating oil supply, resulting in the possibility of further lubricating troubles as it weakens the oil film between rubbing surfaces. In order to prevent these undesirable effects, the upper portion of the distillation curve should exhibit a sufficiently low distillation temperature to ensure that all the gasoline in the cylinder is vaporized.

(d) **Vapour lock characteristics.** When the distillation temperatures of gasoline are too low, the gasoline vaporizes in the fuel system. And where the rate of vaporization is too great, it can upset the carburettor metering or even stop the fuel flow to the engine by setting up a vapour lock in the fuel passage. To avoid these difficulties, it is desirable to leave a high boiling temperature throughout the distillation range. But this requirement is not consistent with the requirement in (a), (b) and (c) above. Thus, any particular gasoline must be a compromise of the desired distillation temperatures. The vapour lock characteristics of a gasoline is especially important in aircraft where high altitudes lower the boiling point of the fuel. The maximum Reid vapour pressure of 380°C is 0.7 bar.

(e) **Winter and summer gasoline.** During summer, the atmospheric temperature is high, and therefore, for warm weather, the volatility of an automotive gasoline should be reduced. For cold weather, the distillation curve for the gasoline must be lowered particularly in the range of 0 to 10 per cent to ensure easy starting.

**2. Sulphur Content.** The gasoline should contain only a permissible amount of sulphur, as sulphur forms corrosive compounds that may attack the various parts of the engine causing serious damage. The maximum sulphur content in per cent by weight is 0.25.

**3. Pour Point.** *Pour point is the temperature at which the fluidity ceases.* Thus, it is an indication of the temperature below which it may not be possible to use gravity feeding from the reservoir to the engine. The pour point is important only when the engine is to be used at low temperature. In such cases, the pour point of the oil should be 5 to 7°C below the operating temperature.

**4. Flash Point.** *The flash point is the lowest temperature of the fluid at which inflammable vapours may be formed.* It is important from the point of view of safety and it used as measure of fire hazard. The fuel should have an optimum value of flash point.

**5. Ash Content.** The ash content should be minimum as it increases engine wear and reduces calorific value of fuel.

**6. Gum Deposits.** Some of the unsaturated hydrocarbons have a tendency to oxidize during storage and form a product known as ***gum***. The presence of gum in the fuel causes undesirable deposits on the intake valve, rings, and other engine parts, as well as clogging in carburettor jets. Therefore, there should be a limit to gum content in the gasoline specifications.

**7. Anti-knock Quality.** *The best fuel for SI engines will be the one which has the highest Octane number of PN.* This is so because of the fact, that with a higher octane number of PN a higher compression may be used. This improves the thermal efficiency and power out of the engine. The minimum value of Octane number by Research Method (RM) is 85.

**8. Viscosity.** The viscosity of the fuel must be optimum. Any excess of viscosity results in bad atomization of the fuel.

Table 6.4 shows the specifications of gasoline.

**Table 6.4** Specification of Gasoline (Average) ($C_8$ $H_{18}$)

| *S.No.* | *Characteristics* | *Requirements* |
|---|---|---|
| 1. | Composition (%) Paraffin<br>Naphthenes<br>Olefins<br>Aromatics | <br>51.55<br>30.6<br>17.9 |
| 2. | Octane number (Research Method), Minimum | 84 |
| 3. | Cetane Number | – |
| 4. | Specific gravity | 0.76 |
| 5. | API gravity at 15°C | 57.2 |
| 6. | Viscosity at 10°C<br>40°C | 0.725<br>0.552 |
| 7. | Distillation range, °C<br>Initial BP<br>10%<br>50%<br>90%<br>B.P. | <br>70<br>84<br>120<br>170<br>212 |
| 8. | Pour point; max, °C | 5 to 7 |
| 9. | Flash point, Abel, °C, min | – |
| 10. | Sulphur, total % by weight, max | 0.25 |
| 11. | Lead content g/L, max | 0.56 |
| 12. | Vapor pressure (Reid) at 380°C, bar | 0.7 |
| 13. | Heating value ; MJ/kg | 41.9 to 44 |
| 14. | Self ignition temperature, °C | 300 – 450 |

## 6.6 KNOCK RATING OF SI FUELS (OCTANE NUMBER)

The knocking tendency of SI fuel is generally rated in terms of ***octane number (ON)***. *Thus, octane rating is a measure of the ability of the fuel to resist detonation during combustion.* The rating of a particular SI fuel is accomplished by comparing its performance with that of standard reference fuel which is usually a combination of *iso-octane and normal* heptanes.

*Iso-octane, being a very good anti-knock fuel is arbitrarily assigned a rating of 100 octane number. Normal heptane on the other hand, has very poor anti-knock qualities and is assigned a rating of 0 octane number.* By mixing these two fuels in varying proportions (by volume) any number of blends is possible which will have anti-knock qualities varying from 0 to 100 octane number. *Thus, the per centage, by volume of iso-octane in a mixture of iso-octane and normal heptanes which exactly matches the knocking intensity of a given fuel, in a standard engine, under the given standard operating conditions termed the octane number of rating of the fuel.* Thus, an octane rating of 70 means that the fuel under test has the same knocking tendency as a mixture of 70% iso-octane and 30% normal heptane. Similarly, an octane number of 50 means that the fuel under test has the same knocking tendency as a mixture of 50% iso-octane ($C_8H_{18}$) and 50% normal heptane ($C_7H_{16}$). ***Iso-Octane and normal heptane*** *are known as primary reference fuels. The fuels which are used to match with these primary fuels are called secondary reference fuels.*

The higher the octane number rating of a fuel with greater will be its resistance to knock and, the higher will be its compression ratio which may be used without knocking. Since, power output and specific fuel consumption are functions of compression ratio, it can be said, that these are also functions of the octane number rating.

Sometimes, *certain anti-knock additives, called* **dopes**, *are added in small proportion to the fuel which results in a considerable improvement of the knocking rating of the fuel. For example, and addition of 5 ml of tetraethyl lead* **(T.E.L.)** to one gallon of 80 octane period increases the octane number to 90. The anti-knock quality of fuel above 100 octane number is measured in terms of millilitres.

The anti-knock effectiveness of tetraethyl lead added to fuel is shown in Fig. 6.3. It is obvious from the diagram that the anti-knock effectiveness of tetraethyl lead for the same quantity of lead added, decreases as the total content of lead in the fuel increases. Putting it in a different way, one millilitre of lead added to pure iso-octane will have considerably greater anti-knock effect in comparison

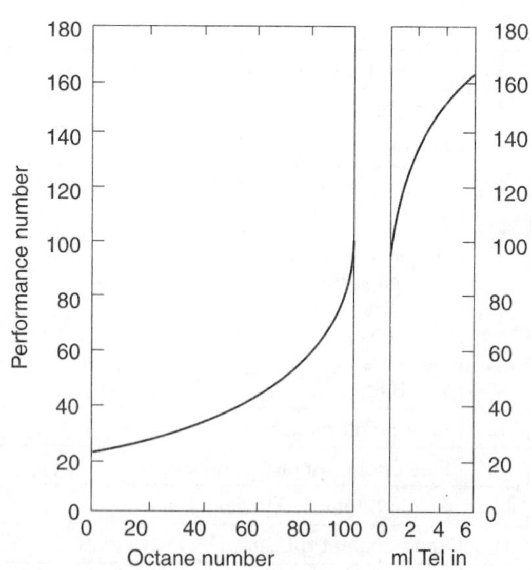


**Fig. 6.3** Peformance number

to the case when the same quantity of lead is added to iso-octane already containing 4 millilitres of lead. Furthermore, one octane number between say 90 and 91 will produce greater anti-knock effect than a similar increase between say 30 and 31 octane numbers. Owing to this non-linear variation a new scale "performance number" **(PN)** has been devised which expresses the approximate relative engine performance.

The **performance number (PN)** *of a fuel is based on the indicated power produced in a supercharged standard engine operating under prescribed conditions.* It represents the maximum power which may be obtained with that fuel, without knock relative to the maximum power which may be obtained using iso-octane, also without knock. Iso-octane is assigned a PN values of 100. *Thus, a supercharged engine having PN value of 130 means that it can produce approximately 1.3 times as much power (without knock) as it can develop with a 100 PN fuel, namely iso-octane (without knock).* The variation of PN with octane number is shown in Fig. 6.3.

Octane numbers, (ON) above 100 can be computed by

$$ON = 100 + \frac{PN - 100}{3}.$$

### 6.6.1 Experimental Method for Octane Number

The octane number of a fuel is determined by testing the fuel in a variable compression CFR (Co-operative Fuel research) engine under specified conditions. There are four methods of testing the octane number–(i) *Research method*, (ii) *Motor method*, (iii) *Aviation method* and (iv) *Supercharging method.*

The engines for the *Motor, Research and Aviation methods* are quite similar, with one cylinder, overhead valves and variables compression ratio (four stroke). The variable compression ratio is obtained by making the head and cylinder in one piece. This cylinder can be raised or lowered by a hand crank thus varying compression volume and thereby compression ratio since the stroke of the piston is constant. The speed is held essentially constant by a synchronous generator that is driven by an engine and connected to an A.C. power lines. The carburettor has three bowls and the fuel feed from each bowl can be varied thus varying the air-fuel ratio. Generally, the test fuel is placed in one of the bowls while blends of the references fuels are placed in the other two bowls. A bouncing pin F

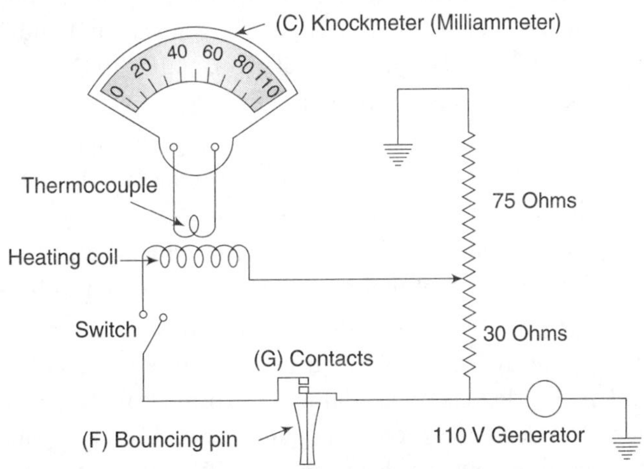


**Fig. 6.4** Knock meter circuit

as shown in Fig. 6.4 records the knock in motor or research method. Pin (P) is mounted on the engine, so that, the diaphragm is directly exposed to combustion pressures. When knock occurs the pin F bounces and closes the contact G, the greater the knock, the longer will be the time that the contacts will be closed.

For *motor and research method* the test procedure is as follows: The engine is run on the test fuel and the fuel-air ratio is adjusted to give the maximum knock intensity. The compression ratio is then varied until the knock is standard (55 units). The compression ratio is now kept fixed at this value and the engine is run, in turn, on two reference fuel blends of known octane number which are kept in the two other bowls. The knock intensity is noted and the test is carried out with different blends of reference fuels till the knock intensity of test fuel is matched with the reference blends. The difference in the octane number of the two reference blends should not exceed 2. Now, by linear interpolation of knockmetre readings the octane number of test fuel is found.

## 6.7 DOPES

*Dopes are substances which when added to fuels in even a very small proportion produces a pronounced anti-knock effect.* The most important dopes (anti-knock additive) is *tetra-ethyl lead (TEL)*, Pb $(C_2H_5)_4$. Commonly, *TEL* is mixed with a proportion of ethylene dibromide, $C_2H_4Br_2$ and is known as *ethyl fluid.* The addition of ethylene dibromide in TEL prevents the deposition of lead inside the engine after the deposition of tetra-ethyl lead which occurs at a temperature above 210°C. Due to the presence of lead and bromine, it forms lead-bromide which moves away with the exhaust gas through discharge valve.

The addition of ethyl fluoride in the petrol, even in the small proportion of 5 cc per gallon, reduces the knocking tendency appreciably and thus increases the value of octane number.

There are other types of dopes also such as $Fe(CO)_5$ and $Ni(CO)_4$ but they are not as effective as TEL. In addition to organic metallic compounds, non-metallic substances also exhibit the phenomenon of dope. They are methyl aniline ($C_7H_9N$), ethyl aniline ($C_8H_{11}N$), toluidine ($C_7H_{11}N$), xylidine ($C_8H_{11}N$), aniline ($C_6H_7N$), phenol ($C_6H_6O$) and benzene ($C_6H_6$). But they are not as effective as TEL.

In some fuels, especially those having an appreciably aromatic content, tetra-ethyl lead, Pb $(CH_3)_4$ has been found to have advantages over tetra-ethyl lead. The superiority of TML to TEL is due to its higher stability.

The mechanism of action of **tetra-ethyl lead is** as follows: The sequence of events leading to hot flame (or fuel ignition) may be considered to be as follows:

(i) *Pre-cool flame reaction, (ii) Appearance of cool flames, (iii) Induction period and (iv) Hot flame.* It has been found that during the pre-cool flame reactions, organic peroxides, aldehydes, ketones and olefins are formed along with small amounts of hydrogen peroxide. The decomposition of organic peroxides takes place during the cool flame region, and the reaction between their decomposition products lead to the formation of formaldehydes, hydrogen peroxide and free radical having tremendous energy. These free radicals give rise to chain-branching reaction, and thus, an explosion.

It has been found experimentally that the addition of TEL has very little effect on the point at which the cool flame appears, but it displaces that auto-igntion (not flame) towards the higher temperature and pressure. It has also been found that TEL tends to eliminate hydrogen peroxide, and thus, its main activity is deactivating the free radicals formed by decomposition of organic peroxides and suppressing chain branching reactions. However, it is believed that it acts primarily as a deactivator of radicals, such as OH, H and $H_2O$. It has also been found that TEL decomposes and oxidizes to form PbO before it becomes active.

## 6.8 INHIBITORS

*Inhibitors are antioxidants which retard the formation of olefin peroxides when added to the fuels of SI engines.* Olefin peroxides are responsible for formation of gum. An example of an *inhibitor is benzyl amino phenol. It prevents gum formation when added to fuels.* Gene-rally, 1 c.c. of inhibitor for 1 litre of the fuel is recommended.

## 6.9 IMPORTANT QUALITIES OF CI ENGINE FUELS ($C_{10}H_{20}$)

Most CI engine fuels (diesel oils) are obtained from fractions of crude petroleum. Note, *the diesel oils are heavier and more viscous than gasoline used in the* SI engines. The following are the important characteristics of CI engines fuels.

**(i) Ignition lag**. Since knock in CI engine can be controlled by decreasing the ignition lag, so CI engine fuel should have short ignition delay. Further, ignition lag controls the starting warm up and smoke. Ignition lag is generally measured in terms of cetane number.

**(ii) Volatility.** Volatility is necessary for good mixing and combustion. Hence, a good fuel should be sufficiently volatile in the operating temperature range.

**(iii) Viscosity.** *The viscosity of fuel in CI engine is more than in SI engines.* They should however, be able to flow through the fuel system and strainers under the lowest operating temperature to which the engine may be subjected.

**(iv) Impurities.** To prevent excessive engine wear, impurities (i.e., solid particles) must be reduced. *Impurities in the CI engine fuel is more than SI fuel.*

**(v) Ash content.** The ash content of the burnt fuel should be minimum.

**(vi) Pour point.** The pour point is important when the engine is used at low temperatures. The fuel should have an optimum value of pour point.

**(vii) Flash point.** There should be an optimum value for flash point as it is important from safety and fire hazard point of view. **Table 6.5** shows the properties of diesel fuels ($C_{10}H_{20}$).

**(viii) Gravity.** The gravity of a fluid is a measure of its density. Specific gravity is defined as the ratio of density of the substance to the density of water, preferably, at the same temperature (15°C). As per American Petroleum Institute (API)

$$\text{API gravity} = \frac{145.5}{\text{specific gravity at } 15°\text{C}} - 131.5$$

As fuel oils are purchased on a volume basis (in litres), API gravity is important for obtaining the mass of purchased fuel.

**(ix) Ignition quality.** *The ignition quality of a diesel fuel is its ability to auto-ignite inside the engine cylinder in the presence of air under different operating conditions and is measured by delay period.* The delay period depends upon self ignition temperature of the oil. Good ignition quality means short delay angle. The ignition quality of diesel oil is measured by its **centane number** (CN) which is discussed in subsequent article.

**Table 6.5** Specifications of Diesel Fuels (Average) ($C_{10}$ $H_{20}$)

| *S.No.* | *Characteristics* | *Requirements* | |
|---|---|---|---|
| | | *High speed diesel (HSD)* | *Light duty diesel (LDD)* |
| 1. | Composition (%), paraffin<br>Naphthenes<br>Olefins<br>Aromatics | <br>66<br>7<br>17 | <br>–<br>–<br>– |
| 2. | Acidity, total mg of KOH/g, max | 0.5 | – |
| 3. | Ash (%) by mass, max | 0.01 | 0.02 |
| 4. | Carbon residue (Ramsbottom) (%) by mass, max | 0.2 | 1.5 |
| 5. | Cetane number, min | 42 | – |
| 6. | Pour point, max | 6°C | 12°C for winter and 18°C for summer |
| 7. | Copper strip corrosion for 3 hr at 100°C | 90 | – |
| 8. | Flash point<br>(a) Abel, °C, min<br>(b) Pensky, °C, min | <br>38<br>66 | <br>–<br>– |
| 9. | Kinematic viscosity CS, at 38°C | 2 to 7.5 | 2.5 to 15.7 |
| 10. | Sediment, (%) by mass, max | 0.05 | 0.10 |
| 11. | Total sulphur, (%) by mass, max | 1.0 | 1.8 |
| 12. | Water content, (%) by vol., max | 0.05 | 0.25 |
| 13. | Total sediments, mg/100 ml, max | 1.0 | – |
| 14. | Heating value (MJ/kg) | 42.9 | 53 |

The ignition qualities of a diesel fuel has a marked influence on cold starting, compression ratio and engine roughness.

**(a) Cold starting.** As diesel fuels are less volatile and more viscous than gasoline, CI engines experience difficulties during cold weather. In order to overcome this problem, a high cetane number fuel (i.e., self-ignition temperature) is recommended. The other alternatives are either use of ether, a volatile fuel with poor anti-knock quality or use of glow plug when engine is cold.

**(b) Compression ratio.** The engine efficiency increases with increase in compression ratio but it will increase the pressure inside the cylinder. As a result, the compression ratio of a diesel engine is kept to a minimum value at which a given fuel assures good starting. If the cetane number is high, the required compression ratio will be lower.

**(c) Engine roughness.** The term engine roughness signifies the intensity of vibration of various parts caused by high rates of pressure rise in the cylinder. The stiffness of the various engine parts decides the intensity of vibration and therefore the effect of cetane number on the rate of pressure rise will vary with the engine design.

**(d) Corrosion and wear.** The fuel for CI engine should not give rise to corrosion and wear before and after the combustion. These requirements can only be met with a minimum amount of sulphur, carbon residue and ash in the fuel. The percentage of sulphur in diesel oils is higher than in gasoline. As a result of combustion, sulphur trioxide is formed due to presence of excess air present inside the cylinder. These trioxides may attack the lubricating oils on the cylinder walls. Further, it may unite with $H_2O$ to form $H_2SO_4$ resulting in corrosion and wear of the engine parts. In view of this, sulphur content in the oil should not exceed 1 per cent.

**(x) Cloud Point.** It is the temperature below which the wax content of the fuel separates out in the form of solid. The waxy solid will clog the fuel line and the fuel filters. Higher the cetane number, lower is the rate of formation of deposits.

**(xi) Cleanliness.** Any dirt is undesirable in diesel oil as it will cause a rapid wear in the injection system. Diesel oil is more viscous than gasoline and it has tendency to hold solid particles in suspension. *This suggests a high degree of filtration is required.*

## 6.10 KNOCK RATING OF CI FUELS

The knock rating of CI fuels is expressed by ***cetane number*** **(CN)**. The rating of a particular CI fuel is accomplished by comparing its performance with reference fuel. These fuels are *n-cetane* with a cetane number of 100, and *alpha methyl naphthalene* with a cetane number of 0. A cetane rating of 60 indicates that fuel has the same ignition characteristics in a standard engine under definite operation conditions as a mixture, by volume of 60 parts of **cetane** and 40 parts of **alphamethyl naphthalene**. It is to be noted, that the fuel is not rated for knock but rather for it ignition characteristics, since ignition delay is major factor that controls the knock in CI engine.

Figure 6.5 shows the effect of cetane number on crank time. It is observed that with the increase in cetane number cranking time reduces.

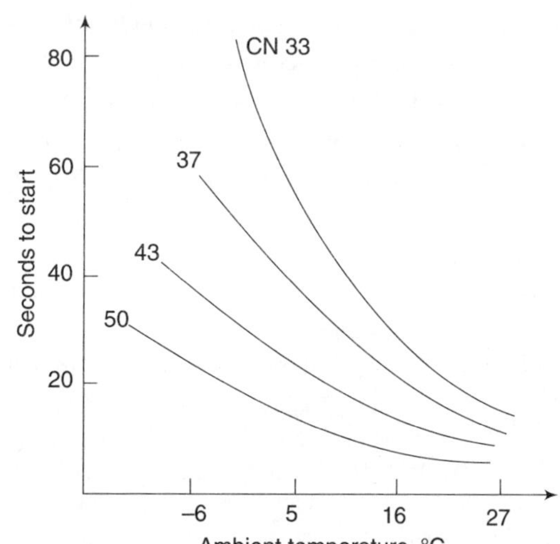


**Fig. 6.5** Effect of cetane number on cranking time

The cetane number of an unknown fuel is determined by testing the fuel in a **variable compression CFR engine operating** under the following conditions.

Engine speed = 900 rpm  Inlet air temperature = 100°C

Jacket water temperature = 65.5°C  Injection timing = 15°b.TDC

Ignition delay period = 13°

The arrangement consists of *three fuel reservoirs*; the test fuel is kept in the central reservoir, and the two other reference fuels are kept in the other two. The engine is first run on the unkonwn fuel with injection timing set at 13° bTDC and the injection pump delivering fuel at specified rate. By varying the compression ratio, the ignition delay is increased or decreased until a portion is found, the test fuel undergoes a 13 degree delay period and, for this condition, the compression ratio is noted. The test is now repeated with the two reference fuel blends, differing in ignition quality by not more than 8 cetane numbers and the compression ratio at which these fuels give 13 degree delay are noted. By interpolation of the compression ratio, the cetane number of the unknown fuel is then determined. Another method for finding the cetane number of fuel is to prepare a graph between cetane number and compression ratio for 13 degree delay period, using fuels of known cetane number. From this, the cetane number corresponding to the compression ratio of the test fuel at 13 degree delay period can be read.

## 6.11 FUELS FOR GAS TURBINES AND JET ENGINES

In gas turbines, the combustion is continuous instead of intermittent which is found in piston-cylinder IC engines. Since burning is continuous, so *properties involving ignition delay (octane and cetane numbers) are not important. This suggests that gas turbine fuels can be made cheaper.*

The main requirements of fuels for power gas turbines is that the fuel and the production of combustion should not cause *corrosion and erosion of nozzles and turbine blades and should not deposit in the combustion chamber and fuel system.* Aviation gas turbines generally confront problems like cold starting, flame blow out at high altitude, evaporation loss at low pressure and boiling due to high stagnation temperature.

*The most common aviation gas turbine fuels are jet propellant —4 (i.e., JP-4) and JP-5, JP-4 is a kerosene/gasoline mixture with high percentage of low octane gasoline whereas JP-5 is also a kerosene/gasoline mixture but with lower gasoline content.* The early jet engine was using kerosene fuel (JP-1).

Table 6.6 shows the characteristics of a fuel of aviation gas turbines.

**Table 6.6** Characteristics of Aviation Turbine Fuels

| *S.No.* | *Characteristics* | *Requirements* | | | |
|---|---|---|---|---|---|
| | | *Jet A* | *JP-1* | *JP-4* | *JP-5* |
| 1. | Flash point, °C, min-max | 43–46 | 43 (min) | – | – |
| 2. | Freezing point, °C max | −40 | −60 | −60 | −48 |
| 3. | Gravity, API (min, – max) | 39–51 | 35 (max) | 45–57 | 36–48 |
| 4. | Reid vapour pressure, bar | – | – | 0.14–0.21 | – |

*Contd...*

*Table 6.6 Contd...*

| | | | | | |
|---|---|---|---|---|---|
| 5. | Distillation, °C | | | | |
| | 10% max | 204 | 210 | – | 204 |
| | 20% max | – | – | 143 | – |
| | 50% max | 232 | – | 188 | – |
| | 90% max | – | 254 | 243 | 288 |
| | EP max | 288 | 300 | – | – |
| 6. | Heating value (lower) MJ/kg, min | 42.92 | 42.7 | 42.92 | 42.7 |
| 7. | Sulphur, % by mass, max | 0.3 | 0.2 | 0.4 | 0.4 |
| 8. | Smoke point, mm (min) | 25 | – | – | 20 |
| 9. | Aromatic, vol (%), max | 20 | 20 | 25 | 25 |
| 10. | Potential gum, mg/100 ml (max) | 14 | 8 | 14 | 14 |

## 6.12 ALTERNATIVE FUELS FOR IC ENGINES

Depletion of petroleum fuels and their rising prices have forced the engineers to think in the direction of non-petroleum fuels as the fuels of IC engines. They are also used for blending purposes for improving the quality of fuels. The following are the non-petroleum or alternative fuels for IC engines fuels.

**(a)** ***Biofuels.*** Biofuels are liquid fuels made from easters, alcohols, ethers and other biomass chemicals. They are renewable fuels that can be produced in any climatic condition using already developed agricultural practices. Common biofuels include **ethanol and biodiesel.** Ethanol is made from sugars or starch, typically from grain or corn. Sugar mills are being encouraged to produce ethanol from sugarcane for blending with petrol, while efforts are on to cultivate such crops as **Jatropha** and **Pongamia**, which yield oil that can either be blended with diesel or even used independently.

While ***Jatropha*** has clearly emerged as the preferred option for cultivation **Pongamia**, a traditional species that has been around ages, too has great potential. Jatropha is a hardy plant which can grow in any soil and under any climatic condition. It grows up to a height of 1.65 m. It needs little water and has a life span 7.45 years. More than 95% of its oil is extractable and can be used as fuel to replace diesel.

Rupees 1200 crore is earmarked for nursing and plantations these plants by Govt. of India. A legislature has been passed to use B5 (5% blend) and successfully $B_2O$ (20% blend) as mandatory all over India.

Biofuels reduce negative environmental effects through lower emissions and alimatic impact. Local production of bio-energy-alcohol, biodiesel, bio-oil is projected to have a broad range of positive economic, social and environmental implications.

The cost of biodiesel is expected to reach between ₹15 to ₹17 per litre of oil.

Indian Oil Corporation, Railway, ONGC, Gujarat Govt. have started using biofuels.

**(b)** ***Benzol.*** It is a mixture of about 70% benzene ($C_6H_6$), 20% toluene ($C_7H_8$), 10% xylene ($C_8H_{10}$) and some sulphur content. Though, it possesses a high anti-knock quality but due to its freezing point (5.5°C), it is not suitable for cold climate. Moreover, the heating value is very low. Generally, it is used as a blending agent with gasoline.

**(c) *Alcohol.*** Methanol ($CH_3OH$) and ethanol ($C_2H_5OH$) are the two forms of alcohol which are used in racing cars due to its high rate of evaporation. This is also used in blending.

**(d) *Gaseous Fuels.*** The gaseous fuels include natural gas, LPG, producer gas, $H_2$, blast furnace gas, bio-gas and coke-oven gas. Among these, natural gas and LPG have the potential to be used in IC engines. *The approximate composition by volume of natural gas is 80% methane, 7% ethanes, 6% propene, 1.5% isobutene, 2.5% butane and 3% pentene plus vapour.* Natural gas is suitable for stationary engine while LPG is suitable for mobile engines.

**(e) *Ammonia.*** A possibility of using ammonia as IC engine fuels has been reported but still research is going on for the concrete results.

### 6.12.1 Alcohol

Alcohol falls under the category of bio-fuel. The pure alcohol is of organic origin and may be produced from a number of different organic sources such as grains, potatoes, corn, sugar, or by fermentation of other products. Alcohol is being considered an engine fuel or more than fifty years and has been used in France and Germany during World Wars I and II.

The main advantages of using alcohols as an engine fuel lies in its high knock resistance characteristics resulting in the use of high compression ratio engine. The main disadvantage is the lower heating value.

*Alcohol is a saturated compound and is designated as ethanol (Ethyl-alcohol, $C_2H_5OH$) or methanol (Methylalcohol, $CH_3$ OH).* Table 6.7 gives a comparison of physical properties of methanol and ethanol with gasoline.

**Table 6.7** Comparison of Physical Properties of Methanol and Ethanol with Gasoline

| *Property* | *Methanol* | *Ethanol* | *Gasoline* |
|---|---|---|---|
| Chemical formula | $CH_3OH$ | $C_2H_5OH$ | $C_8H_{18}$ |
| Molecular weight | 32 | 46 | 114 |
| Density | 0.795 gm/cc | 0.79 gm/cc | 0.69 gm/cc |
| Latent heat of evaporation | 1104 kJ/kg | 853 kJ/kg | 292.7 kJ/kg |
| Lower heating value | 19.64 MJ/kg | 26.76 MJ/kg | 43.91 MJ/kg |
| Mixture heating value (MJ/kg of air) | 3.07 | 2.97 | 2.93 |
| Stoichiometric A/F | 6.4 | 8.96 | 15.3 |
| Ignition limits | 3.15 to 12.8 | 3.5 to 17.0 | 7.0 to 22.0 |
| Self-ignition temperature | 574°C | 537°C | 335°C |
| Octane no. (Research) | > 100 | > 100 | 92 |
| Octane no. (Motor) | 92 | 89 | 80 |

#### 6.12.1.1 Manufacture of Methanol

Methanol or methyl alcohol can be produced from coal. Pulverized lignite or coal is fed to steam/oxygen blown gasifiers (partial combustors) to produce a synthesis gas consisting of CO and $H_2$.

$$H_2O + C = CO + H_2$$

Further, CO and $H_2$ combine to produce $CH_3$ OH in the presence of some catalyst.

$$CO + 2H_2 = CH_3\ OH$$

#### 6.12.1.2 Manufacture of Ethanol

Ethanol or ethyl alcohol can be produced by fermentation of carbohydrates which occur naturally and abundantly in some plants like sugarcane and from starchy materials like corn and potatoes. This means that these fuels can be produced from highly reliable and everlasting raw material sources.

Ethanol is produced by fermentation of carbohydrates by following relation (due to gas)

$$C_6H_{12}O_6 \text{ —— } 2C_2H_5OH \rightarrow 2\ CO$$

#### 6.12.1.3 Advantages and Disadvantages of Alcohol as IC Engine Fuels

Alcohol may be used in both SI and CI engines.

**Advantages**

At the end of combustion of alcohol, greater pressure is generated which results in greater power output and that is why alcohol is used in racing cars.

**Disadvantages**

(a) Due to lower calorific value 1.64 to 2.2 times of alcohol as compared to gasoline is needed to generate equal amount of energy.

(b) The lower vapour pressure and boiling behaviour together with high latent heat of vaporization make vaporization of fuel more difficult and thus put higher demand on the mixture formation system, both in cold starting and warm running condition.

Due to above disadvantages straight alcohols are not normally used in automobile engine except in racing cars. It is the alcohol-gasoline fuel blend which is used in some scale and has a bright future.

### 6.12.2 Hydrogen

*Hydrogen fuel is supposed to be a clean fuel.* Hydrogen powered IC engines came into existence about seventy years ago. Again, there is a revival of interest in the use of hydrogen gas as SI engine fuel mainly due to the following facts:

(i) Hydrogen can be manufactured from an unlimited source.

(ii) It can be burnt easily and products of combustion simply contains $H_2O$ and not HC, $CO_2$, CO and $NO_x$.

Generally, the compression ratio used in hydrogen power engine is around 12:1.

**Advantages**

(i) The thermal efficiency is higher because of higher compression ratio.

(ii) Lean mixture can be burnt more efficiently because of wide flammability limits.

(iii) Higher flame speed allows the spark timing to be closer to TDC resulting in closure to air standard cycle.

(iv) Hydrogen diffuses rapidly to produce a uniform mixture in the combustion resulting in complete combustion.

(v) The calorific value of hydrogen is higher than gasoline.

**Disadvantages**

(i) The fuel system requires expensive and complex auxiliary equipment to inject the gas in the cylinder without backfiring.

(ii) The production of hydrogen is costly.

(iii) The storage and transportation of hydrogen pose great problems.

The **physical properties** of hydrogen fuel are as follows:

| | | |
|---|---|---|
| Boiling point | : | –253°C |
| Heat of vaporization | : | 443.26 kJ/kg at 20°C |
| Lower heating value | : | 119.9 MJ/kg |
| Ignition temperature | : | 585°C |
| Stoichiometric mean ratio | : | 34.3 |

Recently, Ford motors have delivered 30 hydrogen-powered E-450 shuttle buses to select customers.

### 6.12.3 Natural Gas

Recently, natural gas has received much attention as an alterantive fuel to gasoline. Natural gas is a mixture of many gas components, consisting mainly of methane (60 to 95%) with a small amount of other hydrocarbon fuel components. In addition, it contains various small amounts of $N_2$, $CO_2$, He and traces of other gases. An ideal composition of natural gas as an automotive fuel may be as follows:

Methane = 94%; ethane = 4%; propane = 1.7%; $C_4$ = 0.7%; $C_6$ = 0.2; $CO_2 + N_2$ = 0.2%; $H_2$ = 0.1%; CO = 0.1%; $O_2$ = 0.5%; sulphur 10% ppm.

Generally, natural gas is stored as **compressed natural gas** (CNG) at a pressure of 7 to 21 bar and a temperature around –160°C

The following are the **physical properties** of CNG.

(i) Anti-knock quality of CNG is very high. The octane number is 130 which ensures that it burns at high temperature than gasoline.

(ii) It is suitable for high compression ratio up to 12 as compared to 10 in gasoline.

(iii) It is non-toxic and lighter than air.

(iv) The heating value is 50 MJ/kg.

**Advantages**

(i) High octane number makes it suitable as fuel for SI engine.
(ii) Engine emission is low.
(iii) It is cheaper than gasoline, methanol and ethanol.
(iv) It is engine friendly.
(v) It is available worldwide.
(vi) It is easy to tap.
(vii) It is clean and odourless.

**Disadvantages**

(i) Low energy density results in low engine performance.
(ii) Being a gaseous fuel, engine volumetric efficiency is low.
(iii) It needs a large pressured fuel storage tank to cover just 400 km.
(iv) The fuel properties are inconsistent.
(v) Refilling is a slow process.
(vi) Some of its disadvantages may be reduced by using bi-fuel and dual fuel.

## 6.12.4 Liquefied Petroleum Gas (LPG) and Liquefied Natural Gas (LNG)

**LPG** mainly consists of propane (93.1%), ethane (1.8%), propylene (2.8%) and butane (2.3%) by weight. The largest use of LPG at present is as domestic and industrial fuel. However, in recent years, there is an increasing trend to use LPG as motor fuel mainly due to its low cost, clean fuel, high knock resistant and increased engine life. The main drawbacks include lower volumetric efficiency, handling under pressure, difficult to detect leakage, high road sensitively and poor blending characteristics.

Generally, the LPG cylinder is kept in the rear portion of the car and some modification in carburettor is done.

(i) It has a higher octane number, burns more clearly and thus saves maintenance cost.
(ii) It costs around 60% of gasoline.

**LNG** is mainly $CH_4$ with a very small per centage of ethane and propane. It comes from dry natural reservoirs. The boiling temperature is –161°5°C which is the major hurdle in using this as fuel for SI engine.

## 6.12.5 Biogas

*The biogas is generally produced from animal dungs. Some other sources are sewage, crops residue, vegetables wastes, water hyacinch alga, poultry droping ocean kelp, etc.* Biogas is produced by digestion pyrolysis, or hydrogasification. The physical properties of biogas is as follows:

(i) Composition: $CH_4$ = 50 to 60%, $CO_2$ = 30 to 45%; $H_2$ and $N_2$ = 5 to 10% and traces of $H_2S$ and $O_2$.
(iii) Excellent anti-knock properties having octane number 110 with $CO_2$ and 130 without $CO_2$.

(iii) The autoignition temperature is higher than petrol which makes it a safer fuel.
(vi) The mixing properties are good.
(v) The calorific value is less than petrol.

SI engines can be operated on biogas after starting the engines by gasoline. Dual fuel may be used. However, biogas can be used in CI engines. The main advantage of biogas in that is very cheap.

## PROBLEMS

### Things to Think About

6.1 What could be the cause of formation of natural fuels?

6.2 Which region is called the cradle of mankind and why?

6.3 What is the basic principle on which refining process of crude petroleum has been designed?

6.4 What do you mean by cracking in refining process and why it is needed?

6.5 What are the reasons that diesel and gasoline are not single fuel but it is a constituent of many fuels?

### Short and Long Theoretical Questions

6.6 Mention the types of fuels used in IC engines.

6.7 Explain the chemical structure of petroleum.

6.8 Compare the characteristics of the following fuels.

(a) paraffin (b) olefin
(c) diolefin (d) naphthene
(e) aromatic.

6.9 With the help of sketch, briefly explain petroleum refining process.

6.10 Discuss the main properties of SI fuels.

6.11 Discuss the main properties of CI fuels.

6.12 Discuss the knock rating of SI fuels.

6.13 What do you mean by octane number?

6.14 Discuss the experimental method to find octane number.

6.15 What do you mean by dopes and inhibitors?

6.16 Discuss the important qualities of CI fuels.

6.17 Describe the knock rating (cetane number) of CI fuels.

6.18 Discuss the future of LPG as IC engine fuels.

6.19 What will be the consequences if a diesel fuel is used in a SI engine?

6.20 Discus various alternatives fuels for IC engine.

6.21 What is performance number (PN)? How is it found?

### Software Based Question

6.22 Develop a software for the design of petroleum refinery.

### Design and Open Ended Question

6.23 Prepare a comprehensive report on the advance method of petroleum refining process with suitable sketch. Give a report on pollution free IC engines.

### Multiple-Choice Questions

6.24 A SI engine fuel that knocks at low speed will

(a) always knock at high speed (b) have low fire point

(c) not knock at high speed (d) have low volatility

6.25 Octane number of gasoline is obtained by comparing the performance of the engine when running on a mixture of the following hydrocarbons.

(a) ethane and methane (b) iso-octane and TEL

(c) iso-heptane and octane (d) normal heptanes and iso-octane

6.26 Cetane number of diesel oil is obtained by comparing the performance of the engine when the running on a mixture of the following hydrocarbons.

(a) normal heptanes and octane (b) cetane and iso-octane

(c) cetane and alpha methyl naphthalene (d) ethane and cetane

6.27 Performance numbers indicate the

(a) performance rating of engine components.

(b) efficiency of IC engine.

(c) fuels having anti-knock qualities compared to iso-octane.

(d) performance parameters like ip, bp, etc.

6.28 HUCR is the highest compression ratio at which the

(a) engine is most efficient.

(b) engine can be run.

(c) engine gives maximum output.

(d) fuel can be used in a test engine without detonation.

6.29 Performance numbers indicate the

(a) fuels having anti-knock qualities composed of iso-octane.

(b) performance rating of IC engines.

(c) efficiency of a SI engine.

(d) performance parameters like ip, hp, etc.

6.30 The ignition quality of a diesel fuel is measured by its

(1) diesel index (2) delay period

(3) self-ignition temperature (4) cetane number

Of these statements:

(a) 1, 2, 3 and 4 are correct. (b) Only 2 is correct.

(c) 1, 2 and 3 are correct. (d) Only 1 and 2 are correct.

6.31 The flash point is measured for safety of fuel during its storage. The flash point of different fuels in descending order is

(a) petrol, kerosene, diesel
(b) kerosene, diesel, petrol
(c) diesel, kerosene, petrol
(d) diesel, petrol, kerosene

6.32 Alcohol as an engine fuel has

1. high anti-knock characteristics.
2. lower exhaust gas temperature with lower per centage of CO.
3. effective cooling of cylinder walls due to the high value of latent heat of these.

Out of these,

(a) 1 and 2 are correct
(b) 1, 2 and 3 are correct
(c) 1 and 3 are correct
(d) 2 and 3 are correct

## Key to Multiple-Choice Questions

6.24 (c) 6.25 (d) 6.26 (b) 6.27 (c) 6.28 (d) 6.29 (a) 6.30 (b) 6.31 (c)
6.32 (b)

CHAPTER

# 7

# Carburetion and Electronic Gasoline Injection

## OVERVIEW

In SI engines, normally volatile fuel such as gasoline is used. Since the combustion is initiated by spark, so it is essential that preparation of fuel and air mixture should be accomplished outside the engine cylinder. This is achieved by carburettor. *Nowadays, all the gasoline version of passenger cars have switched from carburettor to electronic gasoline injection-multipoint fuel injection (**MPFI**) system. As a result, it is called death of carburettor.* However, all the two-wheeler engines including motorcycles use carburettor. Racing cars are using carburettor due to large volume of air.

This chapter deals with theory and working principle of device involved in carburetion, Electronic gasoline injection system has also been discussed.

## DEFINITION

*The process of formation of a combustible fuel-air mixture by atomizing, vaporizing and mixing the proper amount of fuel with air before admission to engine cylinder according to the requirement of load and speed of the engine is called carburetion and the mechanical device which accomplishes this work is called carburettor.*

## 7.1 THE INTAKE AND EXHAUST SYSTEM AND PROCESSES

The system responsible for preparing the correct mixture of air and fuel as per load and speed, and directing this mixture to each of these cylinders is often termed the *induction or intake system* of SI engine as shown in Fig. 7.1.

*In spark ignition engine, the intake system typically consists of an air filter, a carburettor and throttle or fuel injector and throttle with individual fuel injectors in each intake part and intake manifold.* The **carburettor** is the main element of intake system. A *fuel system* consisting of fuel supply tank, fuel pumps, fuel lines and filters is used to supply the fuel to the carburettor or to the SI injectors. Air is drawn into the carburettor from the atmosphere

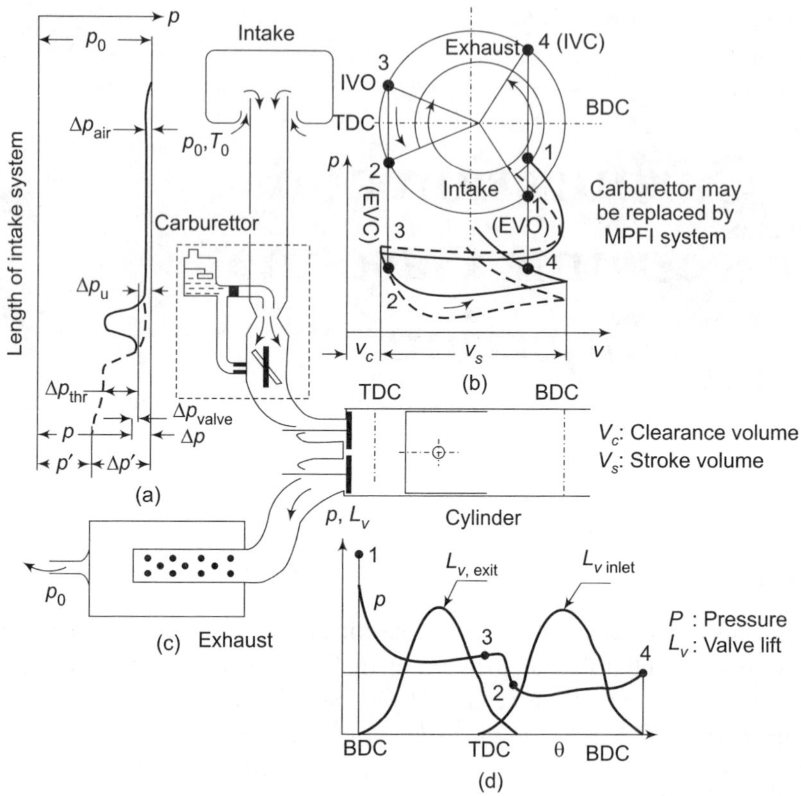


**Fig. 7.1** Intake and exhaust system with processes (a) Intake system and average pressure within it, (b) Valve timing *pressure-volume diagram*, (c) Exhaust system, (d) Cylinder pressure ($p$), valve lift ($L_v$) *vs.* crank angle ($\theta$)

through air filter by the action of engine piston on the intake stroke creating partial vacuum in the cylinder. Some induction system may use supercharger to supply air to the carburettor.

The intake manifold is the ducting or piping through which the fuel-air mixture flows from the carburettor to the cylinder. A **throttle** located in the carburettor regulates the quantity of mixture entering the cylinder as per demand.

The drop is pressure along the intake system depends on engine speed, the flow resistance of the elements in the system, the cross-sectional area through which the fresh charges moves and the charge density. Figure 7.1(a) shows the intake system and time average pressure distribution in it. Here, $p_0$ and $t_0$ are atmospheric conditions, $\Delta p_{\text{air}}$ is pressure loss in air cleaner, $\Delta p_u$ is intake losses upstream of throttle, $\Delta p_{\text{thr}}$ is the pressure losses across throttle, $\Delta p_{\text{valve}}$ is the pressure losses in intake valve.

Figure 7.1(b) depicts the valve-timing and pressure-volume diagram of intake and exhaust processes. Figure 7.1(c) shows the exhaust system which consists of exhaust manifold and a muffler or silencer. Figure 7.1(d) shows the intake and exhaust valve lifts with respects to crank angle.

The usual practice is to extend that valve open phases beyond the intake and exhaust strokes to improve emptying and charging of the cylinders and make the best use of inertia of gases in the intake and exhaust systems. As a usual practice, the exhaust processes usually begins 30° to 60° before BDC. Until about BDC, the burnt cycle gases are discharged due to the pressure difference between the cylinder and exhaust system. After BDC, the cylinder is scavenged by the piston as it moves towards TDC. Typically, the exhaust valve closes 15° to 30° after TDC and the intake valve opens 10° to 20° before TDC. Both valves are open during an overlap period and when $p_i/p_e < 1$ the back flow of exhausted gas into the cylinder and of cylinder gases into the intake will usually occur. The intake valve remains open while 40° to 70° after BDC ensuring the flow of fresh charge into the cylinder after BDC.

## 7.2 PARAMETERS AFFECTING CARBURETION

The following parameters affect significantly the process of carburetion.

(a) The engine speed
(b) The vaporization characteristics of fuel
(c) The temperature and pressure of the incoming air
(d) The design of carburettor and intake manifold

**(a) The engine speed.** Atomization and vaporization of fuel and mixing with air require a finite time to occur. But in high speed engines, the time available for mixture formation is very small, for example, about 0.02 seconds in the case of engine running at 3000 rpm. For this, special care has to be taken in the design of venturi of the carburettor and manifold.

**(b) The vaporization characteristics of fuel.** For efficient vaporization of the fuel, there is need of highly volatile hydrocarbon in the fuel so that within the short period available, the fuel may evaporate.

**(c) The temperature and pressure of the incoming air.** The process of carburetion is greatly influenced by the temperature and pressure of the incoming air. A high temperature results in a high evaporation rate. However, an increase in atmospheric temperature leads to a decrease in power output due to reduction in mass flow.

**(d) The design of the carburettor and induction manifold.** The design of carburettor, the intake system and the combustion chamber affect considerably the uniform distribution of mixture to the various cylinders of the engine. It is only the proper design of carburettor that ensures the supply of desired composition of the mixture under different conditions of operation of the engine. *One of the main disadvantages of carburetion system is that it falls to deliver the equal amount of mixture in all cylinders in a multi-stage engine due to improper distribution in induction manifold.*

## 7.3 AIR-FUEL MIXTURES

In real life situations, an engine runs at different load and speed and as a result, the requirement of air-fuel ratio by the engine varies. There are three different types of mixture of air-fuel which are possible. These are:

(a) Chemically correct (stoichiometric) mixture
(b) Rich mixture
(c) Lean mixture.

### 7.3.1 Chemically Correct Mixture

A chemically or **stoichiometric mixture** is one in which the amount of air present is just enough for complete combustion of fuel. It is computed from the chemical equation for complete combustion of a fuel. For example, complete combustion of octane ($C_8H_8$) is expressed as

$$C_8 H_{18} + (12.5)\ O_2 \rightarrow (8)\ CO_2 + (9)\ H_2O$$

or $$[(8 \times 12) + (18 \times 1)] + (12.5 \times 32) \rightarrow 8\ (12 + 32) + 9\ (2 + 16)$$

or $$1 \text{ kg of fuel} + 3.509 \text{ kg } O_2 \rightarrow \text{for complete combustion}$$

Thus, 1 kg of octane requires 3.509 kg of $O_2$ for complete combustion

Air consists of 23.2 per cent of oxygen and 76.8 per cent of nitrogen by mass, hence the A/F is given by

$$A/F = 3.509/0.232 = 15.12 = 15.12:1$$

For complete combustion, there must be 15.12 kg of air for every kg of $C_3H_8$ fuel. This is called *chemically correct* or *stoichiometric air-fuel ratio.* This chemically correct mixture will vary slightly in numerical value between different hydrocarbons.

### 7.3.2 Rich Mixture

A mixture containing less air than stoichiometric requirement is called a **rich mixture**, e.g., A/F = 10 : 1.

### 7.3.3 Lean Mixture

A mixture containing more air than stoichiometric requirement is called a **lean mixture**, e.g., A/F = 20 : 1.

It is interesting to note that there is a limited range of A/F ratios in a homogeneous mixture, only within which combustion in SI engine will occur. These limits are about A/F = 7 : 1 on rich side and about A/F = 20 : 1 on lean side as shown in Fig. 7.2. Outside this range, that ratio is either too rich or too lean to sustain the flame propagation. This

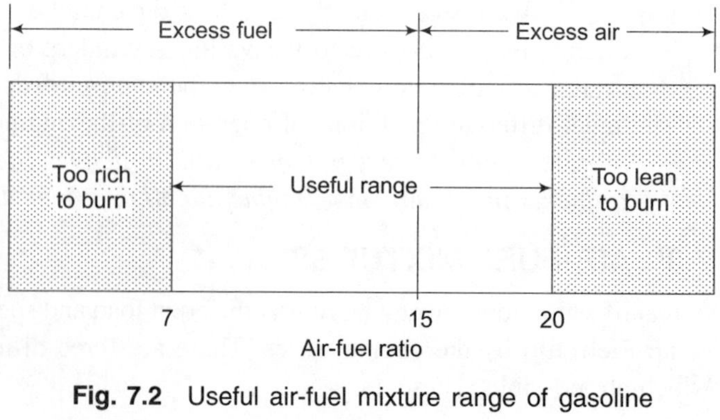


**Fig. 7.2** Useful air-fuel mixture range of gasoline

suggests that the carburettor should provide an A/F ratio as per engine's requirement and that must be within the combustible range.

## 7.4 THEORETICAL CARBURETTOR CONSIDERATIONS

A carburettor is a mechanical device used in SI engines to fulfil the following requirements:

(a) *To meter the liquid fuel in such quantities as to produce the A/F ratio required to meet engine operating conditions.*

(b) *To atomize, and vaporize the fuel and mixing homogeneously with air.*

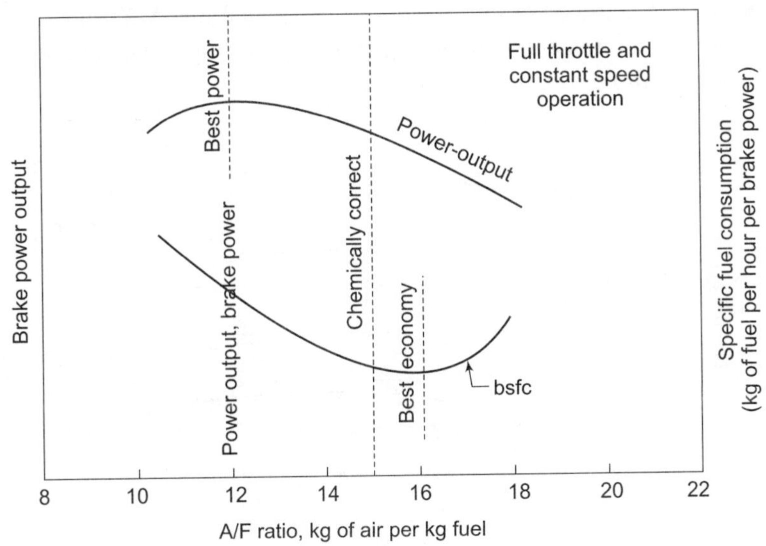


**Fig. 7.3(a)** Effect of A/F ratio on power output and specific fuel consumption

As per engine operating conditions, the A/F ratio required by the engine varies somewhat (7:1 to 20:1) over the engine operating range and the carburettor must be designed to meet this requirement as nearly as possible. As a result of this, carburettor design is complicated.

From the point of view of theoretical consideration for the design and operation, it would simplify the matters if operating range could utilize a carburettor providing a *constant economical A/F ratio throughout* the operating range of the engine rather variable. If it were possible to use such a carburettor, it might operate at a chemically correct A/F ratio as represented by the dashed line *Y-Y′* in Fig. 7.4.

Consider an actual engine operating at full throttle and constant speed. Under these conditions, the A/F ratio will affect both the power output and the brake specific fuel consumption as shown in Fig. 7.3(a).

*The mixture corresponding to the maximum point on the power output curve is called the* ***best power mixture****, with an A/F ratio of approximately* ***12:1****. The mixture corresponding to the minimum point on the sfc curve is called the* ***best economy mixture*** *with an A/F ratio of*

***about 16:1***. *It is interesting to note that the best power mixture is richer than the stoichiometric one and the best mixture is leaner than the stoichiometric one.*

The above facts are based on the full throttle operation. The A/F ratios for the best power and best economy at part throttle are not strictly the same as at full throttle. Ignoring this variation, assume that the A/F ratios for best power and best economy are constant over the full range of throttle operation. Based on this assumption and neglecting the other influencing factors, *the ideal fuel metering* device would be *merely a two position carburettor*. As a result such a carburettor could use set for the best power mixture when maximum performance is desired, and for the best economy mixture when the primary consideration is fuel economy. These two settings are depicted in Fig. 7.4 by the horizontal line *X-X′* and *Z-Z′* respectively. But the actual engine requirements cannot be met by such a simple and convenient carburettor. These requirements will be discussed below.

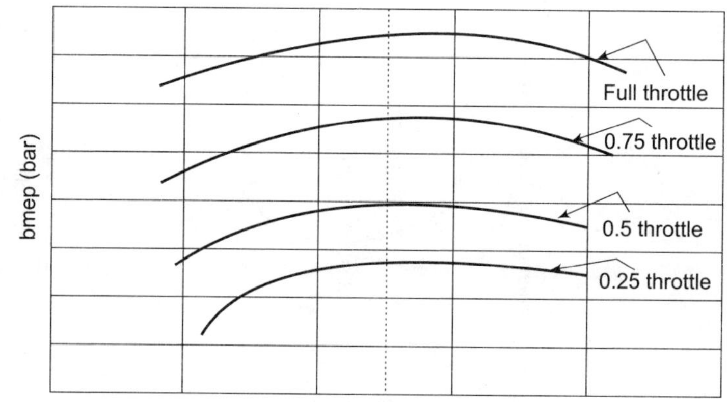


**Fig. 7.3(b)** Effect of equivalence ratio on bmep at constant speed

Figure 7.3(b) the effect of air-fuel-ratio on the engine output under varying throttle opening. It can be observed from the curves that the maximum output at a given throttle opening is nearly the same as that which gives the maximum output at all throttle openings Fig. 7.3(c) depicts the effect of air-fuel ratio on brake specific fuel consumption at various throttle openings. *It is observed that the minimum specific fuel consumption occurs at* $\phi$ = *0.9 at all throttle openings.* Based

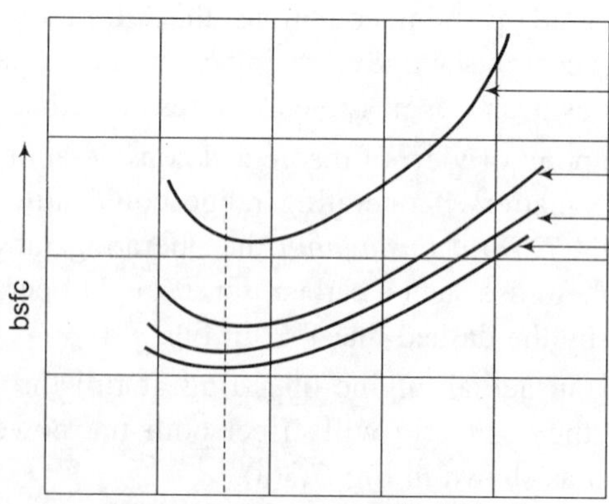


**Fig. 7.3(c)** Effect of equivalence ratio on *bsfc* at constant speed

on these results, and transient conditions requirements the carburettor will require only the following setting:

(a) Starting and idling range
(b) Economy or cruising range
(c) Maximum power range

## 7.5 ACTUAL ENGINE AIR-FUEL MIXTURE REQUIREMENTS

Actual air-fuel mixture requirements in an operating range vary considerably from the ideal conditions discussed in the previous article. The engine requires different ratios of air and fuel under various conditions of load as shown in Fig. 7.4. Under actual conditions of successful operation, the engine requires the carburettor to provide mixtures which follow the general shape of the curve *ABCD*. In general, there are three ranges of throttle operation and they are:

(i) Idling and low load (enriched mixture)
(ii) Economy range with medium load (lean mixture)
(iii) Power range with full load (enriched mixture)

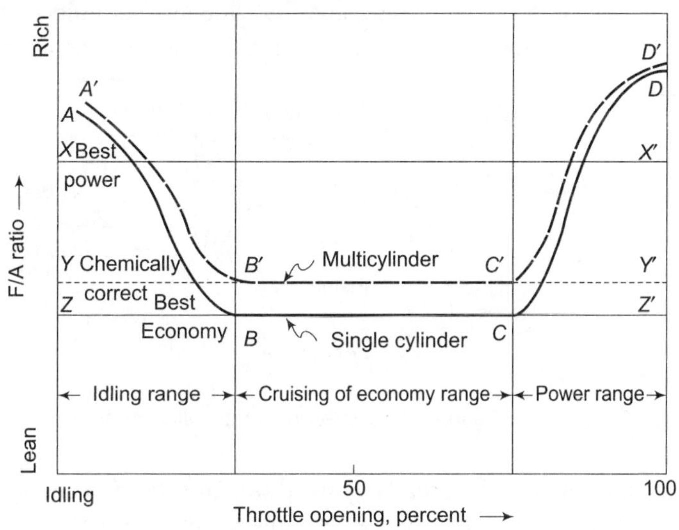


**Fig. 7.4** Mixture requirements

**(i) Idling and low load (0% to 25% of rated power).** *An idling engine is one which operates at no load with nearly closed throttle.* An idling engine demands a rich mixture as indicated by point *A* in Fig. 7.4. This is due to the existing pressure conditions within the combustion chamber and intake manifold which causes exhaust gas dilution of the fresh charge. This may be explained as follows: When the throttle is near the closed position, the pressure in the intake manifold can be far below atmospheric, while the pressure at the end of exhaust stroke in the cylinder is always close to atmospheric pressure. When the intake valve opens, the pressure differential between the combustion chamber and the intake manifold results in a "backward and" initial flow of exhaust gases into the intake manifold. Now, the piston descends on the intake stroke these exhaust gases are drawn back into the cylinder, along with the fresh charge. As a result, the final mixture of fuel and air in the combustion chamber is diluted by the exhaust gas. The presence of this exhaust gas tends to obstruct the mixing fuel and air which is essential for combustion and thus results in a loss of power. To prevent the effects of the exhaust gas dilution and weakening the explosive mixture, the carburettor must furnish a rich mixture (i.e., excess fuel) for idling condition.

Further, the pressure and temperature of the exhaust gas at the end of the exhaust stroke do not vary greatly with changes in load. Since the clearance volume is constant, the mass of the exhaust gas trapped in the clearance space tends to remain constant. At idling, a very small mass of fresh charge is needed because power output is zero, and therefore, the ratio of exhaust gas to fresh charge is high. Figure 7.5 shows the pressure encountered under idling condition.

As the throttling is gradually opened (i.e., as the load increases) from *A* to *B*, the pressure differential becomes smaller and greater amount of fresh charge is inducted which diminishes the exhaust gas dilution of the fresh charge, and thus, the air-fuel ratio is increased to give better economy. Note that a low idling speed demands increasingly richer mixture.

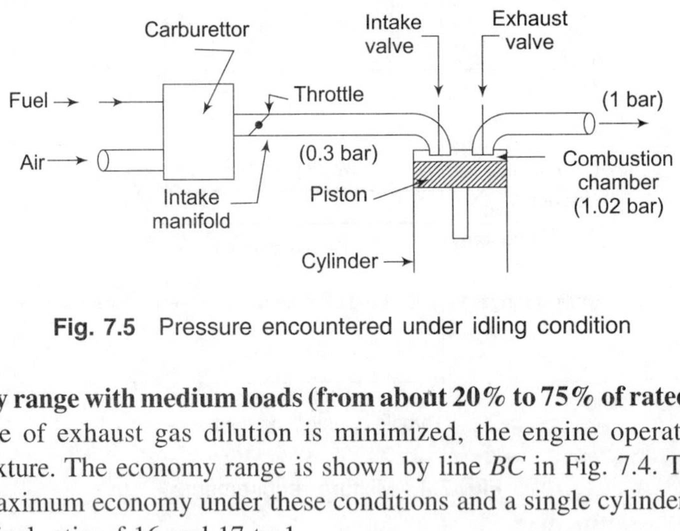


**Fig. 7.5** Pressure encountered under idling condition

**(ii) Economy range with medium loads (from about 20% to 75% of rated power).** Once the interference of exhaust gas dilution is minimized, the engine operates on the most economical mixture. The economy range is shown by line *BC* in Fig. 7.4. The air-fuel ratio will give the maximum economy under these conditions and a single cylinder engine can use highly economical ratio of 16 and 17 to 1.

In the economy range, the lean air-fuel ratio causes the speed of the flame to be slow in travelling across the combustion chamber. This is compensated by advancing the spark for best economy.

**(iii) Power range with full load (from about 75% to 100% of rated power).** During full power range the engine requires a richer mixture as shown by the line *CD* in Fig. 7.4. This is due to the following reasons:

(a) *Provide best power.* Since the maximum power is desired, therefore, it is essential to provide a richer mixture.

(b) *Prevent exhaust valve failure.* When the throttle is opened, the spark must be retarded from the economy setting to avoid knock. But with this spark, setting, the lean mixture will continue to burn as the piston descends on the power stroke (i.e., late burning). Therefore, when the exhaust valve opens, the gases passing around the valve will be at higher temperature than if the mixture had been rich. In addition to this, the excess air in the hot gases will exert a strong oxidizing action. At wide open throttle (i.e., high power), the maximum mass of exhaust gas flows past the valve. Thus, if the mixture is lean, the additional load of higher exhaust temperature and greater mass flow may cause the exhaust valve to burn or melt at wide open throttle and normal speed. This suggests that for power range, the mixture must be rich. Figure 7.6 shows the typical performance curve of an automobile carburettor.

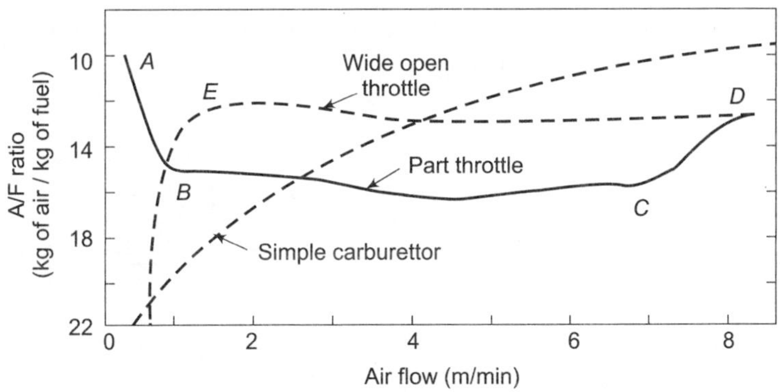


**Fig. 7.6** Typical performance curve of an automobile carburettor

(c) *Inhibit detonation in aircraft engines.* If the mixture is enriched beyond chemically correct proportion, it reduces the flame temperature and thereby tends to reduce detonation. When the mixture is enriched beyond the best power setting, excess fuel is introduced which in vaporizing, tends to further reduce the temperature in the combustion chamber and thereby, helps in reducing the possibility of detonation. *Detonation, especially in aircraft engines is a very serious matter.*

### 7.5.1 Additional Requirements

In addition to the above three requirements, the carburettor should also meet the requirements for ***starting up and acceleration***. To start the engine from cold and even richer mixture than that for idling is required. This is so, because of the fact, the fuel supplied by the carburettor remains liquid (i.e., does not vaporize) or recondenses on the cold cylinder walls and the piston head. It may be possible, that air-fuel ratio at the carburettor, may be rich enough for satisfactory combustion of the mixture but the ratio of evaporated fuel to air in the cylinder may be far too lean to ignite and burn. Therefore, it is necessary to supply, temporarily, as much as 5 or 10 times the normal amount of fuel at the carburettor, to get the correct mixture strength until the manifold and cylinder parts get heated.

To accelerate the engine, it is again necessary to supply an extra quantity of fuel to enrich the mixture temporarily to produce smooth and rapid acceleration with the help of an acceleration pump.

### 7.5.2 Distribution of Mixture in Multi-cylinder Engines

The engine requirements discussed above and depicted by curve *ABCD* in Fig. 7.4 were based on a single-cylinder engine. *In multi-cylinder engine, the distribution of equal amount of mixture becomes a great problem. Also, there exists a variation in the A/F ratio in each cylinder.*

Even the best design of carburettor fails to completely atomize and vaporize the fuel. The mixture passing through the intake manifold generally contains a certain amount of liquid in droplet form. These droplets have greater inertia than the gaseous mixture and as a result, whenever the direction of flow is changed abruptly the droplets tend to continue in the original direction of movement as shown in Fig. 7.7.

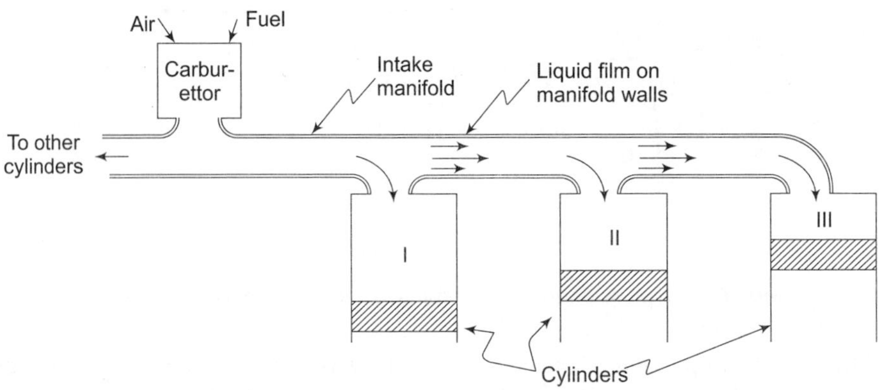


**Fig. 7.7** Intake manifold of a multi-cylinder engine Indicating the flow tendency of liquid fuel droplets

This results in the variation of A/F ratio between the cylinder depending upon the cylinders location and the manifold design.

In addition to the fuel droplets, there exists a thin film of liquid fuel adhering to the inner walls of the intake manifold which results in the uneven distribution.

It is possible to overcome this problem partially by heating the mixture in the intake manifold, causing the less volatile parts of the fuel to vaporize and thus reducing the number of liquid droplets present. But this heating causes the mixture to expand resulting in reduction in the mass of charge and hence reduction in output. The exhaust gas may be used to heat some portion of the manifold.

The other method to overcome this problem partially is to enrich the overall A/F mixture so that the leanest cylinder receives the required air-fuel ratio. Consequently, some of the cylinder receives a richer A/F ratio than necessary, multi-cylinder engines require. This over all enrichment is shown by the curve $A'B'C'D'$ in Fig. 7.4.

## 7.6 A SIMPLE CARBURETTOR

*A simple carburettor consists of a float chamber, a venturi, a throttle and choke* (Fig. 7.8).

**(a) Float Chamber.** *The function of the float chamber is to maintain a constant level of fuel in the carburettor unit.* This is achieved by a float which controls a needle valve placed in the fuel supply line. The fuel is pumped into float chamber by means of a fuel pump. If the amount of fuel in the float chamber falls below the designed value, the float lowers, thereby the needle valve (fuel supply valve) opens and the fuel enters the float chamber at a greater rate than the engine requires. When the designed level has been reached, the float rises up and closes the needle valve, thus, stopping additional fuel flow from the supply system. By this way, a constant level of fuel is maintained in the float chamber.

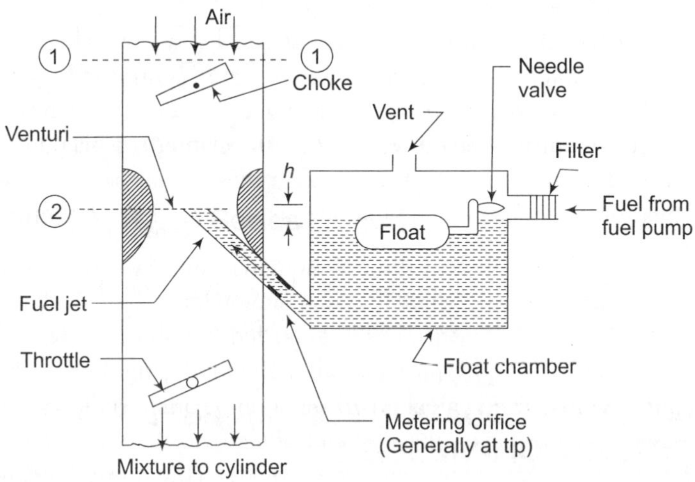


**Fig. 7.8** Simple carburettor (down draught)

A **vent** is provided in the float chamber which communicates the fuel in the chamber with atmospheric pressure. A **filter** is provided in the fuel supply system, so that, no foreign particles should clog the needle valve.

**(b) Venturi.** Air from the atmosphere is drawn through the venturi by the action of the piston on the intake stroke. As the air passes through the venturi, its velocity increases at the throat (due to small area and constant mass flow rate) with a corresponding reduction in pressure. Thus, a partial vacuum is created at the throat called the ***carburettor depression***. Since the float chamber is vented to atmosphere, a pressure differential exists between the float chamber and the tip of fuel discharge nozzle. This pressure differential causes the fuel to discharge (spray) into the air stream in an amount depending upon the magnitude of this pressure differential. The fuel thus sucked is vaporized by the coming air stream. The vaporization of fuel depends upon the pressure at venturi, temperature and velocity of air and nature of the fuel.

Instead of one venturi, some modern carburettors employ **three venturies** for better atomization mixing and maintaining more accurate air-fuel ratio according to load and speed.

**(c) Throttle.** *The function of the throttle is to control the quantity of air-fuel mixture supplied to the engine cylinder, and thus, by this way, control the speed and power of the engine.* It serves as a damper placed between the chamber and induction manifold. The control of the throttle valve is normally with the accelerator pedal through certain linkage. The more the throttle is closed the greater will be the obstacle to flow of mixture and the less will be the quantity of mixture delivered to the cylinder and hence the engine power and speed will reduce till an equilibrium is reached. When the throttle valve is fully open, the engine will suck the maximum amount of mixture and the engine speed will go up till an equilibrium is reached between the engine power and the load on the engine.

**(d) Choke.** *The function of choke is to provide an extra rich mixture during starting and warm up in cold weather to ensure that enough fuel is available in vaporized form, for combustion.* The choke valve is placed in the air intakes system on the upstream side of the venturi. When the choke valve is nearly closed (engine choked), a vacuum is created in the area around the fuel discharge nozzle. The pressure differential between the float chamber and the venturi area forces additional fuel into the air stream, thereby, fulfilling the extra rich engine requirements.

The performance of a simple carburettor may be analyzed on the basis of air-fuel ratio supplied by it. The air-fuel ratio of a simple carburettor may be obtained in subsequent article.

## 7.7 DERIVATION OF EQUATION FOR AIR-FUEL RATIO NEGLECTING COMPRESSIBILITY

Consider a simple carburettor (Fig. 7.8).

Let $p_1$ = pressure of air at section 1–1, i.e., atmospheric pressure, N/m$^2$

$p_2$ = pressure of air at venturi throat at section 2–2, N/m$^2$

$C_2$ = velocity of air at venturi throat, m/s.

$C_f$ = velocity of fuel at discharge from nozzle, m/s

$\rho_a$ = density of air, m$^3$/kg (incompressible)

$\rho_f$ = density of fuel, m$^3$/kg

$h$ = height of the nozzle tip above the fuel level, m

$\dot{m}$ = mass flow rate, kg/s

Considering the air between sections 1–1 and 2–2 and assuming negligible velocity of air at entrance to the carburettor ($C_1 = 0$), we have from Bernoulli's equation

$$\frac{p_1}{\rho_a} = \frac{p_2}{\rho_a} + \frac{C_2^2}{2}$$

or
$$C_2 = \sqrt{\frac{2(p_1 - p_2)}{\rho_a}} \qquad ...(7.1)$$

Similarly, for the flow of fuel

or
$$\frac{p_1}{\rho_f} = \frac{p_2}{\rho_f} + \frac{C_f^2}{2} + gh$$

or
$$C_f = \sqrt{[2\,(p_1 - p_2 - gh \times \rho_f)/\rho_f]} \qquad ...(7.2)$$

If $A_2$ and $A_f$ are the cross-sectional areas of the venturi throat and fuel nozzle respectively in $m^2$, then

$$\text{Mass of air flowing/s} = \dot{m}_a = A_2 \times C_2 \times \rho_a = A_2\sqrt{[2\,(p_1 - p_2)/\rho_a]} \times \rho_a$$

$$= A_2 \times \sqrt{[2\rho_a\,(p_1 - p_2]} \qquad ...(7.3)$$

Similarly, mass of fuel flowing/s

$$\dot{m}_f = A_f \times C_f \times \rho_f = A_f \times \sqrt{2(p_1 - p_2 - gh \times \rho_f)/\rho_f]} \times \rho_f$$

$$= A_f \times \sqrt{2\rho_f\,(p_1 - p_2 - gh \times \rho_f)]} \qquad ...(7.4)$$

From Eqs. (7.3) and (7.4), we have

$$\text{F/A} = \frac{m_f}{m_a} = \frac{A_f \times C_f \times \rho_f}{A_2 \times C_2 \times \rho_a} = \frac{A_f\sqrt{[2p_f(p_1 - p_2 - gh \times \rho_f)]}}{A_2 \times \sqrt{2\rho_a\,(p_1 - p_2)]}}$$

or
$$\text{F/A} = (A_f/A_2)\sqrt{\frac{\rho_f}{\rho_a}}\sqrt{\left(1 - \frac{gh\,\rho_f}{p_1 - p_2}\right)} \qquad ...(7.5)$$

If $h$ is negligible, then F/A is proportional to square root of the ratio of their density.

$$\text{F/A} = \frac{m_f}{m_a}\ \alpha\ \sqrt{\left(\frac{\rho_f}{\rho_a}\right)} \qquad ...(7.6)$$

From Eq. (7.5), the following conclusions may be drawn:

(i) At low air flows where $(p_1 - p_2)$ is less than $gh\rho_f$ there will be no fuel flow.

(ii) With the increase in flow of air, the flow of fuel will start (when $(p_1 - p_2) > gh\rho_f$) and the mixture tends to become richer.

(iii) If flow of air is high where $(p_1 - p_2)$ is larger than $gh\rho_f$, term $gh \times p_f/(p_1 - p_2)$ becomes negligible and thus F/A approaches to $(A_f/A_2)\sqrt{(\rho_f/\rho_a)}$.

(iv) If the density of air decreases, the fuel-air ratio becomes richer. The air density at the throat reduces with increasing flow because of increase in the depression in the venturi. This means, that as the flow of air increases, the mixture becomes richer.

(v) At high altitude, density of air $(\rho_a)$ reduces and the mixture becomes rich.

*In the light of above discussion, it follows that a simple carburettor has the following drawbacks:*

(a) The simple carburettor does not supply a constant fuel-air ratio for the economy range (20–75% load) but supplies a richer mixture as the speed of the engine (or load) increases.

(b) During idling, it does not supply a rich mixture. On the contrary, it supplies a lean mixture or no fuel at all.

(c) There is no provision of delivering extra fuel during acceleration.

(d) At high altitudes, it enriches the mixture.

In order to *overcome these drawbacks,* ***compensating devices*** are used in the actual carburettor.

**EXAMPLE 7.1** A carburettor is tested in a factory without an air cleaner. The main metering system of the carburettor gives an air-fuel ratio of 15.4 : 1 at the seal level condition. The pressure at the venturi throat is found to be 0.8 bar.

The same carburettor is tested again when an air cleaner is fitted at the inlet to the carburettor. The pressure drops due to air cleaner is found to be 0.05 bar when the air flow at sea level condition is 240 kg/h.

Calculate, (a) the throat pressure when the air cleaner is fitted and (b) fuel-air ratio with air cleaner fitted. Assume zero nozzle lip, constant coefficient of flow and the same air flow in both the cases.

***Solution*** When there is no air cleaner

$$p_1 - p_2 = \Delta p_a = 1.013 - 0.8 = 0.213 \text{ bar}$$

With the air cleaner fitted, let the throat pressure be $p_2$ bar

The pressure drop when the air cleaner is fitted

$$= \Delta p_a' = 1.013 - 0.05 - p_2' \text{ bar}$$

Since the air flow and coefficients are the same, hence

$$\Delta p_a = \Delta p_a'$$

$\therefore$ $$0.213 = 1.013 - 0.05 - p_2'$$

or $$p_2' = 0.75 \text{ bar}$$ ***Ans.***

(b) Let $\Delta p_f$ be the pressure difference which causes the fuel flow when there is no air cleaner fitted = $\Delta p_a' = 0.213$ bar. But when the air cleaner is fitted, the new pressure difference for fuel flow will be

$$\Delta p_f' = 1.013 - 0.75 = 0.263 \text{ bar}$$

Hence, more fuel will be sucked by the engine and the mixture will be richer.

$$\text{New air-fuel ratio} = 15.4\sqrt{\left(\frac{\Delta p_a'}{\Delta p_f}\right)} = 15.4\sqrt{\left(\frac{0.213}{0.236}\right)} = 13.859 \qquad \textit{Ans.}$$

## 7.8 EXACT EXPRESSION FOR AIR-FUEL RATIO SUPPLIED BY A SIMPLE CARBURETTOR

The expression for fuel-air ratio as obtained from Eq. (7.5) is approximate, since it does not take into consideration the compressibility of air (i.e., change in density) from inlet to the throat of the carburettor. An exact expression for air-fuel ratio supplied by the carburettor is derived as follows by referring Fig. 7.8.

Let suffixes 1 and 2 denote inlet and throat conditions respectively.

$C_1, C_2$ = velocity of air at inlet and throat, m/s

$p_1, p_2$ = pressure of air at inlet and throat, kN/m$^2$

$A_1, A_2$ = air flow are at inlet and throat, m$^2$

$h_1, h_2$ = specific enthalpy of air at inlet and throat, kJ/kg

$T_1, T_2$ = temperature of air at inlet and throat, K

$v_1, v_2$ = specific volume of air at inlet and throat, m$^3$/kg

$\rho_f$ = density of fuel, $\rho_{f1} = \rho_{f2} = \rho_f$ (i.e., incompressible), kg/m$^3$

$\dot{m}_a$ = mass flow rate of air, kg/s

$\dot{m}_f$ = mass flow rate of fuel, kg/s.

$\rho_{a1}, \rho_{a2}$ = density of air at inlet and throat, kg/m$^3$.

Applying the steady flow energy equation between inlet and throat of the carburettor of air flow (neglecting PE), we have on per kg of air flow:

$$h_1 + \frac{C_1^2}{2} + q_{1-2} = h_2 + \frac{C_2^2}{2} + w_{1-2}$$

But $\quad q_{1-2} = 0, \quad w_{1-2} = 0 \quad$ and $\quad C_1 \simeq 0$, hence

$$C_2 = \sqrt{[2\,(h_1 - h_2)]} = \sqrt{2h_1\,(1 - h_2/h_1)} = \sqrt{2c_p T_1\,[(1 - T_2/T_1)]}$$

Since, $\quad \dfrac{T_2}{T_1} = \left(\dfrac{p_2}{p_1}\right)^{(\gamma-1)/\gamma}$

$$C_2 = \sqrt{2c_p T_1\left[1 - \left(\frac{p_2}{p_1}\right)^{(\gamma - 1/\gamma)}\right]} \qquad ...(7.7)$$

Mass flow of air is given by continuity equation

$$\dot{m}_a = \frac{A_1 C_1}{v_1} = \frac{A_2 C_2}{v_2} = A_2\, C_2\, \rho_{2a} \qquad ...(7.8)$$

But
$$v_2 = v_1 \left(\frac{p_1}{p_2}\right)^{1/y} = \frac{R_a . T_1}{p_1} \left(\frac{p_1}{p_2}\right)^{1/\gamma} \qquad ...(7.9)$$

Putting the value of $C_2$ and $v_2$ into Eq. (7.8), we get

$$\dot{m}_a = \frac{A_2 . p_1}{R_a . T_1} \left(\frac{p_2}{p_1}\right)^{1/\gamma} \sqrt{2 c_p T_1 \left[1 - \left(\frac{p_2}{p_1}\right)^{(\gamma - 1)\gamma}\right]}$$

or
$$\dot{m}_a = \frac{A_2 . p_1}{R_a \sqrt{(T_1)}} \sqrt{2 c_p \left[\left(\frac{p_2}{p_1}\right)^{2/\gamma} - \left(\frac{p_2}{p_1}\right)^{(\gamma + 1)/\gamma}\right]} \qquad ...(7.10a)$$

Here for air,

$c_p$ = 1.005 kJ/kgK,

$R_a$ = 0.287 kJ/kgK and $\gamma$ = 1.4 for air, the mass flow rate in kg/s is

$$\therefore \quad \dot{m}_a = 4.939 \frac{A_2\, p_1}{\sqrt{(T_1)}} \sqrt{\left[\left(\frac{p_2}{p_1}\right)^{1.418} - \left(\frac{p_2}{p_1}\right)^{1.714}\right]} \qquad ...(7.10b)$$

For simplicity, let

$$\varphi = \sqrt{\left(\frac{p_2}{p_1}\right)^{1.418} - \left(\frac{p_2}{p_1}\right)^{1.714}} \qquad ...(7.11)$$

$\therefore$ Eq. (7.10b) becomes

$$\dot{m}_a = 4.939 \frac{A_2\, p_1}{\sqrt{(T1)}} \varphi, \quad \text{kg/s} \qquad ...(7.12)$$

It is to be noted that Eq. (7.12) gives the maximum mass flow rate of air through the venturi throat.

Let $c_{da}$ = coefficient of discharge of venturi throat of air

Then the actual air flow rate will be

$$(\dot{m}_a)_{\text{actual}} = \frac{4.939\, c_{da} . A_2 . p_1}{\sqrt{(T1)}} \varphi \qquad ...(7.13)$$

The fuel flow rate is given by Eq. (7.4)

$$\dot{m}_f = A_f \sqrt{(2\rho_f\, [p_1 - p_2 - gh\, \rho_f])}$$

If coefficient of discharge of fuel orifice is $c_{df}$, the actual fuel flow rate is

$$(\dot{m}_f)_{\text{actual}} = A_f . c_{df} . \sqrt{(2\rho_f [p_1 - p_2 - gh\, \rho_f])} \qquad ...(7.14)$$

If height of the nozzle tip is at the fuel level (i.e., $h = 0$), then

$$(\dot{m}_f)_{\text{actual}} = A_f \cdot c_{df} \cdot \sqrt{(2\rho_f[p_1 - p_2])} \qquad \text{...(7.15)}$$

The pressure drop $(p_1 - p_2)$ in N/m$^2$ is usually measured with a manometer containing water. The manometer depression $\Delta H$ in cm of water at 15°C is related to $(p_1 - p_2)$ is as follows. (Density of water is 1000 kg/m$^3$.)

$$\frac{\Delta H}{100} \times 1000 \times g = p_1 - p_2 = \Delta p_a$$

or

$$p_1 - p_2 = \Delta p_a = 10g\ \Delta H$$

The air-fuel ratio is given by

$$\text{A/F} = \frac{(\dot{m}_a)_{\text{actual}}}{(\dot{m}_f)_{\text{actual}}} = \frac{c_{da} \cdot A_2 \cdot C_2 \cdot \rho_{2a}}{c_{df} \cdot A_f \cdot C_f \cdot \rho_f} \qquad \text{...(7.16a)}$$

From Eqs. (7.13) and (7.14), we have

$$\text{A/F} = 4.939\, \frac{c_{da}}{c_{df}} \cdot \frac{A_2}{A_f} \cdot \frac{p_1}{\sqrt{(T_1)}}\, \varphi\, \frac{1}{\sqrt{(2\rho_f \cdot [p_1 - p_2 - gh\, \rho_f])}} \qquad \text{...(7.16b)}$$

or

$$\text{A/F} = 3.492\, \frac{c_{da}}{c_{df}} \cdot \frac{A_2}{A_f} \cdot \frac{p_1}{\sqrt{(T_1)}}\, \frac{\varphi}{[\rho_f\,(p_1 - p_2 - gh\, \rho_f]^{1/2}} \qquad \text{...(7.17)}$$

It the height of the tip of the nozzle is at the same level as the level of fuel in float chamber (i.e., $h = 0$), then

$$\text{A/F} = 3.492\, \frac{c_{da}}{c_{df}} \cdot \frac{A_2}{A_f} \cdot \frac{p_1}{\sqrt{(T_1)}}\, \frac{\varphi}{[\rho_f\,(p_1 - p_2)]^{1/2}} \qquad \text{...(7.18)}$$

**EXAMPLE 7.2** Design the size of the fuel orifice to give a 13 to 1 air-fuel ratio if the venturi throat is 3.5 cm in diameter and the vacuum at the venturi is 6.5 cm of Hg. Take $c_{da} = 0.92$ and $c_{df} = 0.95$. The air temperature and pressure at the carburettor entrance are 1 bar and 20°C. The fuel orifice is at the same level as that of float chamber fuel. The density of the fuel may be taken as 750 kg/m$^3$.

***Solution*** Refer to Fig. 7.8. Data given are:

A/F = 13:1, $D_2 = 3.5$ cm, $(p_1 - p_2) = \Delta p = 6.5$ cm of Hg, $c_{da} = 0.92$, $c_{df} = 0.95$, $p_1 = 1$ bar, $T_1 = 293$ K, $\rho_f = 750$ m$^3$/kg, $h = 0$, $d_f$ = Required

From continuity equation for air flow

$$\dot{m}_a = c_{da}\, A_2\, C_2\, \rho_{2a}$$

But

$$p_1 - p_2 = 6 \text{ cm of Hg}$$

or

$$p_2 = 1 - \frac{6.5}{76} \times 1.013 = 0.9134 \text{ bar}$$

$$\therefore \quad p_1 - p_2 = \Delta p = 1 - 0.9134 = 0.0866 \text{ bar}$$

$$C_2 = \left[2\, c_p\, T_1 \left\{1 - (p_2/p_1)^{(\gamma - 1/\gamma)}\right\}\right]^{1/2}$$

$$= [2 \times 1.005 \times 10^3 \times 293\, \{1 - 0.9134^{0.2857}\}]^{1/2}$$

$$= 122.72 \text{ m/s}$$

$$\rho_{2a} = \frac{1}{v_2}, \text{ but } v_2 = v_1 (p_1/p_2)^{1/\gamma}$$

or
$$v_2 = \frac{RT_1}{p_1}(p_1/p_2)^{1/1.4} = \frac{0.287 \times 293}{1 \times 10^2}\left(\frac{1}{0.9159}\right)^{1/1.4}$$

or
$$v_2 = 0.851, \text{ so } \rho_{2a} = \frac{1}{0.851} = 1.175 \text{ m}^3/\text{kg}$$

From continuity equation for fuel flow

$$\dot{m}_f = c_{df} \cdot A_f\, C_f \cdot \rho_f$$

But
$$C_f = \left[\frac{2(p_1 - p_2)}{\rho_f}\right]^{1/2}$$

$$= \left[\frac{2(1 - 0.9134) \times 10^5}{750}\right]^{1/2} = 4.8 \text{ m/s}$$

$$\text{A/F} = \frac{\dot{m}_a}{m_f} = 13 = c_{da} A_2\, C_2\, \rho_{2a} / c_{df}\, A_f\, C_f\, \rho_f$$

or
$$13 = \frac{0.92}{0.85} \times \frac{(\pi/4)\,(3.5 \times 10^{-2})^2}{(\pi/4)\, d_f^2} \times \frac{122.72}{4.80} \times \frac{1.175}{750}$$

or
$$13 = \frac{1625 \times 10^{-4}}{3420\, d_f^2}$$

or
$$d_f = 0.191 \times 10^{-2} \text{ m} = 1.91 \text{ mm}$$ ***Ans.***

**EXAMPLE 7.3** A single jet carburettor supplies 6 kg/min of air and 0.4 kg/min of petrol having specific gravity as 0.77. The air is at 1 bar and 20°C. Calculate (a) the diameter of the venturi if the air speed is 90 m/s and the velocity coefficient for venturi is 0.85. Also calculate (b) the diameter of the jet if the pressure drop at the jet is 0.8 times the drop at the venturi and the discharge coefficient for the jet may be taken as 0.68.

***Solution*** Refer to Fig. 7.8. The actual fuel velocity of air at the venturi throat

$$C_2 = c_{da} \sqrt{\left[2 c_p T_1 \left\{1 - (p_2/p_1)^{(\gamma - 1)/\gamma}\right\}\right]}$$

or
$$= 90 = 0.85 \sqrt{2 \times 1000 \times 1.005 \times 293 \left[1 - (p_2/p_1)^{(\gamma - 1)/\gamma}\right]}$$

or
$$1 - \left(\frac{p_2}{p_1}\right)^{(\gamma - 1)/\gamma} = 0.019$$

or $$\left(\frac{p_2}{p_1}\right)^{(\gamma-1)\gamma} = 1 - 0.019 = 0.981$$

or $$\left(\frac{p_2}{p_1}\right) = (0.981)^{\gamma/(\gamma-1)} = 0.9362$$

But $$\frac{v_2}{v_1} = \left(\frac{p_1}{p_2}\right)^{1/\gamma} = \left(\frac{1}{0.9362}\right)^{1/14} = 1.048$$

$$v_2 = 1.048 . R_a \frac{T_1}{p_1} = \frac{1.048 \times 0.287 \times 293}{1 \times 10^2} = 0.881 \text{ m}^3/\text{kg}$$

From the continuity equation at the venturi throat

$$\dot{m}_a = \frac{A_2 C_2}{v_2}$$

or $$A_2 = \frac{\dot{m}_a v_2}{C_2} = \frac{6}{60} \times \frac{0.881}{90} = 0.0009788 \text{ m}^2 = 9.788 \text{ cm}^2 = (\pi/4)\, D_2^2$$

(a) $\therefore$ $$D_2 = 3.53 \text{ cm}$$ ***Ans.***

(b) Drop in pressure at the venturi = $p_1 - p_2 = 1 - 0.9362 = 0.0638$ bar

Pressure drop at the jet = $0.8 \times 0.0638 = 0.05104$ bar

The fuel flow rate = $\dot{m}_f = c_{df} . A_f \sqrt{[2\rho_f (p_1 - p_2)_{\text{jet}}]}$

$$\frac{0.4}{60} = 0.68 \times A_f \sqrt{(2 \times 0.77 \times 1000 \times 0.05104 \times 10^5)}$$

or $$A_f = 0.3496 \times 10^{-5} \text{ m}^2 = 0.03496 \text{ cm}^2 = (\pi/4)\, d_f^2$$

$$d_f = 0.2109 \text{ cm} = 2.109 \text{ mm}$$ ***Ans.***

## 7.9 THE ELEMENTS OF A COMPLETE CARBURETTOR

It is essential to provide the following systems in a simple carburettor to satisfy the demands of an engine under all conditions of operation.

(a) Main metering system (b) Idling system
(c) Power enrichment system (d) Acceleration pump system

## 7.10 MAIN METERING SYSTEM

*The function of the main metering system of a carburettor is to supply a nearly constant fuel-air ratio over a wide range of speed and load (economy range).* We have seen that a simple carburettor does not supply constant fuel-air ratio in the economy range and instead of this it supplies rich mixture as the load increases. Hence, automatic compensating devices are employed to avoid this difficulty.

The following compensating devices are used:

(i) *Compensating jet* which allows an increasing flow of air through a fuel passage as the mixture flow increases.

(ii) *Auxiliary air valve* which automatically admits additional air as mixture flow increases.

(iii) *Tapered metering pin* which reaches into the fuel nozzles and is lifted or lowered, thus, changing the quantity of fuel drawn.

(iv) Changing the position of *jet* in venturi.

(v) *Pressure reduction* in the float chamber.

Each of these is discussed below.

### 7.10.1 Compensation Jet Carburettor

Figure 7.9 shows a compensating jet carburettor. *A* is the main nozzle and *B* is the compensation nozzle (jet). In the compensating nozzle, the fuel flows through the orifice *C* and the chamber *D* which is open to atmosphere is connected with the passage leading to compensating jet. The main nozzle (*A*) supplies a rich mixture as the load increases. The compensating nozzle (*B*) also tends to supply a rich mixture as flow is first increased but this tendency is soon reversed. Note that the fuel from nozzle (*B*) increases with increased suction until delivery equals that obtained from the height of fuel in the flow chamber above the orifice, *C*. When this conditions is reached standpipe is emptied and the pressure on the down stream side of the metering orifice is atmospheric and independent of the pressure in the venturi. The compensating nozzle (*B*) will supply a constant amount of fuel once the standpipe has been emptied and therefore, a leaner mixture is obtained. The lean mixture applied by the compensating jet (*B*) can compensate the rich mixture supplied by the main jet, *A* so that, the over-all mixture ratio can be constant.

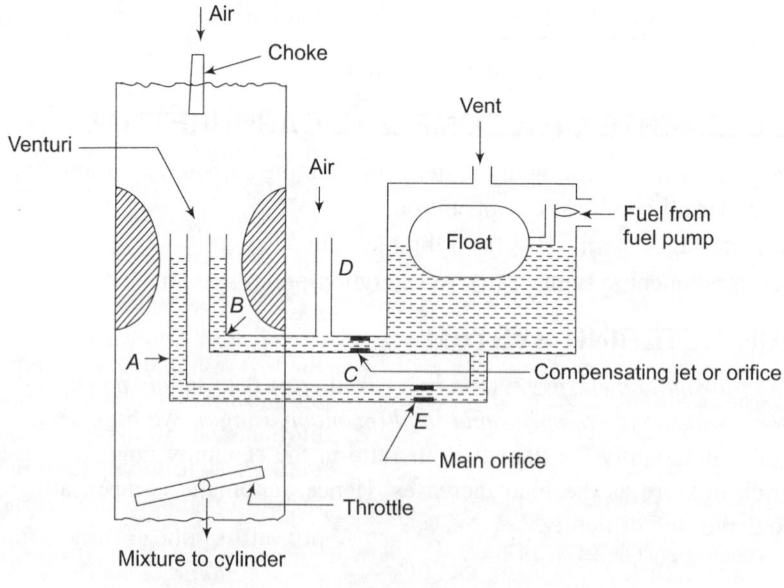


**Fig. 7.9** Compensating jet carburettor

### 7.10.2 Auxiliary Air Valve Carburettor

Figure 7.10 shows an auxiliary air valve carburettor. It consists of a plain nozzle and metering orifice plus an air valve. The air valve admits a regulated quantity of air into the mixing chamber between the venturi and throttle. The operation is as follows. During idling, the engine receives the fuel from the main jet and air from the small air entrance. As the throttle opens, the entrance (*A*) will restrict the flow of air and the decreased pressure in the carburettor will open the auxiliary air valve. One or more springs regulate the position of the air valve so that the mixture rate should first fall and then be held constant. Nowadays air-valve carburettor is not used because of the complicated mechanical adjustment for controlling the A/F. In auxiliary port carburettor, a butterfly valves serves the purpose.

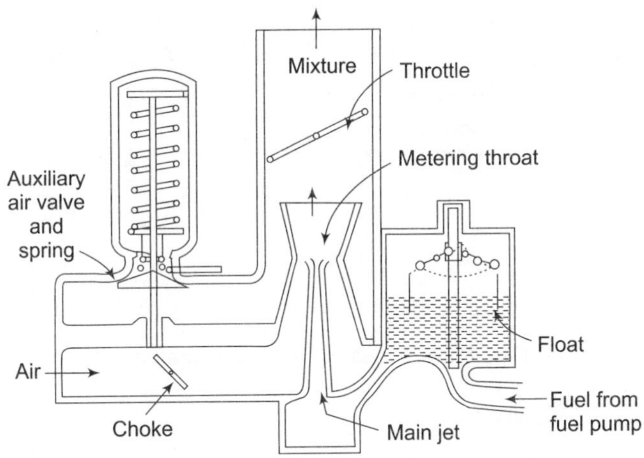


**Fig. 7.10** Auxiliary air valve carburettor

### 7.10.3 Tapered Metering Pin Carburettor

A tapered metering pin can be moved in or out of the fuel orifice, either manually or by means of some automatic mechanism, to control the air-fuel as desired by the operating load and speed. This is because of the fact, that the motion of the pin changes the area of the fuel orifice and thus, the correct amount of fuel flows to the venturi.

### 7.10.4 Position of Jet

By changing the position of jet, the air-fuel ratio is controlled. By raising the venturi position the nozzle moves relatively to provide with smaller suction and thus flow of fuel decreases.

### 7.10.5 Pressure Reduction

Figure 7.11 shows the arrangement of pressure reduction. A large vent line provided with a control valve connects the carburettor entrance with the top of the float chamber and the venturi throat. If the control valve is wide open, the vent line is unrestricted, and thus, the pressure in the float chamber is equal to $p_1$. Under this condition, the pressure difference acting on the fuel oricfice is $(p_1 - p_2)$. When the valve is closed, the float chamber is only in communication with the venturi throat. Under this condition, the pressure acting on the fuel orifice will be $p_2$ only, and the pressure difference on the fuel orifice will be zero so that no fuel will flow. This suggests that it is possible to supply the required quantity of fuel by adjusting the position of control valve.

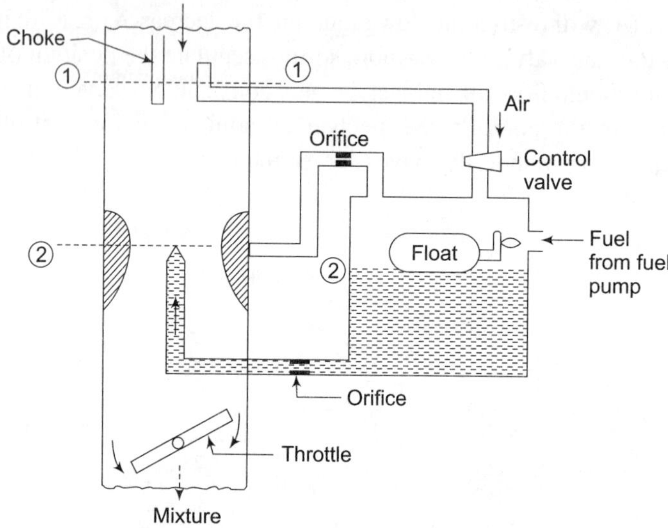


**Fig. 7.11** Pressure reduction method

## 7.11 IDLING SYSTEM

During idling, the intake depression in the throat of the venturi is too small to overcome the surface tension of the fuel and lifts the fuel from the nozzle. In addition to this, during idling, the engine demands a rich mixture while a simple carburettor gives the reverse characteristics. To avoid these drawbacks, an idling jet is provided in the main nozzle.

Figure 7.12 shows the arrangement of an idling system in a carburettor. A small fuel line called idle tube is run from the float chamber to a point on the manifold side of the throttle. To control the mixture ratio a metering orifice is also installed in the idle tube. During idling (throttle

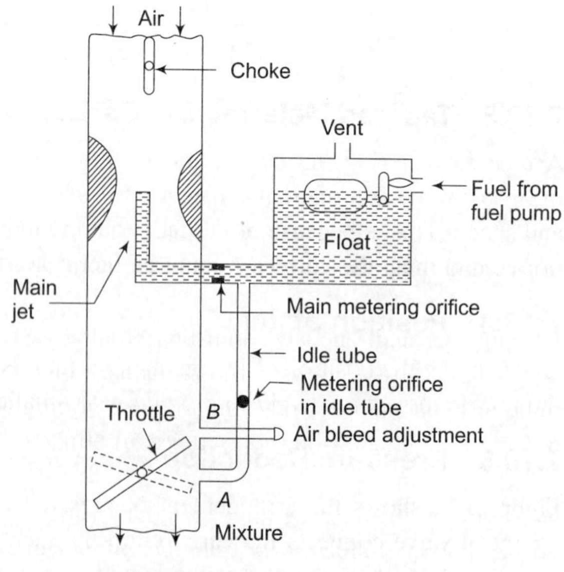


**Fig. 7.12** Idling system

is closed), the manifold pressure decreases, and thus, a strong vacuum is created. Since the pressure in the float is atmospheric, hence, the strong vacuum while idling will pull the fuel through the idling tube and discharges it into the manifold through the holes *A* and *B*. In order to atomize the fuel well, a small amount of air enters the fuel through the idle air bleed. At nearly closed throttle position, the manifold suction is very high and there is always a possibility of delivering a very rich mixture from the idling tube. The small hole (*B*) supplies fuel until the edge of the throttle plate moves past it. Under such a condition, the air pressure on *B* will be much higher than that on *A*, and thus, the air will enter through *B* and reduces the suction of fuel through the idling system.

By this way, the idling fuel air-ratio is controlled from becoming too rich. As the throttle is opened, the pressure in manifold increases and the fuel flow from the idling tube is considerably reduced. The air-fuel ratio rapidly increases with throttle opening until the manifold vacuum is not strong enough to lift the fuel to *A*. From this throttle position to fuel throttle (economy and power range), the idling tube remaining inoperative and the fuel flows only through the main jet. Thus, we see, that idling requirement of the engine is controlled by the idling tube, although the total flow will be made up of the separate flows from both the main nozzle and the idling tube.

An another arrangement of idling system is shown in Fig. 7.13.

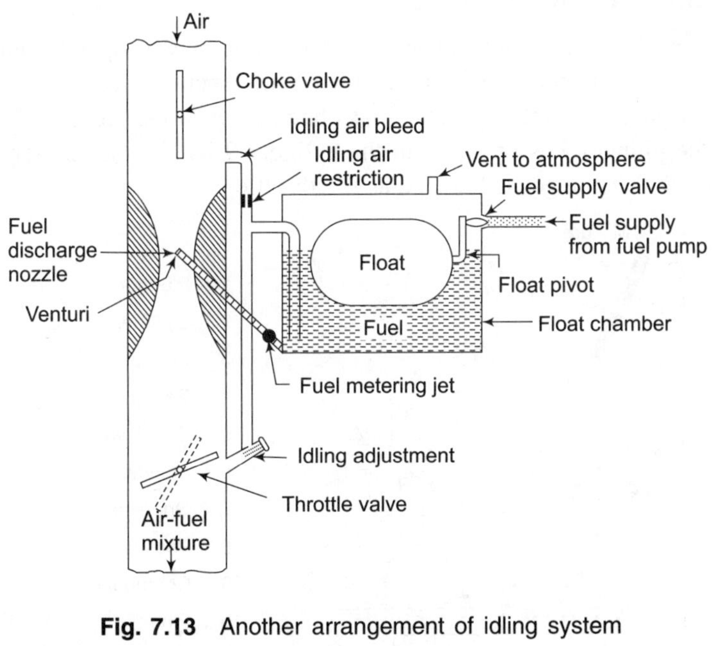


**Fig. 7.13** Another arrangement of idling system

## 7.12 POWER ENRICHMENT SYSTEM

Figure 7.14 shows the power enrichment system. It consists of a power enrichment jet and a power enrichment orifice. The power enrichment orifice is controlled by a needle valve which is operated by the throttle. This means that the operation of power enrichment system depends upon the throttle position. As the needle valve opens, a quantity of fuel is metered in addition to the fuel supplied by the main orifice with greater opening of the throttle, the needle valve gradually opens further and more fuel is delivered through the power enrichment orifice. Note that the fuel passing through the power enrichment orifice has finally to pass through the main jet.

When the needle valves starts to open, almost all the metering is done at the orifice. But at full load, the valve is nearly wide open, hence, most of the metering is done at the power enrichment orifice. In other words, at full load, the maximum amount of fuel added depends on the size of fuel enrichment orifice.

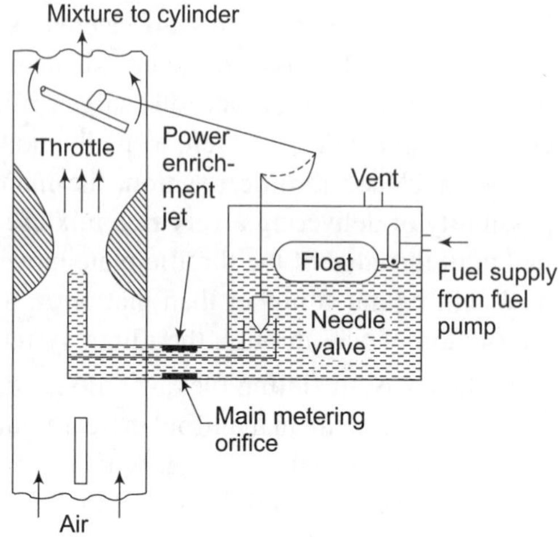


**Fig. 7.14** Power enrichment system (Updraught)

## 7.13 ACCELERATION PUMP SYSTEM

Figure 7.15 shows an acceleration pump system. The pump may be operated either by throttle or by manifold pressure. Here, we will discuss the throttle operated acceleration pump system. The acceleration pump is placed in the float chamber. It is essentially a cylinder-piston arrangement. The plunger 1 is operated in the cylinder 2 by the throttle through a linkage mechanism, so that the plunger is lowered as the throttle is opened.

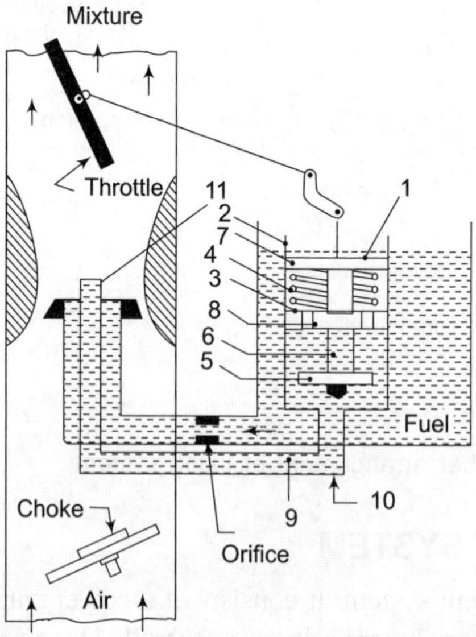


**Fig. 7.15** Acceleration pump system

The free piston 3 is actuated by the plunger inside the cylinder through the spring, 4. A disc type check valve 5 is mounted on the spindle 6 which carries the plunger. If the engine runs under steady state condition, the fuel from the float chamber entering the cylinder through the opening 7 flows through the passage 8 made in the piston, to fill the chamber of 9 and the pipe 10. With the opening of the throttle, there will be a downward movement of the plunger which forces the disc check valve shut and closes the passage 8 in the piston. Thus, the chamber 9 and pipe 10 are cutoff from the float chamber. With further opening of the throttle, the fuel in chamber 9 is compressed and forced out through the accelerating pump discharge nozzle 11. When the throttle is closed, the plunger moves upward opening the passage 8, the float chamber then starts communication with chamber 9 and pipe 10.

## 7.14 AERO-ENGINE CARBURETTORS

Aero-engines operate under conditions very differently from the engines that work at sea level. The peculiar problems that arise in aero-engines are *(i) Altitude effects, (ii) ice-formation and (iii) manoeuvrability.*

### 7.14.1 Altitude Compensation

As the altitude increases the air density reduces and the A/F ratios which is proportional to $\sqrt{\rho_a/\rho_f}$ reduces, i.e., A/F becomes richer with altitude. It is worth to mention that at 7000 m of altitude, the mixture is about 40% richer as the density of air is about one half of the sea level. It is therefore essential to provide an altitude mixture compensation device in aircraft carburettors which should progressively reduce the amount of fuel with altitude.

The enrichment, $E$, due to variation of air density closely follows the relationship

$$E + 1 = \sqrt{\frac{\rho_0}{\rho}} = \sqrt{\frac{p_0 T}{p T_0}}$$

where subscript 0 refers to the calibration condition.

Figure 7.16 shows the flow characteristics of typical carburettor of various altitude levels.

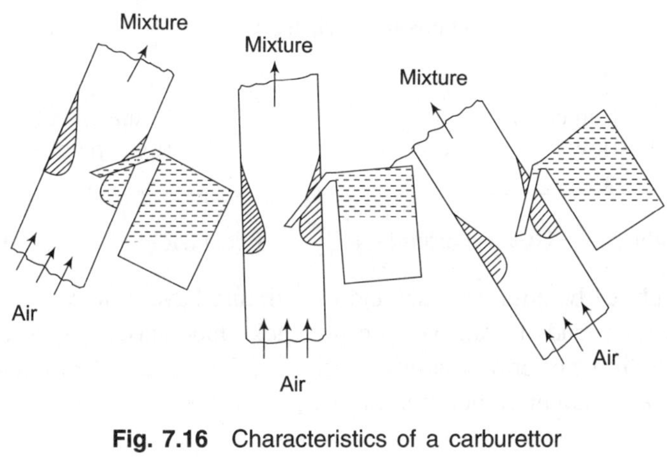


**Fig. 7.16** Characteristics of a carburettor

*One of the methods to maintain a constant air-fuel ratio at all altitudes is to vary the level of the discharge nozzle with respect to the venturi throat.* When the nozzle is in level with the throat, the pressure is the minimum at the venturi throat and fuel discharge in a high pressure at the nozzle and fuel discharge is reduced. In this way, the air-fuel ratio is maintained constant.

*The other method to maintain constant air-fuel ratio at high altitude is to use a back suction control fitted with an automatic control.*

### 7.14.2 Ice Formation

At high altitude, the temperature of the air falls and may be well below the dew point, resulting in the condensation of water vapour present in the air. The freezing of the condensed water may result in the formation of ice. A considerable amount of ice accumulated on the throttle edges reduces the area of flow, which can sometimes prove disastrous. *The formation of ice is avoided either (i) by heating the incoming air or (ii) by introducing alcohol which reduces the freezing point of water vapor.*

### 7.14.3 Manoeuvrability

The military aircraft, in particular, have to make steep dives or inverted flight, etc. Under such conditions, the carburettor may be flooded or cut-off completely which can have disastrous effects. To avoid this, the fuel is injected either into the induction manifold or directly into the cylinder. This is called **injection carburettor**.

## 7.15 TYPES OF CARBURETTORS

The carburettors are classified as following.

### 7.15.1 Upward and Down Draught Carburettor

The carburettors described earlier are all down and upward draught carburettors. In the down draught carburettor, air enters from the top while in the upward, the air enters from the bottom and the mixture goes in upward direction. *In the passenger car, down draught is more common.*

**Types of Carburettor**

- Direction of air flow
  - Down draught
  - Upward draught
- Arrangement of float chambers
  - Eccentric
  - Concentric

In the up-draught carburettor, the fuel and the mixture have to move against gravity, and the fuel being heavier tends to fall back on the choke tube surface. Whereas, in the down draught carburettor, force of gravity assists the flow of the fuel and the mixture in the suction manifold. The down draught carburettor can be installed at a level higher than that of the induction manifold to give greater accessibility. Its main disadvantage is the possibility of the petrol getting into the cylinder through the manifold when the float chamber gets flooded due to some defects.

### 7.15.2 Eccentric and Concentric Types

*When the float chamber is placed on one side of the venturi tube it is called an eccentric type.* This arrangement suffers from the disadvantages of supplying either too rich or too lean a mixture when the engine (carburettor) is inclined, e.g., the carburettor of a car moving on steep gradient shown in Fig. 7.17(a).

The drawback of the above carburettor is eliminated by *arranging the float chamber around the venturi. This arrangement is called concentric tube (Fig. 7.17(c)).*

With this arrangement, the fuel in the float chamber is always at the same level as the fuel nozzle as shown in Fig. 7.17(b).

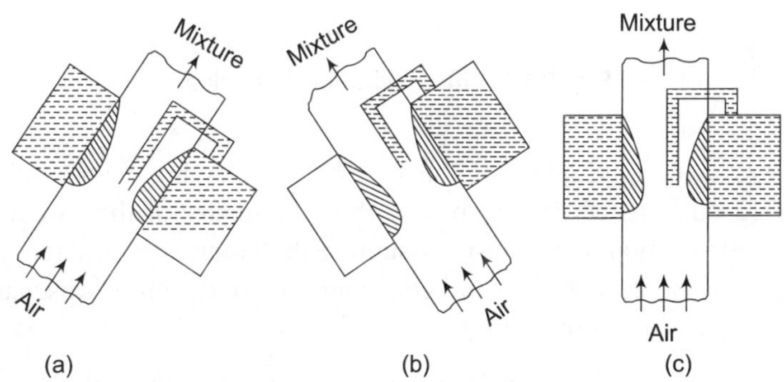


**Fig. 7.17(a, b)** Eccentric carburettor **Fig. 7.17(c)** Concentric carburettor

## 7.16 FUEL PUMP FOR SI ENGINES

Figure 7.18 shows a fuel pump generally used in American passenger cars. *It is actuated by means of a cam or eccentric on the engine camshaft.* As the engine camshaft and the eccentric rotate, they actuate the rocker arm and give a downward movement to it. Due to the downward movement of the rocker arm, the link connected with the diaphragm moves upward against the pressure exerted by the diaphragm spring. Due to upward movement of link, this diaphragm also moves upward which creates a vacuum in the pump chamber, *A*. This (vacuum) suction draws fuel supply tank, causing the fuel to enter the bowl through the inlet passage. From the sediment bowl, the fuel passes through the inlet valve, *B*.

During the upward movement of rocker arm, the spring forces the diaphragm downward forcing the fuel through the outlet valve *D*, and thus to the carburettor through the outlet pipe *G*. If the carburettor float chamber is full and the needle valve is closed, a pressure will be created in the pump chamber *A*. This holds the diaphragm down against the spring pressure, so that, the pump is inoperative in the upward position until the carburettor demands further fuel and, the float needle is open. The pressure produced by this type of pump is of the order of 0.3 to 0.4 bar gauge.

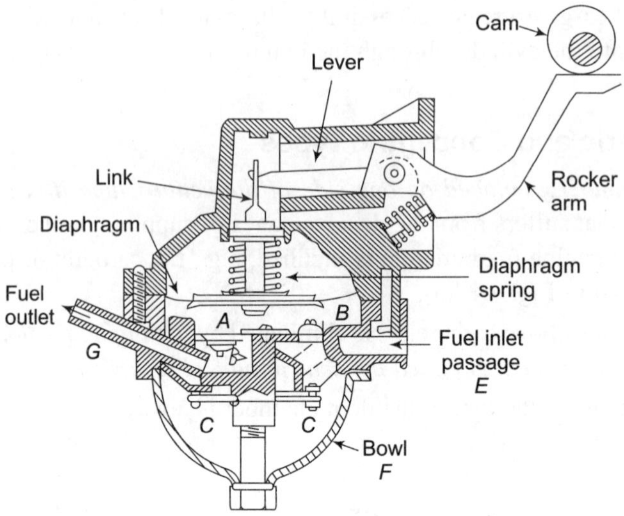


**Fig. 7.18** Fuel pump for SI engine

**EXAMPLE 7.4** A four-cylinder, four-stroke SI engine having cylinder dia of 10 cm and stroke length of 12 cm runs at 2000 rpm. The throat diameter of the carburettor venturi is 3.5 cm. Calculate the suction at the throat assuming the volumetric efficiency of the engine to be 70%. Take density of air to be 1.2 kg/m$^3$ and the coefficient of discharge of air flow 0.8. Neglect the effect of compressibility.

***Solution*** Refer to Fig. 7.8. The actual mass flow rate passing through the venturi throat is given by

$$\dot{m}_a = c_{da}.A_2.C_2.\rho_a\,; \quad \text{kg/s} = c_{da}\,A_2\,\sqrt{2\rho_a\,(p_1 - p_2)}$$

This mass flow rate will be sucked by the engine cylinder. Thus, actual mass flow rate sucked by 4-stroke engine is given by

$$\dot{m}_a = \rho_a\,AL\,\eta_v\,\frac{N}{2 \times 60}.x\,; \text{ kg/s}$$

where $x$ = number of cylinders

or
$$\dot{m}_a = \rho_a.\frac{\pi}{4}\,D^2.L.\frac{N}{2 \times 60}\,\eta_v.x$$

$$= 1.2 \times \frac{\pi}{4}\,(0.1)^2 \times 0.12 \times \frac{2000}{2 \times 60} \times 0.7 \times 4 = 0.0525, \text{ kg/s}$$

Thus, substituting the value of $\dot{m}_a$ in above expression, we get

$$0.0525 = 0.8 \times \frac{\pi}{4}\,(0.035)^2 \times \sqrt{2 \times 1.2(p_1 - p_2)}$$

or $$(p_1 - p_2) = \frac{0.0525^2}{\left[0.8 \times \frac{\pi}{4} \times (0.0353)^2\right]^2 \times 2 \times 1.2} = 1938.54 \text{ N/m}^2$$

or $$\text{Suction} = p_1 - p_2 = 0.0193854 \text{ bar}$$ ***Ans.***

**EXAMPLE 7.5** Design a carburettor for a 4-stroke, 4-cylinder petrol engine for the following data:

Engine capacity = 1500 cc

Speed at maximum power developed = 4200 rpm

Volumetric efficiency of the engine = 0.7

A/F = 13:1

Velocity of air at throat at peak power = 90 m/s.

Coefficient of discharge for venturi and fuel jet = 0.85 and 0.7

The diameter of the emulsion tube = $\frac{1}{25}$ of throat dia.

The level of petrol surface below the venturi throat = 6 mm

The specific gravity of fuel = 0.74

Atmospheric pressure and temperature = 1.013 bar and 20°C.

***Solution*** Refer to Fig. 7.8. The design includes the calculation of main dimensions such as throat dia, fuel jet dia, choke dia, thickness of venturi and tube, etc.

The actual mass flow rate of air sucked by the 4-stroke engine cylinder in

$$\dot{m}_a = V_{\text{cyl}} . \eta_v . \rho_a \frac{N}{2 \times 60}; \text{ kg/s}$$

where $V_{\text{cyl}}$ = engine capacity

$$\rho_a = \frac{p_1}{RT_1} = \frac{p_a}{RT_a} = \frac{1.013 \times 10^2}{0.287 \times (273 + 20)} = 1.204 \text{ kg/m}^3$$

$$\therefore \quad \dot{m}_a = 1500 \times 10^{-6} \times 0.7 \times 1.204 \times \frac{4200}{2 \times 60} = 0.0442 \text{ kg/s}$$

For compressible flow, velocity at the throat (Section 2–2) is given by

$$C_2 = \left[2c_p\, T_1 \left\{1 - \left(\frac{p_2}{p_1}\right)^{(\gamma-1)/\gamma}\right\}\right]^{1/2}$$

or $$90 = \left\{2 \times 1.005 \times 293 \times 10^3 \left[1 - (p_2/p_1)^{(\gamma-1)/\gamma}\right]\right\}^{1/2}$$

or $$1 - (p_2/p_1)^{(\gamma-1)/\gamma} = \left[\frac{90^2}{2 \times 1.005 \times 293 \times 10^3}\right] = 0.01375$$

or $$\frac{p_2}{p_1} = (0.98625)^{\gamma/(\gamma-1)} = (0.98625)^{1.4/0.4} = 0.09526$$

or $$p_2 = 0.9526 \times 1.013 = 0.9649 \text{ bar}$$

For process 1–2, we have, the volume flow rate at throat is given by

$$p_1 \dot{V}_1^\gamma = p_a \dot{V}_a^\gamma = p_2 \dot{V}_2^\gamma$$

or $$\frac{\dot{V}_2}{\dot{V}_1} = (p_1/p_2)^{1/\gamma} = \left(\frac{1}{0.9649}\right)^{1/14} = (1.0363)^{1/1.4} = 1.0257$$

or $$\dot{V}_2 = 1.0257\,\dot{V}_1 = \frac{1.0257 \times \dot{m}_a}{1.204} = \frac{1.0257 \times 0.0442}{1.204} = 0.03765 \text{ m}^3/\text{s}$$

The area at the throat is given by

$$\dot{V}_2 = c_{da} A_2 C_2$$

or $$A_2 = \frac{\dot{V}_2}{c_{da} C_2} = \frac{0.03765}{0.85 \times 90} = 4.9215 \times 10^{-4} \text{ m}^2$$

or $$\frac{\pi}{4}(D_2^2 - d_2^2) = 4.9215 \times 10^{-4}$$

Here, $$d_2 = \frac{D_2}{2.5} \quad \text{where } d_2 = \text{dia of emulsion tube}$$

or $$\frac{\pi}{4}\left(D_2^2 - \frac{D_2^2}{2.5^2}\right) = 4.9215 \times 10^{-4}$$

or $$D_2 = \sqrt{\frac{4.9215 \times 10^{-4} \times 4 \times 0.84}{\pi}} = 0.02294 \text{ m}$$

or $$D_2 = 22.94 \text{ mm}$$ ***Ans.***

The mass flow rate of fuel is given by

$$\dot{m}_f = \frac{\dot{m}_a}{A/F} = \frac{0.0442}{13} = 3.4 \times 10^{-3} \text{ kg/s}$$

The mass flow rate of fuel is also is given by

$$\dot{m}_f = C_{df} \cdot A_f \cdot C_f \rho_f = c_{df} \cdot A_f \cdot [2\rho_f (p_1 - p_2 - gh\rho_f)]^{1/2}$$

or $$3.4 \times 10^{-3} = 0.7 A_f \left[2 \times 0.74 \times 10^3 \left(1.013 - 0.9649 - \frac{9.8 \times 6 \times 0.74 \times 10^3}{10^3 \times 10^5}\right) \times 10^5\right]^{1/2}$$

$$= 0.7\, A_f \times 0.2656 \times 10^4$$

or $$A_f = 3.4 \times \frac{10^{-3}}{0.7 \times 0.2656 \times 10^4} = 18.287 \times 10^{-7}\ \text{m}^2$$

or $$\frac{\pi}{4} d_f^2 = 18.287 \times 10^{-7} \times 10^6\ \text{mm}^2$$

or $$d_f = \left(\frac{18.287 \times 10^{-1} \times 4}{\pi}\right)^{1/2} = 1.525\ \text{mm}$$

Diameter of fuel jet = $d_f$ = 1.525 mm ***Ans.***

The choke dia, i.e., the inlet venturi diameter is calculated by continuity equation

$$\dot{m}_a = \rho_1\, C_1\, A_1$$

But $C_1$ is calculated by energy equation

$$h_1 + \frac{C_1^2}{2} = h_2 + \frac{C_2^2}{2}$$

or $$c_p\, T_1 + \frac{C_1^2}{2} = c_p\, T_2 + \frac{C_2^2}{2}$$

Here, $$\frac{T_2}{T_1} = \left(\frac{p_2}{p_1}\right)^{(\gamma-1)/\gamma}$$

or $$T_2 = 293\ (0.9526)^{0.4/1.4} = 288.96\ \text{K}.$$

Thus, $$1.005 \times 293 + \frac{C_1^2}{2 \times 10^3} = 1.005 \times 288.96 + \frac{90^2}{2 \times 10^3}$$

or $$294.46 + \frac{C_1^2}{2 \times 10^3} = 290.41 + 4.05$$

or $$C_1 = 0$$

Since the aerodynamic losses have not been considered, so inlet velocity is coming about zero. The entering velocity may be assumed to be equal to 20 m/s. Thus,

$$0.0442 = 1.204 \times 20\ A_1$$

or $$A_1 = \frac{\pi}{4} D_1^2 = \frac{0.0442}{1.204 \times 20} = 1.835 \times 10^{-3}$$

or $$D_1 = 0.0483\ \text{m} = 48.3\ \text{mm}$$ ***Ans.***

The thickness of venturi and fuel jet tubes are calculated based on the mechanical design considerations and manufacturing aspects.

This size of float chamber is calculated based on the capacity of a reservoir which may feed the engine for certain period say 30 seconds.

**EXAMPLE 7.6** A 4-stroke SI engine of 1800 cc capacity employed for experimental research work develops maximum power at 5200 rpm. The volumetric efficiency is 75% and the air-fuel ratio is 14:1. Two carburettors are to be fitted in order to achieve the maximum power when the velocity of air at the throat is 100 m/s. The coefficient of the discharge for the venturi is assumed to be 0.8 and that of main jet is 0.7. An allowance should be made for emulsion tube, the diameter of which may be assumed as 1/3 of the throat diameter. The gasoline surface is 6 mm below the throat. Calculate the diameter of the throat and main jet fuel. Assume specific gravity of the gasoline to be 0.75. $p_a$ and $T_a$ may be taken as 1.013 bar and 20°C respectively. Take $\eta_v = 0.75$.

***Solution*** Refer to Fig. 7.8.

The actual total mass of rate of air sucked by the engine cylinder is given by

$$\dot{m}_{at} = V_{\text{cyl}} \cdot \eta_v \cdot \rho_a \frac{N}{2 \times 60} \text{ ; kg/s}$$

Here, $$\rho_a = \frac{p_1}{RT_1} = \frac{p_a}{RT_a} = 1.013 \times \frac{10^2}{0.287 \times (273 + 27)} = 1.176 \text{ kg/m}^3$$

$$\dot{m}_{at} = 1800 \times 10^{-6} \times 0.75 \times 1.176 \times \frac{5200}{2 \times 60} = 0.06879 \text{ kg/s}$$

The mass flow through each carburettor is

$$\dot{m}_a = \frac{\dot{m}_{at}}{2} = \frac{0.06879}{2} = 0.034395 \text{ kg/s}$$

For compressible flow, velocity of the throat (section 2–2) is given by

$$C_2 = \left\{2c_p\, T_1\left[1 - (p_2/p_1)^{(\gamma-1)/\gamma}\right]\right\}^{1/2}$$

or $$100 = \left\{2 \times 1.005 \times 293 \times 10^3\left[1 - (p_2/p_1)^{0.286}\right]\right\}^{1/2}$$

or $$100^2 = 2 \times 1.005 \times 293 \times 10^3\left[1 - (p_2/p_1)^{0.286}\right]$$

or $$1 - (p_2/p_1)^{0.286} = \frac{100^2}{2 \times 1.005 \times 293 \times 10^3} = 1.697 \times 10^{-2}$$

or $$\frac{p_2}{p_1} = 0.941$$

Pressure at throat $= p_2 = 1.013 \times 0.941 = 0.9532$ bar

Volume flow rate at the throat is given by

$$\frac{\dot{V}_2}{\dot{V}_1} = (p_1/p_2)^{1/\gamma} = \left(\frac{1}{0.941}\right)^{1/1.4} = 1.0443$$

or $$\dot{V}_2 = 1.0443 \times \frac{\dot{m}_a}{\rho_a} = 1.0443 \times \frac{0.034395}{1.176} = 0.0305 \text{ m}^3\text{/s}$$

The area at the throat is given by

$$\dot{V}_2 = c_{da} A_2 C_2$$

or $$A_2 = \frac{\pi}{4}(D_2^2 - d_2^2) = \frac{\dot{V}_2}{c_{da} C_2} = \frac{0.0305}{0.7 \times 100} = 4.357 \times 10^{-4}$$

or $$\frac{\pi}{4}\left(D_2^2 - \frac{D_2^2}{9}\right) = 4.357 \times 10^{-4}$$

or $$D_2 = \sqrt{\frac{4.357 \times 10^{-4} \times 9 \times 4}{\pi \times 8}} = 0.02498 \text{ m}$$

or Throat diameter $= D_2 = 24.98$ mm **Ans.**

The mass flow rate of fuel in each carburettor is given by

$$\dot{m}_f = \frac{\dot{m}_a}{A/F} = \frac{0.034395}{14} = 2.456 \times 10^{-3} \text{ kg/s}$$

The mass flow rate is also given by

$$\dot{m}_f = c_{df}.A_f\,[2\rho_f(p_1 - p_2 - gh\rho_f)]^{1/2} = 2456 \times 10^{-3}$$

$$= 0.7A_f\left[2 \times 0.75 \times 10^3 \times 10^5\left(1.013 - 0.9532 - \frac{9.8 \times 6 \times 0.75 \times 10^3}{10^3 \times 10^5}\right)\right]^{1/2}$$

or $$2.456 \times 10^{-3} = 0.7A_f(1.5 \times 10^8 \times 0.05935)^{1/2} = 0.2088 \times 10^4 A_f$$

or $$A_f = \frac{\pi}{4} d_f^2 = \frac{2.456 \times 10^{-3}}{0.2088 \times 10^4} = 11.762 \times 10^{-7}$$

or $$d_f = \sqrt{\frac{11.762 \times 10^{-7} \times 4}{\pi}} = 1.223 \times 10^{-3} \text{ m} = 1.223 \text{ mm}$$ **Ans.**

**EXAMPLE 7.7** A trainer aeroplane SI engine is equipped with a carburettor. Calculate the change of A/F when the plane takes off from sea level to a height of 5000 m. At sea level, the A/F ratio is 15:1 where pressure is 1 bar and temperature is 27°C. The variation of temperature of air with altitude is $T = T_{s1} - 0.0075\,z$ where $z$ is in $m$ and $T_{s1}$ is in $K$. The air pressure variation with altitude $z = 19200 \log_{10}(1/p_z)$ where $p_z$ is in bar. Calculate the A/F at sea level at the altitude of 5000 m and comment on the result if the variation of A/F altitude is linear.

***Solution*** The temperature at an altitude of 5000 m is given by

$$T_z = T_{s1} - 0.0075\,z = 300 - 0.0075 \times 5000 = 262.5 \text{ K}$$

The pressure at an altitude of 5000 m is given by

$$z = 19200 \log_{10} (1/p_z)$$

or
$$5000 = 19200 \log_{10} (1/p_z)$$

or
$$\frac{1}{p_z} = 10^{5000/19200} = 1.821$$

or
$$p_z = 0.549 \text{ bar}$$

Density of air at this altitude

$$= \rho_z = \frac{p_z}{RT_z} = 0.549 \times \frac{10^2}{0.287 \times 262.5} = 0.7287 \text{ kg/m}^3$$

Density of air sea level $= \rho_{sl} = \dfrac{p_{sl}}{RT_{sl}} = \dfrac{1 \times 10^2}{0.287 \times 300} = 1.161 \text{ kg/m}^3$

We know that $\dfrac{A/F_z}{A/F_{sl}} = \sqrt{\dfrac{\rho_z}{\rho_{sl}}} = \sqrt{\dfrac{0.7287}{1.161}} = 0.7922$

or
$$A/F_z = 15 \times 0.7922 = 11.88$$ ***Ans.***

This calculation shows that A/F decreases from sea level of 15 : 1 to 11.88 : 1 at an altitude of 5000 m. Since the variation is linear so this decrease will be linear. This result suggests that there is a need of altitude compensation in carburettor. This may be done by sensing the pressure at the inlet of venturi.

**EXAMPLE 7.8** A 4-stroke, 4-cylinders SI engine has a single carburettor. The throat of the venturi is 20 mm with a coefficient of discharge of 0.8. The fuel orifice has a diameter of 1.15 mm with a coefficient of discharge 0.7. The gasoline surface is 6 mm below the throat. Calculate the following:

(a) the A/F ratio for a pressure drop of 0.1 bar when the nozzle tip is neglected.
(b) the A/F ratio when nozzle tip is taken into account.
(c) the minimum or critical velocity of air required to start the fuel flow when the nozzle tip is neglected.

Take the density of air and fuel to be 1.2 kg/m³ and 760 kg/m³ respectively.

***Solution*** Refer to Fig. 7.8.

(a) When the nozzle tip is neglected

$$\dot{m}_a = c_{da} A_a \sqrt{2\rho_a (p_1 - p_2)}$$

$$\dot{m}_f = c_{df} A_f \sqrt{2\rho_f (p_1 - p_2)}$$

$$\therefore \qquad A/F = \frac{\dot{m}_a}{\dot{m}_f} = \frac{c_{da}}{c_{df}} \cdot \frac{A_a}{A_f} \cdot \sqrt{\frac{\rho_a}{\rho_f}} = \frac{0.8}{0.7} \cdot \left(\frac{20}{1.15}\right)^2 \cdot \sqrt{\frac{1.2}{760}} = 13.73 \qquad \textbf{\textit{Ans.}}$$

(b) When the nozzle tip is taken into account

$$A/F = \frac{\dot{m}_a}{\dot{m}_f} = \frac{c_{da}}{c_{df}} \cdot \frac{A_a}{A_f} \sqrt{\frac{\rho_a}{\rho_f}} \sqrt{\frac{(p_1 - p_2)}{p_1 - p_2 - gh\,\rho_f}}$$

or

$$A/F = \frac{0.8}{0.7} \cdot \left(\frac{20}{1.15}\right)^2 \sqrt{\frac{1.2}{760}} \cdot \sqrt{\frac{0.1}{(0.1 - 6 \times 10^{-3} \times 9.80 \times 760) \times 10^{-5}}}$$

$$= 13.3 \sqrt{\frac{0.1}{(0.1 - 0.006 \times 0.07448)}} = 13.76 \qquad \textbf{\textit{Ans.}}$$

(iii) With the pressure of nozzle tip, the fuel flow will only start when the minimum velocity of air required to create the required pressure difference for the flow of fuel to overcome nozzle tip effect. Thus $p_1 - p_2 = \rho_f\, g\, h_f$. If $C_1 = 0$, then

$$\frac{p_1}{\rho_a} = \frac{p_2}{\rho_a} + \frac{C_2^2}{2}$$

or

$$\frac{C_2^2}{2} = \frac{p_1 - p_2}{\rho_a}$$

For critical velocity, $p_1 - p_2 = \rho_f\, g\, h_f$, thus,

$$C_{\min} = \sqrt{2g\, hf \frac{\rho_f}{\rho_a}} = \sqrt{2 \times 9.8 \times \frac{6}{1000} \times \frac{760}{1.2}} = 8.63 \text{ m/s} \qquad \textbf{\textit{Ans.}}$$

It is to be noted that the nozzle tip is provided in the carburettor to avoid spilling of the fuel due to vibration or inclination of the carburettor.

**EXAMPLE 7.9** A four-cylinder four-stroke spark ignition engine runs at 4200 rpm. The bore and broke of the engine are 80 mm and 75 mm respectively. The engine uses a fuel having 84.21 per cent carbon and 15.79 per cent hydrogen by mass. The volumetric efficiency of the engine is 83% when ambient conditions are 1 bar pressure and 298 K temperature. Calculate the throat diameter of the venturi tube if the air velocity at the throat does not exceed 100 m/s and the actual quantity of air supplied is 0.94 of the stoichiometric volume. Also calculate the fuel flow rate and the pressure drop at the venturi throat. Take coefficient of discharge at venturi throat as 0.98.

***Solution*** The theoretical quantity of air required for complete combustion is given by

$$(\dot{m}_a)_{th} = \left[c \times \frac{32}{12} + H_2 \times 8\right] / 0.23$$

$$= \left[0.8421 \times \frac{32}{12} + 0.1579 \times 8\right]/0.23 = 15.255 \text{ kg/kg of fuel}$$

$$(\dot{m}_a)_{\text{act}} = 0.94 \times (\dot{m}_a)_{th} = 0.94 \times 15.255 = 14.339 \text{ kg/kg of fuel}$$

$$\text{Stroke volume} = V_s = \frac{A.L.x.N}{2 \times 60} \text{ when } x = \text{no. of cylinders}$$

$$= \frac{\pi}{4}(0.08)^2 \times 0.075 \times 4 \times \frac{4200}{2 \times 60} = 0.0528 \text{ m}^3/\text{s}$$

The density of air at ambient condition

$$\rho_a = \frac{p_a}{RT} = \frac{10^5}{287 \times 298} = 1.169 \text{ kg/m}^3$$

Density of fuel vapour entering engine cylinder assuming perfect gas

$$\rho_f = \frac{p}{RT} = \frac{10^5}{(8314/114) \times 298} = 4.6 \text{ kg/m}^3$$

The mixture of air and fuel will acquire the actual volume. Thus,

$$V_{\text{act}} = 0.94 \times 0.0528 = 0.04963 \text{ m}^3/\text{s}$$

$$V_{\text{act}} = 0.04963 = \frac{\dot{m}_a}{\rho_a} + \frac{\dot{m}_f}{\rho_f} = \dot{m}_a\left[\frac{1}{1.169} + \frac{1}{14.339 \times 4.6}\right] = 0.87049\, \dot{m}_a$$

or
$$\dot{m}_a = \frac{0.04963}{0.87049} = 0.057 \text{ kg/s}$$

$$\dot{m}_f = \frac{0.057}{15.255} = 0.03736 \text{ kg/s}$$

The maximum permissible velocity at the throat is 100 m/s, hence

$$C_2 = 100 = \sqrt{2c_pT_1\left[1 - \left(\frac{p_2}{p_1}\right)^{0.286}\right]} = \sqrt{2 \times 1005 \times 298\left[1 - \left(\frac{p_2}{p_1}\right)^{0.286}\right]}$$

or
$$\frac{p_2}{p_1} = 0.9428$$

or
$$p_2 = \text{pressure at throat} = 0.9428 \text{ bar}$$

The depression at the venturi throat

$$\Delta p = 1 - 0.9428 = 0.05272 \text{ bar} = 589 \text{ mm of water}$$

The density of air at the throat

$$\rho_2 = \frac{p_2}{RT_2} = \frac{p_1}{RT_1}\left[\frac{p_2}{p_1}\right)^{1/\gamma} = 1.169 \times (0.9428)^{1/\gamma} = 1.121 \text{ kg/m}^3$$

From continuity equation, at venturi throat

$$\dot{m}_a = c_d.A_2\,\rho_2\,C_2 = 0.057$$

or $$A_2 = \frac{0.057}{0.98 \times 1.121 \times 100} = 0.0004806 = [\pi/4)\,D_2^2$$

Diameter of venturi throat is given by

$$D_2 = \left(\frac{0.0004806 \times 4}{\pi}\right)^{1/2} = 0.0247 \text{ m} = 2.47 \text{ cm}$$ ***Ans.***

## 7.17 BRANDED AUTOMOBILE CARBURETTORS

Various types of branded carburettors are used in automobiles. Some of them are listed below.

(a) Solex carburettor (b) Carter carburettor (c) S.U. carburettor

A brief description of these branded carburettors is given below.

### 7.17.1 Solex Carburettor

It is a famous carburettor for its ease of starting, good performance and reliability. It is available in various models and used in many automobiles such as Fiat, and standared cars and Willey jeep. Now, all use **MPFI system**.

Figure 7.19 shows a schematic diagram of a downdraught Solex carburettor equipped with various provisions. It consists of various fuel circuits such as starting, idling or low speed operation, normal running acceleration, etc. Various circuits are discussed below.

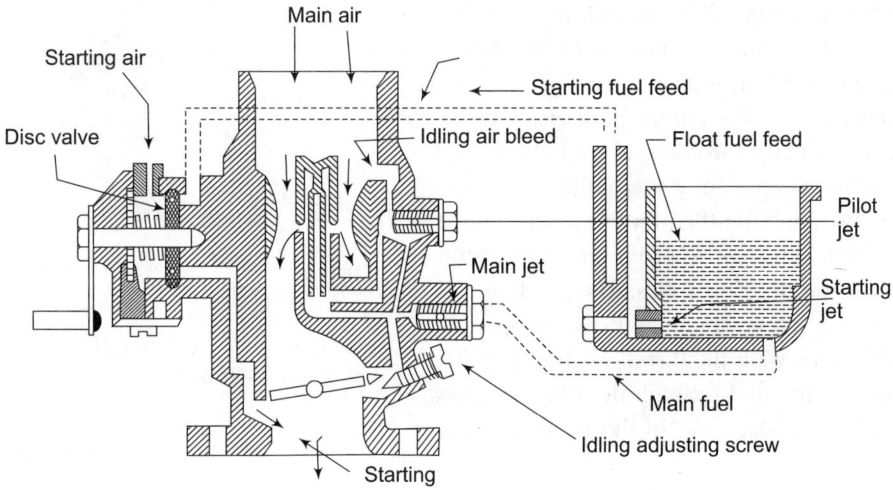


**Fig. 7.19** Solex carburettor

**(a) Normal running.** For normal running, the fuel is provided by the main jet and the air by the choke tube or venturi. Due to depression at the throat, the fuel from the main jet passes into the well of the air bleed emulsion system which has lateral holes (Fig. 7.19). The correct balance of air and fuel is automatically maintained by air entering through and being calibrated by the air correction jet. The metered emulsion of fuel and air is discharged downward through the venturi which is controlled by throttle.

**(b) Starting.** The unique feature of Solex carburettor is the provision of a bi-starter or a progressive starter valve '*C*' which is in the form of flat disc with holes of different sizes (Fig. 7.20). The fuel from the flat chamber enters the starting passage *B* through the starting jet, and finally enters another starting passage *A* through starting valve, *C*. The starter lever brings the starter valve opposite to the fuel supply holes. When the throttle valve is in closed position, the engine suction is applied to the starting passage *A* and thus the air enters from the starting air jet and fuel enters from the small holes provided in the disc valve. This mixture is sufficiently rich to start the engine. As the throttle valve opens, the engine suction is applied to the main jet and the starting lever is turned to the off position thereby stopping the supply of the fuel through the disc valve, *C*.

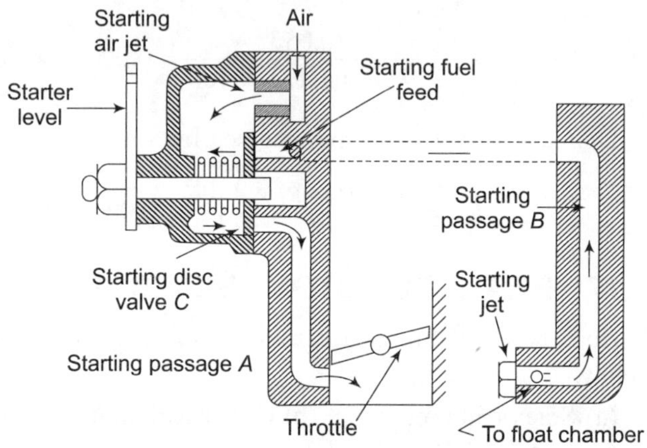


**Fig. 7.20** Starting circuit of the Solex carburettor

**(c) Idling.** Figure 7.21 shows the idling circuit of Solex carburettor. For idling rich mixture is required, which is met by the combined effect of pilot fuel jet, pilot air jet and idling adjusting screw. The fuel from the main jet circuit enters the idle passage through the pilot jet and air enters this passage through the pilot airjet. In the idling condition, the throttle is almost closed and hence suction is applied at the outlet of the idle jets thereby supplying the mixture to the engine through the idle port provided immediately after the throttle on the down stream side. It is the idle adjusting screw which controls the opening of the idle port. With the throttle valve opening, the supply of the fuel through the idle circuit is restricted due to decrease in the suction at the idle passage.

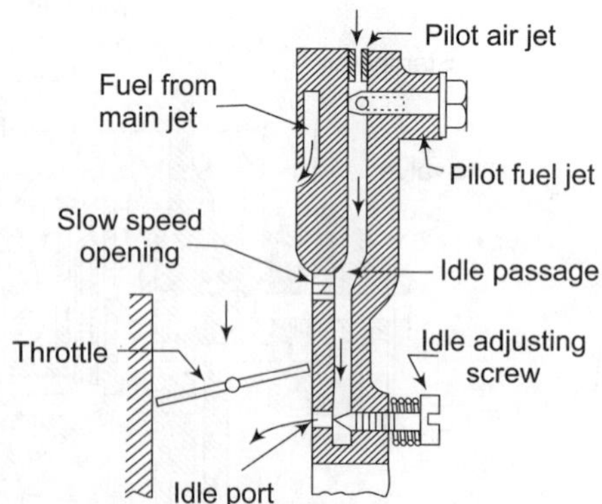


**Fig. 7.21** Idle circuit of the solex carburettor

**(d) Acceleration.** During acceleration, rich mixture of A/F is needed. Figure 7.22 shows accelerating pump system of Solex carburettor which is of diaphragm type. With the pressing of accelerator pedal, the pump lever connected to it is also pressed. This movement of the pump lever pushes the diaphragm towards left with the help of push rod thereby compressing the fuel and forming it into the main jet. The removal of force from the pump lever releases the diaphragm to its original position by means of a spring provided on the left side of the diaphragm. This movement of the diaphragm creates a suction on the left side of the diaphragm which opens the pump valve and admits the fresh fuel into the pump circuit from float chamber.

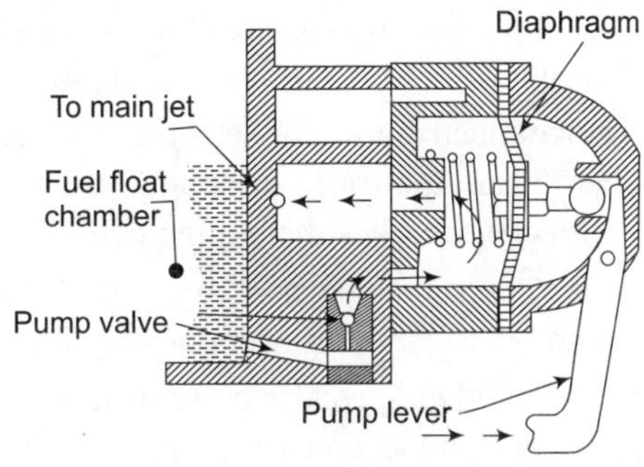


**Fig. 7.22** Diaphragm type acceleration pump system of the solex carburettor

## 7.17.2 Carter Carburettor

Figure 7.23 shows a schematic diagram of downdraught type Carter carburettor. It has conventional type of float chamber in which fuel enters from the fuel system. The air enters from the top to the choke tube, the choke valve of which remains open during running. *The special features of this carburettor is the triple venturi diffusing type of choke which has three venturies,* the smallest lies above the level in the float chamber, other two below the fuel level, one below other.

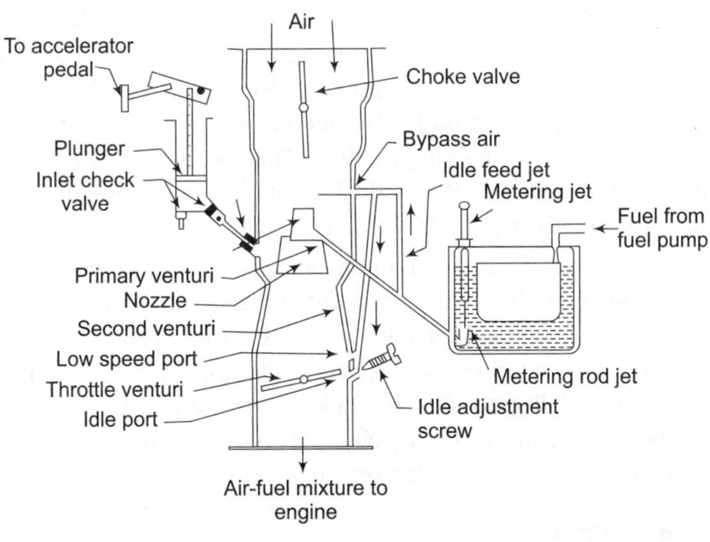


**Fig. 7.23** Carter downdraught type carburettor

**(a) General.** Event at very low speeds, suction in primary venturi is adequate to draw the fuel. The fuel nozzle is placed in the primary venturi at an angle delivering the fuel upwards against the air stream ensuring an even flow of finely divided atomized fuel. The mixture from the primary venturi so produced passes centrally through the second venturi where it is surrounded by a blanket of air stream and finally this leads to the third (main) venturi, where again the fresh air supply insulates the stream from the second venturi. By this way, the mixture reaches the engine cylinder in atomized form.

The main advantages of multiple venturies is better formation of mixture resulting in steady and smooth operation event at very low as well as very high engine speeds.

The another special feature of this carburettor is the use of mechanical metering method. The float chamber consists of a metering rod actuated by a mechanism connected with the main throttle valve. The metering rod has two or more steps of diameter and the area of opening between the metering rod jet and metering rod governs the amount of gasoline drawn into the engine. In the case of high speed, the metering rod is lifted such that the smallest section of the rod is in the jet and the maximum quantity of gasoline flows out to mix with maximum amount of air corresponding to full throttle opening to supply the required mixture.

**(b) Starting.** In the air circuit, a choke is provided for starting. This choke valve is a butterfly valve, one half of which is spring controlled and it is hinged at the centre. For cold starting, the engine is fully choked and as a result the whole engine suction is applied at the main nozzle, which delivers the maximum fuel. The air flow rate is the minimum and hence the mixture so supplied to the engine is very rich. Once the engine is started, the spring controlled half of the choke valve is sucked open to provide correct amount of air during warming up action resulting in very less rich mixture.

**(c) Idling circuit.** Idling requires a rich mixture in small quantity. At idling condition throttle valve is nearly closed and as a result, the whole of the engine suction if applied at the idle port. Consequently, fuel is drawn through idle feed jet and air through the first by-pass and by this way rich mixture is supplied. At low speed, the throttle valve is opened further and as result, the engine suction is also in communications with main fuel jet and so the main fuel nozzle also starts supplying the fuel. This means that at low speed, the fuel is delivered both the main jet and idle jet and air from main venturi and idle air port increasing the quality of mixture and slightly reducing the rich mixture.

**(d) Acceleration.** The acceleration circuit consists of a cylinder plunger working inside a cylinder consisting of inlet check valve and outlet check valve and acceleration fuel jet. The pump plunger is connected to accelerator pedal by throttle control rod. With the sudden opening of the throttle valve by pressing the accelerator pedal, the pump is actuated and a small quantity of petrol is delivered into the choke tube by a jet. As soon as the accelerator pedal is released, the plunger moves up thereby, seeking fuel from the float chamber for the next operation so that the fuel level in the float chamber is maintained.

### 7.17.3 S.U. Carburettor

All the carburettors may be grouped in two categories, either constant choke type a constant vacuum type with automatic variable choke. Zenith solex and Carter fall under the former one while S.U. fall under the latter one. Figure 7.24 shows a simplified sketch of S.U. carburettor of horizontal type.

It consists of a sliding piston, the upper end of which is given a flat form which is known as suction disc. The lower end of the piston is provided with a taper needle which is inserted into the main fuel jet. The up and down movement of the piston and the suction disc is guided by means of a piston rod which reciprocates in piston rod guide. When the piston is moved up and down, the taper needle also moves down and down with the main jet, thus varying the fuel flow. The piston always remains loaded by a helical spring and the movement of the piston controls the main air passage and above thus create an automatic variable choke. The portion above the disc is called suction chamber which is in communication with the air passage by means of a slot provided in the piston.

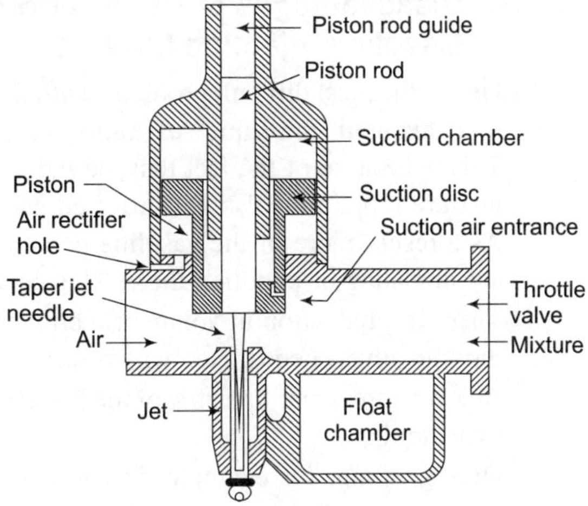


**Fig. 7.24** Horizontal type S.U. carburettor

The main jet of the carburettor can be moved up and down along the taper needle by operating a lever from the dashboard. It is this movement which adjusts the mixture strength throughout the operating range of the carburettor. The lower portion of the suction disc is connected to the atmosphere by means of an air rectifier hole and upper portion to the throttle air passage. The special feature of this carburettor is that it does not have any idling, slow running and accelerating system.

Two forces act on the piston—one downward due to the weight of the piston and other upward due to vacuum in the suction chamber. As a result, at a particular instant the position of the piston is balanced by weight and constant vacuum produced above the piston. During starting, a rich mixture is needed which is supplied by pulling the jet downwards with the lever attached to it. With the opening of throttle, more air is allowed to pass through the inlet due to upward movement of this piston. The taper needle will also move up along with piston which ensures more flow of fuel from the main jet. By this way, the amount of fuel and air is varied with the speed of the engine ensuring the required mixture by the engine.

## 7.18 FUEL SYSTEM FOR SI CARBURETTOR ENGINES

The purpose of both carburettor fuel system and the fuel injection system have the same job, that is, supplying a correct amount of mixture of air and fuel to the engine. The carburettor system consists of a fuel tank, fuel pump, fuel filter, carburettor, intake manifold and fuel lines. The fuel line are tubes connecting the tank fuel pump and carburettor.

## 7.19 FUEL INJECTION SYSTEM FOR SI ENGINES

Most of SI engines had been using modern carburettors in the past, but have certain drawbacks as given below.

### 7.19.1 Disadvantages of Carburettors and Invention of Gasoline Injection System

(i) One of the most difficult problems with the carburettor system of multi-cylinder engine is to maintain the same A/F ratio and same quantity of mixture in each cylinder. This is because of the fact that the induction passages are of unequal lengths and the mixture proportion also gets affected due to fuel condensation in induction manifold. As a result, more of the gasoline particles continue to move to the end of manifold, accumulating or puddling there. This enriches the mixture going to the end cylinder.

(ii) There is a reduction in volumetric efficiency due to pressure of choke tubes, fuel jets throttle valves, bends, etc., which serves as resistance to flow.

(iii) At low temperature, freezing of the fuel takes place unless special device is incorporated to avoid this.

(iv) During titling, for example, in aeroengines, surging of the fuel takes place unless special means are not provided to take care of.

(v) There is a chance of backfiring.

Some of the above disadvantages of carburettor system may be avoided by **gasoline (petrol)** ***fuel injection system***. *However, there is a difference between gasoline injection method and diesel injection method.* In the diesel engine, the fuel is injected into the cylinder and the moment at which fuel injection commences has an important bearing on the performance since the fuel ignites with the highly compressed charge and so it usually commences at about 15° to 25° before TDC during compression stroke. The injection pressure has to be higher than the maximum compression pressure which is about 120 bar and even above 200 bar as the fuel is injected in the cylinder.

*In gasoline, injection, timing is not a critical parameter and hence the fuel may be injected during the suction stroke in the intake manifold of each cylinder* (Fig. 7.25) *at low pressure.* There are two methods of gasoline injections.

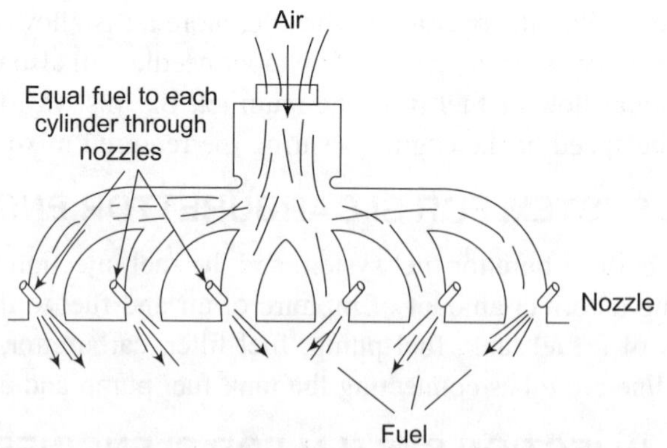


**Fig. 7.25** Basic principle of gasoline injection; (MPFI, near Port)

(a) *Continuous gasoline injection system,* (b) *Timed gasoline injection system*

Both methods may use mechanical or electronic devices. Since it is an electronic age, so electronic fuel injection system is preferred. It is commonly called **multi-point fuel injection (MPFI)** system.

## 7.20 TYPES OF ELECTRONIC FUEL INJECTION (EFI) SYSTEM

The classification of electronic fuel injection (EFI) system may be done as follows:

(i) Mainfold injection

(ii) Port injection

 (a) Timed and (b) continuous

(iii) Gasoline direct injection into cylinders (GDI)

EFI system can be grouped under two heads, namely, ***single point fuel injection (SPFI), and multi-point fuel injection (MPFI)***. In SPFI, one or two fuel injectors are mounted at proper location in the engine to inject fuel into the intake manifold at one point. It is also known as **throttle body injection**. In *MPFI system, one injector is allotted to each cylinder. As the name implies, fuel is injected at more than one location.* For example, in the case of four-cylinder engine, fuel is injected at four points, one at the port of each cylinder. MPFI system is invariably used, and is also known as port injection. Gasoline injection in SI engines can be either **continuous** or **timed injection**.

Figure 7.26 shows the different methods of fuel injection in SI engines: four strokes and two strokes.

### 7.20.1 Continuous Gasoline Injection System

*As the name implies in this system, fuel is sprayed at low pressure continuously into the air supply near inlet manifold.* The amount of fuel sprayed varies with engine speed and power demands. Amount of fuel is governed by air throttle openings, more throttle opening, more is the fuel injected.

*This system has the potential of promoting efficient atomization of the fuel and uniform mixture strength to all cylinders along with providing higher volumetric efficiency.* The system is simpler in many ways and less expensive. **Fuel pressure ranges from 7 bar to 10 bar**.

### 7.20.2 Timed Gasoline Injection System

*The system is similar to the system used in CI engines in which the measured quantity of fuel is injected in the cylinder at a definite time.* Generally, **two types** of system are used in time injection system. One is **multiple *plunger jerk pump system*** which employs a pump having a separate plunger for each cylinder and usually a high injection nozzle pressure. The delivery pressure of pump varies **from 100 bar to 300 bar**. The **other is *low pressure single pump and distributors system*** which employs a *single plunger or gear electric pump delivering fuel at relatively* ***low pressure to a rotating distributor*** *which supplies fuel to each cylinder, in turn, at correct* time during suction stroke. The **pressure ranges from 3.5 bar to 7 bar**.

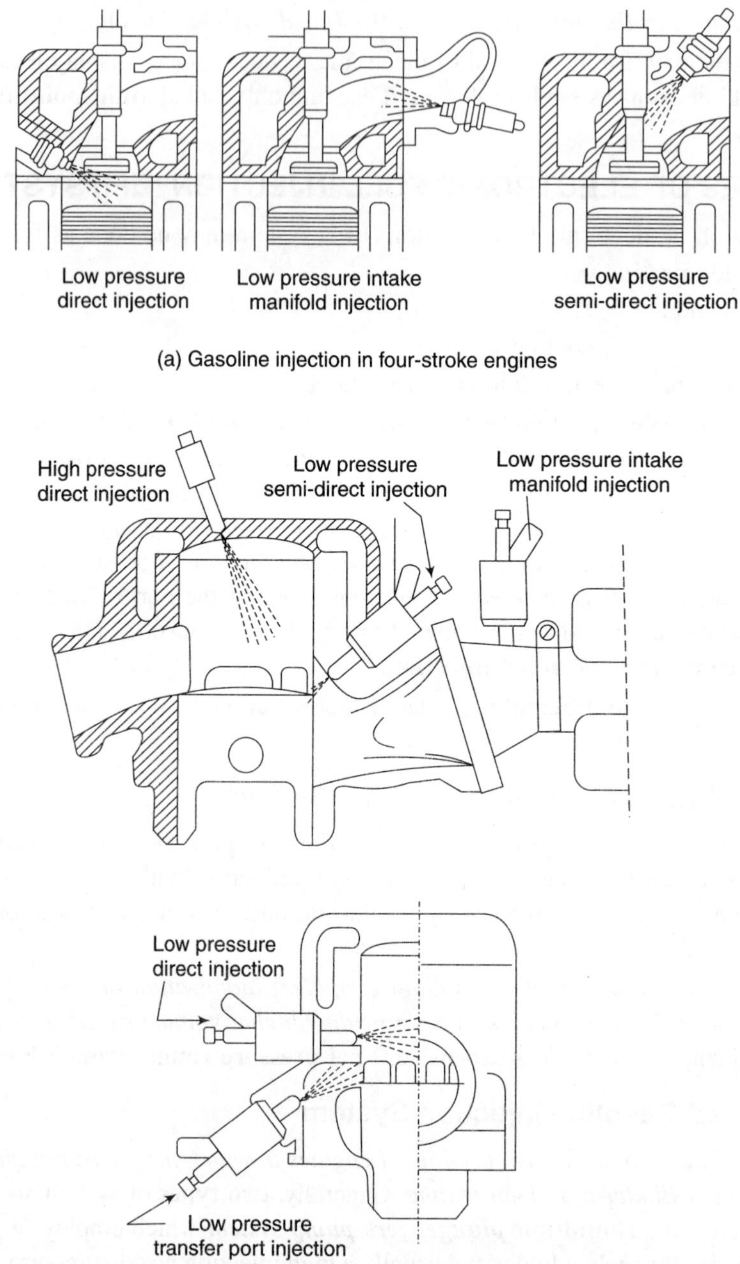


(a) Gasoline injection in four-stroke engines

(b) Fuel injection in two-stroke engines

**Fig. 7.26** Different methods of fuel injection

## 7.21 ADVANTAGES AND DISADVANTAGES OF GASOLINE INJECTION

**Advantages**

(i) It supplies the same A/F ratio and same amount of mixture to each cylinder, resulting in better performance.

(ii) It allows more efficient design of the intake manifold for air only leading to better volumetric efficiency and lower hood height of the car.

(iii) There is no need of manifold heat control valve or heated system is required

(iv) Throttle response is faster because the fuel is under pressure at the injector at all times.

(v) Exhaust emissions during deceleration can be practically eliminated.

(vi) The system can be designed to have a positive or complete fuel shut off when the throttle is released resulting in fuel economy.

(vii) Less pollution from its exhaust as compared to carbureted engine meeting EURO-III or IV or VI emission norms.

(viii) Better atomization and vaporization make less prone to incomplete combustion.

(ix) Formation of ice on throttle plate is eliminated.

**Disadvantages**

(i) High maintence cost

(ii) Possibility of malfunction of some sensors.

With all these many advantages one may wonder why every manufacturer has not gone for fuel injection. The reasons are twofold. First, there is higher cost due to complexity of the system and second the only recently electronics has been brought into the picture.

However, presently, many car engines such as Cello, Volkswagen GM, GmbH, Maruti, Hyundai, Honda, etc., have all introduced **gasoline multipoint fuel injection (MPFI system)**. *It seems that for passenger cars except racing cars, there is a **death of carburettor**.*

## 7.22 ELEMENTS OF FUEL INJECTION IN GENERAL

The function of injection of fuel is the same as that of the carburettor. It meters, atomizes and uniformly distributes the fuel throughout the mass of the air in the cylinders, and supplies the correct amount of air-fuel ratio in each cylinder as per load and speed of the engine. The following are the elements of fuel injection system.

(i) *Pumping element:* It moves the fuel from the fuel tank of the injector through piping, filters, etc.

(ii) *Metering element:* It meters and supplies the fuel at the rate demanded by the speed and load of the engine.

(iii) *Metering control:* It adjusts the rate of metering as per the speed and load of the engine.

(iv) *Mixing element:* It atomizes the fuel and mixes with the air to form a homogeneous mixture for efficient combustion.

(v) *Distributing element:* It divides the metered fuel equally among the cylinders to develop the same power from each cylinder.

(vi) *Timing element:* It allows and stops the flow of fuel-air mixture in the cylinder.

(vii) *Ambient air control:* It affects to compensate for changes in pressure and temperature of either air or fuel that may affect the various elements.

(viii) *Various sensors:* These sense many parameters as per requirement of injection.

## 7.23 SENSORS USED IN EFI SYSTEM

Electronic fuel (gasoline) injection (EFI) system employs a computer, sensors, and a solenoid operated fuel injector to meter and inject the **right amount of fuel either at manifold or port or inside the cylinder as demanded by the speed and load of the engine**. Sometimes, air flow rate is also controlled especially during cold engine starting. The following sensors are used in EFI system.

(i) *Air flow sensor:* It monitors the mass or volume of air flowing into the intake manifold for adjusting the fuel as demanded by load and speed of the engine.

(ii) *Exhaust gas or oxygen content sensor:* It senses the oxygen content in the exhaust of the engine and calculates air-fuel ratio. The output of sensor voltage is an indication of air-fuel ratio inside the cylinders.

(iii) *Engine temperature sensor:* It senses the temperature of engine coolant, and from this data, computer adjusts the mixture strength towards rich side for cold starting.

(iv) *Air inlet temperature sensor:* It senses the inlet temperature of air, the signal of which is used to fine tuning of the mixture strength.

(v) *Manifold pressure sensor:* It senses the manifold pressure (vacuum), the data of which is used by computer for adjusting the mixture strength as per changes in the load.

(vi) *Throttle position sensor:* It senses the movement of throttle plate so that the mixture flow can be adjusted to suit the engine speed and acceleration of the engine.

(vii) *Knock sensor:* Being a mixrophone type sensor, it detects ping or pre-ignition noise that is used that to retard the ignition timing.

(viii) *Camshaft position sensor:* It senses the rotation of engine camshaft/crankshaft for speed and timing of the engine which may be used for required air-fuel ratio.

It is worth to note that the fuel injector of EFI system is basically a ***fuel valve*** equipped with ***solenoid***. In the case of not being energized, its spring pressure makes the injector to remain closed, and there will be no fuel flow in the engine. When the computer sends the required signal, the magnetic field generated attracts the injector armature, and the fuel starts flowing towards manifold or port or in the cylinder.

## 7.24 EFI SYSTEM: LOCATION AND CONTROL

In EFI system, the location of injector has some bearing on the outcome, though minor in nature. There are basically two locations where fuel is injected. They are:

(i) Port injection with MPFI
(ii) Throttle body injection with SPFI

### 7.24.1 Port injection with MPFI Location

Figure 7.27 shows the arrangement of port injection with MPFI. In this system, injector is placed at the port of every cylinder, near the inlet valve. The injector sprays gasoline into the air inside the port of respective cylinder according to the firing order. As a result, the fuel vapour mixes with air throughout resulting in a homogeneous mixture of air and fuel for better combustion. After perfect mixing, the mixture enters the respective cylinder at proper time through inlet valve. In MPFI system, every cylinder is provided with an injector in its inlet port at the same location that ensures the same air-fuel ratio as shown in Fig. 7.28.

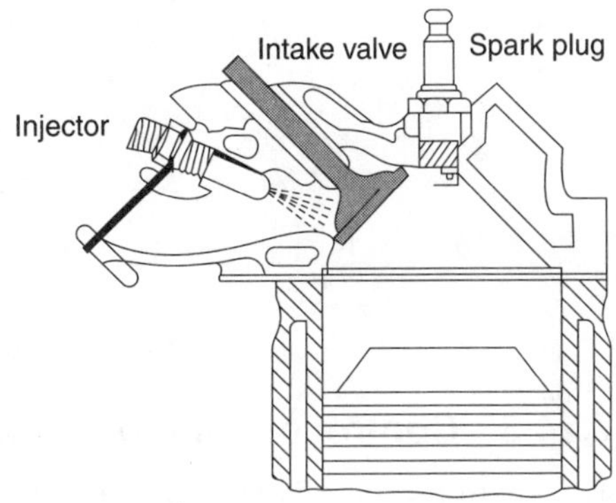


**Fig. 7.27** Port Injection

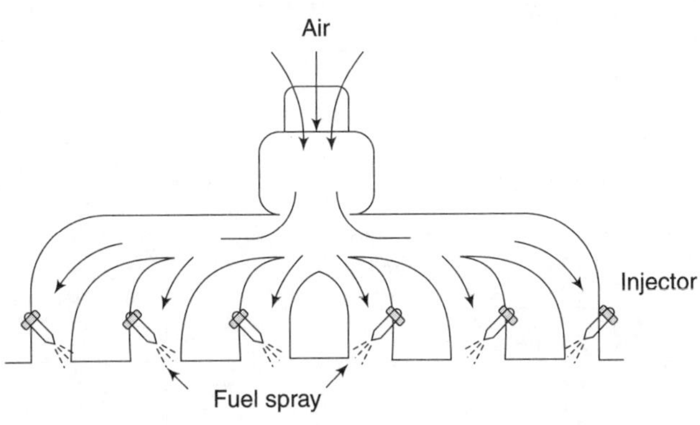


**Fig. 7.28** Multi-point fuel injection (MPFI) near port

### 7.24.2 Throttle Body Injection Location in SPFI

Figure 7.29 shows a simple sketch of throttle body injection location. In fact, it is a single point fuel injection (SPFI). The throttle body is similar to carburettor system. In this system, injector is placed above the throttle body and the injected fuel mixes with air and enters the intake manifold.

**It is worth to note that both systems, port injection and throttle body injection, may be either timed injection or continuous injection.**

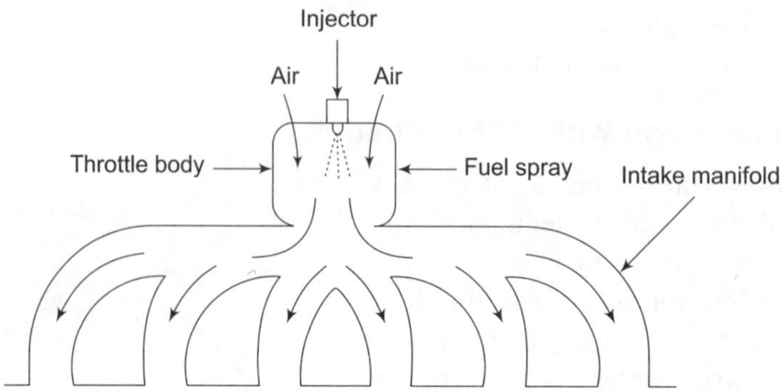


**Fig. 7.29** Throttle body injection (single point)

### 7.24.3 Control of Manifold Injection System

The control system is shown in block diagram in Fig. 7.30. In this system, vacuum in the manifold is first sensed, and the information is given to ECU. Speed sensor senses the speed of the engine, and the information is sent to ECU. All other sensors send the information of ECU. After processing all the information, the ECU sends the signal to the fuel injector to inject the required quantity of fuel into intake manifold as per load and speed of the engine.

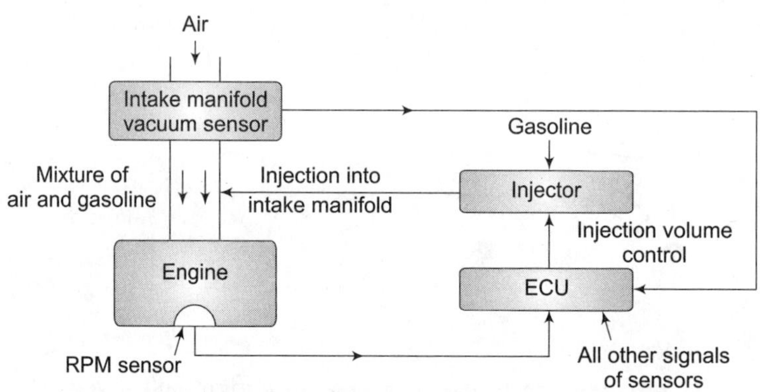


**Fig. 7.30** Control of manifold gasoline injection system

### 7.24.4 Control of Port Injection System

Figure 7.31 shows the block diagram for the control of port injection. In this case also, all the sensors employed send the signals to the ECU that processes all information, and finally sends signal to supply the required quantity of fuel to the injector which injects required quantity of fuel to the port of manifold near inlet valve as per load and speed of the engine.

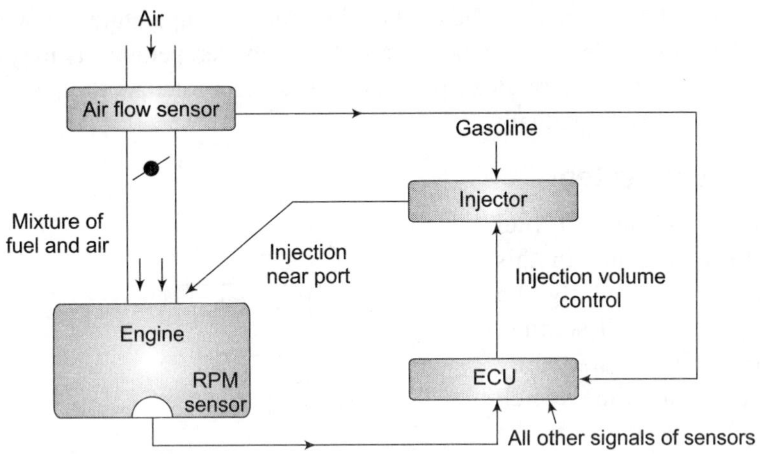


**Fig. 7.31** Control of port injection (MPFI) system

## 7.25 FUNCTIONAL DIVISIONS OF EFI SYSTEM

Sometimes, it is better to divide the whole function into some main division for better understanding. For this purpose, EFI system is divided into major three divisions. They are as follows:

(i) Electronic control unit (ECU)
(ii) Fuel system
(iii) Air induction system

### 7.25.1 EFI Electronic Control System

Electronic control system of EFI is shown in Figs. 7.30 and 7.31. All these sensors, such as oxygen sensor, engine temperature sensor, air flow sensor, air inlet temperature sensor, throttle position sensor, manifolds pressure sensor, speed sensor, knock sensor, etc. send respective signal to ECU which finally processes all these and send the output signal to the injector for injecting required quantity of fuel into intake manifold or port as per load and speed of the engine. When a need arises, the cold start injector timing switches off the ECU and operates the cold start injector as shown in Fig. 7.32.

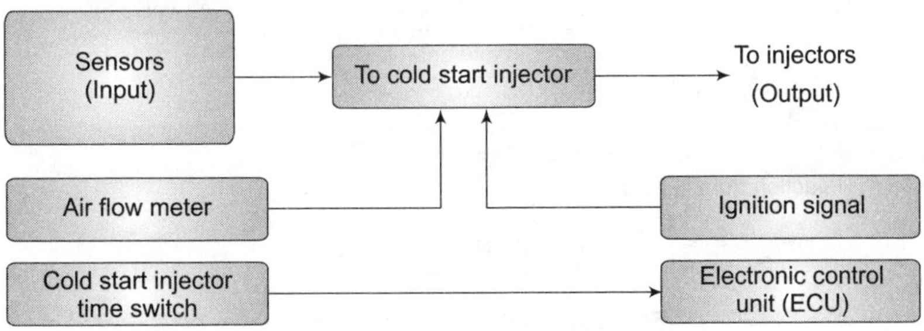


**Fig. 7.32** EFI control system

There happens to be ***air valve*** for the cold engine. As the temperature is low, the air valve speeds up to the engine idle speed to fast idling. When the temperature is low, the air valve opens completely allowing more air to pass into intake manifold. As the temperature rises, the air valve starts closing gradually.

### 7.25.2 EFI Fuel System

Figure 7.33 shows the EFI fuel system in block diagram. In this system, fuel is supplied by the fuel pump. At the time of starting, the cold start injector is started by the cold start injector time switch which injects fuel into the air intake chamber. The main injector also injects fuel. As a result, the air-fuel mixture gets enriched. The pressure regulator regulates the pressure of the fuel. The injector receives signal from the ECU, and injects the fuel into intake manifold.

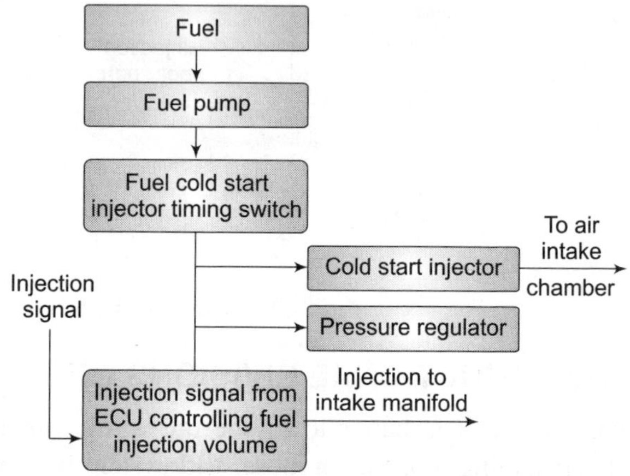


**Fig. 7.33** EFI fuel system

### 7.25.3 EFI Air Induction System

Figure 7.34 shows the block diagram of EFI air induction system. The air cleaner, the air flow meter, the throttle body and the air valve supply a proper amount of air into the intake manifold for the complete combustion of fuel.

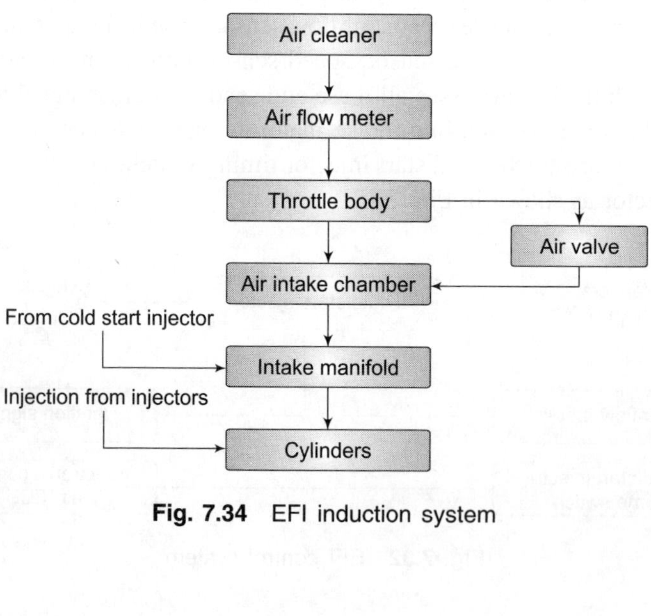


**Fig. 7.34** EFI induction system

## 7.26 INJECTION TIMING

Injection timing plays an important role in electronic gasoline injection system. For example, in the case of a four-stroke engine, the fuel is injected into the intake manifold of each cylinder in equal amount, but at different timings due to the firing order. *The timing at which the injection of fuel takes place inside the intake manifold or port is known as injection timing.*

In order to understand the injection timing, let us consider one cylinder of a four-cylinder engine. In any cylinder, the piston moves up from BDC to TDC position during exhaust stroke. Just before the piston reaches to TDC during exhaust stroke, injection of fuel takes place in the intake manifold or port at about 60-degree crank angle before TDC. In the four-stroke engine, the ECU calculates the appropriate injection timing for each cylinder. The injected fuel mixes with air to give the correct amount of air-fuel ratio as per load and speed of the engine. The duration of opening of injection valve depends on the demand of the engine. In the case of acceleration, the opening of injection valve is controlled longer by ECU.

## 7.27 GROUP MPFI SYSTEM

In group MPFI system, the injectors are not activated individually, but they are activated in groups. In the case of four cylinders engine, there are two groups, each group feeds two injectors, and in six-cylinder engines, there are two groups, each group feeds three injectors.

The advantage of group gasoline injection system lies in two fuel feed pipelines instead of single, so there is no chance of shortage of fuel under some abnormal situation.

The electronic control system of group gasoline injection is shown by block diagram in Fig. 7.35. All the sensors such as exhaust gas, engine temperature, air flow, air inlet temperature, throttle position, manifold pressure, speed of the engine, knock sensor, etc., send respective information to ECU. After receiving all concerned information from sensors, ECU processes all information, and finally sends required signal to the group (i) and group (ii) of injectors and to other parts of the system. After getting signal from ECU, the injectors inject required quantity of fuel either to the manifold or inside cylinder or to the port of respective cylinder.

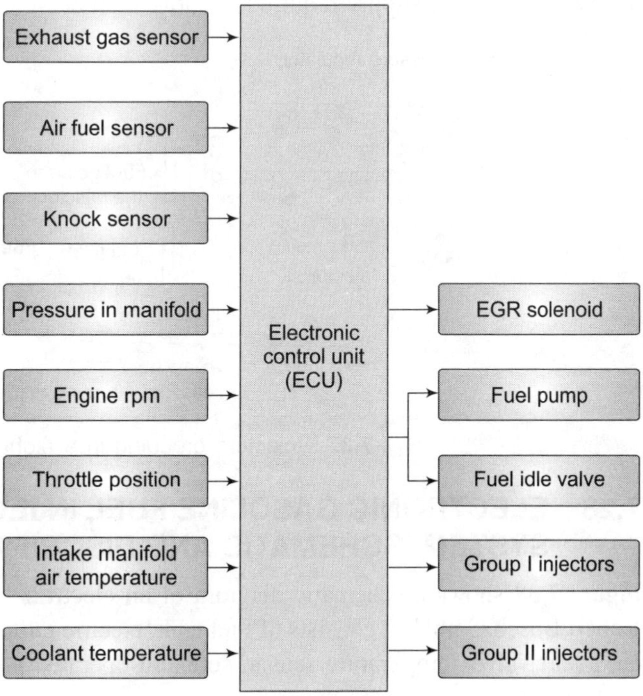


**Fig. 7.35** Block diagram of MPFI control system

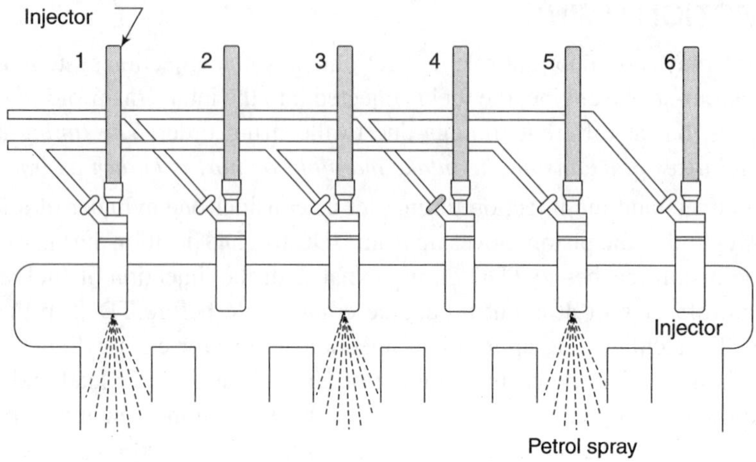


**Fig. 7.36** Injectors grouping in a six-cylinder engine

Figure 7.36 shows the arrangement of grouping of injectors in a six-cylinder engine. In this case, injectors 1, 3 and 5 open at the same time and inject required quantity of fuel either to the manifold or inside cylinder or to the port of the cylinder. By this way, all cylinders receive the same amount of air-fuel ratio, and develop equal power.

Figure 7.37 shows the injectors grouping in an eight-cylinder engine. The ECU sends required signal to the two groups of injectors, each feeds the four injectors and the injectors of respective group supply required quantity of fuel.

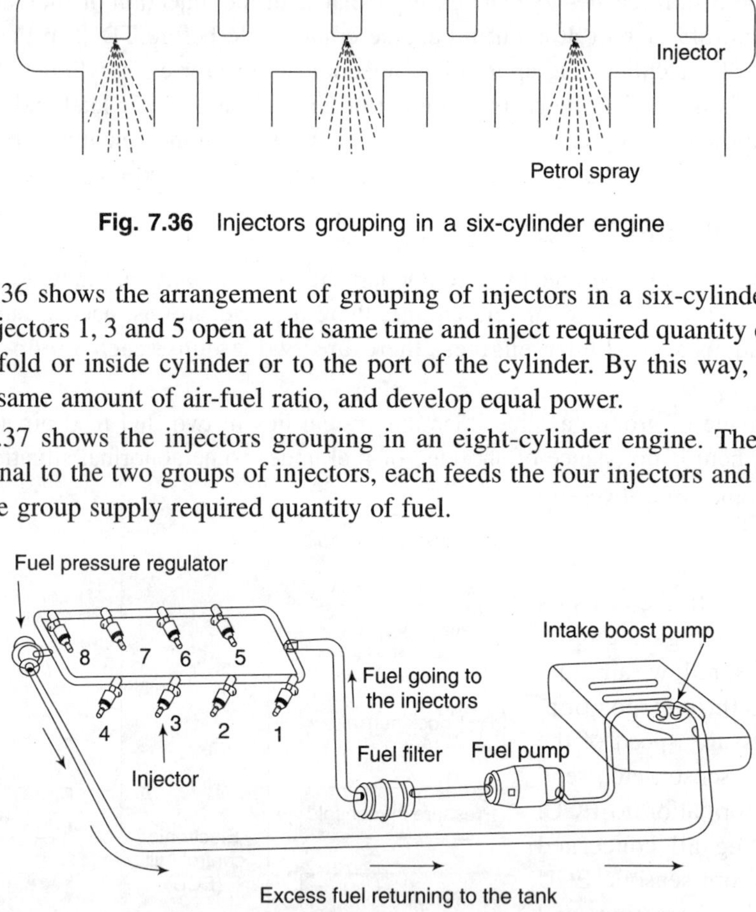


**Fig. 7.37** Injectors grouping in a eight-cylinder engine

## 7.28 ELECTRONIC GASOLINE FUEL INJECTION SYSTEM: SCHEMATIC AND DISCRIPTION

Figure 7.38 shows a schematic diagram of an electronic gasoline injection system used by Robert Bosch GmbH. It consists of fuel tank, electric pump, fuel filter, fuel pressure regulator, cold start valve, temperature sensor, solenoid operated mechanical injection valve, auxiliary

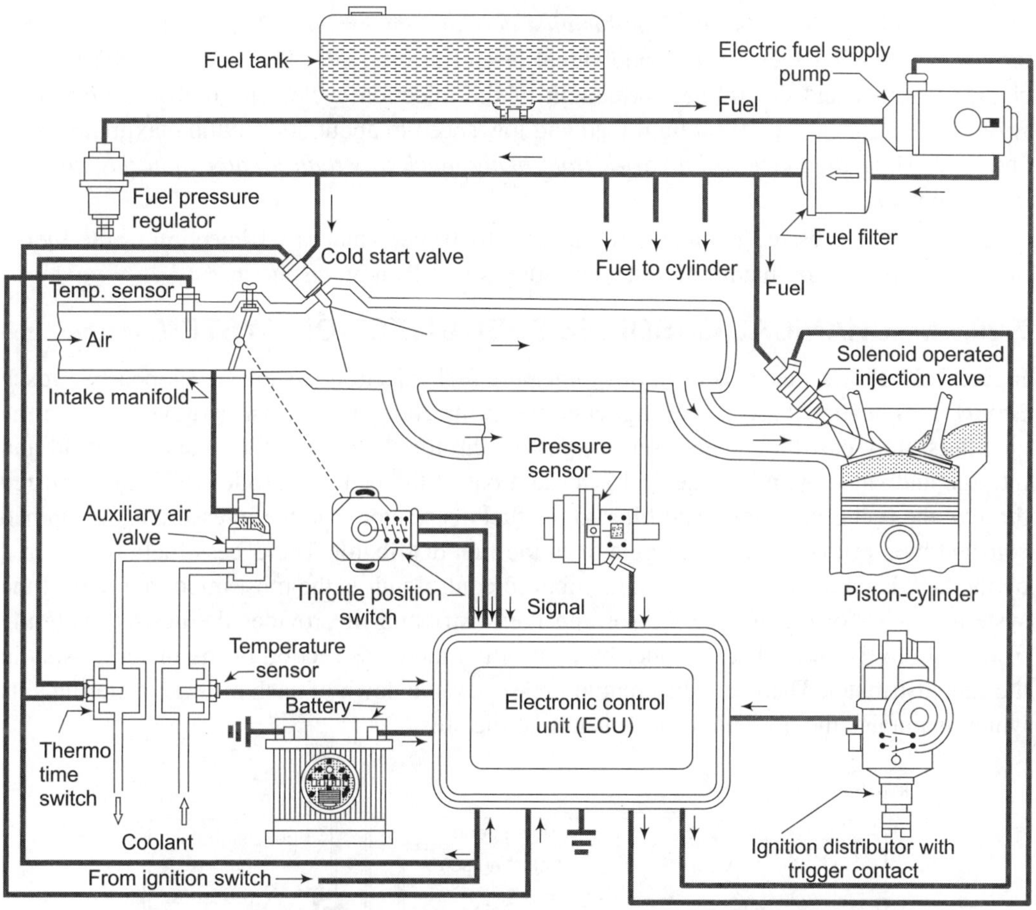


**Fig. 7.38** Schematic diagram of an electronic gasoline-injection system [Robert Bosch GmbH]

air valve, thermo-time switch pressure sensor, throttle position switch, and electronic control unit (ECU). ECU takes the signals from all sources. ECU gets power input from battery. The electric fuel supply pump maintains a high pressure in the fuel line to the solenoid operated injection valves in the intake manifold of each cylinder of a multi-cylinder engine.

When the trigger contacts of ignition distributor close, they send an electric signal to the electronic control unit (ECU) which connects half of the solenoid injection valves to the battery. The important point is that the solenoid injection valves are not individually actuated. Half of them are actuated at a time. Having only two groups of injection valves simplifies the system.

The fuel enters just opposite the intake valves intermittently. Note that individual intake valves open at varying times after injection. *Several factors determine how long an injection valve delivers fuel to the intake manifold. They include the throttle position, intake manifold, vacuum ignoring air temperature and engine coolant temperatures.* Various sensing devices are used to monitor these factors and signal is sent to the ECU. The ECU puts the varying signals together and determines the length of the injection cycle. The fuel injection pulse width varies from about 10° at light load and low speed to about 300° at the maximum speed and load. Thus, *the pulse width varies from being much less than greater than the duration of intake stroke.*

The injection pressure is much less (about 7 to 10 bar) than of CI injection as the fuel is injected at the suction manifold when pressure is less than atmospheric.

## 7.29 MECHANICAL GASOLINE FUEL INJECTION SYSTEM

Figure 7.39 shows mechanical multi-point port fuel injection system developed by Bosch GmbH. It is air-flow based metering, continuous injection system. Air is drawn through the air filters, flow past the air-flow sensor, past of throttle valve, into the intake manifold and into the undivided cylinder. The fuel is sucked out of the tank by a roller cell pump and fed through the fuel accumulator and filter to the fuel distributor. The fuel pressure is maintained constant by a primary pressure regulator in the fuel distributor. The excess fuel not required by the tank is returned to the tank. The mixture-control unit is the most important part of the system consists of the air-flow sensor and fuel distributor. It provides the desired metering of the fuel to the individual cylinder by controlling the cross section of the metering slits in the fuel distributor. There is a differential pressure valve down of each of these metering slits which maintains the pressure drop constant at the slits.

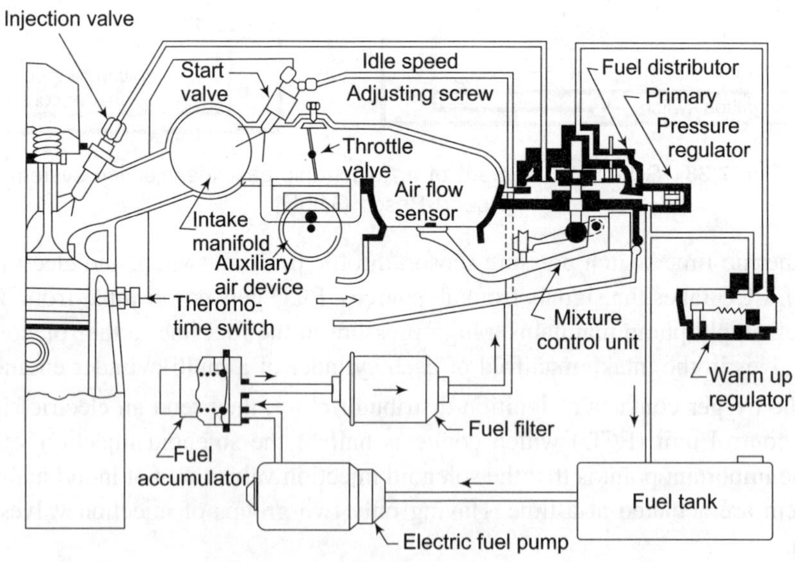


**Fig. 7.39** Mechanical multi-point port fuel-injection system

## 7.30 QUANTITY OF FUEL AND SIZE OF NOZZLE ORIFICE

The theory concerning the quantity of fuel and size of nozzle orifice is the same as that in CI injection system **(section 8.7.6)**. The only difference is in the value of injection timing, injection pressure and cylinder pressure. Both pressures are low in SI engine. Injection pressure may be 7 to 10 bar as it injects in the suction manifold. The injection timing is mainly during suction stroke. The types of nozzle are similar to that of CI injection system (section 8.7).

## 7.31 GOVERNING OF SI ENGINE

Governing of carburetted SI engine is a quantity governing. It is the throttle which controls the quantity of mixture entering into the cylinder as it is under the control of acceleration. In the case of MPFI system, ECU through valve controls the quantity of both fuel and air.

## 7.32 SUMMARY OF FORMULAE

(1) $(\dot{m}_a) = c_{da}\, A_a\, C_a\, \rho_a = c_{da}\, A_2\, C_{2a}\, \rho_{2a}$

(2) $(\dot{m}_f)_{act} = c_{df}\, A_f\, C_f\, \rho_f$

(3) $C_{2a} = c_{da}\sqrt{2c_p T_1\left[1 - \left(\frac{p_2}{p_1}\right)^{(\gamma-1)/\gamma}\right]}$

(4) $C_f = C_{2f} = \sqrt{2\rho_f[p_1 - p_2 - gh\rho_f]}$

(5) $(\dot{m}_a)_{act} = 4.939\, \frac{A_2\, p_1}{\sqrt{T_1}}\, \phi$ where $\phi$

$= \sqrt{\left(\frac{p_2}{p_1}\right)^{2/\gamma} - \left(\frac{p_2}{p_1}\right)^{(\gamma+1)/\gamma}}$

(6) $(\dot{m}_f)_{act} = A_f\, c_{df}\sqrt{2\rho_f\,[p_1 - p_2 - gh\,\rho_f]}$

(7) $\text{A/F} = \frac{\dot{m}_a}{\dot{m}_f} = \frac{c_{da}\, A_2\, C_{2a}\, \rho_{2a}}{c_{df} A_f\, C_f\, \rho_f}$

$= 3.492\, \frac{c_{da}}{c_{df}}\, \frac{A_2}{A_1} \cdot \frac{p_1}{\sqrt{T_1}} \cdot \frac{\phi}{\sqrt{[\rho_f\,(p_1 - p_2)}}$

## PROBLEMS

**Things to Think About**

7.1 With the development of multi-point fuel injection system (MPFI), there is a death of carburettor. Justify this statement.

7.2 What parameters govern the gasoline injection pressure?

7.3 The modern carburettor employs three-venturi system. What is the reason behind it?

7.4 What is the basic philosophy involved in the design of carburettor and gasoline fuel injection system?

### Short and Long Theoretical Questions

7.5 What do you mean by carburettor?

7.6 Give a sketch of a typical induction system of a SI engine and show the pressure of critical point under normal operating carburettor.

7.7 Discuss the various parameters that affect the process of carburetion.

7.8 Explain the following:

(a) Rich mixture, (b) Lean mixture (c) Stoichiometric mixture.

7.9 With suitable sketch, discuss the electronic gasoline-injuction system of Robert Bosch GmbH. List the sensors used in EFI system.

7.10 Discuss the sketch the SI engine air-fuel mixture requirement under all possible conditions of engine operation. Also discuss the distribution of mixture in multi-cylinder engines.

7.11 Explain why a rich mixture is required for the following:

(a) Idling (b) Maximum power

(c) Sudden accelerations

7.12 Explain with block diagram for group MPFI system. List the advantages and disadvantages of EFI System. Discuss the functional divisions of EFI system.

7.13 Derive the exact expression for air-fuel ratio of a simple carburettor.

7.14 List the essential parts of a modern carburettor. Discuss various branded carburettors.

7.15 With the help of suitable sketches, describe the working principle of following system of a modern carburettor.

(a) Main metering system (b) Idling system

(c) Power enrichment system (d) Acceleration pump system

### Numerical Questions

7.16 A carburettor supplies a mixture of A/F of 14:1 when the throttle is set to supply the air at one-half of the full supply. It may be assumed that sixty per cent of the fuel leaving the carburettor nozzle vaporizes in the inductions manifold. Find the air-fuel ratio of the mixture that reaches the cylinder if the throttle is suddenly closed to one-quarter full position, assuming the same air-fuel ratio at the carburettor nozzle and the same percentage of vaporization in the manifold.

[***Ans.*** 11.43 : 1]

7.17 A 7 cm by 8 cm four cylinders, four-stroke cycle SI engine is to have a maximum speed of 3000 rpm and a volumetric efficiency of 80%. If the maximum venturi depression is to be 150 cm of water, what must be the size of the venturi.

7.18 Calculate the size of the fuel orifice of carburettor to give a A/F ratio of 13 : 5 :1 if the venturi throat has 3 cm diameter and the pressure drop in the venturi is 6.5 cm Hg. The fuel orifice is at the same level as that of the float chamber. Take air temperature and pressure at carburettor entrance are 1 bar and 27°C respectively. The density of the gasoline may be taken as 740 kg/m$^3$ and discharge coefficient as unity. [***Ans.*** $d_f$ = 1.63 mm]

7.19 A six-cylinder, four-stroke SI engine works on A/F ratio of 12 to 1. The engine has a bore of 8 cm and a stroke of 11 cm and operates at a speed of 2000 rpm with a volumetric efficiency of 80%. Calculate the diameter of the fuel orifice if the diameter of venturi section is 2.5 cm and density of air may be taken as 1.16 kg/m$^3$. [***Ans.*** 1.875 mm]

7.20 Design the main dimensions of a carburettor for the following data of four-stroke, four-cylinder SI engine.

Bore = 8.255 cm, Stroke = 8.9 cm, Speed = 3000 rpm

Volumetric efficiency = 80%, Venturi depression = 0.1 bar

A/F = 12 to 1, Density of air = 1.2 kg/m$^3$ [***Ans.*** $D$ = 2.97 cm, $d_f$ = 1.46 mm]

7.21 Calculate the A/F ratio at 4600 m altitude in a carburettor adjusted at sea level for a A/F ratio of 15 : 1. Take air temperature and pressure at sea level as 20°C and 1.013 bar.

The temperature of the air decreases with altitude given by the expression

$t = t_s - 0.0065$ h

where $h$ = height in m

$t_s$ = sea level temperature in °C

The air pressure decrease with altitude which is given by relation

$$h = 1922 \log_{10}\left(\frac{1.013}{p}\right)$$

where $p$ = pressure in bar.

7.22 A carburettor of Maruti car is to be designed which has four-stroke SI engine of displacement of 796 cm$^3$ and develops maximum power at 5500 rpm. The volumetric efficiency at this speed is estimated to be 70% and thus A/F is 13.5 : 1. It is estimated that at peak power the theoretical air speed at the choke will be 106 m/s. The coefficient of discharge for the venturi and that of the main petrol jet may be taken as 0.85 and 0.66 respectively. An allowance may be made for the emulsion tube, the diameter of which can be taken as 1/2.5 of the choke at this engine condition. Determine the diameter of suitable choke and main jet of fuel. The specific gravity of petrol may be taken as 0.74 and atmosphere pressure and temperature may be taken as 1 bar and 27°C respectively. [***Ans.*** $D$ = 21.2 mm, $d_f$ = 1.185 mm]

7.23 In order to study the effect of air cleaner on A/F, a carburettor having its float chamber vented to the atmosphere is tested on the test used without an air cleaner at sea level conditions, the main metering system of this carburettor is found to yield a F/A of 0.0655. The venturi throat pressure is 0.842. The carburettor is tested again when the air cleaner is fitted at the inlet of carburettor. The pressure drop due to air cleaner is found to be 0.038 bar when the air flow in the engine at sea level condition is 240 kg/h and pressure at sea level is 1.013 bar. Assuming negligible nozzle lip and constant orifice coefficient the following:

(a) Venturi throat pressure with air cleaner

(b) F/A with air cleaner [***Ans.*** 0.78 bar, 0.0723]

7.24 It is proposed to design a carburettor for Fiat car four-stroke engine of 1090 cc capacity developing maximum poor of 32 kW at 5000 rpm. The volumetric efficiency at this speed is 75% and the A/F ratio is 13:1. At peak power, the theoretical air speed at the choke is 122m/s. The coefficient of discharge for the venturi and main petrol jet may be taken as 0.85 and 0.66 respectively. An allowance may be made for the emulsion tube, the diameter of which can be taken as 1/2.5 of the choke diameter. The petrol surface is 6 mm below the choke. Determine the diameters of choke and main jet. The specific gravity of petrol may be taken as 0.75. The atmospheric pressure and temperature may be taken as 1.03 bar and 300 K respectively.

[***Ans.*** $D$ = 23.3 mm, $d_j$ = 1.297 mm]

7.25 A simple jet carburettor has to supply 5 kg air per minute. The air is at a pressure of 1.013 bar and a temperature of 27°C. Calculate the throat diameter of the choke for air flow velocity of 90 m/s. Take velocity coefficient to be 0.8. Assume isentropic flow. Assume the flow to be compressible. (E.S., 1992)

7.26 An experimental four-stroke petrol engine of 1710 $cm^3$ capacity is to develop maximum power at 5400 rpm. The volumetric efficiency at this speed is assumed to be 70 per cent and air fuel ratio is 13:1. Two carburettors are to be fitted and it is expected that at peak power, the air speed at choke will be 107 m/s. The coefficient of discharge for the venturi is assumed to be 0.85 and that of the main petrol jet 0.66. An allowance should be made for the emulsion tube, the diameter of which can be taken as 1/2.5 of the choke diameter. The petrol surface is 6 mm below the choke at the engine condition. Calculate the size of suitable choke and main fuel jet. The specific gravity of petrol is 0.75. Atmospheric pressure and temperature are 1.013 bar and 27°C respectively. (E.S., 1994)

7.27 Design the main dimensions of fuel pump and injector of four-stroke, four-cylinder SI engine employing MPFI system.

*bp* = 100 kW, *bsfc* = 0.3 kg/kWh, N = 4000 rpm.

Assume other suitable data.

### Software Based Quentions

7.28 Develop a software for the design of modern carburettor for two-wheeler and four-wheeler engines.

7.29 Develop a software for the design of multi-point fuel injection (MPFI) system used in passenger cars.

### Design and Open Ended Question

7.30 Prepare a comprehensive report with complete drawing of a modern carburettor and MPFI system with their features.

### Multiple-Choice Questions

7.31 In a simple carburettor, the air-fuel ratio increases when the

(a) density of air increases (b) density of air decreases

(c) velocity of air increases (d) lip is increased

7.32 In a carburettor, the level of the fuel in the float chamber as compared to the level in the main jet in the venturi is

(a) same (b) lower

(c) higher (d) can be both (b) and (c)

7.33 The Solex carburettor is

(a) downdraft type (b) updraft type

(c) eccentric type (d) multiple-venturi type

7.34. **Assertions (A):** During idling, a carburettor supplies a fuel-rich mixture to engines.

**Reasoning (R):** Because a fixed jet carburettor is provided with slow running and idling management

**Answer Code**

(a) Both A and R are true but R is not the correct explanation of A.

(b) Both A and R are true and R is the correct explanation of A.

(c) A is false but R is true

(d) A is true but R is false

7.35 Electronic fuel injection system:

1. eliminates majority of carburettor pressure losses
2. has better atomization and vaporization of fuel
3. requires more volatile fuels
4. has negligible variation in the air-fuel ratio from one cylinder to another of these statements:

(a) only 1and 2 are correct (b) only 1, 2 and 4 are correct

(c) 1, 2, 3 and 4 are correct (d) 1, 2 and 3 are correct

7.36 Compensating devices are provided in carburettor

(a) to change the quantity of mixture depending upon load

(b) to provide always economy mixture

(c) to modify the mixture strength depending upon requirements under various operating conditions

(d) to supply extra fuel during acceleration only

7.37 The function of a carburettor in SI engine is to control:

(a) A/F ratio

(b) amount of mixture

(c) A/F ratio and amount of mixture

(d) compression ratio and volumetric efficiency

7.38 For petrol engine, the method of governing employed is

(a) quantity governing (b) quality governing

(c) hit and miss governing (d) none of above

## Key to Multiple-Choice Questions

7.31 (c) 7.32 (b) 7.33 (a) 7.34 (a) 7.35 (b) 7.36 (c) 7.37 (c) 7.38 (a)

CHAPTER

# Mechanical and Electronic CI Fuel Injection Systems

# 8

## OVERVIEW

In CI engines, the air necessary for combustion enters the engine during suction stroke through air cleaner and is distributed to various cylinders, either by an intake manifold in a four-stroke cycle engine or by air passage built in the engine block, in a two-stroke engine. Air only is compressed in the cylinder during compression stroke and before the piston reaches TDC position diesel fuel is injected through injection system into the combustion chamber and is ignited automatically due to high temperature of compressed air. The injection system meters, injects and distributes liquid fuel at a pressure considerably much higher than that existing in the cylinder. High injection pressures (110 to 200 bar) are essential not only to overcome the compression pressure, but, also to atomize the fuel and distribute it in a desired manner in the combustion chamber. Higher compression ratio (12 to 20) as compared to those used by SI engine (5 to 10.5) is essential to increase the temperature of compressed air higher than ignition temperature of fuel used so that ignition starts shortly after injection.

From the analysis of air standard cycle, we have seen that thermal efficiency increases with compression ratio. Since actual CI engine operates at considerably higher compression ratio than SI engine and, consequently has higher thermal efficiency. But higher compression ratios results in greater forces acting on the piston, wrist pin, connecting rod, bearing, crankshaft, etc. so in order to counteract these enlarged forces, it is necessary to strengthen engine's components and, hence, a heavier engine results.

It is worth to note that in the CI engine, for a given speed, and irrespective of load, an approximately constant supply of air enters the cylinder. The power produced by the engine directly depends upon the amount of fuel injected into the cylinder. Therefore, the CI engine is often called a *constant supply air engine*.

This chapter deals with the mechanism of mechanical and electronic CI fuel injection systems.

## 8.1 HEAT RELEASE PATTERN IN CI ENGINE

*The period between the start of injection and start of ignition is called* ***ignition delay*** *and is about 0.001 seconds for high speed engines and 0.002 seconds for low speed engines.* After the ignition, temperature and pressure rise rapidly. *The injection period in the CI engine covers about 25 degrees of crank rotation* and thus the process of vaporisation and mixing continues at the increasing rate after the delay period is over. With the advancement of combustion process, the amount of oxygen available for mixing and burning reduces, which in turn, affects the heat release.

Figure 8.1 shows a typical heat release pattern with its various phases of a CI engine. During the ignition delay period, an appreciable amount of vaporized fuel is collected in the combustion chamber and it burns rapidly at the end of ignition delay causing a large amount of heat release achieving the first peak. *It is the chemical kinetics which controls the burning process as the temperature of the mixture remains low.* The burning rate reduces to a minimum and then increases again achieving second peak as more fuel is injected into the combustion chamber. It is to be noted that after the first peak the temperature in the combustion chamber is always high and, therefore, the chemical reaction rate is so high that is cannot be controlled. As a result, the combustion is controlled by the rate of mixing of fuel and air and the availability of oxygen present in the mixture. *The height of second peak is less than that of first peak as it is controlled by the air entrainment rate only.* There is a long-drawn tail after the second peak having downward nature. *This is due to the reduction in diffusivity and turbulent mixing.* The last portion of heat release is again controlled by the chemical kinetics.

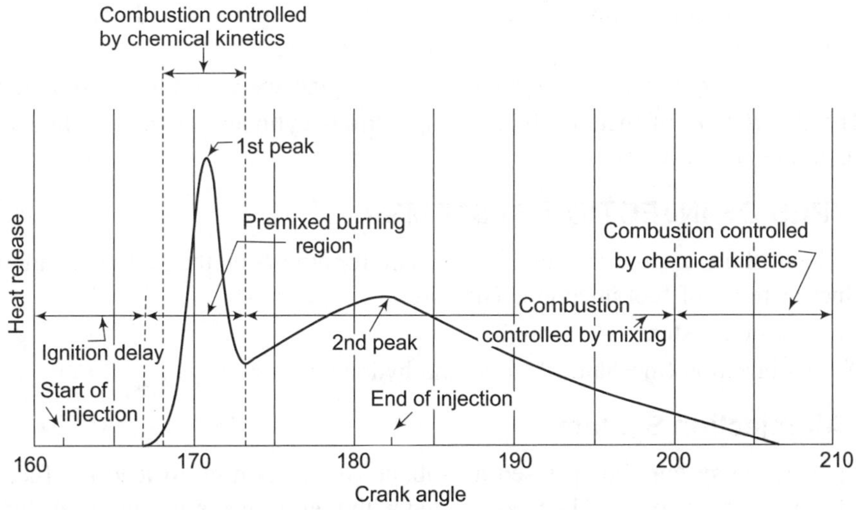


**Fig. 8.1** Typical heat release diagram of CI engine

From the above discussion, it is obvious that the *heat release affects the rate of pressure rise and thus, in turn, noise and vibration of the engine and also the cycle efficiency.*

*The ideal heat release pattern should be a very similar to a* ***double right angled triangles*** *having its peak sometimes after TDC so that the centre of gravity of heat release diagram should be near TDC.* But the actual heat release pattern is *exactly reverse of that* which is desired. This is only possible when the injection rate diagram is similar to as double right angled triangle.

## 8.2 REQUIREMENTS OF AN INJECTION SYSTEM

In order to achieve better performance from the CI engine, the injection system should fulfil the following requirements:

(i) The injection of the fuel in the combustion chamber should be within a precisely *defined timing* of the cycle so that the maximum power is obtained during fuel economy and clean burning.

(ii) *Accurate metering of the fuel injected per cycle* should be achieved as per load and speed of the engine. This aspect is very critical due to the fact that very small quantity of fuel is being handled. *Metering errors result in drastic variation in the desired output. This means that the governing parameters such as clearance between the working parts of pump, fuel orifice and governing mechanism, etc., are to be controlled accurately.*

(iii) The *rate of injection* should be maintained such which should result in desired heat release pattern.

(iv) A high degree of *fuel atomization* should be achieved.

(v) The *spray pattern* of fuel injected should provide rapid mixing of fuel and air in the combustion chamber.

(vi) There should not be any *injection lag* during the beginning and end of injection to avoid any dribbling or after injection of fuel droplets.

(vii) *Injection timing* may be changed to suit the engine speed and load requirements.

(viii) The *distribution* of metered fuel among various cylinders of multi-cylinder engines should be *uniform.*

## 8.3 TYPES OF INJECTION SYSTEMS

Based on the methods to produce required pressure for atomizing the fuel, there are basically two distinct methods of fuel injection. They are:

(a) Air injection system

(b) Airless injection (mechanical, solid and hydraulic) system.

### 8.3.1 Air Injection System

In the air injection system, compressed air (about 70 bar) is used to feed the fuel into the cylinder from the spray nozzle. The nozzle is fed with fuel from a pump atomises the fuel and creates turbulence in the combustion chamber to mix the air and fuel. Though, it is claimed that the air injection does a good mixing of fuel and air with resultant high m.e.p. and has the capability of using high viscous fuel, but it is now almost **obsolete** because it requires a multi-stage air compressor which increases the engine weight and reduces brake power.

### 8.3.2 Airless injection System

Airless injection system is also known as ***solid, (mechanical)*** **and *hydraulic* injection system**. It consists of essentially of two main parts: One is a high pressure pump and the other, a check valve that allows fuel to enter the cylinder, but prevents the cylinder gases from backing up into the fuel system. *The positive displacement pump, which injects the fuel into the combustion chamber, is actuated by mechanical force.* Depending upon the location of the fuel pumps and injectors, the grouping and methods of actuating the pumps, and upon the method used for metering the fuel, the airless or solid injection system may be further classified as:

(a) *Common rail system*
(b) *Distributor system*
(c) *Individual pump system*

All the above three systems consist of the following elements:

(i) Fuel tank in which fuel is stored
(ii) Primary fuel filter
(iii) Low pressure fuel feed pump which delivers the fuel to fine filter.
(iv) Fine filter to filter the find undesirable particles present in the fuel.
(v) High pressure fuel injection pump to meter and pressurise the fuel (110 to 200 bar) for injection.
(vi) Governor to ensure the amount of fuel injected as per load and speed of the engine.
(vii) Injector (nozzle) to receive the high pressure fuel from injection pump and to inject and atomize the fuel in the combustion chamber.
(viii) Pressure relief valve to bypass the fuel to the suction of high pressure pump if pressure is more.

### 8.3.3 Common Rail System

Figure 8.2 shows the arrangement of a common rail injection system. The fuel from the storage tank (*S*) is drawn through the primary fuel filter (*F*) by means of a low pressure fuel feed

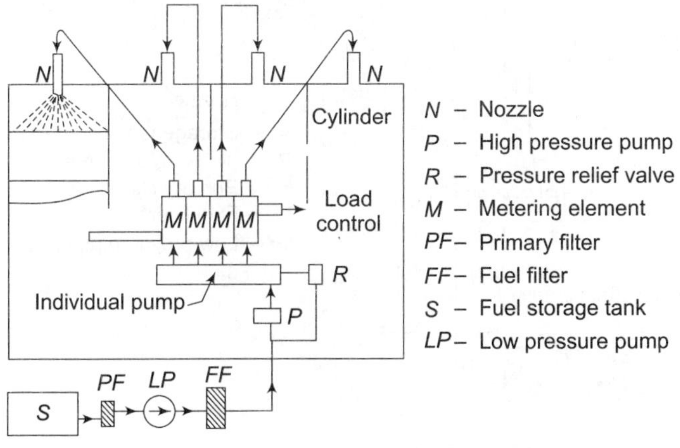


**Fig. 8.2** Common rail system

lift pump (*LP*). The discharge from feed pump enters the main injection pump via fine filter. A single high pressure injection pump delivers the fuel to a fuel header called the **common rail**. The high pressure in the common rail forces the fuel to each of the nozzle located in the cylinder. A pressure relief valve maintains the pressure in the common rail system constant. At the appropriate time, a mechanically operated (by means of push rod and rocker arm) valve allows the fuel to enter the cylinder through the nozzle. It is the fuel valve that controls both the timing of the fuel injection as well as delivery of the required quantity of fuel. *It is mainly used in large and small speed engines.*

### 8.3.4 Distributor System

Figure 8.3 shows the arrangement of distributor system. *In this system, a metering and high pressure pump is used to pump the metered quantity of fuel in the rotating distributor which distributes the fuel to the individual cylinders at the correct time.* The number of injection strokes per cycle for the pump is equal to the number of cylinders. The fuel is fed to the high pressure pump from storage tank through course filter, l.p. pump and fine filter. Since the metering and timing of injection is accomplished by one plunger equal amount of fuel is supplied to each cylinder at the same point in the cycle.

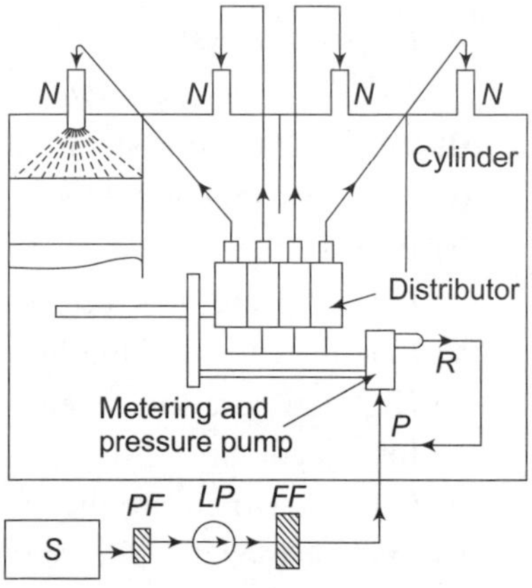


**Fig. 8.3** Distributor system

### 8.3.5 Individual Pump System

Figure 8.4 shows the arrangement of an individual pump system. *This system has an independent high pressure pump for each cylinder which meters, pumps and controls the timing of fuel injection.* Each cylinder is provided with one injector and the pump and injector

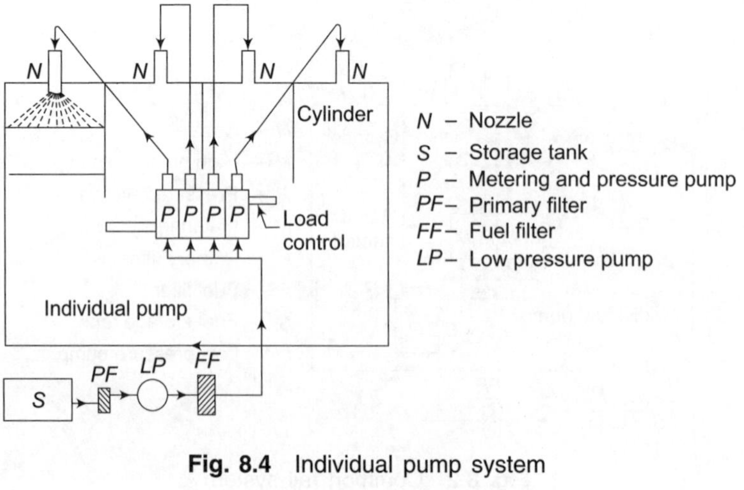


**Fig. 8.4** Individual pump system

may be integrated as one unit. The fuel is brought to the individual pump from storage tank through coarse filter, low pressure pump and fine filter. The high pressure pump is equipped with control mechanism and at the proper time, a rocker arm actuates the plunger and thus injects the fuel into the cylinder. The amount of the fuel injected is regulated by the effective stroke of the plunger. *The most popular fuel injection systems used are* ***individual pump and nozzle type***.

### 8.3.6 Comparison of Various Fuel Injection Systems

Table 8.1 shows the comparison of various fuel injection systems in terms of how the various functions are achieved. Individual pump systems are more common than other solid injection systems.

**Table 8.1** Comparison of Various Fuel Injection Systems

| S.No. | Function | Air Injection System | Air-less (solid) Injection System | | |
|---|---|---|---|---|---|
| | | | Common rail | Distributor | Individual Pump |
| 1. | Metering | Pump | Injection valve | Pump | Pump |
| 2. | Timing | Fuel cam | Fuel cam | Fuel cam | Pump cam |
| 3. | Injection | Spray valve | Spray valve | Fuel cam | Pump cam |
| 4. | Atomization | Spray valve | Spray tip | Spray tip | Spray tip |
| 5. | Distribution | Spray valve | Spray tip | Spray tip | Spray tip |

## 8.4 INJECTION PUMP

*The main aim of the fuel-injection pumps is to deliver accurately metered quantity of fuel under high pressure ranging from 110 bar to 200 bar at the correct instant to the injector fitted on each cylinder of diesel engine.*

Injection pumps are generally classified in two categories, namely.

(a) *Jerk type pump (e.g., Bosch pump).*

(b) *Distributor type pump (e.g., Roosa Master distributor pump)*

### 8.4.1 Jerk Type (Bosch) Fuel Injection Pump

**Working Principle.** *It is a positive displacement pump consisting of a reciprocating plunger inside a barrel.* Figure 8.5 shows a schematic diagram which illustrates the working principle of jerk type pump. Part (a) of this figure depicts the physical configuration of a typical plunger, while the remaining parts of diagram shows its operation in the barrel. The fuel is always available under relatively low pressure, at the port *A* in the side of the barrel. The plunger is driven axially along the barrel by a camshaft and can be rotated about its axis by means of rack *D*, as shown in Fig. 8.5 (b), the movement of which is controlled by the fuel control lever. The port *B* is the orifice through which fuel is delivered to the injector after passing through a spring loaded check valve. Port *C* is the outlet port.

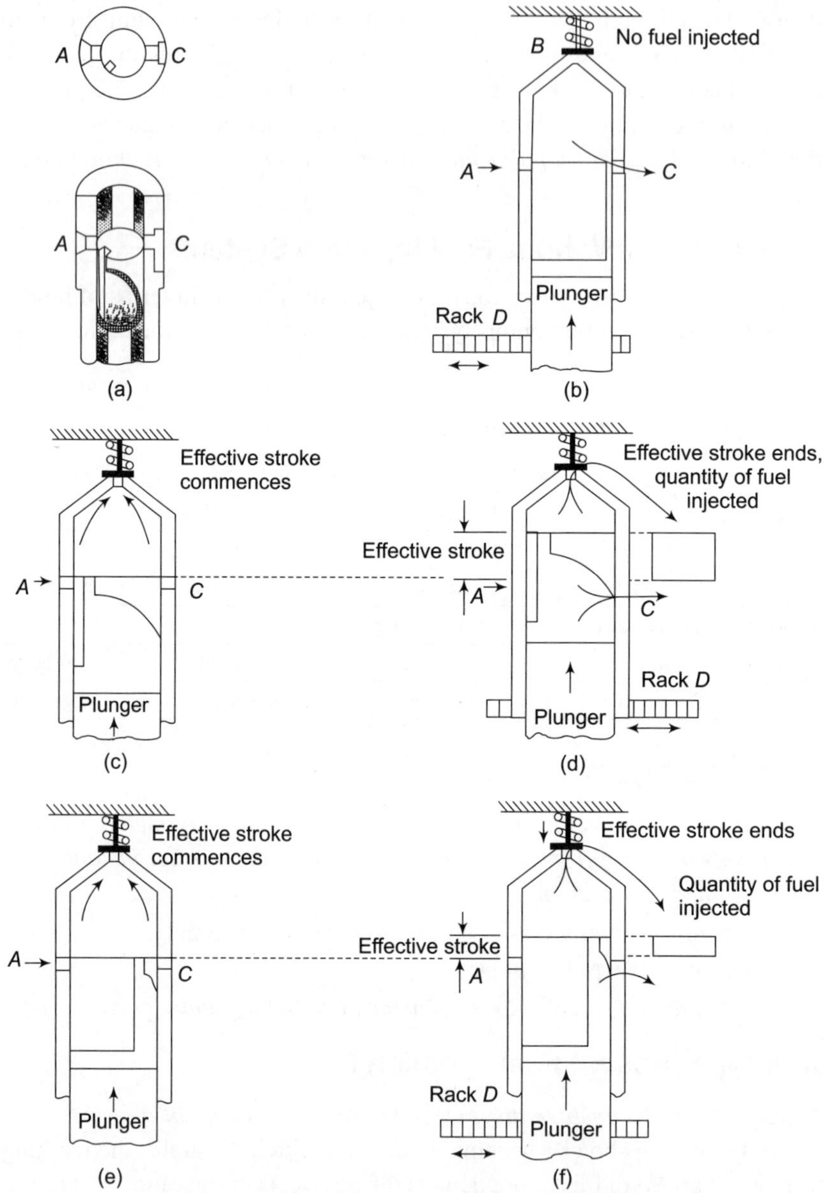


**Fig. 8.5** Working principle of jerk type (Bosch) pump

Figure 8.5(b) shows the position of plunger, rotated by the rack, to a position where no fuel will be injected through the port *B* in the injector and finally to combustion chamber. When the plunger is below port *A*, the barrel above it is fitted with fuel due to well communication with port *A*. As the plunger rises and closes off port *A*, the liquid fuel either may pass through port *B* or port *C*. The fuel takes the easier path to pass out through port *C* and is returned to the supply system instead of through port *B* as it must overcome the force of the spring holding the check valve.

Figures 8.5(c) and (d) represent the position of plunger rotated by the fuel control lever through the rack mechanism unit the plunger is in position to inject the maximum quantity of fuel per stroke. When the plunger moves upward past port *A* (Fig. 8.5c) and traps the fuel in the barrel, it also closes of port *C*. At this stage, the trapped fuel has no alternative except to take on escape route to pass through the check valve by overcoming the spring force and then through port *B* and finally to the injector. This is the beginning of injection.

The fuel injection will continue during the upward movement of plunger until the **helical indentation** on the plunger uncovers port *C*. As this stage (Fig. 8.5d), the fuel will escape through the easier rout out port C, the check valve will close orifice *B* and fuel injection ceases and the effective stroke ends. Here, the effective stroke of the plunger is defined as the axial distance traversed between the time port *A* is closed off by the plunger, and the time port *C* uncovered by the helix (Fig. 8.5d). Both the effective stroke and the relative quantity of fuel injected are depicted in Fig. 8.5 (d).

Figures 8.5(e) and 8.5 (f) shows another positions of plunger rotated by rack to inject less than the maximum quantity of fuel injected following the above sequence of events. At this position, the port *C* uncovers sooner and the effective stroke is shortened resulting in smaller quantity of fuel injection as indicated.

*It is worthwhile to note that the plunger continuously traverses the same axial distance on every stroke as determined by the cam lift and the rotation of the plunger by the rack determines the length of effective stroke and thus the quantity of fuel injected.*

**Constructional details.** Figure 8.6 shows the construction details of individual Bosch injection pump. Here, all the individual high pressure pumps are placed in one housing. High pressure lines start at the pump and terminate at the injector which is located in the cylinder head. The pump is of the positive displacement type, the plungers of which are activated by individual cam (one cam per pump) which are arranged in line. At the proper moment, the cam activates the

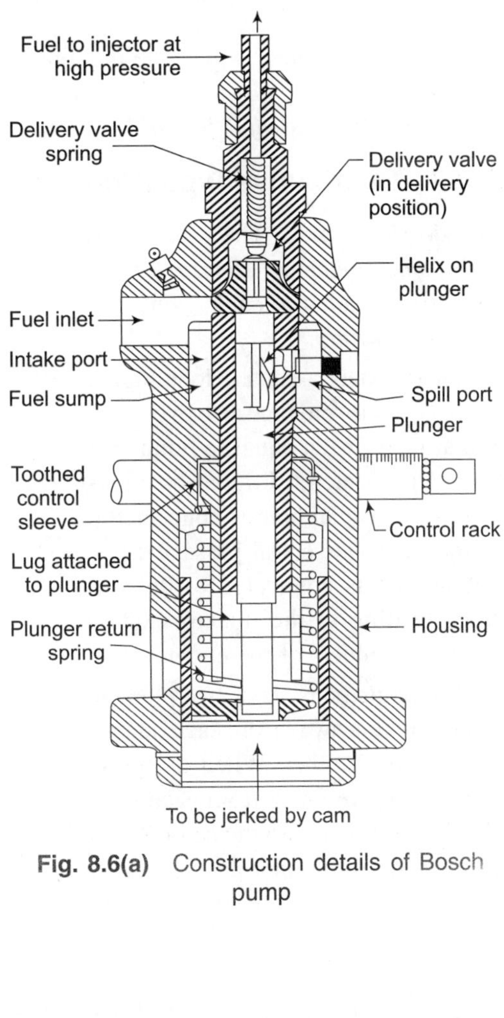


**Fig. 8.6(a)** Construction details of Bosch pump

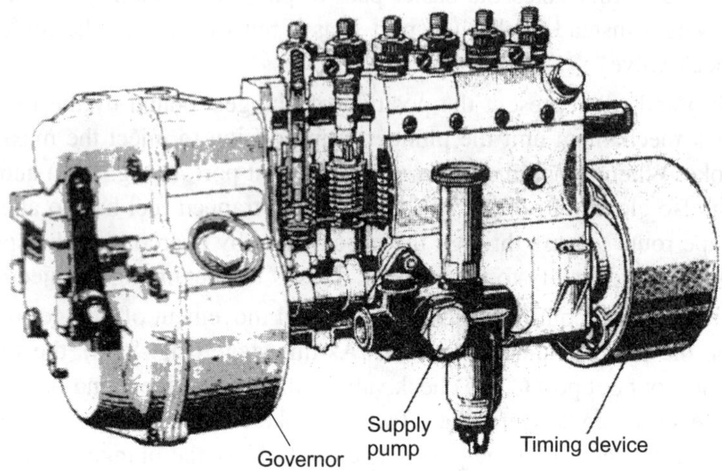


**Fig. 8.6(b)** Diesel fuel system with inline fuel injection pump

plunger axially, which forces the fuel under high pressure to the injector. *The amount of fuel injected is controlled by a horizontal movement of the rack, which in turn, rotates the plunger and thus varies the effective stroke.*

## 8.4.2 Distributive Type (Roosa Master) Pump

**Working Principle.** The working principle of distributive type pump is shown in Fig. 8.7. *It has a single pumping element and the fuel is distributed to each cylinder by means of a rotor.*

There exists a central longitudinal passage in the rotor and also two sets of radial holes, each equal to number of cylinders, located at different positions. One set of radial hole is connected to pump inlet via central passage where second set is connected to delivery lines leading to the injectors of the various cylinders of the engine.

The pump consists of two plungers either moving away from each other or towards each other. When the pump plungers move away from each other the fuel is drawn into the central rotor passage from the inlet port and when the plungers move towards each other

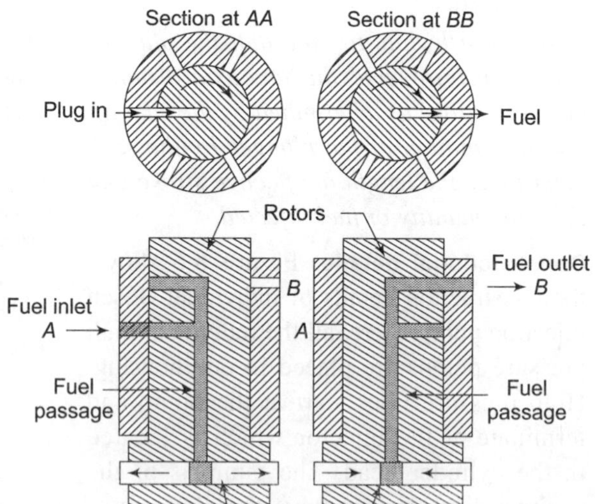


**Fig. 8.7** Working principle of distributor type fuel injector pump

head (pressure) is generated. With the rotation of rotor wherever, the radial delivery passage in the rotor coincides with the delivery port for any cylinder, the fuel is delivered to each injector in turn. *Its small size and light weight are the main advantages of this pump.*

**Constructional Details.** Figure 8.8 shows the construction details of Roosa Master distributor pump. All the elements of pump are housing in one casing. The plungers are actuated either away from each other or towards each other *with the help of roller and shoe and cam ring* (Fig. 8.8a,b,c and d). The rotor distributes the fuel to each cylinder.

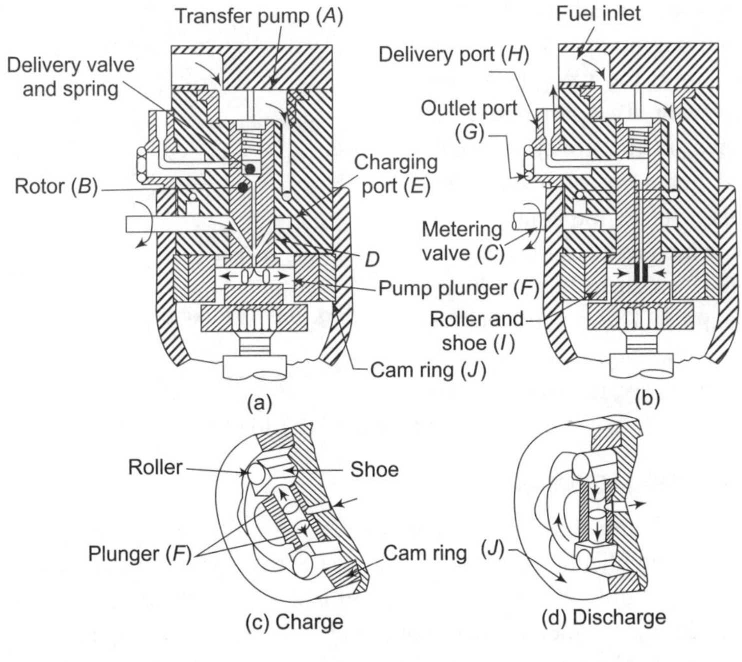


**Fig. 8.8** Construction details of Roosa Master distributor pump

## 8.5 GOVERNING OF CI ENGINES

The governing system employed in CI engine is known as **quality governing**.

Fuel delivered by a ungoverned pump increases with speed whereas the opposite is true about the air intake which results in over-fueling at higher speed.

At idling (low) speed, the engine tends to stall due to insufficiency of fuel. Further, quantity of fuel delivered increases with load resulting in excessive carbon deposits and high exhaust temperature.

As the load on the engine reduces, the speed increases. Drastic reduction in load will cause over-speeding of the engine to a dangerous level leading to failure which suggests that governing of the engine is must. *The function of the speed governing system employed with injection pump is to take care of these limitations by maintaining the speed of the engine fairly constant irrespective of the load.*

There are generally two types of governor employed. They are:
(a) Mechanical governor (b) Pneumatic governor

### 8.5.1 Mechanical Governor

*The purpose of mechanical (centrifugal) governor is to sense the speed of the engine and respond accordingly when the speed limit crosses.* Figure 8.9 shows the working principle of a mechanical speed governor. As the engine speed increases beyond the set limit, the weights of the centrifugal governor fly apart and as a result, the sleeve on the governor shaft moves upward and control lever goes downward. This actuates the control rack in the direction which reduces the amount of fuel delivered, which in turn, reduces the speed of engine. The reverse action takes place when the engine speed tends to decrease. By this way, the speed of the engine is maintained fairly constant irrespective of the load.

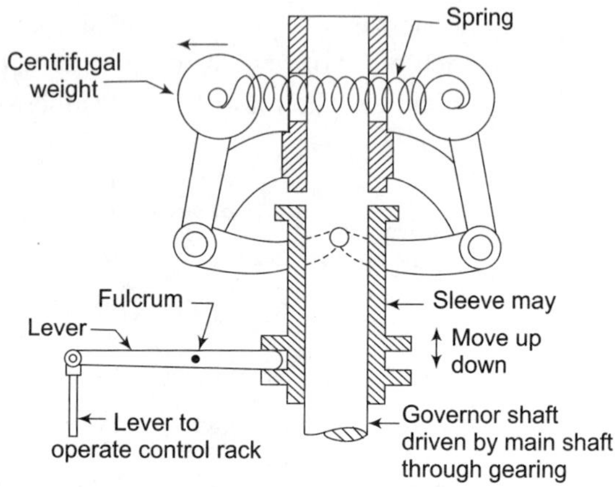


**Fig. 8.9** Working principle of mechanical governor

### 8.5.2 Pneumatic Governor

Figure 8.10 shows pneumatic governor employed with a fuel injection pump of CI engine. The control rack of the injection pump is connected with a spring loaded diaphragm which is housed in a diaphragm unit. A venturi control unit installed on the way to air inlet to cylinder which is linked with the accelerator pedal of the engine. A vacuum pipe connects the venturi unit and vacuum unit. The amount of vacuum applied to the diaphragm is controlled by the accelerator pedal through the position of butterfly valve in the venturi unit. This means the position of accelerator pedal which is a function of speed determines the position of the pump control rack, which in turn, controls the amount of fuel injected to the cylinder.

## 8.6 FUEL INJECTOR

The liquid fuel in the injection system enters the combustion chamber through the injector. *Fuel injector employed in CI engine is of automatic type.* It is mounted on the cylinder body at such a location which yields better performance. Quick and complete combustion is ensured by a well designed fuel injector. The fuel injector assembly consists of the following:

(i) a needle or nozzle valve,
(ii) a compression ring,
(iii) a nozzle, and
(iv) injector body.

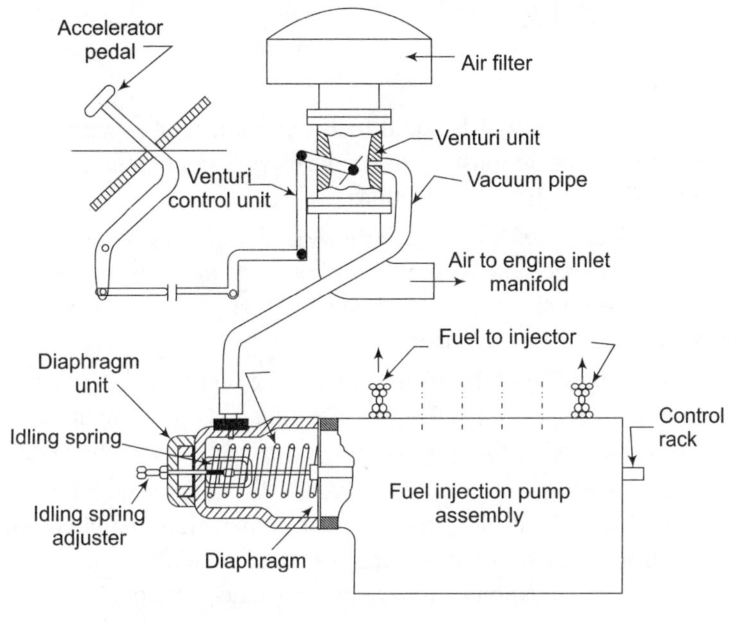


**Fig. 8.10** Working principle of pneumatic governor

Figure 8.11 shows a cross-sectional view of a typical Bosch fuel injector. The fuel from the fuel pump is fed down to the nozzle mouth through long drilling passage. The fuel pressure acts on the differential area of the nozzle valve which lifts against the spring force and, thus allows the fuel to enter the combustion chamber via small hole (orifice) in the form of finally atomized spray. Once, the fuel from the delivery pump gets exhausted, the spring pressure pushes the nozzle valve back on its seat. In order to provide lubrication between nozzle valve and its guide, a small quantity of fuel is allowed to leak through the clearance between them and then drained back to fuel tank through leak off connection provided in the injector body. The adjusting screw provided on the top of spring house is used to adjust the spring tension and hence the valve opening pressure.

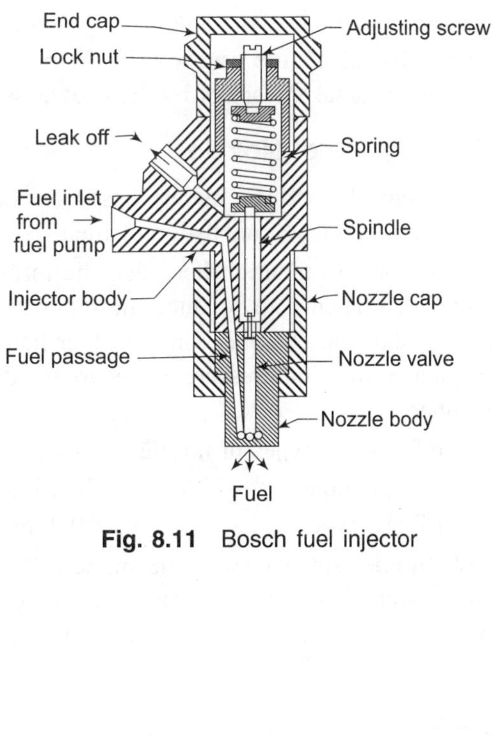


**Fig. 8.11** Bosch fuel injector

## 8.7 INJECTOR NOZZLE

### 8.7.1 Functions

It is the injector nozzle through which liquid fuel in the injection system is sprayed into the combustion chamber. The nozzle must be so designed that, in conjunction with the injection pressure, it should fulfil the following functions.

**(i) Atomization of fuel.** The design of the nozzle should be chosen such that the liquid fuel forced through the nozzle will be broken up into fine droplets, or atomized as soon as it passes into the combustion chamber. This forms the first phase in obtaining proper mixing of fuel and air in the combustion chamber.

**(ii) Proper distribution of fuel.** In the next phase, the fuel must be properly distributed in the desired area of combustion chamber. Factors affecting this phase are the injection pressure, the density of air in the cylinder, the physical properties of fuel and the nozzle design.

It is obvious that higher injection pressure results in better dispersion as well as great penetration of the fuel into all desired locations of combustion chamber. Not only this, it also produces fines droplets resulting in mixing with the air. With the increase in the density of compressed air in the combustion chamber, the resistance offered to the travel of the fuel droplets across the chamber increases which results in better dispersion of the fuel. A very important role is played by the physical properties of the fuel itself such as viscosity, surface tension, etc. in the dispersion aspects of fuel.

**(iii) Prevention of impingement of fuel on walls.** The fuel sprayed by nozzle should not impinge on the surrounding walls of combustion chamber as any fuel striking the walls tends to decompose, producing carbon deposits, unpleasant odour and a smoky exhaust, as well as an increase in fuel consumption.

**(iv) Mixing of fuel and air.** In the non-turbulent type combustion chamber, mixing of fuel and air is taken care of by the nozzle itself.

### 8.7.2 Types of Nozzle

The design of nozzle is mainly based on the types of combustion chamber used ensuring proper and efficient combustion of fuel.

It is worth to mention here that the turbulent type combustion chamber depends upon chamber turbulence to produce the required mixing of the fuel and air. The non-turbulent type combustion chamber, on the other hand, depends almost entirely on both the nozzle design and injection pressure, to secure the desired mixing of fuel and air in the combustion chamber.

The following types of nozzle are used in CI engines (Fig. 8.12).

(a) Single hole    (b) Multi-hole

(c) Pintle type    (d) Pintaux type

**(a) Single Hole Nozzle.** The single orifice nozzle consists of a single hole in the end of nozzle which is closed by nozzle valve (Fig. 8.12a). The hole may be drilled centrally or at an angle to the centre line of nozzle. The size of the hole is around 0.2 mm.

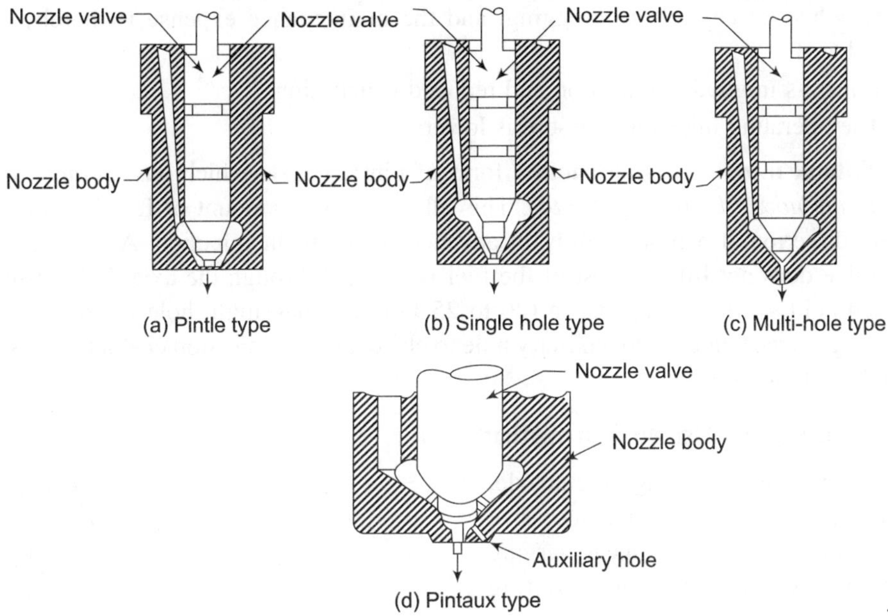


**Fig. 8.12** Types of nozzles

The injection pressure is of the order of 10 to 12 MPa and spray cone angle is about 15°. *The main disadvantages of single hole nozzle include very high injection pressure, tendency to dribble and facilitating poor mixing unless very high air velocities are provided.* Generally, this type of nozzles are used in open combustion chamber.

**(b) Multi-hole Nozzle.** *As the name implies, it consists of a number of holes bored in the tip of nozzle* (Fig. 8.12b). *The number of holes varies from 4 to 18 and the size from 35 μm to 200 μm.* The hole angle is *generally more than 20°*. The operating injection pressure is around 18 MPa. The *main advantage includes the ability to distribute fuel properly.* The main shortcoming is the clogging of hole owing to very smaller size by carbon particles leading to greater maintenance and higher operating expense. These type of nozzles are used in non-turbulent type combustion chamber.

**(c) Pintle Nozzle.** The Pintle nozzle consists essentially of a smaller plunger or pin or pintle in the fuel passage-way, the movement of which controls the flow of fuel in the chamber. This pin or pintle or plunger is the extended form of the stem of the nozzle valve (Fig. 8.12 c). It may be cylindrical or conical in shape. The operating injection pressure is around 8 to 10 MPa. The spray cone angle varies from 0 to 60° by giving tapered to pintle. This type of nozzle is generally employed with turbulent type combustion chamber which utilizes lower injection pressure. The following are the **advantages of pintle nozzles**.

(i) Due to action of pintle, it avoids weak injection and dribbling. The injection rate is very near to the required.

(ii) It is less susceptible to clogging, and the maintenance expense is thereby usually reduced.

(iii) It results in good atomization and reduced penetration.

(iv) The operating injection pressure is lower.

**(d) Pintaux nozzle.** It is an another form of pintle nozzle which has an *auxiliary hole drilled in the nozzle body* (Fig. 8.12d). This allows a small amount of fuel injection in the upstream direction at a time slightly in advance of the main injection. At low speeds the nozzle valve does not lift, so most of the fuel is injected through the axial hole resulting in very good cold starting performance (20 to 25°C lower than multi-hole nozzle). The main disadvantage is the tendency to auxiliary hole to choke and the injection characteristics poorer than multi-hole nozzle.

### 8.7.3 Phases of Spray Formation

As the fuel is injected through the nozzle, spray formed but it consists of its successive phases as shown in Fig. 8.13 due to varying amount of pressure difference existing across the orifice.

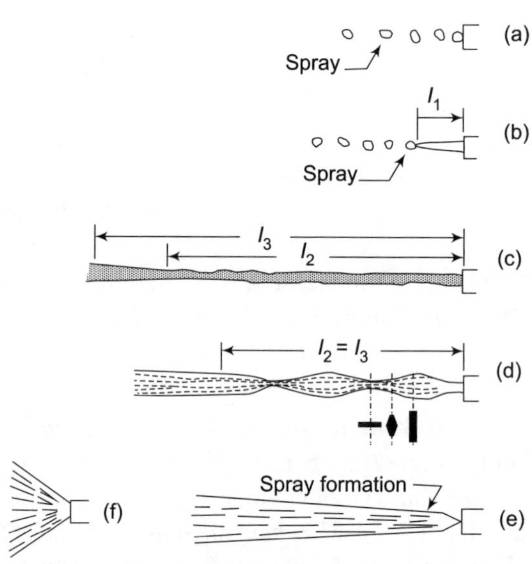


**Fig. 8.13** Phases of spray formation

At the start of fuel injection, single droplets are formed (Fig. 8.13a) as the pressure difference across the orifice is low. With the increase of pressure difference across the orifice, the following phases of spray formation occur successively.

(i) A stream of fuel emerges from the nozzle is shown in Fig. 8.13b.

(ii) The stream encounters aerodynamic resistance from the dense air present in the combustion chamber (Combustion chamber pressure 25 to 35 bar and density 12 to 14 times that of ambient air) and as a result it breaks into a spray, say at a distance of $l_3$ as shown in Fig. 8.13 c. The distance of this breaking point from the orifice is called the *break-up distance*.

(iii) As the pressure difference across the orifice increases, the break-up distance decreases and the *cone angle* increases until the apex of the cone practically coincides with the orifice as shown in Fig. 8.13(d), (e) and (f). All these successive phases of spray formation occurs in a very shortest time.

It is worth to note that the disintegration of the fuel into fine droplets depends upon the relative velocity of fuel and air and the physical characteristics of both fuel and air. The spray angle depends upon the density of the medium in which the fuel is sprayed. The ratio of length to diameter of orifice (L/d) plays an important role on the size of the drop and velocity of the jet. The small value of L/d ratio results in very high jet velocity and thus the nature is turbulent while large value of L/d ratio yields small jet velocity but streamlined flow. A higher L/d ratio results in larger drop size which is due to their high momentum. *The usual range of L/d ratio for diesel engine varies from 2 to 5.*

Large droplets have the capacity for higher penetration but for the point of view of quick mixing and evaporation of fuel, smaller droplets are required. It is the mean droplet diameter which indicates the quality of atomization—**smaller the diameter better is the atomization**. The usual size of the droplets is less than **5 microns**. It is worth to mention that the *droplets size is governed by injection pressure, diameter of the orifice, fuel viscosity of fuel and air density as given below.*

(a) As the injection pressure increases, the mean droplet size decreases.
(b) With increase in air density in combustion chamber mean droplet size decreases.
(c) As the size of the orifice increases, mean droplet size increases.
(d) Higher the fuel viscosity larger is the mean droplet size.

The *penetration* of the spray depends on the momentum of the droplets and the density of the air in the combustion chamber. Higher the momentum greater is the penetration of the spray.

### 8.7.4 Spray Direction

*The relative direction of fuel spray with air plays an important role on atomization of fuel and mixing of air and fuel.*

In the case of down-stream injection (i.e., the direction of the fuel is the same as that of air), the products of the first pass of combustion are swept away from the fuel injected in latter part of the injection process. Upstream injection of fuel provides very good atomization due to very high relative velocity between air and fuel. This result in reduced delay period but at the same time the products of combustion engulf the newly arrived droplets and starve them of necessary oxygen for burning. This results in highly smoky exhaust and poor combustion efficiency.

Keeping in view of the above facts, the upstream injection requires lower injection pressure and large orifice diameter to provide good starting and better part load performance. Further, upstream injection in not suitable in combustion chambers where the air movement is slow. In such cases, downstream injection is generally used for better performance.

### 8.7.5 Injection Timing

Injection timing plays a vital role on the performance of engine. There is an ejection timing control device which would advance or retard the injection timing as the case demands. Figure 8.14 shows a injection timing for a typical diesel engine running at 3300 rpm.

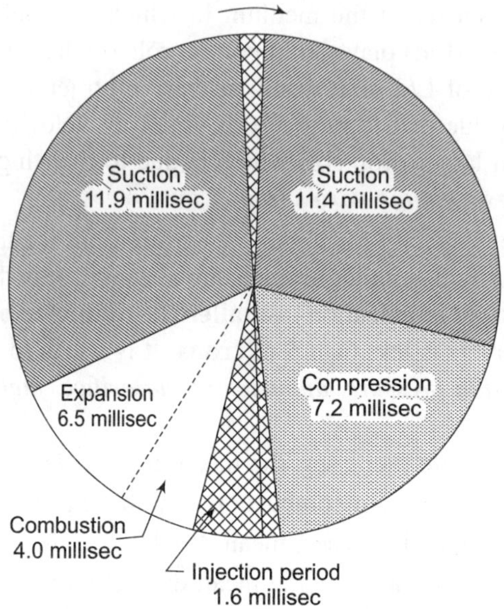


**Fig. 8.14** Injection timing for a typical diesel engine running at 3300 rpm

### 8.7.6 Quantity of Fuel per Cycle and Size of the Nozzle Orifice

The factors affecting the quantity of fuel injected for cycle are the displacement volume and load on the engine. As mentioned earlier, the fuel is injected to the combustion chamber through the nozzle holes. In general, for good atomization, the velocity of the fuel is around 400 m/s. The velocity of fuel through nozzle orifice is expressed by

$$C_f = \sqrt{\frac{2\,(p_{\text{inj}} - p_{\text{cyl}})}{\rho_f}} = \sqrt{\frac{2\,\Delta p}{\rho_f}} \text{ m/s} \qquad ...(8.1)$$

where $p_{\text{inj}}$ = fuel pressure at the inlet of injector, N/m$^2$

$p_{\text{cyl}}$ = pressure of air in cylinder, N/m$^2$

$\Delta p = (p_{\text{inj}} - p_{\text{cyl}})$

$\rho_f$ = density of fuel, kg/m$^3$

The volume of the fuel injected per cycle ($Q_f$) is expressed by

$Q_f$ = coefficient of discharge × Area of all orifices ($A_f$) × Fuel jet velocity ($C_f$) × Time of one injection

or $$Q_f = c_d\, A_f . C_f . \tau$$

or $$Q_f = c_d \left(\frac{\pi}{4}\, d_f^2 . n\right) C_f . \left(\frac{\theta}{360}\, \frac{60}{N}\right) \qquad ...(8.2)$$

where $d_f$ = diameter of one orifice, $m$

$c_d$ = coefficient of discharge

$n$ = number of orifice

$N$ = rpm

$\theta$ = duration of injection in crank angle degrees

With the help of Eqs. (8.1) and (8.2), it is possible to calculate the value $d_f$ if $Q_f$ is known.

The rate of fuel injection is usually expressed in mm$^3$/degree crank angle/litre cylinder displacement volume to avoid effect of engine size.

*Many parameters affect the rate of fuel injected/degree* of crankshaft rotation. These are:

(a) Injector camshaft velocity

(b) The diameter of injector-plunger

(c) The flow area of the tip orifices

As the rate of injection increases the duration of injection decreases for a given fuel input and subsequently introduces a change in injection timing.

This means a higher rate of injection will result in more injection retard from optimum timing and will help in maintaining fuel economy without causing excessive emission of smokes and oxides of nitrogen. Further, the higher injection rate yields increased injection pressure and injector push rod and cam loads which may reduce the engine life.

The quantity of fuel injected per cycle may also be given in terms of *bp*, *bsfc*, speed and number of cylinders

$$Q_f = \frac{(bsfc/60) \times \text{Output}}{(\text{rpm}/2) \,.\, \rho_f . x} = \frac{bsfc \,.\, P}{30\, N \times \rho_f} \qquad \text{...(8.3)}$$

## 8.8 FUEL PRECONDITIONING

In order to obtain a smooth combustion in CI engine, or to use low cetane fuels, fuel preconditioning is one of the approaches.

*By fuel preconditioning, we mean that either all or fraction of fuel– the preliminary injection undergoes a change in state before the main injection takes places.* This change may be either thermal—thermal in temperature or chemical—a change in chemical composition.

In the case of pilot injection, a fraction of fuel is injected into the hot residual gases remaining in the combustion chamber after the exhaust stroke. If the preliminary fuel has the time to develop preflame reactions with air, naturally the ignition delay of the main charge will be reduced resulting in catalyzing ignition.

Experimental results reveal that the rate of pressure rise is considerably reduced due to preconditioning as compared to single injection. This eliminates the knocking tendency of the engine and is a major accomplishment of fuel preconditioning.

## 8.9 NEED OF ELECTRONIC DIESEL INJECTION SYSTEM

Mechanical diesel injection is not able to provide high efficiency and low emission. All parameters related to diesel engine injection process like timing of injection, rate of injection, end of injection, quantity of injected fuel, etc., as per load and speed of the engine have to be perfectly controlled to achieve high efficiency and low emission. Mechanical diesel injection is not able to control all parameters. Electronic fuel (diesel) injection (EFI) system is able to control all parameters and accordingly, it is able to inject required quantity of fuel yielding high efficiency and low emission. EFI system is able to control percisely the following parameters:

(i) Injection timing, (ii) Quantity of fuel injection, (iii) Ignition rate during entire injection period, (iv) Injection pressure, (v) Nozzle opening speed, (vi) Quantity and timing of pilot injection

## 8.10 TYPES AND SENSORS USED IN ELECTRONIC DIESEL INJECTION CONTROL

The following are the various types of control used in electronic diesel injection control.

(i) Electronically controlled injection pump using inline and distributed type.
(ii) Electronically controlled unit injectors.
(iii) Electronically controlled common rail fuel injection system.

The following types of sensors are used in electronic diesel injection control:

(i) Engine speed, (ii) Intake air inlet temperature, (iii) Accelerator pedal position, (iv) Crankshaft position, (v) Lubricating oil temperature, (vi) Intake air mass flow rate, (vii) Turbocharger boost position, (viii) Ambient air temperature, (ix) Knock meter.

## 8.11 ELECTRONICALLY CONTROLLED INLINE INJECTION PUMP

Figure 8.15 shows the block diagram for electronically controlled inline diesel injection pump. About six sensors send the signals to ECU and signals from desired value generators (three in number) are sent to ECU. After processing using inbuilt software, the ECU sends the signals for pumping required injected quantity of fuel to inline pump whose rack is solenoid operated. By this way, the required quantity of diesel fuel is injected through individual injectors in the respective cylinder.

## 8.12 ELECTRONICALLY CONTROLLED COMMON RAIL FUEL INJECTION SYSTEM

In this control system, there are various sensors such as sensors for engine speed, air mass flow, crank speed, camshaft position, turbo air boost pressure, fuel rail pressure, air temperature, coolant temperature, accelerator pedal position, etc. All these sensors send the input signals to ECU (electronic control unit). After processing these signals through software installed, ECU sends outputs to vary the injection timing, fuel quantity, inrition timing, etc. (Fig. 8.16).

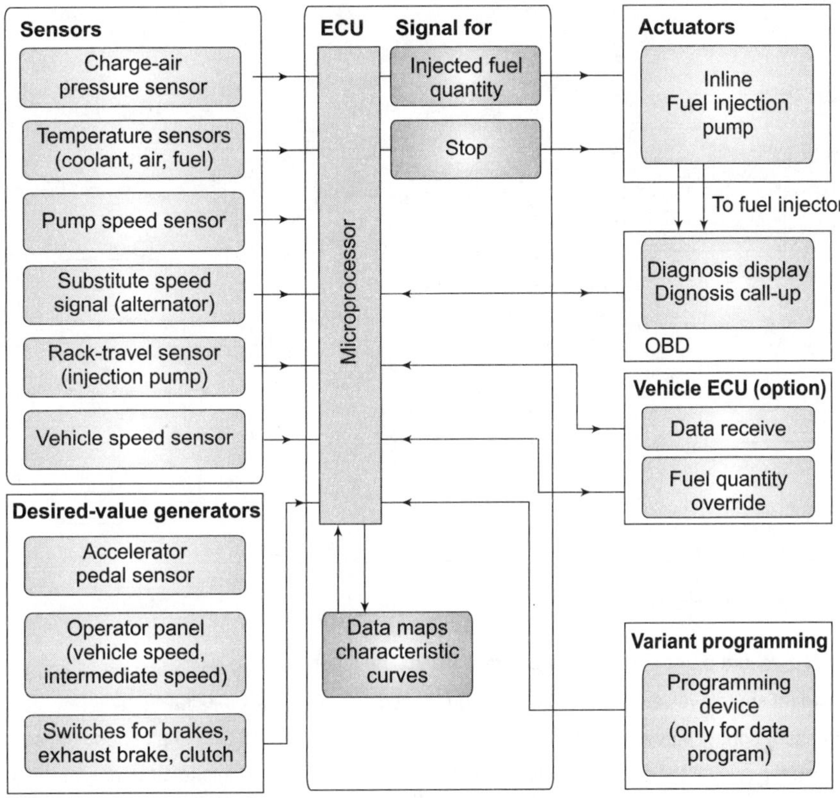


**Fig. 8.15** System block diagram for inline injection pump electronically controlled

ECU also controls systems like fuel injection pump. idle speed control unit, particulate trap regenerator, coolant supply, etc. (Fig. 8.16)

It is worth to note that such management system consists of the following main units: sensors, signal conditioner, analogue to digital converter, ECU, output signal drivers and actuator.

This cotrol system has a wide range of applications, from small to heavy duty engines. The following are some important features:

(i) Very high injection pressure of the order of 1500 bar
(ii) Computer control over start and end of injection
(iii) Injection pressure is independent of speed
(iv) Provision for pilot, main and post injection
(v) Variable injection pressure
(vi) It provides reduced emission level and fuel efficiency

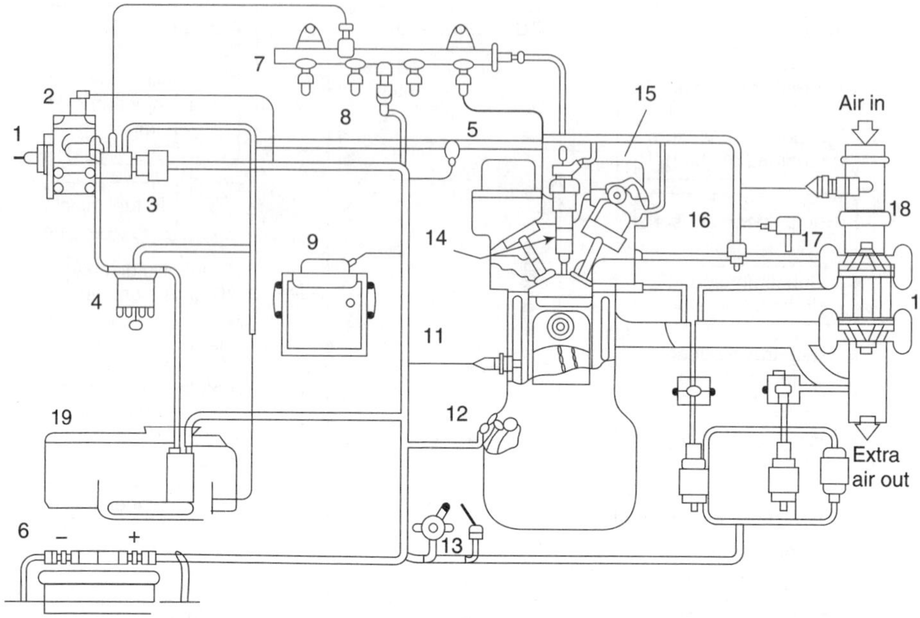


1 High pressure pump
2 Element shutoff valve
3 Pressure control valve
4 Fuel filter
5 Fuel temperature sensor
6 Battery
7 High pressure accumulator
8 Rail pressure sensor
9 ECU
10 Turbocharger
11 Coolant temperature sensor
12 Crankshaft speed
13 Accelerator pedal sensor
14 Injector
15 Camshaft speed sensor
16 Intake air temperature sensor
17 Boost pressure sensor
18 Air mass meter
19 Fuel tank with prefilter and presupply pump

**Fig. 8.16** Bosch common rail electronically controlled diesel injection system

## 8.13 ELECTRONICALLY CONTROLLED UNIT INJECTOR

The injector (Fig. 8.17) used is equipped with solenoid valve. It consists of a multi-hole nozzle, with a spring pressuring the nozzle needle to its sealing seat, a control piston *P*, an orifice *Z* feeding fuel to the control piston and an orifice *A* opened or closed by a solenoid valve. When the valve is de-energized, orifice *A* on the top of control piston is closed. The high pressure fuel flows on the top of control piston through orifice, *Z*. When the solenoid valve is energized, it opens orifice *A* and the fuel on the top of the control piston flows back into the fuel tank via the return line. This reduces the pressure on the control piston because the diameter of the orifice *A* is greater than that of the orifice *Z* and this leads the opening of the injector nozzle. With the de-energisation of solenoid valve, the pressure exerted on the control piston increases and the injector nozzle closes stopping the injection.

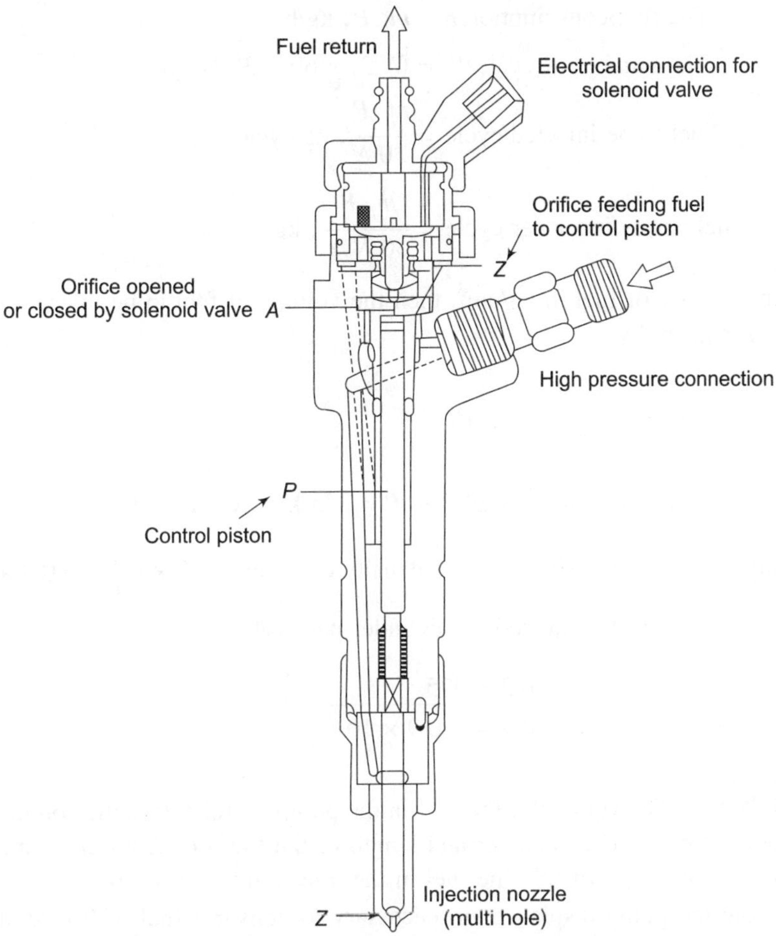


**Fig. 8.17** Electronic injector nozzle

**EXAMPLE 8.1** (a) Derive an expression for the quantity of fuel to be injected per cylinder per cycle for a four-stroke engine is terms of brake specific fuel consumption, b.p. and r.p.m.

(b) Calculate for a four-cylinder, four-stroke CI engine the quantity of fuel to be injected per cylinder per cycle, if the engine consumes 0.3 kg/kWh. The power developed by the engine is 375 kW at a speed of 200 rpm. The specific gravity of the fuel used is 0.9.

***Solution*** (a) Let, $bsfc = \dot{m}_f$; kg/kWh

$bp = P$; kW

$\text{RPM} = N$

No. of cylinder $= x$

$\therefore$ The fuel consumption/$h$ = $\dot{m}_f.P$; kg/h

The number of cycles/$h$ = $(N/2) \times 60 = 30$ N cycles.

$\therefore$ Fuel to be injected/cycle = $\dfrac{\dot{m}_f.P}{30\,N}$; kg/cycle

$\therefore$ Fuel per cylinder per cycle = $\dfrac{\dot{m}_f.P}{30\,N.x}$; kg

If $\rho_f$ be the density of fuel in kg/cm$^3$, then the volume of fuel to be injected per cylinder per cycle ($Q_f$) is given by

$$Q_f = \frac{\dot{m}_f.P}{30\,N\,x\,\rho_f}; \text{ cm}^3$$ ***Ans.***

(b) $m_f = bsfc = 0.3$ kg/kWh, $P = 375$ kW, $N = 200$, $x = 4$

Density of fuel = $\rho_f$ = sp. gravity × density of water = $0.9 \times \dfrac{1}{1000} = 0.9 \times 10^{-3}$ kg/cc.

$\therefore$ The volume of fuel to be injected per cylinder per cycle

$$Q_f = \frac{0.3 \times 375}{30 \times 200 \times 4 \times 0.9 \times 10^{-3}} = 5.2 \text{ cc.}$$ ***Ans.***

**EXAMPLE 8.2** The volume of the fuel in the pump barrel before the commencement of effective smoke is 6 c.c. The diameter and length of the fuel line from the pump to injector is 0.3 cm and 75 cm respectively. The fuel in the injection valve is 2.6 cc.

(a) Calculate the pump displacement necessary to deliver a fuel of 0.1 cc at a pressure 150 bar. Take sump pressure as 1 bar.

(b) Calculate the effective stroke of the plunger if this plunger diameter is 1 cm.

(c) Calculate the power lost in the fuel pumping at an engine speed of 1000 rpm.

Assume coefficient of compressibility of fuel as $80 \times 10^{-6}$ per bar at atmospheric pressure.

***Solution*** Refer to Fig. 8.18.

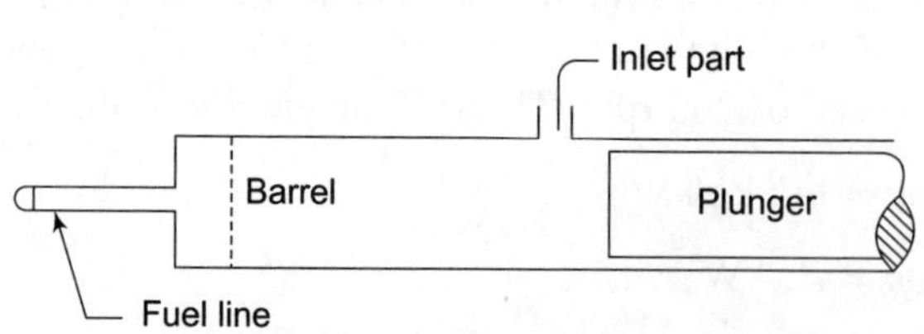


**Fig. 8.18** Example 8.2

Let $V_1$ = total fuel volume

= volume of fuel in the barrel + volume of fuel in the line + vol. of fuel in the injector valve.

$$V_1 = 6 + (\pi/4)\,(0.3)^2 \times 75 + 2.6 = 13.9 \text{ cm}^3$$

Let $V_2$ be the volume when the plunger produces 150 bar of pressure

$\therefore$ Change in volume = $V_1 - V_2 = k\,(p_2 - p_1)\,V_1$

where, $K$ = coeff. of compressibility $= \dfrac{(V_1 - V_2)/V_1}{(p_2 - p_1)}$

$= 80 \times 10^{-6}$ per bar when pressure is taken as atmospheric

$\therefore$ Change in volume $= V_1 - V_2 = 80 \times 10^{-6} \times \dfrac{150 - 1}{1.013} \times 13.9 = 0.16356$ cc.

To deliver 0.26 cc at 150 bar, the total displacement of plunger.

$= (V_1 - V_2) + 0.26 = 0.16356 + 0.26 = 0.32356$ cc. ***Ans.***

(b) Volume $= (\pi/4)\, d^2 \times$ stroke

or $0.32356 = (\pi/4)\,(1)^2 \times$ stroke

$\therefore$ (a) Stroke = 0.4119 cm = 4.119 mm ***Ans.***

(c) Refer to Fig. 8.19.

Suction pressure at *A* is 1 bar and delivery pressure at *B* is 150 bar.

$\therefore$ The work required to build up the delivery pressure

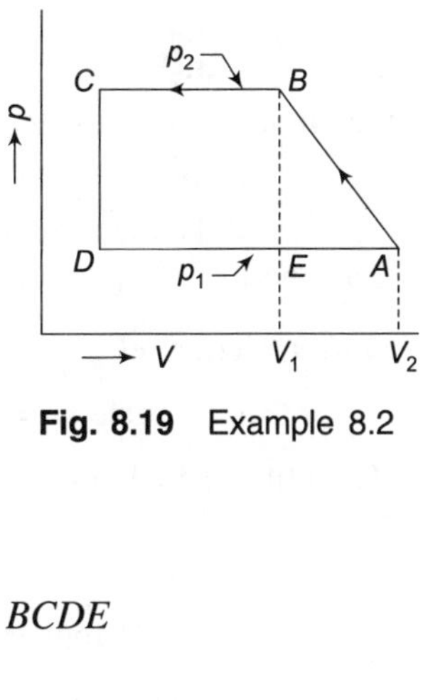


**Fig. 8.19** Example 8.2

$$= \text{Area of } \Delta ABE = \frac{(p_2 - p_1)\,(V_1 - V_2)}{2}$$

$$= \frac{(150 - 1)\,(0.16356)}{2 \times (100)^3} \times 10^2 = 1218.5 \times 10^{-6};\ \text{kJ}$$

The work required to inject the fuel at this pressure = Area of *BCDE*

BC × BE = quantity of fuel injected × $(p_2 - p_1)$

$= 149 \times 10^2 \times 0.16/(100)^3 = 2384 \times 10^{-6}$ ; kJ

$\therefore$ Total work required $= (1218.5 + 2384)\,10^{-6} = 3602.5 \times 10^{-6}$ ; kJ

$\therefore$ Power lost in pumping the fuel $= 3602.5 \times 10^{-6}\,(1000/2) \times (1/60) = 0.03$ ; kW ***Ans.***

**EXAMPLE 8.3** Derive an expression for the diameter of the injector orifice to spray a fuel quantity, $Q$ (m$^3$) per cycle per cylinder in terms of injection pressure, $p_{\text{inj}}$ (kN/m$^2$), combustion chamber pressure $p_{\text{cyl}}$ (kN/m$^2$), density of fuel $\rho_f$ (kg/m$^3$) and period of injection $\tau$ seconds.

***Solution*** Pressure difference causing the fuel flow through the orifice is

$$= (p_{inj} - p_{cyl})\ ;\ \text{N/m}^2$$

∴ Pressure head causing the fuel flow is

$$H_f = p_{inj} - p_{cyl}/\rho_f\ ;\ \text{m of fuel where } \rho_f \text{ in kg/m}^3$$

∴ Velocity of fuel through the orifice from Bernoulli equation = $C_f = \sqrt{(2H_f)}$

or
$$C_f = \sqrt{\frac{2(p_{inj} - p_{cyl})}{\rho_f}}\ ;\ \text{m/s}$$

If $c_d$ is the coefficient of discharge of the orifice, the volume of fuel injected is

$$Q_f = c_d \times \text{Area of the orifice} \times \text{Velocity} \times \text{Injection time}$$
$$= c_d \times (\pi/4)\ d_f^2 \times C_f \times \tau\ ;\ \text{m}^3$$

∴
$$d_f = \sqrt{\left(\frac{4Q_f}{\pi c_d . C_f . \tau}\right)}\ ;\ \text{m}$$

or
$$d_f = \left[\frac{4Q_f}{\pi c_d \tau \sqrt{2(p_{inj} - p_{cyl})/\rho_f}}\right]^{1/2}\ ;\ \text{m}$$ ***Ans.***

**EXAMPLE 8.4** Calculate the diameter of the injector orifice for the following data of a six-cylinder, four-stroke CI engines.

*bp* = 240 kW, *N* = 1300 rpm, *bsfc* = 0.3 kg/kWh, combustion pressure = 36 bar, injection pressure = 200 bar, the specific gravity of the fuel = 0.9, coeff, of discharge of the fuel orifices = 0.92, period of injection = 36 of crank angle.

***Solution*** The fuel consumption = 240 × 0.3 = 72 kg/h

No. of cycles/*h* = (*N*/2) × 60 = (1300/2) × 60 = 39000

∴ Quantity of fuel consumption per cylinder per cycle is

$$= \frac{72}{6 \times 39000} = 0.0003076 \text{ kg}$$

The density of fuel = $\rho_f = 0.9 \times 1000$ kg/m$^3$

∴ Volume of fuel injected per cylinder per cycle

$$Q_f = \frac{0.0003076}{0.9 \times 10^3} = 0.1347 \times 10^{-6} \text{ m}^3$$

$$\text{The period of injection} = \tau = \frac{\theta}{360} \times \frac{60}{N} = \frac{36}{360} \times \frac{60}{1300} = \frac{6}{1300} \text{ s}$$

The velocity of fuel through the orifice

$$= C_f = \sqrt{2(p_{inj} - p_{cyl})/\rho_f} = \frac{\sqrt{2(200 - 36) \times 10^5}}{(0.9 \times 10^3)} = 190.9 \text{ m/s.}$$

$$d_f = \sqrt{\left(\frac{4Q_f}{\pi\, c_d\, C_f\, \tau}\right)} = \sqrt{\left(\frac{4\times 0.3417 \times 1300 \times 10^{-6}}{\pi \times 0.92 \times 190.9 \times 6}\right)}$$

$$= 7.326 \times 10^{-4} \text{ m} = 0.07326 \text{ cm} \qquad \textit{Ans.}$$

**EXAMPLE 8.5** For designing the injector nozzle, calculate the diameter of fuel orifice of a 4-stroke CI engine developing 30 kW per cylinder at 2500 rpm. The specific fuel consumption of fuel 30 API is 0.3 kg/kWh. The fuel is injected at a pressure of 180 bar over a crank travel of 25°. The pressure in the combustion chamber is 40 bar. The coefficient of discharge of orifice is 0.88 and the specific gravity (SG) of the fuel is given by

$$S.G. = \frac{142}{132 + API}$$

***Solution*** Specific gravity of fuel $= \dfrac{142}{132 + 30} = 0.8765$

$$\rho_f = 10^3 \times S.G. = 876.5 \text{ kg/m}^3$$

Velocity of injection of fuel is given by

$$C_f = \sqrt{\frac{2(p_{\text{inj}} - p_{\text{cyl}})}{\rho_f}} = \sqrt{\frac{2(180 - 40) \times 10^5}{876.5}} = 178.73 \text{ m/s}$$

Volume of fuel injected per cycle $= Q_f = \dfrac{(bsfc/60 \times \text{output})}{(\text{rpm}/2)\ \rho_f . x} = \dfrac{bsfc \times P}{30N.x.\rho_f}$.

$$= \frac{0.3 \times 30}{30 \times 2500 \times 1 \times 876.5} = 0.13 \times 10^{-6} \text{ m}^3\text{/cycle}$$

Period of injection, $\tau = \dfrac{\theta}{360} \times \dfrac{60}{N} = \dfrac{25}{360} \times \dfrac{60}{2500} = 1.666 \times 10^{-3}$ sec.

Again $\qquad Q_f = c_d \times \text{Area of orifice} \times \text{Velocity of injection} \times \tau$

or $\qquad A_f = \dfrac{Q_f}{C_d \times C_f . \tau} = \dfrac{0.13 \times 10^{-6}}{0.88 \times 178.73 \times 1.666 \times 10^{-3}} = 0.496 \times 10^{-3} \text{ m}^2$

But $\qquad A_f = \dfrac{\pi}{4} d_f^2$

or $\qquad d_f = \sqrt{\left(0.496 \times 10^{-6} \times \dfrac{4}{\pi}\right)} = 0.7947 \times 10^{-3}\, m = 0.7947 \text{ mm} \qquad$ ***Ans.***

**EXAMPLE 8.6** A spray penetration of 25 cm in 20 milliseconds is obtained at an injection pressure of 180 bar. If the injection pressure of 260 bar had been used, what would have been the time taken to penetrate the same distance assuming the same orifice and combustion chamber design. The cylinder pressure is 30 bar. Use the relation for penetration.

$$S = t\sqrt{\Delta p}$$

where $S$ = penetration in cm

$t$ = time in milliseconds

$\Delta p$ = pressure difference between injection pressure and combustion chamber pressure.

***Solution*** The ratio of penetration under two conditions is given by

$$\frac{S_1}{S_2} = \frac{t_1\sqrt{\Delta p_1}}{t_2\sqrt{\Delta p_2}}$$

where $\Delta p_1 = p_{\text{inj, 1}} - p_{\text{cyl}} = 180 - 30 = 150$ bar

$\Delta p_2 = p_{\text{inj, 2}} - p_{\text{cyl}} = 260 - 30 = 230$ bar

$$\therefore \quad t_2 = \frac{S_2}{S_1} t_1 \sqrt{\frac{\Delta p_1}{\Delta p_2}} = 1 \times 20 \times \sqrt{\frac{150}{230}} = 16.15 \text{ milliseconds}$$ ***Ans.***

**EXAMPLE 8.7** A four-cylinder, 4-stroke diesel engine develops a brake power of 200 kW running at 2500 rpm. The *bsfc* is 0.3 kg/kWh. At the beginning of injection pressure is 40 bar and the maximum cylinder pressure is 60 bar. The injection is expected to be at 200 bar and the maximum pressure at the injector is set to be about 500 bar. Calculate the diameter of the orifice required per injector if the injection takes place over 15 crank angles. Assume the following:

$c_d$ for injector = 0.75, S.G. of fuel = 0.88, atmospheric pressure = 1 bar

Effective pressure difference = Average pressure difference over injection period.

***Solution*** Volume of fuel injected per cycle per cylinder is given by

$$Q_f = c_d . A_f . C_f . \tau$$

$$\text{Brake power/cylinder} = \frac{200}{4} = 50 \text{ kW}$$

$$\text{Fuel consumption per cylinder} = 50\ bsfc = \frac{50 \times 0.3}{3600} = 0.004166 \text{ kg/s}$$

$$\text{Fuel injected per cycle} = m_f = \frac{0.004166 \times 60}{(N/2) \times 60} = \frac{0.004166 \times 60}{(2500/2)} = 1.999 \times 10^{-4} \text{ kg}$$

$$\text{Volume of fuel injected/cycle} = \frac{m_f}{\rho_f} = 1.999 \times 10^{-4}/880 = 0.2271 \times 10^{-6} \text{ m}^3$$

$$\text{Time for fuel injection} = \tau = \frac{\theta}{360} \times \frac{60}{N} = \frac{15}{360} \times \frac{60}{2500} = 0.001 \text{ s}$$

Pressure difference of beginning of injection = 200 – 40 = 160 bar

Pressure difference of end = 500 – 60 = 440 bar

Average pressure difference

$$= p_{\text{inj}} - p_{\text{cyl}} = \frac{440 + 160}{2} = 300 \text{ bar}$$

$$\text{Velocity of injection} = C_f = \sqrt{\frac{2(p_{\text{inj}} - p_{\text{cyl}})}{\rho_t}}$$

$$= \sqrt{\frac{2 \times 300 \times 10^5}{880}} = 261.1 \text{ m/s}$$

$$\therefore \quad A_f = \frac{Q_f}{c_d\, C_f \,.\, \tau} = \frac{0.2771 \times 10^{-6}}{0.75 \times 195.83 \times 0.001} = 1.1596 \times 10^{-6} \text{ m}^2$$

$$\text{or} \quad d_f = \sqrt{\frac{4A_f}{\pi}} = \sqrt{\frac{4 \times 1.1596 \times 10^{-6} \times 10^6}{\pi}} = 1.215 \text{ mm}$$ ***Ans.***

**EXAMPLE 8.8** For the design of individual fuel injection pump and nozzle, calculate the main dimensions of pump and nozzle for a 4-stroke, 4-cylinders diesel engine for the following data:

| | |
|---|---|
| Brake power = 180 kW | *bsfc* = 0.3 kg/kWh |
| Speed = 2500 rpm | Injection pressure = 200 bar |
| Cylinder pressure = 50 bar | S.G. of fuel = 0.875 |

Assume the following for the design:

(i) $c_d$ for injector as 0.75

(ii) Injection takes place over 15° crank angles

(iii) Coefficient of compressibility of oil as $75 \times 10^{-6}$ per bar at atmospheric pressure.

(iv) Coefficient of discharge of orifice as 0.75.

***Solution:*** Volume of fuel injected per cycle = $A_f\, C_f\, \tau$

$$\text{Velocity of fuel injection} = C_f = \sqrt{\frac{2(p_{\text{inj}} - p_{\text{cyl}})}{\rho_f}}$$

$$= \sqrt{\frac{2(180 - 50) \times 10^5}{875}} = 172.37 \text{ m/s}$$

Volume of fuel injected per cycle

$$Q_f = \frac{bsfc \times bp}{30\, N \times \rho_f} = \frac{0.3 \times 180}{30 \times 2500 \times 4 \times 875} = 0.2 \times 10^{-6} \text{ m}^3\text{/cycle.}$$

Change in volume due to compression = $V_1 - V_2 = k \,.\, V_1\,(p_2 - p_1)$

where $V_1$ = total initial fuel volume
= volume of fuel in barrel + volume of fuel in delivery line
+ volume of fuel in injection valve

Here $V_1$ is taken as 100 $Q_f$

$$\therefore \quad V_f = 100 \times 0.2 \times 10^{-6} = 0.2 \times 10^{-4} \text{ ; m}^3$$

Change in volume due to compression

$$= V_1 - V_2 = 75 \times 10^{-6} \times 0.2 \times 10^{-4} (180 - 1) = 0.2685 \times 10^{-6} \text{ m}^3$$

Total displacement of plunger

$$V_d = (V_1 - V_2) + Q_f = 0.2685 \times 10^{-6} + 0.2 \times 10^{-6} = 0.4685 \times 10^{-6} = \frac{\pi}{4} d^2 l$$

Let $l/d = 1$ where $l$ = stroke of plunger, $d$ = dia of plunger

$$\therefore \quad \frac{\pi}{4} d^3 = 0.4685 \times 10^{-6}$$

or

$$d = \left(\frac{0.4685 \times 10^{-6} \times 4}{\pi}\right)^{1/3} = 0.841 \times 10^{-2} \text{ m} = 0.841 \text{ cm} \qquad \textbf{\textit{Ans.}}$$

$$l = d = 0.841 \text{ cm} \qquad \textbf{\textit{Ans.}}$$

$$Q_f = 0.2 \times 10^{-6} = c_d . A_f . C_f . \tau = 0.75 \times \frac{\pi}{4} d_f^2 \times 172.37 \times \left(\frac{15}{360} \times \frac{60}{2500}\right)$$

or

$$d_f = \left(\frac{4 \times 0.2 \times 10^{-6} \times 360 \times 2500}{\pi \times 75 \times 172.37 \times 15 \times 60}\right)^{1/2} = 1.40 \times 10^{-3} \text{ m} = 1.4 \text{ mm} \qquad \textbf{\textit{Ans.}}$$

## 8.14 SUMMARY OF FORMULAE

(1) $\dot{Q}_f = \dfrac{bsfc . P}{32\, N . x . p_f}$ ; $m^3$/cycle for a four-stroke engine

(2) $C_f = \sqrt{\dfrac{(2\rho_{inj} - P_{cycle})}{\rho_f}}$ ; m/s

(3) $Q_f = c_d . A_f . C_f . \tau$

## PROBLEMS

**Things to Think About**

8.1 What is the principle of injection of fuel?

8.2 What is the reason that injection pressure of fuel in CI engine is very high (of the order of 200 bar)?

8.3 Why is the heat release pattern important parameter in CI engine?

8.4 What is the range of ignition delay?

8.5 What is the reason that CI engine is often called a constant supply air engine?

### Viva-voce and Theoretical

8.6 What is the range of injection pressure?

8.7 Discuss heat release pattern in CI engine. How does it affect injection parameters?

8.8 List the requirements of injection system.

8.9 What are the types of injection system?

8.10 What are the various types of airless (mechanical) injection system? Discuss each of them in brief with the help of sketch.

8.11 List the types of injection pump.

8.12 With the help of sketch, discuss the working principle and combustion detail of Bosch (jerk type) fuel injection pump.

8.13 With the help of sketch, discuss the working principle of distributor type fuel injection pump.

8.14 What is the purpose of a using a governor in CI engines? What are the two major types of governor?

8.15 With the help of neat sketch, discuss the working principle of mechanical and pneumatic governor.

8.16 Discuss the function of injector nozzle.

8.17 Mention the types of nozzle and discuss the construction detail of each of them.

8.18 Discuss the phases of spray formation and effect of spray direction.

8.19 Discuss the need of electronic diesel injection system, types of sensors used. With block diagram, show and discuss electronic diesel injection system control-inline and common rail.

8.20 Discuss the working principle of Pintle and Pintaux nozzle.

8.21 Develop the expression for quantity of fuel injected per cycle in terms of *bp*, *bsfc*, *N*, *x* and *f*. Also develop an expression for diameter of orifice.

### Numerical Questions

8.22 A 4-cylinder, 4-stroke CI engine running at 1500 rpm develops 100 kW. The specific fuel consumption is 0.32 kg/kWh. Calculate the diameter of injector orifice if the combustion chamber and injection pressure are 30 bar and 200 bar, respectively. The specific gravity of fuel may be taken as 0.9 and the coefficient of discharge is 0.92. The time of injection is 40 of the crank angle.

8.23 Design the main dimension of fuel pump and injector for four-stroke, four-cylinder diesel engine for the following data:
$bp = 200$ kW, $bsfc = 0.352$ kg/kWh, $N = 3000$ rpm, assume the suitable data.

8.24 A four-cylinder diesel engine develops 120 kW at 2500 rpm. The specific fuel consumption of the engine is 0.3 kg/kWh. Each cylinder has a separate fuel pump, injector and pipe line. At the beginning of effective plunger stroke of one fuel pump, the fuel in the pump barrel is 4 cc the fuel inside the injector is 2 cc and the fuel in the pipe line is 4 cc. The injector pressure is 200 bar and the cylinder charge pressure is 30 bar. Calculate the displacement volume of one plunger per cycle and power lost in pumping fuel to the all cylinders of engine. Specific gravity of fuel is 0.85 and the fuel enter the pump barrel at 1 bar. Coefficient of compressibility of fuel as $80 \times 10^{-6}$ per bar.

8.25 The fuel in the pump barrel of a diesel fuel injection pump before commencement of the effective stroke is 8 cc. The fuel line running from pump outlet to injector is having 3 mm diameter and 600 mm length. The fuel trapped in the injection valve is 3 cc.

(a) Calculate the displacement of plunger to deliver 0.122 cc of fuel at a pressure of 160 bar having pump inlet pressure of 1 bar.

(b) Calculate the effective stroke of a plunger if its diameter is 7 mm. Take coefficient of compressibility of oil as $78 \times 10^{-6}$ at per bar.

8.26 The brake power developed by a six-cylinder, four-stroke diesel engine is 250 kW at 2500 rpm. The *bsfc* is 0.25 kg/kWh. At the beginning of injection, pressure is 40 bar and the maximum cylinder pressure is 60 bar. The injection is expected to be at 190 bar and the maximum pressure at the injection is set to be about 500 bar. Calculate the nozzle orifice diameter if the injection takes place of 160 crank angles. Assume the following:

$c_d$ for injector = 0.75 S.G. of fuel = 0.85 Atmospheric pressure = 1 bar

Effective pressure difference = Average pressure difference over the injection period.

### Software Based Questions

8.27 Develop a computer software for the design of all types of mechanical fuel injection systems for CI engine.

8.28 Develop a computer software for the design of electronic injection system for CI engines.

### Design and Open Ended Question

8.29 Prepare a comprehensive report on Bosch injection system currently used including employing electronic system world over for CI engines. The report should include the drawing, specification and its special features.

### Multiple-Choice Questions

8.30 Solid injection in CI engines means injection of

(a) solid fuel only (b) liquid fuel only

(c) liquid fuel and air (d) solid fuel and air

8.31 A Pintaux nozzle is

(a) simple hole nozzle (b) multihole nozzle

(c) nozzle having an auxiliary hole (d) the same as Pintle hole

8.32 **Assertion (A):** A common rail system for CI fuel injection system is preferred with engines having greater output and higher speeds

**Reasoning (R):** Because the system supplies an increased quantity of fuel whenever the speed of the engine falls.

**Answer Code:**

(a) A is true but R is false (b) A is false but R is true

(c) Both A and R are true (d) Both A and R are false

8.33 The penetration depth of fuel spray in a CI engine

(1) will always increase with the increase in injection pressure

(2) is likely to increase with more viscous fuel

(3) will increase with large length to diameter ratio of the orifice

(4) increases with increase in dispersion of jet of those statements:

(a) 1, 2, 3 and 4 are correct (b) only 2 and 4 are correct

(c) only 1 and 2 are correct (d) only 2 is correct

8.34 The nozzle in a CI engine injects fuel at a sufficiently high pressure so that

(1) fuel penetrates through the high pressure inside the combustion chamber

(2) atomization is better

(3) penetration is not so deep

**Answer Code:**

(a) 1 and 2 are correct (b) 1, 2 and 3 are correct

(c) 1 and 2 are correct (d) 1 and 3 are correct

8.35 Governing of CI engines is called

(a) quantity governing (b) quality governing

(c) quality and quantity governing (d) none of the above

8.36 Heat release pattern in CI engines affects

(a) the rate of pressure rise (b) the noise and vibration of the engine

(c) the cycle efficiency (d) all of them

## Key to Multiple-Choice Questions

8.30 (b) 8.31 (c) 8.32 (c) 8.33 (d) 8.34 (b) 8.35 (b) 8.36 (d)

CHAPTER

# Ignition and Starting Systems

# 9

## OVERVIEW

In SI engine, combustion of the mixture of air and fuel is initiated by producing a spark in the combustion chamber at the appropriate time in the engine. In order to produce spark, an ignition system is needed which forms one of the auxiliary systems of the engine. In this chapter, various types of ignition systems, their details and working principle will be discussed with a brief discussion of starting system. Starting systems of both SI and CI engines are discussed.

## 9.1 IGNITION SYSTEM REQUIREMENTS

The ignition system must provide the following basic requirements:

(i) The system must have a source of electrical energy.

(ii) The system must have a means for boosting the low voltage from the source to the very high potential required to produce a high tension arc across the spark plug gap that ignites the combustible mixture and yields reasonable plug life.

(iii) The system must have a means for timing and distributing the high voltage, i.e., distribute the high potential to each spark plug at the exact instant is required in every cycle for each cylinder.

In order to meet the above requirements the ignition system requires many elements, the basic source of electrical energy is either a battery, a generator or a magnet. The battery and generator normally provide 6 to 12 volts potential direct current, while magneto provides an alternating current of higher voltage but not sufficient to initiate spark. The relatively low voltage produced by any of the three different types of electric sources must be boosted to a very high potential, **10,000 to 20,000** volts, in order to overcome the resistance of the spark plug gap and to release sufficient energy through spark to initiate a self propagating flame front within the combustible mixture. The boosting of the voltage from very low to very high value is achieved by a ignition coil consisting of primary and secondary windings, breaker points and condenser (Fig. 9.1). In order to achieve the timing and distribution of high potential to proper spark plug at the exact instant in each cylinder, distributor and breaker points are used.

## 9.2 TYPES OF IGNITION SYSTEM

Based on the source of low voltage, the two types of basic ignition systems are in use. They are:

(a) Battery Ignition System (b) Magneto Ignition System.

Now, the **electronic ignition system is also coming** up using the battery as the source of low voltage.

## 9.3 BATTERY IGNITION SYSTEM

Figure 9.1 shows a circuit diagram of a battery ignition system for a four-cylinders engine. *It consists of a battery of 6 to 12 volts, ammeter, ignition switch, blast register, ignition coil, breaker points, condenser, distributor and spark plugs.* Ignition coil system is used in passenger cars. Working principle of each element is given below.

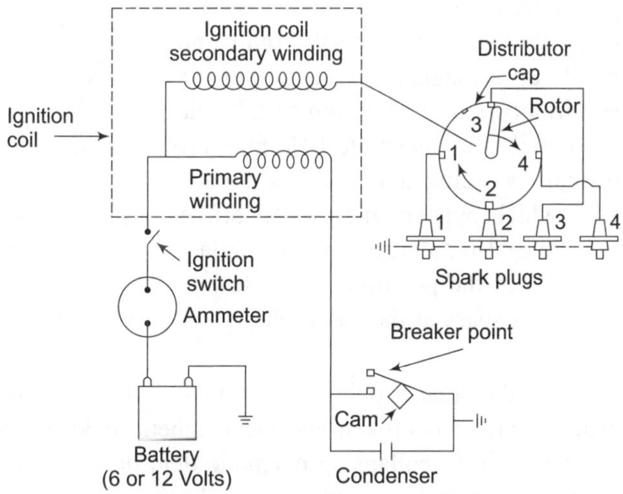


**Fig. 9.1** Battery ignition system

**Battery.** A storage battery (6V to 12V) is used to provide electrical energy for ignition of mixture. It is charged by generator (dynamo) driven by engine (Fig. 9.2). The battery converts the chemical energy into electrical energy. Lead acid battery is used for light commercial vehicles while alkaline is used for heavy duty commercial vehicles. Currently, 12 V battery is universally used due to large capacity to provide higher secondary voltage and better starting.

**Ignition Switch.** With the help of ignition switch, ignition system can be turned on or off.

**Ballast Registor.** A ballast registor is provided in series with the primary winding to regulate the primary current by preventing injury to the spark coil from overheating if the engine should be operated for along time at low speed or should be stalled with the breaker in the closed position.

**Ignition Coil.** The ignition coil consists of a water-proof case containing a primary winding of approximately 200 turns of No. 20 gauge copper wire wound upon an iron core. A secondary wiring of about 18,000 turns of No. 38 gauge copper wire is wound upon the primary with each layer of wire insulated from the next by a breaker strip of oiled paper. From Fig. 9.2, it is obvious that the system has a primary circuit of low voltage current and a secondary circuit of high-voltage current. ***The primary circuit*** *consists of the battery, ammeter, ignition switch, blast register, primary coil winding, condenser and breaker points. The* ***secondary circuit*** *consists of the secondary coil winding, distributor and spark plugs.*

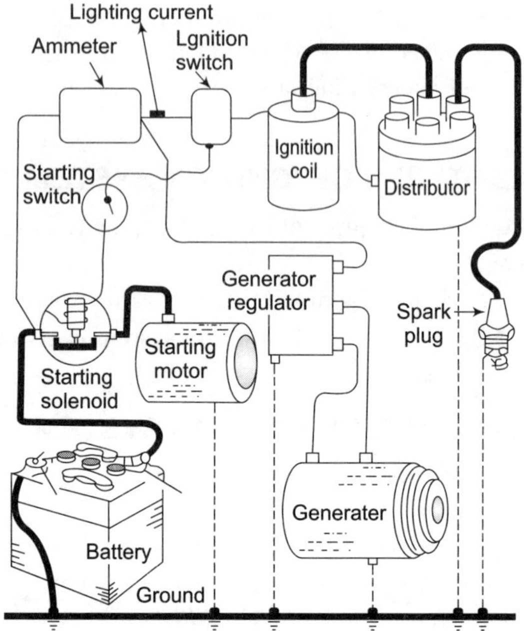


**Fig. 9.2** Block diagram of battery ignition system connected to generator and starting motor

When the ignition switch and breaker points are closed, a low voltage current flows through the primary circuit and builds up a magnetic field around the soft iron core of the ignition coil. While expanding, this magnetic field cuts the primary winding and induces a back emf, which opposes the field itself. A voltage is also induced in the secondary wiring, but the potential is too low to produce a spark at the spark plug gap due to the relatively slow movement of the magnetic flux.

When the breaker points are opened by the action of the cam on the distributor shaft, the primary circuit is broken and the magnetic field begins to collapse. The resulting induced current from the collapsing magnetic field flows in the same direction in the primary circuit as the battery current and charges the condenser which acts as a reservoir for a flowing current. The decrease in flux strength is hastened by discharge from the condenser with the primary current in the opposite direction. Owing to the rapidly collapsing magnetic field, a high voltage is induced in the primary (about 250 volts) and an even higher in the secondary (10,000 to 20,000 volts). The maximum speed of the breakdown of the magnetic field, and the number of turns in the secondary winding determine the maximum voltage that can be induced. The high voltage in the secondary passes through the distributor rotor to one of the spark plug leads and into the spark plug. As soon as sufficient voltage is built up in the secondary windings to overcome the resistance of the gap, i.e., air gap (usually less than 10,000 volts), the arc across the gap ignites the combustible charge of the cylinder.

**Condenser.** The function of the condenser is illustrated in Fig. 9.3. When the breaker points are closed, the strength of the flux field slowly increases from *A* to *B* (constant primary voltage of the battery, say 6V and a maximum primary current of about 4 amp). When the breaker points are open, the field strength falls while the condenser is charging (*BC*) and continues to

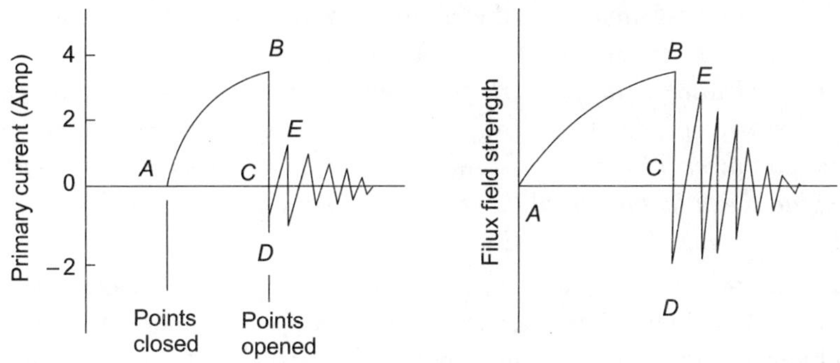


**Fig. 9.3** Magnetic field change in ignition coil

fall sharply while the condenser is discharging (*CD*). An oscillating current (*BCDE*) flows in the resonating circuit with a frequency of about 300 cycles per second. Note, the spark at the plug usually occurs between *B* and *C* and before the rate of change of flux is the maximum at *C* in order to have a safe voltage margin for emergency. The breakdown of the spark gap occurs approximately at 0.0001 sec after the breaker points are opened and represents over 2 crank degrees of lag at 3600 rpm of the engine. The duration of the arc is of the order of 0.001 sec or 22 crank degrees of 3600 rpm engine speed.

**Distributor.** *The breaker points, condenser, distributor rotor and the spark advance mechanism are usually housed in one casing called the* ***ignition distributor.*** The ignition distributor does two types of jobs. It makes and breaks the primary circuit to give a proper timing to spark in each of the cylinder. ***Breaker*** *is generally referred as* ***timer****, because the time or point in the cycle when spark occurs depends upon the opening of the breaker points.* As already pointed out, a cam in used to activate the breaker points. *The breaker cam which is on the distributor shaft is usually driven by a gearing on the engine crankshaft at half the engine speed (in the case of four-stroke engines).* The lobes on the breaker cam correspond to the number of cylinders in the engine. The breaker points are so arranged that the grounded one is stationary, and the insulated one, is attached to a breaker arm that is connected to the primary circuit and, separated from the breaker's cam by a non-conducting rubbing block. By this arrangement, a cam lobe, each time it passes by rubbing block of insulated breaker arm, separates the points.

The second function of the ignition distributor is to lead the high secondary voltage to the proper spark plug to initiate the spark. This is achieved with the help of rotor and cap provided in the ignition distributor. The insulated cap is mounted on the top of the distributor and carries a number of high tension terminals (leads) to the spark plugs. The rotor is mounted on the breaker cam under the cap and rotates with the cam. In this way, the rotor and cap act as a rotating switch that connects the secondary winding to each spark plug, in turn, according to the firing order 1–3–4–2.

To ensure maximum performance of the engine at different speeds and loads, the point in the cycle where the spark occurs, must be regulated. This is achieved automatically by two means, a ***vacuum advance*** and a ***centrifugal advance***. *The vacuum advance on the spark timing is necessary because the lean mixtures in the economy range require an earlier spark timing than do full power mixtures. The centrifugal advance is necessary to compensate for increase in engine speed as the initial flame speed of burning is relatively constant because the spark-plug is near the chamber walls.*

The vacuum advance timing is achieved by attaching the entire breaker point plate to a ***diaphragm*** (Fig. 9.10) held in full retard position by a spring. The other side of the diaphragm, is acted upon by manifold pressure, against the spring resistance. Thus, at part load, the high vacuum of the manifold advances the spark whereas at full load it retards the spark. The centrifugal advance is achieved by attaching the breaker cam to a *weight* (Fig. 9.9). *With the increase in engine speed, the centrifugal force of the weight opens earlier than the points and the spark is thus advanced.*

## 9.4 MAGNETO IGNITION SYSTEM

Figure 9.4 shows a simplified magneto ignition system. It is lighter and more compact than the battery ignition system. It is mostly used in high powered engines, such as, aircraft engine. Magneto system is also used in scooter and motorcycle engines. It *consists of a magneto breaker points, condenser, ignition switch, distributor, spark plug lead and spark plugs.* Here, magneto serves the purpose of a battery and ignition coil. The magneto may have a rotating magnet and a stationary coil or vice versa.

As the rotating magnet turns, the direction of the magnetic flux through the shaft iron core of the coil reverses direction, when breakers are closed. Due to alternating points building up and breaking down of the magnetic flux (four times for each revolution of the rotor with a four-pole magnet), a voltage is induced in the primary and secondary coils. Since the change in flux is not rapid, a high voltage is not induced in the secondary coil to overcome the resistance of the air gap of the spark plug. To overcome this, more rapid breakdown of magnetic flux has to be created and is achieved by the breaker points. When the breaker points are opened by the action of the cam, the condenser is charged followed by a rapid discharge, this produces an almost instantaneous breakdown of magnetic flux. Due to the rapid breakdown of the magnetic flux, a high voltage of the order of 10,000 to 20,000 volts is induced in the secondary windings. This high voltage is fed to the spark plug through the high tension leads and distributor to create the spark. The breaking of the primary circuit, the motion of the rotor and the rotation of the distributor arm must all be

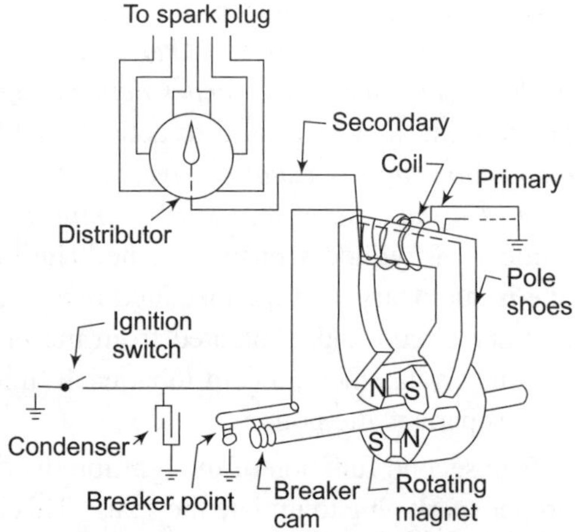


**Fig. 9.4** Magneto ignition system

timed accurately, to produce a spark in each cylinder at the appropriate time in accordance with the firing order.

If the ignition switch is closed, the primary circuit is short circuited, and prevents the building of high voltage in the secondary. This is the reason for the stoppage of the engine when the ignition is closed resulting in grounding of the primary circuit, unlike the battery system.

## 9.5 SPARK PLUGS

*The function of a spark plug is to conduct the high potential from the ignition system into the combustion chamber and then, to allow its discharge across an air gap.*

Figure 9.5 shows spark plug. A spark plug consists essentially of a *terminal, a steel shell, an insulator and two electrodes.* The central electrode connected to the terminal for receiving high voltage from the secondary coil of the ignition system, is well insulated with porcelain, mica or other insulating materials. This central electrode with its insulation is housed in an outer shell made of steel, the lower portion of which is screwed into the cylinder head. The outer electrode is grounded to the cylinder through the steel of the plug. *The material used for electrodes is high nickel alloy to withstand the severe erosion, corrosion and high temperature, to which they are subjected in the combustion chamber.* Since the centre electrode and insulator are subjected to the high temperature of the combustion chamber, heat must flow from the insulator to the steel shell, which is in contact with the relatively cool cylinder head, in order to cool electrodes and prevent pre-ignition.

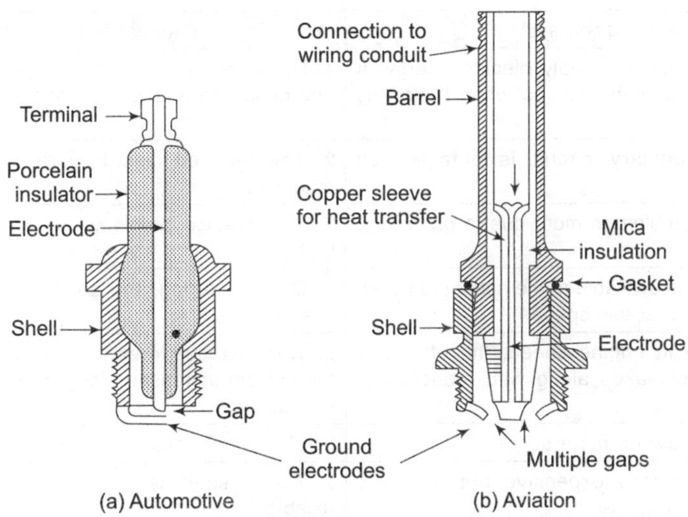


**Fig. 9.5** Spark plugs

Generally, there are two types of spark plugs, the ***hot*** **and the *cold* spark plugs** (Fig. 9.6). They are characterised by their relatively operating temperature range. The operating temperature is governed by the amount of heat transferred, which in turn, depends upon the length of the transfer path and the surface area exposed to the combustion chamber. The spark plug gap ranges from 5 to 8 mm.

The ***cold plug*** *is the one in which the heat path is short and the electrodes remain relatively cool.* Cold plugs are generally used in engines of high output or air cooled engines to avoid pre-ignition. The ***hot plug*** *has a long heat path and is used in the engines which run in cold climate condition.*

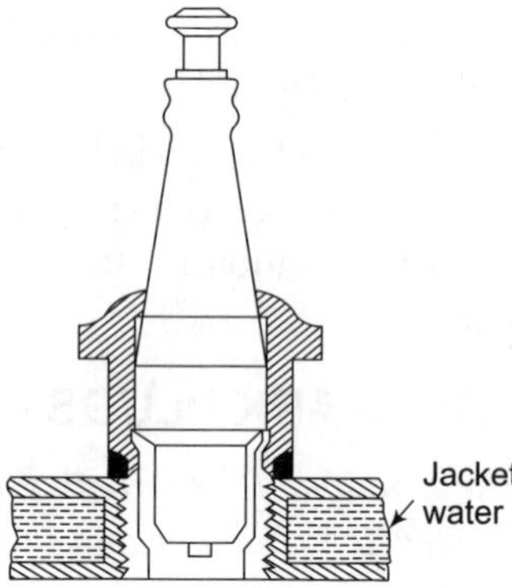


**Fig. 9.6** Hot and cold plugs

The ***establishment of the arc across the electrodes*** *depends on the following factors:* (i) the length of the gap, (ii) the geometry of the gap, (iii) the temperature of the electrodes and enclosed air-fuel mixture, (iv) the density of the mixture, (v) the leakage resistance of the insulator, (vi) the rate of increase of voltage, at the gap, (vii) the presence of ionized gases in the gap, (viii) the air-fuel ratio, and (ix) the electrode material.

## 9.6 COMPARISON BETWEEN BATTERY IGNITION AND MAGNETO IGNITION SYSTEM

Table 9.1 gives the comparison between battery and magneto ignition systems.

**Table 9.1** Comparison between Battery Ignition and Magneto Ignition Systems

| *Battery Ignition system* | *Magneto ignition system* |
|---|---|
| **1.** Battery is a must to supply electric energy. It is impossible to start the engine when battery is discharged. | **1.** No battery is needed to supply electric energy and hence no problem of battery discharge. |
| **2.** Current for primary circuit is obtained by battery. | **2.** The required electric current is generated by magneto. |
| **3.** Maintenance problem is more due to battery. | **3.** Maintenance problems are less as there is no battery. |
| **4.** A strong peak is available at the spark plug even during starting, i.e., at low speed. | **4.** During starting, the spark is weak due to low speed. |
| **5.** With increase in engine speed, the efficiency of the system decreases along with reduction in spark intensity. | **5.** With increase in engine speed, the efficiency of the system increases along with increase in spark intensity. |
| **6.** The system occupies more space. | **6.** The system occupies less space. |
| **7.** The system is less expensive but reliability depends on battery. | **7.** The system is more expensive but highly reliable. |
| **8.** Mostly employed in passenger cars and light commercial vehicles due to good starting characteristics. | **8.** Mostly used in scooters, motorcycles and racing cars due to light weight and less maintenance. |

## 9.7 CONTACT BREAKER AND DWELL ANGLE

### 9.7.1 Contact Breaker

*Contact breaker is a mechanical device (Fig. 9.7) for making and breaking the primary circuit of the ignition coil.* It consists of essentially a fixed metal point against which another metal point makes and breaks contact as it is mounted on spring loaded pivotal arm which is actuated by cam. The material of metal point is one of the *hardest metals, usually tungsten* and each point has a circular flat face of about 3 mm diameter.

Earthing is done to the fixed contact point by mounting it on the base of the contact breaker assembly whereas the arm to which the movable contact point is attached, is electrically insulated. The flow and non-flow of the current through the primary circuit are made possible when points are closed and opened respectively. This is made possible by the pivoted arm which has generally a heel or rounded part of some hard plastic attached in the middle and this heel bears on the cam which is driven by the engine. As a result, every time the cam passes under the heel, the points are forced apart and the circuit is broken.

*When the cam does not pass under the heel, the points are held together by the pressure of spring loaded pivoted arm and thereby closing the primary circuit.*

The contact points are subjected to a severe hammering during their period of service and thus their conditions and adjustment are important and in the case of uneven wear, refacing or replacement is done.

### 9.7.2 Dwell Angle

*The time period, measured in degree of cam rotation during which the contact points remain closed is called the dwell angle or cam angle (Fig. 9.7).*

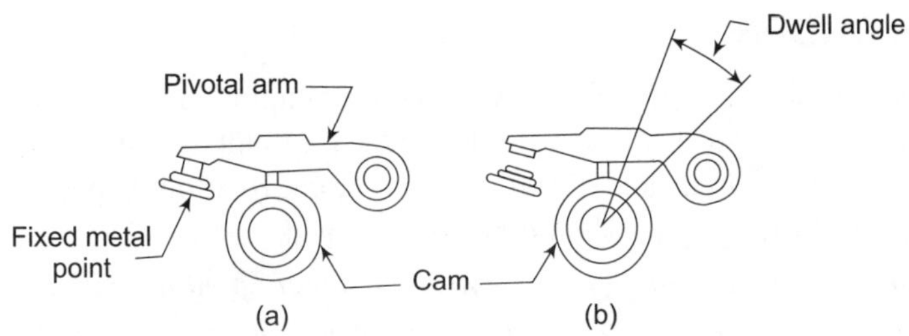


**Fig. 9.7** Contact breaker and dwell angle

The effect of dwell angle on the spark produced is great. It must be sufficiently large to allow magnetic saturation of the primary coil. If the dwell angle is very less it will result in lower secondary voltage and hence poor sparks or even misfiring. If the dwell angle is very large, it over-saturates the winding which results in burning of the condenser and the contact points.

*The dwell angle depends upon the gap between the points and also the angle between the cam lobes.* The range of the gap between the points is of the order of 0.35 mm to 0.55 mm. The smaller the gap larger is the dwell angle. The angle between the cam lobes is a function of number of cylinders. If a single contact breaker is used, the number of lobes on the cam is the same as the number of cylinders. This fact suggests that at a given engine speed, the time available to supply energy to the primary winding decreases as the number of cylinders increases which puts a limit on high speed engine having large number of cylinders. The situation is further worsened due to effects of inertia and spring lag.

*With the increase in number of cylinders, the dwell angle must be reduced as more and more closing and openings operation must be accommodated during every rotation of distributor camshaft. The dwell angle for 4, 6 and 8 cylinders engines are 50°, 38° and 33° respectively.* The dwell period is expressed by the following formula:

$$\text{Dwell period (milliseconds)} = \frac{1000 \times \text{Dwell angle (degrees)}}{6 \times \text{Engine speed (rpm)}}$$

As mentioned, the lower value of dewll angle does not allow sufficient build-up of primary coil energy. To overcome this problem, ***twin contact breakers*** connected in parallel between the primary winding and the earth and a cam having number of lobes half of that of the number of cylinder are used. As a result, the dwell angle becomes almost double, i.e., six cylinder engine will have dwell angle equal to 60°. But this design needs accurate synchronization of the arms of contact breaker which is very difficult to achieve. To overcome this problem, *a **new concept** nowadays has been evolved in which a cam having the same number of lobes as there are number of cylinders is used with **two contact breakers** arranged in parallel.* One contact breaker always opens the points while the other always closes the points, i.e., the dwell angle is obtained thereby allowing sufficient build up of primary coil energy.

## 9.8 FIRING ORDER

In order to initiate combustion of fuel, every cylinder of engine must fire once in every cycle. This means that for a four-stroke, four-cylinder and six-cylinder engines, the ignition system must fire for every 180 degrees and 120 degrees of crank rotation respectively. This further suggests that firing of the cylinder should be well defined.

*The firing order is the order in which various cylinders of a multi-cylinder engine fire. It is the number of cylinders and throws of the crankshaft which decides the number of possibilities of firing order.* For the safeguards of engine, it is desirable to have the power impulses equally spaced and from the point of view of balancing this has led to certain conventional arrangement of crankshaft throws. Over and above of this, there are three following factors which must use considered before deciding the optimum firing order of an engine.

(a) *Engine vibrations*
(b) *Engine cooling*
(c) *Development of back pressure.*

The effect of above parameters may be understood by considering the firing of four-cylinder engine as shown in Fig. 9.8.

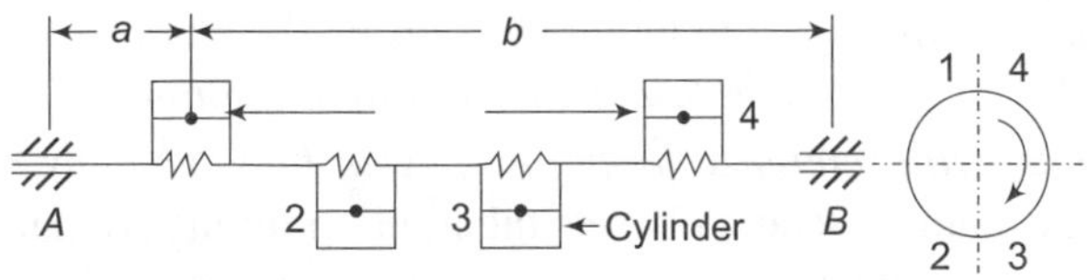


**Fig. 9.8** Firing order

Consider the first cylinder of four cylinder engine is first fired. A pressure, *p* is generated in the cylinder number 1 giving rise to load equal to p.a. and p.b. on the two main bearings *A* and *B* of engine respectively. The load on the bearing *A* is must less than that of on bearing *B*. Now, if the next cylinder fired is cylinder number 2, this imbalance in load on the two bearings would be further increased resulting in server engine vibrations. In order to balance the load, it is desirable to fire the cylinder number 3 instead of 2.

*Firing order that affects the engine cooling greatly.* When the first cylinder is fired, its temperature increases. If the next cylinder that fires is number 2, the portion of the engine between the cylinder number 1 and 2 gets overheated and posing cooling problem. If third cylinder is fired in succession, the overheating is shifted to the portion of the engine between the cylinder number 2 and 3. This localized overheating poses a greater problem on the task of cooling because it requires cooling more at one place than other place. Instead of firing cylinder number 2, if cylinder number 3 is fired, the overheating problem will be minimal.

Further, the *firing order also affects the back pressure of the engine.* After firing the first cylinder, exhaust gases flow out to the exhaust pipe. If the next cylinder fired is the cylinder number 2, we find that before the gases exhausted by the first cylinder go out of the exhaust pipe, the gases exhausted from the second cylinder try to overtake them resulting in increase in back pressure and possibility of arising of back flow, unless and until the exhaust pipe is made bigger in size. Instead of firing cylinder number 2, if cylinder number 3 is fired then by the time the gases exhausted by the cylinder 3 once into exhaust pipe, the gases from cylinder number 1 would have sufficient time to travel the distance between cylinder number 1 and cylinder number 3 and by this way increase in back pressure is prevented.

It is worth to note that all the *three requirements are conflicting to each other to same extent and therefore it is necessary to trade off.* For four-cylinder engine, the possible firing order are as follows.

(a) 1–3–4–2 (b) 1–2–4–3

*The* ***former, i.e., 1–3–4–2 firing order is more common in vertical configuration****.*

For six cylinders engine the possible firing order may be as follows:

(a) 1–5–3–6–2–4 (b) 1–5–4–6–2–3

(c) 1–2–4–6–5–3 (d) 1–2–3–6–5–4

Out of the above, the ***first one,*** **i.e.,** ***1–5–3–6–2–4 is more common***.

## 9.9 SPARK ADVANCE MECHANISM

*The spark timing has a great influence on the performance of engine and it must be regulated to ensure the maximum power and economy at different speeds and loads and this must be done automatically by a mechanism. The purpose of spark advance mechanism is to ensure that under all conditions of operation, ignition takes place at the most favourable instant of time with reference to engine power, fuel economy and the minimum exhaust dilution.* This mechanism accurately sets the ignition timing so that the ignition occurs before TDC position of the piston based on the engine load and speed. In general, advance mechanism is an integral part of the distributor and it automatically regulates the optimum spark advance with reference to load and speed. There are two types of spark advance mechanisms as given below:

(a) *Centrifugal advance mechanism*
(b) *Vacuum advance mechanism*

### 9.9.1 Centrifugal Advance Mechanism

*It controls the ignition timing at full load operation for increase in speed.* As speed increases, the engine requires spark advances as the initial flame speed of the burning mixture remains constant. This is obtained by connecting the contact breaker cam to the driving shaft through a coupling which incorporates *two centrifugal weights* (Fig. 9.9).

With the increase in speed, the weights fly out causing the cam to turn relative to the driving shaft in the direction of rotation and affecting spark advance. The weights are then held in the extended position, in a state of equilibrium corresponding to the shifted timing angle by a retaining spring which exactly balance the centrifugal force. The shifting of the cam by flyweight is either on the basis of rolling contact or sliding contact. The centrifugal force controlled cam is fitted with lower limit stop for purpose of setting the beginning of the adjustment, and also with an upper limit stop to restrict the greatest possible full-load adjustment covering the full range of the speed.

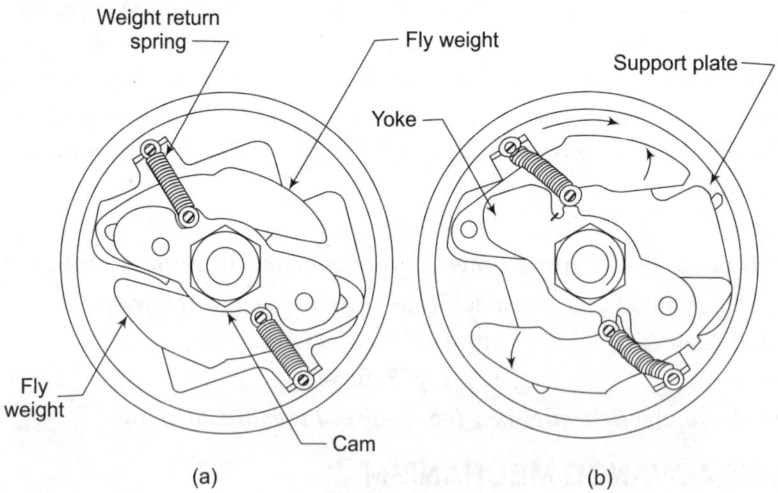


**Fig. 9.9** Centrifugal advance mechanism (sliding contact type)

### 9.9.2 Vacuum Advance Mechanism

At part load condition, i.e., throttle partly closed, the mixture requirement is lean and the lean mixtures require *earlier spark than the rich mixture for maximum power.*

This is achieved by *vacuum advance mechanism which senses the vacuum created at the engine side of the throttle and acting correspondingly by advancing the required spark timing.*

The *ignition advance is accomplished with the help of spring loaded diaphragm connected to the venturi by sensing the vacuum as* shown in Fig. 9.10.

*Under the idling condition, no advance is required.* As soon as the throttle is opened, vacuum is reduced causing the diaphragm to move inwards. Since the diaphragm is connected to movable contact breaker plate, so the inward movement of the diaphragm actuates the contact breaker in such a way angle of ignition advance is achieved with the reduction in induction pressure. The beginning of adjustment is set by the pre-established tension on a compression spring. It is worthwhile to note that the diaphragm area, the spring force and the spring rigidity are all selected according to the partial load advance curve which is to be followed and are all balanced with respect to each other for effective co-ordination.

*The vacuum advance mechanism is strictly an economical device and ensures fuel economy only when the car is properly driven. It operates in conjunction with centrifugal advance mechanism to provide total ignition advance required when the engine is operating under partial load.*

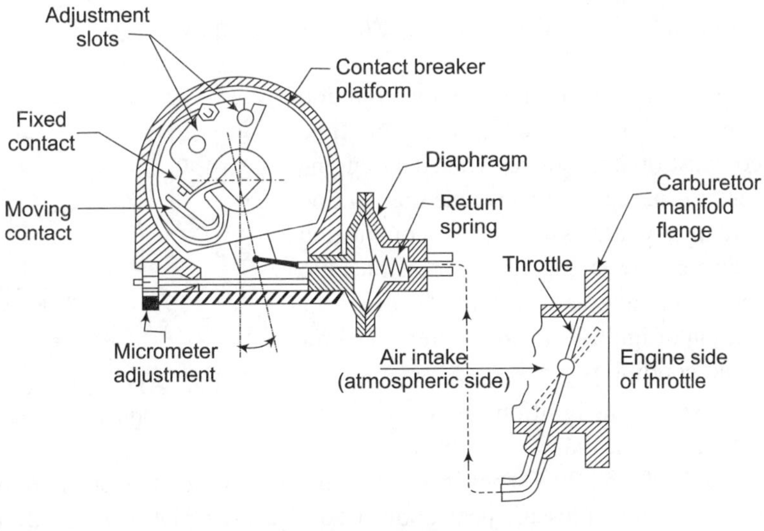


**Fig. 9.10** Vacuum advance mechanism

## 9.10 PARAMETERS AFFECTING IGNITION TIMING

*The power output from an engine is a function of mean effective pressure at a given speed.* This suggests that in order to obtain the maximum power from an engine, the combustion of compressed mixture must deliver its maximum pressure at a time when the piston is about to commence its down stroke, i.e., expansion stroke and the burning of the complete mixture. It means that spark must take place before the piston reaches TDC on its compression stroke. As a general practice, spark should occur at about 15° before TDC position. If the spark occurs too early as compared to TDC position combustion will take place before completion of compression stroke and resulting in rise in pressure will oppose the piston motion and thereby reduces the power developed.

If the spark occurs too late as compared to TDC position, the piston will have already completed a certain part of expansion stroke before the pressure rise occurs due to combustion which will result in corresponding loss of engine power.

*It is the ignition lag which determines the correct instant for the introduction of spark, i.e., ignition timing in the combustion chamber.* Many factors such as compression ratio, mixture strength, part load operation, engine temperature, combustion chamber design and engine speed, the type of the fuel, etc., affect ignition timing. Some of the important engine parameters are discussed below.

## 9.10.1 Engine Speed

The effect of engine speed on ignition advance may be understood by the following equation.

$$\text{Time available for combustion} = \frac{\alpha}{360\,N};\ \text{min}$$

where $\alpha$ = ignition advance in degrees $N$ = speed in rpm

If the engine speed is increased to $2N$, then ignition advance will be $2\propto$ while maintaining the same time available for combustion. This suggests that if the engine speed increases, it is necessary to advance the ignition time correspondingly while assuming the combustion time for a given mixture strength, compression ratio and volumetric efficiency as constant. Figure 9.11 shows the variation of ignition advance with speed for full and part-load conditions.

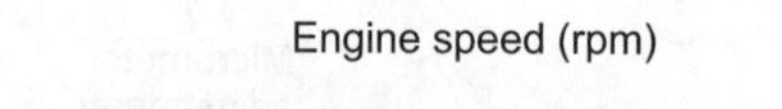


**Fig. 9.11** Effect of speed on ignition advance

The effect of speed on ignition advance can be studied from the indicator diagram taken for a petrol engine running at 1000 and 2000 rpm in which ignition advance has been adjusted in each case to given the maximum pressure peak about 12 degree after TDC (Fig. 9.12). It is obvious from the indicator diagram that at higher speed (2000 rpm), a spark advance of 22 degrees is necessary while at lower speed (1000 rpm), the required advance is only 12.5 degrees. The corresponding delay periods for these two cases are 20° and 10° respectively.

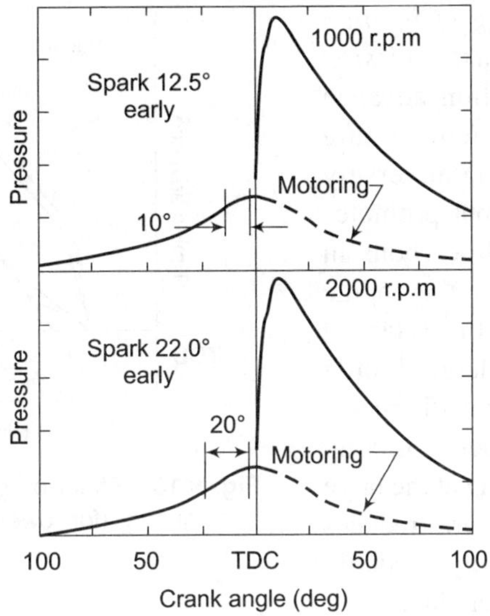


**Fig. 9.12** Effect of spark advance an pressure–crank angle diagram

### 9.10.2 Mixture Strength

It is a well-known fact that rich mixture burns faster. Therefore, it is essential that the spark timing must be retarded if the engine is operating with richer mixture otherwise, the peak pressure will develop before TDC which will oppose the compression process.

### 9.10.3 Part-load Operation

Throttling the incoming charge gives the part-load operation in SI engine. As a result of throttling, a smaller amount of charge enters the cylinder and the dilution of fresh charge due to residual gases is also greater and a result of this, the combustion process becomes slower. In order to compensate the slow combustion process that caused due to part load operation, the spark advance must be increased for the maximum power output and fuel economy.

### 9.10.4 Type of Fuel

Ignition delay is a function of type of fuel. A slow burning fuel needs a higher spark advance than a fast burning of fuel to achieve the maximum power and economy.

## 9.11 IGNITION TIMING AND EXHAUST EMISSIONS

*Environmental pollution is the talk of the day as it is threatening the existence of the mankind. Pollution coming out from the exhaust of automobiles constitutes a large portion of environmental pollution. Some engine operating conditions such as idling, deceleration and rich mixture and closed throttle produce excessive unburnt hydrocarbons and carbon monoxide in exhaust.* The exhaust emission equality in terms of formation of HC and CO is greatly affected by the ignition timing as shown in Fig. 9.13.

*During idling, retarding of ignition timing reduces the exhaust emission in two ways.* As the ignition advance is retarded, exhaust gas temperature increases, which in turn, adversely affects the fuel economy but promotes additional burning of hydrocarbons in the exhaust manifold. Since the engine efficiency is reduced so the retarded timing requires a slightly large throttle opening to maintain the same idle speed. There is a possibility of better mixing and combustion during idling at the large throttle opening, which in turn, reduces exhaust emission appreciably specially during deceleration of the engine.

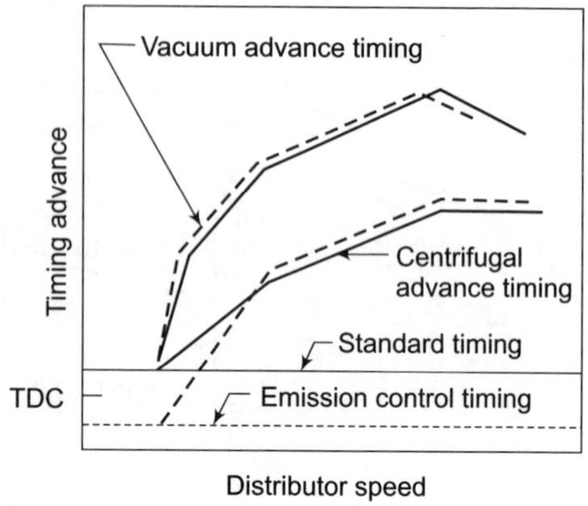


**Fig. 9.13** Effect of ignition advance on emission quality (for lower value of HC and CD)

## 9.12 LIMITATIONS OF CONVENTIONAL IGNITION SYSTEM

The conventional ignition system suffers from the following limitations:

1. *Frequent servicing and replacement of contact point.* During operation, the contact breaker is subjected to high inductive current which results in its arcing, burning and excess wear. This causes error in ignition timing leading to loss of engine power and increased exhaust emissions. To overcome this problem, frequent servicing and replacement of contact points are essential.
2. *Weaker spark, misfiring and reduction in output voltage at high speeds.* At high speed, there is a tendency of the moving arm of the contact breaker to bounce. As a result of this, the effective contact time between the distributor cam and the moving arm is shortened resulting in less energy supply to the primary coil and hence lower voltage of spark plug. This leads to weaker spark and misfiring. Not only this, at high speed, the contact breaker heel fails to follow the cam profile which results in change in ignition timing and possible reduction in output voltage and as a result engine performance is severely affected.
3. *Inefficient system at low speed.* The system becomes inefficient at low speeds because of the fact that relatively high current is drawn from the battery due to the contact remaining closed for larger time.
4. *Poor starting ability.* Slow opening of the breaker points and the cranking speed results in poor starting ability because high tension voltage is available.
5. *Reduced spark duration.* The inductive time constant (L/R ratio) of the primary winding is prohibitively long causing lower rate of primary current build up which results in lower voltage availability across the spark electrodes at moderate to high speeds. As a result of this, the coil current cannot reach its final value in the short intervals between two ignitions resulting in reduced speak duration. Moreover, the ability of ignition

system also reduces to fire fouled plugs because most of the spark energy is lost as current flows through low resistance fouling deposits around the electronic tip.

6. *Poor reproducibility.* The inherent erratic operation of the contact breaker points, the reproducibility of secondary voltage rise peak voltage and firing time is poor. This is one of the reasons of cycle-to-cycle variation in SI engine. Moreover, the system is sensitive to side tracking due to high source impendance.

## 9.13 ELECTRONIC IGNITION SYSTEM

The limitations of convectional ignition system has necessitated the need to void frequent breaker contact servicing and to provide longer ignition system life and better consistency of firing, thereby avoiding the power loss and increased fuel consumption and hydrocarbon emission from the engine. This need has led to the development of ***electronic ignition system.*** *The ability of transistor to interrupt a circuit carrying a relatively high currents makes it an ideal replacement for the breaker points and condenser. The source of electric energy is the battery.*

The following two types of electronic ignition system are used in modern automobiles.

(a) Transistorized coil ignition system (TCI system)
(b) Capacitive discharge ignition system (CDI system)

### 9.13.1 Transistorized Coil Ignition (TCI) System

Transistorized coil ignition system is gaining popularity over conventional ignition system due to the following advantages:

(i) Capability of providing higher output voltage
(ii) Use of electronic triggering maintains required ignition timing
(iii) Reduced ignition system maintenance
(iv) Reduced wear of the components
(v) Increased reliability
(vi) Extended spark plug life
(vii) Improved ignition of lean mixture.

Figure 9.14 shows the circuit diagram of transistorized coil ignition system. The source of electric energy supply to the system is 12 V battery as the conventional system. *The contact breaker and the assembly of the conventional ignition system are replaced by a magnetic pulse generating system which detects the distributor shaft position and sends electrical pulse to an electronic control module.* This module serves many functions. One of the main function of module is to switch off the flow of current to the primary coil inducing a high voltage in the secondary winding which is distributed to the spark plug through distributor similar to the conventional breaker system. The electronic control module is equipped with timing circuit which later closes the primary circuit so that the building up of the primary circuit current can occur for the next cycle. Though, there are many types of pulse generator which may trigger the electronic circuit but *magnetic pulse generator having a gear shaped iron rotor driven by the distributor shaft rotates past the pole of a stationary magnetic pick*

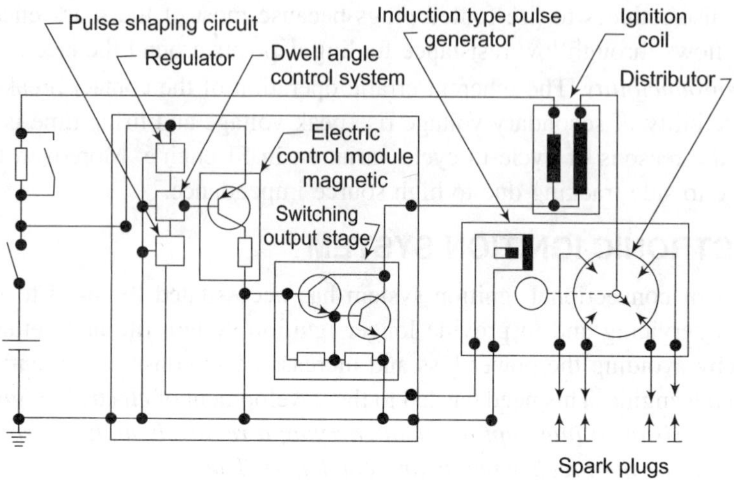


**Fig. 9.14** Transistorized coil ignition system

*up is commonly used.* The number of teeth on the rotor is equal to the number of cylinders and the magnetic field is provided by a permanent magnet.

The magnetic pick-up consists of a coil through which passes the magnetic flux generated by the permanent magnet. As each rotor teeth passes the magnet pole, the magnetic field strength ($\Psi$) first increases and then decreases producing a voltage signal proportional to $d\Psi/dt$. In response to this behaviour the electronic module switches off the primary circuit coil current to produce the spark as the rotor tooth passes through alignment and pick-up coil voltage abruptly reverses and passes through zero. After this voltage reversal, the increasing portion of the voltage waveform is used by electronic module to establish the point at which the primary coil current is switched on for the the ignition pulse for next cycle. *It means that the control unit does the work of contact points, namely, opening and closing of the primary circuit while the rotor does the work of cam. The module is also equipped with dwell angle control system and pulse shaping circuit.*

### 9.13.2 Capacitive Discharge Ignition (CDI) System

Figure 9.15 shows a capacitive discharge ignition system. *The trigger unit of CDI system consists of a capacitor, thyristor power switch, charging device to convert battery voltage to charging voltage of 300 to 500V by means of pulses via voltage transformer, pulse shaping and control unit.*

The function of the capacitor is used to store the ignition energy. The amount of stored energy is determined by the capacitance and changing voltage of the capacitor. The function of ignition transformer is to step up the primary voltage generated at the time of spark by the discharge of the capacitor through the thyristor to the high voltage required at the spark plug.

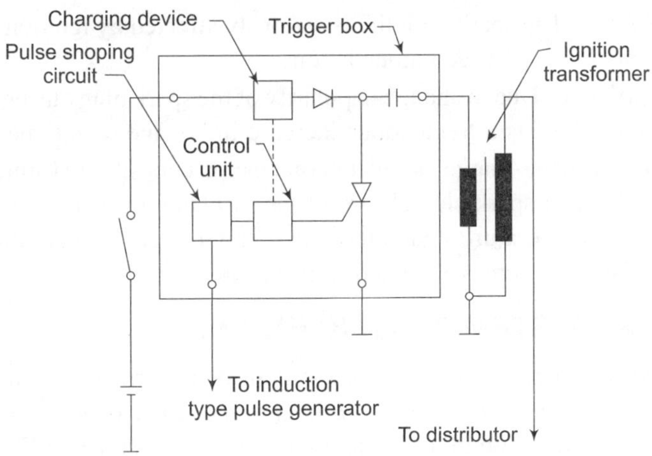


**Fig. 9.15** Capacitive discharge ignition system

The system is insensitive to electrical shunts resulting from spark plug fouling and this forms its main advantage. *The disadvantage includes the chances of ignition failure during lean mixture operating condition owing to the fast capacitive discharge resulting in strong but shorts park.*

## 9.14 FACTORS AFFECTING SPARK PLUG OPERATION

The following factors affect the spark plug operation:

(a) *Voltage requirement and availability.* There is always erosion and corrosion of electrode even during the normal life of the spark plug which causes increase in electrode gap. As a result, the voltage requirement of the spark to given spark increases. In other words, we can say that the reserve voltage decreases. This problem is further augmented by formal wear on the distributor make-and break contacts which reduces the available voltage from the ignition system.

(b) *Mixture strength.* Excessive lean mixtures require higher voltages. Though, the excessive rich mixtures require lower voltage to yield spark but this reduces the spark plug insulator nose operating temperature with the increase in the amount of sooty carbon deposited on the insulator nose. At low speed, the spark plug temperature gets reduced sufficiently so that it is unable to burn off this carbon, which in turn, leads to serious misfiring.

(c) *Misfiring tendency and loss of power.* With the increase in operating life of spark plug, the carbon deposits formed are not continuous and often the spark energy will cause a small track or are on the surface of insulator of discharging across the normal gap. As a result, the spark timing changes which results in loss of power.

(d) *Detonation.* Detonation is caused by shock wave which can excite the electrons of the spark plug and may result in fracture due to vibration. Not only this, the rise in temperature caused by detonation increases the spark plug temperature, which in turn, causes pre-ignition. This pre-ignition may damage the piston crown.

(e) *Ignition advance.* The spark plug life is seriously affected by ignition timing advanced beyond the manufactured recommendations.

(f) *Spark plug polarity.* Due to change in polarity of the spark plug during any replacement of ignition coil, there is a tremendous increase in voltage requirement.

(g) *Combustion deposits.* A large variety of combustion deposits are formed on the walls of combustion chamber spark plug electrode and insulator and each has its own electrical resistance and temperature characteristics. Under high temperature condition, the electrical resistance decreases, which in turn, causes misfire.

## 9.15 STARTING SYSTEM OF AN IC ENGINE

The starting system of an IC engine is a sub-system of electrical system. *It consists of a starting motor driving a pinion. The electrical input to the motor comes from battery through ignition coil.* The wiring diagram of a starting motor system is shown in Fig. 9.16. *The drive*

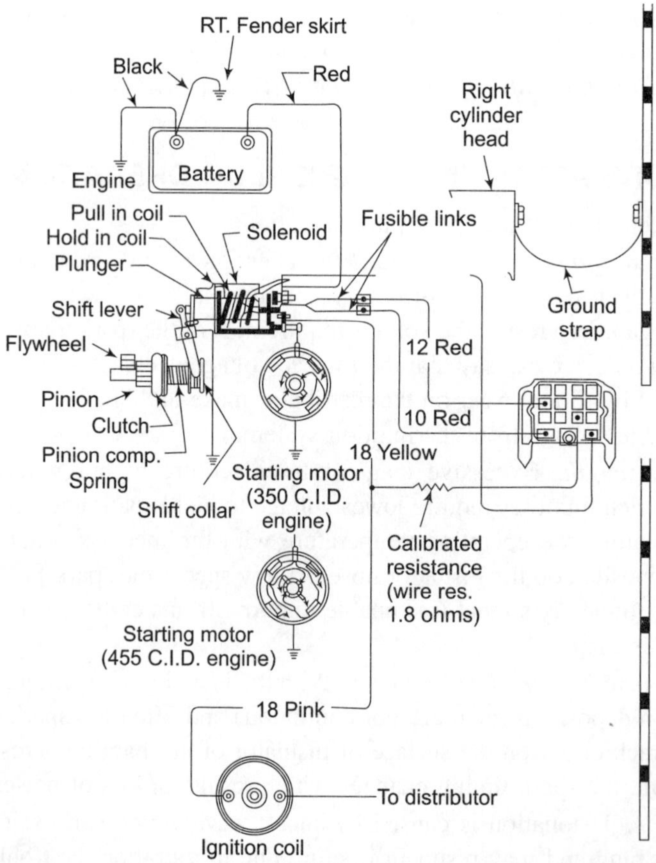


**Fig. 9.16** Wiring diagram of starting motor system for passenger car

*assembly containing a small pinion that, in operation, meshes with teeth on a larger gear on the flywheel* (Fig. 9.16) *or torque convertor.* This provides gear reduction so that the armature must rotate about fifteen times to cause the flywheel to rotate ones. The armature may revolve 2,000 to 3,000 rpm—when starting motor is operated. This makes the flywheel to spin at speeds as high as 300 rpm. This is sufficient for starting the engine.

After the start of the engine, its speed may increase up to 3000 rpm or more. If the starting motor drive pinion remained in mesh with the flywheel, it would be spun at 45,000 rpm which may damage the armature. *In order to prevent such damage, automatic meshing and demeshing devices are used.* In passenger cars, there are two general types **inertia and overhung clutch**. The latter is invariably now used in automobiles. The controls of starting motor have varied from a simple foot operated paddle to automatic device that close the starting motor circuit when the accelerator pedal is depressed. The system presently used in passenger cars and other vehicles has *starting contact in the ignition switch.* When the ignition key is turned against the spring pressure, part on position to START, the starting contacts close. This connects the starting motor solenoid or magnetic switch to the battery. As soon as the engine starts and the ignition key is released, spring pressure returns it to the ON position which opens the starting contacts.

***The starting system is used in both SI and CI engines.***

## PROBLEMS

### Things to Think About

9.1 Why a very high voltage is essential to initiate combustion in a SI engine?

9.2 Why there is a need of firing order?

9.3 Why there is a direct relation between ignition timing and exhaust emission?

9.4 List the various elements of an ignition system.

9.5 Why there is a need of vacuum and centrifugal advance?

### Short and Long Theoretical Questions

9.6 What is the philosophy behind the spark ignition engine?

9.7 Discuss the basic requirements of spark ignition system?

9.8 List the various types of ignition system.

9.9 With the help of a neat sketch discuss the working principle of battery ignition system.

9.10 With the help of neat sketch discuss the working principle of magneto ignition system.

9.11 Differentiate between battery ignition and magneto ignition system.

9.12 With the help of neat sketch, discuss the construction detail of spark plug.

9.13 List the factors on which the establishment of the arc depends.

9.14 Give the constructional details and working principle of contact breaker.

9.15 What do you mean by dwell angle?

9.16 Why there is a need of firing order?

9.17 With the help of a neat sketch discuss the working principle of centrifugal and vacuum advance mechanism.

9.18 Discuss the parameters affecting the ignition timing.

9.19 How does the ignition timing affect exhaust emissions?

9.20 Discuss the limitations of conventional ignition system.

9.21 With the help of a neat sketch, discuss the working principle of electronic transistorized coil ignition system and capacitive discharge ignition system.

9.22 Discuss the factors affecting spark plug operation.

### Software Based Question

9.23 Develop a computer software for the design of various types of ignition systems.

### Design and Open Ended Question

9.24 A real engine requirement for ignition system is enormous. Based on these real requirement, develop an ignition system which will serve all the real requirements. Suggest all the specifications of every element of such system.

### Multiple-Choice Questions

9.25 The sparking voltage across a spark plug ranges from
(a) 6 to 12 V (b) 10 to 1000 V
(c) 1 kW to 2 kV (d) 10 kV to 20 kV

9.26 Ignition timing of a multi-cylinder SI engine can be adjusted by
(a) adjusting the generator pulley (b) rotating the camshaft
(c) adjusting the pump pulley (d) rotating the distributor

9.27 A battery ignition system requires the following element for producing high voltage
(a) ignition coil (b) ignition coil and contact breaker
(c) contact breaker and distributor (d) ignition coil, condenser and contact breaker

9.28 The specific gravity of a battery on being tested by a hydrometer gives a reading of 1180. This signifies that the battery is
(a) half charged (b) three fourths charged
(c) one-fourth charged (d) fully discharged

9.29 The critical spark energy required to ignite a given mixture
(a) decreases with decreasing spark voltage
(b) decreases with increasing spark voltage
(c) increases with decreasing spark voltage
(d) increases with increasing spark voltage

9.30 In a battery ignition system, the intensity of spark is
(a) the maximum at low engine speed and decreases with engine speed
(b) the maximum at low engine speed
(c) the minimum at low engine speed
(d) the minimum at low engine speed and decreases with engine speed

9.31 Cold spark plugs

(1) don't misfire with excess cooling and at low power output

(2) have short heat path and relatively cool

(3) electrodes are used in engines heaving higher output and as air cooled engines. Of these statements:

(a) only 3 is correct (b) only 2 is correct

(c) only 1 and 3 are correct (d) 1, 2 and 3 are correct

9.32 In a 4-cylinder petrol engine, the standard firing order is

(a) 1–2–3–4 (b) 1–4–3–2

(c) 1–3–2–4 (d) 1–3–4–2

## Key to Multiple-Choice Questions

9.25 (d) 9.26 (a) 9.27 (d) 9.28 (b) 9.29 (c) 9.30 (a) 9.31 (d) 9.32 (d)

CHAPTER

# Engine Friction and Lubrication

10

## OVERVIEW

The world is full of friction. Wherever there is a clash of two or more items in any form, friction is produced. Friction may be due to clash of ideas, ego, interest or surfaces in relative motion.

In science and engineering, friction generally refers to force acting between surfaces in relative motion. In internal combustion engines, frictional losses are mainly due to sliding as well as rotating parts. *Usually engine friction is expressed in terms of frictional power and is defined as the difference between the indicated power, (ip) and the brake power, bp.* The mechanical efficiency which is the ratio of *bp* over *ip* is also indicative of mechanical friction.

To improve the performance of engines, it is essential to reduce the frictional loss as minimum as possible. A good engine designer should not allow the total frictional loss more than 10% of energy input. This is achieved by proper lubrication of the engine parts which are either having sliding or rotating motion.

This chapter deals with all aspects of engine frictions and lubrication.

## 10.1 TOTAL ENGINE FRICTION AND MECHANICAL EFFICIENCY

The total engine friction is mainly divided into the following components.

(a) Direct mechanical friction losses
   (i) Crankcase mechanical friction
   (ii) Piston assembly mechanical friction
(b) Blow by losses
(c) Pumping losses
(d) Power loss to drive the components to charge and scavenge (i.e., supercharging and two-stroke scavenging loss)

(e) Power loss to drive other auxiliary components such as water pump, lubricating oil pump, fuel pump, cooling fan, generator, air conditioning, compressor, etc.

The various losses described above can be clubbed into one heading viz., the **mechanical loss** which can be written in terms of **frictional mean effective pressure (fmep)**. *This is defined as the ratio of frictional power divided by engine displacement volume per unit time.* Thus, fmep can be expressed as

$$fmep = mmep + blmep + pmep + cmep + amep \qquad ...(10.1)$$

where

*mmep* = mean effective pressure required to overcome direct mechanical friction

*blmep* = mean effective pressure required to overcome blow-by losses

*pmep* = mean effective pressure required to overcome pumping losses

*cmep* = mean effective pressure required to drive compressor for turbocharging or scavenging pump in two-stroke engines.

*amep* = mean effective pressure required to drive auxiliary components

The mechanical efficiency is defined as the ratio of *bp* or *bmep* to *ip* or *imep*. Thus,

$$\eta_m = \frac{bp}{ip} = \frac{bmep}{imep} \qquad ...(10.2)$$

The total motored friction mep for several four-stroke cycle, four-cylinder SI engines between 845 and 2000 cm$^3$ displacement, at wide open throttle as a function of engine speed is shown in Fig. 10.1. The data are well correlated by an equation of the form

$$fmep = 0.97 + 0.15\ (N/1000) + 0.05\ (N/1000)^2;\ \text{bar} \qquad ...(10.3)$$

where $N$ = rpm

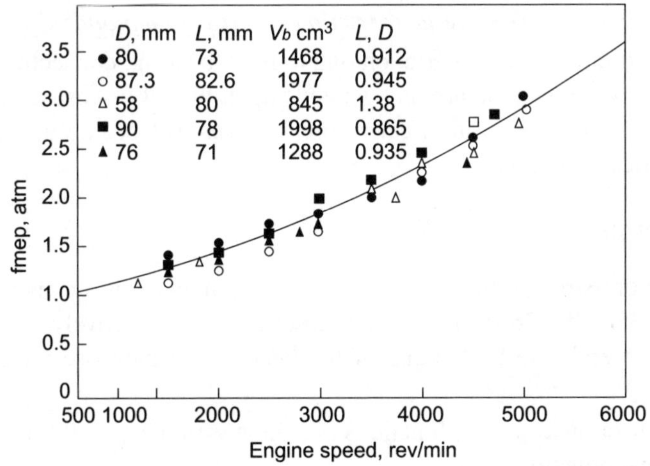


**Fig. 10.1** Motored fmep vs engine speed for SI engine

Figure 10.2 shows the variation of mechanical efficiency and the ratio of *pmep/fmep* as a function of load for a typical SI engine at a fixed mid-speed. *The mechanical efficiency is the highest at full load.*

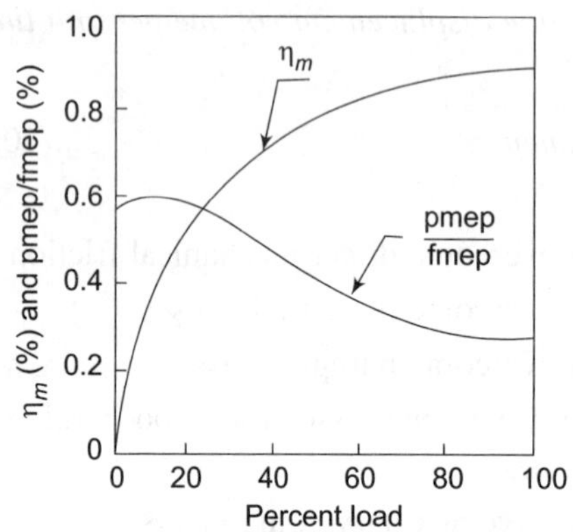


**Fig. 10.2** Mechanical efficiency and pump/fmep vs load

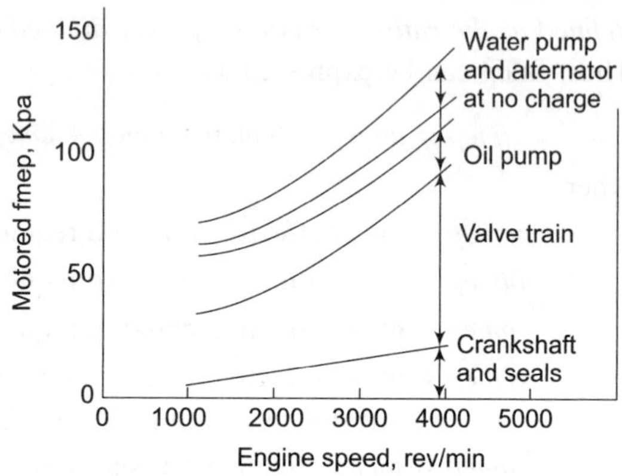


**Fig. 10.3** Motored fmep vs engine speed (a) Four-cylinder, four-stroke SI engines, (b) Six-cylinder diesel engines

Figure 10.3 (a and b) shows the motored (no firing) friction mean effective pressure versus engine speed for four-cylinder SI and six-cylinder diesel engines. These results show the large contribution from the direct mechanical friction.

### 10.1.1 Crankcase Mechanical Friction Losses

The crank case mechanical friction may further be sub-divided into following components.

(i) *Bearing friction* (ii) *Valve gear friction* (iii) *Pump and miscellaneous friction.*

The bearing friction includes the friction due to main bearing, connecting rod bearing and other bearings. It is viscous in nature and depends upon the oil viscosity, the speed, size and geometry of the journal. The main effective pressure (*mep*) lost in journal bearing may be given by an empirical equation.

$$(mep)_{lost} = \frac{D}{L} \cdot \frac{N}{100} \cdot K \qquad ...(10.4)$$

where $D/L$ is the bore-stroke ratio of the engine, $N$ is rpm and $K$ is a constant whose value is generally about 0.85 and 1.76 for petrol and diesel engine respectively. The bearing friction losses are affected very little by loading of the bearing but they rise rapidly with increase in speed.

The valve gear friction losses are functions of engine variables and it is difficult to correlate them with empirical equation.

It is worth to note that the *crankcase mechanical friction is about 15 to 20 per cent of the total engine friction.*

### 10.1.2 Piston Mechanical Friction

*Piston mechanical friction is caused by the movement of piston inside the cylinder.* It can be sub-divided into the following:

(i) Viscous friction

(ii) Non-viscous friction

  (a) Friction due to ring tension

  (b) Friction due to gas pressure forces behind the ring.

The *piston mechanical friction is a function of the piston velocity and is shown* in Fig. 10.4. The viscous friction increases with piston velocity while friction due to gas pressure behind piston ring decreases with increase of piston velocity. Friction loss due to ring tension is more or less constant with piston velocity. The lower part of the piston works more or less under viscous friction conditions, while the degree to which the upper part of the piston can be lubricated also effects the viscous friction. Obviously, the *viscous friction depends upon the viscosity of the lubricating oil and temperature of the various parts of the piston.* The oil film thickness between the piston and cylinder is also affected by the piston side-thrust and the resulting vibrations. The side load acting on the piston due to cylinder gas pressure is shown in Fig. 10.5.

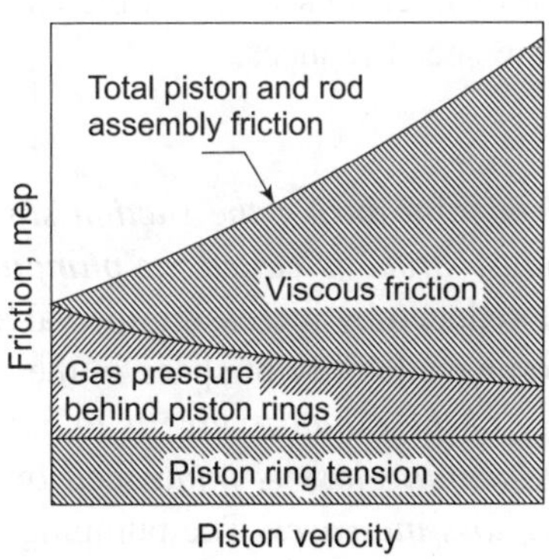


**Fig. 10.4** Variation of piston

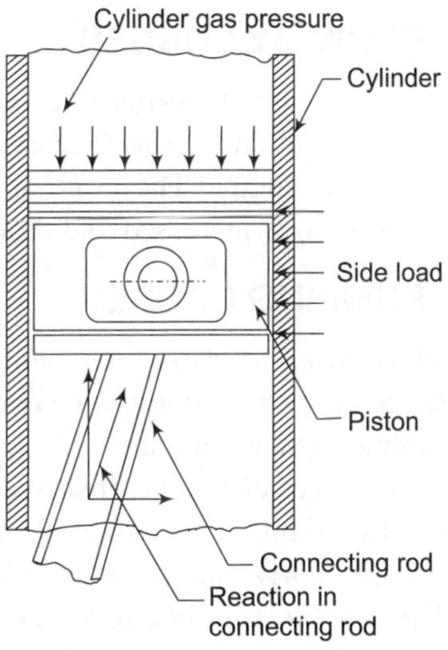


**Fig. 10.5** Side load on piston

The cylinder gas pressure behind the top rings is shown in Fig. 10.6. Owing to the ring tension, the ring presses against the cylinder wall and results in friction losses. Over and above the ring tension, the gas pressure behind the ring also causes friction loss, it is the highest by the top ring but diminishes by the other rings. It is worth to note that no gas pressure can act behind the oil scrapper ring.

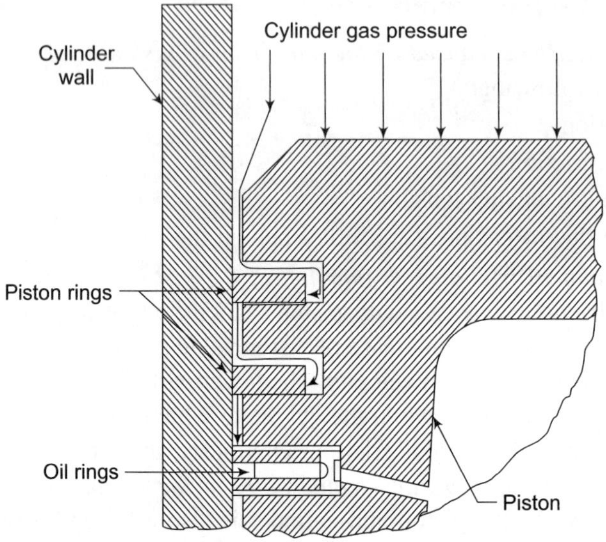


**Fig. 10.6** Cylinder gas pressure acting behind the top piston rings

## 10.2 BLOW-BY LOSSES

*It is the phenomenon of leakage of combustion products past the piston and piston rings from the cylinder to the crankcase.* These losses depend upon inlet pressure, compression ratio and speed of the engine. They vary as the square root of inlet pressure, and increase as the compression ratio is increased and reduce as the engine speed reduces.

## 10.3 PUMPING LOSSES

*The work required to charge the cylinder with fresh mixture during the suction stroke and to discharge the combustion products during the exhaust stroke is called the* ***pumping loss****.* The pumping loss caused mainly by throttling may be reduced by increasing the valve areas. But the valve areas cannot be greatly increased due to unavailability of space in the cylinder head and other limits. *The standard practice for sizing of the exhaust valve is to make them a certain percentage smaller than inlet valves. The pumping loss is influenced by engine speed. With the increase of engine speed, the pumping loss increases.* The pumping loss for a 4-stroke engines may be divided into three parts as shown in Fig. 10.7.

### 10.3.1 Exhaust Blowdown Loss

In order to reduce the work required by the piston to drive out the exhaust gases the exhaust valve is opened before piston reaches TDC on the expansion stroke. During this period, the combustion gases rush out of cylinder due to pressure difference existing between cylinder and atmospheric pressure. This results in a certain loss of power called the ***blowdown loss*** as shown is Fig. 10.7. It is interesting to note that with large valve area and earlier exhaust valve opening, the blowdown loss will be higher whereas with increase in speed, the losses diminish.

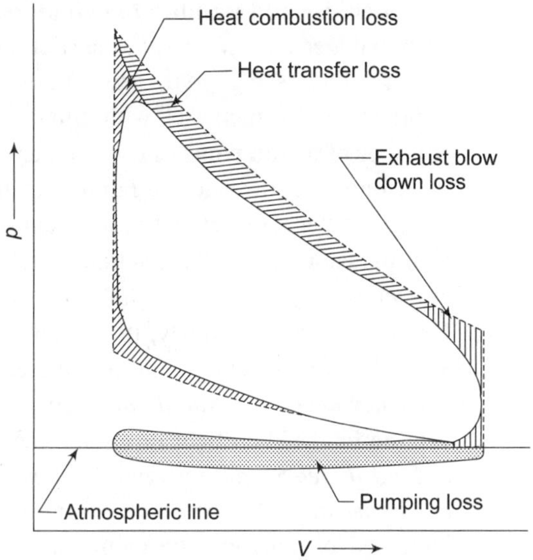


**Fig. 10.7** Pumping losses

### 10.3.2 Exhaust Stroke Loss

The work spent by the piston to force the products of combustion after the blowdown process is called the exhaust stroke loss. In the beginning of blowdown process, the gas pressure inside the cylinder is more than three to four times the gas pressure inside the exhaust pipe, therefore, the gases flow with high velocity. During the end part of exhaust stroke, the cylinder pressure may momentarily drop below the gas pressure in the exhaust pipe and as a result the blowdown is not possible and there is a chance of back flow. But when the piston moves up again, pressure rises and gases are pushed into exhaust pipe. This power required by the piston to drive the exhaust gases out is called the *exhaust stroke loss* as shown in Fig. 10.7. The factors affecting this loss are exhaust valve size, valve timing and valves flow coefficient. *The exhaust stroke loss reduces with the increase in exhaust valve size, early opening of the valve, lower speed and higher valve flow coefficient. But with the increase of size of exhaust valve early opening of valve and high speed exhaust blowdown losses increase and chance of hot spots also increases.* A compromise is made among the conflicting parameters which may result in the minimum overall loss.

### 10.3.3 Intake Stroke Losses

The inlet throttling losses occur due to the restrictions imposed by the elements of intake system such as air cleaner, carburettor venturi throat valve, inlet manifold and inlet valve. To overcome this friction and inertia of gas, energy is supplied to the piston to produce a pressure difference across the inlet valve so that the fresh charge could be drawn into the cylinder. This energy supplied to the piston is intake stroke loss. *Generally, intake and exhaust stroke losses are called* ***pumping losses*** *but exhaust blowdown losses may also be included under this.*

## 10.4 FACTORS AFFECTING ENGINE FRICTION

The following factors affect the engine friction.

(a) **Stroke-bore ratio.** The effect of stroke-bore ratio on engine friction is very small. Obviously, the lower stroke-bore ratio may result in a slight reduction in fmep due to less frictional area.

(b) **Cylinder size and number of cylinders.** It is obvious that larger cylinders have more frictional area and as a result, the friction and economy improve with small number of longer cylinders, e.g., *four-cylinder engine is more efficient than eight-cylinder engine.* Lubrication also increases with number of cylinders.

(c) **Number of piston rings.** Less number of piston rings reduce the friction and improves the economy. However, the number of the piston ring is chosen on the basis of cost, size and other requirement rather than on the basis of cylinders.

(d) **Compression ratio.** The friction increases with compression ratio but the mechanical efficiency may remain constant or increase due to increase in *imep*.

(e) **Engine speed.** Obviously, the engine friction increases with the increase in speed. The lubrication requirement also increases with speed. *At higher speed, mechanical efficiency decreases rapidly and this puts the limit on increasing the engine speed.*

(f) **Engine load.** The effect of engine load on engine friction is nominal. *The increase in load increases the maximum pressure in cylinder which results slight increase in engine friction.* But the decrease in oil viscosity due to higher temperature caused by increased load, reduces, the engine friction and the net result may be very small increase or decrease in engine friction. *The frictional losses in diesel engine are independent of engine load.*

(g) **Oil viscosity.** *The engine friction loss increases with the increase of oil viscosity. The viscosity is affected by oil temperature.* As the oil temperature increases, the viscosity decreases and friction losses reduce up to a certain temperature, generally up to 100°C. If the temperature goes higher than a certain value (45°C), the local oil film is destroyed resulting in metal to metal contact and may lead to failure of the engine.

(h) **Cooling water temperature.** The cooling water (fluid) temperature affects the lubricating oil viscosity. *A rise in cooling water temperature up to a certain limit* (e.g., *95°C) reduces the engine friction through its effect on oil viscosity.* It is worth to note that during starting, the temperature of both oil and the cooling water is low and hence the viscosity is high which results *in high starting friction losses* and rapid engine wear. Determination of engine friction is given in the chapter of engine testing.

## 10.5 LUBRICATION

### 10.5.1 Definition

*Lubrication may be defined as the application of lubricant between two surfaces that are in contact and in a relative motion to one another.* Thus, the maintenance of proper lubrication of all the moving parts is an important function in the operation and maintenance of an engine.

### 10.5.2 Function

The following are the functions of lubrication.

(i) To reduce friction which consequently reduces the power required to overcome the same.

(ii) To reduce wear and tear between the rubbing and wearing surface.
(iii) To cool the surfaces by carrying away heat generated by friction.
(iv) To carry away the particle of worn metal and carbon.
(v) To seal a space adjoining the surfaces such as piston moving in a cylinder.
(vi) To reduce the engine noise and to increase the engine's life.
(vii) To avoid corrosion and deposits.

*If lubrication is not done properly, it may result in a breakdown of lubricating film, which in turn, may cause seizure and other serious damage to the engine. In a real sense, lubrication is the lifeline of the engine. The oils used for lubrication are mineral oils, vegetables oils and animal oil or fat.*

## 10.6 MECHANISM OF LUBRICATION

To understand the mechanism of lubrication, take for example two solid surfaces in contact with each other. If upper surface has to slide with constant velocity across the bottom surface, a ***tangential force*** must be applied. The force due to the weight of the upper block acting perpendicular to the surface is called the ***normal force***. The ratio of the tangential force to the normal force is called the ***dynamic coefficient*** of friction or the ***coefficient of friction*** *(f)*. Thus,

$$\text{Coefficient of friction, } f = \frac{\text{Tangential force}}{\text{Normal force}} = \frac{F_T}{F_N} \qquad ...(10.5)$$

This means that a tangential force is required to overcome the frictional resistance that exists between the surfaces. The question then arises from where this frictional resistance arises.

This frictional resistance arises because the moving surfaces are rough, and hence small irregularities will fit together at the contact area (interface) to give a mechanical lock to the motion. Even the moving surfaces are too smooth, the molecular attraction will be more in the interface and will resist the motion. From the above discussion, we conclude that friction between dry surfaces may arise either from ***surface irregularities*** or from ***molecular attraction*** or from *both.*

*It is interesting to note that coefficient of friction between two dry surfaces tend to be constant and independent of the load (i.e., normal force), relative speed and contact areas of the surfaces but varies with the material and surface finish.*

By providing lubricant between the contacting surfaces, the friction between sliding surfaces can be reduced. *The function of the lubricant is to displace the two surfaces by maintaining an oil film so as to reduce both mechanical interference and molecular attraction.* The lubricant should be such as to adhere to each surface by molecular attraction and at the same time be amenable to molecular displacement. These qualities and conditions indicate that a mobile substance rather than a solid should be used, and therefore fluids and semi-fluids are used as lubricant. *In other words, the effect of lubrication is to replace "solid" friction with a definitely lesser "fluid" friction.*

Consider the case of two parallel plates with viscous oil in between them. The lower one is stationary while the upper one is moving with a constant velocity ($u$) in the direction as shown in Fig. 10.8. The following assumptions are made.

(1) The fluid is incompressible.

(2) The width of the plates in the direction perpendicular to the motion is large enough so that the flow of lubricant in this direction is negligible.

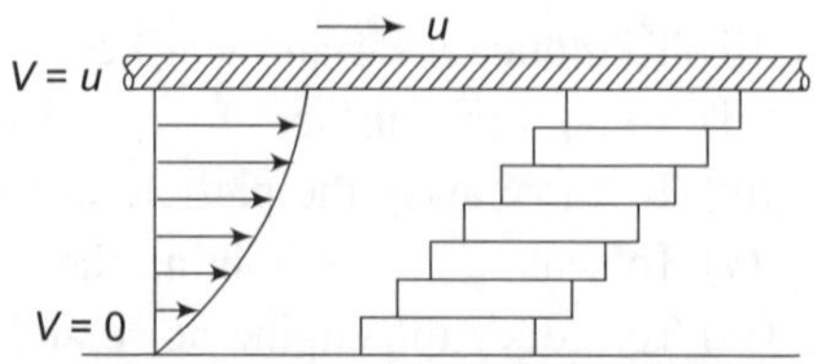


**Fig. 10.8** Mechanism of lubrication in parallel surfaces

Imagine the film as composed of a series of horizontal layers and the force ($F$) causing layers to deform or slide one over the other, just like a deck of card. The layer clinging to the moving surface moves with the same speed as the moving surface. The next layer of fluid moves in the same direction but a slower speed.

The third layer moves at a still slower pace and so on ; ultimately the layer clinging to the stationary surface is not moving at all.

*Thus, the velocity profile across the oil film varies from zero at the stationary plate to the velocity of the plate at the moving surface as shown is Fig. 10.9.* It should however, be noted that each layer of fluid is subjected to a shearing stress and the force required to overcome the stress is the fluid friction. *This fluid friction is due to resistance of the lubricant to the shearing stress. A measure of the resistance to shear is a property called viscosity.* This viscosity of the lubricant helps to drag the fluid into the separating space between the two surfaces and thus retards the escape of the lubricant from the squeezing action of the normal force that tends to bring the surface together. Owing to these effects, a pressure is created in the lubricant. *It is possible to encourage and maintain this pressure by employing a wedge action of the bearing surface.*

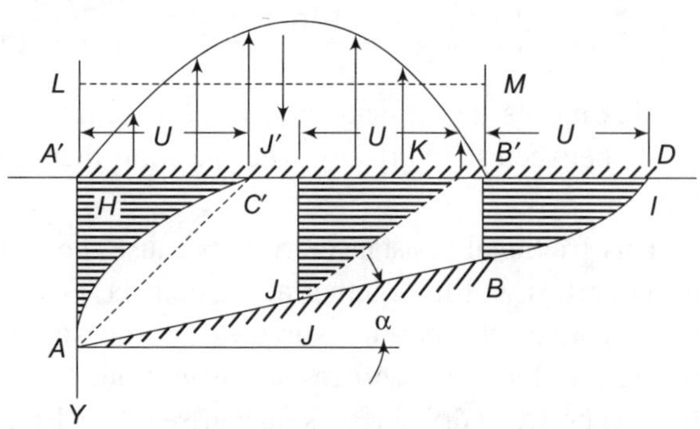


**Fig. 10.9** Mechanism of lubrication on wedge-shaped surfaces

## 10.6.1 Hydrodynamic Lubrication

To understand oil pressure in wedge action, consider a flat surface sliding over an inclined block having an oil film as shown in Fig. 10.9. The cross section of the fluid at the leading edge is less than that at the trailing edge. The velocity distribution across the oil film at the edges are expected to have the shape shown by dotted line but it takes another shape as shown by continuous curved line. This is because of the fact that the lubricating oil between the separating space is contained in a small space and tires to escape. The viscosity of the oil prevents the escape of the oil and pressure is created. The pressure developed tends to cause the flow of oil in all the directions from the point of the maximum pressure, the viscous drag between particles of fluid causes additional oil to enter the separation space. In this way, a constant hydrostatic pressure is built up in the main body of the lubricant which is dependent upon the relative velocity of the surface, geometric configurations, viscosity of the lubricant, and the load. This oil film pressure enables to carry the load. *The type of lubrication where a wedge shaped oil film is formed between two moving surface is called* ***hydrodynamic lubrication***. A special feature of this type of lubrication is that the *load carrying capacity of the bearing increases with increase in relative speed of the moving surface at high speed*, the available time for the oil to squeeze out is less and hence pressure rise tends to be high.

From the above discussion, it follows that a ***positive pressure*** in the oil film is developed in a wedge shaped surface. *The film thickness decreases in the direction of motion and thus it is known as converging film.* It is the ***converging film*** which supports the load.

In the above arrangement, the direction of the motion of the upper plate is important. If it is reversed with the inclination of the lower plate unchanged, the volume of the lubricant that the moving surface tends to drag into the space becomes less than the volume which tends to discharge from the space and as result negative pressure (i.e., less than atmospheric pressure) is developed in the film. Such an oil film is known as ***diverging film*** *and is unable to support any load.*

The *hydrodynamic lubrication exists at place such as thrust bearings, valve tips, cam lifters, cylinder wall, valve guide, main bearings, connecting rod bearings and camshaft bearings.* A detailed discussion will be made later on.

### 10.6.2 Elastohydrodynamic Lubrication

*In the case of very high load acting on the bearings, the bearing material itself deforms elastically against the pressure built up of the oil film. (Fig. 10.10), such type of lubrication is called* ***elastohydrodynamic lubrication****. It occurs between cams and followers, gear teeth and rolling bearings where the contact pressure is very high.*

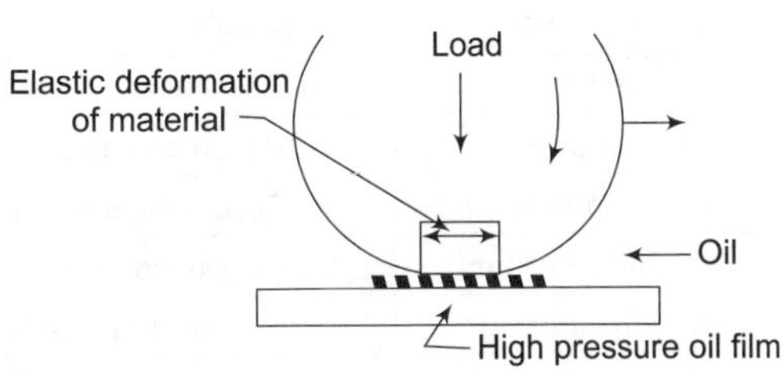


**Fig. 10.10** Elastohydrodynamic lubrication

### 10.6.3 Boundary Lubrication

A situation may arise when the film thickness between the two surfaces in relative motion becomes so thin that formation of hydrodynamic oil film is not possible and the surface high spots or asperities penetrate this thin film to make metal to metal contact, such a lubrication is called ***boundary lubrication*** (Fig. 10.11). *This situation may arise due to insufficient surface area, a drop in relative velocity of the moving surface during the period of starting and stopping, an adequate quantity of lubricant, an increase in the bearing load or a decrease in viscosity of the lubricant due to increase in temperature.* The change from hydrodynamic to boundary lubrication is not an abrupt change, a transition phase may occur in the form of mixed hydrodynamic and boundary type lubrication. *Most of the wear associated with friction occurs during boundary lubrication due to metal to metal contact. The best example of boundary lubrication is the situation where the engine is first started.* During the starting, the shaft is in contact with the bottom of the bearing with only a thin surface film of oil formed on them. The bearing surfaces are not perfectly smooth as they have hills and valleys which tear this film which is constantly formed while the crankshaft is turning slowly. With the increase in speed, the boundary lubrication converts into hydrodynamic lubrication.

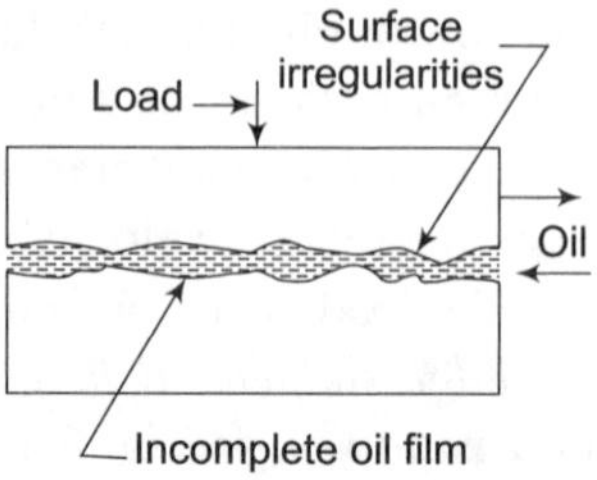


**Fig. 10.11** Boundary lubrication

### 10.6.4 Hydrostatic Lubrication

*Hydrostatic lubrication is one in which a thin oil film resists its instantaneous squeezing out under reversal of loads with relatively slow motions. The oil film acts as a cushion.* In the case of sufficient oil supply, the oil film thickness is restored before next reversal of load.

## 10.7 BEARINGS AND MOTION IN IC ENGINE

A variety of bearings and motions are found in internal combustion engines. They are listed in Table 10.1.

**Table 10.1** Bearings and Motions in IC Engine

| *Motion* | *Bearings* |
|---|---|
| 1. **Sliding contact** | 1. **Journal bearings** |
| (a) Rotating | (a) Crankpins, crankshafts, valve mechanism bearings. |
| (b) Oscillating | (b) Piston pin, knuckle pins, rocker arm bearings. |
| (c) Reciprocating | (c) Slipper bearing—piston, piston rings, valve stems, cross head bearings. |
| 2. **Meshing contact** | 2. Bearing used in worm, bevel, spur and helical gears. |
| 3. **Rolling contact** | 3. Ball, roller and needle bearings. |

## 10.8 JOURNAL BEARING LUBRICATION

Figure 10.12 shows the formation of oil film and oil pressure in a journal bearing. If the shaft (journal) and bearing are at rest shown in Fig. 10.12 (a), the weight of the shaft will permit metal to metal contact between the surface of the journal and bearing on the line of the action of the load due to squeezing out of the oil from under the journal. If the shaft rotates, the slight movement [Fig. 10.12 (b)] permitted by clearance enables the shaft to roll upon an oil film with the speed of rotation pumped oil under the shaft and separating the two surfaces. With increase in speed, the centre of journal moves back from the centre of bearing [Fig. 10.12 (c)] against the resistance of the load, due to increased pressure generated within the oil film. The position shown in Fig. 10.12 (c) is said to be in the region of perfect or film or complete lubrication. In this case, there is no metal to metal contact and load is supported

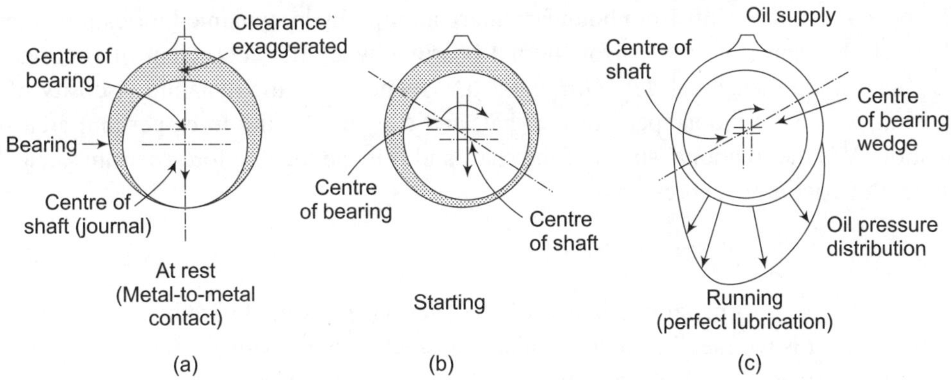


**Fig. 10.12** Fromation of oil film in a journal bearing

by the pressure created in the lubricant (wedge of the oil) by relative motion of shaft and bearing and by the viscosity of the oil.

Oil supplied to bearing should enter in the unload region or diverging section of the oil film. If the oil supply groove is cut in the load region (oil wedge), the oil film pressure is reduced resulting in a large reduction in the load carrying capacity of the bearing.

Under normal operation journal bearings run with a thick film (hydrodynamic) lubrication so that the friction and wear are minimized. However, *it operates in a thin film region when the engine starts. Therefore, nearly all the wear on a journal bearing occurs at the starting period.* Thus, *in order to minimise wear and tear, an engine should be started either at load free or light load and at slow speed.*

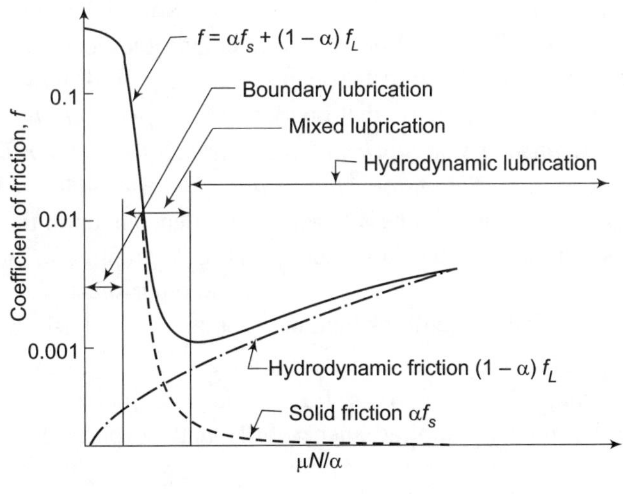


**Fig. 10.13** Stirbeck diagram for journal bearing

The phenomenon of shift from boundary lubrication to hydrodynamic lubrication is given by Stirbeck diagram (Fig. 10.13) for journal bearings when the coefficient of friction, $f$ is plotted against a dimensional *duty parameter* $\mu N/\sigma$ where $\mu$ is the dynamic viscosity of the lubrication, $N$ is the rotation speed of the shaft and $\sigma$ is the loading force per unit area. The lubrication, $N$ is the rotational speed of the shaft and $\sigma$ is the loading force per unit area. The coefficient of friction is expressed as

$$f = \alpha f_s + (1 - \alpha) f_L \quad ...(10.6)$$

where $f_s$ is the metal-to-metal coefficient of dry friction, $F_L$ is the hydrodynamic coefficient of friction, and $\alpha$ is the metal-to-metal contact constant, varying between 0 and 1. As $\alpha \to 1$, $f \to f_s$ and the friction is **called boundary, i.e., close to solid friction**. Under this condition, the lubricating film is reduced to one, a few molecular layers and cannot prevent metal-to-metal contact between asperities. As $\alpha \to 0$, $f \to f_L$ and the friction is called ***hydrodynamic or viscous or thick film***. The lubricant film is sufficiently thick to separate completely the surface in relative motion. In between these regimes, there is a mixed or partial lubrication engine where the transition from boundary to hydrodynamic occurs.

## 10.9 LUBRICANTS

Lubricants are classified under three categories, namely, (a) **Solid lubricants**, (b) **semi-solid lubricants**, and (c) **liquid lubricants or oils**. Normally, *liquid lubricants* (or oils) are mostly used.

**(a)** ***Solid lubricants.*** Mineral substances such as *graphite, talc and mica, mined from* earth, come into the category of solid lubricants. *The solid lubricants are mixed with semi-solid lubricants and the mixture is supplied to the bearings. In some cases, solid lubricants are supplied separately to the bearings.* Solid lubricants have the property of filling the depressions in the bearing surface and they cannot be easily squeezed from the bearings. Though, solid lubricants are rarely used, however, it can be used on bearings and machinery parts that operate at *low speed and low pressure.*

**(b)** ***Semi-solid lubricants.*** *These lubricants are obtained from mineral oils and fatty oils and they do not flow at ordinary room temperature.* Semi-solid lubricants are commonly known as **"greases"**. The thickness to the grease depends upon the percentage of the soap contents. To improve the lubricants properties, graphite or micas is sometimes added. Cheap greases contain clay, wax, resin and other substances which thicken the grease but the lubrication qualities then deteriorate.

Semi-solid lubricants (greases) are normally employed on heavy, slow moving machinery, gear teeth wheel bearings of railroad cars, trucks and automobiles, on revolving parts of machinery where oil cannot be applied successfully and on parts that can not be easily lubricated.

**Liquid lubricants or oils.** Usually, two types of liquid lubricants are used. They are (i) *mineral oils* and (ii) *fatty oils* (vegetable and animal).

**(i) Mineral oils.** *All mineral oils are obtained from crude petroleum by fractional distillation. Mineral oils are mostly used in IC engines, steam engines and machine parts.* They are variable in a wide variety of grades, ranging from an extremely thin oil suitable for lubricating fine mechanical mechanism to a heavy machine like cranes, etc. They have properties that enable them to withstand decomposition, oxidization or gumming.

The specific gravity of mineral lubricating oils range from about 0.86 to 0.94. Their boiling points are high usually above 35°C in the case of refined distilled oils. The flash point varies from 15°C to 35°C. Russian oils have pour point below –15°C, mostly American oils have pour point below –5°C and Scott oils have pour point at about 0°C.

**(ii) Fatty oils.** *Fatty oils are made from vegetable or animal fixed oils.* They decompose in the presence of heat and produce acids that are harmful for metals. They oxidize and tend to produce gum. *But they have two advantages over the mineral oils – (i) on heating, their viscosity is not much lowered and (ii) their adhesiveness is greater.*

The specific gravity of fatty oils varies from about 0.88 to 0.97 at 15°C. *The best known animal oils are* ***lard and sperm oil***. Lard is prepared by separation of solid fat from lard (pig fat) by using pressure. *It alone is a good lubricant and compounded with mineral oils is used for cutting tools.* **Sperm oil** is prepared from sperm whale (large fish-like marine animal). *The advantage of sperm oil is that it possesses a low viscosity among all fatty oils and does not gum and is the best known lubricant for delicate mechanisms, such as watches and mathematical instruments.*

The most common **vegetable oils** used as lubricants are olive oil, cotton-seed oil, linseed oils, caster oil and palm oil. Since they decompose quickly and have a tendency to turn gummy, they are not used alone as lubricants but are compounded with mineral oils.

As an **example**, the lubricants used in **reciprocating compressors** are given below.

Compressor cylinders and stuffing boxes are usually lubricated with mineral oil. The cylinder of oxygen compressor are lubricated with distilled water mixed with 10 glycerol. If the pistons of $CO_2$ compressor are fitted with cup packings, their cylinders are lubricated with glycerol with 10 to 20% distilled water. Chlorine compressors are lubricated with concentrated sulphuric acid (98%) as $Cl_2$ would decompose oil.

In the case of air compressors, oil with a high flash point is required for the lubrication of cylinders and stuffing boxes (for compression temperature of 140°C, flashpoint should be 22°C). The oil should also be resistant to oxidation. In the case of air cooled compressors, oil of still higher flashpoint is selected.

In low pressure compressor (up to 20 ata), the lubricating oil required should have a maximum specific gravity 9.4 N/dm$^3$ and viscosity 4–8° E at 50°C. For high pressure compressors, oil required should have specific gravity 9.8 N/dm$^3$ and a minimum viscosity of 5°E at 50°C. These requirements are met by the permanent oil type V.T. The pour point for both types must be lower than –5°C and the acid number not higher than 0.05 mg KOH/g; the maximum ash content allowable is 0.02 while the maximum water content should not exceed 0.1%. The oils should not contain asphalt or resinous substances.

## 10.10 PROPERTIES OF A GOOD LUBRICANT

A good lubricant should have the following properties:

(i) ***Suitable viscosity*** (*i.e., maximum fluidity*). It should have a suitable viscosity preferably constant which should provide sufficient oil thickness to prevent the wearing of surfaces coming into contact.

(ii) ***High film strength.*** It should have high film strength to prevent metal-to-metal contact and seizure under heavy load.

(iii) ***Higher specific heat.*** The lubricants should have higher value of specific heat so that a large amount of heat may be absorbed and carried away by the lubricant.

(iv) ***Low coefficient of friction.*** The lubricant should have a low coefficient of friction so that only a little oil should be needed to overcome the fluid friction.

(v) ***Oiliness.*** The lubricant should have oiliness to ensure proper adherence to the bearings and lesser friction and wear when lubricant is in the boundary region. It should also give a protective covering against corrosion.

(vi) ***Non-corrosive.*** It should have no tendency to corrosion under any circumstance.

(vii) ***Low pour point.*** It should have a low pour point to ensure the flow of lubricant at low temperature to the oil pump.

(viii) ***No tendency to form deposits.*** It should be devoid of the tendency to form deposits by combining with air, water or the products of combustion in the case of an engine.

(ix) ***Cleaning ability.*** It should be able to clean the deposits of the engines.

(x) ***Dispersing ability.*** It should have dispersing ability to break up and carry, if any foreign material in the oil.

(xi) ***Non-forming characteristics.*** It should have non-forming characteristics to enable to oil to dispel air which encourages oxidation.

(xii) ***Demulsibility.*** Oils when not properly refined from emulsions due to continuous agitation of the oil and water, such an emulsion may prevent the oil to flow freely which may give rise to sludge (a thick muddy substance), formation of which can choke up the whole lubricating system. Demulsibility is the property of the oil to separate itself quickly from an emulsion (milky liquid with only particles suspended in it).

(xiii) ***Low cost.*** It should be easily available at low cost.

(xiv) ***High flashpoint.*** The flashpoint of an oil is the lowest temperature at which it allows inflammable vapours to be formed. *A high flashpoint is required when an oil is to be subjected to high temperatures, for examples, in steam engines, internal combustion engines, and air compressors.*

(xv) **Safe, nontoxic and unexplosive.**

## 10.11 ADDITIVES FOR LUBRICANTS

In order to improve some of the properties of a lubricant such as stability, detergent, oiliness, film strength, anti-corrosiveness, anti-oxidising and anti-foaming, certain oil-soluble organic compounds containing inorganic elements like phosphorous, sulphur, amine derivatives and metals are added to the mineral based lubricating oils. These oil soluble organic compounds are called ***additives***, as given below:

(i) *Oxidation and corrosive inhibitors* (stability additives). The stability additives are used to decrease the oxidation and corrosive characteristics of the oil. The inhibitors are generally in the form of sulphur and phosphorous compounds or amine and phenol derivatives. These compounds have a greater affinity for $O_2$ than the hydrocarbons of the lubricants.

(ii) *Detergent additives. The function of the detergent additives is to disperse the insoluble particles such as carbon, dirt etc., and reduce the tendency to accumulate in the form of sludge.* They are also known as dispersants. The chemical structures of detergent addition are aluminium naphthenate, calcium phenyl stearates, calcium alkyl salicylates, metal salts of acetyl phenol and alkaline earth metal petroleum sulphates. The physical action is similar to soap in keeping dust suspended in oil.

(iii) *Anti-foam agents.* The minute particles of air in a foaming oil increase oxidation and reduce the mass of flow of oil to the bearings. In addition, foaming may cause abnormal loss of oil through the breather. Generally, silicon polymers are used as anti-foam agents.

(iv) *Pour-point improves.* These additives tend to prevent the formation of wax at the low starting temperature. Two of the pour depressants on the commercial market are santopour and acryloid.

(v) *Viscosity index improvers.* High molecule polymers are added to the lubricating oils to increase their viscosity index. A high viscosity index has good starting characteristics together with satisfactory operation under high speed and heavy load conditions.

(vi) *Oiliness and film strength agents.* Many organic substances like sulphur, chlorine and phosphorus compounds are used as additives to improve the oiliness and film strength of the lubricant.

## 10.12 TEST ON LUBRICATING OILS

*Properties of lubricating oils are tested by chemical analysis, by physical tests and by trial in service.* The following are the **important tests** for judging the qualities of the lubricating oils.

**(i) *Viscosity.*** *This is the most important single property of lubricating oils. It is the measure of fluid friction or internal resistance to flow of liquids.* A highly viscous oil is thick and flows slowly. A low viscosity oil is thin and flows readily. Viscosity to some degree is necessary for all lubricants. *A high viscosity oil is used for heavily loaded slow moving journals.* In fast moving machinery, instead of highly viscous oil a low viscous oil is used because a highly viscous oil causes a large loss of power. *The viscosity varies with pressure and temperature of oil.*

To measure the viscosity, **viscometers** are used. The most common viscometers are: (i) **Saybolt viscometer** (used in USA). (ii) **Redword viscometer** (used in UK) and (iii) **Engler viscometer** (used in Germany). The principle of measuring the viscosity is the same in all the cases. *The viscosity is measured by noting the time taken in seconds for the flow of measured quantity of the fluid lubricant through a standardized orifice or short capillary.* The unit of this viscosity is given by *"seconds Saybolts, seconds Redword or seconds Engler"*. It is customary to find the viscosity of oil at 38°C or 50°C.

Sometimes, the ***viscosity index*** is used to represent the degree of viscosity. *It is the measure of time taken for a definite quantity of oil to fall to a certain level in a capillary tube compared with a standard oil.*

**Kinematic viscosity** is an another way of representing the viscosity. It is based on the absolute viscosity and is mathematically calculated for moving liquids. The unit of absolute viscosity is called *poise*.

**(ii) *Specific gravity.*** It is the ratio of weight of an oil to the weight of an equal volume of water, both measured at 15°C.

Sometimes API (American Petroleum Institute) gravity is also used. *It is the comparative weight of the oil to water as read on arbitrary scale.* This scale is commonly used by the petroleum industry API is related with specific gravity by the following equation.

$$\text{API gravity} = \frac{14.5}{\text{Specific gravity}} - 131.5$$

By the definitions, specific gravity of water is 1 while the API gravity is 10. Thus, all oils have API gravity greater than 10, the degrees being given in whole numbers instead of decimal fraction representing their specific gravities on the specific gravity scale.

The specific gravity is measured by using an instrument called a ***hydrometer***. It is graduated in specific gravity and API degrees.

**(iii) *Flash and fire points.*** *The flash point of an oil is the lowest temperature that allows inflammable vapours to be formed. It is found by heating the fuel slowly and then sweeping a flame across the surface of the liquid.* It is measured to determine the type of crude from which oil was made. It is measured by **Pensky–Marten Flash point apparatus.**

*The fire point is the temperature to which the oil must be heated to burn continuously after the test burner has been applied to the escaping vapour.* Commonly, firepoint takes place when the temperature is raised 30 to 40°C above the flashpoint.

**(iv) *Pour point.*** The pour point indicates the flowing characteristics of the oil at low temperature under the force of gravity. If it is required to flow the oil at temperature below the pour point, it is only possible by mechanical agitation of the oil. *For pour point testing, the sample oil is allowed to freeze in test tube until the oil ceases to pour, then adding 30°C to the freezing temperature.*

**(v) *Film strength.*** It is defined according to method of the test. According to **Timken test**, the film strength is defined as the load when incipient seizing of the bearing occurs. According to Almen test, it is defined as the load that the lubricant can carry for 10 seconds without seizure. Commonly, the tests are conducted between steel on steel surfaces at definite speeds and oil temperature but with increasing speed.

Though oiliness and film strength are related but each is an entirely different property. *A lubricant with good oiliness may have a low film strength.*

**(vi) *Corrosiveness.*** The oil should be free from the corrosive property and should afford protection against corrosion. *For testing, a clean strip of pure polished copper is placed in a sample of oil held at 100°C for 3 hours and is then washed in sulphur-free acetone (liquid) and compared with fresh copper strip.* It gives an indication of corrosive effect of the oil under test.

**(vii) *Water test.*** *The presence of water is detected by heating a sample of oil to a few degrees above 100°C.* If the presence of cloudiness in oil is due to water, the oil will not become transparent. The presence of water in oil is undesirable because it retards the capillary action of the wick feed as it wets the wick.

**(viii) *Wick feed test.*** *This test carried for 14 days consists of measuring the quantity of oil lifted by capillary action through a wool wick.* The results are interpreted by comparison of such tests for several oils. The quantity of oil lifted depends upon the diameter of the wick, level of the oil, constitution of the oil and atmospheric conditions.

**(ix) *Specific heat test.*** *This test is important when a lubricating oil is used to cool a bearing or a piston because higher specific heat is required.* This tests is carried out by noting the amount of heat absorbed by 1 kg of oil per kJ per degree rise in temperature.

**(x) *Carbon residue test.*** *This test determines the percentage of carbon residue in a sample of oil after driving out the volatile vapours by applying heat.* This test is important to lubricating oils which are used under high temperature, for example, lubricating oils used in the bearings of internal combustion engines. The high temperature in the IC engines produces chemical changes in the lubricant that causes deposits that clog oil passage and result in greater wear and tear.

**(xi) *Emulsification.*** When oil is agitated at length with water; it forms an emulsion which tends to separate on standing and which is not suitable for lubricating oil. This is found in the case of oil used in steam engine cylinder. *The ability of the oil to separate easily is called demulsibility.* There are two tests for it: (i) steam emulsion test and (ii) water emulsion test.

In steam emulsion test, the time in number of seconds required for an oil to completely separate from an emulsion when steam is passed through the oil in a standard manner. This is known as the steam emulsion number.

In water emulsion test, an emulsion is formed by churning for 5 minutes 27 c.c of oil and 53 c.c of distilled water at either 55°C or 85°C. If complete separation is obtained in one minute, it is termed perfect demulsibility.

**(xii) *Accelerated service test.*** This test is used by operating engineers to know the performance of the oil in actual use on some engine or machines. The operating engineers compare various oils and determine which is the best for a particular purpose by testing each oil, in turn, on same engines or machines under same conditions of operations.

**(xiii) *Cloud point.*** This is the temperature at which paraffin wax or other substances begin to crystallize or separate out from a solution when oil is chilled.

**(xiv) *Neutralization number test.*** (*Acidity*). This test determines the acidity or alkalinity of an oil. It is the weight in milligrammes of potassium required to neutralize the acid content in 1 gramme of oil.

**(xv) *Saponification number test.*** This test is applied only to a compounded oil or to a suspected mineral oil to find out whether it is compounded with an animal fat or vegetables fatty material. This number is the number of milligrammes of KOH required to saponify one gramme of oil.

**(xvi) *Precipitation number.*** *This test is made to determine the amount of oil-insoluble material in an oil. This test is applied to new cylinder and block oils. The lower the precipitation number, the higher the degree of filtration.* This number is obtained by diluting of oil with 20 millilitres of petroleum naptha and centrifuging the solution in a cone shaped tube at 1500 rpm for 10 minutes. *The sediment deposited at the tip of the cone is measured and is known as the precipitation number.*

**(xvii) *Naptha and Bengal Insolubles.*** *This test is carried out on used oils to determine the composition of sludges and solids. This test generally applied to samples taken from the crank cases of the internal combustion engines.*

This test is performed by taking 10 to 20 gm of the sample in a beaker which is heated to drive out any moisture that may be present. The sample is then allowed to cool after which it is diluted with naptha in the proportions of 10 part of oil to 90 parts of naptha. The oil goes into the solution with the naptha but the sludge and solid materials drop out. This insoluble material is removed and filtered in a Gooch filter where the residue is retained by an asbestos mat, washed and then dried. The increase in the weight of Gooch filter gives the weight of insoluble material.

**(xviii) *Functional or mechanical test.*** This test is carried out (i) to test the effects of viscosity (ii) to test the effect on bearing metals. (iii) to test the suitability of locations in the bearings for supply of lubricants, (iv) to test the effect of temperature, pressure and speed on frictional power loss.

The test is performed by a machine known as the *Thurston Oil Testing Machine*.

### 10.12.1 Thurston Oil Testing Machine

Figure 10.14 shows the details of a *Thurston oil testing machine*. It consists of a shaft which is supported on two bearings $B_1$ and $B_2$. To drive the shaft, a pulley $P$ is mounted over this shaft. There are two brasses $C$, upper and lower in the bearing, each of which subtends and arc of contact of 60°. A rectangular housing $D$ holds the brasses in position.

A helical spring ($S$) is provided inside the body of pendulum, $E$. A load is applied by means of a spring. *A* bob, $G$ is fixed at the other end of the pendulum. By means of a screw ($F$) the helical spring is adjusted which applies load to the brasses. The spring is calibrated and the load reading is obtained directly on the scale ($H$) in front of pendulum. The oil is supplied to the bearing through $I$. To measure the temperature of the bearing a thermocouple is provided.

Once the machine is started, the shaft rotates and the load is applied. The friction between the shaft and the brasses causes the pendulum to swing from the vertical position and the reading in *N* obtained on the scale *K* which is proportional to the frictional force. Thereafter, the pressure, temperature and load can be varied to see their effect.

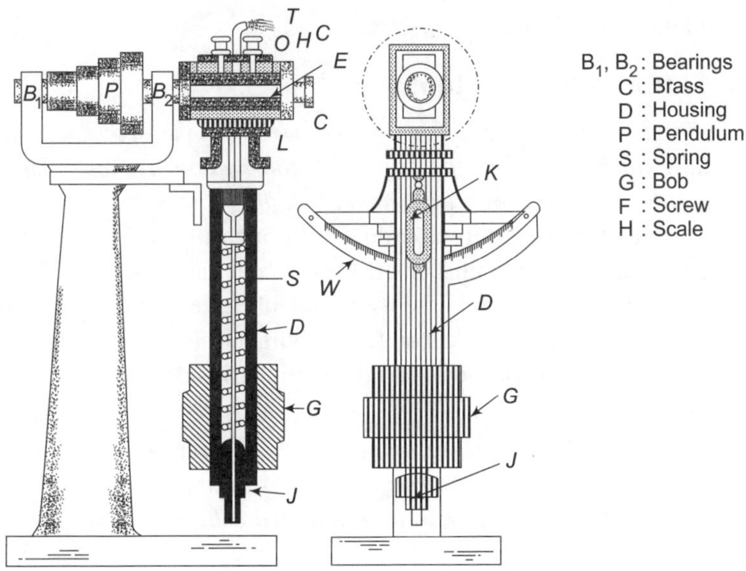


**Fig. 10.14** Thurston oil testing machine

## 10.13 LUBRICATION OF ENGINE COMPONENTS

In internal combustion engines, there are many sliding and moving parts which are to be lubricated to reduce friction. The following parts are to be lubricated:

(a) Piston and cylinders
(b) Crankshaft and their bearings
(c) Crankpin and their bearings
(d) Wristpin and their bearings
(e) Valve gear

These parts are lubricated by lubrication methods which are discussed below.

A typical commercial engine oil is a multigrade engine oil **SAE 20W/50**. Viscometrics manufactured Castrol Ltd. and branded as **"Castrol GTX Extra"**. It meets API SG/CD, MIL-L-46152 D and CCMC G4 performance applications.

## 10.14 LUBRICATION SYSTEM

The function of a lubricating system is to provide sufficient quantity of cool filtered *oil to give positive and adequate lubrication to all the moving parts of the engine*. The lubrication system is classified as the following.

(a) Mist lubrication system
(b) Wet sump lubrication system
  (i) Splash system
  (ii) Pressure feed system
  (iii) Splash and pressure feed system
(c) Dry sump lubricating system

### 10.14.1 Mist Lubrication System

*In two-stroke engines, as the charge is compressed in the crankcase, so it is not possible to have the lubricating oil in the sump for crankcase lubrication. Hence, mist lubrication is adopted in practice. Such engines are lubricated by adding 2 to 3 per cent lubricating oil in the fuel tank.* The oil and fuel mixture is inducted through the carburettor. The gasoline is vapourized, and the oil, in its form of mist enters the cylinder via crankcase. The lubricating oil associated with mist impinges the crankcase walls and lubricates the main and connecting rod bearings. The rest of the oil which passes on to the cylinder during charging and scavenging periods lubricates the piston, piston rings and the cylinder inside wall.

*It is worth to note that the two-stroke engine is very sensitive to particular lubricating oil and fuel combination, as the mixture influences the exhaust smoke, internal corrosion, bearing life, ring and cylinder bore wear, ring sticking exhaust combustion chamber deposits, spark plug fouling, etc. As a result, specially formulated ashless oils are used in two-stroke engines.*

Though, the main advantage of this system lies in its simplicity and low cost because of absence of oil pump, filter and pressure regulating valve, etc. However, there are many ***disadvantages*** which are given below:

(i) As a result of partial or full burning of lubricating oil, heavy exhaust emissions and formation of deposits occur on piston crown, ring grooves and exhaust ports, which in turn, affect the engine efficiency.
(ii) The bearing surfaces may get corroded as a result of close contact with acidic vapours produced during the combustion process.
(iii) The system does not provide effective lubrication as for effective lubrication either separate mixing prior to use or use of some additive to give the oil good mixing characteristics.
(iv) The greater limitation of this system is oil *starvation of the working parts* especially when the throttle is closed in the case of vehicle movement down the hill. This may result in over-heating and piston seizure.
(v) The consumption of lubricating oil is 5 to 15% higher than that of similar size four-stroke engines. This is due to high exhaust temperature and less efficient scavenging.
(vi) The two-stroke engines are over-oiled most of the time due to absence of any control over the lubricating oil.

Some of the problems such as oil starvation, high consumption and corrosion damage are avoided if separate injection pump is used to inject the oil directly into the carburettor.

## 10.14.2 Wet Sump Lubrication

*In wet sump lubrication system, the bottom part of the crankcase contains the lubricating oil from which the oil is supplied to various parts of the engine either by splash or by pressure feed or by combination.*

### *10.14.2.1 Splash System*

The application of this system is limited to only light duty engines. Figure 10.15 shows such a splash system. The system consists of oil in oil sump, strainer, oil pump and oil troughs. The oil in the sump is maintained of a predetermined level. The oil is drawn by a pump through the strainer and is delivered to oil troughs which are located under the big end of all the connecting rods. The oil in troughs is maintained at constant level as they are provided with overflow. As the name implies, a *splasher or dipper* is provided under each connecting rod cap which dips into the oil in the trough at every revolution of the crankshaft and oil is splashed all over the anterior of the crankcase, into the piston rings and on to the exposed portion of the cylinder walls, crankpin bearings, etc. A hole is drilled through the connecting rod cap through which oil will pass to the bearing surface. Oil packets are also provided to catch the splashing oil over all the main bearings and also the camshaft bearings. Excess oil supplied drips back to the sump.

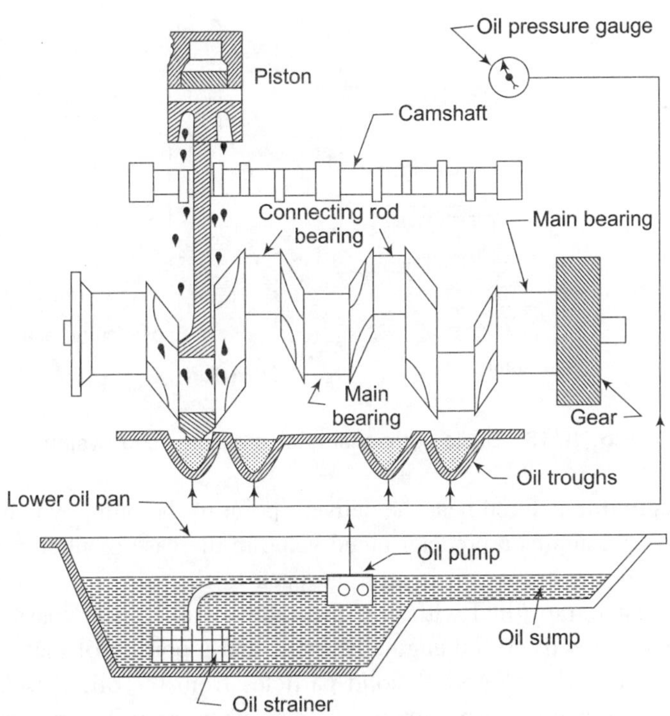


**Fig. 10.15** Splash lubricating system

#### 10.14.2.2 Pressure Feed System

The basic components of pressure feed system is shown in Fig. 10.16. The system along with parts to be lubricated is shown in Fig. 10.17. The main elements of lubricating system consists of oil in crankcase, strainer, pump, pressure regulator, filter, breather and oil galleries. The oil is drawn from sump through strainer which prevents foreign particles and is pumped with the help of gear pump submerged in the oil and driven by camshaft to all the main bearings of the crankshaft through distributing channel. *An* ***oil hole*** *is drilled in the camshaft from the centre of each crankpin to the centre of an adjacent main journal, through which oil can pass from the main bearings to the crankpin bearing. The piston pin receives oil through a hole drilled in the connecting rod. The cylinder walls, tappet rollers, cam piston and piston rings are lubricated by oil spray from around piston pins,* the main and connecting rod bearings.

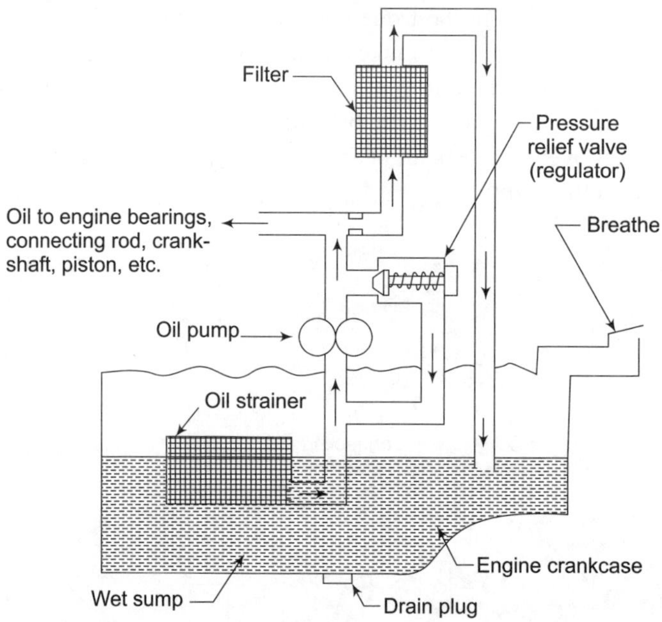


**Fig. 10.16** Basic components of pressure feed system

A **pressure regulator** is fitted near the delivery point of the pump which opens when the pressure in the system attains a predetermined value in the case of filter clogging or the oil cold or sluggish and any excess oil is returned back to the suction of pump. Figure 10.18 shows a typical gear pump filled with pressure relieve valve and bypass. Most of the oil from the pump goes directly to the engine bearings and a portion of the oil passes through a cartridge (fine) filter which removes solid particles from the oil. This filtration reduces the amount of contamination from carbon dust and other impurities present in the oil and is called *bypass filtering* as all the oil does not pass directly to the filter. The advantage lies with bypass filtering is that the clogged filter does not restrict the flow of oil to the engine.

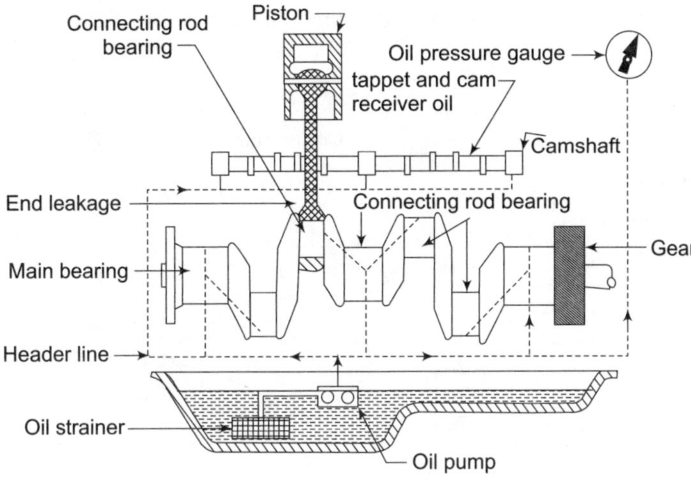


**Fig. 10.17** Pressure feed lubrication system

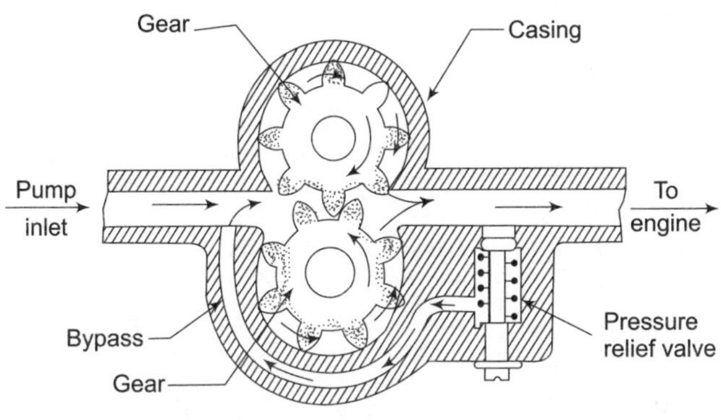


**Fig. 10.18** Gear pump with pressure relief valve

#### *10.14.2.3 Splash and Pressure Feed Lubrication System*

Figure 10.19 shows the schematic diagram of splash and pressure feed lubrication system. *In this case, lubricating oil is supplied under pressure to main and camshaft bearings.* The oil is also supplied under pressure to trough located under the big end of the connecting rod in which dipper dips and thus the crankpin bearings are lubricated by the splash or spray of oil thrown up by dippers.

### 10.14.3 Dry Sump Lubrication

This system is shown in Fig. 10.20. This is also a pressure feed lubrication system but with a slight difference as compared to wet sump lubrication. *In the dry sump, the supply of oil is carried in an external tank – with the help of scavenging pump through strainer and filter.*

*The scavenging pump is placed out of the sump. The capacity of the scavenging pump is always greater than oil feed pump.*

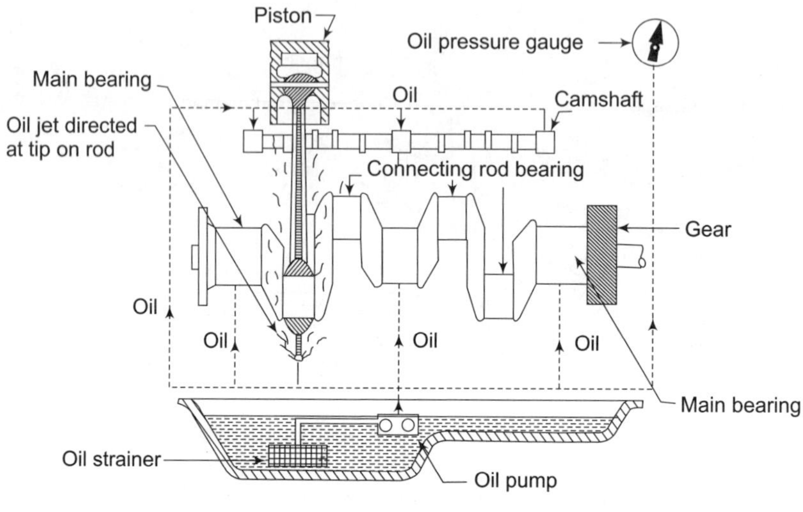


**Fig. 10.19** Splash and pressure feed lubrication system

An oil feed pump draws filtered oil from the supply tank and forces to all the main bearings and camshaft. *Oil is not allowed to get accumulated in the base of the engine.* A filter with a bypass valve is placed in between the scavenge pump and the supply tank. In the case of clogging of the filter, the pressure relief valve opens permitting oil to bypass the filter and reaches the supply tank. To remove the heat from the oil, a separate oil cooler either water or air is usually provided in the dry sump system. The supply tank is usually placed behind the radiator. *Dry sump is generally used in large stationary marine engines.*

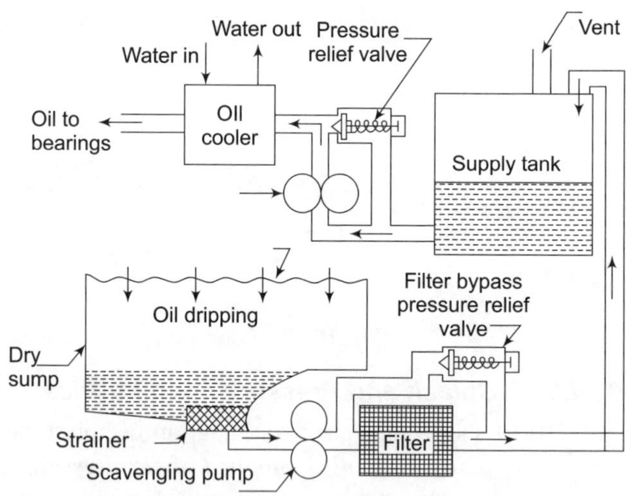


**Fig. 10.20** Dry sump lubricating system

## 10.15 CRANKCASE VENTILATION

The blow-by loss occurs during the compression and expansion strokes due to leakage past the piston. It contains water vapour and sulphuric acid. This causes corrosion of steel parts and sludge formation in lubricating oil of the crankcase. A considerable amount of water condensed may be collected in the sump, and in cold weather this may freeze and may cause damage to the lubricating oil pump. Further, any pressure build-up in the crankcase may cause crankshaft seal leakage. *Therefore, it is essential to remove the blow-by from the crankcase.*

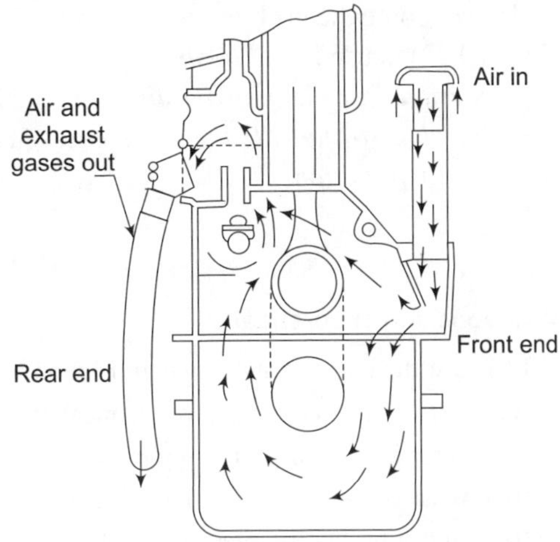


**Fig. 10.21** Open crankcase ventilation

In general, two types of crankcase ventilation system are in use. They are:

(i) *Open crankcase ventilation system*

(ii) *Closed crankcase ventilation system*

In the *open crankcase ventilation system* (Fig. 10.21), a fresh air supply is induced into the crankcase with the help of a *breather.* The air picks up the water vapour before it condenses and also the blow-by gases and flows back to the atmosphere. The outlet from the crankcase should be located at a point where the air is comparatively quiescent and therefore does not hold much oil in suspension. *The main disadvantage of the open system is that the ventilation is quite inadequate when it is most desired such as during idle running at the low speeds. The another disadvantage is increased air pollution.*

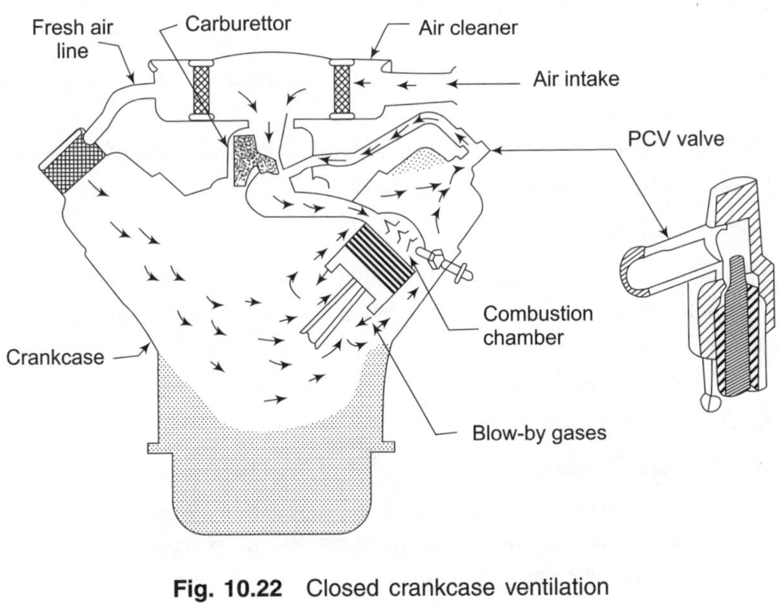


**Fig. 10.22** Closed crankcase ventilation

To overcome the problem of increased air pollution, the closed crankcase ventilation system (Fig. 10.22) is used. *The fresh air supply to the crankcase is taken from the carburettor or air inlet passage. Air cleaner and breather outlets are connected to the intake manifold through a PCV valve so that all the crankcase gases are burnt in the combustion chamber which reduces air pollution.*

## PROBLEMS

### Viva-voce and Theoretical

10.1 List the parts/locations where friction loss is present in IC engines.

10.2 What is the reason that mechanical efficiency increases with engine load?

10.3 What is the cause of pumping losses? How this can be reduced?

10.4 What is the philosophy behind the hydrodynamic lubrication? How is load supported?

10.5 Give a basic difference among hydrodynamic, elastohydrodynamic and boundary lubrication.

### Short and Long Theoretical Questions

10.6 Discuss the various factors that affect engine friction.

10.7 Discuss the function of lubrication.

10.8 Discuss the mechanism of lubrication.

10.9 Discuss the mechanism of various types of lubrication.

10.10 Discuss journal bearing lubrication.

10.11 Describe the types of lubricants.

10.12 Discuss the properties of a good lubricant.

10.13 Discuss the various additives for lubricants.

10.14 Describe various methods of testing of lubricating oil.

10.15 With the help of a sketch, describe the working principle of forced feed lubrication system and other lubrication systems.

10.16 What do you mean by crankcase ventilation? Explain the procedure of the ventilation.

10.17 List the parts to be lubricated in an IC engine.

### Software Based Questions

10.18 Develop a software for the design of lubrication system of an IC engine.

10.19 Develop a software for the design of lubrication system currently used in

(a) Passenger cars

(b) Diesel engine power plants

(c) Diesel rail (loco)engines.

### Design and Open Ended Question

10.20 Prepare a comprehensive report describing the lubrication system used in various types of modern engines used in land, sea and air transport systems with sketch, specifications of each part, merit and demerits of each system.

### Multiple-Choice Questions

10.21 Choose the incorrect answer, a lubricant should have

(a) high film strength to prevent metal-to-metal contact
(b) high pour point
(c) viscosity invariant with temperature
(d) cleansing ability

10.22 The lubricating system incorporated in most of the two-stroke engine is

(a) wet sump lubrication system
(b) splash system for a single cylinder engine
(c) mist lubrication
(d) dry sump lubrication system

10.23 **Assumption (A) :** Crankcase ventilation is employed in all reciprocating IC engines.

**Reasoning (R) :** Because it reduces the blow by losses.

(a) Both A and R are true
(b) A is true but R is false
(c) A is false but R is true
(d) Both A and R are false.

10.24 **Assumption (A) :** Boundary lubrication may not be obtained at the moment of starting a machine from rest.

**Reasoning (R) :** Because it is possible to start and accelerate a machine when the starting loads are low and viscosity is high.

(a) A is true but R is false
(b) A is false but R is true
(c) Both A and R are true
(d) Both A and R are false

10.25 In IC engine, boundary lubrication may occur

(1) when the piston and piston rings are at the beginning and end of stroke
(2) between the piston pin and bushing
(3) between gear teeth

Out of these statements

(a) Only 1 is correct
(b) Only 3 is correct
(c) 1 and 2 are correct
(d) 1, 2 and 3 are correct

10.26 The engine friction is affected by

(a) Cylinder size, number of cylinders and number of piston rings.
(b) Compression ratio, engine load and engine speed
(c) Oil viscosity and cooling water temperature
(d) All of the above

## Key to Multiple-Choice Questions

10.21 (b) 10.22 (c) 10.23 (a) 10.24 (c) 10.25 (d) 10.26 (d)

CHAPTER

# Engine Heat Transfer and Cooling Systems

# 11

## OVERVIEW

*During combustion process, the peak gas temperature in the cylinder of an internal combustion engine is the order of 2500 K. The maximum metal temperature for the inside of the combustion chamber space are limited to much lower values than the gas temperature by a large number of considerations and thus cooling for the cylinder head, cylinder and piston must therefore be provided.* This chapter deals with the engine heat transfer problems and various cooling systems adopted in practice to keep the engine parts safe.

## 11.1 NECESSITY OF ENGINE COOLING

The necessity of engine cooling arises due to the following facts:

(a) **Fatigue cracking due to high thermal stresses.** During combustion period, the heat fluxes to the chamber walls can reach as high as 10 MW/m$^2$. The flux varies substantially with location. The regions of the chamber that are contacted by rapidly moving high temperature burnt gases generally experience the highest fluxes. *In regions of high heat flux, thermal stresses must be kept below levels that would cause fatigue cracking.* So temperature must be less than about 400°C for cast iron and 300°C for aluminium alloy for water cooled engines. For air cooled engines, these values are 270°C and 200°C respectively.

(b) **Piston seize.** The gas side surface temperature of the cylinder wall is limited by the type of lubricating oil used, and this temperature ranges from 160°C to 180°C. *Beyond these temperatures, the properties of lubricating oil deteriorate very rapidly and it might even evaporate and burn damaging piston and cylinder surfaces.* Piston seizure due to overheating resulting from the failure of lubrication is quite common.

(c) **Knock and pre-ignition.** The spark plug and valves may be kept cool to avoid *knock and preignition* problems which results from overheated spark plug electrodes or exhaust valves.

**(d) Decrease in efficiency and power.** The *volumetric and thermal efficiency and power output of the engine decrease* with the increase in cylinder and its head temperature.

All the above factors suggest that there is a need of adequate cooling of the engine.

## 11.2 DISADVANTAGES OF EXCESSIVE COOLING

*A well designed cooling system should provide adequate cooling, but not excessive cooling.* It is worth to note that too much cooling is not as harmful to the engine as overheating, however it is undesirable for the following reasons.

**(i) Poor combustion and fuel economy.** Until the engine is heated to a certain minimum temperature, the fuel is not vapourized sufficiently to give a homogeneous mixture with the air. Such a situation results in poor combustion and fuel economy coupled with a possibility of crankcase dilution of the oil and erratic running of the engine.

**(ii) Increase in friction power.** If the engine temperature is decreased below the set minimum value, there will be an increase in the friction power due to high viscosity of the lubricating oil.

**(iii) Change of valve clearance and setting.** The overcooling may change the valve clearance and setting. The valves are set by the manufacturer to operate above a certain fixed temperature which is the lower limit of the normal operating temperature range of the engine.

**(iv) Different starting and corrosion.** At too low engine temperature, the starting may be difficult. Over and above, at low temperature, corrosion problem may be significant which may reduce the engine life. To avoid the condensation of sulphuric acid formed during combustion, the coolant temperature should be greater than 7°C.

The above facts suggest that the *cooling system should not only cool but also keep the cylinder temperature above a minimum level to avoid poor performance and to avoid corrosion of the engine parts and seizure.*

## 11.3 HEAT DISTRIBUTION IN IC ENGINES

*Heat transfer affects engine performance, efficiency and emissions.* Figure 11.1 shows the typical heat distribution for a reciprocating IC engine obtained from a heat balance test. About 28% of the total heat liberated during combustion is absorbed by the cooling medium. Friction absorbs about 10%. Radiation, exhaust and unaccounted consume about 30%. The remaining about 32% is the useful work.

*The quantity of heat absorbed by the cooling medium depends upon the engine design, size, operating conditions and the type of cooling medium.* Generally, the cooling medium is air, water and organic fluid.

The **functions of cooling system** are as follows:

(i) *To absorb and dissipate the excess heat from the engine in order to prevent the damage to the engine.*

(ii) *To maintain a sufficiently high operating temperature in order to assure smooth and efficient operation of the engine.*

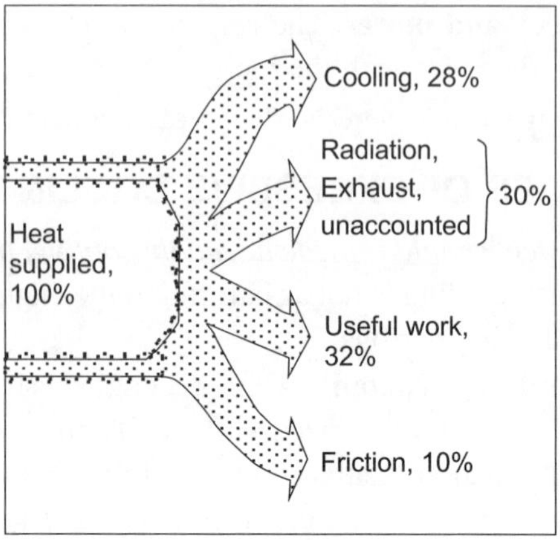


**Fig. 11.1** Typical heat distribution in IC engine

## 11.4 VARIATION OF GAS TEMPERATURE IN CYLINDER

It is the different variables in the engine cylinder which make possible an appreciable variation in the temperature of gases inside the engine cylinder. Figure 11.2 shows the variation of gas temperature inside an engine cylinder. The minimum temperature is observed at the end of suction stroke while the maximum temperature is achieved at the end of combustion. The subsequent expansion and exhaust process reduce the temperature its peak value.

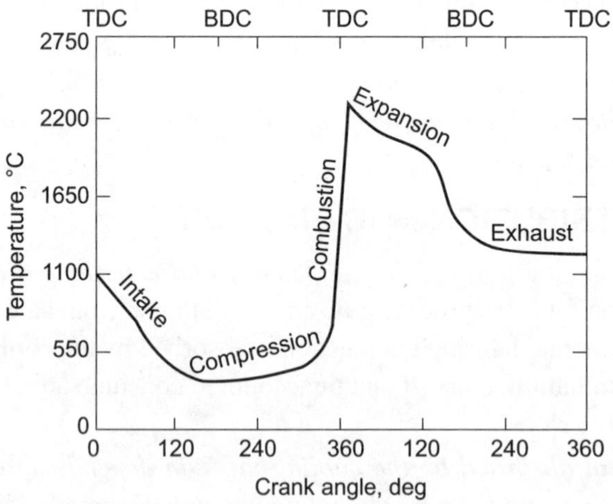


**Fig. 11.2** Gas temperature variation in a cylinder

## 11.5 THERMAL LOADING AND COMPONENT TEMPERATURES

*The variation of heat flux to the combustion chamber walls depend on engine design and operating conditions.* Besides, the heat flux to the various parts of the combustion chamber is not the same. On account of this non-uniform heat flux and different thermal impedances between location of the combustion chamber surfaces and the cooling fluid, the temperature distribution within engine components such as piston, cylinder, cylinder liner, valve, etc., is non-uniform.

### 11.5.1 Piston Temperature Distribution

Figure 11.3 shows typical values of temperature at different parts of a SI engine cast iron piston. Since the piston crown is subjected to very high temperature because this in direct contact with hot combustion gases, so the maximum temperature occurs at the centre of the crown and decreases at the outer edge. *The temperature profile decreases from crown to the bottom of the skirt. There exists a temperature difference between piston outer edge and the centre of the crown which is responsible for flow of heat to the ring belt through the path offered by metal section of the crown. This permits to increase the thickness of the crown from the centre to the outer edge in order to make a path of greater cross section available for the increasing heat quantity. The excessive temperature difference between the crown and outer edge may lead to cracking of the piston.*

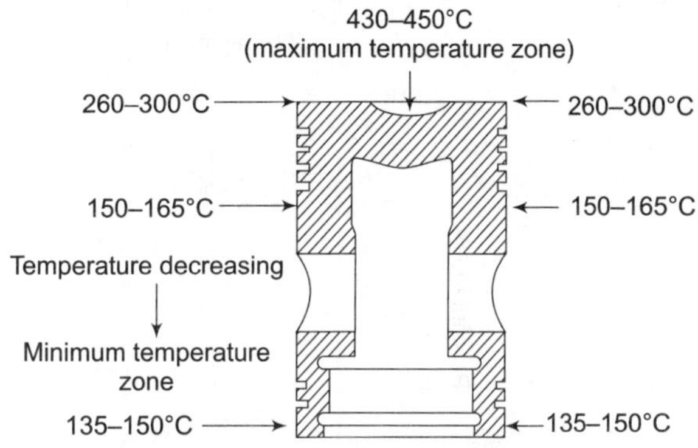


**Fig. 11.3** Typical piston temperature distribution

Figure 11.4 shows the isothermal contour (solid lines) and heat flow path (dashed lines) measured in a high speed diesel engine [2, 73].

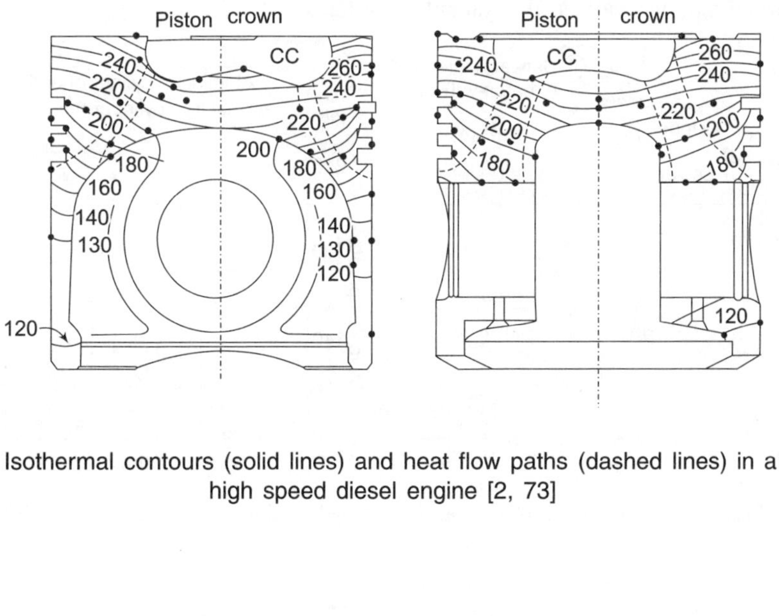


**Fig. 11.4** Isothermal contours (solid lines) and heat flow paths (dashed lines) in a piston of high speed diesel engine [2, 73]

The relative amount of heat flow from various zones of piston as percentage of heat flow in from combustion chamber [2, 73] is shown in Fig. 11.5 for a diesel engine. *It is obvious that about 70 per cent of heat flows out through the ring zone, and much smaller amounts through the pin boss zone underside of the crown and skirt.*

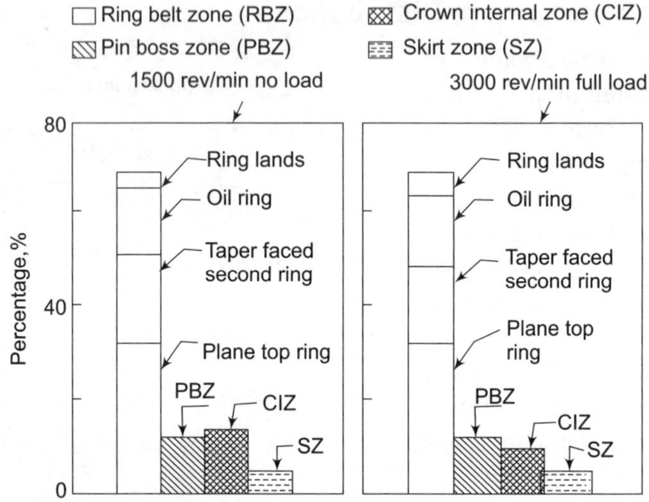


**Fig. 11.5** Heat flows from various zones of piston as a percentage of heat flow in from combustion chamber of diesel Engine [2, 73]

## 11.5.2 Cylinder and Cylinder Liner Temperature Distribution

Figure 11.6 shows the typical temperature of various locations on a four-cylinder SI engine cylinder head [73]. The maximum temperature (hot spots) occurs where the heat flux is high and access for cooling is difficult. Such locations are the bridges between the valves and the region between exhaust valves of adjacent cylinders.

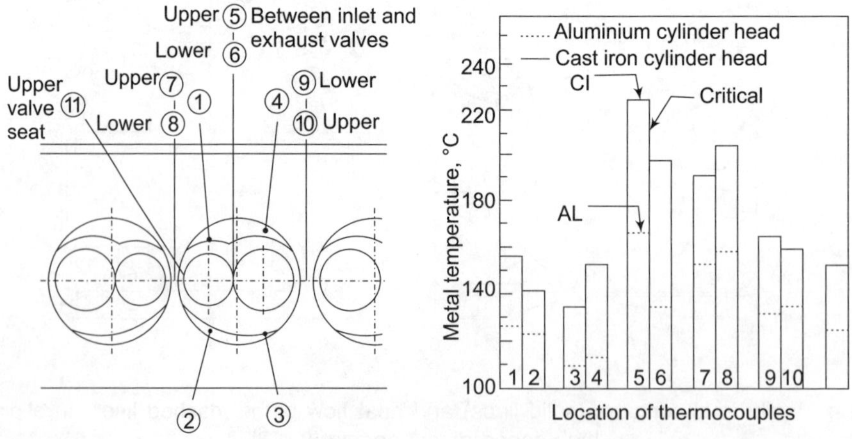


**Fig. 11.6** Cylinder head temperature distribution in SI engine [110, 73]

Figure 11.7 shows temperature and heat flux distribution in the cylinder liner of a high speed direct ignition diesel engine [4, 73]. Owing to the lower regions of the liner are only exposed to combustion products for part of the cycle after significant gas expansion has occurred, the heat flux and temperature decreases significantly with distance from the cylinder head.

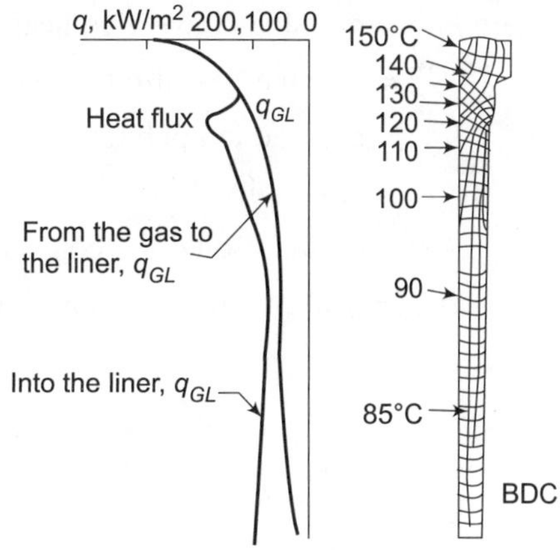


**Fig. 11.7** Temperature and heat flux distribution in the cylinder liner of a high-speed direct injection diesel engine [73]

## 11.5.3 Temperature Distribution in Valve

Figure 11.8 shows the temperature isotherm in a typical engine exhaust valve. In this typical valve, the peak temperature isotherm lies at the surface, the middle of the valve face and on one side of the face-stem joining surface. The temperature variation is large which results in large thermal stresses.

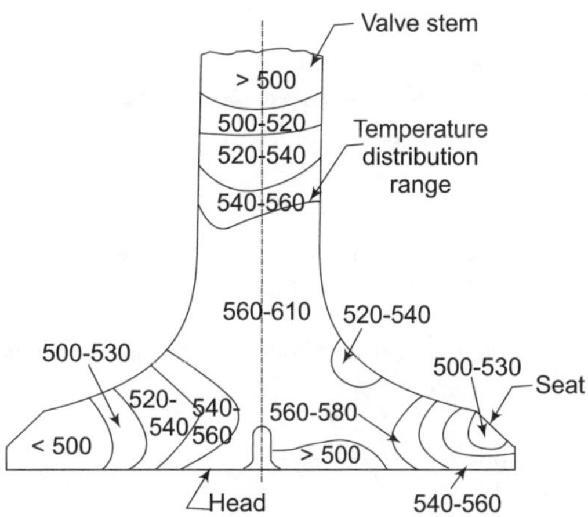


**Fig. 11.8** Exhaust valve temperature isotherm [73]

## 11.6 OVERALL HEAT TRANSFER PROCESS

The overall heat transfer process from the gases within the cylinder through the combustion wall to the coolant flow is shown in Fig. 11.9. The heat flux into the inner wall has, in general, both a convective ($\dot{q}_{CV}$) and a radiation component ($\dot{q}_R$). The heat flux is conducted through the wall ($\dot{q}_{CN}$) and then convected ($\dot{q}_{CV}$) from the coolant.

For a steady one-dimensional heat flow through the wall as shown in Fig. 11.9, the heat flux is given as follows:

**Gas side:** $$\dot{q} = \frac{Q}{A} = \dot{q}_{CV} + \dot{q}_R = h_{c,g}\,(T_g - T_{w,g}) + \sigma\in\,(T_g^4 - T_{w,g}^4) \qquad ...(11.1)$$

where $h_{c,g}$ = gas side convective heat transfer coefficient

$T_g$ = average gas film temperature

$T_{w,g}$ = average temperature of wall, gas side

$\in$ = emissivity

$\sigma$ = Stefan–Boltzman constant

$\dot{q}_{CV}$ = heat flux due to convection (W/m$^2$)

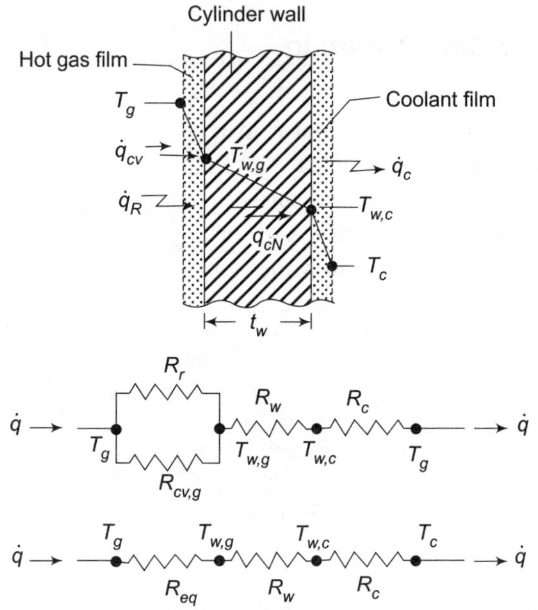


**Fig. 11.9** Temperature distribution and heat flux across combustion chamber wall

**Wall:**
$$\dot{q} = \dot{q}_{CN} = \frac{k(T_{w,g} - T_{w,c})}{t_w} \qquad ...(11.2)$$

where $T_{w,c}$ = average temperature of wall, coolant side

$t_w$ = thickness of wall

$k$ = thermal conductivity of wall material

$\dot{q}_{CN}$ = heat flux due to conduction

$t_w$ = wall thickness

**Coolant side:**
$$\dot{q} = \dot{q}_{CV} = h_{c,c}\,(T_{w,c} - T_c) \qquad ...(11.3)$$

where $h_{c,c}$ = coolant side convective heat transfer coefficient

$T_c$ = coolant temperature

Using the concept of overall heat transfer coefficient

$$\dot{q} = U_0(T_g - T_{w,c}) \qquad ...(11.4)$$

where $U_0$ the overall heat transfer coefficient is given by

$$U_0 = \frac{1}{R_{eq} + R_w + R_c} = \frac{1}{R_{eq} + \frac{t_w}{k} + \frac{1}{h_{c,c}}}\text{; W/m}^2\text{ K} \quad ...(11.5)$$

where $R_{eq}$ = equivalent thermal resistance

$R_w$ = wall thermal resistance

$R_c$ = thermal resistance of coolant film

$\frac{1}{R_{eq}} = \frac{1}{R_r} + \frac{1}{R_{cg}}$ as it acts in parallel

$R_r = \frac{T_g - T_{w,g}}{\sigma\varepsilon (T_g^4 - T_{w,g}^4)}$ = thermal resistance due to radiation

$R_{c,g} = \frac{1}{h_{c,g}}$ = thermal resistance due to convection

*If $U_0$ is known, the heat flux $\dot{q}$ (W/m²) can be calculated and other temperature may be related to each other. After knowing $\dot{q}$, the heat transfer surface area A can be calculated.*

Considering the oil film and scale formation also into consideration the temperature distribution across cylinder wall is shown in Fig. 11.10.

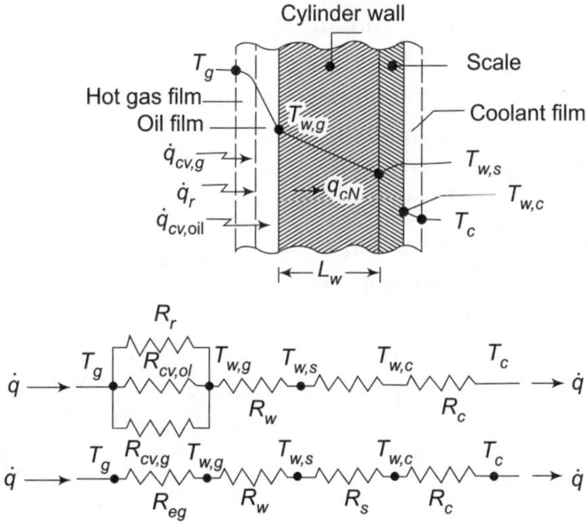


**Fig. 11.10** Temperature distribution

The total thermal resistance is given by

$$R = R_{eq} + R_s + R_w + R_c \quad ...(11.6)$$

where $$\frac{1}{R_{eq}} = \frac{1}{R_r} + \frac{1}{R_{c,g}} + \frac{1}{R_{c,of}}$$

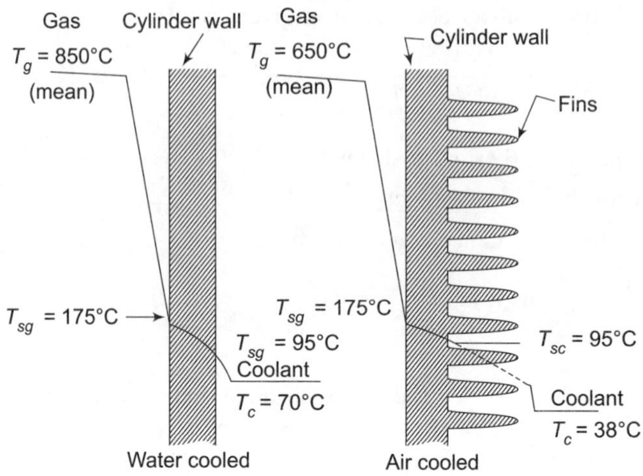


**Fig. 11.11** Typical temperature profile across cylinder wall

$$R_{c,of} = 1/h_{c,of} = \text{thermal resistance due to oil film.}$$

A typical temperature profile across the cylindrical barrel wall for air-cooled and water cooled engine are shown in Fig. 11.11.

## 11.7 HEAT TRANSFER AND ENGINE ENERGY BALANCE

Figure 11.12 depicts the ratio of coolant heat rejection rate to the brake power for a range of engine types and sizes at the maximum power. *It is worth to note that this ratio decreases with increasing engine speed and size.* This is because of the fact that higher gas velocities are used to achieve the desired fuel-air mixing in small diesel engines are compared to bigger size and have less favourable surface-volume ratio.

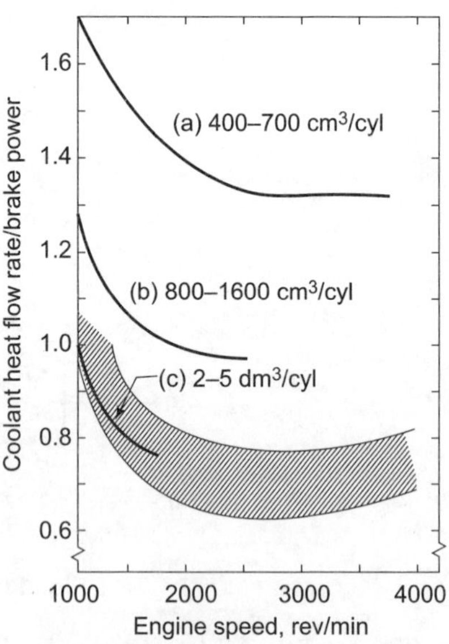


**Fig. 11.12** Ratio of cooling heat flow rate to brake power vs. engine speed. (a) Small automatic diesel (b) Larger automotive diesels, (c) Various diesel, (d) SI engine

The steady-flow energy conservation equation applied to the control volume of an engine (Fig. 11.13) is given by

$$\dot{m}_f (Q_{HCV})_f + \dot{m}_a h_a = P_b + \dot{Q}_{\text{cool}} + \dot{Q}_{\text{misc}} + (\dot{m}_f + \dot{m}_a)h_e \quad ...(11.7)$$

where $P_b$ = brake power

$\dot{Q}_{\text{cool}}$ = heat transfer rate to cooling medium

$\dot{Q}_{\text{misc}}$ = heat rejected oil plus convection and radiation from the engine's external surface

$(Q_{HCV})_f$ = higher heating value of fuel

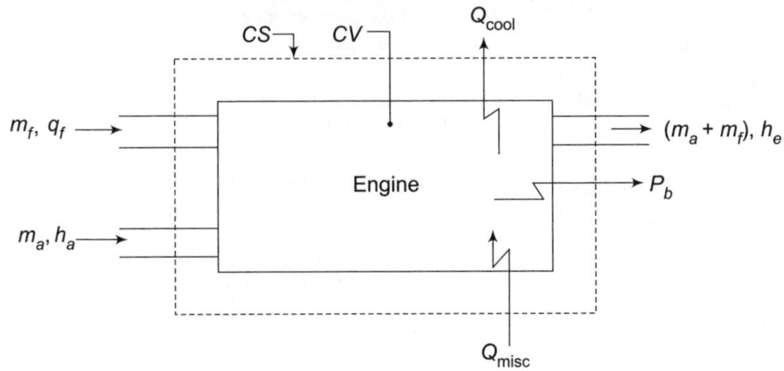


**Fig. 11.13** Control volume of an engine

By breaking exhaust enthalpy ($h_e$) into sensible part ($h_{e,s}$) plus the exhaust reference.

Here, $h_{e,s} = h_e\,(T) - h_e\,(298\text{ K})$

The above equation may be written as

or $$\{P_b + \dot{Q}_{\text{cool}} + \dot{Q}_{\text{misc}} + H_{e,ic} + \dot{m}h_{e,s}\} = \dot{m}_f\, Q_{\text{LHV}} \quad ...(11.8)$$

where $\dot{H}_{e,ic}$ = exhaust enthalpy loss due to incomplete combustion

$Q_{LHV}$ = lower heating value of fuel

$h_{e,s}$ = sensible enthalpy at exit

Table 11.1 shows typical values of these terms as a % of fuel heating value.

**Table 11.1** Energy Balance for Automotive Engine of Maximum Power as a Percentage of Fuel Heating Value

| | $P_b$ (% of heating value) | $\dot{Q}_{cool}$ | $\dot{Q}_{misc}$ | $\dot{H}_{e,ic}$ | $\dot{m}h_{e,sc}$ |
|---|---|---|---|---|---|
| SI Engine | 25–28 | 17–26 | 3–10 | 2–5 | 34–45 |
| Diesel | 34–38 | 16–35 | 2–6 | 1–2 | 22–35 |

Figure 11.14 shows the various energy transfer [5] as a function of vehicle speed. At part load, a greater fraction of the fuel heating value is absorbed into coolant. At low speeds and loads, the coolant heat transfer rate is 2 to 3 times the brake power.

*Computer simulation of SI engine indicates that at typical part-load conditions, a proportional reduction in combustion chamber-wall heat losses of 10 per cent results in a proportional increase in brake fuel conversion efficiency of about 3 per cent [6].*

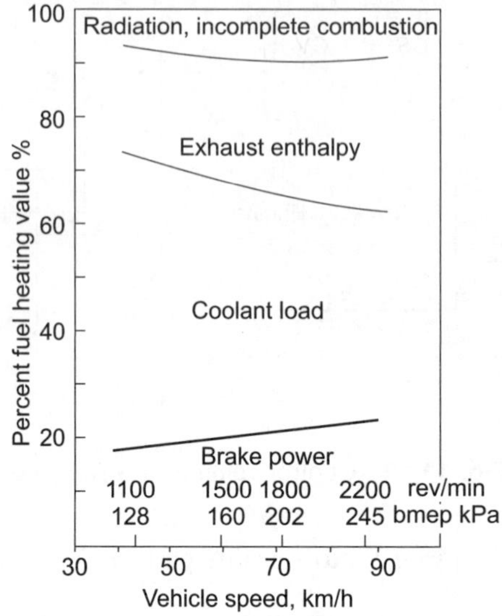


**Fig. 11.14** Energy distribution as a function of vehicle speed

## 11.8 CORRELATION FOR TIME-AVERAGE HEAT FLUX

The correlation of overall heat transfer data for 19 different engines have been made by Taylor and Toong [7] which are given in Fig. 11.15. The Nusselt number (Nu) and Re numbers are defined as

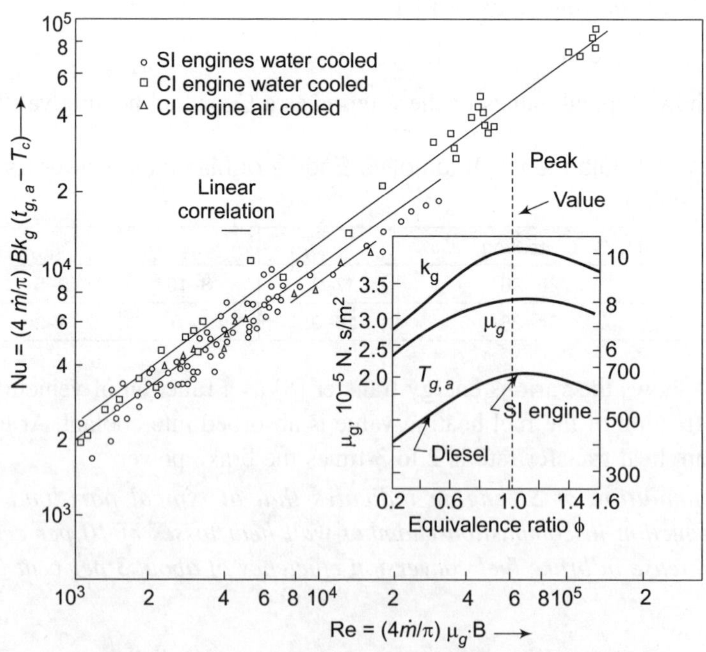


**Fig. 11.15** Overall engine heat transfer correlations [7]

$$\text{Nu} = 4\frac{\dot{Q}}{\pi} Bk_g\,(T_{g,a} - T_c) \qquad ...(11.9)$$

$$\text{Re} = \frac{4\dot{m}}{\pi}\,\mu_g.B \qquad ...(11.10)$$

where $T_{g,a}$ = average effective gas temperature

$B$ = cylinder bore

$\dot{m}$ = charge mass flow rate

## 11.9 CORRELATIONS FOR INSTANTANEOUS SPATIAL AVERAGE HEAT TRANSFER COEFFICIENTS

**Annand** [8] has given the following correlation for convective heat transfer coefficient ($h_c$) by matching the previously published experimental data

$$h_c\frac{B}{k} = a\left(\frac{\rho\, C_m\, B}{\mu}\right)^b \qquad ...(11.11)$$

The value of constant ($a$) varies with intensity of charge motion and engine design. With normal combustion, $a$ = 0.35 to 0.8 with $b$ = 0.7.

where $C_m$ = mean piston speed = LN/30

$B$ = cylinder bore

Gas properties are evaluated at the cylinder-average charge temperature, $T_g$.

**Woschni** [9] has developed a correlation of the form

$$\text{Nu} = 0.035\,Re^m \qquad ...(11.12)$$

where $$\text{Re} = \frac{wB}{\mu}$$

Here $w$ = average cylinder gas velocity (10 m/s)

$B$ = cylinder bore

The above correlation may be written as

$$h_c = 3.26\,B^{-0.2}\,p^{0.8}\,T^{-0.55}\,w^{0.8} \qquad ...(11.13)$$

where $B$ is in metre, $p$ is in kPa, $T$ is in $K$ and $w$ is in m/s.

## 11.10 CORRELATION FOR INSTANTANEOUS LOCAL HEAT TRANSFER COEFFICIENTS

Le **Feuvre *et al.* [9] and Dent [10]** have proposed the forced convective heat transfer coefficient based on the flow over a flat plate

$$\left(h_c\frac{1}{k}\right) = 0.036\left(\frac{\rho v l}{\mu}\right)^{0.8}\left(\frac{\mu c_p}{k}\right)^{0.333} \qquad ...(11.14)$$

where $l$ = length of the plate = $2\pi r$

$v$ = flow velocity over plate = $\omega r$

$\omega$ = solid body angular velocity of the charge

The heat flux at any radius $r$ (with $Pr = 0.73$) is given by

$$\dot{q}(r) = 0.023 \frac{k}{r}\left(\frac{\omega r}{v}\right) 0.8\ [T_g(r) - T_w(r)] \qquad ...(11.15)$$

## 11.11 PARAMETERS AFFECTING ENGINE'S HEAT TRANSFER

The following parameters affect the magnitude of engine's heat transfer and the temperature distribution in the component that comprises the chamber.

**(a) Speed and Load.** *The average engine heat transfer rate increases as speed and load increase; however the cycle heat transfer decreases as speed and load increase.* Figure 11.16 shows the predicted average cycle heat transfer expressed as a percentage of fuel's chemical energy as a function of speed and load [11].

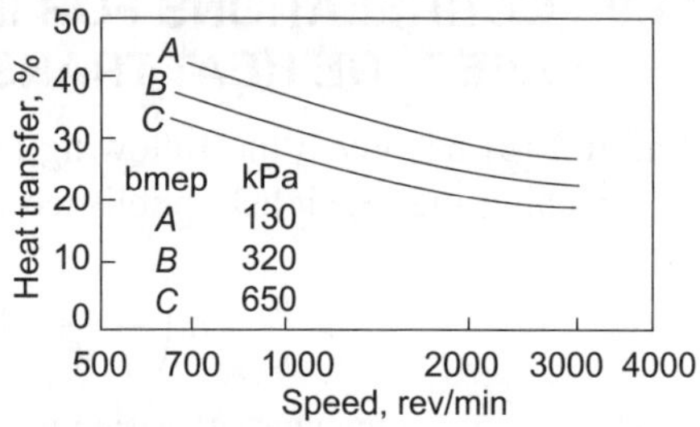


**Fig. 11.16** Predicted average heat transfer per cycle to combustion chamber wall expressed as a percentage of fuel's chemical energy of SI engine as a function of speed and load [11]

*Piston temperature increases with rpm and load, this is because of the fact that the cycle heat transfer decreases, so overheating is expected.*

**(b) Air-fuel (Equivalence) Ratio ($\phi$).** Figure 11.17 shows the effect of A/F (equivalence) ratio or the predicted average heat transfer per cycle expressed as a fraction of fuel's chemical energy in a SI energy for various burn rates ($\Delta\theta_b$ – combustion duration) [11]. The peak heat transfer per cycle occurs at the mixture equivalence ratio, $\phi = 1.0$ for the condition of the maximum power and decreases as mixture is leaned out or enriched from this value. The major effect is through the gas temperature. It can be seen that for $\phi$ greater 1.0 (i.e., leaner mixture), the cycle heat transfer decreases so over-heating of the engine's parts is expected. However, the experimental evidence shows that the peak heat flux occurs at $\phi = 1.1$ and decreases as $\phi$ is leaned out or enriched from this value.

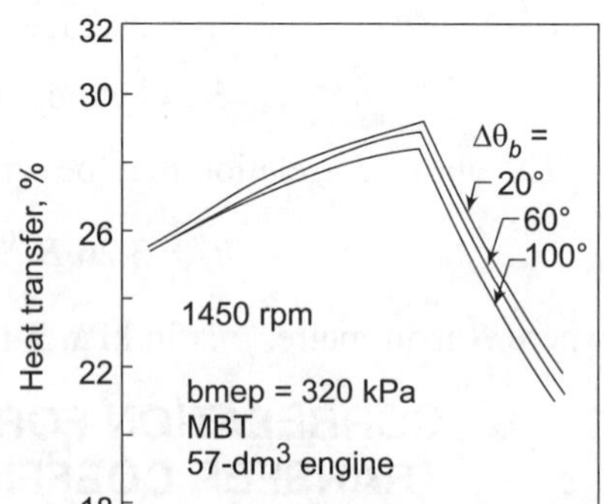


**Fig. 11.17** Effect of equivalence ratio on average heat transfer per cycle for various burn rates in a SI engine [11]

**(c) Compression Ratio (r).** With the increase in compression ratio in the case of SI engine, the total heat flux to the coolant decreases until $r = 10$, thereafter heat flux increases slightly as $r$ increases. The magnitude of this change is modest. The effect of changes in compression ratio on component temperatures depends on location. Generally, head and exhaust valve temperatures decrease with increasing combustion ratio, due to lower expansion and exhaust stroke temperature. The piston and spark plug temperatures increase.

**(d) Spark Timing.** Figure 11.18 shows the effect of spark timing on heat transfer for SI engine for various burn rates. *Retarding the spark timing in as SI engine decreases the heat flux [11].* A similar trend in CI engines with retarding injection timing would be expected. The gas temperature decreases as timing is retarded because combustion occurs later when the cylinder volume is larger. Piston and spark plug electrode temperatures change most with timing variations. Exhaust valve temperature increases as timing is retarded due to higher value of exhaust gas temperature.

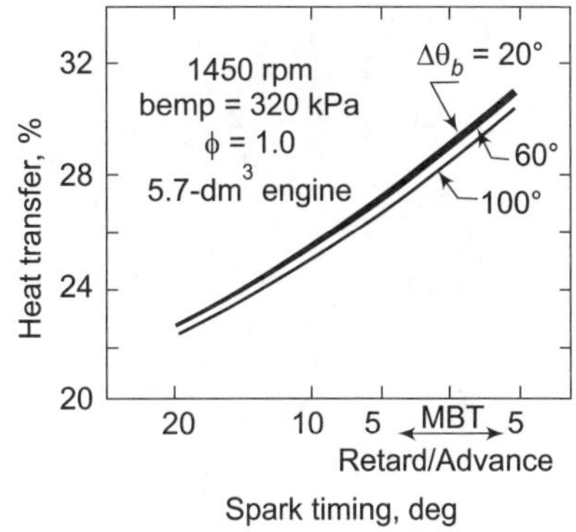


**Fig 11.18** Predicted average heat transfer per cycle at a percentage of fuel's chemical energy as a function of spark timing and burn rate for diesel engine

**(e) Swirl and Squish.** The swirl and squish motion in the combustion chamber increase the gas velocities, which in turn, results in higher heat flux.

**(f) Inlet Temperature.** The heat flux increases linearly with increasing inlet temperature as the gas temperature is increased throughout the cycle. *An increased inlet temperature of 100 K gives a 13 per cent increase in heat flux.*

**(g) Coolant Temperature and Composition.** Figure 11.19 shows the effect of coolant temperature on engine's components of a spark ignition engine [12]. Increasing liquid coolant temperature increases the temperature of coolant directly cooled by the liquid coolant. The exhaust valve and spark plug temperatures are unchanged. *Rapid increase in component temperature with coolant temperature is expected where the heat fluxes are lower (e.g., the cylinder liner) indicating that the heat transfer to the coolant is largely by forced convection.* At higher heat flux locations (e.g., valve bridge), the increase in metal temperature with coolant is less due to the possibility of nucleate boiling regime.

**(h) Wall Material.** *The wall material has a strong influence on the heat losses from the working fluid.* The common metallic component materials of cast iron ($k$ = 54 W/mK) and aluminium ($k$ = 155 W/mK) have substantially different thermal properties and they both operate with combustion chamber surface temperature (200 to 400°C) that are low relatively to burnt gas temperatures. *This creates interest in using material that could operate at much higher temperatures so that the heat losses from the working fluid would be reduced.* ***Ceramic materials**, such as* **silicon nitride** ($k$ = 5 to 10 W/mK) and **zirconia** ($k$ = 1.2) have lower thermal conductivity than cast iron, would operate as higher temperature, and thereby insulate the engine. *For diesel engine, this approach is most feasible.*

**(i) Knock.** Knock results in an increase in gas pressure and temperature above the normal combustion levels. This increases local heat fluxes to regions of the piston, the cylinder head and liner in contact with end gas. *Due to knock, the increase in heat flux in the end gas*

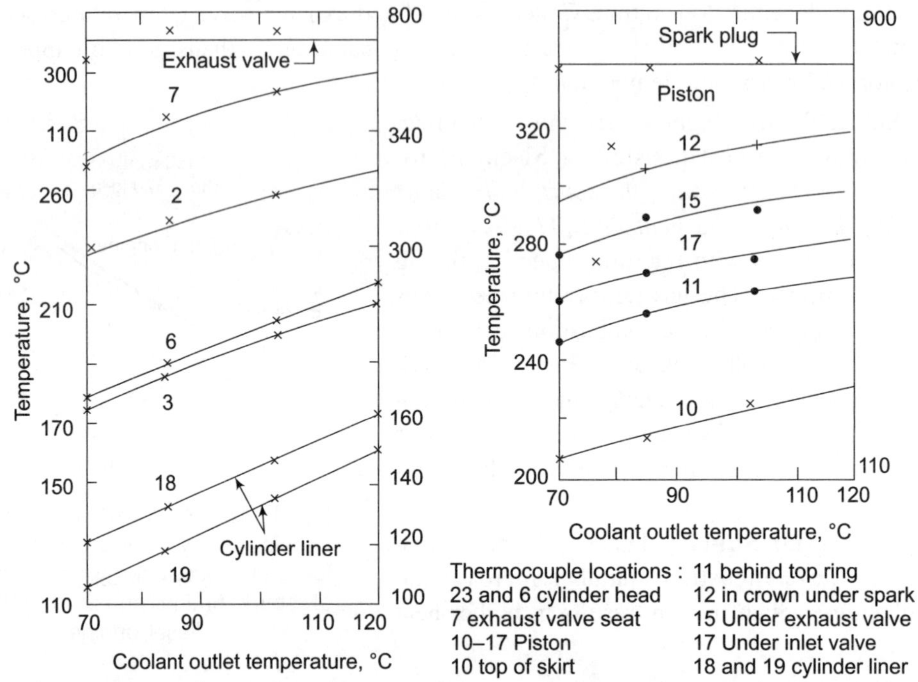


**Fig. 11.19** Effect of coolant temperature on engine's components

*region is even up to three times the normal heat flux. The primary knock damages the piston crown in this region due to the combined effect of extremely high local pressures and higher material temperatures.*

## 11.12 QUANTITY OF WATER AND POWER REQUIRED TO COOL THE ENGINE

The amount of rate of coolant or water ($\dot{m}_c$) required to cool the engine is calculated by

$$\dot{q} = U\,(T_g - T_c) = \dot{m}_c\, c_{pc}\, \Delta T_c$$

or
$$\dot{m}_c = \frac{q}{c_{pc}\,\Delta T_c} \qquad \text{...(11.16)}$$

where $\dot{q}$ = the heat flux rejected from the engine (W/m$^2$)

$\Delta T_c$ = the rise in temperature in coolant (K or °C)

In order to obtain an expression for power required for cooling an engine, the following assumptions are made.

(i) engines run at constant A/F ratio,

(ii) temperature difference between cylinder gases and cylinder wall may be assumed to be constant, and

(iii) the density and temperature of coolant remain constant.

Based on the above assumptions, the power required to cool the engine ($P_c$) is given by (For air **cooled engine**)

$$P_c = \frac{A_e (ip)^{2.25}}{(A_{fin}\ \Delta T_{fin})^{3.75}\ \rho_c^2} \quad ...(11.17)$$

where $A_c$ = effective area; $\rho_c$ = density of coolant; $A_{fin}$ = fin surface area
$\Delta T_{fin}$ = temperature difference between surface temperature and coolant

Power required by cooling water or liquid pump is given by

$$P = \dot{m}_c . gH/\eta_p \quad ...(11.18)$$

where $H$ = head of pump and $\eta_p$ = overall efficiency of pump

## 11.13 TYPES OF COOLING SYSTEM

*A cooling medium is required to cool the engine.* This can be *air* or *liquid.* Based on the cooling medium two types of cooling systems are in general use. They are

(a) Air or direct cooling system
(b) Liquid or indirect cooling system (open or closed system)

**(a)** ***Air cooling system*** *is used for small engines such as scooters, motorcycles, three-wheeler engines, lawn mowers, and for engine whose weight is the main consideration such as aircraft engines. Other areas for air cooled engines are industrial and agriculture engines where liquid cooling is difficult and objectionable. Fins are provided over the cylinder and cylinder heat to affect better cooling.*

**(b)** ***Liquid (water or other coolant) cooling system*** *is used in heavy engines such as passenger cars, trucks, buses, diesel loco, marine engines, some racing motorcycles (Honda, Yamaha), etc. In this system, previously water was used and made to circulate through the jackets provided around the cylinder, cylinder head, valve ports and seats and other hot spots to extract most of the heat.* Now, **synthetic coolant** is gaining popularity.

**Liquid cooling** system is further classified as follows:

(i) Direct or non-return or open system
(ii) Thermo-sphyon or natural circulation system
(iii) Forced circulation system
(iv) Evaporative cooling system
(v) Pressure cooling system

## 11.14 AIR COOLING SYSTEM

In this system, a current of air is made to pass over the outside of the cylinder barrel whose outer surface has been considerably increased by providing cooling fins. Since the heat transfer coefficient between air and metal surface is low, so to increase the heat transfer

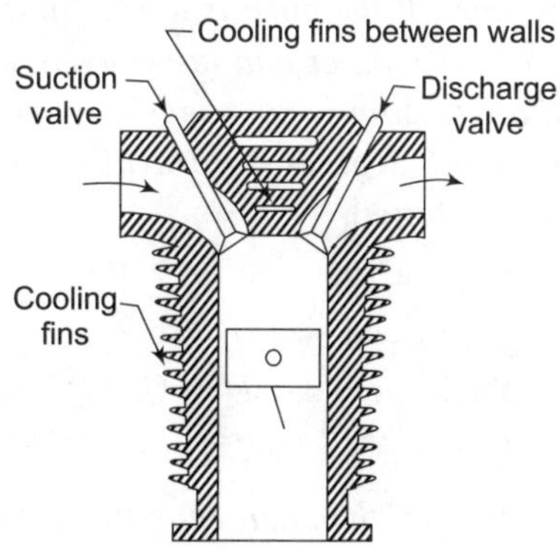


**Fig. 11.20** Air cooled cylinder with fins

rate, the surface area has to be increased by providing fins. Figure 11.20 shows the air cooled cylinder with fins.

Cooling fins are either *cast integral with the cylinder and cylinder head or can be fixed with the cylinder block separately. The former is more common in practice. Various shapes of profile such as rectangular, triangular, trapezoidal, conical, parabolic, etc., are used.* Figure 11.21 shows some of the profiles with their temperature distribution from root to tip. The heat dissipating capacities of the fins are the functions of their cross section, length, shape, material and the spacing between them. Since the heat is dissipated gradually from the fin surface, the temperature of the fin decreases from roots to the tip. *As the temperature difference between tip surface and ambient air is less, so the fin surface nearer to the tip dissipates heat less as compared to the root and this decrease is gradual from root to tip. As a result, the thickness of fin is gradually reduced from root to tip. Among all the four shapes, the least temperature drop is in the rectangular fin (type d) and the maximum temperature drop in triangular (type a) fin.*

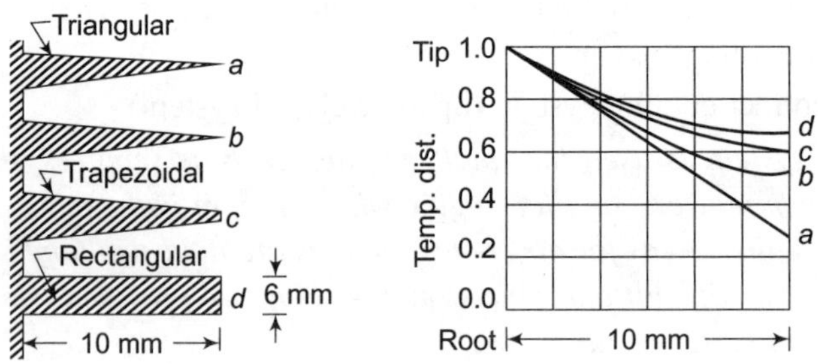


**Fig. 11.21** Shapes of fins and their temperature distribution

*Larger pitch interspacing between the fin offers larger area for cooling air but the heating of air is less, so more cooling air is needed. Smaller pitch results in smaller flow of air for cooling air and hence air is heated more, so less cooling air is required but the pressure drop is more. If the pitch is too small, the laminar boundary layers of the opposite faces interfere with each other and the efficiency of the fins is reduced very much. Therefore, the* ***pitch less than 2.5 mm*** *is not used. The* ***height of the fin*** *varies from* ***15 to 25 mm*** *and the* ***tapering is of 3 to 5 degrees for providing sufficient draft.*** In general, the following dimensions of fins are preferred.

Base or root size ($b$) = thickness of cylinder wall

Tip of end size ($t$) = $b/2$

Pitch ($p$) = $0.06D$ to $0.11D$ where $D$ = cylinder bore

Height (I) = $0.025D$ to $0.35D$.

*It is worth to note that air cooling results in a higher engine temperature.* This necessitates the provision of bigger clearance between the various parts especially between piston and cylinder, and valves and rocker arms.

*The effect of material is also pronounced.* Approximately, 90% of the mass for fins is saved by changing from iron to aluminium.

*Baffles or cowlings are used which forces the air through the space between the fins which are not in the front of the directions of the air.* It is always desired to have negligible kinetic energy loss between the entrance and the exit. Figure 11.22 shows various types of baffles used for air cooled engines. The **type 'b'** is most common for petrol engines where the **type 'd'** is used for diesel engine. The type **'a'** results in the maximum pressure drop.

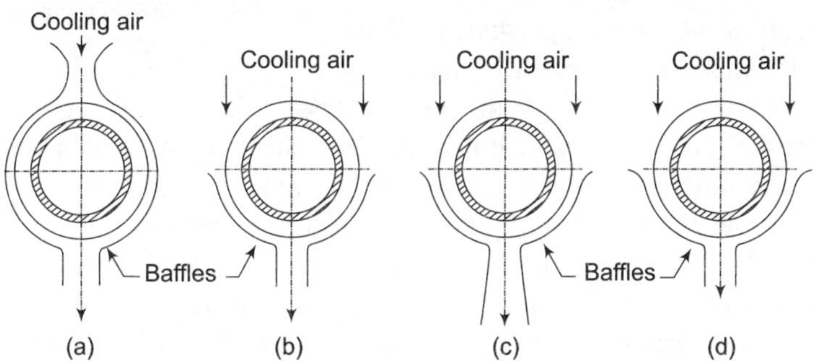


**Fig. 11.22** Types of baffles in air cooling system

## 11.15 LIQUID COOLING SYSTEM

In the liquid cooling system, liquid is circulated through the jacket formed around the cylinder and cylinder head to affect the cooling. *The local velocity of coolant (water), the flow path and the mass flow rate passing through the jacket are important and these should be carefully planned to cool highly thermally stressed areas. The local velocity* **greater than 3 to 4 m/s** *usually* used and sometimes drilled coolant passages are provided to cool highly stressed parts.

As mentioned earlier, **water and synthetic** liquid are used as coolant. *For ease of cold starting, various* **anti-freeze solutions** *are added to water.* These are:

(i) Wool alcohol
(ii) Kerosene
(iii) Denatured alcohol
(iv) Ethylene glycol and propylene glycol
(v) Calcium or magnesium chloride
(vi) Sugar solution
(vii) Glycerine

It is worth to note that some of the anti-freeze solutions form sticky deposits on the water side and reduce the heat transfer and in such case chromates may be used to prevent deposit formation.

**Various liquid cooling systems** used are discussed below.

### 11.15.1 Direct or Non-return Liquid Cooling System

This is an open system and it is applicable only where plenty of water is available. The water from the storage tank is directly supplied through an inlet valve to the engine's cooling water jacket. The hot water coming out from the engine is not cooled for reuse but it is discharged (Fig. 11.23).

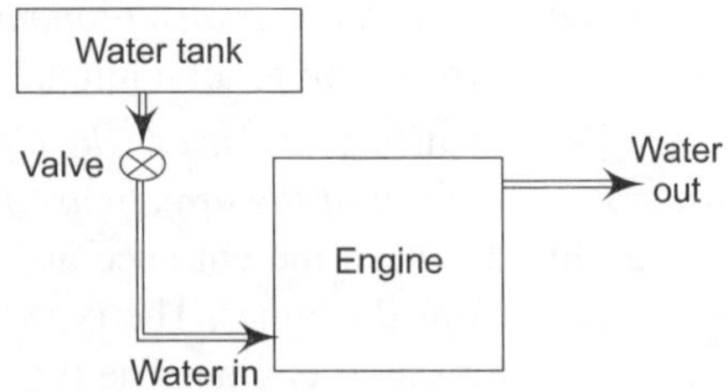


**Fig. 11.23** Direct or non-return cooling system

### 11.15.2 Thermosyphon or Natural Liquid Circulation System

*The system is closed one and designed so that the water may circulate naturally because of the difference in density of water at different temperatures.* Figure 11.24 shows a thermosyphon system. *It consists of water jacket, radiator and fan.* When the water is heated, its density decreases and it tends to rise, while the colder molecules tend to sink. Circulation of water then is obtained as the water heated in the water jacket tends to rise and the water cooled in radiator with the help of air passing over the radiator either by the ram effect or by fan or jointly tends to sink. The direction of natural circulation is shown by arrows. It is worth to note that even under the most favourable conditions the rate of circulation is slow. *This makes the system unfavourable for heavy duty engines that requires high rate of heat transfer. This system may be used on small engines where the requirement of heat rejection rate is small.*

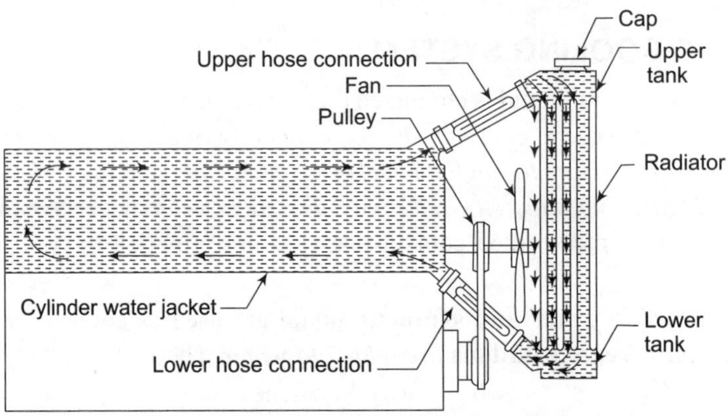


**Fig. 11.24** Thermosyphon cooling system

### 11.15.3 Forced Circulation Liquid Cooling System (Closed System)

Figure 11.25 shows a schematic diagram of a forced circulation closed liquid cooling system. The coolant (water or synthetic coolant) is circulated though the cylinder jacket with the help of pump which is usually a centrifugal type, and driven by the engine through belt and pulley arrangement. A header is usually employed to provide equal distribution of water to all the cylinders. The headers are supplemented by tubes or ducts, to provide a higher rate of flow around the critical sections of the engine such as exhaust valve seats. *The forced circulation system is commonly used in most automotive diesel and petrol engines.* Standby diesel power plant up to 200 kVA use this type cooling system. The system consists of pump, water jacket, radiator, radiator cap, fan and a thermostat, the details of which are explained as following.

**(a) Pump.** The pump which is usually **centrifugal type** is attached at some point on the front end of the cylinder block and is driven by the engine's crankshaft through a friction belt and pulley (Fig. 11.25). The liquid from the bottom of the radiator enters the pump from the lower hose connection and forces the water to circulate through the water jackets. The water leaves the cylinder block through the upper hose connection and enters the radiator where it is cooled by the blowing of cool air through the radiator core by ram effect and fan.

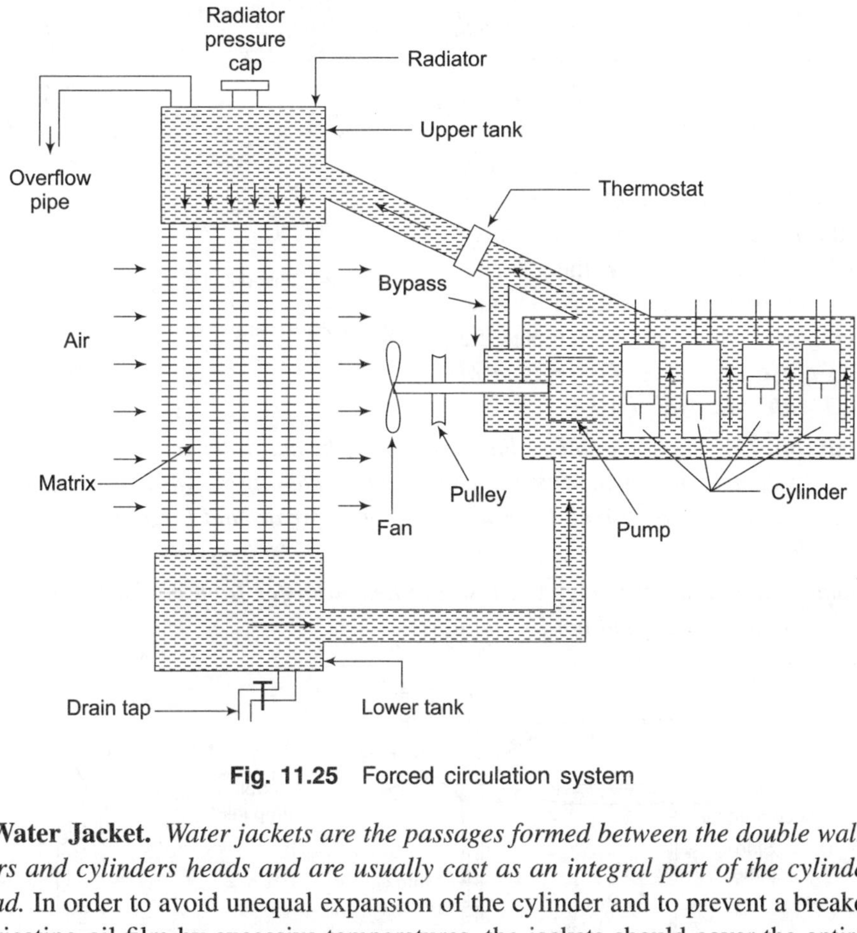


**Fig. 11.25** Forced circulation system

**(b) Water Jacket.** *Water jackets are the passages formed between the double walls of the cylinders and cylinders heads and are usually cast as an integral part of the cylinder block and head.* In order to avoid unequal expansion of the cylinder and to prevent a breakdown of the lubricating oil film by excessive temperatures, the jackets should cover the entire length of the stroke. *Headers are provided to furnish equal distribution of the cooling water to all cylinders.* Special tubes and ducts are provided in the jacket which provides extra cooling around the hotter section.

**(c) Thermostat.** *We know that a too low a cylinder surface temperature may result in severe corrosion damage due to condensation of acid from the gas on the barrel wall.* Such a situation may arise whenever the engine is started from cold. To avoid this situation, a thermostat (usually two) is fitted in the upper hose connection initially prevents the circulation of water below a certain temperature (usually up to 85°C) through the radiator so that water gets heated up quickly. During the warm-up period, the thermostatic valve is closed and the water pump circulates the coolant through the water jacket only. When the normal operating temperature is reached, the thermostatic valve opens and permits the water to circulate through the radiator.

Usually, a **bellows type thermostat** (Fig. 11.26) is used. *It consists of a closed bellows with volatile liquid under reduced pressure.* A link mechanism connected with bellows closes and opens the valve. When the bellows is heated, the liquid vapourizes and creates enough pressure to expand the bellows. The expansion of the bellows operates the linkage which opens the valve under cold condition of engine, the gas gets condensed, the pressure is reduced and the bellows collapses which in turn actuates the linkage to close the valve. *Other type of **valve filled with wax** is also used.*

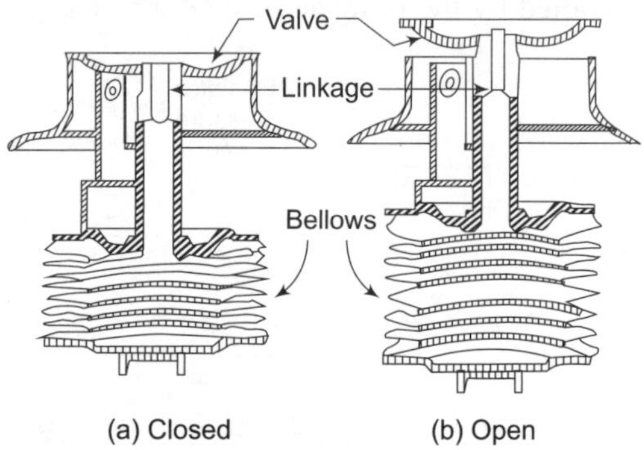


**Fig. 11.26** Details of bellows type thermostat

**(d) Radiator.** *The function of the radiator is to cool the hot water received from the engine. It is a heat exchanger in which water is cooled while passing downward through tubes in the form of fine streams by forced flow of atmospheric air around large metal surface area provided by fins.*

*The name radiator is misnomer because the heat transfer from coolant to the air is by conduction and convection and not by radiation.*

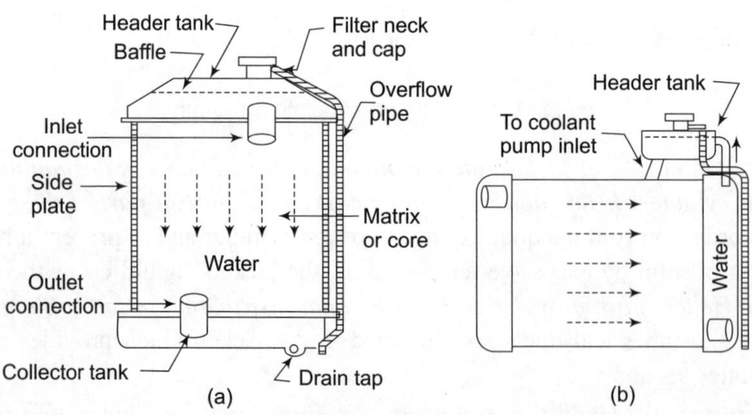


**Fig. 11.27** Radiators (a) Down-flow, (b) Cross-flow

Figure 11.27 shows the two **types of radiators** which are in common use. They are **down flow** [Fig. 11.27 (a)] and **cross-flow** [Fig. 11.27 (b)]. In down-flow type, water flows from top to bottom. *The radiator consists of a header tank with cap, a collector tank at the bottom with a drain tap, matrix or core, side plates and overflow pipe.* In cross-flow type, the hot coolant is supplied to the supply tank, it flows across the radiator and is taken out from the bottom of the collector tank. A separate header tank is usually provided at the upper end of the collector tank to prevent aeration and acts as a coolant reserve. It is the availability of space which mainly decides the choice of the radiator.

Figure 11.28 shows the ***four types of radiator matrix***. They are **(i) honeycomb, mainly (ii) film type, (iii) tube and fin type and (iv) tube and corrugated fin type**. The honeycomb type [Fig. 11.28 (a)] uses a honeycomb block which is made by using a large number of circular tubes upset at each end to a hexagonal shape. Now, it is obsolete due to absence of secondary heat transfer and turbulence.

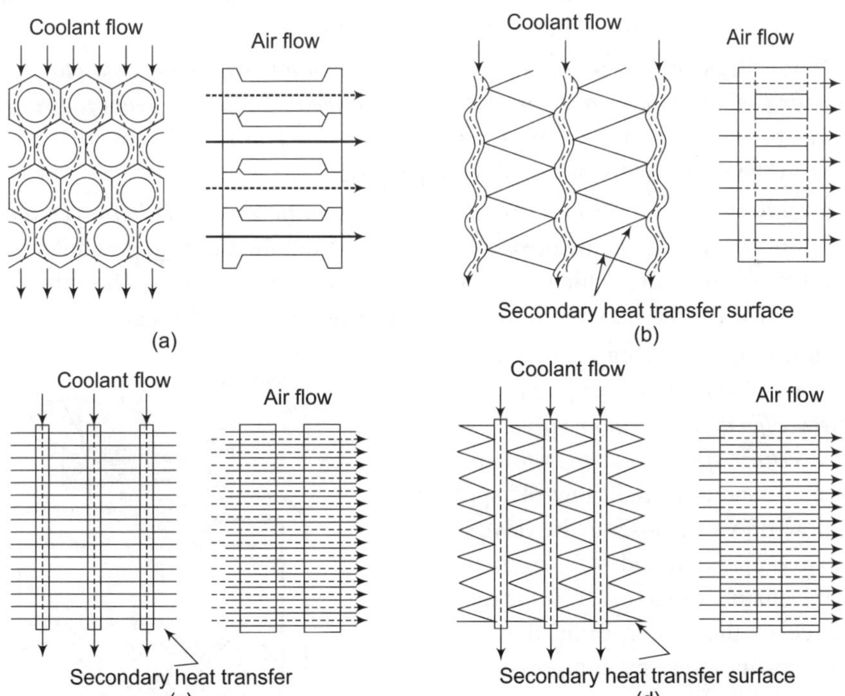


**Fig. 11.28** Radiator matrix (a) Honeycomb, (b) Film type, (c) Tube and fin and (d) Tube and corrugated fins.

**Film type matrix** which is also called the ribbon-cellular matrix is shown in Fig. 11.28(b). It consists of a pair of thin metal ribbon soldered together along their edges so as to form a water way running from header tank to collector tank. Zig-zag cover ribbon formed between two waterways forms the air passage. Copper ribbon acts as air fins. As a result of zig-zag ribbon, the metal surface area is significantly increased and turbulence is also created.

Figure 11.28(c) shows the **tube and fins type matrix**. It consists of a series of long tube extending from top to bottom of the radiator which are surrounded by straight metallic fins forming secondary heat transfer surface. Coolant flow through the tubes and air passes between the fins. This arrangement provides greater structural strength to matrix.

Figure 11.28(d) shows the tube and **corrugated fin type matrix**. This is an improvement over tube and fin type. *The water tubes are made of flattened oval shaped section while zig-zag copper ribbons are used to provide secondary heat transfer areas and turbulence. This type is extensively used.*

*The material chosen for radiators must have good corrosion resistance, good thermal conductivity sufficient strength and easily formable. Copper and fellow brass meet most* of the requirement and are widely used. Where smaller weight is the main consideration, aluminium is also used.

***In many marine diesel and spark ignition engines, the radiator is replaced by a cooler.*** The fresh water leaving the engine is forced through a cooler by a fresh water pump. The sea water absorbs the heat from the fresh water in the cooler.

**(e) Radiator Pressure Cap.** *In order to fill the coolant (water or synthetic coolant), a gate is provided on the top of radiator and there is a need of cap to prevent the surge losses.* This cap will be simple radiator cap. A *radiator pressure cap* is different from an ordinary radiator cap in the sense that it serves double purpose.

(i) *It prevents surge or spill losses and water evaporation and (ii) improves the cooling efficiency of the engine.* The name pressure cap comes from the fact that it maintains pressure above atmospheric pressure which increases the boiling temperature. The pressure cap is generally designed for the maximum inside are pressure of about 1.9 bar.

The radiator pressure cap (Fig. 11.29) consists of a *tightening cap, tightening cap spring, blow-off valve, vacuum valve, valve seats, valve return spring and overflow pipe.* Under normal condition, the blow-off spring holds the blow-off valve closer to the valve seat so the air is compressed and pressure is built up within the chamber marked, *A*. If the inside pressure exceeds the design limit, the blow-off valve comes into play by compressing the pre-calibrated spring, which opens itself and relieves the excess pressure. When the engine is cool, the pressure may drop down slightly than atmospheric pressure. *Care should be taken while removing the pressure cap while the engine is hot as pressure is more and it may hurt the person by spilling hot coolant.*

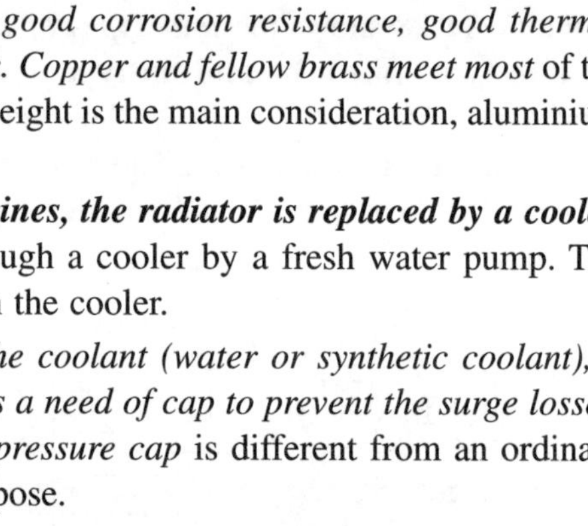


**Fig. 11.29** Radiator pressure cap

**(f) Fan.** Fan at low vehicle speeds with high loads or at idling the cooling, air must be forced through the radiator core in order to provide sufficient cooling. This is done by a belt driven fan. With the increase in the speed of vehicle, a quantity of air is forced through a radiator. *The fan is of axial flow type.* The blades are thin and of aerofoil type. They may be made of metallic sheets but plastic blades are more common due to their height, weight and maintenance free working. Normally, co-polymer moulded polypropylene is used in their construction.

### 11.15.4 Evaporative Liquid Cooling System

*In the evaporative liquid cooling system, the coolant evaporates and this abstracts much more heat from the metal surface equivalent to latent heat of vaporization.*

Figure 11.30 shows evaporative cooling with water cooled condenser. In order to reuse the coolant vapour, it has to be condensed. The vapour so formed goes to the condenser where it gets condensed with the help of secondary water coolant. The condensed vapour flows to the overhead tank by gravity and then to the engine. The coolant is circulated by a pump. The condenser may be cooled by air also.

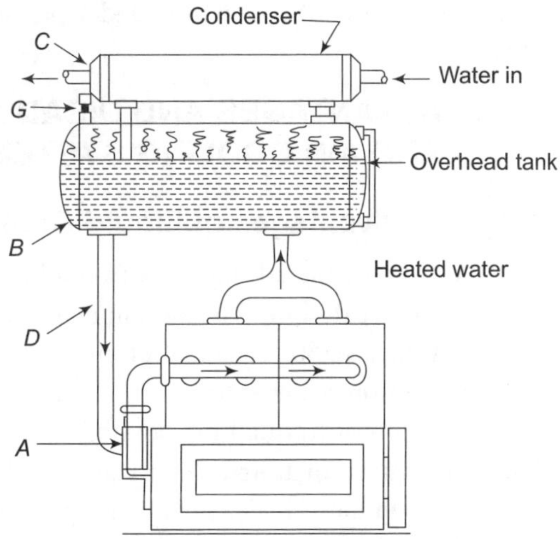


**Fig. 11.30** Evaporative cooling

### 11.15.5 Pressure Cooling System

*In a heat exchanger, the rate of heat transfer depends upon the temperature difference between the two mediums, the area of exposed surface and conductivity of materials.* The pressure cooling system provides the provisions of the larger temperature difference between air and cooling water. *This is achieved by sealing the cooling system from the atmosphere and to allow a certain amount of pressure to occur on the cooling water which increases the boiling point of water.* Generally, the pressure cooling system utilizes pressure equivalent to 2 bar. At 2 bar, the boiling temperature is 120°C.

Figure 11.31 shows the schematic diagram of pressure cooling system. It consists of a cap filled with two valves, one is a safety valve loaded by compression spring and another a vacuum valve. If the coolant is cold, both valves remain closed but as the engine warms up, the coolant temperature rises until it reaches a certain preset value corresponding to the desired pressure. In the event of falling coolant temperature during

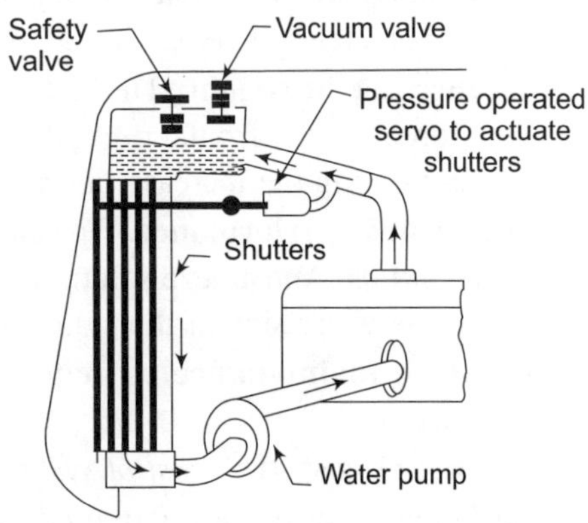


**Fig. 11.31** Pressure cooling system

the engine operation, the valve closes gain until the temperature again rises to the preset temperature corresponding to the desired pressure.

With the switching off, the coolant cools down resulting in formation of vacuum but when the internal pressure falls below the atmospheric pressure, the vacuum valve is opened by the higher outside pressure resulting in the attainment of atmosphere pressure in the system.

In order to prevent the emission of scalding steam or blowing off the cap due to higher internal pressure, a safety device is incorporated in the filter cap so that if an attempt is made to unscrew it while the system is under pressure, the first movement of the cap relieves the pressure.

## 11.16 ADVANTAGES AND DISADVANTAGES OF LIQUID AND AIR COOLING SYSTEM

### 11.16.1 Advantages of Liquid Cooling Systems

The following are the advantages of liquid cooling systems:

(i) **Compact design.** The system enables *compact design* as the heat transfer coefficient of water is 350 times that of air.

(ii) **Minimum overheating.** The chances of *overheating is minimal* as liquid is able to absorb heat from all the possible hot spots.

(iii) **Anywhere installed.** The water cooled engine can be *installed anywhere* (rear or back) in the vehicle. This is not possible in air cooler engine.

(iv) **Higher volumetric efficiency.** The *volumetric efficiency* of liquid cooled engine is *higher* than that of air cooled.

(v) **Lower sfc.** The *specific fuel consumption* of liquid cooled high compression engine is less than that of air cooled.

### 11.16.2 Disadvantages of Liquid Cooled Engines

The following are the disadvantages of liquid cooled engines:

(i) **Dependent.** The system is *dependent on liquid coolant*, if it leaks, it has to be replaced immediately otherwise it may damage the engine.

(ii) **Increase in weight.** The radiator and the pump *increase the weight* of the engine. Moreover, the frontal area of the vehicle gets increased due to presence of radiator resulting in the increase in drag force.

(iii) **Sensitive performance.** The engine's performance becomes *more sensitive to climatic condition*. Anti-freeze solutions are needed during cold weather which may form scale on the water side of the cylinder reducing heat transfer.

(iv) **More maintenance.** It requires *more maintenance* for various parts of cooling system.

(v) **Higher cost.** The *cost of system* is considerably high.

(vi) **Poor warm-up performance.** The warm-up performance is *poor*.

**(vii) Power consumption by pump.** The power consumed by the pump for circulating the water is considerable and hence it reduces the power output of the engine.

### 11.16.3 Advantages of Air Cooling System

(i) The weight of the engine is reduced as much as 20% due to absence of radiator, water jacket and pump. Thus, the weight per kW of the air-cooled engine is less than that of water-cooled.

(ii) Except providing fins, the engine design is much simpler.

(iii) It is less sensitive to climatic condition and there is no need of any freeze solution. The cooling even in hot weather does into deteriorate as greater temperature difference exists.

(iv) The system has a minimum maintenance problem due to the absence of radiator, water pump and thermostat.

(v) The warm-up performance is better resulting in less wear in cylinders.

(vi) The sustained engine performance is achieved due to less chance of carbon deposits and cylinder wall owing to higher mean temperature.

(vii) The installation and control are easy.

(viii) Some degree of damage to cooling fins does not affect the performance.

### 11.16.4 Disadvantages of Air Cooling System

(i) It is only suitable to small and medium output engines.

(ii) Cooling is not uniform due to uneven distribution of air.

(iii) The volumetric efficiency is lower due to higher cylinder temperature.

(iv) It produces more aerodynamic noise.

(v) The specific fuel consumption is slightly higher.

**EXAMPLE 11.1** The brake specific fuel consumption in a 4-cylinder, 100 kW car engine is 0.35 kg/kWh. The calorific value of fuel used in engine is 44 MJ/kg. For efficient engine cooling, calculate the cooling water flow rate. If the water is carried by 40 pipes in the radiator at the mean velocity of 1.5 m/s, calculate the diameter of each pipe. Assume temperature of inlet water as 30°C and outlet temperature of water from engine as 100° C.

***Solution*** Total fuel consumption = $\dot{m}_f \times bsfc \times \text{Power} = 0.35 \times 100 = 35$ kg/h

The heat produced = $Q = 44 \times 10^3 \times 0.35 \times 100 = 1540 \times 10^3$ kJ/h

$$Q = \dot{m}_w\, c_{pw}\, \Delta T_w \quad \text{or} \quad 1540 \times 10^3 = \dot{m}_w \times 4.18 \times (100 - 30)$$

or $$\dot{m}_w = \frac{1540 \times 10^3}{4.18 \times 70} = 5263.15 \text{ kg/h} = 1.461 \text{ kg/s}$$ ***Ans.***

This flow rate is under the closed cycle.

Attain, $\dot{m}_w = v_w . \rho_w = AC . \rho_w = n\ a\ C\ \rho_w$

where $n$ = number of tubes and $a$ = area of one tube

or $\dot{m}_w = 1.461 = 40 \times 1.5 \times 10^3 \times \frac{\pi}{4} d^2$

or $$d = \left[\frac{4 \times 1.461}{\pi \times 40 \times 1.5 \times 10^3}\right]^{1/2} = 0.9869 \times 10^{-2} \text{ m} = 0.9869 \text{ cm}$$ ***Ans.***

**EXAMPLE 11.2** In a 140 kW diesel engine, the temperature of cooling water rises by 50°C while passing through a engine cooling jacket. If the efficiency of diesel engine is 32% and the percentage of energy going to coolant is 28%, calculate the quantity of cooling water required.

***Solution*** $$Q_A = \frac{b_p}{\eta_{th}} = \frac{140}{0.32} = 437.5 \text{ kW.}$$

Heat going to coolant = $\dot{Q}_w = 437.5 \times 0.28 = 122.5$ kW

or $\dot{Q}_w = \dot{m}_w \times c_{pw} \times \Delta T$ or $122.5 = \dot{m}_w \times 4.18 \times (100 - 30)$

or $$\dot{m}_w = \frac{122.5}{418 \times 70} = 0.418 \text{ kg/s}$$ ***Ans.***

## 11.17 SUMMARY OF FORMULAE

(1) $h_c \frac{B}{K} = a \left[\frac{\rho\, cm\, B}{\eta}\right]^b$ where $a = 0.35$ and $b = 0.7$

(2) $\rho_c = \dfrac{A_e\ (ip)^{2.25}}{(A_{fin}\ \Delta T_{fin})^{3.75} \times p_c^2}$

(3) $\rho_p = \dot{m}_c\ gH/\eta_p$.

## PROBLEMS

**Things to Think About**

11.1 For what reasons, spark plugs and valves are to be cooled?

11.2 What is the reason that volumetric efficiency decreases with increase in cylinder and head temperature?

11.3 What is the reason that engine temperature should not go below a certain value?

### Short and Long Theoretical Questions

11.4 Show and comment on the variation of gas temperature in a cycle.

11.5 Why must a reciprocating internal combustion engine be cooled?

11.6 What happens if the engine's temperature is permitted to rise to 200°C, 300°C?

11.7 Will overheating of the cylinder of an engine increase or decrease the tendency to detonate?

11.8 Show the piston, cylinder and valve temperature distribution and mention the hot spot.

11.9 Write the expression for overall heat transfer coefficient.

11.10 Write the energy balance expression.

11.11 Discuss the parameters affecting engine heat transfer.

11.12 Derive the expression for the quantity of water and power required to cool the engine.

11.13 List the types of cooling system.

11.14 Discuss air cooling system.

11.15 List the various anti-freeze solutions used in liquid cooling system.

11.16 With the help of a neat sketch discuss the working principle of forced circulation liquid cooling system.

11.17 Discuss with the help of a sketch the working principle of evapourating cooling system.

11.18 Discuss with the help of a sketch, the working principle of pressure cooling system.

11.19 List the advantages of disadvantages of liquid and air cooling system.

11.20 What is the purpose of baffles in air cooled engines?

11.21 Discuss the working principle of thermostat with sketch.

11.22 Discuss the various types of radiators.

### Software Based Questions

11.23 Develop a software for the design of air and liquid cooling system.

11.24 Develop a software for the design of radiator.

### Design and Open Ended Question

11.25 Prepare a comprehensive report for the design of IC engines cylinder which does not need any cooling by suggesting suitable materials. Give the projection of thermal efficiency of such an engine.

### Multiple-Choice Questions

11.26 The heat flux to the combustion chamber walls during combustion may reach to about

(a) 10 W/m$^2$ (b) 10 kW/m$^2$ (c) 1 MW/m$^2$ (d) 10 MW/m$^2$

11.27 The heat absorbed by coolant in the percentage of energy input is

(a) 5% (b) 15% (c) 25% (d) 45%

11.28 In the liquid cooling system, thermostat valve remains closed till the temperature of coolant reaches to

(a) 65°C (b) 75°C (c) 85°C (d) 95°C

11.29 In the air cooling system, the pitch of the fins should not be less than

(a) 2.5 mm (b) 3.5 mm (c) 4.5 mm (d) 5.5 mm

11.30 In heavy vehicle engines, cooling system used is

(a) open system (b) natural circulation system
(c) evapourating cooling system (d) forced circulation sytem

11.31 For reducing the heat losses to coolant, the cylinder material used may be

(a) cast iron (b) aluminium (c) zirconia (d) none of the above

11.32 The correct sequence in which the water flows through a pressurized water cooling system is

(a) Radiator lower header, pump cylinder jacket, radiator cap, thermostat
(b) Radiator lower header, lower hose, pump, cylinder jacket, thermostat, radiator upper header
(c) Pump, radiator upper header, thermostat, radiator core, radiator cap
(d) Cylinder jacket, pump fan, radiator core, thermostat

## Key to Multiple-Choice Questions

11.26 (d) 11.27 (c) 11.28 (c) 11.29 (a) 11.30 (d) 11.31 (c) 11.32 (b)

CHAPTER

# Intake and Exhaust Systems and Valve Train 12

## OVERVIEW

*The purpose of intake and exhaust systems is to admit the fresh charge in the cylinder during intake stroke and to remove the burnt gases from the cylinder during the exhaust stroke (i.e., at the end of power stroke in each cycle). Inducting the maximum air mass at wide open throttle or full load and retaining that mass within the cylinder is the primary goal of gas exchange process.* The valve mechanism plays an important role for admitting and exhausting the charge.

This chapter deals with fundamentals and description of intake and exhaust systems and valve train. For more details, see Ref. [18 to 29]. *Intake system includes another subsystem such as carburettor or injection systems which have already been discussed.*

## 12.1 INTAKE SYSTEM

In a four-stroke spark ignition engine, the intake system typically consists of an air filter, a carburettor and throttle or fuel injector and throttle or throttle with fuel injectors in each intake port, and intake manifold. During the induction process, pressure losses occur as the mixture/air passes through each element of intake system. There is an additional pressure drop across the intake port and valve. The drop in pressure along the intake systems depends on engine speed, the flow resistance of the elements in the system, the cross-sectional area through which the fresh charge moves, and the charge density.

Figure 12.1 shows the intake system of a four-stroke spark ignition engine. The typical pressure in the intake system is at the start of intake stroke under idling condition.

One of the main elements of intake system is carburettor which has been discussed in the chapter of carburetion. The pressure distribution in the intake system and valve lift has also been discussed in the chapter of carburetion.

### 12.1.1 Design Objectives of Intake Manifolds

The following are the objectives in the design of intake manifolds:

(a) The charge should be *distributed equally among the cylinders* (i.e., mass division) at all speeds and loads.

(b) The charge received by each cylinder *should be well mixed* and has the same physical and chemical characteristics (i.e., quality division).

(c) The volumetric efficiency should be high at full throttle. In order to achieve these objectives, the manifold should have:

(i) Similar passageways should be made to all cylinders (same lengths, diameter and symmetry).

(ii) Turbulence promoters should be provided (high velocities are one method).

(iii) Hot spots may be maintained to reduce large liquid droplets.

(iv) Smooth interior walls should be provided to reduce the thickness of liquid films.

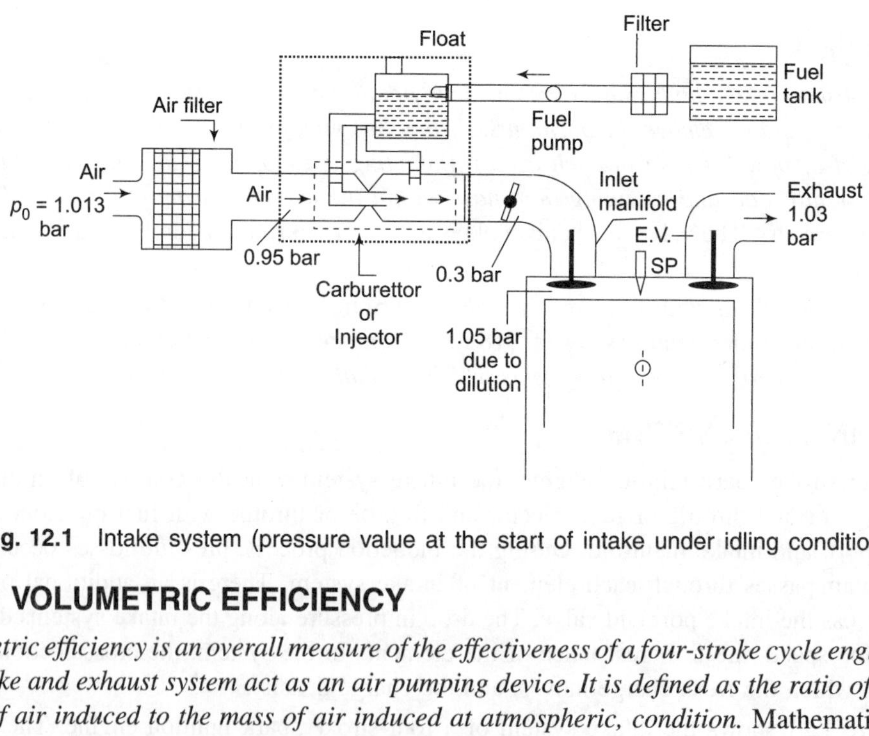


**Fig. 12.1** Intake system (pressure value at the start of intake under idling condition)

## 12.2 VOLUMETRIC EFFICIENCY

*Volumetric efficiency is an overall measure of the effectiveness of a four-stroke cycle engine and its intake and exhaust system act as an air pumping device. It is defined as the ratio of actual mass of air induced to the mass of air induced at atmospheric, condition.* Mathematically,

$$\eta_v = \frac{\dot{m}_a}{\rho_{ao}.V_s.N/2} \qquad \text{...(12.1)}$$

where $\dot{m}_a$ = actual mass of air inducted in cylinder, kg/min

$\rho_{ao}$ = density of air at atmospheric conditions

$V_s$ = stroke volume

$N$ = rpm

*Volumetric efficiency* of IC engine is affected by the *following parameters:*

(a) Fuel type
(b) Fuel-air ratio
(c) Fraction of fuel vapourized in the intake system and fuel heat of vapourization
(d) Mixture temperature
(e) Ratio of exhaust to inlet manifold pressure
(f) Compression ratio
(g) Engine speed
(h) Intake and exhaust manifold and port design
(i) Intake and exhaust valve geometry, size, lift and timings.

It is worth to note that the effects of some of the variables are quasi-static in nature while others are combination of quasi-static and dynamic.

### 12.2.1 Effect of Fuel

*For conventional liquid fuels such as gasoline, the effect of fuel vapour, and therefore, fuel-air ratio, is small. The volumetric efficiency is significantly reduced by the fuel vapour of gaseous fuels in the intake mixture.*

### 12.2.2 Effect of Inlet and Exhaust Pressure Ratio and Compression Ratio

Figure 12.2 shows the effect of inlet and exhaust pressure ratio and compression ratio on volumetric efficiency of an ideal gas. For a fixed compression ratio, the volumetric efficiency decreases with the increase of exhaust to inlet pressure ratio. For any compression ratio, the volumetric efficiency decreases and increases if the ratio of exhaust to inlet pressure increases from 1.0 onward and decreases from 1.0 respectively.

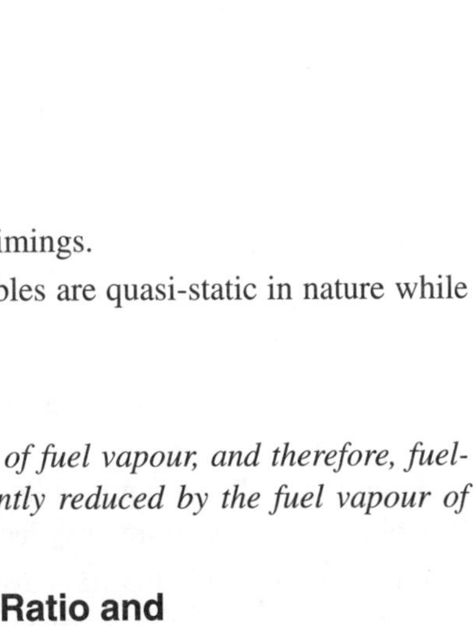


**Fig. 12.2** Effect of exhaust to inlet pressure ratio ($p_e / p_i$) on ideal cycle volumetric efficiency

### 12.2.3 Combined Quasi-static and Dynamic Effects

#### *12.2.3.1 Frictional Losses*

Figure 12.3 shows the pressure loss in the intake system of four-stroke SI engine under steady state condition as a function of engine speed.

This pressure losses are due to friction across the air cleaner, carburettor, throttle and manifold plenum of a

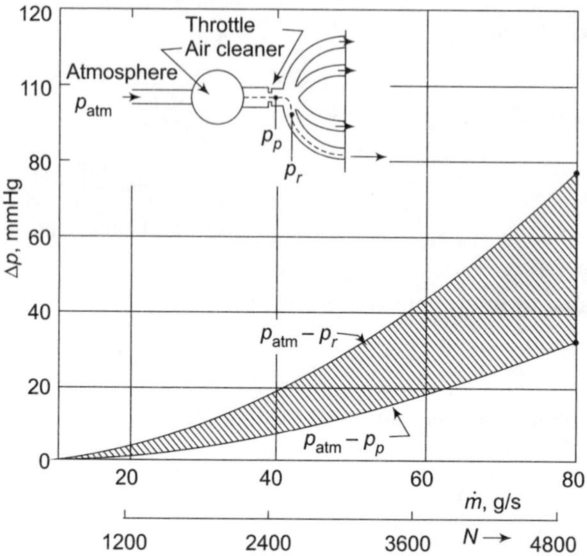


**Fig. 12.3** Pressure losses in the intake system

standard four-cylinder automobile engine intake system. It is obvious that pressure drop increases with speed.

Figure 12.4 shows the time-average manifold gauge pressure as a function of inlet manifold vacuum and speed for a four-cylinder automobile spark-ignition engine. For any speed, the exhaust manifold pressure decreases with increase in inlet manifold pressure. For any inlet manifold pressure, exhaust manifold pressure increases with speed.

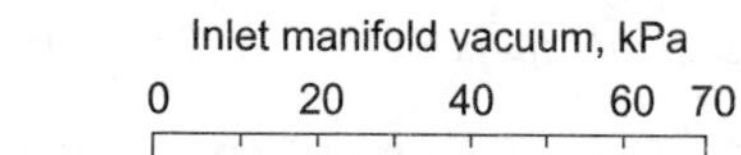


**Fig. 12.4** Exhaust manifold vs. inlet manifold pressure [73]

### *12.2.3.2 Ram Effect*

The pressure in the inlet manifold varies during each cylinder's intake process owing to the piston velocity variation, valve open area variation and the unsteady gas flow effects. Ram effect is caused by inertia and unsteady effect of gas.

### *12.2.3.3 Reverse Flow into the Intake*

Owing to the closing of inlet valve after the start of compression stroke, a reverse flow of fresh charge from the cylinder back into the intake can occur as the cylinder pressure rises due to piston motion towards TDC. *The lowest engine speed causes the largest reverse flow.* It is an inevitable consequence of the inlet valve closing time chosen to take the advantage of the ram effect at high speeds.

*The pressure wave caused by the pulsating flow from each cylinder exhaust may aid or inhibit the gas exchange process in the exhaust system.* These pressure waves propagate at the local sound speed relative to the moving exhaust gas and interact with pipe junctions and ends in the exhaust manifold and pipe. These interactions cause pressure waves to be reflected back towards the engine cylinder and thus air or inhibit the gas exchange process. When the pressure waves aid the gas exchange process by reducing the pressure in the exhaust port towards the end of the exhaust process, the exhaust system is termed as *tuned*.

Figure 12.5 shows instantaneous pressure variations in the intake and exhaust manifold of a four-cylinder SI engine at wide open throttle. The amplitude of the pressure fluctuations increases substantially with increasing engine speed.

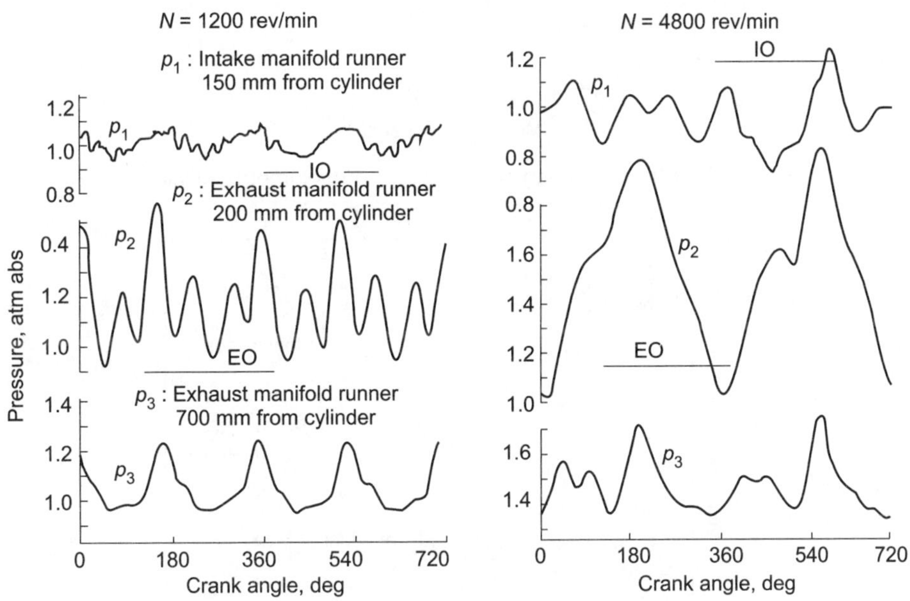


**Fig. 12.5** Instantaneous pressure distribution in the intake and exhaust manifold of 4-stroke SI engine [25, 73]

## 12.2.4 Effect of Speed, Valve Area, Lift and Timing on Volumetric Efficiency

*Volumetric efficiency is affected by the flow in the intake system, which in turn, depends on the velocity of the fresh mixture in the intake manifold, port and valve.* The mixture velocities in the intake system will scale with piston speed. Figure 12.6 shows the variations of volumetric efficiency with mean piston speed of a four-cylinder CI and six-cylinder SI engine. *It is obvious that the volumetric efficiency of SI engine is lower than CI engine due to flow losses in carburettor and throttle, intake manifold heating, the pressure of fuel-vapour and higher residual gas fraction.* The double peak achieved in the diesel curve is due to the intake system turning.

The above results may be explained in a better way by considering the effect of individual phenomena. Figure 12.7 shows the effect of different phenomena on volumetric efficiency. The combined effects result in the curve given in Fig. 12.6.

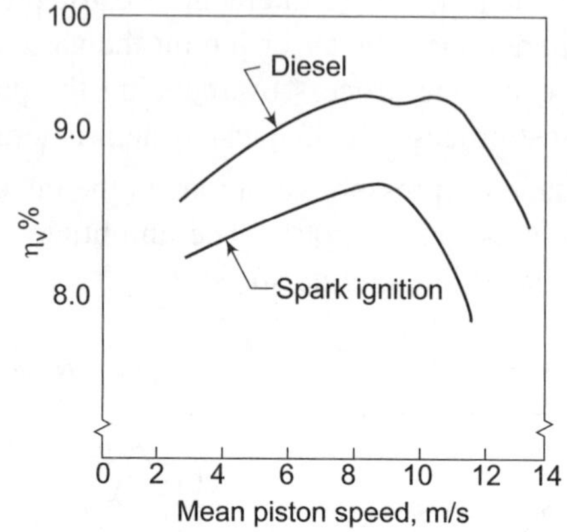


**Fig. 12.6** Effect of mean piston speed on volumetric efficiency

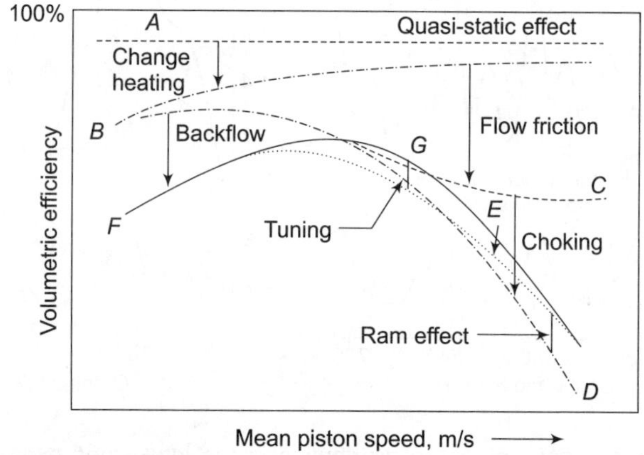


**Fig. 12.7** Effect of volumetric efficiency on different phenomena

Figure 12.8 shows the effect of varying valve timing and valve lift on the volumetric efficiency as a function of speed. It is worth to note that earlier—than normal inlet valve closing reduces back flow losses at low speed and increases volumetric efficiency. But this results in reduced air flow at high speed. Later-than normal inlet valve closing is only advantageous at very high speeds. The valve lift affects the engine breathing considerably. *For any valve lift, there is a specific optimum speeds at which the volumetric efficiency is the maximum.* Low valve lifts significantly reduce the volumetric efficiency over mid speed and high speed operating range while high valve lifts results in higher volumetric efficiency. *Some discussions on volumetric efficiency have been also made in the chapter* **"Measurement, Testing and Performance"**.

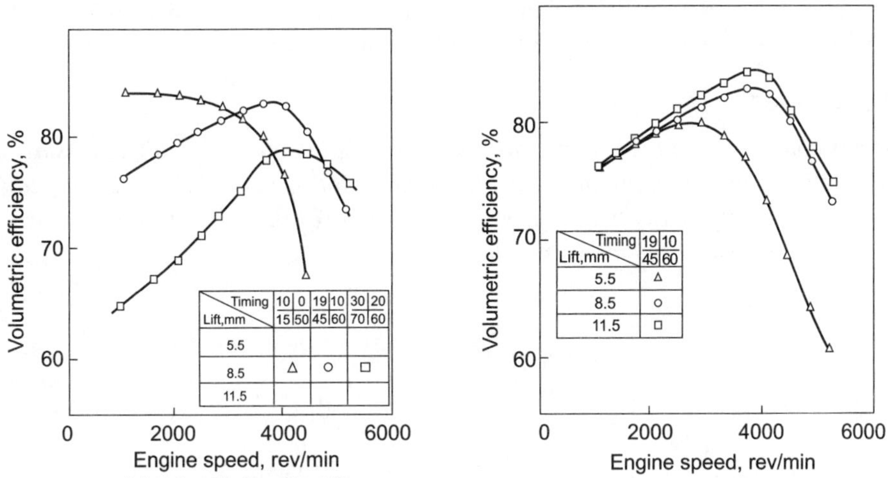


(a) Valve Timing [25, 73]    (b) Valve Lift [25, 73]

**Fig. 12.8** Effect of valve timing and valve lift on volumetric efficiency

## 12.3 EXHAUST SYSTEM

*The exhaust system typically consists of an exhaust manifold, exhaust pipe, often a catalytic converter for emission control and a muffler or silencer.* Figure 12.9 shows the exhaust system. Generally, the muffler is provided at ***two stages** for better results in modern cars.*

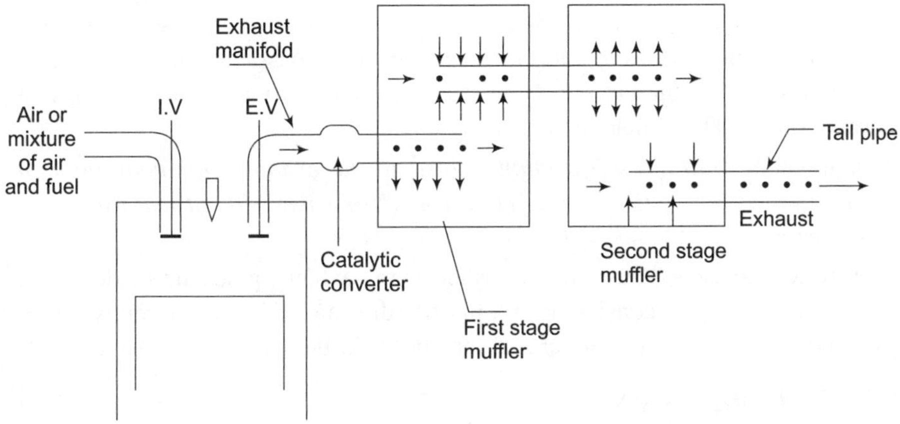


**Fig. 12.9** Exhaust system

*Exhaust system on most modern cars include a* **catalytic converter** *which is discussed in the chapter on* **"Air Pollution"**. Exhaust manifolds have heat control valves. These valves close when the engine is cold and this directs heat to the intake manifold. The heat helps vapourize the in-going gasoline and thus improves cold engine operation.

Some engines have exhaust manifold equipped *with air injection systems*. The system includes an air pump and a series of injection tubes in the exhaust manifold. During operation, the air pump sends a flow of the air into the exhaust manifold opposite the exhaust valve. This extra air helps to burn any HC or CO still left in the exhaust gases.

### 12.3.1 Muffler

*Exhaust gas coming out of the exhaust valve is pulsating in nature with high pressure surges. If it is directly led to the atmospheric, it will create a lot of noise. To reduce the noise and to smooth emission, it is essential to pass the exhaust gases through one stage or two stages muffler where noise will come down drastically.*

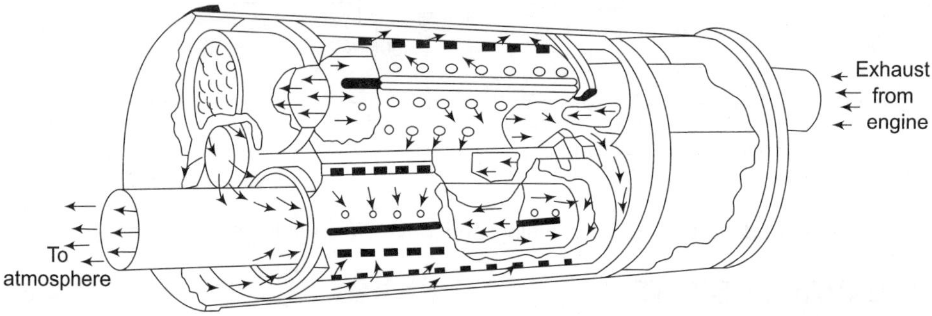


**Fig. 12.10** Muffler

The muffler (Fig. 12.10) consists of a big chamber in which the exhaust pipe terminates and another exhaust (tail) pipe is inserted to carry the exhaust either to second stage or to the atmosphere. It contains a series of holes in the exhaust pipe, passages and resonance chambers to absorbs and damp out the high pressure ranges introduced into the exhaust system as the exhaust valve opens. This quiets the exhaust.

*Some new systems do not use a muffler. Instead the exhaust pipe contains a series of scientifically shaped restrictions that damp the exhaust noise without unduly restricting the flow of exhaust gases.*

In order to reduce the exhaust noises further, many exhaust pipes are made *from* ***a three-ply laminate***. The laminate consisting of a plastic film sandwiched between two metal skins has very good sound-deadening effect and thus holds in the exhaust noises.

## 12.4 THE VALVE TRAIN

A series of parts called the ***"valve train"*** opens and closes the valves at the proper time during the operation. Figure 12.11 shows the main parts of the valve train for only one cylinder. *It consists of camshaft, camshaft sprocket, crankshaft sprocket, timing chain, valve lifter push rod, rocker arm, valve spring retainer, valve guides and suction and discharge valves (generally poppet type).* In a four-cylinder engine, there are at least 8 cams and may go up to 16 cams. To drive the camshaft instead of camshaft sprocket, *crankshaft sprocket and timing chain, pair of gears are used.* The working may be described as follows:

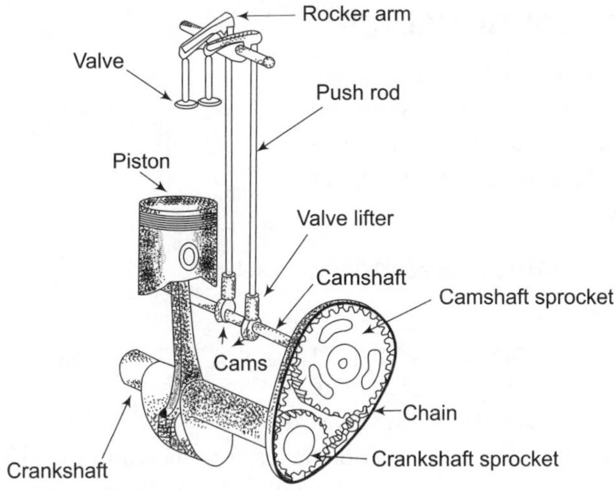


**Fig. 12.11** Valve train parts for a single cylinder engine

The upward movement of the valve lifter causes the push rod to move up and the push rod pushes the rocker arm. The rocker arm is attached to a shaft at its centre. One end of the rocker arm moves up, the other end moves down and pushes down the valve stem which pushes the valve down off its seat and by this way valve opens. At other times, the valve is held closed by a spring the lower end of the valve spring rests against the cylinder head while the upper end of the valve rests against a washer which is locked to the end of the valve stem. When the valve is pushed open the springs gets compressed and stays compressed until the cam lobe moves away from under the valve lifter. As a result, the valve spring pulls the valves back up onto its seat and closes the port. Figure 12.12 shows the action of valve opening.

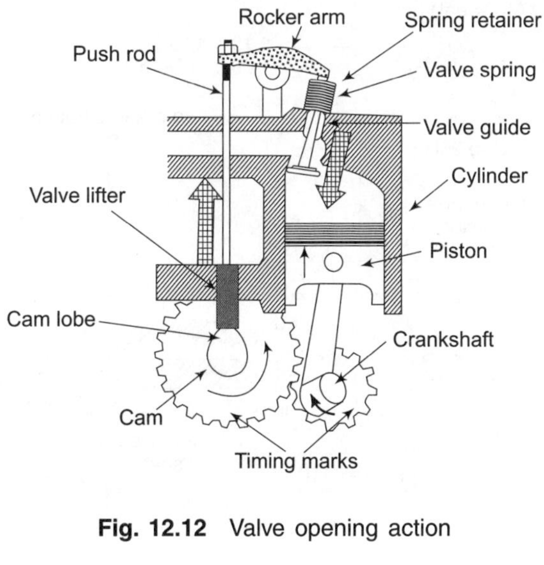


**Fig. 12.12** Valve opening action

## 12.5 FLOW THROUGH VALVES

The most important parts which restrict the flow in the intake and exhaust system are valves or valves and ports together in four-stroke engines. The flow through poppet valves is discussed below.

### 12.5.1 Poppet Valve: Geometry and Timing

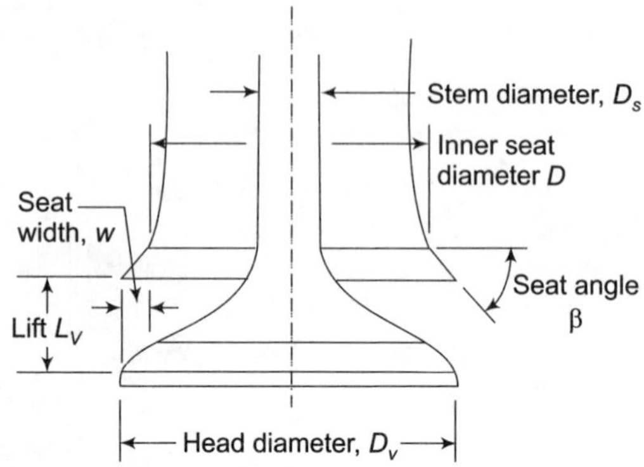


**Fig. 12.13** Poppet valve parameters

The main geometrical parameters of poppet valve are depicted in Fig. 12.13. The proportions of typical inlet and exhaust valves and ports relative to the valve inner seat diameter $D$ is shown in Fig. 12.14. The inlet port is generally circular or so, and the cross-sectional area is no larger than is required to achieve the desired power output.

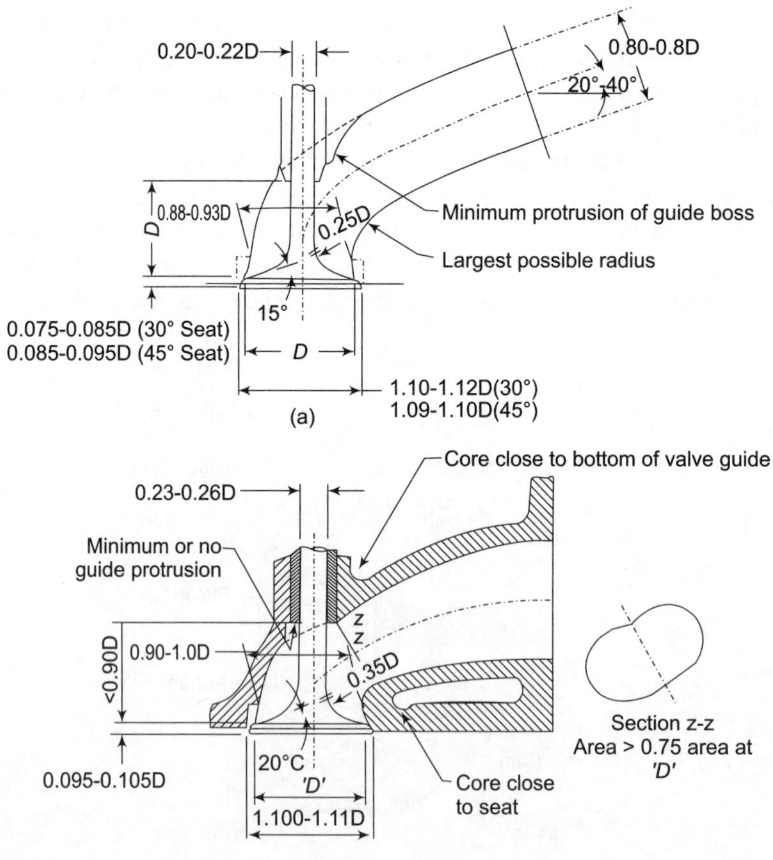


**Fig. 12.14** Proportions of typical (a) Inlet and (b) Exhaust valves [25, 73]

The different requirements of exhaust ports such as better cooling of valve seat and guide and shortest length of exposed valve steam leads to a different design of exhaust port. *Although, a circular cross section is still desirable, a rectangular or oval shape is often essential around the guide boss area.* Table 12.1 shows the typical dimensions of valve in terms of cylinder bore, B [22]. Larger valves sizes or four valves compared to two allow higher maximum air flows for a given cylinder displacement.

Figure 12.15 shows a valve timing, valve-lift profiles and valve open areas for a four-stroke cycle SI engine. There is no universally accepted criterion for defining valve timing points. *SAE defines valve timing events based on reference valve lift position. Another method is based on angular criteria along the lift curve.*

**Table 12.1** Valve Head Diameter in Terms of Cylinder Bore, B

| *Combustion chamber shape* | *Inlet diameter* | *Exhaust diameter* | *Approx. mean piston speed (m/s)* |
|---|---|---|---|
| Wedge or bath tub | 0.43–0.46 B | 0.35–0.37 B | 15 |
| Bowl-in piston | 0.42–0.44 B | 0.34–0.37 B | 14 |
| Hemispherical | 0.48–0.5 B | 0.41–0.43 B | 18 |
| Four-valve port roof | 0.35–0.37 B | 0.28–0.32 B | 20 |

It is worth to discuss the effect of typical valve timing shown in Fig. 12.15 (a). Inlet valve (IV) opens typically 10 to 25° before TDC. *Engine's performance is relatively insensitive to inlet valve opening (IVO).* It should occur sufficiently before TDC so that the cylinder pressure does not dip early in the intake stroke. Inlet valve closing (IVC) usually lies in the range of *40° to 60° after BDC*, to provide more time for cylinder filling under conditions where cylinder pressure is below the intake manifold pressure at BDC. *It is worth to note that IVC is one of the principal factors that determines high-speed volumetric efficiency. Low speed volumetric efficiency is also affected by IVC due to back flows into the intake system.* Exhaust valve opening (EVO) is made to commence 50 to 60° before BDC, well before the end of expansion stroke, so that blowdown can assist in expelling the exhaust gases. *The aim of this setting is to reduce cylinder pressure to close the exhaust manifold pressure as soon as possible after BDC over the full engine speed range. It is worth to note that timing of EVO affects the cycle efficiency since it determines the effective expansion ratio.* Exhaust valve closing (EVC) usually occurring 8 to 12° after TDC ends the exhaust process and determines the duration of valve overlap period. During idling and at light load, in the case of SI engines, EVC regulates the quantity of exhaust gases that flow back into the combustion chamber through the exhaust valve under the influence of intake manifold vacuum.

In the case of high engine speeds and load, it regulates the exhaust of cylinder burnt gases. This discussion leads that EVC timing should occur sufficiently for after TDC so that the cylinder pressure does not rise near the end of the exhaust stroke. Late EVC is favourable to the production of high power at the expense of low-speed torque and idle combustion quality.

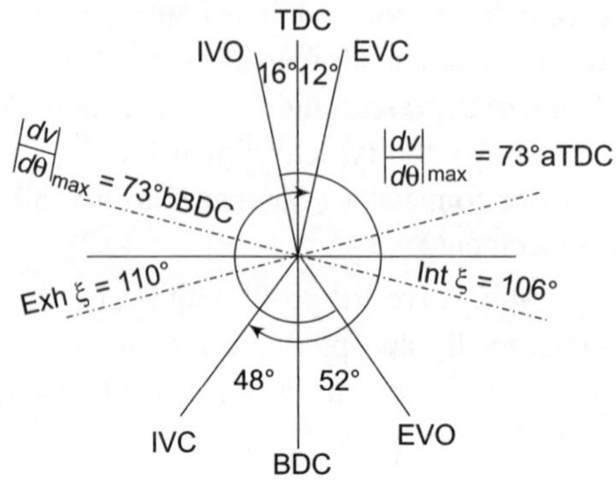

(a) Typical valve timing

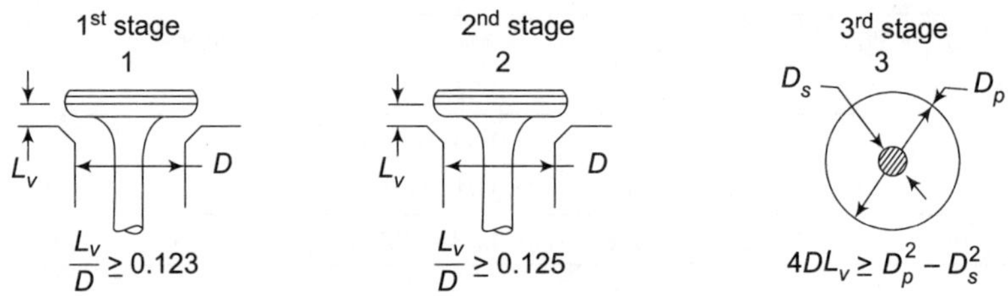

(b) Three stages of valve lift

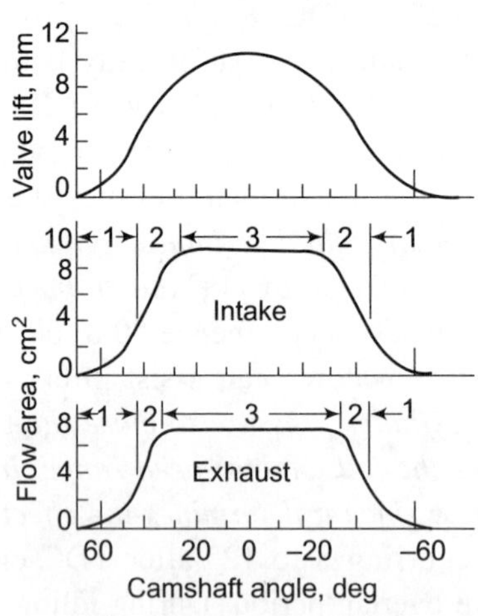

(c) Valve lift and minimum flow area

**Fig. 12.15** (a) Typical valve timing, (b) Three stages of valves lift (c) Valve lift and minimum flow area [25, 27, 73]

The instantaneous valve flow area is governed by valve lift and the geometric details of valve head, seat and steam. Three separate stages may be thought of for the development of flow area as valve lift increases [25]. For low valve lifts (first stage), the minimum flow area corresponds to the frustum of a right circular cone. In this case, the conical face between the valve and the seat, which is perpendicular to the seat defines the flow area. Thus,

For $$\frac{w}{\sin\beta\cos\beta} > L_v > 0$$

the minimum flow area is

$$A_m = \pi L_v \cos\beta\left(D_v - 2w + \frac{L_v}{2}\sin 2\beta\right) \quad ...(12.2)$$

where $\beta$ = valve seat angle

$L_v$ = valve lift

$D_v$ = valve head diameter

$w$ = seat width

For the second stage (i.e., medium valve lifts), the minimum flow area is still the slant surface of a frustum of a right circular cone but with a difference that this surface is no longer perpendicular to the valve seat. In this case, the base angle of the cone increases from (90–β°) toward that of cylinder, 90°. Thus,

For $$\left[\left(\frac{D_p^2 - D_s^2}{4D_m}\right) - w^2\right]^{1/2} + w\tan\beta \geq L_v > \frac{w}{\sin\beta\cos\beta}$$

The minimum flow area, $A_m = \pi D_m [(L_v - w\tan\beta)^2 + w^2]^{1/2}$ ...(12.3)

where $D_p$ = port diameter

$D_s$ = valve steam diameter

$D_m$ = mean seat diameter $(D_v - w)$

For the third stage (i.e., sufficiently large valve lift), the minimum flow area is no longer between the valve head and seat but it is the port flow area minus the sectional area of the valve stem.

Thus, $$L_v > \left[\left(\frac{D_p^2 - D_s^2}{4D_m}\right)^2 - w^2\right]^{1/2} + w\tan\beta$$

the minimum flow area = $A_m = \frac{\pi}{4}(D_p^2 - D_s^2)$ ...(12.4)

Figure 12.15 (c) depicts the intake and exhaust valve opening areas corresponding to a typical valve-lift profile as a function camshaft angle. *It is obvious that the* ***maximum valve lift is normally 12 per cent of cylinder bore.***

### 12.5.2 Flow Rate and Discharge Coefficients

Based on one-dimensional flow analysis, the flow rate through a poppet valve is expressed by

$$\dot{m} = \frac{C_D A_R p_0}{(RT_0)^{1/2}}\left(\frac{p_T}{p_0}\right)^{1/\gamma}\left[\frac{2\gamma}{\gamma - 1}\left\{1 - (p_T/p_0)^{(\gamma-1)/\gamma}\right\}\right]^{1/2} \quad ...(12.5)$$

Under the choked flow condition

$$\dot{m} = \frac{C_D A_R p_0}{(RT_0)^{1/2}} \gamma^{1/2} \left(\frac{2}{\gamma - 1}\right)^{(\gamma+1)/2(\gamma-1)} \quad ...(12.6)$$

where $C_D$ = coefficient of discharge

$p_0, T_0$ = upstream stagnation pressure and temperature

$p_T$ = throat static pressure

$A_R$ = reference area (characteristics of valve design)

$\gamma = c_p/c_v$

The value of $C_D$ and the choice of reference area are linked together, $C_D \cdot A_E$ is called the effective flow area ($A_E$). The most convenient reference area in practice is the valve curtain area ($A_c$)

$$A_c = A_R = \pi D_v L_v \quad ...(12.7)$$

**Inlet Valves.** Discharge coefficient of a typical sharp-cornered inlet poppet valve defined by $C_D = A_E/A_C$ as a function of ratio of valve lift to diameter ($L_v/D_v$) for three segments of flow (*a*, *b*, *c*) is shown in Fig. 12.16. At very low lifts, the flow remains attached to the valve and seat, resulting in high values for the discharge coefficients while at intermediate lifts, the flow separates from the valve head at the inner edge of the valve seat resulting in decreases in discharge coefficients. But with increasing lift, $C_D$ increases. At high lifts, the flow again separates from the inner edge of the valve seat resulting in sharp decrease of $C_D$.

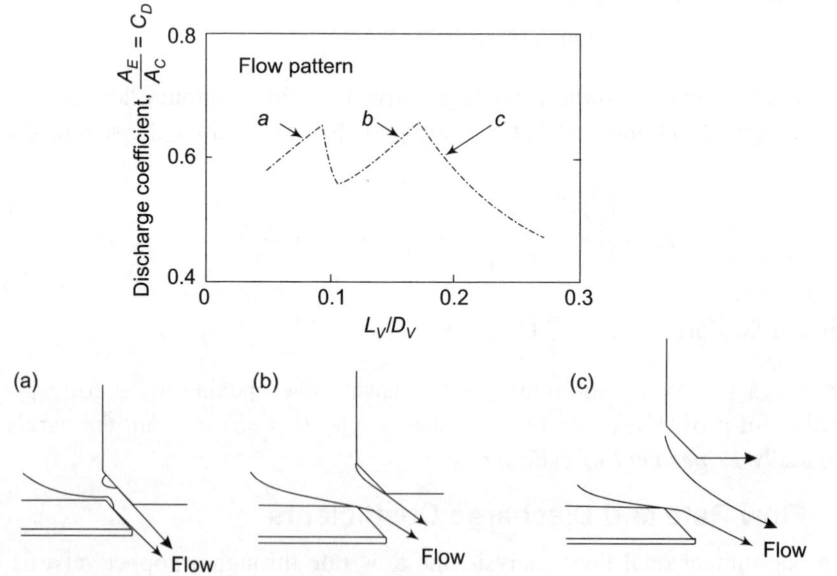


**Fig. 12.16** Discharge coefficient of a typical Inlet valve for different flow regions

The performance of inlet valve assembly not only depends upon the valve lift but is also influenced by many factors such as valve seat width, valve seat angle, rounding of the seat corners, port design, cylinder heat shape, etc.

It is worth to note that at high engine speeds, unless the inlet valve is sufficiently large, the inlet flow during the part of induction process may be choked resulting in lower *volumetric efficiency and this is one of the reasons for making inlet valve larger than discharge valve.*

**Exhaust Valves.** Figure 12.17 shows the flow pattern through exhaust valve at low and high lift. Coefficient of discharge ($C_D$) based on valve's certain area for several different exhaust valves and port combination as a function of $L_v/D_v$ is depicted in Fig. 12.18 [26]. *It is obvious from the result that sharp cornered isolated poppet valve gives highest value of $C_D$. The higher* $L_v/D_v$ ratio results in lower values of $C_D$ *due to breakaway of the flow.*

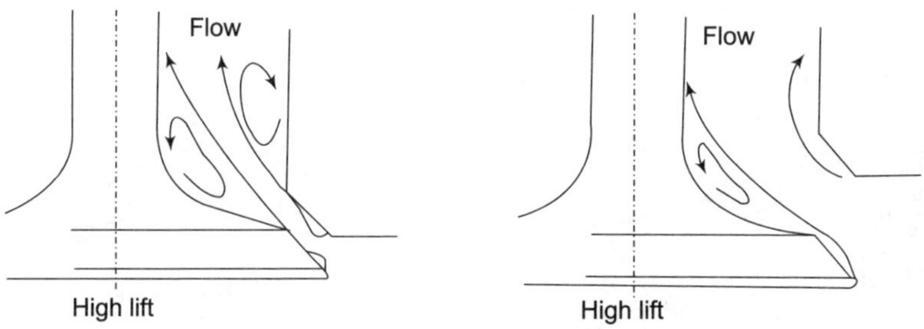


**Fig. 12.17** Flow pattern through exhaust valve

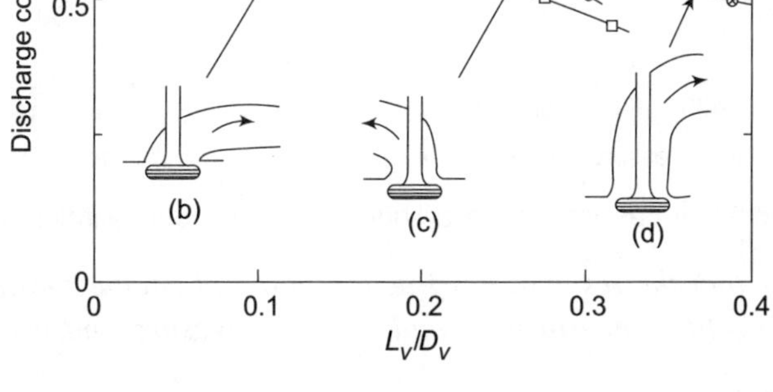


**Fig. 12.18** Discharge coefficient of exhaust valve [25, 26, 73]

## 12.6 RESIDUAL GAS FRACTION IN CYLINDER

*The magnitude of residual gas fraction in the cylinder during compression affects volumetric efficiency engine performance and* $NO_x$ *emission. It is primarily a function of inlet and exhaust pressure, speed, compression ratio, valve timing and exhaust system dynamics.*

Figure 12.19 shows the variation of residual gas fraction for 2- $dm^3$ four-cylinder SI engines as a function of inlet manifold pressure, for a range of speeds, compression ratios and valve overlaps; and also as a function of A/F for a range of volumetric efficiency [26]. *It is obvious from the results that with the increase in inlet manifold pressure, the residual gas fraction decreases. For any value of inlet manifold pressure, the residual gas fraction decreases with increase in speed and pressure ratio but there is a decrease in the angle of overlap.* For any volumetric efficiency that residual gas fraction decreases with the increase in A/F. For any A/F, the residual gas fraction decreases with the increase in volumetric efficiency.

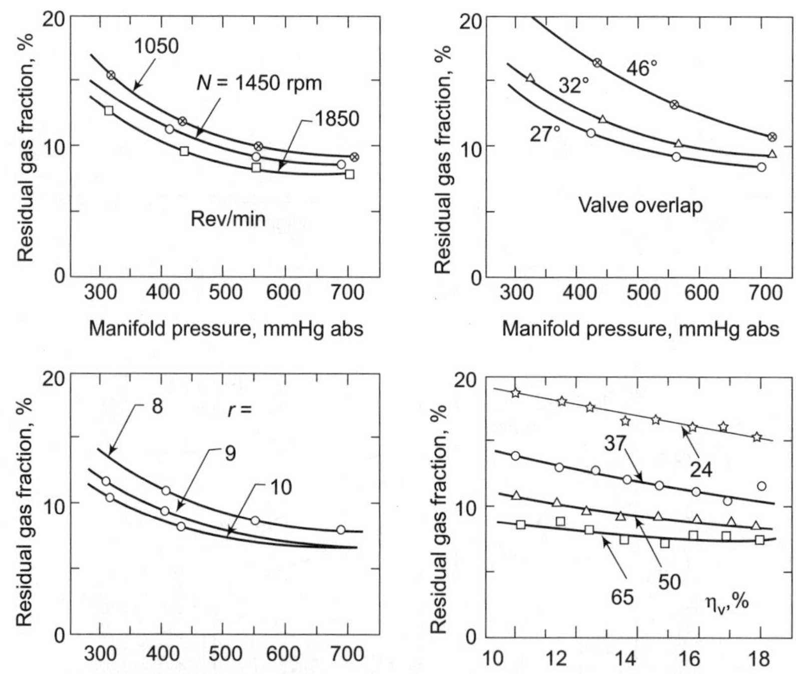


**Fig. 12.19** Residual gas fraction in cylinder for 2-$dm^3$ four-cylinder SI engine [26]

*It is worth to note that the residual gas fraction in diesel engine substantially lower than SI engine due to less difference in inlet and exhaust pressure and much higher cylinder compression ratio.*

## 12.7 EXHAUST GAS FLOW RATE AND TEMPERATURE VARIATION

There is a substantial variation of the mass flow rate and temperature of exhaust gas during the exhaust process. Figure 12.20 shows measured and calculated such as instantaneous variation

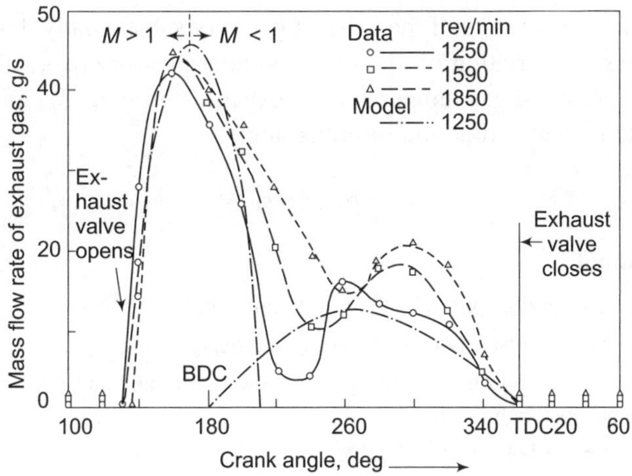


**Fig. 12.20** Instantaneous variation of mass flow rate of exhaust gas through valve as a function of crank angle [25, 73]

of mass flow rate of exhaust gas through valve as a function of crank angle for various speeds. The blow down and displacement phases are clearly evident. There is no substantial variation of rate of exhaust gas during blow-down with speed. The variation of mass flow exhaust gas flow rate along crank angle is sinusoidal but diminishing in nature.

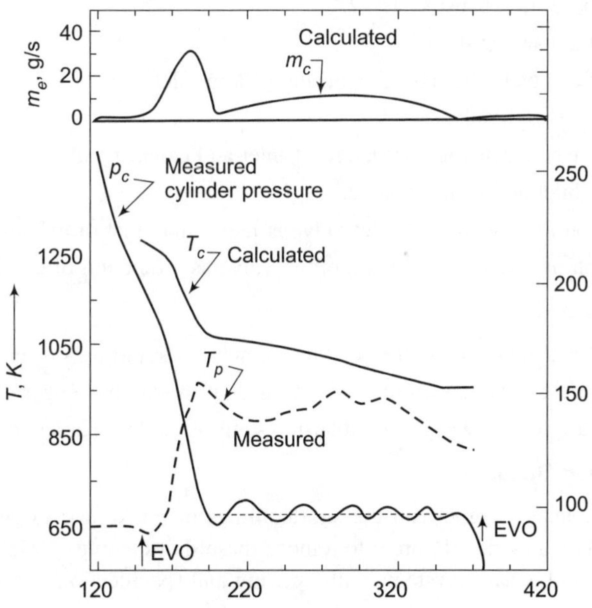


**Fig. 12.21** Exhaust pressure and temperature vs. crank angle [25, 73]

The pressure and temperature of the exhaust gas vary substantially during exhaust process. Figure 12.21 shows the variation of measured cylinder pressure ($p_c$), calculated cylinder gas temperature ($T_c$), measured gas temperature at exhaust port exit ($T_p$) and calculated exhaust gas mass flow rate ($\dot{m}_c$) as a function of crank angle.

## PROBLEMS

### Things to Think About

12.1 What is the main philosophy of intake system design?

12.2 What is the main philosophy of exhaust system design?

12.3 Instead of one inlet and one exhaust valves, now two intake and two exhaust valves are used. What is the purpose behind this?

12.4 What should be the best shape of valve?

12.5 What is the normal valve lift in terms of cylinder bore?

### Short and Long Theoretical Questions

12.6 What is the purpose of intake system?

12.7 Give a sketch of intake system.

12.8 List the design objective of intake manifold.

12.9 Discuss the instantaneous pressure distribution in the intake and exhaust manifolds.

12.10 Discuss the effect of speed, valve area, valve lift and timing on volumetric efficiency.

12.11 What is the purpose of exhaust system?

12.12 Give a sketch of exhaust system

12.13 With the help of a sketch, discuss the working of a muffler.

12.14 Discuss the flow through valves.

12.15 Discuss the nature of discharge coefficient of inlet and exhaust valve.

12.16 Give a layout of inlet and exhaust valve.

12.17 What is the reason that the size of inlet valve is larger than that of exhaust valve?

12.18 Give the variation of exhaust gas flow through valve as a function of crank angle.

### Software Based Questions

12.19 With the help of a sketch discuss the working principle of starting system of an IC engine.

12.20 Develop a software for the design of intake system of motors passenger car engine.

12.21 Develop a software for the design of exhaust system of modern passenger cars engine.

### Design and Open Ended Question

12.22 Prepare a comprehensive report on the shortcomings of intake and exhaust systems used in modern passenger car engine. In order to remove these shortcomings, develop a comprehensive design of intake and exhaust systems with drawing and specifications.

### Multiple-Choice Questions

12.23 The pressure existing in the intake system lies in the range of

(a) 0.4 to 0.5 bar (b) 0.5 to 0.6 bar (c) 0.6 to 0.7 bar (d) 0.8 to 0.9 bar

12.24 The fuel injected in the gasoline MPFI system is at
   (a) in the middle of intake manifold   (b) slightly before the intake valve
   (c) just after the intake valve   (d) in the cylinder

12.25 The magnitude of residual gas fraction in the cylinder during compression process is primarily a function of
   (a) inlet and exhaust pressure
   (b) speed and compression
   (c) valve timing and exhaust system dynamics
   (d) all of these

12.26 The valve head diameter in terms of cylinder bore for bowl-in-piston combustion chamber is
   (a) 0.32–0.34 B suction and 0.24–0.27 B exhaust
   (b) 0.42–0.44 B suction and 0.34–0.37 B exhaust
   (c) 0.24–0.27 B suction and 0.32–0.34 B suction
   (d) 0.34–0.37 B suction and 042–0.44 B exhaust

12.27 The exhaust gas coming out of exhaust valve
   (a) in pulsating in nature   (b) has high pressure surges
   (c) has very high velocity   (d) is having all these

## Key to Multiple-Choice Questions

12.23 (d) 12.24 (b) 12.25 (d) 12.26 (b) 12.27 (d)

CHAPTER

# Supercharging and Turbocharging

# 13

## OVERVIEW

The maximum power developed by a given engine is limited by the amount of fuel that can be burnt efficiently inside the engine cylinder, which in turn, is limited by the amount of air that is introduced into the each cylinder each cycle. *The amount of air inducted per unit cycle can be increased by increasing the engine speed or by increasing the density of air at intake.* If the inducted air is compressed to a higher density than ambient, prior to entry into the cylinder, the maximum power that can be developed by a given engine will be increased as power, torque and mean effective pressure are proportional to inlet air density. *This is the primary purpose of supercharging.*

This chapters deals with the objectives, configurations, types of supercharging and turbocharging.

## 13.1 OBJECTIVES OF SUPERCHARGING

*The term supercharging refers to increasing the air (or mixture) density by increasing its prior pressure to entering the engine cylinder so that amount of air inducted per cycle increases, which in turn, increases the power output of the engine.* The objectives of supercharging may be one or more of the following:

(i) *To increase the power output for a given weight and bulk of the engine.* This is important for aircraft, marine and automobiles engines (passengers and racing cars) where weight and space are important considerations.

(ii) *To compensate for the loss of power due to altitude* especially in aircraft engines where loss of power is about 1% at every 100 m attitude as air density decreases with increases in altitude.

(iii) *To obtain more power from existing engines* specially rail and standby power plants.

(iv) *To obtain better gas-exchange process.*

## 13.2 METHODS OF SUPERCHARGING

Three basic methods are used in supercharging. They are as follows:

**(i)** ***Mechanical Supercharging.*** In this supercharging, a separate pump or blower or compressor, usually driven by power taken from the engine, provides the compressed

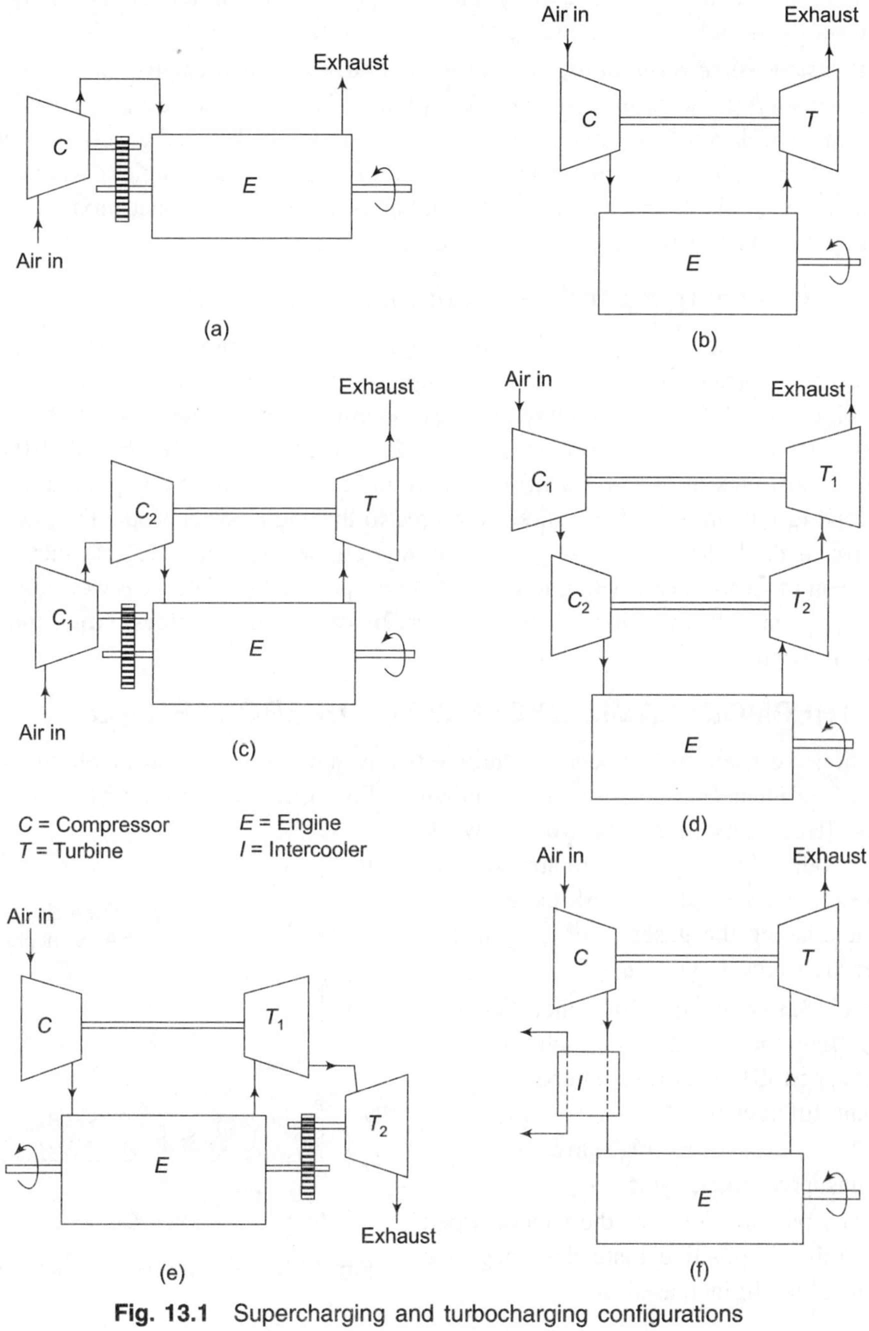


**Fig. 13.1** Supercharging and turbocharging configurations

air. *Centrifugal compressor is mostly used as supercharger. However, in some cases roots blower and gear compressor may also be used.*

**(ii)** ***Turbocharging.*** *Turbocharging employs a turbocharger—a compressor and a turbine on a single shaft—used to boost the inlet air (or mixture) density.* The exhaust gas from the engine cylinder expands through turbocharger turbine and the power developed from the turbine drives the turbocharger compressor.

**(iii)** ***Pressure Wave Supercharging.*** This method uses waves action in the intake and exhaust system to compress the intake mixture. The use of intake and exhaust manifold turning to increase volumetric efficiency is one method of increasing air density. A good example of pressure wave supercharging device is complex, which uses the pressure available in the exhaust gas stream to compress the inlet mixture stream by direct contact of the fluids in narrow flow channels.

### 13.2.1 Supercharging and Turbocharging Configuration

Figure 13.1 shows the various typical configurations of supercharging and turbocharging. The widely accepted arrangement is mechanical supercharger (Fig. 13.1a) or turbocharger (Fig. 13.1b). Figure 13.1(c) shows the combination of engine driven compressor and turbocharger which is generally used **in large marine engines**. Two-stage turbocharger (Fig. 13.1d) provides very high boost pressure **(4 to 7 atm)** to obtain higher *bmep*. Turbo compounding, i.e., use of a second turbine in the exhaust directly geared to the engine drive shaft (Fig. 13.1e) is an alterantive method of increasing engine power and efficiency. Figure 13.1(f) shows another arrangement of turbocharger with intercooler to boost power. Tata Safari car working on diesel fuel is equipped with this type of turbocharger. **In two-stroke engines, supercharging is used for scavenging**.

## 13.3 THERMODYNAMIC CYCLE WITH SUPERCHARGING

As mentioned earlier, the turbocharger turbine is driven by the energy available in the engine exhaust. The ideal energy available is shown in Fig. 13.2 for the case of four-stroke SI engines. It consists of the blowdown work transfer produced by expanding to atmospheric pressure (area *abc*) and the work done by the piston displacing the gases remaining in the cylinder after blowdown (*c d e f*).

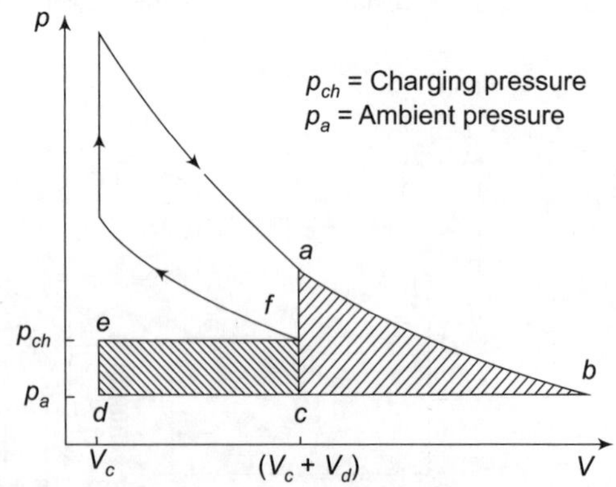


**Fig. 13.2** Ideal available exhaust energy

Figure 13.3 shows the difference between the *p-V* diagrams of an unsupercharged and supercharged SI engines. There are two important differences.

(a) An increase in pressure over the unsupercharged cycle.

(b) The pumping loop of the supercharged engine is positive instead of negative resulting in increased *ip*.

*The gain in the output of the supercharged engine is mainly due to the increase of amount of air inducted for the same swept volume.*

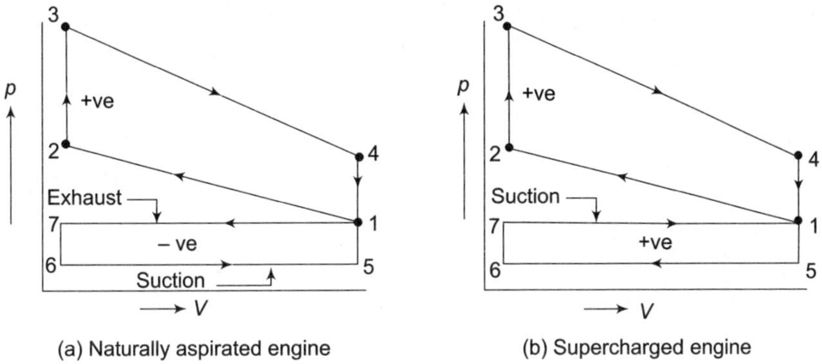


**Fig. 13.3** *p-V* Diagram of unsupercharged and supercharged engine

## 13.4 THERMODYNAMIC RELATIONSHIPS

### 13.4.1 Steady Flow Energy Equation

The steady flow energy rate equation applied to a control volume the turbo-machinery (compressor and turbines) component is given by

$$\dot{Q} - \dot{W} = \dot{m}\left[\left(h + \frac{C^2}{2} + gz\right)_{\text{out}} - \left(h + \frac{C^2}{2} + gz\right)_{\text{in}}\right] \qquad ...(13.1)$$

Neglecting $\dot{Q}$ and changes in KE and PE, the above equation reduces to

$$-\dot{W} = \dot{m}\,(h_{o,\text{out}} - h_{o,\text{in}}) \qquad ...(13.2)$$

If $c_p$ remains constant,

$$-\dot{W} = \dot{m}c_p\,(T_{o,\text{out}} - T_{o,\text{in}}) \qquad ...(13.3)$$

### 13.4.2 Component Efficiencies

The component efficiencies are used to relate actual work transfer rate to that due to isentropic process. The *h-s* diagram for compressor is shown in Fig. 13.4.

For a compressor, the compressor isentropic efficiency is driven by

$$\eta_C = \frac{\text{Isentropic power requirement}}{\text{Actual power requirement}} \qquad ...(13.4)$$

Based on total to total values (01, 02, 03), the compressor isentropic efficiency is given below in which the *suffix s stands for isentropic process (s = constant).*

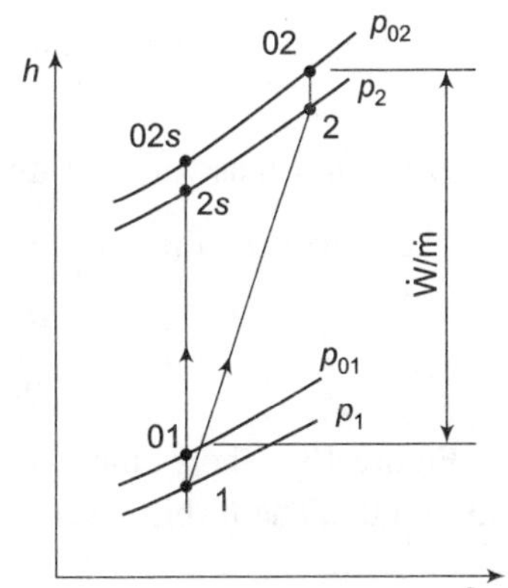


**Fig. 13.4** *h-s* diagram for compressor

$$\eta_{CTT} = \frac{h_{02s} - h_{01}}{h_{02} - h_{01}} \quad ...(13.5)$$

If $c_p$ remains constant, $\eta_{CTT}$ is expressed as

$$\eta_{CTT} = \frac{T_{02s} - T_{01}}{T_{02} - T_{01}} \quad ...(13.6)$$

For isentropic process 01 to 02*s*, we have

$$T_{02s} = T_{01}\left(\frac{p_{02}}{p_{01}}\right)^{(\gamma-1)/\gamma}$$

Thus, Eq. (13.6) is expressed as

$$\eta_{CTT} = \frac{\left(\frac{p_{02}}{p_{01}}\right)^{(\gamma-1)/\gamma} - 1}{\left(\frac{T_{02}}{T_{01}}\right) - 1} \quad ...(13.7)$$

Since the kinetic energy or the gas leaving the compressor is not usually recovered, so a concept of total to static isentropic efficiency may be used to relate actual work. Thus, the total to static isentropic efficiency of compressor is given by

$$\eta_{CTS} = \frac{T_{2s} - T_{01}}{T_{02} - T_{01}} \frac{\left(\frac{p_2}{p_{01}}\right)^{(\gamma-1)/\gamma} - 1}{\left(\frac{T_{02}}{T_{01}}\right) - 1} \quad ...(13.8)$$

The work transfer is now expressed using Eq. (13.6)

$$-\dot{W}_C = \frac{\dot{m}_i . c_{pi} . T_{01}}{\eta_{CTT}}\left[\left(\frac{p_{02}}{p_{01}}\right)^{(\gamma-1)/\gamma} - 1\right] \quad ...(13.9)$$

where $\dot{m}_i$ = mass of inlet mixture

The power required to drive the compressor is given by

$$-\dot{W}_{C,D} = \frac{\dot{W}_c}{\eta_m}$$

Figure 13.5 shows the *h-s* diagram for turbine. Let the inlet and outlet total conditions be 03 and 04. The turbine isentropic efficiency is defined as

$$\eta_T = \frac{\text{Actual power output}}{\text{Isentropic power output}} \quad ...(13.10)$$

Based on total to total values, $\eta_{TTT}$ is expressed as

$$\eta_{TTT} = \frac{h_{03} - h_{04}}{h_{03} - h_{04s}} \quad ...(13.11)$$

$$= \frac{T_{03} - T_{04}}{T_{03} - T_{04s}} = \frac{1 - \dfrac{T_{04}}{T_{03}}}{1 - \left(\dfrac{p_{04}}{p_{03}}\right)^{(\gamma-1)/\gamma}} \quad ...(13.12)$$

Based on total to static condition, the turbine efficiency is expressed as

$$\eta_{TTS} = \frac{h_{03} - h_{04}}{h_{03} - h_{4s}} = \frac{1 - \dfrac{T_{04}}{T_{03}}}{\left(1 - \dfrac{p_4}{p_{03}}\right)^{(\gamma-1)/\gamma}} \quad ...(13.13)$$

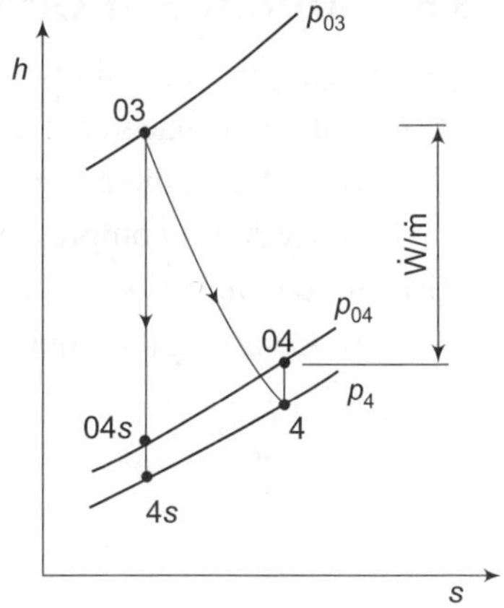


**Fig 13.5** *h-s* diagram for a turbine

The power output from turbine is given by

$$\dot{W}_T = \dot{m}_e (h_{03} - h_{04})$$

$$= \dot{m}_e \, c_{pe} \, \eta_{TTT} \, T_{03} \left[1 - \left(\frac{p_{04}}{p_{03}}\right)^{(\gamma-1)/\gamma}\right] \quad ...(13.14)$$

where $\dot{m}_e$ = mass flow rate of exhaust gas

In the case of turbocharger, the turbine is coupled to the compressor, so

$$-\dot{W}_c = \eta_m W_T \quad ...(13.15)$$

### 13.4.3 Dimensionless Parameters

The operating characteristics of compressors and turbine are expressed by dimensionless parameters. Here

$$\dot{m}, \eta, \Delta T_0 = f(p_{0,\text{in}}, p_{0,\text{out}}, T_{0,\text{in}} N, D, R, \mu, \gamma) \quad ...(13.16)$$

The dimensionless analysis results in

$$\frac{\dot{m}\sqrt{R\,T_{0,\text{in}}}}{p_{0,\text{in}} D^2}, \eta, \frac{\Delta T_0}{T_{0,\text{in}}} = f\left(\frac{ND}{\sqrt{R\,T_{0,\text{in}}}}, \frac{p_{0,\text{out}}}{p_{0,\text{in}}}, \frac{\dot{m}}{\mu D}, \gamma\right) \quad ...(13.17)$$

or

$$\frac{\dot{m}\sqrt{T_{0,\text{in}}}}{p_{0,\text{in}}}, \eta, \frac{\Delta T_0}{T_{0,\text{in}}} = f\left(N/\sqrt{T_{0,\text{in}}}, \frac{p_{0,\text{out}}}{p_{0,\text{in}}}, \frac{\dot{m}}{\mu D}, \gamma\right) \quad ...(13.18)$$

The above non-dimensional parameters are used to plot the characteristics of compressor and turbines.

## 13.5 SUPERCHARGING COMPRESSORS

Supercharging devices can be classified into two main categories. These are:

(a) Positive displacement compressor

- (i) sliding vane compressors
- (ii) Roots blowers
- (iii) Lysholm compressor
- (iv) Serew compressor

(b) Turbocompressor

- (i) Centrifugal compressor

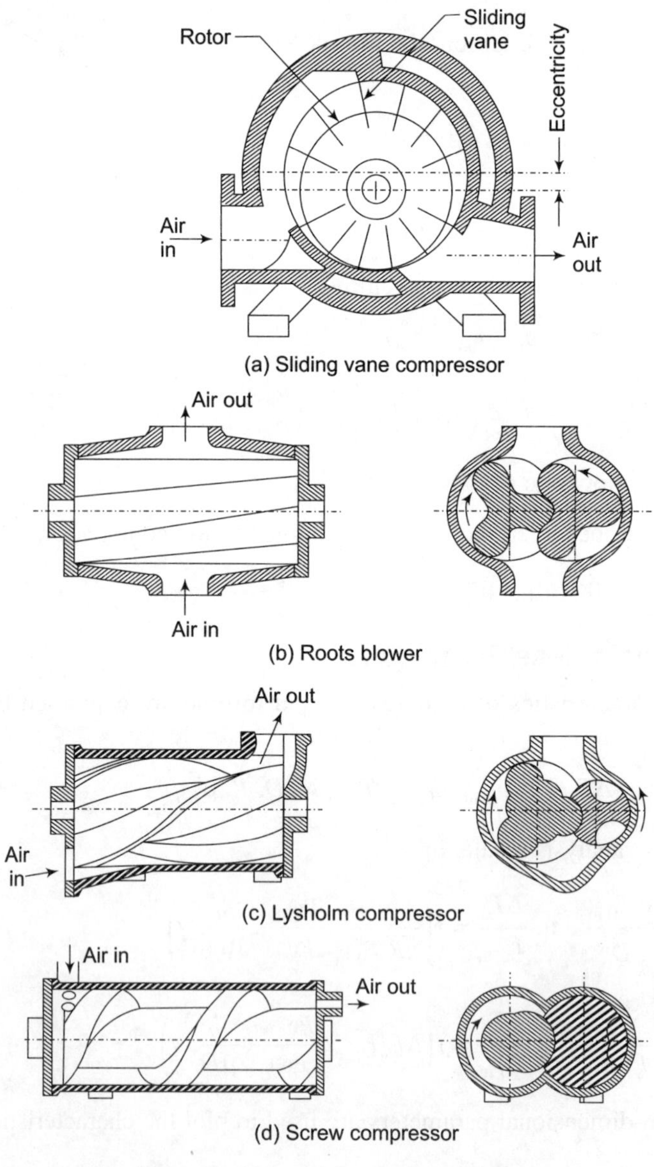


(a) Sliding vane compressor

(b) Roots blower

(c) Lysholm compressor

(d) Screw compressor

**Fig. 13.6** Positive displacement compressors

Figure 13.6 shows the four types positive displacement compressors. In positive displacement compressors, compression is achieved due to push or push and pull of the working fluid.

In turbocompressors, the compression is achieved by the rotation of rotor by external agency which is partly converted into pressure energy due to centrifugal and diffusion action in rotor and rest in diffuser by diffusion action.

***A detailed discussion on compressor is given in chapter 21, in Applied Thermodynamic Book of the author.***

## 13.5.1 Sliding Vane Compressor

*Sliding vane compressor (Fig. 13.6a) consists of deep slots cut into the rotor to accommodate thin vanes varying from 20 to 30 which are free to move radially.* The rotor is mounted eccentrically in the housing. With the rotation of rotor, the centrifugal forces acting on the vanes forces them outward against the housing, thereby dividing the crescent shaped space into several compartments.

Hence, in the initial stage, the space between the rotor, the cylinder and the two adjacent vanes is enlarged and the air is drawn into this space through the suction port. When the rotor reaches the point just opposite to the eccentricity of the rotor, the suction process ends, and further rotation causes the enclosed space to decrease, thus, the compression of air occurs due to decrease in volume and is finally discharged through the outlet port. The compressor is so designed that half of the total pressure rise takes place through internal reversible compression before the entrapped air is released to the delivery port. The air is further compressed in a reversible manner due to momentarily back flow of high pressure from the receiver. The actual flow rate and pressure rise at constant speed will be reduced by leakage. The volumetric efficiency can vary between 0.6 and 0.9 depending on the size of the machine, the quality of designs, method of lubrication and cooling employed.

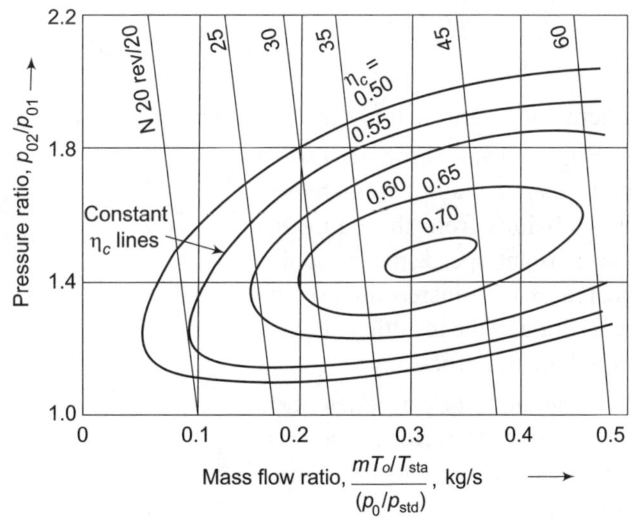


**Fig. 13.7** Sliding vanes compressor performance

The displaced volume $V_D$ is given by

$$V_D = \pi \, \varepsilon \, l \, (D + \varepsilon) \qquad \text{...(13.19)}$$

where $D$ is diameter of rotor, $\varepsilon$ is eccentricity and $l$ is axial length of compressor.

The mass flow rate of is given by

$$\dot{m}\,\frac{\sqrt{T_0/T_{std}}}{p_0/p_{std}} = K \times \rho_i\, \eta_v\, N\varepsilon\, l\,(D + \varepsilon) \qquad \text{...(13.20)}$$

where $K$ is constant, $\eta_v$ is volumetric efficiency.

Figure 13.7 shows the performance characteristics of a typical sliding vane compressor. It is worth to note that the mass flow rate at constant speed depends on pressure ratio. This compressor results in low isentropic efficiency.

### 13.5.2 Roots Blower

Figure 13.6(b) shows schematic diagram of a roots blower. It is another type of positive displacement supercharger. *Roots blowers consists of two rotors connected by gears*. With the rotation of rotors, air trapped in the recesses between the rotor lobes and the housing is carried towards the delivery port without significant change in volume. Since the suction side is closed under this situation, the trapped air is suddenly compressed by the back flow from the higher-pressure delivery line as the recesses are open to the delivery line. This intermittent delivery results in production of non-uniform torque on the rotor and pressure pulses in the delivery line. Such compressors are suitable for low pressure ratio, (up to 1.3).

Parameters such as running clearance, rotor length, rotational speed and pressure ratio affect the volumetric efficiency. The mass flow rate is expressed by

$$\frac{\dot{m}\,\sqrt{T_0/T_{\text{std}}}}{p_0/p_{\text{std}}} = \text{constant} \times \rho_i\, \eta_v\, N\, R^2\, l \qquad \text{...(13.21)}$$

where $R$ = ratio radius and $l$ = blower length

Figure 13.8 shows a typical performance map of roots blower. It is obvious from the figure that the constant speed, the mass flow decreases with increasing pressure ratio through the resulting decrease in volumetric efficiency.

Surge and choking phenomena do not limit its performance range which is supposed to be its main advantage. Its advantages lie in its high noise level, poor efficiency and large size.

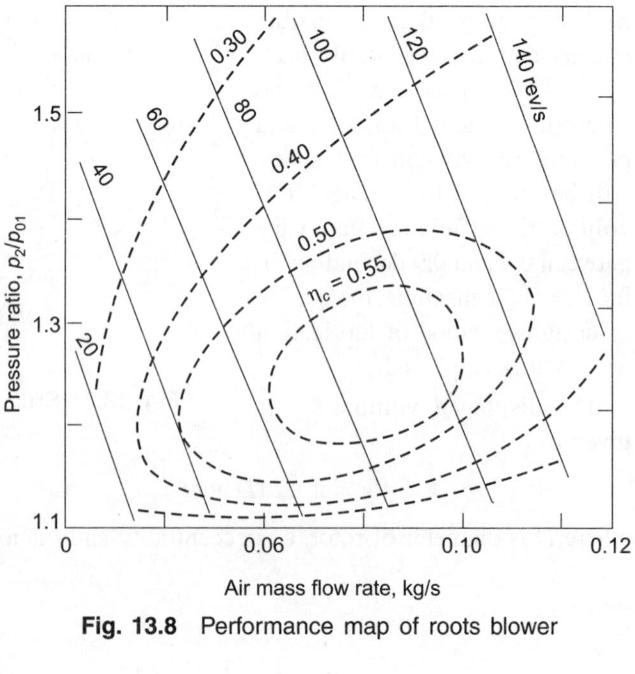


**Fig. 13.8** Performance map of roots blower

### 13.5.3 Lysholm and Screw Compressors

In this compressor (Fig. 13.6 c, d), the lobes are helical having a large pitch that intermesh. As opposed to roots blower, the air is gradually compressed. One rotor (male) has three or four lobes, while the other rotor (female) has four, five or six recesses. The entrapped air is progressively compressed, as it moves through the narrowing passage ways formed by the lobes. These compressors must be precision machined to achieve close tolerance between rotating and stationary elements for satisfactory operation. *They run at very high speeds ranging from 3000 rpm to 30,000 rpm and are generally internally cooled. High values of volumetric and isentropic efficiency are the main advantages of this compressor.*

### 13.5.4 Centrifugal Compressor

A centrifugal compressor is a radial flow turbomachine in which the compression is achieved *due to dynamic action of working fluid (air). Kinetic energy imparted to the air by rotation of rotor is partly changed into pressure energy in the rotor due to centrifugal and diffusion action and the rest in diffuser and volute casing due to diffusion action.* In the case of turbocharger, it is coupled with an exhaust driven turbine. Generally, single stage is used. It is well suited to the high mass flow rates at the relatively low pressure ratios (up to 3.5) required by the engine. For better performance, generally it runs at very high speed.

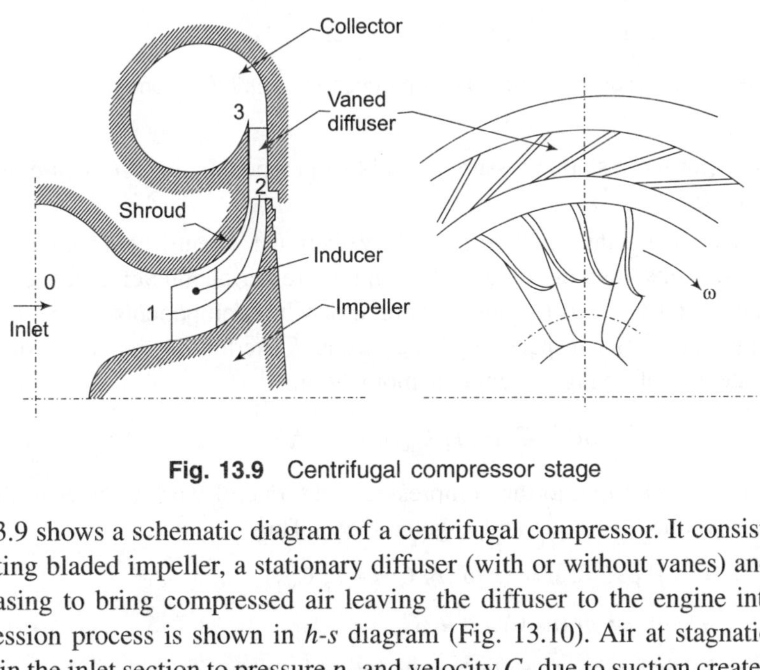


**Fig. 13.9** Centrifugal compressor stage

Figure 13.9 shows a schematic diagram of a centrifugal compressor. It consists of an inlet pipe, a rotating bladed impeller, a stationary diffuser (with or without vanes) and a collector or volute casing to bring compressed air leaving the diffuser to the engine intake system. The compression process is shown in *h-s* diagram (Fig. 13.10). Air at stagnation state 0 is accelerated in the inlet section to pressure $p_1$ and velocity $C_1$ due to suction created at impeller inlet. The enthalpy drop from $h_{01}$ to $h_1$ is equal to $C_1^2/2$. Compression process in the impeller from pressure $p_1$ to $p_2$ which is governed by centrifugal and diffusion action is represented by process 1–2. The stagnation state 02 corresponds to full conversion of kinetic energy $C_2^2/2$ into pressure energy while to the corresponding isentropic state is shown by 2*s*. The process in diffuser is represented by 2–3 in which pressure rises from $p_2$ to $p_3$ due to diffusion action, i.e., slowing down of the kinetic energy in a carefully shaped expanding passage. The final

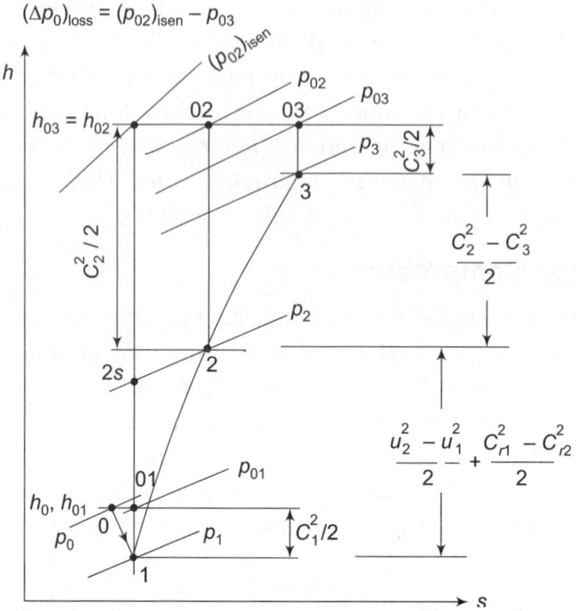


**Fig. 13.10** Representation of compression process in centrifugal compressor on *h-s* diagram

state in the collector has a static pressure $p_3$ and low kinetic energy $C_3^2/2$ while the stagnation state is given by 03.

Figure 13.11 shows the inlet and outlet velocity diagram for centrifugal compressor impeller. Here $C_1$ and $C_2$ are absolute air velocities, $u_1$ and $u_2$ are peripheral velocities, $C_{r1}$ and $C_{r2}$ are relative velocities and $C_{w1}$ and $C_{w2}$ are tangential or whirl components of absolute velocities of impeller inlet and outlet respectively. The torque $T$ exerted on the air by the impeller is expressed as the rate of change of angular momentum.

Thus, $$T = \dot{m}(r_2 C_{w2} - r_1 C_{w1}) \quad ...(13.22)$$

The theoretical work input to the compressor is the rate of work transfer to the air which is given by the product of torque and angular speed. Thus,

$$-\dot{W}_c = T.\omega = \dot{m}.\omega (r_2 C_{w2} - r_1 C_{w1})$$

or $$-\dot{W}_c = m (u_2 C_{w2} - u_1 C_{w1}) \quad ...(13.23)$$

Equation (13.23) is known as Euler's equation of motion.

Since the work transfer to the gas occurs only in impeller, so

$$-\dot{W}_c = \dot{m}(u_2 C_{w2} - u_1 C_{w1}) = \dot{m}(h_{03} - h_{01}) \quad ...(13.24)$$

With the help of velocity triangle, the work done per kg of fluid may be expressed as

$$w_c = \frac{u_2^2 - u_1^2}{2} + \frac{C_{r1}^2 - C_{r1}^2}{2} + \frac{C_2^2 - C_1^2}{2} \quad ...(13.24a)$$

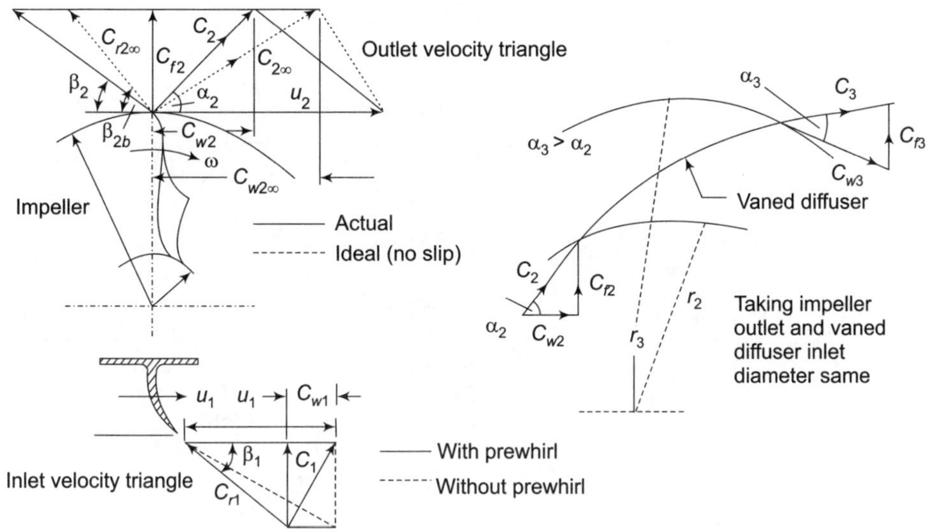


**Fig. 13.11** Velocity diagram for centrifugal compressor

The first term of Eq. (13.24a) signifies the **centrifugal action**, i.e., *the fluid enters the lower diameter and comes out at higher diameter while second term refers to the* **diffusion action**, i.e., *decreases in relative velocity from inlet to outlet.* These two actions, centrifugal and diffusion, are responsible for pressure rise in impeller. The rest K.E. $(C_2^2 - C_1^2)/2$ will be converted into pressure energy in diffuser due to diffusion action. Figures 13.12 and 13.13 show the performance curves of a centrifugal compressor.

The maximum possible pressure energy (i.e., head) developed by a centrifugal machine is given by

$$H_{max} = u_2^2/g \qquad ...(13.24b)$$

## 13.6 TURBOCHARGER TURBINES

As discussed earlier, the *turbocharger turbine (Fig. 13.14) is driven by the energy available in the engine exhaust* (Fig. 13.2). It is a well-known fact that the exhaust from reciprocating internal combustion engine is pulsating in nature. Though, it is possible to design turbines to accept pulsating flow but they operate at low efficiency as compared to steady flow case. There are generally *two methods to recover a fraction of available exhaust energy. They are:*

(a) Constant pressure turbocharging (b) Pulse turbocharging

### 13.6.1 Constant-pressure Turbocharging

In this, an exhaust manifold of sufficiently large volume is used to damp out the mass flow and pressure pulses so that the flow entering the turbine is in steady state condition. The main disadvantages of this approach is that it does not make full use of the high kinetic energy of the gases leaving the exhaust port. It also produces mixing losses in manifold. The advantage of constant-pressure turbocharging lies in high efficient turbine and simple exhaust piping.

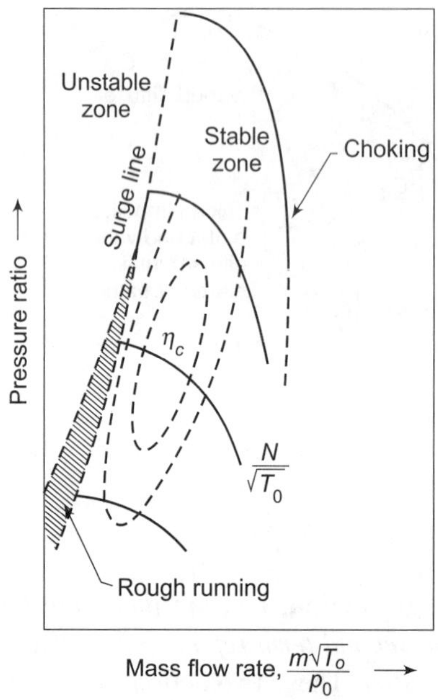


**Fig. 13.12** Centrifugal compressor performance map

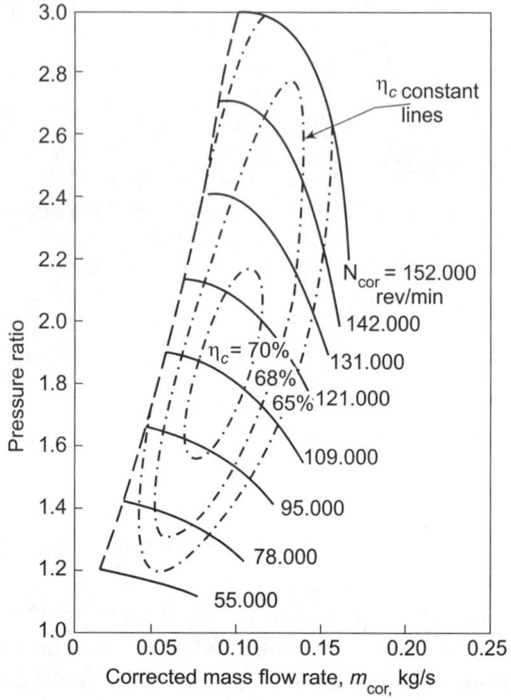


**Fig. 13.13** Correct centrifugal compressor performance map

## 13.6.2 Pulse Turbocharging

It employs short small cross-sectional pipes that connect each exhaust port to the turbine so that the kinetic energy associated with exhaust blowdown can be recovered. The exhaust pulses may be brought to acceptable level by suitably grouping the different cylinder exhaust ports ensuring sequential and minimum overlap of pulses. *The pulse system is commonly used in large diesel engines while constant pressure turbocharging is used in automotive engines.*

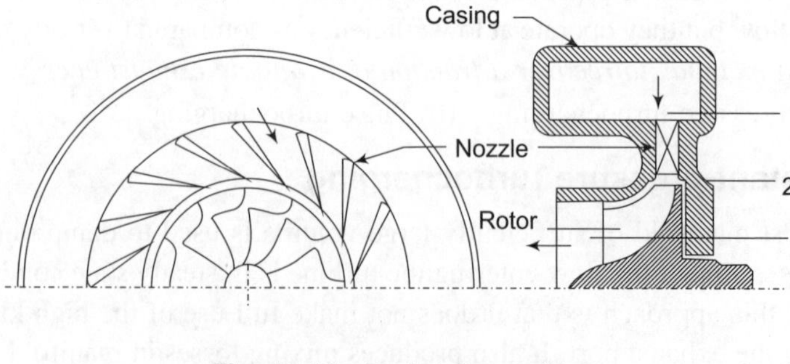


**Fig. 13.14** Radial flow turbine

*Radial and axial flow turbines are used in turbocharger.* In radial flow, the flow is radially inwards. Figure 13.14 shows a schematic diagram of a radial flow inward turbine. It consists of an inlet casing, a set of inlet nozzles, the turbine rotor. Figure 13.15 shows the process in turbine on *h-s* diagram and velocity diagrams. The nozzles (01-2) accelerate the flow, with modest loss in stagnation pressure. Process 2-3 shows the actual expansion process in turbine where the drop in enthalpy results in work transfer. The rotor is designed for the minimum kinetic energy $C^2/2$ at exit. *The work transfer rate from the turbine is a result of change in angular momentum in the working fluid* which is given as

$$\dot{W}_T = T\omega = \dot{m}\omega\,(r_2\,C_{w2} - r_3\,C_{w3})$$
$$= \dot{m}\,(u_2\,C_{w2} - u_3\,C_{w3}) \qquad \text{...(13.25)}$$

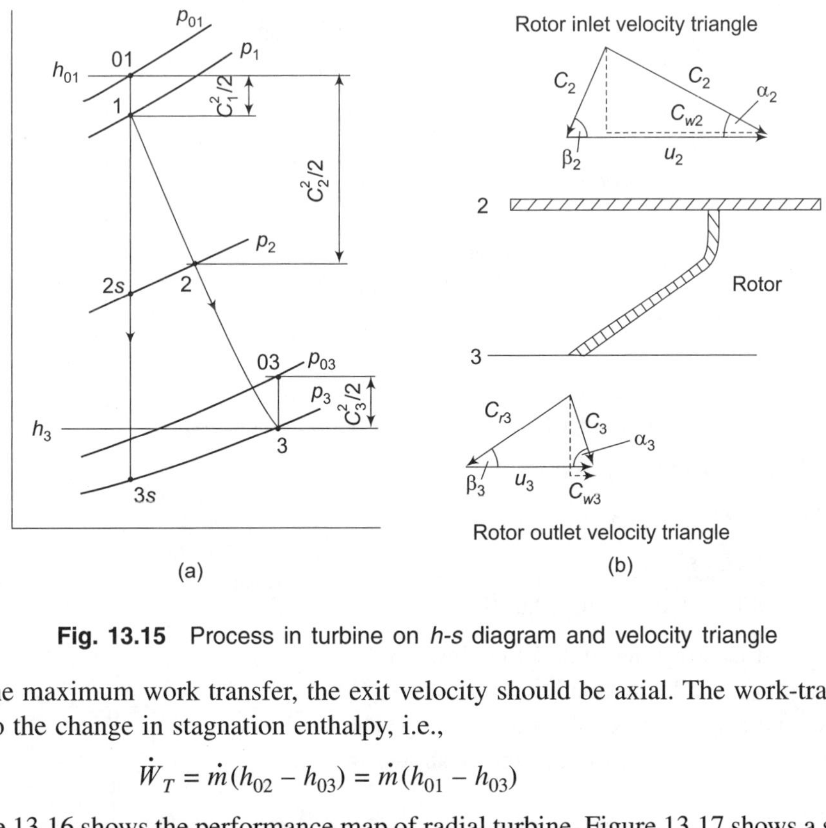


**Fig. 13.15** Process in turbine on *h-s* diagram and velocity triangle

For the maximum work transfer, the exit velocity should be axial. The work-transfer rate relates to the change in stagnation enthalpy, i.e.,

$$\dot{W}_T = \dot{m}(h_{02} - h_{03}) = \dot{m}(h_{01} - h_{03}) \qquad \text{...(13.26)}$$

Figure 13.16 shows the performance map of radial turbine. Figure 13.17 shows a schematic diagram of an axial flow turbine turbocharger. In general, single stage is sufficient to expand the exhaust gas efficiently through the pressure ratios associated with engine turbocharging. It consists of an annular flow passage, a single row of stators (nozzles) and a rotating blade ring. The velocity triangle is shown in Fig. 13.18.

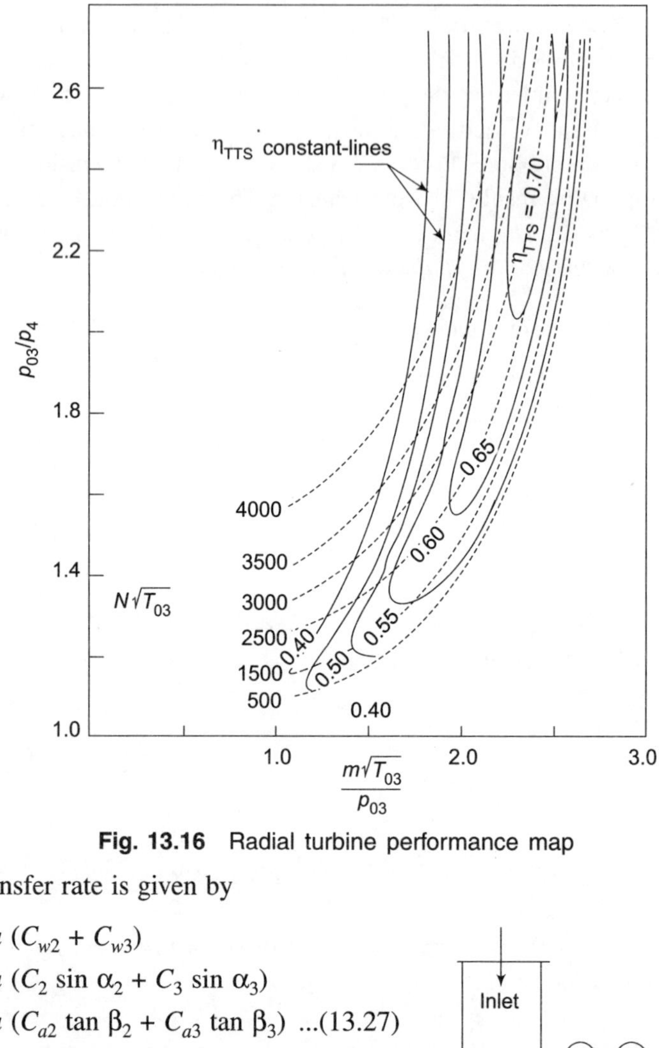


**Fig. 13.16** Radial turbine performance map

The work transfer rate is given by

$$\dot{W}_T = \dot{m}u\,(C_{w2} + C_{w3})$$
$$= \dot{m}u\,(C_2 \sin \alpha_2 + C_3 \sin \alpha_3)$$
$$= \dot{m}u\,(C_{a2} \tan \beta_2 + C_{a3} \tan \beta_3) \quad ...(13.27)$$

Here $\alpha_2$ and $\alpha_3$ are inlet and outlet flow angles where $\beta_1$ and $\beta_2$ are inlet and outlet blade angles. $C_a$ represents axial component of absolute velocity. The performance map of axial flow turbine is shown in Fig. 13.19.

It is essential to check whether the turbine and compressor are matching or not. For this a different type of performance map is shown in Fig. 13.20. It is obvious from the map that the peak efficiency can occur for $0.4 < u/C_s < 0.8$ where $C_s$ is the velocity equivalent of the isentropic enthalpy drop across the turbine stage.

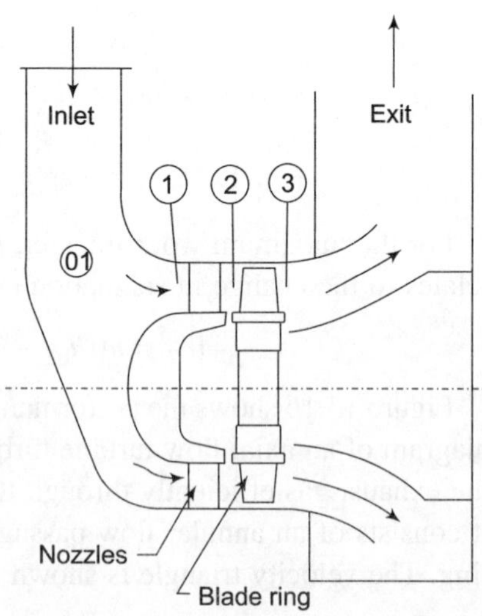


**Fig. 13.17** Single stage axial flow turbine

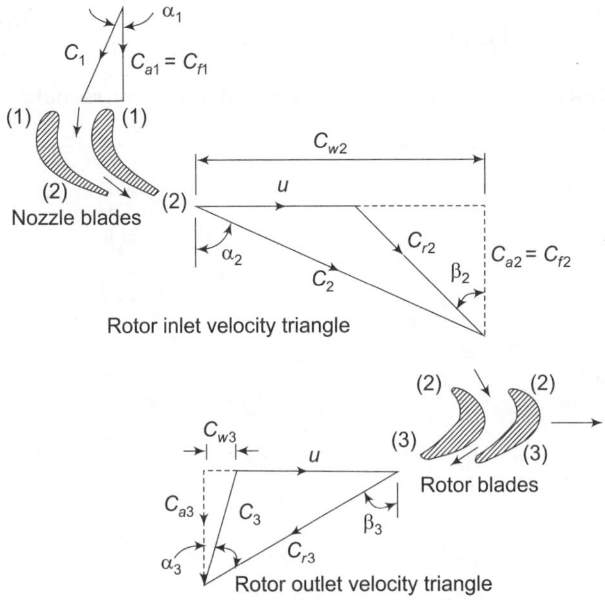


**Fig. 13.18** Velocity triangle of axial flow turbine

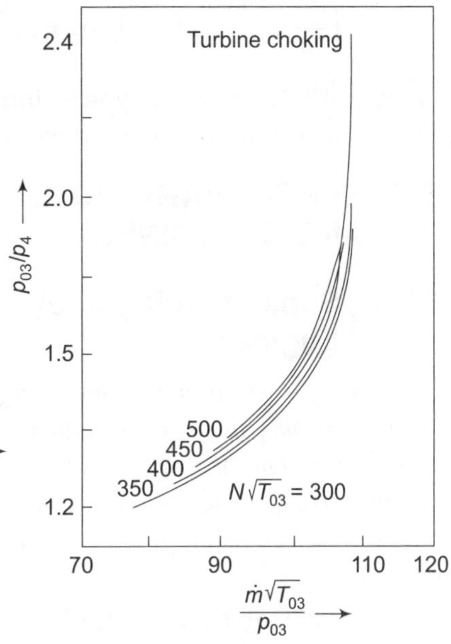


**Fig. 13.19** Performance map of axial flow turbine [73]

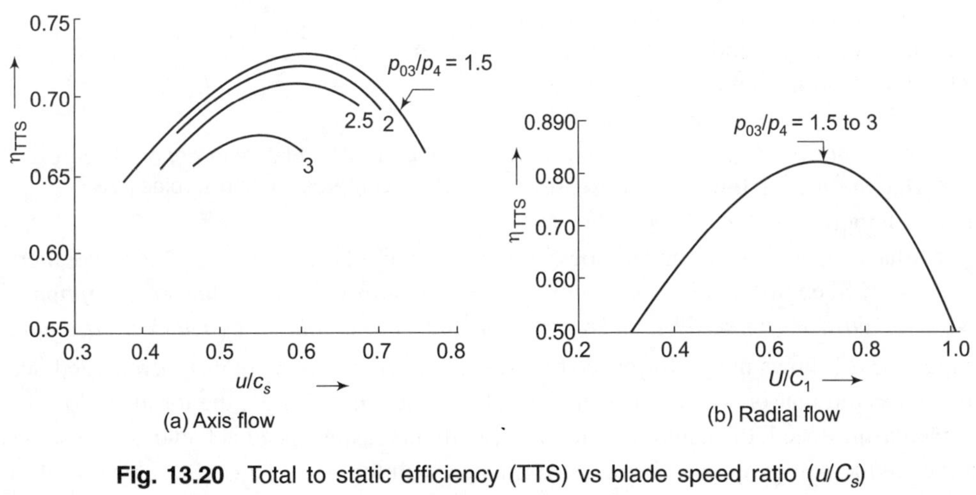


**Fig. 13.20** Total to static efficiency (TTS) vs blade speed ratio ($u/C_s$)

For a given turbocharger, the compressor and turbine characteristics are linked. Since the compressor and turbine are on a common shaft with speed $N$, hence

$$\frac{N}{\sqrt{T_{01}}} = \frac{N}{\sqrt{T_{03}}}\left(\frac{T_{03}}{T_{01}}\right)^{1/2} \qquad ...(13.28)$$

$$\left(\frac{p_{02}}{p_{01}}\right)^{\gamma - 1/\gamma_c} - 1 = \eta_c\,\eta_T\,\eta_m\,\frac{c_{p,T}}{c_{p,c}}\left[1 - \left(\frac{p_4}{p_{03}}\right)^{(\gamma_T - 1)/\gamma_T}\right]\frac{T_{03}}{T_{01}} \quad ...(13.29)$$

Figure 13.21 shows a typical turbocharger characteristics plotted on a turbocharger compressor map for a radial turbine.

## 13.7 SUPERCHARGING OF SI AND CI ENGINES

### 13.7.1 Supercharging of SI Engines

*In general, for SI engines, supercharging is employed only for* **aircraft and racing car engines**. *This is because of the fact that the increase in supercharging pressure increases the tendency to knock and pre-ignition.*

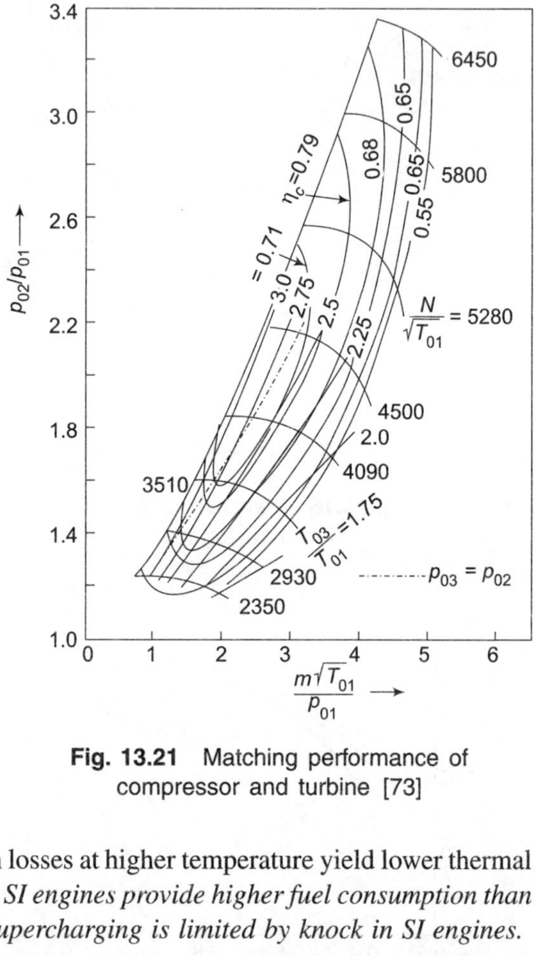


**Fig. 13.21** Matching performance of compressor and turbine [73]

As a result of supercharging, intake pressure and temperature increase which results in increase in volumetric efficiency and thus increased induced mass flow rate, which in turn, increases, power output. Increased intake pressure and temperature reduce ignition delay and increase flame speed, which in turn, increases the tendency to knock. *For this reason, the supercharged SI engines employ lower compression ratios.* But the use of lower compression ratios and increased heat losses due to higher value of specific heats and dissociation losses at higher temperature yield lower thermal efficiency for SI engines. *Thus, supercharged SI engines provide higher fuel consumption than naturally aspirated engines. The degree of supercharging is limited by knock in SI engines.*

Figure 13.22 shows the performance of a supercharged engine with unsupercharged one. It can be observed that brake power is increased over the whole range, the torque is lower and with increasing speed, the nature of torque and with increasing speed is flatter which suggests that the engine will give much better performance with top gear. Thus, the selection of most suitable compression ratio in a supercharged engine, in order to avoid excessive cylinder pressure and temperature, is a matter of great concern.

*It is worth to note that supercharged SI engines using carburettors should have the carburettors before the compressor due to the following reasons.*

(a) The carburettor would be operating at ambient pressure

(b) The entering of charge will suit the system.

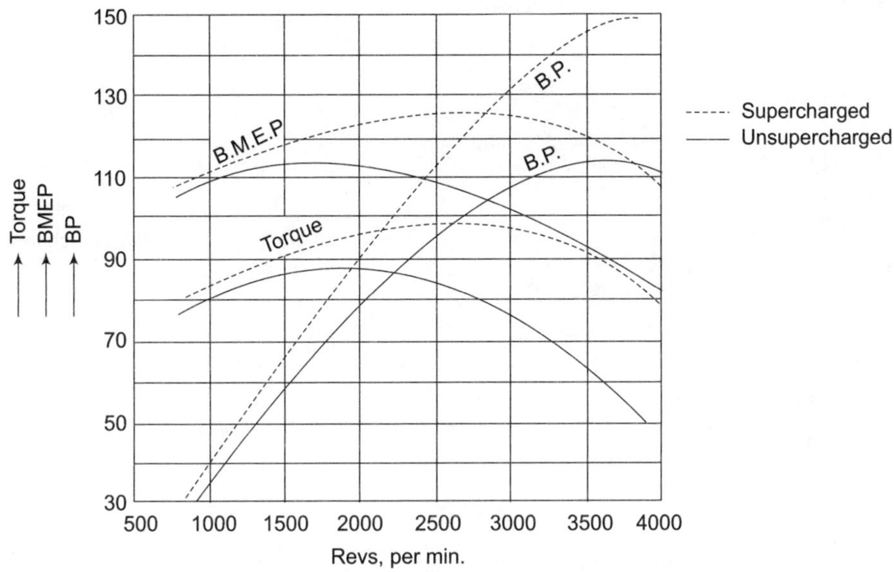


**Fig. 13.22** Comparison of performance of supercharged and unsupercharged engines

(c) The compressor would run away from surge limit
(d) The mixture entering the cylinder would be more homogeneous.
(e) The use of standard fuel pump is possible.

## 13.7.2 Supercharging of CI Engines

*Supercharging in CI engines is employed on marine power plant and* ***automobiles engines.*** It is worth to mention that supercharging does not create any combustion problem unlike SI engines, rather it *improves combustion in CI engines*. This is because of the fact that increase in pressure and temperature of the intake air reduces ignition delay and hence the rate of pressure rise resulting in a better, quieter and smoother combustion. *It is this improved combustion process allows a poor quality of fuel to be used in CI engines.*

Though, supercharging improves combustion process but it *increases mechanical and thermal loading. To overcome that mechanical and thermal overloading problem, the compression ratio may be reduced.* The increased valve overlap also reduces this problem. It is possible to use lower *fuel-air ratios* in a supercharged engine as the increase in fuel flow is less than the increase in air flow. *This results in lower exhaust temperature and thus increased life of the exhaust valve.*

*The degree of supercharging in CI engines is limited by the thermal and mechanical load on the engine and strongly depends on the design of engine and the type of supercharger employed.*

Figure 13.23 shows the fuel consumption for turbocharged and unsupercharged version of six cylinder swirl chamber IDI diesel engine.

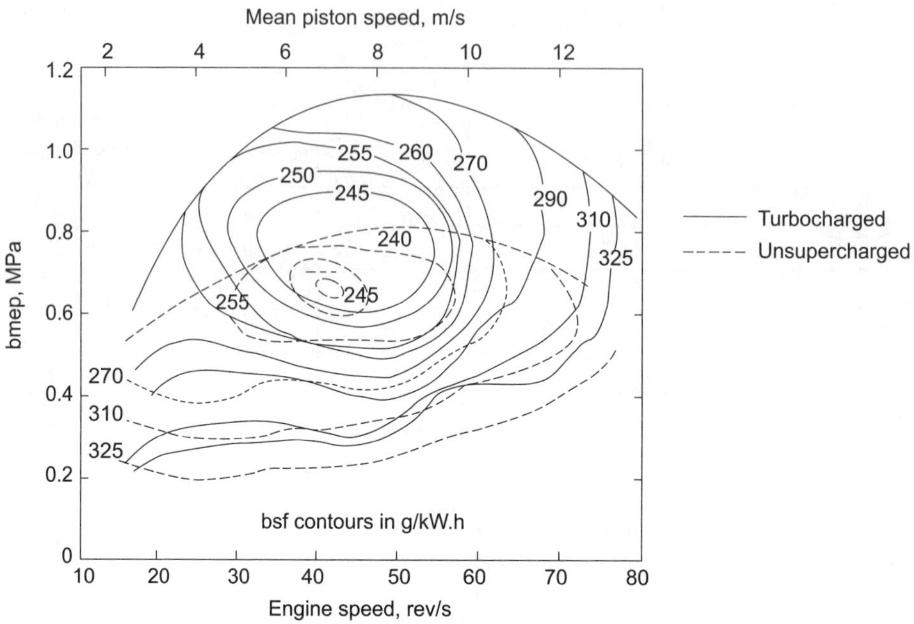


**Fig. 13.23** Fuel consumption map for turbocharged and unsupercharged version of six-cylinder swirl chamber IDI diesel engine [73]

## Comparison between Supercharged SI and CI engines

| *S.No.* | *Parameters* | *Supercharged SI engines* | *Supercharged CI engines* |
|---|---|---|---|
| 1. | Combustion | Problematic | Quiter, smoother and better |
| 2. | Detonation | More | Less |
| 3. | Ignition delay | Decreases | Decreases |
| 4. | Compression ratio | Lower value desirable | Lower value desirable |
| 5. | Volumetric efficiency | Increases | Increases |
| 6. | Thermal efficiency | Decreases | Decreases |
| 7. | sfc | Increases | Decreases |
| 8. | Flame speed | Increases | – |
| 9. | Weak mixture effect | Can run without knock | Can run without knock |
| 10. | Sensitivity to A/F | More | Less |
| 11. | Sensitivity to type of fuel | More | Less |
| 12. | Exhaust temperature and smoke | - | Lower |
| 13. | Application | In aircraft and racing car engines | Automobiles and marine engines |

## 13.8 EFFECTS OF SUPERCHARGING ON THE PERFORMANCE OF THE ENGINE

Performance parameters such as power output, overall efficiency, mechanical efficiency and fuel consumption are affected by supercharging. The following are the effects of supercharging:

(i) **More power output.** The power output of supercharged engine is higher than naturally aspirated one mainly due to higher mass flow rate inducted per cycle.

(ii) **Higher mechanical efficiency.** The mechanical efficiency of supercharged engine is higher than naturally aspirated one. This is because of the fact that increase in *bmep* is much more than the increase in frictional forces due to supercharging.

(iii) **Higher volumetric efficiency.** Supercharging results in better scavenging and reduced exhaust gas temperature in the engine, which in turn, improves volumetric efficiency.

(iv) **Higher sfc in SI and lower in CI.** The specific fuel consumption of supercharged engine is higher in SI and lower in CI as compared to naturally aspirated engine.

(v) **Lower thermals.** The overall efficiency of supercharged engine decreases with the increase of ratio of supercharging.

(vi) **Comparatively cleaner emission.** Supercharged engines furnish comparatively cleaner emission as compound to naturally aspirated engines.

(vii) **Higher cost.** The overall cost is more than naturally aspirated engine.

Figure 13.24 shows the variation of power output and overall efficiency with supercharging ratio. *There exists an optimum supercharging ratio for power output.*

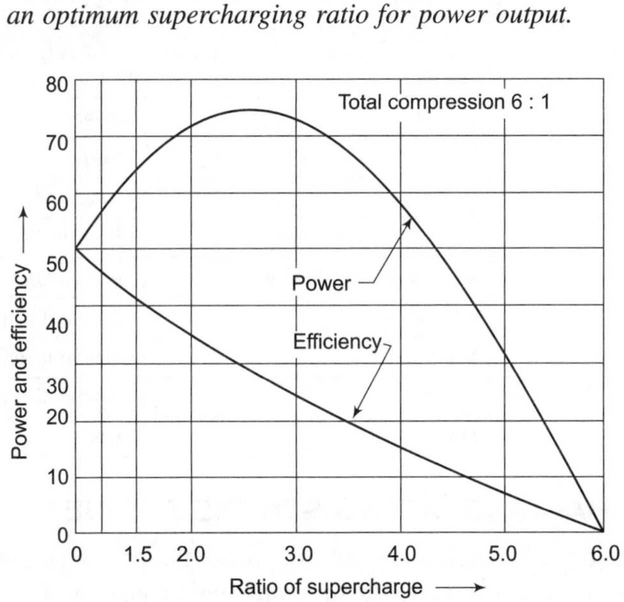


**Fig. 13.24** Supercharging ratio vs power and efficiency

Figure 13.25 shows the effect of supercharging on mechanical efficiency. For a particular speed, mechanical efficiency increases with supercharging ratio.

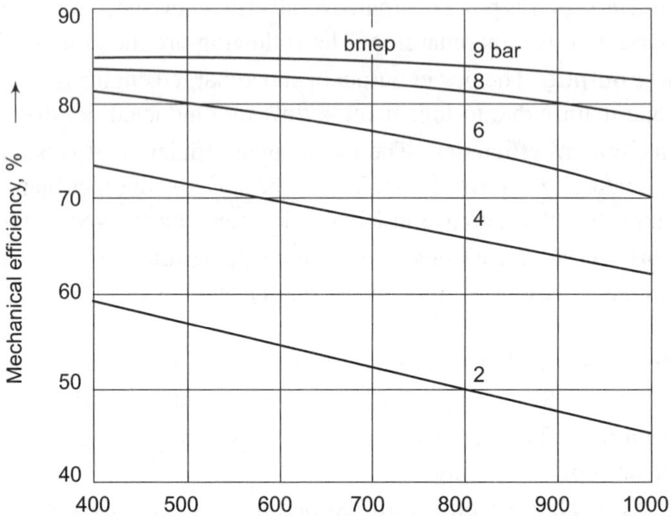


**Fig. 13.25** Effect of supercharging on mechanical efficiency

**Summary of comparison between** supercharged and naturally aspirated engine

| *Parameters* | *Supercharged engines* | *Naturally aspirated engines* |
|---|---|---|
| Volumetric efficiency | Higher | Lower |
| Pressure of inducted air | Higher | Lower |
| Mean effective pressure | Higher | Lower |
| Power produced | Higher | Lower |
| sfc | Higher in SI and lower in CI | Lower in SI and higher in CI |
| Exhaust emission | Comparatively cleaner | Less cleaner |
| Weight of the engine | Heavier | Comparatively higher |
| Load and stress | More | Comparatively less |
| Radiator size | Bigger | Comparatively smaller |
| Cost | More | Less |

## 13.9 TURBOCHARGING OF TWO-STROKE ENGINE

Turbocharging of two-stroke engine not only improves the power output but it also helps in scavenging of the engine. Figure 13.26 shows the turbocharging of a two-stroke engine. It employs mechanically driven blower preceding the turbocharger. The need of crankcase scavenging is thus eliminated.

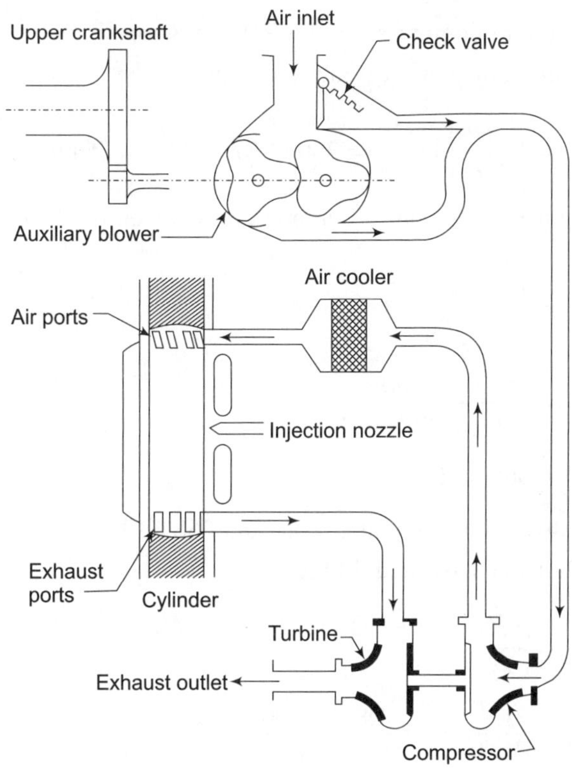


**Fig. 13.26** Two-stroke engines employed with turbocharging

**EXAMPLE 13.1** It is porposed to boost the power of a four-stroke diesel engine of 4000 cc capacity developing 15 kW/m$^3$ of air/minute by a mechanically driven supercharger. The total pressure ratio of turbocharger is 1.8 with total isentropic efficiency of 80%. Assume that at the induction, the cylinder containers volume of charge equal to the swept volume at the pressure and temperature of the delivery from the blower.

Assuming overall mechanical efficiency of 82% of compressor and turbine each, calculate the increase in bp from engine due to turbocharging. The engine running at 3500 rpm has a volumetric efficiency of 78% referred to free air conditions of 1 bar and 300 K.

***Solution*** Refer to Fig. 13.1(a) and 13.4.

$$\text{Swept volume/min} = \dot{V}_s = \text{Engine capacity in cc (cm}^3) \times \frac{N}{2} \times 10^{-6};\ \text{m}^3\text{/min}$$

$$= 4000 \times \frac{3500}{2} \times 10^{-6} = 7\ \text{m}^3\text{/min}$$

Induced volume under unsupercharged condition

$$= \eta_s\, \dot{V}_s = 7 \times 0.78 = 5.46\ \text{m}^3\text{/min}$$

Delivery total pressure of compressor = 1.8 × 1 = 1.8 bar

The total temperature at the end of isentropic compression is given by

$$T_{02s} = T_{01}\left(\frac{p_{02}}{p_{01}}\right)^{(\gamma-1)/\gamma} = 300\ (1.8)^{(1.4-1)/1.4} = 354.91 \text{ K}$$

The actual temperature at the end of compression is given by the concept of compressor efficiency

$$\eta_c = \frac{T_{02s} - T_{01}}{T_{02} - T_{01}}$$

or
$$0.8 = \frac{354.91 - 300}{T_{02} - 300}$$

or
$$T_{02} = 368.63 \text{ K}$$

The compressor delivers 7 $m^3$/min at 1.8 bar and 368.63 K

The equivalent volume at 1 bar and 300 K

$$\frac{p_1\dot{V}_1}{T_1} = \frac{p_2\dot{V}_2}{T_2}$$

or
$$\dot{V}_1 = \frac{p_2}{p_1}\cdot\frac{T_1}{T_2}\cdot\dot{V}_2 = \frac{1.8}{1.0} \times \frac{300}{368.63} \times 7$$

$$= 10.254 \text{ m}^3\text{/min} \quad (\text{assuming } p_0 = p \text{ and } T_0 \text{ and } T)$$

Increase in induced volume due to supercharging

$$= 10.254 - 5.46 = 4.794 \text{ m}^3\text{/min}$$

Thus, increase in indicated power from increased air induced due to supercharging

$$= 15 \times 4.794 = 71.91 \text{ kW}$$

Increase in indicated power due to increased induction pressure due to supercharging

$$= \Delta p \times \dot{V}_s = (1.8 - 1) \times 10^2 \times \frac{7}{60} = 19.83 \text{ kW}$$

Total increase in *ip* = 71.91 + 19.83 = 91.74 kW

Increase in *bp* = 91.74 × 0.82 = 75.22 kW

Mass of air delivered by compressor

$$\dot{m} = \frac{p_2\dot{V}_2}{RT_2} = \frac{1.8 \times 7 \times 10^2}{0.287 \times 368.63 \times 60} = 0.198 \text{ kg/s}$$

Power required to drive the compressor = $\dot{m} c_p\ \Delta T/\eta_m$

$$= 0.198 \times 1.005 \times (368.63 - 300)/0.82 = 16.65 \text{ kW}$$

Net increase in *bp*

$$= 75.22 - 16.65 = 58.57 \text{ kW} \qquad \textbf{\textit{Ans.}}$$

**EXAMPLE 13.2** The following data relate to a diesel engine operating on four stroke cycle at sea level conditions of 1.013 bar and 10°C

$bp = 300$ kW

Speed = 1500 rpm

$bsfc = 0.25$ kg/kWh

A/F = 17.1

$\eta_v = 80\%$

This engine is to be fitted with a supercharger so that it may be operated at an altitude of 2800 m where the atmosphere pressure is 0.7 bar. The power consumed by the supercharger is 8% of the total power produced by the supercharged engine. The temperature of air leaving the supercharger is 35°C. Assume that A/F, $\eta_{th}$ and $\eta_v$ remain the same.

Calculate the following:

(a) Unsupercharged engine capacity and *bmep*

(b) Increase in air pressure required at the supercharger to maintain the same net output of 300 kW.

***Solution***

**(a) Unsupercharged Engine**

$$\text{Fuel consumption} = bsfc \times bp = 0.25 \times 300 = 75 \text{ kg/h}$$

$$\text{Air consumption} = 17 \times 75 = 1275 \text{ kg/h}$$

$$\text{Volumetric efficiency} = \frac{\text{Mass of air induced per cycle}}{\text{Mass of fire air corresponding to swept volume}}$$

or

$$0.8 = \frac{1275 \times 2/1500}{60 \times p_1\ V_s/RT_1}$$

or

$$V_s = \frac{(1275 \times 2/1500) \times 0.287 \times 283}{60 \times 1.013 \times 10^2 \times V_s}$$

or

$$V_s = \frac{1275 \times 2 \times 0.287 \times 283}{60 \times 1500 \times 1.013 \times 10^2 \times 0.8}$$

$$= 0.01419 \text{ m}^3 = 1419 \text{ cc} \qquad \textbf{\textit{Ans.}}$$

But

$$bp = p_{mb}\,(LA)n$$

where $$n = \frac{N}{2} \text{ for 4-strokes}$$

or $$300 = p_{mb} \times 0.01419 \times \frac{1500}{60 \times 2}$$

or $$p_{mb} = \frac{300 \times 60 \times 2}{0.01419 \times 1500}$$

$$= 1691.33 \text{ kN/m}^2 = 16.9133 \text{ bar}$$ ***Ans.***

**(b) Supercharged Engine**

Under supercharged condition, the gross power produced by the engine is

$$= 300 + 0.08 \times 300 = 324 \text{ kW}$$

Mass of air required to produce 324 kW is

$$\dot{m} = \frac{\dot{m}_{unsup}}{\dot{m}_{unsup}} \times P_{super} = \frac{1275}{300 \times 60} \times 324 = 22.95 \text{ kg/s}$$

Since the volumetric efficiency is the same, hence

$$0.8 = \frac{22.95 \times 2 \times 0.287 \times 308}{1500 \times p_2 \times 10^2 \times 0.01419}$$

or $$p_2 = \frac{22.95 \times 2 \times 0.287 \times 308}{0.8 \times 1500 \times 10^2 \times 0.01419} = 2.382 \text{ bar}$$

Increase in pressure required in the supercharger

$$= 2.382 - 0.7 = 1.682 \text{ bar}$$ ***Ans.***

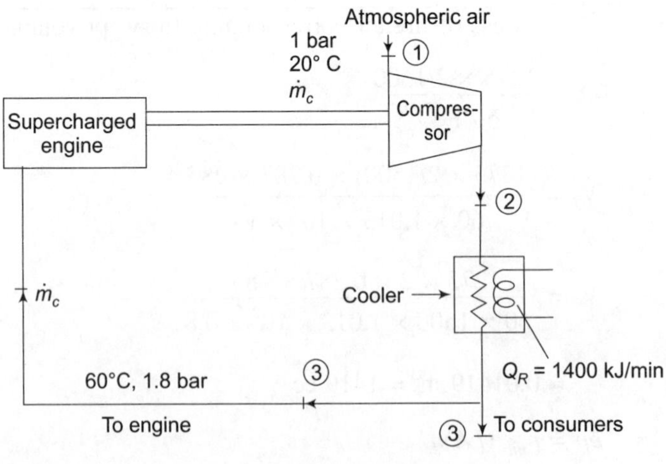


**Fig. 13.27** Example 13.3

**EXAMPLE 13.3** A centrifugal compressor is driven by supercharged four-stroke cycle oil engine. The delivery air does two purposes—firstly it supplies compressed air to engine and secondly it supplies compressed air to consumers. The air enters the compressor at 20°C and 1 bar and the compressed air to consumers. The air enters the compressor at 20°C and 1 bar and the compressed air before being delivered to engine and consumers it first passes through a cooler where heat at the rate of 1400 kJ/min is rejected. The air leaves the cooler at 60°C and 1.8 bar. The supercharged engine has a volumetric efficiency of 0.75 at the condition of 60°C and 1.8 bar. The engine consists of six cylinders having 90 mm bore and 100 mm stroke runs at 2000 rpm and develops 35 kW. Calculate the following if mechanical efficiency is 0.8:

(a) The engine indicated mean effective pressure

(b) The air consumption in kg/min in engine

(c) The air flow rate required into the compressor

***Solution*** Refer to Fig. 13.27.

$$ip = \frac{bp}{\eta_m} = \frac{35}{0.8} = 43.75 \text{ kW}$$

$$ip = p_{mi}\, L\, A\, n\, x = p_{mi}\, LA\, \frac{N}{2 \times 60}\, x$$

or

$$43.75 = p_{mi}\left(\frac{100}{1000}\right)\frac{\pi}{4}\left(\frac{90}{1000}\right)^2 \frac{2000 \times 6}{2 \times 60}$$

or

$$p_{mi} = \frac{43.75 \times 10 \times 4 \times 1000^2 \times 2 \times 60}{\pi \times 90^2 \times 2000 \times 6} = 687.7 \text{ kN/m}^2$$

$$= 6.877 \text{ bar}$$ ***Ans.***

$$\text{Engine swept volume} = V_s = AL\,\frac{N}{2}\,x$$

$$= \frac{\pi}{4}\left(\frac{90}{100}\right)^2 \times \frac{100}{1000} \times \frac{2000}{2} \times 6 = 3.817 \text{ m}^3\text{/min}$$

Actual sucked volume into the engine

$$V_{as} = V_s = 0.75 \times 3.817 = 2.862 \text{ m}^3\text{/min}$$

Actual sucked mass flow rate in the engine

$$\dot{m}_e = \dot{m}_{as} = \frac{p_3\, V_{as}}{RT_3}$$

$$= \frac{1.8 \times 2.862 \times 10^2}{0.287 \times 333} = 5.39 \text{ kg/min}$$ ***Ans.***

Power output of engine = Power consumed by compressor

= Gain in enthalpy of air in compressor

or $$35 = \dot{m}_c \, c_p \, \Delta T = \dot{m}_c \times 1.005 \times (T_2 - 293) \qquad ...(1)$$

Energy balance in the cooler

$$= \dot{m}_c \times 1.005 \, (T_2 - 333) = \frac{1400}{60} = 23.33 \qquad ...(2)$$

Solving Eqs. (1) and (2), we get

$$T_2 = 413 \text{ K}$$

From Eq. (1)

$$35 = \dot{m}_c \times 1.005 \, (413 - 293)$$

or $$\dot{m}_c = 0.29 \text{ kg/s} = 17.41 \text{ kg/min}$$ ***Ans.***

## PROBLEMS

### Things to Think About

13.1 By which principle supercharging enhances the power output of IC engine?

13.2 What is the basic principle involved in using the intercooled turbocharger?

13.3 What is the reason that, for the supercharging, reciprocating compressor is not used?

13.4 What is the basic principle in pressure rise of working fluid through reciprocating centrifugal and axial flow compressor?

13.5 What is the reason that pressure rise per stage in centrifugal compressor more than axial?

### Viva voce and Theoretical

13.6 What do you mean by supercharger?

13.7 List the objectives of supercharging.

13.8 List the method of supercharger.

13.9 Sketch the various configurations of supercharger and turbocharging.

13.10 Show the working process of naturally aspirated and supercharged engine on *p-V* diagram.

13.11 Discuss the working principle of various types of supercharging compressors

13.12 Discuss the performance maps of various types of compressors.

13.13 Differentiate between costant pressure turbocharging and pulse turbocharging.

13.14 Discuss the working principle of radial and axial flow turbine along with performance maps.

13.15 Discuss the turbocharging of two-stroke engines.

13.16 What are the supercharging limits?

13.17 Discuss the effects of supercharging on the performance of engine.

### Numerical Questions

13.18 An eight-cylinder turbocharged intercooled four-stroke cycle diesel engine operates with an inlet pressure of 1.8 bar at a maximum rated speed of 2000 rpm. The bore and stroke of the engine are 130 mm and 140 mm respectively. The volumetric efficiency is 90% based on inlet condition of 1.8 bar and 325 K after cooler. The compressor isentropic efficiency is 72%. Calculate the following:

(a) The power required to drive the turbocharger compressor.

(b) The pressure at the turbine inlet if the exhaust gas temperature is 650°C and turbocharger efficiency is 65%. The turbine exhaust goes to atmosphere.

13.19 What would be the delivery pressure of compressor if the power output of the above engine is further increased by 10%?

### Software Based Question

13.20 Develop a software for the selection and design of a supercharger and turbocharger.

### Design and Open Ended Question

13.21 Prepare a comprehensive report on the type of superchargers and turbochargers used in all types of IC engines and their shortcomings. In order to avoid these shortcomings, suggest the design improvements with drawing and specifications.

### Multiple-Choice Questions

13.22 Supercharging increases power output by

(a) 20% (b) 40% (c) 50% (d) 60%

13.23 The purpose of supercharging is to

(a) increase thermal efficiency (b) increase power output

(c) increase speed of the engine (d) to decrease power output

13.24 An engine is supercharged to get

(a) better mechanical efficiency

(b) better scavenging

(c) reduction in knocking tendency in SI engine

(d) engine operating at higher compression ratio

13.25 A higher boost pressure through supercharging means

1. more air forced into the cylinder
2. increased power output
3. more power required to run the supercharger
4. lower frictional losses

Out of these statements

(a) 1, 2 and 3 are correct (b) 1, 2 and 4 are correct

(c) 1, 2, 3 and 4 are correct (d) only 1 and 2 are correct

13.26 **Assertion (A):** A supercharged IC engine needs a greater overlap of the exhaust and inlet valve timing.

**Reasoning:** As it helps in scavenging of the combustion space and cooling of the exhaust valve and seat.

**Answer Code:**

(a) A is true but R is false

(b) A is false but R is true

(c) Both A and R are true

(d) Both A and R are true but R is not the only reason of A.

13.27 A reciprocating compressor is not suitable for supercharging as

(1) it runs at lower rpm

(2) the pressure ratio is high

(3) the volumetric efficiency is very high

(4) it handles very small quantity of air and it is bulky also.

*Out of these statements:*

(a) All the above are correct (b) Only 2 is correct

(c) Only 1 and 4 are correct (d) Only 1, 2 and 4 are correct

13.28 The ideal head rise per stage in a radial compressor is given by

(a) $u^2/g$ (b) $u_2^2/g$ (c) $u_2^2/2g$ (d) $u_2^3/2g$

13.29 The pressure rise per stage in an axial flow.

(a) 1.1 to 1.2 (b) 1.3 to 1.4 (c) 1.5 to 1.6 (d) 1.7 to 1.8

## Key to Multiple-Choice Questions

13.22 (a) 13.23 (b) 13.24 (b) 13.25 (a) 13.26 (c) 13.27 (a) 13.28 (b) 13.29 (a)

CHAPTER

# 14 Two-stroke Engines

## OVERVIEW

As discussed earlier in two-stroke engines, the cycle is completed in two strokes. In other words, in one revolution of the crankshaft, the cycle is completed involving two strokes. In four-stroke engines, the cycle is completed in two revolutions of crankshaft involving four strokes. *The difference between two-stroke and four-stroke engines lies in the method of filling the cylinder with the fresh charge and removing the burnt gases from the cylinder.* In a four-stroke engine, these operations are performed by the piston during suction and exhaust strokes. In a two-stroke engine, suction is accomplished by air compressed in crankcase or by a separate blower. The induction of compressed air in the cylinder removes the products of combustion through exhaust ports. The engine piston only compresses the fresh charge and expands the products of combustion. *The suction and discharge ports are covered and uncovered by the piston.* In 2-stroke engines, instead of valves, multi-ports are used. Since a two-stroke engine will have twice as many cycles per minute as four-stroke engine operating at the same speed, with the same number of cylinders and same mean effective pressure, theoretically it will develop twice the power than that of four-stroke engine. *In actual practice, it develops 1.6 times power than that of four-stroke engine.*

The two-stroke engines may work on petrol and diesel fuels. The two-stroke SI engines are extensively used in two-wheelers (scooters, mopeds and motorcycles) developing a fraction of 1 kW to 10 kW. Two-stroke SI engines for aircraft applications may develop as high as 2500 kW. *Now, all two-wheeler engines* are using 4-stroke engines instead of 2-stroke engines due to emission norms. *Two-stroke CI engines are used in marine and locomotive engines (2 to 6 MW).*

For more details, See References [14 to 16].

## 14.1 CLASSIFICATION OF TWO-STROKE ENGINES

On the basis of method of scavenging, 2-stroke engines are classified as follows:

(a) Crankcase scavenged engines (b) Separately scavenged engines

Construction details and working principles of above will be discussed in detail.

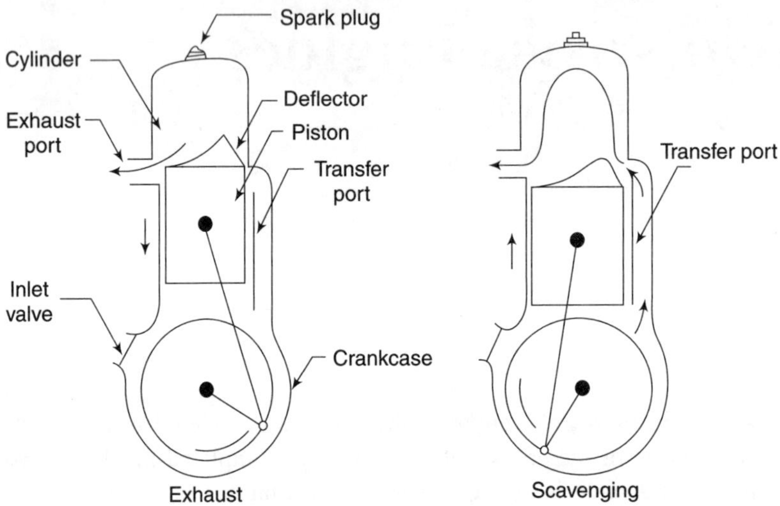


**Fig. 14.1** Crankcase scavenged two-stroke engine

## 14.2 CRANKCASE SCAVENGED ENGINES

Figure 14.1 shows the simplest types of crankcase scavenged two-stroke engines. There are at least ***three ports*** in this type cut in the cylinder wall instead of valves in 4-strokes engines., These are reed valves operated intake port, transfer port and exhaust port. The exhaust and transfer ports are closed and opened by piston itself. The charge (i.e., air-fuel mixture in SI engine) or air (in CI engine) is sucked through the reed valve operated intake port when the pressure in the crankcase reduces due to upward motion of the piston during compression stroke and is compressed in the crankcase by the underside of the piston during the expansion stroke. The compressed charge passes through the transfer port into the engine cylinder flushing the products of combustion out of cylinder through exhaust port. *The process of removing the combustion products from the engine cylinder is called scavenging and this type of engines are called crankcase scavenged engines.*

### 14.2.1 Downward Stroke

Assume the piston is almost at the TDC position when the air or mixture of air an fuel is in compressed form. In the case of SI engine, spark occurs and ignition takes place. In the case of CI engine, the fuel is injected and combustion takes place automatically.

Combustion process is represented by process 2–3 on $p$-$V$ diagram as shown in Fig. 14.2.

As the piston moves down, expansion of gas takes place from state 3 along path 3–4. After completing a portion of a downward stroke, the piston first uncovers the exhaust ports and the cylinder pressure drops to atmospheric level as the combustion products escape through these ports. The exhaust process is shown by process 4–5. With further downward involvement of the piston, the piston uncovers the transfer ports, permitting slightly compressed mixture or air (as the case may be) in the crankcase to enter the engine cylinder. The top of the piston and the ports are given such a shape that the fresh air is directed towards the top of the cylinder rather than to travel towards the exhaust port. This projection of the piston is called ***deflector***. The purpose of this is to scavenge the upper part of the cylinder of the combustion products and also to minimize the flow of fresh air-mixture directly through the exhaust. *Whatever be the shape, a small fraction of fresh air-mixture goes unburnt through the exhaust port.*

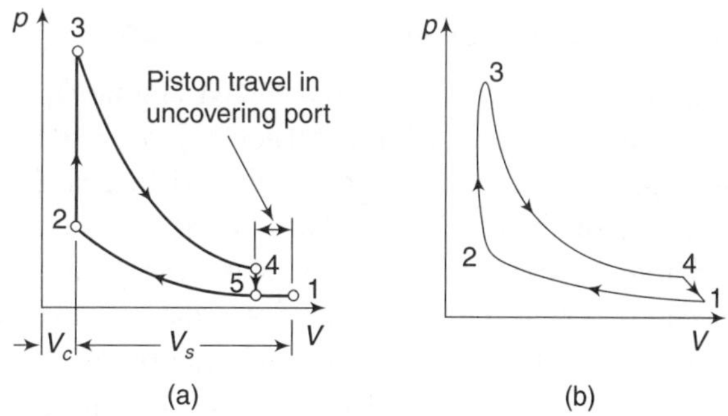


**Fig. 14.2** Ideal and actual *p-V* diagram of two-stroke engine

## 14.2.2 Upward Stroke

As the piston proceeds from bottom centre of upward stroke, the transfer port and then the exhaust ports are closed by the piston enabling compression of the charge. This process is represented as process 1–2. The upward movement of the piston during compression lowers the pressure in the crankcase so that the fresh mixture or air is drawn into the crankcase through the inlet reed valve.

Thus, we can say that compression and induction occur during upward movement of the piston while exhaust and transference occur during downward movement. The injection or ignition occurs when the piston is at the top dead piston. However, in actual engine, ignition occurs 20° before TDC. The actual cycle is slightly different than ideal.

The *valve timing diagram has already been discussed is chapter 1.*

# 14.3 SEPARATELY SCAVENGED ENGINES

In a separately scavenged engine, a external device such as blower is used to scavenge the products of combustion. The following systems fall under the category of separately scavenged engine. These are

(a) Loop scavenged engines
(b) Opposed piston engines
(c) Port and poppet valve engines

### 14.3.1 Loop-scavenged Engines

Figure 14.3 shows a loop-scavenged engine. In this engine, a separate blower (roots type) is used to supply the air or mixture of air and fuel at a pressure of about 1.1 to 1.4 bar into the cylinder. During the expansion stroke, when the piston decends down, the piston uncovers the exhaust ports at about 65° before BDC and the products of combustion begins to leave the cylinder. When the piston reaches the position about 55° before BDC, the inlet ports are uncovered and the scavenging process begins. The inlet ports are so designed, that most of the air flows to the top of the cylinder on the side of the inlet ports, and then downwards on the side of the exhaust ports. This ensures scavenging of the upper part of the cylinder. *Note that the scavenging process in this case is better than that of the crankcase.* **Roots blower system is more common in diesel locomotives.**

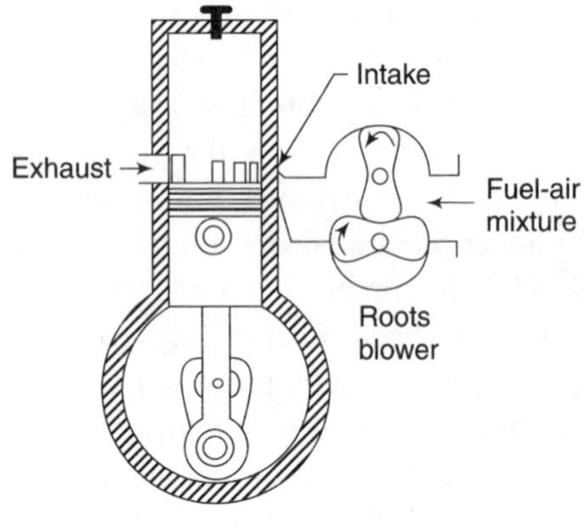


**Fig. 14.3** Loop-scavenged engine

### 14.3.2 Opposed-piston Engines

Figure 14.4 shows an opposed piston engine. It consists of two pistons moving in opposite direction in the same cylinder and connected to the same crankshaft. The ports are so designed that the exhaust ports are covered or uncovered by one piston while the inlet ports are covered or uncovered by the other. During the expansion stroke, each piston moves away from the other from their TDC position, and firstly, the exhaust ports are uncovered for exhaust of combustion products. Some crank degrees later, when the cylinder pressure has been considerably reduced due to further expansion, the inlet ports are uncovered and air (or mixture) compressed by a separate blower begins to enter the cylinder. The shape of the inlet ports should be such as to give the incoming air a swirling motion, in a place normal to the cylinder axis. The function of this swirling motion is to prevent the mixing of fresh charge with products of combustion. During the upward motion of the piston (i.e., when the pistons move towards each other), the exhaust ports close first and the inlet ports later. This results in the filling up of the cylinder with fresh charge up to nearly fully intake manifold pressure during induction period. This type of engine is sometimes, also called, *uniform scavenging engine.*

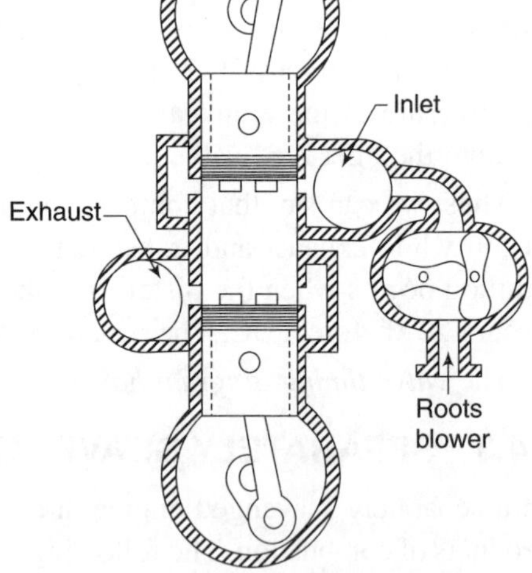


**Fig. 14.4** Opposed-piston engine

### 14.3.3 Port and Poppet Valve Engines (Uniflow Scavenged)

Figure 14.5 shows the details of a port and poppet valve engine. In this case also, the scavenging process is an uniflow one. *The admission of fresh air through inlet ports is controlled by the piston, while the products of combustion are removed through the poppet valves provided in the cylinder head.* Provision for adjusting the timing of the opening and closing of the exhaust valves to suit engines speed and scavenging air pressure, is the main asset of this type of engine.

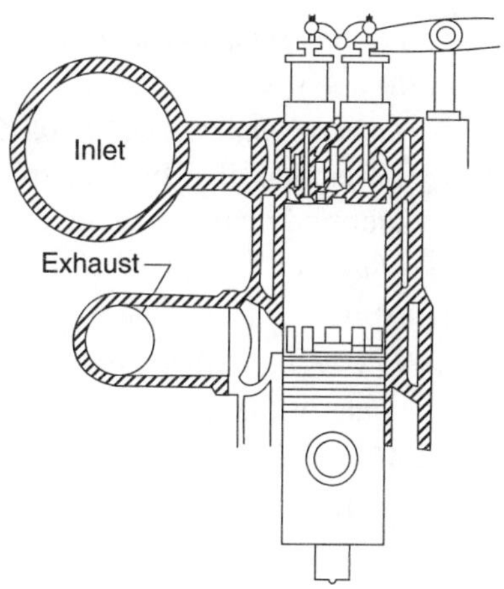


**Fig. 14.5** Port and poppet valve engine

## 14.4 ADVANTAGES OF TWO-STROKE ENGINE OVER FOUR-STROKE ENGINES

The following are the advantages of two-stroke engines:

(i) *Higher power per cycle.* Since the two-stroke engine has twice as much power strokes as a four-stroke engine at the same speed, hence, the two-stroke engine, theoretically develops almost twice the output per revolution of the crankshaft.

(ii) *Simple in design.* For the same output, the two-stroke engines are simple in design and easy to manufacture and operate. Generally, two-stroke engines are employed where small power or very big power is needed, such as auto cycles, scooters, motorcycles, big marine ships, submarines, etc.

(iii) *Easy to start.* It is far easier to start a two-stroke engines than a four-stroke one.

(iv) *Lesser size of flywheel.* Since there is one power stroke for each revolution of the crankshaft, the turning moment is more uniform. Hence, a smaller flywheel is needed.

(v) *Higher mechanical efficiency.* The two-stroke engine has a higher mechanical efficiency as there are no valves, rockers; cams camshafts, which cause friction losses. Also because of its simple construction, the engine's cost is low.

(vi) *Ease of reversing.* The reversing gear mechanism is simpler than the four-stroke engine.

(vii) *Less space and lighter foundation.* It occupies less space, needs lighter foundations, and requires few spare parts.

## 14.5 DISADVANTAGES OF TWO-STROKE ENGINE OVER FOUR-STROKE ENGINES

(i) *Overheating of piston.* Since there is power stroke in every revolution of the crankshaft, the engine has little time for removing heat and the pistons gets overheated. *This imposes a limit on the speed, and the cooling of the piston is done by oil. This is the main reason why two-stroke engines are not used for higher speed and greater output (say passenger cars) even though it develops theoretically twice the power as compared to a 4-stroke engine for the same speed.*

(ii) *Lower compression ratio.* The effective compression ratio in a two-stroke engine is lower than that of four-stroke for the same stroke and clearance volume. Since the thermodynamic efficiency of the engine depends on compression ratio, *hence the efficiency of a two-stroke engine is lower.*

(iii) *Noisy exhaust.* Due to the sudden releases of burnt gases, the exhaust is noisy.

(iv) *Lesser thermal efficiency.* The thermal efficiency of two-stroke engine is lower as some of the unburnt charge escape and also due to poor scavenging.

(v) *More lubrication.* The two-stroke engine requires more lubrication because of high operative temperature. Moreover, petrol vapours contaminate the lubricating oil in the crankcase.

(vi) *Higher fuel consumption.* Due to opening of inlet and exhaust port at the same time, *some of the fresh charge escapes without doing any work.* Hence, the two-stroke engines which are scavenged by fresh charge have higher fuel consumption.

(vii) *Less charge sucked in.* In crankcase scavenged engines, the space occupied by the crank prevents a full charge being sucked in. A two strokes petrol engine may contain only 50 to 55% of the swept volume, filled by a fresh charge, as against 80 to 95% in the case of four strokes engines.

## 14.6 COMPARISON OF TWO-STROKE SI AND CI ENGINES

(a) **Fuel loss.** The two-stroke SI engine suffers from fuel loss due to mixing of fuel-air before entering the cylinder whereas CI engine does not suffer from this due to direct injection of fuel at the end of compression stroke. The problem of two-stroke SI engine may be solved if fuel is injected before the exhaust port close through a fuel injection system instead of carburettor.

(b) **Idling loss.** The idling efficiency of two-stroke SI engine is poor. The two-stroke engine picks up only gradually and may even stop at low speed due to dilution of fresh charge with exhaust gas. At low speeds, there may be backfiring due to slow burning rate. This will be minimized if fuel is injected. *In CI engine, there is no difficulty at idling because the fresh charge is air only and it is not diluted by exhaust gas.*

## 14.7 SCAVENGING PARAMETERS AND MODELS

Various terminologies are used for defining the scavenging parameters and models.

### 14.7.1 Delivery Ratio ($R_{del}$)

The delivery ratio ($R_{del}$) is defined as

$$R_{del} = \frac{\text{Mass of delivered air or mixture per cycle}}{\text{Reference mass}}$$

or
$$R_{del} = \frac{m_{del}}{m_{ref}} \quad ...(14.1)$$

*It compares the actual scavenging air mass (or mixture mass) to that required in an ideal charging process.* The reference mass is defined as displaced volume of ambient air (or mixture density) at *intake conditions. The delivery ratio for diesel engines in the case of uniflow scavenging usually varies from **1.2 to 1.4** except for crankcase scavenged engine where it is less than unity.*

### 14.7.2 Trapping Efficiency ($\eta_{tr}$)

The trapping efficiency is defined as

$$\eta_{tr} = \frac{\text{Mass of delivered air (or mixture) retained}}{\text{Mass of delivered air (or mixture)}}$$

or

$$\eta_{tr} = \frac{m_{ret}}{m_{del}} \qquad \text{(Ranging 80\% to 85\%)} \qquad \text{...(14.2)}$$

*It indicates what fraction of air or mixture supplied to the cylinder is retained in the cylinder.* It is mainly controlled by the geometry of the port and the overlap time.

### 14.7.3 Scavenging Efficiency

The scavenging efficiency is defined as

$$\eta_{sc} = \frac{\text{Mass of delivered air (or mixture) retained}}{\text{Mass of trapped cylinder charge}}$$

or

$$\eta_{sc} = \frac{m_{ret}}{m_{trap}} = \frac{m_{ret}}{m_{ret} + m_{residual}} \qquad \text{...(14.3)}$$

*It indicates to what extent the residual gas in the cylinder has been replaced with fresh charge. The scavenging efficiency is around 85 per cent*

### 14.7.4 Purity of Charge (Purity)

The purity of the charge is defined as

$$\text{Purity} = \frac{\text{Mass of air in trapped cylinder charge}}{\text{Mass of trapped cylinder charge}}$$

or

$$\text{Purity} = \frac{(m_{air})_{trap}}{m_{trap}} \qquad \text{...(14.4)}$$

*It indicates the degree of dilution with burnt gases, of the unburnt mixture in the cylinder.* It is largely controlled by the shape of the combustion chamber and the scavenging arrangement (cross, loop or uniform).

### 14.7.5 Charging Efficiency

The charging efficiency is defined as

$$\eta_{ch} = \frac{\text{Mass of delivered air (or mixture) retained}}{\text{Displaced volume} \times \text{Ambient air density}}$$

or $$\eta_{ch} = \frac{m_{ret}}{V_{disp} \times \rho_{amb}} \qquad ...(14.5)$$

*It indicates how effectively the cylinder volume has been filled with fresh air (or mixture).*

There is an interrelationship between charging efficiency, trapping efficiency and delivery ratio which is expressed as

$$\eta_{ch} = R_{del} \cdot \eta_{tr} \qquad ...(14.6)$$

When the reference mass in the definition of delivery ratio is the trapped cylinder mass ($m_{tr}$), then

$$\eta_{sc} = R_{del} \cdot \eta_{tr} \qquad ...(14.7)$$

In real scavenging processes, the situation is slightly different. Here, mixing occurs as the fresh charge displaces the burnt gases and some of the fresh charge may be expelled. This process may be *modelled* by two limiting cases:

(a) Perfect displacement scavenging model

(b) Complete mixing scavenging model

It the burnt gases are pushed out by the fresh gases without any mxing, it is called **perfect displacement scavenging model**. If entering fresh mixture mixes instantaneously and uniformly with the cylinder contents, it is called **complete mixing scavenging model**.

For perfect displacement model (with $m_{tr}$) as the reference mass in the delivery ratio,

$$\eta_{sc} = R_{del} \quad \text{and} \quad \eta_{tr} = 1 \text{ for } R_{del} \leq 1$$

$$\eta_{sc} = 1 \quad \text{and} \quad \eta_{tr} = R_{del}^{-1} \text{ for } R_{del} > 1 \qquad ...(14.8)$$

For complete mixing model, the mass for air delivered between $t$ and $t + dt$ which is retained is given by

$$dm_{ar} = dm_{ad}\left(1 - \frac{m_{ar}}{m_{tr}}\right) \qquad ...(14.9)$$

where $m_{ar}$ = mass of air retained $\qquad m_{ad}$ = mass of air delivered

$m_{tr}$ = mass of air trapped

Assuming $m_{tr}$ as constant, the integration over the duration of the scavenging process gives

$$\frac{m_{ar}}{m_{tr}} = 1 - \exp\left(\frac{m_{ad}}{m_{tr}}\right)$$

Thus, the scavenging efficiency is given by

$$\eta_{sc} = 1 - e^{-R_{del}}$$

$$\eta_{tr} = \frac{1}{R_{del}}\left(1 - e^{-R_{del}}\right) \qquad \text{...(14.10)}$$

Figure 14.6 shows the variation of $\eta_{sc}$ and $\eta_{tr}$ as a function of delivery ratio for perfect displacement and complete mixing models.

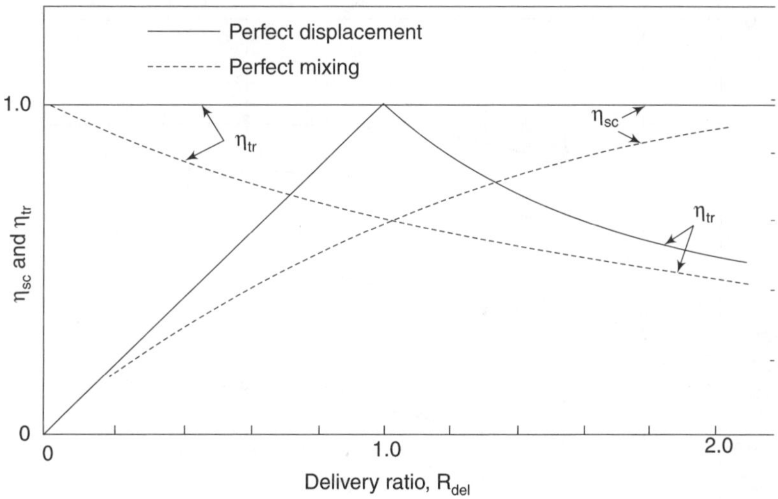


**Fig. 14.6** Scavenging efficiency ($\eta_{sc}$) and trapping efficiency ($\eta_{tr}$) vs. delivery ratio ($R_{del}$)

## 14.8 FLOW THROUGH PORTS

The *port in the cylinder may be rectangular, tapered and may have axial and tangential inclination.*

Figure 14.7 shows the flow patterns expected downstream of piston controlled inlet ports. For small openings, the flow remains attached to the port walls while for fully open ports with sharp corners, the flow detaches at the upstream corners. The flow detachment within the port is prevented by both a rounded entry and converging taper.

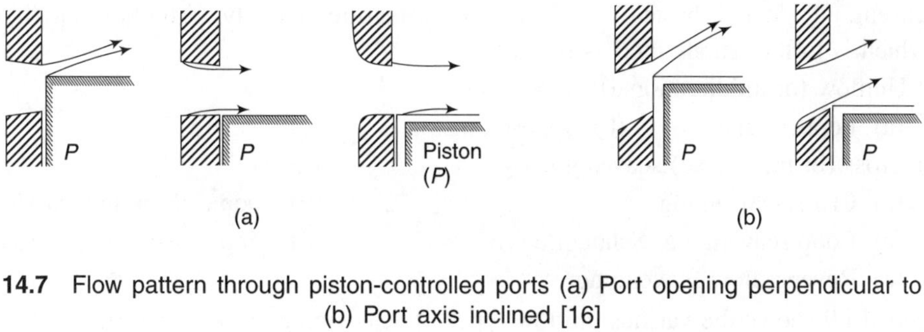


**Fig. 14.7** Flow pattern through piston-controlled ports (a) Port opening perpendicular to wall (b) Port axis inclined [16]

The pressure ratio across the exhaust port varies substantially during the exhaust process. The pressure ratio has a significant effect on the exhaust port discharge coefficient. As the pressure ratio increases, coefficient of discharge also increases at any port open fraction as obvious from Fig. 14.8. For any pressure ratio, the discharge coefficient slightly increases with port open fraction.

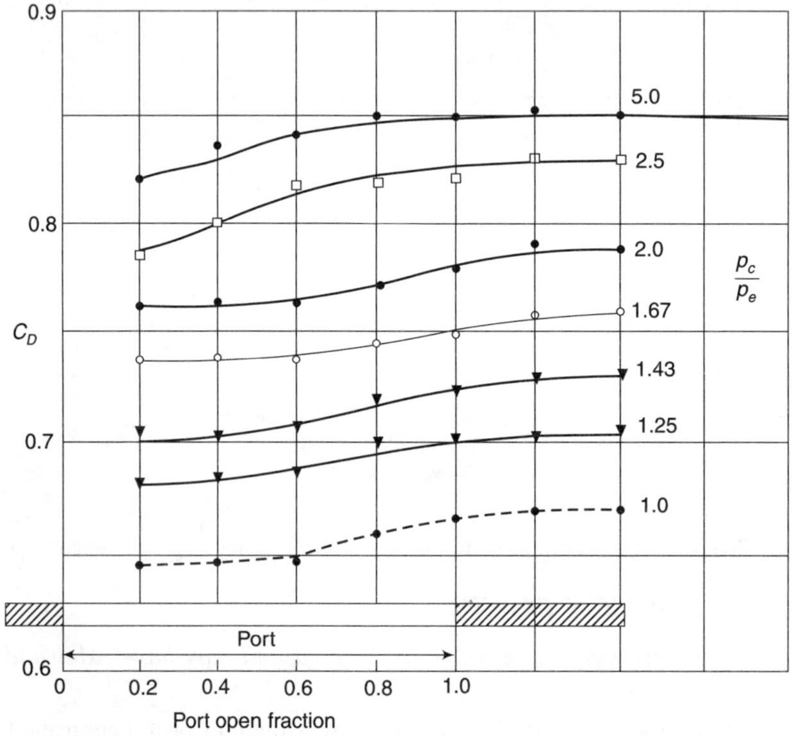


**Fig. 14.8** Discharge coefficient of a single rectangular exhaust port vs. port open fraction

## 14.9 SCAVENGING SYSTEMS

The scavenging system is basically divided in two categories, namely: (i) uniflow and (ii) cross-flow which is further subdivided as follows:

(i) Uniflow (or unidirectional) scavenging
- (a) Poppet valve
- (b) Opposed piston
- (c) Sleeve valve

(ii) Cross (or transverse) scavenging
- (a) Cross scavenging
- (b) Loop scavenging, MAN type
- (c) Loop scavenging, Schnuerle type
- (d) Loop scavenging, Curtis type.
- (e) Reverse flow scavenging.

Figure 14.9 shows the various uniflow types of scavenging system and Fig. 14.10 shows various cross scavenging systems.

### 14.9.1 Uniflow Scavenging

*The uniflow scavenging system is the most perfect method.* In this case, the fresh air charge is admitted at one end of the cylinder and the exhaust escapes at the other end (Fig. 14.9 a, b, c). There is a little short-circuiting between the intake and exhaust openings as the air flow is from end to end. The uniflow scavenging system utilizes three arrangements, namely, ***poppet valve, opposed piston and sleeve valve***. The poppet valve is used either to admit or discharge the charge. Opposed pistons are used to control the inlet and exhaust ports. In sleeve valve arrangement, the inlet and exhaust ports are controlled by the combined motion of piston and sleeve. It is worth to note that all uniflow systems permit unsymmetrical scavenging and supercharging as shown in Fig. 14.9. ***The uniflow system offers highest scavenging efficiency***.

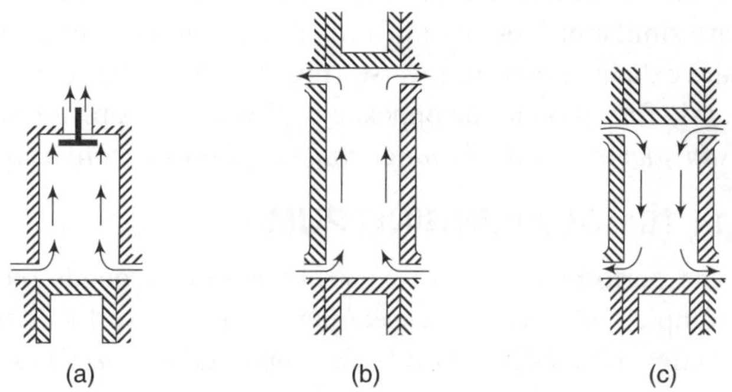


**Fig. 14.9** Uniflow scavenging system

### 14.9.2 Cross Scavenging

*Most small engines are cross scavenged. The cross scavenging system (Fig. 14.10a) incorporates inlet and exhaust ports placed in opposite sides of the cylinder wall.* The air enters from one side, moves up to combustion chamber on one side of the cylinder and then comes down on other side to flow out of the exhaust ports. This is only possible by proper guidance to the flow either by providing deflecting shape of the piston or by inclined ports. *The inlet and exhaust valves are opened and closed by the movement of piston. This system is simpler in construction as compared to uniflow but results in poor scavenging efficiency.*

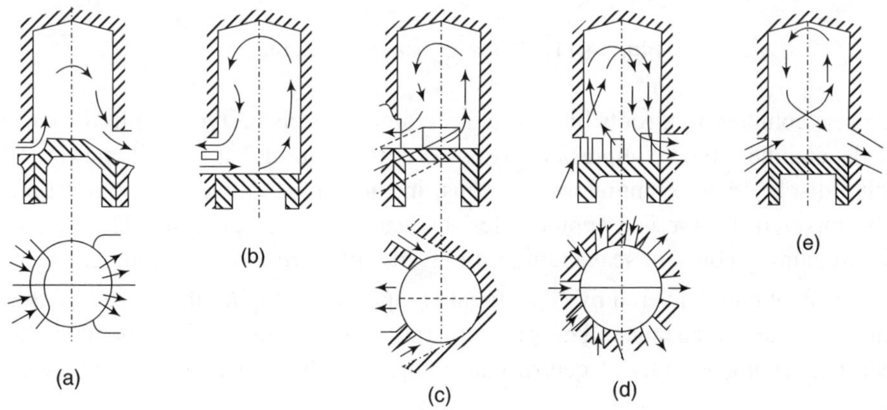


**Fig. 14.10** Cross, loop and reverse scavenging

### 14.9.3 Loop and Reverse Scavenging

Figure 14.10 b, c, d, e show various arrangements of **loop and reverse scavenging**. *These systems avoid the short circulating of the cross-scavenged engine and thus results in better scavenging efficiency. The inlet and exhaust ports are on the same side of the cylinder wall.* The fresh air first sweeps across the piston top, moves up and then down finally out through exhaust ports. In the MAN type loop scavenge (14.10 b), the exhaust port is above the inlet while in Schnuerle type (Fig. 14.10 c) the ports are side by side. The Curtis and Schnuerle are similar except that in Curtis the upwardly directed inlet ports are also placed opposite to the exhaust ports. In reverse (Fig. 14.10 c), the inclined ports are used and the scavenging air is forced on to the opposite wall where it is reversed to the outlet ports. *This system gives satisfactory results in long stroke engines running at low speed.*

## 14.10 SCAVENGING PUMPS

The performance of a two-stroke engine depends on the type of scavenging system and compressor used as a scavenging pump. Figure 14.11 shows the various types of scavenging pumps used such as *crankcase compression, roots blower and centrifugal compressor.*

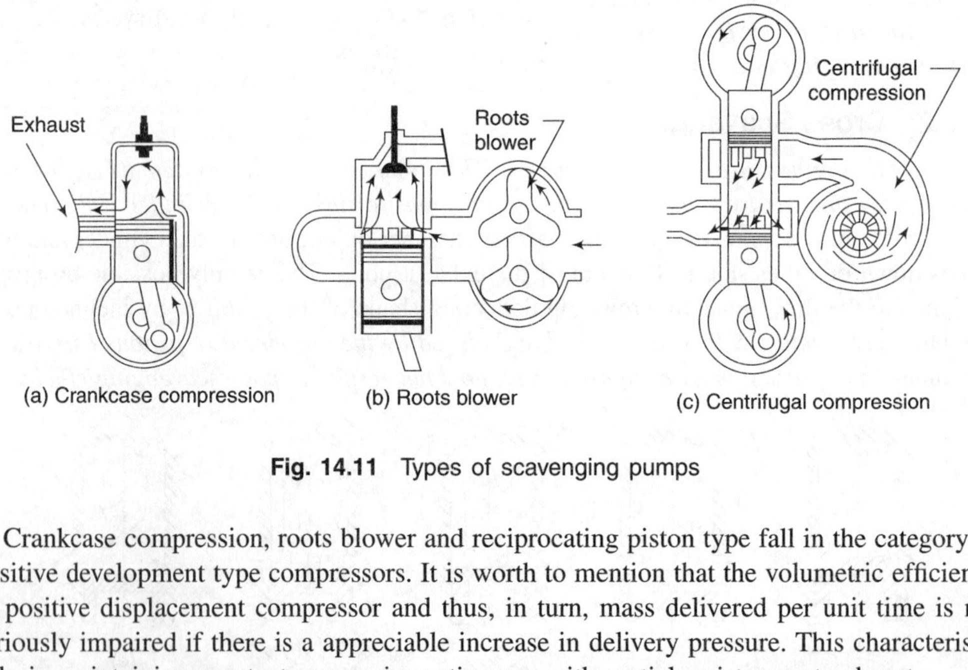


**Fig. 14.11** Types of scavenging pumps

Crankcase compression, roots blower and reciprocating piston type fall in the category of positive development type compressors. It is worth to mention that the volumetric efficiency of positive displacement compressor and thus, in turn, mass delivered per unit time is not seriously impaired if there is a appreciable increase in delivery pressure. This characteristic help in maintaining constant scavenging ratio even with partial resistance at exhaust.

The mass flow rate delivered by centrifugal compressor is highly influenced by the engine and exhaust system resistance. The pressure ratio valves almost linearly with the speed required. This characteristics of centrifugal compressor help in increasing the scavenging efficiency.

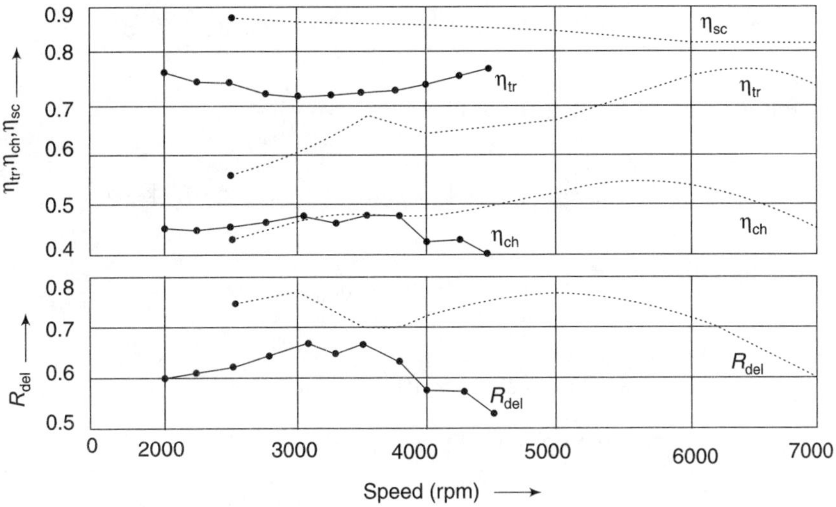


**Fig. 14.12** Delivery ratio ($R_{del}$), trapping efficiency ($\eta_{tr}$) and scavenging efficiency ($\eta_{sc}$) vs. speed for two-stroke SI engines solid line is for 147 cm$^3$ engine, dashed line for loop-scavenged 246 cm$^3$ engine. [14, 15].

## 14.11 PERFORMANCE OF SCAVENGING SYSTEMS

Figure 14.12 shows the delivery ratio, trapping, charging and scavenging efficiencies of two-stroke crankcase scavenged spark ignition engines as a function of engine speed. *The scavenging efficiency decreases with speed while delivery ratio first increases and then decreases as speed increases.*

The comparison of various scavenging system is shown is Fig. 14.13 considering scavenging efficiency and brake mean effective pressure (bmep) as dependent parameters and delivery ratio ($R_{del}$) as independent parameters. *It is obvious from this figure that cross scavenged engine is the least efficient and offers lowest brake mean effective pressure. Uniflow scavenging is the best as it provides better scavenging efficiency and bmep. Loop scavenged engine lies in between.*

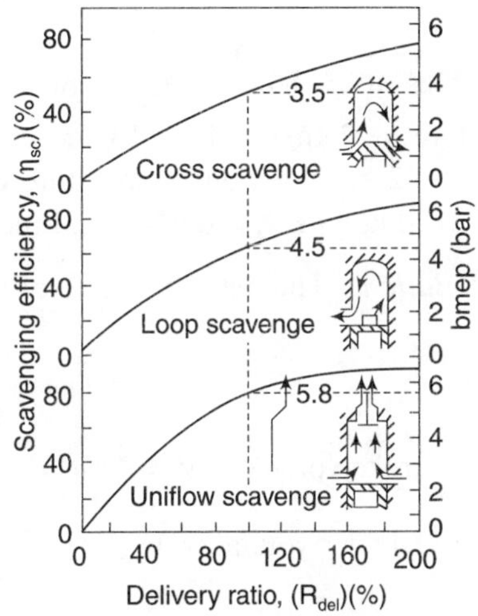


**Fig. 14.13** Comparison of scavenging systems

**EXAMPLE 14.1** A large two-stroke SI engine running at 3000 rpm with 2000 cc and compression ratio 8 consumes 11 kg of fresh charge per minute. The exhaust pressure is 1 bar and the temperature at inlet is 310 K. The scavenging efficiency of the engine is 0.65. Find out the delivery ratio and the trapping efficiency.

***Solution*** Cylinder volume $= V_{cyl} = V_s \times \frac{r}{r-1} = 2000 \times \frac{8}{7} = 2285.7$ cc

The ambient air scavenging density of charge $\rho_{sc} = \frac{p_{ex}}{RT_{inlet}}$

$$= \frac{10^5}{287 \times 310} = 1.124 \text{ kg/m}^3 = 1.124 \times 10^{-6} \text{ kg/cc}$$

The scavenging efficiency $= \eta_{sc} = \frac{m_{ret}}{m_{trap}} = \frac{m_{ret}}{m_{ret} + m_{residual}} = \frac{m_{ret}}{\rho_{sc} \times V_{cyl}}$

or $$0.65 = \frac{m_{ret}}{1.124 \times 2285.7 \times 10^{-6}}$$

or $$m_{ret} = 1669.93 \times 10^{-6} \text{ kg/cycle}$$

The delivery ratio $= R_{del} = \frac{m_{del}}{m_{ref}} = \frac{\text{Mass of fresh charge per cycle}}{V_{cyl} \times \rho_{sc}}$

$$= \frac{11/3000}{1.124 \times 2285.7 \times 10^{-6}} = 1.427$$ ***Ans.***

Trapping efficiency $= \eta_{tr} = \frac{m_{ret}}{m_{del}} = \frac{\eta_{sc}}{R_{del}} = \frac{0.65}{1.427} = 0.455$ ***Ans.***

**EXAMPLE 14.2** The following data relate to a two stroke diesel engine. Bore = 12 cm, stroke = 13 cm, compression ratio = 16, speed = 2000 rpm, exhaust pressure = 1.02 bar, inlet air temperature = 310 K, fuel flow rate = 5 kg/h, air supplied by blower = 130 kg/h, F/A = 0.044. Find the scavenging efficiency and delivery ratio.

***Solution*** The mass flow rate of air supplied to the engine

$$= \dot{m}_{del} = \frac{\dot{m}_f}{F/A} = \frac{5}{0.044} = 113.6 \text{ kg/h}$$

Stroke volume $= V_s = (\pi/4)\, d^2 \times L = (\pi/4) \times (0.12)^2 \times 0.13 = 0.00147 \text{ m}^3$

Cylinder volume $= V_{cyl} = V_s \times \frac{r}{r-1} = 0.00147 \times \frac{16}{15} = 0.001568 \text{ m}^3\text{/cycle}$

The scavenging efficiency $= \eta_s = \frac{m_{del}}{m_{ref}} = \frac{113.6/(60 \times 2000)}{0.001568 \times \rho_{sc}}$

But $\rho_{sc} = \frac{p}{RT} = \frac{1.02 \times 10^5}{287 \times 310} = 1.1464 \text{ kg/m}^3$

Hence, $\eta_{sc} = \frac{113.6/(60 \times 2000)}{0.001568 \times 1.1464} = 0.5264$

$$\text{Delivery ratio} = R_{del} = \frac{\dot{m}_{as}}{V_{syl}} = \frac{13/(60 \times 2000)}{0.001568} = 0.69 \qquad \textbf{Ans.}$$

$$\therefore \text{ Trapping efficiency} = \eta_{trap} = \frac{\eta_{sc}}{R_{del}} = \frac{0.5264}{0.69} = 0.762 \qquad \textbf{Ans.}$$

## 14.12 SOME ASPECTS OF TWO-STROKE SI ENGINES

In general, a two-stroke SI engine employs sealed crankcase suction and compression of the charge before the charge is transferred to the engine cylinder through the transfer ports. This suggests that the fresh charge should be compressed to a pressure higher than the pressure of the burnt gases at the end of expansion stroke to ensure an effective scavenging of burnt gases. During scavenging process, a part of fresh charge goes directly to the exhaust system which is supposed to be one of the major weaknesses of a two-stroke engine. *As a result, the fuel consumption increases and there are excessive emissions of unburnt hydrocarbons. The fuel consumption depends upon charging and trapping efficiency* (Fig. 14.12).

To overcome some of the disadvantages associated with *conventional 2-stroke SI engines, in-cylinder fuel injection system is used. Fuel is supplied at a pressure of about 5.57 bar and air is supplied at 5 bar to atomize the fuel to spray which has a droplet size less than 10 microns.* Obviously, $NO_x$ emissions are significantly lower than from 4-stroke engine owing to high residual gas fraction resulting from low charging efficiency. *But unburnt hydrocarbon emissions are about five times as high as those of equivalent 4-stroke engines.*

## 14.13 SOME ASPECTS OF TWO-STROKE CI ENGINE

*The automotive applications of two-stroke CI engines are rare, however, large two-stroke CI engines are widely used in stationary, locomotives and marine applications. Large marine diesel engine employing cylinder bore ranging from 0.4 m to 1 m are turbocharged to achieve high mean effective pressures and specific output. Obviously, they are low speed engines and employ direct drive arrangement for propulsion of ships.* It is interesting to note that after a continued research work of about 25 years, the output per cylinder gets increased by a factor of about 2 but the fuel consumption also gets decreased by *25% by improving in-cylinder injection system, combustion process, supercharging and scavenging*. In order to obtain higher scavenging efficiency and increased expansion stroke, the *uniflow scavenging system is being preferred to loop scavenging.*

## 14.14 SUMMARY OF FORMULAE

(1) $R_{del} = \dfrac{m_{del}}{m_{ref}}$ (2) $\eta_{tr} = \dfrac{m_{ret}}{m_{del}}$

(3) $\eta_{sc} = \dfrac{m_{ret}}{m_{trap}}$ (4) $\text{Purity} = \dfrac{(m_{air})_{trap}}{m_{trap}}$

(5) $\eta_{ch} = \frac{m_{ret}}{V_{disp} \times \rho_{amb}}$ (6) $\eta_{sc} = R_{del} . \eta_{tr}$

(7) $\eta_{tr} = \frac{1}{R_{del}} (1 - e^{-R_{del}})$

## PROBLEMS

### Things to Think About

14.1 What is the reason that even though two-stroke engine develops theoretically twice power than 4-stroke engine, but 2-stroke engines are not used in passenger car engines?

14.2 What is the reason that two-wheelers motorcycles engines now employ 4-stroke engines instead of 2-stroke enginers?

14.3 What are the critical parameters that need special attention in the further development of 2-stroke engine?

14.4 What do you mean by scavenging? What type scavenging system is used in diesel locomotives?

### Short and Long Theoretical Questions

14.5 Differentiate between two-stroke and four-stroke engines.

14.6 Discuss the advantages and disadvantages of two-stroke engines over four-stroke engines.

14.7 With the help of sketch discuss the working principle of crankcase and separately scavenged two-stroke engines.

14.8 Define the following scavenging parameters:

(a) Delivery ratio (b) Trapping efficiency

(c) Scavenging efficiency (d) Charging efficiency

(e) Purity

14.9 Discuss the modelling of scavenging process.

14.10 Discuss the various scavenging systems.

14.11 Discuss the various scavenging pumps.

14.12 Discuss the comparison between various scavenging systems.

14.13 Discuss the performance of scavenging process as a function of speed.

### Software Based Questions

14.14 Develop a software for the predictions of performance of all types of 2-stroke engines.

14.15 Develop a software for the design of 2-stroke engines used in large ships. Validate the results with existing engines.

### Design and Open Ended Question

14.16 Prepare a comprehensive report on all types of two-stroke engines used in various size of ships with all specifications and design features. Suggest methods to improve the design for further enhancement of performance.

### Multiple-Choice Questions

14.17 In a two-stroke engine with crankcase compression and crankcase valve, the number of ports in the cylinder is

(a) two (b) three (c) one (d) nil

14.18 The delivery ratio for a two-stroke SI engine with crankcase scavenging lies between

(a) 0.1 to 0.4 (b) 0.4 to 0.8 (c) 0.6 to 0.95 (d) 0.5 to 0.9

14.19 As a general rule, a uniflow scavenging system is preferred because it gives higher scavenging efficiency at the condition of

(a) low stroke-bore ratio (b) low bore stroke ratio
(c) higher expression ratio (d) lower expansion ratio

14.20 Scavenging efficiency is defined as

(a) mass of air trapped to mass of air delivered
(b) delivery ratio to trapping efficiency
(c) mass of delivered air retained to reference mass
(d) mass of delivered air retained to mass of the trapped cylinder charge.

14.21 Charging efficiency, trapping efficiency and delivery ratio are related as

(a) $\eta_{ch} = R_{del}/\eta_{trap}$ (b) $\eta_{ch} = \eta_{trap} \times R_{del}$
(c) $\eta_{ch} = R_{del}\ \eta_{trap}$ (d) $\eta_{ch} = \eta_{trap}/R_{del}$

14.22 The scavenging efficiency of a two-stroke SI engine having crankcase scavenging is given by

(a) $0.5 < \eta_{sc} < 0.7$ (b) 0.7 (c) $0.7 < \eta_{sc} < 0.9$ (d) > 0.9

## Key to Multiple-Choice Questions

14.17 (a) 14.18 (c) 14.19 (b)(d) 14.20 (d) 14.21 (b) 14.22 (c)

CHAPTER

# Measurements, Testing and Performance

15

## OVERVIEW

*Testing has been the keyword in the development of human civilization. It signifies the degree of conformation of conceived thought, idea, design, etc., with the actual performance and it covers the whole spectrum involved in getting the performance.* In the field of science and technology, the whole development depends on testing. Nuclear, engines and other tests are their examples.

*Development of IC engines aims to improve the power output, efficiency and reliability of the engines with the minimum cost and the minimum harmful emissions.* In order to achieve this goal, the development engineers tries various design concepts for better results. To know the effect on engine performance of a particular design concepts, one has to resort to testing and based on the results obtained, redesigning is done and testing process is continued. Thus, in the development process, a large number of engine tests are performed by varying the various governing parameters to get better performance. This needs a huge fund for research and development. Nowadays, after obtaining sufficient data from tests, the development work is done by ***computer's simulation (see chapter 17)*** and this saves a lot of money and time. This chapter deals with various types of engine testing and performance evaluation.

## 15.1 PERFORMANCE PARAMETERS

Engine performance in real sense is a relative term. *It indicates how well, one engine is doing its job with relation to the energy supplied in or the degree of success with which it provides useful energy (i.e., conversion of chemical energy to mechanical work) in relation to some comparable engines.* The following parameters are chosen for the evaluation of performance of IC engines.

(a) Engine power (P)

(i) Indicated mean effective pressure (imep); (ii) indicated power (ip); (iii) brake mean effective pressure (bmep); (iv) brake power (bp); (v) torque (T); (vi) friction power (fp)

(b) Engine efficiencies ($\eta$)—(i) air standard efficiency ($\eta_{as}$); (ii) Indicated thermal efficiency ($\eta_{ith}$); (iii) brake thermal efficiency ($\eta_{bth}$); (iv) mechanical efficiency ($\eta_m$); (v) relative efficiency ($\eta_r$); (vi) volumetric efficiency ($\eta_v$); (vii) scavenging efficiency ($\eta_s$), and scavenging ratio ($R_s$); (viii) charge efficiency($\eta_{ch}$); (ix) combustion efficiency ($\eta_{comb}$)

(c) Air-fuel ratio (A/F)

(d) Brake specific fuel consumption (*bsfc*)

(e) Speed and main piston speed ($N$, $c_m$)

(f) Engine load

(g) Specific weight ($w_s$)

(h) Spark or injection timing ($\tau_s$)

(i) Compression ratio ($r$)

(j) Mass of inducted charge ($m_c$)

(k) Heat losses

(l) Emission

(m) Inlet pressure and temperature

The above-mentioned parameters may be grouped in two categories, namely,

(i) **Independent parameters ($N$, $c_m$ load, A/F, $r$, $\tau_s$, etc.)**

(ii) **Dependent parameters ($P$, $\eta$, $sfc$, $m_c$, etc.)**

These are expressed as

**Dependent parameters = *f* (independent parameters)**

or $$(P, \eta, sfc, m_c, \text{etc.}) = f\,(N, c_m \text{ Load}, A/F, r, \tau_s, \text{etc.})$$

*A discussion of above performance parameters has already been given in chapter 1.*

The following points are to be considered in evaluating the performance of an engine.

(i) The maximum power or torque, thermal efficiency and bsfc available at each speed within the useful range of speed **at constant load**.

(ii) The maximum power or torque, thermal efficiency and bsfc available at each load within the useful range of load **at constant speed**.

(iii) Reliability and durability of the engine for the given range of operation.

## 15.2 ENGINE POWER

Basically, three terms are associated with engine power, namely, *indicated power* (ip), *friction power* (fp) and *brake power* (bp) and they are related as

$$ip = bp + fp \qquad ...(15.1)$$

Indicated power can be computed from the measurement of forces in the cylinder (i.e., indicated mean effective pressure). Brake power can be computed from the measurement of forces at the crankshaft (i.e., torque). Motoring of the engine may give frictional power. The various terms associated are discussed below.

### 15.2.1 Indicated Mean Effective Pressure ($p_{mi}$)

It is a well-known fact that pressure in the cylinder varies throughout the cycle and thus it is not possible to develop a simple expression to compute *ip*. Therefore, a concept of indicated mean effective pressure is modelled which is defined as follows.

*The indicated mean effective pressure can be defined as that theoretical constant pressure exerted (which can be imagined) during each power stroke of the engine to produce a work output equal to the indicated power.*

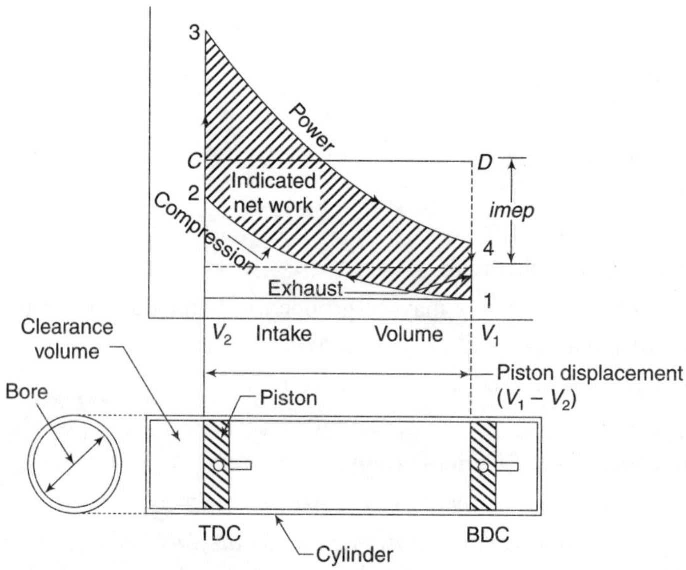


**Fig. 15.1** Concept of mean effective pressure

Referring to *p-V* diagram for an ideal four-stroke SI engine cycle shown in Fig. 15.1, the indicated mean effective pressure ($p_{mi}$) is expressed as

$$p_{mi} \times \text{Stroke volume} = \text{Net work of cycle (i.e., Area 1–2–3–4)}$$

or

$$p_{mi} = \frac{\text{Net work of cycle}}{\text{Stroke volume}} = \frac{W_{net}}{V_1 - V_2}$$

The indicated mean effective pressure is measured with the help of an indicator diagram obtained from an instrument known as ***indicator***. *The indicator diagram is graphical representation of pressure-volume variation during the actual working cycle.* The area enclosed by the actual cycle on indicator card may be measured by a **planimeter** or electronic indicator meter. The indicated mean effective pressure is given by

$$p_{mi} = \frac{\text{Area of indicator diagram in cm}^2 \times \text{Spring scale in bar/cm}}{\text{Length of diagram in cm}} \qquad ...(15.2)$$

### 15.2.2 Indicated Power (*ip*)

*The indicated power is the power actually developed inside the engine cylinder.* The expression for *ip* may be developed as follows

Let $A$ = Area of piston, $m^2$

$L$ = Stroke of piston, m

$p_{mi}$ = Indicated *mep*, bar

The average force on the piston = $100\ p_{mi}.A$ ; kN

∴ Indicated work done per cycle = $100\ p_{mi}.A.L\ \theta$ ; kJ

Indicated work done per minute

$$= 100\ p_{mi}\ A.L \times \text{Working cycles per minute } (n)$$

It is to be noted that most of the engines are single acting. In four-stroke cycle engines, there is one power stroke in two revolutions of crankshaft. In two-stroke engines, there is one power stroke in each revolution of crankshaft.

If $N$ is the speed of the engine is rpm and $n$ is the number of working cycles per minute, then

$n = N$ for single acting two-stroke engines

$n = \frac{N}{2}$ for single acting four-stroke engines

$$\therefore \quad ip = \frac{100\ p_{mi}\ L\ A\ n}{60} \text{ for single cylinder, kW} \qquad ...(15.3)$$

If $x$ is the number of cylinders, then *ip* for single acting and two-stroke engines

$$ip = \frac{100\ p_{mi}\ L\ A\ n\ x}{60}; \text{ kW} \qquad ...(15.4)$$

*ip* for single acting four-stroke engines with $x$ number of cylinders

$$ip = \frac{100\ p_{mi}\ L\ A\ N\ x}{60 \times 2}; \text{ kW} \qquad ...(15.5)$$

### 15.2.3 Brake Power

*The brake power (bp) is the useful power available at the crankshaft or clutch shaft.* The brake power is less than indicated power which is illustrated in energy flow diagram shown in Fig. 15.2.

The fuel is introduced to the combustion chamber (*A*) where it is burnt, converting chemical energy into heat. It would be desirable to utilise all this liberated energy to drive the piston but there are heat losses, primarily to the *exhaust, to the coolant, and to radiation.* The remaining energy, which may be converted to *indicated power (ip)* is utilised *to drive the piston.* The energy acting on the piston passes through the connecting rod and crankshaft (area *B*) to the drive shaft. Again, there are *mechanical energy losses due to friction, pumping power* required to *drive auxiliaries* and unaccounted–the sum of all these, converted to power is called *friction power (fp).* The remaining energy which is delivered to the drive shaft (area *C*) is useful energy for doing mechanical work which is termed as *brake power (bp).* Thus,

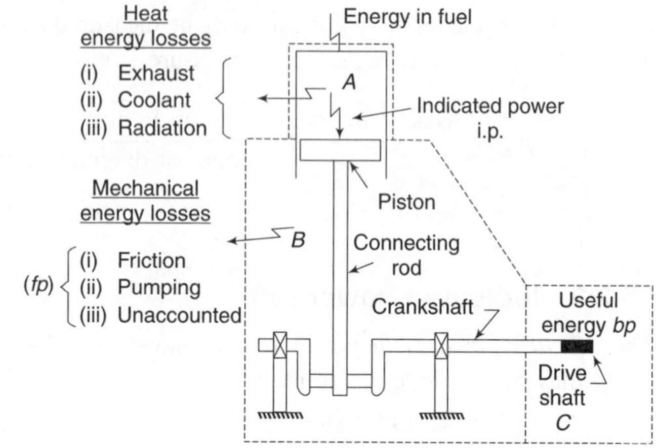


**Fig. 15.2** Energy flow in an IC engine

$$bp = ip - fp \qquad ...(15.6)$$

*For computing brake power, torque is measured by coupling a device* ***called dynamometer*** to the engine output shaft.

The brake power is expressed as

$$bp = \frac{2\pi N T}{60 \times 10^3}\ ;\ \text{kW} \qquad ...(15.7)$$

where $T$ = torque = $W \times R$ ; Nm $\quad W$ = load on the engine; $N$

$R$ = arm radius; m; $\quad N$ = rpm

Sometimes, *bp* as expressed in term of engine constant. Thus,

$$bp = \frac{WN}{C_1} \qquad ...(15.8)$$

where $C_1 = \frac{60 \times 10^3}{2\pi R}$ = an engine constant

*It is worth to note that the torque is the capacity of an engine to do work while power is the rate at which an engine does work.* By citing a simple example of *"a tractor pulling a given load"* may distinguish clearly between the two. *The torque developed by the engine will determine whether or not the tractor is capable of pulling the load, and the power developed will determine how fast the load can be pulled.*

Brake power (bp) is sometimes also expressed in terms of **brake mean effective pressure ($p_{mb}$)**. *The brake mean effective pressure expressed in bar is defined as that theoretical constant pressure (which can be imagined) exerted during each power stroke* of the engine to produce a power output equal to the brake power.

*bp* for single cylinder four-stroke IC engine is

$$bp = \frac{100\, p_{mb}\, L\, A\, N}{60 \times 2}\ ;\ \text{kW} \qquad ...(15.9)$$

*bp* for single cylinder two-stroke IC engine is

$$bp = \frac{100\, p_{mb}\, L\, A\, N}{60}\ ;\ \text{kW} \qquad ...(15.10)$$

Thus, for four stroke, $x$ number of cylinders engines

$$bp = \frac{2\pi\, N\, T}{60 \times 1000} = \frac{100\, p_{mb}\, L\, A\, Nx}{2 \times 60}\ ;\ \text{kW} \qquad ...(15.11)$$

or

$$p_{mb} = KT \qquad ...(15.11a)$$

where $K = \dfrac{4\,\pi}{10^5\, L\, A}$ = a constant for particular engine

Thus, we see that the *bmep (Eq. 18.11a) is directly proportional to the engine torque and is independent of engine speed.*

## 15.3 ENGINE EFFICIENCIES

For computing performance, various types of engine efficiencies have been conceptualised and some of them have been discussed briefly in chapter 1.

### 15.3.1 Air-standard Efficiency ($\eta_{as}$)

The expressions for air-standard efficiency for various type of engines are given in chapter 2. It gives higher value of efficiency as it does not consider losses.

### 15.3.2 Indicated Thermal Efficiency ($\eta_{ith}$)

*It is the ratio of indicated power (ip) to the input fuel energy, i.e., heat added.* Thus,

$$\eta_{ith} = \frac{ip}{q_A} = \frac{ip}{\dot{m}_f \times L.C.V.} \qquad ...(15.12)$$

where $LCV$ = Lower calorific value of fuel

### 15.3.3 Brake Thermal Efficiency ($\eta_{bth}$)

*Brake thermal efficiency is defined as the ratio of brake power to input fuel energy, i.e., heat added*

$$\eta_{bth} = \frac{bp}{q_A} = \frac{bp}{\dot{m}_f \times L.C.V.} \qquad ...(15.13)$$

$\eta_{bth}$ is less than $\eta_{ith}$

Generally, $\eta_{bth}$ varies from 28% to 32%.

### 15.3.4 Mechanical Efficiency ($\eta_m$)

Mechanical efficiency is defined as the ratio of brake power (*bp*) to indicated power (*ip*). Thus,

$$\eta_m = \frac{bp}{ip} = \frac{bp}{bp + fp} = \frac{p_{mb}}{p_{mi}} = \frac{\eta_{bth}}{\eta_{ith}} \qquad ...(15.14)$$

Generally, $\eta_m$ varies from 75% to 85% in IC engines

### 15.3.5 Air Capacity of Two-stroke Engines

Since during the admission period in two-stroke engines, the exhaust ports are opened therefore there is always a chance of some of the fresh charge escaping through the exhaust ports. Thus, *the power output in 2-stroke engine does not depend upon the amount of charge entering the cylinder but the amount of charge retained in the cylinder unlike 4-stroke engines.*

*The ideal air capacity of a two-stroke engine can be defined as the mass of air required to fill the total cylinder volume at inlet temperature and exhaust pressure.*

The ideal air capacity is

$$(\dot{m}_a)_{ideal} = x\,N\,V\,\rho_s \text{ ; kg/min}$$

where $x$ = Number of cylinders $\quad N$ = rpm

$V$ = Total cylinder volume; $\quad \rho_s$ = Scavenging density at $T_i$ and $p_e$

But

$$\rho_s = \frac{p_e}{RT_i} \text{ ; kg/m}^3$$

$\therefore$

$$(\dot{m}_a)_{ideal} = \frac{x\,N\,V\,p_e}{RT_i} \text{ ; kg/min} \qquad ...(15.15)$$

For a single cylinder engine ($x = 1$), we have

$$(\dot{m}_a)_{ideal} = \frac{N\,V\,pe}{RT_i} \text{ ; kg/min} \qquad ...(15.16)$$

### 15.3.6 Relative Efficiency or Efficiency Ratio ($\eta_{rel}$)

*The relative efficiency or efficiency ratio is the ratio of thermal efficiency of an actual cycle to that of ideal cycle. It serves as a useful criterion for indicating the degree of development of the engine.* Thus,

$$\eta_{rel} = \frac{\text{Actual thermal efficiency}}{\text{Air standard efficiency}} = \frac{\eta_{th}}{\eta_{as}} \qquad \text{...(15.17)}$$

$\eta_{rel}$ varies from 75% to 95%.

### 15.3.7 Volumetric Efficiency ($\eta_v$)

*Volumetric efficiency of an engine is defined as the ratio of actual mass of air induced by an engine on the intake stroke to the theoretical mass of air that should have been inducted by filling the piston displacement volume with air at atmospheric temperature and pressure.* Thus,

$$\eta_v = \frac{(\dot{m}_a)_{act}}{(\dot{m}_a)_{th}} = \frac{(\dot{m}_a)_{act}}{\rho_a\, n\, V_s} \qquad \text{...(15.18)}$$

where $n = \frac{N}{2}$ for four-strokes; $n = N$ for two-strokes

$\rho_a$ = density of air at atmospheric condition; $V_s$ = swept volume

*The name volumetric efficiency is a misnomer because it is a mass and not a volume ratio.* A detailed discussion on volumetric efficiency has been discussed in chapter 12.

*Thus, we see that volumetric efficiency is a measure of success with which the air supply and thus the charge is inducted into the engine.* In supercharged engines, the volumetric efficiency may be less or greater than unity and have the denominator of the above equation should be based on intake manifold condition.

The *volumetric efficiency* is affected by many variables such as

(i) *Speed.* As the speed of the engine increases, the volumetric efficiency decreases as high velocity of air flow results in large pressure drop.

(ii) *Inlet pressure and temperature.* At high pressure and low temperature is the fresh charge, the volumetric efficiency gets increased because of the fact that density is increased and a greater mass of charge can be inducted in a given volume.

(iii) *Heat transfer.* As the fresh charge arrives in the hot cylinder, heat is transferred to it from the hot chamber walls and hot residual gases, raising its temperature. This results in a decrease in the mass of the fresh charge admitted and hence a reduction in volumetric efficiency.

(iv) *Clearance ratio.* As the clearance ratio increases, the volumetric efficiency decreases.

(v) *The design of intake and exhaust manifolds.* The exhaust manifold should be so designed as to enable the exhaust products to escape readily, while the intake manifold should be such design so as to bring in the maximum possible fresh charge with the least resistance to flow for better volumetric efficiency.

### 15.3.8 Scavenging Efficiency ($\eta_{sc}$) and Scavenging Ratio ($R_s$)

In the analysis of two-stroke engines, the scavenging efficiency is analogous to volumetric efficiency for 4-stroke engines. The scavenging efficiency $\eta_{sc}$ may be defined as the ratio of mass of fresh mixture actually retained to the ideal mass retained. Hence,

$$\eta_{sc} = \frac{\text{Actual mass of air retained per minute}}{\text{Ideal mass of air retained per minute}}$$

$$\eta_{sc} = \frac{(\dot{m}_a)_{act,\ ret}}{(\dot{m}_a)_{ideal,\ ret}} = \frac{(\dot{m}_a)_{act,\ ret} \times RT_i}{N\,V\,p_e} \qquad ...(15.19)$$

**Scavenging Ratio ($R_s$)** *It may be defined as the ratio of total mass of air entering the cylinder per minute or per stroke to the ideal air capacity per minute or per stroke. This means it is a measure of the excess air used for scavenging.* Hence,

$$\text{Scavenging ratio} = \frac{\text{Total mass of air actually entering cylinder per minute}}{\text{Ideal air capacity per minute}}$$

or
$$R_s = \frac{(\dot{m}_a)_{act}}{(\dot{m}_a)_{ideal}} = \frac{(\dot{m}_a)_{act}}{N\,V} \times \frac{RT_i}{p_e} \qquad ...(15.20)$$

where $(\dot{m}_a)_{act}$ = total mass of air actually entering per minute

With the help of Eq. (15.19) and Eq. (15.20), a new term called *trapping efficiency* may be defined as follows

$$\eta_s = \frac{(\dot{m}_a)_{act,\ ret}}{(\dot{m}_a)_{ideal,\ ret}} = \frac{(\dot{m}_a)_{act}}{(\dot{m}_a)_{ideal,\ ret}} \times \frac{(\dot{m}_a)_{act,\ ret}}{(\dot{m}_a)_{act}}$$

or
$$\eta_s = \eta_{tr} \times R_s \qquad ...(15.21)$$

where $\eta_{tr}$ = Trapping efficiency = $\dfrac{(\dot{m}_a)_{act,\ ret}}{(\dot{m}_a)_{act}}$

A relation between scavenging efficiency and scavenging ratio may be established by

$$\eta_s = 1 - e^{-R_s} \qquad ...(15.22)$$

### 15.3.9 Charge Efficiency ($\eta_{ch}$)

*The charge efficiency indicates how well the piston displacement of a four-stroke engine is utilised.* It is affected by the following:

(i) Compression ratio
(ii) Heat transfer to the charge through manifold
(iii) Valve timing
(iv) Resistance offered to charge in intake manifold

### 15.3.10 Combustion Efficiency ($\eta_{comb}$)

*It is defined as the ratio of actual heat liberated in that combustion chamber to the theoretical heat in fuel.* The actual heat liberated is less than theoretical due to incomplete combustion and dissociation. Combustion efficiency varies from 92% to 97%.

$$\eta_{comb} = \frac{\text{Actual heat liberated}}{\text{Theoretical heat in fuel}}.$$

## 15.4 AIR-FUEL RATIO (A/F)

*Air-fuel ratio is defined as the ratio of mass of air to the mass of fuel.* Thus,

$$\text{A/F} = \frac{\dot{m}_a}{\dot{m}_f} \quad \text{...(15.23)}$$

The stoichiometric air-fuel ratio is around 15 which is a function of fuel.

## 15.5 SPECIFIC FUEL CONSUMPTION (*sfc*)

Specific fuel consumption is one of the most important parameters for the comparison of the engines. *It is defined as the ratio of the amount of kg of fuel used per hour by the engine to the power produced by the engine.* When sfc is based on bp, it is called bsfc.

$$bsfc = \frac{\dot{m}_f \text{ in kg/h}}{bp \text{ in kW}} \text{ ; kg of fuel/kWh} \quad \text{...(15.24)}$$

When *sfc* is based on *ip*, its called *isfc* and is given by

$$isfc = \frac{\dot{m}_f \text{ in kg/h}}{ip \text{ in kW}} \text{ ; kg/kWh} \quad \text{...(15.25)}$$

*bsfc* varies from 0.2 to 0.32 kg/kWh for diesel engines and 0.28 to 0.38 kg/kWh for petrol engines. *sfc is a comparative parameter that shows how efficiently an engine is converting fuel into work.*

## 15.6 BASIC MEASUREMENTS

The following are the basic measurements which should be undertaken to evaluate the performance of an engine on almost all tests.

(i) Speed
(ii) Fuel consumption
(iii) Air consumption
(iv) Friction power
(v) Indicated power
(vi) Brake power
(vii) Noise
(viii) Combustion phenomena
(ix) Emissions

In addition to above, other measurements may be made depending on type of tests.

## 15.7 TYPES OF TESTS

The tests on IC engine can be divided into two types:

1. **Test at variable speed**
2. **Test at constant speed (variable load)**

*Variable speed tests may be at* ***full load*** where the *maximum power and the minimum specific fuel consumption at each different speeds are the objectives.* In variable load tests (constant speed tests), the objective is to determine the variation in specific fuel consumption.

## 15.8 SPEED MEASUREMENT

Speed of the engine is a vital parameter. It is measured by the following instruments:

(i) Mechanical counters and timers
(ii) Mechanical tachometers
(iii) Stroboscope
(iv) Electric counters
(v) Electric generators
(vi) Electric pulse counters using magnetic pickup.

A mechanical or electrical tachometer may be used for measuring speed but both instruments are affected by temperature variation and are not very accurate. The best method of measuring speed is to count the number of revolutions in a given time. *For accurate and continuous measurements of speed, a magnetic pick-up produces a pulse for every revolution and the pulse counter will measure the speed of the engine.*

## 15.9 FUEL CONSUMPTION

Fuel consumption is generally expressed by two ways—by ***volume*** or by ***weight*** during a specific time. *For automobile engines, it is expressed in terms of kilometers per litre (km/litre) while for stationary engines, it is expressed in litre per hour (litre/h). A fuel efficient engine is one which gives the maximum km/h or the minimum litre/h.*

Fuel consumption by an engine is the key parameter and hence its accurate measurement is very essential. Though, its measurement looks to be a simple matter, but its accurate measurement, is confronted by the occurrence of *various undesirable phenomena* given below:

(i) *Vapour bubbles in fuel line.* The engine heat produces vapour bubbles in the fuel line. With the growth of bubbles, the fuel volume increases and back flow of fuel takes place. Some fuel flow meters measure this backflow as if it was forward flow. Some meters do not count backward flow but it counts only when bubble collapses and forward flow takes place. Thus, the measured flow may be less or more than correct one.

(ii) *Vapour bubbles before or inside flow meter.* The formation of vapour bubbles before or inside flowmeter will result in a higher value of measured flow rate than actual one.

(iii) *Presence of swirl.* The presence of swirl in the fuel flow especially in the case of turbine types flowmeter may register a higher flow rate.

(iv) *Density variation.* The density of fuel is a function of temperature which ranges from –10°C to 70°C. Due to density variation, error may crop in measurement.

(v) *Colour of fuel.* The colour of the fuel may affect the measurement of those flowmeters which use light beam.

(vi) *Water hammer type effect.* There is a possibility of formation of water hammer type effect in the float bowl and pipe line of carburettor type SI engine due to periodically opening and closing of fuel by the needle valve of float bowl. This water hammer type effect produces error in the measurement of turbine type flowmeter as it goes on rotating even when the fuel has stopped.

As mentioned above, basically there are two types of fuel measurement methods. They are:

(a) volumetric type (b) Gravimeter type

## 15.9.1 Volumetric Type Flowmeter

### *15.9.1.1 Simplest Method*

The age old simplest method of measuring volumetric fuel consumption is using glass bulbs of known volume and having a mark on each side of the bulb. The stopwatch may measure the time taken by the engine to consume this volume. The volume divided by the time will give the volumetric flow rate.

### *15.9.1.2 Burette Method*

Figure 15.3 shows burette consisting of two spherical glass bulbs having 100 cc and 200 cc capacity respectively for measuring fuel consumption. They are connected by three-way cocks so that one may feed the engine, while the other is being filled. Different capacities of glass bulbs are made to ensure the duration of test approximately constant irrespective of load. In order to avoid the error in sighting the fuel level against the mark on the carburettor, photocells are used. Stop watch is used to ensure the time taken. Volume divided by time gives the volume flow rate.

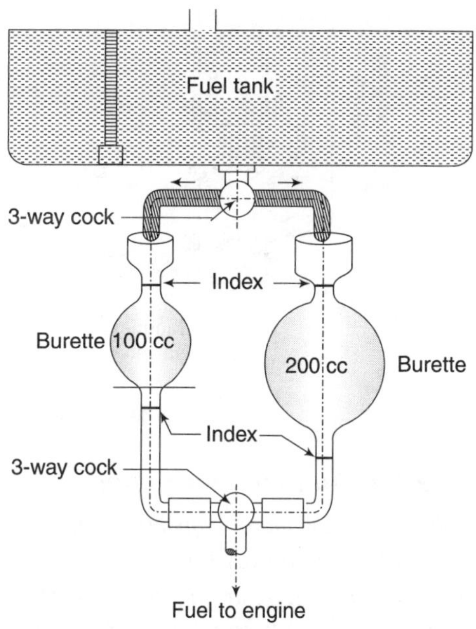


**Fig. 15.3** Burette method for measuring fuel consumption

### *15.9.1.3 Automatic Burette Flowmeter*

Automatic burette flowmeter is commercially available. Figure 15.4 shows the schematic diagram of such an automatic burette flowmeter. It consists of a measuring volume (*A*) which has a photocell (*B*) and a light source (*C*) fitted in a tabular housing. In order to form a point of light on the axis of measuring volume, these housings are put opposite to each other at an angle. It is necessary to provide an air cushion at supply line pressure and to store fuel during measurement, an equalization chamber (*D*) is connected to measuring tube via the air tube (*E*) and magnetic valve (*F*) and equalization pipe (*G*).

As soon as the start button is pressed, the lamps in two photoelectric systems light up and the magnetic valve stops the flow through the instrument. The fuel level in the measuring volume starts falling at a rate depending upon the engine consumption but at the same time an equal amount flows through the equalization tube to the equalization chamber. As soon as the fuel level reaches the upper measuring level (*H*), the focused beam of light from the lamp is reflected on to the opposite photocell and converted into an electrical signal to start a time counter. With the fall of fuel level further and reaching the lower measuring level the new signal generated stops the time counter. The lamps are automatically switched off and valves are opened and the regular flow is restored. By this way, the time period for the consumption of the chosen volume of fuel is thus recorded and thus the volume flow rate is known.

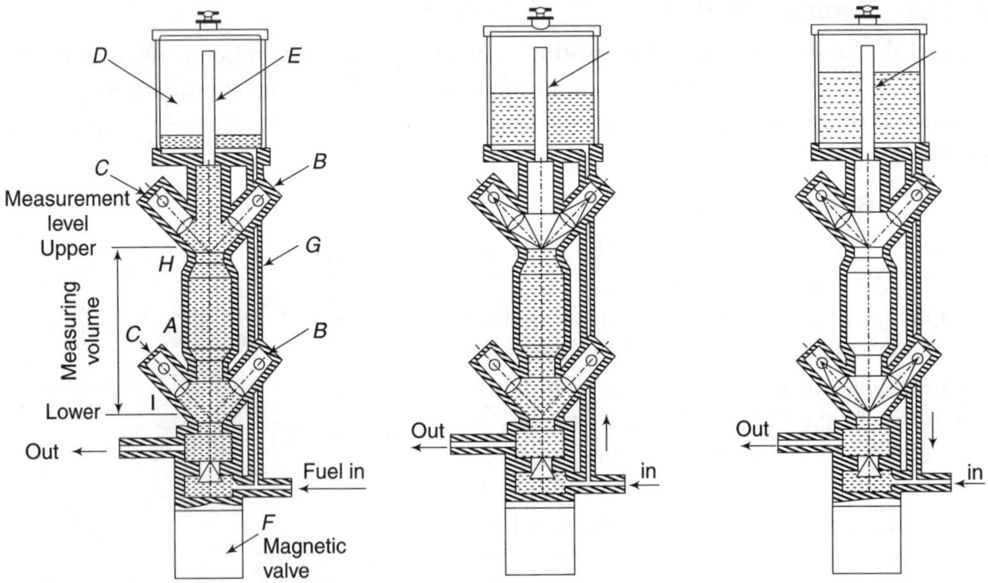


**Fig. 15.4** Automatic volumetric fuel flowmeter

### 15.9.1.4 Turbine Flowmeter

*The turbine flowmeter consists of a turbine which rotates due to the rate of change of momentum caused by fuel flow. A light beam is interrupted by the blades of the turbine rotor every time one goes by. The resulting signals are counted and the volumetric flow is displayed digitally on the screen.* A totalizing mode or a rate mode can be selected to give total flow or flow rate. To reduce the effect of pulsation in the fuel line, a bypass line is used. *It is suitable for large units.*

### 15.9.1.5 Four-Piston Flowmeter

It consists of four radial pistons connected to the same crankshaft. Fuel flows through the port uncovered by the pistons as they move on. The ports are arranged in such a way the fuel flow causes the crank to turn continuously. Two light beams are used to interrupt the crank ten times per revolution to know the magnitude and direction of rotation. The resulting pulses from the two light detectors are combined to determine the direction of rotation and counters are used to count which is displayed in the form of flow rate.

### *15.9.1.6 Orifice Flowmeters*

*Orifice flowmeter is one of the methods to measure level consumption.* Figure 15.5 shows such an orifice flowmeter. It consists of two orifices, *X* and *Y*. Either one or two orifices may be used at a time. *The flowmeters depend on the pressure drop across an orifice. The pressure drop may be calibrated in terms of volume flow rate.*

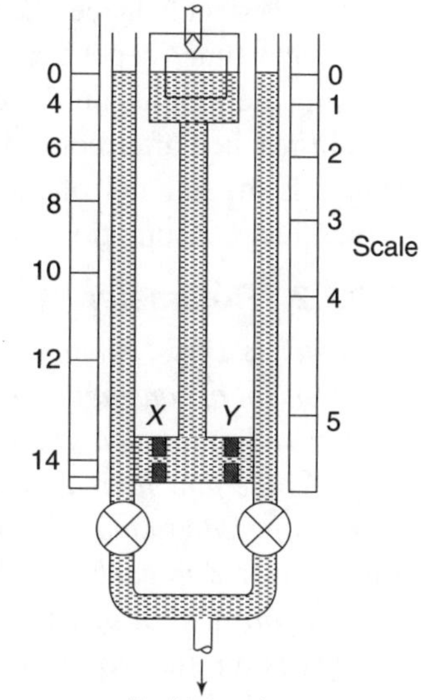


**Fig. 15.5** Orifice flowmeter

## 15.9.2 Gravimetric Fuel Flow Measurement

Though, measured volumetric fuel flow rate may be used to know the fuel consumption in kg/h by multiplying it with its density but it does not give accurate results. So, it is better to measure the consumption by weight directly.

### *15.9.2.1 Fuel Weighing Balance*

One of the accepted methods to measure the amount of fuel used by an engine is to weigh it on a fuel weighing balance as shown in Fig. 15.6. The balance is adjusted until the fuel container is slightly heavier than the balancing weight. As the fuel is consumed by the engine, the scale will gradually approach the balance point. At the

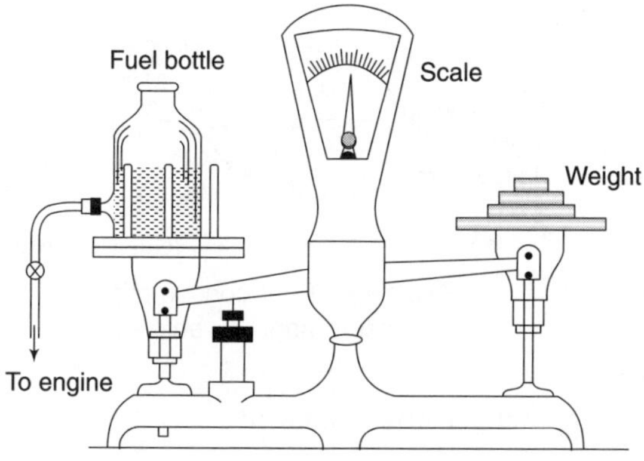


**Fig. 15.6** Fuel weight scales

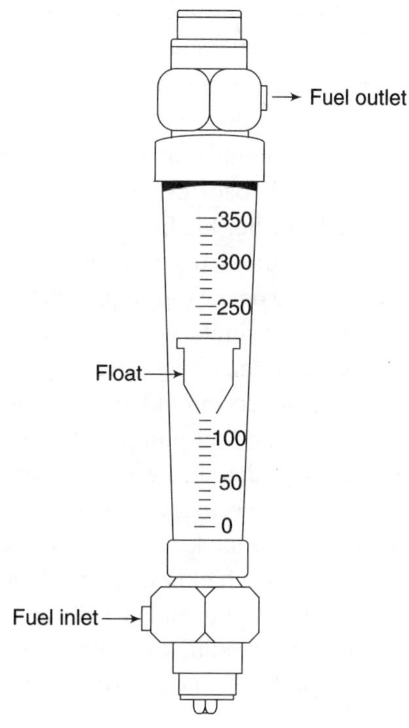


**Fig. 15.7** Rotameter

instant of perfect balance, the stopwatch is started. The pan or beam weight are then recorded. After some time, depending upon the duration of the test, the scale is again adjusted by removing the weights until the fuel is heavier than the balance weights. When perfect balance is again reached after consumption of more fuel, the difference between the two weights at balance is the amount of fuel consumed in the recorded by stopwatch. This method gives average fuel consumption expressed in kg/h.

#### 15.9.2.2 Rotameter

*Rotameter is very common in many automotive laboratories and the aircraft testing for the measurement of consumption of fuel.* Figure 15.7 shows a Fischer–Porter rotameter. *It consists of a tapered graduated glass tube having a float inside it and two connections for inlet and outlet. As the flow increases through the meter, the float rises and the area between the float and the tapered graduated tube proportionately increases. Since flow rate and area of flow are directly related to each other, an advantages of this type of flowmeter is that graduation are linear (calibrated in kg/h) and it can be used for a wide range of flows with good accuracy.* It indicates the fuel consumption at any instant.

### 15.9.3 Fuel Consumption Measurement in Vehicles

As mentioned earlier, the fuel consumption for automobiles is expressed in km/litre. Figure 15.8 shows an arrangement for measuring fuel consumption in km/litre for any vehicle under test. It consists of a glass burette, a needle valve to fill burette, connections to fuel pumps, carburettors and fuel tank.

The glass burette of one litre capacity is connected by tubes and control valves so that the precise number of km/litre is observed on an actual road test involving the effect of speed, traffic, loading, driver's habit and engine conditions on fuel consumption. The tester is hung on the top edge of the right front door glass or similarly suitable location and connection is made by special plastic tubes and the adapters provided. It is the surplus capacity of the fuel pump that fills the glass burette when the needle valve is opened and when a test run is made the control tap directs the fuel to the intake side of the fuel pump. By this way, the fuel is supplied to carburettor or injector at normal working pressure. Speedometer readings in kilometers per 0.5 litre or one litre is noted. By this way the consumption per litre is known.

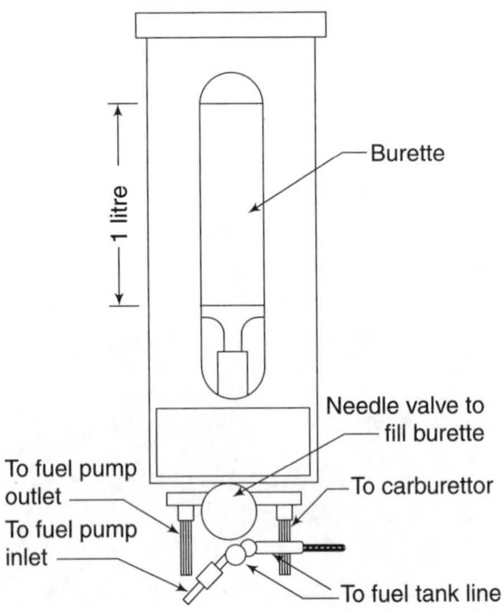


**Fig. 15.8** Fuel consumption measurement in vehicles

## 15.10 AIR CONSUMPTION

After the engine has been designed and constructed, it is essential to measure the air consumption to ensure that restrictions are not present in the intake and exhaust systems that would prevent free breathing of the engine. Moreover, the knowledge of air consumption helps in finding out the air-fuel ratio. *Measuring the air-flow, however is a treacherous problem because of the compressibility effect and pulsating nature.* Therefore, the simple method of using an orifice in the induction pipe is not satisfactory since the reading will be pulsating and unreliable.

It is worth to mention that all kinetic flow measuring systems such as nozzles, orifices and venturies meant to measure flow rates have a square law relationship between flow rate and differential pressure which gives rise to severe errors on unsteady pulsating flow.

Since the major source of errors are caused by pulsating flow, so based on damping out the pulsating flow, two methods are common. They are:

(a) Air box method employing orifice meter; (b) Viscous flow air meter.

### 15.10.1 Air Box Method Employing Orifice Meter

*Orifice meter can be used to measure air consumption in IC engines if pressure pulsations could be damped out by some means.* One of the method of damping out pulsations is to fit an air box of suitable volume (500 to 600 times the swept volume in single cylinder engines and less in the case of multi-cylinder engines) to the engine by flexible pipe. The orifice plate is placed in the side of the box remote from the engine so that maximum pulsation are damped out. Sometimes, internal baffles or diaphragm are provided on one face to reduce the pressure pulsations.

Figure 15.9 shows an air box method to measure air consumption. During engine suction, the air inside the box is at a pressure less than atmospheric pressure. As the box is relatively much too large (500 to 600 times the swept volume), the pressure in the box is not affected by intermittent suction of the engine, thus the pressure in the air box may be assumed to remain constant.

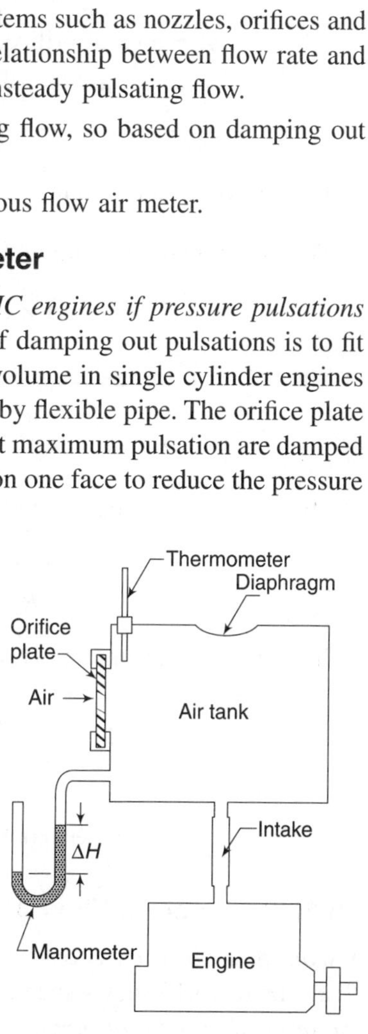


**Fig. 15.9** Air consumption measurement by air box method

The outside air is assumed to flow continuously through the orifice with a constant velocity. This depends on the difference between the pressure difference of air inside and outside the chamber. The pressure difference is measured by means of a manometer, one limb of which is opened to atmosphere and other connected to the inside the box. The pressure difference is usually limited to about 100 mm of water to neglect the compression effect of air. The temperature of the atmospheric air and the barometric reading are taken.

Let $d$ = orifice diameter, $m$

$h_w$ = manometer pressure difference in water column, $m$ of water

$\rho_w$ = density of water, kg/m$^3$

$\rho_a$ = density of air, kg/m$^3$

$c_d$ = coefficient of discharge of orifice

$h_a$ or $(\Delta H)$ = head causing flow through orifice in air column, $m$ of air

$C$ = velocity of flow, m/s

$A$ = area of orifice, m$^2$

$p_a$ = atmosphere pressure of air, bar

$T_a$ = atmosphere temperature of air, K

The equivalent air column $= h_a = h_w \cdot \dfrac{\rho_w}{\rho_a}$

The rate of discharge through the orifice $\dot{V}_a$ = Area × Velocity ; m$^3$/s

$$= A \times c_d \sqrt{2g\, h_a} = A \times c_d \sqrt{2g\, h_w\, \rho_w/\rho_a} \text{ ; m}^3\text{/s} \qquad ...(15.26)$$

Mass of air flow rate $= \dot{m}_a = \dot{V}_a \times \rho_a = A \times c_d \sqrt{2g\,[h_w\, \rho_w/\rho_a]} \cdot \rho_a$

or
$$\dot{m}_a = A \times c_d \sqrt{2g\, h_w\, \rho_w\, \rho_a} \text{ ; kg/s} \qquad ...(15.27)$$

where
$$\rho_w = 1000 \text{ kg/m}^3$$

and
$$\rho_a = \frac{p_a \times 10^2}{R_a\, T_a} = \frac{p_a \times 10^2}{0.287 \times T_a} \text{ ; kg/m}^3$$

If $h_w$ in mm of water and $d$ is in cm, then

$$\dot{m}_a = \frac{\pi}{4}\left(\frac{d}{100}\right)^2 \times c_d \sqrt{2 \times 9.81 \times \frac{h_w}{1000} \times 1000 \times \frac{p_a \times 10^2}{0.287 \times T_a}}$$

or
$$\dot{m}_a = 0.008266\, c_d . d^2 \sqrt{\frac{h_a p_a}{T_a}} \qquad ...(15.28)$$

### 15.10.2 Viscous Flow Air Meter

*Viscous flow air water is one of the best methods to measure air consumption is a pulsating flow.* Figure 15.10 shows such a meter. *It consists of an element where viscous resistance is the principle source of pressure loss and kinetic effects are small.* The meter gives a linear relationship between pressure difference and flow instead of a square law as in the case of nozzles, orifice and venturi. Thus, the flow rate is given by

$$\dot{V} = K_1 \sqrt{h_w}$$

where $h_w$ = pressure differential in water column; $K_1$ = constant

*The viscous element is in the form of a honeycomb passage consisting of a large number of small passages giving Reynolds number less than 200.* The shape of passage is triangular having size approximately equal to 0.5 × 0.5 × 75 mm. At the entrance region, air filter is provided. *Two felt pads, one each side of honeycomb is provided to measure pressure differential across honeycomb by means of a manometer.*

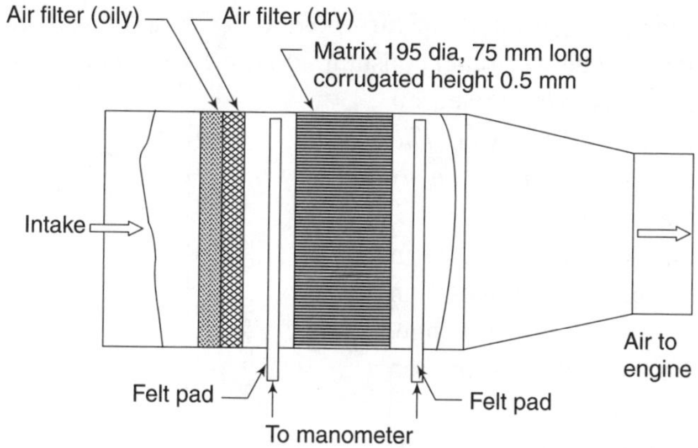


**Fig. 15.10** Akock viscous-flow meter

One of the advantage of viscous flow meter is that large range of flow can be measured without pressure heat being too small.

The viscous-flow meter suffers from the main source of error due to contamination of the small triangular passages by dust. However, by ensuring good filtration at the entry to the meter, and not passing air through the meter unless readings are required, this problem may be minimized.

## 15.11 FRICTION POWER

*The difference between indicated power and brake power is called friction power.* Friction power has a dominating effect on the performance of the engine. Friction loss is mainly attributed to the following mechanical losses:

(i) *Direct friction losses* between piston and cylinder walls, piston rings and cylinder walls and between the crankshaft and their bearings.

(ii) *Pumping losses*

(iii) *Power loss to drive the components* to charge and scavenge

(iv) Power loss to *drive other auxiliary components*

It is the level of friction that decides the maximum output of the engine which can be obtained economically.

*The aim of the designer is to ensure the minimum friction power to have large value of brake power.* The following methods are used to determine friction power.

(i) Willan's line method
(ii) Morse test
(iii) Motoring test
(iv) From difference between *ip* and *bp*.
(v) Retardation test

### 15.11.1 Willan's Line Method

*This method is based on the principle of extrapolation of fuel consumption vs. brake power curve.* A graph is prepared between fuel consumption (on *Y*-axis) and brake power (on *X*-axis) at constant speed and it is extrapolated on the negative axis of the brake power. *The intercept of the negative axis is taken as friction power of the engine at that speed.* This method of extrapolation is shown in Fig. 15.11 by dotted line. The extrapolation is valid because in most of the power range, the relation between the fuel consumption and brake power is linear. Further, at $bp = 0$ the engine consumes some fuel. This means that the energy supplied by fuel at $bp = 0$ is spent in overcoming friction. From the above, we conclude that the extrapolated negative intercept of the X-axis will be the work representing the combined losses due to mechanical friction, pumping and blow-by and as a whole it is known as friction loss of the engine.

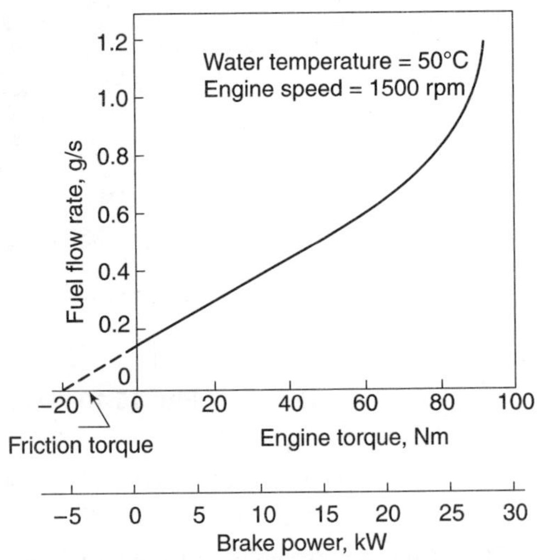


**Fig. 15.11** Willan's line method

The following are the drawbacks of this method:

(i) The long distance is to be extrapolated from data obtained between 5 and 40% load towards the zero line of fuel input. Since the graph is not exactly linear, so the directional margin of curve is rather wide.
(ii) This method holds good mainly for CI engine.

From the above mentioned drawbacks, it is essential that great care should be taken in extrapolating the line and as many readings as possible should be taken at light loads to establish the true nature of the curve. If extrapolated correctly, the accuracy obtained by this method is reasonably good.

### 15.11.2 Morse Test

*The morse test provides a simple, rapid and fairly accurate method of determining ip, fp and mechanical efficiency of a multi-cylinder engine with no more apparatus than a dynamometer (for bp) and tachometer or stroboscope or electric pulse counter (for speed) by making each cylinder inoperative.* This is achieved by short circuiting the spark plug in the case of SI engines and cutting of the fuel supply in the case of CI engines. It is assumed that the friction power remains constant in all the cases. *The test is conducted at* ***constant speed****.*

Suppose that a SI engine has $n$ cylinder (say 4 number; $A$, $B$, $C$ and $D$), then with all cylinders working, measure $bp$ from dynamometer and speed from tachometer and with this data $ip$ is expressed as

$$\sum_{1}^{n} ip_n = \sum_{1}^{n} bp_n + fp \qquad ...(15.29)$$

For 4 cylinders

$$ip_{1234} = bp_{1234} + fp$$

*If first cylinder, say A is cut out from operation by shorting the spark plug, its contribution towards power development is lost and the speed of the engine will decrease.* In order to restore the original speed, the brake load is reduced. *This is done in order to maintain fp constant which is supposed to be independent of load and directly proportional to speed.* With one cylinder cut-off, now only $n$-1 (three) cylinders will develop power but still all the $n$ (four) cylinder will consume friction. Measure the new dynamometer readings for $bp$ and for this

$$\sum_{2}^{n} ip_n = \sum_{2}^{n} bp_n + f_p \qquad ...(15.30)$$

i.e.,

$$ip_{234} = bp_{234} + fp$$

Subtracting Eq. (15.30) from Eq. (15.29), we get

$$ip_1 = ip_A = \sum_{1}^{n} ip_n - \sum_{2}^{n} ip_n = \sum_{1}^{n} bp_n - \sum_{2}^{n} bp_n \qquad ...(15.31)$$

i.e.,

$$ip_1 = ip_A = bp_{1234} - bp_{234}$$

Similarly, we can find the indicated power of the cylinders, viz. $ip_2$, $ip_3$, $ip_4$ .... $ip_n$.

The total indicated power developed by the engine $ip$ is given by

$$ip = \sum_{1}^{n} ip_n \qquad ...(15.32)$$

i.e.,

$$ip = ip_A + ip_B + ip_C + ip_D$$

The frictional power of the engine is given by

$$fp = ip - bp = ip_{1234} - bp_{1234} \qquad ...(15.33)$$

The mechanical efficiency is given by

$$\eta_m = \frac{bp}{ip} \qquad ...(15.33a)$$

The above experiment should be repeated with the same speed but with different setting of throttle. This will show the assumption that $fp$ is independent of load is true.

### 15.11.3 Motoring Test

*The principle of motoring test for finding fp lies in the measuring the brake power when engine is motored to run at the same speed at which it has run by its own power previously.* Based on this principle, the engine is steadily operated at the rated speed by its own power and allowed to remain under the given speed and load condition for sufficient time so that the temperature of the engine components, lubricating oil and cooling water reaches a steady state. In order to absorb the developed power, a swinging field type electric dynamometer is used to absorb the power during this period. The ignition is then cut-off and by suitable electric switching devices, the dynamometer is converted to run as motor so as to crank the engine at the same speed at which it was previously operating. The power supplied by dynamometer to the engine is measured which is the frictional power at the speed.

*The motoring method does not represent closely the real frictional losses due to the following reasons:*

(i) There exists an appreciable difference between the temperature in the motored engine and firing engine.

(ii) The bearings and piston rings are subjected to much higher pressure in firing engine as compared to motored engine.

(iii) Due to cooling of motored engine, the clearance between the piston and cylinder walls is more which results in lower value of piston friction as compared to firing case.

(iv) The air is drawn at a temperature much lower in the case of motored engine as compared to firing.

(v) There exists a difference between the back pressure of motored and firing engine.

This method, however, reasonably gives good results and is very suitable for finding the losses imparted by various engine components.

### 15.11.4 From the Difference between *ip* and *bp*

As mentioned earlier, *fp* may be calculated by the difference between *ip* and *bp*. *ip* is measured by indicator diagram and *bp* is measured by dynamometer.

### 15.11.5 Retardation Test

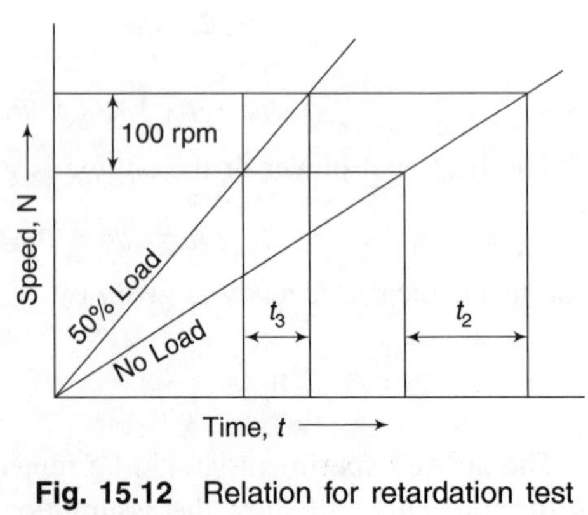


**Fig. 15.12** Relation for retardation test

*The principle of retardation test is based on retarding the engine by cutting the fuel supply and noting the time of fall in speed and the relation between the fall in speed and time yields the friction torque.* Based on this principle, the engine is made to run at no load and rated speed taking into all usual precautions. When the engine is running under steady operating conditions, the supply of fuel is cut-off and simultaneously the time of fall in speeds by say 20%, 40%, 60% and 80% of the rated speed is recorded. Further, this test is again repeated with 50% load on the engine. These values are usually tabulated and plotted in the form of graph for no load and 50% load condition as shown in Fig. 15.12.

From the graph, the time required to fall through the same range (say 1000 rpm) in both, no load and load conditions are found. Let $t_2$ and $t_3$ be the time of fall at no load and load conditions respectively. Let $T$ be the torque at any time, $T_1$ be load torque and $T_f$ be the frictional torque. Thus,

$$T = I\frac{d\omega}{dt} \qquad ...(15.34)$$

where $\omega$ = angular velocity; $I$ = moment of inertia = MK$^2$; $t$ = time

At no load, the torque is friction torque, $T_f$ from Eq. (15.34), we get at no load

$$d\omega = \frac{T}{MK^2}\,dt \qquad ...(15.35)$$

$$\omega_0 - \omega_1 = \frac{T_f}{MK^2}(t_2 - 0)$$

The reference angular velocity $\omega_0$ is that say 1000 rpm, the time of fall for the same range at load $(T_f + T_l)$ is

$$\omega_0 - \omega_1 = \frac{T_f + T_l}{MK^2}(t_3 - 0) \qquad ...(15.36)$$

From Eq. (15.55) and Eq. (15.56), we get

$$T_f = T_l\left(\frac{t_3}{t_2 - t_3}\right) \qquad ...(15.37)$$

### 15.11.6 Comparison of Various Methods of *fp*

Morse tests and Willan's methods are comparatively easy to test, however *Morse tests are more common for classroom laboratories.* Motoring methods result in higher value of *fp*. The most accurate method is *ip* – *bp* if accurately conducted.

## 15.12 INDICATED POWER

Indicated power is the potential output of an engine and it signifies how efficiently the chemical energy is converted into heat energy. The following methods are used to find the indicated power:

(i) by using the indicator diagram obtained from engine indicator.

(ii) by adding measured and *bp* and *fp*.

### 15.12.1 Indicator and Indicator Diagram

*Indicator is a device which measures the variation of pressure in the cylinder over a part or fuel cycle of the engine.* The plot of such variation is called an ***indicator diagram***. It is worth to note that an indicator diagram gives very good indication of the process of combustion and by its analysis, one can study its associated factors such as *rate of pressure rise, ignition lag factors, losses occurring in the suction and exhaust strokes,* knocking and tuning, etc.

**Pressure volume (*p-V*) and pressure-crank angle, (*p, t*)** are the two types of indicator diagrams. It is interesting to note that these two diagrams are mutually convertible. *The area of the indicator diagram represents the indicated power of the engine if accurately measured.*

The following are the main *elements* of an engine indicator—

(i) a pressure sensing device to sense pressure in cylinder.

(ii) a piston displacement or angular position sensing device to sense volume

(iii) a display device to display the variation

There are four *types* of engine indicators:

(a) Piston indicator; (b) Balanced diaphragm type indicator;

(c) Electronic indicator; (d) Optical indicator

Previously, piston indicator was in use but at present, ***electronic indicator*** *is in common use*.

#### *15.12.1.1 Piston Indicator*

*This is an instrument which scales the actual pressure and volume variation in a reciprocating machine.* Figure 15.13 shows a piston indicator. The indicator consists of a removable piston-cylinder arrangement connected to the engine so that the gas pressure is applied against the indicator piston which is attached to the calibrated spring and a drum holding the indicator card. The vertical motion of the indicator piston is multiplied through the pencil arm mechanism and this motion is indicated by the stylus which marks the pressure on a card mounted on a drum. The scale of the vertical displacement of pencil point is expressed in bar which is a measure of cylinder pressure. The drum can be given a swing motion clockwise and anticlockwise proportional to the movement of the engine piston by connecting the rope wound over the drum

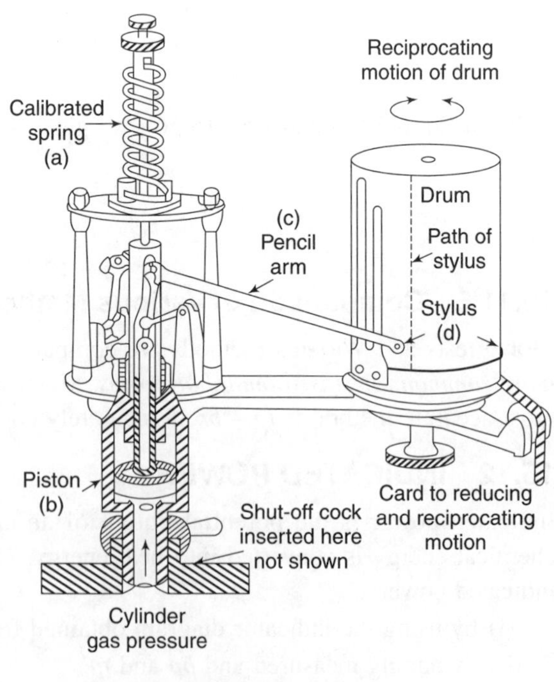


Fig. 15.13 Piston indicator

to a suitable auxiliary mechanism attached to the crankshaft or piston rod. This means that the drum will rotate in synchronism with the engine piston while the stylus moves vertical distances that are proportional to the cylinder pressure. Such a diagram obtained is called a *p-V* diagram.

The drum can also be given, essentially at constant velocity by using coiled spring and the diagram thus obtained is called a pressure-time (*p-t*) diagram.

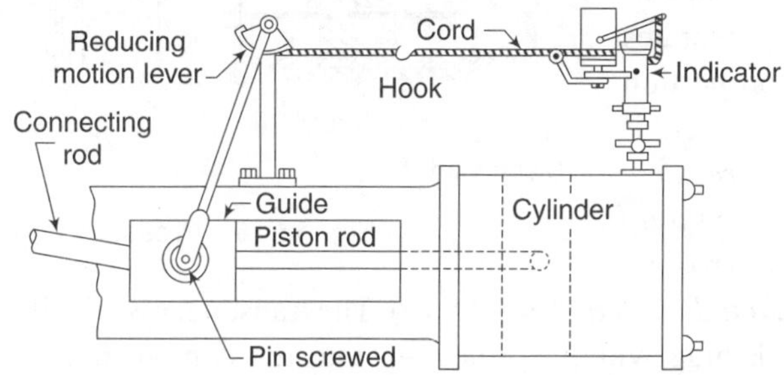


**Fig. 15.14** Another arrangement of piston indicator

To know the pressure by an indicated *p-V* diagram, indicator *spring constant* must be known. This is defined as the change in pressure required to cause one cm of vertical movement of the marking stylus on the indicator diagram. Thus, the scale for the vertical displacement of the pencil point is expressed in bar per cm so that it can be given directly the pressure in bar of the cylinder. Since the stroke is large and cannot be pictured on card, a suitable arrangement as shown in Fig. 15.14 is employed to reduce the stroke. Figure 15.15 shows indicator diagram for working and missed cycle.

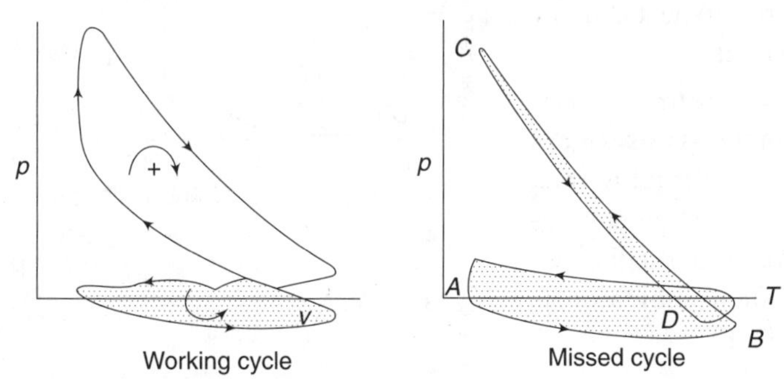


**Fig. 15.15** Indicator diagram for working and missed cycle

### 15.12.1.2 Electronic Indicator

*Electronic indicator is the most accurate method for finding indicator diagram. It consists of the following four main elements.*

(i) a pressure pickup (pressure transducer)
(ii) a pre-amplifier
(iii) a time-base recording device
(iv) a display unit

*The purpose of pressure transducer is to sense the pressure inside the cylinder at an instant of time.* It generates an electric signal in proportion to the pressure to which it is subjected. *The pressure transducer consists of mainly a diaphragm, electrodes, transducer elements, and high lead* (Fig. 15.16). The transducer is usually fitted in the cylinder head just like a spark plug without projecting into the combustion space. The pressure is sensed with respect to displacement of a diaphragm. In order to reduce the displacement and hence the inertia effects, the diaphragm is made very stiff. Displacement of the diaphragm is transmitted to the transducer element which may be any one of the following types:

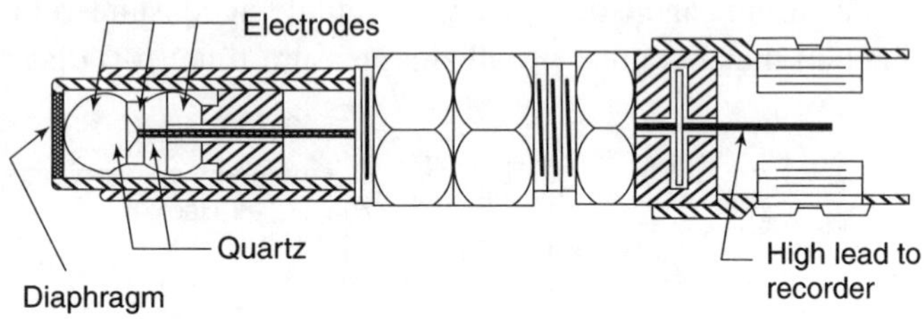


**Fig. 15.16** Pressure transducer

**(i) piezoelectric crystal** **(ii) electromagnetic type**
**(iii) capacitance type** **(iv) strain guage type element**

*This transducer element produces an electrical signal which is amplified before displacement on the oscilloscope or recorded in a recorder. The output of transducer element must be linear and of good frequency response.* It should be insensitive to heat and low acceleration sensitivity.

*Out of the all pressure transducers, the* **electric quartz transducers** are most *satisfactory* for all normal uses in internal combustion engines measurements. A typical *p-t* diagram obtained from such indicator is depicted in Fig. 15.17.

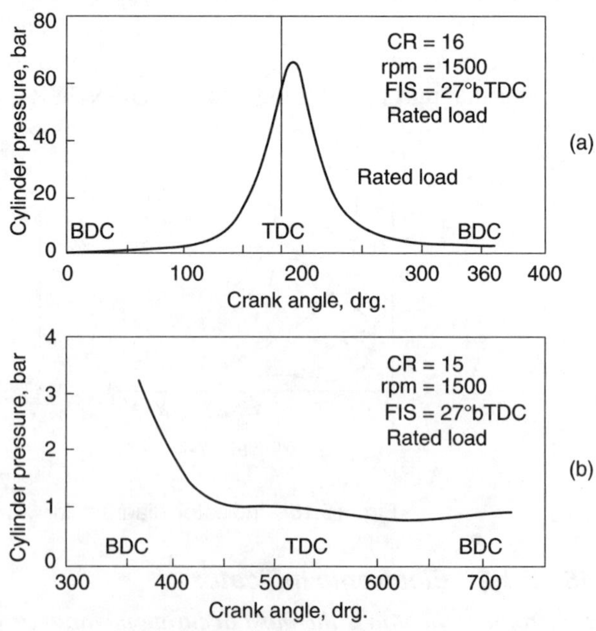


**Fig. 15.17** A typical *p-t* diagram obtained from electronic indicator

## 15.13 BRAKE POWER

*Among all the measurements of internal combustion engines, the measurement of brake power is most important.* It involves the determination of torque and angular speed of the engine output shaft. The torque measuring device is known as ***dynamometer***.

Dynamometer is classified as follows:

(A) **Absorption dynamometer**

(i) Mechanical: (a) Prony braker (b) Rope brake (c) Band brake

(ii) Hydraulic: (a) Water brake (b) Fan brake

(iii) Electric-eddy current dynamometer.

(B) **Transmission dynamometer:** (i) Torsion dynamometer, (ii) Strain gauge dynamometer

(C) **Driving dynamometer**

(i) Electric motor-generator dynamometer

**(a)** ***Absorption Dynamometer.*** *In this type, the power developed by the engine is absorbed in friction. The heat is dissipated to the surrounding where it generally serves no useful purpose.*

**(b)** ***Transmission Dynamometer.*** In this type, *the energy being transmitted to or from dynamometer is not absorbed or dissipated. After measurement, the energy is transmitted to the surroundings in a useful mechanical or electrical form.*

**(c)** ***Driving Dynamometer.*** The unit may be coupled to *either power absorbing or power governing devices; since it may operate as* **motor or generator**. *This is useful to determine the performance characteristics of pumps or compressors.*

### 15.13.1 Prony Brake

Figure 15.18 shows the details of a prony brake. A drum (*a*) is keyed to the shaft of the engines where the output is to be measured. A band (*b*) is tightened around the drum by means of a hand wheel, *c*. The amount of tightening of the band determines the frictional drag, *F* (not shown) acting around the periphery of the drum and resisting the rotation of the shaft. The level-arm frame work, *d*, which is stopped by the platform of the measuring scales prevents the band and its rubbing surface from moving, except through a limited arc.

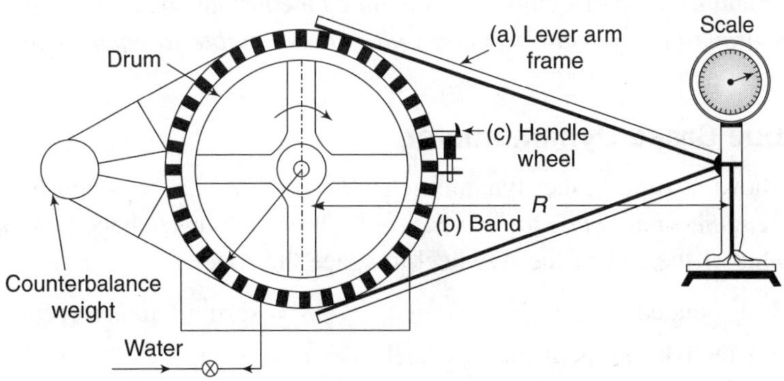


**Fig. 15.18** Prony brake

Let $F$ = frictional force (or drag), N; $W$ = weight recorded by measuring scale, N
$R$ = effective lever arm, m; $N$ = r.p.m of the brake drum
$r$ = brake drum radius, m

*Work is defined as the product of force and distance.* In one revolution of the shaft, the periphery of the drum moves through a distance $2\pi r$ against resisting frictional force, $F$. Thus, work done per revolution = $2 \pi r F$

The total frictional torque acting on the drum = $Fr$. This must be balanced by external moment of the brake which is the product of the reading $W$ and arm $R$. Thus,

$$rF = WR$$

$\therefore$ Work per revolution = $2\pi WR$

Work per minute = $2\pi W RN$; J/min

$$\therefore \quad bp = \frac{2\pi W R N}{60 \times 1000}; \text{ kW} \qquad ...(15.38)$$

$$\text{or} \quad bp = KWN \quad \text{where} \quad K = \frac{2 \pi R}{60 \times 1000} \qquad ...(15.39)$$

Note that $WR$ is called torque, designated as $T$. Hence,

$$bp = \frac{2\pi NT}{60 \times 1000}; \text{ kW} \qquad ...(15.40)$$

$$\text{or} \quad \text{Torque} = T = \frac{60 \times 1000(bp)}{2\pi N}; \text{ kNm} \qquad ...(15.41)$$

If the brake in Fig. 15.18 is not counter-weighed ($W_e = 0$), the force on the scales will be caused in part by the weight of the lever arm frame. The unbalanced weight of the brake is called the tare of the brake. In this case, it is assumed that the brake is counterbalanced and therefore, the tare is zero, thus $W$ *represents the net scale reading*.

*The prony brake is simple in operation, easy to construct and inexpensive. It is extensively used for low speed testing (1000 rpm and 74 kW bp).* At high speed, grabbing and chattering of the band occurs and leads to difficulty in maintaining a constant speed. *Its main disadvantage is its constant torque at any band pressure and is thus, unable to compensate varying load conditions.*

### 15.13.2 Rope Brake Dynamometer

Figure 15.19 shows a rope brake dynamometer. *It consists of a rope wrapped around the flywheel of the engine where bp is to be measured.* Three or four *U*-shaped blocks are located at different points of the rim of the flywheel to space the rope.

Let $W$ = suspended weight, N $S$ = spring balance reading, N
$r$ = effective radius of the flywheel $r = r_f + r_s$
$r_f$ = radius of the flywheel $r_r$ = radius of the rope

The braking torque = $(W - S)_r$; Nm

$$\therefore \qquad bp = \frac{2\pi N(W - S)r}{60 \times 1000} \qquad ...(15.42)$$

$$= K(W - S)N \qquad ...(15.43)$$

where $K = \dfrac{2\pi\, r}{60 \times 1000}$ = Constant for dynamometer

*This dynamometer is extensively used for testing engines for bp because it is inexpensive and easy to fabricate.*

### 15.13.3 Band Brake Dynamometer

Figure 15.20 shows one type of arrangement of band brake dynamometer.

Let $W$ = weight on the right side, N; $\quad w$ = weight on the left side, N

$S$ = spring balance reading, N; $\quad N$ = rpm

$r$ = effective radius of the brake wheel, m

The braking torque = $T = [(W + S) - w]r$; Nm ...(15.44)

$$bp = K\,[(W + S) - w]r;\ \text{kW} \qquad ...(15.45)$$

where $K = \dfrac{2\pi\, N}{60 \times 1000}$ = Constant for dynamometer.

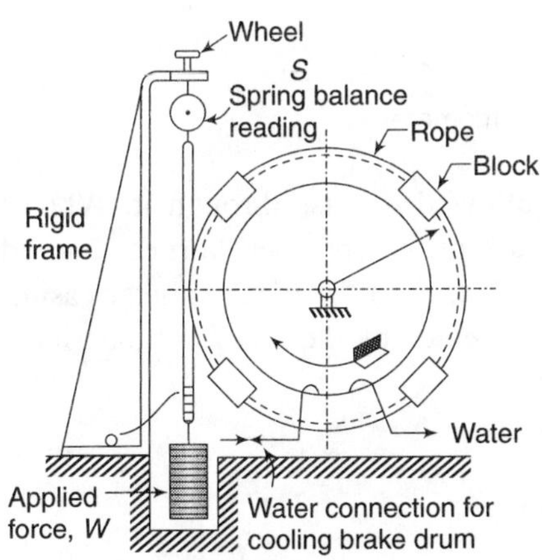


**Fig. 15.19** Rope brake dynamometer

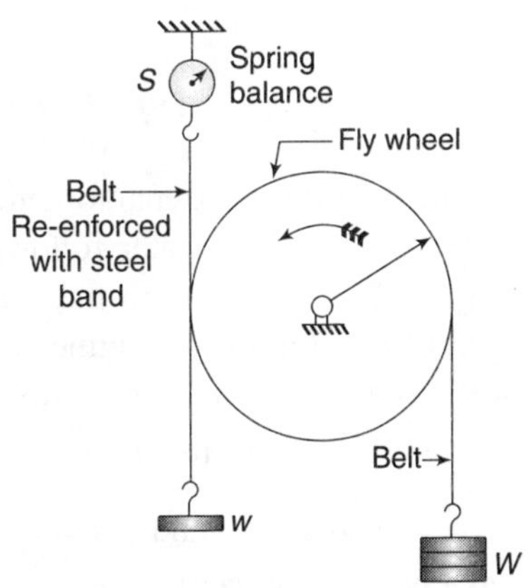


**Fig. 15.20** Band brake dynamometer

### 15.13.4 Hydraulic Dynamometer

*A hydraulic dynamometer uses fluid-friction for dissipating the input energy. It consists of mainly two elements, namely, a rotor and a stationary casing, the rotor being keyed to the driving shaft that rotates inside the casing (Fig. 15.21). The casing is mounted on trunnion (anti-friction) bearings and has a brake arm and balance system attached to its one side (opposite of rotation). The use of trunnion bearing allows the casing to swing freely except for the restrain imposed by the brake drum.* The casing is split in two halves; one of which is placed on either side of the rotor. Semi-elliptical grooves made in the rotor match with corresponding recesses in the casing to form chambers through which a stream of water flow can be maintained.

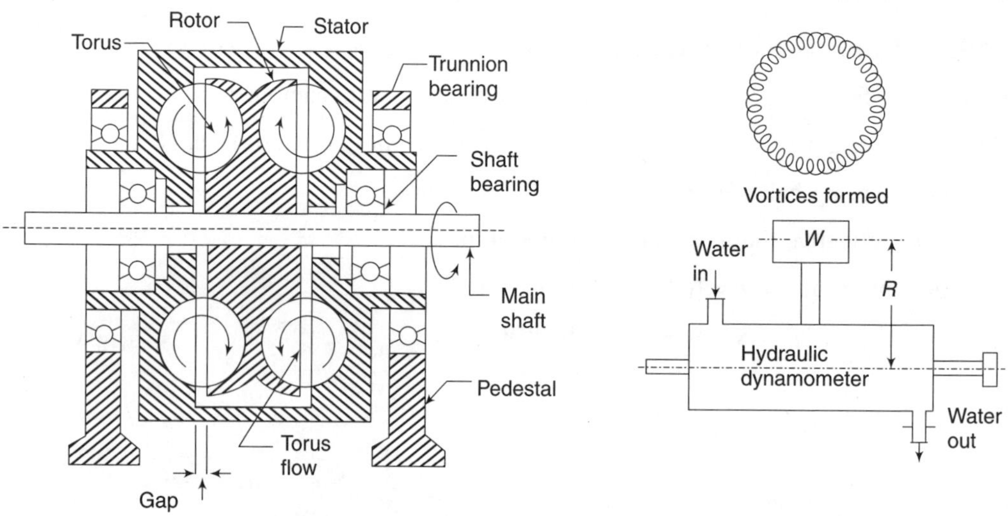


**Fig. 15.21** Hydraulic dynamometer

In order to operate the dynamometer, water is allowed to pass through it. When the dynamometer is operating, water follows a helical path in the chamber. Vortices and eddy current (Fig. 15.21) are set up in the water which tends to turn the dynamometer casing in the direction of rotation of the engine shaft. This is prevented by the brake arm and load on the balance system. The torque measured is given by

$$T = W \cdot R; \text{ Nm}$$

where $W$ = load in $N$ and $R$ = brake arm in m

The brake power is given by

$$bp = \frac{2\pi\, NT}{60 \times 1000}; \text{ kW}$$

The braking action can be controlled by varying either the amount of water and its pressure, or the distance between the rotor and casing. When the casing is full of water, the maximum absorption of power developed is obtained. The minimum absorption of power is obtained with the minimum amount of water required to keep the shaft scales lubricated. Care should be taken so that the amount of water circulated must be sufficient to prevent the formation of steam. *The power absorbed by hydraulic dynamometer varies approximately as the cube of rotational speed and as the fifth power of rotor diameter. It is suitable up to 10,000 rpm and 20,000 kW output engine.*

### 15.13.5 Eddy Current Dynamometer

*It operates on the principle that when an isolated conductor moves through a magnetic flux, voltage is induced and local currents, called eddy currents, flow in a short circular path within the conductor and get dissipated in the form of heat.*

Figure 15.22 shows the basic elements of an eddy current dynamometer. It consists of a toothed steel rotor and a cast iron stator. The rotor is mounted on to the shaft of the engine and rotates inside the smooth stator maintaining a small clearance. An exciting coil fixed on the stator gets energised with a direct current supplied from an external source. The stator is mounted on anti-friction bearings so that it can swing freely and is provided with a brake arm (radius *R*) to which a scale pan or spring balance is attached. The load is controlled by regulating the current in the electromagnets.

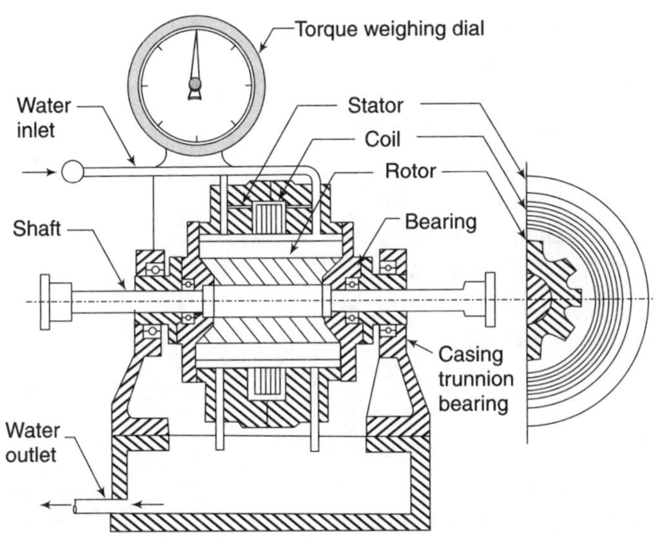


**Fig. 15.22** Eddy current dynamometer

When the dynamometer is under operation, the rotor turns and it causes a constant change in the magnetic flux density at all points in the stator. As a result of this, eddy currents are induced in the stator which opposes the rotation of the rotor.

The torque is measured by

$$T = W \cdot R; \quad \text{Nm}$$

Air circulation by the rotor tooth and water circulation through cooling channel takes away the heat produced in the system. It is suitable for a speed ranging up to 6000 rpm and power up to 250 kW. The dynamometer offers the highest ratio of constant brake power to speed range (upto 5 : 1).

### 15.13.6 Electric Motor-Generator Dynamometer

This dynamometer is essentially a d. c. shunt wound machine which may be operated either as a motor or as an electric generator (Fig. 15.23). This enables to use it as an **absorption** and as a **driving dynamometer**. If consists essentially of two main elements:

(i) a low resistance element armature made of relatively large conductors with a short path for the electric current to flow; and

(ii) a high resistance element field made of many turns of comparatively small diameter wires.

The elementary wiring diagram of this dynamometer is shown in Fig. 15.23. In order to operate the unit as a *motor* to start an engine, the following sequence is adopted.

(a) Open all the switches.

(b) Put the maximum resistance in the circuit by setting the armature rheostat to low speed position in order to limit the current flow in the element to a safe value.

(c) Put no resistance in the field circuit by moving the field rheostat to the low speed position to ensure flow of high current in the field circuit resulting in a high flux and a high torque.

(d) Close the field reverse switch.

(e) Place the armature switch in motor position, *M*. This will connect armature to the main line switch.

(f) Close the circuit breaker manually and then close the main line switch.

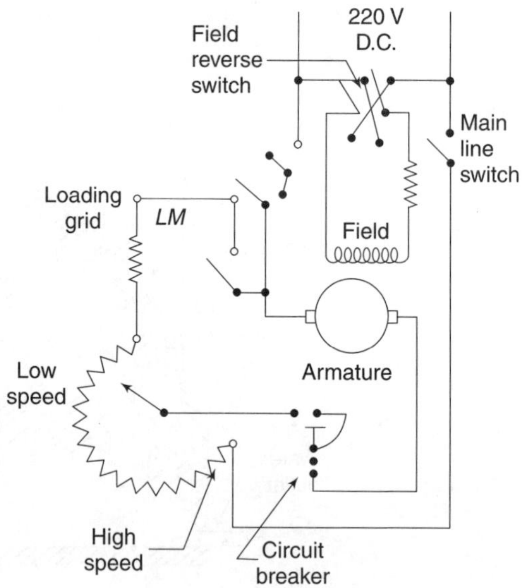


**Fig. 15.23** Wiring diagram of a motor-generator dynamometer

The above sequence of operation starts the dynamometer and runs it as a motor. As the motor speed rises, the back *emf* builds up and this controls the current flow in the circuit. The motor speed is controlled by removing the resistance from the armature circuit enabling supply of full voltage. Low speed rotations are achieved by closing the low speed motoring switch, *L.M.*, and thereby connecting the loading grid across the lines as a voltage divider enabling reduction of armature voltage nearly to zero.

In order to protect the armature from the current overloads, a circuit breaker is incorporated. In case the circuit opens, all controls are returned to the initial starting position before making an attempt to run the dynamometer again as a motor.

The following sequence is adopted to run the *dynamometer as a generator for absorbing* ***energy*** *from an engine.*

(a) Open all the switches.

(b) Set the armature and field rheostats to their respective high position which introduce the maximum resistance in both the armature and field circuits.

As a result of this, a minimum of current would flow and a very little power would be absorbed.

(c) Close the circuit breaker.

(d) Set the armature switch in the generator position $C_l$.

(e) Close the field switch. When the load on the machine increases, the resistance in the armature and field circuit can be reduced.

When the machine is running as a motor, the power absorbed is given by

$$\text{Brake power} = \frac{VI}{1000\,\eta}\ ;\ \text{kW} \qquad ...(15.46)$$

where $V$ p.d. in volts and $I$ is the current in ampere

When running as a generator,

$$\text{Power input} = \frac{VI\,\eta}{1000}\ ;\ \text{kW} \qquad ...(15.47)$$

*This system has the advantages of ease of control, good performance at low speeds. It is suitable upto 4000 kW engine.*

## 15.14 BEVIS GIBSTON FLASH LIGHT TORSION DYNAMOMETER

*This works on the principle that the torque transmitted is directly proportional to the angle of twist.* The torque is calculated from the following relation:

$$\frac{T}{I_P} = \frac{C.\theta}{l} \qquad ...(15.48)$$

where $T$ = torque transmitted by the shaft, $I_p$ = polar moment of inertia

$C$ = modulus of rigidity of the shaft, $\theta$ = angular twist

$l$ = length of the shaft over which the twist is measured

The dynamometer consists of two similar discs having narrow radial slot mounted on the shaft, a light source and an eyepiece (Fig. 15.24). When the shaft transmits power, the disks move relative to each other and the light gets obscured. After moving the eyepiece circumferentially by

**Fig. 15.24** Flash light torsion dynamometer

an amount equal to the twist of the shaft, the flash is projected up. From the angular twist θ, and the shaft constant $I_p$, $C$ and $l$, torque can be determined.

### 15.14.1 Strain Gauge Transmission Dynamometer

Figure 15.25 shows am arrangement of a *strain gauge circuit widely employed for torque measurement from a rotating shaft. It consists of four bonded wire strain gauges mounted on a 45° helix with the axis of rotation, placed in pairs diametrically opposite. The system is temperature compensated and insensitive to bending and thrust.* It is only the torsional deflection that produces a change in the gauge circuit. When shaft is under torsion, gauges 1 and 4 will elongate as a result of tensile component of a pure shear stress on one diagonal axis, while gauges 2 and 3 will contract owing to compressive component on the other diagonal axis. *The tensile and compressive principal strains can be measured and from which shaft torque can be calculated.* Figure 15.26 shows arrangement of strain guage transmission dyanometer.

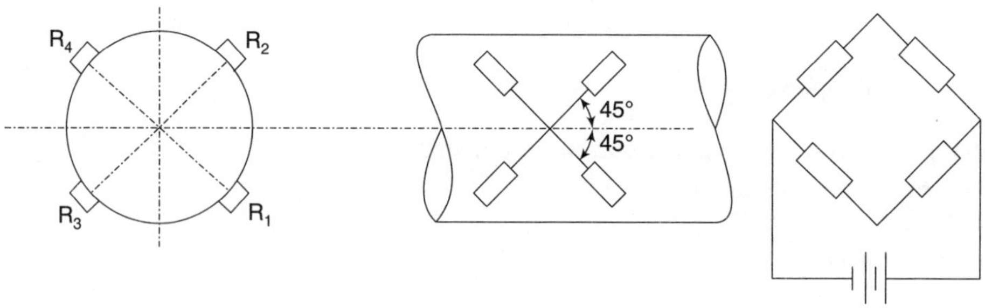


**Fig. 15.25** Strain gauge torsion meter

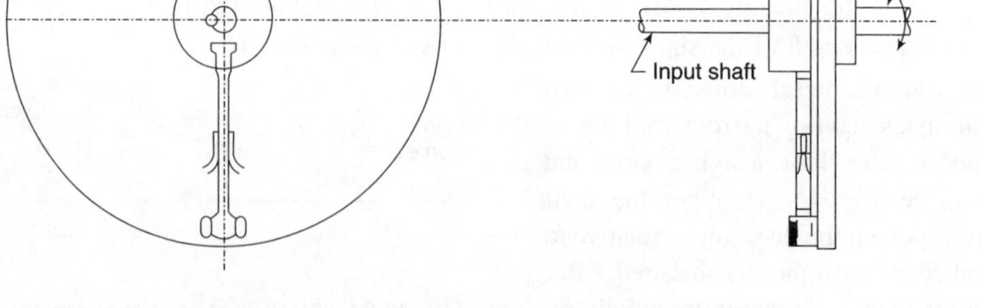


**Fig. 15.26** Arrangement of strain gauge torsion meter

### 15.14.2 Chassis Dynamometer

Chassis dynamometer is used to test the complete vehicle. This dynamometer allows the test engineer to simulate road load conditions in a laboratory and it can be termed as all weather load for the test vehicle. The road load on the vehicle can be put on the engine on a chassis dynamometer and quick results can be obtained. Figure 15.27 shows such dynamometer for tractor. It consists of electric motor, variable speed transmission, hydraulic dynamometer, gearbox, scale and drums. The drums hold the vehicles. Except drums, all other elements are mounted on a pendulum frame.

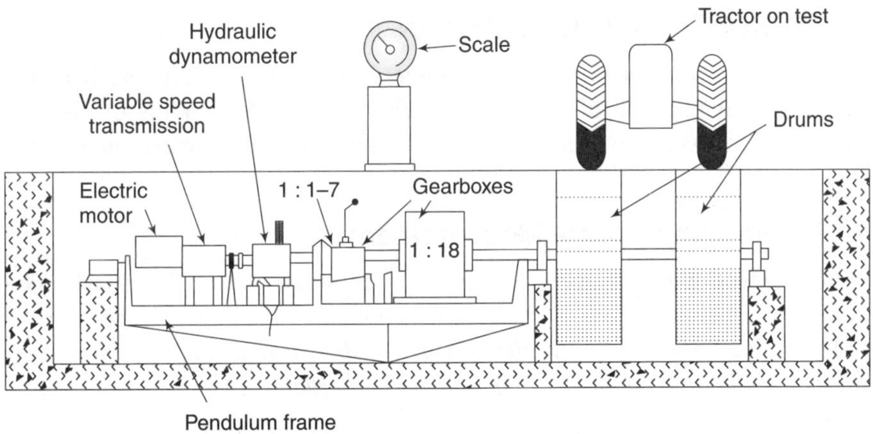


**Fig. 15.27** Chassis dynamometer for vehicle

## 15.15 EXHAUST AND COOLANT TEMPERATURE

One of the simplest methods of measuring the exhaust temperature is thermocouple. *The present technology offers electronic temperature sensitive transducers and ultraviolet radiation analysis for exhaust temperature measurement.*

The coolant temperature is normally measured using suitable thermometers.

## 15.16 NOISE

*A mixture of various types of sound is known as the noise and it is a source of irritation for the listener. A vibrating object creates sound. The vibrations are transmitted to the surrounding air in the form of pressure waves.* If the frequency and intensity of the pressure waves are within specified range (15 to 15000 Hz and 0 to 120 dB intensity), they produce the sensation of sound. *Noise created by motor vehicle is one of the indexes of a quality of vehicle.* The measurement of noise created by motor vehicles is based on a moving vehicles.

*Noise is measured by a sound level meter* (Fig. 15.28). It consists of a microphone, calibrated attenuator, an electronic amplifier and an indicator meter calibrated in *decibels (dB).*

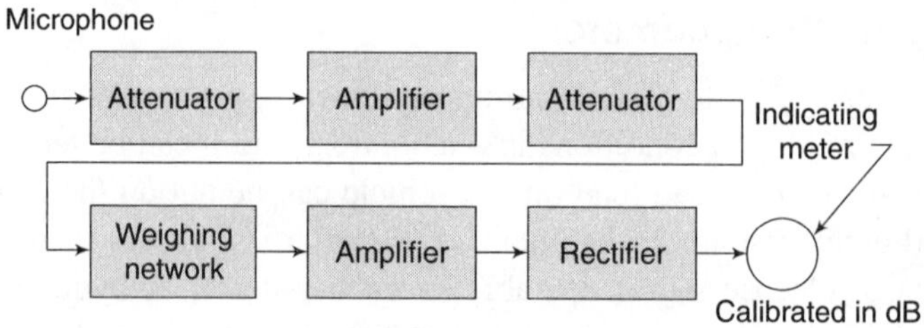


**Fig. 15.28** Sound Level Meter

## 15.17 COMBUSTION PHENOMENON

*Combustion quality is one of the main parameters of an efficient engine.* It is a complex phenomenon and it is not still fully understood. For better understanding of the combustion phenomena, it is necessary to measure the flame temperature, flame propagation, knock, etc.

### 15.17.1 Flame Temperature Measurement

The measurement of flame temperature inside a cylinder is a difficult task. *The recent **technique employed is infrared spectrographic method**.* Figure 15.29 shows a schematic diagram of a infrared spectrographic method. *It consists of detector, lens filter, black body source, thermocouple, potentiometer, amplifier and recorder.*

*The principle of this method is based on the fact that if infrared waves from a black body source pass through a cylinder containing gas at a temperature T, then the black body temperature ($T_b$) equals the cylinder temperature when the rate of absorption of the infrared waves by the gas in the cylinder equals the rate of emission from the gas.* The sensing element is the water vapour, i.e., the detector and optical system is set to record the emission at one of the wavelengths for water vapour –2.6 $\mu m$.

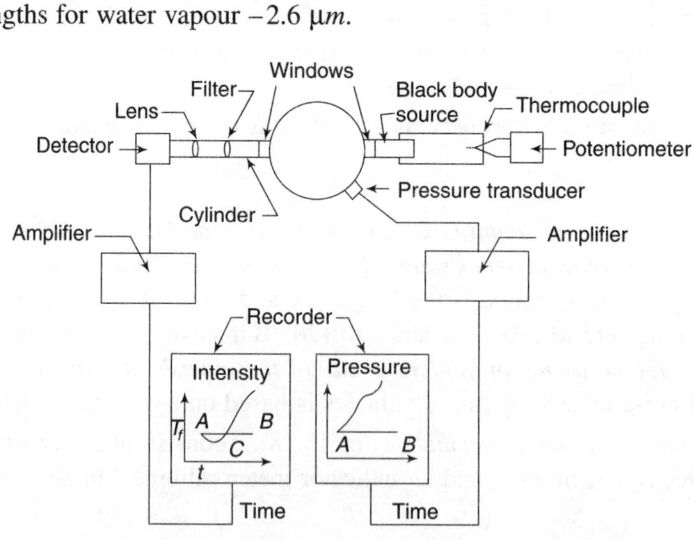


**Fig. 15.29** Infrared spectrographic method

In order to absorb the unwanted wavelength, the black body source for an optical filter is kept ahead of the detector. The cylinder consists of two quartz windows so that there is a direct optical path from the infrared source to the detector. The cylinder gases absorb minimum infrared during the suction stroke. This ensures the passing of infrared waves transmitted from the black body through the cylinder with almost no change in intensity of waves.

The intensity ($I$) corresponds to the black body temperature ($T_b$) measured by thermocouple. It is worth to note that $T_b$ will not be the gas temperature. Suppose if $T_b$ is set above the suction temperature, then at the beginning of compression stroke as the temperature rises, there will be some absorption of the infrared by the cylinder gas and the emission level will be less than from the black body. The intensity recorded by the detector will show a decrease in its value. Further, increase in temperature of cylinder gas will produce a further decrease in recorded intensity until a minimum is reached. With the continuation of stroke, the intensity will now increase until at $C$ (Fig. 15.29 AB horizontal line), the rate of absorption of the infrared waves from the black body source equals the rate of emission from the cylinder gases and thus the cylinder gas temperature equals the black body temperature, $T_b$. The timing (crank angle) at which this situation occurs can be measured by electronic indicators. The above procedure is repeated by selecting various values of $T_b$ which will give flame and time ($T_f - t$) diagram. The accuracy of the result depends on cleanliness of the quartz windows, windows radiation, absorption and homogeneity of the cylinder temperatures, cycle to cycle variation, etc.

### 15.17.2 Flame Propagation

*Flame propagation in the combustion chamber responsible for the efficient combustion and it is affected by many variables such as heat transfer, turbulence, etc.*

*The instrument used to measure the flame location with time is* ***ionization gauges****. Such gauges are made up of a wire in an annulus. The wire is insulated from the body of the gauge.* As soon as the flame reaches the annulus, it becomes conducting since an electrical circuit is completed between the wire and the gauge body. An electronic timing circuit is connecting to the gauge body. A dc voltage is normally applied to the wire as energy input. When the flame arrives at the gauge, the timing circuit is triggered and the time from the reference time is measured. From the location of a number of gauges in the combustion chamber, the flame speed may be calculated.

### 15.17.3 Combustion Process

*For quantitative combustion analysis, high speed colour photography has been successfully used in the past.* Figure 15.30 shows such a system used by Ricardo. The cylinder incorporates either a Quartz or Perspex window. A high speed rotating prism camera operating up to 16,000 frames per second is mounted on a rigid support and mirrors are arranged to record the combustion and crankshaft simultaneously. The dopes are added in the fuel to render low luminosity flames visible. The films are taken and analysed.

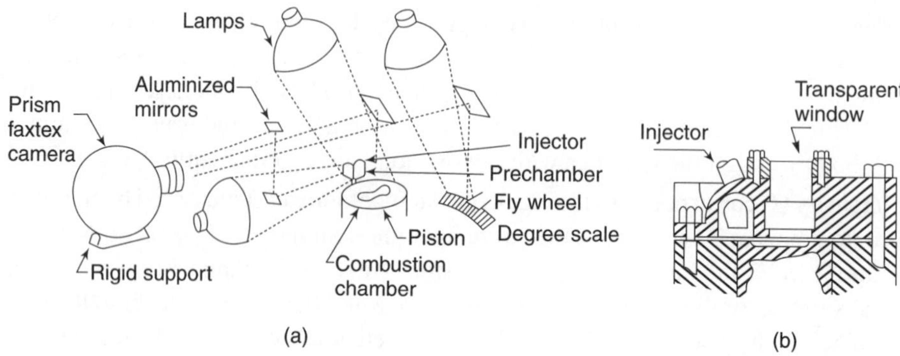


**Fig. 15.30** Recardo high speed photographs technique for combustion analysis

## 15.18 EMISSION MEASUREMENT

*Pollution is one of the great causes imposed on human being due to growth of industrialisation.* Emission from the engine contributes a lot in the air pollution. To control the pollution, its measurement is essential The exhausts of the engine contain the following:

| | |
|---|---|
| (i) Carbon dioxide | (ii) Carbon monoxide |
| (iii) Water vapour | (iv) Oxides of nitrogen |
| (v) Unburnt hydrocarbons (HC) | (vi) Aldehydes |
| (vii) Smoke | (viii) Particulates |

The last two (smoke and particulates) are visible while the rest are invisible. Out of the invisible emission, carbon dioxide and water vapour are harmless compared to others, so their measurements are not described here. However, **$CO_2$ produces greenhouse effect**.

### 15.18.1 Carbon Monoxide

Carbon monoxide is produced due to incomplete combustion either on account of insufficient amount of air or time for combustion. Some CO is always present in the exhaust even of lean mixture. The percentage of CO decreases with speed: *0.5 per cent CO is regarded as the reasonable goal. During idling, CO emissions are high while during deceleration it is the maximum.*

*CO is measured by an instrument known as* ***non-disperse infrared analyser*** **(NDIR)**. The schematic diagram of NDIR is shown in Fig. 15.31. *In the NDIR analyser, the exhaust gas species being measured are used to detect themselves.* ***The direction technique is based on the principle of selective absorption of the infrared energy of a particular wavelength peculiar to a certain gas which will be absorbed by that gas***. This instrument is used for testing and legal certification of automotive exhaust emissions.

### 15.18.2 Unburnt Hydrocarbon

*Unburnt hydrocarbon (HC) is the direct consequence of incomplete combustion in combustion chamber.* Many factors such as combustion chamber shape, L/D ratio and compression ratio

***B*** Receiver
***B1*** Diaphragm
***B2*** Fixed counter electrode
***B3*** Receiver trimmer
***F*** Filter cell
***G*** DC supply
***M*** Measuring cell
***M1*** Sample cell
***M2*** Reference cell
***R*** High impedance resistor
***S1*** I.R. sources
***S2*** Motor
***S3*** Shutter wheel
***V*** Amplifier
***V1*** Mains stabilization

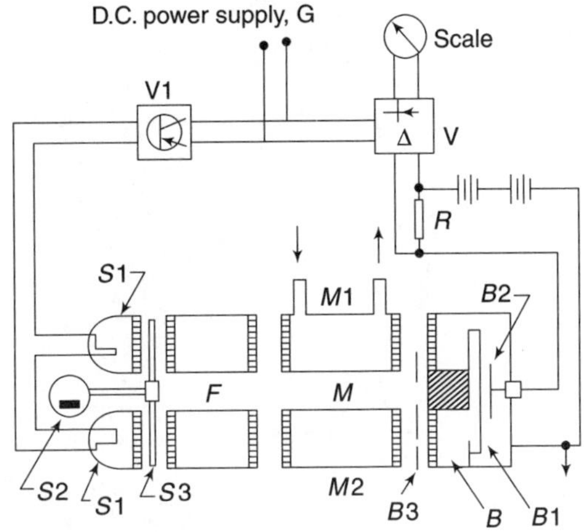


**Fig. 15.31** NDIR method for measuring CO

affect the surface to volume ratio and hence the HC emissions. Lower compression ratio, higher L/D ratio, larger displacement per cylinder and fewer cylinders reduce surface to volume ratio. Lean mixture results in low HC emissions

The well established and accepted method for measuring HC is ***flame ionization detector* (FID)** as shown is Fig. 15.32. The main characteristics of HC compounds is ionization and this principle is employed in FID detector. *Formation of electrically charged particles of ionized carbon atoms from the hydrocarbons in a hydrogen-oxygen flame is achieved in the FID analyser. The current flow in microamperes which is amplified for measurement is a measure of the concentration of hydrocarbons.* The details of various meters used in FID is also given in Fig. 15.32.

## 15.18.3 Aldehydes

*Aldehydes give pungent smell in the engine exhaust and any trained person may identify the degree of aldehydes present in the exhaust.*

There are various methods to measure aldehydes. They are based on wet chemical methods (analytical methods) **like iodine titration technique (ITT)**, **chromotropic acid (CA)** method, **3-methyl**, **2-benzothiazolene hydrozone (MBTH)** method and **2-4**, **Dinitrophenyl hydrazine (DNPH)** method. There are other methods also.

## 15.18.4 Smoke

Smoke of engine exhaust falls under this category of visible emissions and are more irritating and cause nuisance. *Smoke is due to incomplete combustion.* **In diesel engine, smoke may be *blue, white and black***. There are generally two methods employed to measure smoke. They are ***comparison method*** and ***obscuration method***.

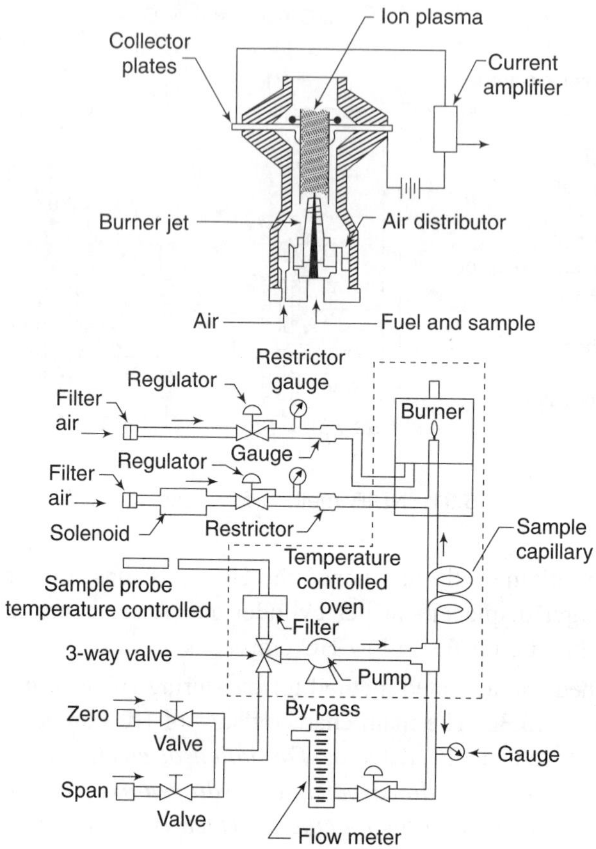


**Fig. 15.32** FID for measuring for unburnt HC

#### 15.18.4.1 Comparison Method

As per pollution ordinances and regulations concerning smokes, most of them are based on estimation of the density of the smoke as it emerges from the exhaust. *One of the most widely used method to find the density of smoke is **Ringlemann chart**.* The chart depicts four shades of grey, *as well as a pure white and an all-black section.* It is very difficult to produce *four shades of grey*. In order to overcome the difficulty of reproducing various shades of grey, the intermediate shades are built from black lines of various widths not shown in Fig. 15.32. United States of Bureau of Mines has printed **four charts on a single 70 cm × 25 cm sheet**. They are given below:

1. All white (means no smoke)
2. Black lines 1 mm thick, 10 mm apart, leaving white space 9 mm square
3. Lines 2.3 mm thick, spaces 7.7 mm square
4. Lines 3.7 mm thick, space 6.3 mm square
5. Lines 5.5 mm thick, space 4.5 mm square
6. All black (means dense black smoke)

So far the use of this chart is concerned, the chart is set up at eye level in line with the stack at such distance (say 10 or more) that the sections appear to be different degree of uniform grey shades. The appearance of the smoke at the exit of silencer is matched against one of the shades on the card and reported as a specific Rigelmann number ranging from 0 (no smoke) to no. 5 (dense black smoke). Any experienced person can estimate the smoke density.

#### *15.18.4.2 Obscuration Method*

Obscuration method for measuring smoke density involves **three basic methods,** namely, **light extinction** type, **continuous filtering** type and **spot filtering type**. They are discussed separately.

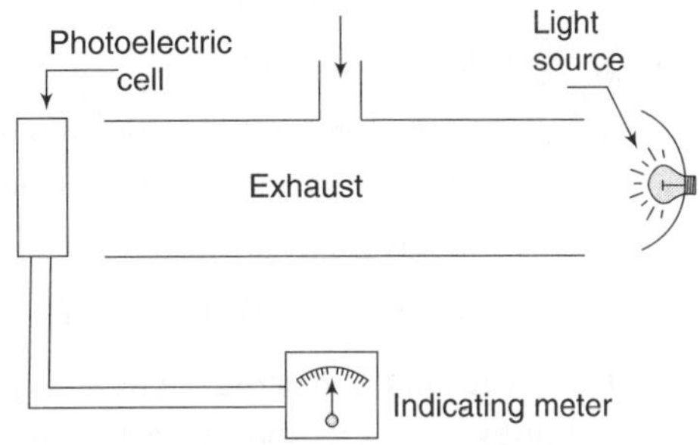


**Fig. 15.33** Basic principle of light extinction type smokemeter

**(a) Light Extinction Type.** *The principle of this method is based on the fact that the intensity of the light beam is reduced by smoke.* This reduction is a measure of smoke intensity. Figure 15.33 shows a schematic diagram of this method. The system consists of a tube (about 45 cm length) through which the exhaust sample passes, a light source at one end and a photocell at the other end.

The amount of light passed through this column of smoke gives the smoke level or smoke density. *For this instrument, the smoke level or smoke density is defined as the ratio of electric output from photocell when sample is passed through the column to the electric output when clean air is passed through it.*

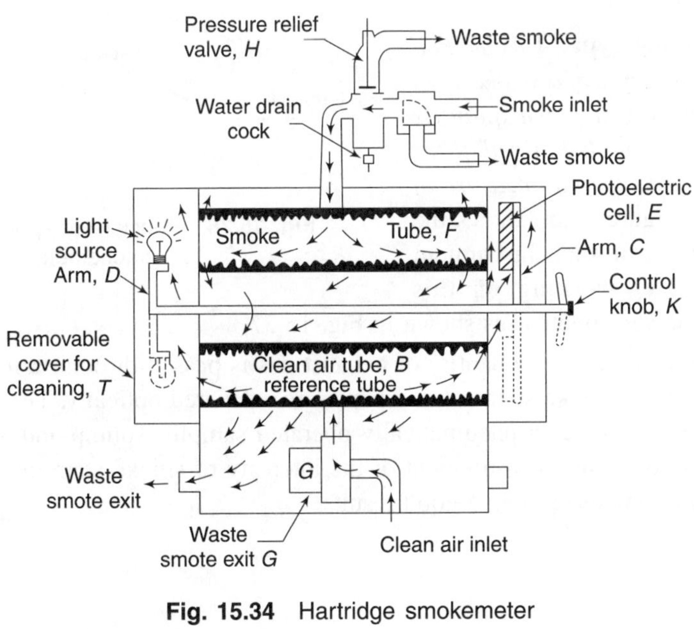


**Fig. 15.34** Hartridge smokemeter

Based on light extinction principle, the **Hartridge smokemeter** has been developed. This smokemeter is shown in Fig. 15.34. *In this smokemeter, light from a source is passed through a standard length tube containing the exhaust gas sample from the engine and at the other end, the transmitted light is measured by a photo electric cell.* With the help of Beer–Lambert law, the fraction of the light transmitted through the smoke ($T$) is estimated as

$$T = e^{-K_{ac}.L_e} \qquad ...(15.49)$$

where $L_e$ = length of light path

$K_{ac}$ = optical absorption coefficient of obscuring matter unit/length = $n\ A\ \theta$

$n$ = number of roots particles per unit volume

$A$ = the average projected area of each particles

$\theta$ = specific absorbance per particle

**(b) Continuous Filtering Type.** As the name implies, this method is based on the principle of continuously passing exhaust gas through a moving strip of filter paper and collecting particles. The schematic diagram of such an instrument is shown in Fig. 15.35. Van Brand smokemeter which is based on this principle is a popular instrument.

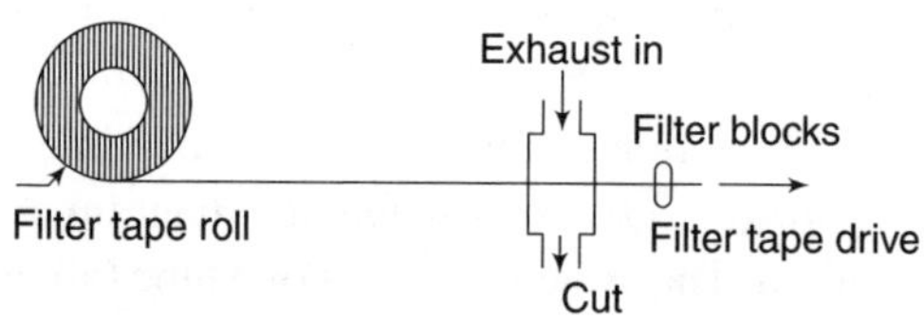


**Fig. 15.35** Continuous filter type smokemeter

In the Van Brand smokemeter, the exhaust sample is passed at a constant rate through a strip of filter paper moving at a preset speed. A strain is imparted to the paper and its intensity is measured by the amount of light which passes through the filter and is an indication of the smoke density of exhaust.

**(c) Spot Filtering Type.** *This method is based on the principle of a smoke stain obtained by filtering a given quantity of exhaust gas through a fixed filter paper which is used for the measure of smoke intensity.* Figure 15.36 shows the schematic diagram of such instrument Bosch smokemeter working on this principle which is very popular as shown in Fig. 15.37.

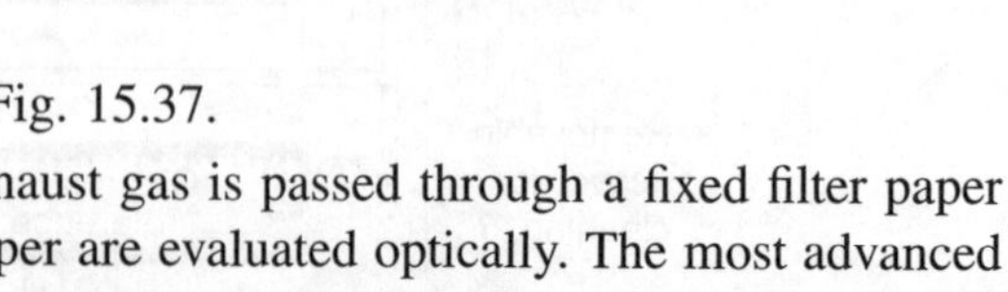


**Fig. 15.36** Principle of spot filtering type smokemeter

In this smokemeter, a fixed quantity of exhaust gas is passed through a fixed filter paper and the density of the smoke stains on the paper are evaluated optically. The most advanced Bosch smokemeter includes a pneumatically operated sampling pump and a photo-electric unit which are used for the measurement of the intensity of smoke stain on the filter paper. This instrument is supposed to give good results.

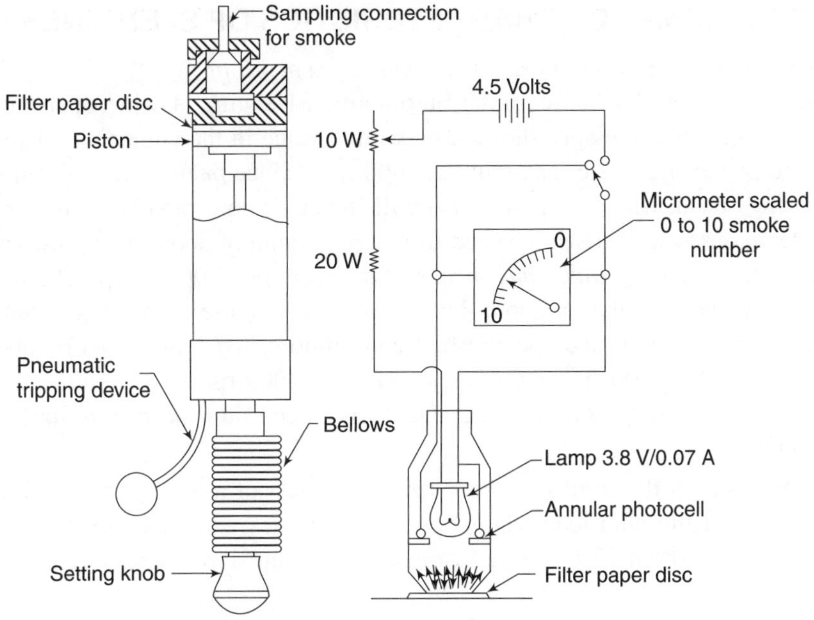


**Fig. 15.37** Bosch smokemeter

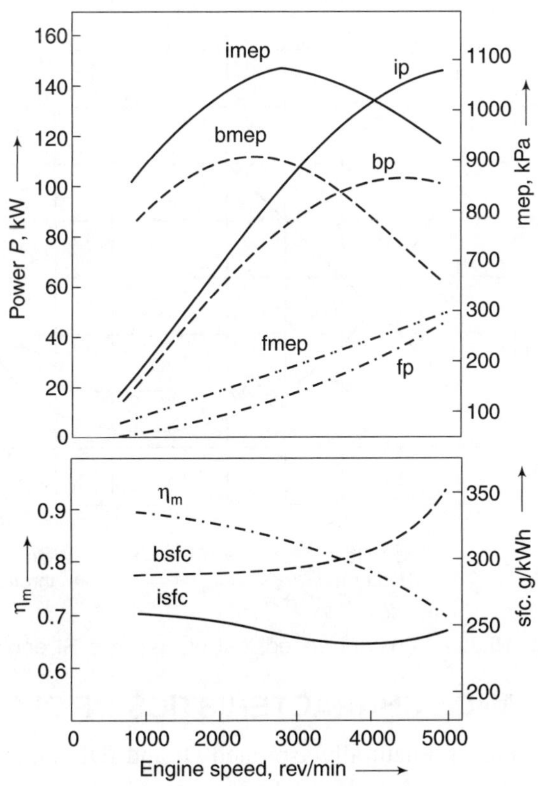


**Fig. 15.38** Variable speed test of a four-stroke six cylinder SI engine

## 15.19 PERFORMANCE CHARACTERISTICS OF SI ENGINES

Figure 15.38 shows the variation of *imep*, *bmep*, *fmep*, *ip*, *bp*, *fp*, $\eta_m$, *bsfc* and *isfc* with engine speed for a 3.8 dm$^3$ six-cylinder automotive SI engine at wide open throttle. The indicated mean effective pressure shows a maximum value in the engine's mid speed range. The indicated power (*ip*) is the maximum at 5000 rpm. Since *fmep* increases almost linearly with increasing speed, frictional power (*fp*) will increase more rapidly. Hence, mechanical efficiency decreases with increasing speed from a maximum of about 0.9 at low speed to 0.7 at 5000 rpm. *As a result of this, bmep peaks at a lower speed than imep.* The brake power (*bp*) shows a maximum value at about 4300 rpm. Any increase above this speed results in a decrease in *bp*. The indicated specific fuel consumption (*isfc*) increases by about 10 per cent from 0.31 to 0.34 over a speed range of 1000 to 4000 rpm. This is due to the fact of decreasing importance of heat transfer per cycle with increasing speed. *bsfc* starts increasing above 3000 rpm.

Figure 15.39 shows the results of a constant speed test of a SI engine. The brake power (*bp*) goes on increasing with load while *bsfc* first decreases with increase in speed, reaches a minimum value at about 75% engine load and then again starts increasing.

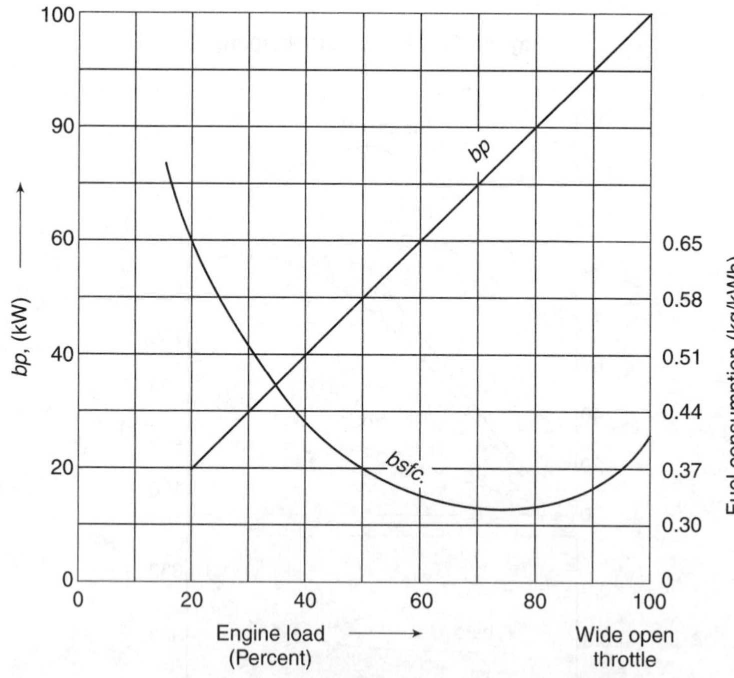


**Fig. 15.39** Constant speed test on 4-stroke SI engine

## 15.20 PERFORMANCE CHARACTERISTICS OF CI ENGINES

Performance characteristics of a naturally aspirated DI and IDI compression ignition engine at full load condition is shown in Fig. 15.40. Except at high engine speeds, brake torque and

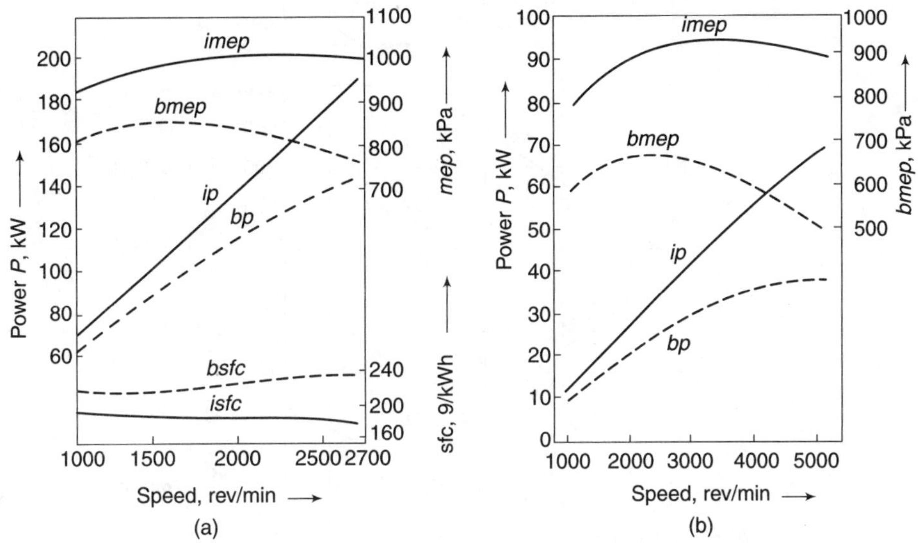


**Fig. 15.40** (a) Variables speed test of (a) 8.4 dm$^3$ six-cylinder naturally aspirated DI engine ($r$ = 16) (b) 1.8 dm$^3$ four-cylinder naturally aspirated IDI engine ($r$ = 22)

*mep* vary only modestly with engine speed. The decrease in torque and *bmep* with increasing engine speed is primarily due to the increase in friction with speed.

Figure 15.41 shows the constant speed test of CI engine. Torque, *bp* and bmep increase with load while *bsfc* first decreases and remains constant between 60 to 80% load.

## 15.21 VARIABLES AFFECTING ENGINE PERFORMANCE

The following are the most important variables which affect the performance of IC engines.

### 15.21.1 Compression Ratio

We have seen that thermal efficiencies of Otto, Diesel and dual cycle increase with compression ratio, $r$. But the highest possible value for the compression ratio depends on the following practical considerations.

(i) In SI engines, too high a value of compression ratio will cause detonation and pre-ignition. In CI engines, too high a value of $r$ will cause pressure above practical possibilities which will need a robust construction of the engine and hence increased structural weight.

(ii) With the increase in compression ratio in SI engines, the tendency of knock increases and it becomes so severe that the power of the engine will commence to decrease due to overheating. The compression ratio at which this occurs is known as the *highest useful compression ratio* (**HUCR**). The value of HUCR depends upon the type of fuel. To suppress detonation in SI engines, anti-knock mixtures known as **dope** is added to the petrol; this **dope** usually consists of *ethyl fluid*.

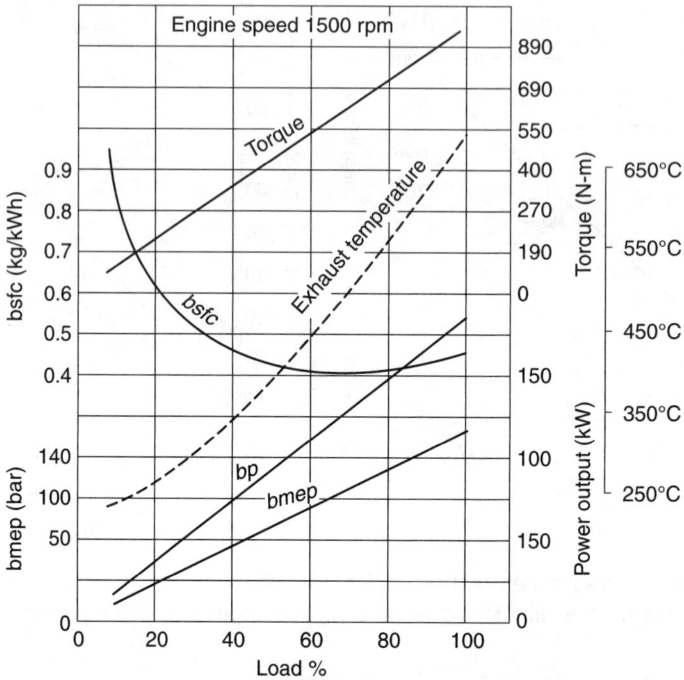


**Fig. 15.41** Constant speed test of 4-stroke CI engine

(iii) Though, theoretically it is true that as the compression ratio increases the efficiency increases but in practice it is not true. Due to various losses, after the optimum value of *r*, the thermal efficiency falls.

The present practice is to use a compression ratio from 5 to 10.5 in SI engines and from 12 to 20 in CI engines.

Figure 15.42 shows the effect of compression ratio on indicated *mep* and thermal efficiency of an eight-cylinder SI engine running at 2000 rpm and at wide open throttle condition. The brake thermal efficiency is the maximum at compression ratio about 14.

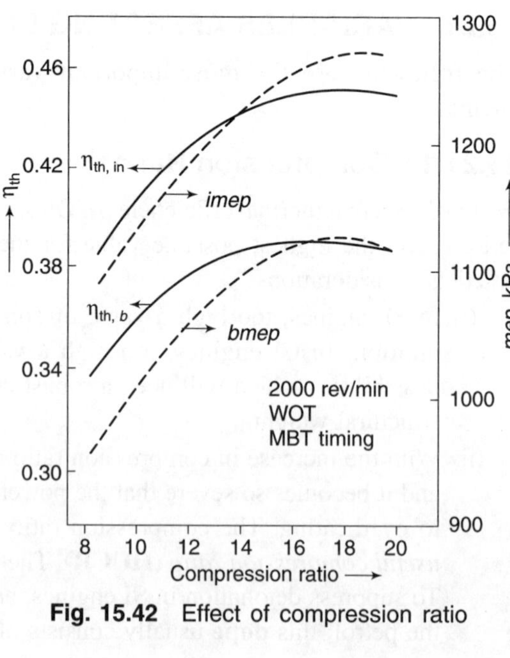


**Fig. 15.42** Effect of compression ratio

## 15.21.2 Air-Fuel Ratio (A/F)

Figure 15.43 shows the variation of *bsfc* and *bmep* with A/F ratio for a diesel engine. It

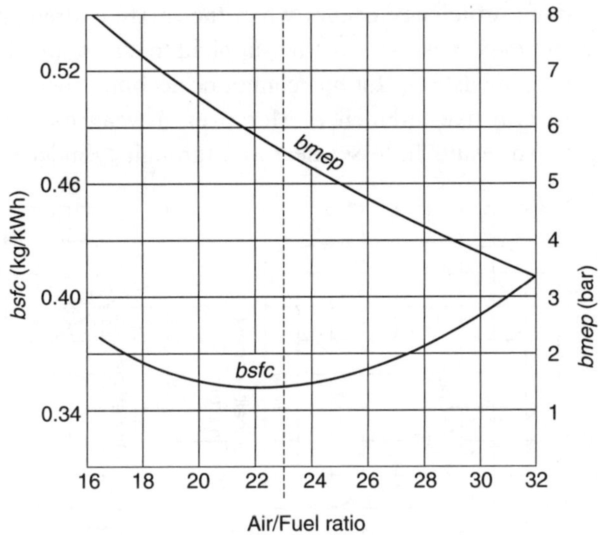


**Fig. 15.43** A/F ratio vs. *bsfc* for a diesel engine

is noticed from the curve that minimum *bsfc* occurs at A/F = 23. There is an increase in *bsfc* at low and high values of A/F. *bmep* goes on decreasing with the increase of A/F.

Figure 15.44 shows the variation of indicated thermal efficiency with mixture strength of a petrol engine. It is obvious that excessive lean mixture causes a rapid reduction in thermal efficiency.

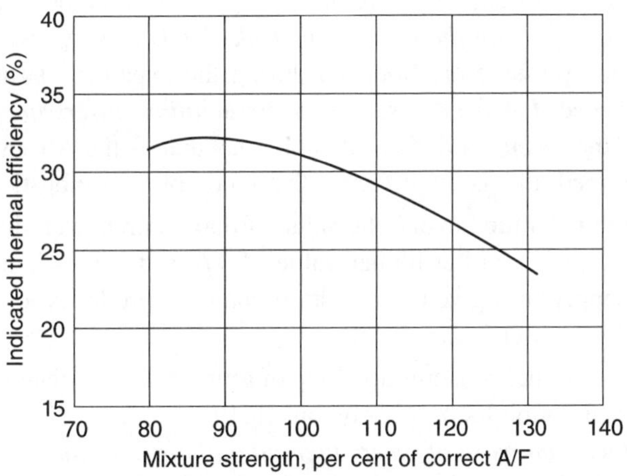


**Fig. 15.44** $\eta_{ith}$ vs mixture strength of a petrol engine

Figure 15.45 shows the variation of *bsfc* and *bmep* with air-fuel ratio of a diesel engine. This curve is a measure of overall efficiency of the engine. *It is observed that the minimum bsfc occurs when the* ***air-fuel ratio is 17****. This is the condition for the maximum efficiency.*

It appears that the maximum efficiency occurs when the mixture strength is 16.4% weak. This is attributed to the fact the excess air must be supplied to ensure that all of the fuel is in contact with the air during combustion. But this cannot be accomplished with correct weight of air even with the most effective turbulence. Moreover, a weak mixture produces lower maximum temperature which results in lesser heat loss through cylinder walls.

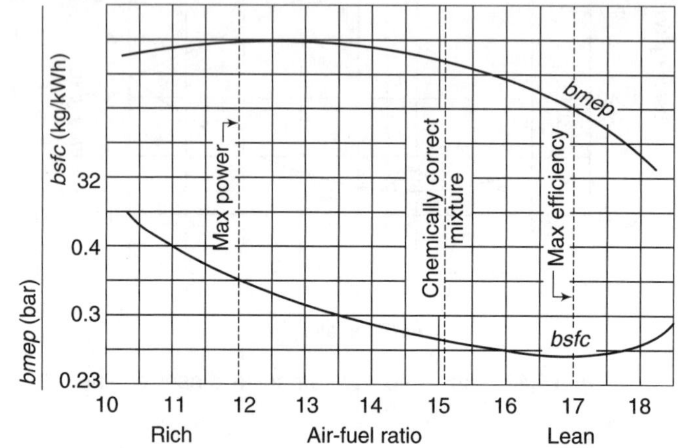


**Fig. 15.45** *bsfc* and *bmep* vs A/F for diesel engines

*If we analyse the bmep curve we find that maximum power occurs with an air-fuel ratio of 12, i.e., when the mixture is 17.8* per cent rich. This is attributed to the fact that a rich mixture gives an incomplete combustion, hence CO may be present in the burnt gases. *The presence of CO in the burnt gases during the expansion stroke tends to suppress dissociation;* hence higher temperature and pressure are obtained during the expansion stroke and more power is consequently produced. *It has been seen that dissociation lowers the temperatures of the* ***gas by 300°C***. From the *bmep* curve, it is also obvious that as the A/F ratio reduces, there is a fall in power developed; this is due to excessive incomplete combustion.

With the increase in A/F ratio beyond the value of maximum power, there is a fall in power developed and this is more rapid at higher value of A/F ratio. This is attributed to the fact that a smaller fuel supply in a given combustion chamber produces lower temperature and pressure and hence less power is developed.

Figure 15.46 shows the fuel consumption loop of four-stroke petrol and diesel engine. From the petrol engine consumption loop, it is obvious that *bsfc reduces (i.e., efficiency increases) up to a certain limit with increase of bmep, thereafter, there is a sudden increase in bsfc and curve loops backwards due to excess supply of fuel which cannot find sufficient oxygen to burn.* Point *A* on the consumption loop refers to the *bmep* at lowest fuel consumption (highest efficiency) while point *B* refers to the maximum *bmep* that can be obtained. Also, for the condition represented on the right of the curve, there are two values of fuel consumption for a given value of *bmep*. The value given by upper portion of the loop is due to an excess supply of fuel; this condition is known as ***choking***.

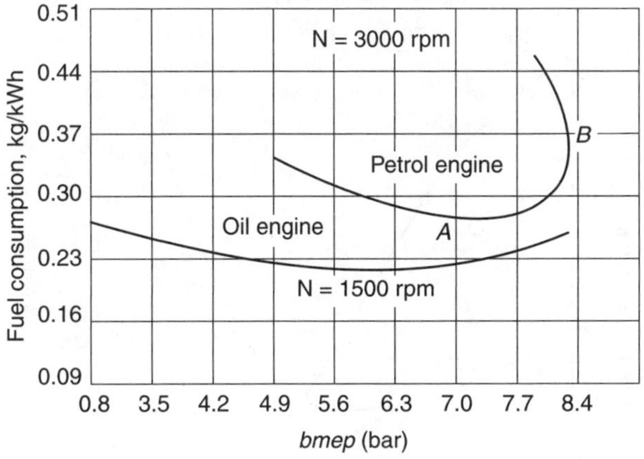


**Fig. 15.46** Fuel consumption loop

*A comparison of the consumption curves for both CI and SI engines reveals that the CI engines maintain high efficiency (i.e., low bsfc) over a large range of bmep (loads) than the SI engines. This fact makes the CI engines more suitable for road vehicles.* The diesel engines usually operate with A/F ratio lying in the *range of 12:1 to 20:1.*

Figure 15.47 shows pressure-volume and pressure-crank diagram for rich, stoichiometric and weak mixture.

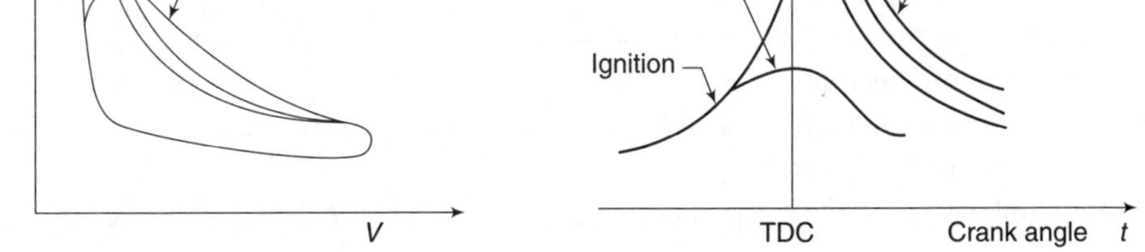


**Fig. 15.47** Mixture strength vs. *p-v* and *p-t* for SI engine

Figure 15.48 shows the variation of brake specific HC and $NO_x$ emission and *bsfc* of fuel with A/F and F/A equivalence ratio of an eight-cylinder SI engine at 1400 rpm. The results show that $NO_x$ decreases with the increase of A/F while HC first increases slowly and later on rapidly; *bsfc* first decreases and then increases rapidly with increase in A/F.

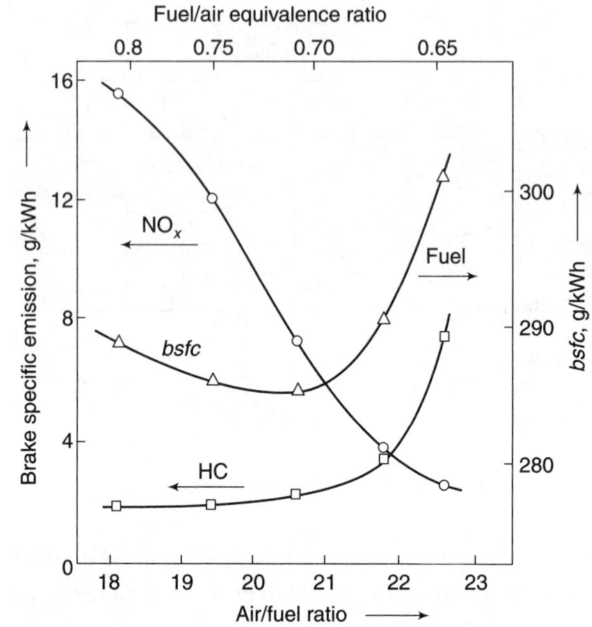


**Fig. 15.48** Variation of HC, $NO_x$ and *bsfc* with A/F

### 15.21.3 Turbulence, Spark or Injection Timing

Figure 15.49 shows the effect of turbulence on *p-t* diagram. Excessive turbulence in the combustion chamber causes greater heat losses to the cylinder walls thus lowering the efficiency. It may also produce high running and prevent complete combustion. Insufficient turbulence also lowers the efficiency due to incomplete combustion of the fuel.

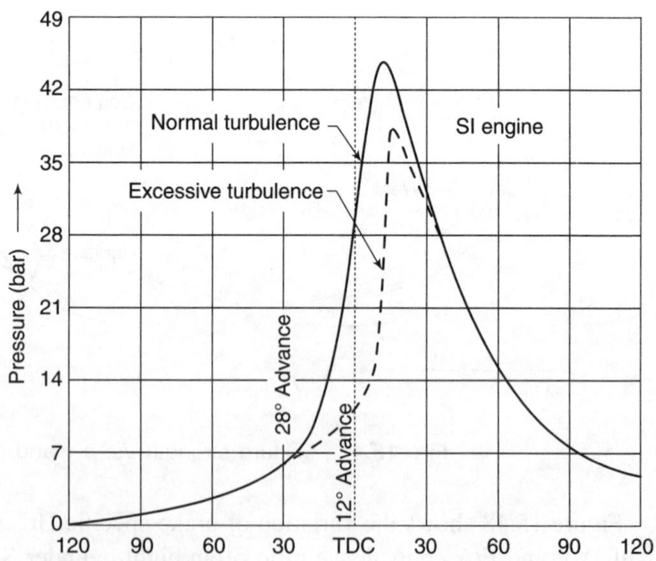


**Fig. 15.49** Effect of turbulence in SI engines

It is obvious from the Fig. 15.49 that the *excessive turbulence lowers maximum pressure.* It also *produces a more rapid pressure rise, even though the ignition is retarded* until 12° before the TDC.

*The spark should be timed and the combustion rate should be controlled so that the maximum pressure occurs as close as possible to the beginning of the power stroke, consistent with a smooth running engine.*

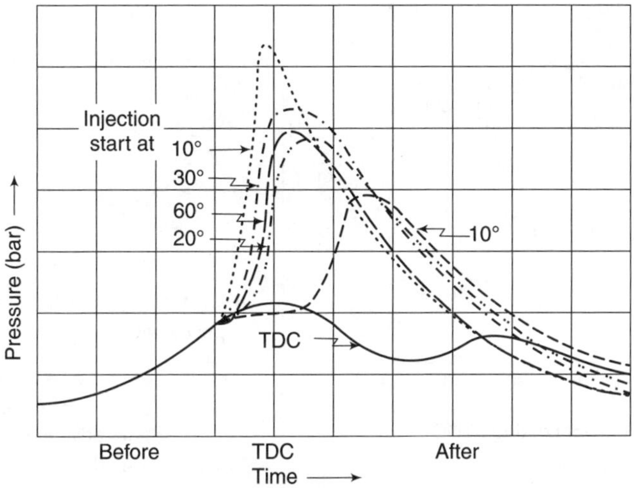


**Fig. 15.50** Effect of injection on *p-t* diagram

So far the turbulence is concerned the same pattern is found in the case of CI engine. Figure 15.50 shows the effect of injection timing on the maximum pressure and rate of pressure rise. It is obvious from *p-t* diagram that when late injection occurs, say at TDC, the maximum pressure decreases appreciably, because by the time combustion takes place the piston is on its way down on the power stroke. As a result, part of the pressure rise due to combustion is utilised in overcoming the expanding volume of the cylinder, thus there is a decrease in maximum pressure of the cylinder. Injection at 40 degree before TDC produces considerably higher maximum pressure but at this condition, the engine may knock. If an early injection at 60° before TDC is used, the maximum pressure again decreases. This is because of the fact that when too early injection takes place, the fuel is not properly prepared for a rapid combustion and thus the combustion is sluggish.

Figure 15.51 shows the variation in brake torque with spark advance for a eight-cylinder automotive SI engine at wide open throttle, at engine speeds varying from 1200 to 4200 rpm. *It is observed that there exists an optimum value of spark advance for each speed at which the brake torque becomes the maximum.*

Fuel injection timing essentially controls the crank angle at which combustion starts. The injection rate depends on the fuel-injector nozzle area and injection pressure. Higher injection rates result in higher fuel-air mixing rates and hence higher heat release rate. Figure 15.52 shows the effect of injection timing and injection rate on *bsfc* for single cylinder naturally aspirated DI diesel engine with swirl. *The higher heat release rates and shorter overall combustion process that result from the increased injection rate decrease the minimum bsfc at optimum injection timing.*

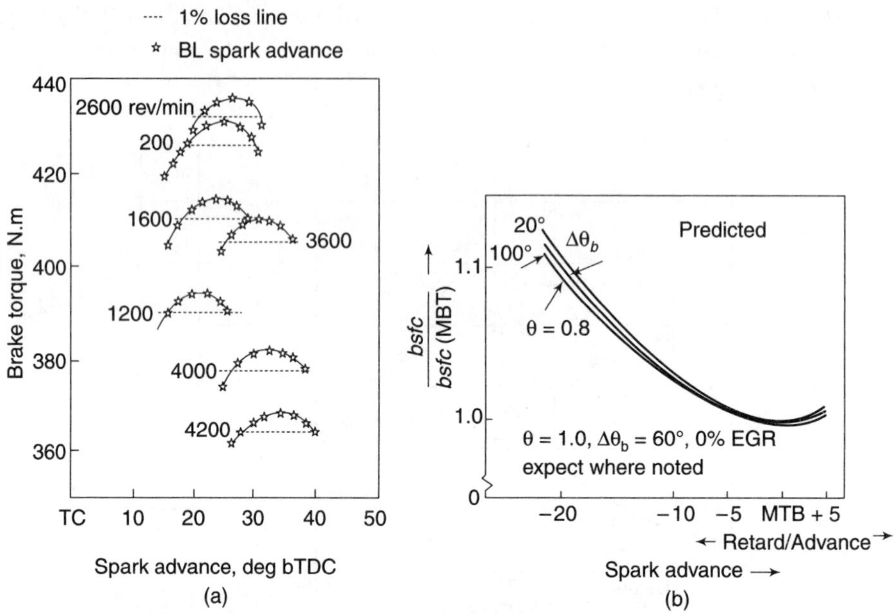


**Fig. 15.51** Variation in brake torque with spark advance, eight-cylinder SI engines at wide open throttle [112, 73]

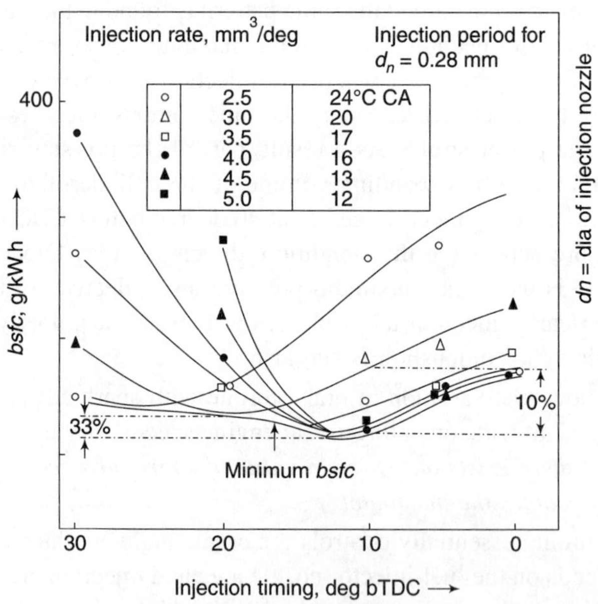


**Fig. 15.52** Effect of injection timing and injection rate on *bsfc* for NADI diesel engine for 0.97 $dm^3$ single-cylinder [111, 73]

Figure 15.53 (a, b) shows effect of injection timing on performance and emission in a diesel engine. *With the increase in injection timing, $NO_x$ decreases, CO first decreases and then increases, HC remains more or less constant as shown in Fig. 15.53(a), smoke first decreases and then increases.*

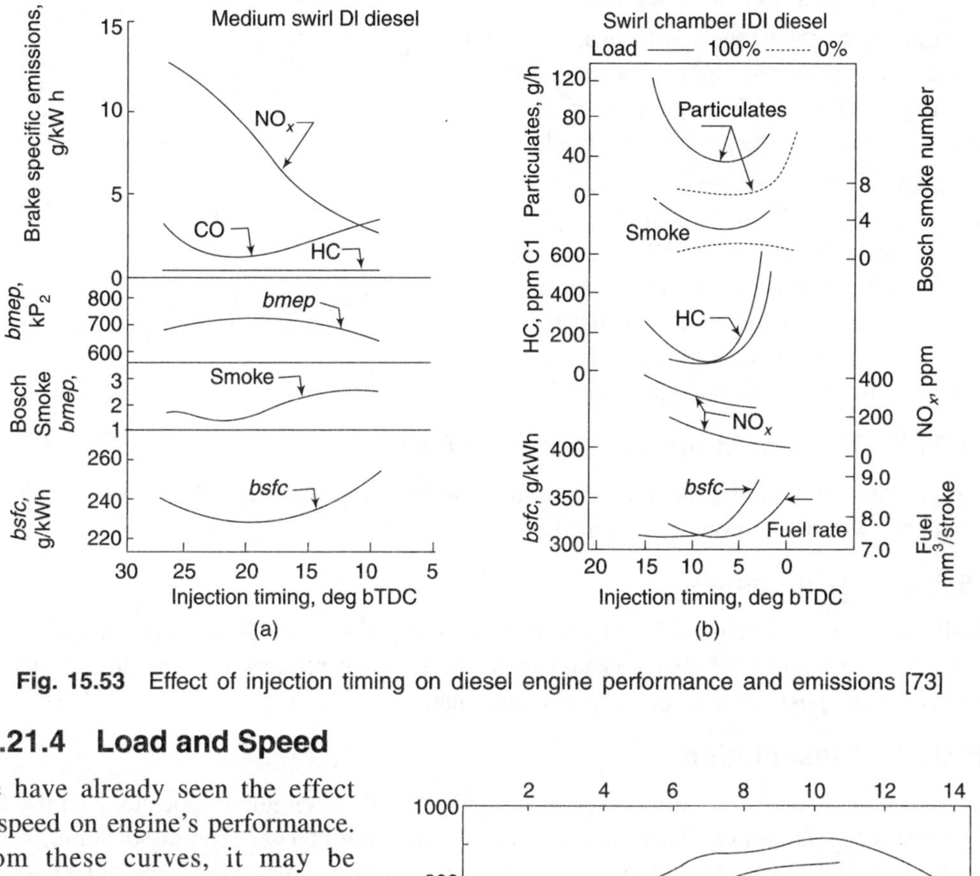


**Fig. 15.53** Effect of injection timing on diesel engine performance and emissions [73]

## 15.21.4 Load and Speed

We have already seen the effect of speed on engine's performance. From these curves, it may be realized that at low speeds, a greater length of time is available for heat transfer to cylinder walls, thus there is a greater proportion of heat loss. At higher speeds, the air consumption is greater, which produces greater *ip*, but along with a greater value of *fp*, consequently the engine speed range must be a compromise. Nowadays, higher speeds are generally preferred.

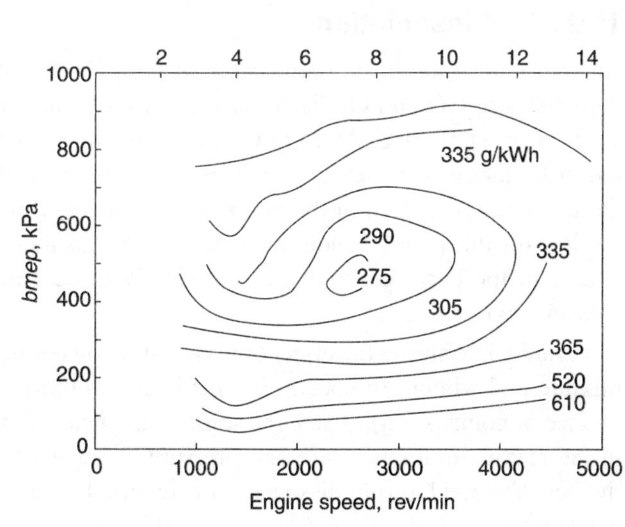


**Fig. 15.54** Performance map for 4-cylinder SI engines

One common way to present the operating characteristics of an internal combustion engine over its full load and speed range is to plot *bsfc* contours on a graph of *bmep* vs engine speed which is known as ***performance map***.

Figure 15.54 shows such performance map for a four-cylinder SI engine. *Maximum bmep occurs in the mid-speed range, the minimum bsfc is located at slightly lower speed and at part load.*

Figure 15.55 shows the performance maps of four-cylinder (1.47 $dm^3$) naturally aspirated DI diesel engine (NADI) with high swirl single hole nozzle. It is obvious from the results that there exists a maximum *bmep* at a particular speed only for any constant *bsfc* curves. There is a minimum value of *bsfc* at a particular speed and bmep.

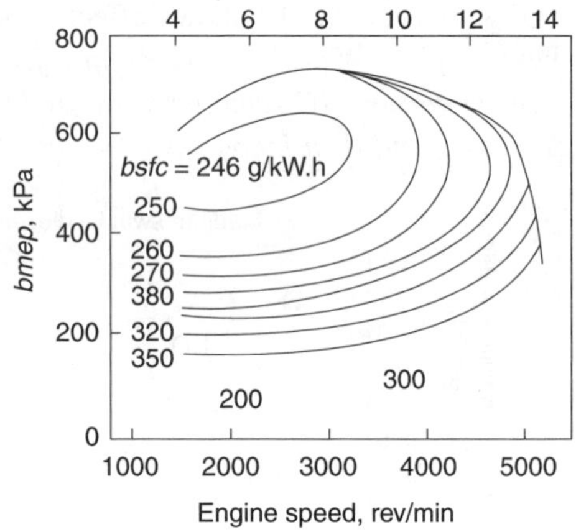


**Fig. 15.55** Performance maps for 4-cylinder NADI diesel engine

### 15.21.5 Mass of Inducted Charge of Air

We have already seen that the greater the mass of the charge (or air) inducted in the cylinder, the higher will be the power developed.

### 15.21.6 Heat Losses

While discussing heat balance, we have seen that higher the heat losses, lower is the efficiency and power developed. Higher constant inlet, temperature reduces the heat losses but it is limited by the possibility of damage to engine parts.

### 15.21.7 Dissociation

At high temperature, the burnt gases in the cylinder of IC engine dissociate. Mainly, $CO_2$ dissociates to CO and $O_2$. There is very little dissociation of $H_2O$. The dissociation of $CO_2$ starts at about 1000°C. At 1500°C, it amounts to 1 per cent. Due to presence of CO and $O_2$, the dissociation is prevented which is noticeable in a rich fuel mixture. On the other hand, there is no dissociation in the burnt gases of a weak mixture as the temperature produced is too low for this phenomenon to occur. This means that the maximum amount of dissociation occurs in the burnt gases of the chemically correct mixture, but decreases with richer or weaker mixtures.

Figure 15.56 shows the effect of dissociation on temperature and *bp* with chemically correct mixture and without dissociation. This is due to the fact that when dissociation occurs (i.e., reverse of combustion), it absorbs some of the heat of the explosion, thus causing a reduction in the maximum temperature and pressure. After the fall of temperature during expansions, the heat absorbed during dissociation is released but it is too late in the stroke to compensate entirely for the lost power. A portion of this heat passes out with the exhaust gases.

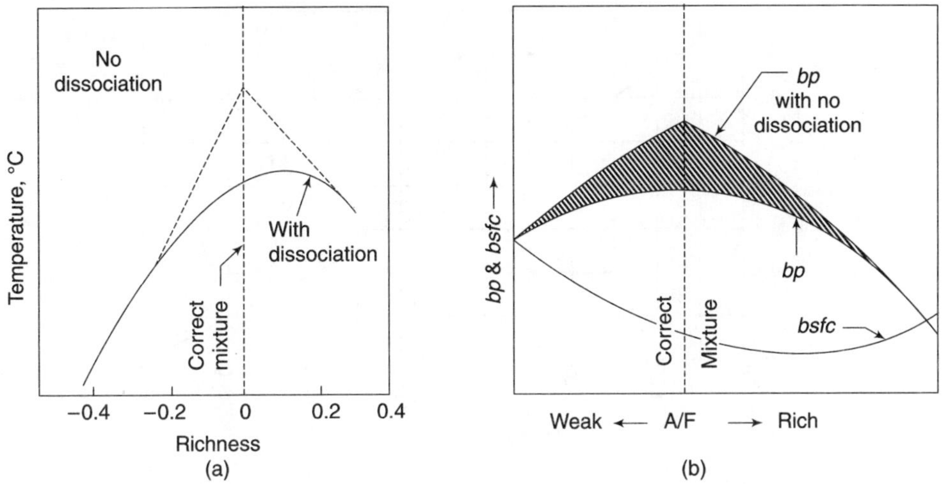


**Fig. 15.56** Effect of dissociation on temperature and *bp*

It is also obvious from the graph that *bp* is the maximum with chemically correct mixture with no dissociation. When a richer mixture is used dissociation rapidly decreases owing to the presence of CO due to incomplete combustion.

## 15.21.8 Effect of Altitude

*The power of the aircraft engine reduces considerably when flying in the rarefied air at high altitudes.* This is mainly due to lack of oxygen owing to the low pressure of the atmosphere. Although, the volume of air drawn in the carburettor remains constant, its *mass is greatly reduced due to its low pressure.* A supercharge corrects this deficiency of air.

At high altitude, the engine power is reduced by the following causes:

(i) There are irregularities of fuel evaporation causing failure to evaporate all the fuel absorbed.

(ii) At high altitude, the pressure of air is excessively low which causes delay in combustion.

(iii) Due to low pressure, the mass of air is greatly reduced, this is corrected by supercharging.

(iv) At low temperature, the viscosity of the lubricating oil increases, which in turn, increases the resistance of bearing and rubbing surfaces.

Though, power of the engine is slightly increased by the effect of reduced back pressure on the exhaust gas; the overall effect at high altitude is to reduce the power output.

Figure 15.57 shows the effect of altitude on power factor. It is noticed from the curve that engine power falls to 60% of its full value at altitude of 4200 m.

In rarefied air of high altitudes, there is a great reduction in frictional resistance of the aircraft, therefore, greater propulsion speeds are obtained with supercharged engines.

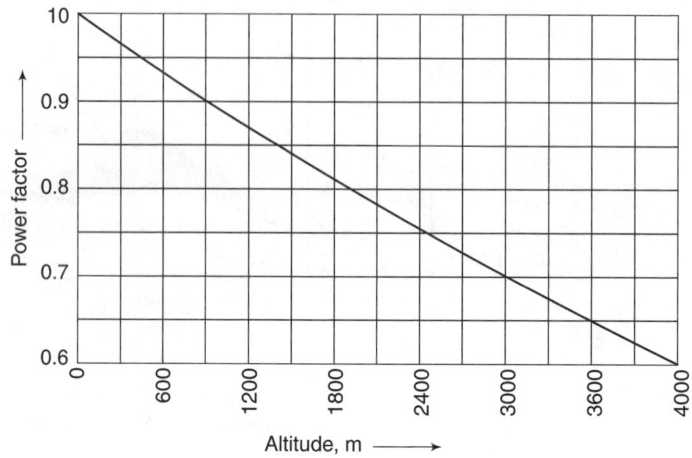


**Fig. 15.57** Effect of altitude

## 15.22 POWER REQUIRED AND POWER AVAILABLE CURVES

If the engine is used to power some vehicle, the *bp* curve represents the power which is available in the engine to drive the vehicle at various engine speeds. The vehicle will require a certain amount of power to drive it at various operating speeds. Thus, we see that the power available curve is a function of engine's performance while power required curve is a function of vehicle's requirement for power to operate at various speeds. Figure 15.58 shows such a power available and power required curve. The maximum speed will be obtained where the curves intersect. *The maximum acceleration will be obtained at the speed where the difference between these curves is the maximum.*

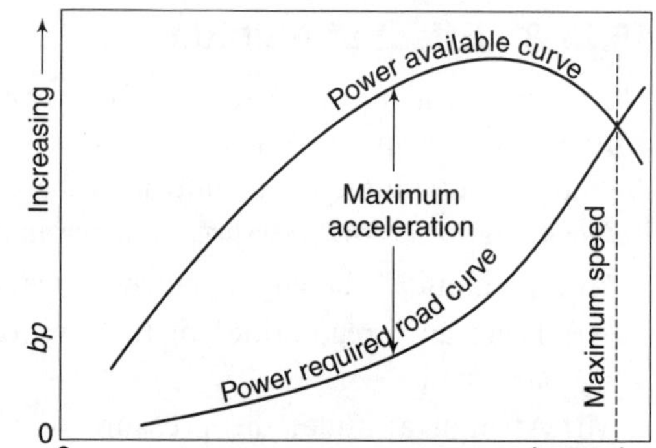


**Fig. 15.58** Power required and power available curve

## 15.23 HEAT BALANCE SHEET

Heat energy is supplied to the engine by the combustion of fuel. *Only a fraction of this heat energy is converted into useful work at the engine's crankshaft and the remainder is lost.* This means, that for the maximum work output, the non-useful energy leaving the system should be as small as possible. The accounting of the energy supplied, the energy losses and the useful energy output may be expressed by which engineer commonly call a ***heat balance,*** which is a determination of the disposition of the energy which is provided to the engine in fuel.

To draw up a heat balance sheet for an internal combustion engine, a complete test must be made on the engine while running at constant speed. The fuel consumption and its calorific value, the quantity and temperature of cooling water, the indicated and brake power, the mass of exhaust gas and its temperature must be known. The various terms used in heat balance sheet must be expressed in heat units.

**(i) *Heat in fuel supplied.*** In oil engines, this will be the mass of oil multiplied by its lower calorific value. If $\dot{m}_f$ is the mass of fuel consumed in kg per sec and LCV is the lower calorific value of fuel in kJ/kg of fuel, then

$$\text{Heat in fuels supplied} = Q_A = \dot{m}_f \times LCV \text{ ; kW} \qquad ...(15.50)$$

For gas engine, the volume of the gas supplied must be measured by a meter. Its pressure and temperature along with barometer reading be taken.

Let $\dot{V}$ = volume of gas supplied m$^3$ per second at NTP

$LCV$ = lower calorific value of fuel in kJ/m$^3$ at NTP

Heat in fuel supplied = $Q_A = \dot{V} \times LCV$ ; kW

**(ii) Heat in *bp*.** Heat equivalent of useful work can be measured by measuring the *bp* of the engine.

$$\text{Heat in } bp = bp = \frac{2\,\pi\, N\, T}{60 \times 1000} \text{ ; kW} \qquad ...(15.51)$$

Sometimes, it is preferred to calculate heat in *ip*.

**(iii) *Heat rejected to cooling water.*** The heat rejected to cooling water is calculated from the equation

$$Q_{cw} = \dot{m}_w \times c_{pw} \times \Delta T_w \text{ ; kW} \qquad ...(15.52)$$

where $\dot{m}_w$ = mass of cooling water, kg/s

$c_{pw}$ = specific heat of water, kJ/kg K

$\Delta T_w$ = temperature rise of cooling water, °C

**(iv) Heat carried away by exhaust gases** (Fig. 15.59)

Let $\dot{m}_g$ = mass of exhaust gases, kg/s

$c_{pg}$ = specific heat of exhaust gases, kJ/kgK

$T_g$ = temperature of exhaust gases, °C

$T_a$ = temperature of ambient air, °C

∴ Heat carried away by the exhaust gas

$$Q_{eg} = m_g . c_{pg} (T_g - T_a) \text{ ; kW} \qquad ...(15.53)$$

In some cases, the energy loss to exhaust gases is determined by passing them through an ***exhaust gas calorimeter***. Figure 15.59 shows such an exhaust gas calorimeter. It is essentially

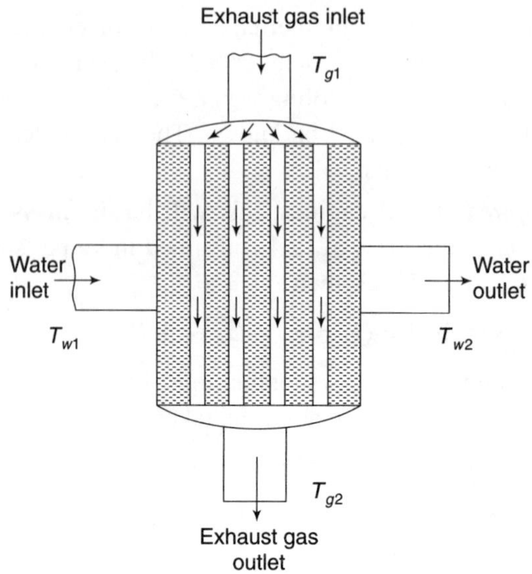


**Fig. 15.59** Exhaust gas calorimeter

a heat exchanger in which the exhaust gases are cooled by circulating water. Assuming 100% heat exchange.

Heat given by exhaust gases = Heat taken by water

If the temperature of water is raised from $T_{w1}$ to $T_{w2}$ while the exhaust gases are cooled from $T_{g1}$ to $T_{g2}$ ; then

$$\dot{m}_g \, c_{pg}(T_{g1} - T_{g2}) = \dot{m}_w \times c_{pw} \, (T_{w2} - T_{w1}) = Q_{eg}$$

If the gas has to fall in temperature to that of ambient temperature, the total energy to exhaust is calculated by

$$Q_{eg} = \dot{m}_w \, c_{pw}(T_{w2} - T_{w1}) \times \frac{T_{g1} - T_a}{T_{g1} - T_{g2}} \; ; \text{kW} \qquad ...(15.54)$$

The heat balance sheet may be prepared as shown in the following table

| *Item* | | *Heat in kW* | *Heat %* |
|---|---|---|---|
| Input | (a) Heat supplied in fuel, kW | | 100 (credit) |
| Expenditure | (b) Heat in *bp*, kW | – | – |
| | (c) Heat carried away by cooling water, kW | – | – |
| | (d) Heat carried away by exhaust gas, kW | | |
| | (e) Heat unaccounted for (by difference), kW | – | – |
| | = [*a* – (*b* + *c* + *d*)] | – | – |
| Total | | – | 100 |

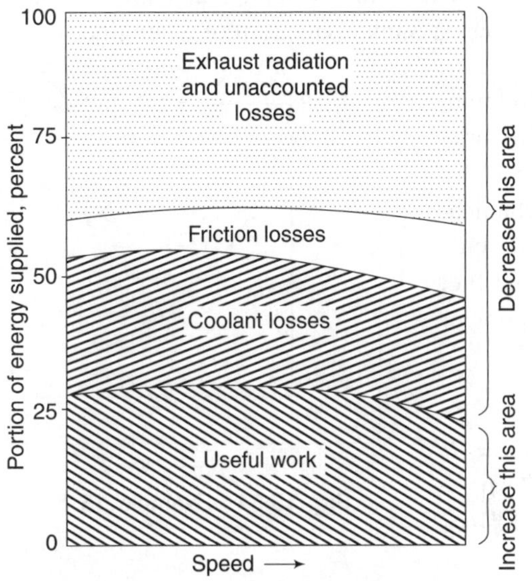


**Fig. 15.60** Heat balance diagram for SI engine at full throttle

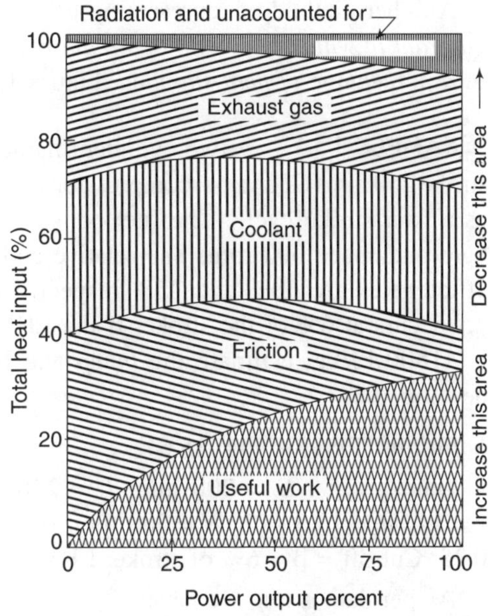


**Fig. 15.61** Heat balance diagram for CI engine

The heat energy distribution may also be expressed in graphical form which is known as **heat balance diagram**. Figure 15.60 shows the heat balance diagram for a SI engine at full throttle for various values of speed. The heat balance remains constant at all speeds except at very high speed. The useful power is only about 30% of the heat supplied. Figure 15.61 shows the heat balance diagram of a CI engine at various loads. The heat balance varies according to the load. At full output, the useful work is about 35% of the heat supplied.

The heat balance may be external or internal. Figure 15.62 shows the external heat balance. Energy absorbed by lubricating oil is very small and is usually neglected.

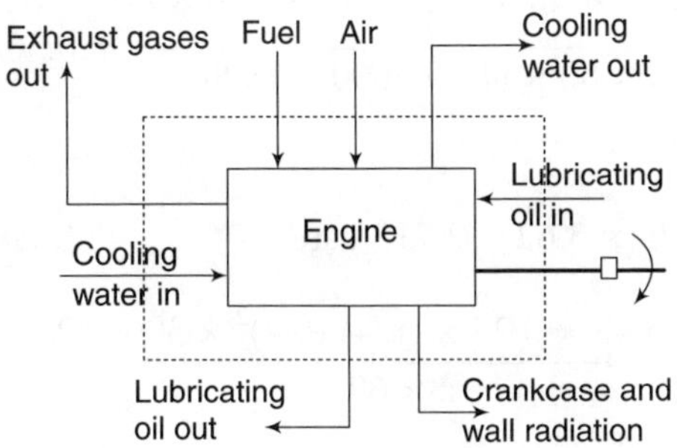


**Fig. 15.62** External heat balance

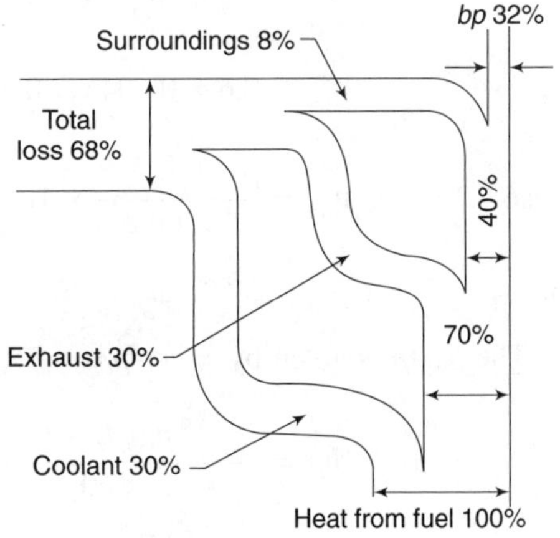


**Fig. 15.63** Sankey diagram

Another method of representing heat balance is **Sankey diagram**. *This is a stream type diagram in which the width of stream represents the heat quantity being considered, usually as a percentage of heat supplied.* Figure 15.63 shows a Sankey diagram of SI engine.

**EXAMPLE 15.1** A two-cylinder (diameter 200 mm, stroke 250 mm) CI engine, compression ratio 14:1 works as a four-stroke cycle. The fuel consumption is 13.5 kg/h and speed 300 rpm. The relative and mechanical efficiencies are 60% and 75% respectively. The cut-off occurs at 6% of the stroke. If the LHV of the fuel is 41800 kJ/kg, estimate the *bmep* and brake power developed.

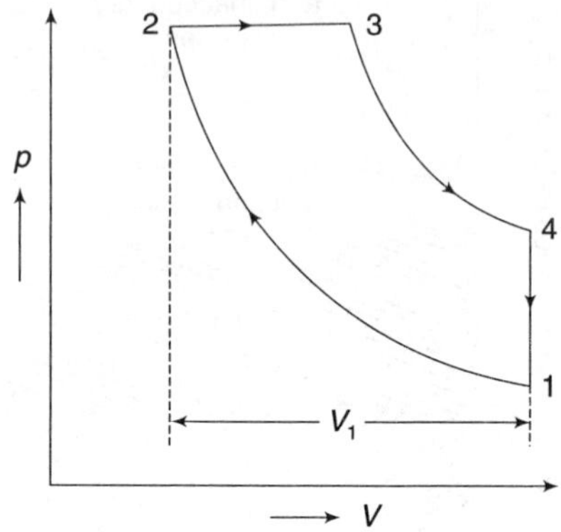


**Fig. 15.64** Example 15.1

***Solution*** Refer to Fig. 15.64.

Data given : $D$ = 200 mm; $L$ = 250 mm; $x$ = 2; $r$ = 14; $\dot{m}_f$ = 13.5 kg/h; $N$ = 300 rpm; $\eta_{rel}$ = 60%; $\eta_m$ = 0.75; Cut-off = $\beta$ = 6% of stroke; LHV = 41800 kJ/kg

As per data given, we have

$$V_3 - V_2 = 0.06V_s = 0.06(V_1 - V_2)$$

or $$\frac{V_3}{V_2} - 1 = 0.06\left(\frac{V_1}{V_2} - 1\right)$$

or $$\beta - 1 = 0.06\,(r - 1) = 0.06\,(14 - 1) \quad \text{where} \quad \beta = \frac{V_3}{V_2}$$

or $$\beta = \text{Cut-off ratio} = 1.78$$

Relative efficiency ($\eta_{rel}$) is given by = $\eta_{rel} = \dfrac{\eta_{ith}}{\eta_{as}}$

But $$\eta_{as} = ASE = 1 - \frac{1}{r^{\gamma-1}}\left[\frac{\beta^\gamma - 1}{\gamma(\beta - 1)}\right] = 1 - \frac{1}{14^{0.4}}\left[\frac{1.78^{1.4} - 1}{1.4(1.78 - 1)}\right] = 0.6045$$

∴ $$\eta_{ith} = 0.6 \times 0.6045 = 0.3627$$

Also, $$\eta_{ith} = \frac{p_{mi}\,L\,A\,N\,x}{2 \times 60\, q_A} \times 10^2 = \frac{10^2 \times p_{mi} \times 0.25 \times \frac{\pi}{4}(0.2)^2 \times 300 \times 2}{2 \times 60 \times (13.5/3600)\,41800} = 0.3627$$

or $$p_{mi} = 3.62 \text{ bar}$$

The *bmep* is given by = $bmep = \eta_m \times p_{mi} = 0.75 \times 3.62 = 2.715$ bar **Ans.**

$$bp = \frac{10^2 \times p_{mb}\,L\,A\,N\,x}{2 \times 60} = \frac{10^2 \times 5.43 \times 0.25 \times (\pi/4)\,(0.2)^2 \times 300 \times 2}{2 \times 60}$$

$$= 21.32 \text{ kW}$$ **Ans.**

**EXAMPLE 15.2** A 6-cylinder, four-stroke 10 cm × 12.5 cm stroke diesel engine develops 50 kW at 1000 rpm. The various efficiencies are mechanical 76%, volumetric 80% under room condition, indicated relative efficiency 88%, theoretical thermal efficiency 52%. The lower calorific value of the liquid fuel is 45000 kJ/kg. Compute (i) *bmep*, (ii) A/F, (iii) *bsfc*. Assume air density as 1.2 kg/m$^3$ under room condition.

***Solution*** Data given : $x = 6$, $s = 4$, $D = 10$ cm, $L = 12.5$ cm, $N = 1000$ rpm, $bp = 50$ kW, $\eta_m = 76\%$, $\eta_v = 80\%$, $\eta_{rel} = 88\%$, $\eta_{ith} = \eta_{as} = 52\%$, LCV = 45000 kJ/kg, $\rho_a = 1.2$ kg/m$^3$

The brake power ($bp$) four-stroke engine is given by

(i) $$bp = \frac{10^2\, p_{mb}\, L\, A\, N\, x}{60 \times 2}$$

or $$50 = \frac{10^2 \times p_{mb} \times 0.125 \times (\pi/4)\,(0.1)^2 \times 1000 \times 6}{60 \times 2}$$

or $$p_{mb} = \frac{50 \times 60 \times 2 \times 4}{10^2 \times 0.125 \times \pi \times (0.1)^2 \times 1000 \times 6} = 10.185 \text{ bar}$$ ***Ans.***

(ii) $$ip = \frac{bp}{\eta_m} = \frac{60}{0.76} = 78.94 \text{ kW}$$

$$\eta_{rel} = \frac{\eta_{ind}}{\eta_{as}} = \frac{\eta_{ind}}{\eta_{ith}}$$

or $$\eta_{ind} = \eta_{as} \times \eta_{irel} = 0.88 \times 0.52 = 0.4576$$

or $$\eta_{ind} = \frac{ip}{q_A}$$

or $$0.4576 = \frac{78.94}{\dot{m}_f \times 45000}$$

or $$\dot{m}_f = 0.0038 \text{ kg/s}$$

Actual air sucked

$$= \dot{V} = A\, L\, N\, x\, \eta_v = \frac{\pi}{4}(0.1)^2 \times 0.125 \times \frac{1000}{60} \times 6 \times 0.8 = 0.0785 \text{ m}^3\text{/s}$$

Mass of air inducted in all cylinder = $\dot{m}_a = 0.0785 \times 1.2 = 0.0942$ kg/s

$$\therefore \quad A/F = \frac{\dot{m}_a}{\dot{m}_f} = \frac{0.0942}{0.0038} = 24.78$$ ***Ans.***

(iii) $$bsfc = \frac{\dot{m}_f}{bp} \times 3600 = \frac{0.0038 \times 3600}{50} = 0.2736 \text{ kg/kWh}$$ ***Ans.***

**EXAMPLE 15.3** During the trial of a single acting oil engine, cylinder diameter 20 cm stroke 28 cm, working on the two-stroke cycle and firing every cycle, the following observations were made:

Duration of trial = 1 hour

Total fuel used = 4.22 kg

Calorific value of fuel = 44670 kJ/kg

Proportion of hydrogen in fuel = 15%

Total number of revolution = 21000

Mean effective pressure = 2.74 bar

Net brake load applied to a drum of 100 cm diameter = 600 N

Total mass of cooling water circulated = 495 kg

Temperature of cooling water, inlet 13°C, outlet = 38°C

Air used = 135 kg

Temperature of air in test room = 20°C

Temperature of exhaust gases = 370°C

Assume $c_{pgas}$ = 1.005 kJ/kgK, $c_{psteam}$ at atmospheric pressure = 2.093 kJ/kg K

Calculate the thermal efficiency and draw up the heat balance.

***Solution*** The indicated power (*ip*) is given by

$$ip = p_{mi}\, L\, A\, N = 2.74 \times 10^2 \times 0.28 \times \frac{\pi}{4} \times 0.22 \times \frac{21000}{3600} = 14.06 \text{ kW}$$

$$bp = 2\pi\, N\, T = 2\pi \times \frac{21000}{3600} \times \frac{600 \times 0.5}{1000} = 11 \text{ kW}$$

$$fp = ip - bp = 14.06 - 11 = 3.06 \text{ kW}$$

$$\text{Heat supplied} = Q_A = \dot{m}_f \times C\,V = \frac{4.22}{3600} \times 44670 = 52.36 \text{ kW}$$

$$\text{Heat taken by cooling water} = \dot{m}_w\, c_{pw}\, \Delta T_w = \frac{495}{3600} \times 4.1868\,(38 - 13) = 14.39 \text{ kW}$$

Water vapour ($H_2O$) formed due to combustion = 0.15 × 9 = 1.35 kg/kg of fuel

$$\text{Air-fuel ratio used} = \frac{135}{4.22} = 31.95 \text{ kg/kg of fuel}$$

Dry flue gases per kg of fuel = $\dot{m}_g$ = (31.95 + 1) – 1.35 = 31.6 kg

Heat carried away by dry flue gases

$$= \dot{m}_g\, c_{pg}\, (T_g - T_a) = \frac{4.22}{3600} \times 31.6 \times 1.005\,(370 - 20) = 13.04 \text{ kW}$$

Heat absorbed by $H_2O = \dot{m}_{H_2O} [h_f + h_{fg} + c_{pg} (T_{sup} - T_s)]$

$$= 1.35 \times \frac{4.22}{3600}[(100 - 20)\ 4.1868 + 2256.9 + 2.093\ (370 - 100)] = 5 \text{ kW}$$

Heat balance sheet

| *Item* | | *Heat in kW* | *Heat %* |
|---|---|---|---|
| Heat input | (a) Heat supplied | 52.36 | 100% (credit) |
| Expenditure | (b) Heat in *bp* | 11.00 | 21.008 |
| | (c) Heat carried away by cooling water | 14.39 | 27.483 |
| | (d) Heat carried away by dry gases | 13.04 | 24.905 |
| | (e) Heat carried away by steam | 5.00 | 9.549 |
| | By difference [*a* – (*b* + *c* + *d*)] | 8.93 | 17.055 |
| Total | | 52.36 | 100% |

**EXAMPLE 15.4** A four-cylinder SI engine running at 1500 rpm delivers 27 kW. The average torque when one cylinder was cut is 120 Nm. The calorific value of the gasoline used is 43 MJ/kg and the brake specific fuel consumption is 0.36 kg/kWh. Calculate mechanical, indicated and brake thermal efficiencies.

***Solution*** The average brake power (*bp*) for 4 cylinders

$$\text{The } \textbf{average } bp = \frac{2\pi N T}{60} = \frac{2\pi \times 1500 \times 120}{60 \times 10^3} = 18.84 \text{ kW}$$

The **average** *ip* with one cylinder $ip = 27 - 18.84 = 8.16$ kW

Total $ip = 4 \times 8.16 = 32.64$ kW

$$\text{Mechanical efficiency} = \eta_m = \frac{bp}{ip} = \frac{27}{32.64} = 0.8272 \qquad \textbf{\textit{Ans.}}$$

The fuel consumption rate is given by

$$\dot{m}_f = \frac{bp \times bsfc}{3600} = \frac{27 \times 0.36}{3600} = 0.0027 \text{ kg/s}$$

$$\therefore \qquad \eta_{ith} = \frac{ip}{\dot{m}_f \times CV} = \frac{32.64}{0.0027 \times 43 \times 10^3} = 28.11\% \qquad \textbf{\textit{Ans.}}$$

$$\therefore \qquad \eta_{bth} = \frac{bp}{\dot{m}_f \times CV} = \frac{27}{0.00027 \times 43 \times 10^3} = 23.25\% \qquad \textbf{\textit{Ans.}}$$

**EXAMPLE 15.5** A six-cylinder, four-stroke gasoline engine operates at 4000 rpm. The bore of each cylinder is 10 cm and the stroke is 12 cm. The clearance volume of each cylinder is 100 cc. The fuel consumption is 20 kg/h and the torque developed is 150 Nm. Calculate (i) the brake power, (ii) the brake mean effective pressure, (iii) the brake thermal efficiency and (iv) the relative efficiency base on brake power. Assume calorific value of fuel as 43 MJ/kg.

***Solution*** The brake power (*bp*) is given by

$$bp = \frac{2\pi N T}{60} = \frac{2\pi \times 4000 \times 150}{60 \times 10^3} = 62.83 \text{ kW} \qquad \textbf{Ans.}$$

Also,
$$bp = \frac{10^2 p_{mb} L A N x}{60 \times 2}$$

or
$$p_{mb} = \frac{60 \times 2 \times bp}{10^2 \times L A N x} = \frac{120 \times 62.83}{10^2 \times 0.12 \times (\pi/4)(0.1)^2 \times 4000 \times 6} = 3.33 \text{ bar} \qquad \textbf{Ans.}$$

$$\eta_{bth} = \frac{bp}{\dot{m}_f \times CV} = \frac{62.83 \times 3600}{20 \times 43 \times 10^3} = 26.3\% \qquad \textbf{Ans.}$$

$$\text{Compression ratio} = r = \frac{V_s + V_c}{V_c} = \frac{(\pi/4)(10)^2 \times 12 + 100}{100} = 10.42 \qquad \textbf{Ans.}$$

$$\text{ASE} = 1 - \frac{1}{r^{\gamma - 1}} = 1 - \frac{1}{10.42^{0.4}} = 0.608 = 60.8\%$$

Relative efficiency based on brake power

$$\eta_{rel} = \frac{\eta_{bth}}{ASE} = \frac{0.263}{0.608} = 43.25\% \qquad \textbf{Ans.}$$

**EXAMPLE 15.6** An eight-cylinder, four-stroke SI engine of 10 cm bore and 11 cm stroke with a compression ratio of 8 is tested at 5000 rpm on a dynamometer which has 55 cm arm. During a 10 minutes test, the dyanometer scale beam reading was 45 kg and the engine consumed 5 kg of gasoline having a calorific value of 44 MJ/kg. The carburettor receives the air at 27°C and 1 bar at the rate of 10.5 kg/min. Calculate (a) the brake power, (b) the brake mean effective pressure, (c) the brake specific fuel consumption, (d) the brake specific air consumption, (e) the brake thermal efficiency, (f) the volumetric efficiency, and (g) the A/F.

***Solution*** The brake power (*bp*) is given by

(a)
$$bp = \frac{2\pi N T}{60} = \frac{2\pi \times 5000 \times (45 \times 0.55 \times 9.81)}{60 \times 10^3} = 127.12 \text{ kW} \qquad \textbf{Ans.}$$

(b)
$$bmep = \frac{60 \times 2 \times bp}{10^2 . LANx} = \frac{60 \times 2 \times 127.12}{10^2 \times 0.11 \times (\pi/4)(0.1)^2 \times 5000 \times 8} = 4.41 \text{ bar} \qquad \textbf{Ans.}$$

(c)
$$bsfc = \frac{\dot{m}_f \times 60}{bp} = \frac{(5/10) \times 60}{127.12} = 0.235 \text{ kg/kWh} \qquad \textbf{Ans.}$$

(d) $$bsac = \frac{\dot{m}_a \times 60}{bp} = \frac{10.5 \times 60}{127.12} = 4.935 \text{ kg/kWh}$$ ***Ans.***

(e) $$\eta_{bth} = \frac{bp}{\dot{m}_f \times CV} = \frac{127.12 \times 60}{(5/10) \times 44 \times 10^3} = 34.66\%$$ ***Ans.***

The volume flow rate of air at intake condition

$$\dot{V}_a = \frac{\dot{m}_a R_a T_a}{p} = \frac{10.5 \times 0.287 \times 300}{1 \times 10^2} = 9.038 \text{ m}^3\text{/min}$$

$$\text{Swept volume} = \dot{V}_s = \frac{\pi}{4} D^2 L \frac{N}{2} x = \frac{\pi}{4} (0.1)^2 \times 0.11 \times \frac{5000}{2} \times 8 = 17.27 \text{ m}^3\text{/min}$$

(f) $$\text{Volumetric efficiency} = \eta_v = \frac{\dot{V}_{act}}{\dot{V}_s} = \frac{9.038}{17.27} = 52.3\%$$ ***Ans.***

(g) $$\text{A/F} = \frac{\dot{m}_a}{\dot{m}_f} = \frac{10.5}{0.5} = 21$$ ***Ans.***

**EXAMPLE 15.7** The power output of a six-cylinder four-stroke CI engine is absorbed by a water brake for which the law is $bp$ = WN/2000 where the brake load, $W$ in newton and the speed, $N$ is rpm. The air consumption is measured by an air box with sharp edged orifice system. The following reading were obtained.

| | |
|---|---|
| Bore = 10 cm | Stroke = 12 cm |
| Speed = 2500 rpm | Orifice diameter = 3 cm |
| Coefficient of discharge = 0.6 | Pressure drop across orifice = 15 cm of Hg |
| Ambient pressure = 1 bar | Ambient temperature = 27°C |
| Brake load = 570 N | C/H ratio by mass = 83/17 |

Time taken for 100 cc of fuel consumption = 20 seconds

Fuel density = 831 kg/m$^3$

Calculate (i) the brake power, (ii) the torque, (iii) the brakes specific fuel consumption, (iv) the volumetric efficiency, (v) A/F, and (vi) the percentage of excess air.

***Solution***

(i) $$bp = \frac{WN}{2000} = \frac{570 \times 2500}{20000} = 71.25 \text{ kW}$$ ***Ans.***

$$bmep = \frac{60 \times 2 \times bp}{10^2 \times LANx} = \frac{60 \times 2 \times 71.25}{10^2 \times 0.12 \times (\pi/4)(0.1)^2 \times 2500 \times 6} = 6.047 \text{ bar}$$ ***Ans.***

(ii) $$T = \frac{60\, bp}{2\pi N} = \frac{60 \times 71.25 \times 10^3}{2\pi \times 2500} = 272 \text{ Nm}$$ ***Ans.***

$$\dot{m}_f = \frac{100}{20} \times 10^{-6} \times 831 \times 3600 = 14.96 \text{ kg/h}$$

(iii) $$bsfc = \frac{\dot{m}_f}{bp} = \frac{14.96}{71.25} = 0.2099 \text{ kg/kWh}$$ ***Ans.***

Density of air= $\rho_a = \dfrac{p}{RT} = \dfrac{1 \times 10^2}{0.287 \times 300} = 1.16 \text{ kg/m}^3$

Volumetric flow rate of air = $\dot{V}_a = A\, c_d \sqrt{2g\, h_a} = A\, c_d \sqrt{2g\, h_{Hg} \cdot \rho_{Hg}/\rho_a}$

$$= \frac{\pi}{4}(0.03)^2 \times 0.6 \sqrt{2 \times 9.81 \times \frac{15}{100} \times \frac{13.6 \times 10^3}{1.16}} = 0.0787 \text{ m}^3\text{/s}$$

Swept volume flow rate

$$= \dot{V}_s = \frac{A\, L\, Nx}{60 \times 2} = \frac{\pi}{4}(0.1)^2 \times 0.12 \times \frac{2500}{60 \times 2} \times 6 = 0.1178 \text{ m}^3\text{/s}$$

(iv) Volumetric efficiency = $\eta_v = \dfrac{\dot{V}_a}{\dot{V}_s} = \dfrac{0.0787}{0.1178} = 66.8\%$ ***Ans.***

$O_2$ required/kg of fuel = $0.83 \times \dfrac{32}{12} + 0.17 \times \dfrac{8}{1} = 3.57$ kg of $O_2$ of fuel

As we know, air contains 23.3% of $O_2$ by weight,

Air required/kg of fuel = $\dfrac{3.57}{0.233} = 15.32$ kg

Actual mass flow rate of air = $\dot{m}_a = \dot{V}_a \times \rho_a = 0.0787 \times 1.16 = 0.0912$ kg/s

(v) Actual A/F = $\dfrac{\dot{m}_a}{\dot{m}_f} = \dfrac{0.0912 \times 3600}{14.96} = 21.94$ ***Ans.***

(vi) % of excess air = $\dfrac{(A/F) - (A/F)_{st}}{(A/F)_{st}} = \dfrac{21.94 - 15.32}{15.32} = 43.25\%$ ***Ans.***

**EXAMPLE 15.8** A four-cylinder, four-stroke gasoline engine working at 3000 rpm develops a brake power of 21 kW. A Morse test was conducted on this engine and the brake power obtained when each cylinder was made inoperative by short circulating the spark plug are 15, 14.2, 14.7 and 14.4 kW respectively. The Morse test was conducted at constant speed. Find the indicated power, mechanical efficiency and bmep when all the cylinders are firing. The bore and stroke of the engine are 8 cm and 9 cm.

***Solution*** For Morse test, we have

$$ip_1 = \sum_1^n bp_n - \sum_2^n bp_n$$

or $$ip_1 = bp_{1234} - bp_{234} = 21 - 15 = 6 \text{ kW}$$

Again, $$ip_2 = bp_{1234} - bp_{134} = 21 - 14.2 = 6.8 \text{ kW}$$

$$ip_3 = bp_{1234} - bp_{124} = 21 - 14.7 = 6.3 \text{ kW}$$

$$ip_4 = bp_{1234} - \text{bp}_{123} = 21 - 14.4 = 6.6 \text{ kW}$$

$\therefore$ $$ip = ip_{1234} = ip_1 + ip_2 + ip_3 + ip_4 = \Sigma ip_n$$

or $$ip = 6 + 6.8 + 6.3 + 6.6 = 25.7 \text{ kW}$$

$\therefore$ $$\eta_m = \frac{bp}{ip} = \frac{21}{25.7} = 81.71\%$$ ***Ans.***

$$p_{mb} = \frac{60 \times 2 \times bp}{10^2 . LANx} = \frac{60 \times 2 \times 21}{10^2 \times 0.09\ (\pi/4)(0.08)^2 \times 3000 \times 4} = 4.64 \text{ bar}$$ ***Ans.***

**EXAMPLE 15.9** During the test of a single-cylinder, four-stroke oil engine, the following results were obtained:

| | |
|---|---|
| Cylinder diameter = 20 cm | Stroke = 25 cm |
| Mean effective pressure = 6 bar | Torque = 410 Nm |
| Speed = 900 rpm | Oil consumption = 10 kg/h |
| Calorific value of fuel = 43 MJ/kg | Cooling water flow rate = 8 kg/min |
| A/F = 30 | Rise in cooling water temperature = 45°C |
| Temp. of exhaust gases = 400°C | Room temperature = 30°C |
| $c_{pg}$ = 1.1 kJ/kg K | |

Find *ip*, *bp*, $\eta_{bth}$ and draw up heat balance sheet.

***Solution***

$$ip = \frac{10^2\, p_{mi}\, L\, A\, Nx}{60 \times 2} = \frac{10^2 \times 6 \times 0.4\ (\pi/4)\ (0.2)^2 \times 900 \times 1}{60 \times 2} = 56.54 \text{ kW}$$ ***Ans.***

$$bp = \frac{2\pi\, NT}{60 \times 10^3} = \frac{2\pi \times 900 \times 410}{60 \times 10^3} = 38.64 \text{ kW}$$ ***Ans.***

$$\text{Heat supplied} = Q_A = \frac{10 \times 43000}{3600} = 119.44 \text{ kW}$$

$$\eta_{bth} = \frac{bp}{Q_A} = \frac{38.64}{119.44} = 32.35\%$$ ***Ans.***

$$\text{Heat carried away by cooling water} = \dot{m}_w\, c_{pw}\, \Delta T_w = \frac{8 \times 4.18 \times 45}{60} = 25.08 \text{ kW}$$

$$\text{Heat carried away by exhaust gases} = \dot{m}_g\, c_{pg}\, \Delta T_g = (\dot{m}_a + \dot{m}_f)\, c_{pg}\, \Delta T_g$$

$$= \frac{10(1 + 30) \times 1.1 \times (400 - 30)}{3600} = 35.04 \text{ kW}$$

Heat accounted (by difference) = 119.44 – (38.04 + 25.08 + 35.04) = 21.28 kW

**Heat Balance Sheet**

| *Item* | | *Heat in kW* | *Percentage* |
|---|---|---|---|
| Heat input | (a) Heat supplied | 119.44 | 100% (credit) |
| Expenditure | (b) Heat to *bp* | 38.64 | 32.35% |
| | (c) Heat to cooling water | 25.08 | 20.99% |
| | (d) Heat to exhaust gas | 35.04 | 29.33% |
| | (e) unaccounted heat by difference [*a* – (*b* + *c* + *d* + *e*)] | 21.28 | 17.33% |
| | Total | 119.44 | 100% |

**EXAMPLE 15.10** Design bore and stroke of a six-cylinder, four-stroke petrol engine for the following data:

Compression ratio = 6; *bp* = 312 kW; *N* = 2500 rpm; stroke = 1.25 × the bore, mechanical efficiency = 0.82; *bmep* = 10 bar, relative efficiency = 0.5 and *C.V.* of fuel = 43960 kJ/kg.

Also calculate the fuel consumption in kg/min and in kg/kWh.

***Solution***

$$ip = \frac{bp}{\eta_m} = \frac{312}{0.82} = 380.48 \text{ kW}$$

But
$$ip = \frac{100\, p_m\, L\, A\, N/2}{60} .x$$

or
$$380.48 = \frac{100 \times 10}{60} \times 1.25 \times D \left(\frac{\pi}{4} D^2\right) \times \frac{2500}{2} \times 6$$

or
$$D = 0.1458 \text{ m} = 14.58 \text{ cm} \qquad \textit{Ans.}$$

$$\text{Stroke} = L = 14.58 \times 1.25 = 18.225 \text{ cm} \qquad \textit{Ans.}$$

Air standard efficiency of the cycle is

$$A.S.E. = 1 - \frac{1}{r^{(\gamma-1)}} = 1 - \frac{1}{6^{0.4}} = 0.5116$$

$\therefore$ Relative efficiency $= \eta_r = \dfrac{\eta_{ith}}{A.S.E.}$

$$\therefore \quad \eta_{ith} = 0.5 \times 0.5116 = 0.2558$$

But
$$\eta_{ith} = \frac{ip \times 60}{(\text{Fuel/min}) \times C.V.}$$

or
$$\text{Fuel/min} = \frac{380.48 \times 60}{0.2528 \times 43960} = 2.05 \text{ kg/min} \qquad \textit{Ans.}$$

$$bsfc = \frac{2.05 \times 60}{312} = 0.394 \text{ kg/kWh}$$ ***Ans.***

**EXAMPLE 15.11** A 4-stroke, four-cylinder petrol engine, bore 9 cm and stroke 12 cm running at 1500 rpm develops brake power equal to 27 kW when using a 20% rich mixture. If the volume of air drawn into the cylinders when measured at 15°C and 76 cm of Hg, is 0.7 of the swept volume, the theoretically correct quantity of air for combustion of 1 kg of petrol is 15 kg. The calorific value of fuel is 43960 kJ/kg. The mechanical efficiency of the engine is 0.9. Determine the indicated thermal efficiency and *bmep*.

***Solution***

$$ip = \frac{bp}{\eta_m} = \frac{27}{0.9} = 30 \text{ kW}$$

$$\text{Swept volume} = \text{A.L.} = (\pi/4)\,\frac{(9)^2}{100^2} \times \frac{12}{100} = 0.763 \times 10^{-3} \text{ m}^3$$

$$\text{Vol. of air drawn in at. S.T.P.} = V = 0.7 \times 0.763 \times 10^{-3}$$

$$= 0.5341 \times 10^{-3} \text{ m}^3\text{/cylinder}$$

∴ Mass of air per cylinder

$$m_a = \frac{pV}{RT} = \frac{1.013 \times 0.5341 \times 10^{-3} \times 10^2}{0.287 \times 288} = 0.6545 \times 10^{-3} \text{ kg}$$

Quantity of petrol associated with this air, for 20% rich mixture

$$\frac{m_a}{A/F} \times 1.2 = \frac{0.6545 \times 10^{-3}}{15} \times 1.2 = 0.05236 \times 10^{-3} \text{ kg/cylinder}$$

It is, of course, impossible for the air to burn more fuel than that at chemically correct mixture strength.

∴ $$\text{Heat liberated} = \frac{0.6545 \times 10^{-3}}{15} \times 43960 = 1918.12 \times 10^{-3} \text{ kJ/cylinder}$$

But heat value in fuel = $1918.12 \times 10^{-3} \times 1.2 = 2.3$ kJ/cylinder charge

∴ $$\eta_{ith} = \frac{(30/4) \times 60}{2.3 \times (1500/2)} = 26.08\%$$ ***Ans.***

We have $$bp = \frac{100\, p_{mb}.L\,A(N/2)\,x}{60}$$

or $$\frac{27}{4} = \frac{100\, p_{mb} \times 0.12\,(\pi/4)\,(9)^2 \times (1500/2)}{60 \times 100^2}$$

or $$p_{mb} = 7.07 \text{ bar}$$ ***Ans.***

**EXAMPLE 15.12** A single-cylinder four-stroke diesel engine running at 350 rpm develops 26.5 kW brake power. The fuel contains 85% of carbon and 15% of hydrogen by weight. The mechanical and brake thermal efficiency may be taken as 86% and 30% respectively. The atmospheric air temperature is 20°C and pressure is 1 bar. The engine uses 25% of excess air and the volumetric efficiency is 82%. Determine the bore and stroke of the engine if bore to stroke ratio is 1: 2 and C.V. of fuel is 43960 kJ/kg of fuel.

***Solution***
$$\eta_{bth} = \frac{bp \times 60}{\dot{m}_f \times C.V.}$$

or
$$0.3 = \frac{26.5 \times 60}{\dot{m}_f \times 43960}$$

or
$$\dot{m}_f = \frac{26.5 \times 60}{0.3 \times 43960} = 0.1205 \text{ kg/min}$$

For complete combustion of fuel, one part of $H_2$ requires 8 parts of $O_2$ and one part of C requires 32/12 parts of $O_2$.

∴ $O_2$ required for complete combustion

$$= 0.15 \times 8 + 0.85 \times \frac{32}{12} = 3.466 \text{ kg/kg of fuel}$$

Since $O_2$ is 23.3% by weight in air

∴ Air required for complete combustion $= \dfrac{3.466}{0.233} = 15.069$ kg/kg of fuel

∴ Air required/min = 0.1205 × 15.069 = 1.815 kg

∴ Actual air supplied with 25% of excess air

$$= \dot{m}_a = 1.25 \times 1.815 = 2.268 \text{ kg/min}$$

$$\text{Volume of air} = \frac{\dot{m}_a R T}{p} = \frac{2.268 \times 0.287 \times 293}{1 \times 10^2} = 1.907 \text{ m}^3/\text{min}$$

Since the speed for the engine is 350 rpm and the engine is of four strokes, hence the number of cycles is 350/2 per minute

∴
$$\text{Vol. per cycle} = \frac{1.907}{350/2} = 0.010897 \text{ m}^3$$

∴ Volumetric efficiency $= \dfrac{0.010897}{\text{Swept volume}} = 0.82$

or
$$\text{Swept volume} = \frac{0.010897}{0.82} = 0.01328 \text{ m}^3/\text{cycle}$$

$$0.01328 = (\pi/4)\, D^2 \times L = (\pi/4)\, D^2 \times 2D = (\pi/2)\, D^3$$

or
$$D = 0.2037 \text{ m or } 20.37 \text{ cm}$$ ***Ans.***

$$L = 20.37 \times 2 = 40.74 \text{ cm}$$ ***Ans.***

**EXAMPLE 15.13** A six-cylinder petrol engine has a bore of 11 cm and a stroke of 12 cm. The engine runs at 2000 rpm using air-fuel ratio of 13.5 to 1. If the air is drawn at atmospheric pressure and 70°C, equivalent to 7/8 of the swept volume per stroke and that the thermal efficiency of the engine is 21%, calculate the power developed at ground level where barometer reads 76 cm of Hg. Assume the C.V. of fuel as 37680 kJ/kg.

What will be the power developed at an altitude of 1500 metres, assuming a drop of 2.5 cm of Hg in the barometer reading for each 300 metres of rise in the altitude?

***Solution***

Swept volume = $(\pi/4)\ D^2\ L = (\pi/4) \times 0.0121 \times 0.12 = 0.0014\ m^3$

$$\text{The mass of air} = m_a = \frac{pV}{RT} = \frac{1.013 \times 10^2 \times 0.00114 \times 7}{0.287 \times 8 \times 343} = 10.26 \times 10^{-6}\ \text{kg/cycle}$$

$$\therefore \text{Mass of petrol per cycle} = \frac{10.26 \times 10^{-4}}{13.5} = 0.76 \times 10^{-4}\ \text{kg/cycle}$$

Heat supplied/minute = $10^{-4} \times 0.76 \times 37680 \times (2000/2) \times 6 = 17182$ kJ/min

$$\text{Power developed} = \frac{\text{Heat supplied per min} \times \eta_{th}}{60} = \frac{17182 \times 0.21}{60} = 60.137\ \text{kW}$$ ***Ans.***

Total drop in barometric reading at 1500 m height = 2.5 × 1500/300 = 12.5 cm of Hg

New barometric reading = 76 – 12.5 = 63.5 cm of Hg

$$= \frac{63.5}{76} \times 1.013 = 0.8463\ \text{bar}$$

Since power is proportional to pressure, hence power at 1500 metres altitude

$$= 60.137 \times \frac{0.8463}{1.013} = 50.24\ \text{kW}$$ ***Ans.***

**EXAMPLE 15.14** Calculate bore, stroke, compression ratio, brake specific fuel consumption and *bmep* of a four-cylinder, four-stroke petrol engine which is supplied with 15 kg of air/kg of octane fuel for the following data:

Air standard efficiency = 52%, relative efficiency = 69%, mechanical efficiency = 84%, volumetric efficiency = 82%, stroke = 1.25 of bore, suction pressure and temperature = 1 bar and 40°C, speed = 2500 rpm, brake power = 75 kW, C.V. = 42000 kJ/kg of fuel.

***Solution*** $\text{ASE} = 1 - (1/r)^{\gamma - 1}$ or $0.52 = 1 - (1/r)^{0.4}$

or $r$ = compression ratio = 6.26 ***Ans.***

$$\eta_{ith} = \text{ASE} \times \eta_r = 0.52 \times 0.69 = 35.88\%$$

$$ip = \frac{bp}{\eta_m} = \frac{75}{0.84} = 89.28\ \text{kW}$$

But $$\eta_{ith} = \frac{ip \times 60}{\dot{m}_f \text{ in kg/min} \times CV} = \frac{ip \text{ (kW)}}{Q \text{ (kJ/s)}}$$

$$\dot{m}_f = \frac{89.28 \times 60}{0.3588 \times 42000} = 0.3554 \text{ kg/min}$$

$$bsfc = \frac{\dot{m}_f}{bp} = \frac{0.3554 \times 60}{75} = 0.2843 \text{ kg/kWh} \qquad \textbf{\textit{Ans.}}$$

$$R_{mix} = \frac{\Sigma(m.R)}{\Sigma m} = \frac{\Sigma(m.R_0/\text{Mol. wt})}{\Sigma m}$$

Since Mol. wt of octane = 114 and Mol. wt of air = 28.96, $R_0 = 8.314$

$$R_{mix} = \frac{(1 \times 8.314/114) + (15 \times 8.314/28.96)}{1 + 15} = 0.2737 \text{ kJ/kg K}$$

For 1 kg of fuel, the mass of mixture = 15 + 1 = 16 kg

For 0.3554 kg of fuel, the mass of mixture = 16 × 0.3354

∴ Volume of the mixture at suction condition is

$$= \frac{16 \times 0.3554 \times 0.2737 \times 313}{1 \times 10^2} = 4.871 \text{ m}^3\text{/min}$$

∴ $$\text{Swept volume} = \frac{4.871}{0.82} = 5.94 \text{ m}^3\text{/min} = \frac{5.94 \times 10^6}{(2500/2)} = 4752 \text{ cm}^3\text{/stroke}$$

$$\text{Swept volume/cylinder} = \frac{4752}{4} = 1188 \text{ cm}^3\text{/stroke}$$

∴ $$(\pi/4)\, D^2 \times L = 1188$$

or $$(\pi/4)\, D^2 \times (1.25D) = 1188$$

or $$D = 10.65 \text{ cm} \qquad \textbf{\textit{Ans.}}$$

$$\text{Stroke} = 1.25 \times 10.65 = 13.31 \text{ cm} \qquad \textbf{\textit{Ans.}}$$

$$bmep = \frac{bp \times 60}{AL(N/2)} = \frac{bp \times 60}{\text{Swept volume}} = \frac{75 \times 60}{100 \times 5.94} = 7.574 \text{ bar.} \qquad \textbf{\textit{Ans.}}$$

**EXAMPLE 15.15** The following observations were taken during the testing of single cylinder four-stroke oil engine:

Brake wheel diameter = 65 cm ; rope dia = 3 cm, speed = 450 rpm; load on band = 270 N; spring balance reading = 40 N; area of indicator diagram = 4.2 cm$^2$; length of indicator diagram = 6.5 cm; spring stiffness = 12 bar/cm; bore = 11 cm; stroke = 16 cm; specific fuel consumption = 0.3129 kWh, calorific value of fuel = 41868 kJ/kg.

Calculate the *bp*, *ip*, mechanical efficiency, indicated thermal efficiency and brake thermal efficiency.

***Solution*** $$bp = \frac{2\pi N T}{60 \times 1000}$$

where torque $= T = (W - S) \times R$

The effective brake radius $= R = \frac{D + d}{2} = \frac{65 + 3}{2} = 34$ cm

$$\therefore \quad bp = \frac{2\pi \times 450 \times (270 - 40) \times 0.34}{60 \times 1000} = 3.685 \text{ kW}$$ ***Ans.***

The mean effective pressure

$$= p_m = \frac{\text{Area of indicator diagram}}{\text{Length of indicator diagram}} \times \text{Spring scale} = \frac{4.2}{6.5} \times 12 = 7.75 \text{ bar}$$

$$ip = \frac{100\, p_m\, LA(N/2)x}{60}$$

$$= \frac{100 \times 7.75 \times 0.16 \times (\pi/4) \times 11^2 \times (450/2) \times 1}{60 \times 10^4} = 4.419 \text{ kW}$$ ***Ans.***

$\therefore$ Mechanical efficiency $= \frac{bp}{ip} = \frac{3.685}{4.419} = 83.38\%$ ***Ans.***

Fuel consumption/$h$ = 0.3129 × 3.685 = 1.153 kg/h

$$\therefore \quad \eta_{ith} = \frac{ip \times 60}{\dot{m}_f \times CV} = \frac{4.419 \times 60}{(1.153/60) \times 41868} = 32.95\%$$ ***Ans.***

Brake thermal efficiency is

$$\eta_{bth} = \eta_m \times \eta_{ith} = 0.8338 \times 32.95 = 27.47\%$$ ***Ans.***

**EXAMPLE 15.16** In an attempt to evaluate the temperature at the end of the suction stroke of petrol engine having a compression ratio of 5, the following assumptions are made:

(i) The clearance is filled with residuals at 1.05 bar and 870°C and that these first expands adiabatically to 0.95 bar behind an imaginary insulating diaphragm separating them from the incoming charge; (ii) for the remainder of the induction stroke, fresh mixture is drawn in at 0.95 bar and 60°C; (iii) the diaphragm is now withdrawn and the gases allowed to mix, while the piston is at the other end of the stroke.

Calculate: (a) the resulting temperature of the charge, (b) the volumetric efficiency based on N.T.P. conditions. Take $c_p$ of residuals and incoming gas as 1.045 and 1.005 respectively and $c_v$ for residual and incoming gas 0.749 and 0.711 kg/kgK respectively.

***Solution*** **For residuals,**

$$R = (c_p - c_v) = (1.045 - 0.749) = 0.296 \text{ kJ/kgK}$$

$$\gamma = \frac{1.045}{0.749} = 1.395$$

Let 1 $m^3$ of clearance volume

$$\text{Mass of residual gas} = m = \frac{pV}{RT} = \frac{1.05 \times 10^2 \times 1}{0.296 \times 1143} = 0.31 \text{ kg/m}^3 \text{ of clearance}$$

This mass of residual gases expands adiabatically behind the imaginary diaphragm so that

$$p_1 V_1^{\gamma} = p_2 V_2^{\gamma}$$

or $$1.05 \times (1)^{1.395} = 0.95\,(V_2)^{1.395}$$

or $$V_2 = \left(\frac{1.05}{0.95}\right)^{1/1.395} = 1.0743 \text{ m}^3$$

During expansion, the temperature falls to

$$T_2 = 1143\left(\frac{0.95}{1.05}\right)^{0.395/1.395} = 1111 \text{ K}$$

Since the residual gases have expanded to 1.0743 $m^3$, hence the

Fresh charge admitted = 5 – 1.0743 = 3.9257 $m^3$

For **incoming gas**: $R = c_p - c_v = (1.005 - 0.711) = 0.294$ kJ/kgK

$$\therefore \text{Mass of fresh charge} = \frac{pV}{RT} = \frac{0.95 \times 10^2 \times 3.9257}{0.294 \times 333} = 3.809 \text{ kg}$$

As the **residuals** and **fresh charge** mix at constant volume, no net external work is being done since piston is at O.D.C. position. If we neglect any loss during the mixing process, we have

Internal energy lost by residuals = Internal energy gained by fresh charge

(a) If $T$ is temperature of the mixture, then

$$0.31 \times 0.749\,(1143 - 333) = 3.809 \times 0.711(T - 333)$$

or $T = 402.44$ K **Ans.**

(b) If (5 – 1) = 4 $m^3$ of fresh charge, i.e., the entire swept volume has been admitted at N.T.P., the mass so admitted is

$$m_s = \frac{1.013 \times 10^2 \times 4}{0.294 \times 273} = 5.048 \text{ kg}$$

$$\eta_v = \frac{3.809}{5.048} = 75.45\%$$ **Ans.**

**EXAMPLE 15.17** A petrol engine having stroke vol. of 0.0012 $m^3$ and compression ratio 5.5 compresses the mixture to 8.5 bar and 350°C. Ignition is started so that the pressure rises along a straight line during explosion and attains its higher value of 28 bar after the piston has travelled 1/30 of the working stroke. The air-fuel ratio is 16:1. Take $R$ for mixture as 0.274 kJ/kg, C.V. of fuel = 43542 kJ/kg and $c_v = 0.962$. Find the heat loss per kg of charge during explosion.

***Solution*** Refer to 15.65.

The stroke volume = $V_1 - V_2 = 0.0012\ m^3 = 1200\ cm^3$

Since the compression ratio is 5.5,

hence, $$r = \frac{V_1}{V_2}$$

$\therefore$ $$V_2 = \frac{V_1 - V_2}{r - 1} = \frac{1200}{4.5} = 266.66\ cm^3$$

and $$V_3 = 266.66 + \frac{1200}{30} = 306.66\ cm^3$$

$$T_2 = 273 + 350 = 623\ K$$

$\therefore$ $$T_3 = T_2 \frac{p_3}{p_2} \frac{V_3}{V_2}$$

$$= 623 \times \frac{28}{8.5} \times \frac{306.56}{266.66} = 2360\ K$$

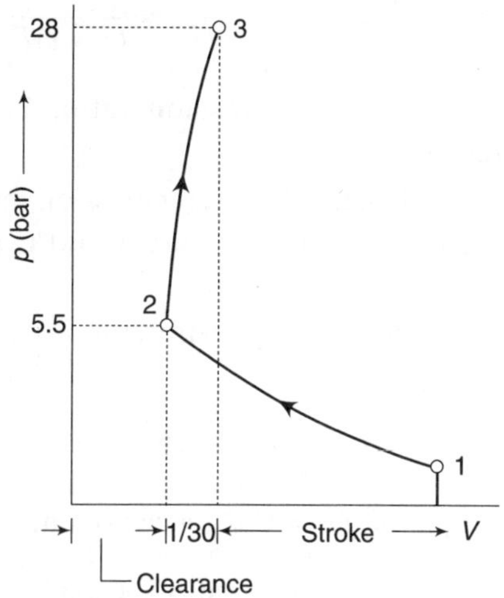


**Fig. 15.65** Example 15.17

Work done during process 2 – 3, we have

$$W_{2-3} = \text{Area under graph} = \frac{10^2\ (28 + 8.5)}{2} \frac{(306.66 - 266.66)}{10^6} = 0.073\ kJ$$

$$\text{Mass of mixture present} = \frac{p_2 V_2}{RT_2} = \frac{8.5 \times 10^2 \times 266.66}{0.274 \times 623 \times 10^6} = 0.00132\ kg$$

Change in internal energy during explosion process, we have

$$\Delta U = U_3 - U_2 = 0.00132 \times 0.962\ (2360 - 623) = 2.205\ kJ$$

$$Q_{2-3} = \Delta U + W_{2-3} = 2.205 + 0.073 = 2.278\ kJ$$

This is the quantity of heat actually given to the mixture in one cycle.

Since the engine admits a mixture of air and petrol of ratio 16 : 1, hence 1 kg of petrol = 1/17 of kg of total charge.

$$\text{Thus, the heat liberated in one cycle} = \frac{1}{17} \times 0.00132 \times 43542 = 3.38\ kJ$$

$\therefore$ Heat lost during explosion = (3.38 – 2.278) = 1.102 kJ

$$= \frac{1.102}{0.00132} = 834.84\ \text{kJ/kg of charge}$$ ***Ans.***

**EXAMPLE 15.18** Prove that the suction temperature in an internal combustion engine is given by the following formula if the density and specific heats remain constant during the suction stroke.

$$T_r = \frac{r\,T_a\,T_g}{(r-1)T_g + T_a}$$

where $r$ = compression ratio, $T_a$ and $T_g$ are atmospheric and exhaust temperature respectively.

A SI engine has a compression ratio of 6. Calculate the temperature at the end of suction if the exhaust temperature is 300°C and atmospheric temperature is at 15°C.

***Solution*** Density of fresh air $= \rho_a = \dfrac{p_a}{287\,T_a}$

Density of exhaust gases $= \rho_g = \dfrac{\rho_g}{287\,T_g}$

Vol. of gas = Clearance volume

$\therefore$ Vol. of air inducted = Clearance volume $\times \left(\dfrac{r-1}{r}\right)$

Therefore, the mass ratio of gases to fresh air is

$$\frac{m_g}{m_a} = \frac{T_a}{T_g(r-1)};\ \text{kg of gases/kg of air}$$

According to heat balance after mixing of fresh air and residual gases,

Heat lost by gases = Heat received by fresh air

$$\therefore \qquad \frac{T_a}{T_g\,(r-1)} \times c_v\,(T_g - T_s) = 1 \times c_v\,(T_s - T_a)$$

where $T_s$ = suction temperature

$$\text{or} \qquad T_s\left[1 + \frac{T_a}{T_g\,(r-1)}\right] = T_a + \frac{T_a\,T_g}{T_g\,(r-1)}$$

$$\text{or} \qquad T_s\left[\frac{T_g(r-1) + T_a}{T_g\,(r-1)}\right] = \frac{T_a\,T_g\,(r-1) + T_a\,T_g}{T_g\,(r-1)} = \frac{r\,T_a\,T_g}{T_g\,(r-1)}$$

$$\text{or} \qquad T_s = \frac{r\,T_a\,T_g}{(r-1)\,T_g + T_a}$$

***Proved***

For the given data,

$$T_s = \frac{6 \times 288 \times 573}{5 \times 573 + 288} = 314 \text{ K} \qquad \textbf{Ans.}$$

**EXAMPLE 15.19** A four-stroke, four-cylinder petrol engine having 62 mm bore and 92 mm stroke when tested at full load with a torque arm of 36 cm gave 170 N net brake load at a rated speed of 3200 rpm. The calorific value of the fuel was 43960 kJ/kg and the fuel consumption was 0.085 kg/min. During Morse test, the spark plug of the cylinder were short-circuited without the change of speed and the corresponding brake loads were 123, 118, 116, and 130 N. Calculate the engine torque, *bmep*, brake thermal efficiency, *ip*, mechanical efficiency, specific fuel consumption and, if A/F = 16.

Also calculate the volumetric efficiency when the the engine was tested in an atmospheric pressure of 1 bar and 15°C and stiochiometic air-fuel ratio 16.

***Solution*** The engine torque = $T = W \times r = 170 \times 0.36 = 61.2$ Nm

$$\therefore \quad bp = \frac{2\pi NT}{60 \times 1000} = \frac{2\pi \times 3200 \times 61.2}{60 \times 1000} = 20.5 \text{ kW}$$

This is *bp* developed at full load in the engine with four cylinders

$$\therefore \quad bp \text{ per cylinder} = \frac{20.5}{4} = 5.125 \text{ kW}$$

$$bmep = \frac{bp \times 60}{(\text{Working strokes/min}) \times \text{Area in m}^2 \times \text{stroke in m}}$$

$$= \frac{bp \times 60}{(N/2)\, AL \times 100} = \frac{5.125 \times 60 \times 10^4}{(3200/2) \times (\pi/4)\,(6.2)^2 \times 0.092 \times 10^2} = 6.919 \text{ bar}$$

**Ans.**

$$\text{Brake thermal efficiency} = \eta_{bth} = \frac{bp \times 60}{m_f \times C.V.} = \frac{20.5 \times 60}{0.085 \times 43960} = 32.91\% \qquad \textbf{Ans.}$$

$$\text{Specific fuel consumption} = \frac{0.085 \times 60}{20.5} = 0.248 \text{ kg/kWh} \qquad \textbf{Ans.}$$

Indicated load = (170 – 123) + (170 – 118) + (170 – 116) + (170 – 130) = 193 N

$$\therefore \quad \eta_m = \frac{\text{Brake load}}{\text{Indicated load}} = \frac{170}{193} = 88.08\% \qquad \textbf{Ans.}$$

$$\therefore \quad ip = \frac{bp}{\eta_m} = \frac{20.5}{0.8808} = 23.27 \text{ kW} \qquad \textbf{Ans.}$$

$$imep = \frac{bmep}{\eta_m} = \frac{6.919}{0.8808} = 7.855 \text{ bar} \qquad \textbf{Ans.}$$

Mass flow rate of air = $\dot{m}_a = 0.085 \times 16 = 1.36$ kg/min

$$\therefore \text{ Volume of air drawn} = \frac{\dot{m}_a RT}{p \times 10^2} = \frac{1.36 \times 0.287 \times 288}{1 \times 10^2} = 1.124 \text{ m}^3/\text{min}$$

$$\text{Swept volume of the engine} = 4 \times \frac{\pi}{4}(0.062)^2 \times 0.092 \times \frac{3200}{2} = 1.777 \text{ m}^3/\text{min}$$

$$\therefore \quad \eta_{vol} = \frac{1.124}{1.777} = 63.25\%$$ **Ans.**

**EXAMPLE 15.20** A four-cylinder, four-stroke petrol engine having 8 cm bore and 10 cm stroke is tested at full throttle at constant speed. The fuel supply is fixed at 0.07 kg/min and the spark plugs of the four cylinders are successively short circuited without change of speed but the brake torque being correspondingly adjusted the power measurements for the all cylinder firings, cylinder No. 1, 2, 3, and 4 short circuited are 12.64, 9.18, 9.29, 9.33 and 9.18 kW *bp* respectively.

Calculate the *ip* of the engine under these conditions. If the calorific value of fuel is 43542 kJ/kg, calculate the indicated thermal efficiency.

Compare this efficiency with that of air standard when the clearance volume of one cylinder being 70 cm$^3$.

***Solution***

$$ip = (12.64 - 9.18) + (12.64 - 9.29) + (12.64 - 9.33) + (12.64 - 9.18) = 13.58 \text{ kW}$$ **Ans.**

$$\eta_{ith} = \frac{ip \times 60}{\dot{m}_f \times C.V.} = \frac{13.58 \times 60}{0.07 \times 43542} = 26.73\%$$ **Ans.**

Volume swept by piston = $(\pi/4)\, 8^2 \times 10 = 502.65$ cm$^3$

$$\text{Compression ratio} = \frac{502.65 + 70}{70} = 8.1807$$

$$\text{A.S.E.} = 1 - \left(\frac{1}{r}\right)^{\gamma-1} = 1 - \left(\frac{1}{8.1807}\right)^{0.4} = 56.85\%$$ **Ans.**

$$\text{Relative efficiency} = \frac{\eta_{ith}}{ASE} = \frac{0.2673}{0.5685} = 47\%$$ **Ans.**

**EXAMPLE 15.21** A single-cylinder, two-stroke engine has 12 cm of bore, 16 cm of stroke and 7 : 1 compression ratio. It runs at 1750 rpm and develops *bmep* 7.2 bar. The mechanical efficiency may be assumed as 0.88. The inlet temperature is 70°C and the exhaust pressure is 1 bar. The amount of air supplied to the engine is 4.7 kg/min. The thermal efficiency of the engine is 0.31, calculate (a) *bp*, (b) *ip*, (c) scavenging ratio, (d) scavenging efficiency, (e) trapping efficiency and (f) thermal efficiency if the fuel is injected after closing of the ports and the fuel air-ratio is 0.07.

***Solution***

(a) $bp = \dfrac{100\, p_{mb}\, L\, A\, Nx}{60}$ (since two strokes)

$$= \frac{100 \times 7.2 \times 0.16 \times (\pi/14)\,(12)^2 \times 1750 \times 1}{60 \times 10^4} = 38 \text{ kW}$$ ***Ans.***

(b) $ip = \dfrac{bp}{\eta_m} = \dfrac{38}{0.88} = 43.18$ kW ***Ans.***

(c) The ideal air capacity per minute = $(\dot{m}_a)_{ideal} = \dfrac{V N p_e \times 10^2}{RT_i}$

But the displacement volume

$$V_d = (\pi/4)\,(12)^2 \times 10^{-4} \times 0.16 = 18.095 \times 10^{-4} \text{ m}^3$$

$\therefore$ Total cylinder volume = $V = \dfrac{r}{r-1} \times V_d$

$$= \frac{7}{6} \times 18.095 \times 10^{-4} = 21.1 \times 10^{-4} \text{ m}^3$$

$$\therefore \quad (\dot{m}_a)_{ideal} = \frac{1750 \times 21.1 \times 10^{-4} \times 1 \times 10^2}{0.287 \times 343} = 3.75 \text{ kg/min}$$

The scavenging ratio = $R_s = \dfrac{\dot{m}_{act}}{(\dot{m}_a)_{ideal}} = \dfrac{4.7}{3.75} = 1.253$ ***Ans.***

(d) Scavenging efficiency = $\eta_{sc} = 1 - e^{-Rs} = 1 - (e)^{-1.253} = 0.7143$ or 71.43% ***Ans.***

(e) $\eta_{tr} = \dfrac{\eta_s}{R_s} = \dfrac{0.7143}{1.253} = 57\%$ ***Ans.***

(f) Fuel injection/min = $(\dot{m}_a)_{ret} \times 0.07$

$$= \eta_{tr} \times (\dot{m}_{act}) \times 0.07 = 0.57 \times 4.7 \times 0.07 = 0.1875 \text{ kg/min}$$

$\therefore$ Fuel carburetted per minute = $(\dot{m}_{act}) \times 0.07 = 4.7 \times 0.07 = 0.329$ kg/min

$$\therefore \quad \eta_{th} = 0.31 \times \frac{0.1875}{0.329} = 17.66\%$$ ***Ans.***

**EXAMPLE 15.22** A six-cylinder, four-stroke direct injection oil engine is tested against a water brake dynamometer for which the *bp* = WN/17280 in kW where *W* is the brake load in newton and *N* is the speed of engine. During the test, the following observations were taken and the air consumption was measured by means of sharp edged orifice: Bore = 9 cm, stroke = 13 cm, speed = 2500 rpm, brake load = 470 N, barometer reading = 76 cm of Hg, orifice diameter = 3.2 cm, coeff. of discharge of orifice = 0.6, pressure drop across orifice = 14 cm of Hg, room temperature = 21°C, fuel consumption = 0.3 kg/min, ratio of carbon to hydrogen by the mass in fuel = 84 : 16.

Calculate (a) the vol. efficiency, (b) *bmep*, (c) the engine torque, (d) *bsfc* and

(e) the ratio of $\dfrac{(A/F)\text{–Stoichiometric ratio}}{\text{Stoichiometric ratio}}$

***Solution:***

$$p = 76 \text{ cm of Hg} = 1.013 \text{ bar}$$

$$\rho_{air} = \frac{p \times 10^2}{R\,T} = \frac{1.013 \times 10^2}{0.287 \times 294} = 1.2 \text{ kg/m}^3$$

$$h = \Delta H \times \rho_{HG} = 14 \times 13.6 = 190.4 \text{ cm} = 1904 \text{ mm of water}$$

$$\text{Volume flow of air/s} = c_d . A \sqrt{\left(\frac{2gh}{\rho_a}\right)}$$

$$= 0.6 \times \frac{\pi}{4}\left(\frac{3.2}{100}\right)^2 \times \sqrt{(2 \times 9.806 \times 1904/1.2)} = 0.0851 \text{ m}^3$$

$$\therefore \text{ Air consumption/stroke} = \frac{0.0851 \times 60}{(2500/2)} = 0.00408 \text{ m}^3$$

$$\text{Swept volume/stroke} = AL \times \text{No. of cylinders}$$

$$= \frac{\pi}{4}\left(\frac{9}{100}\right)^2 \times \frac{13}{100} \times 6 = 0.00496 \text{ m}^3$$

(a) $\therefore$ Volumetric efficiency $= \dfrac{V_{actual}}{V_s} = \dfrac{0.00408}{0.00496} = 82.25\%$ **Ans.**

(b) $bp = \dfrac{W\,N}{17280} = \dfrac{470 \times 2500}{17280} = 67.99 \text{ kW}$

$$bp \text{ of one cylinder} = \frac{67.99}{6} = 11.33 \text{ kW}$$

$$bmep = \frac{bp \times 60}{100\text{A } (m)^2 \times L(m) \times (N/2)}$$

$$= \frac{11.33 \times 60 \times 10^4}{10^2 \times (\pi/4)\,(9)^2 \times (13/100) \times (2500/2)} = 6.575 \text{ bar}$$

(c) $T = \dfrac{bp \times 60}{2\pi\, N} = \dfrac{67.99 \times 60}{2\pi \times 2500} = 0.2597 \text{ kNm}$ **Ans.**

(d) $bsfc = 0.3 \times \dfrac{60}{67.99} = 0.2647 \quad \text{kg/kWh}$ **Ans.**

(e) Oxygen needed/kg of fuel $= 0.84 \times \dfrac{32}{12} + 0.16 \times 8 = 3.52 \text{ kg}$

$\therefore$ Air required/kg of fuel $= \dfrac{3.52}{0.23} = 15.3$ kg

Actual air mass flow rate = 0.0851 × 1.2 = 0.1021 kg/s

$\therefore$ Actual mass of air/fuel $= \dfrac{0.1021 \times 60}{0.3} = 20.42$

$\therefore$ Ratio of $\dfrac{\text{(A/F) – stoichiometric ratio}}{\text{Stoichiometric ratio}} = \dfrac{20.42 - 15.3}{15.3} = 33.46$ ***Ans.***

**EXAMPLE 15.23** Air consumption for a four-stroke petrol engine is measured by means of sharpe edge circular orifice of diameter 3.8 cm having 0.6 as the coefficient of discharge. The drop in pressure through the orifice is equivalent to 145 mm of water. The barometer reads 75.5 cm of Hg. The temperature of air may be taken as 26°C. The compression ratio of the engine is 6. The piston displacement volume is 2000 cm$^3$. The engine develops 29.5 kW bp running at 2600 rpm. The fuel consumption of the engine is 0.14 kg/min and the calorific value of fuel is 43960 kJ/kg. Calculate (a) air-fuel ratio, (b) volumetric efficiency, (c) brake mean effective pressure, and (d) relative efficiency.

***Solution*** The atmospheric pressure $= p = \dfrac{75.5}{76} \times 1.013 = 1.006$ bar

$\therefore$ Density of atmospheric air $= \rho_{air} = \dfrac{p \times 10^2}{RT} = \dfrac{1.006 \times 10^2}{0.287 \times 299} = 1.172$ kg/m$^3$

$h$ = 145 mm of water

Volume of air flowing through the orifice per second $= c_d \times A\sqrt{(2gh/\rho_a)}$

$$= 0.6 \times (\pi/4)\left(\frac{3.8}{100}\right)^2 \sqrt{(2 \times 9.81 \times 145/1.172)} = 0.03352 \text{ m}^3$$

$\therefore$ Air consumption stroke $= \dfrac{0.03352 \times 60}{2600/2} = 0.001547$ m$^3$

The capacity of the engine = 2000 cm$^3$

(b) $\therefore$ Volumetric efficiency $= \dfrac{1547}{2000} = 77.35\%$ ***Ans,***

Mass of air induced $= \dot{m}_a = 0.03552 \times 60 \times 1.172 = 2.357$ kg/min

(a) $A/F = \dfrac{2.357}{0.14} = 16.83$ ***Ans.***

(c) $bmep = \dfrac{bp \times 60}{100\ A\ L\ N/2} = \dfrac{29.5 \times 60 \times 10^6}{(2000 \times 100)\ (2600/2)} = 6.8$ bar ***Ans.***

The air standard efficiency $= 1 - \dfrac{1}{(6)^{(1.4-1)}} = 51.16\%$

$$\text{Brake thermal efficiency} = \eta_{bth} = \frac{bp \times 60}{\dot{m}_f \times C.V.} = \frac{29.5 \times 60}{0.14 \times 43960} = 28.75\%$$

$$\therefore \text{(d) Relative efficiency} = \frac{0.2875}{0.5116} = 0.5619 \text{ or } 56.19\%$$ ***Ans.***

**EXAMPLE 15.24** The following data relate to the testing of a four-cylinder, four-stroke diesel engine:

Bore = 36 cm, stroke = 40 cm, speed = 350 rpm, *bp* = 257 kW, *mep* = 7 bar, fuel consumption = 72 kg/h, C.V. of fuel = 43960 kJ/kg, hydrogen content of fuel = 12%, air consumption = 28.2 kg/min, mass of jacket water = 86 kg/min, rise in temperature of jacket cooling water = 41°C, amount of piston cooling oil = 53 kg/min, temperature rise of cooling oil = 23°C, specific heat of cooling oil = 2.09 kJ/kgK. Room temp. = 20°C, exhaust gas temperature = 325°C, $c_p$ of dry exhaust gas = 1.045 kJ/kgK, sp heat of superheated steam = 2 kJ/kgK, partial pressure of steam in exhaust gases = 0.035 bar.

Draw up the heat balance sheet and calculate the brake specific fuel consumption at half of the load.

***Solution*** Heat supplied in fuel $= \dfrac{72 \times 43960}{60 \times 60} = 879.2$ kW

$$ip = \frac{100\, p_m\, L\, A\, (N/2) x}{60}$$

$$= \frac{100 \times 7 \times 0.4 \times (\pi/4)\,(36)^2 \times (350/2) \times 4}{60 \times 10^4} = 332.5 \text{ kW}$$

$$\therefore \quad fp = 332.5 - 257 = 75.5 \text{ kW}$$

Heat equivalent to $ip$ = 332.5 kW

Heat equivalent to $bp$ = 257 kW

∴ Heat equivalent to $fp$ = 332.5 – 257 = 75.5 kW

Heat carried away by cooling water $= \dfrac{86}{60} \times 4.18 \times 41 = 245.64$ kW

Heat carried away by cooling oil $= 2.09 \times \dfrac{53}{60} \times 23 = 42.46$ kW

Total mass of exhaust gases $= \dfrac{72}{60} + 28.2 = 29.4$ kg/min = 0.49 kg/s

∴ Mass of vapour with exhaust gases $= \dfrac{72}{3600} \times 0.12 \times 9 = 0.0216$ kg/s

∴ Mass of dry exhaust gas = 0.49 – 0.0216 = 0.4684 kg/s

Heat in dry exhaust gases = 0.4684 × (325 – 20) × 1.045 = 149.29 kW

From steam tables, heat lost per kg of steam at given conditions

$$h_g + c_{p\,\text{sup}}\,(t_{\text{sup}} - t_s) = 2730 + 2(325 - 140) = 3100 \text{ kJ/kg}$$

Heat lost in steam = 3100 × 0.0216 = 66.96 kW

$$\text{Indicated thermal efficiency} = \frac{332.5}{879.2} = 37.81\%$$

$$\text{At half load, } bp = \frac{257}{2} = 128.5 \text{ kW}$$

Assuming $f_p$ to remain constant at half load, the *bp* at half load

$$= 128.5 + 75.5 = 205 \text{ kW}$$

$$\text{Heat input at half load} = \frac{204}{\eta_{ith}} = \frac{204}{0.3781} = 539.53 \text{ kW}$$

$$\text{Fuel consumption} = \frac{\text{Heat input}}{C.V.} = \frac{539.53 \times 3600}{43960} = 44.18 \text{ kg/h.}$$

$$bsfc \text{ at half load} = \frac{44.18}{128.5} = 0.3438 \text{ kg/kWh}$$ ***Ans.***

**Heat Balance Sheet**

| | *Item* | *Heat in kW* | *Percentage* |
|---|---|---|---|
| Input | (a) Heat supplied in fuel | 879.2 | 100% (Credit) |
| Expenditure | (b) Heat equivalent to *bp* | 257 | 29.23% |
| | (c) Heat in cooling water | 245.64 | 27.93% |
| | (d) Heat in cooling oil | 42.46 | 4.82% |
| | (e) Heat in dry exhaust gases | 149.29 | 16.96% |
| | (f) Heat in steam in exhaust | 66.96 | 7.61% |
| | (g) Unaccounted heat by difference [*a* – (*b* + *c* + *d* + *e* + *f*)] | 117.85 | 13.44% |
| | Total | 879.2 | 100% |

**EXAMPLE 15.25** During the testing of a four-stroke cycle, eight-cylinder petrol engine, the following data were obtained;

Load on the hydraulic dynamometer = 950 N, speed = 1250 rpm, bore = 70 mm, stroke = 89 mm, fuel consumption = 19.5 kg/h, C.V. = 43542 kJ/kg, jacket water flow = 1400 kg/h, temperature rise in jacket water = 36°C, air-fuel ratio = 32, room temperature = 30°C, exhaust gas temperature = 400°C, fuel analysis = 15% $H_2$ and 85%C, partial pressure of water vapour in the exhaust gas = 0.02 bar, $c_p$ of vapour = 2.09, $c_p$ of dry exhaust gases = 1.005, the constant of hydraulic dynamometer = 17280.

Draw the heat balance sheet.

***Solution*** Heat supplied to the engine $= \dfrac{19.5}{3600} \times 43542 = 235.85$ kW

$$bp = \frac{WN}{K} = \frac{\text{Load} \times \text{rpm}}{\text{hydro. dyn. constant}} = \frac{950 \times 1250}{17280} = 68.72 \text{ kW}$$

$$\text{Heat carried away by jacket cooling water} = \frac{1400}{3600} \times 4.18 \times 36 = 58.519 \text{ kW}$$

Total mass of exhaust gases/kg of fuel = 1 + 32 = 33 kg

Total mass of steam formed = 9 × 0.15 = 1.35 kg

Mass of dry gases = 33 – 1.35 = 31.65 kg/kg of fuel

$$\therefore \text{Mass of dry gases/min} = \frac{31.65 \times 19.5}{60} = 10.286 \text{ kg}$$

$$\text{Mass of water vapour/min} = \frac{1.35 \times 19.5}{60} = 0.4387 \text{ kg}$$

$$\text{Heat carried away by dry gases} = \frac{10.286}{60} \times 1.005(400 - 30) = 63.747 \text{ kW}$$

Since the partial pressure of stream is 0.02 bar, hence the sensible heat and latent heat or enthalpy of saturated steam have to be taken from steam table at this pressure.

∴ Heat carried away by steam in flue gases.

$$= 0.4387\ [h_g + c_p\ (T_{sup} - T_s)] = 0.4387\ [2533.6 + 2.09\ (400 - 17.51)]$$

$$= 1462.18 \text{ kJ/min} = 24.369 \text{ kW}$$

**Heat Balance Sheet**

| | *Item* | *Heat in kW* | *%* |
|---|---|---|---|
| Input | (a) Heat supplied | 235.85 | 100% |
| Expenditure | (b) Heat in *bp*<br>(c) Heat in jacket water<br>(d) Heat in dry exh. gas<br>(e) Heat in water vapour<br>(f) Unaccounted by diff. [*a* – (*b* + *c* + *d* + *e*)] | 68.72<br>58.519<br>63.747<br>24.369<br>20.495 | 29.13%<br>24.8%<br>27%<br>10.33%<br>8.74% |
| | Total | 235.85 | 100% |

**EXAMPLE 15.26** The following data relate to *two-stroke, single-cylinder oil* engine during trial: Room temperature = 20°C, bore = 20 cm, stroke = 26 cm, speed = 400 rpm, brake drum diameter = 120 cm, rope dia = 3 cm, net brake load = 460 N, indicated mean effective pressure = 2.8 bar, oil consumption = 3.7 kg/h, *C.V.* of oil = 42000 kJ/kg of fuel, mass of jacket cooling water = 456 kg/h, rise in temperature of cooling water = 28°C, temperature of exhaust gas leaving the exhaust gas calorimeter = 220°C. Mass flow rate of water passing through exhaust gas calorimeter = 8 kg/min, rise in temperature of calorimeter water = 8°C.

Calculate *ip*, *bp*, mechanical efficiency and brake thermal efficiency. Draw up heat balance sheet.

***Solution*** $bp = \dfrac{2\pi R N W}{60 \times 1000}$; kW

where $R$ = effective drum radius = $\dfrac{120}{2} + \dfrac{3}{2} = 61.5$ cm

$$\therefore \quad bp = \frac{2\pi \times (61.5/100)\ 400 \times 460}{60 \times 1000} = 11.85 \text{ kW} \qquad \textbf{\textit{Ans.}}$$

$$ip = \frac{100\, p_m\, L\, A\, Nx}{60} \text{ since the engine is two stroke, } x = 1$$

$$= \frac{100 \times 2.8 \times 0.26 \times (\pi/4)\ (20)^2 \times 400}{60 \times 10^4} = 15.24 \text{ kW} \qquad \textbf{\textit{Ans.}}$$

$$\therefore \quad \eta_{mech} = \frac{11.85}{15.24} = 77.75\% \qquad \textbf{\textit{Ans.}}$$

$$\eta_{bth} = \frac{bp \times 60}{\dot{m}_f \times CV} = \frac{11.85 \times 60 \times 60}{3.7 \times 42000} = 0.2745 \text{ or } 27.45\% \qquad \textbf{\textit{Ans.}}$$

Heat supplied = $\dfrac{3.7}{3600} \times 42000 = 43.166$ kW

Heat in $bp$ = 11.85 kW

Heat carried away by cooling water = $\dfrac{456}{3600} \times 4.18 \times 28 = 14.825$ kW

Heat carried away by the exhaust gas = $\dot{m}_w . c_{pw} . (T_{w2} - T_{w1}) \times \dfrac{T_{g1} - T_a}{T_{g1} - T_{g2}}$

$$= \frac{8}{60} \times 4.18 \times 8 \times \frac{320 - 20}{320 - 220} = 13.376 \text{ kW}$$

**Heat Balance Sheet**

| | *Item* | *Heat in kW* | *Percentage* |
|---|---|---|---|
| Input | (a) Heat supplied | 43.166 | 100% (Credit) |
| Expenditure | (b) Heat in *bp* | 11.85 | $\dfrac{11.85 \times 100}{43.166} = 27.45\%$ |
| | (c) Heat in cooling water | 14.285 | $\dfrac{14.825 \times 100}{43.166} = 34.34\%$ |
| | (d) Heat in exhaust gas | 13.376 | $\dfrac{13.376 \times 100}{43.166} = 30.98\%$ |
| | (e) Unaccounted heat [*a* – (*b* + *c* + *d*)] | 3.115 | $\dfrac{3.115 \times 100}{43.166} = 7.23\%$ |
| | Total | 43.166 | 100% |

**EXAMPLE 15.27** The following data relate to a four-stroke gas engine trial:

Bore = 25 cm, stroke = 46 cm, clearance volume = 5.65 litres, effective diameter of brake pulley = 1.62 m, total number of revolutions = 8080, total number of explosions = 3520, duration of trial = 40 min, net load on brake = 916 N, mep = 5.76 bar, gas consumption = 7.67 m$^3$ at 13.2 cm of $H_2O$ and 14°C, barometer reading = 74.6 cm Hg, *C.V.* of gas = 1926 kJ/m$^3$ at N.T.P., mass of jacket cooling water = 183 kg, temperature rise in cooling water = 50°C.

Calculate the relative efficiency and draw up the heat balance sheet.

***Solution*** $$bp = \frac{2\pi R N W}{60 \times 1000} = \frac{2\pi \times 0.81 \times 916 \times 8080}{40 \times 60 \times 1000} = 15.69 \text{ kW}$$

$$ip = \frac{100\, p_m\, LAnx}{60}$$

$$= \frac{100 \times 5.76 \times 0.46 \times (\pi/4)\,(25)^2 \times 3520 \times 10^{-4} \times 1}{60 \times 40} = 19.07 \text{ kW}$$

$$\eta_m = \frac{bp}{ip} = \frac{15.69}{19.07} = 82.27\%$$

The absolute pressure at gas meter = $74.6 + \dfrac{13.2}{13.6} = 75.57$ cm of Hg

At gas meter, the absolute temperature = 273 + 14 = 287 K

∴ Volume of gas referred to standard conditions

$$\dot{V}_g = \frac{7.67}{40} \times \frac{75.57}{76} \times \frac{273}{287} = 0.181 \text{ m}^3\text{/min}$$

$$bsfc = \frac{\dot{V}_g}{bp} = \frac{0.181 \times 60}{15.69} = 0.6921 \text{ m}^3\text{/kWh}$$

Heat supplied = $q_A = \dot{V}g \times CV = \dfrac{0.181}{60} \times 19426 = 58.6$ kW

Heat equivalent to $bp$ = 15.69 kW

Heat carried away by jacket water = $m_w\, c_{pw}\, \Delta T_w$

$$= \frac{183}{40 \times 60} \times 4.18\,(50) = 15.936 \text{ kW}$$

$$\eta_{ith} = \frac{ip}{q_A} = \frac{bp}{\eta_m \cdot q_A} = \frac{15.69}{0.8227 \times 58.6} = 0.32$$

Stroke volume = $(\pi/4)\, D^2 L = (\pi/4) \times 0.25^2 \times 0.46 = 0.02258$ m$^3$

Compression ratio = $r$

$$= 1 + \frac{\text{Stroke volume}}{\text{Clearance volume}} = 1 + \frac{0.02258}{0.00565} = 5$$

$$\text{A.S.F.} = 1 - \frac{1}{(r)^{\gamma-1}} = 1 - \frac{1}{(5)^{0.4}} = 0.4746 \text{ or } 47.46\%$$

$$\therefore \text{Relative efficiency} = \frac{\eta_{bth}}{\eta_{as}} = \frac{32 \times 0.8227}{47.4} = \frac{26.32}{47.4} = 55.54\% \qquad \textbf{\textit{Ans.}}$$

**Heat Balance Sheet**

| | *Item* | *Heat in kW* | % |
|---|---|---|---|
| Input | (a) Heat supplied | 58.6 | 100% (credit) |
| Expenditure | (b) Heat in *bp* | 15.69 | 26.77% |
| | (c) Heat loss to cooling water | 15.936 | 27.19% |
| | (d) Exhaust gas and other losses by diff. [*a* – (*b* + *c*)] | 26.97 | 46.04% |
| | Total | 58.6 | 100% |

**EXAMPLE 15.28** A single-cylinder four-stroke gas engine is governed by hit and miss method. The engine has 35 cm of stroke and 20 cm of bore. The *mep* of the working loop and pumping loop are 6.5 bar and 0.5 bar respectively. The engine operating at full load runs at 500 rpm. The *mep* of the dead cycle is 0.650 bar. There are 80 firing strokes/min at full load condition. Calculate the *bp* and mechanical efficiency at full load.

***Solution*** Dead cycles/min $= \frac{500}{2} - 80 = 170$

Net *imep* = 6.5 – 0.5 = 6.0 bar

∴ *fp* = Net *ip* – pumping power of dead cycles

$$= 100 \times 6 \times \frac{\pi}{4} \times \frac{(20)^2}{10^4} \times \frac{0.35 \times 80}{60} - 0.65 \times \frac{\pi}{4}\frac{(20)^2}{10^4} \times \frac{0.35 \times 170}{60} \times 100 = 6.771 \text{ kW}$$

At full load, the engine fires regularly in every two revolutions and so it has 500/2 = 250 fires strokes/min.

$$\therefore \qquad ip = \frac{100 \times 6 \times \pi}{4}\frac{(20)^2}{10^4} \times \frac{0.35 \times 200}{60} = 31.991 \text{ kW}$$

$$bp = ip - fp = 21.991 - 6.771 = 15.228 \text{ kW} \qquad \textbf{\textit{Ans.}}$$

$$\eta_m = \frac{bp}{ip} = \frac{15.228}{21.991} = 69.21\% \qquad \textbf{\textit{Ans.}}$$

**EXAMPLE 15.29** The following data relate to a full load test on a two stroke SI engine:

Speed 500 rpm, brake load 50 kg, *imep* = 3 bar, fuel consumption 5.5 kg/h, rise in jacket water temperature 36°C, jacket water flow rate 450 kg/h, A/F by mass 25, temperature of exhaust gas 360°C, temperature of test room = 27°C, barometric pressure 76 cm Hg, cylinder diameter 22 cm, stroke 25 cm, brake diameter 1.2 m, calorific value of fuel 43 MJ/kg, proportion of hydrogen by mass 15%, $c_{pg}$ = 1.1. kJ/kgK, $c_{pH2O}$ = 2kJ/kg K, $h_{sup}$ = 3180 kJ/kg. Calculate the following:

(i) the indicated thermal efficiency,

(ii) the brake specific fuel consumption, and

(iii) the volumetric efficiency.

Draw up the heat balance.

***Solution***

$$bp = \frac{2\pi NT}{60} = \frac{2\pi \times 500 \times (50 \times 9.81 \times 0.6)}{60 \times 10^3} = 15.66 \text{ kW}$$

$$ip = \frac{10^2 \times p_{mi} LAN}{60} = \frac{10^2 \times 3 \times 0.25 \times (\pi/4)(0.22)^2 \times 500}{60} = 23.75 \text{ kW}$$

(i) $$\eta_{ith} = \frac{ip}{Q_A} = \frac{ip}{\dot{m}_f \times CV} = \frac{23.75}{(5.5/3600) \times 43 \times 10^3} = 36.15\%$$ ***Ans.***

(ii) $$bsfc = \frac{\dot{m}_f}{bp} = \frac{5.5}{15.66} = 0.3512 \text{ kg/kWh}$$ ***Ans.***

Swept volume flow rate = $\dot{V}_s = ALN = \frac{\pi}{4}(0.22)^2 \times 0.25 \times 500 = 4.75$ m$^3$/min

Mass flow rate of air = $\dot{m}_a = (A/F)\,\dot{m}_f = 25 \times 5.5/60 = 2.29$ kg/min

Density of air = $\rho = \frac{p}{RT} = \frac{1 \times 10^2}{0.287 \times 300} = 1.16$ kg/m$^3$

Air consumption rate by engine = $\dot{V}_a = \frac{2.29}{1.16} = 1.974$ m$^3$/min

(iii) Volumetric efficiency = $\eta_v = \frac{\dot{V}_a}{\dot{V}_s} = \frac{1.974}{4.75} = 41.55\%$ ***Ans.***

Heat input = $Q_A = \dot{m}_f \times CV = \frac{5.5}{3600} \times 43 \times 10^3 = 65.69$ kW

Heat carried away by dry exhaust gas = $\dot{m}_{gd}\, c_{pg}\,(T_g - T_a) = (\dot{m}_a + \dot{m}_f)\, c_{pg}\,(T_g - T_a)$

$$= \dot{m}_f \left[(A/F) + 1 - \frac{\dot{m}_{H2O}}{\dot{m}_f} \times 9\right] c_{pg}\,(T_g - T_a)$$

$$= \frac{5.5}{3600}[25 + 1 - 9 \times 0.15] \times 1.1\,(360 - 27) = 13.79 \text{ kW}$$

Heat carried away by the steam in the exhaust gas

$$= \left(\frac{\dot{m}_{H2O}}{\dot{m}_f} \times 9 \times \dot{m}_f\right) [h_{steam} - c_{pw}\, t_R]$$

Enthalpy of steam is given by

$$h_{steam} = c_{pw}\,(100 - t_R) + h_{fg} + c_{pH2O}\,(T_{sup} - T_{\text{sat}})$$

$$= 4.18\,(100 - 27) + 2250 + 2(360 - 100)$$

$$= 3075.14 \text{ kJ/kg}$$

Heat carried away by steam in the exhaust gas

$$= \left(0.15 \times \frac{5.5}{3600} \times 9\right) (3075.14 - 4.18 \times 27)$$

$$= 6.107 \text{ kW}$$

Unaccounted heat loss by difference

$$= 65.69 - (15.66 + 18.81 + 13.79 + 6.109)$$

$$= 11.32 \text{ kW}$$

**Heat Balance Sheet**

| | *Item* | *Heat in kW* | *Percentage* |
|---|---|---|---|
| Input | (a) Heat supplied by fuel | 65.69 | 100% (Credit) |
| Expenditure | (b) Heat to *bp*<br>(c) Heat to cooling water<br>(d) Heat carried by exhaust gas<br>(e) Heat carried by steam<br>(f) Heat accounted by difference [*a* – (*b* + *c* + *d* + *e*)] | 15.66<br>18.81<br>13.79<br>6.109<br>11.32 | 23.83%<br>29.63%<br>20.99%<br>9.29%<br>17.26% |
| | Total | 65.69 kW | 100% |

## 15.24 GOVERNING OF IC ENGINES

There is an inter-relationship between power developed, external load and the engine speed. In automobiles, the speed requirements are variables as per need while in diesel engine power plant, the speed is to be maintained constant irrespective of load. There are three following methods of governing:

(i) Quantity governing

(ii) Quality governing

(iii) Hit and miss governing

**(i) Quantity governing.** *In this method, mass of charge is varied without varying the composition,* i.e., A/F ratio. All the carburettors (i.e., SI) engines employ quantity governing. They are governed by throttling. In stationary engines, the desired speed is maintained by the governor acting upon a butterfly valve, and in automobiles, the desired speed in maintained by foot or manual control of the throttle.

**(ii) Quality governing.** *In this method, mass of fuel is varied resulting in varying the quality (A/F) of the charge.* All the diesel engines and multi-point fuel injection (MPFI) SI engine employ quality governing (see section 8.5). In most oil engines, the amount of fuel injected is regulated by the fuel injection pump whose action is under the control of governor. In diesel engine power plant, the speed is sensed by a centrifugal governor and accordingly the fuel injection is varied to maintain speed irrespective of load.

Gas engine may utilise either quantity governing or quality governing. The quantity governing may be applied to the whole range of loads. The main disadvantage of quantity governing is that the throttling of the charge reduces the suction pressure which reduces the pressure at the end of compression resulting in reduction in thermal efficiency and increases the pumping loss at smaller loads.

*The main advantage of quality governing is fuel economy.* But gas engines working at a small load cannot use quality governing due to possibility of lean mixture which may not burn. *Thus, quality governing is suitable for large gas engines only.*

**(iii) Heat and miss governing.** In this method, the fuel inlet valve remains closed during a particular cycle so that no fuel enters the cylinder and the engine is said to be operating in a dead cycle. The main drawbacks of this method is that it lowers the efficiency and provides uneven turning moment of flywheel. This method is used in small gas engines.

## 15.25 SUMMARY OF FORMULAE

(1) $ip = bp + fp$

(2) $p_{mi} = \dfrac{W_{net}}{V_1 - V_2}$

(3) $p_{mi} = \dfrac{\text{Area of indicator diagram in cm}^2 \times \text{Spring scale in bar per cm}}{\text{Length of diagram in scale}}$

(4) $ip = \dfrac{p_{mi}\, LAnx}{60}$ where $n = \dfrac{n}{2}$ for 4-stroke and $n = N$ for 2-stroke

(5) $bp = \dfrac{2\pi\, NT}{60 \times 10^3} = \dfrac{p_b LAnx}{60}$ where $T$ is Nm and $p_{mb}$ is in kN/m$^2$

(6) $\eta_{ith} = \dfrac{ip}{\dot{m}_f \times (LCV)_f}$

(7) $\eta_{bth} = \dfrac{bp}{\dot{m}_f \times (LCV)_f}$

(8) $\eta_{rel} = \dfrac{\eta_{th}}{\eta_{as}}$

(9) $(\dot{m}_a)_{ideal} = \dfrac{xNVp_e}{RT_i}$

(10) $\eta_{rel} = \dfrac{\eta_{th}}{\eta_{as}}$

(11) $\eta_v = \dfrac{(\dot{m}_a)_{act}}{(\dot{m}_a)_{th}} = \dfrac{(\dot{m}_a)_{ideal}}{\rho_a \, nV_s}$

(12) $\eta_{sc} = \dfrac{(\dot{m}_a)_{act}}{(\dot{m}_a)_{ideal}} = \dfrac{(\dot{m}_a)_{act,\, ret} \times RT_i}{NV\, p_e}$

(13) $R_s = \dfrac{(\dot{m}_a)_{act}}{(\dot{m}_a)_{ideal}} = \dfrac{(\dot{m}_a)_{act}}{NV} \times \dfrac{RT_i}{p_e}$

(14) $\eta_s = \eta_{tr} \times R_s$

(15) $\eta_{sc} = 1 - e^{-R_s}$

(16) $\eta_{comb} = \dfrac{q_{act}}{q_{ideal}}$

(17) $\text{A/F} = \dfrac{\dot{m}_a}{\dot{m}_f}$

(18) $isfc = \dfrac{\dot{m}_f \text{ in kg/h}}{ip \text{ in kW}}$

(19) $bsfc = \dfrac{\dot{m}_f \text{ in kg/h}}{bp \text{ in kW}}$

(20) $\dot{m}_a = Ac_d\sqrt{2g\, h_w \rho_w \rho_a} = 0.008266\ c_d . d^2\sqrt{\dfrac{h_a p_a}{T_a}}$

(21) $ip = \sum\limits_1^n ip_n$

(22) $T = [(W + S) - w]r$ for band brake dynamometer

(23) $Q_f = \dot{m}_f\,(LCV)_f$

(24) $Q_{bp} = 2\pi NT/60 \times 10^3$

(25) $Q_{cw} = \dot{m}_w\, c_{pw}\, \Delta T_w$

(26) $Q_{eg} = \dot{m}_g\, c_{pg}\,(T_g - T_a)$

(27) $Q_{eg} = \dot{m}_w\, c_{pw}\,(T_{w2} - T_{w1})\, \dfrac{T_{g1} - T_{a1}}{T_{g1} - T_{g2}}$

## PROBLEMS

### Things to Think About

15.1 List the performance parameters in IC engines.

15.2 What is the purpose of measurements and testings?

15.3 What is measured in measuring the power output of an IC engine?

### Short and Long Theoretical Questions

15.4 What do you mean by friction power?

15.5 List the various methods to find friction power and discuss them in brief.

15.6 List the independent and dependent parameters.

15.7 What do you understand by indicated mean effective pressure? Derive an expression for indicated power.

15.8 What are the various engine efficiencies? Define each of them.

15.9 List the various basic measurements meant for IC engine.

15.10 List the types of tests usually conducted on IC engines.

15.11 List the various instruments which are used for speed measurements. Which are the instruments that give accurate measurement?

15.12 Discuss the various undesirable phenomena which hamper in the accurate measurement of fuel consumption.

15.13 Discuss the various methods of fuel consumption.

15.14 Discuss an arrangement for fuel consumption of a vehicle under test.

15.15 Discuss the various methods of measuring air consumption of engine. Give the relative merits and demerits.

15.16 Derive the expression for mass flow rate by an air box method employing orifice meter.

15.17 Discuss Morse test.

15.18 Discuss the various methods of measuring indicated power. Give relative merits and demerits.

15.19 Discuss the basic features of electronic indicator. List the various types of transducers used in electronic indicator with working principle.

15.20 What is the function of a dynamometer? Classify the dynamometer.

15.21 Discuss the working principle of various types of dynamometer, such as prony brake, hydraulic, electrical, etc.

15.22 Discuss the working principle of rotameter and turbine flow meter.

15.23 Discuss the method to measure flame temperature and flame propagation. Discuss their limitations.

15.24 List the various constituents of emission from an IC engines.

15.25 Discuss the working principle of various instruments which are used to measure the emissions from IC engines.

15.26 Discuss the performance characteristics of SI and CI engines.

15.27 List the variable affecting engine performance. Discuss them in brief.

15.28 What do you mean by heat balance of an IC engine? Discuss the heat balance diagram of an IC engine at constant speed. Discuss its physical significance.

15.29 Explain the internationally accepted methods of measuring the following invisible emission.

(a) Oxides of nitrogen. (b) Carbon monoxide,

(c) Unburnt hydrocarbon (d) Aldehydes.

15.30 How will you measure noise?

15.31 Explain the effect of the following factors on the performance of an engine

(a) compression ratio (b) A/F

(c) spark timing (d) engine speed

(e) mass of inducted charge (f) heat losses

15.32 Explain the reason of poor part load thermal efficiency of the SI engine compared with CI engine.

15.33 Mention the factors that limit brake power and brake mean effective pressure.

15.34 What are the methods available for improving the performance of an engine?

15.35 Draw a typical performance map for four-stroke SI engine showing contours of constant *bsfc*.

15.36 With the help of a Sankey diagram explain the energy flow through an engine.

15.37 Show the effect of dissociation on temperature and *bp*.

15.38 Discuss the effect of altitude.

15.39 What do you mean by power available and power required curves?

### Numerical Questions

15.40 Design bore and stroke of 4-stroke petrol engine for the following data:

$bp$ = 73.5 kW, $N$ = 400 rpm, $bmep$ = 8.5 bar, $\eta_{mech}$ = 80% $bsfc$ = 0.346 kg/kWh, $CV$ = 44100 kJ/kg, $r$ = 6.

Also calculate the *ip*, *bp*, air standard and relative efficiency and *imep*. Assume bore is equal to stroke.

[***Ans.*** 32 cm, 32 cm, $\eta_{ith}$ = 29.4%, $\eta_{bth}$ = 23.52%, $ASE$ = 51.1%, $\eta_r$ = 57.6% 10.62 bar]

15.41 During the testing of a 4-stroke oil engine fitted with a simple rope brake dyanometer the following readings were taken.

Brake wheel diameter = 60 cm, brake rope dia. 2.5 cm dead load on the brake = 200 N, spring balance reading = 50 N, speed = 450 rpm, area of the indicator diagram = 4.2 cm$^2$, length of the indicator diagram = 6 cm, spring constant = 6 bar/cm, bore 10 cm, stroke = 15 cm, $bsfc$ = 0.3 kg/kWh, $CV$ = 43960 kJ/kg.

Determine the *bp*, *ip* mechanical efficiency and indicated thermal efficiency.

15.42 Determine the mean effective pressure from indicator diagram of a 4-stroke oil engine. Area of the positive loop = 8.2 cm$^2$, base of diagram = 5 cm, spring stiffness = 5 bar per cm, area of the negative loop = 14.7 cm$^2$, base of diagram = 5 cm, weak spring used has stiffness of 0.2 bar/cm. [***Ans.*** 7.6 bar]

15.43 During testing of single-cylinder 2-stroke oil engine, following data were obtained.

Brake torque = 640 N, speed = 350 rpm, mean effective pressure = 5.6 bar, cylinder diameter = 21 cm, stroke = 28 cm, oil consumption = 9.16 kg/h, CV = 42705 kJ/kg. Determine (a) the mechanical efficiency, (b) the indicated thermal efficiency, (c) the brake thermal efficiency, and (d) the specific fuel consumption.

[***Ans.*** $\eta_m$ = 74%, $\eta_{th}$ = 32.2%, $\eta_{bth}$ = 23.75%, 0.353 kg/kWh]

15.44 The following data relate to an oil engine at full load condition:

*ip* = 22.2 kW, *bp* = 18.2 kW, fuel consumption = 5.44 kg/h, *CV* = 43000 kJ/kg, jacket water inlet temperature = 15°C, jacket water outlet temperature = 71°C, mass of cooling water = 4.6 kg/min, water inlet temperature to exhaust gas calorimeter = 55°C, mass flow of water in calorimeter = 8 kg/min, final temperature of exhaust gases = 82°C, room temperature = 16°C, A/F = 20, $c_{pg}$ = 1.005 kJ/kg K.

Calculate the thermal efficiency, mechanical efficiency and prepare heat balance sheet.

[***Ans.*** $\eta_m$ = 82.2%, $\eta_{ith}$ = 34.2%, heat to *bp* = 28.08% heat to jacket water = 27.72% heat to exhaust gas = 37.6%, unaccounted = 6.372%]

15.45 A four-stroke gas engine is governed by hit and miss method of governing. The gas consumption is 3.59 m$^3$/h at 15°C and 76 cm of Hg. The engine runs at 224 rpm in a test in which 60 explosions per minute are recorded. The air consumption was measured by a sharp edged orifice of 2.54 cm dia and the presence difference across the orifice is measured as 9 cm of $H_2O$. The temperature of air is 12°C and barometric reading is 75 cm of Hg. If the coefficient of discharge of the orifice is 0.6, find (a) the mass of air supplied to the engine in kg/min and (b) air-fuel ratio by volume. [***Ans.*** (a) 0.838 kg/min, (b) 6.1]

15.46 During the Morse test over a four-cylinder, four-stroke petrol engine having 6 cm bore and 8 cm storke, the fuel supply to the engine was fixed at 0.635 kg/min and the spark plugs of the cylinders were successively short circuited maintaining speed by changing the brake load. The results are:

*bp* with all cylinder working = 11.9 kW
" " cylinder No. 1 cut-off = 8.45 kW
" " " " 2 cut-off = 8.56 kW
" " " " 3 cut-off = 8.6 kW
" " " " 4 cut-off = 8.48 kW

Calculate the *bp* and indicated thermal efficiency if CV of fuel is 43542 kJ/kg. Also compare the efficiency with the air standard efficiency if the clearance volume in each cylinder is 7.05 cm$^3$. [***Ans.*** 13.52 kW, 29.41% ASE = 55.2%]

15.47 A Wright 14-cylinder, 16 cm × 18 cm four-stroke cycle air plane engine develops 1176 kW $b_p$ at 2400 rpm for a short time operation at take-off, with 661 kW at 1900 rpm as the recommended cruising operation power. Determine for this engine, (a) the piston displacement, (b) the maximum *bmep* (c) the causing bmep (d) the maximum torque, and (e) the cruising torque. [***Ans.*** (c) 9.8 bar]

15.48 A four-cylinder, four-stroke petrol engine has 7.5 cm bore and 10.5 cm stroke. The clearance volume in each cylinder is 1/7$^{th}$ of the swept volume. The speed of the engine taken by a tachometer is 2000 rpm and at this speed the torque is 70 N. The fuel consumption is 4.5 kg/h. Assuming brake thermal efficiency ratio based on air standard cycle is 60, calculate *bmep* and calorific value of fuel. If the engine has a mechanical efficiency of 80%, calculate the indicated thermal and relative efficiency. Prepare heat balance if exhaust gas temperature is 300°C.

[***Ans.*** 7.425 bar, 33913 kJ/kg, 42.38%, 75%]

15.49 The following data were taken during the test of a single-cylinder, 4-stroke oil engine.

Cylinder dia = 28 cm, stroke = 22.5 cm, gross *mep* = 7.24 bar, pumping *mep* = 0.4 bar, speed = 200 rpm, net load on the brake = 1300 N, effective diameter of the brake 1.16 m, fuel used/h = 9 kg, CV of fuel = 42000 kJ/kg.

Find *ip*, *bp*, mechanical efficiency and indicated thermal efficiency of the engine.

*Hint.* Net $p_m$ = gross *mep.* – Pumping *mep* [***Ans.*** 29.25, 21.35, 73%, 28%]

15.50 The following data relate to a 4-stroke cycle, four-cylinder diesel engine having 35 cm bore and 40 cm stroke:

*bp* = 88 kW, rpm = 350, fuel consumption = 32 kg/h, CV = 42705 kJ/kg, air consumption = 30 kg/min, cooling water = 50 kg/min, temperature rise in cooling water 35°C, exhaust gas temperature = 210°C, room air temperature =16°C $c_{pg}$ = 1.01, fuel consists of 15% of $H_2$, $c_p$ of steam = 2.09 kJ/kg K. Draw up the heat balance sheet.

15.51 Find the maximum brake power which can be developed in the cylinder of a four-stroke gas engine which runs at 200 rpm. The diameter of the piston is 300 mm and stroke 400 m. Clearance volume is 25% of the stroke volume. The gas used consists of CO = 19.7 %, $H_2$ = 28.8%, $CO_2$ = 14.4%, $N_2$ 37.1. Assume that the total mixture at NTP admitted per suction stroke is 0.875 of total volume behind the piston at the end of stroke and that the thermal efficiency is 35%. Calorific value of $H_2$ = 121.4 MJ/kg, calorific value of carbon burning from CO to $CO_2$ = 23.4 MJ/kg, Density of air = 1.2 kg/m$^3$. [***Ans.*** 47 kW]

15.52 A nine-cylinder petrol engine of bore 150 mm and stroke 200 mm has a compression ratio of 6:1 and develops 360 kW at 2000 when running on 20 % rich mixture. The fuel used has a calorific value of 43 MJ/kg and contains 85.3 carbon and 14.7%, $H_2$. Assuming volumetric efficiency of 70% at 17°C, and mechanical efficiency of 90%, calculate the indicated thermal efficiency of the engine. [***Ans.*** 30.84%]

15.53 During a test of a four-cylinder four-stroke petrol engine of 75 mm bore and 100 mm stroke, the following results were obtained at full throttle at a constant speed and with a fixed setting of the fuel supply of 0.082 kg/min.

*bp* with all cylinders working = 15.24 kW

*bp* with cylinder number 1 cut-off = 10.45 kW

*bp* with cylinder number 2 cut-off = 10.38 kW

*bp* with cylinder number 3 cut-off = 10.23 kW

*bp* with cylinder number 4 cut-off = 10.45 kW

Calculate the (i) indicated power of the engine under these conditions. If the calorific value of fuel is 44 MJ/kg, find (ii) the indicated thermal efficiency of the engine. (iii) Compare this with air standard efficiency. the clearance volume of one cylinder being 115 cc.

[***Ans.*** (i) 4.79 kW, 4.86 kW, 5.01 kW, 4.79 kW, 19.45 kW, (ii) 32.35%, (iii) 69.1%]

15.54 The following data refer to the trial of a four-stroke mechanical injection diesel engine of 200 mm bore and 400 mm stroke, gross *mep* = 6.2 bar, pumping *mep* = 0.44 bar, speed 270 rpm, brake torque = 468 Nm, fuel used = 3.85 kg of oil per hour of gross calorific value 46600 kJ/kg, cooling water 6 kg/min, temperature rise in cooling water 50°C.

Draw up a heat balance sheet for the trial expressing various quantities in kJ/min and calculate (i) the mechanical efficiency of the engine.

If the fuel contains 13.5% $H_2$ (by mass) and the air supply to the engine 1.71 kg/min at 17°C. Calculate (ii) the heat carried away per minute by the exhaust gases when their temperature is 280°C. Assume a mean specific heat of 1.1 kJ/kg of gas.

[***Ans.*** (i) 16.1 kW, (ii) 0.271 kg/kWh]

15.55 A four-cylinder, four-stroke diesel engine develops 28 kW at 2000 rpm. Its *bsfc* is 0.26 kg/kWh. Find the power output of the engine and its *bsfc* when the fuel rate is reduced by 40% at the same speed. Assume mechanical efficiency 0.8, the indicated thermal efficiency changes linearly with equivalence ratio and there is 1% increase in indicated thermal efficiency for 6% increase in the equivalence ratio. Equivalence ratio at higher flow rate is 0.666.

15.56 A four-stroke cycle naturally aspirated direct-injection diesel engine is being developed to provide 200 kW of power at the engine maximum rated speed. Assuming suitable data estimate the following:

(a) The compression ratio, the number of cylinders, the cylinder bore and stroke and the maximum rated speed of an appropriate engine design that would provide this maximum power.

(b) The brake specific fuel consumption of this engine design at the maximum power operating point.

(c) The approximate increase in brake power that would result if the engine was turbocharged.

15.57 A six-cylinder four-stroke cycle, 10 cm × 12.5 cm stroke diesel engine develops 50 kW at 1000 rpm. The various efficiencies are mechanical 76%, volumetric 80%, theoretical thermal 52%. The lower calorific value of the liquid fuel is 45000 kJ/kg. Compute:

(i) *bmep,*

(ii) A/F,

(iii) *bsfc*

15.58 The following data were obtained during the trial of a single-cylinder 2-stroke cycle diesel engine:

Cylinder bore 23 cm, stroke 25 cm rpm, 350 fuel consumed 0.3 kg/min with a calorific value of 43 MJ/kg, area of indicator diagram 60 cm$^2$, length of diagram 7.8 cm, spring constant 8.5, load on brake drum 1125 N at 1.25 m radius, cooling water used 18 kg/min, temperature of water entering and leaving at 18°C and 62°C respectively, atmospheric temperature 48°C, A/F 28, exhaust gas temperature 410°C, mean sp. heat of exhaust gases 1.08 kJ/kgK. Calculate *ip*, *bp* $\eta_m$ $\eta_{ith}$ and $\eta_{bth}$. Also draw up heat balance sheet.

### Software Based Questions

15.59 Develop a software for the design and performance of a modern car and bus engines.

15.60 Develop a software for predicting the basic engine dimensions, its performance and drawing up heat balance.

### Design and Open Ended Question

15.61 Develop new concepts for measuring, *ip*, *bp*, $\dot{m}_a$, $\dot{m}_f$, noise and emission from engine other than the existing ones with schematic instruments, etc.

**Multiple-Choice Questions**

15.62 Morse test is conducted on the engine to calculate the mechanical efficiency of

(a) single-cylinder SI engine (b) single-cylinder CI engine
(c) multi-cylinder engines (d) only on two-stroke engines

15.63 Motoring test is carried out on an IC engine to determine

(a) brake power (b) indicated power
(c) friction power (d) fuel consumption

15.64 Morse test is carried out on IC engines to determine

(a) thermal efficiency of two-stroke engines
(b) mechanical efficiency of multi-cylinder engines
(c) volumetric efficiency of multi-cylinder engines
(d) brake power produced by individual cylinders.

15.65 Rope brake dynamometer is used to test

(a) small engines (b) large engines
(c) low speed engines (d) high speed engines

15.66 A CI engine is preferred to SI engines in road transportation because

(1) the part load efficiency is better
(2) the fuel is cheaper and thermal efficiency is higher.
(3) the brake power increases with increasing speed of these statements

(a) only 1 and 3 are correct (b) 1 and 2 are correct
(c) only 2 and 3 are correct (d) 1, 2 and 3 are correct

15.67 **Assertion (A):** Qualitative governing is used in large gas engine where fuel economy is important.

**Reasoning (R):** Because this method will ensure a lean mixture at small loads.

**Answer Codes:**

(a) A is true but R is false (b) Both A and R are true
(c) Both A and R are false (d) A is false but R is true

15.68 Flame ionization detector is used for measuring

(a) CO (b) HC
(c) $NO_x$ (d) $CO_2$

## Key to Multiple-Choice Questions

15.62 (c) 15.63 (c) 15.64 (b) 15.65 (b) 15.66 (d) 15.67 (b) 15.68 (b)

CHAPTER

# 16 IC Engines and Alternatives Under Development

## OVERVIEW

Due to various limitations of most developed and widely used petrol and diesel engines, some other types of internal combustion engines are under developing stage and some of them are now commercialized at low scale. *Electric and hydbrid fuel cell vehicles have been commercialized at low scale* (see chapter 22). They include:

(i) Stratified charge engines
(ii) Stirling engines
(iii) Wankel rotary combustion engine
(iv) Variable compression ratio engine
(v) Free piston engines
(vi) Dual fuel engines
(vii) Electric vehicles
(a) battery type
(b) fuel-cell type } **(see chapter 22)**
(viii) Biodiesel powered engines

In this chapter, each of them will be described briefly.

## 16.1 STRATIFIED CHARGE ENGINE

*The petrol engine exhibits very good full load power characteristics, e.g., high degree of air utilization, high speed and flexibility but yields poor part load efficiency, whereas the diesel engine has good part load characteristics but exhibits poor air utilization, low-smoke limited power and higher weight to power ratio.* The other requirements of diesel engine such as use of higher compression ratio for better starting and good combustion over a wide range of engine operation pose additional problems of maintenance and high losses in diesel operation. It is a well-known fact that for an automotive engine, both part load efficiency and full load power are very important because for about 90 per cent of their total operating life, the engines work under part load conditions and it is the maximum power output that controls the speed, acceleration and other characteristics of the vehicle.

The above listed limitations of petrol and diesel engines have impressed upon to develop an engine which combines the advantages of both petrol and diesel engines and avoid their disadvantages as far as possible. *Stratified charge engine is an attempt in this direction as it falls in mid-way between diesel and petrol engines.*

*The purpose of stratified charge engine is to provide different fuel-air mixture strengths at various places in the combustion chamber – a relatively rich mixture at and in the vicinity of spark plug and a lean mixture in the rest of combustion chamber while the overall mixture is lean.* It falls under the category of SI engines.

*Stratified charge engine yields higher thermal efficiency due to higher compression ratio and reduced air-pollution due to lean mixture as compared to conventional petrol engine.*

There are two methods of charge stratification according to the method of formation of heterogeneous mixture in the combustion chamber.

(i) Using fuel injection and positive ignition including swirl production.

(ii) Using carburetion alone.

## 16.1.1 Stratification by Fuel Injection and Positive Ignition

### *16.1.1.1 Recardo First Stratification Charge Engine*

Recardo in 1992 attempted first to design a stratified charge engine which is shown in Fig. 16.1. A relatively rich mixture was formed at the spark plug by an auxiliary spray while another spray injecting fuel along the major axis of the combustion chamber formed a leaner mixture, which in turn, produced overall variable mixture strength and lean mixtures. It yields efficiency as **high as 36%**. The main drawback of this engine was its narrow range.

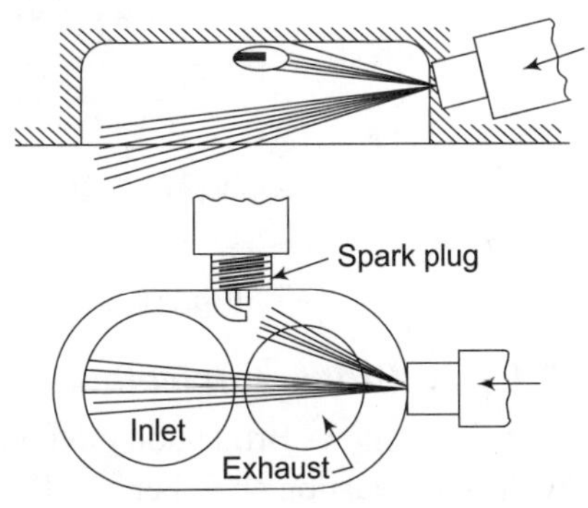


**Fig. 16.1** Recardo first stratified charge engine

### *16.1.1.2 Prechamber Stratified Charge Engine*

Figure 16.2 shows prechamber stratified charge engine developed by Recardo after his first stratfiied engine. It consists of a small prechamber fitted with injector and the spark plug. This ensures rich mixture near spark plug by fuel supplied from injection and lean mixture elsewhere supplied by carburettor. By this way, an overall leaner mixture burns in combustion chamber. *The main drawback of this engine is the use of prechamber which means loss of thermal efficiency due to throttling and costly fuel injection equipment.*

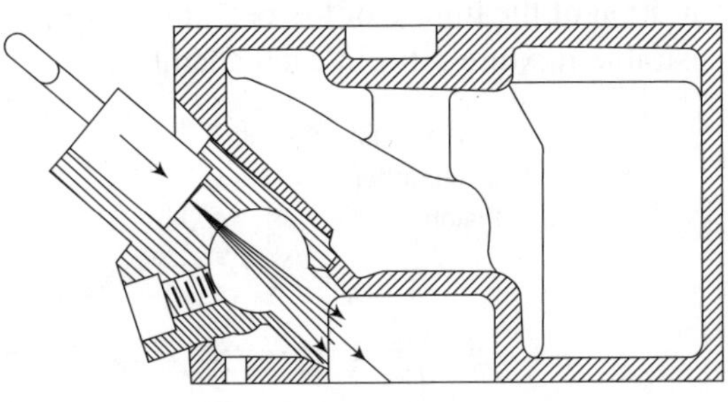

**Fig. 16.2** Recardo's prechamber stratified charge engine

### 16.1.1.3 Volkswagen PCI Stratified Charge Engine

This engine (Fig. 16.3) consists of a spherical unscavenged prechamber comprising of about 25 to 30 per cent of compression volume and is linked to the main combustion chamber by a relatively large flow transfer passage. The main combustion is of disc shape and contains no squish surfaces but a slight swirl is induced by the intake port. In order to avoid overrichment at spark plug, the injection nozzle and spark plug are arranged in sequence in the flow direction. The total volume of fuel is injected partly into the prechamber and partly in the intake manifold, with a relatively rich mixture at spark plug under all operating conditions. The regulation of load is achieved primarily by adjusting the mixture strength introduced into the main combustion chamber. *The main advantages of this system are that it does not need variation of fuel injection timing for stratification and does operate on very lean mixture which is good for air pollution point of view.*

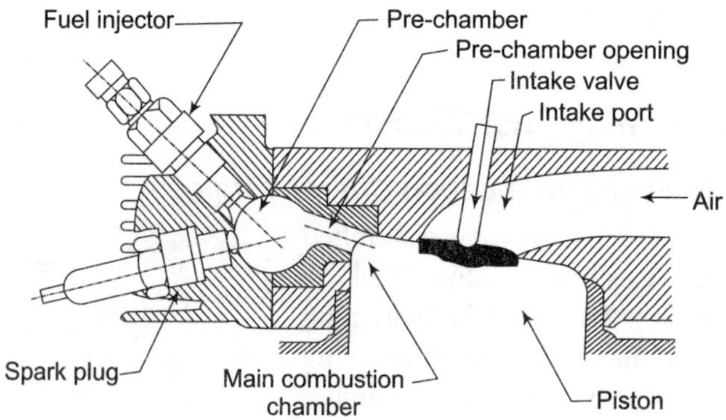


**Fig. 16.3** Volkswagen PCI stratified charge engine

### 16.1.1.4 Broderson Method of Stratification

Figure 16.4 shows Broderson method of charge stratification which consist of an engine with a divided combustion chamber auxiliary or prechamber and the main chamber. The prechamber contains the injector, the intake valve and the spark plug. Whole of the fuel is injected into the prechamber. The direction of the air flow in the combustion chamber changes as the piston passes BDC position and it is this fact which forms the basis of charge stratification. The variation of the timing of the beginning of injection and the duration of injection controls the desirable mixture ratio in each chamber.

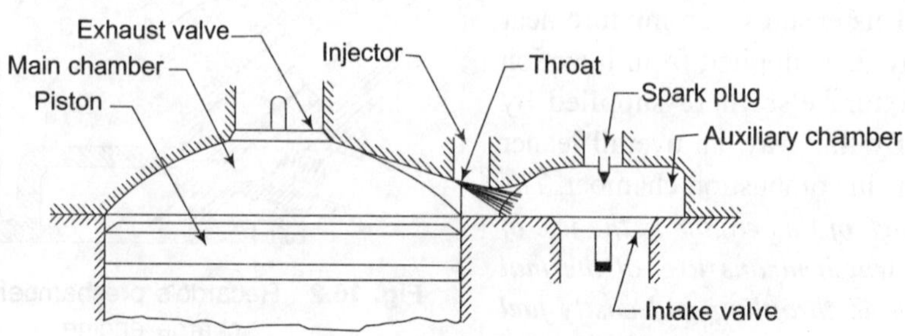


**Fig. 16.4** Broderson method of charge stratification

*This method posseses the advantage of altogether avoiding the pumping losses due to throttling and carburettor venturi, in addition to the high part load efficiency obtained by burning lean mixture and some knock resistance at full load. The other advantage of this engine is to burn propane, kerosene, diesel and gasoline fuel, i.e., it has multifuel capability.* The main drawback includes the need of special attention to be paid for matching the injection and ignition equipment with the design of prechamber and main chamber.

## 16.1.2 Swirl Stratified Charge Engine

The prechamber system does not provide fuel charge stratification. The matching of air swirl and fuel injection provides better charge stratification in an open combustion chamber. Based on this philosophy, many new designs have come. They include:

(i) Witzky swirl stratification process (WSSP)
(ii) Texco combustion process (TCP)
(iii) Ford combustion process (FCP)
(iv) Ford PROCO
(v) New Deutz combustion process (NDCP)

### *16.1.2.1 Witzky and Swirl Stratifications Process (WSSP)*

This engine (Fig. 16.5) is basically unthrottled spark ignition engine with fuel injection and no carburetion. During the suction stroke a high velocity swirl is imparted to the intake air by proper design of intake part, while the fuel is injected during the compression stroke against the swirl direction at some suitable angle. As a result, the swirling air forces the fuel droplets to follow a spark path by virtue of charge forces and directs them towards the centre of combustion chamber. This results in a good degree of stratification — a rich ingnitable mixture near the spark plug over the full load range and leaner mixture a distance from the spark plug.

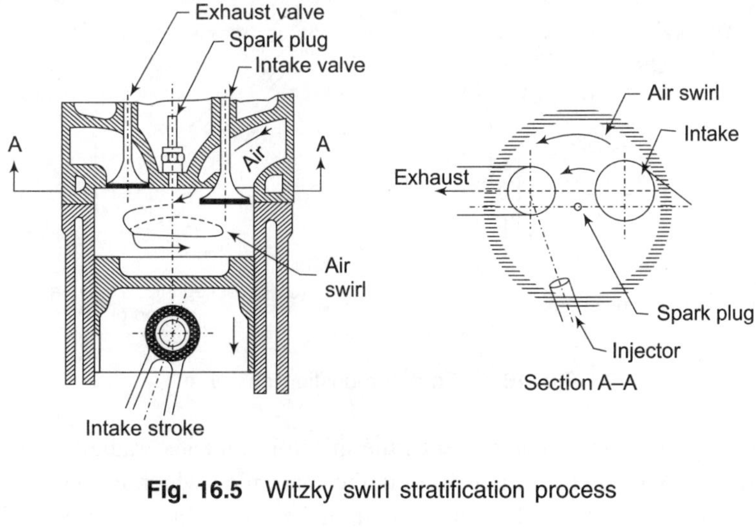


**Fig. 16.5** Witzky swirl stratification process

### 16.1.2.2 Texaco Combustion Process (TCP)

Figure 16.6 shows Texaco combustion process in which a high degree of swirl is imparted to the air and the fuel is injected across the swirling air in down stream direction which ensures full impregnation of air by fuel. The spark plug is positioned directly downstream and as near as possible to injector. For proper operation of TCP, a good coordination of air swirl, injection and positive ignition is needed. The fuel is injected near the end of compression stroke, about 30° bTDC and simultaneously or just after the spark plug is caused so that the first part of the injected fuel which mixes with air during its passage to the spark plug forms a flame front.

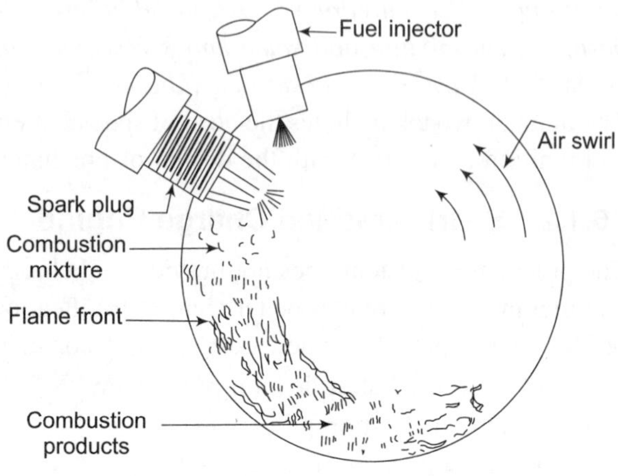


**Fig. 16.6** Texaco combustion process (TCP)

### 16.1.2.3 Ford Combustion Process (FCP)

Figure 16.7 shows Ford combustion chamber. In this case, attempts are made to obtain part load stratification by proper coordination of air swirl, fuel injection and positive ignition and at the same time maintain the fuel load potential of the homogeneous air-fuel mixture.

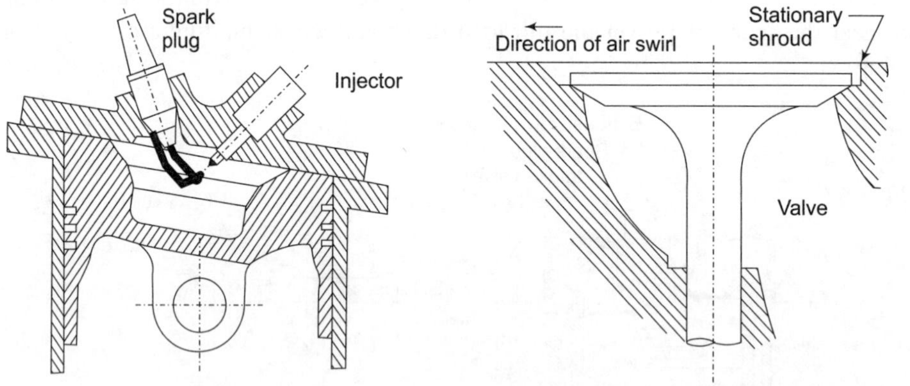


(a) FCP combustion chamber

(b) Recessed intake valve and directional port of FCP

**Fig. 16.7** Ford combustion chamber

In order to meet the above requirements, the injector is located radially and at 56 degree from the head face. A long electrode is used in the spark plug which is so positioned that the spark gas is about 10 mm ahead of the injector along its centre line. To produce the maximum possible turbulence a recessed intake valve with stationary shroud is used to impart the intake air a unidirectional pattern and the stationary shroud reduces anti-swirl air flow.

### *16.1.2.4 Ford PROCO Engine*

This engine employs the second generation Ford combustion process which was developed to give fuel economy and maximum power equal to that of a carbureted engine. *The main aim of Ford PROCO process is to minimize the exhaust emissions with minimum fuel economy loss, using air throttling technique.*

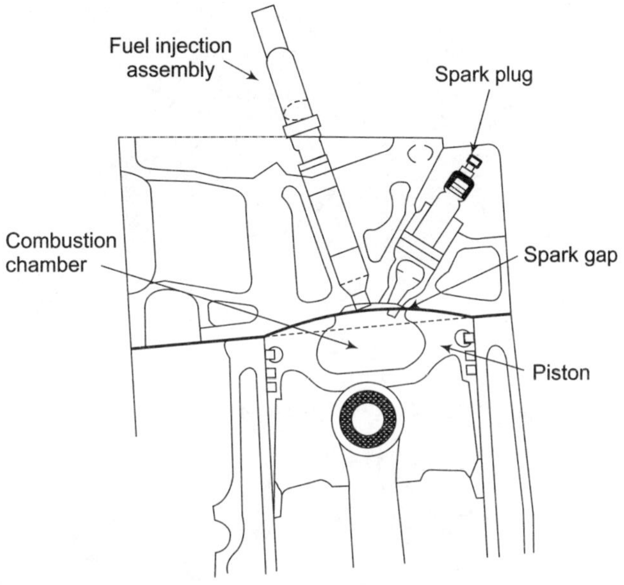


**Fig. 16.8** Ford PROCO engine

Figure 16.8 shows Ford PROCO engine. Its intake part is the most important element. *It is shaped to impart a swirl around the cylinder bore axis.* The swirl speed in the engine is about 3 to 5 times the crankshaft speed. The swirling charge is compressed with a compression ratio of 11:1 into a cup-shaped combustion chamber situated concentrically in the piston with about 65 per cent squish area. The injector is located near the centre of the cylinder bore and provides a soft, low-penetrating, wide angle conical spray resulting in a rich mixture at centre and lean elsewhere. The spark plug located either near the bore centre line or just above the spray. The main *difference between FCP and Ford PROCO is the timing of fuel injection.*

### *16.1.2.5 New Deutz Combustion Process (NDCP)*

This engine has been developed basically as a multi-fuel process and possesses a high resistance to knock. Figure 16.9 shows new Deutz combustion process in which a high degree of swirl is generated during the suction stroke through a swirl port. The swirl velocity is further increased during the compression stroke by a cylindrical piston combustion chamber inclined by angle of 20° from the cylinder axis.

The injector is of two hole nozzle type which supplies atomized fuel across the swirling air. The swirling air causes the fuel drops to form a rotating ring of mixture. The fuel concentration in the ring decreases towards the middle of combustion chamber. *In this process, the fuel droplets vaporize before reaching the wall and due to organised swirl do not spread over the whole chamber but are concentrated in the vicinity of the wall of the form of ring. This ensures rapid combustion.*

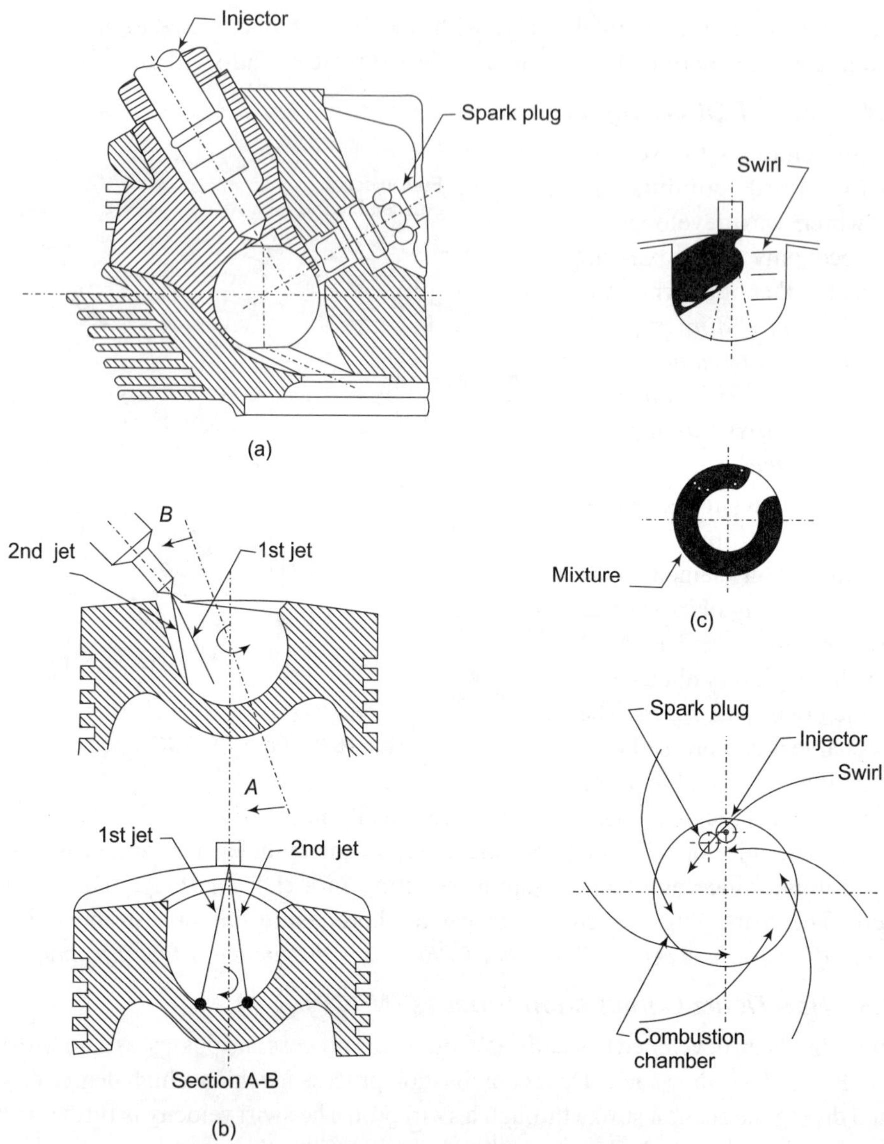


**Fig. 16.9** New Deutz combustion process

### 16.1.3 Stratification by Carburetion

Figure 16.10 shows Honda Compound Vortex Controlled Combustion (CVCC) engine. It consists of a prechamber containing a rich mixture, coupled to the main chamber which contains a lean mixture. Stratification is obtained by carburetion. It consists of an auxiliary combustion chamber around the spark plug and the small intake valve which is fixed to each cylinder. This means two separate intake valves are used on each cylinder— one for auxiliary chamber and other for main chamber. *The carburettor is of three barrels in which the smallest*

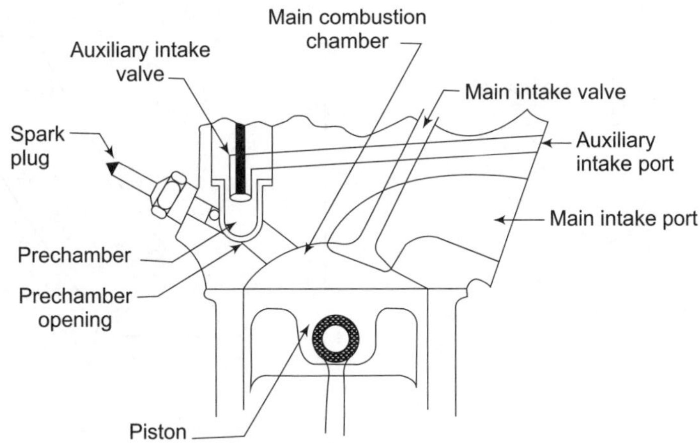


**Fig. 16.10** Honda CVCC engine

*venturi is used to supply rich mixture to each auxiliary chamber and the other two venturies are used to supply a very lean mixture to the main chamber.*

*This engines gives highly leaner mixture than stoichiometric ratio and good emissions.*

## 16.2 STIRLING ENGINE

Stirling enigne has been named after its inventor Scottish Minister Robert Stirling who developed this engine in 1816. Due to inherent advantages but some teething problems research for Stirling engine had been at standstill. In recent years, the research is going on in this field. An efficiency of 40% coupled with a specific power of 82 kW per litre of swept volume has already been achieved and its performance is bound to increase with continued research. This engine has the potential for simultaneously delivering excellent fuel economy, extremely low emissions, low noise level and performance comparable to that of a pre-emission controlled Otto engine with good derivability.

In conventional IC engines, the burning of fuel is internally and intermittently whereas the Stirling engine burns its fuel outside the engine itself and continuously. The engine works on the principle of regeneration. The regenerator employed within the engine itself which would store the rejected heat energy during heat rejection process and supply the same during heat addition process. Figure 16.11 shows the representation of Stirling cycle in *p-V* and *T-s* diagram. It consists of two isothermal and two constant volume processes as given below.

1 – 2 : Reversible isothermal compression process

2 – 3 : Heat addition at constant volume

3 – 4 : Reversible isothermal expansion

4 – 1 : Heat rejection at constant volume

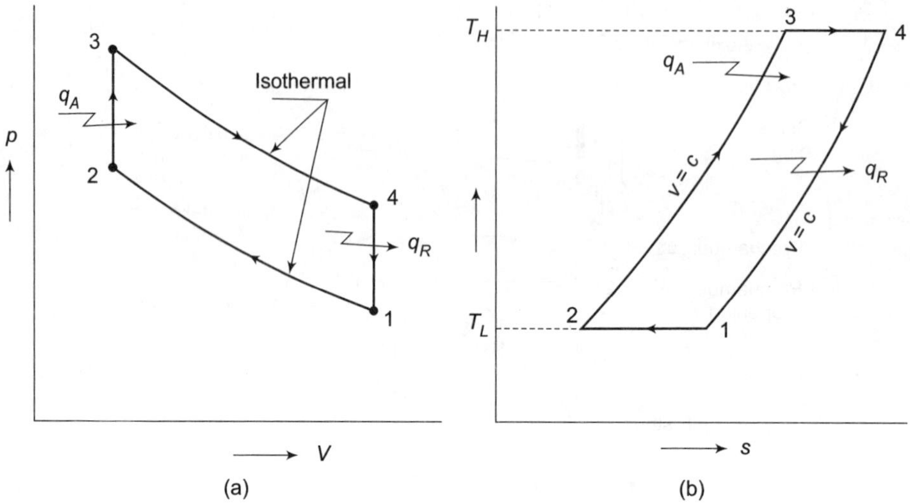


**Fig. 16.11** Stirling cycle on *p-V* and *T-s* diagram

The regenerator used in the Stirling cycle consists of a wire mesh stacked in a duct as shown in Fig. 16.12. The air in the space at the lower end is kept at a high temperature ($T_H$) by heating with a furnace below it. In the upper space, a low temperature ($T_L$) of air is maintained by a supply of cold water passing through the tubes in contact with it. Hence, two heat reservoirs, one at high temperature ($T_H$) and other low temperature ($T_L$) are obtained. The wire mesh interposed ensures uniform temperature gradient from $T_H$ to $T_L$. This provides a large heat transfer area and a high heat capability.

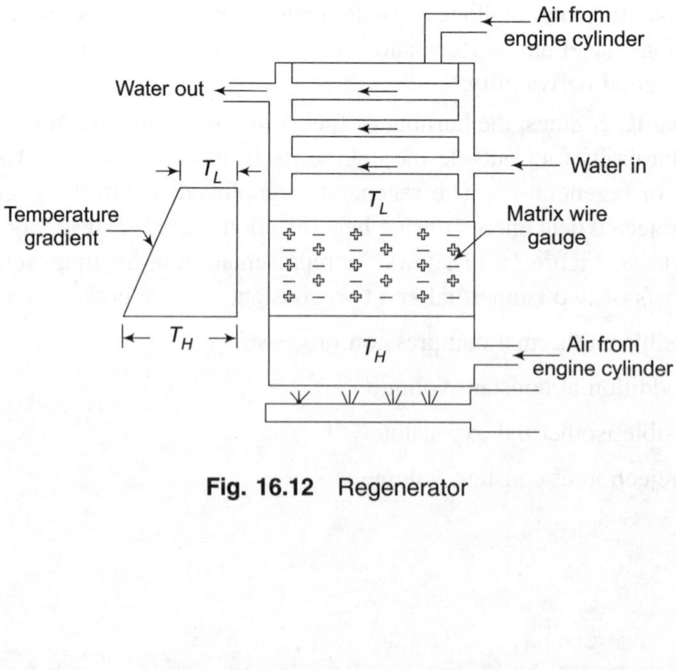


**Fig. 16.12** Regenerator

The thermal efficiency of Stirling cycle is given by

$$\eta_{thermal} = \frac{R(T_H - T_L)\ lnr}{RT_H\ lnr + (1 - \eta_{reg})\ c_v\ [T_H - T_L]} \approx 1 - \frac{T_L}{T_H} \text{ if } \eta_{reg} = 1$$

Figure 16.13 shows the two working models of Stirling engine based on single and double cylinder. In the case of double cylinder, the linkage of the piston is in such a way that the left hand piston does not move when the right hand piston compresses the gas from state 1 to 2. Only a small pressure rise is obtained due to heat removal. The left hand piston now begins its downward movement while the right hand piston further compresses the gas and forces it through generator. The net result is constant volume process in which the gas is transferred to the left hand cylinder and charged to state 3, that is next added to the cycle at a rate which allows the regenerated into cooling cylinder.

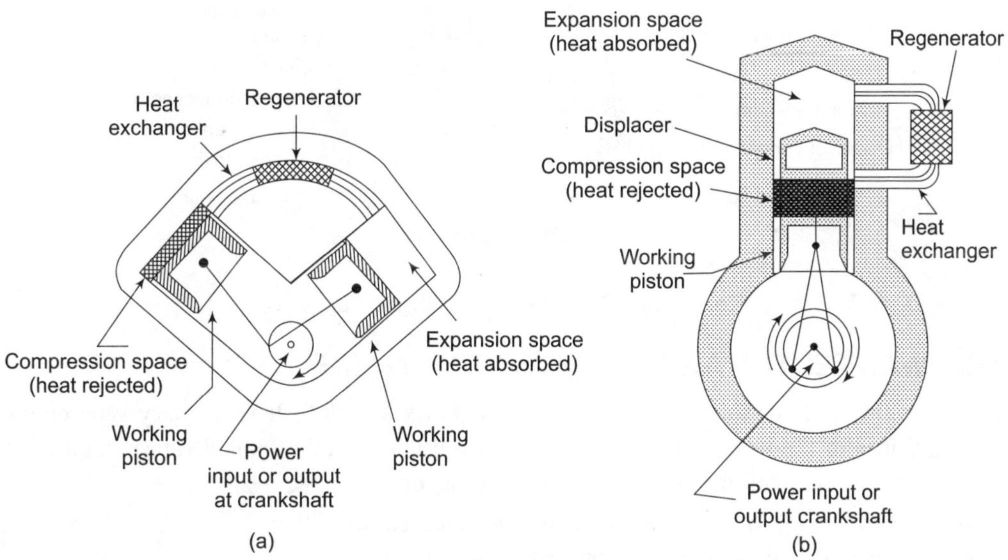


**Fig. 16.13** Double and single cylinder Stirling engine

## 16.2.1 Stirling Engine Using Rhombic Drive

Figure 16.14 shows the construction of Stirling engine using rhombic drive. The rhombic drive because of symmetry, allows even a single cylinder engine to be perfectly balanced. The centre of the gravity of the whole system is kept stationary by means of suitably located and sized counter weight in the crankshaft. The horizontal forces exerted by the connecting rod are exactly balanced at each yoke.

There are five major elements of this engine. They are:

(i) Carburettor and heater head (ii) Regenerator

(iii) Cooler (iv) Power piston and displacer piston

(v) Buffer space and rhombic drive

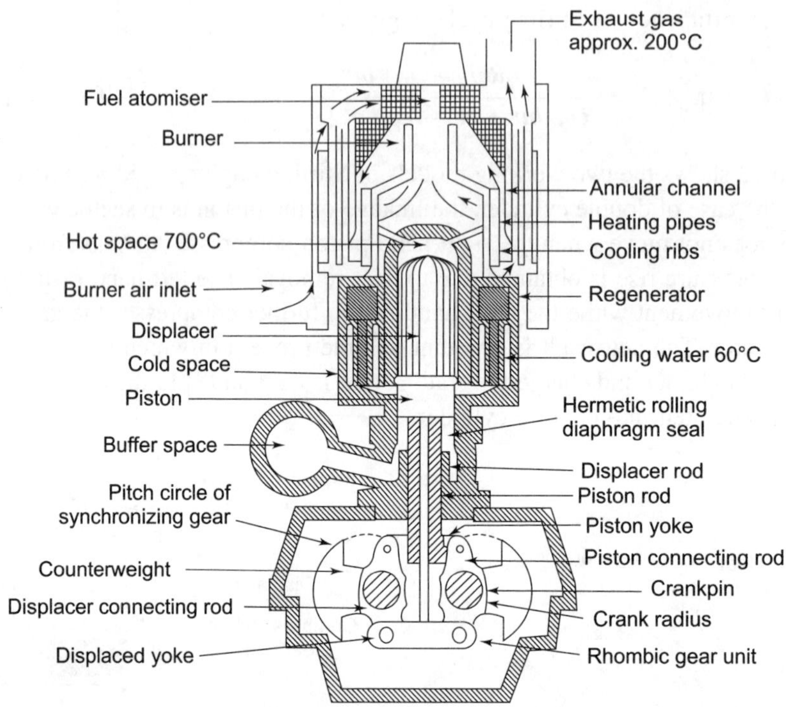


**Fig. 16.14** Stirling engine with rhombic drive

## 16.3 WANKEL ROTARY COMBUSTION ENGINE

Wankel engine is after the name of its inventor **Felix Wankel**. It is a rotary type engine. Although, it is in the process of development at present but this Wankel rotary engine may soon find large application in the field of road transport.

*Wankel engine (Fig. 16.15) resembles with both reciprocating engine and turbine.* Its sequence of operation is exactly the same as in reciprocating piston engines. *The change in volume of the charge is affected by the rotating element (piston) in Wankel engine instead of to and fro motion of the piston is reciprocating engine.* The similarity between Wankel engine and gas turbine lies in the fact there are not reciprocating parts and the motion is purely rotary.

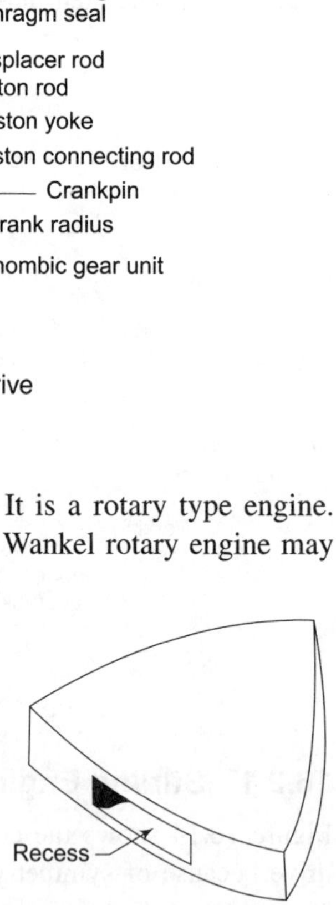


**Fig. 16.15** Wankel pump

**Construction.** This engine consists of two moving parts and a stationary cylinder with an internal bore of *epitrochoidal shape* fitted with inlet and exhaust ports. Out of the two moving parts, one is a *piston of triangular shape* and the other being an output shaft. The output shaft is integral with an eccentric which engages with the inner bore of the piston. The inner bore of the piston also has at one end a wheel with internal teeth which engages with a gear wheel rigidly mounted on the inside face of the cylinder end. The piston rotates about its own axis and also about the axis of the fixed gear.

The outside shape of the piston is triangular having curved on sides. For the correct compression ratio, there is a recess in each curve face of the piston as shown in Fig. 16.16. There is a continuous contact of the apexes of the triangular piston with the inner surface of the cylinder. To check leakage, the apexes of the piston are fitted with special gas seal. The configuration and combination of the cylinder and piston forms three chambers, the position of which vary with the rotation of piston. The combustion of air-fuel charge is initiated by means of a spark plug fitted into the cylinder. The cylinder is water jacketed and carburettor/ injection system supplies the required air-fuel mixture.

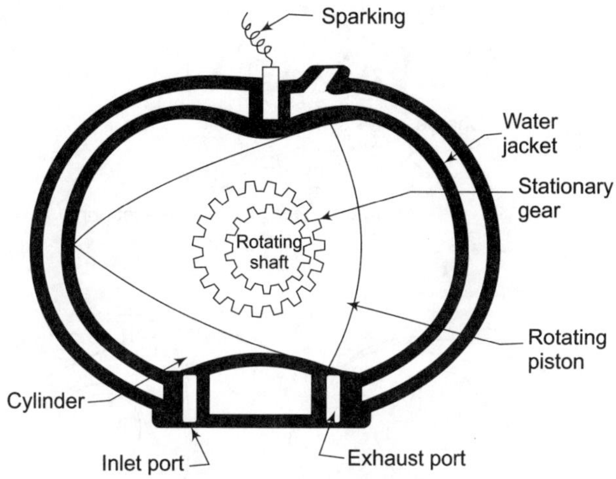


**Fig. 16.16** Wankel engine details

**Working principle.** Figure 16.17 shows the principle of working of Wankel engine. Let the three faces of the piston be 1–2, 2–3 and 3–1. To understand its working, it will be better to know what happens in a particular chamber, say formed by the face 1–2 as it moves round the cylinder.

Figure 16.17(a) shows the induction of the charge into the chamber formed by the piston face 1–2. Figure 16.17(b) represents the compression of the trapped charge. Figure 16.17(c) shows the ignition and the combustion of the charge. Figure 16.17(d) represents the expansion process. Figure 16.17(e) shows the exhaust of the combustion products through exhaust port.

Note, that in turn, exactly similar operation will take place in the chamber formed by the piston faces 2–3 and 3–1 as they rotate. Thus, three explosions occur in each revolution of the piston causing an uniform torque on the output shaft.

The speed of the output shaft depends upon the number of teeth on annulus and rigid gear. For example, output shaft makes three revolutions in every revolution of the piston about its own axis when the number of the teeth on the rigid gear wheel is two-thirds that of annulus. From the concept of epicyclic gear trains, the speed relation may be found out.

**Advantages**

(i) As there is no reciprocating parts, hence engine is free from vibrations.
(ii) Simpler design since there are only two moving parts.
(iii) Weight and space occupied and much lower than reciprocating engines for the same power output.

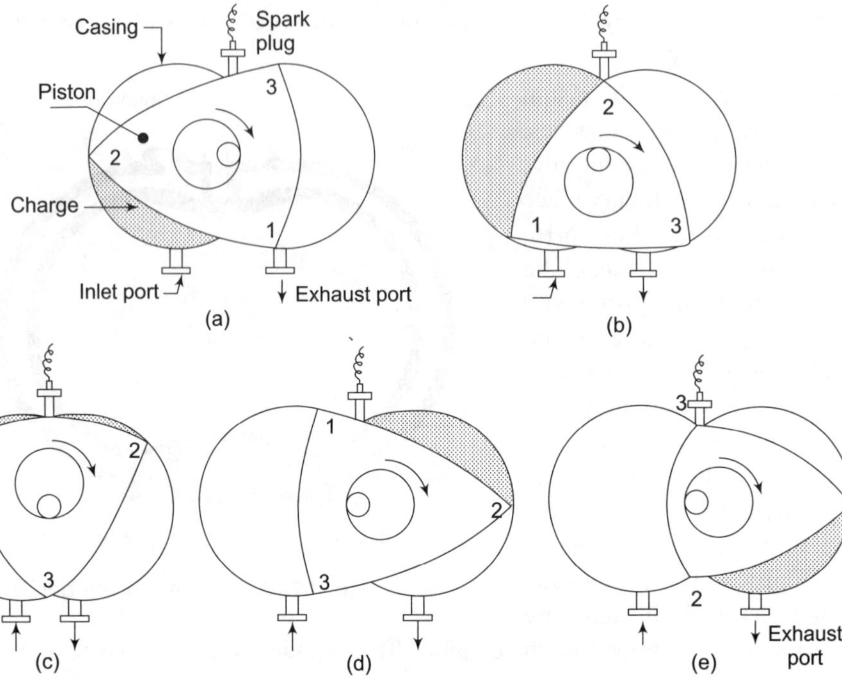


**Fig. 16.17** Sequence of operation in Wankel engine

**Disadvantages**

(i) Due to three explosions per revolution of the piston, the working temperature is very high.

(ii) Quick wear of gas seals fitted at apexes

(iii) Lubrication is difficult

## 16.3.1 Wankel Engine Geometry

Figure 16.18 shows Wankel engine has an inner contour of two-lobed epitrochoid in which rotates a rotor having an equilateral curved sided triangular shape. A trochoid is the shape traced by a point on the radius of extension of a radius of a circle as it falls without shipping on a line, a circle or a curve.

A two-lobed epitrochoid contour (Fig. 16.18a) is formed when the generating circle rolls on its inner circumference around the outer circumference of a fixed cycle. In this case, the ratio of rolling circle radius $R_r$ to the fixed circle radius $R_f$ is 3.2. As the rolling circle, moves around the circumference of the fixed circle, its centre $E_t$ describes a circular path around the centre of fixed circle ($C_f$) at a radius equal, to $R_r - R_f$ which comes to be equal to $R_f/2$ and termed as eccentricity. The trochoid is traced by the generating point $P$ which is at a distance $R$ from the centre of the rolling circle; $2(R+e)$ is the major diameter while $2(R–e)$ is the minor diameter. Since the radii of the two circles are in the ratio of 3:2, so the Wankel

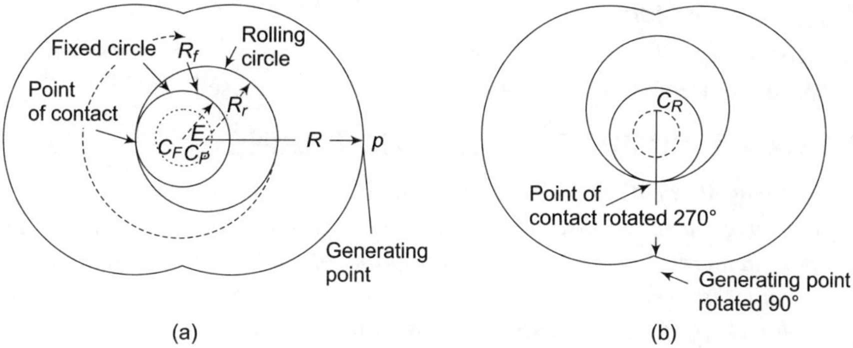


**Fig. 16.18** Wankel engine geometry

engine shows a characteristic of producing a thermodynamic phase over every 270° rotation of the output shaft.

## 16.3.2 Auxiliary Systems of Wankel Engine

The various auxiliary system includes sealing lubrication, cooling, induction and ignition system.

### *16.3.2.1 Sealing System*

The apex seals must seal the straight line contact between the rotor apex and the rotor housing bore and at the same time also seal the sides clearance at their ends. The side gas seals, one each on both sides of the rotor are also used. Oils seals are used to keep the oil to centre of rotor and prevent its loss into the combustion chamber.

### *16.3.2.2 Lubrication System*

Lubrication system of Wankel engine is divided into two parts:

(a) Lubrication of the rotor housing, phasing gear and the shaft

(b) Lubrication of apex seals, side seal and bore.

To meet the above requirements, two separate circuits are formed from one oil sump. An oil pump pumps the oil which goes to main bearing, eccentric bearings, rotor and side plate. A dosage pump pumps small quantity of oil into the fuel pump.

### *16.3.2.3 Cooling System*

Water or air cooling may be used as per requirement. In this engine, rotor housing, rotor seals especially apex seals and the end cover housing require cooling.

### *16.3.2.4 Induction System*

In Wankel engines, ports are used for induction. Three types of intake ports are used:

(i) Peripheral ports (ii) Side ports

(iii) Combination of side and peripheral ports

#### 16.3.2.5 Ignition System

Generally, a condenser discharge system or some other electronic ignitions system is used to meet high speed requirement. One or two spark plugs may be used.

## 16.4 VARIABLE COMPRESSION RATIO ENGINE

*The main requirements of IC engines are high specific power output accompanied by high thermal efficiency, good reliability, longer engine life and pollution free emission. There is a host of problems in increasing the specific power output of the engine. For example, increasing engine speed imposes dynamic loads and increased wear thereby, reducing reliability and engine life with poor part load performance. Use of high pressure using turbocharger results in high peak cycle pressure with higher thermal loads.*

*The use of variable compression ratio (VCR) is one of the methods to solve high peak pressure problem caused by high specific power output engine.* The low compression ratio is used at full load and high compression during starting out at part load. The use of low compression ratio at full load allows the turbocharger to boost the intake pressures without increasing the peak cycle pressure. At starting, the turbocharger output is zero, so a high compression is needed.

*A SI engine may be used as VCR engine but concept wise VCR engine is most suited to turbocharged diesel engines due to two main reasons.* Firstly, VCR concept is beneficial only at part load and the part load efficiency of diesel engine is higher than that of gasoline. Secondly, the diesel engine has better multi-fuel capability.

### 16.4.1 Theoretical Analysis

Figure 16.19 shows the representative of VCR engine on *p-V* diagram. The variation of gas temperature with crank angle is shown in Fig. 16.20. For comparison the diesel engine with constant compression ratio is also depicted. From the representation, we observe the following:

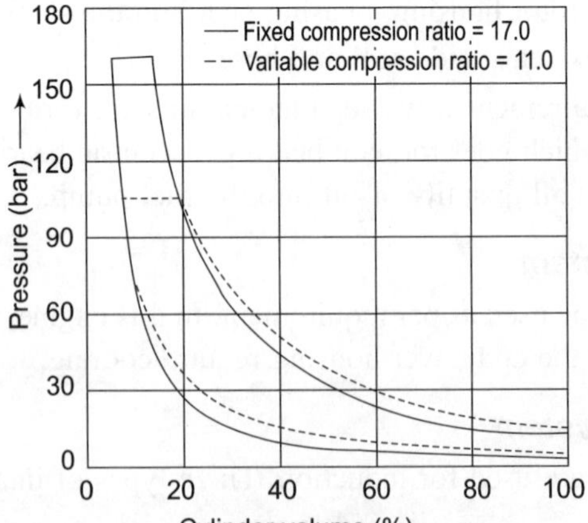


**Fig. 16.19** Theoretical representation of VCR engine

(i) Both the pressure and temperature expansion curves for VCR engine lie above those for constant compression ratio engines. This shows that expansion is slower at low compression ratio.

(ii) During expansion, the temperature of the gas is higher in the case of VCR engine indicating hotter running of exhaust valve of VCR engine.

(iii) The maximum compression is determined by the starting condition and load on the engine.

## 16.4.2 Methods of Obtaining Variable Compression Ratio

There are two basic methods to obtain variable compression ratio. They are:

(i) By altering clearance volume

(ii) By altering both clearance volume and swept volume

The first method requires either raising or lowering the entire crankshaft or raising or lowering the piston crown. The second method requires a variable throw crankshaft for changing the stroke.

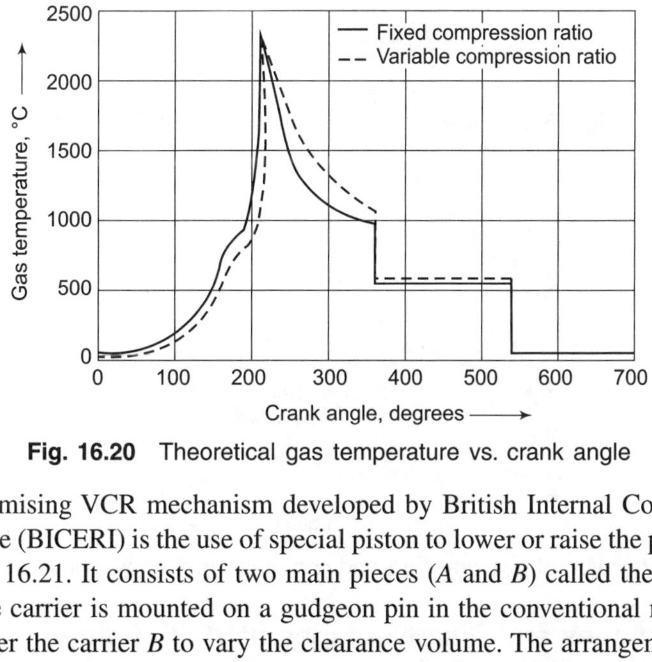


**Fig. 16.20** Theoretical gas temperature vs. crank angle

The most promising VCR mechanism developed by British Internal Combustion Engine Research Institute (BICERI) is the use of special piston to lower or raise the piston skirt which is shown in Fig. 16.21. It consists of two main pieces (*A* and *B*) called the shell and carrier respectively. The carrier is mounted on a gudgeon pin in the conventional manner while the shell *A* slides over the carrier *B* to vary the clearance volume. The arrangement of these two parts of the piston is made such so that it forms two chambers, *C* and *D*. These two chambers are kept full of lubricating oil supplied via a hole in the connecting rod and non-return valve (*F*) from the lubricating system. The gas load is carried by the oil in chamber *C*. With the increase in load, the gas pressure is increased to a present value, the spring loaded relief valve *L* opens and discharges oil to the main sump. It is the relationship between the oil pressure in two chambers and the cylinder gas pressure which slides down the piston shell to a required position and by this way variable compression is affected. Figure 16.22 shows the variation of gas pressure and compression ratio with load in the engine.

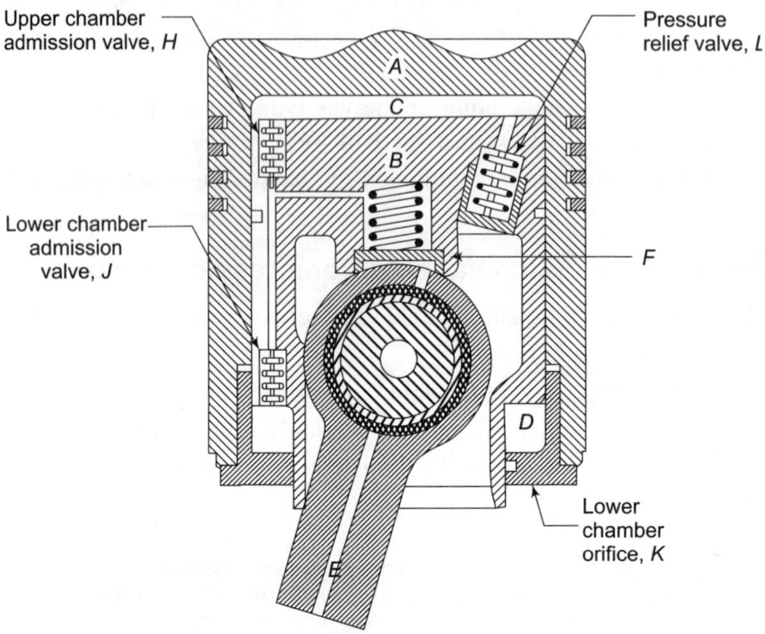


**Fig. 16.21** BICERI variable compression ratio engine

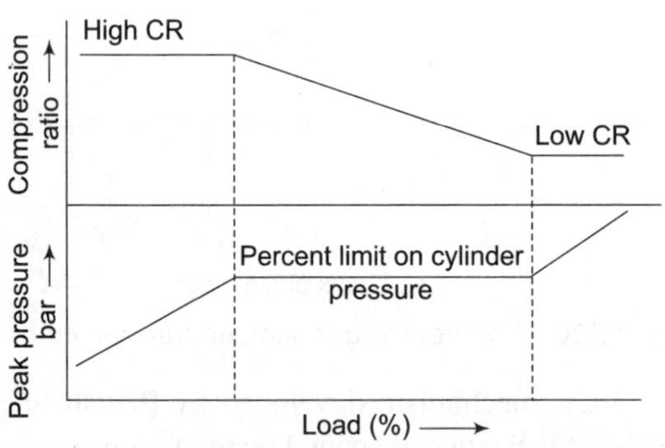


**Fig. 16.22** Variation of gas pressure and compression ratio with load in VCR engine

### 16.4.3 Advantages and Disadvantages of VCR Engine

The following are the parameters on the basis of that we can discuss the advantages and disadvantages of VCR engine.

(a) *Power output and compactness.* VCR engine is very compact and has a high power to weight ratio without any penalty on *sfc*.

(b) *Thermal and Structural Loads.* VCR engine provides smoother combustion at both lower and high compression ratio. It proves the lower thermal and structural loads and a very high specific output.

(c) *Specific Fuel Consumption and Thermal Efficiency.* The specific fuel consumption of VCR engine is slightly higher and thermal efficiency is slower as compared to constant CR engine.

(d) *Engine Noise.* VCR engine by virtue of having a constant peak pressure reduces the low frequency noise.

(e) *Cold Starting and Idling.* Owing to the use of a higher compression ratio at low loads the VCR engine exhibits a good cold starting and idling performance at low ambient temperatures.

(f) *Multifuel Capability.* VCR engine is most suitable to multifuels due to higher compression ratio at starting and part load.

(g) *It is very difficult and complex to obtain variable compression ratio during running and as a result its commercialization has not taken place yet.*

## 16.5 FREE PISTON ENGINE

*Free piston engine is a combination of a reciprocating engine and a rotary turbine in order to achieve the high thermal efficiency with high power/weight ratio.* The exhaust energy from IC engine is appreciable and goes as a waste. It is possible to drive turbine. This idea has culminated as free piston engine which is a highly supercharged diesel cylinder using its exhaust into a gas turbine for producing power, while the whole output of the diesel cylinder is used for compressing the air for its own supercharge. Since no external power is available from the diesel cylinder and as it produces only high pressure gas for a turbine, it is known as a **gasifier** and the whole unit including the turbine is termed as *free-piston engine.*

These are two types of free piston engines. They are:

(i) Inward compression type (ii) Outward compression type

Inward compression gasifiers compress air during the inward motion of the pistons whereas outward compression type compresses air during the outward motion of the pistons. The arrangement of inward type is simple as compared to outward type.

### 16.5.1 Construction and Operation of Free Piston Engine

The schematic diagram of an inward compression type free piston engine is shown in Fig. 16.23. It consists of main two units, namely, a *gasifier* and a *turbine*. The plant may be considered a highly supercharged two-cycle opposed piston diesel engine with hot gases of combustion expanding down to atmospheric pressure in the turbine.

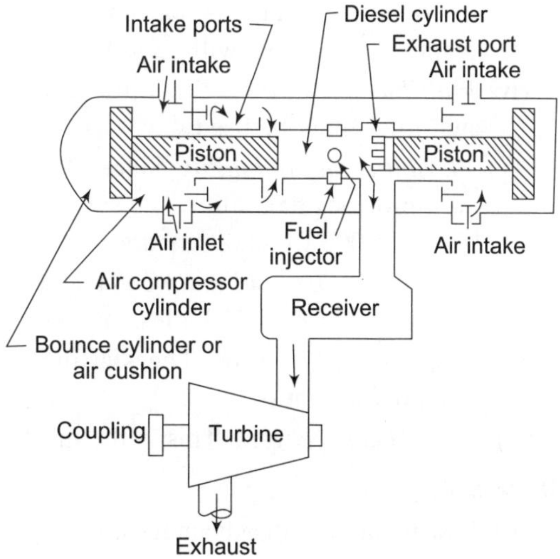


Fig. 16.23 Free piston gas generator

The gas generator consists of a diesel cylinder and combustion chamber in the centre. Fuel is injected in the centre of the diesel cylinder and is burnt and the expansion of the gases of combustion forces the piston outward or apart. Each of the piston is made integral with the other piston which acts as a single stage air compressor on its inner face and acts as an air bounce cushion on its outer face. The energy stored in the bounce or air cushion cylinder due to the outward movement of the pistons is utilised to drive the pistons inward compressing the air in both the air cylinder and diesel cylinder. Air at about 3 to 6.5 bar passes through the valves from the compressor cylinder into a central air space from where it enters the diesel cylinder through intake port on the left. The air scavenges and supercharges the diesel cylinder. The mixture of scavenging air and gases of combustion at a temperature around 340°C pass through the exhaust ports and are expanded down to atmospheric pressure in the turbine which produces power.

## 16.5.2 Advantages and Disadvantages of Free Piston Engines

#### Advantages

The following are the advantages of free piston engines:

(i) *Simplicity.* Due to the absence of crankshaft and connecting rod, the engine becomes more simple mechanically.

(ii) *Power to weight ratio.* The power to weight ratio of free piston engine is surpassed only by simple gas turbine.

(iii) *Lower turbine operating temperature.* The operating temperature for the maximum efficiency and power output is around 400°C to 500°C which is much lower as compared to conventional gas turbine in which the operating temperature is 1000° to 1400°C. This is due to the fact that this turbine has not to supply work input to compressor for compressing air. As a result of lower operating temperature, the need of cooling the blade is not required. The cheap blade materials may be used and the manufacturing cost will be less.

(iv) *Multifuel capability.* Since the compression ratio varies from about 25 to 50, so multifuels may be used.

(v) *Flexibility and reliability.* Since the gasifier and turbine are not mechanically coupled, so it provides flexibility. The system may work as compressor alone. The plant life is about 25,000 to 30,000 hours.

(vi) *Vibration noise and maintenance.* The better balancing of the engine produces less vibration and noise. The less number of bearings require less maintenance.

(vii) *Starting and control.* The starting of free piston engine is easy due to variable compression ratio.

(viii) *Waste heat recovery.* This is the most suitable for waste heat recovery.

#### Disadvantages

The following are the disadvantages:

(i) *Poor fuel economy.* The engine results in higher brake specific fuel consumption as compared to diesel engine.

(ii) *Stability.* The stable operation is possible due to requirement of fuel as per load.

(iii) *Part load efficiency.* The part load efficiency of the engine is poor as the compression ratio is decided by the combustion pressure inside the cylinder.

(iv) *High combustion rate.* The combustion rate per cylinder is 4 to 6 times those encountered in conventional diesel engines. This necessitates more wall lubrication, ring wear and cooling problems.

(v) *Reduction gearing.* The high speed turbine (3000 rpm to 9000 rpm) is to be geared with say marine propeller (100 rpm to 300 rpm). This needs a large reduction gearing ratio.

### 16.5.3 Applications of Free Piston Engines

The following are the applications of free piston engines:

(i) Submarine air compressor

(ii) Medium power range power units

(iii) Ship propulsions

(iv) Driver of high speed centrifugal compressor and water pumps

(v) Mixed gas-steam cycle

## 16.6 DUAL-FUEL ENGINES

*Dual-fuel engines are those engines which run, on two fuels, a liquid fuel (dual or pilot) and a gaseous fuel (primary) or two different liquid fuels.* Portable generating sets of Honda and Birla Yamaha are two-stroke SI engine which run on dual-fuel gasoline and kerosene combination. The engine is started with gasoline (petrol) fuel and after warm-up period, the engines switch over to kerosene for driving the generator. Dual fuel engines are gaining popularity due to rising prices of petroleum products along with easily available and cheaper gaseous fuels like natural gas, sewage gas, gobar gas, etc.

**Advantages.** The following are the advantages of dual-fuel engines:

(i) Dual-fuel engines using gaseous fuels are internally cleaner with longer life of lubricating oil.

(ii) Only a fraction of costly fuel (dual fuel) is needed.

(iii) The performance of dual-fuel engines is better than diesel at full load.

(iv) It proves more economical, if sewage gas is used as primary fuel.

(v) It can generate synthetic gas provided methane is burnt inside the engine.

Dual-fuel engines using gaseous fuel as main fuel proves economical over liquid fuel due to higher self-ignition temperature and higher compression ratio requirement for smooth operation.

### 16.6.1 Working Principle

Figure 16.24 shows the dual-fuel control systems arrangements in a four-stroke diesel engine. During suction stroke, air at a pressure, a little above the atmospheric pressure is inducted into the cylinder. Once the suction stroke is a underway, the gas valve opens and the gas is metered into the relatively high velocity air steam to form a homogeneous mixture. The amount of gas going in it is determined by the pressure existing before the valve which is controlled by the governor.

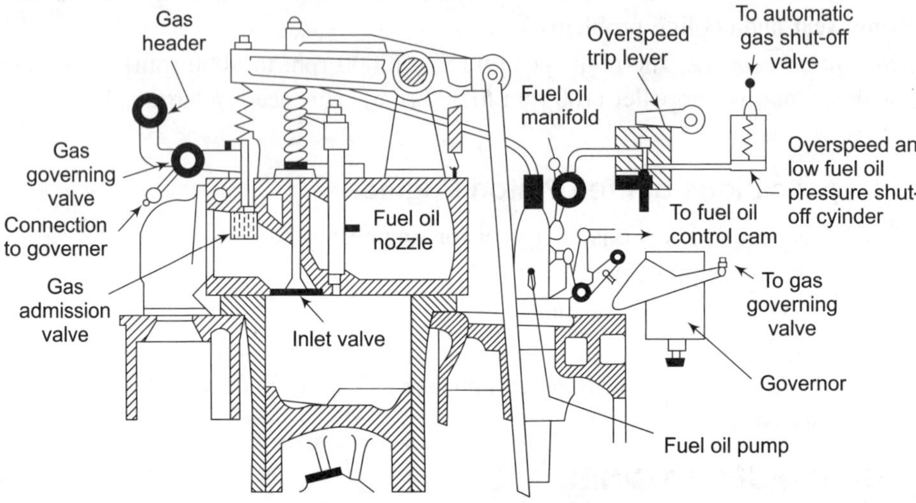


**Fig. 16.24** A dual-fuel engine

The homogeneous and lean mixture is compressed during compression stroke and near the end of top a dead centre, a small amount of liquid fuel called dual fuels is injected through the conventional diesel fuel injection system. The pilot fuel injection acts as a source of ignition. The gas-air mixture in the vicinity of injected spray ignites at a number of spots and a number of flame fronts are established. The combustion gets underway smoothly.

It is worth to note that in a 4-stroke engine, the gas is supplied in the inlet manifold where it mixes with incoming air to form a homogenous mixture. However, in a two-stroke engine, the gas may be injected at low pressure (1.5 to 3.5 bar) directly inside the cylinder after piston covers the exhaust ports.

The output of the engine is controlled by changing the amount of gaseous fuel (primary fuel). The quantity of pilot fuel usually remains constant for a given engine and 10% of engine output at full load is met by pilot fuel. Figure 16.25 shows that *p-t* diagram for dual fuel engine using different fuels.

### 16.6.2 Performance

Figure 16.26 shows the comparison of *sfc* of diesel and dual-fuel engines. It is evident from the curves that the *sfc* (i.e., efficiency) at part load is slightly poor than diesel engine but at part load, dual-fuel performance is better. This is because of increased delay period at low mixture strengths, *Experimental results reveal that the efficiency of dual engines are 2 percentage higher than diesel engine at full load under the same operating conditions.*

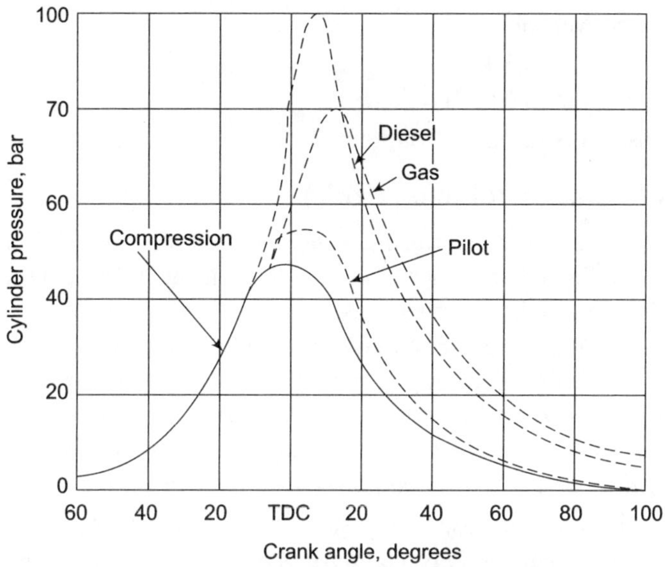


**Fig. 16.25** *p-t* diagram for a dual-fuel engine working on different fuels

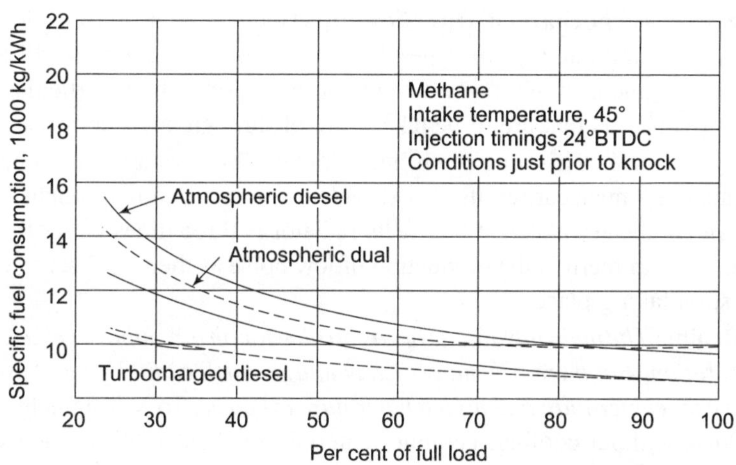


**Fig. 16.26** Comparison of performance of dual-fuel and diesel engines at varying loads

## 16.7 ELECTRIC VEHICLES

Due to emission from IC engines using hydrocarbon as fuels, the development of eco-friendly engines such as electrical vehicles have taken accelerated pace. They have a great impact on energy, environment and transportation and on high technology promotions, new industry creation and economic development worldwide. Currently, the following two types of electric vehicles are in development stage, *now developed and plying on roads*.

(a) Hybrid Electric Vehicles (HEV)
(b) Fuel Cell Vehicles (FCV)
(c) Pure Electric Vehicles (PEV)

The detail discussion is given in **Chapter 22**.

### 16.7.1 Hybrid Electric Vehicles (HEV)

*This falls under the category of intermediate step between the conventional IC engines and electric vehicles.* In India, the brake power of cars varies from 20 kW to 60 kW but the power required during the city driving is much less, around 5 kW, due to lower acceleration rates in Indian traffic conditions. In advanced countries also, acceleration limit in city driving is less. At such part loads, the efficiency of the engine is very poor. *A hybrid vehicle is one which permits the use of a smaller engine running continuously at a fixed load and speed near maximum efficiency and an electric drive consisting of generator-storage system-traction motor, which provides varying loads and power requirements on the road.*

The possible engines for the base load system for hybrid vehicles are the following.

**(i) MPFI SI Engine with 2-Way Catalytic Convertor.** The multi-point fuel injection (MPFI) SI engine with catalytic convertor has become a standard engine for passenger cars all over the world as this engine meets the most stringent emission standards.

**(ii) High Speed Turbocharged Direction-injection Diesel Engines.** Now, high speed turbocharged direction injection diesel engine is competing with MPFI SI engines for HEV. This engine is highly fuel efficient but formation of high particulate and $NO_x$ are the major concerns that are to be taken care of through advance research activity.

**(iii) Stirling Engine.** This engine is still under development though, it has been conceptualised much earlier. Its closed cycle operation requires combustor, regenerator cooler and preheater makes it bulky. Its radiator is three times larger than that of a SI engine. Its main merit is the production of low noise as there is absence of intermittent combustion taking place.

**(iv) Gas Turbine:** *The main advantage of gas turbine for an automotive vehicles is vibration-free operation. But there are many disadvantages such as high fuel consumption, high turbine exit temperature and high temperature exhausts.* To withstand high temperature of 1600 K without cooling, ceramic materials for blades are needed which is very costly affair.

Table 16.1 shows some of important features of these engines.

**Table 16.1** Important Features of Base Load System for Hybrid Engines

| *Prime mover* | *Specific power (W/kg)* | *Brake thermal efficiency (%)* | *Relative emission* |
|---|---|---|---|
| SI engine | | | Low |
| 4-stroke | 650 | 35 | |
| 2-stroke (arbital) | 875 | | |
| DI diesel engine | 525 | 42.5 | High PM, $NO_x$ |

| | | | |
|---|---|---|---|
| Stirling engine (1990) | 800 | 36.5 | Low |
| Gas turbine for automotive application | 500 | 32 | High $NO_x$ |

Over and above the prime mover, requirement *hybrid electric vehicle requires good energy storage devices like batteries of high specific power, flywheel hybridized with battery and ultra-capacity hybridized with battery. Batteries may be of lead-acid, nickel-cadimum, or lithium-ion nickel metal hybrid type.* As the batteries have limited power, so it is desirable to develop advanced type of flywheel hybridized with battery. *Flywheels made of reinforced composite fibre are being developed and they are run in an airtight vessel to reduce friction.* They can store the kinetic energy of the decelerating vehicles and can give back to vehicles during acceleration. The battery may charge the ultra capacitor when the power drawn from the battery is low and during acceleration, support is given by battery.

The US government has entrusted some leading industries to develop a super car under the programme **"Partnership for New Generation Vehicles" (PNGV)** which can give 34 km per litre. ***And an ideal PNGV needs an engine with best fuel economy of diesel, low noise level of stirling engine, vibration free operation of gas turbine and low emission level of a gasoline engine.***

Figure 16.27 shows the comparison of gasoline equivalent fuel economy of current production vehicles with hybrid vehicles under development.

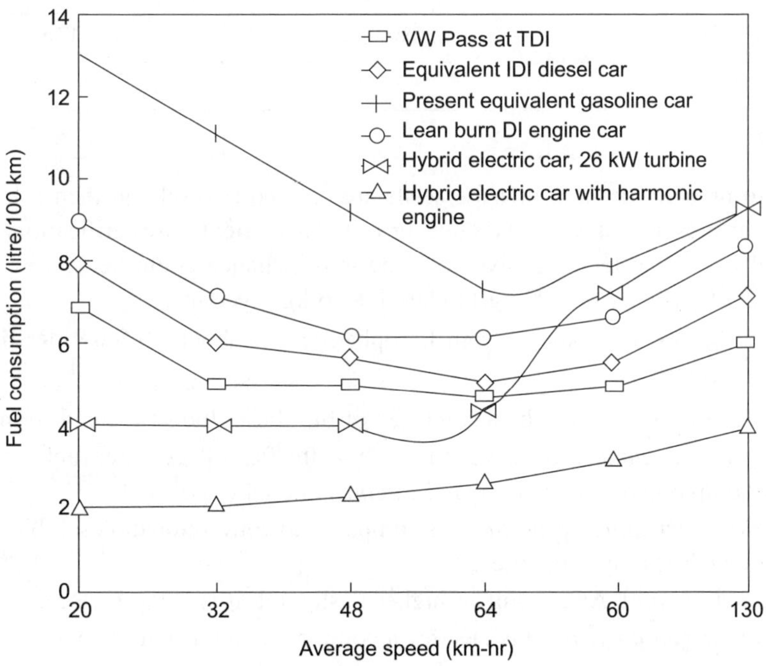


**Fig. 16.27** Comparison of fuel economy of current production vehicles with hybrid vehicles under development

## 16.8 BIODIESEL POWERED ENGINES

The quest for alternative fuels has drawn the attention of researchers towards ***Jatropha curcus*** plants which is now grown worldwide to get biodiesel. The seeds of Jatropha plant contain viscous oils (50% by weight). The raw oil upon transesterification can be used as a fuel in diesel engine without any modification either in design on accessories of engine. It is worth to note that properties of Jatroha oil are similar to the biodiesel extracted from rapseed oil which is widely used in Europe.

Table 16.2 depicts the properties of biodiesel fuel extracted from the seeds of Jatropha plants.

**Table 16.2** Properties of Biodiesel from Jatropha Plants

| *Parameter* | *Jatropha oil (raw)* | *Jatropha oil (trans-esterificated* |
|---|---|---|
| Density (gm/cm$^3$) at 20°C | 0.92 | 0.879 |
| Flashpoint (°C) | 236 | 191 |
| Cetane no. (Pso 5165) | 23 to 41 | 51 |
| Viscosity (mm$^2$/s) at 30°C | 52 | 4.84 |
| Total glycerine (%) | – | 0.088 |
| Free glycerine (%) | – | 0.015 |
| Phosphorus (PPM) | 290 | 17.5 |
| Sulphated ash (%) | – | 0.014 |
| Methanol (%) | – | 0.06 |

In order to prepare the fuel, first the seeds are pressed into oil and then it is mixed with methanol in the presence of catalyst (NaOH or KOH) and after transesterification, the products are biodiesel (methyl ester) and glycerol. The mass balance could be 100 kg oil + 24 kg methanol + 2.5 kg NaOH $\rightarrow$ 100 kg biodiesel + 26 kg glycerine.

At present, the cost of biodiesel from Jatropha seeds is about **US $ 0.9 per litre** which is supposed to be costly.

**Advantages.** The following are the advantages of bio-diesel fuels obtained from Jatropha:

(i) *Higher cetane number.* It has been found that the fuel offers better fuel economy, better power output and torque compared to conventional diesel fuel.

(ii) *Superior lubricating properties.* As compared to conventional diesel, bio-diesel offers superior lubricating properties.

(iii) *Higher flashpoint.* As a result of higher flashpoint, it is safer to store.

(iv) *Higher oxygen content.* This ensures a complete combustion of fuel.

(v) *Better exhaust emission quality.* As a result of this, the fuel is capable to meet Euro III norms.

**Disadvantages.** The following are the disadvantages of bio-diesel fuels.

(i) High cost of production.

(ii) Availability on a commercial scale is low.

(iii) There may be food scarcity as more and more land will be engaged in growing bio-diesel plants which will result in high food prices.

## PROBLEMS

### Things to Think About

16.1 List the limitations of petrol, diesel and gas engines working world over at present and pinpoint these aspects on which now engine may be developed.

16.2 Why the part load performance of an automobile engine is very important consideration?

16.3 What are the reasons that the following research engines have not found for the commercial use?

(a) Stratified charge engine

(b) Stirling engine

(c) Wankel rotary combustion engines

(d) Variable compression ratio engine

(e) Free piston engines

### Short and Long Theoretical Questions

16.4 What do you mean by stratified charge engine?

16.5 What are the important methods of charge stratification?

16.6 With neat sketches, discuss the construction and working principles of various stratified engines.

16.7 With the help of a sketch, discuss the working principle of Stirling engine.

16.8 Discuss the advantages of Stirling engine over conventional type

16.9 With the help of sketch, discuss the working principle of Wankel engine.

16.10 Discuss the geometry of Wankel engine.

16.11 Discuss in brief about auxiliary system of Wankel engine.

16.12 What are the methods to obtain variable compression in IC engines?

16.13 With the help of a sketch, discuss the working principle of a VCR engine.

16.14 Discuss the working principle of hybrid fuel-cell engines.

16.15 With the help of a sketch, discuss the working principle of free piston engine.

16.16 Discuss the working principle of electric vehicle engines.

16.17 Discuss the advantages and disadvantages of free piston engine.

### Software Based Question

16.18 Develop a software for the design of following developing engines as mentioned in question 16.3.

### Design and Open Ended Questions

16.19 Prepare a comprehensive report on any promising developing engines with design features, and limitation.

16.20 Give a totally new concept of engine which will be highly efficient, eco-friendly, economical and reliable for automotive use, which can run on road, run under/over water and fly in the sky.

### Multiple-Choice Questions

16.21 A Stirling engine

(1) can use any heat source
(2) is perfectly balanced and noise free
(3) cannot be safely overloaded even for a brief period
(4) requires very less quantity of lubricating oil

Out of these statements

(a) 1, 2 and 4 are correct (b) 1, 2 and 3 are correct
(c) only 1 and 2 are correct (d) all are correct

16.22 Knock in dual fuel engine can be controlled by

(a) excess air supply (b) increased cooling of piston
(c) reduced pressure of the gaseous fuel (d) the use of cold combustion air
(e) any of the above

16.23 Wankel rotary engine has

(a) higher weight/power ratio (b) lower specific fuel consumption
(c) lower rpm (d) higher volumetric efficiency

16.24 **Assumption (A):** Hybrid electric vehicles do not require a prime mover

**Reasoning (A):** Because they run on fuel cells

**Answer Code:**

(a) Both A and R are false (b) Both A and R are true
(c) A is false but R is true (d) A is true but R is false

## Key to Multiple-Choice Questions

16.21 (a) 16.22 (c) 16.23 (d) 16.24 (a)

CHAPTER

# Modelling of IC Engines 17

## OVERVIEW

*By modelling in engineering problems, we mean developing and using the appropriate combination of assumptions and equations that permit critical features of the process to be analyzed.* Modelling activities can make major contributions to engine engineering at different levels of generality corresponding to different steps of model development. The following are the modelling activities of IC engines.

(i) To develop a more complete understanding of the process under study from the discipline of formulating the model.

(ii) To identify key controlling variables to provide guidelines for more rational and therefore less costly experimental development efforts of engines.

(iii) To predict engine behaviour over a wide range of design and operating variables to screen concepts prior to major hardware programs, to determine trends and tradeoff and, if the model is sufficiently accurate, to optimize design and control engines.

(iv) To provide a rational basis of design innovation of engines.

It is worth to note that each of these contributions is valuable. *It is the accuracy of the model, its testing and validation and time and efforts required to use the models for extensive sets of calculations and to interpret the results which ensure whether a model is ready to pass from one stage to the next.*

There are two basic types of models for predicting performance and emissions.

(i) Thermodynamic model

(ii) Fluid dynamic model (or multi-dimensional models)

At present, it is not possible to construct models that predict engine operation from basic governing equations alone. On account of the complexity of engine processes and our inadequate understanding at a fundamental level, most engine models are incomplete.

*Empirical relations and ad-hoc approximations are often needed to bridge gaps in our models.*

This chapter deals with the types of models and their primary components that are being developed and used to describe engine operating and emissions characteristics.

Though, a large number of literature is available on engine simulation, only some of them are given here in References [53 to 70] which may be consulted for more details.

## 17.1 GOVERNING EQUATIONS FOR OPEN THERMODYNAMIC SYSTEMS

While modelling the engine components, it is often required to model a region of the engine as an open thermodynamic system such as cylinder volume and intake and exhaust manifolds. Such an open system is shown in Fig. 17.1. *The important equations are mass and energy conservation which are supposed to be building blocks for thermodynamic based models.*

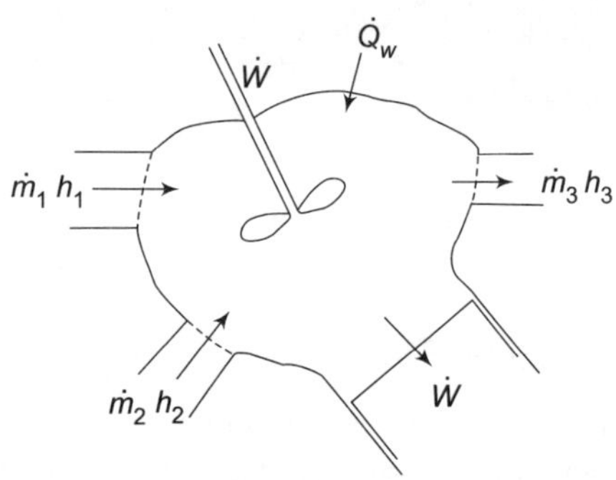


**Fig. 17.1** Open thermodynamic system

### 17.1.1 Conservation of Mass

Rate of change of the total mass $\dot{m}$ of an open system is equal to the sum of the mass flows into and out of the system. Thus,

$$\dot{m} = \sum_j \dot{m}_j \qquad ...(17.1)$$

Here, mass flows into the system are taken positive and mass flows out are taken as negative.

The conservation of fuel chemical element is given by

$$\dot{m}_f = \frac{d}{dt}(m.f) = \sum_j \dot{m}_{f,j} = \sum_j \dot{m}_j.f_j \qquad ...(17.2)$$

where $f$ = fuel fraction = $m_f/m$

$m_f$ = mass of fuel

Differentiating Eq. (17.2) yields the rate of charge of fuel fraction which is expressed as

$$\dot{f} = \sum_j \left(\frac{\dot{m}_f}{m}\right)(f_j - f) \qquad ...(17.3)$$

The fuel/air equivalence ratio (φ) and its rate are expressed as

$$\varphi = \frac{1}{(F/A)_s}\cdot\frac{f}{(1-f)} \text{ and } \dot{\phi} = \frac{1}{(F/A)_s}\cdot\frac{\dot{f}}{(1-f)^2} \qquad ...(17.4)$$

The suffix *s* shows stoichiometric condition.

### 17.1.2 Conservation of Energy

The first law of thermodynamics for the open system (Fig. 17.1) is expressed as

$$\dot{E} = \dot{Q}_w - \dot{W} + \sum_j \dot{m}_j h_j \qquad ...(17.5)$$

There are two approaches to define the rate of change of state of the open system depending upon internal energy ($u$) or enthalpy ($h$) in the term of $\dot{T}$ and $\dot{p}$. Thus, total energy state ($\dot{E}$) is expressed as

$$\dot{E} = \frac{d}{dt}(mu) \qquad ...(17.6a)$$

or

$$\dot{E} = \frac{d}{dt}(mh) - \frac{d}{dt}(pV) \qquad ...(17.6b)$$

We know that

$$u = u(T, p, \varphi);\ h = (T, p, \varphi);\ p = p(T, p, \varphi) \qquad ...(17.7)$$

The rate of change of $u$, $h$ and $p$ can be written in the form of

$$\dot{\alpha} = \left(\frac{\partial \alpha}{\partial T}\right)\dot{T} + \left(\frac{\partial \alpha}{\partial p}\right)\dot{p} + \left(\frac{\partial \alpha}{\partial \varphi}\right)\dot{\varphi} \qquad ...(17.8)$$

where $\alpha$ represented $u$, $h$ or $p$

Using the ideal gas law $p = \rho RT$ and $pV = mRT$, equation for $\dot{p}$ can be expressed as

$$\dot{p} = \frac{\rho}{\partial p/\partial \rho}\left[-\frac{\dot{V}}{V} - \frac{1}{\rho}\frac{\partial \rho}{\partial T}\dot{T} - \frac{1}{\rho}\frac{\partial \rho}{\partial \varphi}\dot{\varphi} + \frac{\dot{m}}{m}\right] \qquad ...(17.9)$$

The equation for $T$ can be expressed as

$$\dot{T} = \frac{\left[B - \frac{p}{D}\frac{\partial u}{\partial p}\left(\frac{\dot{m}}{m} - \frac{\dot{V}}{V} + \frac{\partial R}{\partial \varphi}\frac{\dot{\varphi}}{R}\right) - \frac{\partial u}{\partial \varphi}\dot{\varphi}\right]}{\left[\frac{\partial u}{\partial T} + \frac{C}{D}\frac{p}{T}\frac{\partial u}{\partial p}\right]} \qquad ...(17.10)$$

where

$$B = -RT\frac{\dot{V}}{V} + \frac{1}{m}\left(\dot{Q}_w + \sum_j \dot{m}_j h_j - \dot{m}u.\right)$$

$$C = 1 + \frac{T}{R}\frac{\partial R}{\partial T}$$

$$D = 1 - \frac{p}{R}\frac{\partial R}{\partial p}$$

The cylinder volume can be expressed by

$$\frac{V}{V_c} = 1 + \frac{1}{2}(r - 1)\left[R_{la} + 1 - \cos\theta - (R_{la}^2 - \sin^2\theta)^{1/2}\right] \quad ...(17.11)$$

where $R_{la}$ = ratio of connecting rod length (1) to crank radius ($a$) = $\frac{l}{a}$

$V_c$ = clearance ratio

$r$ = comprehension ratio

$\theta$ = crank angle

The specific heat is expressed in the form of polynomial and as a function of $T$

$$\frac{c_p}{R} = a_{11} + a_{12}\,T + a_{13}\,T^2 + a_{14}\,T^3 + a_{15}\,T^4 \quad ...(17.12)$$

The enthalpy is expressed as

$$h = \int c_p\,dT \quad ...(17.13)$$

Equation (17.1), (17.3), (17.4), (17.9) and (17.10) can now be solved to obtain the state of open system as a function of time with the help of Eqs. (17.11 to 17.13).

## 17.2 INTAKE AND EXHAUST FLOW MODELS

The air flow into the engine's cylinders is governed by the behaviour of the intake and exhaust systems. Inducting the maximum air flow at full load at any given speed and retaining that mass within the engine's cylinders is a primary design goal as higher the air flow greater is the power produced.

Three types of models for intake and exhaust flow are given below:

(a) Quasi-steady models for flows through the restrictions provided by valve and port.

(b) Filing and emptying models to take care the finite volume of critical manifolds components.

(c) Gas dynamic models for finding the spatial variations in flow and pressure throughout the manifolds.

### 17.2.1 Quasi-Steady Flow Models

The intake system consisting of air cleaner, throttle manifold, port and valve may be considered a series of interconnected components which each constitutes a significant flow reatriction.

The gas flow rate through each component is calculated using steady one-dimensional flow equation given below. At throat, the mass flow rate is

$$\dot{m}_{real} = \frac{C_D A_T p_0}{\sqrt{R T_0}} \left(\frac{p_T}{p_0}\right)^{1/\gamma} \left[\frac{2\gamma}{\gamma - 1}\{1 - (p_T/p_0)^{\gamma - 1/\gamma}\}\right]^{1/2} \quad ...(17.14)$$

For a choked flow

$$\dot{m}_{real} = \frac{C_D A_T p_0}{\sqrt{R T_0}} \gamma^{1/2} \left(\frac{2}{\gamma + 1}\right)^{(\gamma+1)/2(\gamma-1)} \quad ...(17.15)$$

Quasi-steady models are often used to calculate the flow into and out of the cylinder through the inlet and exhaust valves. This has been discussed in the chapter of "Intake and Exhaust Systems".

### 17.2.2 Filling and Emptying Models

In this model, the manifolds are represented by finite volumes where the mass of gas can increase or decrease with time. The models treat the whole intake or exhaust system as a single volume or division of these systems into many sections with a flow restrictions such as the air cleaner, throttle valve on inlet valve. Each volume is treated as a control volume which contains gas at a uniform state.

The mass and energy conservation equations given by Eqs. 17.1 to 17.11 coupled with other information given by Eqs. 17.14 and 17.15 may be used to define the gas state in each volume. For intake and exhaust flows these equations can be simplified since the volume are fixed ($\dot{V} = 0$).

A simple filling and emptying model is based on the assumption that at any given time the manifold pressure is uniform. The continuity equation for air flow into and out of the intake manifold is

$$\frac{dm_{a,m}}{dt} = \dot{m}_{a,th} - \Sigma \dot{m}_{a,cyl} \qquad \text{...(17.16)}$$

where $m_{a,m}$ = mass of the air in the manifold

$m_{a,th}$ = mass of the air past the throttle

$m_{a,cyl}$ = mass of the air in the cylinder

The pressure in the manifold is given by ideal gas equation

$$\frac{dp_m}{dt} + \frac{\eta_v V_d N}{2V_m} p_m = \dot{m}_{a,th} \cdot \frac{R.T_m}{V_m} \qquad \text{...(17.17)}$$

The predicted values agree well with the experimental results.

### 17.2.3 Gas Dynamic Models

Gas dynamic models use the mass, momentum and energy conservation equations for the unsteady compressible flow in the intake and exhaust systems. Normally, the one-dimensional unsteady flow equations are used. *Finite difference or finite element techniques are used to solve these equations.*

#### 17.2.3.1 Unsteady Flow Equations

Consider a control volume for unsteady one-dimensional flow within a straight duct shown in Fig. 17.2. It is assumed that the area change over the length $dx$ of the control volume is small so the flow is essentially one-dimensional.

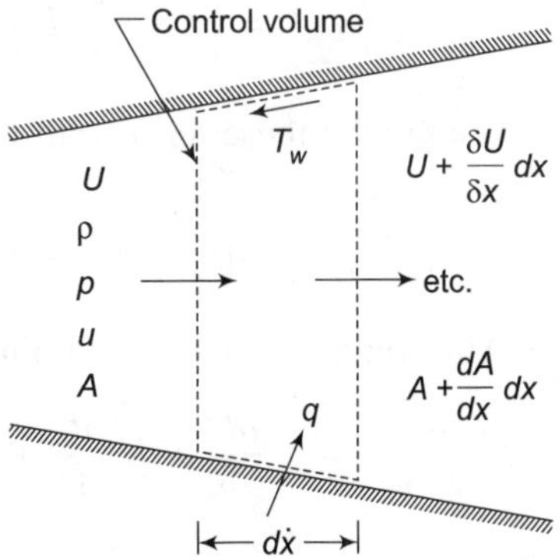


**Fig. 17.2** Control volume for unsteady one-dimensional flow analysis

As per mass conservation, the rate of change of mass within the control volume is equal to the net flow into the control volume. Thus,

$$\frac{\partial}{\partial t}(\rho A\, dx) = \rho A U - \left[\rho A U + \frac{\partial}{dx}(\rho A U)\, dx\right] \qquad ...(17.18)$$

where $U$ = velocity of fluid

After rearrangement and retaining only finite order quantities, we have

$$\frac{\partial \rho}{\partial t} + \frac{\partial}{\partial x}(\rho U) + \frac{\rho U}{A}\frac{dA}{dx} = 0 \qquad ...(17.19)$$

According to momentum conservation equation, the net pressure forces plus the wall shear force acting on the control volume is equal to the rate of change of momentum within the control volume plus the net flow of momentum out of the control volume.

Thus, the net forces and momentum changes are given by:

***Pressure forces:*** $pA - \left(p + \frac{\partial p}{dx}dx\right)\left(A + \frac{dA}{dx}dx\right) + p\frac{dA}{dx}dx = -A\frac{\partial p}{dx}dx$

***Shear forces:*** $-\tau_w \pi D\, dx = -\varepsilon \frac{\rho U^2}{2}\pi D dx$

where $D$ = equivalent diameter = $(4A/\pi)^{1/2}$

$\xi$ = friction coefficient = $\tau_w/[(1/2)\rho U^2]$

The rate of change of moratorium within the control volume is given by

$$= \frac{\partial}{\partial t}(U\rho A\, dx)$$

and the net efflux of momentum within the control volume is given by

$$= \left(\rho + \frac{\partial \rho}{\partial x}dx\right)\left(U + \frac{\partial U}{dx}dx\right)^2\left(A + \frac{dA}{dx}dx\right) - \rho U^2 A = \frac{\partial}{\partial x}(\rho U^2 A)dx$$

Combining these terms into the momentum equation yields

$$-A\left(\frac{\partial p}{\partial x}\right)dx - \xi\frac{\rho U^2}{2}\pi D dx = \frac{\partial}{\partial t}(\rho U A\, dx) + \frac{\partial}{\partial x}(\rho U^2 A)dx \qquad ...(17.20)$$

After rearranging and combining with Eq. (17.19) yields

$$\frac{\partial U}{\partial t} + U\frac{\partial U}{\partial x} + \frac{1}{\rho}\frac{\partial p}{\partial x} + 2\xi\frac{U^2}{D} = 0 \qquad ...(17.21)$$

#### *17.2.3.2 Energy Conservation*

According to the first law of thermodynamics for a control volume, the energy within control volume changes due to heat and shear work transfers across the control volume surface and due to a net efflux of stagnation enthalpy resulting from flow across the control volume surface. The stagnation enthalpy is

$$h_0 = h + \frac{U^2}{2} = u + \frac{p}{\rho} + \frac{U^2}{2}$$

The heat transfer rate $\dot{Q}_W$ is given by

$$\delta \dot{Q}_W = \dot{q}\, \rho\, A\, dx$$

where $\dot{q}$ = heat transfer per unit mass per unit time into the control volume.

The rate of change of energy within the control volume is given by

$$\frac{\partial E}{\partial t} = \frac{\partial}{\partial t}\left[(\rho\, A\, dx)\left(u + \frac{U^2}{2}\right)\right]$$

The net efflux of stagnation enthalpy is given by

$$\frac{\partial}{\partial x}\left[(\rho\, UA)\left(u + \frac{p}{\rho} + \frac{U^2}{2}\right)\right]dx$$

After rearranging, equation for energy conservation becomes

$$\frac{\partial}{\partial t}\left[(\rho A\, dx)\left(u + \frac{p}{\rho} + \frac{U^2}{2}\right)\right] + \frac{\partial}{\partial x}\left[(\rho A\, U)\left(u + \frac{p}{\rho} + \frac{U^2}{2}\right)\right]dx - \dot{q}\rho A\, dx = 0 \quad ...(17.22)$$

Expanding the above and using the mass and momentum conservation equation yields

$$\frac{\partial u}{\partial t} + U\frac{\partial u}{\partial x} = \dot{q} + 2\xi\,\frac{U^3}{D} - \frac{p}{\rho A}\,\frac{\partial\,(UA)}{\partial x} \quad ...(17.23)$$

Since $u = c_v T$ and $R/c_v = \gamma - 1$, hence Eq. (17.23) takes the form

$$\frac{\partial p}{\partial t} + U\frac{\partial p}{\partial x} - a^2\left(\frac{\partial \rho}{\partial t} + U\frac{\partial \rho}{\partial x}\right) - (\gamma - 1)\,\rho\left(\dot{q} + 2\xi\frac{U^3}{D}\right) = 0 \quad ...(17.24)$$

where $a^2 = (\text{sound speed})^2 = \left(\frac{\partial p}{\partial \rho}\right)_s = \gamma\frac{p}{\rho}$

The above one-dimensional unsteady flow equations could be solved by finite difference or finite element methods.

#### *17.2.3.3 Finite Difference Methods*

The conservation Eqs. (17.19), (17.20) and (17.22) can be rearranged and written in matrix form as

$$\frac{\partial}{\partial t}\begin{pmatrix}\rho \\ \rho U \\ \rho u\end{pmatrix} + \frac{\partial}{\partial x}\begin{pmatrix}\rho U \\ \rho U^2 + p \\ \rho U + \rho U u\end{pmatrix} = \begin{bmatrix} -\rho U \dfrac{dA}{dx} \\ -\rho \dfrac{U^2}{A}\dfrac{dA}{dx} - \rho \dfrac{2\xi U|U|}{D} \\ -\dfrac{4h_c(T - T_w)}{D\rho} - \dfrac{1}{A}\dfrac{dA}{dx}\left(\dfrac{1}{2}\rho U^3 + \dfrac{\gamma}{\gamma - 1} Up\right)\end{bmatrix}$$

These equations have the vector form

$$\frac{\partial F}{\partial t} + \frac{\partial G}{\partial x} = H \qquad ...(17.25)$$

where $G$ and $H$ are functions of $F$ only

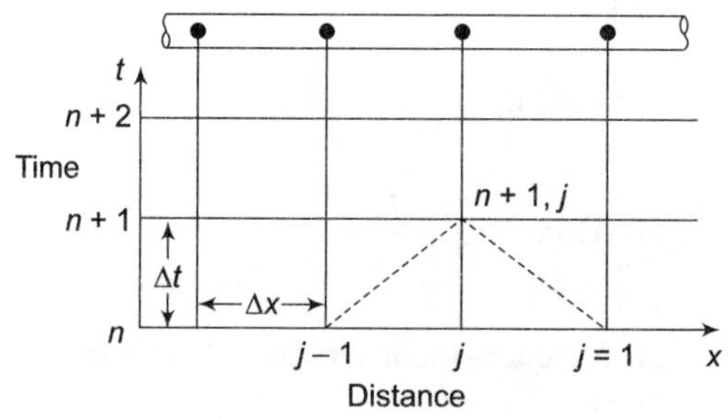


**Fig. 17.3** Lax–Wendroff finite difference method

Several finite difference methods have been used to solve Eq. (17.25). Figure 17.3 shows one-step Lax–Wendroff method which is discussed below.

$$F_j^{n+1} = F_j^n - \frac{1}{2}\frac{\Delta t}{\Delta x}\left(G_{j+1}^n - G_{j-1}^n\right) + \Delta t\, H_j^n$$

$$+\frac{1}{4}\left(\frac{\Delta t}{\Delta x}\right)^2\left[\left(G_{j+1}^{in} + G_j^{in}\right) - \left(G_j^{in} + G_{j-1}^{in}\right)\left(G_j^n - G_{j-1}^{in}\right)\right] \qquad ...(17.26)$$

where $G' = \dfrac{\partial G}{\partial F}$

Equation (17.26) is first order accurate, unless $H$ is small. For stability in the integration process, the time and mesh size must ratify the requirement that

$$C = (|U| + a)\frac{\Delta t}{\Delta x} < 1 \qquad ...(17.27)$$

where $C$ = the Courant number

For the solution, the **boundary conditions** must be satisfied.

Figure 17.4 shows the predicted mass flow rate through the exhaust valve using the above technique.

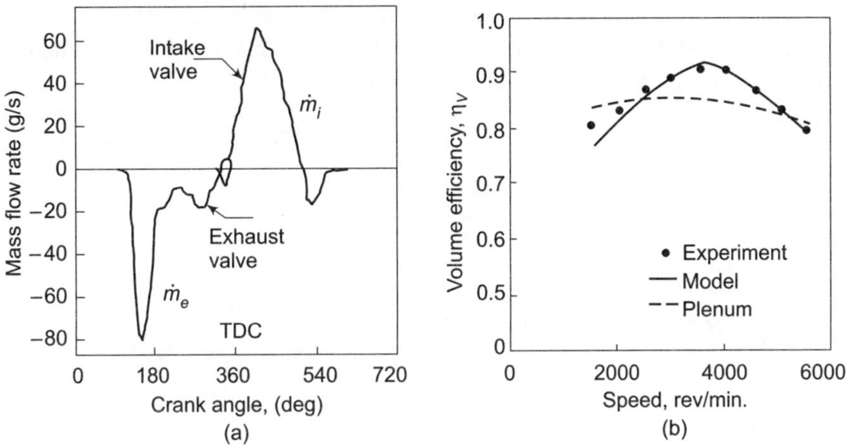


**Fig. 17.4** Predicted mass flow rate through the exhaust valve [103, 73]

## 17.3 THERMODYNAMIC BASED IN-CYLINDER MODELS

### 17.3.1 Overall Model Structure

*Solutions of energy and mass conservation equations permit the cylinder pressure and the work transfer to the piston to be calculated provided the mass transfer into and out of the cylinder during intake and exhaust, the heat transfer between the in-cylinder gases and the cylinder head, piston and cylinder liner and the rate of change of burning are known.* Thermodynamic engine models are used to predict engine operating conditions. These models effectively follow the changing thermodynamic and chemical state of the working fluid through the engine's various processes such as intake, comprehension, combustion, expansion and exhaust processes, they are often called ***engine cycle simulations.***

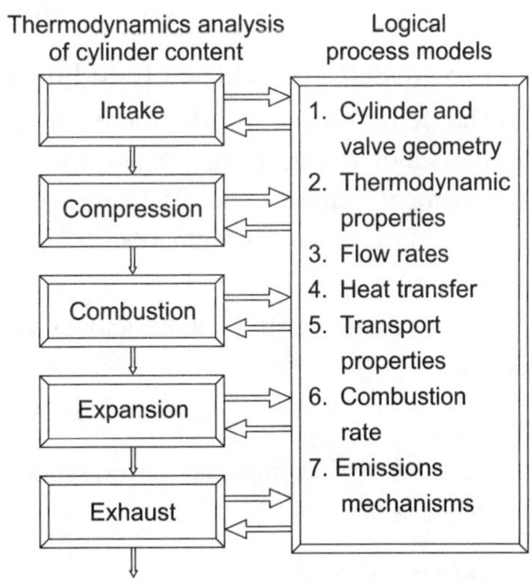


**Fig. 17.5** Structure of engine cycle simulation

*The first step of this engine cycle simulation is to use the first law of thermodynamics for an open system, Eq. 17.5.*

This is applied to the cylinder volume for the intake, compression, combustion, expansion and exhaust processes that in sequence make up the engine's operating cycle. Figure 17.5 shows the structure of this type engine simulation. During each process, submodels are used to describe geometric features of the cylinder and valves or ports, the thermodynamic properties of the unburnt and burnt gases, the mass and energy transfers across the system boundary and the combustion process. Conservation equation is the form of Eqs. (17.1), (17.3) and (17.10) or (17.11) for the intake and then compression process may be used and takes the form as.

***Intake Process***

$$\dot{m} = \dot{m}_i - \dot{m}_e \qquad ...(17.28)$$

$$\dot{f} = \frac{\dot{m}_i}{m}(f_i - f) - \frac{\dot{m}_e}{m}(f_e - f) \qquad ...(17.29)$$

$$\dot{T} = \frac{B}{A}\left[\frac{\dot{m}_i}{\dot{m}}\left(1 - \frac{h}{B}\right) - \frac{\dot{V}}{V} - \frac{C}{B}\varphi + \frac{1}{Bm}(\dot{m}_i h_i - \dot{m}_e h_e - \dot{Q}_w)\right] \qquad ...(17.30)$$

where, $m$ = mass of the gas in the cylinder

$\dot{m}_i$ = mass flow rate through inlet valve

$\dot{m}_e$ = mass flow rate through exhaust valve

$f$ = fuel fraction = $m_f/m$

Subscripts $i$ and $e$ denote properties of flow through intake and exhaust valves.

***Compression Process***

$$\dot{m} = 0 \quad \text{and} \quad \dot{f} = 0 \qquad ...(17.31)$$

$$\dot{T} = \frac{B}{A}\left(-\frac{\dot{V}}{V} - \frac{Q_w}{Bm}\right) \qquad ...(17.32)$$

The pressure is calculated from Eq. (17.9).

The composition and thermodynamic properties can be computed from the relations of properties of working fluid. Mass flows across open valves are usually calculated using one-dimensional compressible flow equation for flow through restriction or filing and emptying models. For more accurate results, unsteady gas dynamic models (Art. 17.2.3) may be used.

Heat transfer during intake and exhaust is calculated using Nusselts–Reynolds equation as

$$\text{Nu} = 0.0483\, Re^{0.783} \qquad ...(17.33)$$

For combustion process, various models have been developed, some of them are given in subsequent articles.

***Exhaust Process***

$$\dot{m} = -\dot{m} \quad \text{and} \quad \dot{f} = -\frac{\dot{m}_e}{m}(f_e - f) \qquad ...(17.34)$$

$$\dot{T} = \frac{B}{A}\left[-\frac{\dot{m}_e}{m}\left(1 - \frac{h}{B} - \frac{\dot{V}}{V} - \frac{C}{B}\varphi + \frac{1}{Bm}(-\dot{m}_e h_e - \dot{Q}_w)\right)\right] \quad ...(17.35)$$

where $h_e$ = enthalpy of the flow through exhaust valve, it is the cylinder average enthalpy for flow out of the cylinder and the exhaust system gas enthalpy if reverse flow occurs.

It is worth to note that the engine operating cycle should end with the working fluid at the same state that it started out. For the first calculations of the sequence of processes in Fig. 17.5, property values defining the initial state of the fluid in the cylinder were assumed. The cycle calculation is repeated with the appropriate new initial values, until the discrepancy is sufficiently small. Only a few interactions bring convergence in the calculation.

The working fluid state is now defined throughout the operating cycle. The work transfer to the piston per cycle.

$$W_c = \oint p dV$$

can now be calculated. From $W_c$, $m_a$, $m_f$ and $N$ we can calculate all performance parameter such as power, torque, *mep*, *isfc*, $\eta_{itc}$, $\eta_v$, residual gas fraction, total heat transfer, etc. using friction model, brake quantities may be calculated.

Other more sophisticated model may be used to predict engine emissions.

Cycle simulations and combustion models for SI and CI engines are discussed in subsequent article.

### 17.3.2 Spark-Ignition Engine Models

SI engine models generally follows the conceptual structure shown in Fig. 17.5. Our focus here is on the *combustion submodels*. The following are the assumptions of *combustion submodels*.

(i) The fuel, air and residual gas charge is essentially uniformly premixed.

(ii) The volume occupied by the reaction zone where the fuel-air oxidation process actually occurs is normally small compared with clearance volume.

(iii) The contents of combustion chamber during combustion can be analyzed as two zones—an unburnt and burnt zone.

(iv) In the absence of strong swirl, the leading edge and mean burnt gas front can be approximated by a sphere.

For a given combustion chamber shape and assumed flame centre location (i.e., the spark plug), the spherical burning area $A_b$, the burnt gas volume ($V_b$) and the combustion chamber surface wetted by the burnt gas can be calculated from flame propagation relations derived below.

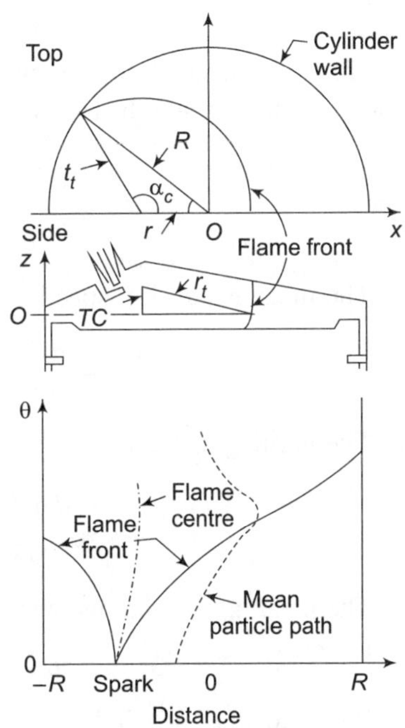


**Fig. 17.6** Flame geometry parameters [73]

Figure 17.6 shows the geometrical parameter which defines the flame surface. Let $r_c$, $\alpha_c$, $z_c$, be the co-ordination of the flame; $r_f$, be the radius of the best fit circle to the flame front silhouette. The flame is initiated at the spark plug. The flame front area $A_f$ is defined as the spherical surface of radius $r_f$, coinciding with the leading edge of the flame contained within the combustion chamber. The enflamed volume ($V_f$) is defined as the volume within the chamber behind this flame front. The burnt gas radius ($r_b$) is defined as the radius of the spherical surface within the combustion chamber which would contain all the burnt gas behind. Thus,

$$V_b\,(r_b, r_c, a_c, z_c) = V_b\,(p, \theta) \qquad ...(17.36)$$

The spherical burning area ($A_b$) is the area of this spherical surface. Thus,

$$A_b = \frac{\partial V_b(r_b, r_c, a_c, z_c)}{\partial r_b} \qquad ...(17.37)$$

The laminar burning area ($A_L$) is defined as the surface area of the flame would have if it burnt at the laminar flame speed. Thus,

$$A_L = \frac{dm_b/dt}{\rho_u S_L} \qquad ...(17.38)$$

The mean expansion speed of the front ($u_f$) is expressed as

$$u_f = \frac{dA_s/dt}{L_s} \qquad ...(17.39)$$

where $A_s$ is the shadow area enclosed by the best fit through the leading edge of the flame and $L_s$ is the arc length within the chamber of this best fit circle.

and $$L_s = \frac{\partial A_s}{\partial r_f} \qquad ...(17.40)$$

The mean expansion speed of the burnt gas ($u_b$) is given as

$$u_b = \frac{\partial V_b/\partial t}{A_b} \qquad ...(17.41)$$

The burning speed ($S_b$) is expressed as

$$S_b = \frac{dm_b/dt}{\rho_u A_b} \qquad ...(17.42)$$

The mean gas speed just ahead of the flame front ($u_g$) is defined as

$$u_g = u_b - S_b \qquad ...(17.43)$$

From Eqs. (17.38) and (17.42), we have

$$S_b\,A_b = S_L\,A_L \qquad ...(17.44)$$

The relationship between the mass fraction burnt, $\alpha_b$ (= $m_b/m$) and the volume fraction occupied by the burnt gas, $y_b$ (= $V_b/V$) is expressed as

$$x_b = \left[1 + \frac{\rho_u}{\rho_b}\left(\frac{1}{y_b} - 1\right)\right]^{-1} \qquad ...(17.45)$$

Combining Eqs. (17.41), (17.42) and (17.45), we get

$$\frac{u_b}{S_b} = \frac{\rho_u/\rho_b}{[(\rho_u/\rho_b) - 1]x_b + 1} \qquad ...(17.46)$$

The heat transfer occurs largely between the burnt gases and the walls and is proportional to chamber surface area wetted by the burnt areas $A_{b,w}$ and it is expressed as follows:

$$\dot{Q}_u = A_{u,w}\, h_{c,u}(T_u - T_w) \qquad ...(17.47)$$

and

$$\dot{Q}_b = A_{b,w}\, h_{c,b}(T_b - T_w) \qquad ...(17.48)$$

Based on the above model, Fig. 17.7 shows the SI engine spherical flame surface area as a function of flame radius and enflamed volume.

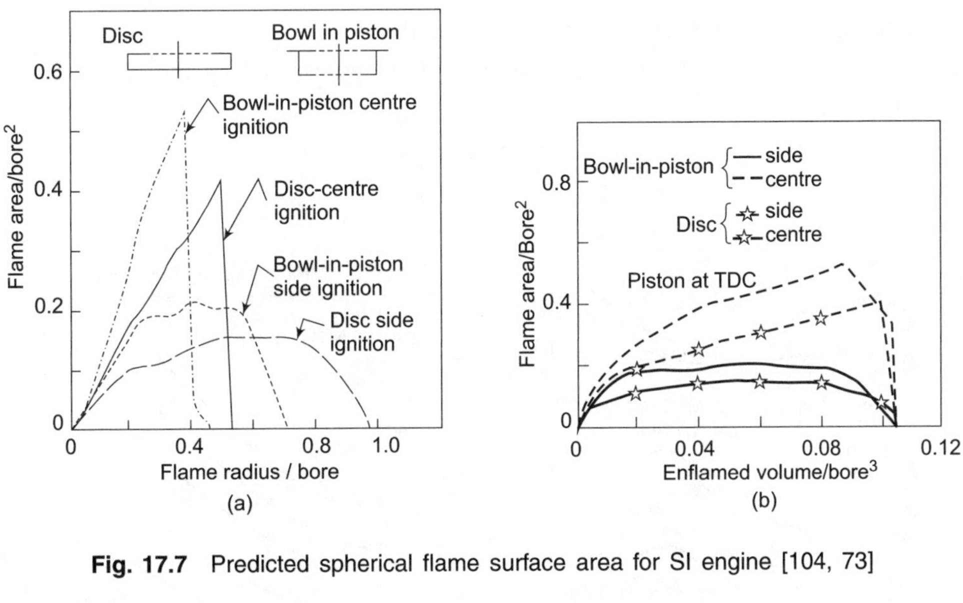


**Fig. 17.7** Predicted spherical flame surface area for SI engine [104, 73]

In order to calculate the effect of heat transfer on the burnt gas state more accurately, the burnt gas zone can be modelled in two parts — an adiabatic and a boundary layer region. Using the open system conservation equation and the combustion model, the predicted cylinder pressure, unburnt and burnt gas tempreture heat transfer rate and boundary layer thickness based on assumed 50° total burnt duration for a dm$^3$ eight-cylinder engine at wide open throttle and at 2500 rpm is shown in Fig. 17.8.

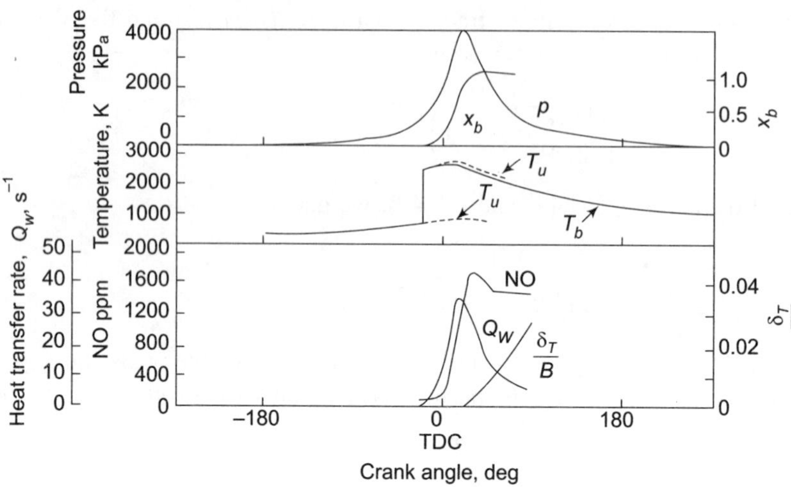


**Fig. 17.8** Predicted cylinder pressure, unburnt and burnt gas temperature, heat transfer rate and boundary layer thickness as a function of crank angle for SI engine [105, 73]

Figure 17.9 shows the predicted and measured *isfc* and exhaust temperature as a function of equivalence ratio for a SI engine operated at 1250 rpm.

The S-shaped mass fraction burnt profile is often represented by the Wiebe function expressed as

$$x_b = 1 - \exp\left[-a\left(\frac{\theta - \theta_o}{\Delta\theta}\right)^{m+1}\right] \quad ...(17.49)$$

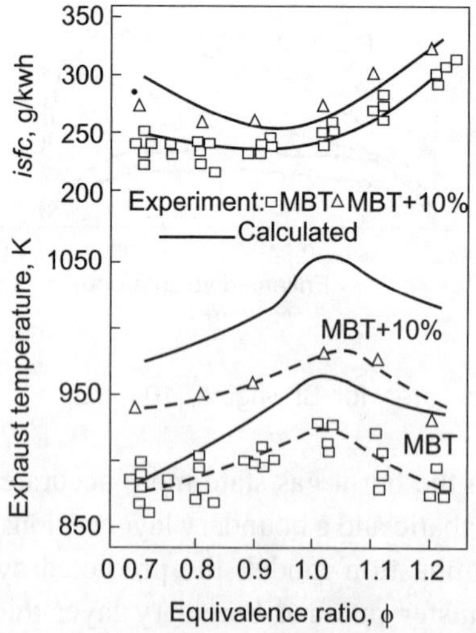


**Fig. 17.9** Predicted and measured *isfc* and exhaust temperature and a function of φ for SI engine [106, 73]

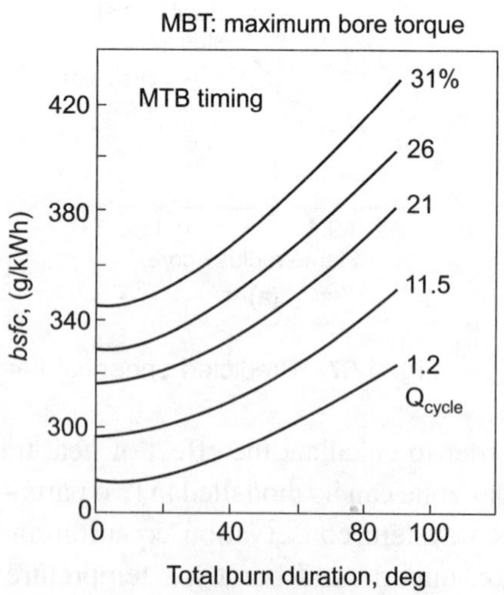


**Fig. 17.10** Predicted *bsfc* vs Total burn duration [107, 73]

Figure 17.10 shows the predicted *bsfc* as a function of heat transfer per cycle to the combustion chamber walls and total burn duration at 1250 rpm of a SI engine.

**Various other advanced models are available including turbulence model.**

## 17.4 DIRECT-INJECTION CI ENGINE MODELS

As discussed earlier, in direct-injection CI engines, the liquid fuel is injected into the cylinder as one or several jets just prior to ignition. *Except large engines, the air flow is usually swirling about the cylinder axis at up to 10 times the speed of the engine.* As a result of swirling, the rate of entrainment of air into the fuel jet increases, which in turn, increases, the fuel-air mixing rate. The unsteady liquid-fuel jet phenomena—atomization, liquid jet and droplet motion, fuel vaporization, air entrainment, fuel-air mixing and the ignition chemistry —also play a vital role in the heat release process. The complex ignition and combustion process make their modelling more complex and it is not yet possible to model all phenomena from a fundamental basis.

The following assumptions are made for DICI engine model:

(i) The cylinder contents are a single open system.

(ii) The only mass flows across the system boundary while the intake and exhaust valves are closed are the fuel and crevice flow.

(iii) Crevice flow effects are omitted.

$$\frac{dQ}{dt} - p\frac{dV}{dt} + \dot{m}_f h_f = \frac{dU}{dt} \quad ...(17.50)$$

There are two common methods to obtain combustion information from pressure data using Eq. (17.50). One approach yields heat release rate while other yields a fuel mass burining rate. In both methods, the cylinder contents are assumed at a uniform temperature at each instant in time during the combustion process.

### 17.4.1 Heat-Release Analysis

In Eq. (17.50), $dQ/dt$ becomes the difference between the chemical energy and the heat transfer from the system if $U$ and $h_f$ are taken to be sensible internal energy of the cylinder contents and sensible enthalpy of injected fuel respectively. Since, $h_{s,f} = 0$, Eq. (17.50) becomes

$$\frac{dQ_n}{dt} = \frac{dQ_{ch}}{dt} - \frac{dQ_{nt}}{dt} = p\frac{dV}{dt} + \frac{dU_s}{dt} \quad ...(17.51)$$

Here, $dQ_n/dt$ is called apparent heat release rate.

If the contents of the cylinder are modelled as ideal gas, then

$$\frac{dQ_n}{dt} = \frac{\gamma}{\gamma - 1}p\frac{dV}{dt} + \frac{1}{\gamma - 1}V\frac{dp}{dt} \quad ...(17.52)$$

Here, $\gamma = c_p/c_v$, which ranges from 1.3 to 1.35 for diesel engine.

$$Q_{ch} = \int_{t_{start}}^{t_{end}} \frac{dQ_{ch}}{dt}\, dt = m_f Q_{LHV} \quad ...(17.53)$$

Equations (17.50) to (17.53) are also true with crank angle (θ) as independent variable instead of time $t$.

### 17.4.2 Fuel Mass Burning Rate Analysis

Equation (17.50) takes the form:

$$\frac{d}{dt}(mu) = -p\frac{dV}{dt} + \frac{dQ}{dt} + h_f\frac{dm}{dt}$$

Hence, $Q$ is the heat transfer to the gas within the combustion chamber and $m$ is the mass within the combustion chamber.

Since $u = (T, p, \varphi)$ and $R = R(T, p, \varphi)$

$$\therefore \qquad \frac{du}{dt} = \frac{\partial u}{\partial T}\frac{dT}{dt} + \frac{\partial u}{\partial p}\frac{dp}{dt} + \frac{\partial u}{\partial \varphi}\frac{d\varphi}{dt} \qquad ...(17.54)$$

$$\frac{dR}{dt} = \frac{\partial R}{\partial T}\frac{dT}{dt} + \frac{\partial R}{\partial p}\frac{dp}{dt} + \frac{\partial R}{d\varphi}\frac{d\varphi}{dt} \qquad ...(17.55)$$

$$\frac{d\varphi}{dt} = \frac{1 + (F/A)_0}{(F/A)_s m_0} \cdot \frac{dm}{dt} \qquad ...(17.56)$$

where suffix $s$ denotes stoichiometric value and $o$ denotes the initial value prior to fuel injection.

Combining above

$$\frac{1}{m}\frac{dm}{dt} = \frac{-\left(\frac{RT}{V}\right)\left(\frac{dV}{dt}\right) - \left(\frac{\partial u}{\partial p}\right)\frac{dp}{dt} + \left(\frac{1}{m}\right)\left(\frac{dQ}{dt}\right) - C\,B}{u - h_f + D\left(\frac{\partial u}{\partial \phi}\right) - C\left[\left(\frac{D}{R}\right)\left(\frac{\partial R}{\partial \varphi}\right)\right]} \qquad ...(17.57)$$

where $$B = \frac{1}{p}\frac{dp}{dt} - \frac{1}{R}\frac{\partial R}{\partial p}\frac{dp}{dt} + \frac{1}{V}\frac{dV}{dt}$$

$$C = \frac{T\left(\frac{\partial u}{\partial T}\right)}{1 + \left(\frac{T}{R}\right)\left(\frac{\partial R}{\partial T}\right)}$$

and $$D = \frac{\left[1 + \left(\frac{F}{A}\right)_o\right]m}{\left(\frac{F}{A}\right)_s m_o}$$

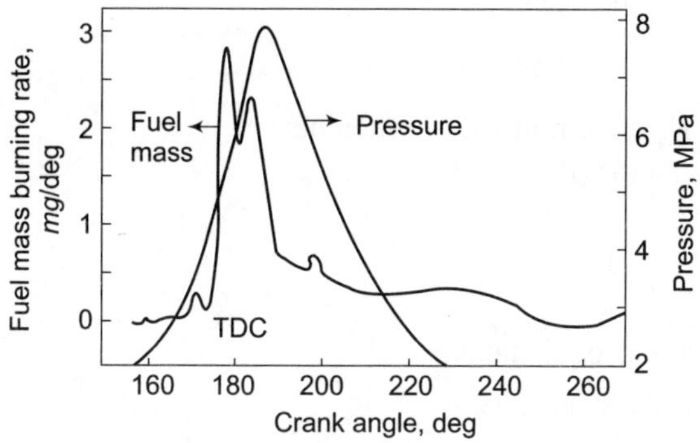


**Fig. 17.11** Cylinder pressure and fuel mass burning rate vs Crank angle

Equation (17.57) may be solved numerically for $m(t)$.

Figure 17.11 shows the predicted cylinder pressure ($p$) and fuel mass burning rate for DICI at 3200 rpm.

Lyn proposed three basic injection, mixing and burning patterns important in diesel engines based on single zone heat release model.

(A) *Fuel injection across the chamber with substantial momentum.* Mixing proceeds immediately as fuel enters the chamber and is little affected by combustion.

(B) *Fuel deposition on the combustion chamber walls.* There is a negligible mixing during the delay period due to limited evaporation. After ignition, evaporation becomes rapid and its rate is controlled by access of hot gases to the surface. This means burning is delayed by ignition lag.

(C) *Fuel distributed near wall.* Mixing proceeds during the delay at a smaller rate. After ignition, mixing is accelerated.

The fraction of the injected fuel that burns in each of these phases is empirically linked by ***Waston et al.*** [53]. One algebraic function is used to describe the mixing controlled heat release. These two functions are weighed with a phase proportionality factor, β, which is largely a function of the ignition delay.

Thus, $$\frac{m_{f,b}(t')}{m_{f,0}} = \beta f_1 + (1 - \beta) f_2 \quad ...(17.58)$$

where $m_{f,b}$ = mass of fuel burnt

$m_{f,s}$ = total fuel mass injected per cycle per cylinder

$t'$ = time from ignition non-dimensional by total time allowed for combustion.

$= (t - t_{ign})/\Delta t_{comb}$

The premixed burning function is

$$f_1 = (t - t'^{K_1})^{K_2} \quad ...(17.59)$$

The mixing controlled function is

$$f_2 = 1 - \exp(-K_3\, t'^{K_4}) \qquad \text{...(17.60)}$$

where $K_1$, $K_2$, $K_3$ and $K_4$ are empirical coefficients.

The proportionality factor ($\beta$) is expressed as

$$\beta = 1 - \frac{a\varphi^b}{\tau_{id}^c} \qquad \text{...(17.61)}$$

where $\varphi$ = overall fuel/air equivalence

$a, b, c$ = empirical constants

For typical turbocharged truck engine

$$K_1 = 2 + 1.25 \times 10^{-8}\,(\tau_{id}N)^{2.4}, \quad K_2 = 5000$$

$$K_3 = \frac{14.2}{\varphi^{0.644}}, \quad K_4 = 0.79\, K_3^{0.25}$$

$\tau_{id}$ = ignition delay in millisecond, $N$ = engine speed, rpm for $a$, $b$ and $c$, range are $0.8 < a < 0.95$, $0.25 < b < 0.45$, $0.25, < c < 0.5$

The overall structure of this particular complete diesel engine performance and emission model is shown in Fig. 17.12. Figure 17.13 shows an example of output from the above model.

**There are various other advanced models available.**

## 17.5 FLUID MECHANICS BASED MULTIDIMENSIONAL MODELS

Fluid mechanics based multi-dimensional models are used to predict the details of the flow field within engines, the heat transfer and combustion processes by solving governing conservation equation (partial differential equations) using numerical methods. The models include one-dimensional, two-dimensional and three-dimensional unsteady flow phenomena.

The principal components of these multidimensional engines are the following:

(i) The mathematical models or equations are used to describe the flow processes. The most important one is the turbulence model.

(ii) The discretization procedure used to transform the differential equations of the mathematical model into algebraic relations between discrete values of velocity, pressure, temperature, etc., on a computing mesh conforming to the geometry of engine concerned engine components.

(iii) The solution algorithm whose function is to solve the algebraic equation.

(iv) The computer codes which translate the numerical algorithm into computer language for computation.

The governing equations are conservation equations for mass, momentum (i.e., Navier–Stokes), energy and species concentration and may be written as

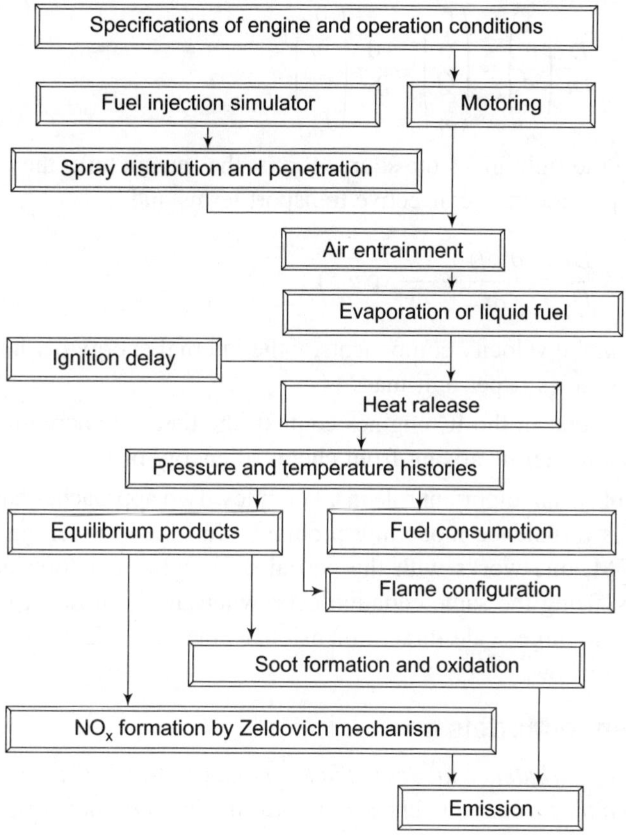


**Fig. 17.12** Structure of thermodynamics based DI diesel engine simulation [108]

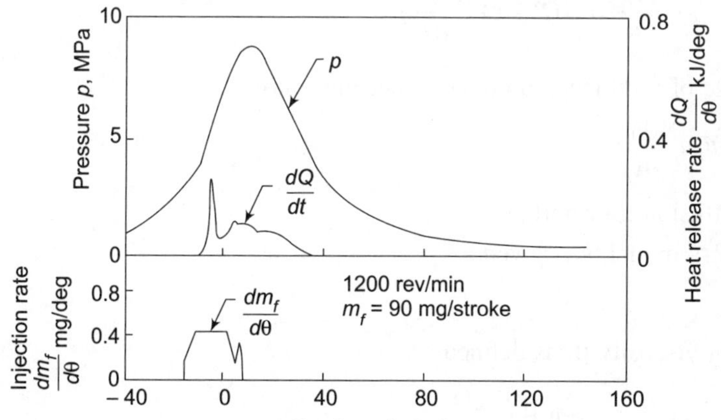


**Fig. 17.13** Predicted fuel injection rate. Heat release rate profile and cylinder pressure vs crank angle. [108, 73]

$$\frac{D}{Dt}\begin{pmatrix}1\\u_i\\e\\Y_a\end{pmatrix}=\begin{pmatrix}0\\0\\Q\\S_a\end{pmatrix}-\frac{\partial}{\partial x_j}\begin{pmatrix}0\\\tau_{ij}\\q_j\\J_{\alpha i}\end{pmatrix} \quad ...(17.62)$$

The first term on the right gives the source terms, the second term the diffusion transport. The *D*/*Dt* operator provides the convective transport terms and is

$$\frac{Df}{Dt}=\frac{\partial(\rho f)}{\partial t}+\frac{\partial}{\partial x_j}(\rho u_j f) \quad ...(17.63)$$

where ρ is density, $u_i$ the velocity components, *e* the internal energy per unit mass and $Y_a$ the concentration of φ species α per unit mass.

It is worth to note that in the IC engines context, the thermal energy term (*Q*) involves a viscous term and source terms arising from chemical energy of fuel.

Turbulent flows plays an important role in CI engines. Two approaches have been developed for dealing with this turbulence modelling problem—full-filed modelling (FFM) and large-eddy (LES). In FFM, one works with the partial differential equation describing suitably averaged quantities, using the same equation everywhere in the flow. In LES approach, one actually calculates the large-scale three-dimensional time-dependent turbulence structure in a single realization of flow.

### 17.5.1 Turbulence Models

*The most widely used turbulence model is the k-ε model [54-56]. This assumes a Newtonian relationship between the turbulent stresses and mean strain rates, and computes the (fictitious) turbulent viscosity appearing in the relationship from the local turbulent kinetic energy k (=* $\overline{u_i u_j}/2$*) and its dissipating rate, ε.* The equation is

$$\frac{\overline{D}\bar{k}}{dt}=\bar{\rho}(P-\varepsilon)-\frac{\partial}{\partial x_j}\bar{j}_k \quad ...(17.64)$$

where $P$ = rate of turbulence production per unit mass

$$=-\overline{u_i u_j}\frac{\partial\bar{u}}{\partial x_j} \quad ...(17.65)$$

$J_k$ = diffusion transport

In the most *k*-ε model, ε is given as

$$\varepsilon \propto k^{3/2} \quad ...(17.66)$$

The turbulent viscosity $\mu_T$ is defined as

$$\mu_T=\frac{C_0\,\bar{\rho}\,k^2}{\varepsilon} \quad ...(17.67)$$

where $C_0$ = model constant

Also

$$\bar{\rho}\,\overline{u_i u_j} = \frac{2}{3}\bar{\rho}\,k\,\delta_{ij} + \frac{2}{3}\mu_T\,\nabla.\bar{u}_{ij} - 2\mu_T\,\bar{S}_{ij} \qquad ...(17.68)$$

where,

$\bar{S}_{ij}$ = strain rate of the $u_i$ filed

$$= \frac{1}{2}\left(\frac{\partial \bar{u}_i}{\partial x_j} + \frac{\partial \bar{u}_j}{\partial x_i}\right) \qquad ...(17.69)$$

The model is completed with a transport equation for ε.

Thus,

$$\frac{\bar{D}\varepsilon}{Dt} = W - \frac{\partial H_i}{\partial x_i} \qquad ...(17.70)$$

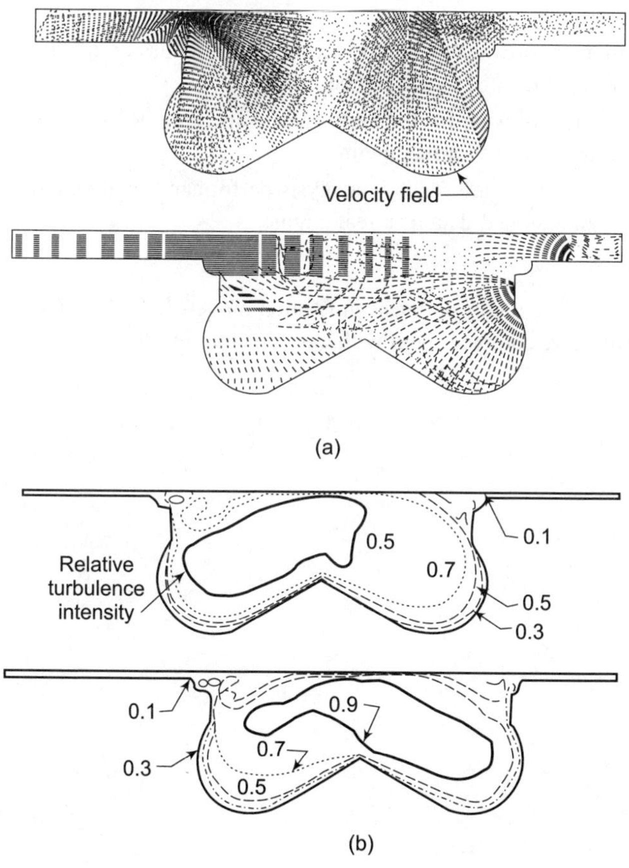


**Fig. 17.14** Predicted velocity field in combustion chamber of DICI engine [109, 73]

where $$W = \left(-C_2 + C_1 \frac{P}{\varepsilon}\right)\frac{\bar{\rho}\varepsilon^2}{k}$$

where $C_1$ and $C_2$ = constant

These equations are solved either by finite element method or finite difference method. Figure 17.14 shows predicted flow field within the offset bowl of DI diesel engines. [73]

## PROBLEMS

### Things to Think About

17.1 What do you mean by modelling?

17.2 What are the two basic types of models for predicting performance and emissions?

17.3 What are the three types of intake and exhaust flow models?

17.4 What should be logical procoss models for thermodynamics based in cylinder models?

### Software Based Questions

17.5 Develop a computer software for the performance and emission of SI engine and give results for an assumed input data of a real engine.

17.6 Develop a computer software for the performance and emission of CI engines and give results for an assumed input data of a real engine.

17.7 Develop a computer software for flow analysis performance and emissions of CI engines and give results for an assumed data of a real engine.

### Open Ended Question

17.8 Prepare a compressive report on latest simulation models for SI and CI engines and give some simulated results and compare them with experimental results.

CHAPTER

# Gas Turbines and Jet Propulsion

# 18

## OVERVIEW

*The gas turbine obtains its power by utilising the energy of burnt gases and air which is at high temperature and pressure by expanding through the several rings of fixed and moving blades and thus causing a rate of change of momentum in working fluid as a result of change in the direction of flow in the moving blade channel.* It thus, resembles a steam turbine. To get a high pressure (of the order of 4 to 40 bar) of working fluid which is essential for expansion, a compressor is required. The quantity of the working fluid and speed required are more, so, in general a centrifugal or an axial compressor is employed. The compressor is driven by the turbine and so it is coupled to the turbine shaft. If after compression, the working fluid were to be expanded in a turbine, then, assuming that there were no losses in either components, the power developed by the turbine would be just equal to that absorbed by the compressor and the work done would be zero. But the power developed by the turbine can be increased by increasing the volume of the working fluid at constant pressure, or alternatively increasing the pressure at constant volume. Either of these may be done by adding heat so that the temperature of the working fluid is increased after the compression. To get a higher temperature of the working fluid, a combustion chamber is required where combustion of air and fuel takes place at constant pressure giving a temperature rise to the working fluid.

*Thus, a simple gas turbine cycle consists of (i) a compressor (in general axial flow compressor), (ii) a combustion chamber, and (iii) a turbine.* Since the compressor is coupled with the turbine shaft, so the compressor absorbs appreciable amount (about 50%) of the power produced by the turbine and hence lowers the efficiency. The net work is the difference between the turbine work and work required by compressor to drive it. *The thermal efficiency of gas turbine plant varies from* ***32% to 47%.****

***For more details, see author's book "Steam & Gas Turbines, Turbomachines, and Power Plant Engineering".**

Gas turbines are constructed to work on the following fuels; oil, coal gas, natural gas, hydrogen, kerosene (JF), producer gas, blast furnace gas and pulverized coal. At present simple, open cycle gas turbine single unit is built up to 400 MW. *In a 135 MW unit built by GE in 1998, the compressor pressure ratio is 13.5 achieved by 18 stages of axial flow compressor. There are 16 annular combustion chambers and 3 stages air cooled turbine in which air is bled from compressor for cooling.* The speed of the turbine is 3600 rpm. The turbine *inlet temperature (TIT) is as high as 1400°C. The plant thermal efficiency is about 35%. Recently (in 2004), GE has developed intercooled (LMS 100) gas turbine working on compressor pressure ratio 42 and TIT = 1653 K yielding 47% efficiency at ISO conditions. Now, gas turbines are supposed to be prime mover of 21$^{st}$ century using natural gas and hydrogen.* This chapter deals with cycle analysis of shaft gas turbine cycle and jet propulsion.

## 18.1 CLASSIFICATION AND APPLICATIONS

### 18.1.1 Classification

(A) On the basis of combustion process, the gas turbine is classified as follows:
- (i) Continuous combustion or constant pressure type. The cycle working on this principle is called *Joule or Brayton cycle*
- (ii) The explosion or constant volume type. The cycle working on this principle is called *Atkinson cycle.*

(B) Another classification, based upon the ***action of expanding gases*** similar to steam turbine is:
- (i) Impulse turbine
- (ii) Impulse reaction turbine,

(C) On the basis of **path of working substance,** the gas turbine is again classified as:
- (i) Open cycle gas turbine (working fluid enters from atmosphere and exhausts to atmosphere).
- (ii) Closed cycle gas turbine (working fluid is confined within the plant).
- (iii) Semi-closed cycle (part of the working fluid is confined within the plant and another part flows from and to the atmosphere).

(D) On the basis of ***direction of flow***
- (i) Axial flow
- (ii) Radial flow (less common, suitable for low power output unit)

### 18.1.2 Applications

Gas turbines are used in (i) central power stations either alone or as combined cycle, (ii) stand-by power plants, (iii) locomotive power plants, (iv) jet engines (commercial jet planes and fighter planes), (v) automotives, (vi) pumping station in long pipeline (oil or gas), (vii) main power source for laboratories, (vii) marine, (viii) as a driver of industrial compressor, pumps, helicopter engines, etc.

## 18.2 SIMPLE OPEN CYCLE GAS TURBINE (CONSTANT PRESSURE HEAT ADDITION) OR AIR STANDARD BRAYTON (OR JOULE CYCLE)

The Joule or Brayton cycle is the most idealized cycle for the simple gas turbine power plant shown Figs. 18.1. and 18.2. Atmospheric air is compressed from $p_1$ to a high pressure $p_2$ in the compressor and delivered to the burner or combustion chamber where fuel is injected and burnt. The combustion process occurs nearly at constant pressure. Due to combustion, heat is added to the working fluid from $T_2$ to $T_3$. The products of combustion from the combustion chamber are expanded in the turbine from $p_3$ to atmospheric pressure $p_1$ and then discharged into the atmosphere. The turbine and compressor are mechanically coupled, so the net work is equal to the difference between the work done by the turbine and work consumed by the compressor. *To first run the compressor, a starter is needed.* When the turbine starts running the starter is cut-off. *In actual plant, turbine bladings are cooled by bleeding the air from compressor to increase its life.* About 10% to 20% of partly/fully compressed air is used to cool the turbine bladings. In the air standard cycle, blade cooling is not considered.

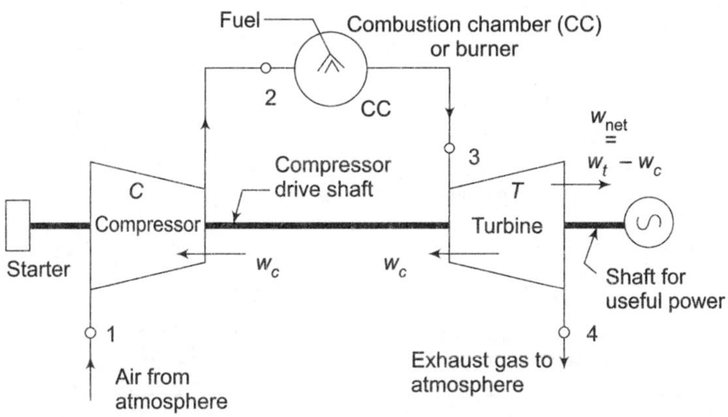


**Fig. 18.1** Simple open cycle gas turbine. *C* = compressor, *CC* = combustion chamber or burner, *T* = turbine, *P* = propulsive power coupling, *S* = starter

Representation of ideal Brayton cycle on *p-V* and *h-s* or (*T-s*) diagrams are shown in Fig. 18.3 (a) and 18.3 (b)

Process 1-2*s* is the isentropic compression in the compressor

Process 2*s*-3 is constant pressure heat addition in the combustion chamber

Process 3-4*s* is isentropic expansion in the turbine

Process 4*s*-1 is constant pressure heat rejection.

Here, *s* stands for isentropic process.

Referring to Fig. 18.3, we may find out the thermal efficiency on the basis of 1 kg of working fluid and $c_p$ remaining constant.

Heat supplied = $q_A = h_3 - h_{2s} = c_p\,(T_3 - T_{2s})$

Heat rejected = $q_R = h_{4s} - h_1 = c_p\,(T_{4s} - T_1)$

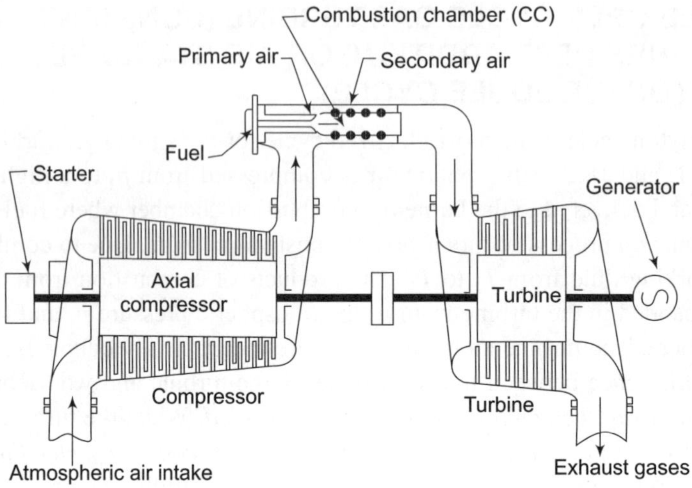


**Fig. 18.2** Gas turbine plant

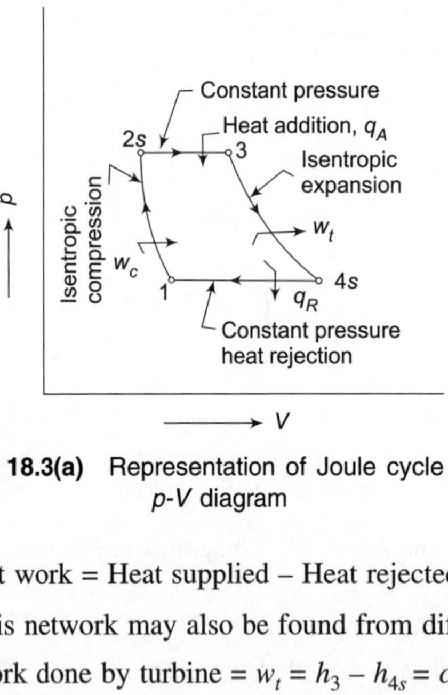


**Fig 18.3(a)** Representation of Joule cycle on *p-V* diagram

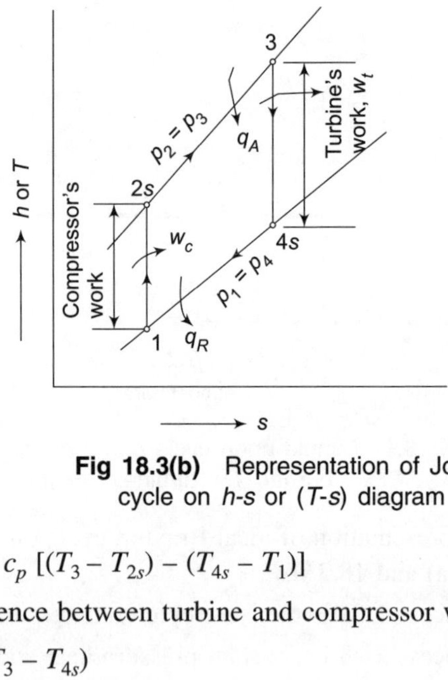


**Fig 18.3(b)** Representation of Joule cycle on *h-s* or (*T-s*) diagram

Net work = Heat supplied – Heat rejected = $c_p\ [(T_3 - T_{2s}) - (T_{4s} - T_1)]$

This network may also be found from difference between turbine and compressor work.

Work done by turbine = $w_t = h_3 - h_{4s} = c_p\ (T_3 - T_{4s})$

Work consumed by compressor which is supplied by turbine

$$w_c = h_{2s} - h_1 = c_p\ (T_{2s} - T_1)$$

Net work $\quad w_{net}$ = Work done by turbine – Work consumed by compressor

$$= w_{net} = c_p\ [(T_3 - T_{4s}) - (T_{2s} - T_1) = c_p\ [(T_3 - T_{2s}) - (T_{4s} - T_1)]$$

The thermal efficiency,

$$\eta_{th} = \frac{\text{Net work}}{\text{Heat supplied}} = \frac{w_t - w_c}{q_A} = \frac{c_p(T_3 - T_{2s}) - (T_{4s} - T_1)}{c_p (T_3 - T_{2s})}$$

or
$$\eta_{th} = 1 - \frac{T_{4s} - T_1}{T_3 - T_{2s}} \quad ...(18.1)$$

We know that $\frac{T_{2s}}{T_1} = \left(\frac{p_2}{p_1}\right)^{(\gamma-1)/\gamma}$ and $\frac{T_3}{T_{4s}} = \left(\frac{p_3}{p_4}\right)^{(\gamma-1)/\gamma}$

since $p_2 = p_3$ and $p_1 = p_4$

$$\therefore \quad \frac{T_2}{T_1} = \frac{T_3}{T_{4s}} = \left(\frac{p_2}{p_1}\right)^{(\gamma-1)/\gamma} \quad \text{or} \quad \frac{T_1}{T_{2s}} = \frac{T_{4s}}{T_3} = \frac{T_{4s} - T_1}{T_3 - T_{2s}}$$

Putting the value of $\frac{T_{4s} - T_1}{T_3 - T_{2s}}$ in Eq. (18.1), we get

$$\eta_{th} = 1 - \frac{T_1}{T_{2s}} = 1 - \frac{T_{4s}}{T_3} \quad ...(18.2)$$

Let. $r_p = \text{pressure ratio} = \frac{p_2}{p_1}$

So $\frac{T_{2s}}{T_1} = \left(\frac{p_2}{p_1}\right)^{(\gamma-1)/\gamma} = (r_p)^{(\gamma-1)/\gamma}$

Therefore, the thermal efficiency,

$$\eta_{th} = \frac{1}{(r_p)^{(\gamma-1)/\gamma}} \quad ...(18.3)$$

From the above equation, we see that the thermal efficiency of the Brayton cycle is the same as that of the Otto cycle with a difference that compression ratio is replaced by compressor pressure ratio. Note that for turbine and compressor work, we have neglected the change in K.E. and P.E. energies.

We may work on **stagnation values** of pressure ($p_o$), temperature ($T_o$) and enthalpy ($h_o$) also. The **static values** may be replaced by **stagnation values**.

A plot of $\eta_{th}$ as a function of $r_p$ is shown in Fig. 18.4. It is obvious from Fig. 18.4 that the *efficiency progressively increases with increasing value of pressure ratio.*

*When gas turbine is used in electricity generating plant, whole of the expansion of gas takes place in turbine and the generator mounted on the turbine shaft generates electricity. Such turbine is known as* ***power turbine***.

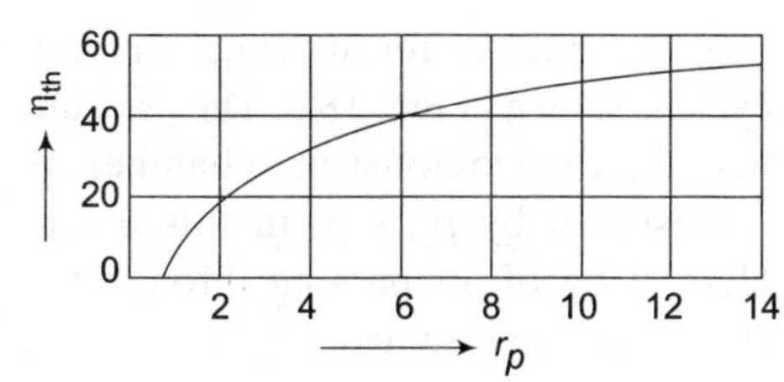


**Fig. 18.4** $\eta_{th}$ vs. $r_p$

*When it is used in* ***jet engines****, the gas is expanded partially in the turbine to drive the compressor and accessories, and the rest of the expansion takes place in a nozzle put just immediate in the exhaust part of the turbine to produce a high velocity jet (i.e., thrust) for the propulsion of the aircraft in the opposite direction of jet according to Newton's third law of motion. Such turbine is known as* ***aviation or aircraft turbine.*** In the case of **helicopters**, there are two trubines, one turbine drives the compressor, called **compressor turbine** as both are coupled on the same shaft. Another turbine, called **power turbine** which develops power to drive **propeller**.

## 18.3 ACTUAL BRAYTON CYCLE

The actual gas turbine cycle differs from the theoretical cycle in the following main respects:

(1) *Due to frictional losses* in the compressor and turbine, the compression and expansion processes are not frictionless and take place with some increase in entropy (i.e., the processes are irreversible adiabatic). In the ideal case, the compressor and turbine efficiency are 100 per cent but in practice, turbine and compressor efficiencies are less than 100 per cent.

(2) *A small pressure loss* occurs in the combustion chamber. This loss is so little that it can be neglected for the sake of simplification of the problem wherever necessary. A small pressure loss also occurs in the exhaust hood of the turbine.

(3) The mass of the gas flowing through the turbine is $(1 + F/A)$ times the mass of air flowing through the compressor where F/A represents the fuel-air ratio. ($F/A = \dot{m}_f/\dot{m}_a$).

(4) The specific heat of combustion gas is slightly higher than that of the air. This increase is so little that the specific heat of combustion gas may be taken as that of air for simplicity wherever necessary.

(5) The turbine bladings are cooled by air bled from compressor at appropriate points, however, it will not be considered in the analysis.

A *T-s* diagram for an actual Brayton cycle is shown in Fig. 18.5. The pressure loss in the combustion chamber is represented by $p_3 - p_2$ in this cycle. Here, 1 kg of air passing through the combustion is assumed.

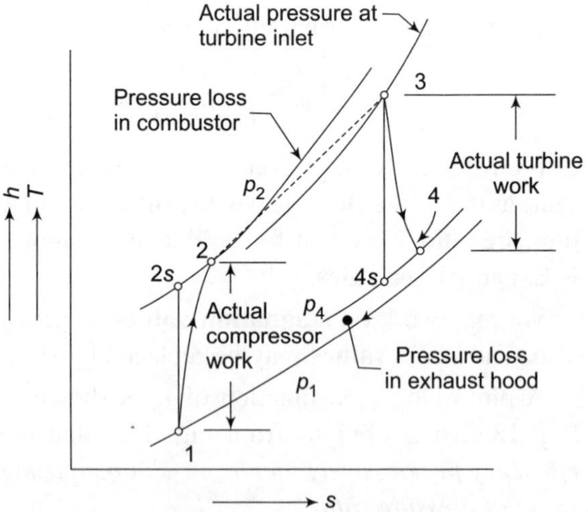


**Fig. 18.5** Actual brayton cycle

1 – 2*s* is isentropic compression. 1 – 2 is actual compression.

3 – 4*s* is isentropic expansion. 3 – 4 is actual expansion.

The compressor efficiency,

$$\eta_c = \frac{\text{Isentropic compression work}}{\text{Actual compression work}} = \frac{w_c}{w_{ca}}$$

Based on static values, $\eta_c = \dfrac{h_{2s} - h_1}{h_2 - h_1} = \dfrac{c_p\,(T_{2s} - T_1)}{c_p\,(T_2 - T_1)} = \dfrac{T_{2s} - T_1}{T_2 - T_1}$ ...(18.4)

**Based on total or stagnation values,** $\eta_{oc} = \dfrac{h_{02s} - h_{01}}{h_{02} - h_{01}} = \dfrac{T_{02s} - T_{01}}{T_{02} - T_{01}}$

The turbine efficiency,

$$\eta_t = \frac{\text{Actual turbine work}}{\text{Isentropic turbine work}} = \frac{w_{ta}}{w_t}$$

Actual turbine work = $w_{ta} = h_3 - h_4$ kJ/kg

$\eta_t$ and $\eta_c$ are generally in the neighbourhood of 85%.

If specific heat of combustion gas $c_{pa}$ may be taken as that of air, then

Actual turbine work = $w_{ca} = c_p\,(T_3 - T_4)$ kJ/kg

Isentropic turbine work = $w_t = h_3 - h_{4s} = c_p(T_3 - T_{4s})$ kJ/kg

So, turbine efficiency based on static values is

$$\eta_t = \frac{h_3 - h_4}{h_3 - h_{4s}} = \frac{c_p(T_3 - T_4)}{c_p(T_3 - T_{4s})} = \frac{T_3 - T_4}{T_3 - T_{4s}} \quad ...(18.5)$$

Based on total values, $\eta_{ot} = \dfrac{T_{03} - T_{04}}{T_{03} - T_{04s}}$

The thermal efficiency of the cycle is computed as follows

Actual net work = $w_{neta}$ = Work done by turbine – Work supplied to compressor

$$w_{neta} = \left[\left(1 + \frac{F}{A}\right)(h_3 - h_4) - (h_2 - h)\right]; \text{ kJ/kg of air}$$

where $F/A$ = Fuel-air ratio = $\dot{m}_f/\dot{m}_a$

Actual heat supplied = $q_{Aa} = \left[\left(1 + \dfrac{F}{A}\right) h_3 - h_2\right]$; kJ/kg of air

So, thermal efficiency = $\eta_{tha} = \dfrac{\left(1 + \dfrac{F}{A}\right)(h_3 - h_4) - (h_2 - h_1)}{\left(1 + \dfrac{F}{A}\right) h_3 - h_2}$

or

$$\eta_{tha} = \frac{\left(1 + \dfrac{F}{A}\right)c_{pg}\,(T_3 - T_4) - c_{pa}(T_2 - T_1)}{\left(1 + \dfrac{F}{A}\right)c_{pg}.T_3 - c_{pa}.T_2} \quad ...(18.6)$$

If we neglect the mass of fuel in comparison to the mass of air used, then

$$\eta_{tha} = \frac{(h_3 - h_4) - (h_2 - h_1)}{(h_3 - h_2)}$$

If we assume that the specific heat of air and combustion gas be the same, then

$$\eta_{tha} = \frac{(T_3 - T_4) - (T_2 - T_1)}{(T_3 - T_2)} \quad \text{...(18.7)}$$

or $$\eta_{tha} = \frac{(T_3 - T_{4s})\, \eta_t . \eta_c - (T_{2s} - T_1)}{(T_3 - T_2)\, \eta_c}$$

or $$\eta_{tha} = \frac{w_t\, \eta_t . \eta_c - w_c}{q_{Aa} . \eta_c} \quad \text{...(18.8)}$$

From the Eq. (18.8), it is clear that *the thermal efficiency is increased by improving either $\eta_t$ or $\eta_c$ or both.* All the above expressions may be based on total values.

Here, $q_{Aa}$ = actual heat supplied = $\dot{m}_f \times$ C.V. $\times\ \eta_{cc}$

where $\eta_{cc}$ is the combustion efficiency, in the range of 0.98 to 0.99.

The above expression depends on sense where it is applied.

### 18.3.1 Polytropic or Small Stage Efficiency

The isentropic efficiency deals with the overall performance of a dynamic compressor (centrifugal and axial).

Sometimes, it is desirable to establish the *true aerodynamic efficiency* of compressor based on the concept of *small stages within a compressor or turbine* (Figs. 18.6 and 18.7).

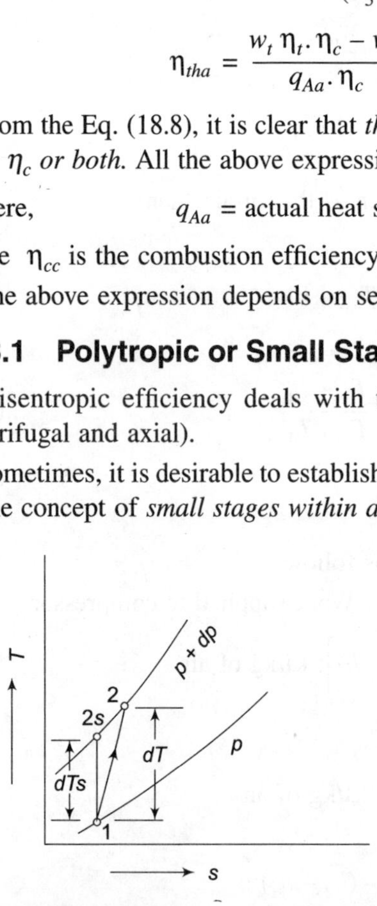


**Fig. 18.6** Polytropic compression

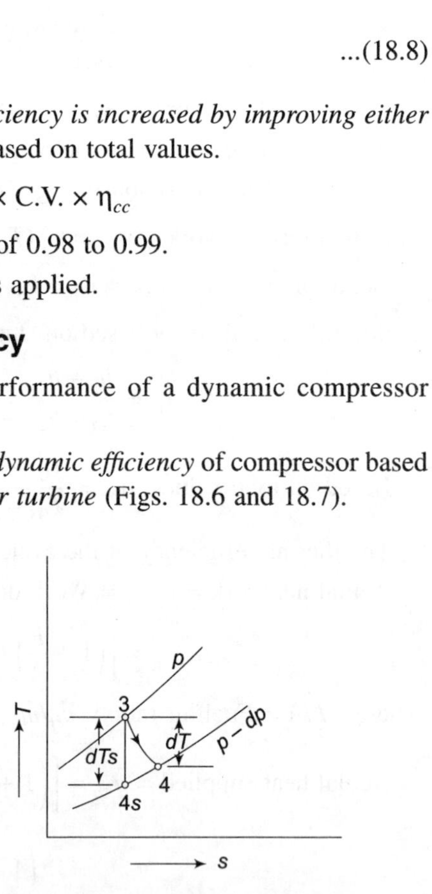


**Fig. 18.7** Polytropic expansion

The polytropic efficiency for compressor is

$$\eta_{pc} = \frac{\gamma - 1}{\gamma} \times \frac{n}{n-1} \quad \text{...(18.9)}$$

Similarly, for polytropic expansion process,

$$\eta_{pt} = \frac{n-1}{n} \times \frac{\gamma}{\gamma - 1} \qquad \text{...(18.10)}$$

where $\eta_{pt}$ is polytropic efficiency for turbine

***Note:*** The clue for solving any problem on gas turbines is to know the values of temperature at all states in the cycle by possible methods.

## 18.4 THE CYCLE AIR RATE AND WORK RATIO

### 18.4.1 The Cycle Air Rate

In addition to the thermal efficiency which is a measure of fuel economy, the size of the plant is equally important in so many applications, particularly in the fields of aviation. The size of the plant depends on the rate of air in relationship to the power output.

*Air rate, (AR), is defined as the air flow required in kg per kWh output.* Thus,

$$AR = \frac{\dot{m}_a}{\text{kWh output}} = \frac{\text{Mass of air required}}{\text{kWh output}}$$

$$AR = \frac{\dot{m}_a}{\dot{m}_a\,(w_{net}/3600)} = \frac{3600}{w_{net} \text{ in kJ/kg}}; \quad \text{kg/kWh} \qquad \text{...(18.11a)}$$

The reciprocal of air rate is termed **specific power**. Air rate is actually the criterion of the size of the plant, i.e., *the lower the air rate, the smaller the plant.*

### 18.4.2 The Cycle Work Ratio

*A good design of gas turbine not only requires a high thermal efficiency and a low air rate, but it must also have a high work ratio.*

*The work ratio, (WR), is defined as the ratio of net work to the turbine work.* Thus,

$$WR = \frac{w_{net}}{w_t} \qquad \text{...(18.11b)}$$

From Eq. (18.11b), it is also clear that for a good gas turbine, the work ratio should be high. The work ratio may also be used as a guide in the determination of the size of the gas turbine and as an indicator of the sensitivity of the plant to the decrease of the component efficiency.

## 18.5 THE SPECIFIC FUEL CONSUMPTION

The specific fuel consumption (*sfc*) is a measure of fuel aconomy. *It is defined as the mass of fuel per power output and is expressed in kg/kWh.* Thus,

$$sfc = \frac{3600\ \dot{m}_f \text{ in kg/s}}{W_{net} \text{ in kW}} = \frac{3600\ (\dot{m}_f/\dot{m}_a)}{w_{net}(\text{kJ/kg})}\ ; \text{kg/kWh} \qquad \text{...(18.12)}$$

where $\dot{m}_f$ = mass flow ratio of fuel, kg/s and $\dot{m}_f/\dot{m}_a$ = fuel-air ratio

*sfc* is around 0.3 kg/kWh

**EXAMPLE 18.1** An open cycle gas turbine plant works between the fixed absolute temperature limits 300K and 1500K, the absolute pressure limits begin 1 bar and 14 bar. The isentropic efficiency of compressor 0.85 and that of turbine is 0.86. Estimate the actual thermal efficiency of the plant and the power available for driving generator, if the calorific value of fuel is 42000 kJ/kg. Assume $\eta_{comb.} = 99\%$, $\eta_m = 0.98$ for compressor and turbine each, $\eta_{gen} = 0.985$ and $\dot{m}_a = 500$ kg/s. Also calculate air standard efficiency.

***Solution*** Refer to Fig. 18.1 and Fig. 18.8.

Given : $T_1 = 300$ K, $p_1 = 1$ bar, $p_2 = 14$ bar, $T_3 = 15000$ K, $\eta_{comb.} = 99\%$, $\eta_{mc} = \eta_{mt} = 0.98$, $\eta_{gen} = 0.985$, $\eta_c = 0.85$ and $\eta_t = 0.86$, $\dot{m}_a = 500$ kg/s,

$$\text{C.V.} = 42000 \text{ kJ/kg. Let us take } \gamma = 1.4$$

$$\frac{T_{2s}}{T_1} = \left(\frac{p_2}{p_1}\right)^{(\gamma-1)/\gamma} = (14)^{0.4/1.4} = 2.125$$

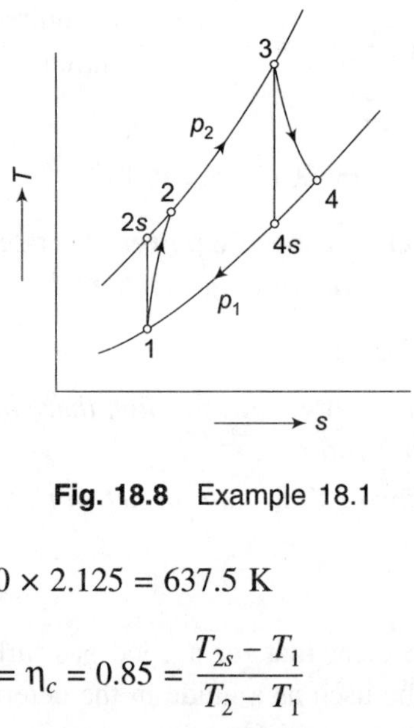


**Fig. 18.8** Example 18.1

or $$T_{2s} = 300 \times 2.125 = 637.5 \text{ K}$$

The compressor efficiency $= \eta_c = 0.85 = \dfrac{T_{2s} - T_1}{T_2 - T_1}$

or $$T_2 = 300 + \frac{637.5 - 300}{0.85} = 697.05 \text{ K}$$

We have $$\frac{T_3}{T_{4s}} = \left(\frac{p_3}{p_4}\right)^{(\gamma-1)/\gamma} = (14)^{0.4/1.4} = 2.125$$

or $$T_{4s} = \frac{T_3}{2.125} = \frac{1500}{2.125} = 705.88 \text{ K}$$

The turbine efficiency $= \eta_t = \dfrac{T_3 - T_4}{T_3 - T_{4s}}$

or $$T_3 - T_4 = 0.86\ (1500 - 705.88) = 682.94\ \text{K}$$

or $$T_4 = T_3 - 682.94 = 1500 - 682.94 = 817.06\ \text{K}$$

The efficiency of combustion chamber = $\eta_{comb}$ = 99%

C.V. of fuel = 42000 kJ/kg of fuel

Let $\dot{m}_a$ and $\dot{m}_f$ be the kg of air and fuel per second and $c_p$ for air and gas be the same,

then $$(\dot{m}_a + \dot{m}_f)\ h_3 - \dot{m}_a\ h_2 = \dot{m}_f \times \text{C.V.} \times \eta_{comb}$$

or $$(\dot{m}_a + \dot{m}_f)\ c_{pg}\ T_3 - \dot{m}_a \cdot c_{pa}\ T_2 = \dot{m}_f \times \text{C.V.} \times \eta_{comb}$$

or $$\frac{\dot{m}_a}{\dot{m}_f} = \frac{\text{C.V.} \times \eta_{comb} - c_{pg}\ T_3}{c_{pg}\ T_3 - c_{pa}\ T_2} = \frac{\text{C.V.} \times \eta_{comb} - c_p\ T_3}{c_p(T_3 - T_2)}$$

or $$\frac{\dot{m}_a}{\dot{m}_f} = \frac{42000 \times 0.99 - 1.005 \times 1500}{1.005\ (1500 - 697.05)} = 49.65 \text{ kg of air per kg of fuel}$$

$$\text{Fuel-air ratio} = \frac{\dot{m}_f}{\dot{m}_a} = 1 : 49.65$$

Actual work consumed by the compressor/kg of air

$$w_{ca} = c_p(T_2 - T_1)/\eta_{mc} = 1.005\ (697.05 - 300)/0.98 = 407.17\ \text{kJ/kg}$$

Actual work done obtained by the turbine/kg of air *assuming* $c_p$ = 1.005 is given by

$$w_{ta} = c_p\left(1 + \frac{\dot{m}_f}{\dot{m}_a}\right)(T_3 - T_4) \times \eta_{mt}$$

$$= 1.005\left[\left(1 + \frac{1}{49.65}\right)(1500 - 817.06)\right] \times 0.98 = 686.17\ \text{kJ/kg}$$

Without considering $\dot{m}_f$, $w_{ta}$ = 672.62 kJ/kg

So, we may neglect the mass of fuel for simplification

Network = Work obtained by the turbine – Work consumed by the compressor

$$w_{net} = 686.17 - 407.17 = 279 \text{ kJ/kg of air}$$

Here, $\dot{m}_a$ = 500 kg/s

So, mass of fuel, $\dot{m}_f = \dfrac{500}{49.65}$ = 10.07 kg/s (This is actual $\dot{m}_f$)

Total net power available = $W_{neta} = W_{ele} = \dot{m}_a \times w_{net} \times \eta_{gen}$

$$= 500 \times 279 \times 0.985/10^3 = 137.4\ \text{MW}.$$ ***Ans.***

Thermal efficiency of the plant

$$= \frac{W_{neta}}{q_{Aa}} = \frac{W_{neta}}{\dot{m}_f \times \text{C.V.}} = \frac{137.4 \times 10^3}{10.07 \times 42000} = 32.48\%. \qquad \textbf{\textit{Ans.}}$$

***Note:*** It is to be noted that in $\dot{m}_f$ calculation, $\eta_{comb}$ has been considered.

$$ASE = 1 - \frac{1}{(r_p)^{\gamma-1/\gamma}} = 1 - \frac{1}{(14)^{0.4/1.4}} = 52.96\% \qquad \textbf{\textit{Ans.}}$$

**EXAMPLE 18.2** A gas turbine plant consists of one turbine for compressor drive and another for output and both have their own combustion chambers which are served by air directly from the compressor. Air enters the compressor at 1 bar and 15°C and is compressed with an isentropic efficiency of 0.86. Gas inlet temperature and pressure in both turbines are 1300°C and 30 bar. The isentropic efficiency of turbine is 0.86. The mass flow rate of air at the compressor inlet is 130 kg/s. The calorific value of fuel is 42000 kJ/kg. Calculate the output of the plant and thermal efficiency assuming $c_p$ = 1.00.5 and $\gamma$ = 1.4 for air and $c_p$ = 1.128 kJ/kg K and $\gamma$ = 1.34 for gases. Also find air standard efficiency.

Assume $\eta_m$ = 0.98 for compressor and turbine each, $\eta_{comb}$ = 0.995

***Solution*** Refer to Fig. 18.8 but system sketch will be different. Here, $r_p$ = 30, $T_1$ = 273 + 15 = 288 K

$$T_{2s} = T_1 \times (30)^{0.286} = 761.81 \text{ K}$$

$$T_2 = T_1 + \frac{T_{2s} - T_1}{0.86} = 288 + \frac{761.81 - 288}{0.86} = 838.94 \text{ K}$$

$$T_3 = 1300 + 273 = 1573 \text{ K}$$

$$\frac{T_3}{T_{4s}} = (30)^{0.34/1.34} = (30)^{0.2537} = 2.369$$

$$T_{4s} = \frac{1573}{2.369} = 663.99 \text{ K}$$

$$\therefore \quad T_4 = T_3 - (T_3 - T_{4s}) \times 0.86 = 1573 - (1573 - 663.99) \times 0.86 = 791.25 \text{ K}$$

Work consumed by compressor = $w_{ca} = c_p\,(T_2 - T_1)/\eta_{mc}$

$$= 1.005\,(838.94 - 288)/0.98 = 564.99 \text{ kJ/kg of air}$$

Work done by both turbines = $w_{ta} = c_{pg}\,(T_3 - T_4) \times \eta_{mt}$

$$= 1.128\,(1573 - 791.35) \times 0.98 = 864.17 \text{ kJ/kg of air}$$

Gas required per kg of air compressed in compressor drive turbine to supply power to compressor

$$\frac{\dot{m}_{gc}}{\dot{m}_a} = \frac{w_{ca}}{w_{ta}} = \frac{564.99}{864.17} = 0.653 \text{ kg}$$

Let $\dot{m}_f$ be the total amount of fuel supplied per kg of air in the combustion chambers.

The heat supplied = $q_{Aa} = (\dot{m}_a + \dot{m}_f)\, h_3 - \dot{m}_a\, h_2 = \dot{m}_f \times \text{C.V.} \times \eta_{comb}$

or $$(1 + \dot{m}_f)\, c_{pg} \times T_3 - 1 \times c_{pa}\, T_2 = \dot{m}_f \times 42000 \times 0.995$$

or $$(1 + \dot{m}_f)\, 1.128 \times 1573 - 1.005 \times 838.94 = \dot{m}_f \times 42000 \times 0.995$$

or $$\dot{m}_f = 0.0232 \text{ kg of fuel/kg of air}$$

Total amount of gases per kg of air supplied in power turbines

$$= 1 + 0.0232 = 1.0232 \text{ kg}$$

Since gases required per kg of air in compressor driven turbine = 0.653 kg,

hence, gases supplied to output turbines = 1.0232 – 0.653 = 0.3702 kg/kg of air

Power developed = 0.3702 × 130 × 864.27 = 41589 kW = 41.5891 MW **_Ans._**

Thermal efficiency = $\dfrac{864.17 \times 0.3702}{0.0232 \times 42000} = 32.83\%.$ **_Ans._**

$$ASE = 1 - \frac{1}{(r_p)^{(\gamma-1)/\gamma}} = 1 - \frac{1}{(14)^{0.2887}} = 52.967\%$$ **_Ans._**

**EXAMPLE 18.3** Air at temperature of 15°C enters a gas turbine plant working at pressure ratio of 15. Turbine inlet temperature is 1250°C. Polytropic efficiency (i.e., small stage efficiency) of compressor and turbine is 0.91. Assume $c_p$ = 1.005 for air and 1.128 kJ/kgK for gases and calorific value of fuel used = 42000 kJ/kg of fuel. Calculate (a) overall efficiency, (b) specific output, (c) fuel to air ratio, and (d) specific fuel consumption.

***Solution*** Refer to Fig. 18.8. Given $\dfrac{p_2}{p_1} = 15$, $T_1$ = 288 K, $T_3$ = 1250 + 273 = 1523 K

$$c_p = 1.005 \text{ kJ/kgK, C.V.} = 42000 \text{ kJ/kg},$$

$$c_{pg} = 1.128 \text{ kJ/kgK}, \eta_{pc} = \eta_{pt} = 0.91$$

Polytropic efficiency = $\eta_{pc} = \left(\dfrac{n}{n-1}\right)\left(\dfrac{\gamma - 1}{\gamma}\right)$

where $n$ = index of polytropic compression ($n > \gamma$)

$$\therefore \quad \left(\frac{n-1}{n}\right) = \left(\frac{\gamma-1}{\gamma}\right) \cdot \frac{1}{\eta_{pc}}$$

Now, from gas laws, the actual temperature is

$$\frac{T_2}{T_1} = \left(\frac{p_2}{p_1}\right)^{(n-1)/n} = \left(\frac{p_2}{p_1}\right)^{(\gamma-1)/\gamma\eta_{pc}}$$

$$T_2 = 288(15)^{0.4/1.4 \times 0.91} = 673.98 \text{ K}$$

Again, for real expansion processes,

$$\frac{n-1}{n} = \frac{\gamma - 1}{\gamma}.\eta_{pt}$$

Hence, $$\frac{T_3}{T_4} = \left(\frac{p_2}{p_1}\right)^{(\gamma-1)\eta_{pt}/\gamma} = (15)^{0.4 \times 0.91/1.4} = 2.022$$

$\therefore$ $$T_4 = \frac{1523}{2.022} = 753.21 \text{ K}$$

(b) Net work done = Turbine work – Compressor work = $w_{ta} - w_{ca}$

$$= c_{pg}\,(T_3 - T_4) - c_{pa}\,(T_2 - T_1)$$

$$= 1.128\,(1523 - 753.21) - 1.005\,(673.98 - 288) = 480.42 \text{ kJ/kg}$$ ***Ans.***

Heat supplied = $(c_{pg}\,T_3 - c_{pa}\,T_2) = 1.128 \times 1523 - 1.005 \times 673.98 = 1040.59$ kJ/kg

(a) Overall efficiency $= \frac{\text{Net work}}{\text{Heat supplied}} = \frac{480.42}{1040.59} = 46.16\%$ ***Ans.***

(c) Fuel-air ratio = Fuel used per kg of air

$$= \frac{\text{External heat supplied in kJ/kg of air}}{\text{Calorific value of fuel in kJ/kg of fuel}} = \frac{1040.59}{42000} = 0.0247$$ ***Ans.***

(d) Specific air consumption $= \frac{1}{\text{Specific output}} = \frac{1}{480.42}$ kg/kWs

Specific fuel consumption = Specific air consumption $\times \frac{F}{A}$

$$= \frac{3600 \times 0.0247}{480.42} = 0.185 \text{ kg/kWh}$$ ***Ans.***

**EXAMPLE 18.4** A simple gas turbine takes in air at atmospheric pressure and 15°C and compresses air in the compressor up to 6 bar. The air enters the combustion chamber and is heated to a maximum temperature of 750°C, then it enters the turbine and expands to atmospheric pressure. The isentropic efficiency of compressor and turbine each is 0.8, combustion efficiency 0.9, fall of pressure through the combustion system is 0.1 bar, $c_p$ for both air and gases 1.00.5 kJ/kgK, $\gamma = 1.4$. Determine the flow of air and gases for net power of 1100 kW developed. Calculate also the heat supplied per kg of air and work ratio.

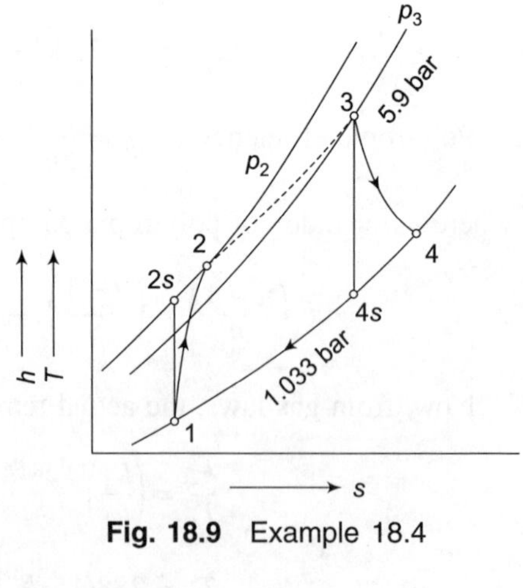


**Fig. 18.9** Example 18.4

***Solution*** Refer to Fig. 18.9. Data given as: $p_1$ = 1.013 bar, $p_4$ = 1.033 bar, $T_1$ = 288 K, $p_2$ = 6 bar, $p_3$ = 5.9 bar, $\eta_c = \eta_t = 0.8$, $\eta_{com} = 0.9$, power = 1100 kW.

$$\frac{T_{2s}}{T_1} = \left(\frac{p_2}{p_1}\right)^{(\gamma-1)/\gamma} = \left(\frac{6}{1.013}\right)^{(0.4)/1.4} = 1.662$$

$$T_{2s} = 288 \times 1.662 = 478.65 \text{ K}$$

$$T_2 - T_1 = \frac{(T_{2s} - T_1)}{\eta_c} = \frac{478.65 - 288}{0.8} = 238.31 \text{ K}$$

Also
$$T_{4s} = \frac{T_3}{(p_3/p_4)^{(\gamma-1)/\gamma}} = \frac{1023}{(5.9/1.033)^{(0.4)/1.4}} = 623.78 \text{ K}$$

$$(T_4 - T_3) = (T_3 - T_{4s}) \times \eta_t = (1023 - 623.78) \times 0.8 = 319.37 \text{ K}$$

Net work = Turbine work – Compressor work = $c_p\,(T_3 - T_4) - c_p\,(T_2 - T_1)$

= 1.005 (319.37 – 238.31) = 81.46 kJ/kg of air

Power developed = 1100 = $\dot{m}_a \times 81.46$

or $\dot{m}_a$ = 13.50 kg/s ***Ans.***

Effective heat supplied = $c_p\,(T_3 - T_2)$ kJ/kg of air

= 1.005 (1023 – 526.31) = 499.17 kJ/kg or air

Actual heat supplied = $\frac{449.17}{0.9}$ = 554.63 kJ/kg or air ***Ans.***

Work ratio = $\frac{\text{Actual work}}{\text{Turbine work}} = \frac{81.46}{320.96}$ = 25.38% ***Ans.***

## 18.6 OPTIMUM PRESSURE RATIO FOR THE MAXIMUM SPECIFIC OUTPUT IN ACTUAL SIMPLE GAS TURBINE CYCLE

Referring to Fig. 18.5, we have

$$\frac{T_{2s}}{T_1} = \frac{T_3}{T_{4s}} = \left(\frac{p_2}{p_1}\right)^{(\gamma-1)/\gamma} = \left(\frac{p_3}{p_4}\right)^{(\gamma-1)/\gamma} = (r_p)^{(\gamma-1)/\gamma} = x$$

Actual compressor work = $w_{ca} = (h_2 - h_1) = \frac{(h_{2s} - h_1)}{\eta_c} = \frac{c_p(T_{2s} - T_1)}{\eta_c}$; kJ/kg

Actual turbine work = $w_{ta} = (h_3 - h_4) = (h_3 - h_{4s})\,\eta_t = c_p \times (T_3 - T_{4s})\,\eta_t$, kJ/kg

Thus, the net output = $w_{neta} = c_p\,(T_3 - T_{4s})\eta_t - \frac{c_p(T_{2s} - T_1)}{\eta_c}$

$$= c_p\,\eta_t . T_3\left(1 - \frac{T_{4s}}{T_3}\right) - \frac{c_p\,T_1}{\eta_c}\left(\frac{T_{2s}}{T_1} - 1\right)$$

$$= c_P \eta_t T_3 \left(1 - \frac{1}{x}\right) - \frac{c_p T_1}{\eta_c} (x - 1) \quad ...(18.13)$$

For the maximum output, differentiating $w_{net}$. w.r.t. to $x$ and equating to zero

$$\frac{dw_{neta}}{dx} = 0 = c_p . \eta_t . T_3 \left(\frac{1}{x^2}\right) - \frac{c_p . T_1}{\eta_c}$$

So $$x = \sqrt{\left(\eta_t . \eta_c . \frac{T_3}{T_1}\right)} = \left(\eta_t . \eta_c . \frac{T_{max}}{T_{min}}\right)^{1/2}$$

or $$(r_p)^{(\gamma - 1)/\gamma} = \left(\eta_t . \eta_c . \frac{T_3}{T_1}\right)^{1/2}$$

or $$(r_p)_{opt} = \left(\eta_t . \eta_c . \frac{T_3}{T_1}\right)^{\gamma / 2(\gamma - 1)} = \left(\eta_t . \eta_c . \frac{T_{max}}{T_{min}}\right)^{\gamma / 2(\gamma - 1)} \quad ...(18.14)$$

## 18.7 OPTIMUM PRESSURE RATIO FOR THE MAXIMUM CYCLE THERMAL EFFICIENCY

From the above article, the net output is from Eq. (18.13)

$$w_{neta} = c_p \eta_t T_3\left(1 - \frac{1}{x}\right) - \frac{c_p T_1}{\eta_c} (x - 1)$$

Heat supplied $= c_p (T_3 - T_2)$

This approximation is only for simplification of problem without any appreciable error,

Let $$\frac{T_3}{T_1} = y \quad \text{and} \quad \frac{T_3}{T_4} = \frac{T_2}{T_1} = (r_p)^{(\gamma - 1)/\gamma} = x$$

So, the heat supplied $= c_p(T_3 - T_1 . x) = c_p T_1(y - x)$

The actual thermal efficiency of the cycle is

$$\eta_{th_a} = \frac{\eta_t T_3(1 - 1/x) - T_1(x - 1)/\eta_c}{T_1(y - x)} = \frac{\eta_t y(1 - 1/x) - (x - 1)/\eta_c}{(y - x)}$$

For the maximum thermal efficiency

$$\frac{d\eta_{tha}}{dx} = 0 = \left[\eta_t y(1/x^2)(y - x) - \frac{1}{\eta_c}(y - x)\right]$$
$$- \left[\eta_t y\left(1 - \frac{1}{x}\right) - \frac{1}{\eta_c} (x - 1)\right] (-1)$$

or $$\left[\frac{\eta_t(y - x)}{x^2} - \frac{y - x}{\eta_c}\right] + \left[\frac{\eta_t y(x - 1)}{x} - \frac{(x - 1)}{\eta_c}\right] = 0$$

or $$\left[\frac{\eta_t\,\eta_t.y - x^2}{x^2\eta_c}\right](y - x) + (x - 1)\left[\frac{\eta_c.\eta_t.y - x}{x.\eta_c}\right] = 0$$

or $$\eta_c.\eta_t.y^2 - yx^2 - \eta_c.\eta_t\,y\,x - x^3 = x^3 - x^2\,\eta_c\,\eta_t.y - x^3 + x.y\,\eta_c\,\eta_t$$

Multiplying throughout by $\frac{1}{x^2}$, we have

$$[y\,(\eta_c\,\eta_t - 1) + 1] - \frac{2\eta_c\,\eta_t\,y}{x} + \frac{\eta_c\,\eta_t\,y^2}{x^2} = 0$$

or $$\frac{1}{x^2}(\eta_c\eta_t\,y^2) - \frac{1}{x}.(2\eta_c\,\eta_t\,y) + [y\,(\eta_c\,\eta_t - 1) + 1] = 0$$

This is quadratic equation in $x$

$$\frac{1}{x} = \frac{2\eta_c\,\eta_t y \pm \sqrt{4\{[(\eta_c\eta_t\,y)^2 - 4(\eta_c\eta_t\,y^2)]\,[y(\eta_c\eta_t - 1) + 1]\}}}{2\eta_c\eta_t\,y^2}$$

Taking only positive roots, we have

$$x = \frac{2\eta_c\eta_t y^2}{2\eta_c\left[\eta_t\,y + 2\sqrt{(\eta_c\eta_t y)^2 - (\eta_c\eta_t\,y^2)\,[y(\eta_c\eta_t - 1) + 1]}\right]}$$

or $$x = \frac{2\eta_c\eta_t\,y^2}{2\eta_c\eta_t\,y[1 + \sqrt{\{(y - 1)\{(1/\eta_c\eta_t) - 1\}\}}}$$

Therefore,

$$x = \frac{y}{\left[1 + \sqrt{[(y - 1)\{(1/\eta_c\eta_t) - 1\}}\right]}$$

Thus, $$(r_p)_{opt} = \left[\frac{T_3/T_1}{1 + \sqrt{\{[(T_3/T_1) - 1][(1/\eta_c\eta_t) - 1]\}}}\right]^{\gamma/\gamma - 1}$$

## 18.8 MEANS OF IMPROVING THE EFFICIENCY AND THE SPECIFIC OUTPUT OF SIMPLE CYCLE

The improvement of the simple cycle plant may be done by the following methods:

**(1) Regeneration.** This is done by preheating the air with the turbine exhaust, thus saving the fuel consumption.

**(2) Improving turbine output.** This may be done by

- **(i) Reheating.** The whole expansion in the turbine is achieved in two or more stages and reheating is done after each stage.
- **(ii)** Increasing the value of the maximum cycle temperature, i.e., turbine inlet temperature. This requires:

(i) Better quality of fuel.

(ii) New materials which can withstand high temperatures.

**(iii) Improved turbine efficiency.** It depends on design improvements.

**(iv) Improved turbine blade cooling techniques.** This requires less cooling air bled from compressor.

**(3) Reducing compressor input.** This may be done by

**(i) *Intercooling.*** Compressor work is reduced by intercooling the air between compressor stages.

**(ii) *By lowering inlet temperature to compressor.*** It is done by inlet refrigeration or fogging. The mass flow rate increases due to higher density.

**(iii) *By increasing the compressor efficiency.*** This depends upon the design improvement.

**(iv) *Water injection.*** By injection water at the inlet to compressor, the work output and efficiency are increased due to the extra mass of water injection and cooling of air.

## 18.9 OPEN CYCLE GAS TURBINE (OCGT) WITH REGENERATION (OR RECUPERATION)

In this method, the air delivered by the compressor passes through a heat exchanger utilizing the gases exhausted from the turbine; the heated air then passes into the combustion chamber and already partly heated air is employed to burn the fuel. Since some heat is added to the air in the heat exchanger itself, so the same turbine gas inlet temperature is attained as that when no heat exchanger is employed; with a lower fuel consumption, hence thermal efficiency will be higher accordingly. The compressor, turbine and net work are not affected by the addition of the regenerator (or recuperator) to the cycle, only the heat required to be supplied in the combustion chamber is decreased which gives the gain in thermal efficiency.

**Ideal Cycle.** Process 1–2*s*–5–3–4*s*–6–1 shows the ideal open cycle gas turbine with regeneration as shown in Fig. 18.10 (a, b). The high pressure air at state 2 is passed through a regenerator (or heat-exchanger) where it is heated at constant pressure from $T_2$ *to* $T_5$ due to which the exhaust gas from the turbine at state 4 is cooled from $T_4$ to $T_6$. With perfect heat exchange in the heat exchanger, in an ideal case, the temperature of air leaving the heat exchanger will be equal to the temperature of exhaust gas entering the heat exchanger and the temperature of exhaust gas leaving heat exchanger will be equal to the temperature of air entering heat exchanger. Thus,

$$T_5 = T_{4s} \quad \text{and} \quad T_6 = T_{2s} \quad \text{i.e.,} \quad h_5 = h_{4s} \quad \text{and} \quad h_6 = h_{2s}$$

The air from the heat exchanger at temperature $T_5$ then passes through the combustion chamber where it is heated from temperature $T_5$ to $T_3$ ( i.e., from $h_5$ to $h_3$). The advantage of exhaust heat exchange is that the amount of heat supplied in the combustion chamber is reduced from $(h_3 - h_2)$ to $(h_3 - h_5)$, i.e., from $c_p(T_3 - T_2)$ to $c_p\,(T_3 - T_5)$. Thus, the thermal efficiency is increased because the heat supplied is reduced.

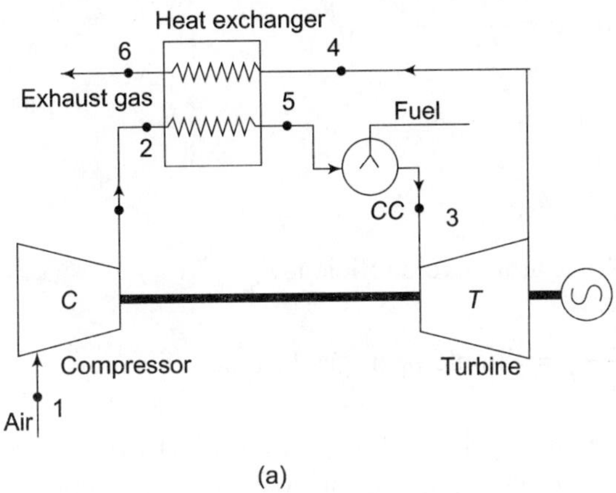


**Fig. 18.10** (a) OCGT with regeneration

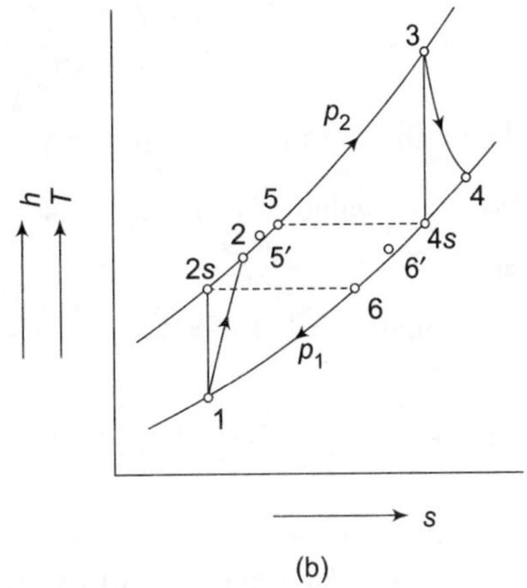


**Fig. 18.10** (b) Process on *T-s* chart for OCGT with regeneration

Assuming 1 kg of air flowing through the compressor.

Turbine work = $w_t = h_3 - h_{4s} = c_p\,(T_3 - T_{4s})$ ; kJ/kg

Compressor work = $w_c = h_{2s} - h_1 = c_p\,(T_{2s} - T_1)$ ; kJ/kg

Net work = $w_{net} = c_p[(T_3 - T_{4s}) - (T_{2s} - T_1)]$

Heat supplied = $q_A = h_3 - h_5 = c_p(T_3 - T_5) = c_p(T_3 - T_{4s})$

So the thermal efficiency,

$$\eta_{th} = \frac{c_p[T_3 - T_{4s}) - (T_{2s} - T_1)]}{c_p(T_3 - T_{4s})} = 1 - \frac{T_{2s} - T_1}{T_3 - T_{4s}} \qquad ...(18.15)$$

But we know that $\dfrac{T_{2s}}{T_1} = \left(\dfrac{p_2}{p_1}\right)^{(\gamma-1)/\gamma} = (r_p)^{(\gamma-1)/\gamma}$

and

$$\frac{T_3}{T_{4s}} = \left(\frac{p_3}{p_2}\right)^{(\gamma-1)/\gamma} = \left(\frac{p_2}{p_1}\right)^{(\gamma-1)/\gamma} = (r_p)^{(\gamma-1)/\gamma}$$

So

$$\frac{T_{2s}}{T_1} = \frac{T_3}{T_{4s}} \quad \text{or} \quad \frac{T_{2s}}{T_3} = \frac{T_1}{T_{4s}} = \frac{T_{2s} - T_1}{T_3 - T_{4s}}$$

By putting the value of $T_{2s} - T_1$ in Eq, (18.15), we get

$$\eta_{th} = 1 - \frac{T_{2s}}{T_3} = 1 - \frac{T_{2s}}{T_1}\cdot\frac{T_1}{T_3}$$

So

$$\eta_{th} = 1 - \frac{T_1}{T_3}(r_p)^{(\gamma-1)/\gamma} = 1 - \frac{T_{min}}{T_{max}}(r_p)^{(\gamma-1)/\gamma} \qquad ...(18.16)$$

Let $$t = \frac{T_3}{T_1} = \frac{T_{max}}{T_{min}} \quad \text{and} \quad c = \frac{T_{2s}}{T_1} = \frac{T_3}{T_{4s}} = (r_p)^{(\gamma-1)/\gamma}$$

Then, the thermal efficiency $= \eta_{th} = 1 - \frac{c}{t}$ ...(18.16a)

For fixed value of $c$, $\eta_{th}$ increases with $t$.

**Cases:**

(1) When $c = 1$, $\eta_{th} = 1 - \frac{1}{t} = 1 - \frac{T_{min}}{T_{max}}$ = Carnot cycle efficiency

(2) When $t = c^2$, $\eta = 1 - \frac{1}{c} = 1 - \frac{1}{(r_p)^{(\gamma-1)/\gamma}}$ = Simple open gas turbine efficiency.

**Actual Cycle.** 1–2–5′–3–4–6′–1 shows the actual cycle as shown in the Fig.18.10. In actual practice, the compression and expansion process are not isentropic. Assume 1 kg of air flowing through the compressor. If we neglect the mass of the fuel and assume the same value of $c_p$ for air and exhaust gas, turbine and compressor work may be computed per kg of air as follows:

Actual turbine work $= w_{ta} = h_3 - h_4 = c_p(T_3 - T_4)$

Actual compressor work $= w_{ca} = h_2 - h_1 = c_p(T_2 - T_1)$

Actual net work $= w_{neta} = c_p[(T_3 - T_4) - (T_2 - T_1)]$

Actual heat supplied $= q_{Aa} = h_3 - h'_5 = c_p(T_3 - T'_5)$

Actual thermal efficiency, $(\eta_{tha})$,

$$\eta_{tha} = \frac{\text{Actual net work}}{\text{Actual heat supplied}} = \frac{(T_3 - T_4) - (T_2 - T_1)}{T_3 - T_5'}$$

or $$\eta_{tha} = \frac{(T_3 - T_{4s})\eta_t - [(T_{2s} - T_1)/\eta_c]}{T_3 - T_5'}$$

or $$\eta_{th} = \frac{(T_3 - T_{4s})\,\eta_t\,\eta_c - (T_{2s} - T_1)}{\eta_c\,(T_3 - T_5')}$$ ...(18.17)

In actual practice, perfect heat exchange is never possible in the regenerator. As a result, $h'_5 < h_4$ and $h'_6 > h_2$ (i.e., $T_5' < T_4$ and $T_6' > T_2$) in the actual cycle. The regenerator effectiveness, $\varepsilon_r$, is defined as

$$\varepsilon_r = \frac{\text{Actual heat exchange}}{\text{Maximum theoretical heat exchange}}$$

The actual heat exchange $= (h'_5 - h_2) = (h_4 - h'_6)$

The maximum theoretical heat exchange $= (h_4 - h_2) = h'_5 - h'_6$

Hence, $$\varepsilon_r = \frac{h'_5 - h_2}{h_4 - h_2} = \frac{h_4 - h'_6}{h'_5 - h'_6}$$

or $$\varepsilon_r = \frac{T_5' - T_2}{T_4 - T_2} = \frac{T_4 - T_6'}{T_5' - T_6'},$$ ...(18.18)

If $\varepsilon_r$ is known, $T_5'$ can be found out and hence thermal efficiency can be evaluated.

*A drawback of such a heat exchanger, acting as a regenerator or recuperator, is the low overall heat transfer coefficient when both the heating and heated fluids are gases.* The maximum effectiveness is about 75 per cent. *This, in turn, requires a large heat exchange area.* Another drawback is the resistance to flow of the gases, which causes a *pressure* drop in the compressed air, and increase the back pressure on the turbine. However, by careful design these losses can be kept low.

Figure 18.11 shows the effect of regeneration on the *specific fuel consumption* for various values of compression ratio. It is obvious from the figure that up to a certain value of compression ratio, the specific fuel consumption is less except for much lower value of regenerator effectiveness. After a certain value of pressure ratio, i.e., when the temperature of compressed air is more than the exhaust temperature of gas, the specific fuel consumption increases. Thus, we can say that ***use of regenerator is limited by pressure ratio, for a fixed value of the maximum temperature in the cycle.*** *Note that the specific fuel consumption is the minimum at lower value of pressure ratio, i.e., a lower value of compressed air temperature.*

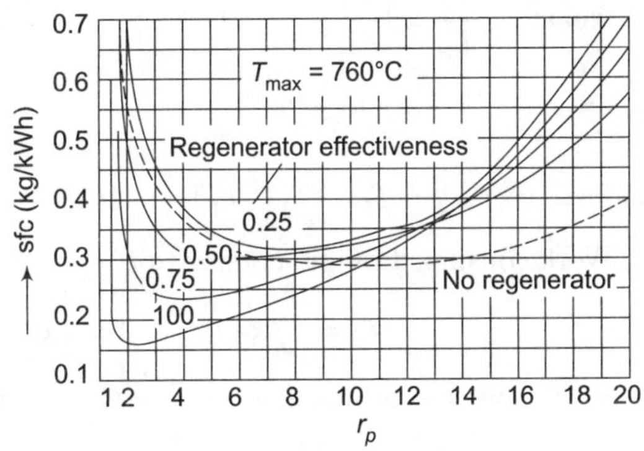


**Fig. 18.11** Effect of regeneration (or recuperation)

**EXAMPLE 18.5** In a gas turbine plant, the air at 10°C and 1 bar is compressed to 12 bar with compression efficiency of 80%. The air is heated in the regenerator and the combustion chamber till its temperature is raised to 1400°C, and during the process, the pressure falls by 0.2 bar. The air is then expanded in the turbine and passed to regenerator which has 75% effectiveness and causes a pressure drop of 0.2 bar. If the isentropic efficiency of the turbine is 85%, determine the thermal efficiency of the plant.

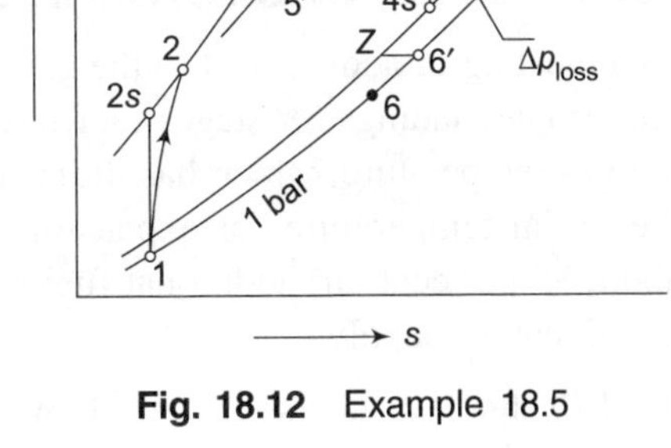


**Fig. 18.12** Example 18.5

***Solution*** Refer to Fig. 18.12. Though, there is a fall in pressure in regenerator and combustion chamber but for simplicity, addition of heat will be assumed at constant pressure, i.e., the process 2–5–3 will be assumed as 2–*x*–5–3 and process 4–6 will be assumed as 4-*z*-6.

**Given:** $T_1 = 273 + 10 = 283$ K

$$\therefore \quad T_{2s} = T_1 \left(\frac{p_2}{p_1}\right)^{(\gamma-1)/\gamma} = 283 \times (12)^{0.286} = 576 \text{ K}$$

Now $$T_2 = T_1 + (T_{2s} - T_1)\eta_c = 283 + \frac{576 - 283}{0.8} = 649.25 \text{ K}$$

$$p_3 = 12 - 0.2 = 11.8 \text{ bar and } p_5 = 1 + 0.2 = 1.2 \text{ bar}$$

Again, $$\frac{T_3}{T_{4s}} = \left(\frac{p_3}{p_{4s}}\right)^{0.286} = \left(\frac{11.8}{1.2}\right)^{0.286} = 1.922$$

or $$T_{4s} = 870.44 \text{ K}$$

and $$T_4 = T_3 - (T_3 - T_{4s})\eta_t = 1673 - (1673 - 870.44) \times 0.85 = 990.82 \text{ K}$$

Work consumed by the compressor

$$= w_{ca} = c_p \times (T_2 - T_1) = 1.005\,(649.25 - 283) = 368.08 \text{ kJ/kg}$$

Work done by turbine $= w_{ta} = c_p \times (T_3 - T_4) = 1.005(1673 - 990.82) = 686.09$ kJ/kg

Net work $= w_{neta} = w_{ta} - w_{ca} = 686.09 - 368.08 = 318.01$ kJ/kg

The regenerator effectiveness is given by

$$\varepsilon_r = \frac{T_5' - T_2}{T_4 - T_2} = 0.75$$

or $$T_5' = 649.25 + (990.82 - 649.25) \times 0.75 = 905.42 \text{ K}$$

Heat supplied $= q_{Aa} = c_p(T_3 - T_5') = 1.005\,(1673 - 905.42) = 771.41$ kJ/kg

$$\eta_{th} = \frac{318.01}{771.41} = 41.22\%.$$ ***Ans.***

## 18.10 OPEN GAS TURBINE CYCLE WITH REHEAT

By reheating or adding heat to the gases after they have passed through a part of the rows of the turbine bladings (or stages), a **further increase in work done is obtained**. In reheating, the gas temperature, which has dropped due to expansion is brought back to approximately the initial temperature for expansion in the next stage. Since the working fluid contains about 85 per cent air, additional fuel can be burnt by injecting it into the gases without any additional air supply.

**Ideal Cycle.** 1–2*s*–3–4*s*–5–6*s*–1 shows the ideal cycle gas turbine with reheat as shown in Figs.18.13 and 18.14 The combustion gas from the combustor CC1 at temperature $T_3$ is partially expanded in the turbine number 1 from pressure $p_2$ to intermediate pressure $p_x$. After this, it is then passed through combustion chamber ($CC2$) where it is reheated or additional heat is added at constant pressure ($p_x$) so that the temperature of the gas is raised from $T_4$ to $T_5 = T_3$. After this, the gas is expanded in the second stage turbine from pressure $p_x$ to $p_1$.

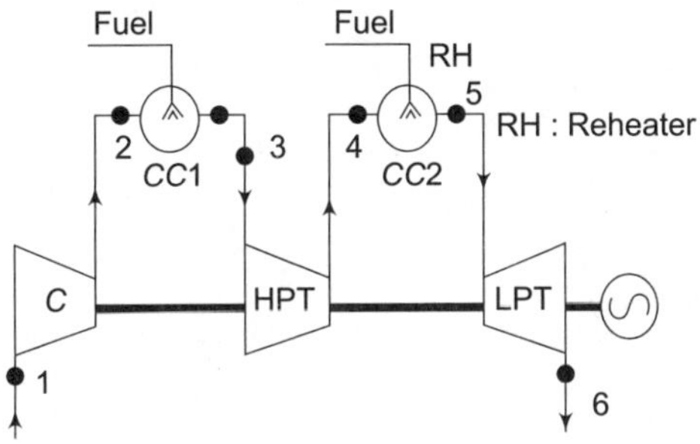


**Fig. 18.13** Open gas turbine cycle with reheat

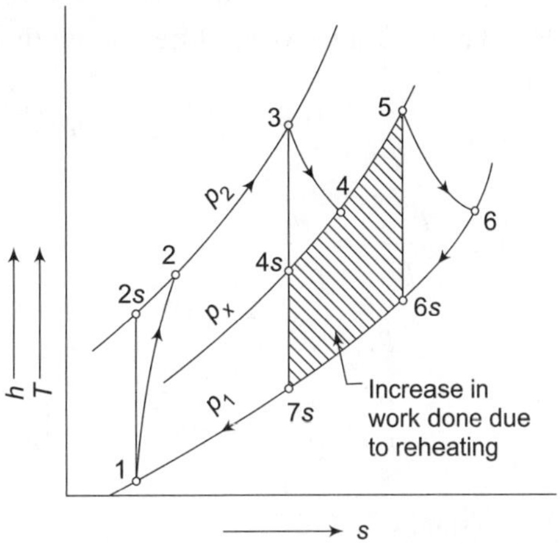


**Fig. 18.14** Reheat on *T-s* diagram

The net work done is more with reheat compared to the ideal cycle (without reheat). The efficiency with which the heat is added to the gas in the combustion chamber ($CC2$) at lower pressure ($p_x$) is lower than the efficiency when this heat is added to the gas in the combustion chamber ($CC1$) at higher pressure ($p_x$), so efficiency decreases as a result of reheating, only the net work is more by hatched area $4s$–5–$6s$–$7s$ shown in Fig. 18.14.

Net work = $w_{net} = (h_3 - h_{4s}) + (h_5 - h_{6s}) - (h_{2s} - h_1)$

Heat supplied = $(h_3 - h_{2s}) + (h_5 - h_{4s})$

So, the thermal efficiency,

$$\eta_{th} = \frac{(h_3 - h_{4s}) + (h_5 - h_{6s}) - (h_{2s} - h_1)}{(h_3 - h_{2s}) + (h_5 - h_{4s})} = \frac{(T_3 - T_{4s}) + (T_5 - T_{6s}) - (T_{2s} - T_1)}{(T_3 - T_{2s}) + (T_5 - T_{4s})}$$

For the maximum output, there must be an optimum pressure at which reheating should be done. As we know, the compressor work ($w_c$) is not affected by reheating, so for a maximum output, we have to find the condition where turbine work ($w_t$) is the maximum.

$$w_t = c_p[(T_3 - T_{4s}) + (T_5 - T_{6s})] = c_p\left[T_3\left(1 - \frac{T_{4s}}{T_3}\right) + T_5\left(1 - \frac{T_{6s}}{T_5}\right)\right]$$

$$= c_p\left[T_3\left\{1 - (p_x/p_2)^{(\gamma-1)/\gamma}\right\} + T_5\left\{1 - (p_1/p_x)^{(\gamma-1)/\gamma}\right\}\right]$$

Here, $T_3 = T_5$ and $c_p$, $T_3$, $p_1$, $p_2$ are constants. The only variable is $p_x$.

Let $c_p T_3 = C = \text{constant}$ and $\dfrac{\gamma - 1}{\gamma} = a$

Then $$w_t = C\left[1 - \left(\frac{p_x}{p_2}\right)^a + \left(\frac{p_1}{p_x}\right)^a\right]$$

For the maximum work, differentiate the above equation w.r.t. $p_x$ and equate to zero.

$$\frac{dw_t}{dp_x} = 0 = ap_x^{a-1}\, p_2^{-a} - ap_x^{-a-1}\, p_1^a$$

Hence, $$\frac{p_x^{a-1}}{p_2^a} = \frac{p_1^a}{p_x^{a+1}}$$

or $$p_x = \sqrt{(p_1 p_2)} \qquad \text{...(18.19)}$$

or $$\frac{p_x}{p_1} = \frac{p_2}{p_x} = \sqrt{(p_2/p_1)} \quad \text{or} \quad r_{p1} = r_{p2} = \sqrt{(r_p)}$$

For $n$ stages,

$$r_{p1} = r_{p2} = r_{p3} = \sqrt[n]{(r_p)}$$

*So, we can say that the optimum intermediate pressure of reheating is the square root of initial and final pressure, i.e.,* the pressure ratio in each stage of turbine is equal to the square root of total pressure ratio.

*For finding out thermal efficiency for the maximum output in this cycle, this value of intermediate pressure should be used.*

**Actual Cycle.** 1–2–3–4–5–6–1 shows actual cycle with reheat as shown in Fig. 18.14.

Actual turbine work = $w_{ta} = (h_3 - h_4) + (h_5 - h_6) = c_p[(T_3 - T_4) + (T_5 - T_6)]$

Actual compressor work = $w_{ca} = h_2 - h_1 = c_p(T_2 - T_1)$

Actual heat supplied = $(h_3 - h_2) + (h_5 - h_4) = c_p[(T_3 - T_2) + (T_5 - T_4)]$

$$\therefore \quad \eta_{tha} = \frac{(T_3 - T_4) + (T_5 - T_6) - (T_2 - T_1)}{(T_3 - T_2) + (T_5 - T_4)}$$

or $$\eta_{tha} = \frac{(T_3 - T_{4s})\eta_t + (T_5 - T_{6s})\eta_t - [(T_{2s} - T_1)/\eta_c]}{(T_3 - T_2) + (T_5 - T_4)} \qquad \text{...(18.20)}$$

In this case also, the condition for the maximum output will be the same; i.e., $r_{p1} = r_{p2}$.

## 18.11 OPEN GAS TURBINE CYCLE (OGTC) WITH INTERCOOLING

*The net work of the gas turbine cycle may be increased by saving some compression work.* This is done by using several stages of compression with intercooling of air between stages.

The air from the first stage compressor is cooled in a intercooler approximately to initial temperature before entering the second stage compressor. *The effect of intercooling is to increase the network and decrease efficiency as compared to the simple ideal cycle without intercooling. But there is one value of optimum intercooling pressure where there may be also small gain in thermal efficiency along with appreciable increase in power output (LMS 100 model).*

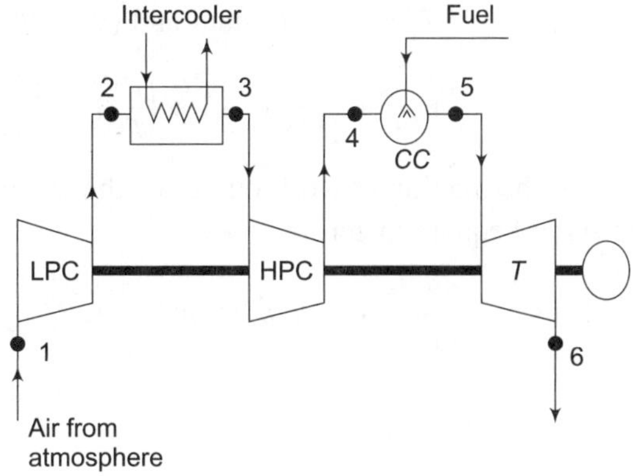


**Fig. 18.15** OGTC with intercooling

**Ideal Cycle.** 1–2*s*–3–4*s*–5–6*s*–1 shows the ideal cycle gas turbine with intercooling as shown in Figs. 18.15 and 18.16. In the first stage compressor, atmospheric air is compressed from $p_1$ to $p_{2'}$ it is then cooled from temperature $T_{2s}$ to $T_3 = T_1$ in the intercooling at constant intermediate pressure $p_x$, and finally compressed from $p_x$ to $p_2$ in the second stage of compressor. At constant pressure $p_2$ heat is added in the combustion chamber up to $T_5$ and after that it is expanded in the turbine from pressure $p_2$ and temperature $T_5$ to pressure $p_x$. It is evident from Fig. 18.16 that the turbine work is the same as for the simple cycle but the work supplied to the compressor is reduced because the constant pressure lines converge towards the left. The work supplied to compressor is saved due to intercooling which may be shown as area 2*s*–3–4*s*–a–2*s*, so this is the amount by which net work of the cycle is increased as compared to ideal cycle without intercooling. But the heat supplied is increased by amount $(h_a - h_{4s})$ because the compressed air delivered by second stage compressor is at lower temperature $(T_{4s})$ than in the simple cycle. So the effect of the intercooling is to decrease the thermal efficiency of the cycle. This result is to be expected because the heat supplied in the intercooler is a complete loss, no part of it is converted into useful work.

Turbine work = $w_t = h_5 - h_{6s} = c_p\,(T_5 - T_{6s})$

Compressor work = $w_c = (h_{2s} - h_1) + (h_{4s} - h_3) = c_p\,(T_{2s} - T_1) + c_p[(T_{4s} - T_3)]$

Heat supplied = $h_5 - h_{4s} = c_p\,(T_5 - T_{4s})$

So, the thermal efficiency = $n_{th} = \dfrac{(T_5 - T_{6s}) - (T_{4s} - T_3) - (T_{2s} - T_1)}{T_5 - T_{4s}}$ ...(18.21)

For the minimum compression work in the compressor or the maximum work output in this cycle, only the compressor work is responsible.

$$w_c = c_p\,[(T_{2s} - T_1) + (T_{4s} - T_3)] = c_p\left[\left\{T_1\left(\frac{T_{2s}}{T_1} - 1\right) + T_3\left(\frac{T_{4s}}{T_3} - 1\right)\right\}\right]$$

$$= c_p\,T_1\left[\left(\frac{p_x}{p_1}\right)^{(\gamma-1)/\gamma} - 1 + \left(\frac{p_2}{p_x}\right)^{(\gamma-1)/\gamma} - 1\right] \text{ as } T_3 = T_1$$

Let $c_p T_1 = C = \text{constant and } (\gamma - 1)/\gamma = a$

$$w_c = C\left[\left(\frac{p_x}{p_1}\right)^a - 1 + \left(\frac{p_2}{p_x}\right)^a - 1\right]$$

For the maximum work output or the minimum compression work, differentiate $w_c$ w.r.t. to $p_x$ and equate to zero.

Thus, $$\frac{dw_c}{dp_x} = 0 = a.p_x^{a-1}.p_1^{-a} - a.p_2^a.p_x^{-a-1}$$

or $$\frac{p_x^{a-1}}{p_1^a} = \frac{p_2^a}{p_x^{a+1}}$$

or $$p_x = \sqrt{(p_1 p_2)} \qquad ...(18.22)$$

or $$\frac{p_x}{p_1} = \frac{p_2}{p_x} = \sqrt{\left(\frac{p_2}{p_1}\right)} \quad \text{or} \quad r_{p1} = r_{p2} = r_{p2} = \sqrt{(r_p)} \qquad ...(18.23)$$

For $n$ number of stages, $r_{p1} = r_{p2} = r_{p3} \,\ldots\ldots\ldots \sqrt[n]{(r_p)}$

*Thus, we see that for the minimum compression work, the optimum intermediate pressure ($p_x$) is the square root of initial and final pressure.* For the maximum work output, this value of intermediate pressure, $p_x$ is used.

**Actual Cycle.** 1–2–3–4–5–6–1 shows the actual cycle gas turbine with intercooling as shown in Fig. 18.16.

$$\text{Intercooling effectiveness} = \varepsilon_{int} = \frac{h_2 - h_3}{h_2 - h_1} = \frac{T_2 - T_3}{T_2 - T_1}$$

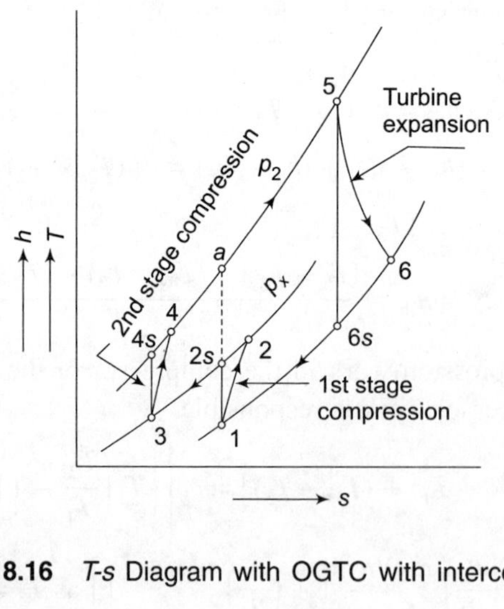


**Fig. 18.16** *T-s* Diagram with OGTC with intercooling

$$\text{Hence, } \eta_{tha} = \frac{(T_5 - T_6) - (T_4 - T_3) - (T_2 - T_1)}{(T_5 - T_4)}$$

$$\eta_{tha} = \frac{(T_5 - T_{6s})\eta_t - (T_{4s} - T_3)/\eta_c - (T_{2s} - T_1)/\eta_c}{(T_5 - T_4)} \quad \text{...(18.24)}$$

It this case also, we can show that for the minimum work input to the compressor

$$r_{p1} = r_{p2} = r_{p3} \ldots\ldots \sqrt[n]{(r_p)}$$

## 18.12 OPEN GAS TURBINE CYCLE WITH INTERCOOLING, REHEAT AND REGENERATION

The power output as well as the thermal efficiency of this cycle is much greater than that of the simple cycle. It is to be noted that the power output is increased due to reheat and intercooling and thermal efficiency is due to regeneration.

**Ideal Cycle.** 1–2*s*–3–4*s*–5–6–7*s*–8–9*s*–10–1 shows such type of ideal cycle as shown in Figs. 18.17 and 18.18 the following are the various processes.

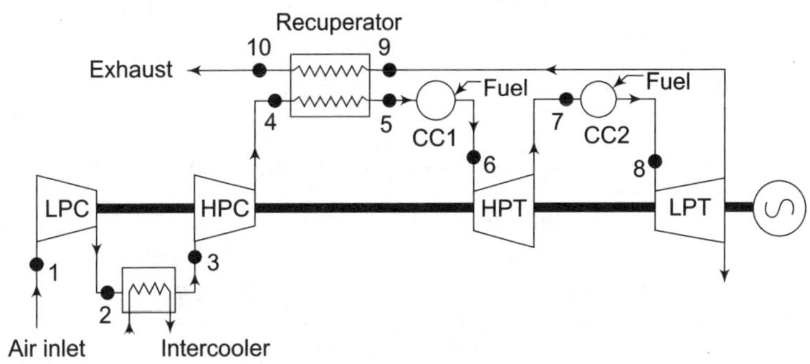


**Fig. 18.17** OGTC with intercooling. reheat and regeneration

**Ideal Cycles**

***Process 1–2s.*** Isentropic compression of air in first stage compressor $C_1$. For the minimum compression work, the pressure ratio $(r_{p1})$ of compressor $C_1$ should be equal to $\sqrt{r_p}$.

***Process 2–3.*** Cooling of air in the intercooler at constant pressure $(p_x)$ from temperature $T_2$ to $T_3 = T_1$

***Process 3–4s.*** Isentropic compression of air in second stage compressor, $C_2$. For the minimum compression work, the pressure ratio $(r_{p2})$ of compressor $C_2$ should be equal to $\sqrt{r_p}$.

***Process 4s–5$_p$.*** Heating of air in the regenerator at constant pressure $(p_4)$ from temperature $T_4$ to $T_{5p}$. For perfect heat exchange in the intercooler, $T_{5p}$ should be equal to $T_{9s}$.

***Process 5p–6.*** Addition of heat at constant pressure ($p_2$) in the combustion chamber from the temperature $T_{5p}$ to $T_6$.

***Process 6–7s.*** Isentropic expansion of gas from pressure $p_2$ to $p_x$ in the first stage of turbine. For the maximum turbine work pressure ratio $p_2/p_x = \sqrt{(r_p)}$.

***Process 7s–8.*** Reheating of gas at constant pressure ($p_x$) in the combustion chamber $CC_2$ from temperature $T_{7s}$ to $T_8 = T_6$.

***Process 8–9s.*** Isentropic expansion of gas from pressure $p_x$ to $p_1$ in the second stage of turbine. For the maximum turbine work, pressure ratio $p_x/p_1 = \sqrt{(r_p)}$.

***Process 9s–10p.*** Cooling of exhaust gas in the regenerator from temperature $T_{9s}$ to $T_{10p}$. If there is a perfect heat exchange in the regenerator then $T_{10p} = T_{4s}$.

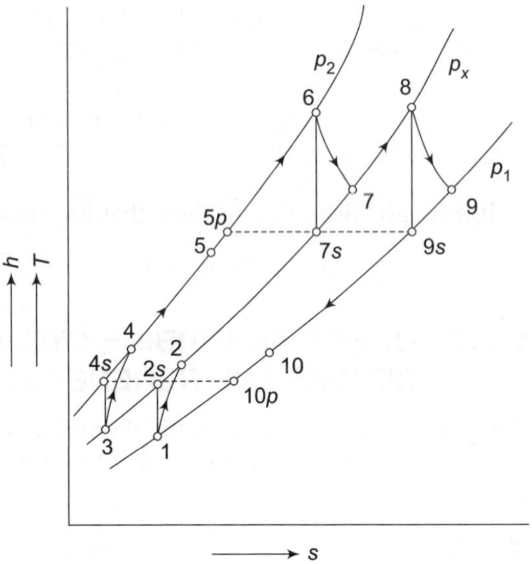


**Fig. 18.18** Intercooling reheating and regeneration on *T–s* diagram

Thermal efficiency is given by

$$\eta_{th} = \frac{(h_6 - h_{7s}) + (h_8 - h_{9s}) - (h_{2s} - h_1) - (h_{4s} - h_3)}{(h_6 - h_{5p}) + (h_8 - h_{7s})}$$

$$\eta_{th} = \frac{(T_6 - T_{7s}) + (T_8 - T_{9s}) - (T_{2s} - T_1) - (T_{4s} - T_3)}{(T_6 - T_{5p}) + (T_8 - T_{7s})} \quad \text{...(18.25)}$$

**Actual Cycle.** 1–2–3–4–5–6–7–8–9–10 shows such an actual cycle as shown in Fig. 18.18.

Actual thermal efficiency,

$$\eta_{tha} = \frac{(h_6 - h_7) + (h_8 - h_9) - (h_2 - h_1) - (h_4 - h_3)}{(h_6 - h_5) + (h_8 - h_7)}$$

or

$$\eta_{tha} = \frac{(T_6 - T_7) + (T_8 - T_9) - (T_2 - T_1) - (T_4 - T_3)}{(T_6 - T_5) + (T_8 - T_7)}$$

$$= \frac{(T_6 - T_{7s})\eta_t + (T_8 - T_{9s})\eta_t - (T_{2s} - T_1)/\eta_c - (T_{4s} - T_3)/\eta_c}{(T_6 - T_5) + (T_8 - T_7)} \quad \text{...(18.25a)}$$

Regeneration effectiveness is given by

$$\eta_r = \frac{h_5 - h_4}{h_9 - h_4} = \frac{h_9 - h_{10}}{h_5 - h_{10}}$$

or

$$\varepsilon_r = \frac{T_5 - T_4}{T_9 - T_4} \quad \text{...(18.26)}$$

If $\varepsilon_r$ is known, then $T_5$ can be found out. Once $T_5$ is known, we can find out the thermal efficiency.

Similarly, intercooler effectiveness is given by

$$\varepsilon_{int} = \frac{h_2 - h_3}{h_2 - h_1} = \frac{T_2 - T_3}{T_2 - T_1}$$

If $\varepsilon_{int}$ is known, $T_3$ can be calculated.

## 18.13 EFFECT OF VARIOUS MODIFICATIONS

The effects of various modifications in the simple cycle on the work output and thermal efficiency are given here. The following are the results for air standard cycle having $r_p = 4$, $T_{max} = 864$ K, and $p_1 = 1$ bar, $T_{min} = 288$ K.

| *Addition to simple cycle* | | *Effect on efficiency* | *Effect on output* |
|---|---|---|---|
| (a) Regeneration | | +50% | Nil |
| (b) Intercooling | | −6.5%* | +10.2% |
| (c) Reheat | | −10.4% | +24.5% |
| (d) Reheat | + Regeneration | +66.7% | +24.0% |
| (e) Intercooling | + Regeneration | +68% | +10.2% |
| (f) Reheat | + Intercooling | −18.2% | +34.7% |
| (g) Reheat | + Intercooling | | |
| | + Regeneration | +80% | +34.7% |

* At the certain optimum pressure there may also be small gain in thermal efficiency.

*So we see that the thermal efficiency increases compared to simple cycle by addition of the following in the simple cycle turbine.*

(a) Regeneration
(b) Reheat and regeneration.
(c) Intercooling and regeneration.
(d) Reheat, intercooling and regeneration
(e) Water injection.

**EXAMPLE 18.6** The following data refer to a gas turbine using intercooling, regeneration and reheating arrangement.

Pressure ratio = 49, compressor inlet temperature = 26°C, turbine inlet temperature = 1400°C, compressor efficiency = 0.88, turbine efficiency = 0.88, regenerator effectiveness = 0.8, inlet pressure to compressor = 1 bar. Determine, (a) cycle thermal efficiency, (b) cycle work ratio, (c) cycle air rate, and (d) net work output.

***Solution*** Refer to Fig. 18.18.

Given, $r_p = \frac{p_2}{p_1} = 49, \quad \eta_c = \eta_t = \varepsilon_t = 0.88, \quad T_1 = 299 \text{ K}, \quad p_1 = 1 \text{ bar}, \quad T_6 = 1673 \text{ K}$

The optimum pressure for the minimum work of compressor and the maximum work of turbine is given by

$$p_x = \sqrt{(p_1 p_2)} = \sqrt{49} = 7 \text{ bar}$$

$$T_{2s} = T_1\left(\frac{p_x}{p_1}\right)^{(\gamma-1)/\gamma} = 299(7)^{0.4/1.4} = 521.33 \text{ K}$$

Now $T_2 - T_1 = \frac{(T_{2s} - T_1)}{\eta_c} = \frac{521.33 - 299}{0.85} = 261.56$

$$T_2 = 299 + 261.56 = 560.56 \text{ K} = T_4$$

Work consumed by first stage compressor

$$= c_p (T_2 - T_1) = 1.005 (560.56 - 299) = 262.86 \text{ kJ/kg}$$

For perfect intercooling and the minimum work to compressor, the work consumed in both stages will be the same. Hence, total work done on two stages of compressor

$$= 262.86 + 262.86 = 525.72 \text{ kJ/kg}$$

**Note:** If there is no perfect intercooling, $T_3$ will be calculated from intercooler effectiveness ($\varepsilon_{int}$) and work inputs to both compressors will not be the same. It has to be calculated separately.

$$T_{7s} = \frac{T_6}{(7)^{(\gamma-1)/\gamma}} = \frac{1673}{1.743} = 959.83$$

Again $T_6 - T_7 = \eta_t(T_6 - T_{7s}) = 0.88(1673 - 959.83) = 627.58$

$$T_7 = 1673 - 627.588 = 1045.41 \text{ K} = T_9$$

Work done by first stage of turbine = $c_p (T_6 - T_7) = 1.005 \times 627.58 = 630.7$ kJ/kg

Assume that reheating is done at the optimum pressure up to the intitial inlet turbine temperature, i.e., $T_6 = T_8$ and $T_7 = T_9$, *hence the work done by both stages will be the same.*

Total work done by the turbine = 630.7 + 630.7 = 1261.4 kJ/kg

Regenerator effectiveness

$$\varepsilon_r = \frac{T_5 - T_4}{T_9 - T_4} = \frac{T_5 - T_2}{T_9 - T_2} = 0.8$$

or $\quad T_5 - 560.56 = 0.8\ (1045.41 - 560.56) = 387.88$

or $\quad T_5 = 387.88 + 560.56 = 948.44\ \text{K}$

Heat supplied $= c_p[(T_6 - T_5) + (T_8 - T_7)$

$= 1.005[(1673 - 948.44) + 1673 - 1045.41)] = 1358.91\ \text{kJ/kg}$

(a) Cycle thermal efficiency

$$= \frac{\text{Turbine work} - \text{Compressor work}}{\text{Heat supplied}} = \frac{1261.4 - 525.72}{1358.91} = 54.13\% \qquad \textbf{\textit{Ans.}}$$

(b) Cycle work ratio $= \dfrac{\text{Net work}}{\text{Turbine work}} = \dfrac{735.42}{1261.4} = 0.5831$ ***Ans.***

(c) Cycle air rate $= \dfrac{3600}{w_{net}} = \dfrac{3600}{735.42} = 4.89\ \text{kg/kWh}$ ***Ans.***

(d) Net work done = 735.42 kJ/kg ***Ans.***

**EXAMPLE 18.7** An open cycle gas turbine is equipped with intercooling, regeneration and reheat arrangements. The H.P. turbine is used for compressor drive and L.P. turbine for output. The specific output of the installation is 132 kW/kg/s and overall thermal efficiency of 32%. The following data refer to this installation:

Pressure ratio in each compressor stage = 2.5

Isentropic efficiency in each compressor stage = 0.86

Isentropic efficiency in H.P. turbine = 0.87

Isentropic efficiency in L.P. turbine = 0.84

Pressure loss in air side of heat exchanger = 6% of entry pressure

Pressure loss in intercooler = 3% of entry pressure

Pressure loss in reheater = 6% of entry pressure

Turbine inlet temperature (for both) = 1020 K

Atmospheric temperature = 288 K

Temperature after intercooling = 298 K

Atmospheric pressure = 1 atm = 1.013 bar

Calculate permissible pressure drop on gas side and effectiveness of heat exchanger.

***Solution*** Refer to Fig. 18.19. $p_1 = 1.013$ bar

$$p_2 = 2.5 \times 1.013 = 2.5325 \text{ bar} \quad \text{and} \quad p_3 = p_2 \times 0.97 = 2.4565 \text{ bar}$$

$$p_4 = 2.5 \times p_3 = 2.5 \times 2.4565 = 6.141 \text{ bar}$$

$$p_5 = 6.141 \times 0.94 = 5.772 \text{ bar} = p_6$$

$$T_1 = 288 \text{ K} \quad \text{and} \quad T_{2s} = T_1(2.5)^{0.286} = 374.28 \text{ K}$$

$$\therefore \quad T_2 = 288 + \frac{374.28 - 288}{0.86} = 388.32 \text{ K}$$

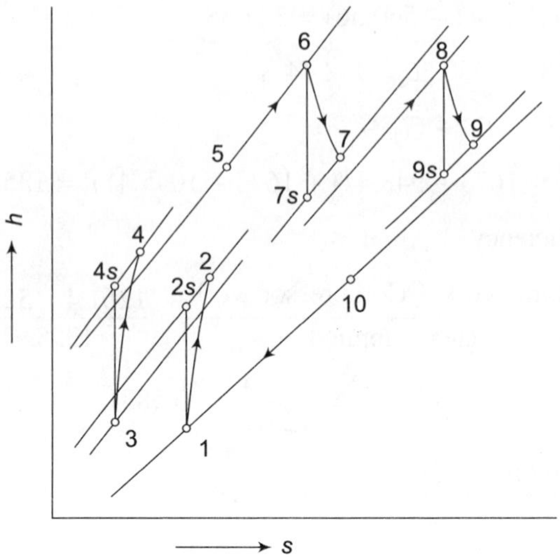


**Fig. 18.19** Example 18.7

$$T_{4s} = T_3\,(2.5)^{0.286} = 298(2.5)^{0.286} = 387.28 \text{ K}$$

$$T_4 = 298 + \frac{387.28 - 298}{0.86} = 401.8 \text{ K} \quad \text{and} \quad T_6 = 1020 \text{ K}$$

Since compressor is driven by H.P turbine, thus

$$c_p(T_6 - T_7) = c_p(T_2 - T_1) + c_p(T_4 - T_3)$$

or $$(1020 - T_7) = (388 - 288) + (401.8 - 298) = 203.8$$

or $$T_7 = 1020 - 203.8 = 816.2 \text{ K}$$

We know that $$(\eta_t)_{hp} = \frac{T_6 - T_7}{T_6 - T_{7s}}$$

or $$T_{7s} = T_6 - \frac{T_6 - T_7}{(\eta_t)_{hp}} = 1020 - \frac{1020 - 816.2}{0.87} = 785.74 \text{ K}$$

We have $$\frac{T_6}{T_{7s}} = \left(\frac{p_6}{p_7}\right)^{(\gamma-1)/\gamma} = \left(\frac{p_6}{p_7}\right)^{0.286}$$

$$\frac{p_6}{p_7} = \left(\frac{1020}{785.74}\right)^{3.496} = 2.4898 \quad \text{or} \quad p_7 = \frac{5.772}{2.4898} = 2.31 \text{ bar}$$

$$p_8 = 0.94 \times 2.31 = 2.17 \text{ bar} \quad \text{and} \quad T_8 = 1020 \text{ K}$$

The net output is 132 kW/kg/s = 132 kJ/kg

This output is developed by L.P. turbine, so

$$c_p(T_8 - T_9) = 132 \quad \text{or} \quad T_8 - T_9 = \frac{132}{1.005} = 131.34$$

$$T_9 = T_8 - \frac{132}{1.005} = 1020 - \frac{132}{1.005} = 888.65 \text{ K}$$

Also $$(\eta_t)_{lp} = \frac{T_8 - T_9}{T_8 - T_{9s}}$$

$$T_{9s} = T_8 - \frac{T_8 - T_9}{0.84} = 1020 - \frac{131.34}{0.84} = 863.65 \text{ K}$$

Now $$\frac{T_8}{T_{9s}} = \left(\frac{p_8}{p_9}\right)^{0.286}$$

or $$\frac{p_8}{p_9} = \left(\frac{1020}{863.5}\right)^{3.496} = 1.7890 \quad \text{or} \quad p_9 = \frac{2.17}{1.7890} = 1.212 \text{ bar}$$

But $p_{10} = p_1 = 1.013$ bar

Hence, permissible pressure drop in heat exchanger on gas side

$$= 1.212 - 1.013 = 0.199 \text{ bar}$$ ***Ans.***

Gas at temperature $T_9$ enters the heat exchanger and comes out at temperature $T_{10}$ with a pressure drop 0.199 bar.

Heat supplied in the cycle = $c_p[(T_6 - T_5) + (T_8 - T_7)]$ kJ/kg

Net work = 132 kJ/kg

$$\eta_{th} = 0.32 = \frac{132}{1.005[(1020 - T_5) + (1020 - 816.2)]}$$

or $T_5 = 812.34$ K

The effectiveness of heat exchanger is

$$\varepsilon_r = \frac{T_5 - T_4}{T_9 - T_4} = \frac{812.34 - 401.8}{888.65 - 401.8} = 84.32\%$$ ***Ans.***

## 18.14 EFFECT OF REGENERATION, INTERCOOLING AND REHEATING ON EFFICIENCY

It is evident from the curves shown in Figs. 18.20 and 18.21 that higher thermal efficiency is possible from simple cycle with regeneration, intercooling and reheat. It is also obvious (from the curves) that *due to addition of regeneration, intercooling and reheat, the peak value of thermal efficiency shifts towards lower pressure ratio compared to simple cycle.*

*In Fig. 18.21, A is simple open cycle, B is simple cycle with regeneration, C is open cycle with regenerator and intercooler, D is open cycle with regenerator, intercooler and reheater, E is ideal cycle.*

*In Fig. 18.21, A is simple open cycle, B is open cycle with regenerator, C is open cycle with regenerator and intercooler, D is open cycle with regenerator and reheater, E is open cycle with regenerator, intercooler and reheater.*

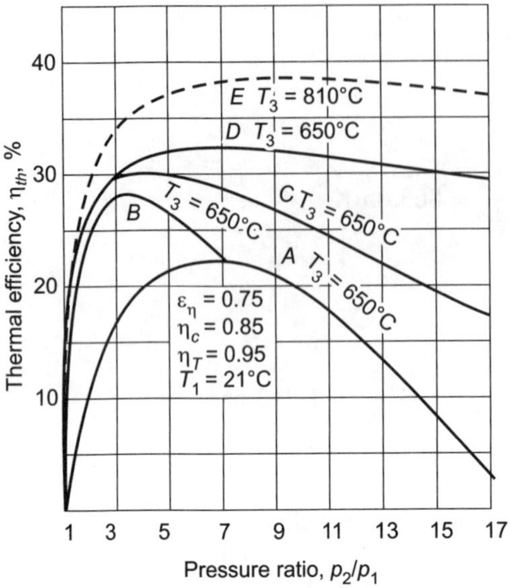


Fig. 18.20 $\eta_{th}$ *vs.* $r_p$ for various $T_3$

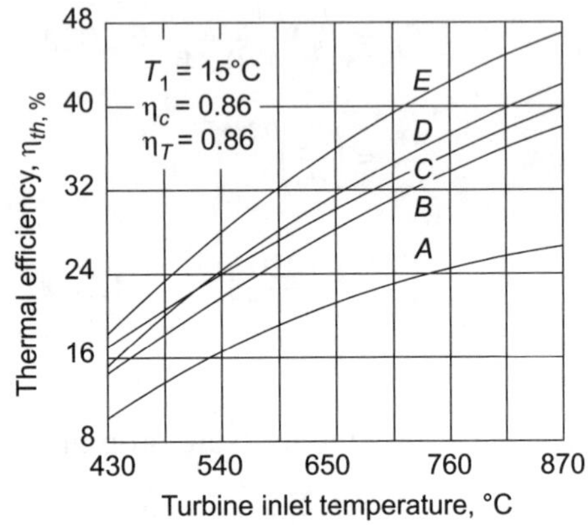


Fig. 18.21 $\eta_{th}$ *vs.* TIT

## 18.15 EFFECT OF OPERATING VARIABLES ON THERMAL EFFICIENCY

The thermal efficiency of an actual simple cycle gas turbine depends upon the following operating variables.

(a) Pressure ratio, $r_p = p_2/p_1$.
(b) Turbine efficiency, $\eta_t$
(c) Turbine inlet temperature, $T_3$
(d) Compressor efficiency, $\eta_c$
(e) Compressor inlet temperature, $T_1$
(f) Specific fuel consumption, kg/kWh

### 18.15.1 Effect of Variation of Turbine Inlet Temperature and Pressure Ratio

Figure 18.22 shows that as the *turbine inlet temperature increases, thermal efficiency increases with the other variables being held constant. For each value of turbine inlet temperature, there is an optimum pressure ratio for the maximum thermal efficiency.* For lower value of turbine inlet temperature, the thermal efficiency first increases and after reaching a maximum value, it decreases rapidly.

But for higher value of turbine inlet temperature, the peaks of the curves are flatter giving a greater range of optimum value of pressure ratio. *The turbine inlet temperature is limited by turbine blading material.* In the present tecnology, TIT is 1700°K while turbine safe blade temperature is 1123K. These values may increase in future.

### 18.15.2 Effect of Variation of Turbine and Compressor Efficiency

Figure 18.23 shows that the thermal efficiency is very much sensitive to the variations of turbine and compressor efficiency. The dotted line shows the thermal efficiency for ideal simple open cycle. As the efficiency of turbine and compressor increase, the thermal efficiency increases. There is an optimum pressure ratio for each set of component efficiency.

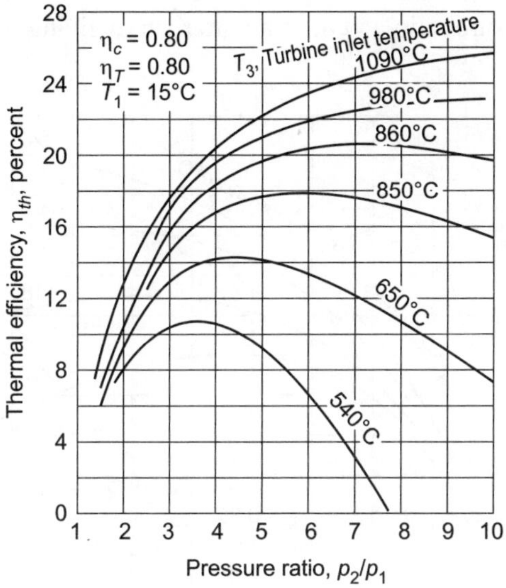


**Fig. 18.22** $\eta_{th}$ vs. $r_p$ and $T_3$

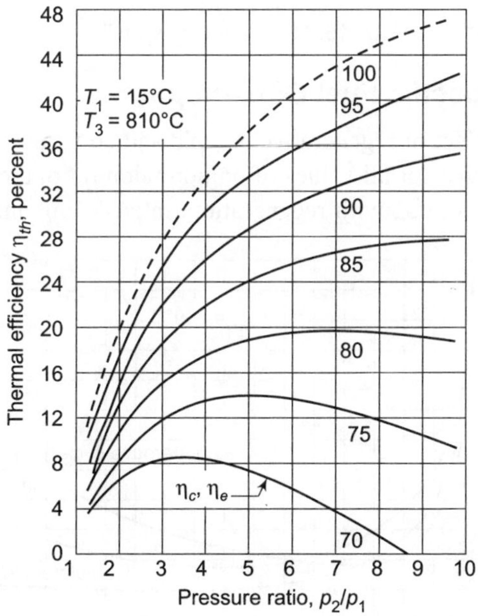


**Fig. 18.23** $\eta_{th}$ vs. $r_p$ and $\eta_t$ and $\eta_c$

### 18.15.3 Effect of Variation of Compressor Inlet Temperature or Atmospheric Air

Figure 18.24 shows that as the air inlet temperature is lowered, the thermal efficiency increases. Peaks are at a higher pressure ratio and have a flatter curve giving optimum value of pressure ratio at a greater range.

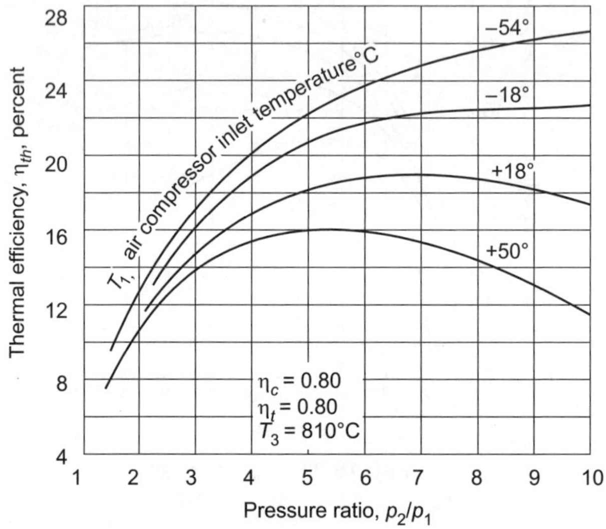


**Fig. 18.24** $\eta_{th}$ vs. $r_p$ and $T_1$

### 18.15.4 Effect of Specific Fuel Consumption

Figure 18.25 shows the effect of regenerative, simple and complete cycle on the specific fuel consumption. It suggests that for all values of compression ratio, the fuel consumption is the minimum in complete cycle, i.e., with regeneration, intercooling and reheat.

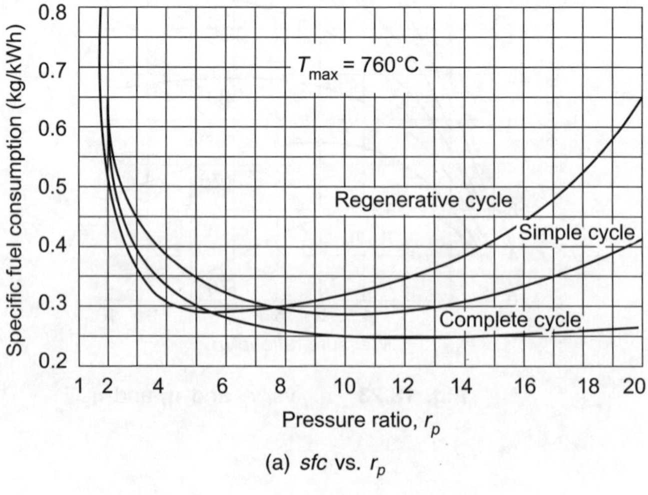


(a) *sfc* vs. $r_p$

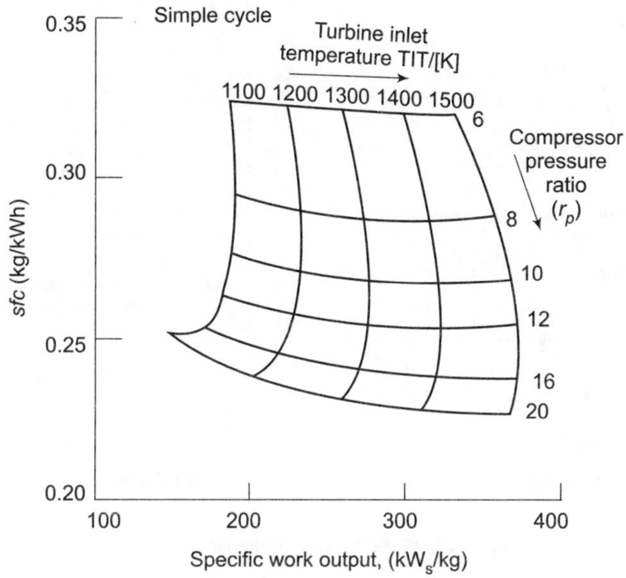


(b) *sfc* vs. $r_p$, TIT and $w_{sp}$

**Fig. 18.25** Sprcific work output for various $r_p$

In the case of simple cycle for any TIT, as the compressor pressure ratio increases, specific fuel consumption decreases, while specific work output first increases, and then slightly (Fig. 18.25b) decreases.

## 18.16 CLOSED CYCLE GAS TURBINE

As it has been seen that in the open cycle, the exhaust gas from the turbine is rejected into the atmosphere but *in the case of closed cycle gas turbine, the same working fluid continuously circulates. The closed cycle gas turbine is similar to steam turbine because as in the latter case, the working substance is confined within the plant.*

Since the closed cycle gas turbine continuously circulates the same working fluid, gases of higher density than air must be used as working fluid, the heat added must be supplied through a heat exchanger (generally called **air heater**) from an external source and the heat rejected from the system must be through another heat exchanger with a cooling medium generally called **precooler**.

Figure 18.26 shows a closed cycle gas turbine. The gas is compressed from pressure $p_1$ to $p_2$ in the compressor. The compressed gas is heated in a air heater where fuel and air are supplied from outside, and the temperature of the gas is increased from $T_2$ to $T_3$ at constant pressure. After this, the compressed and heated gas expands in the turbine. The exhaust gas from the turbine is cooled in a precooler from temperature $T_4$ to $T_1$ and is recirculated to the compressor. Note that all the modifications such as regenerative, intercooling and reheating are possible.

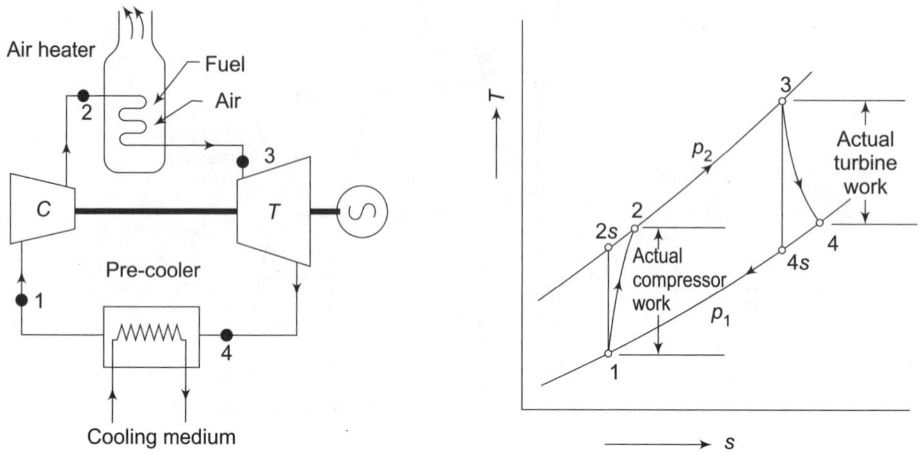


(a) Closed cycle

(b) Closed cycle on *T-s* diagram

**Fig. 18.26** Closed cycle gas turbine

## 18.17 THE ADVANTAGES AND DISADVANTAGES OF CLOSED CYCLE GAS TURBINE OVER THE OPEN CYCLE

### 18.17.1 Advantages

(i) ***Higher thermal efficiency.*** Appreciably higher thermal efficiencies are obtainable for the same maximum and minimum temperature limits as with the open cycle gas turbine.

(ii) ***Reduced size.*** The density of the working fluid is increased in the closed cycle by placing the system under an initial overall high pressure.

Since the working medium is not required to support combustion, so it is possible to use a gas of higher density and higher specific heat than air, such as the monatomic gases like krypton, argon, xenon and mercury vapour as the working fluid. The increase in the density of the working fluid reduces the size of all components and ducts of the system for the same power output. It also permits the use of higher temperature for a given stress limit.

(iii) ***Improved part load efficiency.*** Part load efficiency of a closed cycle is improved as compared to the open cycle because the method of control of the output in the closed cycle does not vary the temperature and pressure ratio of the system, and also, the efficiency during the load change.

(iv) ***Improved heat transmission.*** The rate of heat transmission is improved.

(v) ***Lesser fluid friction loss.*** The fluid friction loss is reduced because of higher Reynolds number.

(vi) ***No contamination.*** No corrosive gases circulate through the closed system. As the working fluid does not contain the combustion gases, the turbine and regenerator are not subjected to carbon deposits and remain relatively clean. The compressor also remains free of dust and other foreign deposits because the working medium is cleaned before being put into the system. This advantage plays an important role in increasing the life of the plant as well as its efficiency compared to the open cycle plant.

(vii) ***Greater output.*** Much greater outputs are possible and these do not involve excessive weight.

(viii) ***Inexpensive fuel.*** Because the closed cycle utilises external heating, any inexpensive solid fuel such as coal can be used.

(ix) ***No loss of working medium.*** Since the working circuit is closed, there is no loss of working medium to make good.

### 18.17.2 Disadvantages

(i) ***Dependent system.*** The closed cycle depends on cooling water which must be provided for the precooler. This eliminates the use of this system as an aeronautical engine. It is best suited for marine propulsion where the provision of cooling water is not a problem.

(ii) ***Complexity.*** The load control of the closed system is complex and costly compared to open system. Since the initial pressure, i.e., the compressor inlet pressure is high and the working fluid is other than air, the system has to be gas tight against high pressure which increases its cost and also poses engineering problems.

(iii) ***Air heater.*** A heavy and longer air heater is required which is relatively inefficient compared to the internal combustion chamber of the open cycle.

**EXAMPLE 18.8** The following particulars relate to closed cycle, gas turbine using air as the working medium:

| | |
|---|---|
| Atmospheric temperature = 26°C. | Maximum temperature = 870°C |
| Initial pressure at compressor = 1 bar | Final pressure of compressor = 5 bar |
| Turbine efficiency = 0.84 | Compressor efficiency = 0.8 |
| Calorific value of fuel = 41840 kJ/kg | Heater loss = 10% heating value |
| $c_p$ = 1.005 kJ/kg K | $\gamma$ = 1.4 |

Calculate (a) compressor work, (b) heat supplied, (c) turbine work, (d) net work, (e) fuel-air ratio, (f) thermal efficiency, (g) air-rate, (h) work ratio, and (i) specific fuel consumption.

***Solutions.*** Refer to Fig. 18.26.

Since the cycle is closed one, so there is no chance of mixing of air and fuel.

$$T_{2s} = T_1(r_p)^{(\gamma-1)/\gamma} = 299 \times (5)^{0.286} = 473.56 \text{ K}$$

$$\text{(a) Actual compressor work} = \frac{c_p\,(T_{2s} - T_1)}{\eta_c} = c_p\,(T_2 - T_1)$$

$$= \frac{1.005(473.56 - 299)}{0.8} = 219.29 \text{ kJ/kg} \qquad \textbf{\textit{Ans.}}$$

(b) $$c_p(T_2 - T_1) = 219.29$$

or $$T_2 = \frac{219.29}{1.005} + 299 = 517.199$$

Heat supplied = $c_p\,(T_3 - T_2) = 1.005\,(870 + 273 - 517.199) = 628.93$ kJ/kg ***Ans.***

(c) Now $T_{4s} = \dfrac{T_3}{(r_p)^{(\gamma-1)/\gamma}} = \dfrac{1143}{1.5838} = 721.68$ K

Actual turbine work = $c_p(T_3 - T_{4s}) \times \eta_t$

$$= 1.005(1143 - 721.68) \times 0.84 = 355.67 \text{ kJ/kg} \qquad \textbf{\textit{Ans.}}$$

(d) Net work = $w_{net} = 355.67 - 219.29 = 136.38$ kJ/kg ***Ans.***

(e) As 10% of heating value is lost, i.e., the efficiency of heater is 90%, each kg of fuel contributes $0.9 \times 41840 = 37656$ kJ/kg of fuel.

Heat supplied = 628.93 kJ/kg of air

Therefore, air-fuel ratio = $\dfrac{37656}{628.93} = 59.87$ kg of air/kg of fuel ***Ans.***

(f) Thermal efficiency = $\dfrac{\text{Net work}}{\text{Heat supplied}} = \dfrac{136.38}{628.93} = 21.68\%$ ***Ans.***

(g) Air rate = $\dfrac{3600}{w_{net}} = \dfrac{3600}{136.38} = 26.39$ kg/kWh ***Ans.***

(h) Work ratio = $\dfrac{\text{Net work}}{\text{Turbine work}} = \dfrac{136.38}{355.67} = 38.34\%$ ***Ans.***

(i) Specific fuel consumption = $\dfrac{\text{Air rate}}{\text{Air – fuel ratio}} = \dfrac{26.39}{59.87} = 0.4407$ kg/kWh ***Ans.***

## 18.18 SEMI-CLOSED CYCLE GAS TURBINE

When some part of the working fluid is confined to the plant and another part flows into the atmosphere, it is called semi-closed cycle. Figure 18.27 shows such a cycle. It may be placed between the open and closed cycles so far the advantages and disadvantages are concerned. It is basically a high pressure system, and the component parts are smaller than an open cycle for the same power output. It gives a better part-load performance compared to open system. The air heater is smaller than that of closed cycle.

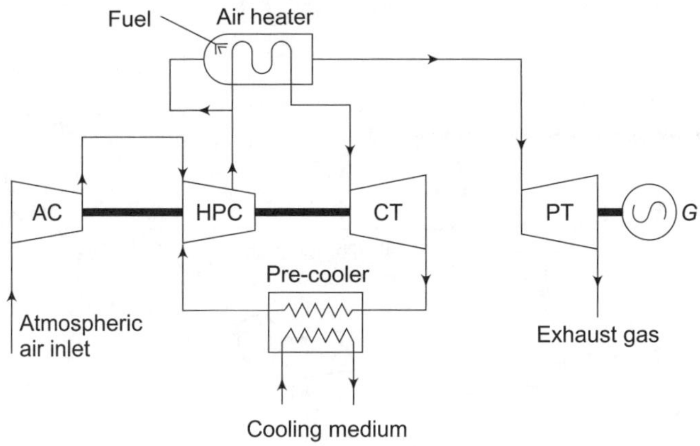


**Fig. 18.27** Semi-closed cycle

The basic working medium is air. Compressed air from auxiliary compressor and exhaust air of turbine driving compressor, passing through the precooler enters the high pressure compressor and is compressed. The high pressure air before entering the air heater in two parts, one part serving the power turbine is used for internal combustion in the air heater and another part which does not mix with the fuel is heated by the heat of external combustion so that all the time this part of air may be circulated in a closed system. The exhaust of power turbine goes to atmosphere.

**EXAMPLE 18.9** In a semi-closed circuit gas turbine plant, air is taken by l.p. compressor at 1 bar, 290K and is compressed to 2 bar, and passes into h.p. compressor through an intercooler with perfect intercooling and is again compressed to 4 bar. A part of compressed air enters the turbine driving the compressor after receiving heat from heat exchanger and the combustion chamber at 1073 K and rest part of air enters the power turbine at 1123K. The plant produces 1837 kW. Find (a) the temperature of the power turbine exhaust, (b) ratio of mass flow of air in both the circuits, (c) the overall efficiency, and (d) the power required to drive the compressor. Assume effectiveness of heat exchanger = 0.75. Isentropic efficiency of compressors = 0.84, isentropic efficiency of turbine = 0.86, efficiency of combustion = 0.98, mechanical efficiency of power turbine = 0.99

$c_p$ = 1.003, kJ/kg K, $\gamma$ = 1/4 for air, $c_p$ = 1.127. kJ/kg K, $\gamma$ = 1.34 for gases

***Solution.*** Refer to Fig. 18.28.

$$\frac{T_9}{T_{10s}} = \left(\frac{p_9}{p_{10}}\right)^{(\gamma-1)/\gamma} \quad \text{or} \quad T_{10s} = \frac{T_9}{(p_9/p_{10})^{(\gamma-1)/\gamma}} = \frac{1123}{1.421} = 790 \text{ K}$$

$$T_{10} = T_9 - (T_9 - T_{10s})\eta_t = 1123 - (1123 - 790)\, 0.86$$

$$T_{10} = 836.62 \text{ K}$$

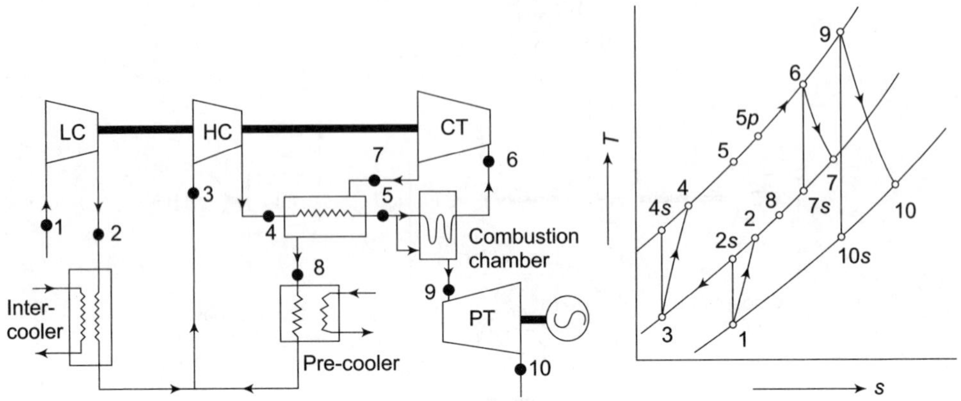


**Fig. 18.28** Example 18.9

$$\frac{T_{2s}}{T_1} = \left(\frac{p_2}{p_2}\right)^{\gamma-1)/\gamma} = (2)^{0.286} = 1.219 \quad \text{or} \quad T_{2s} = 290 \times 1.219 = 353.5 \text{ K}$$

Now $$T_2 = 290 + \frac{T_{2s} - T_1}{\eta_c} = 290 + \frac{(353.5 - 290)}{0.84} = 365.59 \text{ K}$$

Since the intercooling is perfect and pressure ratio is the same in both states, therefore, $T_4 = T_2 = 365.59$ K.

Since air serving the compressor turbine does not mix with fuel, $\gamma$ of air will be considered here

$$\frac{T_6}{T_{7s}} = \left(\frac{p_6}{p_7}\right)^{0.4/1.4} \quad \text{or} \quad T_{7S} = \frac{1073}{1.219} = 880.22 \text{ K}$$

$$T_7 = T_6 - (T_6 - T_{7s}) \times 0.86 = 1073 - (1073 - 880.22) \times 0.86 = 907.2 \text{ K}$$

Net output = 1837 kW.

Let $\dot{m}_1$ be the mass flow rate of air through power turbine

$$1837 = \dot{m}_1 \, c_{pg} \, (T_9 - T_{10}) \times \eta_m$$

or $$\dot{m}_1 = \frac{1837}{1.127 \times (1123 - 836.62) \times 0.99} = 5.749 \text{ kg/s}$$

Let $\dot{m}_2$ be the kg of air flowing the closed circuit. According to question

Work done by h.p. turbine = Work consumed by compressor

$$\dot{m}_2 \, c_p \, (T_6 - T_7) = c_p[\dot{m}_1 \, (T_2 - T_1) + (m_1 + m_2) \, (T_4 - T_3)]$$

or $$\dot{m}_2 \, (1073 - 907.2) = 5.749 \times 75.59 + (5.749 + \dot{m}_2) \times 75.59$$

or $$165.8\,\dot{m}_2 = 869 + 75.59\,\dot{m}_2$$

$$\dot{m}_2 = \frac{869.12}{90.21} = 9.63 \text{ kg/s}$$ ***Ans.***

(b) Hence, the ratio of mass flow of air in both circuits

$$\frac{\dot{m}_2}{\dot{m}_1} = \frac{9.63}{5.749} = 1.675$$ ***Ans.***

Since the air serving the compressor turbine does not mix with fuel so there is indirect heating of this amount of air and that is why the temperature of air serving both turbines is not the same. Thus, heating of $\dot{m}_2$ kg of air will be treated as a heating in heat exchanger (shell and tube type)

Also, $$\varepsilon_r = 0.75 = \frac{T_5 - T_4}{T_7 - T_4} = \frac{T_5 - 365.59}{907.2 - 365.59}$$

or $$T_5 = 771.79 \text{ K}$$

$$\text{Heat supplied} = \frac{\dot{m}_1\, c_{pg}\,(T_9 - T_5) + \dot{m}_2\, c_{pa}\,(T_6 - T_5)}{0.98}$$

$$= \frac{5.749 \times 1.127\,(1123 - 771.79) + 9.63 \times 1.005\,(1073 - 771.79)}{0.98}$$

$$= 5296.62 \text{ kJ/kg}$$

(c) $$\text{Overall thermal} = \eta = \frac{\text{Net output}}{\text{Heat supplied}} = \frac{1837}{5296.62} = 34.68\%$$ ***Ans.***

(d) Power required to drive the compressor = power developed by h.p. turbine

$$= 9.63 \times 1.005\,(1073 - 907.2) = 1604.63 \text{ kW}$$ ***Ans.***

## 18.19 ADVANTAGES AND DISADVANTAGES OF GAS TURBINE OVER STEAM TURBINE POWER PLANTS

### 18.19.1 Advantages

The ***advantages*** of a gas turbine are:

(i) No feed water supply system is required

(ii) There is no need of a condensing plant.

(iii) No boiler is required

(iv) There are less components to be designed

(v) The maintenance cost is low because working fluid is usually air. Installation time is less.

(vi) It is quick to start and there is no banking loss. This is important in aviation, locomotive and emergency power plants.

(vii) It has lower specific weight and size, i.e., lower weight and size per power output; this is advantageous in aviation.

(viii) It has lower operating pressure.

(ix) It is an independent system except closed cycle plant. It is best suited for aviation purpose.

(x) The initial cost and operating costs are lower.

### 18.19.2 Disadvantages

(i) Previously, the power developed was less than steam turbine because of the lower pressure ratio and turbine inlet temperature, therefore, for electrical power generation gas turbine was rarely used. *But now with the development of larger unit (100 to 400 MW) gas turbine are used as base power or captive power. This is extensively used in combined cycle plant.*

(ii) Previously, the thermal efficiency was slightly lower (1 or 2% lower). *But with new development, the thermal efficiency of gas turbine has reached upto to 47%. So now, it is an advantage.*

(iii) Air rate is high, so that the open cycle gas turbine is a great disadvantages in marine applications.

(iv) The part load efficiency is poor.

(v) The life of turbine (even though it is cooled) is less due to high temperature.

## 18.20 ADVANTAGES AND DISADVANTAGES OF GAS TURBINES OVER DIESEL OR PETROL ENGINES

### 18.20.1 Advantages

The advantage are as follows:

(i) **Better balance** due to absence of reciprocating masses.

(ii) **Better mechanical efficiency** due to absence of numerous sliding or bearing members.

(iii) **High power plant** of greater capacity than diesel engines can be built.

(iv) The **specific weight is lower** and the size is smaller. This is important for aviation purpose.

(v) It can operate with **cheaper fuels** like any of hydrocarbon fuels from high octane gasoline down to heavy diesel oil, even solid fuels (pulverized coal).

(vi) It requires less lubrication and is easy and inexpensive to operate.

(vii) It has a much **higher operating speed.**

(viii) **Smokeless exhaust.**

(ix) Comparatively **silent operation** due to continuous exhaust.

(x) It can be used in a **jet propulsion** unit because of continuous exhaust at high temperature and pressure.

(xi) The system is **flexible** and thus reheating, regenerative and intercooling arrangements can be added to increase efficiency and output.

(xii) It has **lower inlet temperature** of air at high attitude which increases the thermal efficiency.

### 18.20.2 Disadvantages

The disadvantages are as follows:

(i) It has poor part load efficiency.
(ii) It has a high air rate, so that open cycle is not suitable in marine applications.
(iii) It is sensitive to changes in compressor and turbine efficiencies.
(iv) It is not suitable for marine applications. Closed cycle gas turbine is rarely used for marine purpose.

## 18.21 JET PROPULSION

*The principle of jet propulsion is based on Newton's second and third laws of motion. Momentum is imparted to a mass of fluid in such a manner that the reaction of the imparted momentun gives propulsive force.* This is done by expanding the gas which is at high temperature and pressure through a nozzle so that the gas with markedly increased velocity in the form of jet comes out into the atmosphere and its reaction in the opposite direction gives the propulsive force. *The open cycle gas turbine is most suited for the jet propulsion.* If the working fluid is expanded in the turbine such that the power developed is only sufficient to drive the compressor and accessories and the rest of the expansion is achieved in a nozzle which is placed just after the turbine, then this unit will serve as a jet propulsion. The gas from the turbine while passing through the nozzle will be accelerated and will come out in the from of the jet with a tremendous velocity The reaction of this jet propels the unit in a forward direction (i.e., opposite direction of the jet). Jet engines are shown in Fig. 18.29.

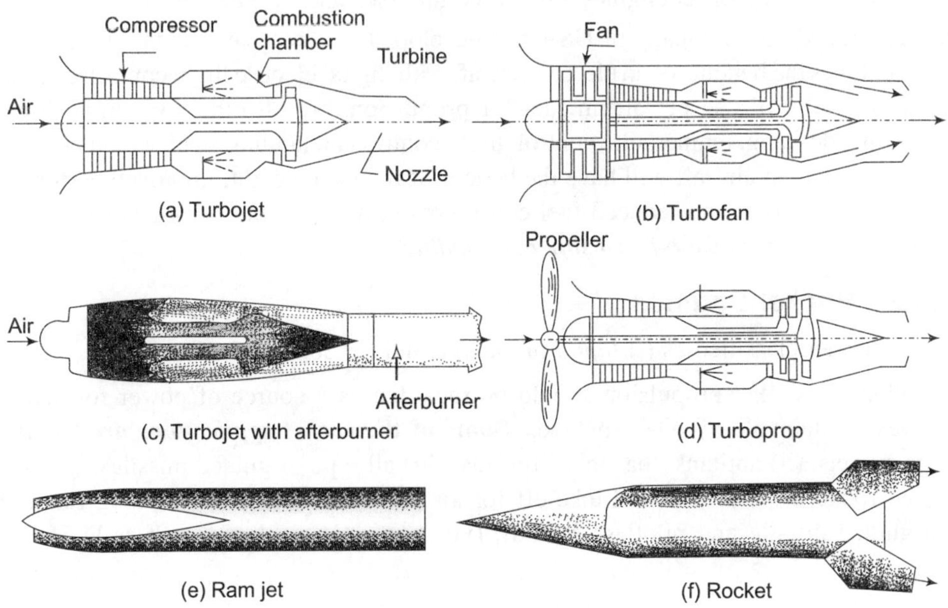


**Fig. 18.29** Jet engines

The jet propulsion engines are classified as follows:

**(i) Atmospheric jet engines. (Breathing engines)**

**(ii) Rocket engines. (Non-breathing engines)**

***Atmospheric jet engines*** *require oxygen from atmospheric air for combustion of the fuel.* As a result, their performance depends to a great extent on the forward speed of the engine and on the atmospheric pressure and temperature.

*The* ***rocket engine carries*** *its own oxidizer for the combustion of the fuel and is, therefore, independent of atmospheric air. It requires the maximum specific fuel consumption among all types of propulsion.*

## 18.21.1 Atmospheric Jet Engines

They are further **classified** as:

**(a) Steady combustion system; continuous air flow**

(i) *Turbo jet.*

(ii) *Turbo jet with after burner* (also known as turbo ram jet, turbojet with tail-pipe burning, and turbojet with reheater).

(iii) Turbofan.

(iv) *Turboprop* (also known as prop jet).

(v) *Ram jet* (also known as Athodyds and Lorin tube).

**(b) Intermittent combustion system; intermittent air flow**

(i) *Pulse jet* (also known as aeropulse, resojet, Schmidt tube, and intermittent jet).

*The turbojet, with after burner, turbofan and turboprop are all modified forms of the simple open cycle gas turbine.* The ram jet and pulse jet are athodyds (*aero-thermo-dynamic-ducts*), i.e., straight duct type of jet engines without compressor and turbine wheels.

***Note:*** The Lockheed Company of USA had developed an integrated digital electric aircraft which combines the benefits of an electric aircraft with digital electric flight control technology. Such an aircraft will use petroleum fuel for propulsion, but all other system such as air-conditioning, de-icing, temperature control, flight control and landing gears would be powered electrically. Such an aircraft will have the benefit of reduced weight, improved performance, a lower lifecycle costs and reduced fuel consumption. *Now, all the aeroplanes use state-of-the-art technology for take-off, cruising and landing.*

## 18.21.2 Rocket Engines

They are further classified as (i) *liquid propellant* and (ii) *solid propellant.*

**Applications.** Rocket propulsion should be regarded as a source of power for achieving objectives unattainable by other methods. **Some of the *rocket applications are*** (i) artillery barrage rockets, (ii) antitank "bazooka" rockets, (iii) all type of guided missiles, (iv) aircraft launched rockets. (v) jet assisted take-off for air planes, (vi) engines for long range, high speed guided missile and pilotless aircraft, (vii) transonic airplanes such as D-558, (viii) space vehicles. *Space vehicles helped to explore the Moon, the Mars and other planets. Space satellites are now the lifeline.*

## 18.22 THE TURBOJET

*The turbojet engine consists of a diffuser at the entrance which slows down the entrance air and thereby compresses it, called the ramming effect; a simple open cycle gas turbine and an exit nozzle which expands the gas and converts the thermal energy of the exit gas into kinetic energy which is emitted out in the form of a jet. The increased velocity of air thereby produces thrust* and based on Newton's third law of motion, the plane propels in the opposite direction of thrust. Such a turbojet engine is shown in Fig. 18.30 with the arrangement of the diffuser, compressor, combustion chamber, turbine and exhaust nozzle. The basic cycle for the turbojet engine is the Joule or Brayton cycle as shown on *T-s* diagram in the Fig. 18.31.

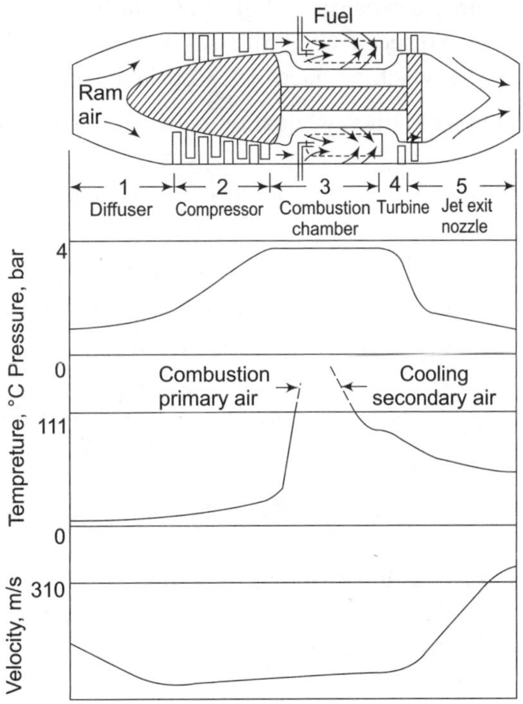


**Fig. 18.30** Turbojet

In process 0 to 1*s*, the air entering from atmosphere is diffused isentropically from velocity $C_a$ down to zero (i.e., $C_1 = 0$). This means that the diffuser has an efficiency of

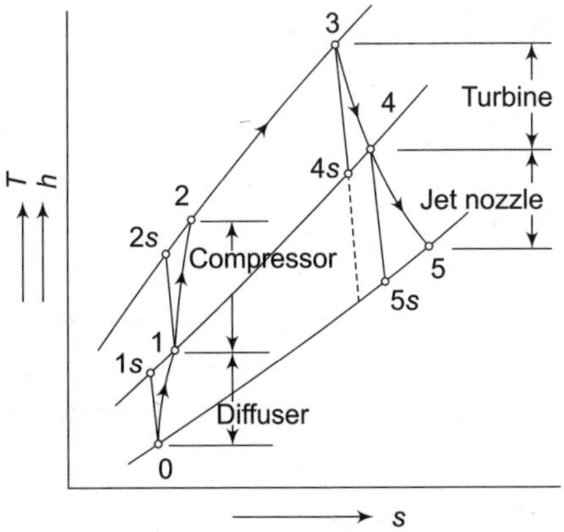


**Fig. 18.31** T-s representation of turbojet

100 per cent. This is termed *ram compression*. Process 0–1 is the actual ramming process. Process 1 to 2s shows the isentropic compression of air. And 1 and 2 shows the actual compression of air, 2*s* to 3 shows the ideal addition of heat at constant pressure $p_2 = p_3$ and 2 to 3 shows the actual heat addition at constant pressure $p_2 = p_3$. Process 3 to 4*s* shows the isentropic expansion of gas in the turbine and 3 to 4 shows the actual expansion of gas in the turbine respectively. Processes 4–5*s* and 4–5 show the isentropic and actual expansions in the nozzle which produces jet. Consider 1 kg of working fluid flowing through the system.

**Diffuser.** Energy equation between 0 and 1*s* gives

$$\frac{C_a^2}{2} + h_0 + q_{01} = \frac{C_{1s}^2}{2} + h_{1s} + w_{01}$$

where $C_a$ = velocity of entering air from atmosphere.

In an ideal diffuser, $C_{1s} = 0$, $q_{o1} = 0$ and $w_{01} = 0$

So enthalpy at state 1*s* is

$$h_{1s} = h_o + \frac{C_a^2}{2 \times 1000} \text{ ; kJ/kg}$$

or
$$T_{1s} = T_o + \frac{C_a^2}{2 \times 1000 \times c_p} \qquad \text{...(18.27)}$$

**Process 0 – 1,** shows the actual process in diffuser

$$\text{Diffuser efficiency} = \eta_d = \frac{h_{1s} - h_o}{h_1 - h_o} = \frac{T_{1s} - T_o}{T_1 - T_o}$$

∴
$$h_1 = h_o + \frac{C_a^2}{2 \times 1000 \times \eta_d}$$

or
$$T_1 = T_o + \frac{C_a^2}{2 \times 1000 c_p \cdot \eta_d} \qquad \text{...(18.28)}$$

**Compressor.** Using energy equation

$$h_1 + \frac{C_1^2}{2} + q_{12} = h_{2s} + \frac{C_{2s}^2}{2} + w_c$$

Neglecting the change in kinetic and potential energy and heat loss, the ideal work consumed by compressor

$w_c = h_{2s} - h_1 = c_p(T_{2s} - T_1)$, Here negative sign shows work input.

$$\text{The actual compressor work} = w_{ca} = h_2 - h_1 = \frac{(h_{2s} - h_1)}{\eta_c} = \frac{c_p\,(T_{2s} - T_1)}{\eta_c}$$

This compressor work has to be supplied by the turbine.

**Combustion Chamber.** The ideal heat supplied

$$= q_A = h_3 - h_{2s} = c_p(T_3 - T_{2s})$$

The actual heat supplied $= q_{Aa} = (1 - F/A)h_3 - h_2 = c_{pg}\,(1 + F/A)T_3 - c_{pa}\,T_2$

**Turbine.** Using energy equation between 3–4, we get

$$h_3 + \frac{C_3^2}{2} + q_{34} = h_{4s} + \frac{C_{4s}^2}{2} + w_t$$

If $q_{34} = 0$, then the turbine work

$$w_t = (h_3 - h_{4s}) + \frac{C_3^2 - C_{4s}^3}{2 \times 1000}\ ;\ \text{kJ/kg}$$

Neglected the change in K.E., we have

$$w_t = h_3 - h_{4s} = c_p(T_3 - T_{4s})$$

Actual turbine work $= w_{ta} = h_3 - h_4 = c_p(T_3 - T_4) = c_p(T_3 - T_{4s})\eta_t$

For the simplification, turbine work is equal to compressor work

$$c_p(T_3 - T_4) = c_p(T_3 - T_{4s})\eta_t = \frac{(T_{2S} - T_1)\,c_p}{\eta_c}$$

or $$T_4 = T_3 - (T_3 - T_{4s})\eta_t = T_3 - \frac{T_{2s} - T_1}{\eta_c}$$

**Jet nozzle.** Using energy equation between states 4 and 5, neglecting $w$ and $q_{4\text{–}5}$

$$h_4 + \frac{C_{4s}^2}{2} = h_{5s} + \frac{C_{5s}^2}{2}$$

In actual case,

$$h_4 + \frac{C_4^2}{2} = h_5 + \frac{C_5^2}{2}$$

If $C_4$ is very much less compared to $C_5$, so

$$h_4 = h_5 + \frac{C_5^2}{2}$$

or $$C_5 = \sqrt{(2 \times 100\,(h_4 - h_5))} = 44.72\,\sqrt{(h_4 - h_5)}\ \text{m/s}$$

or $$C_5 = 44.72\sqrt{(c_p\,\eta_n\,(T_4 - T_{5s})}$$

Thermal efficiency is given by

$$\eta_{th} = \frac{(h_3 - h_4) - (h_2 - h_0)}{h_3 - h_2} = \frac{(T_3 - T_4) - (T_2 - T_o)}{T_3 - T_2} \qquad \text{...(18.29)}$$

## 18.23 THRUST, THRUST POWER, PROPULSIVE EFFICIENCY AND THERMAL EFFICIENCY

### 18.23.1 Thrust, T

Let $C_a$ = vehicle velocity through the air, m/s. Since the atmospheric air is assumed to be at rest, the velocity of the air entering the engine, relative to the engine, is the velocity of the vehicle $C_a$. It is also called approach of air.

Let $C_j$ = velocity of the jet relative to the exit nozzle, m/s

$\left(1 + \frac{\text{Fuel}}{\text{Air}}\right)$ = mass of products leaving the nozzle for 1 kg of air

So, the thrust of jet engine is

$$T = \left(1 + \frac{\text{Fuel}}{\text{Air}}\right)C_j - C_a, \quad \text{N/kg of air} \qquad ...(18.30)$$

Neglecting the mass of fuel

$$T = (C_j - C_a) \quad \text{N/kg of air} \qquad ...(18.31)$$

It is assumed that the vehicles is flying in quiescent air,

### 18.23.2 Thrust Power, TP

*It is the time rate of development of the useful work achieved by the engine, i.e., it is the product of the thrust time the flight velocity of the vehicle, or*

$$TP = T.C_a = [(1 + \text{Fuel/Air})\, C_j - C_a]C_a . \quad \text{W/kg} \qquad ...(18.32)$$

Neglecting the mass of the fuel, it reduces to

$$TP = (C_j - C_a)C_a \quad , \quad \text{W/kg} \qquad ...(18.33)$$

or

$$TP = \frac{(C_j - C_a)C_a}{1000} \quad , \quad \text{kW/kg} \qquad ...(18.33a)$$

### 18.23.3 Propulsive Power, PP

*This represents the energy required to change the momentum of the mass flow of gas and may be expressed as the difference between the rate of kinetic energies of the entering air and the exit gases,* i.e.,

$$PP = \Delta KE = \frac{(1 + \text{Fuel/Air})\, C_j^2}{2} - \frac{C_a^2}{2}\, ; \text{W/kg} \qquad ...(18.34)$$

Neglecting mass of fuel, it reduces to

$$PP = \frac{C_j^2 - C_a^2}{2}\, ; \quad \text{W/kg}$$

or $$PP = \frac{C_j^2 - C_a^2}{2 \times 1000}; \quad \text{kW/kg} \qquad ...(18.35)$$

### 18.23.4 Propulsive Efficiency, $\eta_p$

$$\eta_p = \frac{TP}{PP} = \frac{2[1 + \text{Fuel/Air}) C_j - C_a]C_a}{[(1 + \text{Fuel/Air}) C_j^2 - C_a^2]} \qquad ...(18.36)$$

Neglecting the mass of the fuel,

$$\eta_p = \frac{2(C_j - C_a)C_a}{C_j^2 - C_a^2} = \frac{2(C_j - C_a)C_a}{(C_j + C_a)(C_j - C_a)}$$

or $$\eta_p = \frac{2C_a}{C_j + C_a} = \frac{2}{1 + \left(\frac{C_j}{C_a}\right)} \qquad ...(18.37)$$

It is obvious from Eq. (18.37), *that propulsive efficiency increases as the aeroplane velocity approaches jet velocity but the thrust decreases.*

### 18.23.5 Thermal Efficiency of the Plant, $\eta_{th}$

$$\eta_{th} = \frac{\Delta \text{KE}}{(\text{Fuel/air}) \times \text{Calorific value}}$$

$$= \frac{(1 + \text{Fuel/Air})C_j^2 - C_a^2}{2 \times 1000 \times (\text{Fuel/Air}) \times \text{Calorific value}} \qquad ...(18.38)$$

Neglecting the mass of the fuel,

$$\eta_{th} = \frac{(C_j^2 - C_a^2)}{2 \times 1000 \times (\text{Fuel/Air}) \times \text{Calorific value}} \qquad ...(18.39)$$

The overall efficiency is given by

$$\eta_{overall} = \eta_{th} \times \eta_p = \frac{[(1 + \text{Fuel/Air})(C_j - C_a)C_a}{1000 \times (\text{Fuel/Air}) \times \text{Calorific value}} \qquad ...(18.40)$$

Neglecting mass of the fuel,

$$\eta_{ov} = \frac{(C_j - C_a)C_a}{1000 \times (\text{Fuel/Air}) \times C.V.} \qquad ...(18.41)$$

**The Jet Efficiency ($\eta_{jet}$).**

$$\eta_{jet} = \frac{\text{Final K.E. in the jet}}{\text{Isentropic heat drop in jet pipe + Carryover from turbine}} \quad ...(18.42)$$

## 18.24 THE TURBOFAN ENGINE

*The turbofan engines, which are invariably used in commercial and fighter planes are basically a modified version of turbojet engine with an aim to reduce jet noise and to improve propulsive efficiency.* In this engine, unlike turbojet engines, a portion of total flow bypasses part of compressor, combustion chamber, turbine and hot nozzle before using being ejected through a separate cold nozzle (Fig. 18.32). *This means thrust is made up of two components, the cold stream or fan thrust and hot stream thrust. In some cases,* it is sometimes desirable to mix the two streams and eject them as a single jet having reduced velocity.

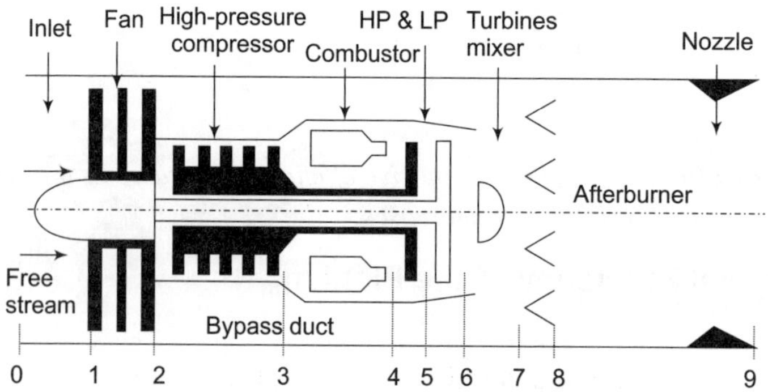


**Fig. 18.32(a)** Two-spool turbofan engine with mixed exhaust and afterburner

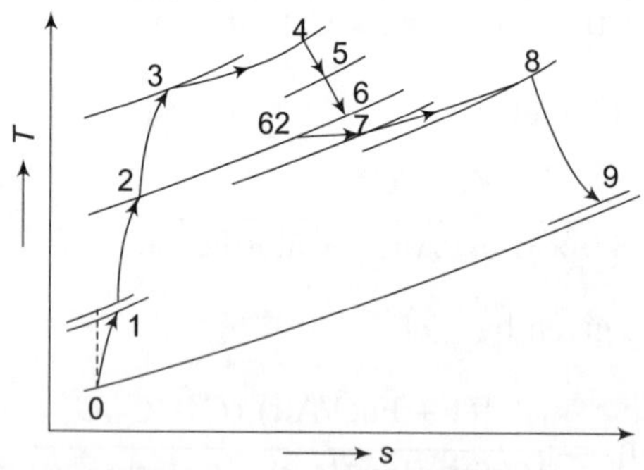


**Fig. 18.32(b)** *T-s* representation of two-spool turbofan engine

In the analysis of turbofan engines, the term **bypass ratio** plays an important role. *The bypass ratio is defined as the ratio of the flow through the bypass duct (i.e., cold stream) to flow at entry to the high pressure compressor (i.e., hot stream). Thus, the bypass ratio (B) is*

$$B = \frac{\dot{m}_c}{\dot{m}_h} \qquad ...(18.43)$$

or
$$\dot{m}_c = \frac{\dot{m}B}{B+1}\ ; \quad \dot{m}_h = \frac{\dot{m}}{B+1}$$

and
$$\dot{m} = \dot{m}_c + \dot{m}_h \qquad ...(18.44)$$

$$\text{Thrust} = T = (\dot{m}_c\, C_{jc} + \dot{m}_h\, C_{jh}) - \dot{m}\, C_a \qquad ...(18.45)$$

*The value of bypass ratio varies from 0.3 to 8 or even more. Turbofan engines may be double or triple spool engines with or without mixer and afterburner.* Figure 18.32(a, b) show the schematic of two-spool turbofan engine with mixed exhaust and afterburner with its *T-s* representation. The schematic and *T-s* representation of three-spool turbofan engine with separate exhaust and no afterburner are shown in Fig. 18.33(a, b). The condition after mixing will be found by enthalpy and momentum balance. In the two-spool engine, the fan is driven by LP turbine while HP compressor is driven by HP turbine (Fig. 18.32a). In the three-spool (14.33a) engine, LP turbine drives fan, IP turbine drives LP compressor while HP turbine drives HP compressor.

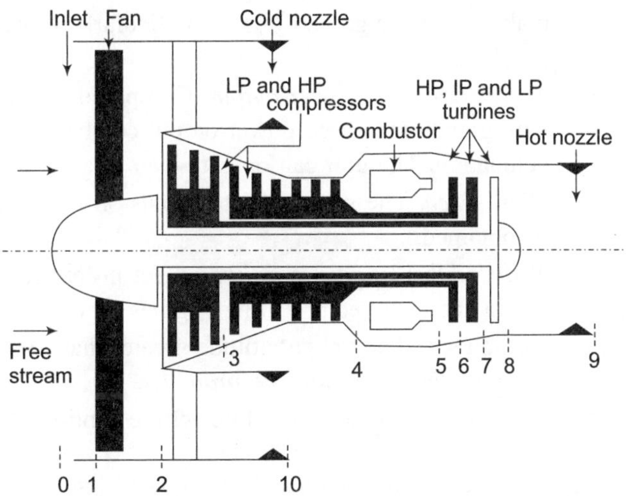


**Fig. 18.33(a)** Three-spool turbofan engine, with separate exhaust and no afterburner

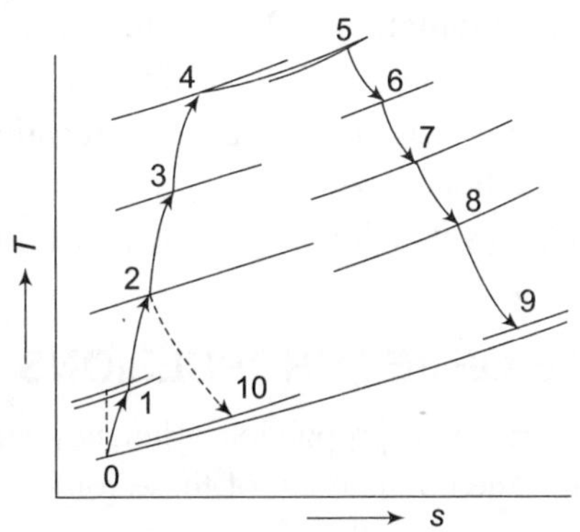


**Fig. 18.33(b)** *T-s* representation of three-spool turbofan engine with separate exhaust and no afterburner

## 18.25 ADVANTAGES AND DISADVANTAGES OF JET PROPULSION OVER THE OTHER SYSTEMS

**Advantages**

(i) *Low specific weight.* The specific weight of jet propulsion is one-fourth to one-half of the reciprocating engine.

(ii) *Not unbalanced force.* There is no reciprocating parts and so jet propulsion is free from unbalanced forces. Greater reliability is thus achieved.

(iii) *Small frontal area.* The frontal area of the propulsion is less than one-fourth the frontal area of the reciprocating engines which decreases nacelle drag greatly and hence makes available greater power, particularly at high loads. This reduces the air cooling problem.

(iv) *No restriction in power output.* Compared to reciprocating engine, the jet propulsion with greatly increased power output can be built because the power is not limited by detonation. The unit can operate over a large range of mixture strength.

(v) *High speed.* The speed of jet propulsion is not limited by propeller. High speed can be obtained.

(vi) *Neither lubrication not radiators.* Jet propulsion requires neither internal lubrication nor radiators as reciprocating engines require.

(vii) At high speed and at an altitude greater than 10,000 metres, the *efficiency of the jet is much higher than that of a propeller.*

(viii) Combustion and delivery of power are continuous, while peak and fluctuating pressure do not occur.

(ix) There is *no slip* stream loss, the drag is reduced, and warm compressed air is available for cabin heating.

(x) The unit permits of a *better position for the pilot* whilst the absence of a propeller permits of a smaller undercarriage.

**Disadvantages**

(i) Particularly at low pressure, the thermal efficiency is lower. At low attitude and speed up to 148 m/s, the fuel consumption is 2 to 3 times that of a reciprocating engine.

(ii) The plant is very noisy, materials costly and life short.

(iii) The compression ratio is not constant as in the reciprocating but various approximately with the square of the rotational speed.

(iv) There are certain difficulties which are encountered in the running of the propulsive unit.

## 18.26 PERFORMANCE OF JET PROPULSIONS

Figure 18.34 shows the comparison of propulsion efficiency among reciprocating, turbojet and rocket. Figure 18.35 shows the comparison of these for specific fuel consumption.

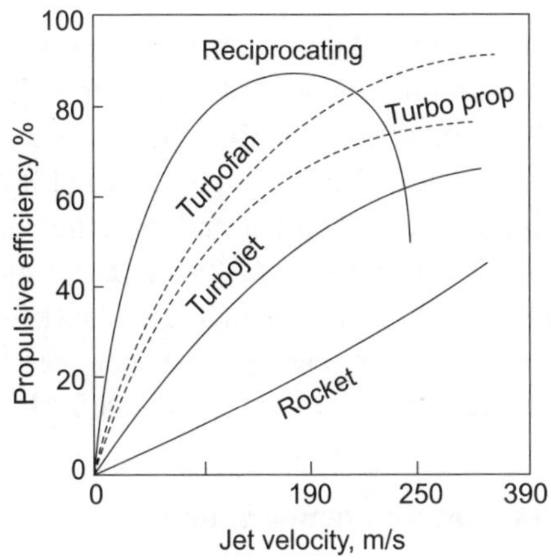


**Fig. 18.34** Performance of jet propulsion

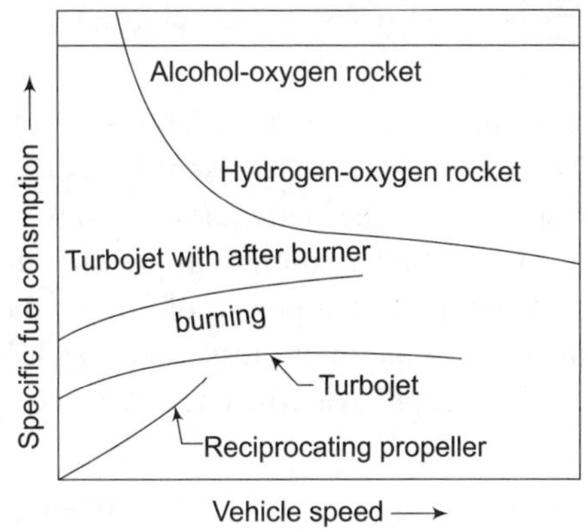


**Fig. 18.35** Comparison of *sfc*

Comparison of the relative performance characteristics of the various propulsion power plants is given below in Table 18.1.

**Table 18.1** Relative Performance Characteristics of Propulsions

| *Engine* | *Relative fuel consumption per power output* | *Relative weight of the engine/power output* | *Relative frontal area per power output* | *Probable best speed range*, m/s |
|---|---|---|---|---|
| Reciprocating | | | | 48.3 to 240 |
| Turboprop | | | | 97 to 290 |
| Turbojet | | | | 240 to 270 |
| Ramjet | | | | 480 to 1/80 |
| Pulse jet | | | | 48.5 to 1180 |
| Liquid rocket | | | | Above 965 |

**EXAMPLE 18.10** In a jet propulsion unit, the total pressure and total temperature at intake to the compressor are 0.6 bar and 0°C. The speed of the propulsion unit is 190 m/s. The total temperature and total pressure of gases after the combustion entering the turbine are 750°C and 3.1 bar. The isentropic efficiencies of compressor and turbine are 85% and 80% respectively. The static back pressure of the propulsion nozzle is 0.52 bar and the efficiency of the nozzle based on total pressure drop available is 90%. Determine (a) power consumed by the compressor per kg of air, (b) the air-fuel if the calorific value of fuel is 41840 kJ/kg of fuel, (c) the total pressure of gas leaving the turbine, (d) thrust per kg of air per second, and (e) propulsive efficiency. Assume $c_p$ of gases = 1.1296, $\gamma_g$ = 1.33 m, $c_{pa}$ = 1.005 kJ/kg, $\gamma_a$ = 1.4

***Solution*** Refer to Fig. 18.36. **Working in total pressures and temperatures**, we have

$$T_{2s} = T_1\left(\frac{p_2}{p_1}\right)^{(\gamma - 1)/\gamma} = 273\left(\frac{3.1}{0.6}\right)^{0.286} = 436.65 \text{ K}$$

From the definition of compressor efficiency, $\eta_c = \dfrac{T_{2s} - T_1}{T_2 - T_1}$

$$T_2 = \frac{(T_{2s} - T_1)}{\eta_c} + T_1 = \frac{(436.65 - 273)}{0.85} + 273 = 465.52 \text{ K}$$

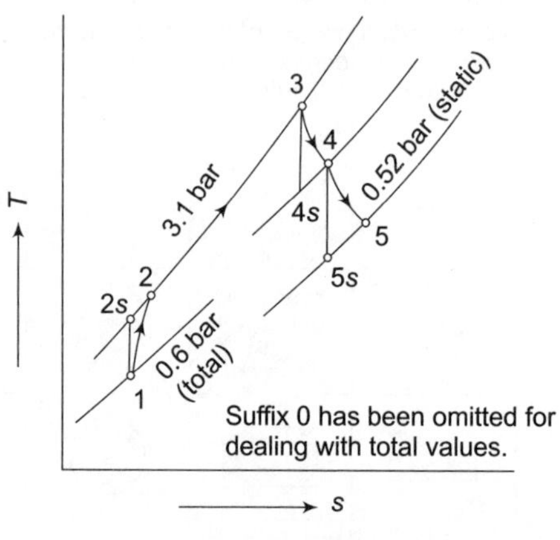


**Fig. 18.36** Example 18.11

(a) Power consumed by compressor = $c_p(T_2 - T_1)$

$$= 1.005(465.52 - 273) = 193.48 \text{ kW/kg}$$ ***Ans.***

(b) $(\dot{m}_a + \dot{m}_f)\, c_{pg}\, T_3 - \dot{m}_a\, c_{pa}\, T_2 = \dot{m}_f \times \text{C.V.} \times \eta_{cc}$ ; here $\eta_{cc} = 1$

or
$$\frac{\dot{m}_a}{\dot{m}_f} = \frac{\text{C.V.} - c_{pg}.T_3}{c_{pg}.T_3 - c_{pa}\,T_2}$$

$$= \frac{41840 - 1.1296 \times 1023}{1.1296 \times 1023 - 1.005 \times 465.52} = 59.166$$

Air-fuel ratio = 59.166:1 ***Ans.***

(c) The power developed by the turbine is consumed by the compressor

$$c_{pa}(T_2 - T_1) = c_{pg}(T_3 - T_4)\ (1 + \dot{m}_f/\dot{m}_a)$$

or $$1.005(465.52 - 273) = 1.1296(1023 - T_4)\left(1 + \frac{1}{59.166}\right)$$

$$T_4 = 1023 - \frac{1.005(465.52 - 273)}{1.1296 \times 1.0169} = 854.56 \text{ K}$$

$$T_3 - T_{4s} = \frac{(T_3 - T_4)}{\eta_t} = \frac{1023 - 854.56}{0.80} = 210.55 \text{ K}$$

$$T_{4s} = 1023 - 210.55 = 812.45 \text{ K}$$

Now $$p_4 = p_3\left(\frac{T_{4s}}{T_3}\right)^{\gamma/(\gamma - 1)} = 3.1\left(\frac{812.45}{1023}\right)^{1.33/0.33} = 1.2246 \text{ bar}$$

Total pressure of gas leaving the turbine = 1.2246 bar ***Ans.***

For the first approximation, $T_{4s}$ and $p_4$ may be assumed as static values

Now $$T_{5s} = T_4\left(\frac{p_5}{p_4}\right)^{(\gamma - 1)/\gamma} = 854.56\left(\frac{0.52}{1.2246}\right)^{0.33/1.33} = 690.45 \text{ K}$$

From the definition of nozzle efficiency,

$$\eta_n = \frac{(T_4 - T_5)}{(T_4 - T_{5s})}$$

$$T_5 = T_4 - (T_4 - T_{5s})\eta_n = 854.56 - (854.56 - 690.945) \times 0.9 = 698.31 \text{ K}$$

Heat drop in the nozzle = $\Delta h_{nozzle} = h_4 - h_5 = c_{pg}(T_4 - T_5)$

$$= 1.1296(845.56 - 698.31) = 166.33 \text{ kJ/kg}$$

$$C_5 = C_j = 44.72\sqrt{(\Delta h_{nozzle})} = 44.72\sqrt{(166.33)} = 576.74 \text{ m/s}$$

Thrust, $T = (C_j - C_a) = (576.74 - 190) = 386.74$ N/kg/s. ***Ans.***

(e) Propulsive efficiency $= \dfrac{2\,C_a}{C_j + C_a} = \dfrac{2 \times 190}{576.74 + 190} = 49.56\%$ ***Ans.***

**EXAMPLE 18.11** A jet propelled unit travels at 180 m/s in air at 0.65 bar and –6°C. Air first enters diffuser in which it is brought to rest relative to the unit and it is then compressed in a compressor through a pressure ratio of 5.8 and fed to a turbine at 925°C. The gas expands through the turbine and then through nozzle to atmospheric pressure (i.e., 0.65 bar). The efficiency of diffuser and nozzle are 0.9 each. The compressor and turbine efficiencies are 0.8 each. Pressure drop in the combustion chamber is 0.14 bar. Find the fuel-air ratio and the specific thrust of the unit. If the inlet cross section of the diffuser is 0.1 m$^2$, calculate the total thrust and propulsive efficiency. Assume calorific value of fuel as 44141 kJ/kg of fuel.

***Solution*** Refer to Fig. 18.37.

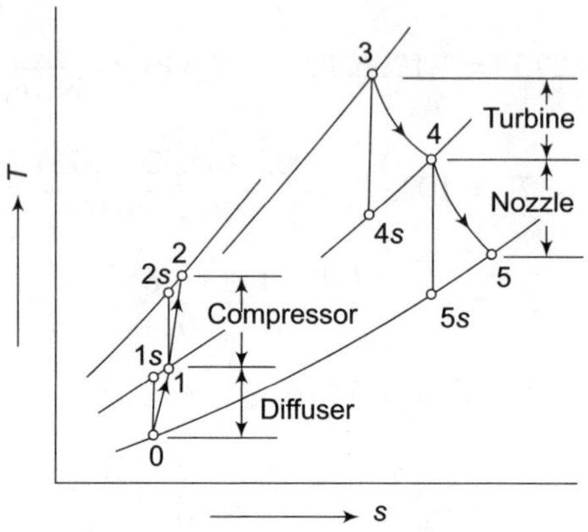


**Fig. 18.37** Example 18.12

For ideal diffuser i.e., for process 0-1*s*, from energy equation

$$h_{1s} = h_0 + \frac{C_a^2}{2} \quad \text{or} \quad h_{1s} - h_0 = \frac{C_a^2}{2}$$

$$T_{1s} = T_0 + \frac{C_a^2}{2c_p} = 267 + \frac{180^2}{2 \times 1000 \times 1.005} = 283.11 \text{ K}$$

For actual diffuser, $\eta_d = \dfrac{h_{1s} - h_0}{h_1 - h_0}$

or $$h_1 = h_0 + \frac{(h_{1s} - h_a)}{\eta_d} \quad \text{or} \quad T_1 = T_0 + \frac{C_a^2}{2c_p\eta_d}$$

$$T_1 = 267 + \frac{180^2}{2 \times 1000 \times 1.005 \times 0.9} = 284.95 \text{ K}$$

Now $$\frac{T_{1s}}{T_o} = \left(\frac{p_1}{p_0}\right)^{(\gamma - 1)/\gamma} = \frac{283.11}{267} = 1.06$$

or $$p_1 = 0.65(1.06)^{3.5} = 0.797 \text{ bar}$$

$$T_{2s} = T_1\left(\frac{p_2}{p_1}\right)^{0.286} = 284.95\left(\frac{5.8}{0.797}\right)^{0.286} = 502.68 \text{ K}$$

again $$T_2 = T_1 + \frac{T_{2s} - T_1}{\eta_c} = 284.95 + \frac{(502.68 - 284.95)}{0.9} = 526.6 \text{ K}$$

$$T_3 = 925 + 273 = 1198 \text{ K}$$

Let $\dot{m}_f$ be kg of fuel per sec and $c_{pa} = c_{pg} = c_p$

$$(\dot{m}_a + \dot{m}_f)\, c_p\, T_3 - \dot{m}_a\, c_p\, T_2 = \dot{m}_f \times C.V. \times \eta_{cc} \text{ ; here } \eta_{cc} = 1$$

or
$$\frac{\dot{m}_a}{\dot{m}_f} = \frac{C.V. - c_p . T_3}{c_p(T_3 - T_2)} = \frac{44141 - 1.005 \times 1198}{1.005(1198 - 527.17)} = 63.68 \qquad \textbf{\textit{Ans.}}$$

$$p_3 = p_2 - 0.14 = 5.8 \times 0.797 - 0.14 = 4.4826 \text{ bar}$$

In jet engines, the function of turbine is to drive the compressor and accessories. Here, assume that the turbine drives only compressor, so

$$c_p(T_2 - T_1) = c_p(T_3 - T_4) \quad \text{or} \quad 527.17 - 284.90 = 1198 - T_4$$

or
$$T_4 = 995.73 \text{ K}$$

Also
$$T_{4s} = T_3 - \frac{(T_3 - T_4)}{\eta_t} = 1198 - \frac{(1198 - 995.73)}{0.8} = 945.17 \text{ K}$$

$$\frac{T_3}{T_{4s}} = \left(\frac{p_3}{p_4}\right)^{0.286}$$

or
$$\frac{p_3}{p_4} = \left(\frac{1198}{945.17}\right)^{3.5} = 2.292 \quad \text{or} \quad p_4 = \frac{4.4826}{2.292} = 1.955 \text{ bar}$$

Now,
$$\frac{T_4}{T_{5s}} = \left(\frac{p_4}{p_5}\right)^{0.286} = \left(\frac{1.955}{0.65}\right)^{0.286} = 1.37$$

or
$$T_{5s} = \frac{995.73}{1.37} = 726.81 \text{ K}$$

and
$$T_5 = T_4 - 0.9(T_4 - T_{5s}) = 995.73 - 0.9(955.73 - 726.81) = 789.7 \text{ K}$$

Velocity at the exit of nozzle is

$$C_j = 44.72\sqrt{(h_4 - h_5)} = 44.72\sqrt{c_p(T_4 - T_5)}$$

$$= 44.72\sqrt{[1.005 \times (955.73 - 789.7)]} = 577.66 \text{ m/s}$$

$$\text{Specific thrust} = (1 + \dot{m}_f) \times C_j = \left(1 + \frac{1}{63.88}\right) 577.66 = 586.67 \text{ N/kg of air} \qquad \textbf{\textit{Ans.}}$$

Volume of flowing air = $0.1 \times 180 = 18 \text{ m}^3/\text{s}$

$$\text{Mass flow} = \dot{m} = \frac{p.V}{RT} = \frac{0.65 \times 10^2 \times 18}{0.287 \times 267} = 15.268 \text{ kg/s}$$

Total thrust = 586.67 × 15.268 = 8957.27 N. ***Ans.***

$$\text{Propulsive efficiency} = \frac{2\,C_a}{C_j + C_a} = \frac{2 \times 180}{577.66 + 180} = 47.51\% \qquad \textbf{\textit{Ans.}}$$

**EXAMPLE 18.12** For the consumption of 430 kg of petrol, a flying bomb had a range of 260 km, an average speed 174 m/s and a thrust of 2678 N. Assuming the calorific value of fuel 42676 kJ/kg, the maximum temperature rise in the combustion chamber as 820°C, the diameter of the discharge nozzle as 0.304 metre, the altitude of the flight as 606 metres, and the $c_p$ of exhaust gases as 1.05 kJ/kg/K calculate.

(a) The fuel-air ratio, (b) the approximate temperature of the exhaust gases and their relative velocity to the bomb, and (c) the propulsive efficiency and overall thermal efficiency of the power unit.

***Solution*** Let $\dot{m}_f$ be the mass of fuel per kg of air compressed and $\eta_{cc}$ then

$$42676\, \dot{m}_f = \dot{m}_a\left(1 + \frac{\dot{m}_f}{\dot{m}_a}\right) \times 1.05 \times 820$$

or

$$\frac{\dot{m}_f}{\dot{m}_a} = \frac{\text{Fuel}}{\text{Air}} = 0.0206 \text{ kg of fuel/kg of air} \qquad \textit{Ans.}$$

$$\text{Air-fuel ratio} = \frac{\dot{m}_a}{\dot{m}_f} = \frac{1}{0.0206} = 48.54$$

Let $C$ be the velocity of flying bomb and $C_j$ be the velocity of the jet relative to machine. With this value of $C$, the air enters the machine at inlet, i.e.; $C_a = C$.

$$\text{Thrust} = [(1 + \dot{m}_f)\, C_j - C)]\dot{m}_a = [(1 + \dot{m}_f)C_j - C)];\ \text{J/kg of air}$$

$$\text{Duration of flight} = \frac{260000}{174} = 1494.25 \text{ sec.}$$

$$\text{Fuel consumption} = \frac{430}{1494.25} = 0.2877 \text{ kg/s}$$

$$\text{Air consumption} = 0.2877 \times 48.54 = 13.96 \text{ kg/s}$$

Now

$$\text{Thrust} = 2678 = [(1 + 0.0206)C_j - 174] \times 13.96$$

or

$$C_j = 358.44 \text{ m/s.} \qquad \textit{Ans.}$$

$$\text{Specific volume of the exhaust gas} = \frac{(\pi/4)\,(0.304)^2 \times 358.44}{13.96 + 0.2877} = 1.79 \text{ m}^3\text{/kg}$$

At an altitude of 606 metres, the barometric pressure from gas table is 0.864 bar. Therefore, temperature of exhaust gases

$$T_{eg} = \frac{pv}{R} = \frac{0.864 \times 10^2 \times 1.79}{0.287} = 538.87 \text{ K} \qquad \textit{Ans.}$$

Hence, propulsion efficiency, neglecting mass of fuel.

$$\eta_p = \frac{2C_a}{C_j + C_a} = \frac{2C}{C_j + C} = \frac{2 \times 174}{358.44 + 174} = 65.35\%. \qquad \textit{Ans}$$

Considering the amount of fuel

$$\eta_p = \frac{2[(1 + \dot{m}_f)C_j - C_a]C_a}{(1 + \dot{m}_f)C_j^2 - C_a^2}$$

$$= \frac{2[(1 + 0.0206) \times 358.44 - 174]174}{(1 + 0.0206) \times 358.02^2 - (174)^2} = 66.19\%. \qquad \textbf{Ans.}$$

$$\text{Overall thermal efficiency} = \eta_{oth} = \eta_{th} \times \eta_p = \frac{[1 + (\text{Fuel/Air})C_j - C_a]C_a}{1000 \times (\text{Fuel/Air}) \times C.V.}$$

$$= \frac{[(1 + 0.0206)358.44 - 174] \times 174}{1000 \times 0.0206 \times 42676} = 3.79\%. \qquad \textbf{Ans.}$$

**EXAMPLE 18.13** The following particulars related to a jet engine which develops a thrust power 770 kW when flying at an attitude of 9120 metres with a speed of 217.5 m/s —internal efficiency of turbine 0.85; efficiency of compressor = 0.85; efficiency of jet tube = 0.9; inlet pressure and temperature = 0.036 bar and –45° C; temperature of gas velocity in ducts constant = 670°C; pressure ratio $r_p$ = 5; C.V. of fuel = 42676 kJ/kg; velocity in ducts constant = 197 m/s; for air, $c_p$ = 1.005, $\gamma$ = 1.4; $R$ = 0.287; for combustion gases $c_p$ = 1.087 kJ/kg K gases during expansion, $\gamma$ = 1.33.

Calculate (a) overall thermal efficiency of the machine, (b) rate of air consumption, (c) the power developed by the turbine, (d) the outlet area of the jet tube, and (e) specific consumption in kg per kg thrust.

***Solution*** Refer to Fig. 18.37. Let $\dot{m}_f$ kg *of fuel per kg of air* be required,

$$42676\, \dot{m}_f \times \eta_{cc} = (1 + \dot{m}_f) \times 1.087\, T_3 - 1.005 \times T_2\ ;\ \text{Here } \eta_{cc} = 1$$

$$T_1 = 227.5 \text{ K} \quad \text{and} \quad T_{2S} = T_1\, (r_p)^{0.286} = 360.48 \text{ K}$$

$$T_2 = T_1 + \frac{(T_{2s} - T_1)}{\eta_c} = 227.5 + \frac{(360.48 - 227.5)}{0.85} = 383.94$$

or $$T_3 = 670 + 273 = 943 \text{ K}$$

Hence, the heat supplied/kg of air

$$= 42676\, \dot{m}_f = (1 + \dot{m}_f) \times 1.087 \times 943 - 1.005 \times 383.94$$

or $$\dot{m}_f = 0.01535 = \text{Fuel-air ratio}$$

$\therefore$ $$\text{Air-fuel ratio} = \frac{1}{0.01535} = 65.14$$

The discharge velocity $C_j = C_5$ cannot be determined from the thrust equation because rate of airflow in unknown. It may be determined from expression of jet efficiency.

$$\eta_{jet} = \frac{\text{Final K.E. in the jet}}{\text{Isentropic heat drop in jet pipe + carry over from turbine}}$$

$$= \frac{C_j^2/2}{c_p(T_4 - T_5) + C_4^2/2}$$

Here, $C_4 = 197$ m/s

Since the turbine work is to drive the compressor only, so

$$c_{pa}(T_2 - T_1) = c_{pg}(1 + F/A)(T_3 - T_4)$$

or $$T_4 = 943 - \frac{1.005 \times 156.44}{1.087(1 + 0.01535)} = 800.54 \text{ K}$$

Let $r_{pt}$ = expansion pressure ratio in turbine, i.e. $r_{pt} = p_3/p_4$

$r_{pj}$ = expansion pressure ratio in jet tube, $= p_4/p_5$

$$\therefore \quad r_{pt} \times r_{pj} = \frac{p_3}{p_4} \times \frac{p_4}{p_5} = \frac{p_2}{p_5} = r_p = 5$$

We also know that $\eta_t = \dfrac{T_3 - T_4}{T_3 - T_{4s}}$

or $$T_{4s} = T_3 - \frac{(T_3 - T_4)}{\eta_t} = 943 - \frac{(943 - 800.54)}{0.85} = 775.4 \text{ K}$$

Now, $$\frac{T_3}{T_{4s}} = \left(\frac{p_3}{p_4}\right)^{(\gamma - 1)/\gamma} = \left(\frac{p_3}{p_4}\right)^{0.33/1.33}$$

or $$\frac{p_3}{p_4} = r_{pt} = \left(\frac{943}{775.4}\right)^{4.03} = 2.2161$$

Hence, $$r_{pj} = \frac{p_4}{p_5} = \frac{5}{2.2161} = 2.2562$$

Thus, $$T_{5s} = \frac{T_4}{(r_{pj})^{0.33/1.33}} = \frac{800.54}{(2.2562)^{0.248}} = 654.30 \text{ K}$$

Putting all these values in jet efficiency equation, we have

$$0.9 = \frac{C_j^2/2}{1.087 \times 1000(800.54 - 654.30) + (197)^2/2}$$

$$C_j = 566.62 \text{ m/s.}$$

(a) Overall thermal efficiency $= \eta_{ov} = \dfrac{(1 + \text{(Fuel/Air)})(C_j - C_a)C_a}{1000 \times \text{(Fuel/Air)} \times \text{Calorific value}}$

$$= \frac{[(1 + 0.01535)(566.62 - 217.5)] \times 217.5}{1000 \times 0.01535 \times 42676} = 11.76\%.$$

**Ans.**

Thermal power = Thrust × Velocity of machine

$$\text{Thrust} = \frac{770}{217.5} = \frac{[(1 + \dot{m}_f/\dot{m}_a)C_j - C_a)]\,\dot{m}_a}{1000}$$

$$\therefore \qquad \dot{m}_a = \frac{770 \times 1000}{[1.01535 \times 566.62 - 217.5]\,217.5} = 9.89 \text{ kg/s}$$

Power developed by the turbine = $\dot{m}_a\,(1 + \dot{m}_f/\dot{m}_a)c_{pg}\,(T_3 - T_4)$

$$= 9.89 \times 1.01535 \times 1.087\,(943 - 800.54) = 1555.01 \text{ kW}$$

Now
$$\frac{C_j^2 - C_4^2}{2} = c_{pg}\,(T_4 - T_5)$$

or
$$T_5 = T_4 - \frac{C_j^2 - C_4^2}{2 \times 1000\,c_{pg}} = 800.54 - \frac{566.62^2 - 197^2}{2 \times 1000 \times 1.087} = 652.91 \text{ K}$$

The atmospheric pressure is 0.306 bar and let the exit pressure be equal to the atmospheric.

$$\text{Density of exhaust gases} = \rho = \frac{0.306 \times 10^2}{0.287 \times 652.91} = 0.1633 \text{ m}^3/\text{kg}$$

Discharge of jet area is given by continuity equation

$$= A \times C_j \times \rho = \dot{m}_a(1 + \dot{m}_f/\dot{m}_a)$$

or
$$A = \frac{9.89 \times 1.01535}{566.62 \times 0.1633} = 0.1085 \text{ m}^2$$ **Ans.**

$$\text{Specific fuel consumption} = \frac{0.01535 \times 9.89 \times 3600}{1000 \times 770/217.5} = 0.1543 \text{ kg/thrust–hr}$$ **Ans.**

## 18.27 TURBOPROP

Figure 18.38 shows a turboprop or propjet engine. *It consists of a geared propeller attached to a turbojet engine. To add more power to the propeller shaft, the turbine is modified by adding additional stages. In this case, nearly 80 to 90 per cent of the power developed is used to drive the shaft and the remaining 10 to 20 per cent is used to obtain thrust from the jet.* The control of the engine is governed by changing the propeller pitch and the quantity of fuel burnt in the combustion chamber.

*The turboprop engine cycle is the same as that of the turbojet engine cycle except the turbine expansion process is greater.* Regenerators, intercoolers and reheaters may be added to the turboprop cycle for the improvement in power and efficiency.

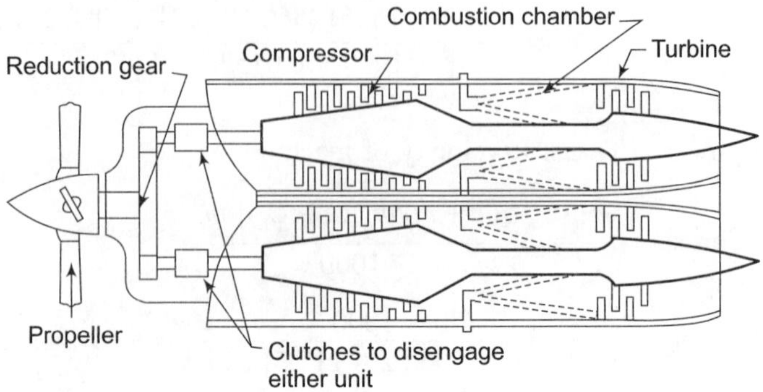


**Fig. 18.38** Dual arrangement of turboprop engine

The turboprop engine combines in the merit of turbojet engines, i.e., *low specific weight, small frontal area, simplicity and lower vibration and the merit of the propeller are high power for take-off and climb, high propulsive efficiency at speeds below 650 km/h at sea condition, high thrust per unit frontal area and good fuel economy.*

**EXAMPLE 18.14** The following data relate to a twin-spool turbofan engine where the fan is driven by the l.p. turbine and the compressor by the h.p. turbine. Separate cold and hot stream nozzles are used in the turbofan.

Overall pressure ratio = 30.0, fan pressure ratio = 1.7, bypass ratio, $\dot{m}_c/\dot{m}_h = 5.5$, stagnation turbine inlet temperature = 1600 K, mechanical efficiency = 0.98, Combustion pressure loss = 2 bar, mass flow rate of air = 220 kg/s, lower calorific value of fuel = 43000 kJ/kg, $c_{pa}$, $c_{pg}$ and $\gamma$ for gas = 1.005, 1.148 kg/kgK, 1.33 respectively, polytropic efficiency for fan, compressor and turbine each = 0.9, ambient condition = 1.0 bar and 288 K, nozzle efficiency = 0.95.

Calculate the specific thrust, air-fuel ratio, and specific fuel consumption at sea level condition.

***Solution*** Refer to Fig. 18.32.

The values of $(n - 1)/n$ for polytrophic compression and expansion are given by

$$\text{For compression} = \frac{n-1}{n} = \frac{1}{\eta_p}\left(\frac{\gamma-1}{\gamma}\right) = \frac{1}{0.9 \times 3.5} = 0.3175$$

$$\text{For expansion : } \frac{n-1}{n} = \eta_p\left(\frac{\gamma-1}{\gamma}\right) = \frac{0.9}{3.5} = 0.257$$

Under static condition assume $T_{01} = T_a$ and $p_{01} = p_a$ as nothing is mentioned about the ramming action. The actual condition will be directly available.

$$T_{02} = T_{01}\left(\frac{p_{02}}{p_{01}}\right)^{(n-1)/n} = 288 \times (1.7)^{0.3175} = 340.84 \text{ K}$$

$$T_{03} = T_{02}\left(\frac{p_{03}}{p_{02}}\right)^{(n-1)/n} = 340.84\ (17.647)^{0.3175} = 847.94\ \text{K}$$

The cold nozzle pressure ratio is fan pressure ratio (FPR) $= \frac{p_{02}}{p_0}$ = FPR = 1.7

The critical pressure for cold nozzle $= \frac{p_c}{p_{02}} = \left(\frac{2}{\gamma+1}\right)^{\gamma/(\gamma-1)} = \left(\frac{2}{2.33}\right)^{1.3/0.33} = 0.54$

or $$\frac{p_{02}}{p_c} = \frac{1}{0.54} = 1.851$$

or $\frac{p_{02}}{p_a} < \frac{p_{02}}{p_c}$, so the nozzle is not choking. Thus, $p_8 = p_a$

The cold thrust $= T_c = \dot{m}_c\, C_8 = \dot{m}_c\, C_j$

where $$C_8 = C_j = [2c_{pg}\,(T_{02} - T_8)]^{1/2}$$

But $$T_{02} - T_8 = \eta_j\, T_{02}\left[1 - \left(\frac{p_a}{p_{02}}\right)^{(\gamma-1)/\gamma}\right] = 0.95 \times 340.84\left[1 - \left(\frac{1}{1.7}\right)^{0.2481}\right] = 39.95\ \text{K}$$

Thus, $$C_8 = C_j = [2 \times 1.005 \times 10^3 \times 39.95]^{1/2} = 283.37\ \text{m/s}$$

The cold mass flow rate $= \dot{m}_c = \frac{\dot{m}\, B}{B+1} = \frac{220 \times 5.5}{5.5+1} = 186.15$ kg/s

The cold thrust $= T_c = 186.15 \times 283.37 = 52.749$ kN

Heat balance around the combustor gives : $\left(1 + \frac{\dot{m}_f}{\dot{m}_h}\right) c_{pg}\, T_{04} - c_{pa}\, T_{03} = \frac{\dot{m}_f}{\dot{m}_h}(LCV)_f$

or $$\left(1 + \frac{\dot{m}_f}{\dot{m}_h}\right) \times 1.148 \times 1600 - 1.005 \times 847.94 = \frac{\dot{m}_f}{\dot{m}_h} \times 43000$$

or $$\frac{\dot{m}_f}{\dot{m}_h} = 0.0239.$$ ***Ans.***

Since h.p. turbine drives h.p. compressor, so

$$\left(1 + \frac{\dot{m}_f}{\dot{m}_h}\right) c_{pg}\,(T_{04} - T_{05}) = \frac{c_{pa}(T_{03} - T_{02})}{\eta_m}$$

or $$T_{04} - T_{05} = \frac{1.005}{1.148} \times \frac{1}{1.0239} \times \frac{(847.94 - 340.84)}{0.98} = 442.41$$

or $$T_{05} = 1600 - 442.41 = 1157.59\ \text{K}$$

Since 1.p. turbine drives fan, so $(T_{05} - T_{06}) = (B + 1)\dfrac{c_{pa}}{c_{pg} \cdot \eta_m}(T_{02} - T_{01})$

$$= (5.5 + 1) \times \frac{1.005}{1.148 \times 0.98}(340.84 - 288) = 306.81 \text{ K}$$

or $$T_{06} = 1157.59 - 306.81 = 850.78 \text{ K}$$

Now, $p_{06}$ may be found as follows: $\dfrac{p_{04}}{p_{05}} = \left(\dfrac{T_{04}}{T_{05}}\right)^{n/n-1} = \left(\dfrac{1600}{1157.59}\right)^{1/0.257} = 3.532$

or $$p_{05} = \frac{p_{04}}{3.523} = \frac{p_{03} - \Delta p}{3.523} = \frac{30 - 2}{3.523} = 7.947 \text{ bar}$$

$$\frac{p_{05}}{p_{06}} = \left(\frac{T_{05}}{T_{06}}\right)^{n/n-1} = \left(\frac{1157.59}{850.78}\right)^{1/0.257} = 3.314$$

or $$p_{06} = \frac{p_{05}}{3.314} = \frac{7.947}{3.314} = 2.398$$

The hot nozzle pressure ratio $= \dfrac{p_{06}}{p_a} = \dfrac{2.398}{1} = 2.398$

The critical pressure ratio in hot nozzle is given by

$$\frac{p_c}{p_{06}} = \left(\frac{2}{\gamma + 1}\right)^{\gamma/\gamma - 1} = \left(\frac{2}{2.33}\right)^{1.33/0.33} = 0.54$$

or $$\frac{p_{06}}{p_c} = \frac{1}{0.54} = 1.851$$

Since $\dfrac{p_{06}}{p_a} < \dfrac{p_{06}}{p_c}$, hence nozzle is unchoked. Thus, $p_7 = p_a$

$$T_{06} - T_7 = \eta_j\, T_{06}\left[1 - \left(\frac{p_a}{p_{06}}\right)^{(\gamma - 1)/\gamma}\right]$$

$$= 0.95 \times 850.78\left[1 - \left(\frac{1}{2.398}\right)^{0.33/1.33}\right] = 157.68 \text{ K}$$

$$C_7 = C_{jh} = [2c_{pg}(T_{06} - T_7)]^{1/2} = [2 \times 1.148 \times 157.68 \times 10^3]^{0.5} = 601.69 \text{ m/s}$$

$$\dot{m}_h = \frac{\dot{m}}{B + 1} = \frac{220}{5.5 + 1} = 33.84 \text{ kg/s}$$

The hot gas thrust $= T_h = \dot{m}_h\, C_7 = 33.84 \times 601.69 = 20.361$ kN

The total thrust $= T = T_c + T_h = 52.749 + 20.361 = 73.11$ kN ***Ans.***

The specific fuel consumption (*sfc*),

$$sfc = \frac{(\dot{m}_f/\dot{m}_h)\,\dot{m}_h \times 3600}{T} = \frac{0.0239 \times 33.84 \times 3600}{73.11 \times 10^3} = 0.039 \text{ kg/hN}$$

***Ans.***

## 18.28 RAM JET

Ram jet was invented by a French engineer, Rene Lorin in 1913. It is also called **athodyd Lorin tube or flying stovepipe.** It is a steady combustion or continuous flow engine and has the simplest construction of any propulsion engine. Figure 18.39 shows a schematic diagram of a ram jet engine. It consists of an inlet diffuser, a combustion chamber and an exit nozzle or tailpipe.

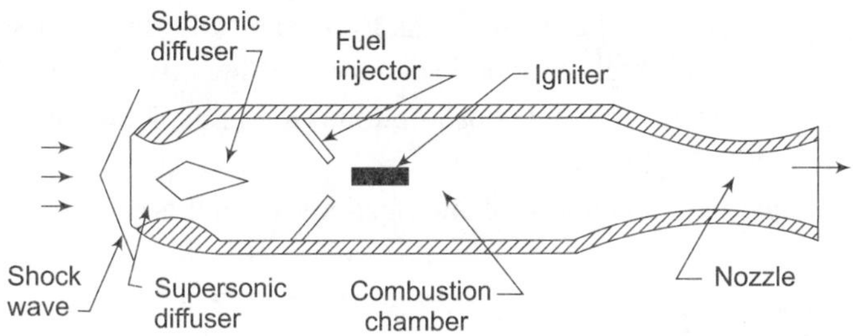


**Fig. 18.39** Ram jet

*As the ram jet has no compressor, hence the entire compression depends upon the ram compression.* The ram pressure ratio increases very slowly in the subsonic speed range. *That is why ram jet is boosted up to a speed of 290 km/h by suitable means such as a turbojet or a rocket before the ram jet will produce any thrust and must be boosted to even higher speeds before the thrust produced exceeds the drag.*

After the boosting of ram jet, velocity of air passing through the diffuser decreases and hence pressure increases. This is called **ram compression** and a ***"pressure barrier"*** is created after the end of the diffuser. The fuel is injeced through injection nozzles into the combustion chamber where it is ignited by means of a spark plug. The expansion of gases towards the diffuser entrance is restricted by thc pressure barrier at the end of the diffuser and as a result the gases are constrained to expand through the tail pipe and out of the exit nozzle at a high velocity.

As the ram jet engine has no turbine, the temperature of the gases of combustion is not limited to a relatively low value as in the turbojet engine. The air-fuel ratio is 15:1, the exhaust temperature ranges 1800°C to 2100°C. The jet action gives the necessary forward thrust to the engine.

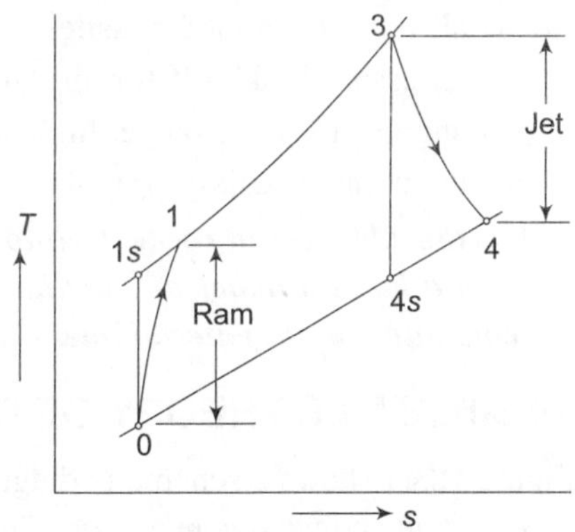


**Fig. 18.40** *T-s* diagram of Ram jet cycle

Figure 18.40 shows the representation of ram jet cycle on *T-s* diagram. The basic characteristics of the ram jet engines are:

(i) Simple in construction.

(ii) No moving parts and hence free from unbalancing.

(iii) Greater thrust per unit engine weight than any other propulsion engine at supersonic speed except rocket.
(iv) The thrust per unit frontal area increases both with the efficiency and the air flow through the engine, therefore, much greater thrust per unit area is obtainable at high supersonic speeds.
(v) The best performance of ram jet engine is obtained at flight speed of 1700 km/h to 2200 km/h.
(vi) For successful operation, the diffuser has to be carefully designed so that the kinetic energy associated with high entrance velocity is efficiently converted into pressure.
(vii) At low moderate speeds, the fuel consumption is too large. However, the fuel consumption decreases with flight speed.
(viii) The performance is independent of fuel technology and a wide range of fuels can be burned.

## 18.29 ROCKET ENGINES

*A rocket engine has its own oxidiser and does not depend on the surrounding air. Hence, it can operate at any place even in vacuum. The fuel and the oxidiser are carried into the body of the unit which is to be propelled. The two substances together are termed as* **propellants.** The propellants may be of two types:

(i) Solid propellants, and (ii) liquid propellants

The following are the requirements of **an ideal rocket propellant:**

(i) high heat value, (ii) low toxicity and corrosiveness, (iii) highest possible density so that it occupies less space, (iv) reliable and smooth ignition, and (v) stability and ease of handling and storing.

Rocket propulsion at this time should not be regarded as a competitor of the existing means for propelling aeroplanes but as a source of power for reaching objectives unattainable by other methods. Some of the applications of the rockets engines are:

(i) artillery barrage rockets. (ii) anti-tank "Bazooka" rockets.
(iii) all types of guided missiles. (vi) aircraft launched rockets.
(v) jet-assisted take-off for airplanes.
(vi) engines for long range, high speed guided missile and pilotless aircraft.
(vii) main and auxiliary propulsion engines on transonic aeroplanes such as the D-558.

***It is the advances in rocket technology which landed the man on the moon long back and sending a large number of satellites in space orbiting the earth and serving the mankind for communication, entertainment, video-conferencing, etc.***

## 18.30 BASIC THEORY OF OPERATION OF ROCKET ENGINES

Figure 18.41 shows a schematic diagram of a liquid bi-propellant rocket engine. *It consists of an injection system, a combustion chamber and an exit nozzle. In the combustion chamber, the oxidizer and fuel burn which produces high pressure. The pressure thus produced depends on mass flow rate of propellants, the cross-sectional area of the nozzle throat and the chemical characteristics of the propellant.* The products of combustion are ejected to the atmosphere at supersonic speed through the nozzle. In the nozzle, the pressure energy is converted into kinetic energy. The reaction of this high velocity jet emerging from the nozzle produces the thrust of the rocket engine.

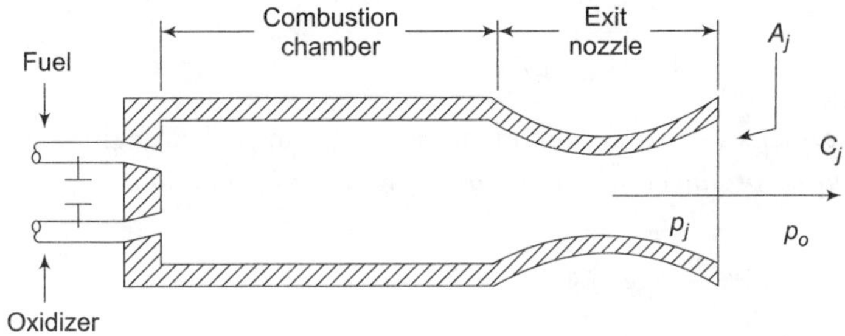


**Fig. 18.41** Components of a liquid bi-propellant rocket

**Thrust.** The thrust produced is due to the resultant of the pressure forces acting upon the inner and outer surfaces of the rocket motor.

Hence, resultant internal forces $= \dot{m}_p\, C_j + p_j A_j$ ...(14.46)

where $\dot{m}_p$ = mass flow rate of propellants, (kg/s), $C_j$ = exit or jet velocity relative to nozzle, (m/s), $p_j$ = exit static pressure, (N/m$^2$), $A_j$ = exit area, m$^2$

The resultant external forces acting on the rocket motor. $= p_o A_j$

where $p_0$ = atmospheric pressure, N/m$^2$

$\therefore$ Thrust $= T$ = Resultant of total pressure forces $= \dot{m}_p\, C_j + A_j\,(p_j - p_0)$ ...(14.47)

The above equation shows the effect of atmospheric pressure on the thrust of a rocket engine, *The lower is the atmospheric pressure the higher is the thrust. This means when* $p_0 = 0$, i.e., *the rocket engine will produce the maximum thrust in vacuum.*

Sometimes, the thrust is expressed in terms of mass flow rate of propellants and an 'effective jet velocity', Thus,

$$T = \dot{m}_p\, C_{je} \qquad ...(14.48)$$

*The effective jet exit velocity* is a hypothetical velocity and it is expressed as

$$C_{je} = C_j + \frac{A_j}{\dot{m}_p}(p_j - p_0)\ ; \quad \text{m/s}$$

$C_{je}$ is an important parameter in rocket engine performance.

**Thrust Power** (TP) *It is defined as the thrust multiplied by the flight velocity,*

or $$TP = TC_a = \dot{m}_a\, C_{je}\, C_0\ ; \quad \text{Watt} \qquad ...(14.49)$$

**Propulsive Efficiency.** It is expressed as

$$\eta_{pro} = \frac{TP}{TP + \text{K.E. loss}} = \frac{\dot{m}_p\, C_{je}\, C_a}{(m_p\, C_{je}\, C_a) + [\dot{m}_p(C_{je} - C_a)^2/2}$$

$$= \frac{2(C_a/C_{je})}{1 + (C_a/C_{je})^2} \qquad \text{...(14.50)}$$

**Specific Impulse.** This is also an important parameter in rocket engine performance and is defined as *the thrust produced per unit mass flow rate of propellant consumption.*

$$I_{sp} = \frac{T}{\dot{m}_p} = C_{je}\ ;\ \text{N–s/kg} \qquad \text{...(14.51)}$$

## 18.31 SUMMARY OF FORMULAE

(1) $\eta_{th} = 1 - \frac{1}{(r_p)^{(\gamma - 1)/\gamma}}$

(2) $\eta_{pc} = \frac{\gamma - 1}{\gamma} \times \frac{n}{n - 1}$

(3) $\eta_{pt} = \frac{n - 1}{n} \times \frac{\gamma}{\gamma - 1}$

(4) $AR = \frac{3600}{w_{net} \text{ is kJ/kg}}$

(5) $WR = \frac{w_{net}}{w_t}$

(6) $(r_p)_{opt} = \left(\eta_t . \eta_c . \frac{T_{max}}{T_{min}}\right)^{\gamma/2(\gamma - 1)}$

(7) $\varepsilon_r = \frac{\text{Actual heat exchange}}{\text{Maximum theoretical heat exchange}}$

(8) $r_p = \sqrt{p_1\, p_3}$

(9) $T = (1 + \dot{m}_f/\dot{m}_a)C_j - C_a$

(10) $TP = \frac{(C_j - C_a)C_a}{1000}$, kW/kg

(11) $PP = \frac{(1 + \dot{m}_f/m_a)C_j^2 - C_a^2}{2 \times 10^3}$

(12) $\eta_p = \frac{2(1 + \dot{m}_f/\dot{m}_a)C_j - C_a]C_a}{[1 + \dot{m}_f/(\dot{m}_a)C_j^2 - C_a^2]}$

(13) $\eta_{th} = \frac{C_j^2 - C_a^2}{2 \times 10^3\, (\dot{m}_f/\dot{m}_a) \times (LCV)_f}$

(14) $\eta_{ov} = \eta_{th} \times \eta_p = \frac{[(1 + \dot{m}_f/\dot{m}_a)C_j - C_a]C_a}{10^3 \times (\dot{m}_f/\dot{m}_a) \times (LCV)_f}$

(15) $\eta_{jet} = \frac{\text{Final KE in the jet}}{\text{Isentropic heat drop in jet pipe + carry-over from turbine}}$.

## PROBLEMS

### Things to Think About

18.1 What is the philosophy involved in conversation of energy from a gas turbine plant?

18.2 What are the main parameters responsible for improving thermal efficiency and power output from gas turbine?

18.3 What is the maximum efficiency and power output from a single unit of efficient gas turbine at present?

18.4 What is the effect of inlet air temperature compressor on the gas turbine performance?

18.5 Why turbofan jet engines are used in commercial planes as compared to jet engines?

### Short and Long Theoretical Questions

18.6 What is the principle or jet propulsion?

18.7 It is true that rocket is independent of atmospheric air. Give reasons.

18.8 What do you mean by propulsive power and propulsive efficiency?

18.9 What are the advantages of jet propulsion over the other systems?

18.10 What will happen if there is no combustion chamber in the gas turbine plant?

[ ***Ans.*** Even the compressor may not be driven by the turbine and what to talk of output (also obvious from *p-V* or *T-s* chart)]

18.11 Why is thermal efficiency of the gas turbine less, even though the turbine develops greater power?

[***Ans.*** A big amount of (about 50%) of power is consumed by compressor]

18.12 What are the methods of improving the thermal efficiency of simple open cycle gas turbine?

18.13 Why is the thermal efficiency of gas turbine high at very high altitude?

[***Ans.*** Due to lower value of inlet air temperature]

18.14 Is the use of regenerator in a simple cycle gas turbine limited?

[***Ans.*** Yes, by the pressure ratio]

18.15 Axial compressor is more preferred than centrifugal compressor in aviation. Give reasons.

18.16 What are the various methods which are used to improve the efficiency and output of a gas turbine? Give a neat sketch, in each of the cycle and show the variation of efficiency and output with pressure ratio.

18.17 State merits and demerits of a closed cycle gas turbine over open cycle.

18.18 Derive an expression for the optimum pressure ratio giving the maximum cycle thermal efficiency of gas turbine cycle, if the compressor efficiency is $\eta_c$ and turbine efficiency is $\eta_t$. The maximum cycle temperature is $T_3$ and the minimum cycle temperature, $T_1$.

18.19 Derive an expression for the optimum pressure ratio giving the maximum specific output in simple cycle gas turbine.

18.20 Prove the following expression of thermal efficiency of a gas turbine with ideal regenerative and complete reheat, using reheat pressure for yielding the maximum output.

$$\eta_{th} = 1 - \frac{T_1(r_p)^{(\gamma - 1)/2\gamma}\,[(r_p)^{(\gamma - 1)/2\gamma} - 1]}{2\eta_c\,\eta_t \cdot T_3}$$

where $T_1$ = compressor inlet temperature; $T_3$ = turbine inlet temperature

$r_p$ = overall pressure ratio; $\eta_c$, $\eta_t$ = compressor and turbine efficiencies

18.21 Prove that the work done by the turbine per kg of gas is less by the following quantity, when pressure drops are considered as compared to the case when they are neglected in an open cycle of turbine using a heat exchanger

$$\frac{\gamma - 1}{\gamma} \times \frac{c_p T_3 \times \eta_t}{(\gamma - 1)/\gamma}\left(\Delta p_2 + \frac{\Delta p_1}{r_p}\right)$$

where $\Delta p_1$ is pressure drop in combustion chamber and air side of heat exchanger and $\Delta p_2$ is pressure drop in gas side of heat exchanger; $T_3$ = turbine inlet temperature, $\eta_t$ = turbine efficiency; $p_2$ = exhaust pressure; $r_p$ = pressure ratio.

18.22 List the advantages and disadvantages of a gas turbine over steam turbines and diesel or petrol engines.

18.23 Prove that the output of a simple gas turbine plant is positive only when the product of compressor and turbine efficiencies is greater than $\left(\frac{T_1}{T_2}\right)(r_p)$ when $r_p$ is pressure ratio, $T_I$ and $T_3$ are the compressor and turbine inlet temperatures.

## Numerical Questions

18.24 Air enters a compressor at 10°C and at atmospheric pressure and is compressed through a pressure ratio of 5. The temperature of gaseous mixture enters the turbine at 850°C and expands to atmospheric pressure. Determine the thermal efficiency of the cycle under the following conditions.

(a) The system operates on ideal Brayton cycle.

(b) The isentropic efficiency of compressor and turbine is each 0.85.

(c) With condition (b) a regenerator of 75% effectiveness is introduced.

(d) With condition (c) an intercooler which cools to the 10°C is introduced at optimum pressure ratio.

(e) Repeat the calculation for pressure ratio 20 and turbine inlet temperature of 1400°C.

[***Ans.*** (a) 36.65% (b) 24.15%,(c) 36.6% (d) 35.9%]

18.25 Sketch diagramatically the arrangement of an internal combustion turbine to work on a open cycle with two stages of compression and expansion of the working substance and regenerative after compression.

18.26 If the total pressure ratio is 25, the maximum cycle temperature is 1400°C and the minimum 20°C, compression and expansion are isentropic and intercooler and regenerator are each 75% efficient. Find (a) the thermal efficiency of the cycle. If the efficiency of the turbine and compressor units are each 0.85, (b) what is the maximum thermal efficiency that may be expected from the turbine.

18.27 The air enters the compressor of a gas turbine at temperature 15°C and the atmospheric pressure operating at pressure ratio of 5:1. The isentropic efficiency of compressor is 0.8. The inlet temperature to turbine is 600°C. What must be the isentropic efficiency of turbine in order that the overall cycle efficiency may be 18.%. Assume $c_p$ = 1.006 for air and $c_p$ = 1.09 for gases and $R$ = 0.287 for both air and gases.

**Hints:** Find $\gamma$ air and gas. [*Ans.* 89.4%]

18.28 A gas turbine is designed to operate under the following conditions:

Maximum temperature = 1350°C, $r_p$ = 20

Inlet pressure and temperature = 1.05 bar and 15°C

Turbine isentropic efficiency = 88%

Compressor isentropic efficiency = 87%

Mechanical efficiencies of the compressor and turbine each = 99%

Efficiency of combustion = 98%

(a) Determine the improvement in plant thermal efficiency that would result from the addition of heat exchanger of 75% effectiveness and (b) with reheater with the reheated temperature of 1300°C. Allow a pressure loss of 0.905 bar in the heat exchanger, for air $c_p$ = 1.005 and $c_p$ = 1.13 kJ/kg K for gas. Assume that in the heat exchanger, the mean specific heat of the exhaust products and air is the same. Pressure loss in combustion chamber = 0.2 bar.

18.29 The air enters the compressor of a gas turbine at atmospheric pressure and 30°C and is compressed to 22 bar with an isentropic efficiency of 89%. The turbine inlet temperature is 1400°C. The isentropic efficiency of the turbine is 88% Estimate the flow of air in kg/s for a net power of 200 MW. Assume $\eta_{cc}$ = 0.98, $\eta_m$ = 0.95, $\eta_{gen}$ = 0.985.

18.30 The atmospheric air in which a jet travels at the speed of 223 m/s is 0.562 bar. The gas enters the nozzle at 2 bar and 833K and leaves the nozzle exit at pressure 1 bar. Calculate the velocity of jet at nozzle exit, nozzle exit area, propelling thrust and thrust power if the nozzle efficiency is 0.95 and the mass flow 22.5 kg/s. [*Ans.* 533 m/s, 852.1 cm$^2$,12240 N, 2668 kW]

18.31 Air enters at the rate of 900 kg/min a compressor jet aircraft travelling at 241 m/s. The air-fuel ratio is 60:1 and the compression pressure ratio is 6:1. The colorific value of fuel is 42,000 kJ/kg. Neglecting all losses, calculate the thrust, the specific fuel consumption and the propulsive efficiency. [*Ans.* 10300 N, 40%]

18.32 Atmospheric air enters at pressure 1.013 bar and temperature 15°C compressor of a turbojet engine. The isentropic efficiency of compressor is 0.8. Calculate the pressure at the turbine exhaust if gas enters the turbine at 1060°C and leaves the turbine at negligible velocity. The isentropic efficiency of turbine is 0.8. Assume $c_p$ = 1.005 and $\gamma$ = 1.4 for air and $c_p$ = 1.13 and $\gamma$ = 1.33 for gases.

If the mass flow rate of air through the compressor is 22.5kg/s and the nozzle efficiency is 0.94, calculate the nozzle exit area and the thrust of the jet leaving nozzle when the jet is stationary. [*Ans.* 1.792 bar, 979 cm$^2$, 11700 N]

18.33 An aircraft travels at 241 m/s at an altitude at which the pressure is 0.46 bar and the temperature is –17°C, the calorific value of fuel burnt being 42000 kJ/kg. Assuming that the inlet duct (diffuser) converts the entire kinetic energy of the air into pressure energy and that the isentropic efficiency of compressor is 80% when the pressure ratio is 5, calculate the air-fuel ratio if the maximum temperature must not exceed 874°C. Assume $c_p$ as 1.005 kJ/kg. [*Ans.* 61.4]

18.34 A jet engine requires 1544 kW of thrust power at 193 m/s. The fuel consumption is 0.4299 kg per propulsive kW per her and the C.V. of fuel is 42000 kJ/kg of fuel. The temperature rise in the combustion chambers is limited to 510°C. Calculate the air-fuel ratio, and the plant efficiency.

[*Ans.* 63.3, 1.910 m./s., relative to machine, 7500 N. 327%. 19.63%]

### Software Based Questions

18.35 Develop a software for predicting the performance of a gas turbine and jet propulsion system. Validate the results with latest systems.

18.36 Develop a software for the design of gas turbine and jet propulsion system. Validate the results with latest system.

### Design and Open Ended Questions

18.37 Prepare a comprehensive report on all models of gas turbine plants and commercial aircraft manufactured by various firms covering all aspects such as compressor ratio, turbine output, turbine inlet temperature, turbine exhaust temperature, speed, stages of compressor and turbines, types and number of combustion chambers, plant efficiency, price per MW, per thrust (N), Mach number, thrust, propulsive efficiency, seating capacity, sfc, BPR, etc., for the use of consumers.

18.38 At present, one of the main problems in gas turbine is the cooling of blades which causes some losses. Develop a gas turbine and jet propulsion systems free from cooling of blades so that the maximum temperature does not remain the limitation. For this, complete design report is to be prepared.

### Multiple-Choice Questions

18.39 Intercooling in gas turbines at an optimum pressure
(a) increases net output (b) increases thermal efficiency
(c) both (a) and (b) (c) decreases fuel consumption

18.40 The blades of gas turbine are subjected to the maximum temperature as per current metallurgical limit of about
(a) 500°C (b) 900°C (c) 1400°C (d) 2000°C

18.41 Regenerator in a gas turbine
(a) increases thermal efficiency (b) decreases heat loss to exhaust
(c) permits to use higher compression ratio (d) increases fuel consumption

18.42 The actual thermal efficiency, of a modern gas turbine ranges from
(a) 20 to 25% (b) 25 to 30% (c) 30 to 35% (d) 35 to 40% or higher

18.43 In the analysis of axial flow gas turbine, the assumptions are
(1) Degree of reaction is usually 0.5
(2) The rotor blades are twisted
(3) Free vortex flow is assumed for radial equilibrium
(4) The rotor blades are equiangular at a given radius.
Out of this,
(a) 1, 2 and 3 are correct (b) 1, 2 and 4 are correct
(c) 1, 3 and 4 are correct (d) 1, 2, 3 and 4 are correct

18.44 If $C_a$ is the aircraft velocity and $C_j$ represents the jet velocity, the propulsive power is given by

(a) $\frac{C_j^2 - C_a^2}{2}$ (b) $\frac{2C_a}{C_j + C_a}$ (c) $\frac{C_j^2}{2}$ (d) $\frac{(C_j - C_a)C_a}{2}$

18.45 In a rocket, the air velocity relative to the aircraft is

(a) equal to the velocity of aircraft (b) equal to the jet velocity
(c) half of the velocity of aircraft (d) zero

18.46 In gas turbine, theoretically combustion take place

(a) at constant volume (b) at constant pressure
(c) at entropy remain constant (d) at constant temperature

18.47 In a gas turbine, combustion takes place at constant pressure due to

(a) open system (b) closed system only
(c) constant velocity (d) constant enthalpy

18.48 Restricted burning is adopted in rockets when it is required to

(a) produce a small thrust for a small duration
(b) produce a small thrust for a large duration
(c) produce a large thrust for a small duration
(d) all of the above

18.49 **Assertion (A):** The propulsion efficiency of a jet propulsion system will always be lower than its maximum value.

**Reasoning (R):** Because the thrust power at the maximum propulsive efficiency will be zero.

**Answer Code:**

(a) Both A and R are true (b) A is true but R is false
(c) A is false but R is true (d) Both A and R are false

## Key to Multiple-Choice Questions

18.39 (a) 18.40 (b) 18.41 (a) 18.42 (d) 18.43 (a) 18.44 (a) 18.45 (d) 18.46 (b)
18.47 (a) 18.48 (b) 18.49 (a)

CHAPTER

# Air Pollution and Control 19

## OVERVIEW

*The problem of air pollution came into existence from the day the first fire was lit*. Basically, air pollution is the presence of foreign substances in air. The requirement of a good environment is that the air should be pure and free from undesired ingradients, viz., harmful gases, suspended particles (SPMs), etc. In general, the accepted air quality is:

Carbon monoxide (CO) = $10^3$ μg/m$^3$, sulphur dioxide ($SO_2$) = 30 μg/m$^3$, nitrogen oxide ($NO_2$) = 30 μg/m$^3$, suspended particles matter (SPM) = 100 μg/m$^3$.

But the rapid industrialization and use of large number of automobiles have adversely affected the environment: **Some specific definitions of air pollution** are given below.

(i) *Air pollution means the presence in the outdoor atmosphere of one or more contaminants such as dust, fumes gas, mist, odour, smoke or vapour, in quantities, with characteristics, and of duration such as to be injurious to human, plant or animal life or to property, or which unreasonably interferes with the comfortable to enjoyment of life and property.* ***—Engineers Joint Council, USA***

(ii) *Air pollution is the excessive concentration of foreign matter in the air which adversely affects the well being of the individual or causes damage to property.*
***—American Medical Association***

(iii) *Air pollution is the presence in ambient atmosphere of substances, generally resulting from the activity of man, in sufficient concentration, present for sufficient time and under circumstances which interferes significantly with the comfort, health or welfare of precious or with the full use or enjoyment of property.*
***—Indian Standards Institutions, IS 4167, 1996***

**Greenhouse Effect.** A phenomenon called the green house effect is caused by the increasing amounts of $CO_2$ formed in the atmosphere. The primary source of $CO_2$ is the consumption

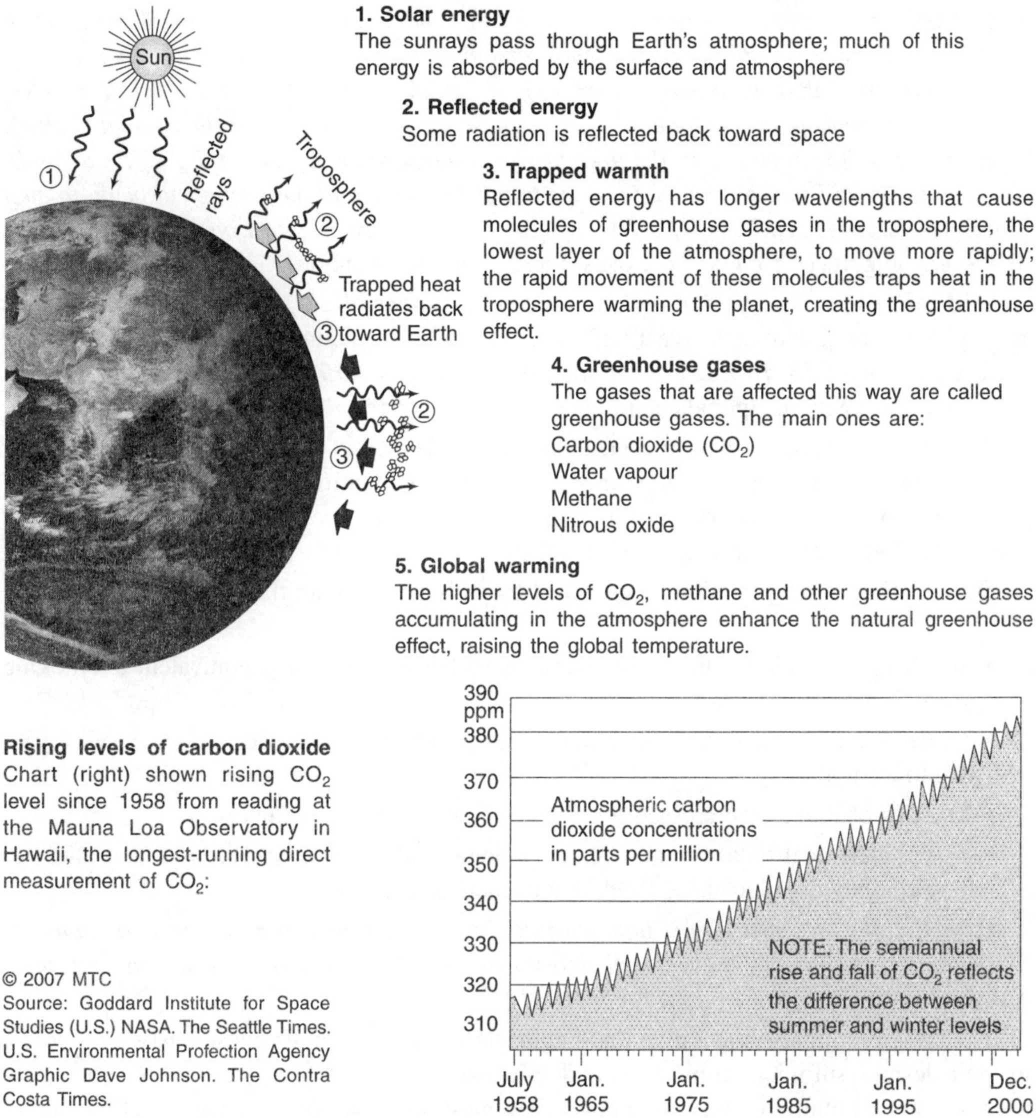


**Fig. 19.1** The greenhouse effect [Adapted from *Times of India*]

of fossil fuels. *The production of $CO_2$ is much more than the sinks (e.g., plants and sea) can absorb them and therefore the level of $CO_2$ in the atmosphere is increasing.* The greenhouse effect as the name suggests is analogous to the phenomena used to keep the ambient condition of plants warmer in a green house. In the botanist's greenhouse, the sunrays while passing through the glass walls suffer scattering which changes (increases) its wavelength. The glass pans are opaque to this longer wavelength, and so the radiant heat is retained. In the earth's case, what are referred to as "greenhouse gases" which means *primarily $CO_2$ act as the glass* *wall and the earth itself as the greenhouse. The sun's radiations can pass through the layer of $CO_2$ and heats up the earth but it stops that same heat being re-emitted by the earth. The net result of all this is that there is a danger of the earth gradually being heated up. If the $CO_2$ content of the atmosphere is doubled which can occur by the middle of the present century, then it will result in increase in the present world temperature by about 3.6°C. As a result, all the glaciers will recede and the ice caps in the Antarctic and the Arctic will begin to melt which will increase sea level by a few metres and thus most of cities on sea-shore may be submerged.* Figure 19.1 depicts the causes and effect of greenhouse gases.

**Acid rain** *is another effect of pollution.* Large quantities of $SO_2$ and $NO_2$ emitted into the atmosphere through chimneys react with moisture in the air and produce *sulphuric acid and nitric acid which falls as acid precipitation on the earth's surface.* Acidification of streams and lakes affect aquatic animals and plants leading to *destruction of ecosystems.*

In general, some actions of people, viz., *rapid industrialization, transportation, deforestation, etc., are the primary causes of pollution* and as the population increases, the problem of pollution also will increase proportionally. Air pollution means different things to different people, Further, it varies from place to place.

The situation of air pollution has reached to such a grim state that it presents a horrible scene as be given below:

- In metropolitan cities, the air we breath is so harmful that, it is equivalent to smoking about 20 cigarettes a day
- As per World Bank Report, in India about 40,000 premature deaths occur every year due to air pollution
- 21 cities of India have crossed the dangerous limits of air pollution.
- As per statistical data, the number of patients with respiratory diseases and allergies have roughly doubled since 1990, due to increased air pollution.

*Air pollution can cause death, impair health, reduce visibility bring about vast economic losses and contribute to the general deteriorations of ecology of both urban and rural regions.*

It is therefore highly imperative that engineers of all disciplines consciously incorporate in their designs sufficient constraints and safeguards to ensure that they do not contribute to atmospheric pollution. Over and above, they must apply various methods to eliminate air pollution so that the natural environment is restored.

Air pollution involves a sequence of events, namely, the generation of pollutants and their release from a source; their transport, transformation and removal from the atmosphere; and their effects on human beings, materials and ecosystems.

Now world over, *emissions norms are adhered for automobiles and other polluting agencies EURO VI or Bharat stage VI* emission norm is followed in India from 2020 for automobiles (see Table 19.14).

This chapter deals with the air pollution caused by IC engines (automobiles) and its control. The other sources of it are out of the above scope of this book.

## 19.1 SOURCE AND CLASSIFICATION OF AIR POLLUTANTS

In order to control the air pollution, an inventory of air contaminants is a must. Air pollution can either be natural or may be the consequences of activities of man like industrial operation, deforestation, transportation, etc. The industrial contaminants may include by-products of *external combustion* like smoke, dust and $SO_2$ or by-products of *internal combustion engines* like reaction in petrol and diesel engines. The pollutants can either by *primary* or *secondary* in nature. The sources of pollutants may be *stationary* or *mobile*. The following passage deals with details of source and their classifications.

**Air pollutants** can be **classified** mainly in ***three categories*** as given below:

(i) ***Natural contaminants*** (e.g., natural fog, pollen grains, bacteria and products of volcanic eruption. Pollen is matter of interest among all natural contaminants as it causes allergy, asthma and fever to some individuals).

(ii) **Aerosols (particulate)** (e.g., dust, smoke, mists, fog and fumes in size varying from 0.01 μm to 100 μm in diameter).

(iii) **Gases and vapours**

Table 19.1 shows the sources of air pollution indicating type of aerosols, gases and vapours.

**Table 19.1** Sources of Air Pollution

| *S.No.* | *Sources* | *Aerosols* | *Gases and vapours* |
|---|---|---|---|
| 1. | Combustion processes (domestic burning, thermal power plants, automobiles, aeroplanes, railways and refuge burning) | Dust fume, smoke | $SO_2$, $NO_2$, $CO_2$, CO, organic vapours and odours |
| 2. | Chemical processes (paper mills, cement, fertilizers, etc.) | Dust, fume, mist | Process-dependent—$SO_2$, CO, $NH_3$, $NO_2$ organic vapours and odours |
| 3. | Petroleum operation | Dust, mist | $SO_2$, $H_2S$, $NH_2$, CO, hydro-carbons |
| 4. | Mineral processing | Dust fume | Process dependent – $SO_2$, CO, fluorides and organic vapours |
| 5. | Metallurgical processes (aluminium, refineries, steel plants) | Dust, fume | $SO_2$, CO, fluorides organic vapours |
| 6. | Agricultural activities (a) crop spraying (b) field burning | Dust, mist, smoke, fly ash | Organic phosphates, chlorinated hydrocarbons, sulphur oxides, organic vapours |
| 7. | Food and feed operation | Dust, mist | Odourous materials |
| 8. | Nuclear energy programmes (a) fuel fabrication (b) ore preparations (c) bomb explosions | Dust | Fluorides, Iodine - 131 and Aegas - 41, Radioactive gases (Sr - 90 Cs - 137, C - 14 etc.) |
| 9. | Natural contaminants | Fog, pollen grains, bacteria | All types of gases due to volcanic eruption. |

*Contd.*

*Table 19.1 Contd.*

| | | | |
|---|---|---|---|
| 10. | Earth moving operations (construction such as road, buildings, dams) | Dust | – |
| 11. | Materials handling and processing such as loading and unloading of sand, gravel, coal ores, lime cement and crushing and grinding of ores | Dust | – |

### 19.1.1 Primary and Secondary Air Pollutants

Air pollutants may also be classified as primary and secondary pollutants.

Primary air pollutants as those emitted directly from identifiable sources.

**Examples of primary air pollutants include**

(i) Finer particles (less than 100 μm in diameter); (ii) coarse particles (greater than 100 μm in diameter); (iii) sulphur compounds; (iv) oxides of nitrogen; (v) carbon monoxide; (vi) halogen compounds; (vii) organic compounds; and (viii) radioactive compounds

**Example of secondary air pollutants include**

(i) Ozone (ii) Formaldehyde (iii) Peroxy acetyl nitrate.
(iv) Photochemical smog (v) Acid mists ($H_2SO_4$)

### 19.1.2 Stationary and Mobile Sources

Air pollutants (emission) may also be classified by source, as stationary or mobile.

Another method of classifying air pollutants is based on the following:

**(i) Point source** (large stationary sources such as power plants)

**(ii) Area source** (small stationary sources and mobile sources with indefinite routes such as residential heating, commercial heating).

**(iii) Live source** (mobile sources with definite routes such as high vehicles, railroad, etc.).

## 19.2 METEOROLOGICAL FACTORS INFLUENCING AIR POLLUTION

Concentration of air pollutants in a particular area largely depends on meteorological conditions. The important meteorological parameters that influence air pollution can be classified into primary and secondary parameters. **Primary parameters include:**

(i) Wind direction and speed (ii) Temperature
(iii) Atmospheric stability (iv) Mixing height

**Secondary parameters include**

(i) Precipitation (ii) Humidity
(iii) Solar radiation (iv) Visibility

## 19.3 EFFECT OF AIR POLLUTION ON HUMAN HEALTH

*The growth of civilization through industrialization has produced the greatest environmental evils known as air pollution.* The air we breathe has life supporting properties till it is unpolluted. Once it is polluted, it has life damaging properties.

An average man breathes **22,000 times a day and takes in 16 kg air each day.** As per estimate, a man can live for five weeks without food and five days without water, but only for five minutes without air. The following are the factors of pollutants that affect human health.

(i) Nautre of the pollutants
(ii) Concentration of pollutants
(iii) Duration of exposure of pollutants
(iv) State of health of the receptor
(v) Age group of the receptor

### 19.3.1 Mechanism of Action of Air Pollutants on Health

The human health gets affected by air pollution as a result of contact between the pollutants and the body at the surface of skin and exposed membranes. Contact with exposed membranes surfaces is of utmost importance because of their high absorptive capacity compared to that of the skin. It is *worth to note that air-borne gases, vapours, fumes, mist and dust may cause irritation of the membranes of the eyes, nose, throat, larynx, tracho-bronchial tree and lungs.*

### 19.3.2 Effects on Health

The following are the effects of air pollution on health.

(i) Eye, nose, throat and respiratory track irritation.
(ii) Gases such as $H_2S$, $NH_3$ and mercaptans cause odour nuisance even at low concentration.
(iii) Mortality and morbidity rates increase.
(iv) Pollen causes asthmatic attacks to some.
(v) Chronic pulmonary diseases like bronchitis and asthma, are aggravated by a high concentration of $SO_2$, $NO_2$, particulate matter below 10 μm in diameter and photochemical smog.
(vi) CO produces stress on those suffering from cardiovascular and pulmonary diseases by combining with the haemoglobin in the blood.
(vii) Hydrogen fluoride causes diseases of bone (osteofluorosis) and mottling of teeth.
(viii) Carcinogenic agents in the pollutants cause cancer.
(ix) Respiratory diseases like silicosis, asbestosis, etc., are caused by dust particles.
(x) Certain heavy metals such as lead may cause poisoning through inhalation.

## 19.4 POLLUTANTS FROM IC ENGINES

Population of passenger cars and other auto vehicles in the world were more than 1.4 billion in 2020. The growth rate is about 15% every year. The major pollutants being emitted by IC engines are:

(i) Carbon compounds (CO and $CO_2$)
(ii) Sulphur compounds ($SO_2$ and $H_2S$)
(iii) Nitrogen oxides (NO, $NO_2$, $N_2O$)
(iv) Hydrocarbon (benzene and benzpyrene)

(v) Metals (Ph, As, Cd, Ni)
(vi) Photochemical products (PAN and $PB_2N$)
(vii) Particulate matter (soot and smoke)
(viii) Acid droplets ($HNO_3$ and $H_2SO_4$)
(ix) Complex organic substances (ether and acetic acid)

*Spark-ignition and diesel engine are a major source of urban air pollution. Diesel engines create relatively minor pollution problems compared to gasoline powered ones.* In advanced countries like USA and UK, the air pollutants by automobiles are above 50% of total air-pollutants. *Emissions from gasoline powered vehicles are generally classified as following:*

**SI Engines**

(a) Exhaust emissions, (b) Crankcase emissions, (c) Evaporative emissions

The amount of pollutants, that an automobile emits depend on a number of factors, including the design and operation (idle, acceleration, deceleration, etc.).

The approximate contribution of the pollutants from the various sources of *SI engines* are given below.

(i) Exhaust emissions : 65% HC and almost all CO and $NO_x$
(ii) Evaporative emission : 20% unburnt HC
(iii) Evaporation emission : 15% unburnt HC
(tank and carburettor). (In MPFI, there is no carburettor.)

SI engine exhaust gases contain the following pollutants:-

(a) $NO_x$ : 500 to 1000 ppm (or 20 g/kg of fuel)
(b) CO : 1 to 2% (or 200 g/kg of fuel)
(c) HC : 1000 to 3000 ppm (or 25 g/kg of fuel)
(d) Particulates : 05% of total emission
(e) Traces of $SO_2$ and compounds of lead

Fuel evaporation and piston blow by gases are effectively controlled in most modern engines by returning the blow-by gases from the crankcase to engine intake system and by venting the fuel-tank and carburettor float bowl through a vapour absorbing carbon cannister which is purged by some of the engine intake air during normal engine operation.

**CI Engines**

**Main emission from CI engine exhausts are the following:**

(a) $NO_x$ : About 500–1000 ppm, i.e., 20 g/kg of fuel
(b) HC : About 200–600 ppm, i.e., 5 g/kg of fuel
(c) CO : About 0.05–1%, i.e., 10 g/kg of fuel
(d) Particulates : About 15% of total emission
(e) Smoke : Contains about 0.1μ diameter soot particles
(f) Odour
(g) $SO_2$, $SO_3$, $H_2SO_4$

*The* ***diesel engine exhaust*** *contains only about a tenth of the amount of CO exhausted by a gasoline engine, although, its HC emission may be one-fifth of gasoline engines.* $NO_x$ *emission of diesel engines is comparable to those of gasoline engines. Blow-by is negligible in the diesel, since the cylinder contains only air on the compression stroke.* Evaporative emissions are also low as the diesel engine employs a closed injection fuel system and also because the fuel is less volatile as compared to gasoline. ***The major problems of diesel engines are smoke, particulate and odour***. Specific hydrocarbon compounds in the exhaust gases are the source of diesel odour. It is a matter of importance that diesel engines are important source of ***particulate emissions;*** between about **0.2 and 0.5 per cent of the fuel mass** is emitted as small as **0.1 μm diameter** particles which consist primarily of soot with some additional absorbed hydrocarbon material. ***The particulate emission from diesel engines amounts to about 15% of total emission, while that of SI engines amounts to 5% of total emission.***

If alcohol fuels are used in SI and CI engines aldehyde emissions increase. Petrol and diesel fuels contain sulphur. Sulphur is oxidized to produce sulphur dioxide ($SO_2$) of which a fraction can be oxidized to sulphur trioxide, $SO_3$, which combines with water to form a sulphuric acid aerosol.

## 19.5 MECHANISM OF FORMATION OF POLLUTANTS IN SI ENGINES

### 19.5.1 Exhaust Emission

Exhaust emissions are the major source of automotive emissions. *If the perfect combustion takes place, the exhaust gas would consist only of carbon dioxide and water vapour plus the excess oxygen.* But owing to several reasons, the combustion is incomplete and as a result, the exhaust consists of carbon monoxide, a deadly poisonous gas, unburnt hydrocarbon, $NO_x$, $SO_2$, and compounds of lead. In addition to these pollutants, there are large number of organic compounds such as ketones, aldehydes, etc., which form **smog** in the presence of sunlight.

Figure 19.2 shows the processes by which pollutants form within the cylinder of a SI engine during four different phases of the engine operating cycle – compression, combustion, expansion and exhaust. ***Nitric oxide (NO)*** *forms throughout the high-temperature burnt gases behind the flame as a result of chemical reactions involving nitrogen and oxygen atoms and molecules which do not attain chemical equilibrium (Fig. 19.2b). The higher the burnt gas temperature, the higher the rate of formation of NO.* During expansion stroke, the burnt gases cool which results in freezing of reactions involving NO and leaving NO concentrations far in excess of levels corresponding to equilibrium at exhaust conditions.

***CO*** *is formed during combustion process.* This is due to the fact that with rich fuel-air mixtures, there is insufficient oxygen to burn fully all the carbon in the fuel to $CO_2$. Further, even with lean mixtures, dissociation ensures significant production of CO. But during expansion stroke, CO oxidation process freezes due to cooling of burnt gases (Fig. 19.2c).

**Emission of *unburnt hydrocarbon*** through several different sources are given below.

(i) ***Unburnt mixture forced into crevices.*** During compression and combustion (Fig. 19.2a and b), the increasing cylinder pressure forces some of the gas in the cylinder into crevices or narrow volumes, connected to the combustion chamber. Most of this gas

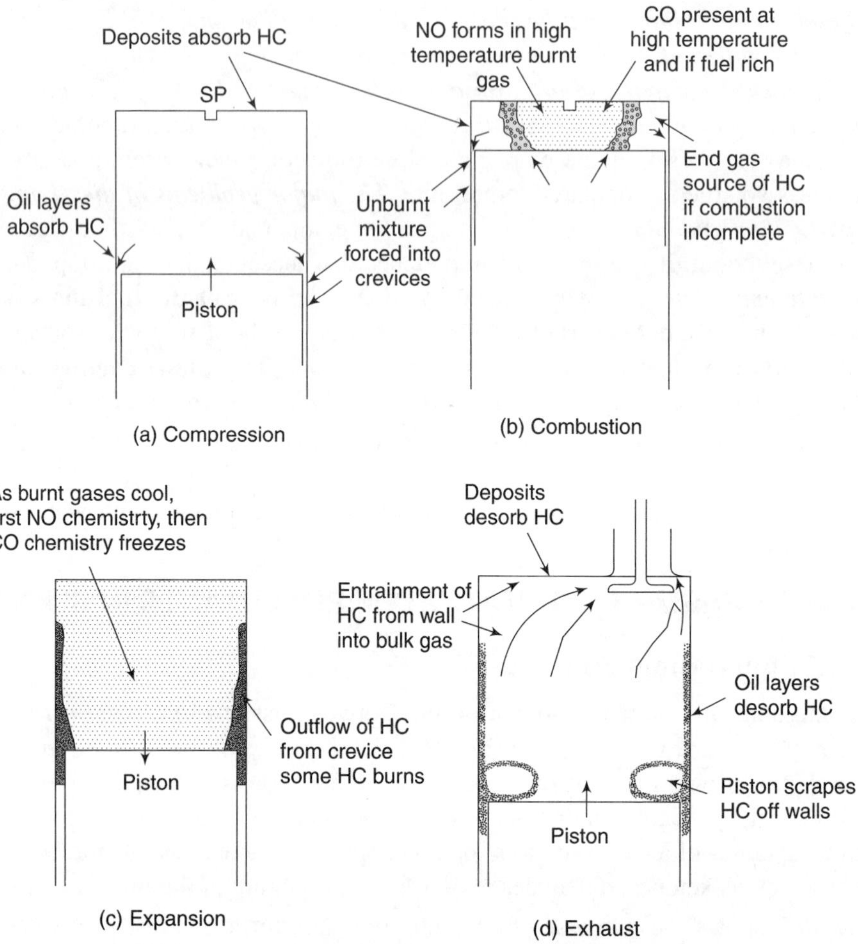


**Fig. 19.2** Pollutants formation mechanism in a SI engines

is unburnt fuel-air mixture. It is worth to note that much of this escapes the primary combustion process because the entrance to these crevices is too narrow for the flame to enter and hence it does not burn. During expansion and exhaust processes, the gas which leaves these crevices is one source of unburnt HC emissions.

**(ii)** ***Combustion chamber walls.*** Combustion chamber walls are the another source of HC emissions. A quenched layer containing unburnt and partially burnt fuel-air mixture is left at the wall when the flame is extinguished as it approaches the wall. It has been observed that HC layers in these thin layers is $\leq 0.1$ mm burn up rapidly when the combustion chamber walls are clean but porous deposits on the walls of the engine in actual operation increase engine HC emissions.

**(iii)** ***Engine oil left in a thin film on wall.*** The third source of unburnt hydrocarbons in any engine oil left as a thin film on the cylinder wall, piston and cylinder head. These oil layers are capable of absorbing and desorbing fuel hydrocarbon components before and after combustion respectively, thus permitting a fraction of the fuel to escape the primary combustion process unburned and hence the HC emissions.

**(iv)** ***Incomplete combustion due to bulk quenching of the flame.*** The fourth source of unburnt HC is done to bulk quenching of the flame. Such conditions arise during transient engine operation when the fuel air-fuel ratio, spark timing and the fraction of the exhaust recycled for emission control may not be properly matched.

It is noticeable that substantial oxidation of hydrocarbon which escapes the primary combustion process can occur during expansion and exhaust depending upon the temperature and oxygen concentration time histories of the this HC as they mix with bulk gases.

*Exhaust emissions are greatly affected by fuel-air equivalence ratio, (Φ), ignition timing and design of engine.*

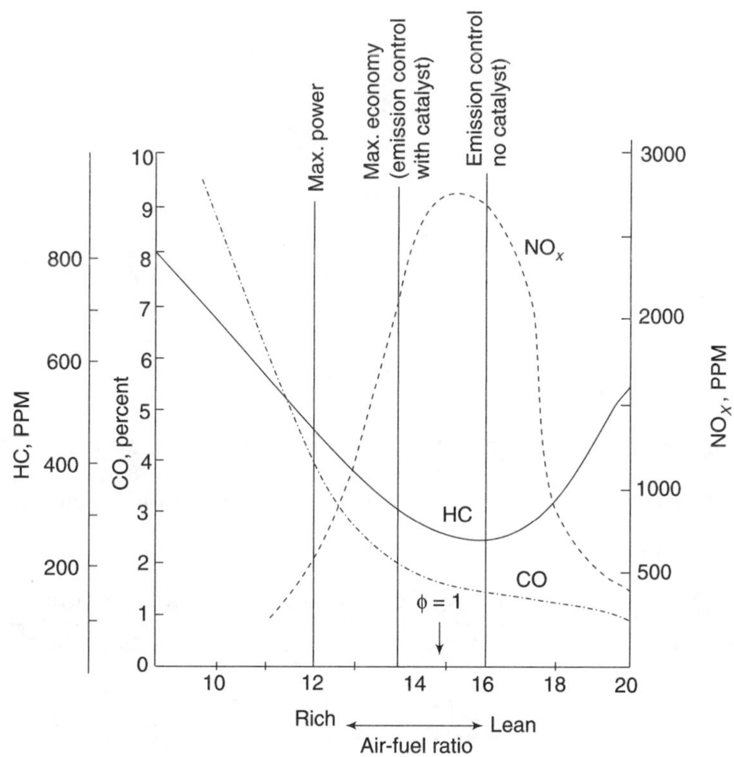


**Fig. 19.3** Variation of HC, CO and NO concentration in the exhaust of a SI engine

*Figure 19.3 shows the variation of HC, CO and $NO_x$ concentration in the exhaust of a conventional SI engine with fuel-air equivalence ratio. SI engine is normally operated close to stoichiometric, or slightly fuel-rich,* to ensure smooth and reliable operation. From the figure, it is obvious that leaner mixtures give lower emission until the combustion quality becomes poor (i.e., leading to misfiring), when HC emissions rise sharply and engine operation becomes erratic. During starting from cold, fuel vapourization is slow and hence rich mixture is provided for easy combustion. As a result, until the engine warms up and this enrichment is removed, CO and HC emissions are high but NO emissions are lower.

At part load conditions, lean mixtures are used which result in lower HC and CO emissions. The maximum NO emissions are obtained in the case of stoichiometric mixture.

It is worth to note that use of recycled exhaust to dilute the engine intake mixture lowers the NO levels, but also deteriorates combustion quality.

The combustion of HC and $NO_x$ in the presence of sunlight and certain atmospheric conditions produce ***photochemical smog***. The *mechanism of smog formation* is shown in Fig. 19.4.

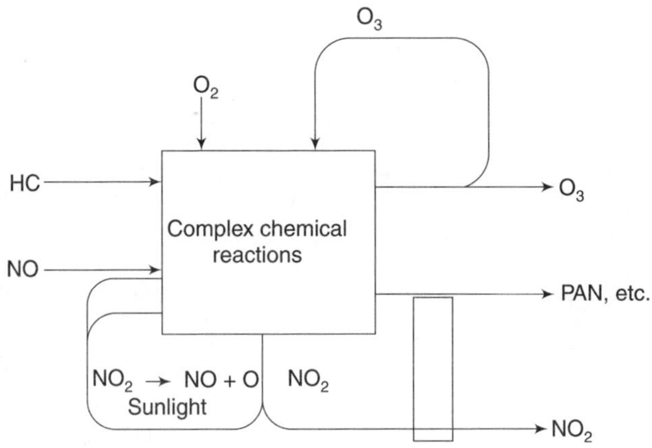


**Fig. 19.4** Mechanism of photochemical smog formation

Table 19.2 shows the effect of engine operating conditions on exhaust emission (from uncontrolled car).

**Table 19.2** Effect of Engine Operating Conditions on Exhaust Emissions

| *Conditions* | *Idle* | *Acceleration* | *Cruising* | *Decelerations* |
|---|---|---|---|---|
| *A/F ratio* | *11–12.5* | *11–13* | *13–15* | *11–12.5* |
| Engine speed, rpm | 400–500 | 400–500 | 1000–3000 | 3000–1000 |
| Air flow, cfm | 6–8 | 30–35 | 15–35 | 6–8 |
| Cylinder vacuum, Hg | 16–20 | 0–7 | 7–19 | 20–25 |
| Exhaust emissions CO, % | 4–6 | 0–6 | 1–4 | 2–4 |
| NO, ppm | 10–50 | 1000–4000 | 1000–3000 | 10–50 |
| HC, ppm | 50–1000 | 50–500 | 200–300 | 4000–2000 |
| Unburnt fuel% of supplied | 4–6 | 2–4 | 2–4 | 20–60 |

From Table 19.2, it may be observed that the *mode of operation greatly affects total emissions. A stop-and-go driving which means longer periods of high emission transient modes (acceleration and deceleration) will result in more emissions than sustained steady speed driving.*

*Organic and inorganic compounds of higher molecular weights and lead compounds resulting from the use of TEL are exhausted in the form of particulates of sizes varying from 0.02 μm to 0.06 μm* (**$PM_{2.5}$** - particulate matter less than 2.5 μm). About 75% of the lead burnt in the engine is exhausted into the atmosphere in the form of **$PM_{2.5}$** and rest is deposited on the engine parts.

*Traces of products of partial oxidation consisting of formaldehyde, acetaldehyde phenols, acids, ketones, ethers, etc., are also found due to incomplete combustion of fuels.*

### 19.5.2 Crankcase Emissions

*Crankcase emission refers to the phenomenon of leakage past the piston and piston rings from the cylinder to the crankcase. HC emissions due to blow-by amounts to about 20 per cent of the total HC emission for the engine. If the piston rings are worn, HC emissions increase to about 30%.*

The mechanism of crankcase emission is as follows.

Air-fuel mixtures trapped in the top land clearance and behind the top rings are unable to burn due to wall quenching effect. The cylinder forces this quenched gas (HC) past the piston ring and into the crankcase along with some burnt gases, in the ratio of 85:15. The blow-by rate depends on the top land clearance and the position of the top ring because some of the quenched gas is recycled in the combustion chamber and the ability of this to burn is controlled on the nearness to spark plug and the flame speed, otherwise it will go in the form of HC. Table 19.3 shows the volume percentage of exhaust and blow-by gases measured by Orsat analysis.

**Table 19.3** Volume Percentage of Exhaust and Blow-by Gas Emissions

| *A/F Ratio* | *Type of Emission* | *Volume%* | | |
|---|---|---|---|---|
| | | *$CO_2$* | *$O_2$* | *CO* |
| Rich | Exhaust gas | 11.0 | 0.7 | 4.6 |
| A/F = 12.6 | Blow-by gas | 1.5 | 17.9 | 0.6 |
| Lean | Exhaust gas | 13.0 | 0.6 | 0.5 |
| A/F = 16.4 | Blow-by gas | 1.8 | 18.4 | 0.3 |

### 19.5.3 Evaporative Emissions

*Evaporative emissions contribute about 15 to 25 per cent of total HC emission from a gasoline engine.* There are two main sources of evaporative emissions. They are ***fuel tank*** and ***carburettor*** (now fuel injection, where evaporation chance is more or less nil).

#### 19.5.3.1 Fuel Tank Emissions

*Fuel tank emission occurs by displacement of vapour during filling of petrol tank or by evaporation of fuel in the tank, forcing the vapour through a breather vent to the atmosphere.* Temperature of the tank varies according to engine operation. When the temperature is low, the fuel tank "breathes in" air but when the tank temperature goes high it "breathes out" air plus gasoline vapour.

The mechanism of fuel tank emission is given below.

A partially filled fuel tank under the condition of open to atmosphere is subjected to two forces, namely, the partial pressure of vapour phase of hydrocarbon and vapour pressure of liquid phase of hydrocarbons and they remain in equilibrium till temperature of both phases is the same. If the temperature of liquid is increased, say by engine operation, the vapour pressure of the liquid will increase and it will vapourize in an attempt to restore equilibrium. Due to vapourization of additional liquid, the total pressure of the tank increases and since the tank is open to atmosphere, the vapour will flow out of the tank. There will be a further outflow of vapour if vapour temperature also increases.

*A large number of variables such as ambient temperature, fuel tank temperature, mode of vehicle operation, the amount of fuel in the tank, volatility of the fuel, the capacity of the tank, design and location of fuel tank with aspect to exhaust system and flow pattern of the heated air underneath the vehicle affect the evapourative emissions.*

*Less the tank fill, greater is the evaporative emission.* Table 19.4 shows the effects of tank fill and temperature on evaporative emissions. It is obvious from the table as the temperature of the tank increases, emission due to evaporation increases.

The operational modes substantially affect the evaporation loss. About 75% of HC losses from tank are $C_4$ and $C_5$ hydrocarbons.

**Table 19.4** Effect of Tank Fill on Evaporative Emissions

| *Tank fill* | *Ambient Temperature (°C)* | *Temperature Rise during Test (°C)* | *Evaporative Emission (%)* |
|---|---|---|---|
| 1/4 | 19 | 7 | 5.7 |
| 1/2 | 16 | 4 | 1.2 |
| 3/4 | 18 | 2 | 0.1 |
| Full | 22 | 3 | 0.0 |

#### 19.5.3.2 Carburettor Loss

Carburettor losses result from the following.

(a) *External venting of float-bowl relieving internal pressure as the carburettor heats.*

(b) *Hot soak losses which occur after the engine has been stopped, as a result of evaporation of petrol stored in the bowl.* The loss is effected through vent pipe or through the air cleaner.

Most of the loss from the carburettor occurs due to direct boiling of the fuel in the carburettor bowl during hot soak. **This loss is absent now due to gasoline injection (MPFI).**

The type of emission are $C_5$ and $C_6$ hydrocarbon. The bowl volume and bowl temperature affect the evaporation.

## 19.6 MECHANISM OF FORMATION OF POLLUTANTS IN CI ENGINES

*CI engines suffer from the main drawback that the fuel distribution is non-uniform in most of the critical parts of cycle. The processes of pollutant formation are* strongly dependent on fuel distribution and how distribution changes with time due to mixing.

Figure 19.5 depicts how various parts of the fuel jet and the flame affect the formation of NO, unburnt HC, and soot or (particulates) during the 'pre-mixed' and mixing controlled phases of diesel combustion in a direct injection (DI) engine with the presence of swirl. The formation of nitric oxide occurs in the high-temperature burnt gas region as before, but temperature and F/A ratio distributions within the burnt gas are non-uniform which results in the highest formation rate of nitric oxides at close to stoichiometric regions.

*The formation of soots (or particulates) occurs in the rich unburnt fuel containing core of the fuel spray, within the flame region, where the fuel vapour is heated by mixing with burnt gases.* The oxidation of soots then starts in the flame vapours zone when it contacts unburnt oxygen giving rise to luminous character of the flame.

The regions where the flame quenches both on the walls and where excessive dilution with the air prevents the combustion process from either starting or going to completion, give rise to the formation of hydrocarbons and aldehydes. *Fuel that vapourizes from the nozzle during later stages of combustion is also source of HC.*

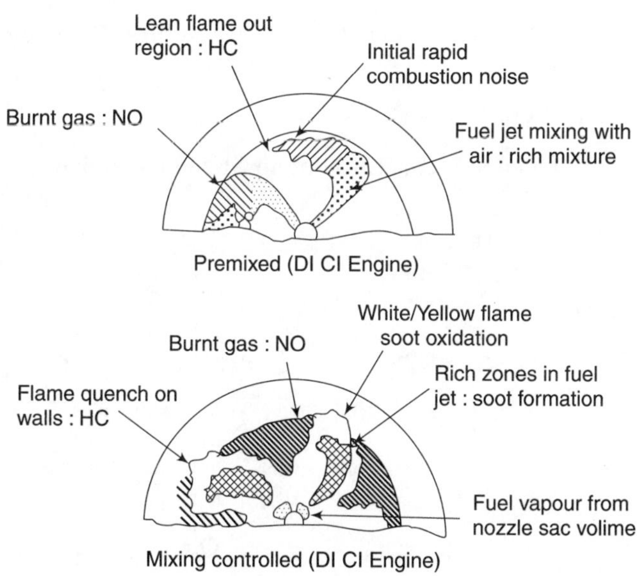


**Fig. 19.5** Pollution formation mechanism in CI engine

## 19.7 KINETICS AND FORMATION OF NITROGEN OXIDES

### 19.7.1 Kinetics of NO Formation

*Nitric oxide (NO) is the predominant oxide of nitrogen produced inside the engine cylinder. The principal source of NO is the oxidation of atmospheric (molecular) nitrogen at high flame temperature.* Moreover, the oxidation of nitrogen present in the fuel is an additional source of NO formation. Gasoline contains negligible amount of nitrogen while diesel fuels contain more nitrogen. *It is worth to note NO and $NO_2$ are grouped together as $NO_x$ emissions.*

In combustion of near-stoichiometric fuel-air mixtures, the principal reactions governing the formation of NO from molecular nitrogen as given below.

$$O + N_2 = NO + N \qquad ...(19.1)$$

$$N + O_2 = NO + O \qquad ...(19.2)$$

$$N + OH = NO + H \qquad ...(19.3)$$

### 19.7.2 Kinetics of $NO_2$ Formation

Based on the consideration of chemical equilibrium, the ratio of $NO_2/NO$ in the burnt gases at typical flame temperature should be negligibly small. Experimental evidences confirm this for SI engines but in CI engines, $NO_2$ can be 10 to 30 per cent of total exhaust oxides of nitrogen emissions. *A possible mechanism of* $NO_2$ *formation is as follows.* NO *formed in the flame zone can be rapidly converted to* $NO_2$ *via the following reactions.*

$$NO + HO_2 \longrightarrow NO_2 + OH \qquad ...(19.4)$$

Conversion of this $NO_2$ to NO occurs unless the $NO_2$ formed in the flame gets quenched by mixing with cooler fluid via

$$NO_2 + O_2 \longrightarrow NO + O_2 \qquad ...(19.5)$$

Figure 19.6 shows NO and $NO_2$ concentration in SI and CI engines [34].

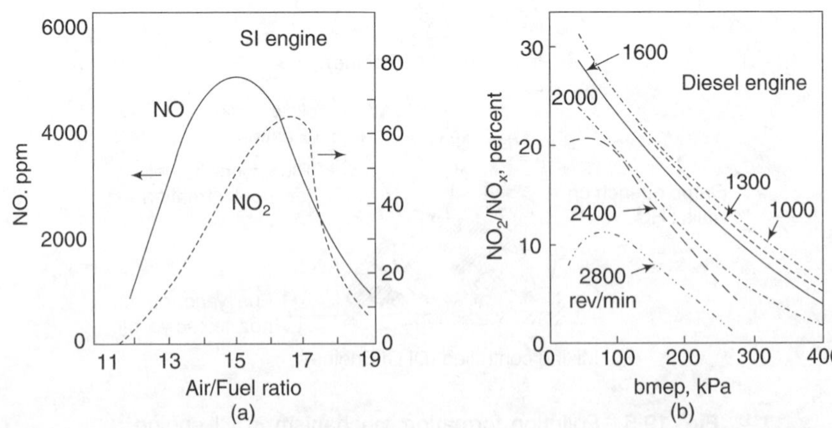


**Fig. 19.6** (a) NO and $NO_2$ concentration in SI engines (b) $NO_2/NO_x$ concentration in diesel engines [34]

Figure 19.6 (a) and (b) indicates that the maximum value for the ratio $NO_2/NO$ for the SI engine is about 2 per cent at equivalent ratio of about 0.85 while for diesel engines, this ratio is higher especially at lightloads.

## 19.7.3 NO Formation in SI Engines

The most important SI engine variables that affect NO emission are the following:

(a) Fuel-air equivalence ratio ($\phi$)

(b) Burnt gas fraction of the in-cylinder unburnt mixture

(c) Spark timing

*The burnt gas fraction depends on the amount of diluent such a recycled exhaust gas (EGR) used for $NO_x$ emissions control as well as the residual gas fraction.*

### 19.7.3.1 F/A Equivalence Ratio

The effect of variation of F/A equivalence ratio on NO emission is depicted in Fig. 19.7. *The maximum burnt gas temperature occurs around $\Phi = 1.1$. But at this oxygen concentration are low. Burnt gas temperature falls with the enrichment of mixture. As the mixture is leaned out, increasing oxygen concentration initially offsets the falling gas temperature and NO emission reaches its peak value at about $\Phi = 0.9$.* In rich mixtures, NO concentration freezes easily in the expansion process and little NO decomposition occurs while in rich mixtures, substantial NO decomposition occurs.

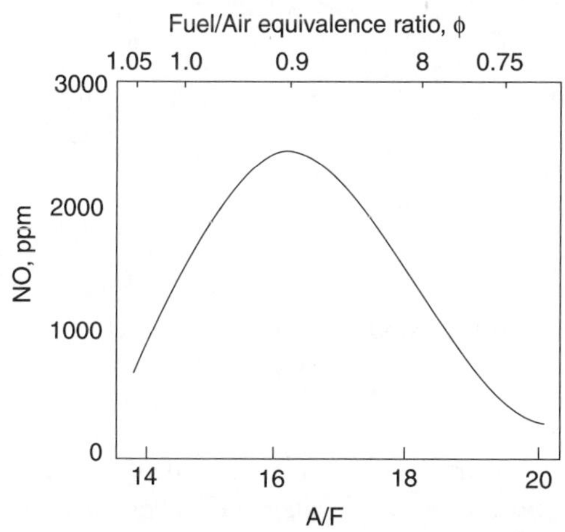


**Fig. 19.7** Effect of equivalence ratio on NO emission in a SI engine [34, 35]

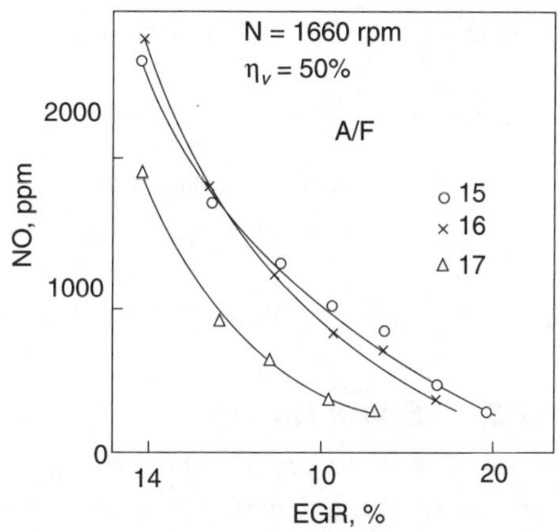


**Fig. 19.8** Effect of EGR on NO concentrations [34, 35]

### 19.7.3.2 Burnt Gas Fraction

The unburnt mixture in the cylinder of SI engine consists of fuel vapour, air and burnt gas. The burnt gas are residual gases from the pervious cycle and any exhaust gas is recycled to the intake for $NO_x$ emissions control. *It is worth to note that residual gas fraction is a function of load, valve timing, speed, A/F and compression ratio.* Burnt gases act as a diluent in the unburnt mixture. The gas temperature reached after combustion varies inversely with burnt gas mass fraction. *This suggests that NO emission levels may be reduced by increasing the burnt gas mass fraction.* Side effect of this is to reduce the combustion rate which results in unstable combustion.

Figure 19.8 depicts the effect of increasing the burnt gas fraction by recycling exhaust gas (EGR) to the intake system just below the throttle plate of a SI engine [35]. There is a substantial reduction in NO concentrations with increasing value of EGR but not more than 15% to 25%. EGR is tolerable by any engine. *This is due to the fact that the burnt gas diluents in the unburnt mixtures reduces flame temperature by increasing the heat capacity of cylinder charge per unit mass of the fuel.* Figure 19.9 shows the effect of different diluents gases added to the engine intake flow [36]. It is obvious from Fig. 19.9 (a) that equal volume percentage of different diluents produce different reductions in NO emission but correlation with diluent heat capacity results in more or less single curve.

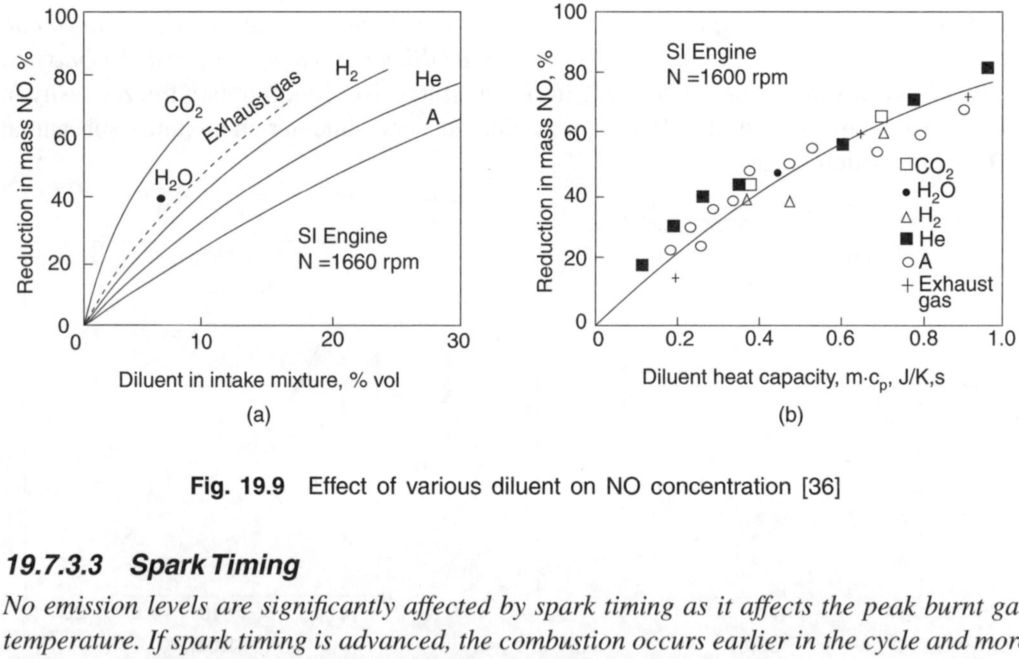


**Fig. 19.9** Effect of various diluent on NO concentration [36]

### 19.7.3.3 Spark Timing

*No emission levels are significantly affected by spark timing as it affects the peak burnt gas temperature. If spark timing is advanced, the combustion occurs earlier in the cycle and more fuel is burnt before TDC which results in higher peak cylinder pressure. This results in higher peak burnt gas temperature, hence higher NO formation rates. If spark timing is retarded, peak cylinder pressure decreases because more fuel burns after TDC. Lower the peak pressure results in lower peak burnt gas temperature and lower NO formation rates.*

Figure 19.10 shows the effect of spark timing on NO emissions [35]. *It is obvious from the figure that NO emission levels steadily decrease as spark timing is retarded form MBT timing and moved closer to TDC.*

### 19.7.4 $NO_x$ Formation in CI Engines

*$NO_x$ formation in diesel engine arises mainly due to non-uniform fuel distribution during combustion.* Mixture which burns early in the combustion process is of special importance since it gives rise to a higher temperature of burnt gas which results in higher NO formation rate. After the time of peak pressure, there is a decrease in burnt gas temperature due to expansion and due to mixing of high temperature gas with air or cooler burnt gas, which in turn, freezes the NO chemistry. *It is worth to note that freezing occurs more rapidly in the diesel engine as compared to SI engine and hence there is very less decomposition of NO.*

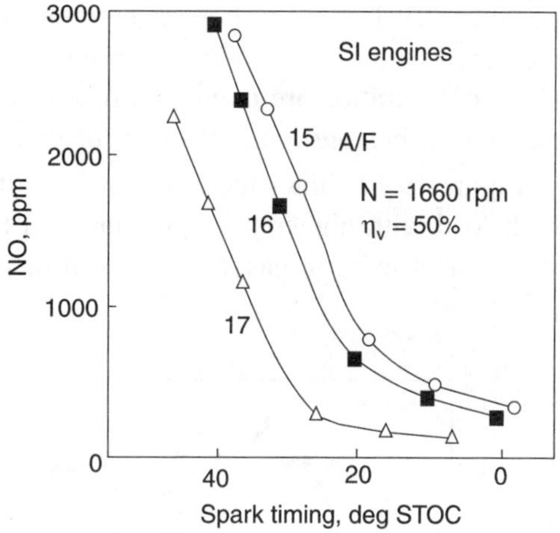


**Fig. 19.10** Effect of spark timing on NO emission [35]

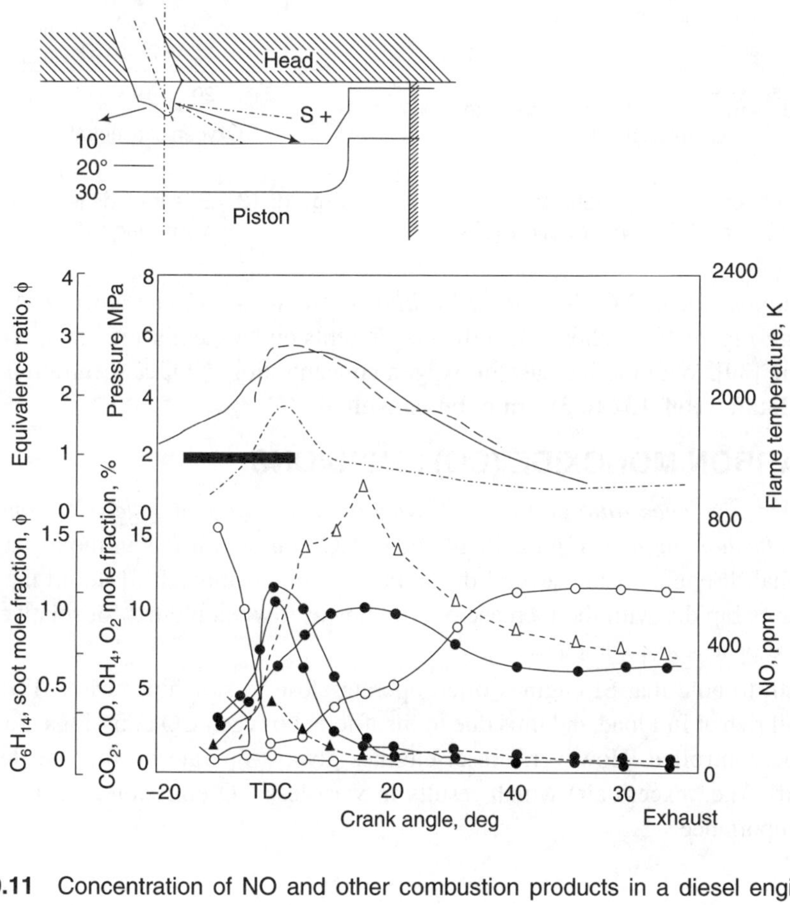


**Fig. 19.11** Concentration of NO and other combustion products in a diesel engine [37]

Figure 19.11 shows the concentration of NO, CO, soot and other combustion products measured with a rapid acting sampling valve in a direct injection diesel engine [37]. NO and $CO_2$ concentration are maximum *at about 15° after TDC. The 20 crank angle degrees after the start of combustion is the critical time period during which all of the NO forms.*

Figure 19.12 shows the effect of overall equivalence ratio on $NO_x$ concentration [38]. At high load with higher peak pressure and hence temperature, and larger regions of close to stoichiometric burnt gas, NO levels in increase.

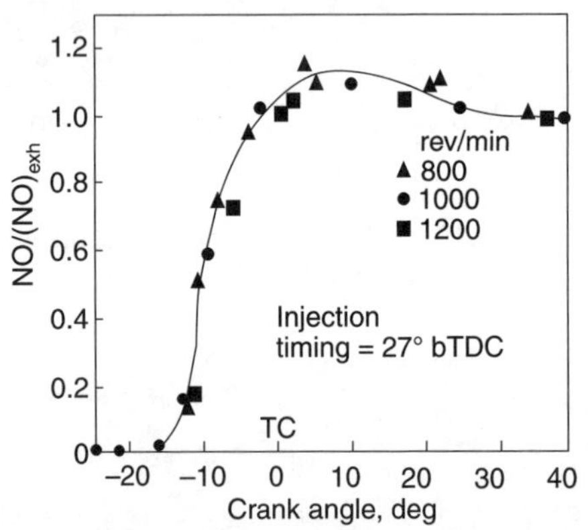


**Fig. 19.12** Effect of overall equivalence ratio on $NO_x$ and NO concentration [39]

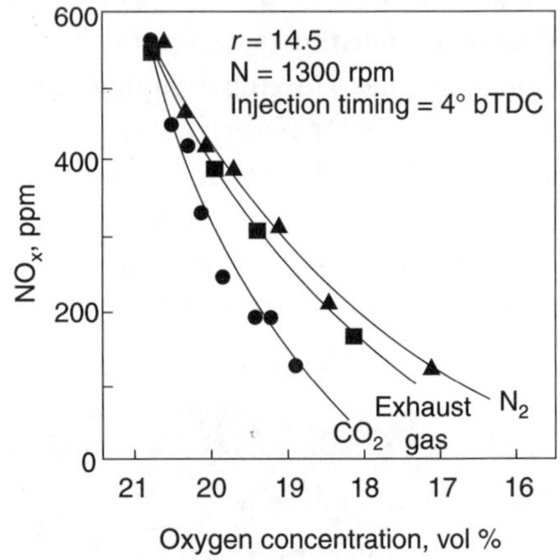


**Fig. 19.13** Effect of diluents on $NO_x$ concentration [39]

*The formation rate of $NO_x$ is reduced by diluents (recycled exhaust gas, $CO_2$ $N_2$) added to the intake air.* Figure 19.13 shows the effect of diluents on $NO_x$ emission in a direct injection diesel engine [39]. With the increase in oxygen concentration, $NO_x$ concentration decreases. **For more details, Ref. [34 to 39] may be consulted**.

## 19.8 CARBON MONOXIDE (CO) EMISSIONS

*It is the F/A equivalence ratio that primarily controls carbon monoxide (CO) emission from internal combustion engines.* Figure 19.14 shows CO emission levels in the exhaust hood of a conventional SI engine using several different fuel compositions [48]. In all the cases, CO level decreases rapidly with the increase of A/F, however when plotted for relative A/F ratio, it gives more or less a single curve.

It is worth to note that SI engines often operate close to stoichiometric A/F ratio at part load and fuel rich at full load and thus due to insufficient oxygen CO emissions are significant and must be controlled. However, diesel engines always operate on the fuel lean side of stoichiometric (i.e., excess air) which results in very less CO emission and thus it is not a matter of importance.

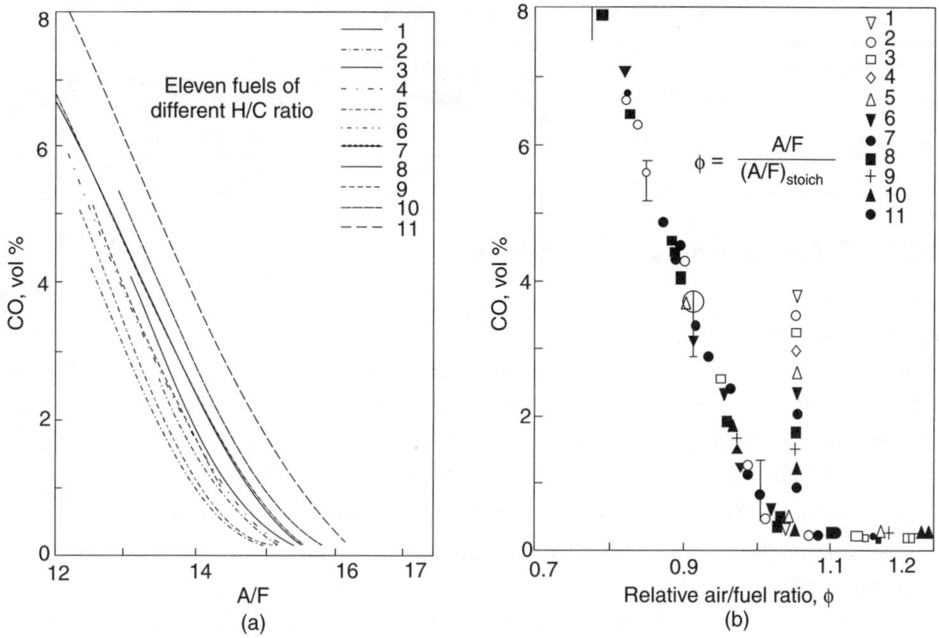


**Fig. 19.14** CO emission from SI engines using different H/C ratio [48, 73]

It multi-cylinder engines, because CO increases rapidly as the inlet mixture becomes richer than stoichiometric, cylinder to cylinder variations in equivalence ratio about the mean value are important parameters. This suggests that improved cylinder to cylinder F/A distribution is essential to reduce CO emissions. Further, CO emission during engine warm-up are much higher than emissions in the fully warmed up state because of use of rich mixture.

Additional reductions in CO beyond what can be achieved in the engine are only possible with the exhaust gas treatment device.

## 19.9 UNBURNT HYDROCARBON EMISSIONS

*Incomplete combustion of hydrocarbon fuels results in the hydrocarbon or organic emission.* In general, the level of unburnt hydrocarbons (HC) in the exhaust gases is generally specified in terms of total hydrocarbons concentration expressed in parts per million carbon atoms ($C_1$). Engine exhaust gases contain a large number of hydrocarbons. Table 19.5 shows the average breakdown by class of the hydrocarbons in SI engine exhaust gases.

It is worth to note that fuel decomposition can significantly influence the composition and magnitude of the organic emissions. Fuel containing high proportions of aromatics and olefins produce relatively higher concentration of reactive carbons. It is interesting to note that many of the organic compounds formed in the exhaust are not present in the fuel which suggests that significant pyrolysis and synthesis occurs during the combustion process.

**Table 19.5** HC Comparison of SI Engine Exhaust (by class)

| *Conditions* | *Carbon, per cent of total HC* | | | |
|---|---|---|---|---|
| | *Parafin* | *Olefin* | *Acetylene* | *Aromatics* |
| Without catalyst | 33 | 27 | 8 | 32 |
| With catalyst | 57 | 15 | 12 | 26 |

### 19.9.1 HC Emission from SI Engines

*HC emission from the exhaust of SI engine under normal conditions ranges from 1000 to 3000 ppm which corresponds to about 1 to 2.5 per cent of the fuel flow under the conditions of the maximum combustion efficiency.* As discussed earlier, HC emission rises with richness of fuel in mixture as compared to stoichiometric. Moreover, HC emission also rises with very lean mixture due to incomplete combustion or misfire. Further, any HC escaping the primary combustion process may oxidize in the expansion and exhaust process.

There are **five possible HC emission formation mechanism** *for SI engines.*

(i) *Flame quenching* at the combustion chamber walls, leaving a layer of unburnt fuel-air mixture adjacent to the wall.

(ii) *Filling of crevic volumes* with unburnt mixture which, since the flame quenches at the crevic entrance, escapes the primary combustion process.

(iii) *Absorption of fuel vapour into oil layers* on the cylinder wall during intake and compression. Followed by desorption of fuel vapour into the cylinder during expansion and exhaust.

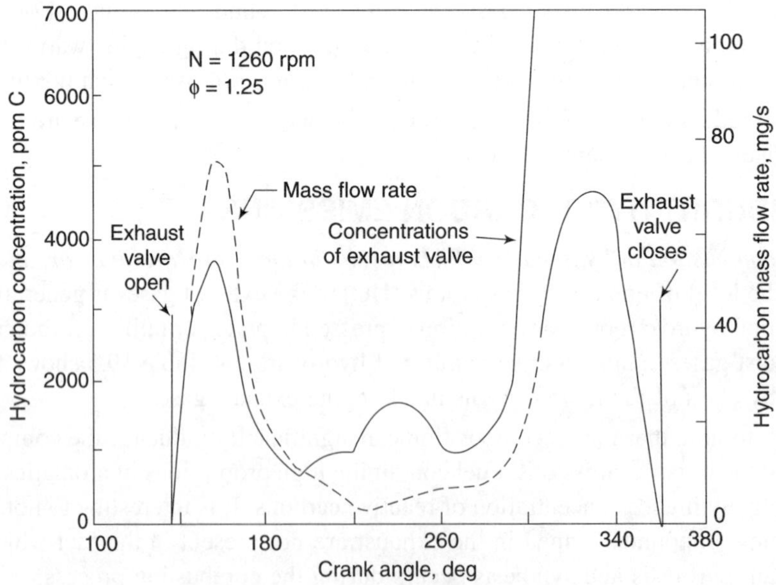


**Fig. 19.15** Variation of HC concentration and HC mass flow rate at the exhaust valve during exhaust process [41]

(iv) *Incomplete combustion* is a fraction of the engine operation cycle when combustion quality is poor.

(v) *Deposit built up* on the combustion chamber wall.

It is interesting to note that all these mechanisms except misfire result in unburnt hydrocarbons close to the combustion chamber walls and not in the bulk of cylinder gases resulting in non-uniform distribution of HC in the exhaust gases. Figure 19.15 shows variation of HC concentration and mass flow rate along crank angle during exhaust process [41]. It is obvious that unburnt HCs are exhausted in two peaks of approximately equal mass. The first of these peaks coincides with the exhaust blow down just after opening of exhaust valve and second occurs towards the end of exhaust stroke where HC concentration are very high and the mass flow rate is relatively low.

#### 19.9.1.1 Flame Quenching at the Walls

*There is a quench layer on the cold combustion chamber walls of a SI engine.* The quench layer thickness measured were in the range of 0.05 to 0.4 mm (thinnest at high load). These quench layers contain unburnt and partially burnt fuel-air mixture. When the walls are clean, these unburnt HC burns rapidly but there is always a chance of porous deposits on the wall which does not help in burning of HC and so there is HC emissions.

Figure 19.16 shows the variation in concentration of HC species through the combustion, expansion and exhaust processes [42]. The fuel ($C_3H_8$) concentration drops rapidly to a low value when flame arrives at value but at the same time intermediate hydrocarbon product concentration rises and then fall sharply below 1 ppm. Onward 60 aTDC, all HC concentrations rise and vary on engine operation conditions including flame quenching.

#### 19.9.1.2 Crevic HC Mechanism

*As discussed earlier, crevices in the combustion chamber walls—small volumes with narrow entrances have been the major sources of unburnt HC emissions. The largest of these crevice regions is the volumes between the piston rings and the cylinder wall.* The other type of crevice volumes *include threads around the spark plug, the space around the plug centre electrode, crevices around the intake and exhaust valve heads and the head gasket crevice.* The following are the important crevices processes occurring during the engine cycle.

During compression process, cylinder pressure rises and hence the unburnt mixture is forced into the crevice regions. As these volumes are thin, they have a large surface-volume ratio, which helps in cooling down the gas following into each crevice by heat transfer up to the wall temperature. During combustion, pressure rise takes place which further forces unburnt mixture into the crevice volumes. With the approach of flame at each crevice, it can either propagate into the crevice and fully or partially burn the fuel and air within the crevice or it can quench at the crevice entrance depending upon crevice entrance geometry, the composition of the unburnt mixture and its thermodynamic state. Until the cylinder pressure starts to decrease, burnt gases will flow into each crevice after the flame arrival and quenching. With the decrease of cylinder pressure as compared to crevice gas pressure, gas flows back from each crevice into the cylinder. This happens during expansion process. Depending on spark plug location in relation to the top ring gap, well above 50 per cent of this gas can be unburnt fuel-air mixture.

Figure 19.17 shows the effect of top land clearance on HC emissions. It is obvious from Fig. 19.17 that as the top land clearance increases until the clearance equals about 0.18 mm, when emission drops to zero clearance level.

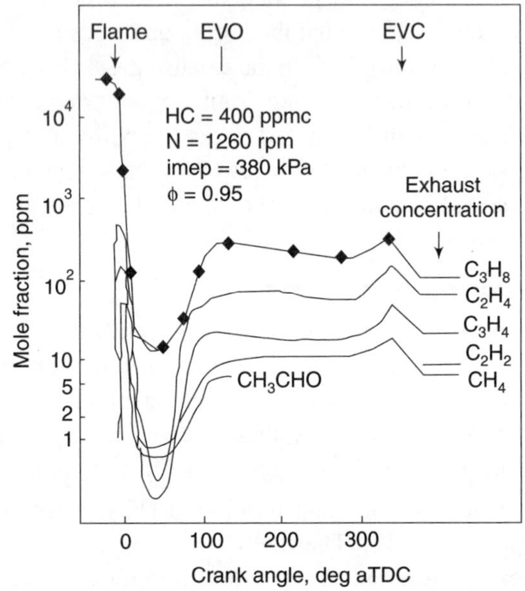


**Fig. 19.16** Variation of HC adjacent to combustion chamber walls as a function of crank angle [42]

It has been observed that the ring back crevices produces approximately 80 per cent of total scaled HC emissions, the heat gasket crevice about 13 per cent and the spark plug threads 5 per cent.

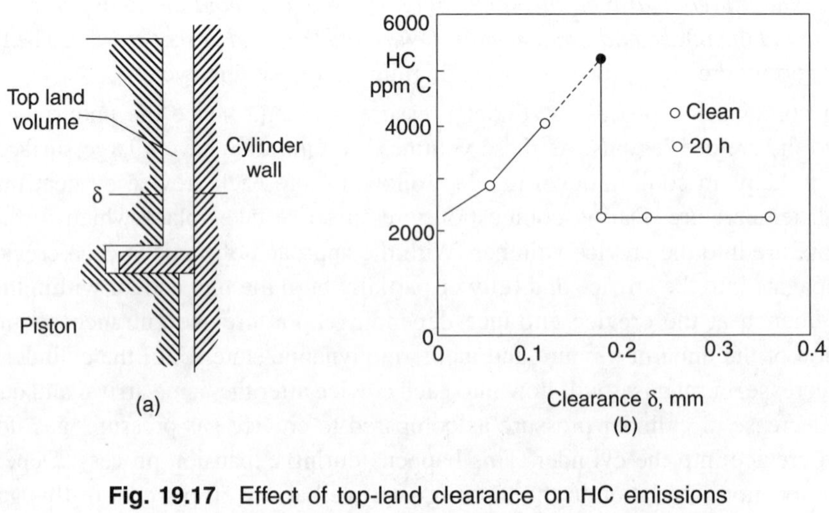


**Fig. 19.17** Effect of top-land clearance on HC emissions

Table 19.6 depicts a summary of estimate of the total mass of gas and mass of unburnt mixture flowing into and out of crevice regions of a typical SI engine.

**Table 19.6** Amount of Gas Flowing into and out of Crevic Region

| S.No. | Description | % mass ppm c |
|---|---|---|
| 1. | Total gas in all crevice regions | 8.2 |
| 2. | Total gas back to combustion chamber | 7.0 |
| 3. | Unburnt back to combustion chamber | 3.7-7.05000-9400 |
| 4. | Unburnt to blow-by | 0.5-1.2 |
| 5. | Total unburnt escaping primary combustion | 4.2-8.2 |

### 19.9.1.3 Absorption and Desorption in Engine Oil

The presence of lubricating oil in the fuel or on the walls of combustion chamber causes HC emission in the exhaust. *Exhaust HC levels from a clean engine can be double or triple when operated on a fuel containing 5 per cent lubricating oil over a period of about 10 minutes.* The phenomenon of HC emission by the presence of engine oil could be explained as follows by absorption and desorption mechanism.

During intake and exhaust processes, the fuel vapour concentration within the cylinder is close to the inlet manifold concentration. This means, for one crankshaft revolution, any oil film on the walls will absorb fuel vapour. During latter part of the compression, even if the oil was saturated during intake, absorption will continue according to Henry's law. During combustion process, the fuel vapour concentration in the bulk gases goes essentially to zero so that the absorbed fuel vapour will desorb from the liquid oil film into the gaseous combustion products. It is interesting to note that desorption could continue throughout the expansion and exhaust stroke. Some of the desorbed fuel vapour will mix the high-temperature combustion products and oxidize; but desorbed vapour that remains in the cool boundary layer or mixes with the cooler bulk gases late in the cycle may escape full oxidation and contribute to unburnt HC emission in the exhaust.

### 19.9.1.4 Poor Combustion Quality

Poor combustion quality causing flame extinction in the bulk gas, before all of the flame front reaches the wall causes HC emission under certain engine operating condition. This is explained as follows.

During expansion stroke, with the fall of cylinder pressure, the temperature of the unburnt mixture ahead of the flame decreases, which results in slowing the burning rate. If the pressure and temperature fall too rapidly the flame can be extinguished. Such a bulk quenching of flame has been observed in SI engines resulting in a very high HC concentration for the particular cycle. Engine operating condition where bulk quenching is most likely to occur are at idle and light load where engine speed is low and the residual gas fraction is high, with high dilution with excessive EGR or very lean mixture.

Figure 19.18 shows a comparison of HC emissions with moderate and faster burn rate as a function of EGR. Here, $(COV)_{imep}$ *is the coefficient of variation in imep.* As the percentage of EGR increases, $(COV)_{imep}$ and HC emission increase lesser for fast burn rate and more for moderate burn rate and $NO_x$ decreases.

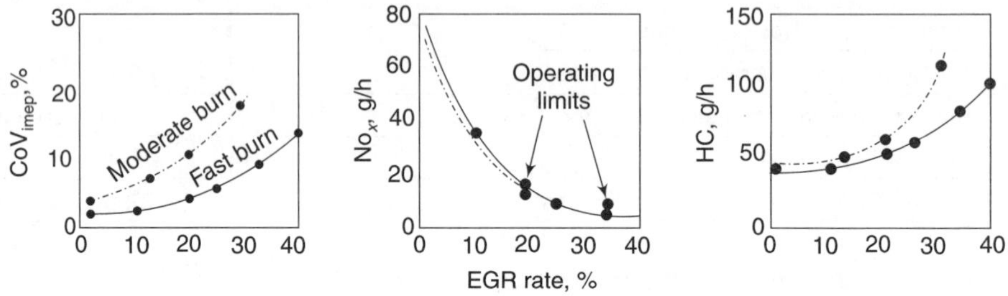


**Fig. 19.18** Effect of burn rate and EGR or $(COV)_{imep}$, $NO_x$ and emission

### *19.9.1.5 Effect of Deposits*

*It has been observed that when a vehicle has run over several thousands kilometre, deposit builds up on the combustion chamber walls which causes unburnt HC.* If the gasoline contains lead, the increase in HC emissions varies between 7 to 20 per cent. The removal of deposits results typically in a reduction in HC emissions to close to clean engine levels. Soft sooty deposits, such as those which accumulate after running the engine on a rich mixture, also cause an increase in HC.

The mechanism of deposit emission is still unknown. It is believed that absorption and desorption of hydrocarbon by these surface deposits may be the mechanism of emissions by these deposits. However, deposits can also build up in the piston ring crevice regions. Due to reduction in volume of crevice regions, HC emissions may decrease but due to changes in piston-cylinder wall clearance due to deposits may cause flame quenching resulting in increased emissions.

### *19.9.1.6 Hydrocarbon Oxidation*

Figure 19.19 shows the flowchart of complete SI engine HC formation and oxidation mechanism within the cylinder and exhaust system. There is a considerable evidence that substantial oxidation of HC does occur. The oxidation of unburnt HC in the thin quench layer formed on the combustion chamber walls oxidation occurs quickly. An average value for passenger car vehicle exhausts in that fuel compounds comprise 40 per cent of the total HC. *There is a direct evidence that HC oxidation in the exhaust systems occurs.*

An experimental study of HC existing from a simulated crevice volume has exhibited that complete HC oxidation only occurs when cylinder gas temperature is above 1400 K.

Overall, probably about half of the unburnt HC formed by the source mechanisms described above will oxidize within the engine's cylinder depending upon operating conditions and engine design.

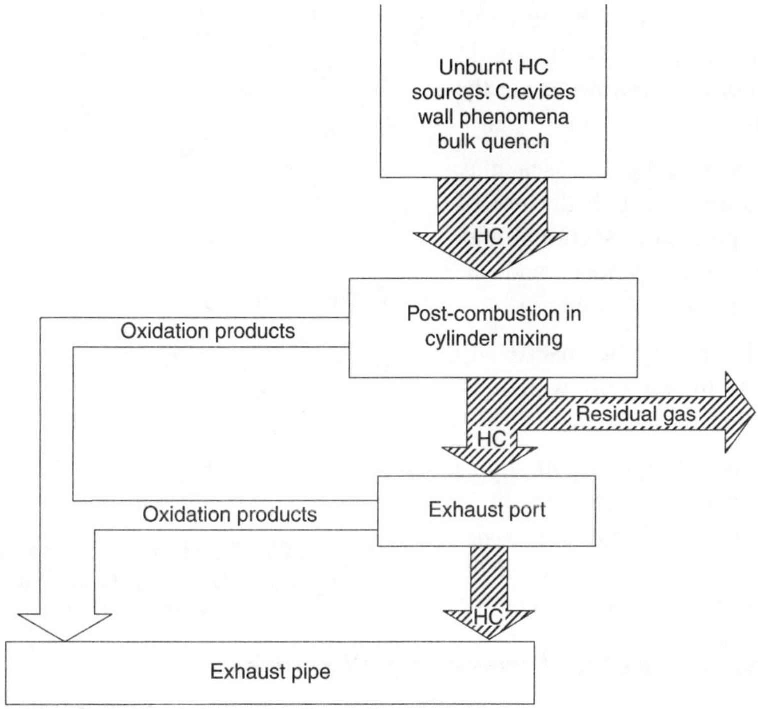


**Fig. 19.19** Flowchart of Complete SI Engine HC Formation and Oxidation Mechanism within the Cylinder and Exhaust System

It is interesting to note that to oxidize the hydrocarbons in the gas phase, or residence time of order of 50 ms or larger at temperature in excess of 600°C are required. *To oxidize carbon monoxide temperature in excess of 700°C are required. The average gas temperature at the cylinder exit are about 800°C and the average gas temperature at the exhaust port is about 600°C.* It is obvious with these temperature data that only in the exhaust port and upstream end of the manifold can have any significant gas-phase HC oxidation occurs.

Figure 19.20 shows the effect of exhaust gas temperature on HC and CO burn up in the exhaust of SI engines as a function of residence time. It is obvious from the results that rate of oxidation of HC and CO increases rapidly with increasing temperature of exhaust gas ($T_e$) up to a certain value of residence time (100 ms in the case of HC and 200 ms in the case of CO). With the increase of residence time (higher load) HC emission decreases rapidly up to a certain value and after that there is no effect of residence time.

The per cent of unburnt HC existing in the cylinder which reacts in the exhaust system with most of the reaction occurring in the port varies between a few and 40 per cent. *Engine operating conditions that give highest exhaust temperature such as stoichiometric operation, higher speeds, retarded spark plug, timing, lower compression ratio, etc., and longest residence time such as lighter load gives relatively higher percentage reduction in HC emissions due to oxidation of HC.*

In order to increase the degree of reaction for oxidation occurring in the exhaust port, the following are the possible methods.

(i) The exhaust heat losses may be reduced through the use of larger port cross-sectional area (to reduce flow velocity and surface area per unit volume).

(ii) Port liner may be inserted to provide higher port wall temperature.

(iii) Better port design may minimize hot exhaust gas impingement on the walls. *Further, exhaust reactors (catalyst) may be used for oxidation in exhaust system.*

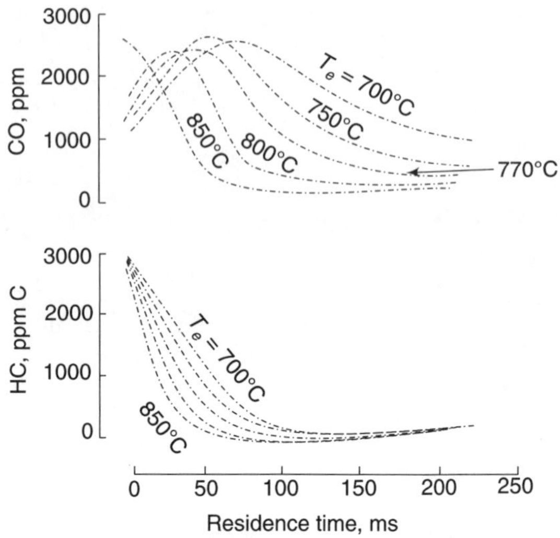


**Fig. 19.20** Effect of exhaust gas temperature or HC and CO burnup in the exhaust of a SI engine [46, 73]

### *19.9.1.7 Summary of HC Emissions in SI Engines*

Table 19.7 depicts the summary of critical factors and engine variable in HC emission mechanism. *The total process is divided into four sequential steps as given below.*

(i) Formation of HC

(ii) In-cylinder mixing and oxidation

(iii) Fraction of HC flowing out of cylinder

(iv) Oxidation in exhaust system.

For details on HC emissions, Ref. [40 to 46] may be consulted.

**Table 19.7** Critical Factors and Variables for HC Emissions Mechanisms in SI Engines

| | | |
|---|---|---|
| ***I. Formation of HC*** | | |
| (A) **Crevices** | | |
| (1) Crevice volume | (2) Crevice location | (3) Engine load |
| (4) Crevice wall temperature | (5) A/F (composition) | |
| (B) **Oil layers** | | |
| (1) Oil consumption | (2) Wall temperature | (3) Engine speed |
| (C) **incomplete combustion (poor combustion quality)** | | |
| (1) Burn-rate and variability | (2) A/F (Composition) | (3) Engine load |
| (4) Spark timing | | |
| (D) **Combustion chamber walls (flame quenching absorption and resorption)** | | |
| (1) Deposits | (2) Wall roughness | |

*Contd.*

***II. In-cylinder Mixing and Oxidation***

- (A) **Mixing rate with bulk gases**
  - (1) Engine speed (2) Swirl in combustion chamber
  - (3) Combustion chamber shape
- (B) **Bulk gas temperature during expansion and exhaust**
  - (1) Engine speed (2) Sparks timing (3) A/F (composition)
  - (4) Compression ratio (5) Heat loses to walls
- (C) **Bulk gas oxygen concentration–** A/F (or equivalence ratio)
- (D) **Wall temperature**
  - (1) Important parameters if HC source near wall
  - (2) For cervices; importance depends on geometry.

***III. Fraction of HC flowing out of cylinder***

- (A) **Residual fraction**
  - (1) Engine load (2) Exhaust pressure (3) Valve overlap
  - (4) Compression ratio (5) Engine speed
- (B) **In-cylinder flow during exhaust stroke**
  - (1) Valve overlap (2) Exhaust valve size and location
  - (3) Combustion chamber shape (4) Compression ratio (5) Engine speed

***IV. Oxidation in exhaust system***

- (A) *Exhaust gas temperature*
  - (1) Speed (2) Spark timing
  - (3) A/F (composition) (4) Compression ratio
  - (5) Secondary air flow (6) Heat losses in cylinder and exhaust
- (B) **Oxygen concentration**
  - (1) A/F (equivalence ratio) (2) Secondary air flow
- (C) **Residence time**
  - (1) Speed (2) Load
  - (3) Volume of critical exhaust system component
- (D) **Exhaust reactors**
  - (1) Oxidation catalyst (2) Three-way catalyst (3) Thermal reactor

## 19.9.2 HC Emission Mechanisms in Diesel Engines

The difference in the composition of diesel and gasoline fuel and methods of introducing the fuels in the engines make the distinction between the HC emission mechanisms of diesel and gasoline engines. *Diesel fuel contains hydrocarbon compounds with higher boiling points (i.e., higher molecular weights) as compared to gasoline.* Further, there is an occurrence of substantial pyrolysis of fuel components within the fuel sprays during the diesel combustion process. As a result, the composition of the unburnt and partially oxidized hydrocarbons in the diesel exhaust is much more complex than in the SI engine.

It is worth to note that the complex heterogeneous nature of diesel combustion, where fuel evaporation, fuel air and burnt-unburnt gas mixing and combustion can occur simultaneously and affect diesel engine HC emissions. The following are the two primary paths by which fuels can escape the normal combustion process unburnt.

(i) the fuel-air mixture can become too lean to autoginite or to support a propagating flame at the conditions prevailing inside the combustion chamber, or

(ii) during the primary combustion process, the fuel-air mixture may be too rich to ignite or support a flame.

The fuel that escapes combustion can then be consumed only by slower thermal oxidation reactions later in the expansion process after mixing with additional air. This suggests that some HC will remain unconsumed.

Figure 19.21 shows the schematic representation of diesel HC formation mechanism for two cases—one for fuel injected during delay period and other for fuel injected while combustion is occurring [47]. Fuel injected during the ignition delay period (Fig. 19.21) will mix with air to produce a wide range of A/F ratios such as locally overlean, locally overreach and combustible mixture. Due to slow reaction or no ignition or improper flame propagation, over-lean and over-reach mixture will not autoignite and will give product of incomplete combustion consisting of unburnt HC. Even some of the combustible mixture due to bulk quenching will result in incomplete combustion consisting of unburnt HC. *If the ignition delay increases from 12° to 25° exhaust HC emission increases from 200 ppm C to 2000 ppm C.*

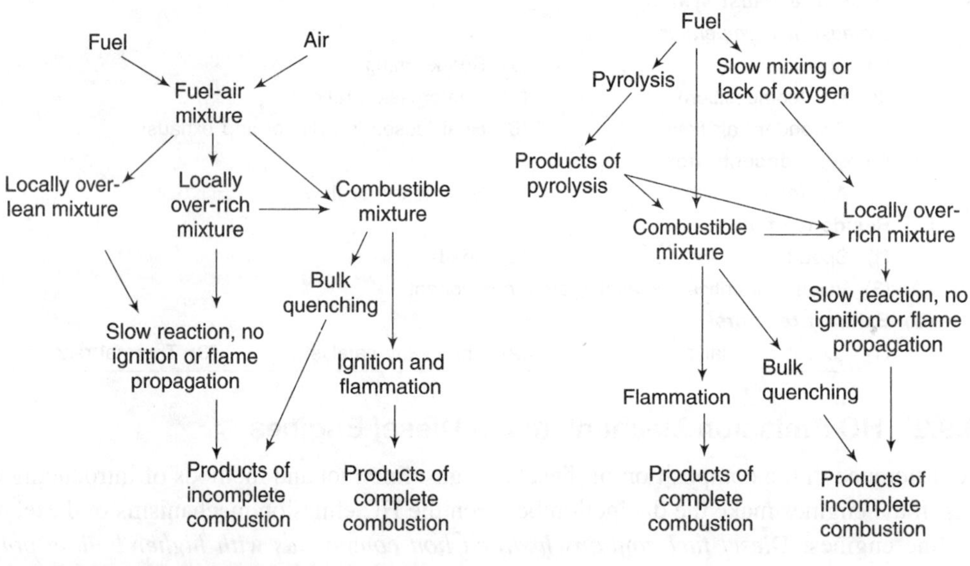


**Fig. 19.21** Diesel HC formation mechanisms (a) fuel injected during delay (b) after delay

The fuel injected while combustion is occurring (after delay period) either goes through pyrolysis process or slow mixing with air (undermining) which results in complete combustion. Due to bulk quenching, some of the combustion mixture produces incomplete combustion.

It is interesting to note that HC emission level from diesel vary widely with operating conditions. Engine idling and high load operation produce significantly higher HC emission than full load operation. Over-fuelling increases rapidly HC emissions (if $\phi$ increases from 1.0 to 1.1, exhaust HC emission increases sharply from 200 to 600 ppmC).

In concluding remarks, we can say that the following three main parameters affect the diesel HC emissions.

(i) Overleaning, (ii) under-mixing, (iii) quenching and misfire

## 19.10 PARTICULATE EMISSIONS

### 19.10.1 SI Engine Particulate

*The particulate emission from SI engines consists of three classes, namely,* ***lead****,* ***organic particulates*** *(including soot) and* ***sulphates.***

*The automobiles engines working over regular and premium leaded gasoline (which contains about 0.15 g $P_b$/ litre) emit particulates at the rate of typically 100 to 150 mg/km, out of which 25 to 60% is the lead.* When the engine is cold, following start-up, the particulate emissions are considerably higher. The exhaust temperature affects the emissions significantly. The particle size distribution with leaded fuel is as follows:

- *80% by mass below 2 μm diameter*
- *40% by volume below 2 μm in diameter*

*It is believed that most of these particles form and grow in the exhaust system as a result of vapour phase condensation enhanced by coagulation.* Some of the particles, either from or are deposited on the walls where agglomeration may occur while some of the particles are emitted directly without settling. If the exhaust flow rate is, suddenly increased many of these particulates are removed. Only a small fraction between 10 to 50 per cent of the lead consumed in the fuel is exhausted and remainder gets deposited within the cylinder and exhaust system. *Overly rich mixtures produce soot.*

Sulphate emission levels depend on the fuel suffer content, the operating conditions of the engine and the type of catalyst system used in the exhaust effect. Unleaded gasoline contains 150 to 600 ppm by weight sulphur. This gets oxidized within the engine cylinder to $SO_2$ which can be oxidized further to $SO_3$ by exhaust catalyst. This $SO_3$ combines with water at ambient temperature to form a sulphuric acid aerosol. The typical sulphate emission rates from automobiles are about 20 mg/km.

*Automobile engines operating on unleaded gasoline reduce particulate emissions to about 20 mg/km (on-fifth of leaded) with catalyst. It is interesting to note that this particulate is primarily a soluble (condensed) organic material.*

### 19.10.2 Diesel Engine Particulates, Soot and Smoke

#### *19.10.2.1 Characteristics of Diesel Engine Particulates, Soot and Smoke*

*The particulate emission from diesel engines consists of combustion generated carbonaceous material* ***(soot)*** *on which some organic compounds get absorbed. Incomplete combustion of the fuel is the main source of most of particulates, however, some are contributed by the lubricating oils.*

*Light-duty diesel automobiles give particulate emission rates in the range of 0.2–0.6 kg/km while for larger direct-injection engines emission rate are 0.5–1.5 g/bkWh. The parameters that govern the composition of the particulate material are engine exhaust conditions and particulate collections and particulate collections system.* For example, when exhaust temperature is above 500°C, the individual particles are principally clusters of many small spheres or spherules of carbon of diameter ranging from 15–30 nm. But at temperature below 500°C, particles get coated with absorbed and condensed high molecular weight organic compounds including ketones, ethers, organic acids and polynuclear aromatic hydrocarbons.

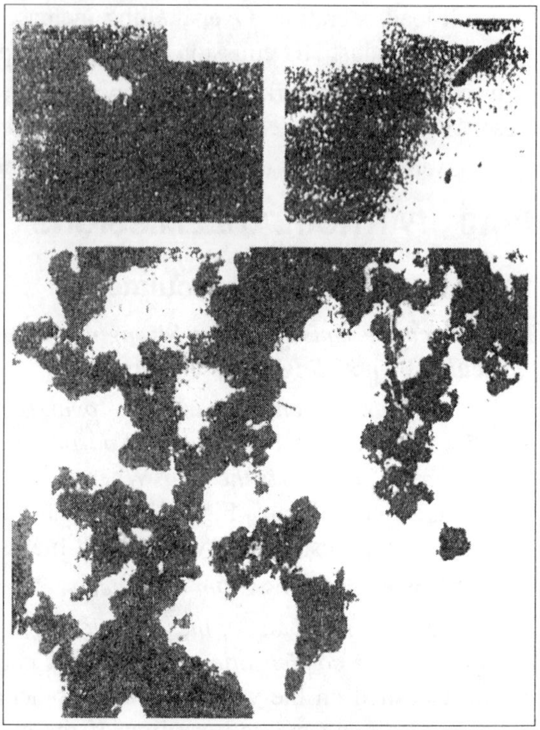

**Fig. 19.22** Photomicrography of diesel particulates [48, 73]

Figure 19.22 shows the structure of *diesel particulate material obtained* from the **photomicrography** [48] *from the exhaust of an IDI diesel engine.* The observation of the same reveals that it consists of collection of primary particles (spherules) agglomerated into aggregates (particles). The appearance of individual particles ranges from clusters of spherules to chains of spherules. It is interesting to note that clusters may contain as many as 400 spherules. The spherules are combustion generated soot particles which range from 10 to 80 nm in diameter.

For the analysis the exhaust particulate is usually partitioned with an extraction solvent into a soluble fraction and a dry soot fraction. Table 19.8 shows the chemical composition of particulate matter.

**Table 19.8** Chemical Composition of Particulate Matter

| *Type* | *Idle* | *40 km/h* |
|---|---|---|
| Extractable composition | $C_{23}\ H_{29}\ O_{4.7}\ N_{0.21}\ C_{24}$ | $H_{30}\ O_{2.6}\ N_{0.18}$ |
| H/C | 1.26 | 1.63 |
| Dry soot composition | $CH_{0.27}\ O_{0.22}\ N_{0.01}\ CH_{0.21}$ | $CH_{0.21}\ O_{0.15}\ N_{0.01}$ |
| H/C | 0.27 | 0.21 |

Figure 19.23 shows a lattice image of diesel particle [49]. It is a concentric lamellate structure arranged around the centre of each spherule similar to the structure of carbon black. A surface area of about 200 $m^2/g$ has been measured for diesel soot.

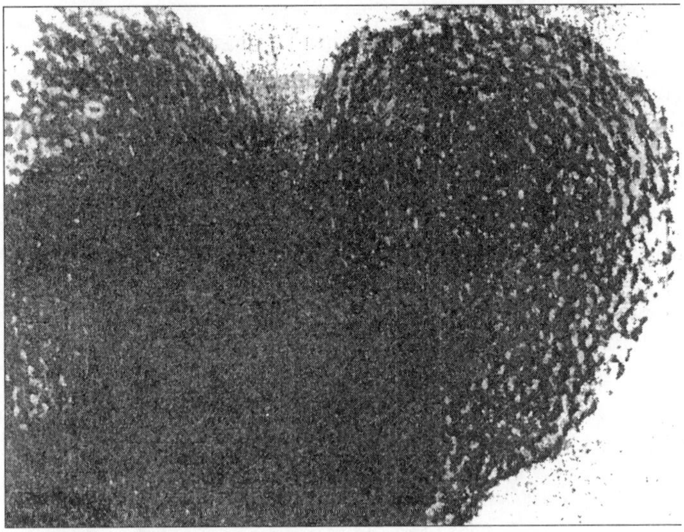

**Fig. 19.23** Lattice-imaging micrograph of a diesel particulate [49, 73]

#### 19.10.2.2 Formation of Particulate, Soot and Smoke

As discussed earlier, *soot particles are primarily from the carbon in the diesel fuel.* The mechanism of soot formation is normally attributed to the following reaction:

$$2CO \rightarrow CO_2 + C$$

It suggests that the formation of CO is catalyzed by carbon, therefore they build-up rapidly and polymerize. It is worth to note that soot formation process starts with a fuel molecule containing 12 to 22 carbons atoms and H/C ratio of about 2 and ends up with particles typically a few hundred nanometers in diameter, composed of spherules 20 to 30 nm in diameter each containing some 105 carbon atoms and having an H/C ratio of about 0.1. Diesel combustion environment at temperatures between about 1000 K to 2800 K at pressure of 50 to 100 atm and with sufficient air to burn fully at full load causes the formation of soot.

There are two stages of soot formation. They are:

(i) **Particle formation.** *This occurs when the first condensed phase material arises from the fuel molecules via their oxidation and or pyrolysis products which include various unsaturated HC particularly acetylene and its higher analogues ($C_{2n}$ $H_2$) and polycyclic aromatic hydrocarbon (PAH) are considered the most likely precursors of soot in flames. The first recognizable soot particles in the form of nuclei having diameter less than 2 nm appears as a result of condensation reactions of gas-phase species.*

(ii) **Particle growth.** This includes both surface growth coagulation, and aggregation. Surface growth gives rise to the generation of bulk of the solid-phase material and it involves the attachment of gas-phase species to the surface of particles and their incorporation into the particulate phase. Figure 19.24 shows the paths to soot formation plotted between species of molecular weight (M) vs. hydrogen mole fraction ($\overline{x}_H$) [48].

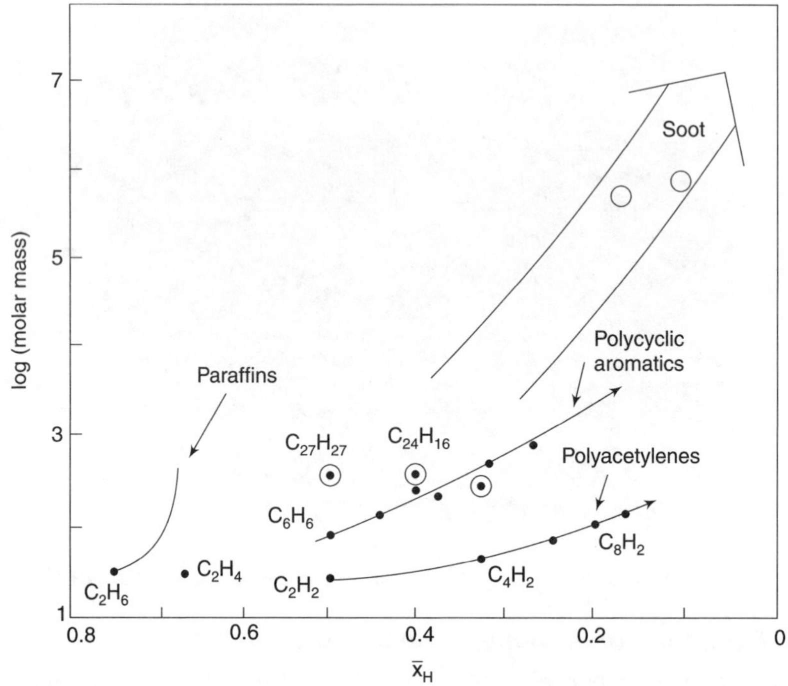


**Fig. 19.24** Paths to soot formation [48]

It is obvious that neither purely polyacetylene chain growth nor purely PAH growth would lead to soot particles which have $\bar{x}_H$ in the range of 0.1 to 0.2. The stages of particle generation and growth constitute the soot formation which is a non-equilibrium process. Figure 19.25 *shows the processes leading to net* production of diesel particulates [50].

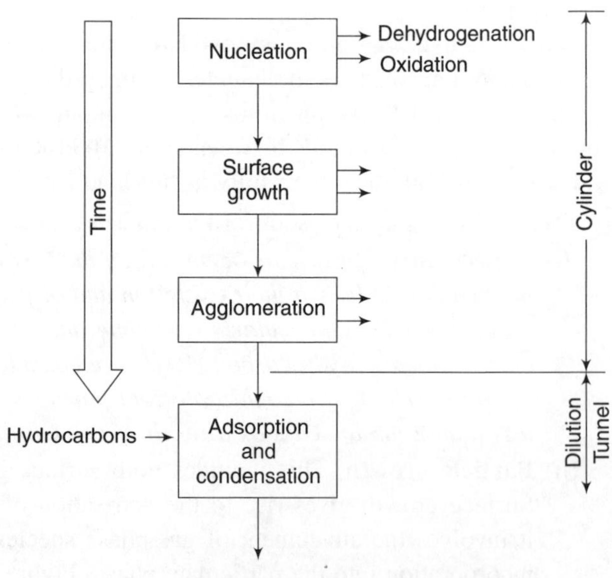


**Fig. 19.25** Processes leading to net production of diesel particulates

A simple mechanistic model for formation of soot from aromatic and aliphatic hydrocarbon compounds has been proposed by Amanu [48] as shown in Fig. 19.26. At low temperature, an aromatic hydrocarbon can produce soot via direct (fast) condensation reactions producing graphite like structure.

While at high temperature (above 1800 K), aliphatic hydrocarbon can produce soot via indirect (slow) route through fragmentation reactions.

It has been discussed earlier, oxidation of soot at the precursor, nuclei and particle stages can occur. It has been also discussed that a large fraction of soot formed in oxidized within the cylinder before the exhaust process commences. Oxidation of soot depends on the diffusion of reactants and products from surface as well as the kinetics of the reaction. Agglomeration will have an indirect influence on the amount of soot oxidized through its effect on surface area. For more details on particulate emission, Ref. [48 to 51] may be consulted.

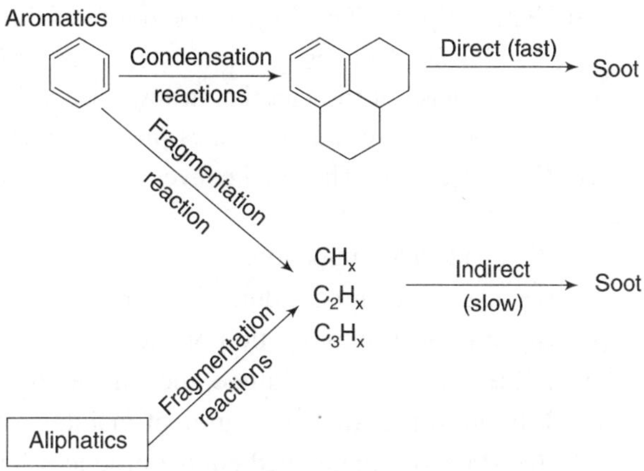


**Fig. 19.26** Mechanistics model for soot formation

### *19.10.2.3 Factors Affecting Formation of Smoke*

The following factors affect the formation of smoke.

**(a) Quality of Fuel.** The different qualities of fuel affect the smoke in the following manner:

(i) More volatile fuels produce less smoke

(ii) Heavier fuels produce more smoke

(iii) High cetane number fuels produce less white smoke but black smoke is not affected by cetane number

(iv) 45 cetane number fuel may act as optimum value to produce white smoke within acceptable limits.

**(b) Engine Load.** The variation in engine load affects the smoke as follows:

(i) Overloading of the engine produces black smoke.

(ii) The intensity of smoke increases rapidly from no load position to fuel load then rapidly from no load position to fuel load position.

**(c) F/A Ratio.** The variation of F/A affect the smoke as follows:

(i) A richer mixture produces more smoke due to less arailability of oxygen.

(ii) The intensity of smoke increases with F/A.

(iii) The smoke density decreases with increase in F/A for turbocharged engine but increase for naturally-aspirated engine.

**(d) Types of Engine.** The following are the effects of types of engine on smoke formation

(i) Turbocharged engines produce smoke of lower level than naturally aspirated engines due to ample oxygen available.

**(e) Engine Speed.** The engine speed affects the smoke level as follows:

(i) The smoke density is too much at extreme engine speed, i.e., low and high. This is because of the fact that at low speed, there is a heating of charge while at higher speed, wire drawing takes place at inlet valve.

**(f) Fuel Injection.** The smoke level increases, with the following occurrence of injection system:

(i) inadequate injection,

(ii) excess for longer duration, and

(iii) improper automization of fuel.

**(g) Altitude.** At higher altitude, the smoke level is more.

**(h) Engine Maintenance.** Degree of maintenance of engine affects smoke as follows:

(i) Properly maintained engine produces low level of smoke.

(ii) Engine with worn piston rings produces smoke of higher level due to the burning of lubricating oil at with fuel.

## 19.11 EMISSION CONTROL IN SI AND CI ENGINES

Engine emissions may be controlled *at two stages by checking or removing the pollutants formed.* They are:

(i) ***Inside engine cylinder*** (ii) ***Engine exhaust system***

The pollutants formed inside the engine cylinder depends on many factors such as (i) *engine design,* (ii) *operating conditions* and (iii) *fuel used. Even after the best design, most favourable operating conditions and best fuel used, there will be sufficient pollutants formed inside the engine cylinder.* Any further reduction can be obtained by removing pollutants from the exhaust gases in the engine's exhaust system. **In totality, the following devices are used to reduce exhaust emissions.**

(1) *Engine design modifications*

(2) *Treatment of exhaust gases (through the following three methods):*

(a) *Catalytic convertors*

(i) *Oxidizing catalysts for HC and CO*

(ii) *Reducing catalysts for $NO_x$*

(iii) *Three-way catalysts for all three pollutants (HC, CO and $NO_x$).*

(b) *Thermal reactors (for HC and CO)*

(c) *Traps or filters for particulates*

(3) Evaporative emission control

(4) Blow-by or crankcase ventilation (PCV)

(5) Various other control systems

(6) Preventive maintenance

The various emission control methods are given below:

### 19.11.1 Engine Design and Modification of Operating Variables for Reducing Exhaust Emissions

**SI Engines**

**(a) Use of Leaner A/F.** The carburettor now injection system, may be modified to provide relatively lean and stable A/F mixtures during idling and cruise operation. *It is well known that burning leaner mixture greatly reduces HC and CO emissions* (Fig. 19.3). However, this reduction will reduce emissions only if there is a good mixture formation and distribution.

One of the methods of producing a uniform mixture and refined fuel metering is the adoption of a **triple venturi** (three bands) carburettor incorporating a single primary barrel for idle and light load operation and two large secondary barrels having the air capacity required for good performance of the engine. It has been experimented that a dual manifold system ensures better fuel distribution. **Use of MPFI system helps in this regard.**

**(b) Reduced Valve Overlap.** Increased valve overlap allows some mixture to escape directly without burning and thus increases emissions levels. This can be controlled by reducing valve overlap.

**(c) Pretreatment of Mixture.** It is done to improve vapourization and mixing of fuel and air. The following are the methods to accomplish this.

(i) Use of narrower venturies to produce higher air speeds.

(ii) Provision of exhaust heated hot spots to vapourize the heavier liquid droplets in the air stream

(iii) Use of exhaust heat to pre-heat the mixture at part loads.

(iv) Use of automatic transmission.

(v) Special devices for reducing fuel supply during deceleration.

**(d) Retarding Ignition Timing.** Retarding ignition time provides increased time for fuel burning. The controls are designed to retard the spark timing at idle but provides normal spark advance during acceleration and cruising. *Retarding decreases the maximum temperature, which in turn, reduces $NO_x$ emission. At the same time, it also reduces HC emission causing higher exhaust temperature. However, there is a loss in power and fuel economy.*

**(e) Lower Compression Ratio.** The use of lower compression ratio reduces the quenching effect by reducing the quenching area in the combustion chamber, which in turn, reduces the HC emissions. *$NO_x$ emissions also get reduced due to lower maximum temperature achieved as a result of low compression ratio.* However, there is some reduction in power and fuel economy.

**(f) Low Quench Combustion Chamber.** There is a need of modification of combustion chamber design using attempts *to avoid flame quenching zones* which promote incomplete combustion resulting in HC emissions. *The low quench combustion chambers include reducing surface to volume ratio, reduced squash area, reduced ideal space around piston ring and reduced distance of the top piston rings from the top of the piston.*

**(g) Piston and Ring Variables.** *To overcome partly the crevice effect, a system known as sealed ring orifice system may be used which reduces HC emissions. This system has been put in practice in many American automotive engines.* This system minimises the amount of gas volume between the top land and bore and to block the leakage path through the gas into the crankcase, as well as the passage of gas into the ring back. Table 19.9 depicts the exhaust emission levels of various modified and unmodified engines.

**(h) Mode of Operation (Speed and Load).** Cruising and acceleration will reduce CO and HC formation and $NO_x$ formation will be unaffected. Operating at design or medium load will reduce $NO_x$ foruration.

**(i) Modification in Fuel Comoposition.** Use of methane and propane result in much lower CO.

**Note:** *The control mechanism given in items 3 to 6 may be clubbed in Design Modifications Heading.*

**Table 19.9** Exhaust Emission Levels of Various Modified and Unmodified Engines

| *S.No.* | *SI Engines* | *Exhaust Emissions, g/km* | | |
|---|---|---|---|---|
| | | *HC* | *CO* | *$NO_x$* |
| 1. | Catalytic reactor | 1.1 | 7.5 | 0.375 |
| 2. | Modified carburettor, ignition timing and heat conserving exhaust manifold | <0.44 | <6.5 | <1.65 |
| 3. | Unmodified engines | 6.8 | 50.0 | 2.5 |
| 4. | Exhaust reactor and air injection | 0.44 | 7.5 | 0.375 |
| 5. | Rich mixture, exhaust reactor and recycled exhaust | <0.94 | <12.5 | <0.81 |
| 6. | Gas turbine | 0.53 | 3.12 | 0.63 |

### Desirable Design Characteristics

Table 19.10 may serve a *concluding remark for desirable design characteristics of a SI engine for low exhaust emissions.*

**Table 19.10** Engine Design Characteristics for Low Emission

| *S.No.* | *Engine variables* | *Exhaust Emission* | | |
|---|---|---|---|---|
| | | *HC* | *CO* | *$NO_x$* |
| 1. | A/F | >15:1 | >15:1 | <13:1 or >17:1 |
| 2. | Manifold depression | – | <380 mmHg | High |
| 3. | Air inlet temperature | – | High | 70 to 190°C |
| 4. | Ignition timing | Retard | Retard | Retard |
| 5. | Coolant temperature | – | High | Low |
| 6. | Compression ratio | Low | Low | Low |
| 7. | Combustion chamber surface to volume ratio | High | Low | High |
| 8. | Exhaust temperature | – | High | High |
| 9. | Combustion chamber deposit | Avoid | Avoid | Avoid |
| 10. | Valve overlap | Minimum | Minimum | – |
| 11. | Mode of operation (speed and load) | | Cruising and Acceleration | Design load or medium load |

Figure 19.27 shows the exhaust emission and effect with gasoline, LPG, CNG and diesel, Empirical values taken from five SI and five CI vehicles, **status 1993 in Europe** including indirect emission (manufacture and transport)

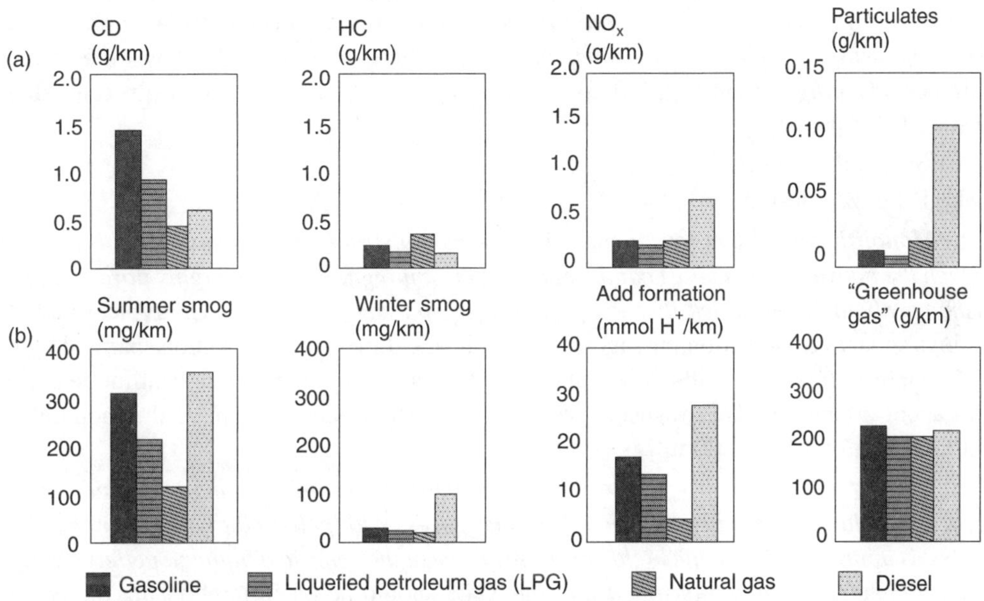


**Fig. 19.27** Comparison of exhaust emission with gasoline LPG, CNG and diesel

## 19.11.2 Treatment of Exhaust Gases for Reducing Exhaust Emission in SI Engines

In a SI engine, the exhaust gas temperature can vary from 300 to 400°C during idling operation and can reach as high as 900°C during high power operation, however, the *most common range is* **400 to 600°C**. The exhaust gas may contain modest amount of oxygen when operating at lean mixture or more substantial amount of CO when operating at rich mixture. The story in the case of diesel engine is different as it operates always at lean and ensures substantial amount of air in the exhaust having temperature range *varying from* ***200 to 500°C***.

The above discussion suggests that *the removal of the gaseous pollutants from the exhaust gases after they leave the engine cylinder can be either* **thermal** or **catalytic**. *To oxidize HC in gaseous phase without a catalyst, a residence time of about 50 ms and temperature in excess of 600°C are required. In order to oxidize CO, temperature in excess of 700°C are required. Catalytic oxidization of CO and HC in the exhaust can be achieved at a temperature as low as 250°C. Catalytic oxidization of NO requires temperature 400°C.* ***For removal of CO, HC and NO from the exhaust of SI engine, use of catalyst is widespread. Particulates in the exhaust gas may be removed by trap.***

#### 19.11.2.1 Catalytic Converters of SI Engines

*It consists of an active catalytic material in a specially designed metal casing which directs the exhaust gas flow through catalyst used.* For the oxidization of CO and HC or for reduction of NO, the active material generally employed is ***nobel metals***. Though, base metal oxides can also use be used. *The active materials must be distributed over a large surface area to enable mass transfer characteristics efficiency of catalyst.* Figure 19.28 shows the common ***two types of configuration*** [52, 73] commonly used as **catalytic converters in the exhaust stream.** They are

(a) Monolith design

(b) Pelletized design

**(a) *Monolith design*** *employs a ceramic honeycomb structure or monolith held in a metal in whch the active (nobel metal) catalyst material is impregnated into a highly porous alumina washcoat about 20 μm thick that is applied to the passageway walls.* The typical monolith employs square cross-section passageways with inside dimensions of 1 mm separated by thin (0.15 to 0.3 mm) porous walls having number of passageways per square centimetre varying from about 30 and 60. The washcoat weighing 5 to 15 per cent weight of the monolith has a surface area of 100 to 200 $m^2/g$.

**(b) Pelletized Design.** *It employs a bed of spherical ceramic pellets to provide a large surface area in contact with the flow of exhaust gases. With pellet catalysts, the noble metal catalyst is impregnated into the highly porous surface of the spherical alumina pellets (typically 3 mm diameter) to a depth of about 250 μm.* Care should be taken in choosing the pellet to have good crush and abrasion resistance after exposure to temperature of order of 1000°C. In order to achieve close to 100 per cent conversion efficiency, the gas flow is directed down through the bed to provide a large flow area and low pressure drop. *In the pelletized type design, the gas flow is turbulent resulting in high mass transfer rate while in the monolith catalyst passageways, it is laminar in nature.*

**Types of Catalysts:** The **various types of catalysts** used in the exhaust gas treatment are as follows:

**(i) Oxidation Catalysts for CO and HC.** The exhaust gas stream of SI engine typically contains the following:

| | | | |
|---|---|---|---|
| $CO_2$ and $H_2O$ | : 12% | NO | : 100 to 2000 ppm |
| $SO_2$ | : 20 ppm | $O_2$ | : 1 to 5 percent |
| CO | : 0.2 to 5 perent | HC (as $C_1$) | : 1000–6000 ppm (small amounts of lead and phosphorus) |

*The function of an oxidation catalyst is to oxidize CO and HC to $CO_2$ and water, respectively, The oxidation of saturated HC which comprises about 20 to 30 per cent is most difficult to take place. It is a well-known fact that the case of oxidation increases with increasing molecular weight. In order to oxidize the CO and HC, sufficient oxygen must be present and this may be supplied by the engine itself operating at lean mixture or with the help of pump that introduces air into the exhaust ports just downstream of the valve.* The required air can also

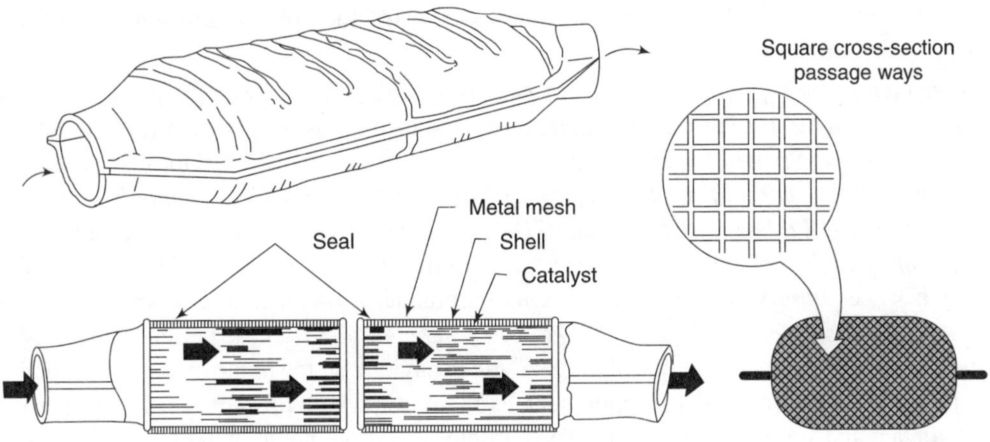


(a) Monolith Design

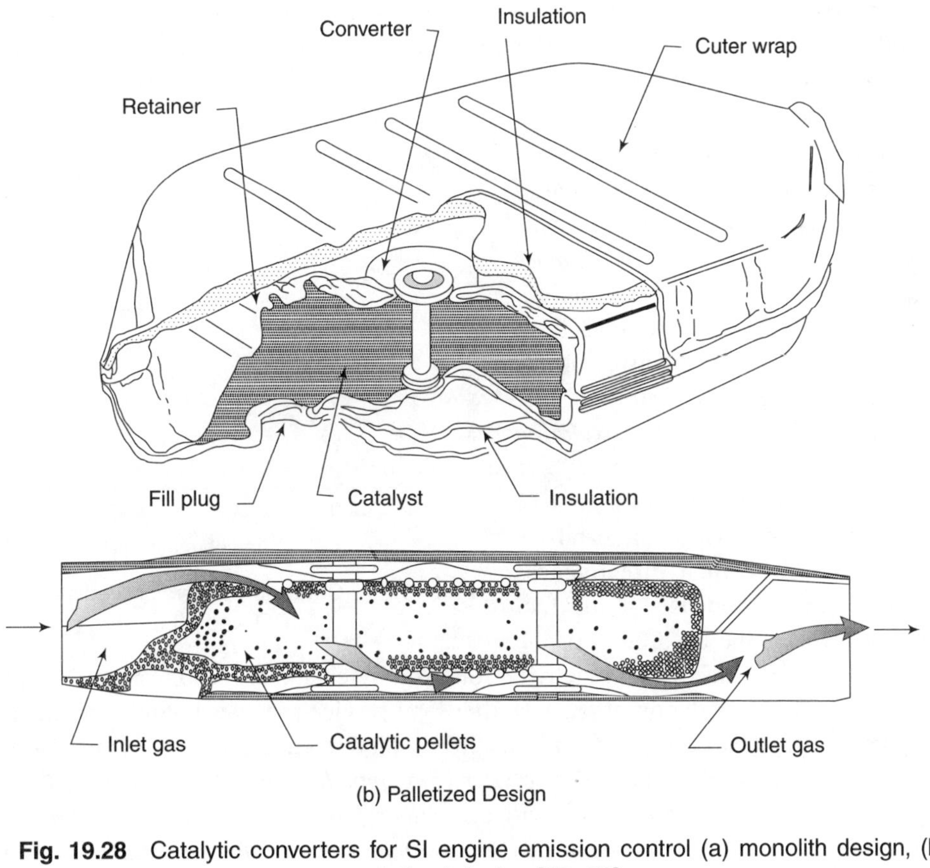


(b) Palletized Design

**Fig. 19.28** Catalytic converters for SI engine emission control (a) monolith design, (b) pelletized design [52, 73] be used by venturi air addition into the exhaust port using the pressure pulsations generated by the exhaust process.

***Noble metals*** *are supposed to be the most suitable as the catalytic materials* owing to their high intrinsic activity for HC oxidation, thermal resistance to loss of low-temperature activity and much less deactivated by the sulphur in the fuel than base metal oxides. *A mixture of* ***platinum (Pt) and palladium (Pd) is one of noble metals*** *which is commonly used. The specific activity of Pd is higher than that of Pt for the oxidization of CO, olefins and methane but Pt is more active than Pd for the oxidization of paraffin hydrocarbons.* However, similar activity has been observed for the oxidization of aromatic compounds. *It is worth to note that pure noble metals sinter rapidly in the 500 to 900°C temperature range experienced by exhaust catalysts.* Since surface atoms control the catalytic behaviour, so the noble metals are dispersed as finally as possible on an inert support such as $Y - A_2\,O_3$ which prevents particle-to-particle metal contact and suppresses sintering. The particle size of the noble metal particles depends upon temperature to which it is exposed, the fresh catalyst numbers less than 50 nm while exposed one due to operation honeycombs catalyst varies between 1 and 2 g/dm$^3$ of honeycombs volume with Pt/Pd = 2 on a weight basis. *As per thumb rule, the ceramic honeycomb volume required is about half of the piston displacement volume so that the ratio of volume flow rate of exhaust to the converter volume over the normal operating range should vary from 5-30/second.*

*The conversion efficiency of a catalyst which is defined as the ratio of the rate of mass removal in the catalyst of the particular constituents of interest (say HC) to the mass flow rate of the constituent into the catalyst, is given in Fig. 19.29.* It is obvious from the figure that catalyst is ineffective until its temperature has risen above 250 to 300°C. *The conversion efficiency of a new oxidation catalyst at high temperature is around 89 to 99 per cent for CO and 95 per cent for HC.* It is worth to mention that as the catalysts spend time in service their effectiveness deteriorates. This is because of the fact that catalyst involves the absorption of the reactants on to surface sites of high activity, followed by chemical reaction then desorption of the products. The degradation of catalyst after long service involves both the deactivation of these sites by catalyst poisons and a reduction in the effective area of this sites through the process of sintering. *Poisoning of catalyst means impeding of catalytic activity through prolonged contact with interfering elements that either physically block the active sites or interact chemically with the active material the most important poisons for catalytic converter are the lead in fuel antiknock agents and the phosphorus in oil additives.* Figure 19.30 shows

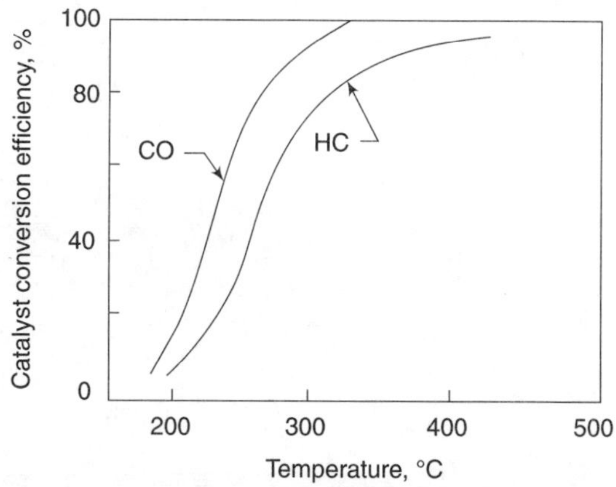


**Fig. 19.29** Catalytic conversions efficiency for CO and HC

the effect of lead concentration on HC conversion efficiency of catalyst [52]. It is obvious from the figure that lower the lead concentration, higher is the HC conversion efficiency.

***Oxidation-catalyst** equipped vehicles may emit sulphuric acid aerosols. This is because of the fact that unleaded gasoline contains 150 to 600 ppm by weight of sulphur which leaves combustion chamber as $CO_2$ but gets converted into $SO_3$ due to oxidation by catalyst.* This $SO_3$ *combines with water at ambient conditions to form an $H_2SO_4$ aerosol.* Figure 19.31 shows the conversion efficiency of $SO_2$. At low temperature, the conversion is very poor while at high temperature it is as high as 90% [52].

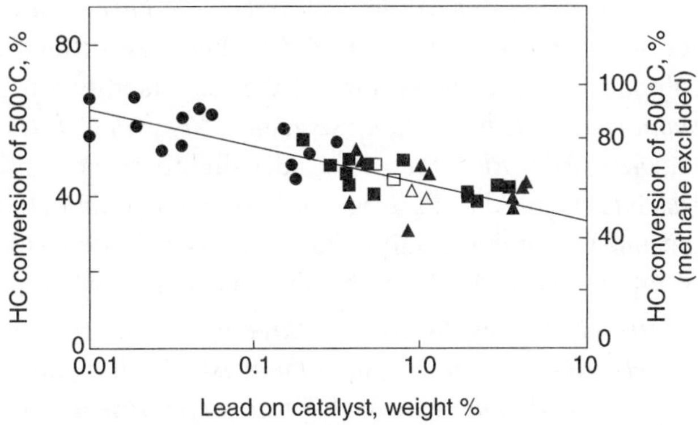


**Fig. 19.30** Effect of lead concentration on conversion efficiency

**(ii) NO catalysis.** *In the present technology, no catalyst is available for the decomposition of NO to $N_2$ and $O_2$. Thus, NO is removed by the* ***process of reduction*** *using the CO, HC and $H_2$ in the exhaust as per the possible NO reactions in* ***six ways*** *given below:*

(i) $NO + CO \longrightarrow \frac{1}{2} N_2 + CO_2$

(ii) $2\ NO + 5\ CO + 3\ H_2O \longrightarrow 2\ NH_3 + 5\ CO_2$

(iii) $2\ NO + CO \longrightarrow N_2O + CO_2$

(iv) $NO + H_2 \longrightarrow \frac{1}{2} N_2 + H_2O$

(v) $2\ NO + 5\ H_2 \longrightarrow 2\ NH_3 + 2\ H_2O$

(vi) $2\ NO + H_2 \longrightarrow N_2O + H_2O$

*This reduction can be carried out under rich conditions where there is an excess of reducing species over oxidizing species catalyst. This system requires a follow-up oxidation catalyst together with addition of air from an air pump before the oxidation catalyst to remove the remaining CO and HC. The advantage of this two-bed system is that it can remove all three pollutants, namely, NO, CO and HC.* The disadvantage lies in more fuel consumption and formation of $NH_3$ rather than $N_2$.

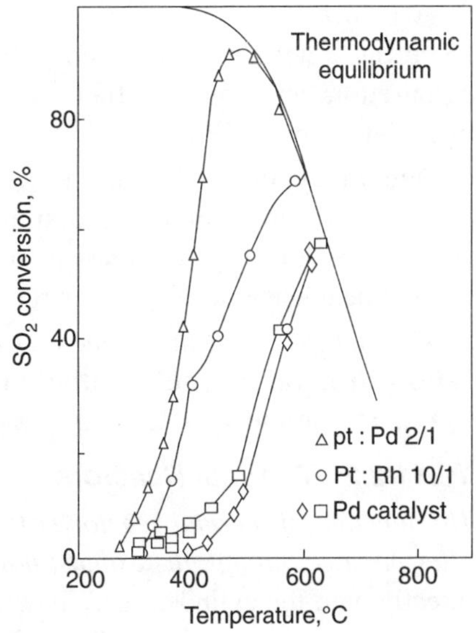


**Fig. 19.31** $SO_2$ conversion efficiency

*For reduction of NO by CO or $H_2$, the* **base metal catalyst may be CuO or NiO in** *the temperature range of 350 to 600°C* but they suffer from the drawback of deactivation by sulphur and have shown limited thermal stability when used in vehicles exhausts. *Alumina supported noble metal catalysts reduce NO with CO- $H_2$ mixture and their reduction activity in the order Ru > Rh > Pd > Pt.* Under slightly rich conditions, Ru and Rh produce considerably less $NH_3$ than Pd or Pt. It is worth to note that though these properties make ruthenium (Ru) the most desirable catalyst but it forms volatile oxides under oxidizing conditions which results in loss of Ru from the alumina support when put in service.

**(iii) Three-way catalyst.** *A three-way catalyst is one which removes all three pollutants namely, NO, CO and HC simultaneously, Such a catalyst is suitable only when the engine is operated at all times with an A/F ratio at or close to stoichiometric in which both NO reduction and CO and HC oxidation can be done in a single catalyst bed.*

Figure 19.32 depicts the conversion efficiency for NO, CO and HC as a function of A/F. It is obvious from the figure that there is a narrow range (window) **of A/F ratios (14.5 to 14.6) near stoichiometric in which high conversion efficiencies for all three pollutants are achieved.**

It is impossible for any ordinary carburettor to control A/F so precisely to run the engine within the window. It can be achieved sometimes with sophisticated carburettors and gasoline fuel-injection system (MPFI).

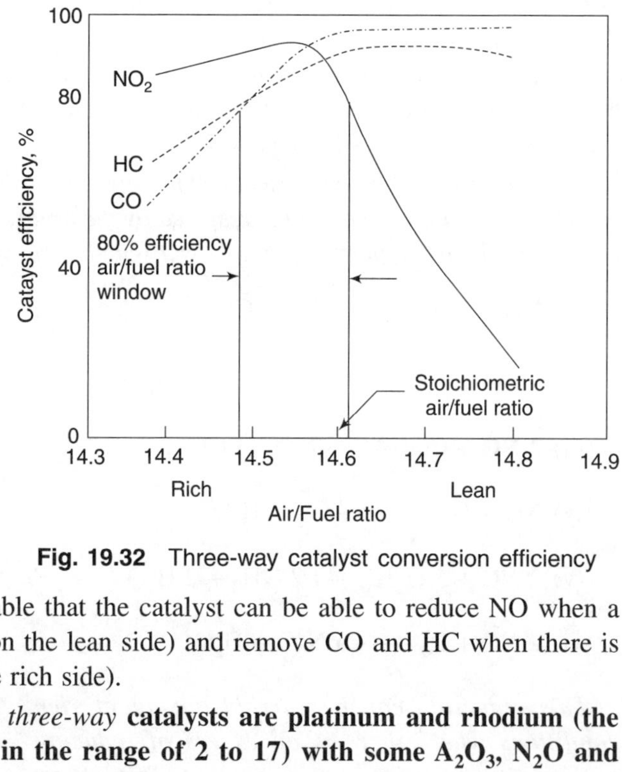


**Fig. 19.32** Three-way catalyst conversion efficiency

Due to cyclic variation in exhaust gas composition, it is desirable that the catalyst can be able to reduce NO when a slight excess of oxygen is present (on the lean side) and remove CO and HC when there is a slight deficiency of oxygen (on the rich side).

*The material used in commercial three-way* **catalysts are platinum and rhodium (the ratio Pt/Rh varying substantially in the range of 2 to 17) with some $A_2O_3$, $N_2O$ and $CeO_2$. The support material is generally preferred as alumina.**

### *19.11.2.2 Thermal Reactors*

*The function of thermal reactor is to enhance the oxidization of products of exhaust gases after passage through the exhaust port.* A thermal reactor is an enlarged manifold that bolts directly onto the cylinder head. Its aim is to promote rapid mixing of the hot exhaust gases with any secondary air injected into the exhaust port to remove nonuniformities in temperature and composition in the exhaust gases and to retain the gases at a high enough temperature for sufficient time to oxidize much of the HC and CO which exists in the cylinder.

Figure 19.33 shows a thermal reactor usually used. *The requirements of temperature levels in a thermal reactor for bulk gas oxidation of HC and CO are 600°C and 700°C respectively which are higher than that required by catalytic converters.* In order to achieve greater reduction in emissions, the reactor must be designed to reduce heat losses and increase residence time. Further, for rapid warm-up after engine start, a low thermal inertia reactor is desirable. *In general, a thermal reactor consists of a thin steel liner which acts as a core of the reactor inside a cast iron outer casing with suitably arranged flow paths. By this way, the heat losses by thermally isolating the core will be a minimum.*

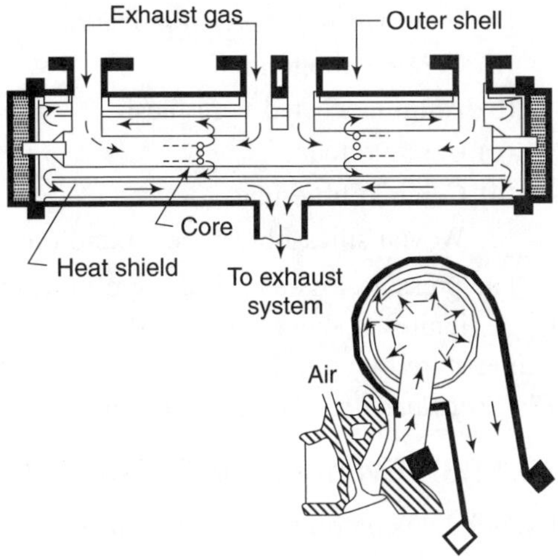


**Fig. 19.33** Thermal reactors for HC and CO oxidation

The parameters that affect the effectiveness of the reactor are its operating temperature, the availability of excess oxygen mixed throughout the reacting gases and the reactor volume. The operating temperature depends on the reactor inlet gas temperature, heat losses and the amount of HC, CO and $H_2$ burnt up in the reactor—which results in temperature rise. As a result of above influencing parameters, reactors with fuel-rich cylinder exhaust gas and secondary air exhibit greater fractional reductions in HC and CO emission than reactor with fuel-lean cylinder exhaust.

The reactor effectiveness with fuel-rich engine operation is limited by the mixing of secondary air and engine exhaust gases in the exhaust port and the reactor core. However, mixing in the reactor itself is promoted by suitably arranging the reactor inlet and exit ports and by using baffles. It is the improved control of secondary air that increases significantly the burning up of CO emissions.

#### *19.11.2.3 Particulate Traps for SI and CI Engines*

***Diesel engine particulate emissions*** *are reduced by trap oxidizer which is a temperature-tolerant filter. After serving some period, the filter is cleaned off by oxidizing the accumulated particulates.* The technology suffers from the following drawbacks:

(i) The filter, even when clean, increases the pressure in the exhaust system.

(ii) The exhaust pressure increases steadily as the filter collects particulate matters.

(iii) Collected particulate matter does not ignite and oxidize under normal operation.

(iv) The burn-up process after ignition of particulate is to be controlled carefully to prevent excessively high temperature and trap damage or destruction.

There are *various types of* ***particulate filters*** as given below:

(a) Ceramic monoliths
(b) Alumina coated wire mesh
(c) Ceramic foam
(d) Ceramic fibre mat
(e) Woven silica fibre rope wound on a porous tube

The pressure loss and filtering efficiency of each of these are different. Diesel particulate matter ignites at about 500 to 600°C but the catalytic coatings on the traps reduce the ignition temperature by up to 200°C. The particulate oxidation rate depends on the trap temperature. ***The traps are able to reduce the particulate emissions by about 70%.***

### 19.11.3 Evaporative Emission Control for Fuel Vapour

Evaporative control system minimizes the amount of vapour escaping into the atmosphere. It consists of a charcoal canister, a vapour purge control system, a control vapour system and a fuel tank vapour control system. Figure 19.34 shows such a system in Honda car.

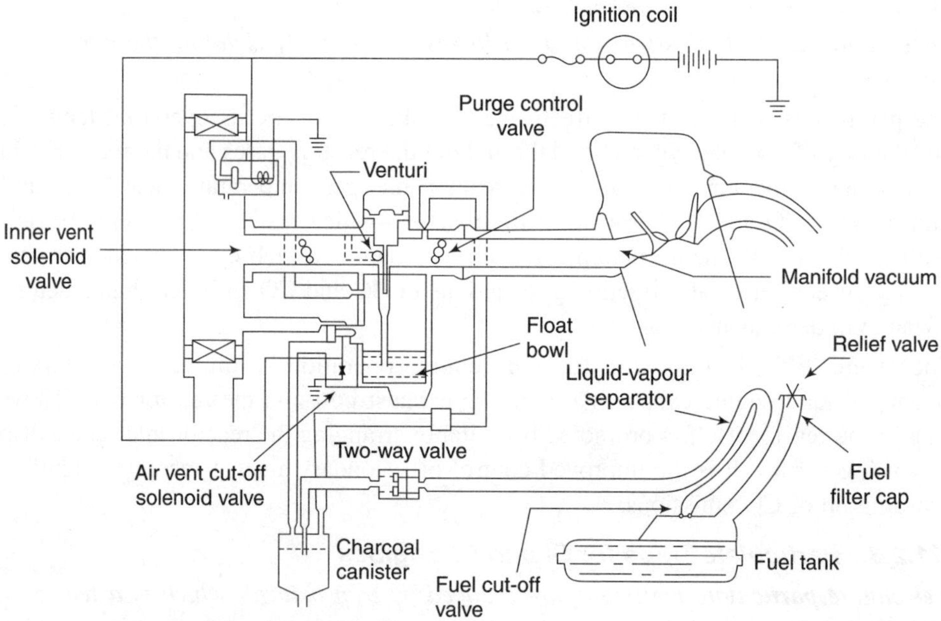


**Fig. 19.34** Evaporative emission control [*Courtesy.* Honda Car Co]

The charcoal canister is a container for temporary storage of fuel vapour until it is purged from the canister to engine and burnt. *The vapour purge system* will come into operation when the canister is to be purged. In the event of intake air temperature below 65°C and coolant temperature above 63°C, manifold vacuum is directed to the purge control valve, which allows venturi vacuum to draw fresh air through the bottom of canister into the engine, purging the fuel vapour from the canister.

The carburettor vapour control system utilises the vapour formed in carburettor fuel float bowl through canister. The air vent cut-off solenoid valve regulates air flow to the carburettor float bowl. In the event of non-running of engine, the outer air vent passage opens, so that the fuel vapour in the float bowl can be vented to the charcoal canister. Where the engine speed is above 400 rpm, the air vent passage is delayed momentarily to ease engine start-up. When the engine speed is above 400 rpm, inner vent solenoid valve normally opens the sub-inner air vent passage, but will only allow fuel vapour to be vented to the air cleaner. ***In the case of gasoline fuel injection, there is no use of carburettor vapour control system.***

The fuel tank vapour control system consists of a fuel filter cap, a two-way valve liquid vapour separator and a fuel cut-off valve. All the fuel vapour inside the fuel tank is directed to the charcoal canister through the fuel cut-off and the liquid/vapour separator to prevent the building up of vapour in the tank which might otherwise escape when the fuel filter cap is removed. The pressure and the vacuum in the tank are regulated by a two-way valve. Over and above, there is a relief valve in the fuel filter cap in case there is an excessive build-up of pressure vacuum.

### 19.11.4 Blow-by or Positive Crankcase Ventilation (PCV) System

The blow-by or positive crankcase ventilation (PCV) system prevents blow-by vapour from escaping into the atmosphere by diverting it into inlet manifold. It consists of a PCV valve with a spring loaded plunger as shown in Fig. 19.35. As soon as the engine starts, the plunger is lifted in proportion to manifold vacuum and blow-by vapour is drawn directly into the intake manifold.

### 19.11.5 Other Control Systems for Exhaust Gas Treatment

The other control systems based on many factors to control the pollution are given below.

(i) Feedback control
(ii) Air injection control
(iii) Exhaust gas recirculation (EGR)
(iv) Throttle control
(v) Ignition timing control
(vi) Antiafterburn valve control
(vii) PGM-CARB control unit

**(i) Feedback Control.** This control is meant of correct amount of air-fuel ratio. This is performed by oxygen sensor, the EACV and the control unit. The oxygen sensor (Fig. 19.36) sends the signal to the control unit to access where the air-fuel ratio is richer or leaner as compared to stoichiometric air-fuel ratio. The control unit receives other signals from speed sensor, TW sensor to the electronic air control valve (EACV) which opens the required air passage from the air cleaner to the intake manifold to maintain the required A/F as demanded by speed, load and emission reduction.

**(ii) Air Injection Control.** In order to promote oxidation of unburnt hydrocarbon, this controls makes use of vacuum pulses in the exhaust manifold to draw air from the air cleaner in the exhaust manifold. The control unit energizes the air suction control solenoid valve, to apply manifold vacuum to the air suction cut-off diaphragm valve which opens the secondary air passage by which fresh air enters the exhaust manifold.

**(iii) Exhaust Gas Recirculation (EGR) System for SI and CI Engines.** It is designed to control oxides of nitrogen emission by recirculation a part of exhaust gas into the intake manifold through EGR valve mounted into the intake manifold (Fig. 19.35). The control system provides the correct amount of exhaust gas recirculation as per load and speed.

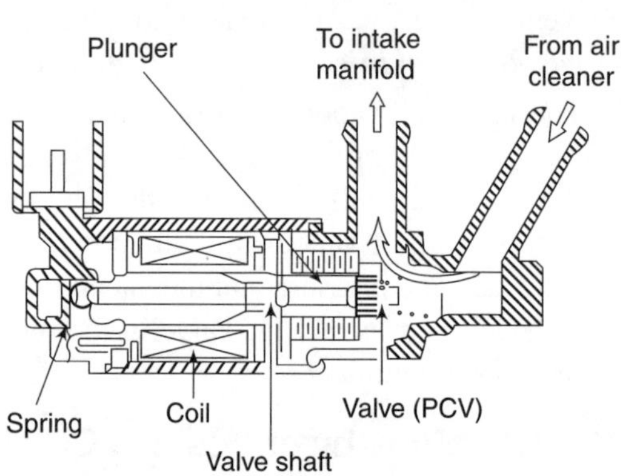


**Fig. 19.35** Positive crankcase ventilation

**(iv) Throttle Control System.** Throttle control system controls the emissions during vehicle decelerations or during shifting of gears with the help of dashpot and a cranking opener, which prevent the throttle valve from shutting rapidly (Fig. 19.37). Under idle condition of the engine, intake manifold vacuum applied to the diaphragm of the throttle controller through the orifice of the dashpot check valve pulls up the diaphragm rod ensuring throttle valve at the idle position.

**(v) Ignition Timing Control.** Ignition timing control, combined with the vacuum advance affects the time at which each spark plug ignites the total air-fuel mixture as per load, speed and constant temperature.

**(vi) Anti-afterburn Valve.** The anti-afterburn valve allows fresh air into the intake manifold when manifold vacuum suddenly increases such as during deceleration.

**(vii) PGM-CARB Control.** This unit consists of logic circuits that sense the inputs and supply outputs as required to control exhaust emissions and provide smooth engine operation through various valves and contents.

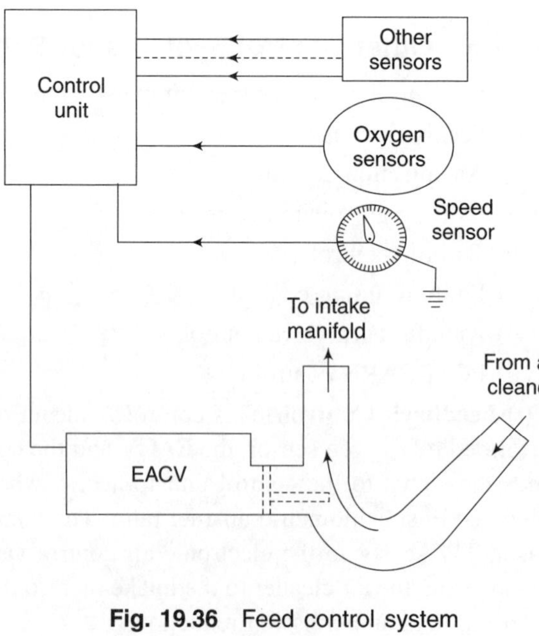


**Fig. 19.36** Feed control system

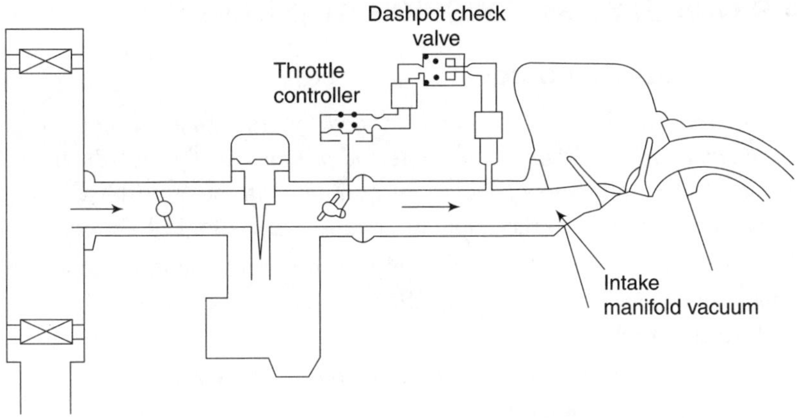


**Fig. 19.37** Throttle control

### 19.11.6 Emission Control through Preventive Maintenance

During running of engine, many problems crop up in the proper working of engine to yield more CO such as: (i) rich-fuel mixture, (ii) improper idle speed, (iii) defective ignition (iv) clogged silencer, (v) dirty/clogged air cleaner, (vi) defective choke, and (vii) clogged injector.

So a proper preventive maintenance is essential to check the improper combustion leading to formation of CO. A check on CO should be made during maintenance to reduce pollution.

## 19.12 ACID RAIN

The natural rain in the absence of pollution-free atmosphere is a clean water and its **pH value ranges from 5.6 to 7**. However, it is slightly acidic due to the formation of carbonic acid $H_2CO_3$ by the following chemical reaction

$$CO_2 + H_2O = H_2CO_3$$

The exhaust emissions from various industries, automobiles, aircraft, power plants, etc. contain acidic oxides such as $NO_2$, $SO_2$, HCl, etc. When these oxides dissolve in rain water particles, or the moisture present in the atmosphere, they form acids and when it falls on the earth with rain water it is called **acid rain**. The following is the mechanism of acid rain formation

$$N_2O + O_2 + H_2O \rightarrow 2HNO_2$$

$$4NO_2 + O_2 + 2H_2O \rightarrow HNO_3$$

$$SO_2 + H_2O \rightarrow H_2SO_3$$

$$2SO_2 + O_2 \rightarrow 2H_2O \rightarrow 2H_2SO_4$$

$$HCl + H_2O \rightarrow HCl$$

**The pH value of acid rain is less than 5.6.**

## 19.13 AIR QUALITY AND EMISSION STANDARDS

### 19.13.1 Air Quality Standards

*Air quality standards (AQS) are legal limits framed by government agency on levels of air pollutants in the ambient air driving a given period of time.* In other words, they characterise the allowable level of a pollutants or a class of pollutants in the atmosphere and thus set the amount of exposure permitted to the population and/or to ecological systems.

Air quality standards have evolved differently in different countries depending on many considerations such as exposure conditions, the socio-economic situation and the importance of other health related problems.

There are basically two types of air quality standards. They are:

(a) Ambient air quality standards

(b) Other air quality standards

(i) Quasi-emission standards

(ii) Soiling index

(iii) Odour standards

(iv) Visibility standards

(v) Standards for particulate matter deposited

Table 19.11 shows the air quality standards adopted by EPA and Table 19.12 depicts Indian air quality standards. Table 19.13 depicts the air pollutions in major Indian cities.

**Table 19.11** Air Quality Standards Adapted by EPA

| *S.No.* | *Parameter* | *Standard concentration* | | *Remarks* |
|---|---|---|---|---|
| | | *μg/m³* | *ppm* | |
| 1. | Suspended particulate matter | 75 | – | Annual geometric mean |
| 2. | Sulphur dioxide | 260 | – | 24h |
| | | 80 | 0.83 | Annual mean |
| 3. | Nitrogen oxides | 100 | 0.05 | Annual mean |
| 4. | Carbon monoxides | 1000 | 9 | 8 h, once a year |
| | | 4000 | 35 | 1 h, not more than once per year |
| 5. | Ozone | 235 | 0.12 | 1 h daily (max) |
| 6. | Non-methane hydrocarbons | 160 | 0.24 | 6-9 h, annual mean |
| 7. | Lead | 1.5 | | 3 months average |

**Table 19.12** Indian Air Quality Standards

| *Category Area* | | *Concentration in μg/m³* | | | |
|---|---|---|---|---|---|
| | | *SPM* | *$SO_2$* | *$NO_x$* | *CO* |
| A | Industrial and mixed use | 500 | 120 | 120 | 5000 |
| B | Residential and rural | 200 | 80 | 80 | 2000 |
| C | Sensitive | 100 | 30 | 30 | 1000 |

**Table 19.13** Annual Summary of Air quality in Major Cities in India for 1981 by NEERI

| City | SPM ($\mu g/m^3$) | $SO_2$ ($\mu g/m^3$) 100 $cm^2$/day | SR (sulphation rate) mg $SO_3$ | DF (dust fall MT/ $km^2$/month |
|---|---|---|---|---|
| Ahmedabad | 217 | 31 | 0.10 | 91 |
| Mumbai | 184 | 38 | 0.45 | 11 |
| Kolkata | 522 | 90 | 0.45 | 34 |
| Delhi | 433 | 31 | 0.2 | 31 |
| Hyderabad | 143 | 30 | – | – |
| Jaipur | 308 | 06 | 0.18 | 19 |
| Kanpur | 432 | 13 | 0.19 | 34 |
| Chennai | 155 | 08 | 0.33 | – |
| Nagpur | 199 | 12 | 0.22 | 34 |

*Table 19.13 shows the situation is alarming in cities like Delhi, Kolkata, Jaipur and Kanpur in 1981 itself. In the present 2020, the situation in much more worse as obvious from TV news daily on air pollution in major cities of India.*

## 19.13.2 Emission Standards

Emission standards are legal limits which establish permitted emission levels for specific group of emitters and requires that all members of these groups emit no more than these permitted emission levels.

There are separate emission standards for mobile and stationary sources.

Table 19.14 gives the emissive norms adopted in India from European norms (EURO).

**Table 19.14** Emission (EURO) Norms Adopted in India

| Norms | CO gm/km | | HC + NO gm/km | | PM gm/km |
|---|---|---|---|---|---|
| | Petrol | Diesel | Petrol | Diesel | Petrol/Diesel |
| Euro I | 3.16 | 3.16 | 1.13 | 1.13 | –0.18 |
| Euro II | 2.20 | 1.00 | 0.50 | 0.70-0.90 | –0.08-0.10 |
| Euro III | 2.3 | | 0.2 + 0.15 | | Followed from 2000 |
| Euro IV | 1.0 | | 0.1 + 0.08 | | Followed from 2005 |
| Euro VI | - | | 43% reduction of HC + $NO_x$ for diesel engines 25% reduction in $NO_x$ for petrol engines | | 82% reduction in PM for diesel engines |

Now EURO VI or Bharat Stage VI has beeen adopted in India from 2020.

The emission limits prescribed are expressed as concentration of pollutants per unit volume of air under *standard or normal* conditions written in $Nm^3$. India has adopted 25°C, 760 mm Hg and zero per cent moisture as standard conditions.

The standard volume is expressed as

$$V_{std} = V_m Y\left(\frac{298}{273 + T_m}\right)\left(\frac{p_{bar} - p_m}{760}\right)$$

where $Y$ is calibration factor of the dry gas meter

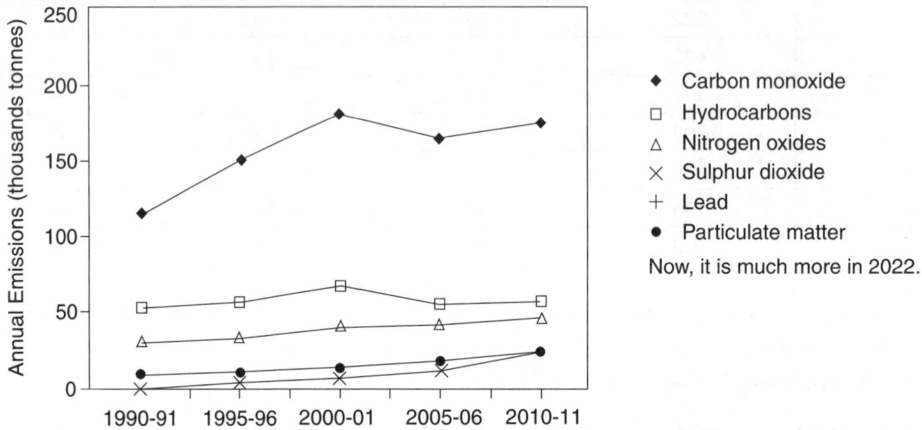


**Fig. 19.38** Projection of annual vehicular pollution load in Delhi

## PROBLEMS

### Things to Think About

19.1 What do you mean by greenhouse effect and acid rain?

19.2 Discuss the causes of greenhouse and acid rain.

19.3 What are the effects of greenhouse on human health and ecology?

19.4 What measure should be adopted to avoid greenhouse effect?

### Short and Long Theoretical Questions

19.5 What do you mean by air pollution?

19.6 Discuss sources and classification of air pollutants.

19.7 What do you mean by primary and secondary air pollutants?

19.8 Discuss the meteorological factors which affect air pollution.

19.9 Discuss the effects of air pollution on human health.

19.10 Give a brief summary of classification of emission and emission contribution from gasoline engines.

19.11 Discuss the mechanism of formation of pollutants in SI engines.

19.12 Discuss crankcase and evaporative emission from SI engines.

19.13 Discuss the mechanism of formation of pollutants from CI engines.

19.14 Discuss the kinetics and formation of $NO_x$.

19.15 Discuss the mechanism of $NO_x$ formation in CI engines.

19.16 Discuss the CO emission from SI engines.

19.17 Discuss HC emission mechanism from SI engines in detail.

19.18 Give a list of critical factors and engine variables for HC emission mechanism in SI engines.

19.19 Discuss HC emission mechanism in diesel engines.

19.20 Discuss about particulate emission from IC engines.

19.21 Discuss the various methods for exhaust gas treatment of IC engines.

19.22 Discuss the engine design and operating variables modification to reduce emission from SI engines.

19.23 Discuss the various types of catalytic converters used in SI engines.

19.24 Discuss thermal reactors used for exhaust gas treatment.

19.25 Discuss trap used in exhaust gas treatment.

19.26 Discuss air quality and emission standards.

### Software Based Questions

19.27 Develop a software for predicting the air pollution in big cities of world.

19.28 Develop a software for predicting the quality and quantity of emission from SI and CI engines based on their engine capacity.

19.29 Develop a software for the design of equipment for treatment of exhaust gases from IC engines to conform the EURO IV and EURO VI standards

### Design and Open Ended Question

19.30 Prepare a comprehensive report on present greenhouse and acid rain effects prevailing all over the world. Suggest measures and corresponding design in improvement of equipment/machines to suppress greenhouse and acid rain effects.

### Multiple-Choice Questions

19.31 The exhaust emission level from SI engine can be reduced by

(a) burning a lean mixture

(b) using thermal reactor in exhaust port

(c) use of catalyst in silencer

(d) all of the above

19.32 The HC emission from a SI engines is due to

(a) incomplete combustion

(b) crevice effects

(c) oil deposits on combustion chamber walls

(d) all of the above

19.33 **Assertion (A):** In SI engines, HC emission can be reduced by having an additional spark plug at an appropriate place inside the combustion chamber.

**Reasoning (R):** As a result of combustion at two points, the length of flame travel gets reduces leading to less expansion quenching

**Answer Code:**

(a) Both A and R are true but R is the correct explanation of A

(b) Both A and R are true but R is not the only reasoning of A

(c) A is true but R is false

(d) A and R both are false

19.34 **Assertion (A):** A part of exhaust gas if circulated through the inlet manifold, the concentration of HC gets reduced.

**Reasoning (R):** Because it reduces the maximum temperature in the cycle.

**Answer Code:**

(a) A is false but R is true (b) A is true but R is false

(c) Both A and R are true (d) Both A and R false

19.35 A three-way catalytic converter used in SI engine

(i) uses leaded petrol

(ii) first converts CO into $CO_2$

(iii) A/F ratio near to stoichiometric

Among these statements

(a) Only (i) is correct (b) Only (ii) is correct

(c) both (ii) and (iii) are correct (d)both (i) and (ii) are correct

## Key to Multiple-Choice Questions

19.31 (d) 19.32 (d) 19.33 (b) 19.34 (a) 19.35 (c)

CHAPTER

# Design of IC Engine Parts

20

## OVERVIEW

*Ultimate aim of any theory is to design and a ultimate aim of any design is to produce which will have minimum cost, and the maximum reliability and will need the minimum supervision.* This chapter deals with the design of IC engine parts. The design procedure of the following principal parts is dealt in this chapter.

- Cylinder and cylinder head
- Piston assembly
- Valve gear mechanism
- Connecting rod
- Crankshaft

*The basic mechanical design philosophy is to draw the loading diagram of the part to be designed; select the materials according to loading and failure criteria and then use relevant governing equation/empirical relations for design. Finally, check for safety factor from strength point of view, such as yield stress, etc. Based on this design dimensions, 3D stress analysis is used; the problem is solved by* ***FEM by varying*** *the dimensions and shapes for the desired stress distribution. If 3D flow/heat analysis is needed, this should be done before stress analysis.*

## 20.1 SELECTION OF TYPE OF ENGINES

The power ouput needed and the type of service required mainly govern the selection of the type of IC engine. The consideration of the *following parameters* is a must while selecting the type of the engine.

**(a) Working Cycle:** There are mainly two cycles on which IC engines work. They are *Otto cycle and Diesel cycle*. For small output engines, both the cycles may be used but the first cost always favours the Otto cycle. At present, Diesel cycle is usually used for large engines such a marine and stationary engine (power plant) and also for heavy duty automobiles such as trucks, buses, and car engines. The Otto cycle is used in passenger cars and every large gas engine. Fuel cost also plays the main role in deciding the cycle.

**(b) Working Stroke.** There are distinct advantages and disadvantages of two-stroke and four-stroke engines, however, in general, where weight and space are main considerations two-stroke engines are preferred and for low fuel consumption, low emission and long run at full load, four-stroke engines are preferred. Small engine where simplicity is more important than fuel economy, such as two-wheelers, two-stroke engines are generally preferred. But two-stroke engines pollute air more and as a result **all the new generation two-wheelers are now for stroke engines**. Further, the large marine and locomotive engines are two-stroke engines due to less weight power output considerations. Light weight high speed petrol engines such as passenger cars are generally four-stroke engines. Medium size oil and petrol engines are built on both types.

**(c) Number of Cylinders.** *The governing parameters for the selection of number of cylinders are torsional moment in crankshaft, engine weight, power of the engine, speed, and efficiency. For smoother torsinal moment on crankshaft, lower engine weight and higher efficiency, it is better to use a greater number of cylinders instead of less number of large cylinders.* The small cylinders run at high speed and larger cylinders run at low speed (e.g., marine engines). From the above discussion, it follows that the *desirable crankshaft speeds is an important factor in the selection of size and number of cylinders. It is always beneficial to increase the power of engine by a greater number of cylinders instead of increasing its dimensions. The number of cylinders in horizontal engines may vary from one to four while* in vertical engines including V-type it may vary from one to sixteen.

**(d) Arrangement of Cylinder.** The arrangement of cylinders could be in-line, horizontal, vertical, *V shape*, and radial. *The balance and degree of compactness to the engine depending on service govern the selection of arrangement of cylinder.* Low speed engines and for which floor consideration is no consideration are built with horizontal cylinders. For higher speeds, vertical in-line arrangement is more suitable. For compact, shorter and lighter constructions, V-type and radial engines are most suitable.

**(e) Single and Double Acting.** The great majority of IC engines are built single acting due to less complexity in construction and operation. Double acting engines are used only for very large power output with large cylinders.

**(f) Engine speed (N).** *Engine speed is one of the most important parameters as it is responsible for engine dynamic proportion.* In the past, there was a trend of increasing the engine speed which resulted in the reduction of engine size, weight and overall dimensions. As the engine speed increases, the inertial forces grow, cylinder filling becomes worse, the exhaust toxicity increases, wear of the engine parts increases and the engine service life drops. Because of this, in the last decade the engine speed practically became stable, and in certain type of US automobiles, the engine speed even reduced.

*In the present technology, the engine speed of cars ranges from 4000 to 8000 rpm and only in racing cars, the speed is kept above 8000 rpm.* In order to reduce inertia stresses and increase the service life, the engine speed for trucks and tractors has been reduced to 1500 to 2500 rpm, however there are models which run at 3000 to 4000 rpm.

**(g) Mean Piston Speed ($C_m$).** *The main piston speed is a criterion of the engine speed.* The type of the engine based on speed may be classified with the help of mean piston speed as follows:

*Low speed engine:* $C_m < 6.5$ m/s   *High speed engine:* $C_m > 6.5$ m/s

*In general, the mean piston speed of IC engines varies from 3 to 18 m/s.* The mean piston speed selected for the following engines are given below:

- Carburettor, now injection engine for cars: 12 to 15
- Carburettor engine for trucks: 9 to 12
- Automobile diesel engine: 6.5 to 12
- Tractor diesel engine: 5.5 to 10.5

*With increase in piston speed, mechanical losses rise, thermal stresses of the parts increase, the engine service life grows shorter.* In this connection, an increase in the mean piston peed involves the necessity of improving the life of parts, use of more durable materials in the engine building industry and improving the quality of oils in use. *Due to this reason, the aluminium pistons are used to reduce the inertia stressed in the case of use of higher piston speed.*

**(h) Compression Ratio ($r$).** The compression ratio is one of the most important characteristics of the engine. Its choice mainly depends on the type of mixing method, type of fuels and combustion process in the engine, whether supercharged or not, the engine speed, cooling system and other factors.

The compression ratios of IC engines are within the following limits:

| | | |
|---|---|---|
| SI engines | : | 5 to 11 |
| Diesel engines with open chambers | : | 14 to 17 |
| Swirl chamber diesel engines | : | 16 to 20 |
| Precombustion chamber diesel engines | : | 16.5 to 21 |
| Supercharged diesel engines | : | 11 to 17. |

**(i) Stroke-bore ratio (L/D)** *of any engine affects the engine's performance, hence its proper selection is a must. From the consideration of combustion chamber design, $L/D = 1.5$ gives the best results but for high surface volume ratio, $L/D$ ratio should be unity. In practice, $L/D$ ratio varies from 1 to 2 but the most commonly used value is unity.* For air-cooled aircraft engines, $L/D = 1$ while for others, it varies from 1.2–1.35.

## 20.2 PRINCIPLE OF SIMILITUDE

Following the principle of similitude, the similar engine should have—

(i) The same $L/D$ ratio
(ii) The same volumetric efficiency
(iii) The same compression ratio
(iv) The same mean effective pressure
(v) The same speed factor.

## 20.3 GENERAL DESIGN CONSIDERATION

*The size of the engine (bore and stroke) as a whole is determined on the basis of mean effective pressure of gas in the cylinder and the speed. The mean effective pressure is determined by the detailed heat analysis of actual cycle. The design of individual parts depends on the maximum cylinder pressure and the acceleration of moving parts.* The following loads develop the stresses in the various parts of the engine.

(i) Gas pressure in the cylinder (ii) Inertia and centrifugal forces
(iii) Frictional forces (iv) Torsional moment reaction and weight
(v) Vibrations of crankshaft (vi) Thermal loads

The loading caused by gas pressure continuously varies during the working cycle and reaches its maximum within a comparatively small portion of the piston stroke. *It is worth to note that loading due to inertial force varies periodically and sometimes reaches in high-speed engines the values exceeding the load due to gas pressure and may be dangerous during resonance.*

Forces, because of thermal loads resulting from heat liberation due to combustion of mixture and friction affect strength of materials and cause extra stress in mating parts when they are differently heated and have different linear/volumetric expansion causing severe thermal stress.

## 20.4 DESIGN CONDITIONS

*It is the operating conditions of the engine which change the basic loads acting on the engine parts.* Figure 20.1 shows the choice of design operating conditions. Generally, the engine parts are designed for most server operating conditions.

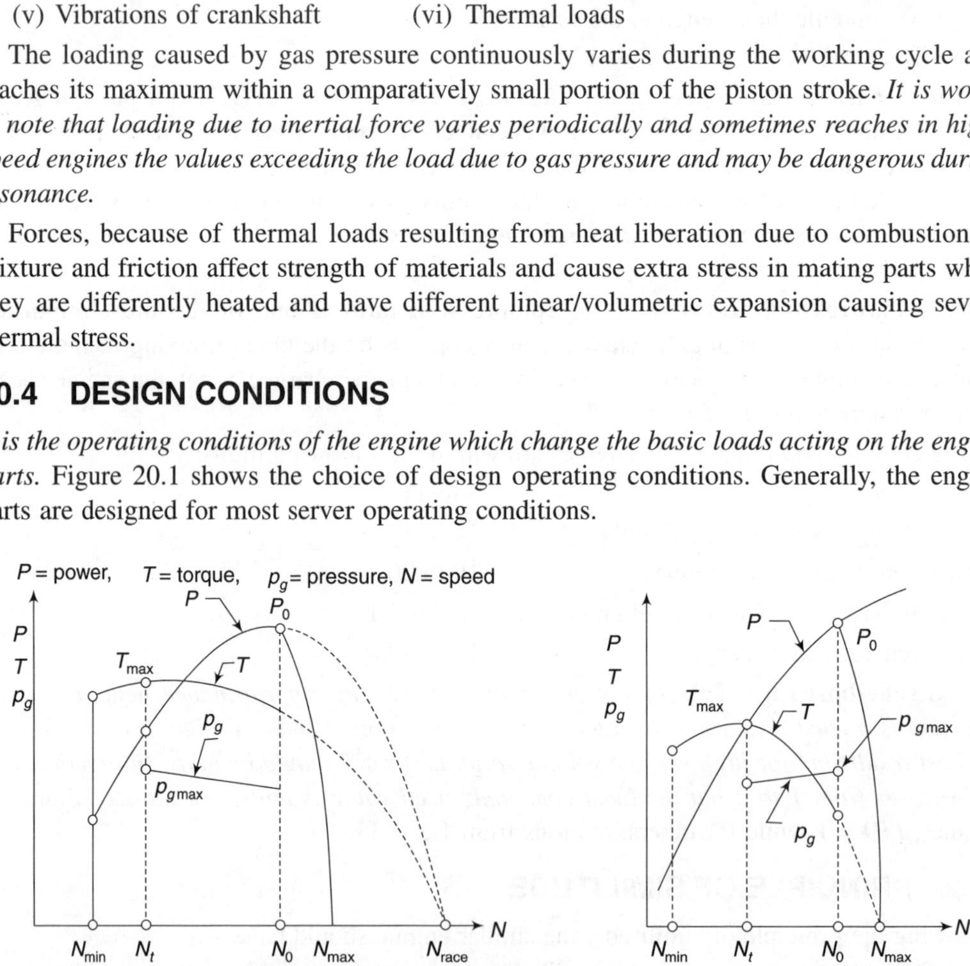


**Fig. 20.1** Choice of design operating conditions (a) Carburettor (now MPFI) engine (b) Supercharged diesel engine

For **carburettors (now MPFI) engines**, the basic design operating conditions include following data:

(i) The maximum torque ($T_{max}$) at the engine speed ($N$) = (0.4 to 0.6) $N_0$ when the gas pressure ($p_z$) reaches its maximum ($p_{z\ max}$), while the inertial forces are comparatively small. Here, $N_0$ is nominal speed.

(ii) The nominal speed output ($P_0$) of speed $N_0$, when all analyses of parts are made with taking into account the joint effect of gas and inertia loadings.

(iii) The maximum speed at idling ($N_{id\,max}$) = (0.5 to 1.20) $N_0$, when inertial forces reach their maximum, while the gas pressure is small.

For **high-speed diesel engine,** we use the following design operating conditions.

(i) The nominal power output ($P_0$) at nominal engine speed ($N_0$), when the pressure reaches its maximum ($p_{z,\ max}$), while the parts are designed to challenge the joint effect of gas and inertial loads.

(ii) The maximum speed at idling, $N_{id\ max}$ = (1.04 to 1.07) $N_0$, at which inertial forces reach their maximum.

*It is worth to note that while designing the parts of a carburettor (now MPFI) engine, the maximum gas pressure ($p_{z,\ max}$) is determined by the heat analysis made for the maximum torque operation.* Alternatively, it is assumed as approximately equal to the designed maximum combustion pressure ($p_g$) obtained from the heat analysis for the nominal power operation. The inertial forces in the maximum torque computations are neglected.

Further, while making nominal power operation computations, we assume that gas force ($p_z$) acts together with the maximum inertial force at TDC. It is the heat analysis which gives the value of the maximum gas force. The gas pressure is neglected in making the maximum speed computation for idling operation.

## DESIGN OF PRINCIPLE PARTS

## 20.5 DESIGN OF CYLINDER AND CYLINDER HEAD

### 20.5.1 Function

***The primary function of an IC engine cylinder is to contain the working fluid and the secondary function is to guide the trunk piston.***

### 20.5.2 Construction

The construction of cylinder depends upon the type of engine. If the engine is very small, the cylinder, the water jacket and the frame may be manufactured as one piece, while for small engines, cylinder and jacket can be cast as one piece. In the case of larger engine and most of the high speed engines, the cylinders have separate cylinder liner or sleeves which are more economical as these can be easily replaced after wear and tear. Further, the cylinder jacket may be of cheaper grade cast iron having wear resistance. This arrangement takes care of longitudinal expansion in better way. Portable small engines are air cooled and fins are provided over cylinder head and cylinder.

There are two types of liner—***wet liner and dry liner***. (Fig. 20.2). *As the name implies, the wet liner is one in which water in jacket is in direct contact with the outer wall of the liner, whereas the dry liner is pressed into the cylinder proper.* In practice, the engines having bore more than 13 cm use wet type liner.

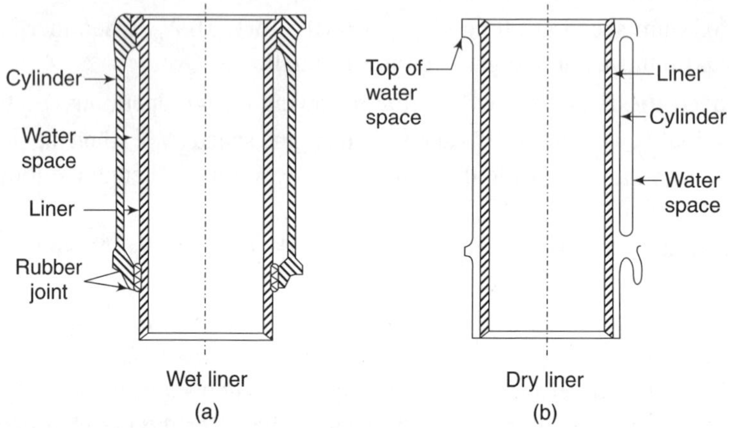


**Fig. 20.2** (a) Wet and (b) dry type of liner

The following are the **advantages of dry type liner:**

(i) Replacement of dry liner is simple.

(ii) There is no danger of water leakage either into crankcase or combustion chamber.

(iii) The upper part of the liner is better cooled.

(iv) Cylinder centre gets reduced due to absence of a heavy flange at the top of liner.

The **disadvantages of dry type liner** include the following:

(i) The casting is complicated

(ii) The heat flow through the composite wall gets reduced

The following are the **advantages of wet type liner:**

(i) Owing to the longitudinal expansion of the liner, the cylinder block is relieved of the stresses.

(ii) There is a considerable reduction of foundry problem since the large internal cores of the cylinder block and be properly supported.

The **disadvantages of wet liner** include the following:

(i) The replacement is slightly difficult as compared to dry type liner.

(ii) There is a danger of water (coolant) leakage into the crankcase and the combustion space if the casting is defective.

### 20.5.3 Qualities of a Good Liner

The following are the qualities of a good liner:

(i) It should have sufficient strength to resist the gas pressure

(ii) The material should be sufficiently hard to resist wear. This also ensures less wear of piston and piston rings.

(iii) It should be capable of resisting the thermal stresses caused due to the heat flow through the liner wall.

(iv) The material should be corrosion resistant.

(v) It should be capable of taking good bearing surface.

(vi) The shape of the liner should be symmetrical to avoid unequal deflection due to gas load and unequal expansion due to thermal load causing distortion.

(vii) There should not be any distortion of inner surface due to restraining fixings.

**Liner Materials.** Based on the qualities needed for a good liner, the **following liner materials** are used in order of preference.

(i) A good grade grey cast iron with homogeneous and closed grained structure, i.e., perlitic and other similar cast iron. It may be centrifugally cast.

(ii) Nickel cast iron and nickel chromium cast iron (generally 2% nickel and 10.65% chromium).

(iii) Nickel chromium cast steel with molybdenum.

(iv) For aircraft engines, the liners are often used of alloy steel forgings to get very light cylinders.

### 20.5.4 Surface Treatment of Cylinder Liner

It has already been stated that the *cylinder liner should be quite hard to resist wear, so the surface treatment is a must.* The following surface treatment methods are used:

(i) In one method, after initially machining, the liner is heated in an oven to about 840°C (limit is a function of material). After heating, the liner is quenched in oil and then tempered for about 180°C. After tempering the surface of the liner is finished by a honing operation ensuring the bore within close limits of the desired roundness and straightness. The hardness is maintained around Rock-well C50.

(ii) In the second method, the cast iron liners are heated from inside with the help of a high frequency induction current. After heating, the liners are quenched by water ensuring a surface hardness of Rock-well C 52 to 55. The liners are then tempered to lower the hardness of about C50. After minor tempering, the liner bore is honed to the final finish ensuring roundness and closeness.

(iii) In special cases, the cylinder bore is chromium electroplated to have a very hard and porous surface. A good oil film is formed and retained on this surface, which in turn, reduces surface wear appreciably.

### 20.5.5 Stresses Induced in Liner

Cylinder liner is subjected to two types of stresses, namely, ***thermal stresses*** *and* ***pressure stresses****. The thermal stresses induce compressive stresses in the inner fibres of the wall and tensile stresses in the outer fibres of the wall. The pressure stresses due to gas pressure induce tensile stresses throughout the outer wall.* As a result of opposing nature of stresses on the inside of wall, the compressive stresses due to temperature will be relieved by the pressure stresses, while on the outside of the wall, the tensile stresses due to temperature will be increased by the pressure stresses acting in the same direction. It is worth to note that the temperature stress is proportional to temperature difference, to the wall thickness and, to the modulus of elasticity of the wall material. The pressure stress is also a function of the wall thickness and go on decreasing with the increase in wall thickness.

From the above discussion, it follows that ***temperature stresses*** can be reduced by the following methods:

(i) by making the wall thickness thinner using material of high tensile strength, and

(ii) by lowering the temperature difference between inside and outside of the wall by increasing the velocity of coolant (water) through the jacket.

Failure of liner due to temperature and pressure stresses is unknown. *Two **main reasons of liner failure are distortion and wear**. Wear is taken cares by surface treatment.*

### 20.5.6 Distortion of Liner

As discussed, *the liner fails due to distortion, so the cylinders, cylinder block and liners should be designed to reduce distortion to a minimum.* Distortion of liner is caused due to the following reasons:

(i) *Temperature stresses.* Any defect in the casting on the inner wall will change the temperature stress distribution and the liner may get distorted. Similarly, any ribs or bosses on the outer wall of the liner will impose local restraint on the wall causing distortion. This suggests that ribs or bosses on the outer wall should be avoided as far as possible.

(ii) *Axial expansion.* Axial expansion of liner may cause distortion. Due to this reason one end is made free.

(iii) *Gas pressure* also causes distortion.

(iv) Distrotion may also be caused by *bolting on the cylinder* head or any other part.

(v) Distrotion may be caused due to *fixing*, especially at the lower water joint. To overcome this problem, rubber rings are used as it is not compressible.

### 20.5.7 Cylinder Lubrication

*Even with the most careful arrangement of cylinder lubrication,* ***about 3 to 4% of indicated power of the engine is lost due to friction between the cylinder liner and piston rings to maintain a satisfactory gas seal.*** In the beginning, when the engine starts from cold, the oil will just ooze from the bearing being quite thick resulting in oil starving and thus high wear of the liner. To avoid this problem, sometimes jacket wall and liner are bolted with a bolt having an oil passage throughout (Fig. 20.3) to provide the lubrication of liner.

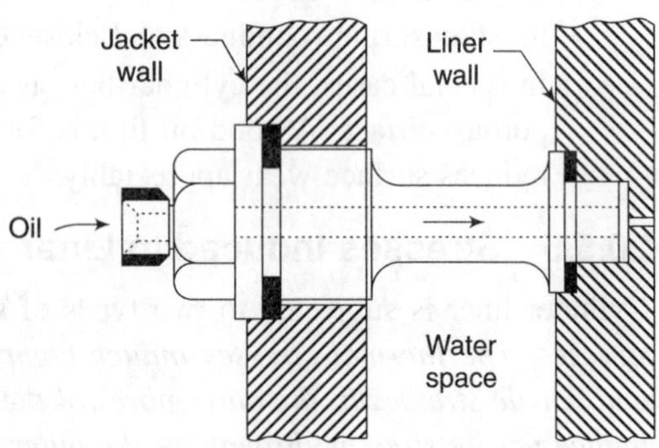


**Fig. 20.3** Typical liner lubrication

### 20.5.8 Liner Joints

*Joining of liner with cylinder water jacket should be done with utmost care.* Figures 20.4 and 20.5 show the various arrangements of upper liner and lower liner water joints. Care should be taken to ensure that the upper joint should not touch at the corners. *The stress concentration should be avoided.* The joint may be metal to metal or copper joint may be adopted. In order to prevent water joints, expansion joints, may be provided at the lower or bottom free end of the liner.

**Fig. 20.4** Upper liner water joint

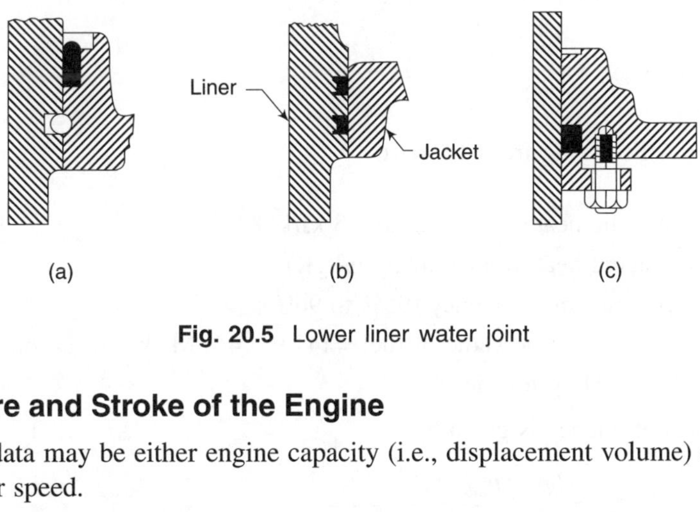


**Fig. 20.5** Lower liner water joint

### 20.5.9 Bore and Stroke of the Engine

Design input data may be either engine capacity (i.e., displacement volume) or brake power for a particular speed.

Engine size or power output may be obtained from heat analysis.

The indicated power developed from an IC engine is given by

$$ip = \frac{p_{mi}\, L\, A\, n\, x}{60} \qquad ...(20.1)$$

where $p_{mi}$ = indicated mean effective pressure, kN/m$^2$; $L$ = stroke, m;
$A$ = piston area, m$^2$; $x$ = number of cylinder; $n$ = number of revolutions
= N/2 for 4-stroke = N for 2-stroke; $ALx$ = engine capacity

The indicated mep ($p_{mi}$) is expressed as

$$p_{mi} = K_{lr}\,(p_{mi})_{as} \qquad ...(20.2)$$

where $K_{lr}$ = coefficient to take care of losses in real cycle ranging from 0.8 to 0.85.

$(p_{mi})_{as}$ = indicated mep based on air standard cycle expressed in (Ref. chapter 2)

$$(p_{mi})_{as} = \frac{p_1\, r\,(\alpha - 1)\,(r^{\gamma-1} - 1)}{(\gamma - 1)(r - 1)} \text{ for SI engine} \qquad ...(20.3)$$

and

$$(p_{mi})_{as} = \frac{p_1\, r\,[\gamma r^{\gamma-1}\,(\beta - 1) - (\beta^{\gamma} - 1)]}{(\gamma - 1)\,(r - 1)} \text{ for CI engine} \qquad ...(20.4)$$

where $\alpha = \dfrac{p_3}{p_2} = \dfrac{T_3}{T_2}$; $\qquad \beta = \dfrac{T_3}{T_2}$

$r$ = compression ratio $= \dfrac{V_1}{V_2}$

$T_2 = T_1\, r^{\gamma - 1}$

The maximum temperature of gas ($T_3$) in the cycle is given by heat release expression

$$(\text{LCV})_f \,.\, \eta_{comb} = \left(\frac{\dot{m}_a}{\dot{m}_f} + 1\right) c_{pg}\, T_3 - \frac{\dot{m}_a}{\dot{m}_f}\, c_{pa}\, T_2 \qquad ...(20.5)$$

where $\dfrac{\dot{m}_a}{\dot{m}_f}$ = A/F varying from 14 to 16

$c_{pg}$ = specific heat of gas (1.2 to 1.3 kJ/kgK)

$c_{pa}$ = specific heat of air (1.005 kg/kgK)

$\eta_{comb}$ = combustion efficiency (95% to 96%)

$(\text{LCV})_f$ = lower calorific value of fuel = (41.9 – 44) MJ/kg for gasoline

= 42.9 MJ/kg for diesel

The mechanical efficiency is given by

$$\eta_m = \frac{bp}{ip} = \frac{p_{mb}}{p_{mi}} = \frac{p_{mi} - p_{mf}}{p_{mi}} \qquad ...(20.6)$$

where $p_{mf}$ = 0.034 + 0.0113 $C_m$ ; MPa

$C_m = \dfrac{LN}{30}$, m/s

*The mechanical efficiency is a function of speed.* It ranges from: $\eta_m$ = 0.8 to 0.9. The higher value of $\eta_m$ is for low speed.

The sizing of the cylinder is based on the assumption of L/D ratio.

Engine volume (capacity) = $V = ALx$

Let $\frac{L}{D} = K_1$, so

Engine volume $= V = \frac{\pi}{4} D^3 K_1 x$ where $x$ = no. of cylinders ...(20.7)

Clearance on both sides may be taken as 10 to 15% of the stroke volume to calculate the length of cylinder. Thus,

$$L_{cyl} = \text{Stroke} + \text{Clearance on both sides}$$

Fuel consumed by engine is expressed as specific fuel consumption which is given by

$$bsfc = \frac{3600\, \dot{m}_f}{bp}\ ;\ \text{kg /kWh} \quad ...(20.8)$$

where $\dot{m}_f$ = mass of fuel flow rate, kg/s

The $\dot{m}_f$ is calculated from air consumption rate $\dot{m}_a$ and A/F. Thus,

$$\dot{m}_a = \frac{p_{ind} \cdot V}{R_a\, T_{ind}}\ ;\ \text{kg/h} \quad ...(20.9)$$

where $p_{ind} = p_1 - \Delta p = 0.9$ bar = average induction pressure

$T_{ind} = T_1 + \Delta T = 330$ K = average induction temperature

$\dot{V}_a = V.n \times 60 = \frac{60\,\text{VN}}{2}$ ; m$^3$/h for 4 stroke

$= 60$ VN; m$^3$/h for two-stroke

The actual maximum gas pressure achieved inside the engine due to losses is given by,

$$p_{g,\max} = 0.8\, p_3 \quad ...(20.9a)$$

### 20.5.10 Thickness of Cylinder and Liner Wall

The cylinder wall or liner thickness is stressed due to gas pressure and the piston's side thrust. *The piston's side thrust may cause bending of the wall but at the point of the maximum thrust, the gas pressure has decreased sufficiently, so the stress caused by side thrust is negligible. Gas pressure causes two types of stresses, namely,* ***longitudinal and circumferential (hoop)*** *acting perpendicular to each other, resulting in reduction in net stress in each direction.* The longitudinal stress is negligible.

The wall thickness of cylinder or liner is usually calculated considering it as a thin cylinder which is expressed by

$$t_c = \frac{p_{g,\max} \cdot D}{2\,\sigma_h} + K_r + K_c \quad ...(20.10)$$

where $t_c$ = wall thickness of cylinder or lines in mm, $D$ = cylinder bore, mm

$p_{g,\max}$ = actual maximum gas pressure = $0.85 \times p_3$

$\sigma_h$ = maximum hoop (circumferential) stress = 35 to 105 N/mm$^2$ depending on the size and material, larger values are used for smaller bore

$K_r$ = reboring factor (Table 20.1), $K_c$ = casting allowance (1 to 1.5 mm)

**Table 20.1** Reboring Factor

| Cylinder bore in mm | 50 | 100 | 150 | 200 | 250 | 300 | 350 | 400 | 450 | 500 |
|---|---|---|---|---|---|---|---|---|---|---|
| $K_r$, in mm | 1.5 | 2.3 | 4.0 | 6.0 | 7.5 | 9.5 | 10.5 | 12.5 | 12.5 | 12.5 |

The thickness of the liner wall is also given by

$$t_c = 0.5\ D\ [\sqrt{\{(\sigma_h + 0.4\ p_{g,\ max})/(\sigma_h - 1.3\ p_{max})\} - 1}] \qquad ...(20.11)$$

The thickness of cylinder wall is also expressed by **empirical equation**. *The thickness usually varies from about* ***4.5 mm to more than 25 mm*** *depending upon the bore of the cylinder.* The liner of oil engine may be

$$t_c \geq \frac{D}{15} \qquad ...(20.12)$$

For dry liners, the total thickness ($t'_c$) is the thickness of liner and that of cylinder wall. The thickness of the dry liner is given as

$$t'_c = 0.03D \text{ to } 0.04D \qquad ...(20.13)$$

The thickness of the inner wall of the automobile engine may be expressed by

$$t_c = 0.04D + 1.6 \text{ mm} \qquad ...(20.14)$$

The thickness of the jacket wall may be given by

$$t_c'' = 0.032D + 0.6 \text{ mm} \qquad ...(20.15)$$

Water space between outer cylinder wall and the inner jacket wall may be given as

$$t_c''' = 0.08D + 6.5 \text{ mm} \qquad ...(20.16)$$

### 20.5.11 Cylinder Head

In practice, a separate cylinder cover or head is provided in all engines except the smallest ones. *Cylinder head is very complicated in configuration and is the most difficult part to be designed and manufactured.*

The *following factors govern* the design of cylinder head:

(i) Air and gas passage

(ii) Accommodation of valves and their gears

(iii) Accommodation of spark plug or fuel injectors

(iv) Subjected to severe alternating loads and high temperature causing drastic stresses in cylinder head.

The ***materials*** selected for cylinder head are the following:

(i) Close grained cast iron

(ii) Alloy cast iron containing nickel; chromium and molybdenum for small and medium sized engines.

(iii) Low carbon steel for large engines

(iv) Aluminium alloy.

*The cylinder block heads must be rigid enough to prevent distortion of the valve seats and other parts of the cylinder head in the engine operation. The cylinder head rigidity is ensured on the proper selection basic dimensions.* In this connection, experimental data are widely utilised in practice.

Thickness of head lower support wall ($\delta_h$) and thickness of water jacket walls ($\delta_j$) for engines with a cylinder bore up to 150 mm can be taken by the following thumb rules.

- Carburettor (MPFI) engines : $\delta_h = 0.09$ D
- Diesel engines : $\delta_h = (1.5 + 0.09$ D) mm
- All engines : $\delta_j = (2.2 + 0.03$ D) mm

If the cylinder head is approximately a flat circular plate, the thickness can be calculated by the following equation

$$t_h = D\sqrt{\frac{C . p_{gmax}}{\sigma_t}} \qquad ...(20.17)$$

where $C = 0.1$; $\sigma_t$ = allowable stress (35 to 56 N/mm$^2$)

*Low value of allowable stress is desirable to take care of highly stressed cylinder head.*

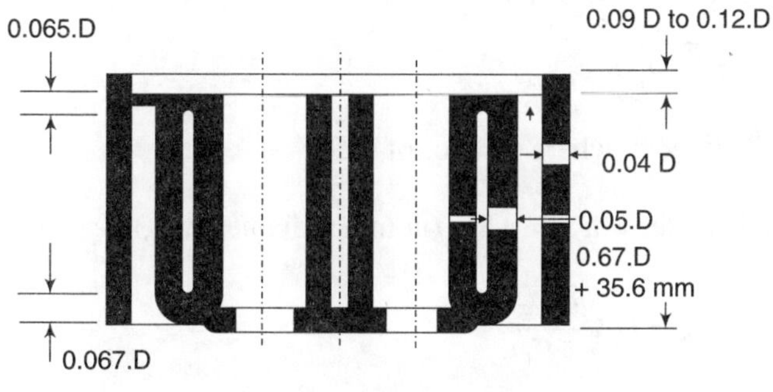


(a) Typical cylinder head proportions

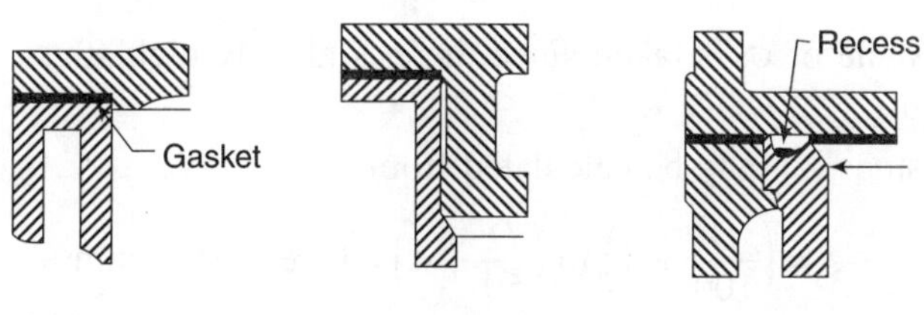


(b) Typical cylinder head joints

**Fig. 20.6**

Figure 20.6(a) shows a typical proportion of cylinder head for four-stroke engines having bore varying from $D = 0.1$ m to 0.45 m.

Figure 20.6(b) shows typical methods for securing a tight joint between the cylinder and cylinder head.

### 20.5.12 Cylinder Head Studs

*Studs are used to join the head to the monoblock structure (Fig. 20.7) of the engine. They resist to the action of preloading forces; gas pressure and loads arising due to temperature difference and coefficient of liner expansion of the cylinder head, monoblock and stud materials. It is essential that the number of stud (z), their dimensions and preloading must provide reliable sealing of the joint between the cylinder head and the cylinder block under all operational conditions.*

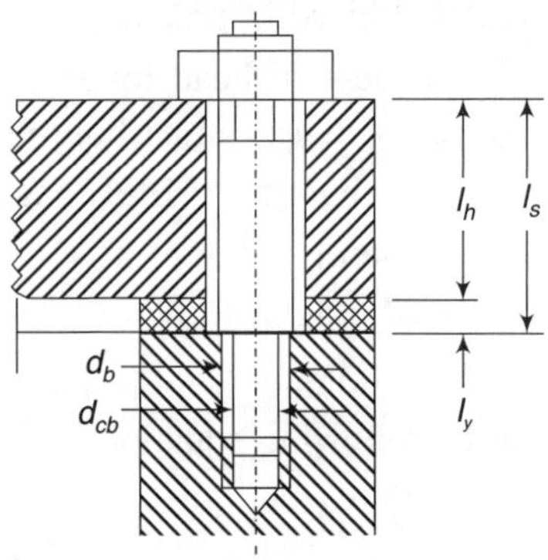


**Fig. 20.7** Design diagram of a stud

The materials used for studs are carbons *steels* of high limit of elasticity, *high alloy steels and nickel alloy steel having yield* point ranging from 630 to 950 MPa.

The gas pressure actually acts upon the pitch circle diameter of the studs but as the stud diameter ($d_b$) is not known initially, the pressure may be assumed to be acting on the cylinder diameter. The centre of the stud should be 1.25 $d_b$ to 1.5 $d_b$ from the inner wall of the cylinder.

Pitch circle diameter = $D_p = D + (1.25\ d_b \text{ to } 1.5\ d_b)$

Load on the stud = $\frac{\pi}{4} D_p^2 \times p_{gmax}$

But load = $z \times \frac{\pi}{4} d_{cb}^2 \times \sigma_t$ where $z$ = no. of studs

Generally, core diameter = $d_{cb} = 0.8\ x$ (nominal diameter, $d_b$)

Thus, $$d_{cb} = D\sqrt{\frac{p_{gmax}}{z\,\sigma_t}} \qquad ...(20.18)$$

where $\sigma_t$ = allowable fibre stress (35 to 70 N/mm$^2$)

$d_{cb}$ = core diameter of stud; $D$ = cylinder bore

Generally, low value of $\sigma_t$ is taken since there is already high stress in the studs due to tightening of the nuts.

The number of studs ($z$) may be calculated from

$$z = \left(\frac{D}{100} + 4\right) \text{ to } \left(\frac{D}{50} + 4\right) \text{ where } D \text{ is in mm} \qquad ...(20.19)$$

The pitch circle diameter ($D_P$) of stud is given by

$$D_p = D + 1.4\, d_b$$

The pitch of the stud may be calculated from $19\sqrt{d_b}$ to $29\sqrt{d_b}$.

*In general, the engine producing* ***more than 10 kW liquid cooling*** *is used and less than 10 kW,* ***air cooling*** *by fins is used.* Air cooled fins are provided over the cylinder and head and major part of cylinder. The liquid cooled engine employs jacket passage. *Fins or jacket passage are designed on the basis of heat transfer to coolant which may be assumed around 25 to 30%.* The heat taken by cooling medium is expressed as

$$Q_{cooling} = (25\% \text{ to } 30\%) \times \dot{m}_f \times LCV \times \eta_{comb} = A_0\, U_0\, (T_s - T_\infty)$$

where $T_\infty$ = cooling fluid temperature; $U_0$ = overall heat transfer coefficient

and $T_s$ = surface temperature of cylinder

### Cooling Fins

Overall heat transfer coefficient from a cylinder with fins (Fig. 20.8) is expressed as

$$U_0 = \frac{h_t}{s+b}\left[\frac{2}{a}\left(1+\frac{w}{2r_0}\right)\tan h(a\,(w+0.5)] + s_0\right]$$

where $s$ = mean spacing of tapered fins

$b$ = thickness of the fin (0.5 mm to 1.5 mm)

$w$ = fin width (13 mm to 35 mm)

$a = \sqrt{2\, h_i/kb}$

$h_i$ = special heat transfer coefficient,

W/m$^2$ K = $f(s$, air velocity) = 40 W/m$^2$ K

$r_0$ = outer radius of cylinder, mm

$s_0$ = spacing of tapered fins on the cylinder surface, mm

$k$ = thermal conductivity, W/m K

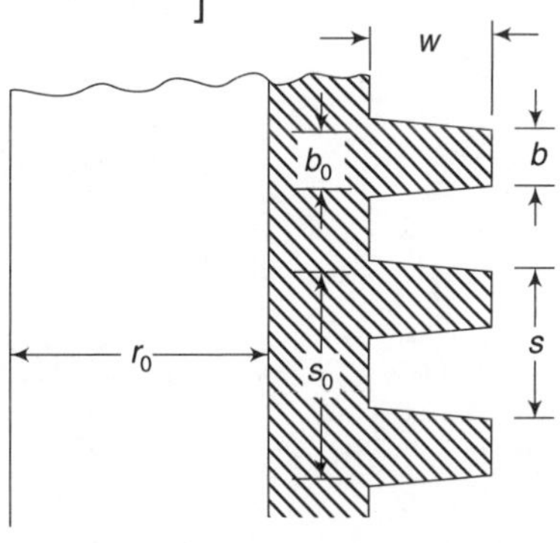


**Fig. 20.8** Fin dimensions

The other details of the fins are: the total taper of the fins are from 3.5 to 5 or about 1.5 mm to 2.3 mm per cm length. The spacing of the fins $p$ = 1.5 to 3 per cm.

### Water jacketing

$$Q_{cooling} = \dot{m}_w\, c_{pw}\, (T_{w2} - T_{w1}) = U_0\, A_0\, (T_s - T_w)$$

where $T_s$ = surface temperature of cylinder, $U_0$ = overall heat transfer coefficient

For finding $U_0$, the following correlation for convective heat transfer coefficient (local) may be used

$$\left(\frac{h_c l}{k}\right) = 0.036\left(\frac{\rho\, v\, l}{\mu}\right)^{0.8}\left(\frac{\mu\, c_p}{k}\right)^{0.333}$$

where $l = 2\pi\, r,\quad v = r\,\omega$

$h_c$ = convective heat transfer coefficient

The other correlation (spatial average) is

$$h_c\ (\mathrm{W/m^2}\ K) = 3.26\ D(m)^{-0.2}.\ p(kPa)^{0.8}\ T(K)^{-0.55}.\ v_g\,(\mathrm{m/s})^{0.8}$$

## 20.6 PISTON ASSEMBLY

### 20.6.1 Function

As discussed earlier, *the main function of the piston is to receive the impulse from the expanding gases and transmit the energy to the crankshaft through the connecting rod.* Further, the piston must also disperse a large amount of heat resulting from the combustion of fuel and air in the combustion chamber to the cylinder wall. In the internal combustion engines, the piston used is usually of **trunk type** (Fig. 20.9).

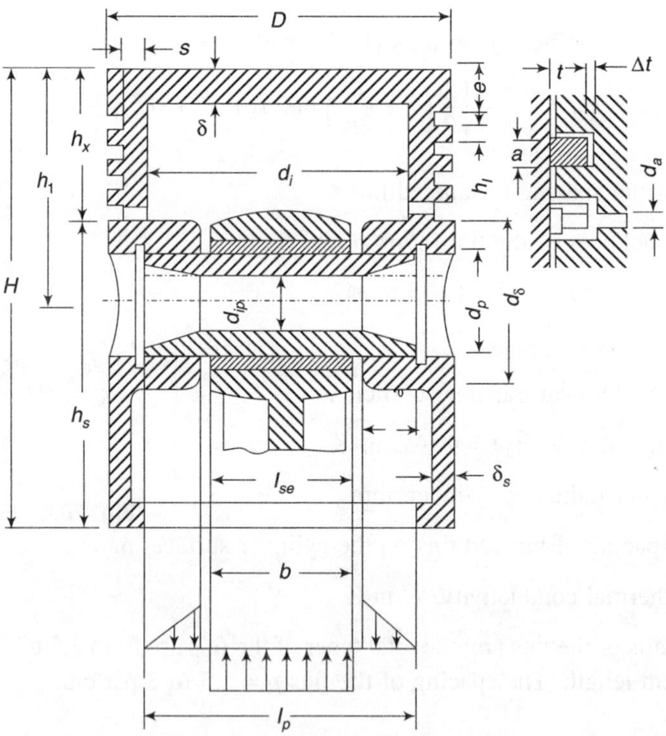


**Fig. 20.9** Trunk type piston

A trunk type piston consists of:

(i) crown or head to carry the cylinder pressure,

(ii) skirt acts as a bearing surface for connecting rod side thrust,

(iii) a piston pin to connect it to the connecting rod, and

(iv) piston rings to seal cylinder and to scrap the excess oil on cylinder (liner) wall.

*It is the design of combustion chamber that decides the profile of the piston crown.* The engine power rating and thermal efficiency decide the thermal stresses to be developed in the piston. *The connecting rod/stroke ratio and the gas pressure decide the amount of side thrust on the cylinder wall.* Many factors including the bottom and design govern the number and type of piston rings.

### 20.6.2 Design Objectives

*The present trend is to develop internal combustion engines of increased power capacity, highly efficient, compact and releasing minimum pollution,* This is made possible by improved engine design including increased compression ratio and engine speed. The following are the design objectives:

(i) The piston should be made of enormous strength and heat resistant to withstand high pressures and inertia forces.

(ii) The piston should have sufficient rigid construction to withstand thermal and mechanical distortion.

(iii) It should dissipate heat quickly from the crown to the ring and bearing area and then to cylinder walls.

(iv) It should form an effective gas and oil seal.

(v) It should have minimum weight to minimise the inertia forces.

(vi) It should have sufficient bearing area to prevent undue wear.

(vii) It should be symmetrical in design to allow even expansion under thermal load.

(viii) It should provide the minimum noise during reciprocating motion.

(ix) The power consumed by friction should be as minimum as possible.

(x) The material of thc piston must possess good wearing qualities to provide sufficient hardness at the operating temperature.

(xi) The material should be anti-corrosive.

### 20.6.3 Materials

The following is the list of materials that are commonly used for the piston of IC engines.

(i) Cast iron
(ii) Cast aluminium
(iii) Forged aluminium
(iv) Aluminium alloy
(v) Cast steel
(vi) Forged steel

Pistons are seldom made of cast iron. However, cast iron may be used for mean piston speed below 6 m/s. ***Aluminium alloys are the ideal materials for pistons and all modern pistons are made of aluminium alloys***.

### 20.6.4 Piston Design Proportions

Based on experimental investigation, piston design proportions (Fig. 20.9) have been prepared. Table 20.2 gives such a design proportions mostly in terms of cylinder bore ($D$). *After adopting these proportions, stresses in various parts may be checked whether they are in the safe limit or not.*

**Table 20.2** Piston Design Proportions

| S. No. | Descriptions | SI Engine | Diesel Engine |
|---|---|---|---|
| 1. | Piston crown thickness, $\delta$ | (0.05–0.10) $D$ | (0.12–0.20) $D$ |
| 2. | Piston height, $H$ | (0.8–1.3) $D$ | (1.0–1.7) $D$ |
| 3. | Height of piston top part, $h_t$ | (0.45–0.75) $D$ | (0.6–1.0) $D$ |
| 4. | Piston skirt height, $h_s$ | (0.6–0.8) $D$ | (0.6–1.1) $D$ |
| 5. | Boss diameter, $d_0$ | (0.3–0.5) $D$ | (0.3–0.5) $D$ |
| 6. | Distance between boss and faces, $b$ | (0.3–0.5) $D$ | (0.3–0.5) $D$ |
| 7. | Thickness of skirt wall, $\delta_s$, mm | 1.5–4.5 | 2.0–5.0 |
| 8. | Thickness of piston crown wall, $s$ | (0.05–0.10) $D$ | (0.05–0.10) $D$ |
| 9. | Distance to the first piston groove, $e$ | (0.06–0.12) $D$ | (0.11–0.2) $D$ |
| 10. | Thickness of first piston ring land, $h_1$ | (0.03–0.05) $D$ | (0.04–0.07) $D$ |
| 11. | Radial thickness of piston ring, $t$, compression ring<br>Oil ring | (0.04–0.045) $D$<br>(0.038–0.043) $D$ | (0.04–0.045) $D$<br>(0.038–0.043) $D$ |
| 12. | Number of piston rings (width 4 mm) | 2–4 | 3–5 |
| 13. | Difference between free gap and compressed gap of piston ring, $A_0$ | (2.5–4.0) $t$ | (3.2–4.0) $t$ |
| 14. | Radial clearance of ring in piston groove, $\Delta t$, mm<br>Compression ring<br>Oil ring | <br>0.70–0.95<br>0.91–1.1 | <br>0.7–0.95<br>0.9–1.1 |
| 15. | Piston inner diameter, $d_i$ | $D$–2 ($s$ + $t$ + $\Delta t$) | $D$–2 ($s$ + $t$ + $\Delta t$) |
| 16. | Number of oil holes in piston, $n'_0$ | 6–12 | 6–12 |
| 17. | Oil passage diameter, $d_0$ | (0.3–0.5) $a$ | (0.3–0.5) $a$ |
| 18. | Pin's outer diameter, $d_p$ | (0.22–0.28) $D$ | (0.3–0.38) $D$ |
| 19. | Pin's inner diameter $d_{ip}$ | (0.65–0.75) $d_p$ | (0.5–0.70) $d_p$ |
| 20. | Pin length, $l_p$<br>Retained pin<br>Floating pin | <br>(0.88–0.93) $D$<br>(0.78–0.88) $D$ | <br>(0.88–0.93) $D$<br>(0.80–0.90) $D$ |
| 21. | Connecting rod bushing length, $l_{se}$<br>Retained pin<br>Floating pin | <br>(0.28–0.32) $D$<br>(0.33–0.45) $D$ | <br>(0.28–0.32) $D$<br>(0.33–0.45) $D$ |

*Though the piston design proportions have been given but dimension of its critical part may be calculated with the help of governing equation and may be compared from the proportions as a check.*

### 20.6.5 Piston Crown

*The piston crown is designed for bending by the maximum gas forces ($p_{g\,max}$) as uniformly loaded round plate freely supported by a cylinder.* In the case of carburettor (SI) engines, maximum gas pressure occurs when operating at maximum torque while in diesel engines a maximum gas pressure takes place usually when operating at maximum power. The piston crown may also be designed for heat dissipation but design based on strength is more preferred.

In some cases, the piston crown is made of different material from that of the lower part and joined to it with the help of studs. This is done in view of using material of high strength (i.e. cast steel) for the crown in large oil engines. Such a construction helps replacing the crown only if it is damaged due to huge pressure and temperature. In such case, it is essential that the upper part should never touch the cylinder walls.

#### 20.6.5.1 Strength Basis Crown Thickness

*As mentioned above, the crown thickness on the basis of strength can be calculated by considering it to be a flat plate of uniform thickness fixed at the edges and uniformly loaded by gas load ($p_{gmax}$).* The crown thickness ($\delta$) is given by

$$\delta = \sqrt{\frac{3\,p_{gmax}\,D^2}{16\,\sigma_t}} = 0.433\,D \times \sqrt{\frac{p_{g\,max}}{\sigma_1}} \qquad \text{...(20.20)}$$

where $\delta$ = crown thickness, mm; $D$ = cylinder bore, mm

$p_{gmax}$ = max. gas pressure, N/mm$^2$; $\sigma_t$ = allowable stress in bending, N/mm$^2$

The allowable stresses in bending for various materials are given below:

(i) Grey cast iron (ultimate strength 150 N/mm$^2$) ; $\sigma_t$ = 40 to 45 N/mm$^2$

(ii) Nickel cast iron, aluminium alloy ; $\sigma_t$ = 80 to 110 N/mm$^2$

(iii) Forged steel (ultimate strength 380 N/mm$^2$); $\sigma_t$ = 90 to 120 N/mm$^2$

#### 20.6.5.2 Heat Dissipation Basis Crown Thickness

The thickness ($\delta$) may be calculated by assuming the steady state heat transfer and crown to be flat and of constant thickness by energy balance.

Heat absorbed by the piston from the gases

= heat flow through the head from the centre to the cylinder wall

$$\pi r^2 q = 2\pi r \delta k \frac{dT}{dr} \qquad \text{...(20.21)}$$

where $r$ = piston radius; $q$ = heat flux, W/m$^2$; $T$ = temperature of crown, $K$

$k$ = conductivity of piston crown

If $H$ be the heat flow gases in watts, then

$$H = Q_A = q\frac{\pi}{4}D^2$$

Using the boundary conditions, $T = T_c$ at $r = 0$

and $T = T_e$ at $r = D/2$, after integration, we get ...(20.22)

$$\delta = \frac{H}{12.56k\,(T_c - T_e)} = \frac{Q_A}{12.56\,k\,(T_c - T_e)}$$

The heat flow, $H$ through head may be expressed as

$$H = Q_A = C \times \dot{m}_f \times HCV \times bp/3600 \quad ...(20.23)$$

and $$bp = \frac{p_{mb}\,L\,A\,n}{60} \quad ...(20.24)$$

where The conductivity, $k$ may be taken as

$k = 46.6$ W/mK for gray cast iron $= 174.5$ W/mK for aluminium alloy

Temperature difference may be taken as

$T_c - T_e = 220°C$ for cast iron with $T_c = 425°C$

and $T_c - T_e = 75°C$ for aluminium with $T_c = 260°C$

$\dot{m}_f$ = mass of fuel, kg/kWh = 0.24 to 0.3 kg/kWh

$C$ = constant = 0.05 = portion of heat absorbed by piston

## 20.6.6 Ribs

*In general, a piston consists of a number of ribs to meet the various requirements.* Side thrust from the connecting rod is transmitted to the skirt through the piston pin boss. In order to distribute this side thrust properly to avoid the distortion of the skirt, a stiffening ribs are provided at the centre line of this boss and extending around the skirt. *Triangular ribs are provided just below the ring section and extending to the pin bosses to carry the gas load. In certain cases, these ribs also extend across the crown. These ribs not only take the gas load but also dissipating the heat to the cylinder wall through piston pin.*

It is worth to note that the skirt expansion may cause the skirt out of round and according to some designs, it is a bad practice to provide ribs on a surface subjected to expansion, and in design without ribs the piston pin boss in extended upto the crown.

*In practice, four to six ribs are provided radially to the crown and their thickness varies from $\delta/3$ to $\delta/2$ where $\delta$ is the crown thickness.*

## 20.6.7 Piston Rings

*As discussed earlier, the main function of the piston rings is to prevent the leakage of gas past the piston.* There are two types of piston rings as given:

(a) Compression rings or pressure rings (b) Oil scrapper or oil central rings.

The compression rings are inserted at the top portion of the piston.

The functions of compression rings are the following:

(i) *To seal the space between the piston and cylinder and thus prevent the leakage of high pressure gases or air past the piston.*

(ii) *To transfer the heat from the piston crown to the cylinder walls.*

(iii) *To absorb partly side thrust.*

So far the number of piston rings is concerned, there is no hard and fast rules. Automobiles and aircraft pistons may have three to four rings, however stationary compression ignition engines may use five to seven piston rings. *According to one design philosophy, it is better to use a large number of narrow rings than a few wide shallow rings as the heat transfer is better from rings to the liner than from the piston wall to the rings.*

The function of the oil scraper rings are the following:

(i) *To scrap the lubricating oil from the surface of the liner so as to minimise the flow of oil into the combustion chamber.*

(ii) *To allow sufficient oil to go upwards during the upward stroke for the proper lubrication of liner.*

In small engines, the piston may use one oil scraper ring but in large engines, it may use two or three. *The oil rings are inserted below the compression rings.* In the case of more than one oil ring, one may be filled below the gudgeon pin and the remaining above the gudgeon pin. Figure 20.10 shows the various types of piston rings and their grooves.

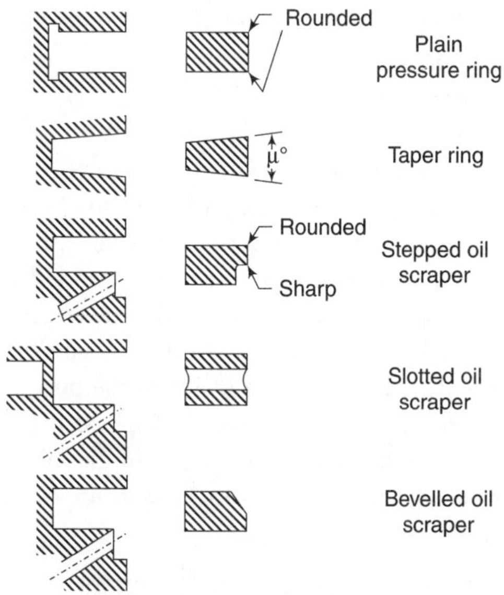


**Fig. 20.10** Types of piston rings and their grooves

*So far the cross section of the compression rings are concerned, the reactangular cross section is widely used. The diameter of the ring is slightly made larger than the cylinder bore to produce pressure against the liner wall.* In order to insert the piston with rings in cylinder, a part of the ring is ***cut-off***. Over and above, this cut allows **thermal expansion** at highest temperature without the buckling of the ring. Figure 20.11 shows the various types of cuts in the ring. Care should be taken in giving sufficient gap between the end of the cut so that even at the highest temperature, the ends will not touch each other during expansion of rings to avoid their buckling. *Square cut ends are more common.*

The story in the case of two-stroke engines is different as thin rings are avoided to reduce the possibility of their fracture when riding over the parts. Further, ring ends must never come over the parts due to rotation of the piston ring in the grooves. To overcome this problem,

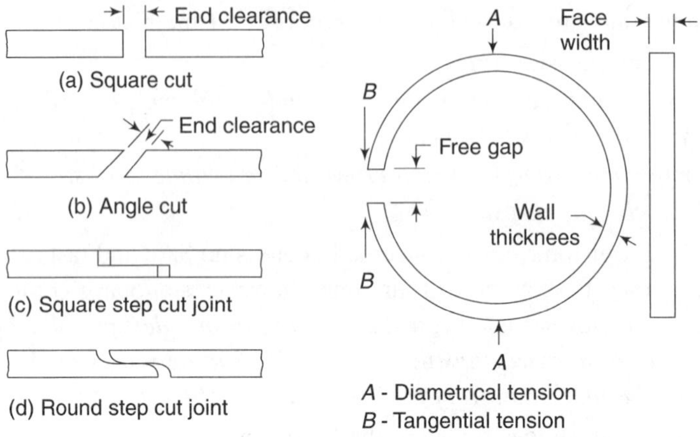


**Fig. 20.11** Various types of cuts in piston rings

the piston rings are pinned or anchored to avoid the rotation, so that the ring gap is always confined to the plain part of the liner. *It is very important to ensure that ring joints should be spaced equally around the piston and should not be one below the other to avoid the leakage of the gas.* The location of the ring by pin is necessary in the case horizontal engines to stop blow-by.

*Prevention of oil pumping in the case of excess oil consumption is essential.* Figure 20.12 shows the various arrangements of preventing oil pumping with the help of piston rings. In one case, the upper edges of the piston rings are rounded off (Fig. 20.12a). In the second case, the lower groove edge of the last piston ring is beveled off and small holes are drilled towards the inside of the piston to assist in removal of the excess oil (Fig. 20.12b). In third case, an under-cut is provided in the last ring with small holes drilled towards the inside of the piston (Fig. 20.12c).

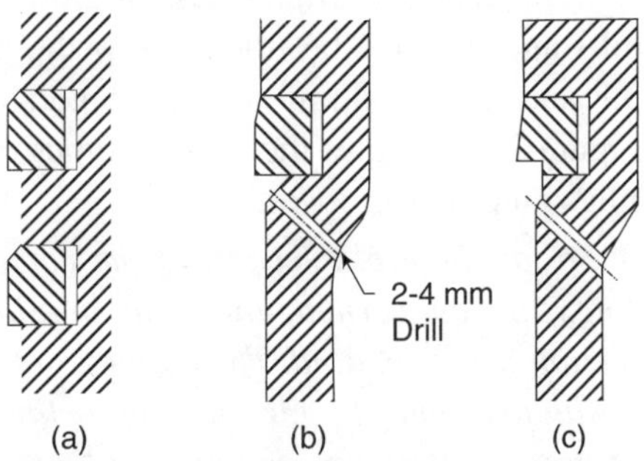


**Fig. 20.12** Methods of preventing oil pumping by piston rings

### *20.6.7.1 Piston Ring Materials*

*The piston ring materials must possess good wearing qualities and spring characteristics even at high temperature.* Following are the common materials for piston rings:

(i) Cast iron

(ii) Alloy cast iron having ultimate strength varying from 175 to 350 N/mm$^2$ depending on the composition.

(iii) For high output aircraft engine, heat treated high carbon cast iron alloy containing nickel and molybdenum having tensile strength varying from 425–700 N/mm$^2$.

(iv) To have gas-tight seal, high stresses are used in piston ring design to obtain high wall pressure, especially for high engine speed. Such high ring tension increases the metal to metal friction. To avoid friction, composite rings are used in which soft metal inserts made of bronze are used on the outer surface of ring.

*In general, to avoid wear of the liner and the ring, the rings are* ***chrome plated.*** In some engines, bimetallic rings are used to increase the strength and to reduce probability of ring fracture. The steel ring is brazed to the inner surface of the cast iron ring (Fig. 20.13).

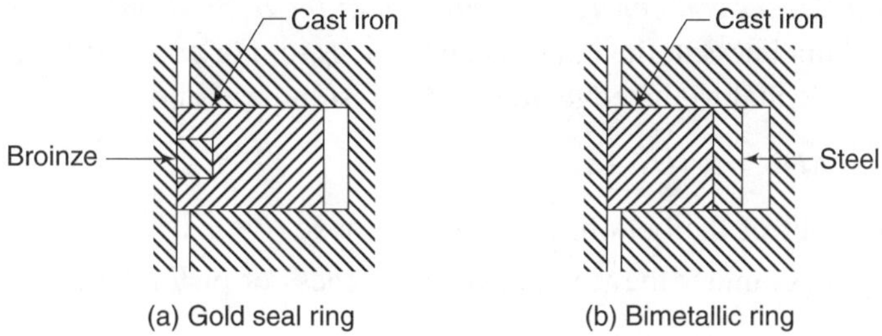


**Fig. 20.13** Composite and bimetallic rings

### *20.6.7.2 Design Procedure of Piston Rings*

The radial thickness of the ring is a function of designed wall pressure. The following equation gives the radial thickness

$$t = D\sqrt{\frac{3\,p_w}{\sigma_t}} \qquad ...(20.25)$$

where $t$ = radial thickness of the ring, mm

$p_w$ = wall pressure, N/mm$^2$

$\sigma_t$ = allowable stress in bending, N/mm$^2$

The desired wall pressure exerted by the ring may vary from 0.0245 to 0.042 N/mm$^2$.

In order to avoid excessive stress where slipping the ring into the groove, the allowable stress is taken very high in the range of 84 N/mm$^2$ to 112 N/mm$^2$ for cast iron piston rings

The axial width ($a$) may be taken as

$$a = 0.7\,t \text{ to } t \qquad ...(20.26)$$

The axial width ($a$) is also calculated by empirical relation

$$a = \frac{D}{10\,z} \text{ where } z = \text{number of rings} \qquad ...(20.27)$$

The width of the ring land ($h_1$) may be taken as

$$h_1 = 0.75\,a \text{ to } a \qquad ...(20.28)$$

The width of the top land is made larger than the other ring lands to protect the top land from the high temperature conditions existing at the top of the piston. Thus,

Width of the top ring land = $h_i = \delta$ to $1.2\delta$

$$= 0.03\,D \text{ to } 0.05\,D \quad \text{for SI engines} \qquad \text{...(20.29)}$$

### 20.6.8 Piston Barrel

*Based on the ideal heat transfer condition, the thickness of the piston barrel should be equal to the crown thickness tapering down to zero at the bottom end.* But based on strength consideration, wall thickness at the bottom cannot be zero. For equal heat transfer area under the crown should be equal to the edge area of the piston head and, thus,

$$\pi D\delta = \pi\,(D_1 - s)s$$

where $D_1$ = diameter of the bottom of the ring groove

$s$ = thickness under the ring above = thickness of piston barrel

$$\therefore \qquad s = \frac{D_1}{2} + \sqrt{D_1^2 - 4\,D\delta} \qquad \text{...(20.30)}$$

Here $D_1 = D - (2t + 0.006\,D + 0.5 \text{ mm})$ at compression rings

and $D_1 = D - (2t + 0.006\,D + 1.5 \text{ mm})$ at oil grooves

It is also possible to express '$s$' empirically,

$$s = (0.05 \text{ to } 0.10)\,D \text{ for SI and CI engines} \qquad \text{...(20.31)}$$

Take $b$ = depth of the ring groove, about 0.4 mm larger than $t$.

### 20.6.9 Piston Skirt

As the general terminology of skirt implies, it refers to the bottom portion. Thus, the position of the piston barrel below the ring section up to the open end is known as **skirt**. *It is the skirt which takes the side thrust of the connecting rod. This suggests that its length is an important parameter to take side thrust and its length should be such that the side thrust pressure should not exceed 0.21 to 0.28 N/mm² for low speed and upto 0.49 N/mm² for high speeds.*

The normal side thrust ($R$) is given by

$$R = (0.03 \text{ to } 0.10)\,F_{max} = D\,h_5\,p_R \qquad \text{...(20.32)}$$

where $F_{max} = \frac{\pi}{4} D^2 \times p_{max}$

$h_s$ = length of skirt, mm

$p_R$ = side pressure N/mm²

Length of piston, $H$ = Length of skirt + Length of ring section + Top land

As per empirical equation,

$$H = (0.8 \text{ to } 1.3)\, D \text{ for SI engines}$$
$$(1.0 \text{ to } 1.7)\, D \text{ for CI engines}$$

It is worth to note that length of position determines the side thrust on liner. A longer piston gives less piston slap but increases the reciprocating weight, so a compromise is made.

Thickness of the skirt wall = $\delta_s$ = 1.5 to 4.5 mm for SI engines

= 2.0 to 5.0 mm for CI engines ...(20.33)

*The shape of combustion chamber may be selected from the Articles of Types of combustion chambers for SI and CI engines discussed earlier.*

### 20.6.10 Piston Pin or Gudgeon Pin

*Piston pin (or gudgeon pin or wrist pin) is the most important link which links between piston and connecting rod. Whole of the gas load developed in the cylinder as a result of combustion passes through the piston pin.* There are three types of connections between piston pin and connecting rod namely ***full floating, semi-floating and fixed***.

- **(i) Full Floating.** In this arrangement, pin floats both in the piston boss and the small end bush against end movements (Fig. 20.14a) to ensure that the ends of the pin do not touch and score the liner.
- **(ii) Semi-floating.** In this arrangement, the pin is free to move either in the piston boss but clamped in the small end of the connecting rod (Fig. 20.14b) or is free to move in the small end of the connecting rod and secured to the piston bosses by means of a screw.
- **(iii) Fixed.** As the name implies, in this arrangement, the pin is locked in the piston of both the ends. The disadvantage of this arrangement is the confinement of whole of the swinging motion to the small end and uneven wear of piston pin.

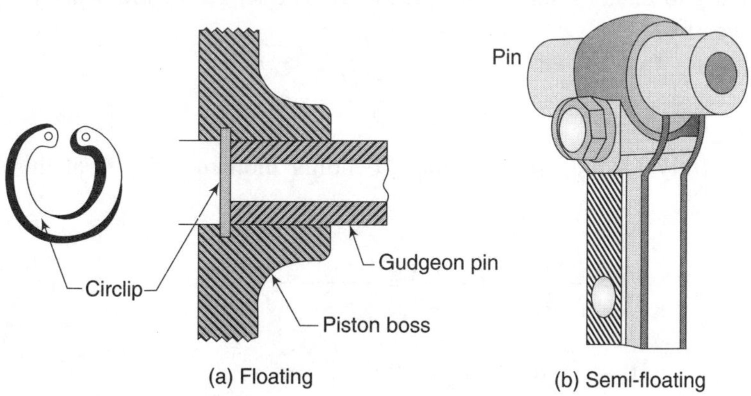


**Fig. 20.14** Floating and semi-floating piston pin

#### 20.6.10.1 Piston Pin Materials

*The piston pin is subjected to the effect of* ***alternating loads*** *resulting in stresses of* ***bending shear, bearing and ovalization***. Because of this high strength and toughness requirements, the following materials are used.

(i) Case hardened or nitrogen hardened steel alloy containing nickel, chromium, molybdenum or vanadium having tensile strength varying from 700 to 910 N/mm$^2$.

#### 20.6.10.2 Piston Pin Design Procedure

**Direct Method**

*Since the piston pin is heavily loaded by alternating loads so it should be designed for the maximum gas load or the inertia force of piston whichever is larger. The* maximum stresses in the piston pins of carburettor (now MPFI) engines occur when engines are operating at a maximum torque while in diesel engines the maximum stress occur when running at rated conditions. The piston pin bearing area should be roughly equally divided between the piston pin boss and the connecting rod bearing. The design criterion is the bearing pressure and the diameter of the pin is determined by *equating the gas load and bearing load.*

Thus, $$l_{cr}\, d_p\, p_b = \frac{\pi}{4} D^2\, p_{g\max} \qquad ...(20.34)$$

where $l_{cr} = l_{se}$ length of the piston pin in the connecting rod small end bearing (about 0.45 $D$)

$d_p$ = diameter of the piston pin; $p_b$ = bearing pressure

Generally, a high value of bearing pressure is used about 15 N/mm$^2$ for large pistons to about 31.5 N/mm$^2$ for smaller pistons.

It is essential to check in bending with the gas load theoretically taken as uniformly distributed (though in actual, it is not) over the length $l_p$, and the pin supported at the centre of the bosses at two ends as shown in Fig. 20.15. The length of the beam ($l_p$) between the supports is given by

$$l_p = l_{cr} + \frac{D - l_{cr}}{2} = \frac{l_{cr} + D}{2}$$

Due to uniformly loaded, the maximum bending moment will be at the centre of the pin.

$$M_b = \frac{F_{g\max}}{2}\left(\frac{l_p}{2} - \frac{l_{cr}}{2}\right) = \frac{F_{g\,max} \cdot D}{2} \qquad ...(20.35)$$

Also, $$M_b = \frac{\pi}{32} d_p^3\, \sigma_b \qquad ...(20.36)$$

*In order to reduce the weight of the pin, it is usually made hollow, with outside ($d_p$) and inside diameter, $d_{ip}$*

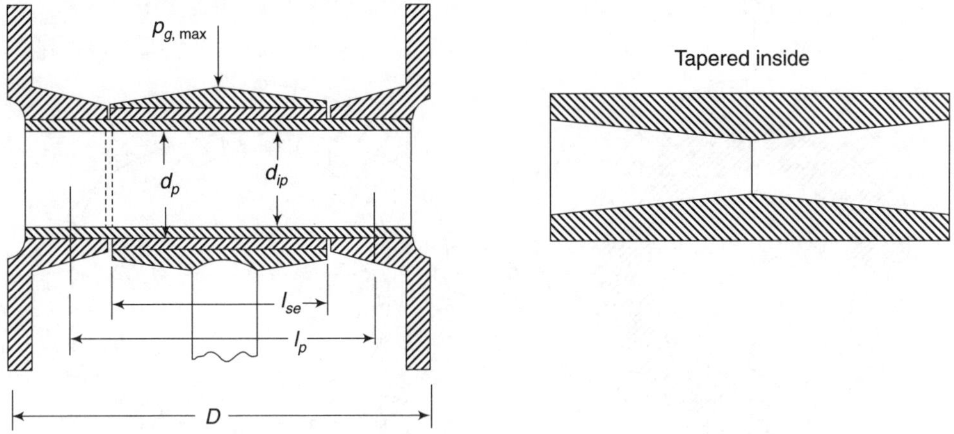


**Fig. 20.15** Piston pin

$$M_b = \frac{\pi}{32} \times \frac{d_p^4 - d_{ip}^4}{d_p} \times \sigma_b \qquad ...(20.37)$$

where $\sigma_b$ = allowable stress in bending = 84 N/mm$^2$ for case hardened carbon steel
= 140 N/mm$^2$ for heat treated alloy steel, higher value for special alloy steel.

*It is worth to note that the connecting rod bearing is usually of high grade bronze. Piston pin holes in all aluminium alloy pistons are super finished while all piston pins are lapped to a mirror finish. These surface finish requirements for piston pins and their holes are a must.*

**Indirect Method**

*Figure 20.16 shows the load and stress distribution on the piston pin. The indirect method of design includes the selection of the piston pin dimension from Table 20.2 and then to check the stresses whether they are safe or not.*

Piston pin analysis includes determination of the pin specific pressures on the small end bushing and on bosses, and also the stresses caused by bending shear and ovalization.

The total force ($F$) acting on the piston pin is given by

$$F = p_{z,\, max} . Ap + k . F_i \text{ ; MN} \qquad ...(20.38)$$

where for carburettor engine (now MPFI), $p_{z\ max}$ is the maximum gas pressure when operating at the maximum torque in MPa

$k$ = 0.76 to 0.86, a factor accounting for the mass of the piston pin

$F_i$ = Inertia force of the piston assembly

$= - m_p\, \omega^2\, r\, (1 + \lambda) \times 10^{-6}$ at $N = N_t$ in MN

$A_p$ = area of the piston pin

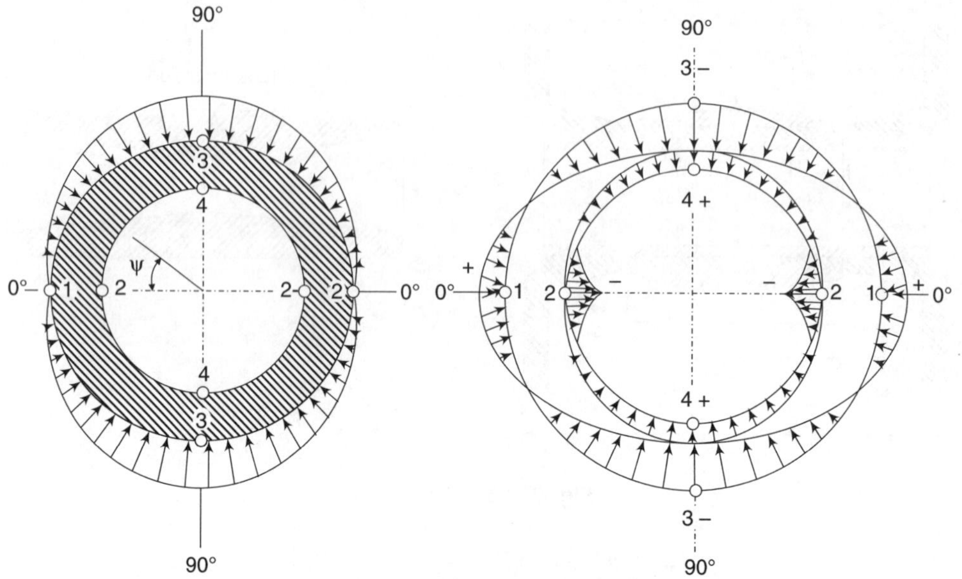

**Fig. 20.16** Load and stress distribution on piston pin

For diesel engine, $p_{z\,max}$ is the maximum gas pressure at rated condition in MPa

$k$ = 0.68 to 0.81, a factor according the mass of the piston pin

$F_i$ = inertia forces of the piston assembly

$= - m_p\, \omega^2\, r\,(1 + \lambda) \times 10^{-6}$ at $N = N_0$ in MN

The specific pressure exerted by the piston pin in (MPa) on the small end bushing of connecting rod is given by

$$q_{cr} = \frac{F}{d_p\, l_b} \qquad ...(20.39)$$

where $d_p$ = outer diameter of the pin, m

$l_b$ = length of the pin bearing surface at the small end, m

The specific pressure exerted by a floating piston pin on the bosses is given by

$$q_b = \frac{F}{d_p\,(l_p - b)} \qquad ...(20.40)$$

where $l_p$ = overall length of the piston pin, m

$b$ = distance between the boss end faces, m

$l_p - b$ = length of pin bearing surface in the bosses, m

In modern automobiles and tractor engines,

$q_{cr}$ = 20 to 60 MPa

$q_b$ = 15 to 50 MPa

The lower limits are meant for tractor engines.

The bending stress expressed in MPa in the pistion pin, consequent upon the loading (Fig. 20.16) is given by

$$\sigma_b = \frac{F(l_p + 2b - 1.5\, l_b)}{1.2\,(1 - \alpha^4)\, d^3 p} \qquad ...(20.41)$$

where $\alpha = \dfrac{d_{ip}}{d_p}$

For automobile and tractor engines, $\sigma_b$ = 100 to 250 MPa.

The tangential stress expressed in MPa due to the pin shear in the section planes between the bosses and the connecting rod small end are given by

$$\sigma_t = \frac{0.85\, F(1 + \alpha + \alpha^2)}{[(1 - \alpha^4)\, d_p^2]} \qquad ...(20.42)$$

For automobile and tractor engines, $\sigma_t$ = 60 to 250 MPa for piston pin made of alloy steel. The lower values of $\sigma_t$ are meant for tractor engines.

On account of non-uniform distribution of forces to the piston pin, it is strained in operation in its cross section (ovalisation) resulting in different stresses along length and section.

It is worth to note that the maximum ovalisation of the piston takes place in middle which is the most strained portion. The maximum increase in the horizontal diameter, ($\Delta d_{p\max}$) is given by

$$\Delta d_{p\max} = \frac{1.35\, F}{E . l_p} \left(\frac{1 + \alpha}{1 - \alpha}\right)^3 [0.1 - (\alpha - 0.4)^3] \qquad ...(20.43)$$

where $E$ = modulus of elasticity of the pin materials

The value of $\Delta d_{p\max}$ should not exceed 0.02 to 0.05 mm.

The maximum ovalisation stress occurs on the pin's internal surface in a horizontal plane.

$$\sigma_{\psi = 0} = -\frac{15\, F}{l_p\, d_p}\left[0.19\, \frac{(1 + 2\alpha)(1 + \alpha)}{(1 - \alpha)^2\, \alpha} + \frac{1}{1 - \alpha}\right] [0.1 - (\alpha - 0.4)^3];\ \text{MPa} \qquad ...(20.44)$$

The $\sigma_{\psi = 0}$ should not exceed = 300 – 350 MPa

### 20.6.11 Piston Clearance

*Due to thermal expansion and distortion of piston, there is need of clearance between piston and the liner. The thermal expansion may be counteracted by making the piston more or less tapered and the amount of taper depends on the following factors:*

(i) Operating temperature

(ii) Relative expansion between piston and liner

(iii) Running clearance adopted

*Care should be taken in providing proper clearance as excessive clearance leads to piston slap and too less clearance leads to piston seizure.*

For aluminium alloy piston, the following clearances are recommended

(i) Automobiles : 0.0375 to 0.15 mm
(ii) Two- and three-wheelers : 0.0375 to 0.075 mm
(iii) Stationary oil engines : 0.0625 to 0.125 mm
(iv) Tractor : 0.075 to 0.185 mm

### 20.6.12 Troubles Encountered by Piston

*The pistons have to operate under severe conditions such as high pressure, high temperature, inertia forces, side thrust, etc. so the following troubles are encountered by pistons:*

(i) Seizure of piston in cylinder
(ii) Burning of the piston crown
(iii) Cracking of piston crown and side walls
(iv) Sticking of piston ring
(v) Breaking of piston rings
(vi) Excessive lubricating oil consumption

## 20.7 CONNECTING ROD ASSEMBLY

### 20.7.1 Function

*As mentioned, the connecting rod connects the piston and crankshaft. The main function of the connecting rod is to transmit the forces acting on the piston pin to the crank pin. Sometimes, in certain cases, its secondary function is to transmit the lubricating oil from crankpin to piston pin and then for splash of jet cooling of the piston crown.*

Figure 20.17 shows the usual form of connecting rod assembly used in internal combustion engines. *It consists of an eye at the small end for the piston in bearing, a long sank, a big end opening which is usually split to accommodate the crankpin bearing shells and connecting rod bolts.*

### 20.7.2 Cross-sectional Shape of the Connecting Rod

The following are the usual cross-sectional shap of the connecting rod:

(i) Rectangular
(ii) Circular
(iii) Tubular
(iv) I-section
(v) H-section

For low speed engines, usually rectangular or circular with flattened ends are used while for high speed engine, slightness of the connecting rod is a major criteria and for this either I-sections or H-sections are used.

### 20.7.3 Length of Connecting Rod

The parameters that decide the length of the connecting rod is its angularity with respect to the cylinder centre line, *which in turn, depends upon the ratio of connecting rod length and stroke (l/r).* The smaller the ratio, l/r, the greater is the angularity. This angularity produces a side thrust of the piston against the liner. This means side thrust and resulting wear increase

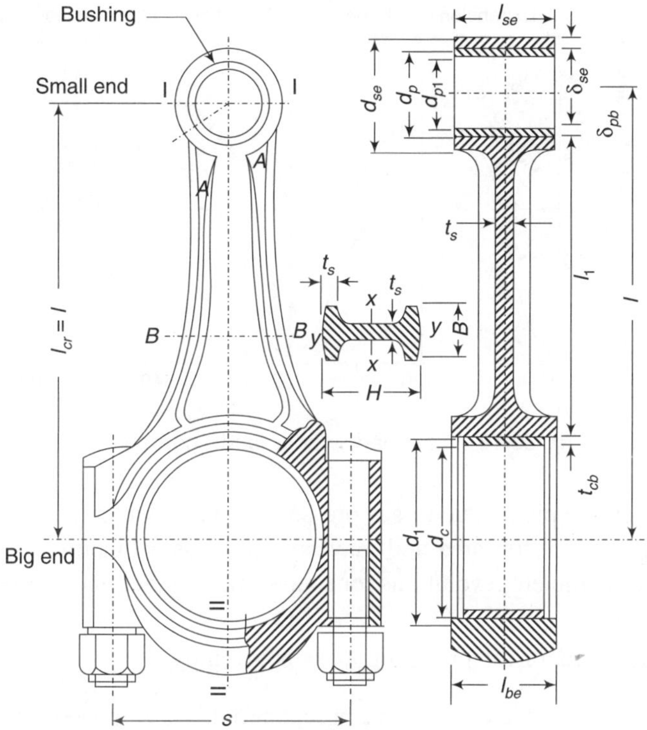


**Fig. 20.17** Connecting rod assembly

with angularity (i.e., with a decrease of $l/r$). However, the decrease of $l/r$ ratio decreases the height of the engine. ***In practice, l/r ratio is taken between 4 to 5.***

## 20.7.4 Stresses in Connecting Rod

The connecting rod is subjected to the following forces:

(a) *The combined effect of gas pressure on the piston and the inertia of the reciprocating masses*

(b) *The inertia of the connecting rod*

(c) *The friction of the piston rings and of the piston*

(d) *The friction of the piston and the crankpin bearing*

The combination of the above forces set up the stresses in the connecting rod.

### *20.7.4.1 The Load Due to Gas Pressure and Inertia of Reciprocating Mass*

The gas pressure is a function of type of the cycle (SI or CI), the crank angle, and other cycle parameters.

The gas pressure can be calculated with the help of an indicator diagram of the engine as a function of crank angle.

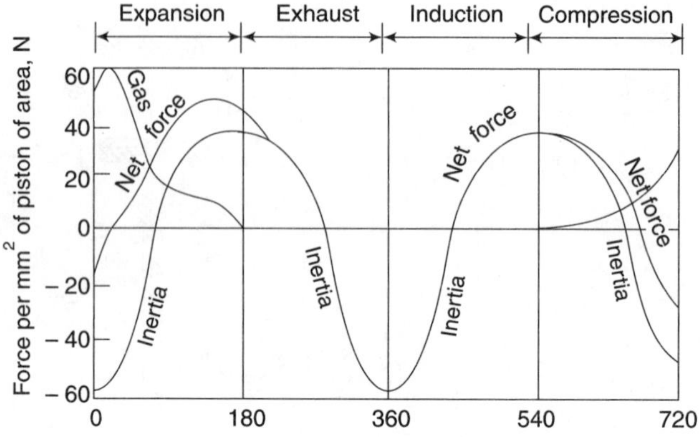


**Fig. 20.18** Gas load, inertia load and new load

The direct load on the piston due to gas pressure ($F_g$) is calculated by multiplying the gas pressure so obtained by piston area and may be plotted as a function of crank angle (Fig. 20.18) covering two complete revolutions for four-stroke engines and one revolution for two-stroke engine.

The load due to inertia of reciprocating mass is given by

$$F_i = \text{Mass of the reciprocating masses} \times \text{Acceleration}$$

or
$$F_i = m_r \, \omega^2 \, r \left( \cos \theta + \frac{r}{l} \cos 2\theta \right) \qquad ...(20.45)$$

where $m_r$ = mass of reciprocating masses, kg = $F_i/g$

= mass of piston including that of rings + mass of piston pin + mass of one-third portion of connecting rod (small end portion)

**For aluminium piston,**

Mass of reciprocating masses = $0.1245\ D^2N$

where $D$ = bore diameter, cm, $\omega$ = angular velocity of crank, rad/s

$\theta$ = crank angle from TDC $r$ = crank radius, m

$l$ = connecting rod length, m

A curve between inertia forces ($F_i$) and the crank angle ($\theta$) with the help of the above equation may be drawn preferably on the same ordinate as the gas load and then the net load is plotted as shown in Fig. 20.18.

It is worth to note that net vertical load ($F_{net}$) on the connecting rod will be the algebraic sum of the gas load and the inertia load.

$$F_{net} = F_g + F_i \quad \text{and} \quad \sigma_{net} = \sigma_g + \sigma_i$$

Due to angularity of rod, the actual load ($Q$) along the connecting rod will be more than the net load (Fig. 20.19) and this load is the maximum when the crank angle, $\theta = 90°$ as obvious from the following expression.

$$Q = \frac{F_{net}}{\cos\varphi} = \frac{F_{net}}{\sqrt{1 - \sin^2\varphi}} = \frac{F_{net}}{\frac{r}{l}\sqrt{(l/r)^2 - \sin^2\theta}} \qquad ...(20.46)$$

But at $\theta = 90°$, the gas load gets decreased considerably due to expansion of the gas. Therefore, *as a practice, the load on the connecting rod is taken corresponding to the maximum gas load.*

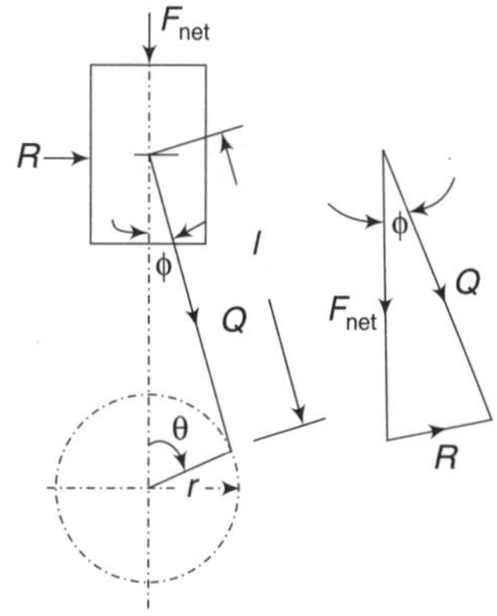


**Fig. 20.19** Force acting on connecting rod

### 20.7.4.2 Force due to the Friction of Piston Rings and of Piston

The friction force ($F_f$) due to piston ring may be given by

$$F_f = \pi\, D\, a\, z . p_w . \mu \qquad ...(20.47)$$

where $D$ = cylinder bore $\quad$ $z$ = number of piston rings

$p_w$ = wall pressure of rings (0.0245 to 0.042 N/mm$^2$)

$a$ = axial width of rings $\quad$ $\mu$ = coefficients of friction (0.1)

*The friction force produced by the piston is caused by the normal component of the piston pressure which varies from 3 to 10% of piston pressure.*

Due to lubrication, the *coefficient of friction may be taken about 0.06.* Thus, the friction force caused by piston is at the most 0.6 per cent of piston pressure which is negligible. *Thus, the force caused by friction due to piston rings and piston may be neglected for the design.*

### 20.7.4.3 Inertia of Connecting Rod

*The connecting rod is subjected to inertia force which has two components along the rod (i.e., longitudinal component) and normal to the rod (i.e., transverse component). The longitudinal component is taken into account by considering about one-third portion of connecting rod on the small end side as reciprocating mass and remaining two-thirds mass as rotating with the crank.* The inertia of reciprocating portion of connecting rod has been already considered with piston mass.

*The normal component of the inertia force will cause a centrifugal force to act on every part of the rod. This force tends to bend the connecting rod outwards, away from the centre line and thus causes bending stress which is alternating in nature.* For analysis purposes, the following assumptions are made.

(i) Connecting rod may be assumed to be of uniform section throughout its length

(ii) The bending force is zero at the piston pin and the maximum at the crank pin

(iii) The variation of bending force is assumed to be triangular (Fig. 20.20).

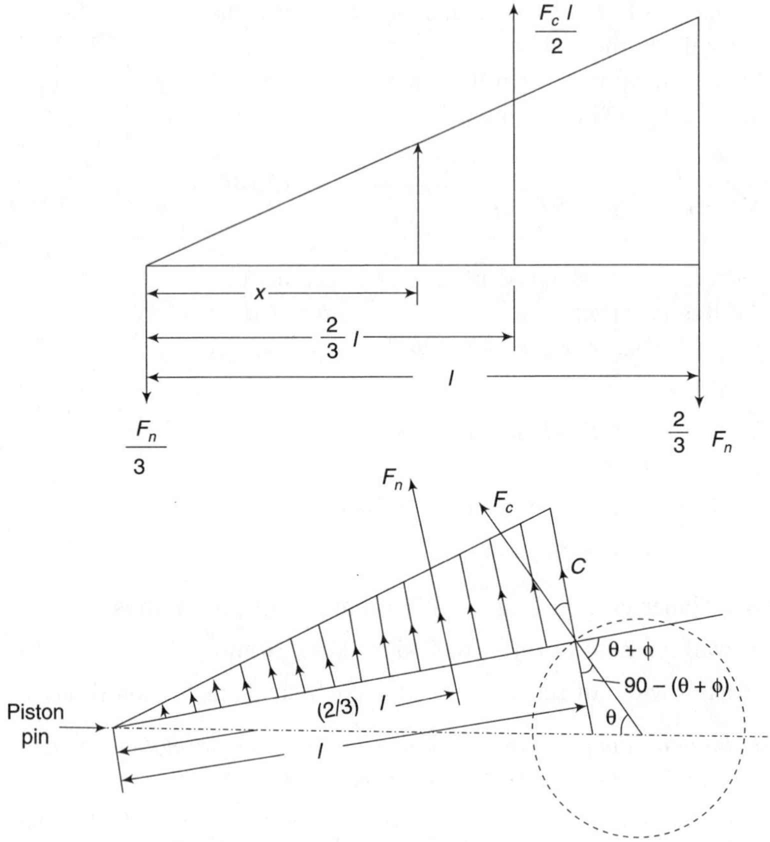


**Fig. 20.20** Force acting on connecting rod

From the above assumption, it is evident that the *bending stress will be the maximum when the crank and the connecting rod are at right angles.*

The centrifugal forces ($F_c$) acting on a unit length at the crankpin is given by

$$F_c = \rho\, A\, \omega^2\, r \sin (\theta + \phi) \qquad ...(20.48)$$

where $\rho$ = density of the materials, kg/m$^3$ $\quad A$ = cross-sectional area of rod, m$^2$

$\omega$ = angular speed, rad/s $\quad r$ = crank radius, m

$F_c$ will be the maximum when crank and connecting rod are at right angle to each other (i.e., $\theta + \varphi = 90°$) and its maximum value will be

$$(F_c)_{max} = \rho\, A\, \omega^2\, r \qquad ...(20.49)$$

The resultant normal inertia force ($F_n$) acting on the connecting rod is given by

$$F_n = (1/2)\, (F_c)_{\max} \cdot l \qquad ...(20.50)$$

and it will act at a distance of (2/3) $l$ from the piston pin.

*As per assumptions, one-third of the mass of the connecting rod is concentrated at the piston pin and two-thirds at the crankpin, so the reactions at the two bearings due to $F_n$ will be in the same ratio, i.e., (2/3) $F_n$ at crank pin and (1/3) $F_n$ at the piston pin (Fig. 20.20).*

Following the loading on the connecting rod, the bending moment on the connecting rod at a distance $x$ from the piston pin

$$M_x = \frac{F_n}{3}.x - \frac{1}{2}.\frac{(F_c)_{\max}.x}{l}.x.\frac{x}{3}$$

or

$$M_x = \frac{F_n.x}{3l^2}(l^2 - x^2)$$

For the maximum bending moment,

$$\frac{dM_x}{d_x} = 0 = \frac{F_n}{3} - \frac{3F_n.x^2}{3l^2} \quad \text{or} \quad x = \frac{l}{\sqrt{3}}$$

*This means that the maximum bending moment occurs at a distance $\frac{l}{\sqrt{3}}$ from piston pin.*

Now, the maximum bending moment is expressed as

$$M_{\max} = \frac{F_n}{3l^2}\frac{l}{\sqrt{3}}\left(l - \frac{l^2}{3}\right) = 0.128\ F_n.l$$

The maximum transverse bending stress is given by

$$(\sigma_b)_{\max} = \frac{M_{\max}}{Z} = \frac{0.128\ F_n.l}{Z} \qquad ...(20.51)$$

The crank angle, θ at which the maximum bending moment occurs is given by an empirical equation.

$$\theta = 90 - \frac{3500}{\left(\frac{l}{r} + 7.82\right)^2}$$

Thus, for $\frac{l}{r} = 4$, $\theta = 65°$ from TDC and for $\frac{l}{r} = 5$, $\theta = 68.7°$ from TDC

#### 20.7.4.4 Friction of the Two-End Bearing

Even with good lubrication, friction is produced in the bearing. *The effect of friction force at the two-end bearings is to bend the connecting rod into an "S" shape.* As a result of this tendency of friction force, the thrust along the rod gets displaced from the axis of section resulting in an eccentric loading. This loading will further increase the compressive stress due to the direct load. *The bending caused by bearing forces will be the maximum at the two ends where the shank of the connecting rod joins.*

### 20.7.5 Materials of Connecting Rod

As discussed above, during the engine's operation, the connecting rod is subjected to the alternating gas and inertial forces and sometimes these forces produce impact loads. *Thus, the materials chosen for connecting rod must be highly resistant to fatigue and should have adequate strength and stiffness with the minimum weight.* In general, the connecting rods of internal combustion engines are manufactured by **drop forging**. The following materials may be used for connecting rods.

(i) For industrial engines, **carbon steel** with ultimate tensile strength ranging from 550 to 670 MPa may be used.

(ii) For automobiles, **alloy steel** having a strength of about 750 to 940 MPa may be used. Manganese steel is more common.

(iii) For aero-engines, **nickel chrome steel** having ultimate tensile strength of about 940 to 1350 MPa may be used.

(iv) For low speed horizontal gas engines, **steel cast** may be used.

(v) For high speed engines, **duralumin and aluminium alloy** may be used

*In general, after drop forging, the outer surfaces are kept unfinished but in the case of high duty aero-engines, the outer surfaces are highly finished. In order to increase the fatigue strength after drop or press forming, the connecting rods may undergo for thermal treatment such as polishing, shot blasting, normalizing, hardening and tempering.*

### 20.7.6 Design Procedure of Connecting Rod

The above discussed stress analysis reveals that the maximum compressive load on the rod is the *algebraic sum of the gas load ($F_g$) and the inertia load due to reciprocating masses ($F_i$)*. From the plot of these loads on crank angle, it is also obvious that the effect of inertia load is to decrease the maximum compressive load. This load gets further reduced slightly by the piston ring and piston friction (though, it is negligible).

Further, there is a transverse (normal) component of bending stress on the connecting rod due to its inertia. The total stress in the connecting rod is the direct compressive stress due to net piston load ($\sigma_{net}$) and the maximum transverse bending stress, $\sigma_{b\ max}$ (i.e., whipping stress) and is given by

$$\sigma = \sigma_{net} + (\sigma_b)_{max}$$

*It has also been established that the transverse bending stress is the maximum at about 65° to 68° from TDC where the gas load decreases considerably as it is maximum at TDC.* A plot of net and $(\sigma_b)_{max}$ against crank angle ($\theta$) reveals that the maximum stress due to combined effect occurs at about $\theta = 90°$ and its value is quite low as compared to the direct stress due to maximum gas load. Under this scenario, the usual procedure of designing connecting rod is to **design the connecting rod for the maximum gas load** with a reasonable **factor of safety (say 5)**. This wil give robust section which can withstand the maximum bending stress due to inertia of the connecting rod and bearing friction.

As discussed above, the maximum gas load produces alternating direct compression and tensile stress to compression and expansion strokes, so it is numerically much larger than tensile. This suggests that ***connecting rod should be designed as a strut***. Ranking Gordon formula for the buckling load as given below may be used in a section where radius to gyration with respect to rod length is comparably larger and I-section satisfies this.

$$\text{Buckling load} = \frac{\sigma_{cru} . A}{1 + a\left(\frac{l}{k}\right)^2}; \quad \text{N} \qquad ...(20.52)$$

where $\sigma_{cru}$ = ultimate crushing stress; N/m$^2$; $A$ = sectional area of strut, m$^2$

$l$ = equivalent length of strut, m;

$k$ = radius of gyration about axis of buckling, m

$a$ = constant = $\frac{1}{7500}$ for m.s., $\frac{1}{9000}$ for WI and $\frac{1}{1600}$ for CI

The buckling load is given by

Buckling load = Maximum gas load × Factor of safety

$$= \frac{\pi}{4} D^2 . p_{g\max} . \text{FOS} \qquad ...(20.53)$$

Factor of safety (FoS) for strut may be taken as follows:

- Light shock, FoS = 4.2 to 7.5
- Heavy shock, FoS = 6.75 to 15
- Steady load, FoS = 3.375 to 5

### 20.7.6.1 Sectional Area

With the help of Ranking Gordon formula, sectional area ($A$) of the rod may be calculated.

### 20.7.6.2 Section

As discussed, *I-section is the most widely used section for connecting rod* (Fig. 20.21). For buckling about the neutral axis (*x-x*), the strut is freely hinged because in the plane of motion the ends of the rod are direction free and so freely hinged at the piston pin and the crankpin. But in the plane perpendicular to the plane of motion for buckling about the axis *y-y*, the strut is fix ended due to the constraining effect of the bearings at the piston and crankpins. This means that for buckling about axis *y-y*, the rod is stronger (say four times as strong as for buckling about axis *x–x*). As a rule, the rod should be equally strong in both planes, i.e.,

$$4\, I_{yy} = I_{xx} \quad \text{or} \quad K_{yy}^2 = \frac{1}{4} K_{xx}^2 \qquad ...(20.54)$$

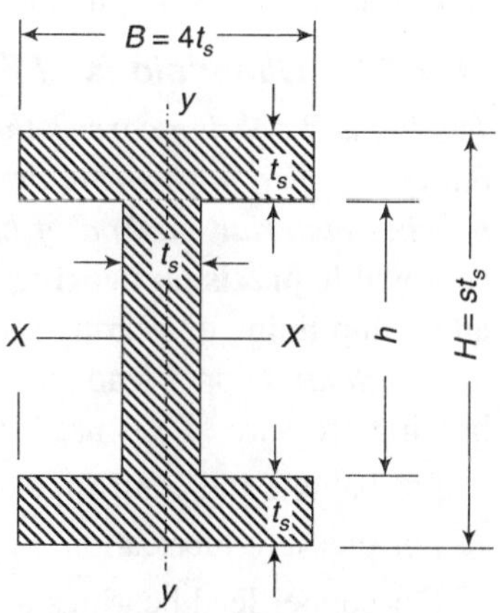


**Fig. 20.21** *I*-section of connecting rod

Let the flange and web thickness of section be $t_x$

Depth of the section = 5 $t_s$

Width of the section = 4 $t_s$

Area of the section = 11 $t_s^2$

$$I_{xx} = \frac{1}{12}(BH^3 - bh^3) = \frac{419}{12}t_s^4$$

$$I_{yy} = \frac{1}{12}(2t_s B^2 + ht_s^3) = \frac{131}{12}t_s^4$$

$$\therefore \quad K_{xx}^2 = \frac{I_{xx}}{A} = \frac{419\, t_s^4}{12 \times 11\, t_s^2} = 3.18\, t_s^2$$

$$\therefore \quad K_{yy}^2 = \frac{I_{yy}}{A} = \frac{131\, t_s^4}{12 \times 11\, t_s^2} = 0.955\, t_s^2$$

$$\therefore \quad \frac{K_{yy}^2}{K_{xx}^2} = \frac{1}{3.2}$$

The ratio $\frac{K_{yy}^2}{K_{xx}^2} = \frac{1}{3.2}$ is slightly higher than 1/4, so the proportion of the section, chosen are satisfactory. However, this section is to be modified avoiding sharp corners by the following way.

*Section so obtained by above-mentioned procedure is at the middle of the rod. Width is kept constant throughout the length connecting rod but the depth varies as follows:*

Depth at crank end = 1.1 to 1.25 × Depth of mid-point

Depth at the piston pin end = 0.75 to 0.9 × Depth at mid-point

### *20.7.6.3 Dimensions of Big End*

*The big end bearing carries the greatest load, i.e., the maximum gas load and thus the crank pin is designed for bearing pressure. The parameters that affect the bearing pressure are the bearing material, method of lubrication and space limitations.* In practice, the big end has **removable precision bearing shells** and the shells may be made of **bronze or steel or brass** and a thin lining of bearing metal (0.03 $d_c$ to 0.05 $d_c$). The *bearing metal may be tin or lead base babbitt,* copper lead, cadmium silver, etc. Owing to space limitations, a high allowable bearing pressure ($p_b$) is used. Following are the value of bearing pressure:

For babbitt and centrifugal force lubrication, $p_b \leq 7$ N/mm$^2$

For pressure lubrication, $p_b = 8.4$ to 10.5 N/mm$^2$

For copper lead bearings and pressure lubrication, $p_b = 12.5$ N/mm$^2$

Since the maximum gas load $(F_g)_{max}$ is carried by big end bearing, so

$$F_{g\max} = l_c . d_c . p_b \qquad \text{...(20.55)}$$

where $l_c$ = length of the crank pin, mm

$d_c$ = diameter of the crank pin, mm = inner diameter of the big end

$p_b$ = allowable bearing pressure, N/mm$^2$

Generally, $\frac{l_c}{d_c}$ ranges from 1.25 to 1.5

The values of $l_c$ and $d_c$ may also be calculated by the following

- $l_c$ = (0.6 to 0.85) $D$
- $d_c$ = (0.56 to 0.75) $D$

#### 20.7.6.4 Design of Small End

The design of piston pin has already been discussed. *The piston pin bearing is a high quality bronze bushing but it may also be made of aluminium alloy and its thickness 0.055 to 0.085 times the cylinder bore.* In the case of larger cylinder bore ($D$ > 0.15 m), the usual practice is to lock the bush against rotation by means of a setscrew. However, in some heavy duty engines, the *antifriction needle bearing is used.*

The small end bearing consists of an inner and outer race between which small diameter rollers are located and the piston pin acts as inner race.

The maximum gas force is given by

$$F_{g\max} = l_p . d_p . p_b \qquad ...(20.56)$$

where $= \frac{l_p}{d_p}$ = 1.5 to 2, $d_p$ = diameter of inner side

$p_b$ = 15 to 32 N/mm$^2$

*The dimensions of $d_p$ (= $d_{p0}$) and $l_p$ are also available from piston assembly design and any one may be selected.*

#### 20.7.6.5 Bolts for Big End Cap

*The maximum load acting on the bolts of big end cap is the inertia force at TDC on the exhaust stroke and it should form the design load. The compression and explosion loads due to gas pressure will be compressive in nature and the bolts and cap are unaffected by these.* In practice, the number of bolts ($z_b$) used for fastening the crank in bearing to the connecting rod may be two or four. The material used for connecting rod bolts should be nickel alloy steel having a tensile strength of 700 to 1050 MPa. *The factor of safety may be taken as 6 because it is subjected to repeated but not alternating stresses.* The allowable stress taken usually varies from 120 to 175 MPa for nickel alloy steel.

The inertia force is given by

$$F_i = m_r . \omega^2 \, r \left(\cos\theta + \frac{r}{l}\cos 2\theta\right) \qquad ...(20.57)$$

This $F_i$ is taken care by bolts, thus,

$$F_i = z_b . A . \sigma_t \qquad ...(20.58)$$

where $z_b$ = number of bolts; $A = \frac{\pi}{4} d_b^2$

$d_b$ = bolt core diameter; $m_r$ = mass of reciprocating masses

### 20.7.6.6 Connecting Rod Big End Bearing Cap

*The maximum load acting on the big end bearing cap is the inertia force at TDC on the exhaust stroke, so the bearing cap should be strong and as rigid as possible.* The cap acts as a beam freely supported at the bolt centres and may be assumed to be loaded in between uniformly distributed load and centrally.

The maximum bending moment is

$$M_b = \frac{F_i . s}{6}\ ;\ \text{Nm} \qquad ...(20.59)$$

where $F_i$ = inertia load

$s$ = distance between bolt centres

= diameter of the bearing + twice the thickness of bearing liner + diameter of the bolt + some clearance

Thickness of bearing liner = Thickness of the shell + Thickness of bearing metal

Thickness of bearing shell = 0.05 $D$

Thickness of bearing metal ≈ 1 to 1.5 mm

Also,

$$M_b = \sigma_b . Z \qquad ...(20.60)$$

where $Z = \frac{bc^2}{6}$

$b$ = width of cap = length of bearing, mm

$c$ = thickness of cap, mm

$\sigma_b$ = 120 N/mm$^2$ to 300 N/mm$^2$ depending upon speed

Tables 20.3 to 20.5 give the proportional dimension of connecting rod.

**Table 20.3** Small End Proportions

| S.No. | Description | SI Engine | CI Engine |
|---|---|---|---|
| 1. | Inner diameter of small end, $d_p$<br>without bushing<br>with bushing | <br>$d_p = d_{p0}$<br>(1.10-1.25) $d_p$ | <br>$d_p = d_{p0}$<br>(1.10–1.25) $d_p$ |
| 2. | Outer diameter of end, $d_{se}$ | (1.25-1.65) $d_p$ | (1.3–1.7) $d_p$ |
| 3. | Length of small end, $l_{se}$<br>retained pin<br>floating pin | <br>(0.28-0.32) $D$<br>(0.33-0.45) $D$ | <br>(0.28–0.32) $D$<br>(0.33–0.45) $D$ |
| 4. | Minimum radial thickness of end wall, $\delta_{se}$ | (0.16-0.27) $d_p$ | (0.16–0.27) $d_p$ |
| 5. | Radial thickness of bushing wall, $\delta_{pb}$ | (0.055-0.085) $d_p$ | (0.07–0.085) $d_p$ |

**Table 20.4** Big End Proportions

| S.No. | Descriptions | Proportions |
|---|---|---|
| 1. | Crank pin diameter, $d_c$ | (0.56-0.75) $D$ |
| 2. | Shell wall thickness, $t_s$<br>thin-walled<br>thick-walled | <br>(0.03-0.05) $d_c$<br>0.1 $d_c$ |
| 3. | Distance between connecting rod bolts, $s$ | (1.30-1.75) $d_c$ |
| 4. | Big end length, $l_{be}$ | (0.45-0.95) $d_c$ |

**Table 20.5** Connecting Rod Shank Proportions

| S.No. | Dimensions of connecting rod shank | SI Engine | CI Engine |
|---|---|---|---|
| 1. | $t_{smin}$ | (0.5-0.55) $d_{p0}$ | (0.5-0.55) $d_{p0}$ |
| 2. | $t_s$ | (1.2-1.4) $t_{min}$ | (1.2-1.4) $t_{min}$ |
| 3. | B | (0.5-0.6) $t_s$ | (0.55-0.75) $t_s$ |
| 4. | H | (2.5-4.0) mm | (4.0-7.5) mm |

These Tables may be used to check design dimensions calculated from governing equations.

## 20.8 CRANKSHAFT

### 20.8.1 Function

As discussed earlier, *the function of the crankshaft is to transform a reciprocating motion of the piston into rotary motion or vice versa.* Crankshaft consists of the following:

(a) Shaft part revolving in the main bearings
(b) Crank pins to which the big ends of the connecting rod are connecting
(c) Crank webs or checks which connects the crank pins and the shaft part.

### 20.8.2 Classification of Crankshaft

The crankshafts may be classified on the basis of the following:

(a) *Position of crank.* On the basis of crank position, the crankshaft may be classified as:
   (i) Crankshaft with side crank or overhung crank (Fig. 20.22)
   (ii) Crankshaft with centre crank (Fig. 20.22)

(b) *Number of cylinders.* On the basis of number of cylinders, crankshafts may be classified as:
   (i) Single cylinder crankshaft
   (ii) Multi-cylinder crankshaft

A crankshaft can be manufactured with two side cranks on each end or with two or more centre cranks. A crankshaft with only one side crank is called a ***single throw crankshaft*** and

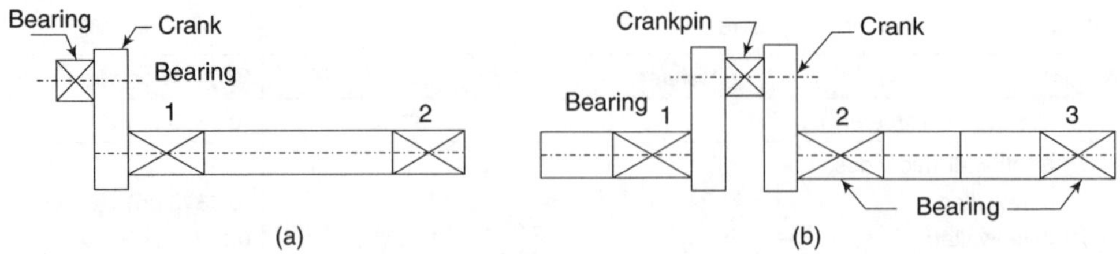


**Fig. 20.22** Types of crankshaft

it is used in *single cylinder engines*. The one with two side cranks or two or more centre cranks is known as ***multi-throw crankshaft*** and is used in *multi-cylinder engines*.

So far the application of overhung is concerned, it is mainly used for medium size and large horizontal engines. The need of only two bearings is its main advantage. The main disadvantage is the danger of its failure, however, the chance of its failure in less in three bearings. *This suggests that the number of bearings is one of the main parameters in the design of crankshaft. Crankshaft of large engines generally have a bearing between each pair of cranks and a bearing on each end shaft. In automobiles, the number of bearings should be less to make engine light in weight and short in length.*

Further, the throws, orientation and number of bearings are decided by the arrangement of the engine cylinder such as in line, *V* type or *W*, etc.

### 20.8.3 Materials and Their Requirements

*The crankshaft is the most complicated and strained engine part subjected to cyclic loads due to gas pressure, inertial forces and their couples. These forces and moment cause considerable stresses of torsion, bending, tension and compression in the crankshaft materials. The type of loading is fatigue one.*

The materials of crankshafts should fulfil the following requirements:

(i) It should have sufficient strength to withstand the bending and twisting moments.

(ii) It should have sufficient rigidity to keep the distortion a minimum.

(iii) It must possess enough stiffness to resist the stresses due to torsional vibrations of the shaft.

(iv) It should have properly distributed mass to reduce vibration well below the critical limit of the shaft.

(v) It should have sufficient projected area of crankpins and journals to keep the bearing pressure.

(vi) It should have minimum weight especially for aero-engines.

(vii) It should be made much heavier and stronger than necessary from the strength point of view to meet the requirements of rigidity and vibrations.

The following **materials are** *used for crankshaft.*

(i) For industrial engines either **0.35 carbon steel** of ultimate tensile strength of about 500 to 525 MPa or **0.45 carbon steel** of ultimate tensile strength of about 625 to 750 MPa is commonly used.

(ii) For transport engines, ***alloy steel*** (i.e., *manganese)* having ultimate strength of about 785 to 940 MPa may be used.

(iii) For aero-engines, ***nickel, chromium steel*** having ultimate tensile strength of about 940 to 1100 MPa may be generally used.

Sometimes, for low speed engines, heavy duty cast iron or cast steel may be used as the materials of crankshaft.

### 20.8.4 Manufacturing of Crankshaft

*Since the crankshaft is the most critical part of an engine, so great care must be taken in its manufacturing.* Small crankshafts are manufactured from **drop forging** *while larger crankshafts are* ***first forged and then machined to shape.*** In the case of cast crankshaft, casting is done in permanent moulds with maximum accuracy and minimum of machining. During machining of shaft, special care should be taken to avoid springing.

*The size of the journal and crankpin need accuracy and so they are grounded to exact size.*

*After completion of manufacturing, the crankshafts are statically and dynamically balanced. During balancing either extra weight is embedded to or removed from the crankshaft as per need.*

### 20.8.5 Stresses in Crankshaft

*The crankshafts are subjected to cyclic loads due to gas pressure, inertia forces and their couples which induce bending and also shear stresses due to torsional moments.* Failure of the crankshafts have been reported due to progressive fracture caused by repeated bending or reversed torsional stresses. This shows that the type of loading on the crankshaft is **fatigue loading** and therefore *endurance limit of the material should form the basis for design.*

In order to avoid the stress concentration and fatigue, abrupt changes in the section of crankshaft should be avoided and two different cross sections must be blended with a **large fillet radius**, $r$ ($r \geq 0.2d$, Fig. 20.23). In the case of space limitations, the crank web should be under cut to obtain the fillet but this makes the web weak and to compensate this weakness, the width of the web is increased. For the safety of crankshaft, a high value of ***safety factor***, *of the order of* ***4 to 5****, based on endurance limit* consisting of reversal of bending and torsional stresses may be taken.

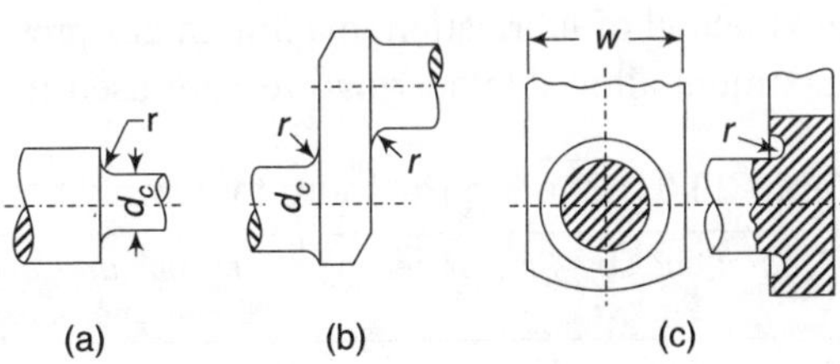


**Fig. 20.23** Provision of fillets in crankshaft

The endurance limit ($EL$) of the following materials is as follows:

(i) Carbon steel and cast steel:

$EL$ = 225 MPa in bending and 124 MPa in shear

(ii) Chrome nickel and other alloy steels:

$EL$ = 525 MPa in bending and 290 in shear

(iii) Alloy cast iron:

$EL$ = 140 MPa in bending and 140 MPa in shear

The allowable stresses of the following materials used in the design of crankshaft are given below:

(i) Carbon steel :

Bending ($\sigma_b$) = 56.0 to 75.0 MPa

Shear ($\sigma_s$) = 31 to 42 MPa.

Combined stress = $\left(\frac{1}{2}\right)$ × Elastic limit in tension

(ii) Alloy cast iron :

Bending ($\sigma_b$) = 35 to 47 MPa

Shear ($\sigma_s$) = 35 to 47 MPa.

Combined stress = Elastic limit in pure tension

(iii) Chrome nickel and other alloy steels:

Bending ($\sigma_b$) = 130 to 175 MPa

Shear ($\sigma_s$) = 72 to 97 MPa.

Combined stress = (1/2) × Elastic limit in pure tension

### 20.8.6 Bearing Pressure

*Bearings, such as main, crankpin and piston pin are needed in the crankshaft, so the allowable bearing pressures are important for their design.* The parameters that affect the allowable bearing pressure are the journal velocity, change of direction of the bearing pressure amount and method of lubrication, maximum gas pressure and space limitation. Table 20.6 shows the maximum allowable bearing pressure used in the design of crankshaft.

**Table 20.6** Bearing Pressure for Crankshafts

| *Type of engine* | *Kind of bearing* | *Maximum allowable bearing pressure (MPa)* | *Minimum* $\frac{Zn}{p}$ | *Clearance c/d* | *l/d ratio* |
|---|---|---|---|---|---|
| Automotive and aircraft engines | Main | 5.62-11.95 | 2.13 | 0.001 | 0.6-1.8 |
| | Crankpin | 10.5-24.5 | 1.42 | 0.001 | 0.6-1.4 |
| | Piston pin | 26.0-35.0 | 1.14 | 0.001 | 1.5-2.0 |
| Four stroke gas and oil engines | Main | 5.0-8.5 | 2.85 | 0.001 | 0.6-2.0 |
| | Crankpin | 10.0-14.0 | 1.42 | < 0.001 | 0.6-1.5 |
| | Piston pin | 12.5-17.0 | 0.86 | < 0.001 | 1.5-2.0 |
| Two stroke gas and oil engines | Main | 3.5-5.5 | 3.6 | 0.001 | 0.6-2.0 |
| | Crankpin | 7.0-14.0 | 1.72 | < 0.001 | 0.6-1.5 |
| | Piston pin | 8.5-12.5 | 1.14 | < 0.001 | 1.5-2.0 |

## 20.8.7 Design Procedure

*Design philosophy of the crankshaft assumes it as a beam with two or more supports. It is essential to check the design against failures at least for two crank positions, one when the bending moment is maximum and the other when the twisting moment is maximum. Over and above, the additional moment due to the flywheel weights, belt tension and other forces must be considered.*

The design procedure involved include the determination of the magnitudes of the various loads acting on the crankshaft, distance between supports, and other necessary dimensions of the crankshaft.

It is worth to discuss the dynamic analysis of crankshaft which will help in its designing. Dynamic analysis of crankshaft for multi-cylinders are out of scope of this book.

### *20.8.7.1 Analysis of Centre Crankshaft for Single Cylinder*

**20.8.7.1.1 Crank at Angle of Maximum Bending (Dead Centre)** *When crank is on dead centre, the maximum bending moment will act on the crankshaft due to the maximum gas load and the resulting force diagram* is shown in Fig. 20.24.

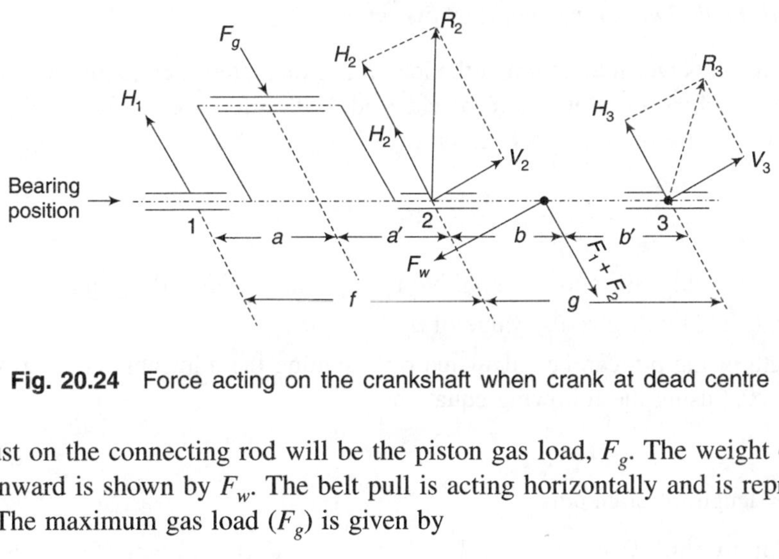


**Fig. 20.24** Force acting on the crankshaft when crank at dead centre

The thrust on the connecting rod will be the piston gas load, $F_g$. The weight of flywheel acting downward is shown by $F_w$. The belt pull is acting horizontally and is represented by $(F_1 + F_2)$. The maximum gas load $(F_g)$ is given by

$$F_g = \frac{\pi}{4} D^2 . p_{g\max} \quad ...(20.61)$$

This $F_g$ will cause two horizontal reactions $H_1$ at bearing 1 and $H_2$ at bearing 2, so that

$$H_1 = H_2 = \frac{F_g}{2} \quad \text{if} \quad a = a'$$

and

$$H_1 = \frac{F_g . a}{f} \quad \text{and} \quad H_2 = \frac{F_g . a'}{f} \text{ if } a \neq a' \quad ...(20.62)$$

where $f$ = the centre distance between two main bearings.

$a = a'$ = the centre distance between main bearing and big end journal bearing

The weight of the flywheel, $F_w$ will result in two vertical reactions, namely, $V_2$ and $V_3$ at bearings 2 and 3 respectively and given by

$$V_2 = V_3 = \frac{F_w}{2} \quad \text{if} \quad b = b' = \frac{g}{2}$$

and $$V_2 = \frac{F_w}{g} b' \quad \text{and} \quad V_3 = \frac{F_w}{g}.b \quad \text{if} \quad b \neq b' \qquad ...(20.63)$$

Further, the belt pull $(F_1 + F_2)$ will result in two horizontal reactions, $H_2$ and $H_3$ at bearing 2 and 3.

$$H_2' = H_3 = \frac{F_1 + F_2}{2} \quad \text{if} \quad b = b' = \frac{g}{2}$$

and $$H_2' = \frac{F_1 + F_2}{g}.b' \quad \text{and} \quad H_3 = \frac{F_1 + F_2}{g}.b \qquad ...(20.64)$$

*It is worth to note that when the crank at dead centre, twisting moment will be absent and the shaft will be designed for bending stress only.*

**(i) Crankpin.** It is worth to note that the length and diameter of crankpin may be calculated by connecting rod theory. But it is to be checked. With reference to Fig. 20.23, the bending moment at the centre of crankpin is given by

$$M_b = H_1.\frac{f}{2} = \frac{\pi}{32} d_c^3.\sigma_b \qquad ...(20.65)$$

where $\sigma_b$ = allowable stress in bending, N/m$^2$ $\quad d_c$ = diameter of the crankpin

Equation (20.65) will give the value of $d_c$.

The length of the pin can be calculated by assuming the allowable bearing pressure $(p_b)$ from Table 20.3 using the following equation

$$F_g = l_c.d_c.p_b$$

where $l_c$ = length of crankpin

**(ii) Left Hand Crank Web.** *The crank web is subjected to two types of stresses—one direct compressive stress and other bending stress due to gas load, $F_g$.* Generally, the thickness and width of web are calculated empirically and the stresses are checked whether these are in safe limit or not.

The thickness $(t)$ of the crank web can be given empirically by

$$t = 0.22\,D \quad \text{to} \quad 0.32\,D \qquad ...(20.66)$$

or $$t = 0.5\,d_c \quad \text{to} \quad 0.6\,d_c$$

or $$t = 0.65\,d_c + 6.35 \text{ mm}$$

The width ($w$) of the crank web may be given by

$$w = \frac{9}{8} d_c + 12.7 \text{ mm} \qquad ...(20.67)$$

The total stress ($\sigma_{\text{total}}$) on the web is given by

$$\sigma_{\text{total}} = \text{direct stress } (\sigma_d) + \text{Bending stress } (\sigma_b)$$

$$= \frac{H_1}{w.t} + \frac{H_1\left(\frac{f}{2} - \frac{t_c}{2} - \frac{t}{2}\right)}{\frac{1}{6} wt^4} \qquad ...(20.68)$$

As a check, the total stress, ($\sigma_{\text{total}}$) should not exceed the allowable stress in bending.

**(iii) Right Hand Crank Web.** In general, the bearings, 1 and 2 have the same length and are symmetrical to the cylinder centre line, therefore, $H_1 = H_2$. Thus, the dimensions of right hand crank web will be the same as that of left hand web.

**(iv) Shaft Under the Flywheel.** *The shaft under the flywheel is subjected to bending stress. The bending moment is caused due to flywheel weight and belt pull.* The bending moment due to flywheel weight ($F_w$) (Fig. 20.23) is given by

$$M_f = V_3 . \frac{g}{2} = \frac{F_w}{2} . \frac{g}{2} = \frac{F_w \cdot g}{2} \qquad ...(20.69)$$

The bending moment due to the belt pull is given by

$$M_b = H_3 . \frac{g}{2} = (F_1 + F_2) \frac{g}{4} \qquad ...(20.70)$$

It is worth to note that these bending moments act at right angles to each other, so the resultant bending moment is given by

$$M_b = \sqrt{M_f^2 + M_b^2} = \frac{\pi}{32} d_{sfw}^3 . \sigma_b \qquad ...(20.71)$$

where $d_{sfw}$ = diameter of the shaft under flywheel

$\sigma_b$ = allowable stress in bending

**20.8.7.1.2 Crank at Angle of Maximum Twisting Moment** *The maximum twisting moment on the crankshaft will occur when the tangential force, ($F_t$) is the maximum and this can be calculated graphically by taking pressure from the net effect diagram for different crank angle.* This angle usually falls between 25° and 30° from the dead centre for Otto cycle engine and between 30° to 40° for diesel engines. At this angle, the gas pressure is not the maximum.

Figure 20.25 shows the force diagram when the crank at an angle of the maximum twisting moment. If $F_g$ is the gas load acting along the cylinder centre line, then the thrust $Q$ along the connecting rod is given by

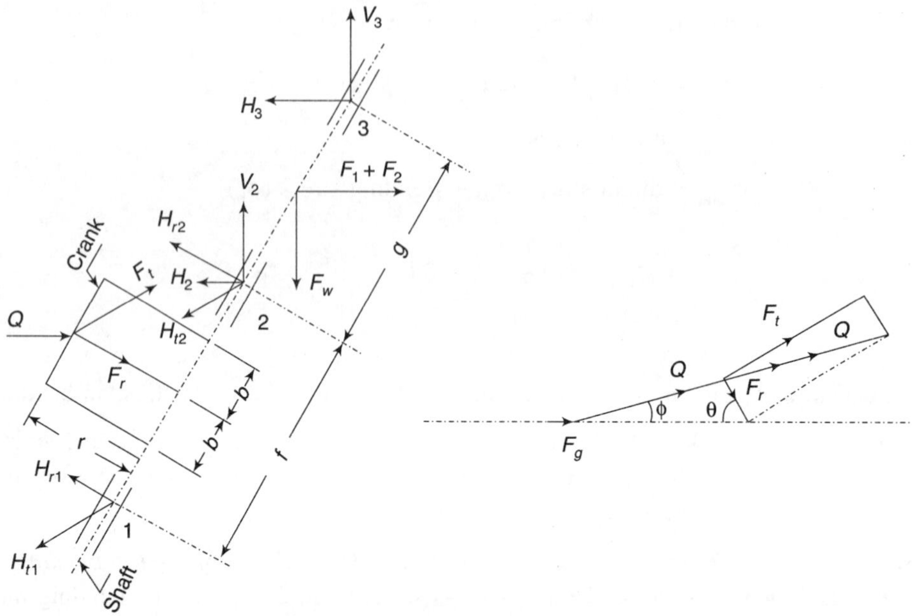


**Fig. 20.25** Force acting on the crankshaft when crank is at an angle of the maximum twisting moment

$$Q = \frac{F_g}{\cos \phi} \tag{20.72}$$

where $\phi = \sin^{-1}\left(\frac{\sin \theta}{l/r}\right)$

$l$ = length of connecting rod

$r$ = crank radius

The thrust, ($Q$) is resolved into two components, namely, tangential force ($F_t$) and the radial force ($F_r$) along the crank and are expressed as

$$F_t = Q \sin (\theta + \phi)$$

and $$F_r = Q \cos (\theta + \phi) \quad ...(20.73)$$

Further, $F_t$ will produce two reactions, $H_{r1}$ and $H_{r2}$ at bearings 1 and 2 respectively.

$F_r$ will also produce two reactions $H_{r1}$ and $H_{r2}$ at bearings 1 and 2 respectively.

But, the reactions at the bearing due to $F_w$ and ($F_1 + F_2$) will be the same as before, i.e., $H_2'$ and $H_3$.

At the position of crank at angle of the maximum twisting moment, the different sections of the crankshaft will be subjected to both bending and torsional moments and these must be checked for combined stress. *In practice, the failure due to shear stress may form the criteria for designing.*

**(i) Crankpin.** Referring to Fig. 20.24, the bending moment at the centre of the crankpin is given by

$$M_b = H_{r1} \cdot \frac{f}{2} \quad ...(20.74)$$

Twisting moment is given by

$$M_t = H_{r1} \cdot r \quad ...(20.75)$$

The equivalent twisting moment is given by

$$M_{te} = \sqrt{M_b^2 + M_t^2} = \frac{\pi}{16} d_c^3 . \sigma_s \quad ...(20.76)$$

where $d_c$ = crankpin diameter

$\sigma_s$ = allowable shear stress, N/m$^2$

Equation (20.76) gives the value of crankpin diameter ($d_c$) if other terms are known.

**(ii) Shaft under the Flywheel.** Referring to Fig. 20.24, the resultant bending moment on the crankshaft due to flywheel and the belt pull will be the same as when the crank is at the dead centre. This will give the diameter of shaft under flywheel ($d_{sfw}$).

The twisting moment is given by

$$M_t = F_t . r$$

∴ Equivalent twisting moment is given by

$$M_{te} = \sqrt{M_t^2 + M_r^2} = \frac{\pi}{16} d_{sfw}^3 . \sigma_s \quad ...(20.77)$$

Equation (20.77) will give the value of $d_{sfw}$ if other terms are known.

**(iii) Shaft at Juncture of Right Hand Crank Arm.** Referring to Fig. 20.24, the bending moment is given by

$$M_b = H_1' \left(\frac{f}{2} + \frac{l}{2} + \frac{t}{2}\right) - Q\left(\frac{l}{2} + \frac{t}{2}\right) \quad ...(20.78)$$

For symmetrical bearings, $H_1' = \dfrac{Q}{2}$

where $H_1'$ = total reaction at bearing 1 due to $Q$.

The twisting moment is given by

$$M_t = F_t . r \quad ...(20.79)$$

The equivalent twisting moment is given by

$$M_{te} = \sqrt{M_t^2 + M_b^2} = \frac{\pi}{16} d_2^3 \, \sigma_s \quad ...(20.80)$$

where $d_2$ = diameter of shaft at the junction of right hand crank.

Equation (20.80) gives the value of $d_2$ if other terms are known.

**(iv) Right Hand Crank Web.** Radial and tangential components of force $Q$, direct compression and torsion cause the bending stress in two planes, normal to each in the right hand crank web. This suggests that the design of the right hand crank web is to be checked against the total combined stress. Bending moment caused on right hand crank web due to radial component of $Q$ is given by

$$M_{br} = H_{r2}\left(\frac{f}{2} - \frac{l}{2} - \frac{t}{2}\right) \qquad ...(20.81)$$

Also
$$M_{br} = \sigma_{br}.Z = \sigma_{br}.\frac{1}{6}\,w.r^2 \qquad ...(20.82)$$

Equations (20.81) and (20.82) give the value of $\sigma_{br}$ if other terms are known.

The maximum bending moment due to tangential components of $Q$ occurs at the juncture of crank shaft and shaft and is expressed by

$$M_{bt} = F_t\left(r - \frac{d_2}{2}\right) = \sigma_{bt}\,Z = \sigma_{bt}.\frac{1}{6}\,tw^2 \qquad ...(20.83)$$

Stress developed in the right hand crank web due to direct compression is given by

$$\sigma_d = \frac{F_r}{2.w.t} \qquad ...(20.84)$$

The total compressive stress caused by $\sigma_{br}$, $\sigma_{bt}$ and $\sigma_d$ will be the maximum at the upper left corner to the cross section of crank and is given by

$$\sigma_{\text{comp}} = \sigma_{br} + \sigma_{bt} + \sigma_d \qquad ...(20.85)$$

The twisting moment on the area is given by

$$M_t = H_{t1}\left(\frac{f}{2} + \frac{l}{2}\right) - F_r.\frac{l}{2}$$

or
$$M_t = H_{t2}\left(\frac{f}{2} - \frac{l}{2}\right)$$

The shear stress caused by twisting moment is expressed as

$$\sigma_s = \frac{M_t}{Z_p} = \frac{M_t}{w.t^2/4.5} \qquad ...(20.86)$$

where $Z_p$ = polar section modulus = $wt^2/4.5$

The total combined stress is given by

$$\sigma_{c\max} = \frac{\sigma_{\text{comp}}}{2} + \sqrt{\sigma_s^2 + \left(\frac{\sigma_{\text{comb}}}{2}\right)^2} \qquad ...(20.87)$$

For safe design, it is essential that $\sigma_{c\max}$ must be within safe limits otherwise $w$ can be increased as other dimensions are not affected by $\sigma_{c\max}$.

**(v) Left Hand Crank.** Dimensions of the left hand crank may be safely taken as that of right hand crank as it is not severely stressed as that of right hand crank. This is because of the fact that right hand crank has to transmit power to flywheel while left hand does not transmit any power.

**(vi) Crankshaft Bearings.** Bearing 2 is heavily loaded as compared to 1, so bearing 2 is to be checked for the safe bearing pressure. The total pressure or bearing 2 is given by

$$p_{bt} = \frac{F_g}{2} + \frac{F_w}{2} + \left(\frac{F_1 + F_2}{2}\right) \qquad \text{...(20.88)}$$

In practice, distance between bearings 1 and 2 may be taken equal to twice the cylinder diameter.

### *20.8.7.2 Analysis of Simple Side Crankshaft for Single Cylinder*

Figure 20.26 shows the nomenclature used for the design of simple side crankshaft. The design procedure is on the same lines as for centre crankshaft. It is possible to estimate the dimension for the crankpin and the journal before checking the design of crankshaft for the position of the maximum bending moment and for the maximum twisting moment.

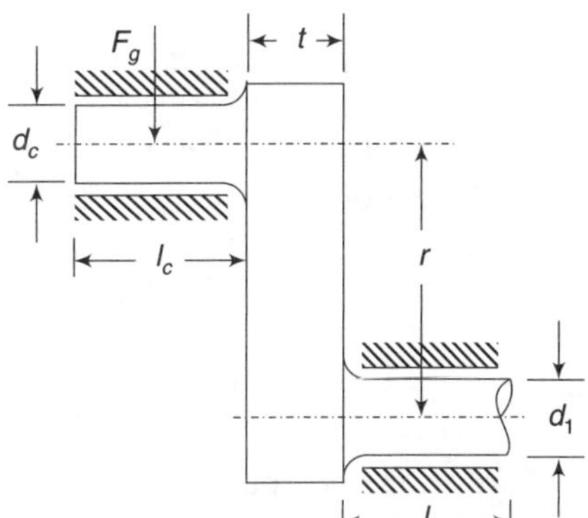


**Fig. 20.26** Nomenclature used in simple side crankshaft

**(i) Crankpin.** In normal practice, the dimensions of the crankpin are determined from the bearing and lubrication requirements and then checked for strength and deflection.

*The crankpins of the side crank engines are modelled as cantilever beams subjected to gas load, $F_g$.* It is possible that the load may not be evenly distributed over the crankpin length, the bending moment may be taken as $0.75\ F_g.l$ rather $0.5\ F_g.l$. The maximum bending stress will act at the junction of the pin and crankweb. Thus,

$$0.75\ F_g.l_c = \frac{\pi}{32}\, d_c^3.\sigma_b \qquad \text{...(20.89)}$$

where $\sigma_b$ = allowable bending stress

By assuming $l/d$ ratio ranging from 0.6 to 1.4, the diameter of the pin ($d_c$) and its length ($l_c$) may be calculated by Eq. (20.89) provided $\sigma_b$ is known. The bearing pressure, $p_b = F_g/l_c\ d_c$ may be checked.

**(ii) Journal.** *The journal is subjected to bending.* The bending moment at the middle of the journal is given by

$$M_b = F_g(0.75\, l_c + t + 0.54\, l_1) = \frac{\pi}{32} d_1^3 . \sigma_b \qquad ...(20.90)$$

where $t$ = crank web thickness = 0.45 $d$ to 0.75 $d$

$l_1 = 1.5\, d_c$ to $2d_c$

Equation (20.90) will give the value of the journal diameter, $d_1$

**20.8.7.2.1 Crank at Angle of the Maximum Bending (Dead Centre).** The forces acting on the crankshaft at an angle of the maximum bending which occurs at dead centre is shown in Fig. 20.27.

The gas load ($F_g$) causes two horizontal reactions at bearings 1 and 2 which are expressed as

$$H_1 = F_g . \frac{(a+b)}{b}$$

and $$H_2 = -F_g . \frac{a}{b} \qquad ...(20.91)$$

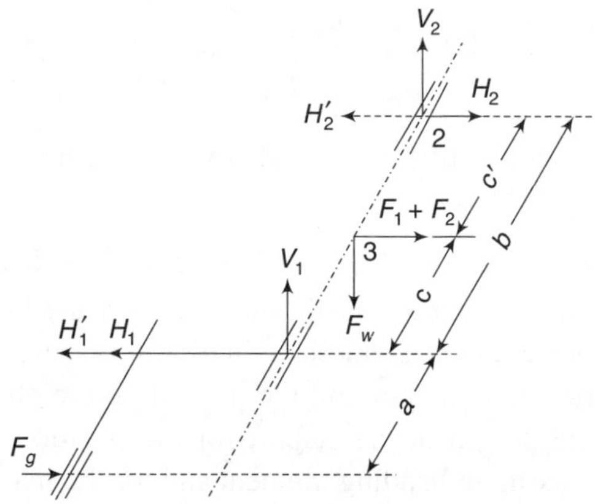


**Fig. 20.27** Forces at the side crankshaft on dead centre

The weight of flywheel ($F_w$) results in two vertical reactions which are given by

$$V_1 = F_w . \frac{c'}{b}$$

and $$V_2 = F_w . \frac{c}{b} \qquad ...(20.92)$$

Further, the belt pull ($F_1 + F_2$) causes two horizontal reactions which are expressed as

$$H_1' = (F_1 + F_2) . \frac{c'}{b}$$

and $$H_2' = (F_1 + F_2) . \frac{c}{b} \qquad ...(20.93)$$

**(i) Crank Web.** *When the crank is at dead centre, the crank arm is subjected to bending moment and to a direct compression.* Thus

$$\text{Bending moment} = M_b = F_g \left(0.75\, l_c + \frac{t}{2}\right)$$

The bending stress caused by this bending moment is given by

$$\sigma_b = \frac{M_b}{Z} \quad \text{where} \quad Z = \frac{1}{6} w t^2$$

The direct stress, ($\sigma_d$) is expressed by

$$\sigma_d = \frac{F_g}{w.t}$$

Thus, the total stress acting on the crank web is given by

$$\sigma_t = \sigma_b + \sigma_d \qquad ...(20.94)$$

The total stress ($\sigma_t$) should not exceed the safe limit for safe design.

**(ii) Shaft Under Flywheel.** This is subjected to bending moment caused by horizontal bending moment due to gas load and belt pull and the vertical bending moment due to the flywheel weight.

The bending moment due to gas load = $M_{bg} = F_g\,(a + c) - (H_1 + H_1')c$

The bending moment due to belt pull = $M_{bp} = (F_1 + F_2).b/4$

The total horizontal bending = $M_h = M_{bg} + M_{bp}$

The vertical bending moment = $M_v = F_g.b/4$

The effective bending is given by

$$M_c = \sqrt{M_h^2 + M_v^2} = \frac{\pi}{32}\,d_{sfw}^3\,w\,\sigma_b \qquad ...(20.95)$$

Equation (20.95) will give the diameter of the shaft under flywheel ($d_{sfw}$).

**20.8.7.2.2 Crank at Angle of Maximum Moment.** Figure 20.28 shows the various forces acting on the crankshaft when it is an angle of the maximum torsional moment. The tangential component, ($F_t$) and the radial component ($F_r$) of the connecting rod thrust ($Q$) can be determined like the case of centre crankshaft.

Radial component, $F_r$ of $Q$ gives two reactions at bearings 1 and 2 which are given as

$$H_{r1} = F_r.\frac{a+b}{b} \text{ and } H_{r2} = -F_r.\frac{a}{b} \qquad ...(20.96)$$

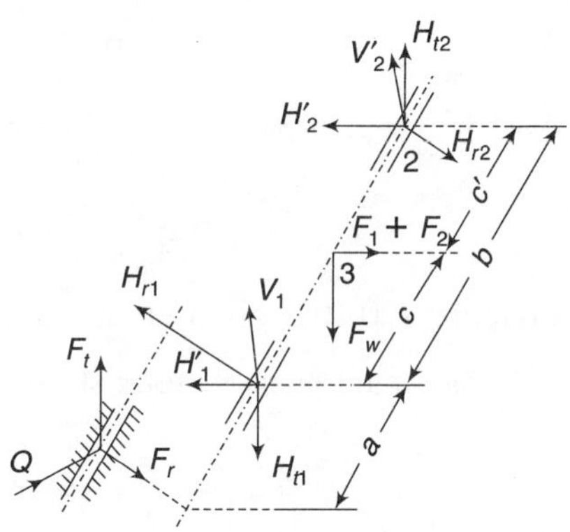

**Fig. 20.28** Forces on the crankshaft. At an angle of maximum torsional moment

The tangential component, $F_t$ of $Q$ gives two tangential reactions at bearings 1 and 2 which are expressed as

$$H_{t1} = F_t.\frac{a+b}{b} \quad \text{and} \quad H_{t2} = -F_t.\frac{a}{b} \qquad ...(20.97)$$

**(i) Crank Web.** *The crank web is the dangerous part of crankshaft and its most dangerous sections is where the web joins the shaft.* This sections is subjected to the following forces.

(a) bending stress due to $F_t$

(b) bending stress due to $F_r$

(c) direct compression stress due to $F_r$

(d) shear stress due to torsion

The bending moment due to $F_t$ is given by

$$M_{bt} = F_t\left(r - \frac{d_1}{2}\right) = \sigma_{bt}.Z = \sigma_{bt}.\left(\frac{1}{6}\,t.w^2\right) \quad ...(20.98)$$

Equation (20.98) will give the value of $\sigma_{bt}$.

The bending moment due to $F_r$ is given by

$$M_{bt} = F_t\left(0.75\,l_c + \frac{t}{2}\right) = \sigma_{bt}.Z = \sigma_{bt}.\left(\frac{1}{6}\,t.wt^2\right) \quad ...(20.99)$$

Equation (20.99) gives the value of $\sigma_{bt}$

The direct compression ($\sigma_d$) is caused by $F_r$ and is expressed by

$$\sigma_d = \frac{F_r}{w.t} \quad ...(20.100)$$

Thus, the total compression stress is given by

$$\sigma_c = \sigma_{bt} + \sigma_{br} + \sigma_d \quad ...(20.101)$$

The torsional moment is given by

$$M_t = F_t\left(0.75\,l_c + \frac{t}{2}\right)$$

And the shear stress, ($\sigma_s$) is given by

$$\sigma_s = \frac{M_t}{Z_p} \quad \text{where } Z_p = \text{polar section modulus} = \frac{w.t^2}{4.5} \quad ...(20.102)$$

Equation (20.102) gives the value of $\sigma_s$.

The maximum possible total stress caused by $\sigma_c$ and $\sigma_s$ is given by

$$\sigma_{max} = \frac{\sigma_c}{2} + \sqrt{\sigma_s^2 + \left(\frac{\sigma_c}{2}\right)^2} \quad ...(20.103)$$

It is to be checked here that $\sigma_{max}$ should not exceed the safe limit.

**(ii) Shaft at Junction of Crank.** The shaft at junction of crank is subjected to bending and torsional moment.

The bending moment is given by

$$M_b = Q\,(0.75\,l_c + t) \quad ...(20.104)$$

The torsional moment is expressed as

$$M_t = F_t . r \quad ...(20.105)$$

The equivalent twisting moment is given by

$$M_{te} = \sqrt{M_t^2 + M_b^2} = \frac{\pi}{16} d_1^3 . \sigma_s \quad ...(20.106)$$

Equation (20.106) gives the value of $d_1$

**(iii) Shaft Under Flywheel.** The shaft under flywheel is subjected to bending. The twisting moment is given by

$$M_t = F_t . r \quad ...(20.107)$$

The equivalent bending moment is the same as dead centre case.

The equivalent twisting moment is expressed as

$$M_{te} = \sqrt{M_t^2 + M_e^2} = \frac{\pi}{16} d_{sfw}^3 \sigma_s \quad ...(20.108)$$

Equation (20.107) gives the value of $d_{sfw}$.

**(iv) Journals.** In general, the diameter of the two journals is kept the same. Their lengths are determined on the basis of bearing pressure and the maximum reactions at the journal as the previous case.

## 20.9 VALVE GEAR MECHANISM

### 20.9.1 Introduction

The valve gear mechanism of internal combustion engines includes the combination of those parts of the engine which actuate the various intake, exhaust, fuel and air starter valves, open and close them at the proper time with respect to the position of piston and crankpin and keep them open or closed during the required time of crankshaft rotation. The construction of valve gear mechanism varies considerably depending upon type, speed and size of the engines. A simple arrangement of valve gear mechanism for vertical, over heated valve four-stroke engines is shown in Fig. 20.29. while for horizontal engines, the mechanism is shown in Fig. 20.30. The working principle of this mechanism is discussed below.

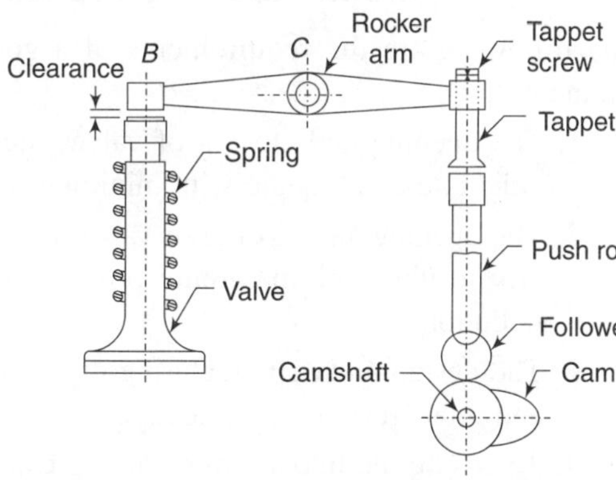


**Fig. 20.29** Valve gear mechanism for vertical engines

The cam of the rotating camshaft of vertical engines pushes the cam follower and pushes rod upward and thus transmits the cam action to the rocker arm which moves on a fulcrum pin and changes the above motion into a downward motion of the valve and thus the valve opens. The closing of the valve is affected when the cam follower moves over the circular position of the cam which releases the pushing action of the valve and the valve returns to its seat by the action of valve spring.

Another mechanisms of the valve gear for vertical engines locates the camshaft at the cylinder head level and thus eliminates the push rod by using a roller type cam follower as an integral part of the rocker arm.

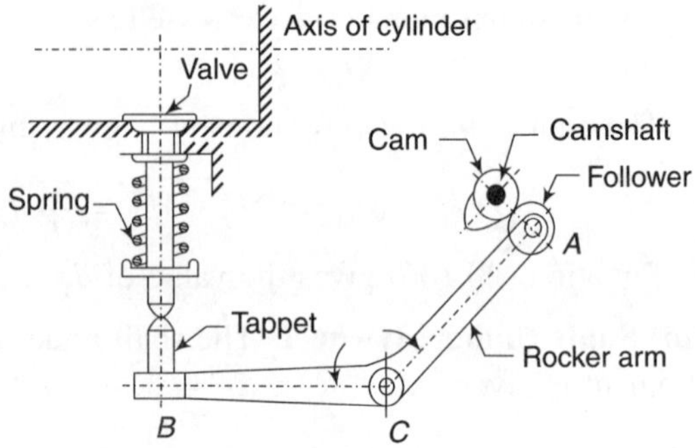

**Fig. 20.30** Valve gear mechanism for horizontal engines

**Some aspects of valve gear have been discussed in the chapter on Intake and Exhaust System and Value Gear.**

### 20.9.2 Types of Valve

There are two types of valve—poppet or **mushroom valve** and **sleeve valves**. The most common type is the **poppet valve**. Figure 20.31 shows a poppet type valve. The poppet valve geometry has already been discussed in the chapter on ***Intake and Exhaust System and Valve Gear***. It consists of a head or disk at the end of rod or stem. A margin is provided to avoid this edge.

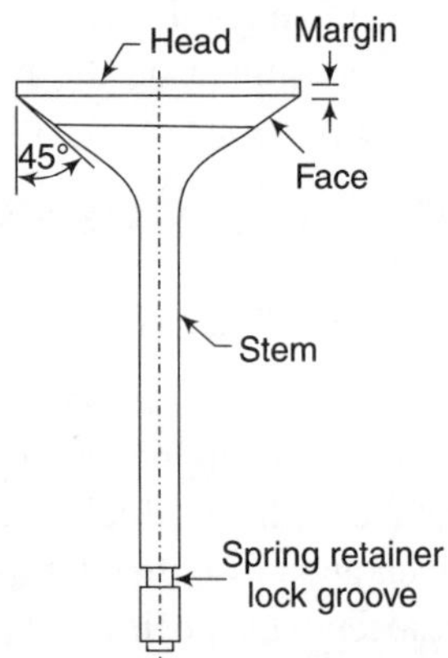

**Fig. 20.31** Poppet valve

### 20.9.3 Requirements of a Good Valve

The following are the requirements of a good valve of IC engine.

1. The opening and closing of valves must take place at a definite crank angle with minimum noise and wear.
2. The opening and closing of valve must be timed properly to yield the maximum power and efficiency of the engine with the minimum pollution.
3. The design and manufacturing of the valves should be such to ensure tightly seating and thus preventing the escape of the gases during the working process.
4. To ensure the maximum breathing capacity, the valve opening should be sufficient.
5. To reduce inertia forces, the valve should be as light as possible.
6. The valve must be strong enough to withstand the continued pounding action caused by continued opening and closing of valves.
7. The materials chosen for valves should be such so as to withstand the high temperature, erosive and corrosive condition prevailing inside the cylinder.

### 20.9.4 Determination of Principal Parts of Valves

**(a) Materials for Valves.** Exhaust valves have to withstand very high temperature while inlet valves run cooler.

The following are the materials for inlet valves:

(i) Carbon steel (ii) Nickel steel

(iii) Chrome nickel steel (iv) Chrome molybdenum alloys

All these are hardened to withstand the repeated high stress.

*Since the exhaust valves have to withstand high temperature, so their materials must be able to maintain their strength at high temperatures and offer resistance to air hardening, deformation, oxidation, wear, pitting and etching.* The following are the materials used for exhaust valves:

(i) austenitic type chrome nickel steel (ii) standard chrome nickel steel

(iii) cobalt nickel steel (iv) high speed steel

(v) stainless steel

*The materials of the valve seat inserts may be cast iron high speed steels, monel metal and various alloys and bronzes.*

**(b) The Size of Valve Ports.** The size of the valve port is very important as it determines the breathing capacity of the engines. It depends upon the cylinder diameter and the cylinder centre distance. Larger the port area, greater is the breathing capacity of the engine. The size of valve port may be calculated by the following equation

$$A_{piston} \cdot C_{mp} = A_{port} \cdot C_{port}\, x \cdot c_d$$

where $A_{piston}$ = area of piston, $C_{mp}$ = mean piston speed, $A_{port}$ = area of port

$C_{port}$ = average velocity of gas through port, $x$ = number of similar valves

$C_d$ = coefficient of discharge.

The gas velocity should be moderate to maintain the pressure difference. Generally,

$C_{port}$ = 35 to 50 m/s for stationary engines

= 50 to 75 m/s for automotive and aircraft engines

*It is worthy to note that better charging by inlet valve and better cooling of the exhaust valves are required for high engine output.*

*As an example if the cylinder pressure decreases by 0.007 MPa due to restricted air* intake, the output decreases by about 7 per cent whereas for 0.007 MPa increase of back pressure due to restricted gas outlet system, the output decreases merely about 1.5 per cent. ***Therefore, for better charging, the inlet port is made 20 to 40 per cent larger than exhaust valve.*** The following points must be considered to obtain maximum flow with the minimum pressure drop.

(i) Small seat angle provides the maximum opening with a given lift.

(ii) Rounding of all the edges of the valves and seats reduces resistance to flow.

(iii) Large and smooth passage ways to and from valves reduce resistance to flow.
(iv) Never allow the pressure drop to exceed the critical value, i.e., 0.53 times the manifold pressure.
(v) The passage between the valves and combustion chamber should be such that it should not restrict the flow.

**(c) Valve Seat Angles, Lifts, Valve Seats and Inserts.** Figure 20.32 shows nomenclature used for the design of valves. The head of the valve (i.e., the seating surface of valves) may be flat or bevelled. The bevelled valves are generally conically shaped heads to get a self-centring action and due to this advantage these are universally used. The bevelled angle may be kept 30° to 45°. Generally $\alpha = 45°$ is more common. However, to obtain a greater opening area of inlet valve for a limited life it may be made with $\alpha = 30°$ but for exhaust valve, $\alpha$ cannot be less than 45° otherwise the head will become thin and cannot withstand the high temperature of burnt gas.

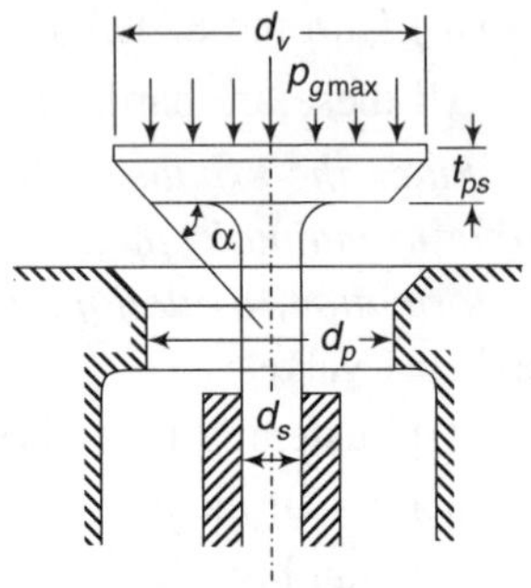


**Fig. 20.32** Nomenclature of poppet valve

The valve lift ($l_v$) is calculated by equation the area across the valve seat to the port area. Thus, by neglecting the valve stem area, the valve lift is given by

$$\pi\, d_{vp}\, l_v \cos \alpha = \frac{\pi}{4} d_{vp}^2$$

or

$$l_v = \frac{d_{vp}}{4 \cos \alpha} \qquad \text{...(20.109)}$$

For flat headed valve, $\alpha = 0$

**As a practice, the angle of the valve seat is made 1/2 to 1 degree larger than valve angle ($\alpha$) for providing more effective seal.**

*The valve seats are eroded due to continuous pounding of the valve under high temperature conditions. To overcome this problem, separate replaceable valve seats of hard and heat resisting alloys are provided.*

**(d) Thickness of Valve Head ($t_h$).** The valve head may be considered a flat plate subjected to the maximum gas load, ($F_g$) as uniformly distributed. The thickness ($t_h$) of the valve head or disk may be calculated as

$$t_h = k\, d_{vp} \sqrt{\frac{p_{g,\max}}{\sigma_b}} \qquad \text{...(20.110)}$$

where $d_{vp}$ = valve port diameter (mm), $p_{g\max}$ = the maximum gas pressure (N/mm²)

$\sigma_b$ = allowable stress is bending, $k$ = constant, = 0.54 for CI ($\sigma_b$ = 28 N/mm²)

= 0.42 for carbon steel ($\sigma_b$ = 56 N/mm²)

**(e) The Valve Stem Diameter ($d_s$)**

The valve stem diameter is given by empirical equation

$$d_s = \frac{d_{vp}}{8} + 6.33 \text{ mm} \quad \text{to} \quad \frac{d_{vp}}{8} + 11 \text{ mm}$$

Proportion of valve is shown in Fig. 20.33 in terms of valve port diameter. ***The proportions are also given in terms of cylinder bore in the chapter "Intake and Exhaust Systems and Valve Trains".***

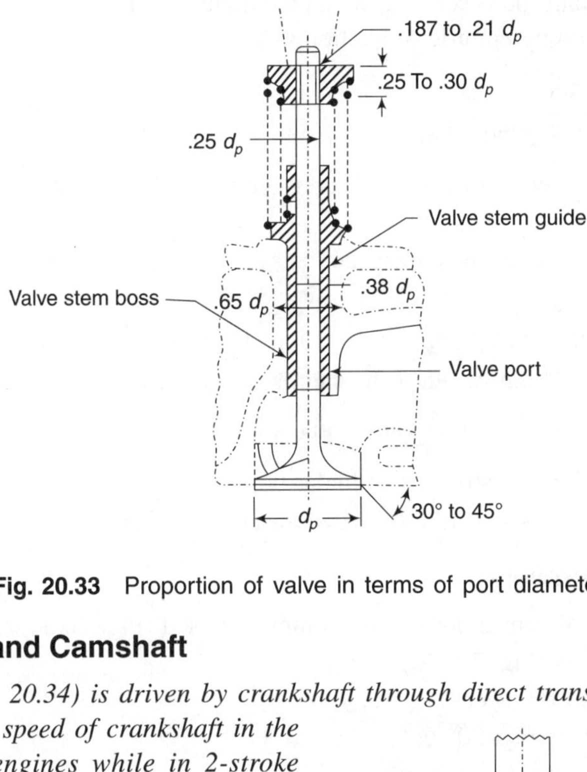


**Fig. 20.33** Proportion of valve in terms of port diameter

## 20.9.5 Cams and Camshaft

*The camshaft (Fig. 20.34) is driven by crankshaft through direct transmission system and runs at half of the speed of crankshaft in the case of 4-stroke engines while in 2-stroke engines, the camshaft speed is the same as of crankshaft.* The cams are integral part of the camshaft. *Generally, the cams are forged as one piece. The cam is an eccentric projection on a revolving disc used for the opening and closing of valves.*

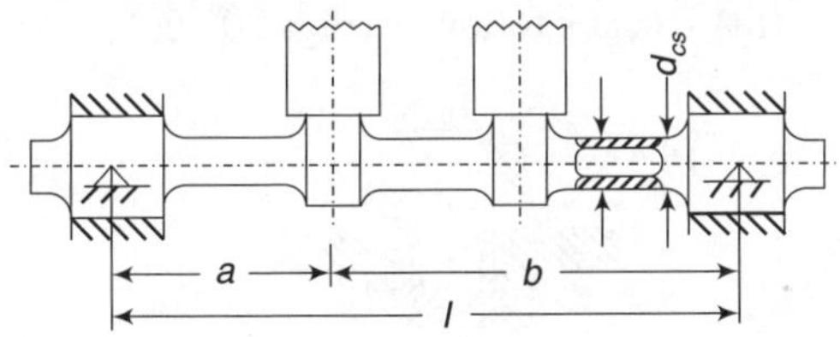


**Fig. 20.34** Camshaft

The cams are of the following four types:

(i) Tangent (ii) Concave flank cams

(iii) Convex flank cams (iv) Proportional cams

Figure 20.35 shows the tangent cam with approach ramp. The tangent cam is constructed of a circle, a noise circle and two flanks tangent to both circle.

*The material for the camshaft is forged steel, usually of nickel chromium alloy steel. These are heat treated and are carried in plain bearings.* In the case of larger camshafts, these are made often bored hollow.

In order to avoid excessive accelerations, the base circle of cam should be kept as small as possible and the roller or follower should be as large as possible.

Though, the shaft diameter depends upon its length to take care of stiffness of shaft, however, in general, the diameter of cam shaft is calculated by empirical equation.

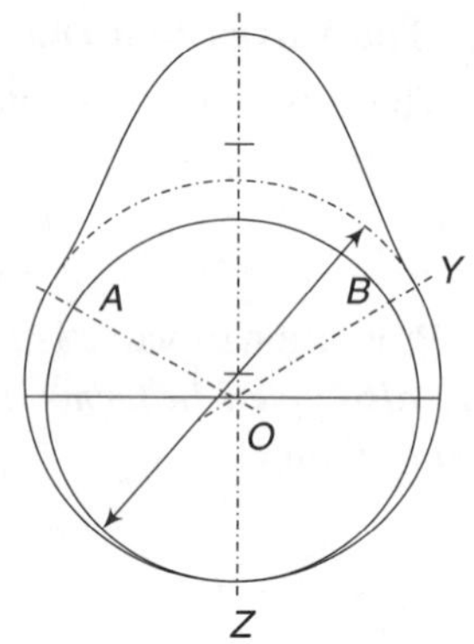


**Fig. 20.35** Tangent cam with approach ramp.

Diameter of camshaft

$$= d_{cs} = 0.16 \times \text{Cylinder bore} + 12.7 \text{ mm} \quad ...(20.111)$$

The base circle diameter of the cam is about 3 mm greater than the camshaft diameter for solid cams.

The width of the cam depends upon the load to be carried. The loads for well lubricated roller followers are:

Load = 215 to 250 N/mm width if cam is face hardened

= 110 to 140 N/mm width if the cam is of plain cast iron or steel.

For roller follower, the width of cam is given by

$$\text{Cam width} = w_c = 0.09 \times \text{Cylinder bore} + 6 \text{ mm to } 0.11 \text{ cylinder bore} + 12 \text{ mm} \quad ...(20.112)$$

### 20.9.6 Cam Followers

The following types of cam followers are commonly used (Fig. 20.36).

(i) Roller followers (Fig. 20.36a)
(ii) Flat mushroom follower (Fig. 20.36b)
(iii) Rocker follower (Fig. 20.36c)

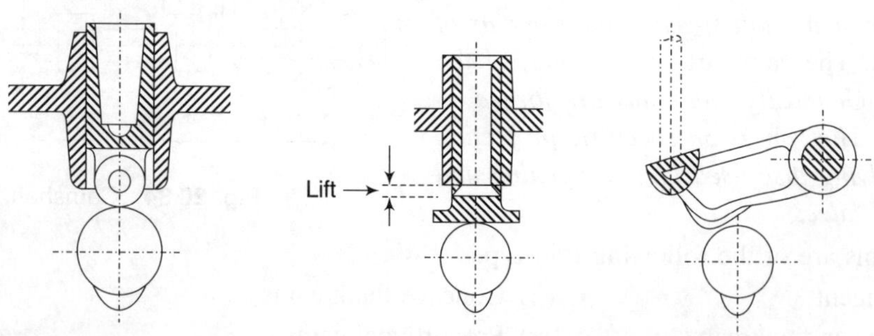


**Fig. 20.36** Cam followers

Generally, use of roller followers is universal for bore above 12.5 cm but below this size bore, either mushroom or rocker followers are chosen.

The diameter of roller is expressed as

$$\text{Diameter of roller} = \left(\frac{1}{2} \text{ to } \frac{2}{3}\right) \times \text{Cam base circle diameter} \quad ...(20.113)$$

The forced lubrication may be used on roller pin.

### 20.9.7 Rocker Arm

The force required to operate the suction valve is relatively small as compared to exhaust valve. However, in general practice the dimensions of rocker arm are made same for both valves. The following forces act on the valve.

(i) Weight of the valve
(ii) Inertia of valve
(iii) Vacuum on the suction stroke for suction valve and high pressure on the start of exhaust stroke.

These forces are taken care of by fitting spring, the normal load of which is equivalent to a pressure of about 0.07 N/mm$^2$ of valve area in the case of slow speed engines while for high speed engines, it may upto 0.14 N/mm$^2$.

Refer to Fig. 20.37. The force at $B$ ($F_B$) during periods of positive acceleration is given by

$$F_B = F_g + F_s + F_a \quad ...(20.114)$$

where $F_g$ = gas load; $F_s$ = spring force; $F_a$ = force due to valve acceleration

The gas load ($F_g$) is given by

$$F_g = A_v \times p_{exhaust} \quad \text{where} \quad A_v = \text{area of valve}$$

$$p_{exhaust} = \text{pressure of gas at the time of exhaust } (0.35 \text{ to } 0.42 \text{ N/mm}^2)$$

The spring force ($F_s$) is given by $F_s = A_v \times p_{suction}$

where $p_{suction}$ = negative pressure at suction stroke

The force due to valve acceleration ($F_a$) is given by $F_a = m_v . a_v$

where $a_v$ = acceleration of valve

The force at $A$ ($F_A$) is given by $F_A = F_B + F_{ra}$

where $F_{ra}$ = force due to rocker acceleration = $m_r . a_r$

The force at $C$ ($F_c$) is given by

$$F_C = F_A + F_B \quad ...(20.115)$$

*The design philosophy of the rocker arm may be based on treating this as a simply supported beam subjected to a point load at fulcrum, C.* The section of the arm adjacent to fulcrum ($C$) (Fig. 20.37) is calculated by the following equation.

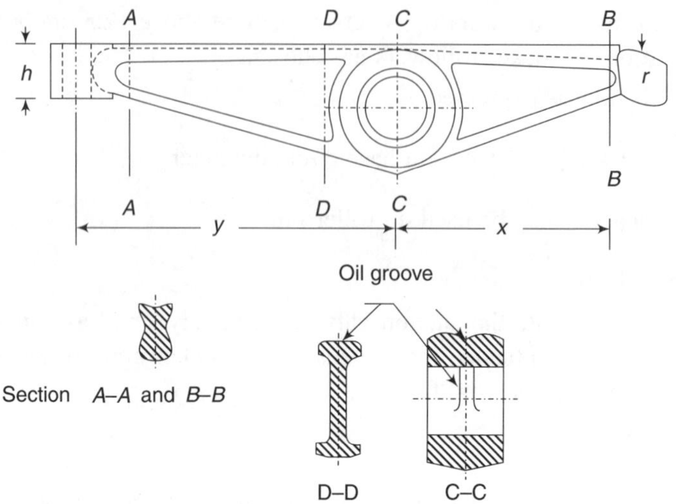


**Fig. 20.37** Rocker arm with I-section

$$M_b = \sigma_b . Z \qquad ...(20.116)$$

where $\sigma_b$ = safe stress in bending, $Z$ = section modulus, $M_b$ = bending moment

*The rocker arm may be made of cast iron, cast steel or mild steel stamping. The factor of safety may be taken high (up to 10) to take care of varying stress and shock.* In order to reduce the valve inertia, its section may be made of **I-section** for high speed engine and of **rectangular section** for low speed engines. The fulcrum pin is to be designed in bearing and so be checked in double shear. In the case of rocker pin with its own support, the bending may be neglected but if the number of rockers mounted on the same spindle, the bending stress must be considered for its design.

The load on the rocker arm is the total reaction at fulcrum, $C$. Thus, reaction at $C$ is expressed as

$$F_c = l_{rp} . d_{rp} . p_b = 2 \times \frac{\pi}{4} d_{rp}^2 . \sigma_s \qquad ...(20.117)$$

where $l_{rp}$ = length of rocker arm pin, mm

$d$ = diameter of rocker arm pin, mm $l_{rp}/d$ = 1.5 to 2

$p_b$ = bearing pressure, N/mm$^2$ = 4.55 to 10.5 N/mm$^2$

$\sigma_s$ = safe shear stress, N/mm$^2$ = 42 N/mm$^2$

Equation (20.117) gives the value of diameter ($d_{rp}$) and length ($l_{rp}$) of the pin.

The material of the bearing bush may be bronze and aluminium alloy. In order to find outer diameter of boss, the bending at CC may be considered.

It is worthy to note that for equal strength, the rocker arm is tapered towards the ends both in width and depth keeping the thickness of flange and web as constant. As a general practice, the height ($h$) is given by

$h$ = (1.0 to 1.4) + Nominal diameter of tappet screw.

The radius ($r$) should be chosen such that the contact between the rocker and valve or push rod is full tangential.

The lever ratio $\left(\frac{y}{x}\right)$ should range from $\frac{1}{1.5}$ to $\frac{1}{1.6}$. The distance ($x$) should not be less than five times the valve lift. The distance between the fulcrum pin centre and the top end of the valve in closed position affects the side thrust on the valve and its valve may be taken equal to one-third of valve life.

### 20.9.8 Valve Springs

The valve spring is one of the most important elements of valve assembly. *The objective of the valve spring is to provide sufficient force during the valve lifting process to overcome the inertia of valve gear and to keep it in contact with cam without bouncing and chattering. This is achieved by putting the spring under compression during its installing.*

The spring is subjected to the maximum load when it is at full lift and this may be treated as design load. Thus,

Total load = Initial load + Load at full lift

But the initial load is expressed as

Initial load = Force on the valve to draw it into the cylinder on the suction stroke

Load at full lift = Full lift × Stiffness of spring

The shear stress ($\sigma_s$) induced in the spring is given as

$$\sigma_s = \frac{8\,F\,D\,K}{\pi\,d^3} \qquad ...(20.118)$$

where $F$ = total spring load, N; $D$ = coil diameter, mm

$d$ = wire diameter, mm; $K$ = stress increase coefficient = $\frac{4\,S-1}{4\,S-4} + \frac{0.615}{S}$

$S$ = spring index = $\frac{D}{d} \geq 5$

The deflection of the spring is given by

$$y = \frac{8\,F.S^2.z_c}{C.d} \qquad ...(20.119)$$

where $z_c$ = number of effective coils; $C$ = modulus of rigidity

Since the spring is to be used as a compression ring, two more coils should be provided to act as bearing at the ends.

Length of wire ($L$) required is given by

$$L = \pi.D.z_c \qquad ...(20.120)$$

Minimum free length of spring is given by

= Length of spring when fully compressed + Initial compression + Full lift

Free length of spring = (Minimum free length) × 1.07 (20.121)

Sometimes two concentric rings may be provided for the safety purpose.

### 20.9.9 Push Rods

*Push rods of the valve assembly are subjected to buckling and therefore Ranking Gordon expression of buckling given below may* be helpful for its design.

$$F = \frac{\sigma_c . A_{pr}}{1 + a\left(\frac{l_{pr}}{k}\right)^2} \quad ...(20.122)$$

where $F$ = Load on the push rod at point $A$. $\sigma_c$ = safe buckling stress (70 N/mm$^2$)

$A_{pr}$ = area of push rod = $\frac{\pi}{4}(d_0^2 - d_i^2)$, $a = \frac{1}{7500}$ for steel

$l_{pr}$ = length of push rod depends upon the height of engine.

The body of the rod may be solid or tubular. The end condition of push rod will depend upon the general condition of gear mechanism to drive the camshaft.

The materials chosen for push rods may be bright drawn mild steel with 0.4% content having yield stress about 420 N/mm$^2$. Safety factor of 3 to 4 may be used.

**EXAMPLE 20.1** Design cylinder, cylinder head, piston assembly, connecting rod and crankshaft of a 4-stroke vertical SI engine to be used in a two-wheeler for the following data:

Engine capacity = $V$ = 100 cc

Speed = $N$ = 8500 rpm

***Solution*** **(A) Cylinder and Cylinder Head**

Based on the cycle heat analysis, the power output of the engine may be calculated. Since the capacity of the engine is less, so one cylinder is sufficient ($x$ = 1).

Assume compression ratio, $r$ = 9

The indicated mep based on air-standard cycle given by Eq. 20.8 may be given as

$$(p_{mi})_{as} = \frac{p_1 . r(\alpha - 1)(r^{\gamma-1} - 1)}{(r-1)(\gamma - 1)} \quad \text{where} \quad \alpha = \frac{p_3}{p_2} = \frac{T_3}{T_2}$$

$$T_2 = T_1 \, r^{\gamma-1} = 288 \, (9)^{0.4} = 693.5 \text{ K}$$

The maximum temperature, $T_3$ in the cycle is given by Eq. (20.5).

$$\text{LCV} \times \eta_{comb} = \left(\frac{\dot{m}_a}{\dot{m}_f}\right) c_{pg}\, T_3 - \frac{\dot{m}_a}{\dot{m}_f}\, c_{pa}\, T_2$$

or $$44 \times 10^3 \times 0.97 = (15 + 1) \times 1.27\, T_3 - 15 \times 1.005 \times 693.5$$

or $$T_3 = 2600 \text{ K}$$

$$\alpha = \frac{T_3}{T_2} = \frac{2600}{693.5} = 3.749$$

$$p_2 = p_1\, r^{\gamma} = 1.013 \times 9^{1.4} = 21.95 \text{ bar}$$

Thus, $$(p_{mi})_{as} = \frac{1.013 \times 9\,(3.749 - 1)\,(9^{0.4} - 1)}{(9 - 1)\,(0.4)} = 12.3 \text{ bar}$$

The actual indicated mep is less than $(p_{mi})_{as}$ due to various losses in real cycle.

$$p_{mi} = K(p_{mi})_{as} = 0.85 \times 12.3 = 10.45 \text{ bar}$$

The indicated power is given by

$$ip = \frac{p_{mi}\, L\, A\, N\, x}{2 \times 60} = \frac{10^2 \times 10.45 \times 100 \times 8500 \times 1}{2 \times 60 \times 10^6} = 7.4 \text{ kW}$$

The brake power = $bp = \eta_m . ip = 0.82 \times 7.4 = 6$ kW **_Ans._**

This means that a 100 cc engine produces 6 kW of brake power at 8500 rpm.

For cylinder dimensions, assume $\frac{L}{D} = 1$.

Engine volume = $V = ALx = 100$ cc

or $$\frac{\pi}{4} D^3 \times 1 = 100 \times 10^3$$

or $$D = \left(\frac{4 \times 10^5}{\pi}\right)^{1/3} = 50.2 \text{ mm} = 50 \text{ mm}$$

Thus, $$L = D = 50 \text{ mm}$$ **_Ans._**

The length of cylinder is given by

$$L_{cy} = \text{Stroke} + \text{Clearance on both sides}$$

or $$L_{cy} = 50 + 2 \times 0.1 \times 50 = 60 \text{ mm}$$ **_Ans._**

***Note:*** **The rounding off dimension has been done without following any standard**

The mass flow rate of air inside cylinder at mean induction condition

$$\dot{m}_a = \frac{p_{ind} . \dot{V}_a}{R_a . T_{ind}} = \frac{p_{ind} \times 60\, V\, N}{R_a \times T_{ind} \times 2}\ ; \text{kg/h}$$

$$= \frac{0.9 \times 10^2 \times 60 \times 100 \times 10^{-6} \times 8500}{0.287 \times 330 \times 2} = 23.63 \text{ kg/h}$$

$$\therefore \quad \dot{m}_f = \frac{23.63}{15} = 1.57 \text{ kg/h, hence } bsfc = \frac{1.57}{6} = 0.26 \text{ kg/kWh} \qquad \textbf{\textit{Ans.}}$$

***Material:*** The material of cylinder is grey cast iron.

The **thickness of cylinder wall** is given by Eq. (20.10)

$$t_c = \frac{p_{g,\max} \cdot D}{2\sigma_h} + K_r + K_c$$

Here, $p_{g,\max} = 0.85 \times p_3 = 0.85 \times 82.29 = 69.94 = 70 \text{ bar} = 7 \text{ N/mm}^2$

$$t_c = \frac{7 \times 50}{2 \times 50} + 1.5 + 1 = 5 \text{ mm}$$

The thickness may be checked by empirical Eq. (20.14)

$$t_c = 0.045\,D + 1.6 \text{ mm} = 3.85 \text{ mm}$$

Thus, the design is safe. *This may also be checked from strength point of view.*

The final cylinder thickness depends upon the conceptual design arrangement of cylinder and cylinder head. If the cylinder is having flange to bolt the cylinder head, then 5 mm thickness is sufficient. If no flange, no cylinder head is provided, then thickness of the cylinder will accommodate the stud. In this case, we adopt the later case and as a result the thickness will be

$$t_c = 5 \text{ mm} + 2 \times d_b \approx 5 + 2 \times 3 = 11 \text{ mm}$$

The thickness of the head lower support wall

$$\delta_h = 0.09\,D = 0.09 \times 50 = 4.5 \text{ mm} \quad \textbf{\textit{Ans.}}$$

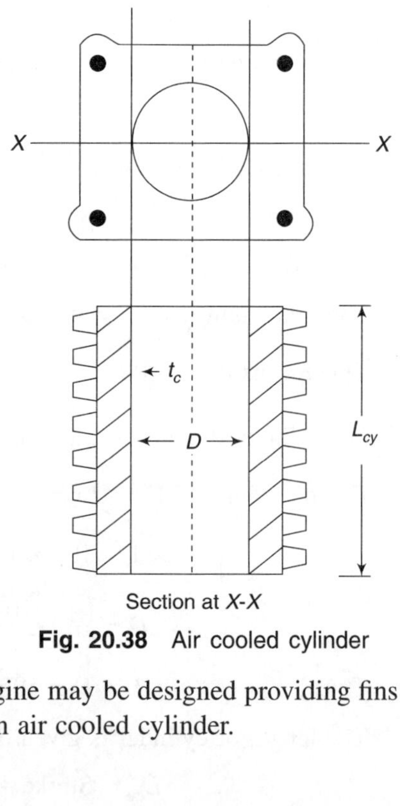


**Fig. 20.38** Air cooled cylinder

Since the size of the engine is small, so air cooled engine may be designed providing fins over the outer surface of cylinder. Figure 20.38 shows an air cooled cylinder.

### Cylinder Head

Since the temperature is very high, so aluminium alloy cylinder head may be designed.

The thickness of cylinder head is given by Eq. (20.17)

$$t_h = D \times \sqrt{\frac{C \times p_{g,\max}}{\sigma_t}} = 50 \times \sqrt{\frac{0.1 \times 7}{45}} = 6.23 \text{ mm} = 7 \text{ mm}$$

Since, stud is to be accommodated so $t_h$ is extended to

$$t_h = 12 \text{ mm}$$

**Cylinder Head Stud**

The number of studs ($z$) may be calculated from

$$z = \frac{D}{50} + 4 = \frac{50}{50} + 4 = 5, \text{ rounded to 6 number} \qquad \textit{Ans.}$$

The core diameter ($d_{bc}$) of the stud is given by

$$d_{bc} = D\left(\frac{p_{g,\max}}{z.\sigma_t}\right)^{1/2} = 50 \times \left(\frac{7}{6 \times 70}\right)^{1/2} = 6.45 \text{ mm} = 7 \text{ mm} \qquad \textit{Ans.}$$

The nominal diameter ($d_b$) of the stud is given by

$$d_b = \frac{d_{bc}}{0.8} = \frac{7}{0.8} = 8.75 = 9 \text{ mm} \qquad \textit{Ans.}$$

The pitch circle diameter ($D_p$) of the stud is given by

$$D_p = D + 1.4\, d_b = 50 + 1.4 \times 9 = 62.6 = 63 \text{ mm} \qquad \textit{Ans.}$$

**Cooling Fins**

Since the engine size in small, so instead of liquid cooled, it may be made air cooled by providing cooling fins on cylinder and cylinder head surfaces. It is assumed that the heat rejected to cooling air is 25% of the energy input.

$$Q_{cooling} = 0.25 \times \dot{m}_f \times \text{LCV} \times \eta_{com}$$

$$= 0.25 \times \frac{1.57}{3600} \times 44 \times 10^3 \times 0.97 = 4.65 \text{ kW}$$

The heat transfer through cooling fins (Fig. 20.8) is expressed by

$$Q_{cooling} = A_0\, U\, (T_s - T_\infty)$$

where $$U = \frac{h_i}{s+b}\left[\frac{2}{a}\left(1 + \frac{w}{2r_0}\right)\tan h[a\,(w + 0.5b)] + s_0\right]$$

Let, $A_0$ = outer surface area of cylinder and cylinder head; $w = 30$ mm
$b = 1$ mm; $r_0 = 25$ mm; spacing is 1.5 to 3 per cm; $s = 3$ mm
$s_0 = 2.5$ mm; $k = 100$ W/m$^2$ K; $h_i = 50$ W/m$^2$ K at $s = 3$ mm
$a = \sqrt{2}\, h_i / k_b$

$Q_{cooling}$ calculated matches fairly the value of 4.65 kW for the selected temperature of cylinder materials ($T_s$) to withstand the high temperature.

**(B) Piston Assembly**

The piston proportions given in Table 20.2 and Fig. 20.9 may be used to calculate the dimension of various parts of piston assembly. However, some of the dimensions may be calculated from governing equation and its validity will be checked through the proportions given.

The **material chosen for piston is aluminium alloy**. The crown thickness ($\delta$) of piston is calculated by Eq. (20.20).

(1) $\delta = 0.433\, D \times \sqrt{\dfrac{p_{g\max}}{\sigma_t}} = 0.433 \times 50 \times \dfrac{\sqrt{7}}{110} = 5.46 \approx 5.5$ mm **Ans.**

As per proportion thumb rule

$$\delta = (0.05 \text{ to } 0.01)\, D = 0.05\, D = 0.05 \times 50 = 2.5 \text{ mm (lower range)}$$

Also, $\delta = 0.1\, D = 0.1 \times 50 = 5$ mm (upper range)

Crown thickness calculated from equation is comparable with proportion of upper range.

It has been observed that it is easy to select the piston dimensions from proportions given in Table 20.2. The various values (Fig. 20.9) are:

(2) Piston height, $H = 1.2\, D = 1.2 \times 50 = 60$ mm

(3) Height of piston top part, $h_t = 0.7 \times D = 0.7 \times 50 = 35$ mm

(4) Piston skirt height, $h_s = 0.7\, D = 0.7 \times 50 = 35$ mm

(5) Boss diameter, $d_0 = 0.4\, D = 0.4 \times 50 = 20$ mm

(6) Distance between boss and faces, $b = 0.4 \times 50 = 20$ mm

(7) Thickness of skirt wall, $\delta_s = 4$ mm

(8) Thickness of piston crown wall, $s = 0.08 \times D = 0.08 \times 50 = 4$ mm

(9) Distance to the first portion groove, $e = 0.12\, D = 0.12 \times 50 = 6$ mm

(10) Thickness of first piston ring land, $h_1 = 0.05\, D = 0.05 \times 50 = 2.5$ mm

(11) Radial thickness of piston ring, $t$

Compression ring $= 0.04\, D = 0.04 \times 50 = 2$ mm

Oil ring $= 0.04\, D = 0.04 \times 50 = 2$ mm

(12) Piston ring width, $a = 4$ mm

(13) Difference between free gap and compressed gap of piston ring,

$A_0 = 4t = 4 \times 2 = 8$ mm

(14) Radial clearing of ring in piston groove, $\Delta r$

Compression ring = 0.95 mm

Oil ring = 1.1 mm

(15) Piston inner diameter, $d_i = 50 - 2\,(4 + 2 + 0.95) = 32.1$ mm

(16) Number of oil holes in piston, $n_0' = 8$

(17) Oil passage diameter, $d_0 = 0.5\, a = 0.5 \times 4 = 2$ mm

(18) Pin's outer diameter, $d_p = 0.28\, D = 0.28 \times 50 = 14$ mm

(19) Pin inner diameter, $d_{ip} = 0.75\ d_p = 0.75 \times 14 = 10.5$ mm

(20) Pin length, floating, $l_p = 0.88\ D = 0.88 \times 50 = 44$ mm

(21) Connecting rod bushing length, $l_{se} = 0.45\ D = 0.45 \times 50 = 22.5$ mm

***Note:*** Sr. 11, 12, 18, 19, 20, 21 are to be checked or reused later on while designing piston ring and piston pin.

#### Piston Rings

*Four rings are chosen in all—three compression rings and one oil ring.* The material chosen for piston rings is cast iron.

(11) The radial thickness of the ring is given by (Eq. 20.25)

$$t = D \times \sqrt{3\frac{p_w}{\sigma_t}} = 50 \times \sqrt{3 \times \frac{0.04}{85}} = 1.87 \text{ mm} = 2 \text{ mm} \qquad \textbf{\textit{Ans.}}$$

The calculated value is the same as from proportion, at Serial No. 11.

#### Piston Pin

The *material chosen for the piston pin is* **heat treated alloy steel**. The piston pin diameter (Fig. 20.15) is calculated by Eq. (20.37).

$$M_b = \frac{\pi}{32} \times \frac{d_p^4 - d_{ip}^4}{d_p} \times \sigma_b$$

where
$$M_b = \frac{F_{g,\max} \cdot D}{8}$$

Here,
$$F_{g,\max} = \frac{\pi}{4} D^2 \times p_{g,\max} = \frac{\pi}{4}(50)^2 \times 7 = 13.7 \times 10^3 \text{ N}$$

Thus,
$$M_b = \frac{13.7 \times 10^3 \times 50}{8} = 85.6 \times 10^3 \text{ N·mm}$$

$$d_p = 14 \text{ mm} \quad \text{and} \quad d_{ip} = 10.5 \text{ mm}$$

Substituting the values in Eq. 20.37,

$$85.6 \times 10^3 = \frac{\pi}{32} \cdot \frac{14^4 - 10.5^4}{14} \times \sigma_b$$

or
$$\sigma_b = \frac{85.6 \times 10^3}{184.15} = 464 \text{ N/mm}^2$$

The value is within the safe limit as its value is 540 N/mm$^2$ for heat treated special allow steel. The safety factor is 540/74.39 = 1.16, $d_p$ may be increased.

### (C) Connecting Rod Assembly

Refer to Fig. 20.16. The section chosen for connecting rod is I-section (Fig. 20.21) as it is light in weight and keep the inertia forces small and it can withstand the high pressure also. *Material chosen for connected rod is **carbon steel**.*

### Section of Rod

Refer to Fig. 20.21.

Flange and web thickness of I-section = $t_{sh}$ ; Depth of section = 5 $t_s$

Width of section of 4 $t_s$; Area = $A = 11\ t_s^2$ ; $K_{xx}^2 = 3.18\ t_s^2$; $K_{yy}^2 = 0.995\ t_s^2$

The section of the rod is determined by Rankine-Gordon formula (Eq. 20.52)

$$\text{Buckling load} = \frac{\sigma_{\text{cru}} \cdot A}{1 + a\left(\frac{l}{k}\right)^2}\ ;\ \text{N}$$

The buckling load = $F_{g,\,\max}$ × FOS = 13.7 × $10^3$ × 5 = 68.5 × $10^3$ N

The ratio of length of connecting rod and stroke ($l/r$) is chosen as 5

Thus, length of connecting rod = $l = 5 \times \frac{\text{Stroke}}{2} = 5 \times 25 = 125$ mm **Ans.**

The crushing stress, $\sigma_{cru}$ for carbon steel = 330 MPa

$$K^2 = K_{xx}^2 = 3.18\ t_s^2$$

Thus, the buckling load is

$$68.5 \times 10^3 = \frac{330 \times 10^6 \times 11\ t_s^2}{1 + \frac{1}{7500} \times \left(\frac{0.125^2}{3.18\ t_s^2}\right)}$$

or $$68.5 \times 10^3 = \frac{68.5 \times 10^3}{7500} \times \frac{0.125^2}{3.18\ t_s^2} = 330 \times 10^6 \times 11\ t_s^2$$

or $$= 86.57 \times 10^7\ t^4{}_s - 0.16337 \times 10^5\ t^2{}_s - 0.01 = 0$$

Let $$t_s^2 = T$$

or $$8.657 \times 108\ T^2 - 1.6337 \times 10^4\ T - 0.041 = 0$$

By trial and error, $T = 0.2 \times 10^{-4} = t_s^2$

or $$t_s = 0.4472 \times 10^{-2}\ \text{m} = 4.47\ \text{mm} \approx 4.5\ \text{mm}$$ **Ans.**

Depth of section = 5 $t_s$ = 5 × 4.5 = 22.5 mm **Ans.**

Width of section = 4 $t_s$ = 4 × 4.5 = 18 mm **Ans.**

The section obtained here is of middle section. The modified section (Fig. 20.22) to take of rounding of sharp corner is

Depth at crank end = 1.2 × Depth at midpoint = 1.2 × 22.5 = 27 mm **Ans.**

Depth at piston end = 0.9 × Depth at midpoint = 0.9 × 22.5 = 21 mm **Ans.**

Thus the section at the crank end = 27 × 21 mm

The section at small end = 18 × 21 mm

The modified form of section is given in Fig. 20.39.

### Big End

*The big end shell may be made of brass.* The thin lining of bearing metal (lead chosen) is of 1 mm. Centrifugal force lubrication is chosen as the engine is small. The bearing pressure selected as 10 N/mm$^2$.

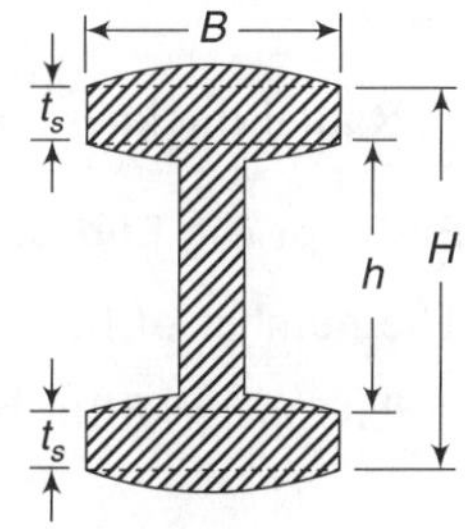


**Fig. 20.39** Modified section of C.R.

The length ($l_c$) and diameter of crank pin ($d_c$) may be taken as:

$$F_{g,\max} = l_c \cdot d_c \cdot p_b$$

Here, the ratio, $\dfrac{l_c}{d_c}$ selected as 1.3

$$\therefore \quad 13.7 \times 10^3 = 1.3\, d_c^2 \times 10$$

or $$d_c = \left[\frac{13.7 \times 10^3}{1.3 \times 10}\right]^{1/2} = 32.6 = 33 \text{ mm}$$ **Ans.**

$$\therefore \quad l_c = 1.3 \times 33 = 42.9 = 43 \text{ mm}$$ **Ans.**

These values may be checked by empirical equation

$$d_c = (0.56 \text{ to } 0.75)\, D = 27.8 \text{ mm to } 37.5 \text{ mm}$$

$$l_c = (0.6 \text{ to } 0.85)\, D = 30 \text{ mm to } 42.5 \text{ mm}$$

The values calculated by governing equation lies in the range:

Distance between connecting rod bolts, $s = 1.5\, d_c = 1.5 \times 33 = 49.5$ mm

Big end length, $l_{bc} = 0.8\, d_c = 0.8 \times 33 = 26.5 \text{ mm} = b$

Shell was thickness = $0.04\, d_c = 0.04 \times 33 = 1.3$ mm

### Small End

The thickness of bushing = $\delta_{pb} = 0.06\, D = 3$ mm **Ans.**

The piston pin's outer diameter, $d_p$

= inner diameter of small end, $d_{p0} = 0.28\, D = 14$ mm

*The material chosen for piston pin bushing is of high quality bronze.*

Piston in diameter which is also inner diameter of small end is given by Eq. 20.56.

$$F_{g,\,max} = l_p . d_p . p_b$$

Assume $\quad l_p/d_p = 2.0 \quad$ and $\quad p_b = 32$ N/mm$^2$

or $\quad 13.7 \times 103 = 1.8\, d^2_p \times 32$

or $\quad d_p = 14$ mm (as calculated above)

The outer diameter of end, $d_{se} = 1.5\, d_p = 21$ mm

Length of small end, $l_{se} = 0.4\, D = 0.4 \times 50 = 20$ mm

Minimum radial thickness of end wall, $\delta_{se}$

$$= 0.25\, d_p = 0.25 \times 14 = 3.5 \text{ mm}$$

Radial thickness of bushing wall, $\delta_{pb} = 0.08\, d_p = 0.08 \times 14 = 1.12$ mm $\approx 1.2$ mm

#### Bolts for Big End Cap

The number of bolts, $z_b = 2$ assumed.

The material for bolts is nickel alloy. Inertia force taken by bolts is given by Eq. (20.57)

$$F_i = F_r . \omega^2 . r \frac{\left(\cos\theta + \frac{r}{l}\cos 2\theta\right)}{g} = z_b . A . \sigma_t$$

Here, $\quad \sigma_t = 170$ MPa

$$\omega = \frac{2\,\pi\,N}{60} = \frac{2\,\pi \times 8500}{60} = 890.1 \text{ rad/s}$$

$$r/l = 1/5 \text{ and } r = \frac{L}{2} = 25 \text{ mm}$$

$$F_r = 10 \text{ N (after estimation)}$$

$\therefore$

$$F_i = 10\,(890.1)^2\; 2 \times 0.025 \left(\cos\theta + \frac{1}{5}\cos 2\theta\right)/9.81 = 24.22 \times 10^3 \text{ N}$$

Thus, $\quad 24.22 \times 10^3 = 2 \times \frac{\pi}{4} \times d_b^2 \times 170 \times 10^6$

or $\quad d_b = 0.00952 = 9.52$ mm $\approx 10$ mm

$$\text{The outer diameter} = d_{b0} = \frac{d_b}{0.8} = \frac{10}{0.8} = 12.5 \text{ mm}$$

*If the crankpin is not an internal part to crankshaft, then big end is not a split one and then there is no need of any bolt.*

#### Connecting Rod Big End Bearing Cap

The cap dimensions may be calculated by Eq. (20.58)

$$M_b = \frac{F_i . s}{6} = \sigma_b . z = \sigma_b \times \frac{b c^2}{6}$$

$s$ = distance between bolt centre

= diameter of bearing + twice of thickness of bearing liner

+ diameter of bolt + some clearance

= 33 + 2 × 3.5 + 12.5 + 2 = 54.5 mm

The bending moment is

$$M_b = \frac{F_i \cdot s}{6} = \frac{24.22 \times 10^3 \times 54.5}{6} = 220 \times 10^3 \text{ N.mm}$$

Also, $$M_b = \frac{\sigma_b \cdot bc^2}{6}$$

Here, $\sigma_b = 270 \text{ N/mm}^2$ and $b = l_{se}$ = width of cap = 26.5 mm

Thus, $$220 \times 10^3 = \frac{270 \times 26.5\, c^2}{6}$$

or $$c = \text{Thickness of cap} = \left(\frac{220 \times 10^3 \times 6}{390 \times 26.5}\right)^{1/2} = 12.88 \approx 13 \text{ mm}$$

Now, $\sigma_b$ is to checked.

**(D) Crankshaft**

Since the number of cylinders chosen is one so centre crankshaft may be taken with one crank for its obvious advantages. In this case, the main bearings are of the same length. The material selected for crankshaft is plain *carbon steel.*

The flywheel is used as pulley. The estimated weight of the flywheel is 10 kN and the total belt pull is about 3 kN.

**(i) Crank at Dead Centre.** Refer to Fig. 20.24 showing forces acting on the crankshaft. Equations (20.62 to 20.64) will give the various force components

$$F_g = F_{g\,\max} = 13.7 \times 10^3 \text{ N} = 13.7 \text{ kN}$$

Let $f = 2D = 2 \times 50 = 100$ mm and $a = a' = \frac{f}{2} = 50$ mm

Let $b = b' = \frac{g}{2}$

$$H_1 = H_2 = \frac{F_g \cdot a}{f} = \frac{13.7 \times 10^3 \times 50}{100} = 6.85 \text{ kN}$$

$$V_2 = V_3 = \frac{F_w}{2} = \frac{10}{2} = 5 \text{ kN} \quad \text{if} \quad b = b' = \frac{g}{2}$$

$$H_2' = H_3 = \frac{F_1 + F_2}{2} = \frac{3}{2} = 1.5 \text{ kN} \quad \text{if} \quad b = b' = \frac{g}{2}$$

**(ii) Crankpin.** The dimensions of crankpin have already been calculated but it is to be checked. The bending moment at the centre of crankpin is given by Eq. (20.65)

$$M_b = H_1 . \frac{f}{2} = \frac{\pi}{32} d_c^3 \sigma_b$$

$$M_b = 6.85 \times 10^3 \times 0.05 = 0.3425 \times 10^3 \text{ Nm}$$

Taking $\sigma_b$ for carbon steel as 75 MPa, we have

$$0.3425 \times 10^3 = \frac{\pi}{32} d_c^3 \times 75 \times 10^6$$

$$wd_c = \left( \frac{0.3425 \times 10^3 \times 32}{\pi \times 75 \times 10^6} \right)^{1/3} = 0.0359 \text{ m} = 36 \text{ mm}$$

The allowable bearing pressure already assumed is10 N/mm$^2$

$$F_g = l_c \, d_c \, p_b \quad \text{or} \quad 13.7 \times 10^3 = l_c \times 36 \times 10$$

or $$l_c = 38 \text{ mm}$$

The value of $l_c$ and $d_c$ are comparable, so already calculated values, i.e., $d_c$ = 33 mm, $l_c$ = 43 mm may be taken.

**(iii) Left Hand Crank Web.** The thickness of left hand crank web is calculated by any one of the Eq. 20.66.

$$t = 0.65 \, d_c + 6.35 \text{ mm} = 0.65 \times 33 + 6.35 = 27.8 \text{ mm} = 29 \text{ mm}$$ ***Ans.***

The width ($w$) may be calculated by Eq. 20.67

$$w = \frac{9}{8} d_c + 12.7 \text{ mm} = \frac{9}{8} \times 33 + 12.7 = 49.8 \text{ mm} = 50 \text{ mm}$$ ***Ans.***

$\sigma_{total}$ given by Eq. (20.68) may be checked

$$\sigma_{total} = \frac{H_1}{w.t} + \frac{H_1 \left( \frac{f}{2} - \frac{l_c}{2} - \frac{t}{2} \right)}{\frac{1}{6} w \, t^4}$$

$$= \frac{6.85 \times 10^3}{50 \times 29} + \frac{6.85 \times 10^3 \left( \frac{100}{2} - \frac{43}{2} - \frac{29}{2} \right)}{\frac{1}{6} \times 50 \times 29^4}$$

$= 4.72$ N/mm$^2$ which is safe

**(iv) Right Hand Crank Web.** Same dimensions are taken as that of left hand crank web

**(v) Shat under Flywheel.** As already mentioned the main bearings are of the same length. Thus,

$$l_1 = l_2 = l_3 = \left(\frac{f}{2} - \frac{l_c}{c} - \frac{t}{2}\right) \times 2 = 2\left(\frac{100}{2} - \frac{45}{2} - \frac{29}{2}\right) = 28 \text{ mm}$$

Assuming width of flywheel = 30 mm, hence, $g = 28 + 30 = 58$ mm

Giving allowance for gearing and clearance, $g$ may be taken as

$$g = 70 \text{ mm}$$

Bending moment due to flywheel weight, $M_f = \dfrac{F_w \cdot g}{4} = \dfrac{10 \times 0.07}{4} = 0.175$ kNm

Bending moment due to the belt pull

$$M_b = (F_1 + F_2)\frac{g}{4} = \frac{3 \times 0.07}{4} = 0.0525 \text{ kNm}$$

Resultant bending moment,

$$M_r = \sqrt{M_f^2 + M_b^2} = \sqrt{(0.175^2 + 0.0525^2)} = 0.0915 \text{ kNm}$$

But

$$M_r = \frac{\pi}{32} d_{sf}^3 \sigma_b \qquad \text{...(20.123)}$$

$$0.0915 \times 10^3 = \frac{\pi}{32} d_{sf}^3 \times 40 \times 10^6$$

or

$$d_{sf} = \left(\frac{0.0915 \times 10^3 \times 32}{\pi \times 40 \times 10^6}\right)^{1/3} = 0.0285 \text{ m} = 28.5 \text{ mm}$$

Design may be checked for crank at angle of the maximum moment.

Students are advised to draw the part to be designed first manually and then by using standard package of graphics such as **AUTOCAD or PROE or CATIA**.

## PROBLEMS

### Things to Think About

20.1 What are the basic philosophies involved in the design of following parts?

(a) cylinder (b) cylinder head (c) piston (d) piston pin

(e) connecting rod (f) crankshaft (g) camshaft (h) valve assembly

20.2 Draw the loading diagram on the elements in Question No. 20.1.

20.3 List the design considerations for elements in Question No. 20.1.

20.4 What are the basis for a material selection for the design of any element given in Question No. 20.1

20.5 What are the basis for selection of manufacturing method element given in Question No. 20.1.

### Manual Design Questions

20.6 Design the main parts of Maruti 1000 cc, Zen, WagonR SI engines. Draw the parts manually.

20.7 Design the main parts of Hero Honda motorcycle engine and draw the parts manually.

### Software Based Questions

20.8 Develop a software for the design of two-wheeler scooter and motorcycle engines. Print the design data including graphics for 150 cc scooter engine.

20.9 Develop a software for the design of fuel efficient 4-stroke passenger car engine. Print the design data with graphics for WagonR engine.

20.10 Develop a software to design the main parts of racing car engine. Draw the graphics by standard package.

20.11 Develop a software for the design of following system of IC engine:

(a) Intake system
(b) Exhaust system
(c) Ignition system
(d) Injection system
(e) Lubrication system
(f) Cooling system

Print the design data for passenger car engines of your choice.

### Design and Open Ended Question

20.12 Prepare a comprehensive report on design features of Rolls-Royce car engine of high end segments. Give the layout of whole design layout of this engine.

### Multiple-Choice Questions

20.13 Choose the incorrect assumptions in the design of connecting rod

(a) uniform section through the length
(b) bending force variation is triangular
(c) bending force variation is constant
(d) bending force is zero at the piston pin and the maximum at the crankpin

20.14 Piston pin is designed for

(a) maximum gas load
(b) maximum inertia force on piston
(c) which is larger in (a) and (b)
(d) average gas load

20.15 The design criteria for piston pin is

(a) bearing pressure
(b) bearing temperature
(c) maximum torque
(d) maximum bending stress

20.16 Which factor/factors govern the design of cylinder head

(a) air/gas passage and accommodation of valve and their gear
(b) accommodation of spark plug or fuel injectors
(c) subjected to severe alternating loads
(d) all of them

## Key to Multiple-Choice Questions

20.13 (c) 20.14 (c) 20.15 (a) 20.16 (d)

CHAPTER

# Compressors

21

## OVERVIEW

*Compressors are power consuming thermodynamic devices which convert mechanical energy into head or pressure energy.* The uses of compressed gases and vapours have very wide applications in engineering, petrochemicals, fertilizers, day-to-day life and many other fields. This chapter deals with thermodynamics aspect of compressors.

## 21.1 FUNCTION OF A COMPRESSOR

The *function of a compressor is to compress the gases and vapours from low pressure to high pressure.* According to the second law of thermodynamics, this is only possible when the work is done on the gas by an external agency, such as prime movers, electric motors, etc., using direct and indirect transmission as shown in Fig. 21.1. Out of the work received by the compressor, some is absorbed in overcoming friction, some is lost as radiation, some is lost in cooling the fluid employed to cool the compressor and the rest is utilised in delivering

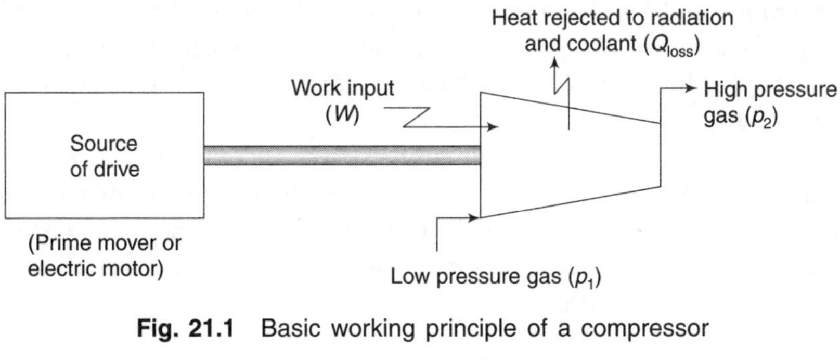


**Fig. 21.1** Basic working principle of a compressor

* **Note:** For more details are Author's book **Applied Thermodynamics.**

gas at high pressure. Thus, a compressor sucks in a gas at low pressure (atmospheric air in the case of air compressor), compresses it and delivers it at high pressure to a storage vessel called receiver (only in reciprocating type) from where it may be carried by a pipe line to wherever it is desired.

## 21.2 USE OF COMPRESSED GASES AND VAPOURS

The use of compressed gases and vapours in industry and in day-to-day life are many. Some of them are given below:

**(i)** ***Mining.*** In mines, owing to danger of fire, no electrical or fuel motor can be used. The air is compressed above the mines and passed down to mines after cooling it to its initial temperature for operating air motors, pneumatic drills, spades, picks, haulage pumps, motor generators, drill sharpeners, coal cutters, etc.

**(ii)** ***Engineering.*** In engineering for operating hammers, drills, hoists, tube scalars, paint sprayers, chucks, pile drivers, fuel atomizers, blast furnaces, Bessemer converters, torpedoes, *starting IC engines*, **supercharging**, pneumatic brakes for locomotives, gas turbines, etc.

**(iii)** ***General.*** In general, for pneumatic tyres, pneumatic tubes for conveyance of messages, lift gates, liquid air and separation of gases, compression of lighting gas, air lift pumps, air dash pots, glass blowing, conveying materials like sand, concrete and powdered fuel along pipeline, etc.

**(iv)** ***Chemical Industries.*** Compressed gases also play an important role in the heavy chemical industry, for example, in the manufacture of soda, in hydrogenation of brown coal with pressure of about 300 atm; in the manufacture of synthetic ammonia where pressures of up to 1000 atm are applied; in polymerization of ethylene where pressure of up to 2500 atm is employed, in fertilizers such as synthesis gas compressors for ammonia plants, ammonia refrigeration, compressors for ammonia plants, carbon dioxide, compressor for urea plants; oxo-syn, gas make up and recycle services for methanol plants, natural gas compressor for urea plants and many other similar processes of manufacturing such as refineries, oil and exploration, of petrochemicals, etc.

**(v)** ***Refrigeration and Air Conditioning.*** Compressed vapours play a vital role in refrigerating plants such as freeze, cold storage and air conditioners which enable us to raise our standard of living by economic handling of perishable foods and maintaining room air temperature.

**(vi)** ***Other Applications.*** Compressed or rarefied gases are utilised in numerous other field of human activity, such as agriculture, military engineering, medicine, transport and physical research. In fact, it would be difficult to find a branch of industry where utilization of compressed gases and vapours would not afford a materials rationalization of the manufacturing processes.

## 21.3 CLASSIFICATION OF COMPRESSORS

The compressors are classified on the basis of the following:

**(A) Design and principles of operation.** On the basis of design and principles of operation, compressors are classified in two main groups. They are further subdivided into many sub-groups as follows and given in branched form.

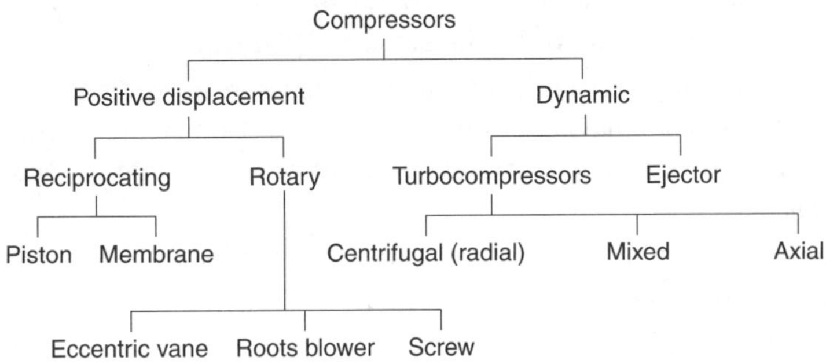


Turbocompressors are also sometimes called rotary compressor.

**(i) *Positive displacement compressors.*** *In the positive displacement compressors, the rise in pressure is achieved by positive displacement of gas caused by displacement of one or more elements,* i.e., *the pressure of gas is increased by decreasing its volume due to positive displacement.*

In piston compressors (reciprocating compressors), the gas volume decreases and pressure increases due to action of one or more reciprocating pistons moving axially in the cylinder. In membrane compressors, the volume variations are affected by deflection of the elastic partition. For rotary compressors, it will be more correct to say that pressure increases due to **push and pull of gas** caused by the rotary movement of the rotor.

**(ii) *Dynamic compressors.*** *In dynamic compressors, the rise in pressure is achieved by the dynamic action of gas.* In ***turbocompressors***, the kinetic energy imparted to the gas by the rotation of rotor is changed into pressure energy, partly in rotor and rest in a diffuser. *The energy level of the working fluid goes on increasing along the flow stream.*

*In centrifugal, mixed and axial flow compressors, the kinetic energy imparted to the gas by the rotation of the rotor is partly changed into pressure energy in the rotor due to change in moment of momentum of the flowing fluid passing through the rotor and the rest in the diffuser.* These are also called **turbocompressors**. In centrifugal compressors, the flow is *radial*. The kinetic energy imparted to the gas in centrifugal compressor is partly changed into pressure energy due to *diffusion and centrifugal action in rotor and the rest in diffuser due to diffusion action.* In mixed flow compressor, the flow is neither radial nor axial but mixed. In axial flow compressors, the flow is axial and there is an *absence of centrifugal action.* In the ejector, the driving vapour first attains a high velocity passing through a nozzle, mixed with the gas to be compressed in a mixing chamber, it is then compressed in a diffuser due to the change of kinetic energy into pressure energy.

*Reciprocating compressors are suited to small and medium discharge and low to high pressure. Centrifugal compressors are suited to medium discharge and low to medium pressure, while axial compressors are suited to high discharge and low pressure.*

**(B) Pressure.** On the basis of final pressure, the compressors are classified as:

(i) *Low pressure compressor* whose final pressure does not exceed 10 bar.

(ii) *Medium pressure compressor* with a range 10 bar to 80 bar.

(iii) *High pressure compressor* with a range 80 to 1000 bar.

Multistage reciprocating compressors for delivery pressure exceeding 1000 bar are termed as **hypercompressors.** They have as many as 10 stages. Compressors with a high suction pressure (e.g., 270 atm) and low compression ratio (e.g., 1.2) are termed ***circulation*** or ***circulation pumps***.

Further, depending upon the pressure rise limit, (the multiblades system) may be classified as (according to ASME Code):

(i) ***Fan,*** in which the pressure ratio < 1.1.

(ii) ***Blower,*** in which the pressure ratio > 1.1 < 2.3.

(iii) ***Compressor,*** in which the pressure ratio > 2.3.

**(C) Capacity.** A classification of compressor according to capacity based on the quantity of free air delivered is as under:

(i) *Small compressor* handling up to 9 $m^3$/min

(ii) *Medium compressor* handling 9 to 3000 $m^3$/min

(iii) *Large compressor* handling more than 3000 $m^3$/min

**(D) Number of stages.** The compressor may be of *single stage or multistage.*

**(E) Medium to be compressed.** Compressors are also classified on the basis of medium to be compressed. It may be air, *oxygen, hydrogen, nitrogen, acetylene, lightening gas and vapours and gases used* in refrigeration cycle such as *methyl chloride, ammonia, Fr-12, etc.* Their physical and chemical properties exert a considerable influence upon the design of the compressor.

**(F) Method of cooling.** On the method of cooling, compressors are classified such as *water cooled* or *air cooled.*

**(G) Arrangement of cylinders.** On the basis of the arrangement of cylinders, compressors are classified as being *vertical, horizontal, V type, W type, radial, horizontal balance opposed, etc.*

**(H) Types of drives.** On the basis of drives, compressors are classified as *motor driven, IC engine driven, turbine driven,* etc. **(See chapter 13).**

Note that each type of compressor has its own operational characteristics which limit its efficient application.

*The selection of a compressor for a particular type of job depends mainly upon the delivery pressure, free air delivered and speed requirements.*

# (A) RECIPROCATING COMPRESSORS

## 21.4 RECIPROCATING COMPRESSORS (PISTON TYPE)

*In this compressor, the gas volume decreases and pressure increases due to the action of one or more reciprocating pistons moving axially in one or more cylinders.* It may be single acting or double acting, single cylinder or multi-cylinder, single stage or multi-stage. The compressor is used to *meet the need for low, medium and high compression ratio and low and medium gas volumes. For efficient operation, a compressor's maximum working limit is 300* $m^3/min$.

In single-stage reciprocating compressor, pressure ratio is kept between 5 to 8. Till recently, multi-stage reciprocating compressors have been built for delivering gas at 5000 atm pressure. In these compressors, the speed varies from 100 to 1500 rpm. Reciprocating compressors can be driven directly or indirectly by internal combustion engine, or steam engine or electric motors, etc. It cannot be driven by steam or gas turbines because these have high speed. The delivery from the compressor is intermittent, so the gas is stored in a **receiver** from where continuous flow of compressed gas can be supplied.

The reciprocating compressors have a wide use in comparison to the other types of compressors. It is widely used in the refrigeration system such as freeze, air conditioner and cold storage, mining works, chemical factories, fertilizer factories, garages, etc. It is being gradually replaced in some industries like fertilizer industries, large cold storages, central air conditioning system in large hotels, etc., by the centrifugal compressors.

Figure 21.2 shows a single stage, single acting and single cylinder reciprocating compressor in simplified form. The crankshaft is rotated by an external agency which may be any prime

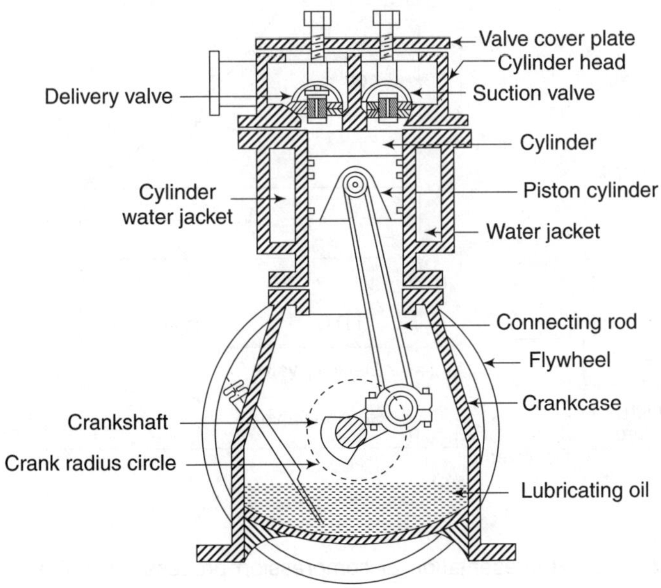


**Fig. 21.2** Details of reciprocating compressor

mover, or electric motor which produces reciprocating motion of piston in the cylinder. If the air is admitted to one side of the piston only, the compressor is said to be *single acting*. A *double acting* compressor admits air to each side of the piston alternatively, i.e., one face of the piston performs the suction stroke while the other compresses the air. The output of a double acting compressor is almost double that of a single acting compressor.

On the downward stroke of the piston, the pressure in the cylinder falls and the atmospheric air enters through the suction valve (delivery valve closed) till the piston reaches bottom dead centre position. When the piston goes up, the suction and delivery valves are both closed and the air is compressed till the delivery valve opens due to the difference in pressure of cylinder and delivery manifold. As the piston descends on the next downward stroke, the air trapped in a clearance volume expands and the pressure falls to suction pressure; the inlet valve then opens and the cycle is repeated.

## 21.5 SINGLE-STAGE RECIPROCATING COMPRESSOR WITHOUT CLEARANCE

Consider a single-stage, single acting, ideal reciprocating compressor having no flow resistance at suction or delivery valve, no friction losses and no clearance. The working fluid is a perfect gas (say air).

Let $p_1$, $V_1$ and $p_2$, $V_2$ be the pressure (kN/m$^2$) and volume (m$^3$/s) before and after the compression respectively. Representation of compression process on *p-V* and *T-s* diagrams are shown in Fig. 21.3 and Fig. 21.4 respectively.

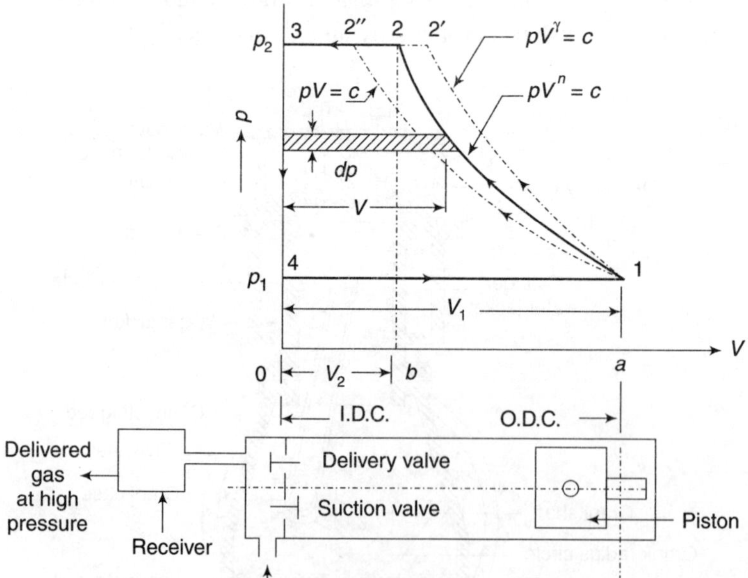


**Fig. 21.3** Representation of compression process on *p-V* diagram

Now, consider the piston at the inner dead centre position at point 4, at this instant the suction and delivery valves are closed. When the piston moves outwards, a vacuum is created. Because of the pressure difference across the suction valve, the suction valve opens and the fluid is drawn into the cylinder at constant pressure ($p_1$) through the suction manifold. *The suction process is represented* by 4-1. In actual practice, the suction pressure in the cylinder is less than pressure ($p_1$) in the manifold due to the resistance offered to the fluid by the suction valve. When the piston reaches the outer dead centre position, the suction valve closes at point 1 and the piston then starts moving towards inner dead centre position. Since both valves are closed, the induced fluid enclosed in the cylinder volume diminishes, causing a pressure rise. *The compression process is shown by* 1-2. The compression process will stop only when the pressure rise in the cylinder exceeds the pressure in the delivery manifold by an amount equal to the resistance offered to the passage of the fluid through the delivery valve. At point 2, the delivery valve opens and the compressed air goes to the receiver.

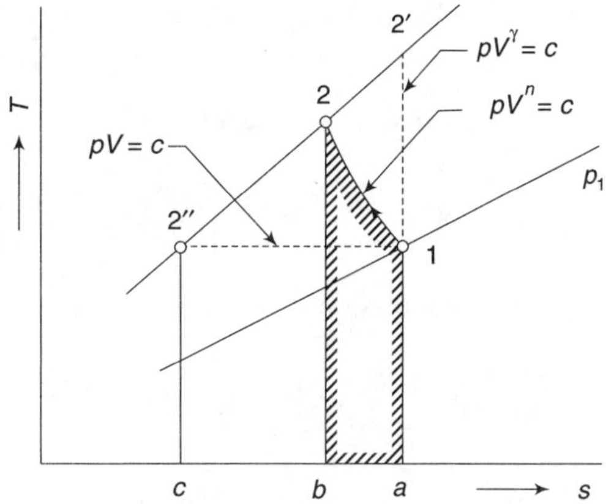


**Fig. 21.4** Representation of compression process on *T-s* diagram

**First Method:**

Wok done per cycle = $W$ = area 12341 = area $ba12b$ + area $ob230$ – area $oa140$

If the work done on the fluid is assumed to be negative and work done by the fluid assumed to be positive, then

$$-W = \left(\frac{p_2 V_2 - p_1 V_2}{n-1} + p_2 V_2 - P_1 V_1\right) = \frac{n}{n-1}(p_2 V_2 - p_1 V_1) \qquad ...(21.1)$$

**Second Method:**

The above equation of work done per cycle can also be obtained from the concept of steady flow through the machine neglecting change in KE and PE.

$$-W = \int_{p_1}^{p_2} V\,dp \qquad ...(21.2)$$

Assume $pV^n = c$ **for the compression process**

As $\quad pV^n = p_1 V_1^n = p_2 V_2^n = c$ thus, $V = V_1 (p_i/p)^{1/n}$

$$\therefore \quad -W = \frac{n}{n-1}(p_2 V_2 - p_1 V_1)$$

or
$$-W = \frac{n}{n-1} p_1 V_1 \left[\left(\frac{p_2}{p_1}\right)^{(n-1)/n} - 1\right]; \text{ kW}$$

We know that $p_1 V_1 = \dot{m}RT_1$

Thus, $-W = \frac{n}{n-1} \dot{m}RT_1 \left[\left(\frac{p_2}{p_1}\right)^{(n-1/n} - 1\right]$; kW where $\dot{m}$ is in kg/s ...(21.3)

If the compression process is adiabatic, i.e., $n = \gamma$, then

$$-W = \frac{\gamma}{\gamma-1} p_1 V_1 \left[\left(\frac{p_2}{p_1}\right)^{(\gamma-1)/\gamma} - 1\right]; \text{ kW}$$

or
$$-W = \frac{\gamma}{\gamma-1} \dot{m}RT_1 \left[\left(\frac{p_2}{p_1}\right)^{(\gamma-1)/\gamma} - 1\right]; \text{ kW} \quad ...(21.4)$$

If the compression process is isothermal ($pV = c$), i.e., all the heat generated during compression is withdrawn, so that the temperature of the fluid remains constant, less work will be required to be done on per kg of air to raise its pressure from $p_1$ to $p_2$. The change in internal energy will be zero, this is also clear from $p$-$V$ and $T$-$s$ diagrams. The areas 1-2–2″– 1 on $p$–$V$ and $T$–$s$ diagrams show the saving in work and heat withdrawn due to isothermal compression. In actual compressor, this is not possible to achieve but it is attempted to reach as close as possible to isothermal compression by efficient cooling of cylinder. In other words, attempts are made to reduce the value of polytropic exponent, $n$.

Work done on gas per cycle in **isothermal compression** is

$$-W = \int_{p_1}^{p_2} V\,dp$$

But $pV = p_1 V_1 = p_2 V_2 = c$ or $V = c/p$, so

$$-W = c \int_{p_1}^{p_2} \frac{dp}{p} = p_1 V_2 \ln \frac{p_2}{p_1} = p_1 V_2 \ln \frac{V_1}{V_2} = \dot{m}RT_1 \ln \frac{p_2}{p_1} \quad ...(21.5)$$

It is to be noted that in isothermal compression ($n = 1$) from $p_1$ to $p_2$ the work input is equal to the heat rejected as is obvious from $T$-$s$ diagram. But in polytropic ($n < \gamma$), a portion of the heat is rejected and remaining is used to increase the temperature of air. The polytropic exponent ($n$) can be calculated from

$$n = \frac{\ln p_1 - \ln p_2}{\ln V_2 - \ln V_1} = \frac{\ln (p_1/p_2)}{\ln (V_2/V_1)} \quad ...(21.6)$$

Thus, for fixed values of $p_1$, $p_2$ and $V_1$, $n$ depends on $V_2$.

## 21.6 SINGLE-STAGE COMPRESSOR WITH CLEARANCE AND VOLUMETRIC EFFICIENCY

*In actual compressor, there is a small clearance between the cylinder head and piston for thermal expansion, machine tolerance and for preventing the piston from striking the cylinder head.* It is desirable that the clearance should be as minimum as possible and as a rule this clearance length is 0.005 L + 0.5 mm where L is the stroke of the piston in mm. This space together with all other space between the closed valve and the piston at inner dead centre position is called the ***clearance volume***. The *displacement volume* or the stroke volume ($V_s$) is defined as the *volume swept by the piston in one stroke.*

*The clearance ratio, (∈) is defined as the ratio of clearance volume, ($V_c$) to the piston displacement volume (or stroke volume), $V_s$.* Thus,

$$\in = \frac{V_c}{V_s} \quad \text{...(21.7)}$$

*The clearance ratio depends on the arrangement of the valves in the cylinder and on the mean piston velocity ($C_m = LN/30$). It should be as minimum as possible and it ranges: ∈ ≤ 5% for well designed and constructed compressor having moderate piston velocity, ∈ = 5 to 13% for compressor with high piston velocity ($C_m$ > 3 m/s) with group valves.*

Figure 21.5 shows *p-V* and *T-s* diagrams of a compressor with clearance. Let $V_{as}$ be the actual sucked volume. Suction process is represented by 4-1, compression by 1-2 and delivery by 2-3. Let the law of compression be $pV^n = c$. The fluid trapped in the clearance volume ($V_c = V_3$) expands during suction stroke from the delivery pressure ($p_2$) at point 3 to the suction pressure ($p_1$) and occupied volume ($V_4$) at point 4. The expansion of the trapped volume for gas is represented by a process 3-4. It is clear from *p-V* diagram that suction volume has decreased

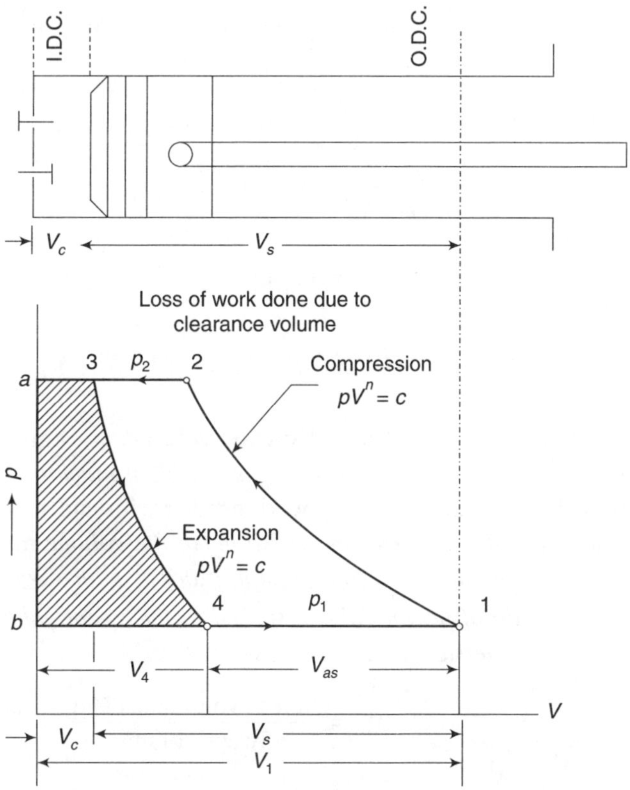


**Fig. 21.5** *p-V* and *T-s* diagrams of a compressor with clearance

from $V_s$ to $(V_1 - V_4)$. So, *we see, that the effect of clearance is to reduce the volume of the gas actually sucked in per working cycle.*

*The cylinder is normally cooled by water jacketing or by surrounding air.* The small volume of fluid expanding from the clearance volume ($V_c$) contacts a large cooling surface and, consequently, the re-expansion, curve at the beginning is steeper than the adiabatic curve. With further expansion, the temperature of the fluid falls below the temperature of the piston surface and the cylinder wall, and heat flows from the piston and the cylinder to the fluid. Thus, the *exponent of re-expansion curve is variable.* But for simplicity, it is assumed constant. The exponent of the *re-expansion curve may not be equal to that of compression, so n may be taken as the exponent of re-expansion curve to distinguish it from the compression curve.*

Work done = (Work done needed for compression without clearance – Work done by the gas in expanding from volume $V_3$ to $V_4$)

$$-W = \frac{n}{n-1} p_1 V_1 \left[\left(\frac{p_2}{p_1}\right)^{(n-1/n} - 1\right] - \frac{n'}{n'-1} p_2 V_3 \left[1 - \left(\frac{p_1}{p_2}\right)^{(n'-1/n'}\right]$$

$$= \frac{n}{n-1} p_1 V_1 \left[\left(\frac{p_2}{p_1}\right)^{(n-1)/n} - 1\right] - \frac{n'}{n'-1} p_1 V_4 \left[\left(\frac{p_2}{p_1}\right)^{(n'-1)/n} - 1\right]$$

If $n' = n$, then

or $$-W = \frac{n}{n-1} p_1 (V_1 - V_4) \left[(p_2/p_1)^{(n-1)/n} - 1\right]$$

or $$-W = \frac{n}{n-1} p_1 (V_{as}) \left[(p_2/p_1)^{(n-1/n)} - 1\right]$$

$$-W = \frac{n}{n-1} \dot{m}_{as} RT_1 \left[(p_2/p_1)^{(n-1/n)} - 1\right] \qquad ...(21.8)$$

where $V_{as} = V_1 - V_4$ = **actual sucked volume**, $m^3$. Equations (21.3) and (21.8) reveal that the *work to be done to compress a gas in actual machine is independent of the clearance ratio* ($\in = V_c/V_s$). Comparing the work needed with and without clearance. $V_1$ is replaced by $V_{as}$. Thus, the *effect of clearance is to make the volume taken in per stroke less than the swept volume. The maximum compression process is also controlled by the clearance volume.*

*The ratio of actual volume sucked to the swept volume* per cycle by the piston is known as the *volumetric efficiency*. Thus,

$$\eta_v = \frac{\text{Actual volume sucked}}{\text{Swept volume}} = \frac{V_{as}}{V_s} = \frac{V_s + V_c - V_4}{V_s}$$

But $$V_4 = V_c (p_2/p_1)^{1/n} ; \in = V_c/V_s\} \quad \text{and} \quad V_{as} = V_s + \in V_3 (p_2/p_1)^{1/n}$$

or $$\eta_v = 1 + \in - \in (p_2/p_1)^{1/n} \qquad ...(21.9)$$

## 21.7 MULTI-STAGE RECIPROCATING COMPRESSORS

In a single-stage compressor, if the compression ratio is increased, the final temperature increases, and the volumetric efficiency decreases to such an extent, that at a compression ratio $p_2/p_1 = [(1/\in) + 1]^n$, it becomes zero as is obvious from Fig. 21.6.

*Leakage past the piston starts and high compression requires robust cylinder construction. A high compression, temperature affects the operation of delivery valves, diminishes the lubricating properties of the oil and increases the risk of ignition in piping and receiver.*

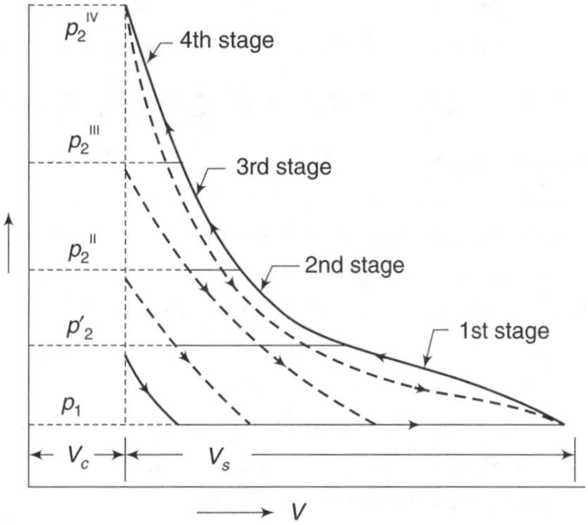


**Fig. 21.6** Showing the effect of increase in compression ratio in a single stage

It is for these reasons that if a higher compression above (6 to 8 atm) is needed, the overall compression ratio is sub-divided into *two or more stages with lower compression ratios,* usually the maximum compression pressure ratio for small single-stage compressors is 8 and for large machines it is 5. In practice, delivery pressure up to 5.5 bar in single stage, 5.5 to 35 bar in two stages and from 35 to 84 bar in three stages. An intercooler is used between two stages so that compressed air from the first stage cools down before it enters the second stage with low temperature nearly about equal to initial temperature in the first stage.

If pressure ratio is assumed to be constant, then the overall pressure ratio is given by

$$r_p = (r_{p\,stage})^Z \quad \text{where } Z = \text{number of stages} \qquad ...(21.10)$$

or

$$Z = \text{number of stages} = \frac{\ln r_p}{\ln (r_{p,\,stage})} \qquad ...(21.11)$$

*Though, in actual practice, pressure ratio in each stage is not kept constant.* Equation (21.11) is **valid for all types of compressors**.

## 21.8 ADVANTAGES OF MULTI-STAGE COMPRESSION

The following are the advantages of multi-stage compression.

(i) ***Less power.*** Less power is required to run a multi-stage compressor as compared to a single-stage compressor for the same delivery pressure and the same quantity of free air due to intercooling. For a large pressure range, the power reduction for drive is about 20%.

(ii) ***Better mechanical balance.*** Due to the phasing of operation into two or more stages, a better mechanical balance for machine is attained.

(iii) ***Increased volumetric efficiency.*** Volumetric efficiency increases due to lower temperature and lower pressure range.

(iv) ***Better lubrication.*** Better lubrication is possible due to lower working temperature.

(v) ***Less leakage loss.*** Due to lower pressure range, there is less leakage loss past the piston.

(vi) ***Lighter cylinder.*** Due to lower pressure range, a lighter cylinder serves the purpose.

## 21.9 TWO-STAGE AIR COMPRESSORS

Figure 21.7 shows schematic diagram of a two-stage reciprocating air compressor. The source of drive is the same for both the stages. Air is admitted at point 1 in the cylinder or the L.P. compressor and is compressed to some intermediate pressure between the intake and delivery pressure. Air at high temperature and pressure is brought to an **intercooler (heat exchanger)** where the air is cooled down approximately to the temperature of intake air to L.P. cylinder by properly regulating the supply of cooling water in the intercooler. It is to be noted that cooling of air by water jacketing is not sufficient. The cylinder cooling is a must to take out heat of cylinder. The air cooled from intercooler enters the H.P. cylinder. The compressed air from H.P. cylinder goes to receiver or storage vessel from where it can be used. In most of the cases, compressed air from H.P. cylinder is made to pass through a aftercooler. The aftercooler cools the air before going to receiver.

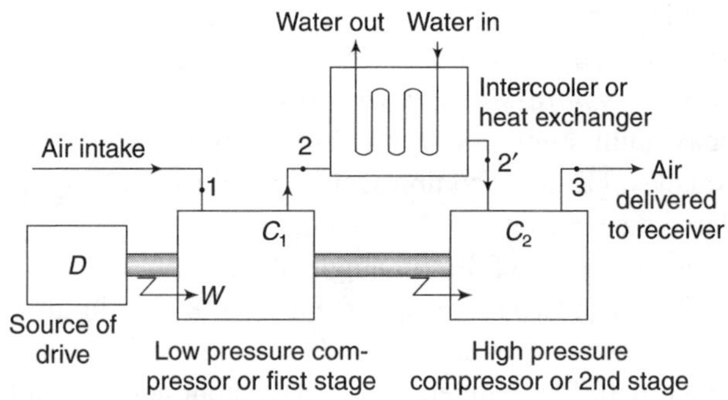


**Fig. 21.7** Two-stage air compressor

In a multistage compressor, the air leaving the second stage is again cooled in second intercooler before being passed to the third stage and this process is repeated until the required pressure is reached.

Figure 21.8 shows a two-stage compressor having six cylinders, four in the first stage and two in the second stage.

## 21.10 MINIMUM WORK INPUT IN MULTI-STAGE COMPRESSION

Let us take the case of a two-stage air compressor with intercooler and assume that the compressor is without clearance. The *p-V* and *T-s* diagrams of the compression process of this compressor are shown in Figs. 21.9 and 21.10 respectively. Now, in process 1-2, the compression of air is achieved in the L.P. stage compressor. At point 2, the air is cooled in an intercooler to its initial suction temperature (i.e., $T_2' = T_1$), though in actual practice, the intercooling is not perfect, i.e., temperature of air from the intercooler is slightly higher than the suction air temperature but for simplicity we will take it as perfect intercooling and at constant pressure in the intercooler, though in actual practice this is not the case as there is

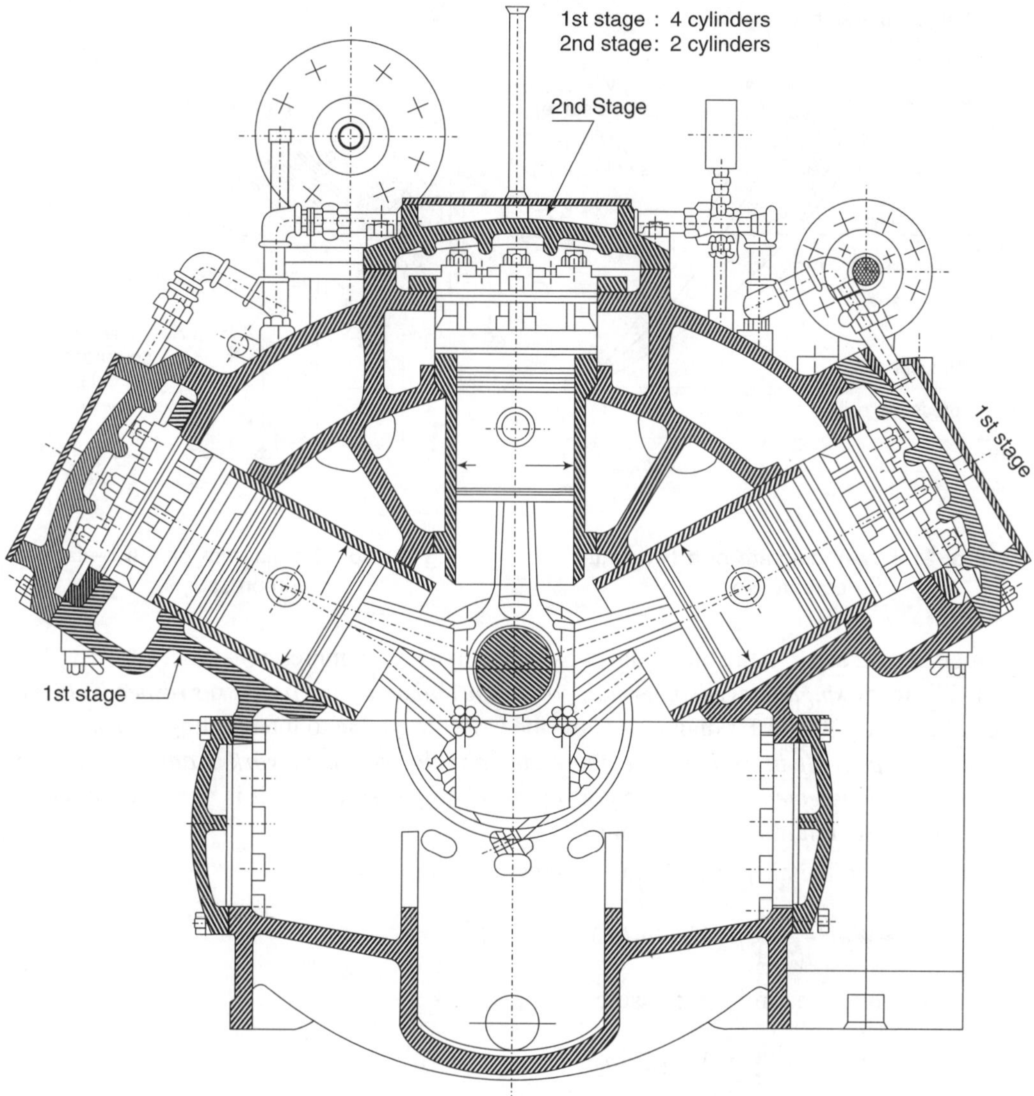


**Fig. 21.8** Details of a two-stage compressor

a drop of pressure in the intercooler. We will also take for granted that the pressure during suction and delivery in each stage remains constant.

Now, the air coming out from intercooler is compressed in the H.P. stage compressor and this process is represented by 2′–3.

At point 3, the air delivered at constant pressure is represented by 3–4. This area 1–2– 2′–3–4–5–6 shows the work done on the two-stage compressor with intercooler. The index of compression, ($n$) is taken as the same in both stages along with the mass flow rate.

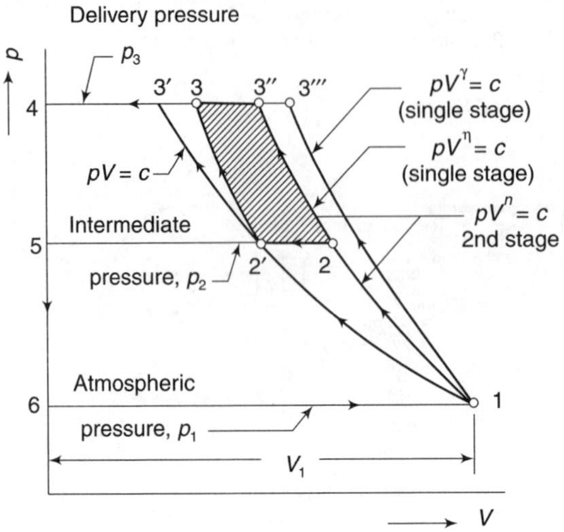


**Fig. 21.9** *p-V* diagram of a two-state compressor

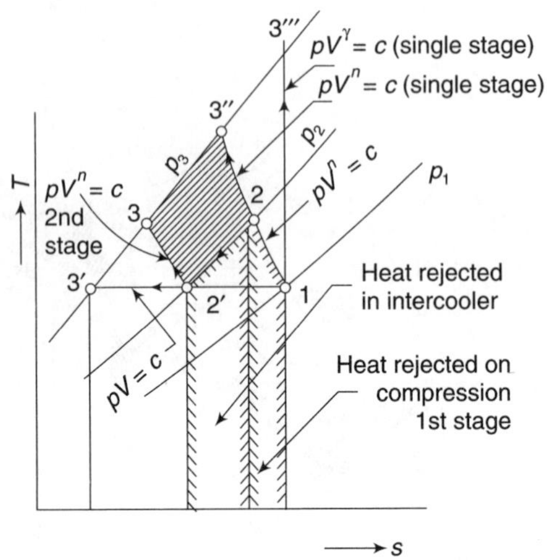


**Fig. 21.10** *T-s* diagram of a two-stage compressor

Processes 1-2 and 2′-3 are the actual two-stage polytropic compression. Processes 1-3′, 1-3″ and 1-3‴ are the *single-stage isothermal, polytropic and adiabatic processes respectively.* The hatched area on *p-V* and *T-s* diagrams show the work saved due to intercooling. *The minimum work input is possible in isothermal compression and the maximum work in adiabatic process since no heat is rejected.* By a choice of intermediate pressure ($p_2$), it is possible to find the condition for the minimum work input.

Work done on the air per cycle in low pressure compressor is

$$-W_2 = \frac{n}{n-1} p_1 V_1 \left[(p_2/p_1)^{(n-1)/n} - 1\right]$$

And the work done in high pressure compressor is

$$-W_2 = \frac{n}{n-1} p_2 V_2' \,[(p_3/p_2)^{(n-1/n} - 1]$$

$$-W = (-W_1) + (-W_2)$$

$$= \frac{n}{n-1} p_1 V_1 \left[(p_2/p_1)^{(n-1)/n} - 1\right] + \frac{n}{n-1} p_2 V_2' \left[(p_3/p_2)^{(n-1)/n} - 1\right]$$

Since the intercooling is assumed perfect and there is no pressure drop of air in the intercooler, so

$$p_1 V_2 = p_2 V_2'$$

Thus, $$-W = \frac{n}{n-1} p_1 V_1 \left[(p_2/p_1)^{(n-1)/n} + (p_3/p_2)^{(n-1/n)} - 2\right] \tag{21.12}$$

In Eq. (21.12), $n$, $p_1$, $V_1$, $p_3$ are fixed. *The only variable is the intermediate pressure, $p_2$.* The minimum work input, (i.e., maximum efficiency) can be found by differentiating Eq. (21.12) with respect to $p_2$ and equated to zero. Thus,

$$\frac{dW}{dp_2} = 0$$

Let $(n - 1)/n = a$ where $a$ is constant

$$\therefore \qquad -W = a\, p_1\, V_1\left[(p_2/p_1)^a + (p_3/p_2)^a - 2\right]$$

$$\frac{dW}{dp_2} = 0 = ap_1^{-a} \cdot p_2^{a-1} - ap_3^a \cdot p_2^{-a-1}$$

$$\text{or} \qquad \frac{p_1^{-a} \cdot p_2^a}{p_2} = \frac{p_3^a}{p_2^a \cdot p_2} \quad \text{or} \quad (p_2/p_1)^a = (p_3/p_2)^a$$

$$\text{or} \qquad p_2/p_1 = p_3/p_2$$

$$\text{or} \qquad p_2 = \sqrt{(p_1 \cdot p_3)} \qquad \text{...(21.13)}$$

Thus, we see, that for the *minimum work input, the intermediate pressure is the geometrical mean of initial and final pressure.* Substituting the value of $p_2$.

$$-W = \frac{n}{n-1} p_1\, V_1\left[\left(\sqrt{(p_1\, p_3)}/p_1)\right)^{(n-1)} + \left(p_2/\sqrt{(p_1\, p_3)}\right)^{(n-1)/n} - 2\right]$$

$$= \frac{2n}{n-1} p_1\, V_1\left[(p_3/p_1)^{(n-1/2n} - 1\right] \qquad \text{...(21.14)}$$

$$\frac{2n}{n-1} mRT_1\left[(p_3/p_1)^{(n-1)/2n} - 1\right] \qquad \text{...(21.15)}$$

In general, for *N number of stages*, from Eq. (21.13), we have

$$r_{p1} = r_{p2} = r_{pN}$$

$$\text{or} \qquad p_2/p_1 = p_3/p_2 = p_4/p_3 = p_{(N+1)}/p_N = N\sqrt{\left[p_{(N+1)}/p_1\right]} \qquad \text{...(21.16)}$$

Hence, the minimum work input for $N$ stages is

$$-W = \frac{Nn}{n-1} p_1\, V_1\left[(p_{N+1}/p_1)^{(n-1)/Nn} - 1\right] \qquad \text{...(21.17)}$$

For a three-stage compressor, the minimum work input will be

$$-W = \frac{3n}{n-1} p_1\, V_1\left[(p_4/p_1)^{(n-1)/3n} - 1\right] \qquad \text{...(21.18)}$$

In conclusion, we may say it is necessary that for the minimum work input for compression air in a multi-stage compressor with intercooler, the following must be true.

(i) *Work done in all stages is equal.*

(ii) *The compression ratio in each state is the same.*

(iii) *The air is cooled in the intercooler to the initial suction temperature after each stage of compression.*

In the event of intercooling, being imperfect, we must treat each stage as a separate compressor and the total power will be the sum of the powers per stage

$$-W = \frac{n_1}{n_1 - 1} p_1 V_1 \left[\left(\frac{p_2}{p_1}\right)^{(n_1 - 1)/n_1} - 1\right] + \frac{n_2}{n_2 - 1} p_2 V_2 \left[\left(\frac{p_2}{p_1}\right)^{(n_2 - 1)/n_2} - 1\right] \quad \text{...(21.19)}$$

In multi-stage refrigerating compressors, if the ratio between the condenser pressure and evaporator pressure is greater than 7, we normally select two-stage compression unless very small installations are concerned.

*The main factors involved in the selection of the number of stages are combination of economic and technical considerations.*

## 21.11 INDICATOR DIAGRAM

*The indicator diagram obtained from an indicator is a tool to study the performance of a reciprocating machine. It is a picture record of the simultaneous variation of volume and pressure of the working fluid in a cylinder as the piston reciprocates.* In other words, it tells the actual events in the cylinder of the compressor. Figure 21.11 shows such an indicator diagram of single-stage reciprocating compressor.

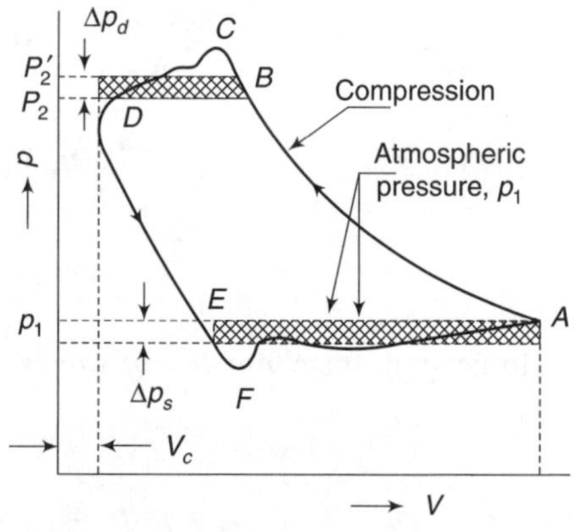


**Fig. 21.11** Indicator diagram

The piston at *A*, with cylinder fully charged with gas at the suction pressure ($p_1$) will, during the compression stroke, compresses the gas along the curve *ABC*. When the pressure in the cylinder exceeds the ordinate $p_2$, at point *B*, the energy of the piston is directed to unseating the discharge volume. At this point, the curve slope breaks into the cap at point *C*, which releases the gas in the cylinder into the cylinder discharge passage. But due to valve flutter, the gas is discharged in series of pressure waves, diminishing in amplitude and general slope, until the piston reaches the end of the stroke at *D*. Thereafter, the direction of piston travel is reversed. On the return stroke, the trapped gas in the clearance volume ($V_c$) expands along the curve *DEF*. A further reduction of pressure occurs below point *E* at ordinate $p_1$ to point *F*, the differential pressure ($\Delta p_s$) is required to unseat the suction valve. *Due to flutter, the suction line is wavy.* So it is clear that the suction takes place only when the pressure in the cylinder is below atmospheric pressure, $p_1$.

The areas above line *BD* and below *AE* represent the work required to unseat the valve element and overcome the valve friction. The sharp bursts of flow at points *C* and *F* undoubtedly are the source of menacing pulsation forces that on occasions violently vibrate the piping system. The cross-hatched area ($\Delta p_s$), the strip below *AE* represents the average suction valves frictional resistance.

## 21.12 MEAN EFFECTIVE PRESSURE AND INDICATED POWER

It has already been seen that the height on the *p-V* diagram is proportional to the pressure on the compressor piston. The average height of the diagram in proper pressure units is the *indicated mean effective pressure* and can be easily determined by measuring the enclosed area with a planimeter and dividing the area by the length of the diagram. Thus, by definition of the spring constant,

Indicated mean effective pressure (imep)

$$= p_m = \text{Mean diagram height} \times \text{Spring constant}$$

$$= \frac{\text{Area of indicator card (cm}^2) \times \text{Spring scale (bar/cm)}}{\text{Length of diagram (cm)}}\ ; \text{bar}$$

Indicated power, (*ip*) is based on the indicated mean effective pressure

Let $A$ = piston area, m$^2$ $\qquad$ $L$ = length of stroke, m

$N$ = rpm $\qquad$ $p_m$ = *mep*, bar

$x$ = number of cylinders in a stage

$i$ = 1 for single acting, $\qquad$ $i$ = 2 for double acting.

Hence, work done per cycle = $100\ p_m\ (\text{kN/m}^2) \times A \times (\text{m}^2) \times L\ (m) \times (x \times i)$

Work done per minute = $100\ p_m\ ALN\ (x.i)$; kg/min

$$\therefore \quad \text{Indicated power} = ip = \frac{100\ p_m\ ALN}{60}\ (x.i);\ \ \text{kW} \qquad ...(21.20)$$

For single acting and single cylinder

$$ip = \frac{100\ p_m\ ALN}{60}\ ;\ \text{kW} \qquad ...(21.21)$$

## 21.13 COMPRESSOR POWER EFFICIENCIES

**(i) Mechanical Efficiency.** *It is defined as the ratio of indicated power to the brake power.*

$$\eta_m = \frac{ip \text{ of compressor}}{bp \text{ of compressor}} \quad \text{where } \eta_m \text{ varies from 0.85 to 0.96.}$$

**(ii) Isothermal Efficiency.** *It is defined as the ratio of isothermal work input to actual work input (i.e., polytropic work input)*

$$\eta_{iso} = \frac{\text{Isothermal work input}}{\text{Actual work input}} = \frac{p_1 V_1 \ln (p_2/p_1)}{[n/(n-1)]\, p_1 V_1 \, [(p_2/p_1)^{(n-1)} - 1]}$$

or
$$\eta_{iso} = \frac{\ln (p_2/p_1)}{[n/(n-1)] \left[(p_2/p_1)^{(n-1)/n} - 1\right]} \qquad \text{...(21.22)}$$

As we have seen, **isothermal compression is the best because it needs the minimum work input**. But it has some drawbacks, for instance, atmospheric air always contains some water vapour in it, so that isothermal compression would lead to condensation of air in the cylinder. By circulating chilled water through the jacket isothermal efficiency greater than 100% is obtained resulting in $n < 1$. Though, this will require the minimum work input, but in practice, it creates many problems and **so it is seldom done**.

**(iii) Adiabatic Efficiency, $\eta_{ad}$:** *It is defined as actual work input to adiabatic work input.*

$$\eta_{ad} = \frac{\text{Actual work input}}{\text{Adiabatic work input}} = \frac{[n/(n-1)] \left[(p_2/p_1)^{(n-1)/n} - 1\right]}{[\gamma/(\gamma-1)] \left[(p_2/p_1)^{(\gamma-1)/\gamma} - 1\right]} \qquad \text{...(21.23)}$$

## (B) ROTARY COMPRESSOR

## 21.14 SLIDING VANE COMPRESSORS

*Sliding vane compressors (Fig. 21.12) (a) consist of deep slots cut into the rotor to accommodate thin vanes varying from 20 to 30 which are free to move radially. The rotor is mounted eccentrically in the housing. With the rotation of rotor, the centrifugal forces acting on the vanes force them outward against the housing thereby dividing the cresent shaped space into several compartments.*

Hence, in the initial stage, the space between the rotor, the cylinder and the two adjacent vanes is enlarged and the air is drawn into this space through the suction port. When the rotor reaches the point just opposite to the eccentricity of the rotor, the suction process ends, and further rotation causes the enclosed space to decrease thus the compression of air occurs due to decrease in volume and is finally discharged through the outlet port. *The compressor is so designed that half of the total pressure rise takes place through internal reversible compression before the entrapped air is released to the delivery port. The air is further compressed in an irreversible manner due to momentarily back flow of high pressure from the receiver.* The actual flow rate and pressure rise at constant speed will be reduced by leakage. The volumetric efficiency can vary between 0.6 and 0.9 depending on the size of the machine, the quality of design, method of lubrication and cooling employed.

The displaced volume $V_D$ is given by

$$V_D = \pi\, \varepsilon\, l\, (D + \varepsilon) \qquad \text{...(21.24)}$$

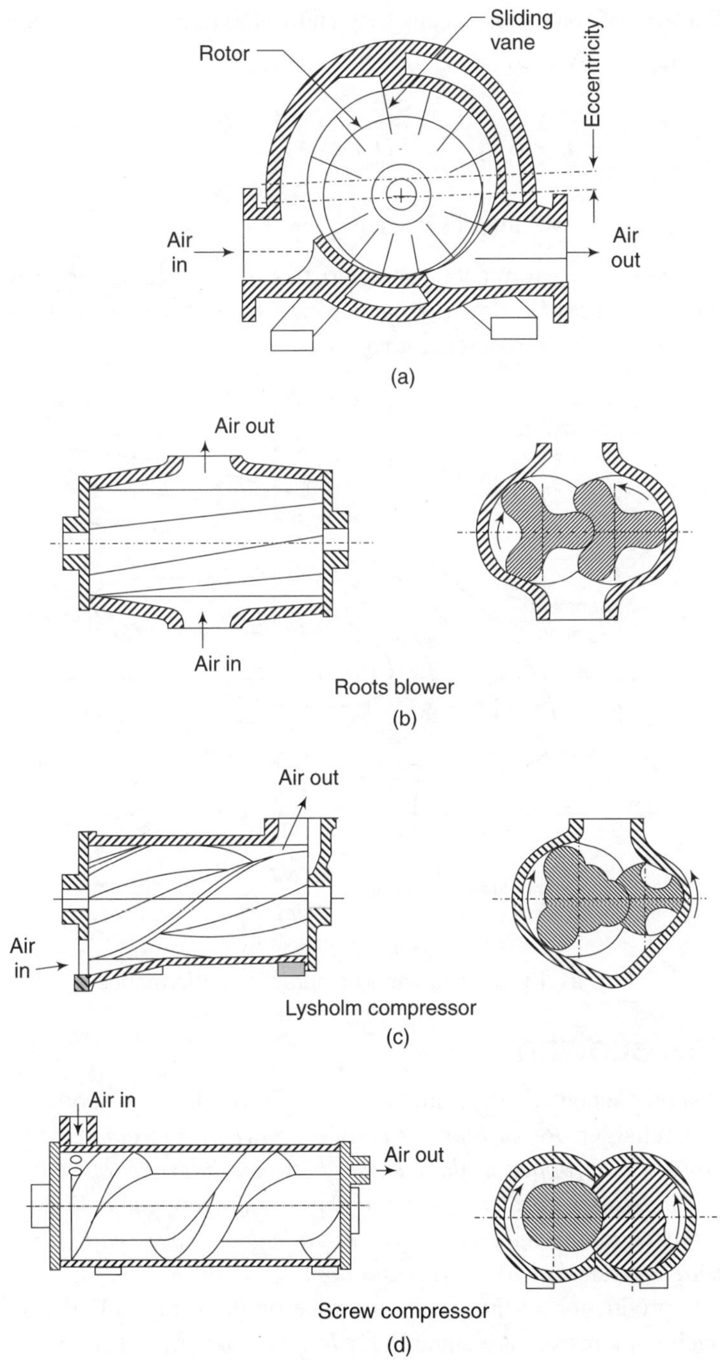


**Fig. 21.12** Rotary compressor

where $D$ is a diameter of rotor, $\varepsilon$ is eccentricity and $l$ is axial length of compressor.

The mass flow rate is given by

$$\frac{\dot{m}\sqrt{T_0/T_{std.}}}{p_0/p_{std.}} = K \times \rho_i\, \eta_v\, N\, \varepsilon\, l(D + \varepsilon) \qquad ...(21.25)$$

where $K$ is constant, $\eta_v$ is volumetric efficiency.

Figure 21.13 shows the performance characteristics of a typical sliding vane compressor. It is worthy to note that the mass flow ratio at constant speed depends on pressure ratio. This compressor results in low isentropic efficiency.

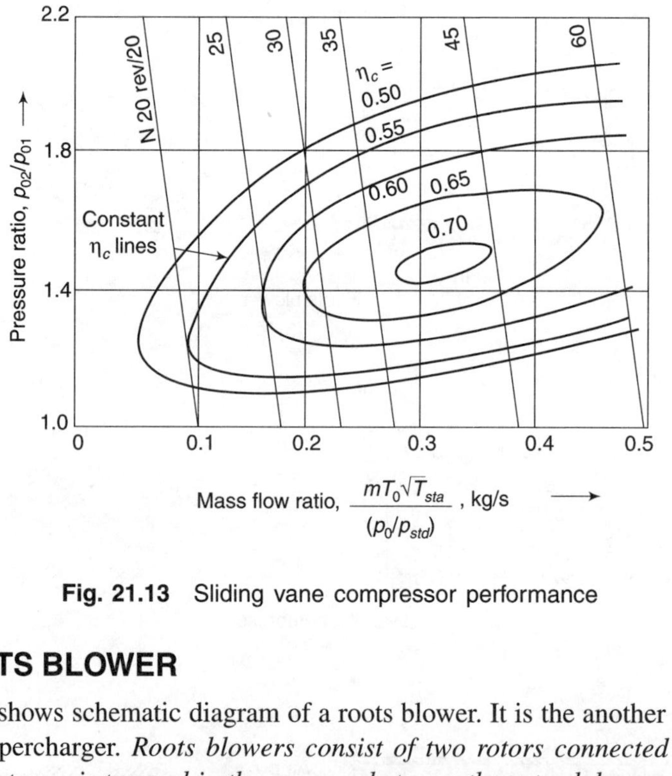


**Fig. 21.13** Sliding vane compressor performance

## 21.15 ROOTS BLOWER

Figure 21.12(b) shows schematic diagram of a roots blower. It is the another type of positive displacement supercharger. *Roots blowers consist of two rotors connected by gears. With the rotation of rotors, air trapped in the recesses between the rotor lobes and the housing is carried towards the delivery port without significant change in volume. Since the suction side is closed under this situation, the trapped air is suddenly compressed by the back flow from the higher pressure delivery line as the recesses are open to the delivery line.* This intermittent delivery results in production of non-uniform torque on the rotor and pressure pulses in the delivery line. Such compressors are suitable for *low pressure ratio* (up to 1.3).

Parameters such as running clearances, rotor length, rotational speed and pressure ratio affect the volumetric efficiency. The mass flow rate is expressed by

$$\frac{\dot{m}\sqrt{T_0/T_{std.}}}{p_0/p_{std.}} = \text{constant} \times \rho_i\, \eta_v\, N\, R^2\, l \qquad ...(21.26)$$

where $R$ = rotor radius and $l$ = blower length

Figure 21.14 shows a typical performance map of a roots blower. It is obvious from the figure that at constant speed, the mass flow decreases with increasing pressure ratio through resulting decrease in volumetric efficiency.

*Surge and choking phenomena do not limit its performance range which is supposed to be its main advantage. Its disadvantages lie in its high noise level, poor efficiency and large size.*

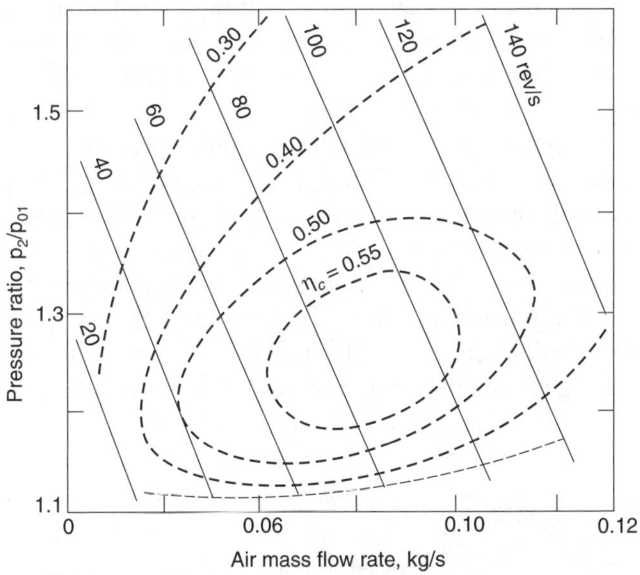


**Fig. 21.14** Performance map of roots blower

## 21.16 LYSHOLM AND SCREW COMPRESSORS

In this compressor (Fig. 21.12 c, d), the lobes are helical having a large pitch that intermesh. As opposed to roots blower, the air is gradually compressed. *One rotor (male) has three or four lobes, while the other rotor (female) has four, five or six recesses.* The entrapped air is progressively compressed as it moves through the narrowing passage ways formed by the lobes. These compressors must be precision machined to achieve close to tolerances between rotating and stationary elements for satisfactory operation. *They run at very high speeds ranging from 3000 rpm to 30,000 rpm and are generally internally cooled. High values of volumetric and isentropic efficiency are the main advantages of this compressor.*

## (C) TURBOCOMPRESSORS

## 21.17 CENTRIFUGAL COMPRESSOR*

*A centrifugal compressor is a radial flow turbomachine in which the compression is achieved due to dynamic action of working fluid (air). Kinetic energy imparted to the air by rotation of rotor is partly changed into pressure energy in the rotor due to centrifugal and diffusion action and the rest in diffuser and volute casing due to diffusion action.* In the case of turbocharger, it is coupled with an exhaust driven turbine. Centrifugal compressor may be single or multi-stage up to 8. It is well suited to the high mass flow rates at the relatively low pressure ratios (up to 3.5) required by the engine. For better performance, generally it runs at very high speed.

Figure 21.15 shows a schematic diagram of a centrifugal compressor. *It consists of a inlet pipe, a rotating bladed impeller, a stationary diffuser (with or without vanes) and a collector or volute casing—to bring the compressed air leaving the diffuser to the engine intake system* or to the other users. The compression process is shown on *h-s* diagram (Fig. 21.16). Air at stagnation state 0 is accelerated in the inlet section to pressure ($p_1$) and velocity ($C_1$) due to suction created at impeller inlet. The enthalpy drop from $h_{01}$ to $h_1$ is equal to $C_1^2/2$. Compression process in the impeller from pressure $p_1$ to $p_2$ which is governed by centrifugal and diffusion action is represented by process 1-2. The stagnation state 02 corresponds to full conversion of kinetic energy $C_2^2/2$ into pressure energy while to the corresponding isentropic state is shown by 2*s*. The process in diffuser in represented by 2-3 in which pressure rises from $p_2$ to $p_3$ due to diffusion action, i.e., slowing down of the kinetic energy in a carefully shaped expanding passage. The final state in the collector has a static pressure ($p_3$) and low kinetic energy ($C_3^2/2$) while the stagnation state is given by 03.

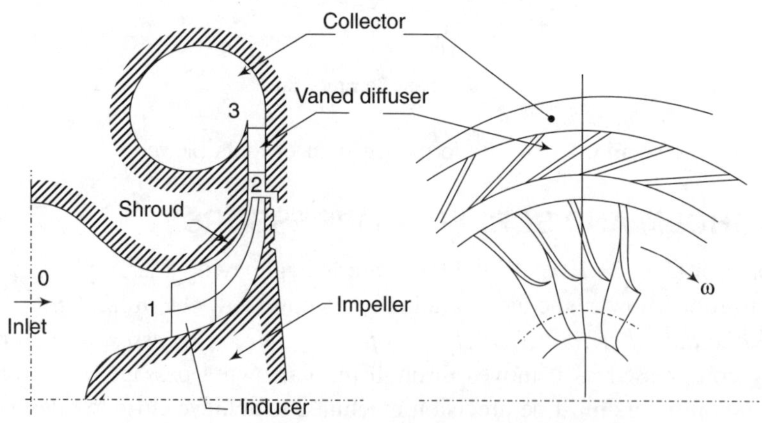


**Fig. 21.15** Centrifugal compressor stage

* For more detail see Author's book *"Applied Thermodynamics"*.

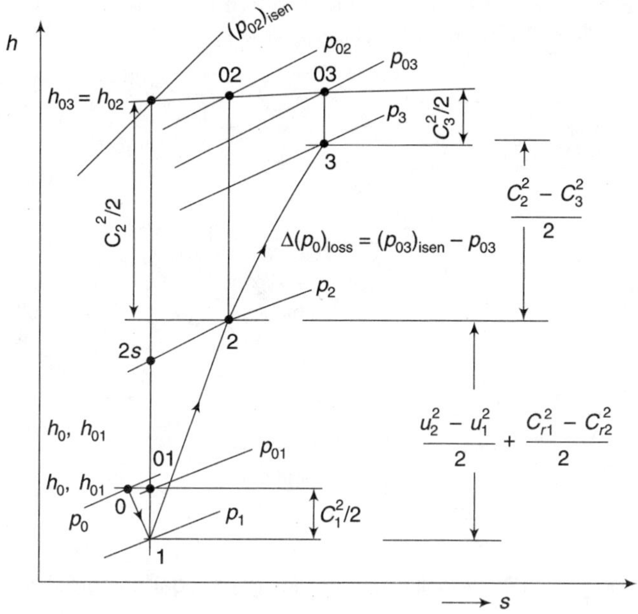


**Fig. 21.16** Representation of compression process in centrifugal compressor on *h-s* diagram

Figure 21.17 shows the inlet and outlet velocity diagrams for centrifugal compressor impeller. Here $C_1$ and $C_2$ are absolute air velocities, $u_1$ and $u_2$ are peripheral velocities, $C_{r1}$ and $C_{r2}$ are relative velocities and $C_{w1}$ and $C_{w2}$ are tangential components of absolute velocities of impeller inlet and outlet respectively. The torque ($T$) exerted on the air by the impeller is expressed as the rate of change of angular momentum.

Thus, $$T = \dot{m}(r_2\, C_{w2} - r_1\, C_{w1}) \qquad ...(21.27)$$

The theoretical work input to the compressor is the rate of work transfer to the air which is given by the product of torque and angular speed. Thus,

$$-W_c = T.\omega = \dot{m}\,\omega(r_2\, C_{w2} - r_1\, C_{w1})$$

or $$-W_c = m(u_2\, C_{w2} - u_1\, C_{w1}) \qquad ...(21.28)$$

Equation (13.28) is know as **Euler's equation of motion**

Since the work transfer to the gas occurs only in impeller, so

$$-\dot{W}_c = \dot{m}(u_2\, C_{w2} - u_1\, C_{w1}) = \dot{m}(h_{03} - h_{01}) = \dot{m}c_p\,(T_{03} - T_{01}) \qquad ...(21.29)$$

With the help of velocity triangles, the work done per kg of fluid may be expressed as

$$-w_c = \frac{u_2^2 - u_1^2}{2} + \frac{C_{r1}^2 - Cr_2^2}{2} + \frac{C_2^2 - C_1^2}{2} \qquad ...(21.29a)$$

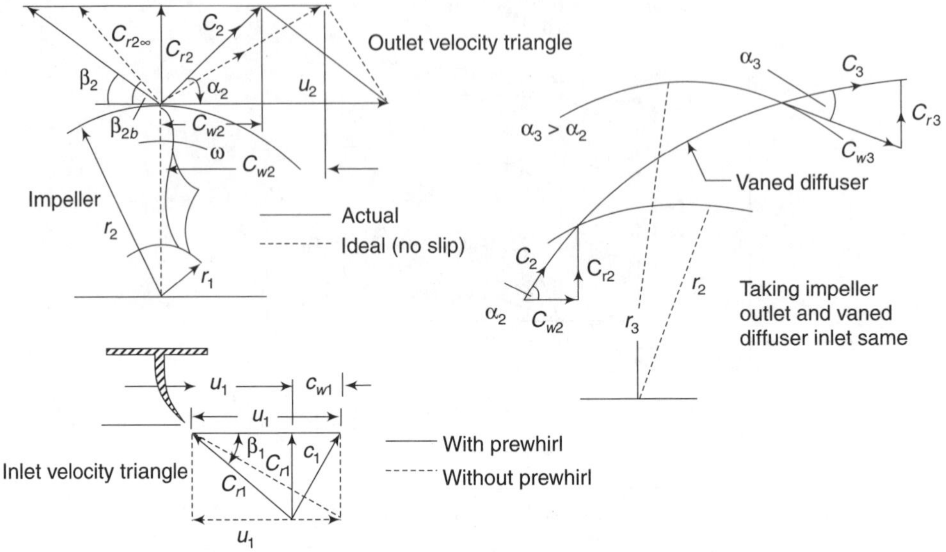


**Fig. 21.17** Velocity diagram for centrifugal compressor

*The first term of Eq. (21.29a) signifies the* **centrifugal action,** i.e., *the fuel enters at lower diameter and comes out at higher diameter while second term refers to the* **diffusion action**, i.e., decrease *in relative velocity from inlet to outlet. These two actions centrifugal and diffusion are responsible for pressure rise in impeller.* The **rest K.E.** $(C_2^2 - C_1^2)/2$ will be converted into pressure energy in diffuser and volute casing.

*The maximum possible pressure energy (i.e., head) developed by a centrifugal machine is given by*

$$H = \frac{u_2^2}{g}$$

The pressure ratio per stage varies from 2 to 5. It is possible to achieve a delivery pressure of up to 400 bar in a multi-stage (about 8) centrifugal compressor stage with an isentropic efficiency varying from 84% to 86%.

## 21.18 SLIP FACTOR OF CENTRIFUGAL COMPRESSOR

*Under ideal conditions, fluid particles follow exactly the path of blades when the blades are infinite in number* without any thickness resulting in $\beta_{2\infty}$ equal to the impeller blade angle $\beta_{2b}$ at the tip. But in actual practice, when the number of blades is finite, the air is trapped between the impeller vanes and the fluid is reluctant to move around the impeller (Fig. 21.18) due to inertia. Generally, the *slip factor is defined as the ratio of work done under actual conditions to the work done under ideal conditions (i.e., infinite number of blades)*

$$\mu = \frac{u_2 C_{w2} - u_1 C_{w1}}{u_2 C_{w2\infty} - u_1 C_{w1\infty}} \qquad ...(21.30)$$

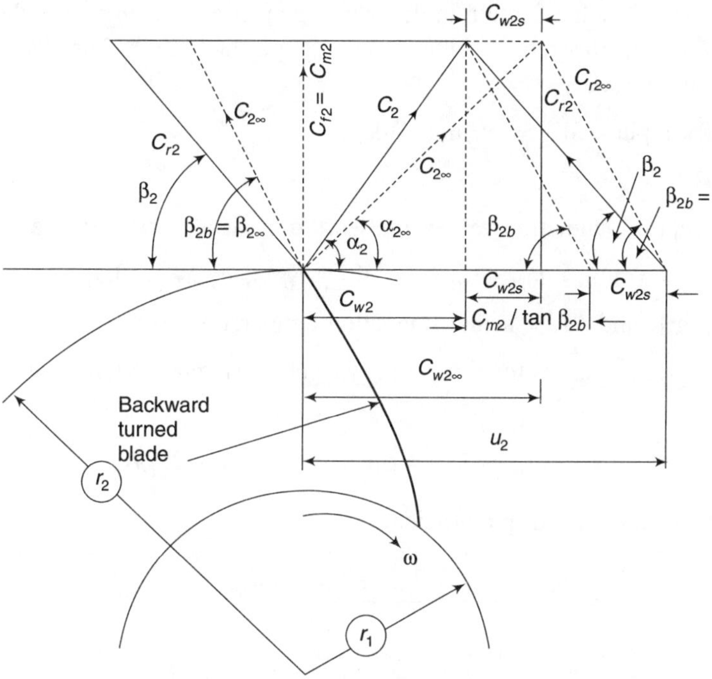


**Fig. 21.18** Showing slip

If $\quad C_{w1} = 0 = C_{w1\infty}$, hence

$$\mu = \frac{C_{w2}}{C_{wu2\infty}} \quad \text{...(21.31)}$$

For radial blades, $C_{w2\infty} = u_2$, as $\beta_{2b} = \beta_{2\infty} = 90°$, hence $\mu = \dfrac{C_{w2}}{u_2} = 1 - \dfrac{C_{w2s}}{u_2}$ ...(21.32)

where $C_{w2s}$ is slip velocity

The value of μ varies from 0.9 to 0.92

## 21.19 PRESSURE RATIO IN A CENTRIFUGAL COMPRESSOR STAGE

The theoretical work input to the compressor may be given by the Euler's work assuming $C_{w1}$ to be zero

Theoretical work input = $w_{th} = C_{w2} . u_2$ ; J/kg ...(21.33)

Maximum head developed = $(H_{th})_{max} = u_2^2/g$ as in the ideal case, $C_{w2} = u_2$. ...(21.34)

From the concept of slip factor, the theoretical work input for radial bladed impeller is

$$w_{th} = \mu\, u_2^2; \quad \text{J/kg} \quad \text{...(21.35)}$$

Due to other losses, such a friction in the blade channel and disc friction, the actual work input is greater than this theoretical value. A power input factor, $\sigma$ is introduced here to take into account these losses.

$\therefore$ Actual work input = $W_{act} = \sigma\mu\, u_2^2$; J/kg ...(21.36)

where $\sigma = 1.04$

From the concept of stagnation values, the work done is also expressed as

Actual work done = $w_{act} = h_{02} - h_{01} = c_p\,(T_{02} - T_{01}) = c_p\,\Delta T_0$; J/kg ...(21.37)

where $c_p$ is in J/kg K and $\Delta T_0$ = actual stagnation temperature rise.

Equating the above equations for work done, we get the stagnation temperature rise through the impeller

$$w = c_p\,(T_{02} - T_{01}) = \sigma\mu\, u_2^2 = c_p\,(T_{02s} - T_{01})/\eta_{0is} \qquad \text{...(21.38)}$$

where $\eta_{0is}$ = total to total isentropic efficiency

$$= \frac{h_{02s} - h_{01}}{h_{02} - h_{01}} = \frac{T_{02s} - T_{01}}{T_{02} - T_{01}} = \frac{\Delta T_{0s}}{\Delta T_0}$$

Thus, $$\frac{T_{02s}}{T_{01}} = 1 + \frac{\eta_{0is}\,(T_{02} - T_{01})}{T_{01}} = 1 + \frac{\eta_{0is}\,\Delta T_0}{T_{01}} = 1 + \eta_{0is}\,\frac{\sigma\mu\, u_2^2}{c_p\,T_{01}} \qquad \text{...(21.39)}$$

But $$\frac{p_{02}}{p_{01}} = \left[\frac{T_{02s}}{T_{01}}\right]^{\gamma/\gamma - 1} \qquad \text{...(21.40)}$$

Hence, the pressure ratio achieved in a centrifugal compressor stage is given by

$$\frac{p_{02}}{p_{01}} = \left[1 + \eta_{ios}\,\frac{\sigma\mu\, u_2^2}{c_p\,T_{01}}\right]^{\gamma/\gamma - 1} = \left[1 + \eta_{ise}\,\frac{\Delta T_0}{T_{01}}\right]^{\gamma/(\gamma - 1)} \qquad \text{...(21.41)}$$

If $u_2$ or $\Delta T_0$ and $\eta_{ise}$ are known, we can calculate stagnation pressure ratio in a centrifugal compressor stage.

## 21.20 INFLUENCE OF IMPELLER BLADE SHAPE

Let us suppose, for simplicity, that the number of impeller blades is infinitely large, i.e., the flow direction at impeller outlet is identical with the blades outlet angle. The various shapes of impeller blades on the same impeller are schematically represented in Fig. 21.19. On the basis of outlet angles ($\beta_2$), the shape of blades can be divided into three basic groups.

(i) Backward turned blades; $\beta_2 < 90°$

(ii) Radial blades, $\beta_2 = 90°$

(iii) Forward turned blades, $\beta_2 > 90°$

The velocity triangle in each case is shown. In all the cases, the valve of $u_2$ is the same. From Fig. 21.19 it is obvious that for backward turned blades, the tangential component ($C_{w2}$) is much reduced (lowest in all) and consequently for a given speed, the impeller will need a low energy input as is obvious from Euler's equation. It the case of forward turned blades $C_{w2}$ is increased (maximum) and consequently, the energy input in forward turned blades is the maximum. In addition to this, the value of absolute velocity ($C_2$) is the maximum. In the case of radial blades, the valve of $C_{w2}$ lies between backward and forward curved blades. Hence, from the *efficiency point of view, the backward turned blades are the best.*

Generally, the blades of an impeller are radial ($\beta_{2b}$ = 90°) which are used as turbocharger and heavy duty system. However, recent trend is to use backward turned ($\beta_{2b}$ < 90°) blades which results in higher compression efficiency. Forward turned blades are used in exhaust fan systems where use of K.E. is important.

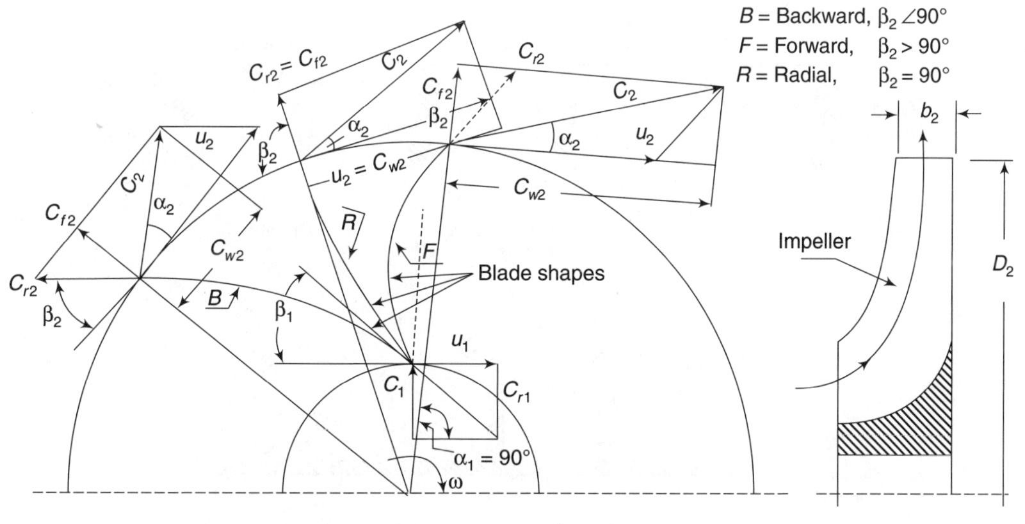


**Fig. 21.19** Influence of blade shape

## 21.21 PREWHIRL

The centrifugal compressors employed in the gas turbines or jet engines run at very high speed and there is always a likelihood of the formation of shock wave in the flow passage. To avoid this, the Mach number at any point in the flow passage should not exceed more than unity. The maximum value of Mach number corresponding to relative velocity is found at the inlet. Thus,

$$\text{Mach number at inlet} = \frac{\text{Velocity of air relative to vanes at inlet}}{\text{Sound velocity}}$$

or

$$M_{r1} = \frac{C_{r1}}{\sqrt{(\gamma R\, T_1)}} \quad \text{where } T_1 = \text{static temperature at inlet.} \qquad ...(21.42)$$

If $C_{r1}$ for a certain fixed flow rate (i.e., $C_{f1}$) and the rotational speed (i.e., $u_1$) are too high for satisfactory operation of compressors, the fluid is given an **initial pre-rotation** called **prewhirl** by means of fixed guide vanes attached to the casing at inlet of compressors. When the air passes into the impeller eye through these vanes, the air is given the necessary prewhirl. *The relative velocity $C_{r1}$ (i.e., $M_{r1}$) is reduced without affecting the flow velocity $C_{f1}$, i.e., mass flow rate.* This is shown in Fig. 21.20. Here *ABC* is the inlet velocity triangle showing the axial; energy of air without any prewhirl ($\alpha_1 = 90°$) and *ABC′* is the inlet velocity with prewhirl. The work input with prewhirl is

$$w = C_{w2}\, u_2 - C_{w1} . u_1 \qquad ...(21.43)$$

From Eq. (21.43), it follows *that the prewhirl reduces the work input by the amount $C_{w1} . u_1$ at the cost of additional construction* and also reduces the Mach number at inlet with reference to relative velocity to avoid the shock formation.

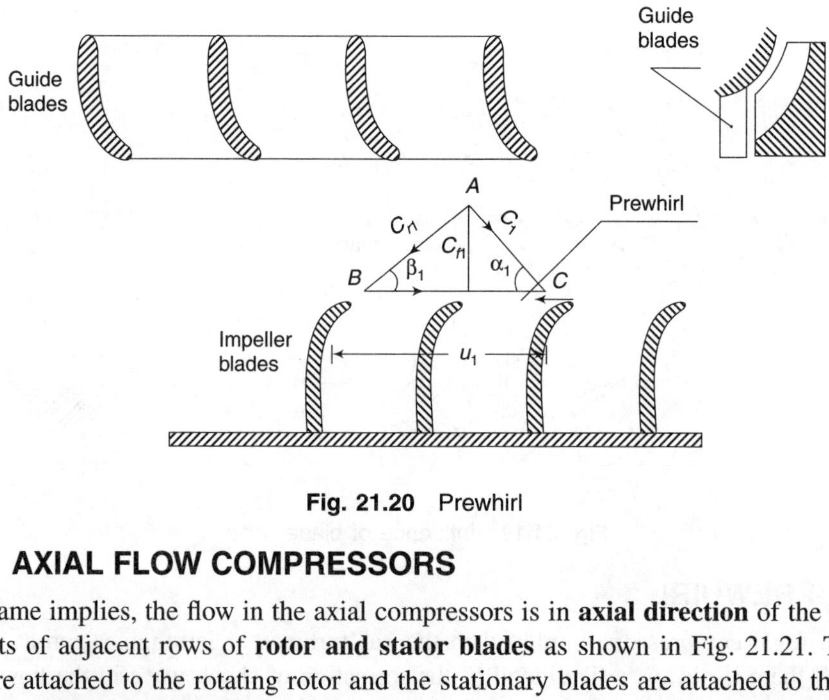


**Fig. 21.20** Prewhirl

## 21.22 AXIAL FLOW COMPRESSORS

As the name implies, the flow in the axial compressors is in **axial direction** of the machine. It consists of adjacent rows of **rotor and stator blades** as shown in Fig. 21.21. The rotor blades are attached to the rotating rotor and the stationary blades are attached to the casing. Some part of the kinetic energy imparted to the air by the rotor is converted into pressure energy in the rotor due to *diffusion action* because of decrease in relative velocity and the rest is converted into pressure in the stator due to *diffusion action* on account of decrease in absolute velocity. The stator blades also redirect the air into angle suitable for entry to the succeeding set of moving blades. *The rotor as well as the stator blade channels area of diverging type* (i.e., *area increases from inlet to outlet*). which is responsible for diffusion action. *A row of moving blades with a succeeding set of stationary blades is* called **a stage of axial flow compressor**. Generally, the maximum pressure ratio achieved in a stage of axial compressor is about 1.12 to 1.2, hence to obtain pressure ratio of 14, attainable by axial compressor, 15 to 20 stages are required. GE of USA has built a 18 stages compressor having delivery pressure equal to 13.5 bar in 1988. *Presently (in 2020), axial compressor has been built having pressure ratio up to 40 through multistaging.*

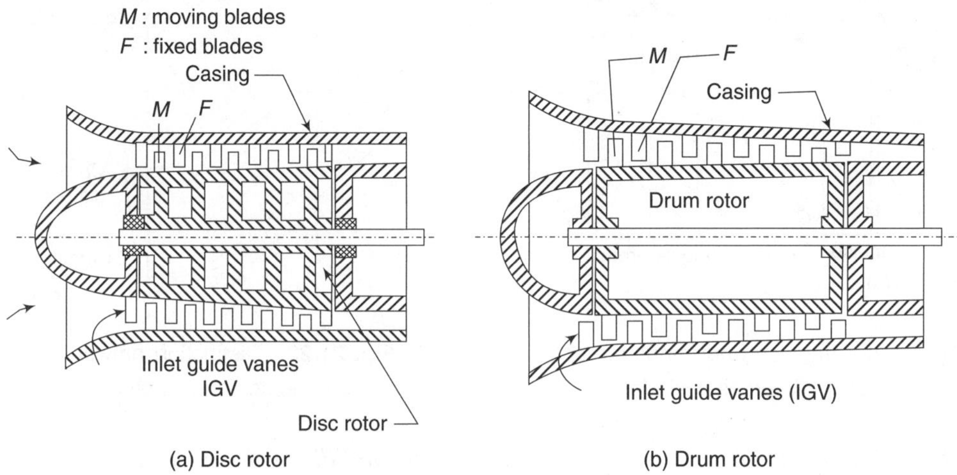


**Fig. 21.21** (a, b) Axial flow compressor with two forms of rotor

Two forms of rotor have been used, namely, the **drum and disc types** (Fig. 21.21a, b). *The **disc type** is used where consideration of low weight is more important than cost as in aircraft applications.* The drum type is more suitable for static industrial application, however, ***bolted disc type** is also common for industrial gas turbine.* In some applications, combination of both types have been used.

The other important constructional aspect is the ***contraction of the flow annulus*** from the low to the high pressure end of the compressor. This practice is followed to maintain the axial velocity at a reasonably constant level throughout the length of the compressor despite the increasing density of the working fluid. *It is worth to note that most of the axial flow compressor are designed on the basis of constant axial velocity throughout the stage.* Contraction of annulus is also possible by contracting the casing and diverging the rotor.

According to aerodynamic conditions, the *blades are of aerofoil section*. The compressors may be of *impulse or reaction type*. The blade profile, pressure and velocity distribution are shown in Fig. 21.22 and Fig. 21.23 showing the difference between axial compressor and turbine blades.

### Materials

***Rotor bladings*** The following compressor rotor blade materials are in common use.

*(a) Fibrous composites, (b) Aluminium (c) Titanium (d) Steel and (e) Nickel alloy. The above listings are in increasing order of weight and their ability to withstand high temperature.*

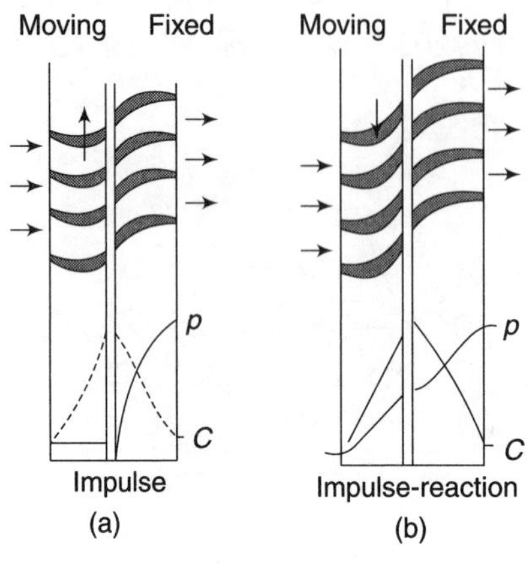


**Fig. 21.22** Impulse and impulse-reaction axial compressor blades

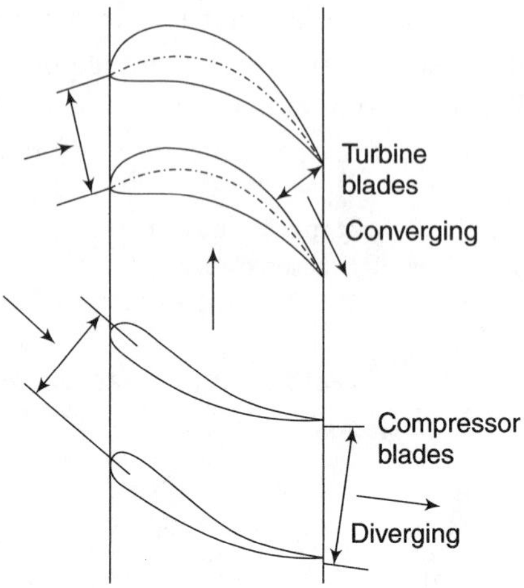


**Fig. 21.23** Distinction netween axial compressor and turbine blades

***Rotor.*** *The rotor material includes steel for shafts and disc but aircraft engines may use titanium at the front stages and nickel alloy in the rest.*

***Stator bladings.*** *Stator bladings may be of the same materials as that of rotor but steel is the most common.*

***Casings.*** *Casings may be of cast magnesium, aluminium steel or iron or fabricated from titanium or steel.*

Numerically controlled machine makes dies and the blades are manufactured by **precision forging.** The blades are also manufactured by **investment casting.** The blade root and tips are machined on NC or CNC machine. Blades are also machined by CNC copying machine.

## 21.23 VELOCITY TRIANGLE AND WORK DONE OF A STAGE OF AXIAL FLOW COMPRESSORS

Figure 21.24 shows the velocity triangle of axial flow compressor. Assume that air approaches the rotor blade with an absolute velocity ($C_1$) at an absolute angle ($\alpha_1$) with axial direction. Combining with the peripheral velocity ($u$) of the blades, the relative velocity of the air will be $C_{r1}$ at angle of $\beta_1$ with the axial direction. The axial component of the absolute velocity $C_1$ is $C_a$. Air after passing from the diverging passage formed between the rotor blades which does work on air and increases its absolute velocity, will emerge with relative velocity ($C_{r2}$) at an angle ($\beta_2$) less than $\beta_1$ due to the camber of the blades. Due to diffusion action, the relative velocity ($C_{r2}$) at rotor outlet is less than ($C_{r1}$). Therefore, some pressure rise will be accomplished in the rotor. The velocity ($C_{r2}$) in combination with $u$ gives the absolute velocity ($C_2$) at the exit from the rotating blades making an angle ($\alpha_2$) with the axial direction velocity, $C_a$. Here, the axial component of velocity $C_a$ (or flow velocity $C_f$) is constant throughout the stage.

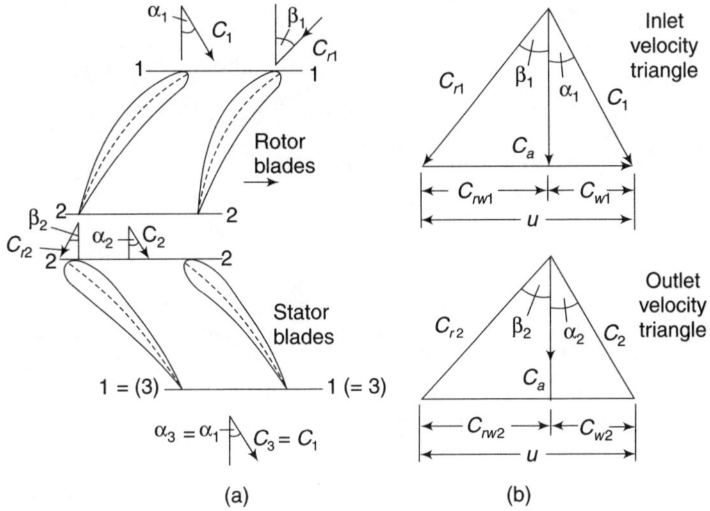


**Fig. 21.24** Axial flow compressor stage

The air then passes through the passages formed by the stator blades where its pressure is further increased due to diffusion and the air comes out with the velocity, $C_3$ (less than $C_2$) inclined at angle of $\alpha_3$ with the axial direction. The stator blades are designed such that $C_3 = C_1$ and $\alpha_3 = \alpha_1$ so that the air enters the next stage without shock. Note that unlike centrifugal machines, the angles here are measured from the axial and not from the tangential direction. From the velocity triangle

$$u/C_a = \tan \alpha_1 + \tan \beta_1 \qquad \text{...(21.44)}$$

$$u/C_a = \tan \beta_2 + \tan \alpha_2 \qquad \text{...(21.45)}$$

Assume 1 kg of flow of air through the compressor stage. From Newton's second law of motion.

Tangential force per kg = $C_{w2} - C_{w1}$

Work absorbed the stage per kg of air

$$w_{st} = u\,(C_{w1} - C_{w2}) = c_p\,(T_{02} - T_{01}) \qquad \text{...(21.46)}$$

where $C_{w1}$ and $C_{w2}$ are the peripheral or whirl components of absolute velocity at inlet and exit of rotating blades respectively.

The expression for work done may be put in terms of axial velocity and air angles.

$$w_{st} = u\,C_a\,(\tan \alpha_2 - \tan \alpha_1) \qquad \text{...(21.47)}$$

From Eqs. (21.44) and (21.46) we have

$$w_{st} = u\,C_a\,(\tan \beta_1 - \tan \beta_2) \qquad \text{...(21.48)}$$

The work input to the axial flow compressor may also be obtained by the Euler's equation. We get for each kg of air delivered.

$$w_{st} = u_2\, C_{w2} - u_1\, C_{w1}$$

Since $C_{w2} - C_{w1} = C_{rw1} - C_{rw2}$ and $u_1 = u_2 = u$, the above equation is modified as

$$w_{st} = u(Cr_{w1} - Cr_{w2}) = u\,.\,\Delta C_{rw} = c_p\,(\Delta T_0)$$

By the use of velocity triangle and cosine theorem.

$$w_{st} = \frac{C_{r1}^2 - C_{r2}^2}{2} + \frac{C_2^2 - C_3^2}{2} \qquad \text{...(21.49)}$$

Here, $C_3 = C_1$

The above equation is graphically represented on *h-s* or *T-s* diagram shown in Fig. 21.25. *The first term on the right side of the above equation introduces the part of the work supplied by a rotating cascade which is inverted into pressures due to diffusion action in rotating cascade itself. The second term represents the increments of kinetic energy in rotating cascade that has to be converted into pressure energy in stationary cascade.* Comparing this equation to the work input to centrifugal compressor, we find that the term centrifugal action $[(u_2^2 - u_1^2)/2]$ *is missing in axial flow compressors. Due to this reason the pressure ratio per stage in axial flow compressor is much less than of centrifugal compressor.* Regardless of efficiency of compression, the stage temperature rise will be given by the equation

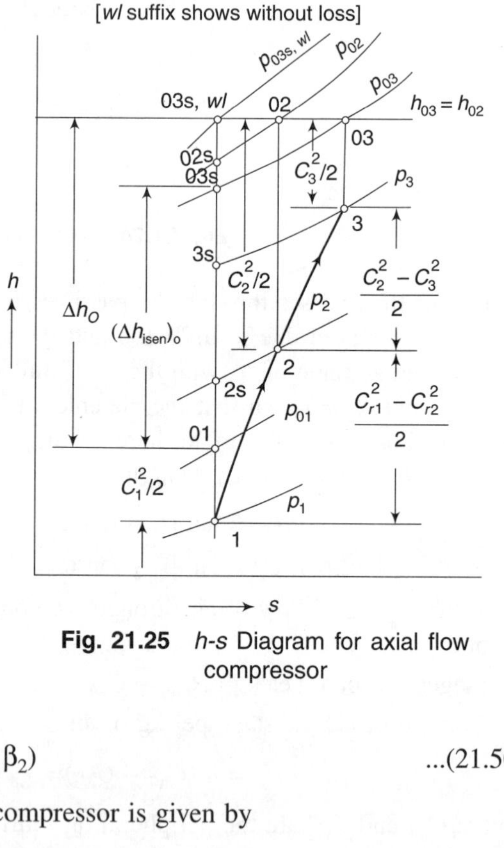


**Fig. 21.25** *h-s* Diagram for axial flow compressor

$$(\Delta T_0)_{act} = \frac{uC_a}{c_p}(\tan\beta_1 - \tan\beta_2) \qquad \text{...(21.50)}$$

The pressure ratio achieved in axial flow compressor is given by

$$\frac{p_{02}}{p_{01}} = \left[1 + \eta_{ise}\frac{\Delta T_0}{T_{01}}\right]^{\gamma/(\gamma-1)} \qquad \text{...(21.51)}$$

## 21.24 LOSSES IN AXIAL FLOW COMPRESSOR STAGE

In actual practice, various losses occur while the fluid through a compressor stage. The total pressure loss arises in three ways as given below.

(i) Profile losses on the surface of the blades

(ii) Skin friction on the annulus walls

(iii) Secondary flow losses

Figure 21.26 shows the various losses represented on the graph between stage efficiency and flow coefficient. Some performance curves are also given in the chapter on **Supercharging and Turbocharging.**

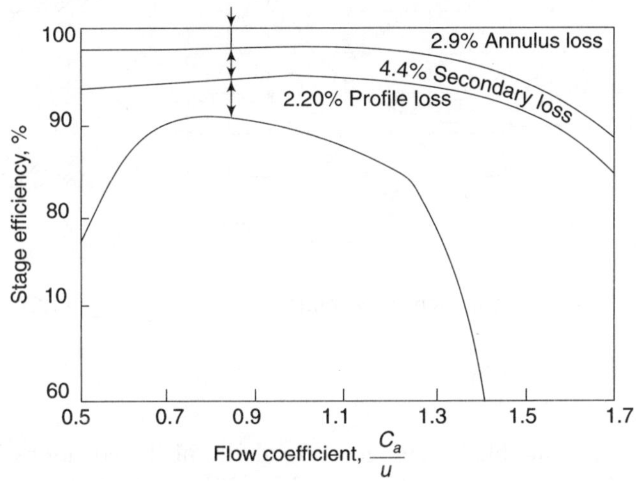


**Fig. 21.26** Losses in compressor stage

### 21.24.1 Profile Losses

*By profile losses, we mean the total pressure loss of two-dimensional rectilinear cascade arising from the skin friction on the surface and due to the mixing of flow particles after the blades.* This is mostly determined experimentally. According to **Howell** (Fig. 21.27a)

$$C_{D_p} = f\left(\frac{i - i^*}{\in^*} . \frac{s}{c}\right) \quad ...(21.52)$$

where $i$ = angle of incidence and $i^*$, $\in^*$ = angle of incidence and deflection angle under design conditions respectively.

This is unique curve used for all cascades. The off-design performance can be calculated from this curve.

### 21.24.2 Skin Friction Losses on the Annulus Walls

*The wall friction total pressure losses arising from the skin friction on the annulus walls and the secondary losses are difficult to analyse as the boundary layer growth on these walls is a complex three-dimensional phenomenon.* According to Howell, the annulus drag coefficient (Fig. 21.27b) may be calculated from the formula:

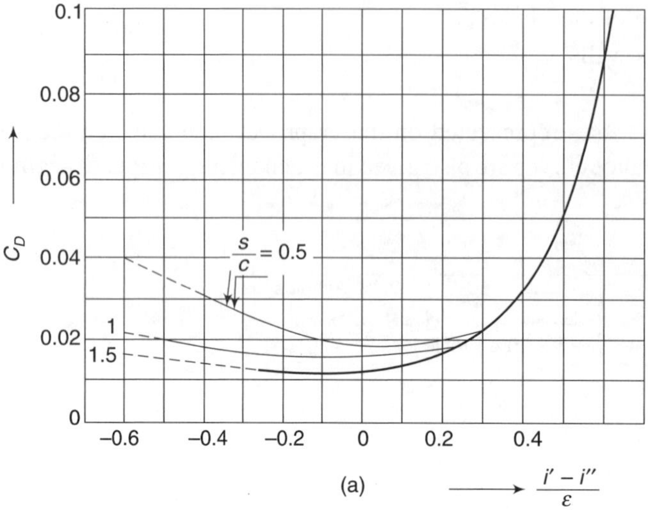


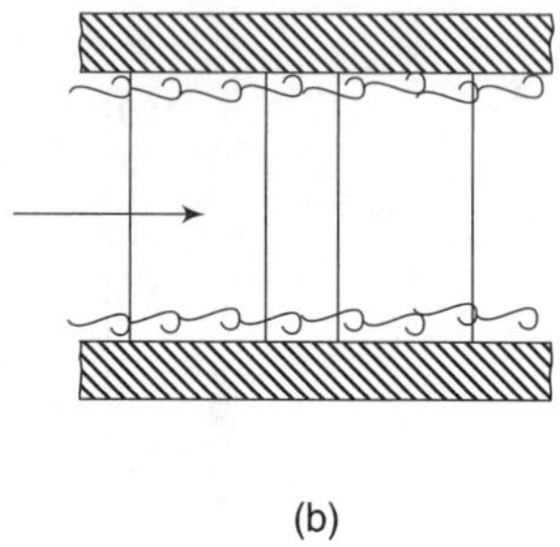


**Fig. 21.27** (a) Off-design performance of cascades

**Fig. 21.27** (b) Annulus drag

$$C_{Da} = \frac{0.02s}{b} \qquad ...(21.53)$$

where $b$ is the height of the blade. The general validity of the equations is limited because according to the relation, the drag coefficient ($C_D$) will increase if fewer blades are arranged along the periphery. Actually, two parameters, s/c and b/c are mainly responsible for this loss. According to **Haller**:

$$C_{Da} = 0.018\,\frac{c}{b} \qquad ...(21.54)$$

This equation is more suitable for calculating $C_D$. It takes better account of losses because of the fact that the frictional losses on the walls depend to a greater extent on the circumference effects between the wall and the profile boundary layers than on the wall boundary layer alone.

### 21.24.3 Secondary Flow Losses

*Generally, secondary flows in an axial flow compressor blade channel are produced by the combined effects of curvature and boundary layer. In other words, secondary flow is produced when a streamwise component of velocity is developed from the deflection of an initially sheared flow. Such secondary flow occurs when a developed pipe flow enters bend, when a sheared flow passes over an aerofoil of finite thickness or an aerofoil or finite lift or when a boundary layer meets an obstacle normal to the surface over which it is flowing (e.g., a wind blowing past a telegraph pole).*

*One of the most important engineering aspects of secondary flow occurs in axial turbomachinery aerodynamics where boundary layers growing on the casing and hub walls of the machines are deflected by rows of blades, stationary and rotating.* Consider the curved blade passage shown in Fig. 21.28(a) and in particular the section *X–X*. As a result of curved flow path, the pressures on the concave side of the passage at *A* will be higher than those of the convex side at *B*. This variation in pressure is shown by a thick line in Fig. 21.28(a). *Due to boundary layer (i.e., friction effect), the flow velocities at the inner and outer radii, i.e., at the upper and lower boundary walls of the passage are lower than those in the centre or mid height of the blade. This means that due to boundary layer, pressures at the inner and outer radii will be higher than those in the mid-height. As a result, the pressure gradients across the section X–X are reduced and the pressure changes at the upper and lower ends of the passage is shown by the dotted line. Hence, we see* $p_A > p_C > p_D > p_B$, *with the result that two secondary flows shown by the dotted lines are set up in the blade channels.* If the secondary flow is superimposed on the main flow, the central part of the issuing jet will follow a path such as *EF* in which the jet is underdeflected whereas the parts of the jet near the roof and floor of the blade will follow a path such *EG* and will tend to be overdeflected. A view of secondary losses at the exit of row of blades is shown in Fig. 21.28(b).

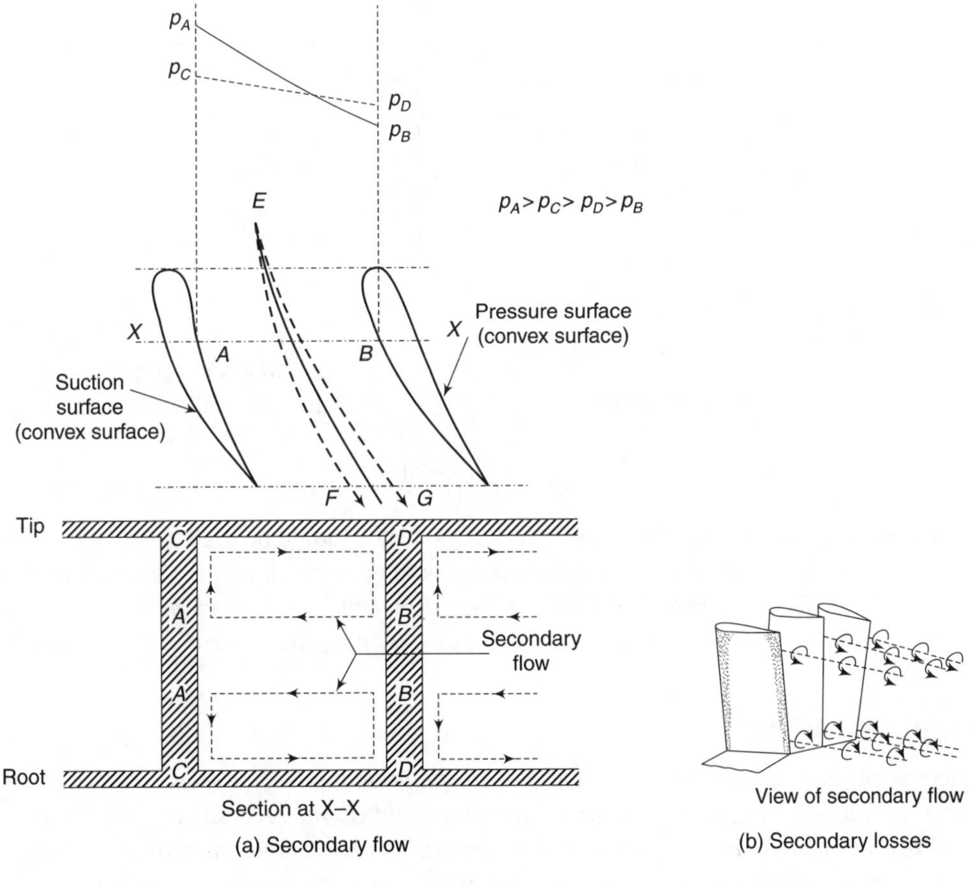


(a) Secondary flow

(b) Secondary losses

**Fig. 21.28**

## 21.25 WORK DONE FACTOR

It is difficult to draw the actual velocity triangles at root, mean and tip because the contraction of main stream cannot be estimated accurately. *The axial velocity at the mean section is known to be greater than the mean axial velocity, and the mean blade section (and most of the span) will do less work than is estimated from velocity triangles based on the mean axial velocity. The tip and root sections might be expected to do more work because of the low values of axial velocity at entry to these sections, but local stalling of the blades will reduce this work. This means that the average temperature rise across the stage will be less than as estimated from velocity triangles based on the mean through flow velocity.* Figure 21.29(a) shows the axial velocity distribution along blade height. The variation of work done factor with the number of stages are shown in Fig. 21.29(b). **Howell** suggested that the stagnation enthalpy rise across a stage may be given by

$$c_p \Delta T_0 = \lambda u (C_{a2} \tan \alpha_2 - C_{a1} \tan \alpha_1) \qquad ...(21.55)$$

where $\lambda$ is a work done factor which may be taken as approximately 0.85.

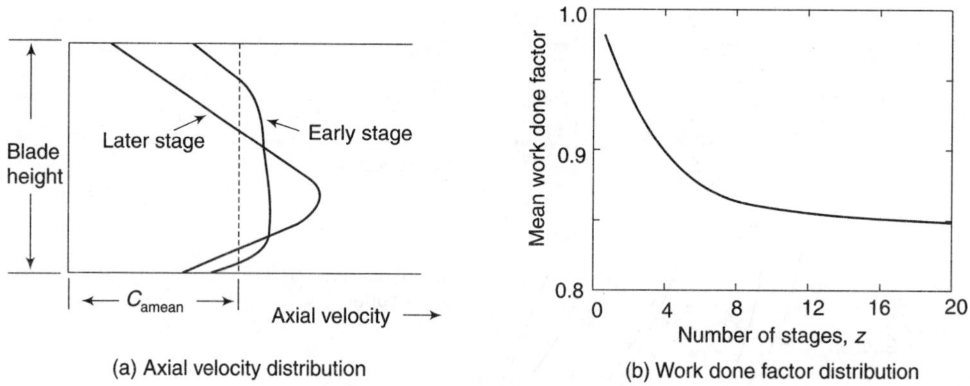


**Fig. 21.29**

According to other investigators, the work done factor should be high (0.96) at entry where the annulus wall boundary layers are thin, but should be reduced progressively to 0.85 in the later stages of the compressor. **Howell** *suggested that the mean work done factor depends upon the number of stages. As the number of stages increases, it first reduces rapidly and then slowly.*

## 21.26 SURGING

There is an *unstable limit of operation of centrifugal and axial flow compressors, known as surging. Surging is caused due to unsteady, periodic and reversal of flow through the compressor when the compressor has to operate at less mass flow rate than a predetermined value (a value corresponding to the maximum pressure). As the flow is drastically reduced than this predetermined value, this surge can reach such a magnitude as to endanger the compressor and in many cases mechanical failure may result.* The alternating stresses to which the rotor of the machine is subjected during this irregular working condition, may damage compressors, rotor blading and seals. *Severe surges have been known to bend the rotor shaft.*

To understand the causes of surging, consider the head capacity curve of a centrifugal or axial flow compressor as shown in Fig. 21.30. This curve consists of **two portions**, portion *AB* having a **positive slope** *and portion BC having a* **negative slope**. *A* and *C* are the two extreme points of the curve. Point *A* corresponds to the flow condition when the discharge valve is fully closed and point *C* corresponds to the flow condition when the discharge valve is fully open. In the case of fully closed discharge valve, the flow rate is zero and the static pressure developed is termed shut-off head. In the case of fully open discharge valve, the mass flow rate is the maximum but the pressure developed is zero. This means that the compression efficiency is zero.

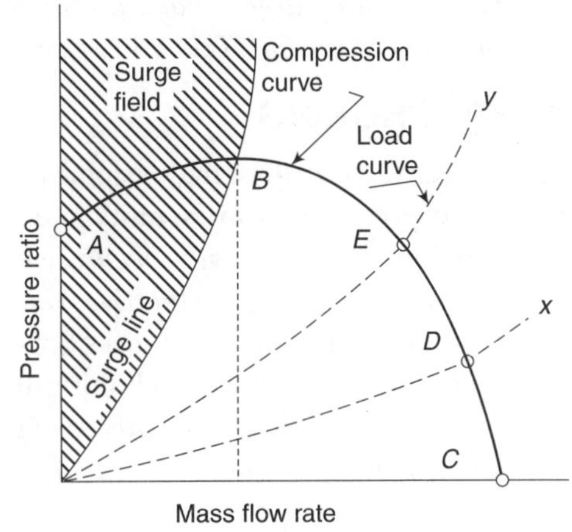


**Fig. 21.30** Surging

Now, consider the case when the discharge valve is opened slightly, the delivery of fluid will commence and the static pressure will rise due to contribution of diffuser towards pressure rise. As the discharge valve is further opened, the pressure goes on increasing till the maximum pressure rise represented by point *B* is attained. This is the point of maximum efficiency for a given speed, inlet pressure and temperature. With the further opening of the discharge valve beyond point *B*, the mass flow rate increases but the pressure rise and efficiency decreases. Finally, when the pressure ratio approaches unity, the mass flow rate would be the maximum but the compression efficiency would be zero.

*The opening and closing of the discharge valve can be treated as the external load on the compressor.* This means that the compressor has to operate in conjunction with external load on the compressor. *Thus, the point of the compressor curve and load curve will give the* operating point. For example, *D* shows such an operating point. Now, suppose that external load is increased due to the partial closing of the discharge valve. This will increase the back pressures in the discharge line. As a result, the point of operation will shift to point *E* which is the point of intersection of the new load line (*OY*) and the compression curve. The new operating point (*E*) is possible and stable because the compressor will produce more head which will take care of increased back pressure in the discharge line. With the further increase in external load due to closing of the valve, the operating point shifts say between the region *AB*. Under this condition of operation, the mass flow rate will be less than the predetermined value, a value corresponding to point *B*. The compressor produces less head and consequently the delivery will not be *possible since high pressure would be existing in the discharge line.*

*This will result momentarily in the reversal of flow.* After a lapse of time, the fluid from the delivery pipe would leave and the back pressure would fall. The compressor would again start delivering fluid and the *cycle would be repeated* with instability occurring again. *Thus, when the flow through the compressor is less than a predetermined value, a surge or pulsation begins and the air surges to and fro through the whole compressor instead of giving a steady stream in one direction. This gives rise to* **surging**.

## 21.27 CHOKING

Theoretically, mass flow rate becomes the maximum when the pressure ratio is unity, i.e., there is no compression. *This generally occurs when the Mach number corresponding to relative velocity at inlet becomes sonic.* The maximum mass flow rate possible in a compressor is known as **choking flow**. *Choking means fixed mass flow rate regardless of pressure ratio i.e., the characteristic becomes vertical as shown in Fig. 21.31.*

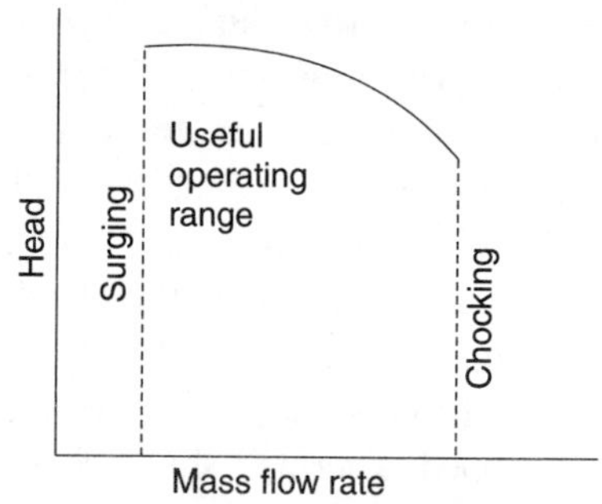


**Fig. 21.31** Choking

## 21.28 STALLING

*Stalling of a stage of axial flow compressor is defined as the aerodynamic stall or the breakaway of the flow from suction side of the blade aerofoil.* The breakaway of the flow from the suction side may be due to lesser mass flow rate than designed value due to non-uniformly in the blade profile. Thus, stalling is ahead phenomenon of surging. A multi-stage compressor may operate stable in the unsurged region with one or more of the stages *stalled and the rest of stages unstalled.* In other words, *stalling is a local phenomenon whereas surging is a complete system phenomenon.*

Many researchers have studied the operation of single-stage fans and compressors at flow coefficients low enough to cause stalling. The most detailed survey of various types of stalling has been given by **Smith** and **Fletcher**. Their results have been summarised by Carter as follows:

(i) ***A rotating stall** which may rotate in the opposite direction of rotation of the compressor at approximately one-half the compressor rotational speed.* Such a stall cell includes a region of very small through flow velocity or even reversed flow. This mechanism of stalling appears to be the basic stalled flow; as the flow is reduced, other stall cell appears.

(ii) **A band of reversed flow located by spiders.**

(iii) A centrifuging effect—a flow inwards at the root and outwards at the tip.

(iv) A flow in and out at all radii which Carter calls a **surge**.

According to Carter, it is possible that type (iii) and (iv) may include type (i).

Since the **rotating stall** appears to be the basic, hence it is necessary to discuss its mechanism in detail.

***Rotating stall.*** Consider flow through a cascade as shown in Fig. 21.32. Suppose that some non-uniformity in the approaching flow or in a blade profile causes blade $B$ to stall. The air now flows on the blade $A$ at an increased incidence due to the blockage of channel $AB$. The blade $A$ stall as a result but the flow on the blade $C$ is now at a lower incidence, and blade $C$ may unstall. In this way, a stall cell may move along the cascade in the direction of the lift of the blades. In the relative system, stalls appear to move in direction opposite to that of the rotor blades. In other words, *the **phenomenon of reduction in lift force at high angle of incidence is known as stalling***.

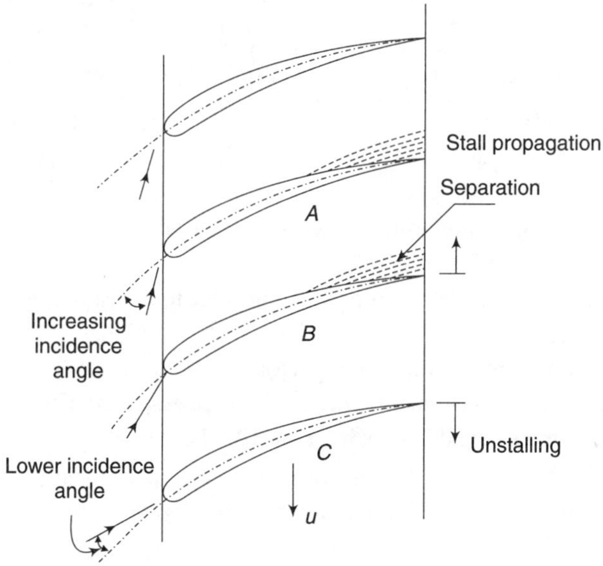


**Fig. 21.32** Mechanism of stall propagation

*If the natural frequency of vibration of the blades coincides with the frequency at which the stall cells pass a blade; the resonance and possible mechanical failure of the blades may occur.* Carter suggests that since stall cell has a substantially square wave from higher harmonics may be present that may excite a blade.

## 21.29 PERFORMANCE OF CENTRIFUGAL AND AXIAL FLOW COMPRESSORS

Figure 21.33(a, b) shows the relationship between pressures ratio, power and efficiency versus rate of flow for various values of speed such as $N_1$, $N_2$, etc., for centrifugal and axial flow compressor respectively. The pressure curve is flatter in the case of centrifugal compressor. At a certain speed, efficiency increases as the flow rate increases and reaches a maximum value after which it decreases. Accordingly, as the flow rate increases the power consumed also increases.

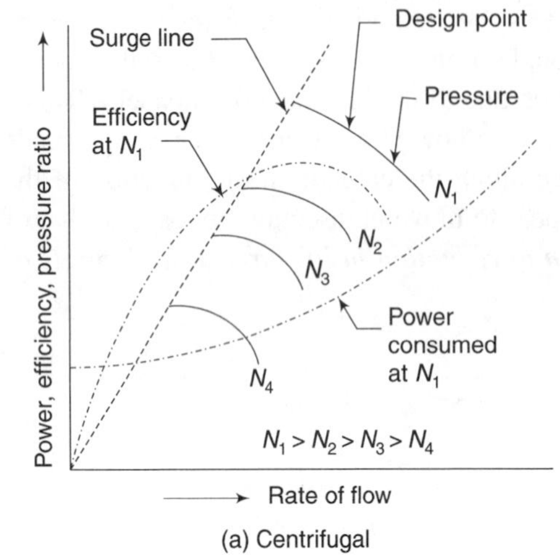


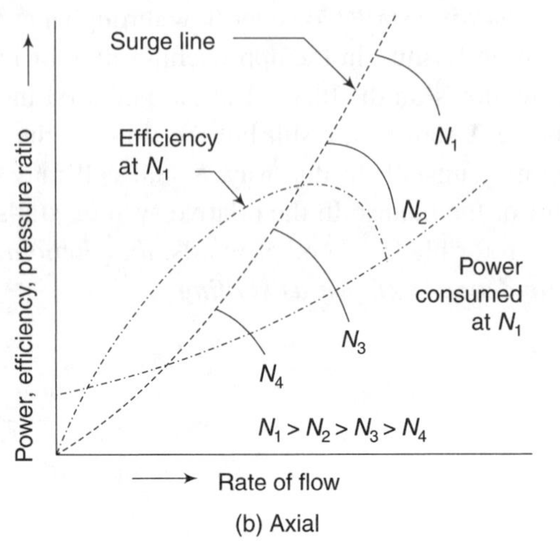


**Fig. 21.33** (a, b) Performance of centrifugal and axial flow compressors

Such a plot makes no consideration of varying inlet temperature and pressure. In addition to this, these plots cannot show the comparison of performance for similar compressors of different sizes. To account for all these, the performance curves are plotted with dimensionless parameters. **These dimensionless parameters are pressure ratio,** $p_2/p_1$, speed parameter, $N_1/\sqrt{T_1}$ and flow parameters $\dot{m}\sqrt{T_1}/p_1$ **(See Fig. 13.12).**

Figures 21.34 and 21.35 show the performance and the constant efficiency curves in the case of centrifugal flow and axial compressors respectively.

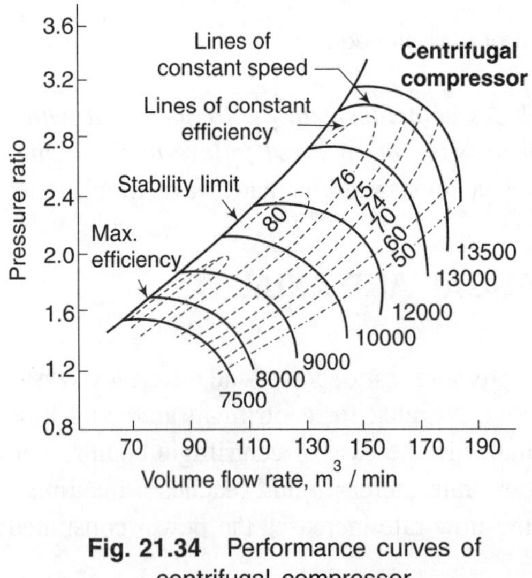


**Fig. 21.34** Performance curves of centrifugal compressor

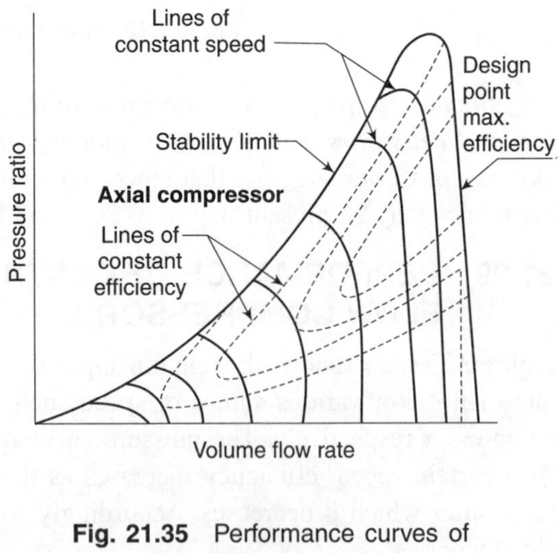


**Fig. 21.35** Performance curves of axial flow compressor

## 21.30 COMPARISON BETWEEN CENTRIFUGAL AND RECIPROCATING COMPRESSORS

The following salient features distinguish between centrifugal and reciprocating compressors shown in tablular form

| S.No. | Parameters | Reciprocating | Centrifugal |
|---|---|---|---|
| (i) | Balancing efficiency | Presence of reciprocating masses makes the machine *poorly balanced* and hence vibration problems are greater. | Absence of reciprocating masses makes the machine *better balanced.* |
| (ii) | Mechanical | Presence of numerous sliding or bearing members *lower its mechanical efficiency.* | Absence of numerous sliding or bearing members *improvers its mechanical efficiency.* |
| (iii) | Cost | *Higher* installed *first cost.* | *Lower* installed *first cost* where pressure and volume conditions are favourable. |
| (iv) | $r_p$ per stage | Pressure ratio per stage is *high,* about 5 to 8. | Pressure ratio per stage is about 2 to 4.5. |
| (v) | Delivery pressure | Capability of delivering high pressure. By multi-staging, high delivery pressure upto 5000 atm may be achieved. | Capable to delivering medium pressure. By multi-staging, the delivery pressure up to 450 atm may be achieved. |
| (vi) | Volume | Capable of delivering small volume. By using multi-cylinders, the volume may be increased. | Capable of delivering greater volumes per unit of delivering space. |
| (vii) | Flexibility | Greater flexibility in capacity and pressure range. | No flexibility in capacity and pressure range. |
| (viii) | Maintenance | Higher maintenance expense. | Lower maintenance expense. |
| (ix) | Continuity of service | Lesser continuity of service. | Greater continuity of service and dependability. |
| (x) | Compression efficiency | Higher compression efficiency at compression ratio above 2. | Higher compression efficiency at compression ratio less than 2. |
| (xi) | Drive | Adaptability to low speed drive. | Adaptability to high speed, low maintenance cost drivers like turbines. |
| (xii) | Attention | More *operating attention.* | Less operating attention. |
| (xiii) | Mixing of lub. of fluid | There is always a chance of *mixing* of working fluid with lubricating oil. | No chance of *mixing* of working fluid with lubricating oil. |
| (xiv) | Suitability | Suitable for low, medium and high pressure and low and medium gas volumes. | Suitable for low and medium pressure and large gas volumes. |

## 21.31 COMPARISON BETWEEN CENTRIFUGAL AND AXIAL FLOW COMPRESSORS

The following main features give the comparison between centrifugal and axial flow compressors shown in tabular form. The comparison helps us to form some criteria for the choice between these two compressors.

| *S.No.* | *Parameters* | *Centrifugal* | *Axial* |
|---|---|---|---|
| (i) | Flow direction | The flow is *radial.* | The flow is parallel to the direction axis of the machine (*i.e., axial*). |
| (ii) | $r_p$ per stage | The pressure ratio per stage is high about 4.5:1. Thus, the unit is compact. In supersonic compressors, the pressure ratio per stage is about 10 but at the cost of efficiency. Operation is not so difficult and risky. | The pressure ratio per stage is *low*, about 1.2:1. This is due to absence of centrifugal action. To achieve the pressure ratio equal to that of per stage in centrifugal above 10 stages are required. Thus, the unit is less compact and less rugged. The pressure ratio per stage in supersonic compressor is about 10 but the efficiency drops rapidly. The compression is achieved due to shock wave but the operation is risky. |
| (iii) | Efficiency | The *isentropic efficiency* of centrifugal compressor is about 80 to 84%. | The *isentropic efficiency* of axial flow compressor is about 86 to 88%. |
| (iv) | Range of operation | Centrifugal compressors have a *wide range of operation* between surging and choking limit, The head capacity curve is flat. The part load performance is better. | Axial compressors have a *narrow range of operation* between surging and choking limit. The part load performance is poor. |
| (v) | Frontal area | Centrifugal compressors have a *larger frontal area*, so more drag force if used in jet engines. | Axial flow compressors have a *smaller frontal area* for the same mass flow rate and pressure value than that of centrifugal. This makes the axial flow compressors more suitable for jet engine due to less drag. |
| (vi) | Effect of deposit | When working with the contaminating fluids, the accumulation of deposits on the surface of flow passage *do not adversely affect the performance.* | The *accumulation of deposits* on the surface of flow passages *affect* adversely the performance of axial flow compressors. |
| (vii) | Starting torque | Centrifugal compressors need *low starting torque.* | Axial flow compressors need *high starting torque.* |
| (viii) | Cost | Its *construction is simple, rigid, relatively cheap.* At high altitude, it is less sensitive to icing troubles. | Its construction *is complex and costly.* It is sensitive to icing troubles at high altitudes. |
| (ix) | Multistaging | *Multistaging is slightly difficult* and up to 450 bar delivery pressure is possible. | It is *most suitable for multi-staging* and delivery pressure up to 50 bar is possible in a single casing. |
| (x) | Application | It is used in blowing engines in steel mills low pressure refrigeration, big central air-conditioning plants, fertilizer industry. Supercharging internal combustion engines, gas pumping in long distance pipe line, petrochemical industries, etc. Previously it was used in jet engines. | Due to higher efficiency and smaller frontal area, axial flow compressor are mostly used in *jet engines.* Axial flow compressor are also preferred in *power plant gas turbines* and steel mills. |

**EXAMPLE 21.1** A single cylinder, single acting reciprocating compressor takes in air 6 m$^3$/min at 1 bar 15°C and compresses air to a delivery pressures of 5 bar guage. The possible compression processes are:

(a) Isentropic compression, $n = 1.4$

(b) Polytropic compression, $n = 1.25$

(c) Isothermal compression, $n = 1$

Find for each case neglecting clearance, calculate the following:

(i) Temperature at the end of compression, (ii) work done during suction by air, (iii) work done on air during compression, (iv) work done on air during delivery, (v) net work done on air, (vi) heat transfer, and (vii) change in internal energy.

Also, compare the brake power of the compressor for the index of compression process $n = 1.25$ and 1.35 (due to partial failure of water cooling system) if the mechanical efficiency is 0.86. Find isothermal efficiency for $n = 1.25$. Take atmospheric pressure 1 bar. Comment on the results.

***Solution*** Refer to Figs. 21.3 and 21.4.

**(a) Isentropic Compression**

(i) Temperature at the end of compression,

$$T_2 = T_1 \left(\frac{p_2}{p_1}\right)^{(\gamma - 1)/\gamma} = 288(6)^{0.4/1.4} = 480.53 \text{ K} \qquad \textbf{\textit{Ans.}}$$

(ii) Work done by air during suction

$$= p_1 V_1 = \frac{1 \times 10^2 \times 6}{60} = \frac{6}{60} \times 10^2 = 10 \text{ kW} \qquad \textbf{\textit{Ans.}}$$

$$\text{Mass of air} = \dot{m} = \frac{p_1 V_1}{RT_1} = \frac{1 \times 10^2 \times 6}{0.287 \times 288} = 7.259 \text{ kg/min}$$

(iii) Work done on air during compression,

$$= \frac{\dot{m}R}{\gamma - 1}(T_2 - T_1) = \frac{7.259 \times 0.287}{0.4 \times 60}(480.53 - 288) = 16.71 \text{ kW} \qquad \textbf{\textit{Ans.}}$$

(iv) Work done on air during delivery

$$= p_2 V_2 = \dot{m}RT_2 = \frac{7.259}{60} \times 0.287 \times 480.53 = 16.68 \text{ kW} \qquad \textbf{\textit{Ans.}}$$

(v) Net work done on air $= W = 16.88 + 16.71 - 10 = 23.39$ kW

This may also be calculated by

$$W = \frac{\gamma}{\gamma - 1} p_1 V_1 \left[\left(\frac{p_2}{p_1}\right)^{(\gamma - 1)/\gamma} - 1\right] = 23.39 \text{ kW}$$

(vi) Since the process is isentropic, so the heat transfer $Q$ is zero. ***Ans.***

(vii) Since the heat transfer is zero, the change in internal energy is equal to work done on air = $\Delta U$ = 23.39 kW

**(b) Polytropic Compression; *n* = 1.25**

(i) $T_2 = 288\ (6)^{0.25/1.25} = 412.1\text{K}$ ***Ans.***

(ii) Work done by air during suction = $p_1 V_1 = \dfrac{1 \times 10^2 \times 6}{60} = 10$ kW

(iii) Work done on air during compression

$$\frac{\dot{m}R}{n-1}(T_2 - T_1) = \frac{7.259 \times 0.287}{0.25 \times 60}(412.1 - 288) = 17.23 \text{ kW}$$ ***Ans.***

(iv) Heat transferred to cylinder walls = $\dfrac{\gamma - n}{\gamma - 1}$ × Work done during polytropic compression

$$= \frac{1.4 - 1.25}{0.4} \times 17.23 = 6.461 \text{ kW}$$ ***Ans.***

(v) Work done during delivery

$$= \dot{m}RT_2 = (7.259/60) \times 0.287 \times 412.1 = 14.309 \text{ kW}$$ ***Ans.***

(vi) Net work done on air = $W$ = 17.23 + 14.309 – 10 = 21.539 kW ***Ans.***

This may also be found from the equation

$$W = \frac{n}{n-1} p_1 V_1 \left[\left(\frac{p_2}{p_1}\right)^{(n-1)/n} - 1\right] = 21.539 \text{ kW}$$

(vii) Change in internal energy $\Delta U = (Q - W)$ = 17.23 – 6.461 = 10.769 kW ***Ans.***

**(c) Isothermal compression**

(i) Due to isothermal compression temperature remains constant, thus

$$T_1 = T_2 = 288 \text{ K}$$

(ii) Work done by air during suction = $p_1 V_1 = 1 \times 10^2 \times \dfrac{6}{60} = 10$ kW ***Ans.***

Work done during compression

$$= p_1 V_1 \ln(p_2/p_1) = 1 \times 10^2 \times (6/60) \ln 6 = 17.91 \text{ kW}$$ ***Ans.***

There is no change in internal energy because the temperature remains constant in isothermal compression.

(iii) Since the internal energy is zero, hence heat transferred will be equal to the work done during compression.

Heat transfer = 17.91 kW ***Ans.***

(iv) Work done on air during delivery = $V_2\,p_2 = p_1\,V_1 = \dfrac{6}{60} \times 10^2 = 10$ kW **Ans.**

(v) Net work done = 17.91 + 10 – 10 = 17.91 kW

This may also be found from $W = p_1\,V_1 \ln (p_2/p_1) = 17.91$ kW

$$\text{Now } \eta_{iso} = \frac{\text{Work done due to isothermal compression}}{\text{Actual work done}} = \frac{17.91}{21.539} = 83.15\% \qquad \textbf{Ans.}$$

Brake power for ***n* = 1.25**,

$$bp = \frac{n}{n-1} \times \frac{p_1\,V_1}{60 \times \eta_m}\left[\left(\frac{p_2}{p_1}\right)^{(n-1)/n} - 1\right]$$

$$= \frac{1.25}{0.25} \times \frac{1 \times 10^2 \times 6}{60 \times 0.86}\,[6^{0.25/1.25} - 1] = 25.05 \text{ kW} \qquad \textbf{Ans.}$$

For $n = 1.35$,

$$bp = \frac{1.35}{0.35} \times \frac{1 \times 10^2 \times 6}{60 \times 0.86}\,[6^{0.35/1.35} - 1] = 26.51 \text{ kW} \qquad \textbf{Ans.}$$

**Comment: Brake power (bp) needed increases with index of polytropic compression**

**EXAMPLE 21.2** A single acting reciprocating air compressor having 8 cm bore and 10 cm stroke compresses air from 1 bar to a delivery pressure of 6 bar. The compressed gas is stored in a receiver having a capacity of 1.2 m$^3$. The speed of the compressor is 500 rpm and consumes 1.175 kW brake power. (a) The indicated mean effective pressure is 2.24 bar. The initial condition in the reservoir were 1 bar and 15°C. After 2 minutes of pumping, the pressure rise in the receiver is 6 bar at 52°C. Find (b) free air delivered per minutc at 15°C, (c) volumetric efficiency at atmospheric conditions, and (d) mechanical and isothermal efficiencies.

***Solution*** Refer to Fig. 21.5 mass of air initially in the receiver

$$m_1 = \frac{p_1 V_1}{RT_1} = \frac{1 \times 10^2 \times 1.2}{0.287 \times 288} = 1.451 \text{ kg}$$

$$\text{Mass of air finally in the receiver} = m_3 = \frac{p_3 V_3}{RT_3} = \frac{6 \times 10^2 \times 1.2}{0.287 \times 325} = 7.719 \text{ kg}$$

∴ Mass delivered to reservoir in 25 minutes = $m_3 - m_1 = 7.719 - 1.451 = 6.268$ kg

(a) Free air delivered/min at 1 bar and 15°C

$$\dot{V}_{del} = \frac{6.268 \times 0.287 \times 288}{25 \times 1 \times 10^2} = 0.2072 \text{ m}^3/\text{min} \qquad \textbf{Ans.}$$

(b) Swept volume = $V_s = (\pi/4)\,(8)^2 \times 10 = 502.65$ cm$^3$ = $502.65 \times 10^{-6}$ m$^3$

Volume delivered/stroke = $\dfrac{\dot{V}_{del}}{N} = \dfrac{0.2072}{500} = 0.4144 \times 10^{-3}$ m$^3$

(c) ∴ Volumetric efficiency = $\eta_v = \dfrac{V_{as}}{V_s} = \dfrac{0.4144 \times 10^{-3}}{502.65 \times 10^{-6}} = 82.36\%$ **Ans.**

$$ip = \frac{100\, p_m\, LAN}{60} = \frac{100 \times 2.24 \times (\pi/4)8^2 \times 500 \times 0.1}{60 \times 10^4} = 0.938 \text{ kW}$$

(d) Mechanical efficiency = $\dfrac{ip}{bp} = \dfrac{0.938}{1.175} = 79.82\%$ **Ans.**

Isothermal power = $p_1\, \dot{V}_{del} \cdot \ln \dfrac{p_2}{p_1} = \dfrac{1 \times 10^2 \times 0.2072 \times \ln(6)}{60} = 0.618$ kW

$$\eta_{iso} = \frac{\text{Isothermal power}}{\text{Indicated power}} = \frac{0.618}{0.938} = 65.56\% \quad \textbf{Ans.}$$

**EXAMPLE 21.3** A single-stage, double acting air compressor delivers at 6 bar. It handles 16 m$^3$/min of air measured at free condition 1 bar and 15°C. The pressure and temperature at the end of suction are 0.96 bar and 30°C. The clearance ratio is 4% and the index of compression and expansion are $n = 1.3$. The speed of the compressor is 500 rpm. Calculate the stroke volume of the compressor. Also calculate the *ip* of the compressor and the temperature of air delivered. The effect of piston rod may be neglected.

***Solution*** Refer to Fig. 21.5 Due to double acting compressor, the indicator diagram on the other side of the piston will be similar to one shown when the effect of piston rod is neglected.

Let the stroke volume be 100 m$^3$ and then the clearance volume = 4 m$^3$

Hence, the total volume = 104 m$^3$

The actual quantity of air handled = $\dot{m} = \dfrac{p\dot{V}}{RT} = \dfrac{1 \times 10^2 \times 16}{0.287 \times 288} = 19.35$ kg/min

The temperature at the end of compression,

$$T_2 = T_1\,(p_2/p_1)^{(n-1)/n} = 303\left(\frac{6}{0.96}\right)^{0.3/1.3} = 462.49 \text{ K} \quad \textbf{Ans.}$$

The total mass of air handled by the double acting compressor for 100 m$^3$ of stroke volume,

$$M = 2\,(m_1 - m_2) \text{ kg/revolution}$$

$$= 2\left[\frac{p_1V_1}{RT_1} - \frac{p_3V_3}{RT_3}\right] = 2\left[\frac{0.96 \times 10^5 \times 104}{287 \times 303} - \frac{6 \times 10^5 \times 4}{287 \times 462.49}\right]$$

$$= 193.45 \text{ kg/revolution}$$

Hence, the actual stroke volume,

$$V_s = 100 \frac{\dot{m}}{M \times N} = 100 \times \frac{19.35}{193.45 \times 500} = 0.02 \text{ m}^3 \qquad \textit{Ans.}$$

$$ip = \frac{\dot{m} \times \left(\frac{n}{n-1}\right) R\,(T_2 - T_1)}{60}$$

$$= \frac{19.35 \times (1.3/0.3) \times 0.287\,(462.49 - 303)}{60} = 63.96 \text{ kW} \qquad \textit{Ans.}$$

**EXAMPLE 21.4** A single acting 14 cm × 10 cm reciprocating air compressor having 4 per cent clearance gives the following data obtained from a performance test. Suction pressure, 0 bar gauge, suction temperature 20°C, barometer 76 cm Hg, discharge pressure 5 bar gauge, discharge temperature 180°C, speed 1200 rpm, shaft power 6.247 kW, mass of air delivered 1.7 kg/min. Calculate (a) the actual volumetric efficiency, (b) *ip*, (c) isothermal efficiency, (d) mechanical efficiency and (e) overall isothermal efficiency.

***Solution*** Refer to Fig. 21.5. Barometric reading = 76 cm of Hg

Hence, atmospheric pressure = 76 cm of Hg = 1.013 bar

Therefore, $p_1 = 0 + 1.013 = 1.1013$ bar and $p_2 = 5 + 1.013 = 6.013$ bar

Actual volume of air sucked $= \dot{V}_a = \dfrac{\dot{m}RT_1}{p_1} = \dfrac{1.7 \times 0.287 \times 293}{1.013 \times 10^2} = 1.411 \text{ m}^3/\text{min}$

Stroke volume $= \dot{V}_s = (\pi/4) \times \left(\dfrac{14}{100}\right)^2 \times \dfrac{10}{100} \times 1200 = 1.847 \text{ m}^3/\text{min}$

(a) ∴ Volumetric efficiency $= \dfrac{\dot{V}_{as}}{\dot{V}_s} = \dfrac{1.411}{1.847} = 76.39\%$ ***Ans.***

(b) $\dfrac{T_2}{T_1} = \left(\dfrac{p_2}{p_1}\right)^{n-1/n}$ or $\dfrac{453}{293} = \left(\dfrac{6.013}{1.013}\right)^{(n-1)/n} = 1.546$

or $\dfrac{n-1}{n} \ln 5.935 = \ln 1.546$

or $n = 1.32$

$$ip = \frac{n}{n-1} \times \frac{\dot{m}R\,(T_2 - T_1)}{60}$$

$$= \frac{1.32}{0.32} \times \frac{1.7 \times 0.287\,(453 - 293)}{60} = 5.366 \text{ kW}$$ ***Ans.***

(c) $iso.\ p.\ \frac{\dot{m}RT_1}{60} \ln (p_2/p_1) = \frac{1.7 \times 0.287 \times 293}{60} \ln \frac{6.013}{1.013} = 4.243 \text{ kW}$

Isothermal efficiency $= \eta_{iso} = \frac{iso.p}{ip} = \frac{4.243}{5.366} = 79.07\%$ ***Ans.***

$\therefore$ Mechanical efficiency $= \eta_m = \frac{ip}{bp} = \frac{5.366}{6.247} = 85.89$ ***Ans.***

(e) Overall isothermal efficiency $= \eta_{iso} \times \eta_m = 0.7907 \times 0.8589 = 67.9\%$ ***Ans.***

**EXAMPLE 21.5** Design the main dimensions of a two-stage, single acting reciprocating air compressor for the following data:

Free air delivered = 4 m$^3$/min, pressure ratio = 16, atmospheric pressure = 1 bar, atmospheric temperature = 293 K, speed = 600 rpm, clearance = 5%. Index of compression and re-expansion $= n = 1.3$.

Assume stroke to be equal to L.P. cylinder diameter, two cylinders in first stage and one cylinder in second stage, overall volumetric efficiency to be 5% less than apparent volumetric efficiency and intercooling perfect.

Also calculate the motor power required to drive the compressor if mechanical and transmission efficiency are 95% and 98% respectively and no change in apparent volumetric efficiency and intercooling perfect.

***Solution*** The intermediate pressure $= p_2 = \sqrt{(p_1\, p_3)} = \sqrt{(1 \times 16)} = 4$ bar

For each cylinder, the apparent volumetric efficiency is

$$\eta_v = 1 + \in - \in (p_2/p_1)^{1/n} = 1 + 0.05 - 0.05\,(4)^{1/1.3} = 0.9047$$

$\therefore$ Overall volume efficiency $= \eta_{ov} = 0.9047 - 0.05 = 0.8547$

In the first stage, the volume sucked is the free air delivered, hence

$$\dot{V} = x.A.L.N.\eta_{ov} = 4 = 2 \times (\pi/4)\, D_1^2 \times D_1 \times 600 \times 0.8547$$

or $$D_1 = \left[\frac{4 \times 4}{2\pi \times 600 \times 0.8547}\right]^{1/3} = 0.1706 \text{ m} = 17.06 \text{ cm}$$ ***Ans.***

Since the mass flow rate is constant, so for perfect intercooling

$x_1\, A_1\, p_1 = x_2.A_2\, p_2$ where $x_1$ and $x_2$ are number of cylinders in stage 1 and 2 repectively.

or $$2 \times \frac{\pi}{4}(17.06)^2 \times 1 = 1 \times \frac{\pi}{4} D_2^2 \times 4$$

or $$D_2 = \left(\frac{2 \times 291.04}{4}\right)^{1/2} = 0.1206 = 12.06 \text{ cm}$$ ***Ans.***

Stroke = $D_1$ = 17.06 cm ***Ans.***

The shaft power in a stage,

$$sp = \frac{100\, p_m\, LAN}{60 \times \eta_m} = \frac{n}{n-1}\, p_1\, V_1\, [(p_2/p_1)^{(n-1)/n} - 1] \times \frac{1}{60 \times \eta_m}$$

Due to perfect intercooling, the shaft power in the first stage will be equal to second stage.

$$sp = 2 \times \frac{1.3}{0.3} \times 1 \times 10^2 \times 4\left[(4)^{0.3/1.3} - 1\right] \times \frac{1}{60 \times 0.95} = 22.929 \text{ kW}$$

Since efficiency of electric motor is 85% at full load, so

$$\text{Motor power} = \frac{1.15 \times sp}{\eta_t} = \frac{1.15 \times 22.929}{0.98} = 26.9 \text{ kW}$$ ***Ans.***

**EXAMPLE 21.6** The following data relate to a two-stage single acting reciprocating air compressor.

Pressure during suction stroke = 0.96 bar, temperature at the beginning of compression = 25°C, delivery pressure = 18 bar, clearance ratio of low pressure cylinder = 5%, index of compression and expansion = 1.25, speed = 200 rpm.

Capacity is 4.6 m$^3$/min measured under free air condition of 15°C and 1.013 bar. The intercooler pressure chosen is such that theoretically the work is shared equally between the two cylinders. Calculate (a) the *ip* and (b) dimension of the low pressure cylinder if the bore is equal to the stroke, and (c) the volumetric efficiency and the mass of cooling water necessary for abstracting heat in the intercooler if the temperature rise of the cooling water is 20°C.

***Solution*** Refer to Fig. 21.36. Since the work is equally shared between the two cylinders so

$$p_2 = \sqrt{(p_s\, p_d)} = \sqrt{(0.96 \times 18)} = 4.15 \text{ bar}$$

We have $$T_B = T_A\left(\frac{p_i}{p_s}\right)^{(n-1)/n} = 299\left(\frac{4.15}{0.96}\right)^{(0.25/1.25)} = 400.7 \text{ K}$$

The total work done per kg for two stages with equally shared work

$$= 2\,\frac{n}{n-1}\, R(T_B - T_A) = 2 \times \frac{1.25}{0.25} \times 0.287\,(400.7 - 299) = 291.879 \text{ kJ/kg of air}$$

$$\therefore \quad ip = \frac{\text{Work done in kJ/kg} \times \text{Mass of air per min in kg}}{60}$$

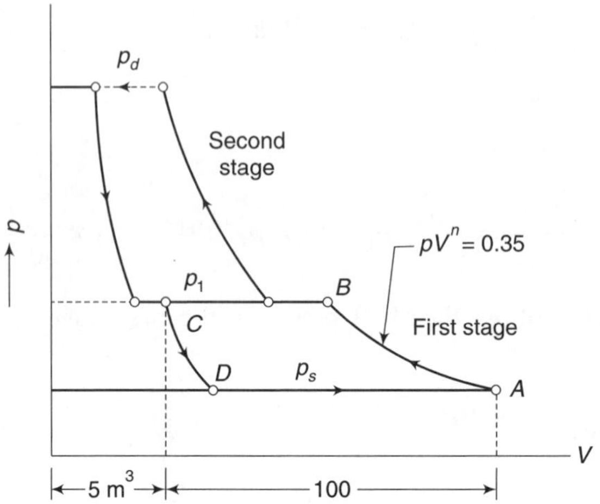


**Fig. 21.36** Example 21.6

$$\text{Mass of air handled} = \dot{m} = \frac{pv}{RT} = \frac{1.013 \times 10^2 \times 4.6}{0.287 \times (273 + 15)} = 5.63 \text{ kg/min}$$

(a) Hence, $ip = \dfrac{291.879 \times 5.63}{60} = 27.38$ kW **Ans.**

(b) Let the stroke volume of L.P. cylinder be equal to 100 m$^3$, then clearance volume will be 5 m$^3$ and total volume = 100 + 5 = 105 m$^3$. Now, mass of the air handled with stroke volume of 100 m$^3$ is given by

$$m = m_A - m_C = \frac{0.96 \times 10^2 \times 105}{0.287 \times 299} - \frac{4.15 \times 10^2 \times 5}{0.287 \times 400.7} = 99.42 \text{ kg}$$

$\therefore$ Actual volume = $\dot{V}_a = \dfrac{100}{99.42} \times 5.63 = 5.662$ m$^3$/min

$\therefore$ $\dot{V}_a = 5.662 = (\pi/4)\, D^2 \times L \times N = (\pi/4)\, D^3 \times 200$

or $D = \left(\dfrac{5.662 \times 4)}{\pi \times 200}\right)^{1/3} = 0.33\ m = L = 33\ CM$ **Ans.**

(c) Since the volumetric efficiency is always referred to as L.P. stage, thus the volume delivered by L.P. cylinder at 4.15 bar 400.7 K is

$$V_B = V_A\,(p_1/p_2)^{1/n} = 105\left(\frac{0.96}{4.15}\right)^{1/1.25} = 32.55 \text{ m}^3$$

$\therefore$ Volume delivered by L.P. cylinder at 4.15 bar and 400.7 K = 32.55 – 5 = 27.55 m$^3$

∴ Volume delivered under free air conditions

$$= \frac{27.55 \times 4.15 \times 10^2 \times 288}{1.013 \times 10^2 \times 400.7} = 81.12\% \text{ m}^3$$

∴ Volume efficiency $= \dfrac{81.12}{100} \times 100 = 81.12\%$ ***Ans.***

(d) Loss of heat in the intercooler per kg of air cooled

$$q_{loss} = c_p(T_B - T_A) = 1.005\ (400.7 - 299) = 102.2 \text{ kJ/kg}$$

Hence, mass of cooling water $= m_w \times C_{pw} \times \Delta t_w = 102.2$

or $$m_w = \frac{102.2}{20 \times 4.18} = 1.222 \text{ kg of water/kg of air.}$$ ***Ans.***

**EXAMPLE 21.7** A centrifugal compressor running at 9000 rpm delivers 600 m$^3$/min of free air. The air is compressed from 1 bar and 20°C to a pressure ratio of 4 with an isentropic efficiency of 0.82. Blades are radial at outlet of impeller and the flow velocity of 62 m/s may be assumed throughout constant. The outer radius of the impeller is twice the inner and the slip factor may be assumed as 0.9. The blade area coefficient equal to 0.9 may be assumed at inlet. Calculate (a) final temperature of air, (b) theoretical power, (c) impeller diameters at inlet and outlet, (d) breadth of impeller at inlet, (e) impeller blade angle at inlet, and (f) diffuser blade angle at inlet.

***Solution*** Refer to Figs. 21.16 and Fig. 21.17. The isentropic temperature of air at the exit of compressor is

$$T_{2s} = T_1\ (r_p)^{(\gamma - 1)/\gamma} = 293 \times (4)^{0.4/1.4} = 435.39 \text{ K}$$

Actual temperature rise $= T_2 - T_1 = \dfrac{435.39 - 293}{0.82} = 173.64$ K

∴ (a) $T_2 = 293 + 173.85 = 466.85$ K ***Ans.***

Mass flow rate $= \dot{m} = \dfrac{p_1 V_1}{RT_1} = \dfrac{1 \times 10^2 \times 600}{0.287 \times 293} = 713.5$ kg/min

(b) ∴ Work done $= \dot{m} c_p\ (T_2 - T_1) = \dfrac{713.5}{60} \times 1.005 \times 173.64 = 2075.19$ kW ***Ans.***

(c) For radial blades, work input to the compressor is given by

Work done $= \dfrac{\mu u_2^2}{1000} = c_p\ (T_2 - T_1)$

Here, $T_2$ is the final temperature of air from the exit of compressor

$$u_2 = \sqrt{\left(\frac{1000\ c_p(T_2 - T_1)}{\mu}\right)} = \sqrt{\left(\frac{1.005 \times 1000 \times 173.64}{0.9}\right)} = 440.33 \text{ m/s}$$

$\therefore$ $$u_2 = 440.33 = \pi D_2 \text{ N/60}$$

$\therefore$ $$D_2 = \frac{440.33 \times 60}{\pi \times 9000} = 0.9344 \text{ m} = 93.44 \text{ cm}$$ ***Ans.***

$$D_1 = \frac{D_2}{2} = \frac{93.44}{2} = 46.72 \text{ cm}$$ ***Ans.***

(d) Volume flow rate $= \pi D_1 b_1 C_{f1} . k_a$ where $k_a$ = blades are coefficient

or Blade inlet width $= b_1 = \dfrac{600}{60 \times \pi \times 0.4672 \times 62 \times 0.9} = 0.122 \text{ m} = 12.2 \text{ cm}$ ***Ans.***

(e) $$\tan \beta_1 = \frac{C_{f1}}{u_1} = \frac{62 \times 2}{440.33} = 0.2816$$

or $$\beta_1 = 15.72°$$

(f) $$\tan \alpha_2 = \frac{C_{f2}}{\mu . u_2} = \frac{62}{0.9 \times 440.33} = 0.1564$$

or $$\alpha_2 = 8.9°$$ ***Ans.***

***Note:*** If temperature and pressure based on stagnation values, then $p$ and $T$ will be replaced by $T_0$ and $p_0$.

**EXAMPLE 21.8** In an eight-stage flow compressor, the overall stagnation pressure ratio achieved is 5:1 with an overall isentropic efficiency of 90%. The temperature and pressure at inlet area 20°C and 1 bar. The work is divided equally between the stages. The means blade speed is 175 m/s and 50% reaction design is used. The axial velocity through the compressor is constant and equal to 100 m/s. Calculate the power required and blade angles.

***Solution*** Refer to Fig. 21.37 for velocity diagrams. In the case of 50% reaction, the blades are symmetrical and hence the velocity diagrams are identical. Thus,

$$\alpha_1 = \beta_2 \text{ and } \alpha_2 = \beta_1$$

Let suffix $z$ denote the number of stages

With isentropic compression, the temperature of the air leaving the compressor stages is

But $$T_{02s} = T_{01}\left(\frac{p_{0z}}{p_{01}}\right)^{(\gamma - 1)/\gamma} = 293\ (5)^{0.4/1.4} = 464 \text{ K}$$

But $$\eta_{0ise} = \frac{T_{0zs} - T_{01}}{T_{0z} - T_{01}} = 0.9 = \frac{464 - 293}{T_{0z} - 293}$$

or $$T_{0z} = 483 \text{ K}$$

The work consumed by the compressor,

$$= c_p(T_{0z} - T_{01}) = (C_{w2} - C_{w1})\, u.z = C_a\,(\tan\alpha_2 - \tan\alpha_1)\, u.z$$

$$\therefore \quad (\tan\alpha_2 - \tan\alpha_1) = \frac{1.005 \times (483 - 293) \times 10^3}{100 \times 175 \times 8} = 1.3639 \qquad ...(1)$$

From velocity triangle, we have, $\frac{u}{C_a} = \tan\alpha_1 + \tan\beta_1 = \frac{175}{100} = 1.75$ ...(2)

From Eqs. (1) and (2), we have $\tan\beta_1 = \frac{1.3669 + 1.75}{2} = 1.5584$

$$\therefore \quad \beta_1 = 57.3° = \alpha_2$$ ***Ans.***

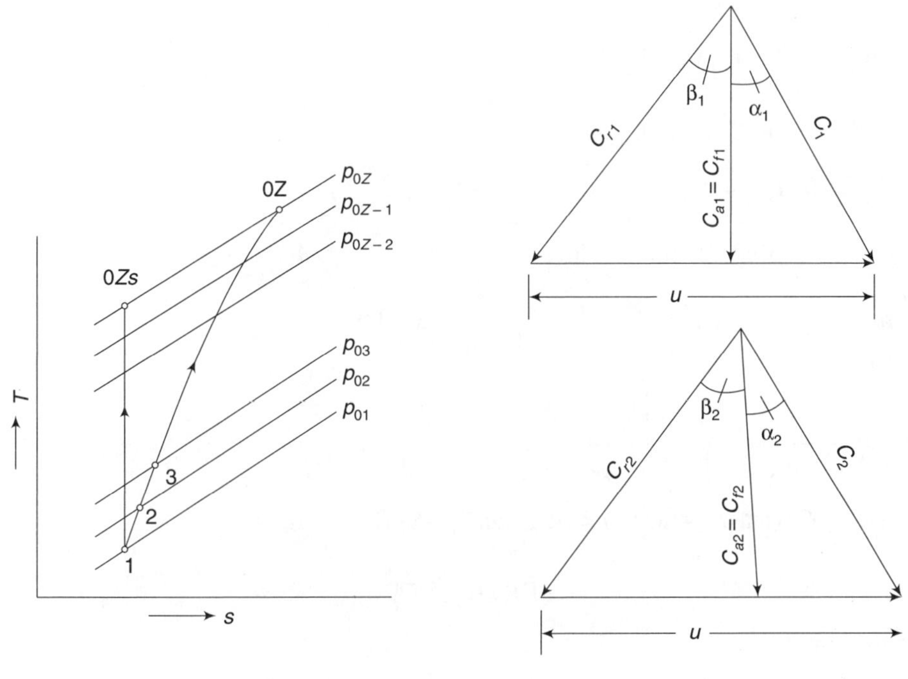


**Fig. 21.37** Example 21.8

Putting the value of $\tan\beta_1$ in Eq. (1), we get

$$\tan\alpha_1 = 1.75 - 1.5584 = 0.1916 \qquad \therefore \quad \alpha_1 = \beta_2 = 10.84°$$ ***Ans.***

Power required $= \dot{m}c_p\,(T_{0z} - T_{01}) = 1 \times 1.005\,(483 - 293) = 190.95$ kW ***Ans.***

## 21.32 SUMMARY OF FORMULAE

(1) $-\dot{W}_c = \dot{m} c_p (T_{0,\text{out}} - T_{0,in})$

(2) $T_{02s} = T_{01}\left(\frac{p_{02}}{p_{01}}\right)^{(\gamma-1)/\gamma}$

(3) $\eta_{cTT} = \frac{h_{02s} - h_{01}}{h_{02} - h_{01}}$

(4) $-\dot{W}_{C,D} = \frac{\dot{W}_c}{\eta_m}$

(5) $\eta_{TTT} = \frac{h_{03} - h_{04}}{h_{03} - h_{04s}}$

(6) $\dot{W}_T = \dot{m}(h_{03} - h_{04})\,\eta_m$

(7) $-\dot{W}_C = \frac{n}{n-1}(p_1 V_1 - p_2 V_2)$

(8) $-\dot{W}_C = \dot{m}(u_2 C_{w2} - u_1 C_{w1})$

(9) $\dot{W}_T = \dot{m}u(C_{w2} + C_{w1})$

(10) $(p_2)_{opt} = \sqrt{p_1 p_3}$ for intercooling

(11) $\frac{p_{02}}{p_{01}} = \left[1 + \eta_{ise}\frac{\sigma\mu\, u_2^2}{c_p\, T_{01}}\right]^{\gamma/(\gamma-1)} = \left[1 + \eta_{ise}\frac{\Delta T_0}{T_{01}}\right]^{\gamma/(\gamma-1)}$

(12) $Z = \frac{\Delta T_0}{(\Delta T_0)_{stage}} = \frac{\ln r_{p0}}{\ln (r_{p0})_{stage}}$

(13) $w_{st} = uC_a(\tan\alpha_2 - \tan\alpha_1) = uC_a(\tan\beta_1 - \tan\beta_2)$

## PROBLEMS

### Things to Think About

21.1 Think of the physics behind a new compression process other than that in positive displacement and dynamic compressors.

21.2 Explain the work input to compressor in the light of second law of thermodynamics.

21.3 Under what cylinder arrangement, the compression and re-expansion take place?

21.4 What factors affect the index of compression and re-expansion?

21.5 What will happen to compressor if the supply of cooling water to cylinder jacket fails?

21.6 What do you mean by turbocompressor?

21.7 What parameters limit the pressure rise in a stage of centrifugal compressor stage?

21.8 What parameters limit the pressure rise in a stage of axial flow compressor?

21.9 What is the reason that pressure ratio achieved in a supersonic turbcocompressor is much higher than that of subsonic ones?

21.10 What is the reason that pressure rise per stage in a centrifugal compressor is more than axial?

21.11 What is the reason that axial flow compressor is more suitable for jet engines?

### Short and Long Theoretical Questions

#### Positive Displacement Compressors

21.12 Why is an external power input needed to drive a compressor? What are the uses of compressed gases?

21.13 How will you classify the compressor? Differentiate between the working principles of positive displacement and dynamic compressors.

21.14 Is it essential to maintain clearance between cylinder head and piston? Give reasons.

21.15 What are the effects of clearance on the performance of reciprocating compressors?

21.16 What is the effect of throttling of gas due to the presence of valves?

21.17 What is the condition for the minimum work input in multi-stage reciprocating compression?

21.18 What is the principle of operation of the reciprocating compressor valves?

21.19 Define volumetric efficiency and state what are the factors on which it depends.

21.20 Is it essential to have a receiver in a reciprocating compressor? Give reasons.

21.21 Why is multi-staging essential for high compression ratio?

21.22 What are the effects of atmospheric conditions on output?

21.23 For which type of works, reciprocating and rotary compressor are suitable?

21.24 Derive expression for volumetric efficiency and discuss its physical significance.

21.25 Derive an expression of the minimum work input to compress the gas in a two- and three-stage reciprocating air compressor with perfect and imperfect intercooling.

21.26 List the advantages of multi-stage compression.

21.27 Derive an expression showing the effect of throttling on volumetric efficiency.

21.28 Show on what factors the volumetric efficiency depends.

21.29 Prove that the heat rejected per stage per kg with perfect inercooling is given by

$$\left[c_p + c_v\left(\frac{\gamma - n}{n - 1}\right)\right](T_2 - T_1)$$

21.30 Show the throttling low pressure suction increase the temperature range in successive stages.

21.31 Show the increasing clearance in one stage causes more work to be done in the lower stages.

21.32 Explain the function of (a) intercoolers and aftercooler in multi-stage compressors.

21.33 Explain the working principles of different types of compressor valves generally used in practice.

21.34 Discuss the working principle of rotary vane, screw and lobe type compressors. Discuss their advantages and disadvantages.

### Turbocompressors

21.35 Define the dynamic compressor.

21.36 Distinguish between the working principles of centrifugal, mixed and axial flow compressors.

21.37 Explain the principle of working of centrifugal compressor with the help of *h-s* chart.

21.38 Derive the expression:

$$w = \frac{C_2^2 - C_1^2}{2} + \frac{C_{r1}^2 - C_{r2}^2}{2} + \frac{u_2^2 - u_1^2}{2}$$

in the case of centrifugal compressor and explain its physical significance.

21.39 Discuss the slip factor of centrifugal compressor.

21.40 Discuss the influence of impeller blade shape on the performance.

21.41 What do you mean by prewhirl?

21.42 Distinguish between axial compressor and turbine blades.

21.43 Discuss the various losses in axial flow compressor stage. Discuss secondary flow losses.

21.44 What do you mean by work done factor?

21.45 Discuss surging in the case of centrifugal and axial flow compressors.

21.46 Discuss choking and stalling phenomena.

21.47 Discuss the performance of centrifugal and axial flow compressors.

21.48 Give a comparison between centrifugal and reciprocating compressors.

21.49 Give a comparison between centrifugal and axial flow compressors.

## Numerical Questions

### Positive Displacement Compressors

21.50 (a) A single-stage reciprocating compressor takes in 1 $m^3$ of air per minute at 1.013 bar and 15°C and delivers it at 7 bar. If the law of comparison is $pV^{1.3}$ = constant, and the clearance is negligible, determine the indicated power. (b) If the speed of the compressor is 300 rpm and the compressor is single acting and single-cylinder machine, determine the cylinder bore assuming L/D = 1.5. Also calculate the electric motor to drive the compressor if the mechanical efficiency of the compressor is 85% and the transmission efficiency is 90%

[***Ans.*** (a) *ip* 4.23 kW, (b) *D* = 14.15 cm, *Pe* = 5.53 kW]

21.51 The free air delivered by a single-stage, double acting air compressor measured at 1.013 bar and 15°C is 14 $m^3$/min. The pressure and temperature in the cylinder during induction are 0.95 bar and 32°C. The delivery pressure is 7 bar and the index of compression and expansion, $n = 1.3$. If the clearance volume is 5% of the swept volume, calculate the indicated power and volumetric efficiency. [***Ans.*** *ip* = 63.65 kW, $\eta_v$ = 81.7%]

21.52 A single acting two-stage reciprocating air compressor compresses 4.5 kg of air per minute from 1.013 bar and 15°C through a pressure ratio of 9. The intercooling is perfect and the law of compression and expansion, $pV^{1.3}$ = constant. Assuming the clearance volume of both stages 5% of their swept volume and the speed of compressor 300 rpm, calculate the indicated power and the cylinder swept volume. Also calculate the heat loss to the cylinder jacket cooling water and the heat loss to the intercooler circulating water.

[***Ans.*** 15.5 kW; 0.0131 $m^3$, 0.00436 $m^3$, 90 kJ/min 375 kJ/min]

21.53 A 23 kW electric motor drive a single-cylinder, single-acting air compressor running at 300 rpm. The mechanical efficiency is 87%. The air inlet conditions are 1.013 bar and 15°C and the delivery pressure is 8 bar. Assuming that the index of compression and expansion is $n = 1$ that the clearance volume is 7% of the swept volume and $L = D$, calculate the free air delivered in $m^3$/min, the volumetric efficiency and the bore and stroke of the compressor.

[*Ans.* FAD = 4.47 $m^3$/min, $\eta_v$ = 73% $D$ = 20.6 cm]

21.54 A single-acting reciprocating air compressor delivers air at 70 bar from an induction pressure of 1 bar. The rate of air compressed is 2.4 $m^3$/.min measured at free air conditions of 1.013 bar and 15°C. Temperature at the end of induction stroke is 32°C.

(a) Calculate the indicated power if the compression is carried out in two stages with an ideal intermediate pressure and perfect intercooling and the index of compression and expansion for both stages is 1.25. Also calculate the saving in power over single-stage compression.

(b) If the clearance volume is 3% of swept volume in each cylinder, calculate the swept volumes of the cylinder for the compressor speed of 750 rpm.

(c) If the mechanical efficiency of the compressor is 85%, determine the electric motor power.

(d) Determine the heat rejected per minute to the jacket cooling water of each stage and the heat rejected per minute to the intercooler. Assume that 50% of the friction power in each stage is transferred to the jacket cooling water.

[*Ans.* (a) 22.7 kW, 6 kW, (b) 0.00396 $m^3$, 0.000474 $m^3$ (c) 26.75 kW, (d) 325 kJ/min]

21.55 A single cylinder, single acting air compressor has a cylinder diameter of 15.25 cm and a stroke of 22.8 cm. Air is drawn into the cylinder at a pressure of 1.013 bar and a temperature of 15.6°C. It is compressed adiabatically to 6.1 bar. Calculate the theoretical power required to drive the compressor if it runs at 100 rpm and the mass of air compressed per minute.

[*Ans.* 1648 W, 0.58 kg/min]

21.56 The pressure in the mains of a compressed air service is to be 8.1 bar and a supply of 28 $m^3$/h at 25°C and that pressure is to be maintained. The air is compressed from an initial pressure of 1.063 bar by a two-stage compressor in which the compression is adiabatic and the air is cooled to its initial temperature of 15°C in the intercooler. What is the minimum power required to compress the air. [*Ans.* 15.5 kW]

**Turbocompressor**

21.57 All the stages of an axial flow compressor have the following conditions at the mean diameter: degree of reaction is half, blade peripheral speed 200 m/s, angles of absolute velocity at inlet and exit from the rotor measured from the axial direction 15° and 45° respectively. A stage efficiency of 88% and work done factor of 0.86 are to be taken as constant throughout the compressor. The combined efficiency of intake and inlet guide vanes is 95% based on static conditions. If the ambient air pressure and temperature are 1 bar and 288 K respectively, find

(a) the static temperature of air at inlet to the first stage,

(b) the pressure ratio of the first stage, and

(c) the static pressure ratio of the air exit from the tenth stage.

[*Ans.* 274.73 K, 1.24, 6.23]

21.58 An axial flow compressor compresses air from an inlet condition of 1 bar and 290 K to a delivery pressure of 5 bar with an overall isentropic efficiency of 87 per cent. The degree of reaction is 0.5 and the blade angles at inlet and outlet are 44° and 13° respectively. The mean blade speed and axial velocity are constant throughout the compressor. Assuming a blade velocity of 180 m/s, and work done factor 0.85, calculate the number of stages and the change in entropy.

[***Ans.*** 12, 0.40388 kJ/kg K]

21.59 The following particulates relate to a centrifugal compressor : inlet diameter of impeller = 61.4 cm, outlet diameter of impeller = 123 cm; speed = 5000 rpm, velocity of flow = 61.6 m/s free air delivered = 1000 $m^3$/min, pressure ratio = 1.33, index of compression = 1.6. Assuming that all pressure rise takes place in the impeller, find the angles at which air from impeller enters the casing, breadth of the impeller blade inlet and outlet.

[***Ans.*** $\beta_1$ = 20.97°C, $\beta_2$ = 15.67°, $\alpha_3$ = 31°, $b_1$ = 14 cm $b_2$ = 7 cm]

21.60 The following particulates relate to a double sided centrifugal compressor having radial bladed impeller. Speed of compressor = 15,000 rpm. Mass flow rate = 10 kg/s, total head pressure at inlet = 1.12 bar, total temperature at inlet = 295 K, isentropic efficiency = 80%, eye root diameter = 15 cm, eye tip diameter = 20 cm, impeller tip diameter = 50 cm Assuming the slip factor of 0.9, a power input factor of 1.04 and axial entry, calculate total head pressures ratio and compressor power. Also calculate angles at impeller root and tip assuming axial component of initial velocity to remain constant.

[***Ans.*** $r_p$ = 3.134, 1438 kW, 35.4°, 77.3°]

21.61 A single-stage, single sided centrifugal compressor delivers 18 kg/s of air with a pressure ratio 4 when running at 15,000 rpm. The inlet conditions are 1 bar and 20° C. The slip factor is 0.9. The power input factor is 1.04 and the isentropic efficiency of compressor is 75 per cent. Calculate (a) the final temperature of air, (b) power required, (c) the tip diameter of the impeller, (d) the blade angles at impeller eye which has root and tip diameter 15 cm and 20 cm respectively.

[***Ans.*** 482.85 K, 3434. 6 kW, 57.5 cm]

21.62 The impeller of a radial bladed centrifugal compressor has an outside diameter of 33 cm and rotates at 18600 rpm. The inlet stagnation temperature is 289 K and the isentropic efficiency is 0.85. Neglecting heat losses, calculate the rise in stagnation temperature through the machine? Also calculate the compression ratio. Take $\sigma$ =1.04 $\mu$ = 0.9. [***Ans.*** 96.7°C, 24]

21.63 A multistage axial compressor is required for compressing air at 293 K through a pressure ratio of 5 to 1. Each stage is to be 50% reaction and the mean blade speed 275 m/s, flow coefficient 0.5 and stage loading factor 0.3 are taken, for simplicity, as constant for all. Determine the flow angles and the number of stages required if the stage efficiency is 88.8%. Assume $c_p$ = 1.005 kJ/kg K and $\gamma$ = 1.4 for air. ***Ans.*** 52.45°, 25°, 9

21.64 An axial flow compressor is required to deliver 50 kg/s of air to stagnation pressures of 5 bar. At inlet to the first stage the stagnations pressure is 1 bar and the stagnation temperature is 23°C. The hub and tip diameters at this location are 0.436 m and 0.728 m. At the mean radius, the reaction is 0.5 for all stages and the absolute air angle at stator exit is 28.8° for all stages The speed of the rotor is 8000 rpm. Calculate the number of stages needed assuming that the polytropic efficiency 0.89 and that the axial velocity at the mean radius is constant through the stages and equal to 1.05 times the average axial velocity.

21.65 Design a centrifugal compressor for the following data:

Total pressure ratio = 4.5, mass flow rate = 30 kg/s speed = 16,000 rpm, $p_1$ and $T_1$ = 1 bar and 300 K.

21.66 Design an axial flow compressor for the following data:

Total pressure ratio = 30, mass flow rate = 500 kg/s, speed = 3000 rpm, $p_1$ and $T_1$ = 1 bar and 3000 K.

### Software Based Questions

21.67 Develop a computer software for the design of main dimensions of reciprocating compressor including the required power of electric motor to drive it.

21.68 Develop a software for the design of a piston, cylinder and valves of a reciprocating compressor.

21.69 Develop a software for the design of rotary positive displacement compressors.

21.70 Develop a software for the complete design of reciprocating compressor covering all types of cylinder arrangement for all types of gases to be compressed up to 1000 bar.

21.71 Develop a software for the design of centrifugal and axial flow compressors and validate the results with an existing modern compressors.

21.72 Develop a software for predicting the performance of centrifugal and axial compressors.

### Design and Open Ended Questions

21.73 Prepare a report on HH frame model of reciprocating compressor manufactured by Nuovo Pignone, Italy discussing the cylinder layout, features, specifications, capacity control, mechanism and performance. Find the main drawback in these compressors and suggests measures to improve this.

21.74 Prepare a report on LPG gas bottling plant discussing layout features, specification of an existing plant. Find its shortcomings and suggest improvements. Includes the complete design of your suggested bottling plant.

21.75 Prcpare a report on the manufacturing process of urea plant with complete layout and feature of a modern working plant. Show the employment of various reciprocating compressor in the layout with specifications. Can you suggest some improvement to save the power input and to maximize the production in the plant?

21.76 In the present technology (2020), multi-stage (8 stages) centrifugal compressor in a single casing is able to produce delivery pressure up to 450 bar. If it is required to produce delivery pressure upto 1000 bar in a single casing for the use in petrochemical industries, what type of design and manufacturing challenges will be imposed by this project. Prepare report discussing these challenges including the conceptual layout of compressor, materials seals, bearings and lubrication system to be selected, and the projected isentropic efficiency.

21.77 In the present technology (2020), multi-stage (22 stages) axial flow compressor in a single casing is able to achieve pressure ratio up on 40 handling about 600 kg/s of air. Such a compressor is being used in gas turbine power plant. If a project is undertaken to achieve pressure ratio up to 60 handling about 800 kg/s of air. What type of design and manufacturing challenges will be imposed by this project. Prepare a report discussing these challenges including the conceptual layout of compressor, materials for all parts, seals bearing and lubrication system to be selected and the projected polytropic efficiency.

### Multiple-Choice Questions

21.78 The work input to compressor handling a particular gas depends on
(a) mass flow rate of gas (b) initial temperature of gas
(c) pressure ratio and index of compression (d) all of them

21.79 The volumetric efficiency of a reciprocating compressor depends on
(a) clearance ratio (b) pressure ratio
(c) index of re-expansion (d) all of them

21.80 The work input in a multistage reciprocating compressor is given by

(a) $\frac{Nn}{n-1} p_1 V_1 \left[(p_{n+1}/p_i)^{(n-1/Nn} - 1\right]$

(b) $\frac{Nn}{n-1} p_1 V_1 \left[(p_n/p_1)^{(n-1/Nn} - 1\right]$

(c) $\frac{(N+1)n}{n-1} p_1 V_1 \left[(p_{n+1}/p_1)^{(n-1/Nn} - 1\right]$

(d) $\frac{(N+1)n}{n-1} p_1 V_1 \left[(p_n + 1/p_1)^{(n-1)/(N+1)n} - 1\right]$

21.81 The volumetric efficiency with flow resistance is given by

(a) $(1-\varepsilon)\left(\frac{p_s}{p_1}\right)^{1/n} - \varepsilon\left(\frac{p_2}{p_1}\right)^{1/n}$ (b) $(1-\varepsilon)\left(\frac{p_s}{p_1}\right)^{1/n} - \varepsilon\left(\frac{p_2}{p_1}\right)^{1/n}$

(c) $(1-\varepsilon)\left(\frac{p_s}{p_1}\right)^{1/n} + \varepsilon\left(\frac{p_2}{p_1}\right)^{1/n}$ (d) none of the above

21.82 The pressure ratio per stage achieved in a centrifugal compressor is
(a) 1.5 (b) 2.5 (c) 3.5 (d) 4.5

21.83 The pressure ratio per stage achieved in an axial flow compressor is
(a) 1.05 (b) 1.2 (c) 1.3 (d) 1.4

21.84 The pressure rise in an axial flow compressor is due to
(a) diffusion action (b) centrifugal action
(c) both centrifugal and diffusion action (d) none of the above

21.85 The pressure rise in a centrifugal compressor is due to
(a) diffusion action (b) centrifugal action
(c) both centrifugal and diffusion action (d) none of the above

### Key to Multiple-Choice Questions (All the answers may vary due to fast development)

21.78 (d) 21.79 (d) 21.80 (a) 21.81 (d) 21.82 (d) 21.83 (b) 21.84 (a) 21.85 (c)

CHAPTER

# Electric Vehicles 22

## OVERVIEW

The pollution created by hydrocarbon fuels (petrol/diesel/natural gas/kerosene) powered vehicles on road, rail and sea produce immense harm to the environment, the ecology, human health and cause greenhouse gas effect. It is said that about 40 per cent pollution is from automobiles run by hydrocarbon fuels. This forced the scientists and engineers to develop green technology vehicles, which should be pollution free. In this attempt, electric vehicles (EV) came in the market. So far the early history is concerned, the invention of the first model of electric vehicle is attributed to various people. In 1828, Anyos Jedik invented an early type of electric motor, and created a small model car powered by his new motor. Between 1832 to 1839, Scottish inventor Robert Andersion also invented a crude electric car. After 1990, the maximum progress took place in the development of electric vehicles.

*The electric vehicle has numerous* ***advantages*** *over the traditional internal combustion engines, such as zero emission, superior efficiency (about 90 per cent), no dependence of fossil fuels, quieter and smooth operation, less moving parts, high mechanical efficiency, fully balanced rotational motion, and very less maintenance cost. The main* ***disadvantages*** *include the charging of battery at about every four to eight hundred kilometre, high cost, about 10 years battery pack life, at least one hour in charging the battery, and wider vehicle size to accommodate huge size battery pack.* Many companies of the world have now involved in developing and manufacturing electric cars, buses, and two-wheelers. The companies like Tesla, Nissan, Toyota, Volkswagen, Tata Motors, Mahindra Electric, Hyundai, Kona Electric, Ford, General Motors, Mercedes-Benz AG, etc., are manufacturing electric vehicles. Tesla is the biggest electric vehicle manufacturer in the world, and its revenue was USD 53.82 billion in 2021.

Based on the **energy resources**, three types of electric vehicles are being developed and manufactured. They are:

(i) Battery based system (no emission)
(ii) Hybrid (gasoline and electric) based system (partial emission)
(iii) $H_2$-$O_2$ fuel cells system (no emission)

Among these three, battery based energy resource system is more common. Battery based electric cars are plying on the road in most of the countries. Fuel cells ($H_2$-$O_2$) based electric vehicles have been developed by Toyota. Their cars and buses are running in many countries. UK and Japan have declared that by 2030, only electric cars will run in these two countries. Later on, many countries will follow this policy to mitigate pollution. Recently Mercedes-Benz electric cars rolled out from Pune factory (India) go up to 850 km in one charge in October, 2022.

This chapter deals with all the three systems of electric vehicles in brief only.

## 22.1 GENERAL DISCRIPTION OF VEHICLE MOVEMENT

Figure 22.1 shows the forces acting on a vehicle moving on a gradient. The tractive effort, $F_t$ on the contact area between tyres of the driven wheels and the road surface propels the vehicle in forward direction. It is produced by the engine torque and is transferred through vehicle transmission and finally to the drive wheels. While the vehicle is moving, there is a resistance that attempts to stop its forward movement. The resistance usually includes the tyre rolling resistance, aerodynamic-drag, and uphill gradient resistance. According to Newton's second law of motion, the vehicle acceleration can be represented as

$$\frac{dV}{dt} = \frac{\Sigma F_t - \Sigma F_{tr}}{\delta . M_v} \quad \text{...(22.1)}$$

where $V$ the vehicle speed, $\Sigma F_t$ is the total tractive effort of the vehicle, $\Sigma F_{tr}$ is the total resistance, $M_v$ is the total mass of the vehicle, and $\delta$ is the mass factor, which is an effect of rotating components in the power train.

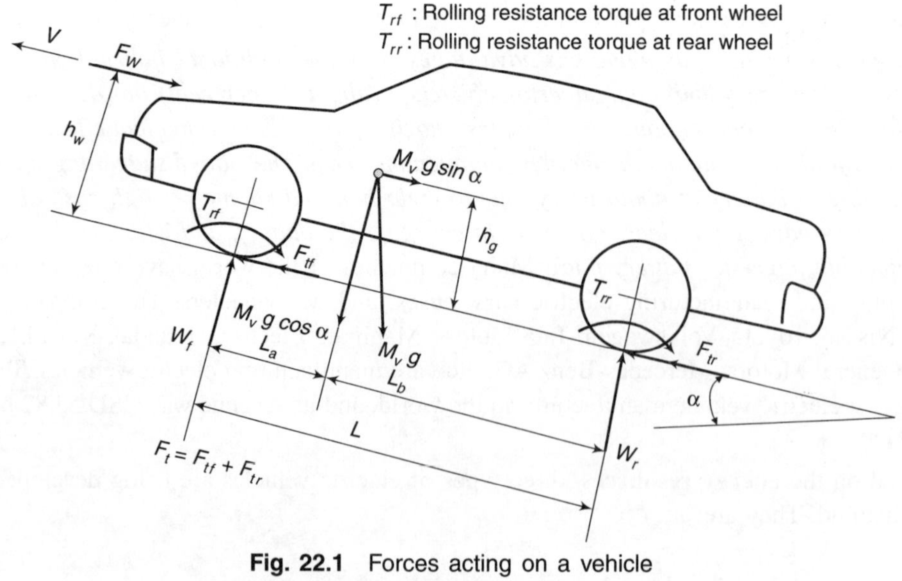


**Fig. 22.1** Forces acting on a vehicle

Equation (22.1) points out that speed and acceleration depend on tractive effort, resistance and vehicle mass. The vehicle resistance opposing its movement includes rolling resistance of tires, rolling resistance torque $T_{rf}$, aerodynamic drag $F_w$, and grading resistance ($M_v$.g.sin $\alpha$).

## 22.2 CONFIGURATION OF ELECTRIC VEHICLES (EVs)

Figure 22.2 shows the simplest configuration of an electric vehicle. It consists of mainly three subsystems. They are electric energy source, electric motor drive, and mechanical transmission. The electric energy source may be chemical battery, $H_2$-$O_2$ fuel cells, ultra capacitor and flywheel. At present, ultra capacitor and flywheel are not used alone. They may be supplement to battery. Electric motor may be DC/AC three-phase induction motor to supply power for converting into tractive force. Mechanical transmission includes gearbox and other necessary elements.

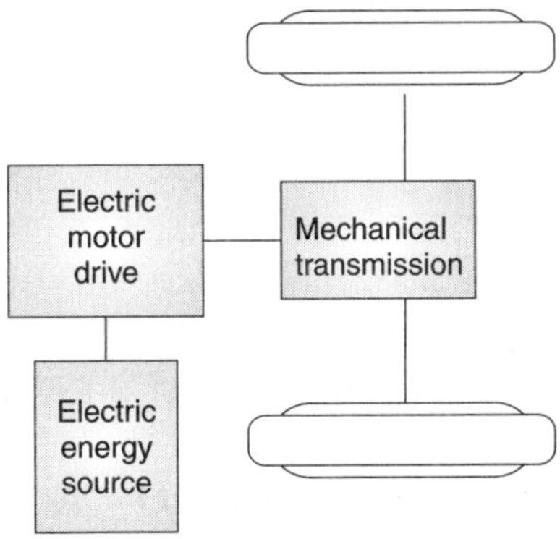


**Fig. 22.2** Layout of an electric vehicle power train

Figure 22.3 shows a generalized system level layout of an electric vehicle. The modern EV is designed and built around a specially designed and developed chassis and frame. The especially designed chassis and frame helps in meeting the structural requirements unique to EVs and making use of greater flexibility potential of electric propulsion system. A modern electric vehicle drive train is conceptually illustrated in Fig. 22.3. The EV drive train comprises three major subsystems: ***Electric motor propulsion assembly, energy resource, and associated auxiliaries***. The electric propulsion subsystem assembly includes a vehicle controller, power electronic converter (to provide required voltage and current, either in DC or AC form), electric motor, mechanical transmission and driving wheel. The energy source subsystem includes the energy source ***(battery,/$H_2$-$O_2$ fuel cells, etc.)***, the energy management unit, and the energy refuelling/recharging unit (in the case of fuel cells, the compressed $H_2$ bottle will be replaced or refuelled, or in the case of battery, it will be recharged by recharging point). The auxiliary subsystem consists of the power steering unit, the cabin climate control (air conditioner/heater), and the auxiliary power supply unit (may be ultracapacitor).

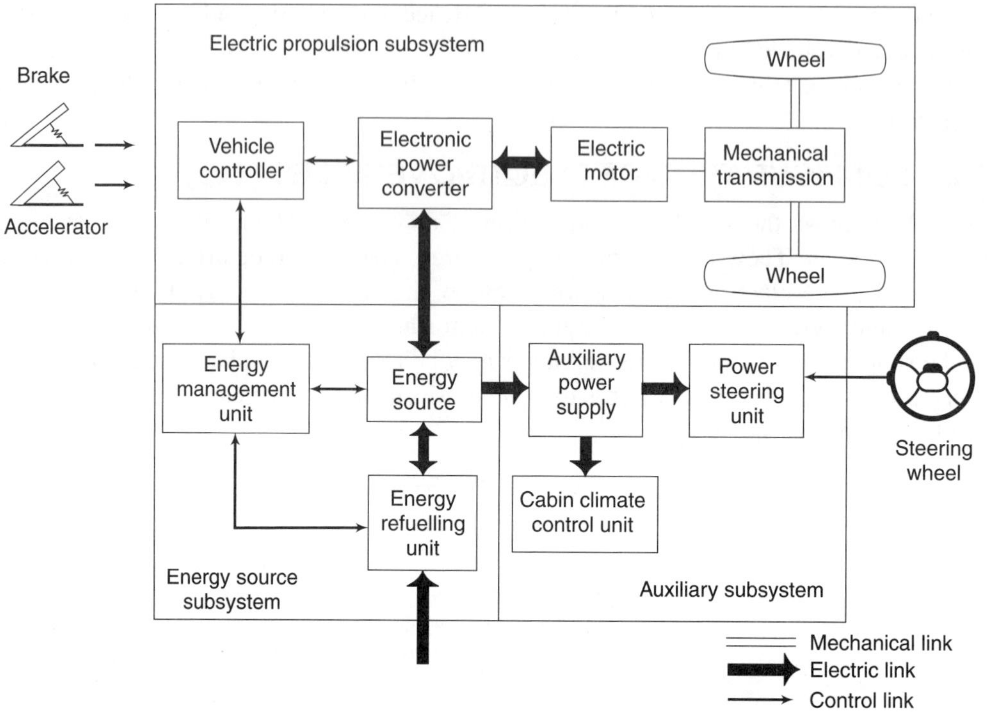


**Fig. 22.3** Generalised system level layout of electric vehicle [102a]

Figure 22.4 (a, b, c, d, e, f) shows the ***possibility of numerous configurations*** on account of variation in electric propulsion characteristics and energy sources. Figure 22.4 (a) depicts the very basic configuration, in which an electric drive train replaces the IC engines of a conventional vehicle drive train. The replacing EV drive consists of an electric motor, a clutch, a gearbox, and a differential. The clutch and gearbox may further be replaced by automatic transmission. The clutch is used to connect or disconnect the power of the electric motor from the driven wheels. A set of gear ratios in the gearbox are used to modify the speed power characteristics to meet the load requirements of vehicle. A set of planetary gears of the differential are utilized to drive both sides of driven wheels to rotate at differential speeds when the vehicle moves on a curved path. Figure 22.4 (b) depicts a fixed gearing system that can replace the multispeed gearbox and clutch of a traditional vehicle as the electric motor develops constant power with high efficiency over a large speed range. This configuration offers a great advantage of a simplified drive train due to the absence of gear shaft as well as reduced size/weight of mechanical transmission, which is eliminated. The arrangement shown in Fig. 22.4(c) is similar to that of Fig. 22.4(b), the electric motor, fixed gearing and differential can be further integrated into a single assembly while both axles point at both driving wheels. The whole drive train can be further simplified and compacted in the direction of the most simple drive train. Figure 22.4(d) illustrates a configuration in which the mechanical differential is replaced by using two independent traction motors. Each of them drives a separate wheel and operates at different speeds when a vehicle is running along a curved path. In order to further simplify the drive train, the traction motor can be integrated inside the wheel. This arrangement is so called wheel drive (Fig. 22.4(e)). A thin planetary gear set may be used to reduce the motor speed and enhance the motor torque. This thin planetary gear set offers the advantage of high-speed reduction ratio as well as an inline arrangement of the input and output shaft. The arrangement in Fig. 22.4(f) offers full abandoning of any mechanical gearing between the electric motor and the driving wheel, by the use of the **out-rotor of low speed electromotor in the in-wheel drive**, that can be directly connected to the driving wheel. The speed control of the electric motor is equivalent to the control of wheel speed, and hence the vehicle speed. However, this arrangement requires the electric motor to have a higher torque to start and accelerate the vehicle. ***A lot of research work is needed to achieve these possibilities.***

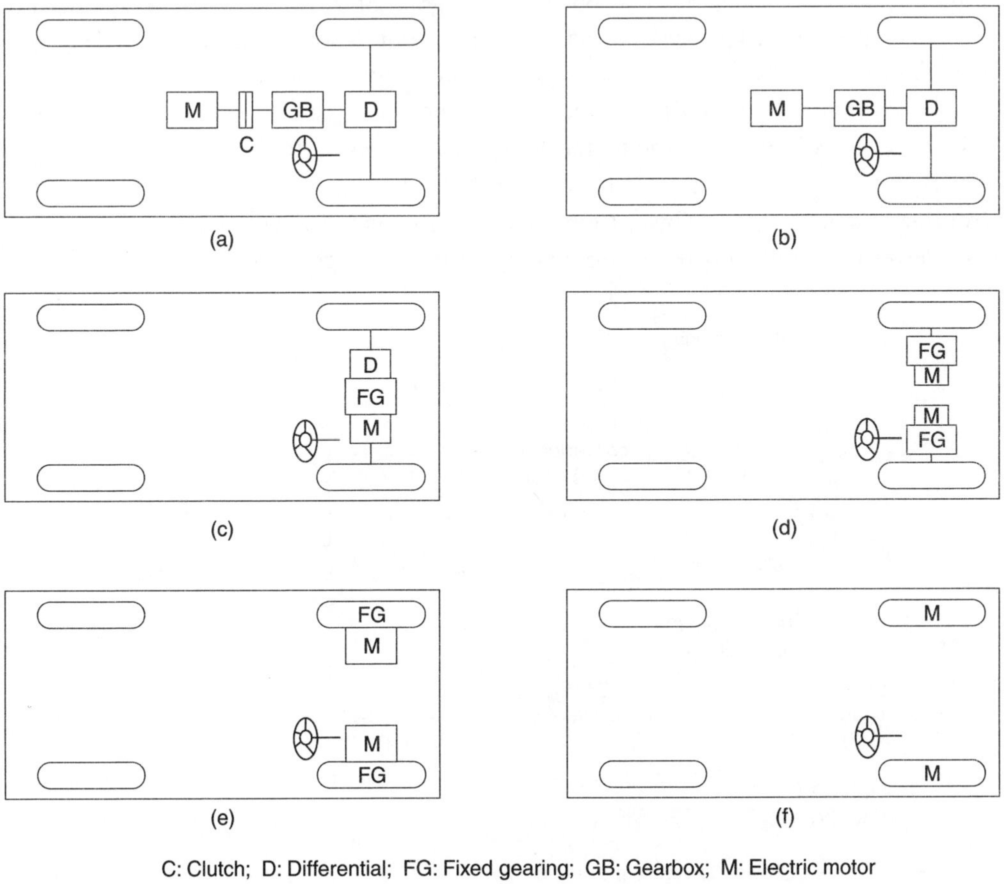


**Fig. 22.4** Possible EV configurations for future development [101, 102a]

## 22.3 CONCEPT OF HYBRID ELECTRIC VEHICLES (HEVs)

Hybrid electric vehicles (HEVs) are powered by two independent power sources—an internal combustion engine and an electric motor, which uses energy stored in batteries or fuel cells. One may be called *primary source* and other may be called *secondary source. A hybrid electric vehicle cannot be plugged into charge the battery. Instead, the battery is charged through regenerative braking and by internal combustion engines.* There may be many types of HEVs. In submarine, when it comes on the surface, it is powered by diesel engines, and when submerged, it is powered by batteries driven electric motor. HEVs have the advantages of both ICE vehicles and EV and overcome their individual disadvantages. In general, a vehicle that has two or more energy sources and energy converter is called hybrid vehicle, and when one of the sources is electric power train, it is called HEV. *Now, battery of HEVs can also be charged by charging point.*

Figure 22.5 shows a generic conceptual flow diagram of a HEV drive train. Hybrid drive trains supply the required power by an adapted power train. There are many available patterns of combining the power flows to meet requirements as given below:

(i) Power train 1 (e.g., ICE) alone delivers power to the load.
(ii) Power train 2 (e.g., electric motor) along delivers power to the load.
(iii) Both power trains 1 and 2 deliver power to load at the same time.
(iv) Power train 2 obtains power from load through regenerative braking
(v) Power train 2 obtains power from power train 1 through charging

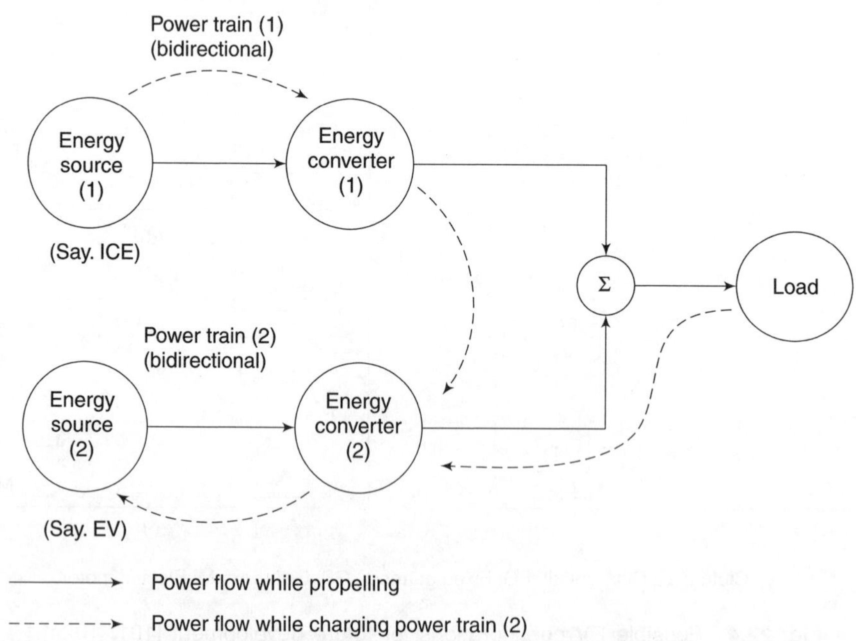


**Fig. 22.5** Conceptual illustration of a hybrid electric train

(vi) Power train 2 obtains power from power train 1 and load simultaneously.
(vii) Power train 1 delivers power to load and power train 2 at the same time.
(viii) Power train 1 delivers power to power train 2 by recharging, and power train 2 delivers power to load.
(ix) Power train 1 delivers power to load, and load delivers power to power train 2.

Therefore, HEVs are now classified into four kinds as given below:

(a) Series hybrid
(b) Parallel hybrid
(c) Series-parallel hybrid
(d) Complex hybrid

Figure 22.6 shows the classification of various hybrid vehicles. In Fig. 22.6(a), a fuel tank-IC engine, and a battery-electric motor are taken, respectively, an example of primary power source (steady power source), and secondary power source (dynamic power source). Of course, the ICE can be replaced by other types of power sources, such as fuel cells. Similarly, the battery can be replaced by ultracapacitor or flywheels and their combinations. Similarly, Figs. 22.6 (b), (c), (d) are other forms of classification.

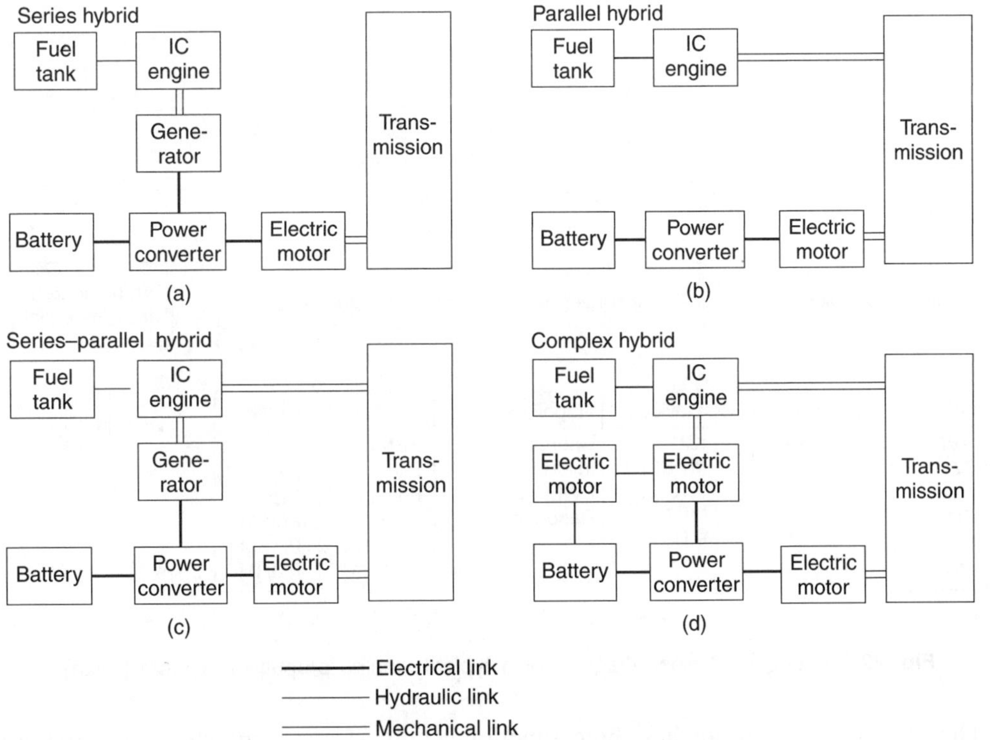


**Fig. 22.6** Classification of possible HEVs [101, 102a]

*The* ***advantages*** *of HEV include less pollution due to electric power, no need of charging battery, and can cover long distance till refuelling of petrol/diesel/gas is needed, an increased overall efficiency, and some saving on fuel consumption. The* ***disadvantages*** *include complexity of the vehicle, heavy and bigger size vehicle to accommodate two power packs, not fully zero emission, and increase in cost.*

## 22.4 ELECTRIC PROPULSION SYSTEMS OF EVs AND HEVs

Electric propulsion system is the heart of electric vehicles (EVs) and hybrid electric vehicles (HEVs). Figure 22.7 shows the functional block diagram of an electric propulsion system. *The propulsion system consists of electric motor, power converter, electronic controllers, and transmission and differential.* The electric motor converts the electrical energy to propel the vehicle through transmission and differential, or, vice versa, and enable regenerative braking and/or to generate electricity for charging the onboard energy storage. The power converter is used to supply the electric motor with proper voltage and current as per requirement of the motor and vehicle. The electronic controllers direct the power converter by providing control signal to it, and thereafter control the operation of the electric motor to develop proper torque and speed, according to the control signal from the drive.

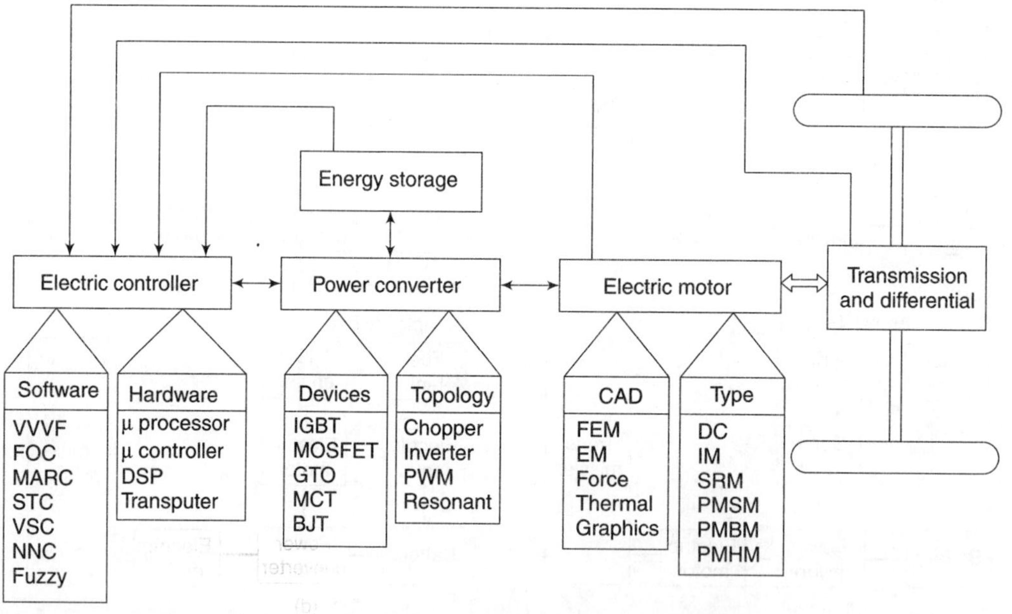


**Fig. 22.7** Functional block diagram of a typical electric propulsion system [102a]

The electronic controller has three functional units: ***sensor, interface circuitry and processor***. The sensor is used to translate measurable parameters such as current, voltage, temperature, speed torque, and flux into electrical signals through the interface circuitry.

These signals after required conditioning to the appropriate level are fed into the processor. The processor output signals are usually amplified via the interface circuitry to drive power convertor.

The power packs of EV and HEVs can be classified into two main groups, namely, the ***commutator motors and commutator-less motors.*** The classification of which is given in Fig. 22.8. Commutator motors are traditionally DC motors, with features such as series excited, shunt excited, compound excited, separately excited, and permanent magnet (PM) excited. DC motors essentially need commutators and brushes to feed current into the armature, thus making them less reliable and unsuitable for maintenance free and high speed operation (up to 18,000 rpm) which is requirement of vehicles. It is also called brushed DC motor. Additionally, winding excited DC motors have rather low specific power density. Nevertheless, because of the maturity of this technology and simple control mechanism, DC motor drives have been prominent in electric propulsion systems. *However, today's automakers use three different types of electric motors in electric (green) vehicles: the brushless DC (BLDC) motors, brushed DC motor, and AC induction motor. Tesla uses AC three-phase induction motor.*

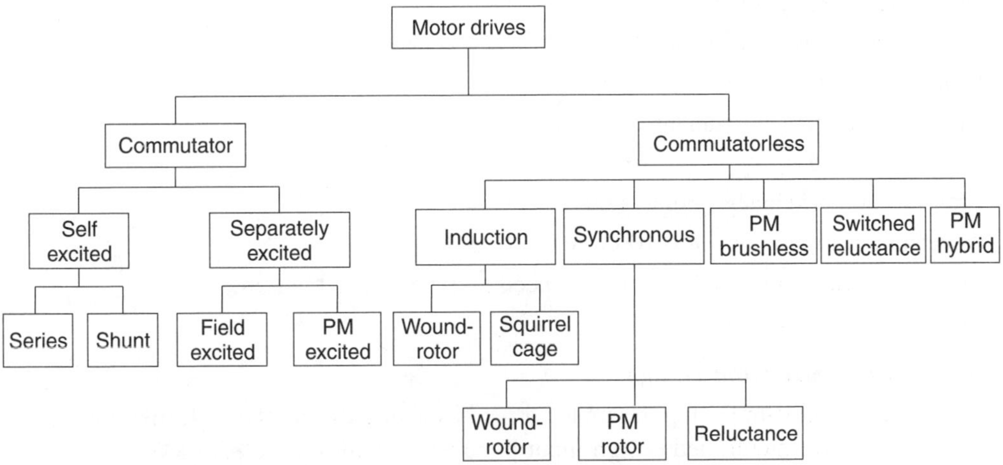


**Fig. 22.8** Classification of electric motor drives for EVs and HEVs applications

## 22.5 POWER RATING DESIGN OF THE TRACTION MOTOR

Similar to the pure electric vehicle, power rating of HEV motor is also calculated in the same way. The power rating depends on total vehicle mass, vehicle acceleration performance requirements, motor characteristics, and transmission characteristics. At the start of the design process, the power rating of the motor drive has to be estimated using the following equation.

$$P_t = \frac{\delta \cdot Mv}{2\, t_a}\,(V_f^2 + V_b^2) + \frac{2}{3}\, M_v \cdot g \cdot f_r \cdot V_f + \frac{1}{5}\, \rho_a\, C_D \cdot A_f \cdot V^3{}_f \qquad \ldots(22.2)$$

where $M_v$ = total vehicle mass in kg

$t_a$ = expected acceleration time in sec.

$V_b$ = Vehicle speed in m/s, corresponding to the motor based speed.

$V_f$ = final speed of the vehicle accelerating in m/s

$g$ = acceleration due to gravity, 9.81 m/s$^2$

$f_r$ = tyre rolling resistance coefficient

$\rho_a$ = air density, 1.202 kg/m$^3$

$A_f$ = frontal area of the vehicle in m$^2$

$C_D$ = aerodynamic drag coeffieient

The first term in Eq. (22.2) represents the power used to accelerate the vehicle mass, and second and third terms represent the average power for overcoming the tyre rolling resistance and aerodynamic drag.

### DESIGN EXAMPLE

**Design input parameters:**

Vehicle total mass = 1500 kg

Rolling resistance coefficient = 0.01

Aerodynamic drag coefficient = 0.3

Frontal area = 2.0 m$^2$

Transmission efficiency (single gear) = 0.9

Acceleration time (from 0 to 100 km/h = 10 ± 1 sec.)

Maximum gradeability > 30% at low speed and 7.5% at 100 km/h

Maximum speed = 150 km/h

**Traction Motor Size from Design**

Using Eq. (22.2) and supposing the motor drive has a speed ratio of $x = 4$, the motor power rating for the above-mentioned design input parameters comes to be **82.5 kW**.

## 22.6 ENERGY STORAGE TECHNOLOGIES

Energy storage is defined as the devices that store energy to systems outside (discharge), and accept energy for storage from outside (charge). There are several types of energy storage technologies that have been proposed for electric vehicles (EVs) and hybrid electric vehicles (HEVs) applications. The energy storage, till date, mainly includes the following devices:

(i) Chemical batteries

(ii) Ultracapacitors or supercapacitors

(iii) Ultrahigh-speed flywheels

Table 22.1 shows the status of battery systems potentially available for electric vehicles. It is obvious from table that although the specific energy is high in advanced batteries, but the specific power has to be improved. **About 300 W/kg might be the optimistic estimate of specific power.**

**Table 22.1** Status of Battery Systems for Electric Vehicles [102a]

| *System* | *Specific power (Wh/kg)* | *Peak power (W/kg)* | *Energy efficiency (%)* | *Cycle life* | *Self discharge (% per 48 hr)* | *Cost USD/kWh* |
|---|---|---|---|---|---|---|
| Acidic ap. Solution Lead/Acid | 35–50 | 150–40 | ≥ 80 | 500–1000 | 0.6 | 120–150 |
| Alkaline aqua sol Nickel/cadmium | 50–60 | 80–150 | 75 | 800 | 1.0 | 350–350 |
| Nickel/Iron | 50–60 | 80–150 | 75 | 1500–2000 | 3 | 200–400 |
| Nickel/Zinc | 55–75 | 170–260 | 65 | 300 | 1.6 | 100–300 |
| Nickel/Metal hyd | 70–95 | 200–300 | 70 | 750–1200 | 6 | 200–350 |
| Al/Air | 200–300 | 160 | ≤50 | - | - | - |
| Iron/Air | 80–120 | 90 | 60 | 500+ | - | 50 |
| Zinc/Air | 100–220 | 30–80 | 60 | 600+ | - | 90–120 |
| Zinc/Bromine | 70–85 | 90–110 | 65–70 | 500–2000 | - | 200–250 |
| Vanadium-redox | 20–30 | 110 | 75–85 | - | - | 400–450 |
| Molten Sodium/sulphur | 150–240 | 230 | 80 | 800+ | - | 250–450 |
| Sodium/nickel | 90–120 | 130–160 | 80 | 1200+ | - | 250–345 |
| Chloride Lithium/Iron | 100–130 | 150–250 | 80 | 1000+ | - | 110 |
| **Lithium-ion** | **80–130** | **200–300** | **>95** | **1000+** | **0.7** | **200** |

## 22.6.1 Lithium-ion Chargeable Battery

Among all the possible batteries, lithium-ion chargeable battery is mostly preferred in electric vehicles and hybrid electric vehicles. Lithium is the lightest of metals, has the greatest electrochemical potential and provides the largest specific energy per weight. The working principle of lithium-ion battery is as follows:

**Basics:** *A battery is made up of an anode (negative electrode), cathode (positive electrode), separator, electrolyte, and two current collectors (positive and negative).* Generally, the negative electrode of a conventional lithium cell is made from carbon, while the positive electrode is typically a lithium metal oxide. *The anode and cathode store the lithium. The electrolyte carries positively charged lithium ions from the anode to the cathode and vice versa through the separator.* The most commonly used electrolyte is comprised of lithium salt, such as $LIPF_6$ in an organic solution. The material of the separator is polyolefin, it has an excellent mechanical properties, good chemical stability and is low cost. The movement of the lithium ions creates free electrons in the anode which creates a charge at the positive current collector. The electrical current then flows from the current collector through a device being powered (say motor) to the negative current collector. The separator blocks the flow of electrons inside the battery.

**Charge/Discharge:** While the battery is discharging and providing an electric current, the anode releases lithium ions to the cathode, generating a flow of electrons from one side to the other. When plugging in the device, the opposite happens. Lithium ions are released by the cathode and received by the anode.

**Energy Density vs Power Density:** In battery technology, the two most common parameters associated with batteries are *energy density and power density.* Energy density is measured in watt-hours per kilogram (Wh/kg). It signifies the amount of energy can store with respect to its mass. Whereas, power density is measured in watts per kilogram (W/kg), and it signifies the amount of power that can be generated by the battery with respect to its mass. In order to understand these two terms clearly, think of *draining a pool. Energy density is similar to the size of the pool, while power density is comparable to draining the pool as quickly as possible. The emphasis on research is to increase the energy density of batteries, while reducing the cost, and maintaining acceptable power density.*

**Battery Pack Architectures:** The voltage level of the battery (in cell form) determines the maximum electrical power which can be delivered continuously.

$$\text{Power} = V.I.$$

The higher the current ($I$), the bigger the diameter of high voltage wires, and higher the thermal losses. For this reason, the current should be limited to a maximum and the nominal power is obtained by having a higher voltage. *For a nominal voltage of 400 V, the battery pack may be estimated. For the auxiliary devices such as headlights, multimedia, heaters, brake lights, wipers, etc., the average consumption could be about 430 W of electrical power. The average energy consumption for the auxiliary loads is about 9.241 Wh/km.*

The direct current (DC) supplied by the battery pack is converted into alternated current (AC) by the inverter. This conversion takes place with an associated loss. The average propulsion efficiency of 90 per cent may be considered from the battery to the wheel. The battery pack may be designed for an average energy consumption of **161.7451 Wh/km**.

All high voltage battery packs are made up from **battery cells** arranged **in strings and modules**. *A battery cell can be regarded as the smallest division of voltage. Individual battery cells may be grouped in parallel and/or series as* ***modules****. Further, battery modules can be connected in parallel to create* ***battery pack****. Depending upon the battery parameters, there may be several levels of modularity.*

**The total battery pack voltage is determined by the number of sells in series.** For example, the *total (string) voltage of 6 cells will be the sum of their individual voltage.* In order to increase the current capability, battery capacity and current capability, more strings have to be connected in parallel. *For example, three strings connected in parallel will triple the capacity and current capability of the battery pack.*

**CASE STUDY:** The high voltage battery pack of **Miysubishi i-MIEV** consists of the following:

- 22 modules made up from 88 cells connected in series. Each module contains 4 prismatic cells. The voltage of each cell is 3.6 volts, and the total voltage of battery pack is 330 volts.

Another example is the high voltage battery pack of **Tesla Model S**, which has the following architecture:

- 74 cells in a parallel group
- 6 groups in series for a module
- 16 modules in series
- **7104 cells in total, cylindrical in shape, 4.2 volts per cell, and 25.2 volts per module.**
- **85 kWh battery pack weighs 540 kg**
- 375 volts generation.

As per estimate of battery pack manufactured by **Panasonic**, some parameters are as follows:

- Number of cells in string: 112
- String energy (Wh): 1290
- Number of strings: 32
- Battery pack energy (kWh): 42.29
- Battery pack capacity: 102.4
- Number of total cells: **3584**
- Battery pack mass (kg): **173.8**
- Battery pack volume (litre): 63
- Battery pack peak current (A): 102.4
- Battery pack peak power (kW): 40.96
- Battery pack continuous current (A): 102.4
- Battery pack continuous power (kW): **40.96**

This battery pack is arranged on a strong aluminium plate whose size lies between four wheels. Constant research and development is going on to develop better battery pack, better battery cells, and other better storaged devices.

### 22.6.2 Features of Ultracapacitors

*The ultracapacitor is characterized by much higher specific power, but much lower specific energy compared to the chemical batteries. Its specific energy is in the range of a few watt-hours per kilogram (1 Wh/kg to 30 Wh/kg), 10 to 50 times lesser than lithium-ion battery. However, its specific power can reach to 10 kW/kg, about 3 to 4 times greater than lithium-ion battery, and which is higher than any type of battery. Due to low specific energy density, and dependent of voltage on the SOC, it is challenging to use ultracapacitor as an energy storage for EVs and HEVs*. However, there are number of advantages that can result from using the ultracapacitor as an auxiliary power source. One promising application is the so called hybrid battery and ultracapacitor energy storage for EVs and HEVs.

One thought process is under active research that the high specific energy and high specific power requirements can be decoupled, thus giving a scope to design a battery which is optimized for the specific energy as well cycle life with lesser attention being given to specific power. Due to the load levelling effect of the ultracapacitor, the high current discharging from the battery and the high current charging to the battery by regenerative breaking is minimized so that the available energy, endurance, and life of the battery can be significantly increased. Double layer capacitor technology is the novel approach for achieving the ultracapacitor concept. Figure 22.9 shows the basic principle of a double-layer capacitor.

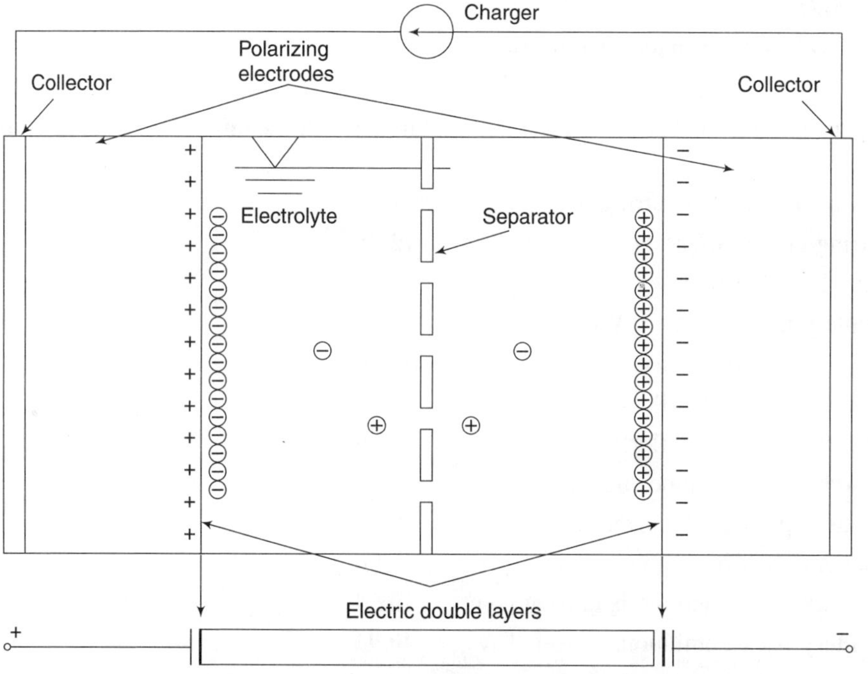


**Fig. 22.9** Basic principle of a double layer capacitor

#### 22.6.2.1 Technology Development for High Performance Ultracapacitor

*The goals set by the U.S. Department of Energy for the inclusion of ultracapacitor in EVs and HEVs at the near-term goals for specific energy and specific power should be better* **than 500 Wh/kg, and 1600 W/kg**. *These goals by the U.S. Department of Energy are very challenging, as these will change the complete scenario of EVs and HEVs*. However, uptill now even near-term goals have not been achieved. Nevertheless, some companies are actively engaged in the research and developemt of ultracapacitors for EVs and HEVs applications.

**Maxwell Technologies** has claimed that is power **BOOSCAP ultracapacitor cells** (2600 farad (F) at 2.5 volt) and integrated modules (**145F and 42 V and 435 F** at 14 V) are in production.

## 22.6.3 Ultrahigh-Speed Flywheel

Flywheels have been used over centuries for storing energy in mechanical form. Oerlion Engineering Company in Switzerland in 1990s made the first passenger bus primarily powered by a massive flywheel. This particular flywheel is a massive steel rotor with hundreds of kilograms that spins in the order of a couple of thousand rpm. In the new technology, the advanced flywheel is a lightweight composite material rotor weighing tens of kilograms and rotating being in the order of 10,000 rpm, and the these are called ***ultra-speed flywheel***.

### 22.6.3.1 Operation Principles of Flywheels

A rotating flywheel stores energy in the form of kinetic energy as given below:

$$E_f = \frac{1}{2} J_f \cdot \omega_f^2 \qquad \ldots(22.3)$$

where $J_f$ is the moment of inertia of the flywheel in kg.m$^2$/sec, and $\omega_f$ is the angular velocity of flywheel in radian/sec. On analysing the above equation, it suggests that the angular velocity of flywheel is the key means of improving its energy capacity and reducing its weight and volume. In the present technology, a **speed of over 60,000 rpm has been achieved**.

However, with available current technology status, it is challenging to directly use the mechanical energy stored in a flywheel to propel a vehicle. This is due to the need for continuous variation transmission (CVT) with a wide gear ratio variation. *The most commonly used approach is to couple an electric machine to the flywheel directly or through a transmission so as constitute the so-called* ***mechanical battery****. Figure 22.10 shows the basic layout of flywheel energy conversion.* Here, this mechanical battery, functions as the energy input and output port, converting the mechanical energy or vice versa.

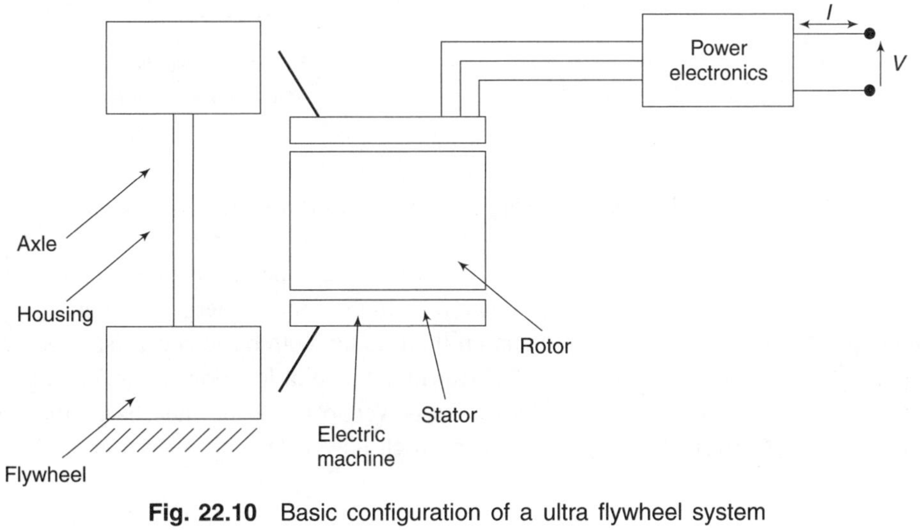


**Fig. 22.10** Basic configuration of a ultra flywheel system

## 22.7 HYBRIDIZATION OF ENERGY STORAGE TECHNOLOGIES

Hybridization offers more options. Hybridization of energy storage is the amalgamtion of two or more storage technologies together so that the advantages of each one can be compensated by others. *For example, hybridization of a **chemical battery with an ultracapacitor** can overcome such challenges as low specific power of electrochemical batteries and low specific energy of ultracapacitor. This enables to achieve high specific energy and high specific power.*

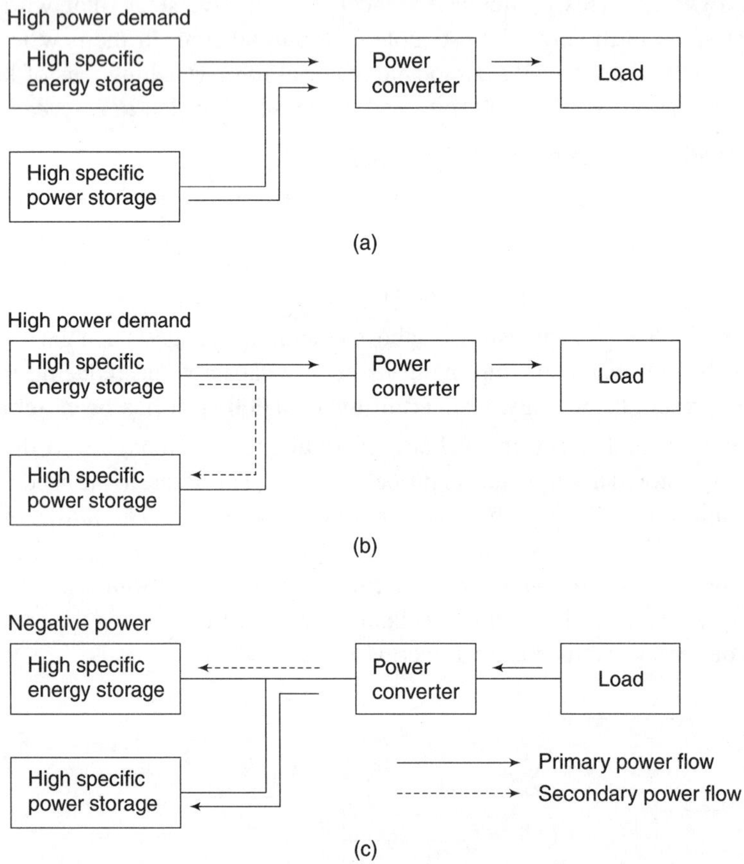


**Fig. 22.11 (a, b, c, d)** Working principle of a hybrid energy operation

Figure 22.11 shows the concept of hybrid energy storage operations, in which one storage energy is with high specific power, and other with high specific energy. During high power demand operations, such as acceleration and hill climbing, both basic energy storages deliver their power to the load as shown in Fig. 22.11(a) In the case of low power demand, operation, the high specific energy storage will deliver its power to the load, and charge the specific power storage to recover the charge lost during high power demand operation as illustrated in Fig. 22.11(b).

During regenerative braking, operations, the peak power is absorbed by the high specific power storage, and only a limited part is absorbed by the high specific energy. As a result, the whole system would be much smaller in weight and size if any one of them alone was the energy storage system.

There are numerous viable hybridization schemes for EVs and HEVs, typically, battery and battery hybrids, and battery and ultracapacitor. The latter being more natural since the ultracapacitor can offer more high power than batteries. *It is expected that in the long run, an ultrahigh-speed flywheel would replace the batteries in hybrid energy storage to obtain a high efficiency, compact, and long-life storage system for EVs and HEVs.*

## 22.8 $H_2$-$O_2$ FUEL CELL BASED ELECTRIC VECHICLE

Hydrogen-oxygen fuel cells produces electricity by electrochemical process. Hydrogen is produced from natural fuel, compressed to high pressure, around 70 bar and above and stored in a separate suitably designed metallic bottle with leakproof valve arrangement to increase the mass of hydrogen. There are other methods of hydrogen production but from natural gas, it is cheaper. The clean hydrogen is produced by electrolysis process which is very costly. At present, in the world there are four main sources for the commercial productions of hydrogen: natural gas, oil, coal, and electrolysis; which account for 48%, 30%, 18% and 04%, respectivaly. Fossil fuels are the dominant souce of industrial hydrogen. Oxygen is taken from air. The direct current electricity so produced is managed to achieve proper voltage and current to drive the three-phase DC/AC motor through using suitable devices (Figs. 22.2, 22.3). The emission from this cell is just water and water vapour. This vehicle is pollution free, as hydrogen is a clean energy. The efficiency of fuel cell conversion is around 60 per cent, though theoretically, conversion efficiency at low load is 80 per cent.

Toyota motors manufactures electric vehicles based on hydrogen-oxygen fuel cells. *Hydrogen compressed bottles are kept on fuel stations, when the hydrogen gas gets exhausted, it is replaced by new one within ten minutes. The other method is refuelling of hydrogen from the gas station like gasoline refuelling. It takes just five minutes for refuelling unlike at least one hour in battery based vehicles.* Hydrogen gas filling stations in compressed form above 70 bar are less even in Japan and USA. Such electric vehicles can cover distance up to 400 km. The cost of fuel cell based car is more costly than battery based electric cars, and hydrogen gas is at least three times costly than gasoline. Though, some cars and buses are plying on the road, but the number of vehicles built is much less than battery based electric cars. In spite of all hurdles, Toyota says that it is the future green vehicles. Hydrogen energy is supposed to be the future green energy.

### 22.8.1 Working Principle of $H_2$-$O_2$ Fuel Cell

#### *22.8.1.1 Fuel Cell Vehicles (PCV)*

*A fuel cell is an electrochemical device which converts the chemical energy of conventional fuels, such as hydrogen, hydrocarbons, etc., directly into electrical energy.* Broadly speaking, it consists of a *non-consumable anode, cathode and an electrolyte* (invariant) fuel, oxidant supply steam and suitable control. In rural areas, $H_2$ may be obtained from biomass whereas in urban areas it may be obtained from sewage waste. By this way, fuel cell may be cheaper. Figure 22.12 shows a schematic diagram of $H_2$-$O_2$ fuel cell. The hydrogen gas serves as a fuel while $O_2$ serves as an oxidizer. Each gas is introduced at high pressure into separate chambers. $H_2$ is supplied to the anode side and oxygen to the cathode side. *Both electrodes are made of porous material to permit passage of gas. The electrolyte is placed between the electrodes. The electrolyte may be solid or liquid.* Porous ceramic (zirconia oxide) polymers and alkaline solution (potassium hydroxide) have been successfully used as electrolytes. When a solid electrolyte is used, its surface is first impregnated with a catalyst to facilitate the ion migration process. Impregnation allows the two sides of the electrolyte to be used as electrodes. In the case of liquid electrolyte, separate electrodes are necessary. When the electrodes are connected to a resistive load, current flows.

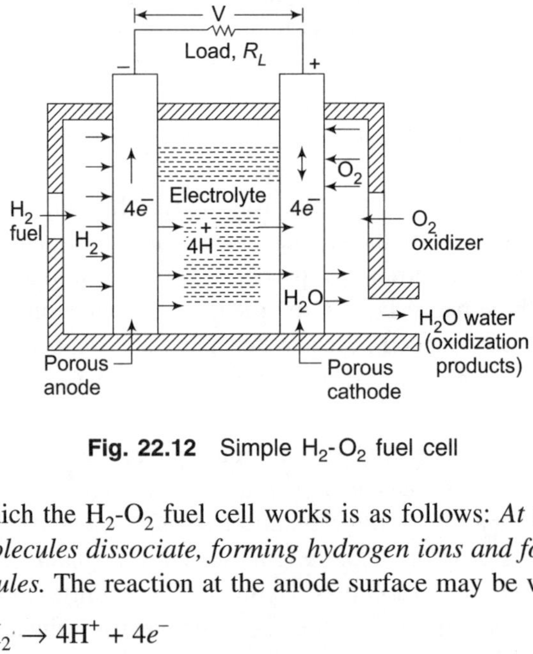


**Fig. 22.12** Simple $H_2$-$O_2$ fuel cell

The principle on which the $H_2$-$O_2$ fuel cell works is as follows: *At the anode-electrolyte interface, hydrogen, molecules dissociate, forming hydrogen ions and four free electrons per two moles of $H_2$ molecules.* The reaction at the anode surface may be written as

$$2H_2 \rightarrow 4H^+ + 4e^- \quad \text{...(22.4)}$$

These electrons flow through the external circuit where they do electrical work and return to the fuel cell at the anode. The hydrogen ions ($4H^+$) migrate through the electrolyte to their cathode interface, where they combine with the returning electrons and with oxygen to form water. The reaction at the cathode is written as

$$4e^- + 4H^+ + O_2 \rightarrow 2H_2O \quad \text{...(22.5)}$$

The overall reaction for the fuel cell is just the sum of the above two equations

$$2H_2 + O_2 \rightarrow 2H_2O$$

On account of the chemical reaction, a quantity of energy representing the enthalpy of combustion of the fuel $\bar{h}_f^\circ$ is released. A part of this energy is available for conversion to electrical energy, the maximum amount is equal to the charge in Gibbs function ($w_e = \bar{g}_1 - \bar{g}_2$).

*The remaining part of the energy remains as heat. The electrical energy output per mole of* $O_2$ or per 2 mole of $H_2$ is expressed as

$$w_e = 4\ N_0 . e . V_r \qquad ...(22.6)$$

where $V_r$ = voltage developed by the reversible cell

$N_0$ = Avogadro's number

$e$ = electronic charge

The maximum value of electrical work under reversible steady-state conditions will be equal to the change in Gibbs free energy,

$$w_e = \overline{g}_1 - \overline{g}_2 \qquad ...(22.7)$$

where $\overline{g}_1$ and $\overline{g}_2$ are the Gibbs free energies of the reactants and products respectively. Hence, the maximum voltage that can be developed in the reversible fuel cell is

$$V_r = \frac{\overline{g}_1 - \overline{g}_2}{n.N_0 . e} \qquad ...(22.8)$$

where $n$ = number of electrons released per molecule of $O_2$. The maximum ideal efficiency of the fuel cell is

$$\eta_{fc} = \frac{w_e}{(-\overline{h}_f^{\circ})} = \frac{\overline{g}_1 - \overline{g}_2}{(-\overline{h}_f^{\circ})} = \frac{nN_0 . e . V_r}{(-\overline{h}_f^{\circ})} \qquad ...(22.9)$$

It is to be noted that in the $H_2$-$O_2$ fuel cell, the enthalpy of formation $\overline{h}_f^{\circ}$ (i.e., $\overline{h}_f^{\circ}$ is – 286 kJ/gm mole. This gives the efficiency of $H_2$-$O_2$ fuel cell equal to 83%.

*Fuel cells are not limited by the Carnot efficiency* and theoretical efficiency as high as can be expected of it. But high efficiencies are possible only at relatively light load and very low voltage (1 volt). *The main advantages of the fuel cells are its simplicity, its high power to weight ratio, emission free and higher efficiency. The disadvantage are its high cost, need for equipment to control the flow of reactants to match the reaction rates, problems of selection of materials and short life, especially at high temperatures.*

The performance of a fuel cell depends primarily on the current density (amperes per unit area) and temperature. Figure 22.13 shows a typical voltage current density for a fixed temperature. This is known as *polarization curve*. Due to the interference at the electrodes, the measured no-load voltage $V_0$ is less than that for the ideal reversible case $V_r$.

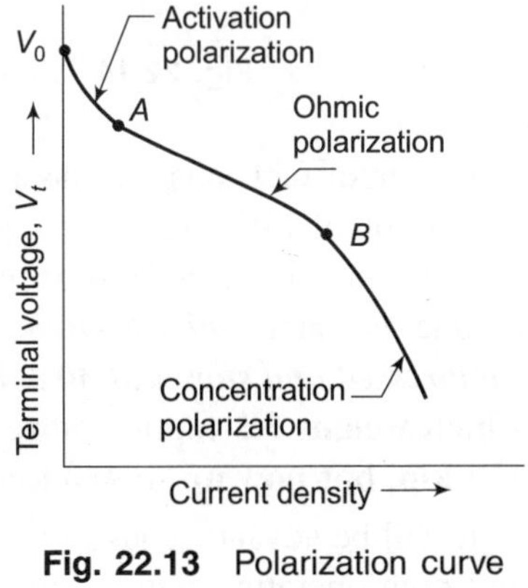

**Fig. 22.13** Polarization curve for a fuel cell

It is obvious from Fig. 22.13 that the cell voltage drops rapidly with light loads. This is due to the irreversible effects at the electrodes and is known as an *activation polarization*. The main reason for the linear drop in the operating range *AB* is the ohmic resistance in the electrolyte. Beyond the point *B* electrostatic effects, due to concentration gradients in the electrolyte, begins to limits the currents, hence the term concentration polarization for this regions of operation.

For an actual fuel cycle, the actual voltage $V_a$ is less than $V_r$

Hence, the efficiency is

$$\eta_0 = \frac{V_a\, n . N_0 . e}{-\bar{h}_f^{\circ}} \qquad \text{...(22.10)}$$

**With $V_a$ = 0.6 volt, $\eta_0$ = 60% at a current density near 1075 amp/m$^2$.** These are the fuel cells which have been used in *Apollo system.*

Figure 22.14 shows the comparison of fuel cell efficiency with conventional IC engines. It is obvious from the results that at part loads, fuel cells are much more efficient (about 60%) IC engines (about 35%). But at high loads, fuel cells slightly suffer from losses on account of internal cell resistance over potential of cathodic and anode reactions and mass transfer problems.

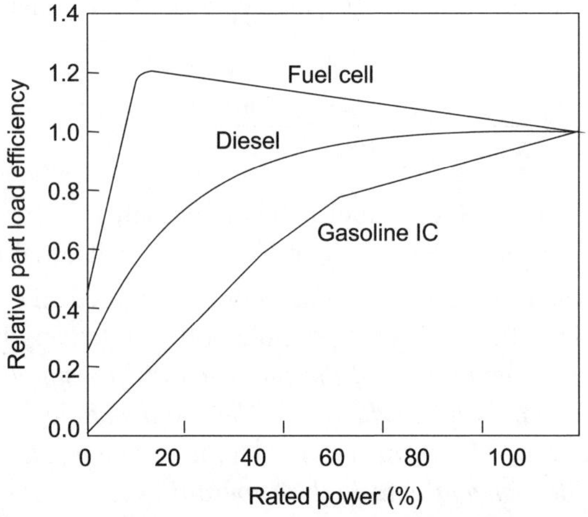


**Fig. 22.14** Comparison of fuel cell efficiency with IC engine

Design of fuel cells are classified on the basis of electrolytes used such as phosphoric acid doped polymer electrolyte, molten carbonate, solid doped *zr*-oxide. These cells operate on hydrogen and oxygen. Hydrogen is fed to the anode and oxygen to the cathode and the final product is water. ***Daimler–Benz has built before 2010 a bus with 45000 litres of hydrogen compressed and stored at 30 MPa pressure and employing a 250 kW fuel cell power plant*** **which weighs 3.5 tonnes more than diesel bus having a city operation range of 140 to 170 km, but now up to 460 km in latest system (2020).**

It will be advantageous to use on board hydrogen producing plant as it offers low weight and high operating range. Methanol, natural gas, LPG, gasoline or other petroleum fuels can be used on board to produce $H_2$ and $CO_2$ by multi-stage chemical reactions processes. *Daimler–Benz vehicles used on board methanol—to hydrogen reforming fuel cell car, NECAR III has a 40-litre and a range of over 400 km.*

*The main drawbacks of fuel cells are low power density and high cost, about three times costly, than diesel. At present, the power density is about 1 kW/litre which is very near to that of diesel engine.* But more improvement in power density is needed, by concentrating on efficient electrolyte materials. Now, cost is coming down.

## PROBLEMS

### Things to Think About

22.1 What is the scope of electric vehicles?

22.2 Will be a death of ICE?

22.3 List the energy storage systems.

### Short and Long Theoretical Questions

22.4 Discuss with a sketch of general layout of EV and HEV.

22.5 Discuss the various possibilities of power train of EV.

22.6 List the advantages and disadvantages of EVs over ICE.

22.7 In how many ways, HEVs may be operated.

22.8 Discuss the general layout of power train of EV with sketch.

22.9 What is the efficiency of EV?

22.10 What factors limit the large size of EV either using battery or fuel cells?

### Software Based Question

22.11 Develop a software for the design and operation of power train based on lithium-ion battery, hybrid system and $H_2$-$O_2$ fuel cells.

### Design and Open Ended Question

22.12 Prepare a report about the specifications of E-vehicles power trains manufactured by all leading manufacturers.

### Multiple Choice Questions

22.13 At present, efficiency of electric vehicle is

(a) 40% (b) 60% (c) 70% (d) 90%

22.14 At present, specific power (Wh/kg) of battery of EVs is

(a) 130 (b) 150 (c) 170 (d) 190

22.15 At present, the peak power (W/kg) of battery of EVs is

(a) 300 (b) 400 (c) 500 (d) 600

## Key to Multiple-Choice Questions

22.13 (d) 22.14 (a) 22.15 (a)

# Appendices

## APPENDIX A

### SI Units

In 1960, the C.G.P.M (Conference Generale des Poids et Measures) recommended the universal adoption of units based on metre (m), kilogram (kg), second (s), degree Kelvin (K) and candela (cd), to cover the entire system of science and engineering and give it the name "System Internationale of Units" abbreviated SI.

Most of the countries including India have switched over to these units.

The fundamental and supplementary units of the SI system are given in Table A1. The derived units from the base and supplementary units are presented in Table A2. Table A3 shows multiples and sub-multiples.

**Table A1** Fundamental and Supplementary SI Units.

| *Quantity* | *Unit* | *Symbol* |
|---|---|---|
| | **Fundamental Units** | |
| Length | metre | m |
| Mass | kilogram | kg |
| Time | second | s |
| Temperature | degree Kelvin | K |
| Current | ampere | amp |
| Luminous intensity | candela | cd |
| Amount of substance | mole | mol |
| | **Supplementary Units** | |
| Plane angle | radian | rad |
| Solid angle | steradian | sr |

**Note:** In addition to thermodynamics temperature scale degree, Kelvin (K), the degree Celsius (°C) is also in use.

**Table A2** Derived Units

| *Quantity* | *Unit* | *Symbol* |
|---|---|---|
| area | square metre | $m^2$ |
| volume | cubic metre | $m^3$ |
| velocity | metre per second | m/s |
| acceleration | metre per second per second | $m/s^2$ |
| density | kilogram per cubic metre | $kg/m^3$ |
| force | newton | N, $kgm/s^2$ |
| pressure, stress | pascal or newton per sq. metre | Pa; $N/m^2$ (IPa = 1 $N/m^2$) |
| work, heat energy | joule | J; Nm |
| power | watt | W, Js(1 W = J/s) |
| specific entropy | joule per kilogram per degree Kelvin | J/kg K |
| enthalpy | joule per kilogram | J/kg |
| rate of heat flow | watt | W |
| heat flux rate | watt per metre square | $W/m^2$ |
| thermal conductivity | watt per metre per degree Kelvin | W/m K |
| heat transfer coefficient | watt per sq. metre per degree kelvin | $W/m^2$ K |
| radiation factor | watt per square metre per degree K to the power four | $W/m^2 K^4$ |
| frequency | hertz | Hz = c/s |
| dynamic viscosity | pascal-second or new newton per second per sq metre | Pa's; $Ns/m^2$ |
| kinematic viscosity | square metre per second | $m^2/s$ |

**Note:** In addition to pressure expressed in pascal or $N/m^2$, bar is also in common use. 1 bar = $10^5$ $N/m^2$, 1 atm pressure = 101.3 $kN/m^2$ = 1.013 bar.

**Table A3** Multiples and Sub-multiples

| | | | | | |
|---|---|---|---|---|---|
| tetra | T | $10^{12}$ | milli | m | $10^{-3}$ |
| giga | G | $10^9$ | micro | μ | $10^{-6}$ |
| mega | M | $10^6$ | nano | n | $10^{-9}$ |
| kilo | k | $10^3$ | pico | p | $10^{-12}$ |
| deci | d | $10^{-1}$ | femto | f | $10^{-15}$ |
| centi | c | $10^{-2}$ | atto | a | $10^{-18}$ |

## A.4 Conversion Factor

**(i)** Factor

IN = $10^5$ dyne = 0.012 kgf

1 $kg_f$ = 9.81 N

**(ii)** Pressure

1 $N/m^2$ = 1 Pa

1 bar = $10^5$ $N/m^2$ = $10^5$ Pa = $10^2$ kPa

1 atm = 1.01325 bar = 101.325 $kN/m^2$ = 760 mm of Hg = 1.03 $kg_f/cm^2$

**(iii)** Energy

| | | | |
|---|---|---|---|
| 1 J | = 1 Nm | = 1 Ws | = $10^7$ erg |
| 1 kcal | = 4.1868 kJ | | |

**(iv)** Power

| | | | |
|---|---|---|---|
| 1 W | = 1 J/s | = 0.86 kcal/h | |
| 1 h.p. | = 75 m.kgf/s | = 733.5 W | = 0.7355 kW |

**(v)** Heat Transfer Rate

1 W/m$^2$ = 0.86 kcal/m$^2$h
= 0.317 BTU/hft$^2$

1 tonne of refrigeration = 50 kcal/min = 3.517 kW = 200 BTU/min

**(vi)** Temperature

K = °C + 273 = $\left(\frac{5}{9}\right)$ (°F + 460)

Temperature difference = $\Delta T = \Delta t$ (°C) or (K)

**(vii)** Latent Heat

1 kJ/kg = 0.2349 kcal/kg

**(viii)** Thermal conductivity and heat transfer coefficient

1 W/mK = 0.86 kcal/mh K = 0.5778 BTU/fhR

1 W/m$^2$K = 0.08598 kcal/m$^2$hK

1 K/W = 1.163 K/kcal

**(ix) Physical Constant**

Speed of light = $2.998 \times 10^8$ m/s

Universal gas constant (Ro) = 8314 J/kg mole K

Characteristics gas constant (R) = Ro/Molecular weight of a gas

# APPENDIX B

**Table B** Comparison among Different Types of Power Plants

| *S.No.* | *Parameter* | *Diesel Engine* | *Gas Turbine* | *Steam Turbine Plant* |
|---|---|---|---|---|
| 1. | Usual power range | 10 kW to 10 MW | 100 kW to 260 MW | 1 MW to 1000 MW |
| 2. | Approx. capital cost | ₹1100 – 1800/kW | ₹800 – 1500/kW (aero) (aero) – 280/kW (heavy) | ₹1400 – 2400/kW (turbine only) |
| 3. | Approx. installed cost | – | ₹1500 – 3000/kW | ₹8000 – 12000/kW |
| 4. | Size and weight/kW | Medium | Small | Large |
| 5. | Specific installation requirements | Heavy foundations, vibration isolation, quietening enclosure | Intake and filters intake and exhaust mufflers | Ash handling facility copious supplies of water exhaust de-pollution |
| 6. | Typical plant life | 10 to 20 years | 15 to 30 years | 25 to 35 years |
| 7. | Inspection/overhaul periods | Annual | Several years | Annual |
| 8. | Efficiency (typical) | 35% to 47% | 30% to 47% | 30% to 40% |
| 9. | Efficiency factor at half load | 1.00 | 0.8 to 0.85 | 0.91 to 0.96. |
| 10. | Fuel grade and type | High liquid (diesel grade) | High to medium liquid, gas | Medium to low gas, liquid solid |
| 11. | Staffing requirements | A few | None (extra to plant) | Many |
| 12. | Lubrication needs | Frequent oil and filter | Oil topping up | Negligible |
| 13. | Maintenance period | Weekly and Monthly | Monthly and annually | Daily |
| 14. | Site selection | No special site | No special site | Special site to meet water and pollution requirement |

# APPENDIX C

**Table C.1** The Properties of Some Gaseous Fuels

| *Gases* | *Chemical formula* | *Molecular weight* | *Density (kg/m$^3$)* | *Lower Calorific (MJ/kg)* | *Value MJ/m$^3$* |
|---|---|---|---|---|---|
| n-Butane | $C_4H_{10}$ | 58.17 | 2.458 | 45.71 | 112.41 |
| Carbon monoxide | CO | 28.01 | 1.185 | 10.1 | 11.97 |
| Ethane | $C_2H_6$ | 30.09 | 1.272 | 47.48 | 60.43 |
| Methane | $CH_4$ | 16.04 | 0.678 | 50.01 | 33.95 |
| Propane | $C_3H_8$ . | 44.14 | 1.865 | 46.35 | 86.42 |
| Hydrogen | $H_2$ | 2.016 | 0.085 | 119.95 | 10.22 |

**Table C.2** The Properties of Some Solid Fuels

| *Solid Fuels* | *Density (No pores) kg/m$^3$* | *Higher Calorific Value (MJ/kg)* |
|---|---|---|
| Charcoal | 1400 to 1700 | 34.30 |
| Coal | 1250 to 1700 | 32 to 36 |
| Coke | 1750 to 2000 | 33.60 |
| Peat | 1150 to 1250 | 19.70 |
| Wood | 500 to 1100 | 17.30 |

**Table C.3** The Properties of Some Liquid Fuels

| *Pure Liquids* | *Chemical Formula* | *Molecule Weight at 1 atm (K)* | *Boiling Point at 298 K* | *Density (kg/litre)* | *Lower Calorific Value at 298 K (MJ/kg)* |
|---|---|---|---|---|---|
| Benzene | $C_6H_6$ | 78.11 | 353 | 0.879 | 40.14 |
| n-Butane (LPG) | $C_4H_{10}$ | 58.17 | 273 | 0.575 | 45.34 |
| n-Octane | $C_3H_{18}$ | 114.23 | 399 | 0.703 | 44.43 |
| n-Heptane | $C_7H_{16}$ | 100.21 | 371 | 0.684 | 44.56 |
| Ethanol | $CH_5O_H$ | 34.06 | 352 | 0.794 | 27.23 |
| Methanol | $CH_3O_H$ | 32.04 | 338 | 0.796 | 19.94 |
| Propane (LPG) | $C_3H_8$ | 44.14 | 231 | 0.505 | 45.98 |
| Liquid fuels | *Density at 188 K (kg/litre)* | *Boiling range (K)* | *Kinematic Viscosity (cSt)* | *Lower. Calorific Value at 188K MJ/kg)* | |
| Diesel | 0.84 | 453 to 613 | 5 | 42.9 | |
| Fuel oil | 0.8 to 0.97 | 453 | 1.4 to 70 | 53.1 | |
| Gasoline | 0.72 to 0.76 | 301 to 473 | 0.5 to 0.65 | 41.9 – 44.0 | |
| Kerosene | 0.73 to 0.83 | – | 2.1 to 3.0 | 43.1 | |

# APPENDIX D

**Table D.1** Enthalpy of Formation, Gibbs Functions of Formation, and Absolute Entropy of Various Substances at 25°C, 1 atm Pressure

| *Substance* | *Molecular Weight M* | $\bar{h}_f^0$ *kJ/kg mole* | $\bar{h}_f^0$ *kJ/kg mole* | $\bar{h}_0$ *kJ/kg mole* |
|---|---|---|---|---|
| CO(g) | 28.001 | −119529 | −137150 | 197.653 |
| $CO_2$(g) | 44.001 | −393522 | −394374 | 213.795 |
| $H_2O$(g) | 18.015 | −241827 | −228583 | 188.883 |
| $H_2O$(1) | 18.025 | −28538 | −237178 | 70.049 |
| $CH_4$(g) | 16.025 | −74873 | −50751 | 186.256 |
| $C_2H_2$(g) | 26.038 | +226731 | +209234 | 200.958 |
| $C_2H_4$(g) | 28.054 | +52283 | +68207 | 219.548 |
| $C_2H_6$(g) | 30.070 | −84667 | −32777 | 229.602 |
| $C_3H_8$(g) | 44.097 | −103847 | −23316 | 270.019 |
| $C_4H_{10}$(g) | 58.124 | −126148 | +16914 | 310.227 |
| $C_8H_{18}$(g) | 114.23 | −208447 | +16859 | 466.835 |
| $C_8H_{18}$(1) | 114.23 | −249952 | +6940 | 360.896 |

**Table D.2** Enthalpy of Combustion of Some Hydrocarbons at 25°C.

| *Hydrocarbon* | *Formula* | *Liquid $H_2O$ in products (Negative of higher heating value)* | | *Vapour $H_2O$ in products (Negative of lower heating value)* | |
|---|---|---|---|---|---|
| | | *Liquid Hydrocarbons kJ/kg fuel* | *Gaseous Hydrocarbons kJ/kg fuel* | *Liquid Hydrocarbon kJ/kg fuel* | *Gaseous Hydrocarbon kJ/kg fuel* |
| **Paraffin family** | | | | | |
| Methane | $CH_4$ | | −55496 | | −50010 |
| Ethane | $C_3H_6$ | | −51875 | | −47484 |
| Propane | $C_3H_8$ | −49975 | −50345 | −45983 | −46353 |
| Butane | $C_3H_{10}$ | −49130 | −49500 | −45344 | −45714 |
| Pentane | $C_5H_{12}$ | −48643 | −49011 | −44983 | −45351 |
| Hexane | $C_6H_{14}$ | −48308 | −48676 | −44033 | −45101 |
| Heptane | $C_7H_{16}$ | −48071 | −48436 | −44557 | −44922 |
| Octane | $C_8H_{18}$ | −47893 | −48256 | −44425 | −44788 |
| Decane | $C_{10}H_{22}$ | −47641 | −48000 | −44239 | −44598 |
| Dodecane | $C_{12}H_{24}$ | −47470 | −47282 | −44110 | −44467 |
| **Olefin family** | | | | | |
| Ethene | $C_2H_4$ | | −50296 | | −47158 |
| Propene | $C_3H_6$ | | −48917 | | −45780 |

*Contd.*

| | | | | | |
|---|---|---|---|---|---|
| Butene | $C_4H_8$ | | –48453 | | –45316 |
| Pentene | $C_5H_{10}$ | | –48034 | | –44996 |
| Hexene | $C_6H_{12}$ | | –47937 | | –44800 |
| Heptene | $C_7H_{14}$ | | –47800 | | –446612 |
| Octene | $C_8H_{16}$ | | –47693 | | –44556 |
| Nonene | $C_9H_{18}$ | | –47612 | | –44475 |
| Decene | $C_{10}H_{20}$ | | –47547 | | –44410 |
| **Alkyl-benzene family** | | | | | |
| Benzene | $C_6H_6$ | –41831 | –42266 | –40141 | –40576 |
| Methyl-benzene | $C_7H_8$ | –42473 | –42847 | –40527 | –40937 |
| Ethyl-benzene | $C_8H_{10}$ | –42997 | –43395 | –40924 | –41322 |
| Propyl-benzene | $C_9H_{12}$ | –43416 | –43800 | –41219 | –41603 |
| Butyl-benzene | $C_{10}H_{14}$ | –43748 | –44123 | –41453 | –41828 |

# References

## Research Papers


[1] Li. C.H., *'Piston Thermal Deformation and Friction Consideration'* SAE Paper 820086. 1982.

[2] Furuhama, S. and Suzuki, H., *'Temperature Distribution of Piston Rings and Piston in High Speed Diesel Engine'* Bull., JSME, Vol. 22, No. 174, pp. 1788–1795, 1979.

[3] Finlay, IC et al., *'Factors Influencing Combustion Chamber Wall Temperature in a Liquid Cooled Automotive SI Engine'*. Proc. Instn. Mech. Engn., Vol. 199, No. D3, pp. 207-214, 1985.

[4] Woschni, G. *'Prediction of Thermal Loading of Supercharged Diesel Engine'*, SAE Paper 790821, 1979

[5] Ament, F., et al., *'Heat Balance Provides Insight into Modern Engine Fuel Utilization'*, SAE Paper 770221, 1977.

[6] Novak, J.M. and Blumberg, P.N. *'Parametric Simulation of Significant Design and Operating Alternative Affecting the Fuel Economy and Emission of SI Engine,* SAE Paper 780943, SAE Trans, Vol. 87, 1978.

[7] Taylor, C.F. and Toon, T.Y., *'Heat Transfer in Internal Combustion Engines,* ASME Paper 57-HT -17, 1957.

[8] Annand, W.J.D., *'Heat Transfer in the Cylinder of Reciprocating Internal Combustion Engines'*, Proc. Inst. Mech Engrs., Vol 177, No. 36, pp. 973-990, 1963.

[9] LeFeuvre, T., et al., *'Experimental Instantaneous Heat Fluxes in a Diesel Engine and Their Correlation'*, SAE Paper 690464, SAE Trans, Vol. 78, 1969.

[10] Dent, J.C. and Sulaiman, S.J., *'Convective and Radiative Heat Transfer in a High Swirl Direct Injection Diesel Engine'*, SAE Paper 770407, SAE Trans, Vol. 56, 1977.

[11] Watts, P.A. and Heywood, J.B., *'Simulation Studies of the Effects of Turbocharging and Reduced Heat Transfer on SI Engine Operation'*, SAE Paper 800289, 1980.

[12] French, C.C.J. and Atkins, K.A. *'Thermal Loading of a Petrol Engine'*, Proc. Instn. Mech. Engrs., Vol. 187, 49/73, pp. 561-573, 1973.

[13] Nakajma, Y., et al. *'Effects of Exhaust gas Recirculation on Fuel Consumption'*, Proc. of Inst of Mech. Engrs., Automobile Div , Vol. 195, No. 30, pp. 369-376, 1981.

[14] Hubes, E.W., *'Measuring the Trapping Efficiency of Internal Combustion Engine through Continuous Exhaust Gas Analysis'*, SAE Paper 710144, SAE Trans., Vol. 80, 1971.

[15] Blair, G.P. and Kenny, R.G., *'Further Developments in Scavenging Analysis for Two-stroke Cycle Engines'*, SAE Paper 800038, SAE Trans., Vol. 89, 1980.

[16] Annod, W.J.D. and Roe, G.E., *'Gas Flow in the Internal Combustion Engine'*, Haessner Publishing, Newfoundland, J.J., 1974.

[17] Benson, R.S. *'Experiments on a Piston Controlled Port'*, The Engineer, Vol. 210, pp. 875-880, 1960.

[18] Takizawa, M. et al., *'A Study of Gas Exchange Process Simulation of an Automotive Multi-cylinder Internal Combustion Engines'*, SAE Paper 820410, SAE Trans, Vol. 91, 1982.

[19] Kay. I.W., *'Manifold Fuel Film Effects in SI Engine'*, SAE Paper 780944, 1978.

[20] Hofbauer. P. and Stator, K. *'Advanced Automotive Power Systems, Part 2. A Diesel for a Subcompact Car'*, SAE Paper 770113, SAE Trans. Vol. 86, 1977.

[21] Armstrong, D.L. and Stirrat, G.F., Ford's 1982, *'3.8 Litre V6 Engine'*, SAE Paper 820112, SAE Trans., Vol. 91, 1982.

[22] Barnes-Moss, H.W., *A Designers Viewpoint' in Passenger Car Engines, Conference Proceedings*, pp. 133-197, Inst. of Mech. Engr, London, 1975.

[23] "SAE Recommended Practice", *Engine Terminology and Nomenclature–General, SI Hand Book, J604d.*

[24] Asmus, T.W., *"Valve Events and Engine Operation"* SAE Paper 830749, SAE Trans Vol. 91, 1982

[25] Kastner, L., et al. *'Poppet Inlet Valve Characteristics and their Influence on the Induction Process'*, Proc. Inst. Mech Engrs., Vol. 178, pt 1, No. 36, pp. 955-978, 1963-1964.

[26] Annand, W.J.D. and Roe, G.E. *"Gas Flows in Internal Combustion Engines"*, Haessner Publishing Newfoundland, N.J., 1974.

[27] Toda, T., et al. *'Evaluation of Burned Gas Ratio (BGR) as a Predominant Factor to $NO_x$', SAE* Paper 760765, SAE Trans., Vol. 85, 1976.

[28] Tabaczynski, et al., *'Time Resolved Measurement of Hydrocarbon Mass Flow Rate in The Exhaust of a Spark-Igntiion Engine'*, SAE Paper 720112, SAE Trans, Vol. 81, 1972.

[29] Caton, J.A. and Heywood, J.B., *'Experimental and Analytical Study of Heat Transfer in an Engine Exhaust Port'*, Int. J. of Heat & Mass Transfer, Vol. 24, No. 4, pp. 581-595, 1981.

[30] Watson, N and Janota, M.S., *'Turbocharging the Internal Combustion Engines'*, Wiley-Interscience Publication, John Wiley, New York, 1982.

[31] Bhinder, F.S., *'Supercharging Compressor–Problems and Potential of the Various Alternatives'*, SAE paper 840243, 1984.

[32] Bhinder, F.S., *'Some Fundamental Consideration–Concerning the Pressure Charging of Small Diesel Engines'*, Paper 830145, 1983.

[33] Brandsletter. W. and Dziggle, R., *'The 4 and 5-cylinder Turbocharged Diesel Engines for Volkswagen and Audi'*, SAE paper 830441, SAE Trans. Vol. 91, 1982.

[34] Newhall, H.K. and Shahed S.M., *"Kinetics of Nitric Oxide Formation of High Pressure Flames"*, Proc. of 13th Int. Symp. on Combustion, pp. 381-390, The Combustion Institute, 1971.

[35] Sakai, Y., et al., *'The Effect of Combustion Chamber Shape on Nitrogen Oxides,'* SAE Paper 730154, 1973.

[36] Quader, A, A., *'Why Intake Change Dilution Nitric Oxide Emission from SI Engines'*, SAE Paper, 710009, Trans, Vol. 80, 1971.

[37] Aoyagi, Y. Kamimoto, et al., *'A Gas Sampling Study on the Formation of Process of Soot and NO in a DI Diesel Engine'*, SAE Paper 800254, SAE Trans. Vol. 89, 1980.

[38] Viocnlescu, F.A. and Borman, G.L. *'An Experimental Study of Diesel Engine Cylinder Averaged $NO_x$ Histories'*, SAE Paper 780228, SAE Trans., Vol. 87, 1978.

[39] Yu, R.C. and Shahed, S.M., *'Effect of Injection Timing and Exhaust Gas Recirculation as Emission from a DI Diesel Engine'*, SAE Paper 811–234, SAE Trans, Vol. 90, 1981.

[40] Harrington, J.A. and Shishu, R.C., *'A Single-Cylinder Engine Study of the Effects of Fuel Type, Fuel Stoichiometry and Hydrogen to Carbon Ratio and CO, NO, and HC Exhaust Emissions,* SAE Paper 730476, 1973.

[41] Takczynski, R.J., et al., *'Time Resolved Measurement of Hydrocarbon Mass Flow Rate in the Exhaust of Spark Ignition Engines'*, SAE Paper 720112, SAE Trans, Vol. 81, 1972.

[42] LoRusso, J.A., et al., *'An Electrohydraulic Gas Sampling Valve with Application of Hydrocarbon Emission Studies'*, SAE Paper 800045,' SAE Trans., Vol. 89, 1980.

[43] Haskell, W.W, and Legate, C.E., *'Exhaust Hydrocarbon Emissions from Gasoline Engines-Surface Phenomena'* SAE Paper 720225, 1972.

[44] Namazian, M. and Heywood, J.B., *'Flow in the Piston-cylinder Ring Crevices of a Spark-Ignition Engine: Effect on Hydrocarbon Emissions, Efficiency and Power'*, SAE Paper 820088, SAE Trans., Vol. 91, 1982.

[45] Kuroda, H., et al., *'The Fast Burn with Heavy EGR, New Approach for Low $NO_x$ and Improved Fuel Economy'*, SAE Paper 780006, SAE Trans., Vol. 87, 1978.

[46] Nakagawa, Y, et al., *'A Fundamental Analysis of HC and CO Oxidation Reaction in the Exhaust System'*, JSAE, Rev. No. 1, pp. 98–106, 1977.

[47] Yu, R.C. and Shahed, S.M., *Effects of Injection Timing and Exhaust Gas Circulation on Emission from a D.I. Diesel Engine'*, SAE Paper 811234, SAE Trans., Vol. 90, 1981.

[48] Amann, C.A. and Siegla, D.C., *Diesel Particulates What they are and Why' Aerosol.* Sci. Technol., Vol. 1, pp. 73-101, 1982.

[49] Lahaya, J. and Prado, G., *'Morphology and Internal Structure of Soot and Carbon Blacks' Particulate Carbon Formation during Combustion,* pp. 33–55, Plenum Press, 1981.

[50] Haynes, B.S. and Wagner, H.G., *'Soot Formation, Prog. Energy Combustion'* Sci., Vol. 7, pp. 229-273, 1981.

[51] *Diesel Technology, Impacts of Diesel-Powered Light-duty Vehicles'*, Reports of the Technology Panel of the Diesel impacts Study Committee National Research Council, National Academy Press, D.C., 1982.

[52] Amann, C.A., *'Control of the Homogeneous-Charge Passenger-car Engine-Defining the Problem'*, SAE Paper 801440, 1980.

[53] Watson, N., et al., *'A Combustion Correlation for Diesel Engine Simulation'*, SAE Paper 800029, 1980.

[54] Reynolds, W.C. *'Modeling of Fluid Motino in Engines–An Introductory Overview'*, in J.N. Mattavi and C.A. Amann (eds). *Combustion Modeling in Reciprocating Engine*, pp. 41-68, Plenum Press, 1980.

[55] El Tahry, S.H. *'K-$\varepsilon$ Equation for Compressible Reciprocating Engines'*, SAE Paper 820040, 1982.

[56] Morel, T. and Mansour, N.N., *'Modeling of Turbulence in Internal Combustion Engines'*, SAE Paper 820040., 1982.

[57] Strcit, E. and Bozman. G.L., *'Mathematical Simulation of a large Turbocharged Two-stroke Diesel Engine'*, SAE Paper 710176, SAE Trans., Vol. 80, 1971.

[58] Assanis, D.N. and Heywood, D.N., *'Development and Use of Computer Simulation of the Turbocompounded Diesel System for Engine Performance and Component Heat Transfer Studies'*, 860329, 1986.

[59] Danota, M.S., et al. *'The Prediction of Diesel Engine Performance and Combustion Chamber Component Temperature using Digital Computers'*, Proc. Inst. of Mech Engrs., Vol. 182, PL/3L., pp. 58-70, 1967-1968.

[60] Heywood. J.B. and et al., *'Development and Use of a Cycle Simulation to Predict SI Engine Efficiency and $NO_x$ Emissions'*, SAE Paper 790291, 1979.

[61] Lavoie, G.A. and Blumberg, P.N., *'A Fundamental Model for Preictidng Fuel Consumption, $NO_x$ and HC Emissions of the Conventional SI Engines'*, Combustion Science an Technology, Vol. 21, pp. 225-258, 1978.

[62] Beretta, G.P. and et al., *'Turbulent Flame Propagation and Combustion in SI Engines,' Combustion Flame*, Vol. 52, pp. 217–245, 1983.

[63] Borgnakke, C., et al., *'A Model for the instantaneous Heat Transfer and Turbulence in SI Engine'*, SAE Paper 800287, 1980.

[64] Whitehouse, N.D. and Way, R.J.B., *'Simple Method for the Calculation of Heat Release Rates in Diesel Engine Bases on the Fuel Injection Rules'*, SAE Paper 710134, 1971.

[65] Cartellieri, W and Johns, R.J.R., *'Multidimensional Modelling of Engine Process : Progress and Prospects'*, Paper presented at 15th CIMAC, Congress, Paris, June 1, 1983.

[66] Amsden, A., et al., *'A Comprehensive Model for 2-D and 3-D Engine Simulations'*, SAE Paper 850554, 1985.

[67] Basso, A., *'Optimization of Combustion Chamber Design for SI Engines'* SAE Paper 840231, 1984.

[68] Schopertons, H. and Lee, W., *'Multidimentional Modeling of Knocking Combustion in SI Engine'*, SAE Paper 850502, 1985.

[69] Cheng, W.K. and Theobald, M.A., *'A Numerical Study of Diesel Engines'*, Paper 87-FE-2, Presented at the ASME, Energy-Source Technology Conference, Dallas, Feb. 1987.

[70] Schapertons, H and Thiele F., *'Three Dimensional Computation for Flow Fields in DI Piston Bowls'*, SAE Paper 860463, 1986.

**Books**

[71] Obert, E.F., *'Internal Combustion Engines'*, Harper & Row, Publishers, N.Y. 3rd Edition, 1973.

[72] Gill, P.W. and Smith (Jr), J.H., *'Fundamental of Internal Combustion Engines'*, Oxford & IBH Publishing Co., New Delhi, 1967.

[73] Heywood, J.B., *'Internal Combustion Engines Fundamentals'*, McGraw Hill Book Co., New York, 1988, 2011.

[74] Rogowki, A.R., *'Elements of Internal Combustion Engines'*, McGraw Hill, 1953.

[75] Taylor, C.F. and Taylor E.S., *The Internal Combustion Engine in Theory and Practice*, M.I.T. Press, 1968.

[76] Popoinch, M., *Fuels and Lubricants*, John Wiley, 1959.

[77] Litchy, L.C. *Combustion Engine Processes*, McGraw Hill, 1967.

[78] Ricardo, H.R., *The Internal Combustion Engines*, McGraw Hill, 1967.

[79] Mathur, M.L. and Sharma, R.P., *Internal Combustion Engine*, Dhanpat Rai & Sons, Delhi 5th Edn., 1990.

[80] Ganesan, V., *Internal Combustion Engines*, Tata McGraw Hill Publishing Co. Ltd., New Delhi, 1984, 4th Edn., 2012

[81] Yadav, R. *"Fundamental of Engineering Thermodynamics,* 7th Edn. 2004; and *Applied Thermodynamics"*, 6th Edn., 2005, Central Publishing House, Allahabad. (now changed)

[82] Ricardo, H.R. and Hempson, J.G.G., *'The High Speed Internal Combustion Engines'* Blackis, 1968.

[83] Cohen, H., Rogers, G.F.C. and Saravanmutto, H.H., *'Gas Turbine Theory'*, Longmans, 3rd Edn., 1992.

[84] Eastop, T.D. and A. McConky, *'Applied Thermodynamics for Engineeriong Technologists'* 3rd Edn., Longmans, 1978.

[85] Yadav, R., *'Stream and Gas Turbines and Power Plant Engineering.'* Central Publishing House, Allahabad, 6th Edn., 1988.

[86] Holeman, J.P., *'Thermodynamics'* 3rd Edn., McGraw Hill Book Co., N.Y., 1988.

[87] Sharma, P.C.; Agrawal, D.K., *'Machine Design'*, S.K. Kataria & Sons, 1986. New Delhi.

[88] Crouse, W.H., *'Automotive Engines'*, Tata McGraw-Hill Publishing Co. Ltd., New Delhi, 8th Edn., 1981.

[89] Spotts, M.F., *Design of Machine Elements*, Prentice-Hall of India Pvt. Ltd., New Delhi 5th Edn. 1981.

[90] Harman, R.T.C. *'Gas Turbine Engineering'*, The Macmillan Press Ltd. London, 1981.

[91] Dixon, S.L., *'Fluid Mechanics, Thermodynamics of Turbomachinery'*, Pergamon Press, N.Y., 2nd Edn., 1975.

[92] Yadav, R., et al. *"Heat and Mass Transfer"*, Central Publishing House, 4th Edn., 2000.

[93] Kern, D.Q. and Kraus, A.D., *'Extended Surface Heat Transfer'*, McGraw Hill, N.Y., 1972.

[94] Frass, A.P. and Ozisik, M.N., *"Heat Exchangers Design"*, Wiley, N.Y., 1965.

[95] Incorpera, F.P. and Dewitt, D.P., *"Fundamental of Heat Transfer"*, John Wiley & Sons N.Y., 1981.

[96] Ferguson, T.B., *"The Centrifugal Compressor Stage"*, Butterworth, London, 1963.

[97] Maleev, V.L., *"Internal Combustion Engine-Theory and Design"*, McGraw Hill, N.Y; 2nd Edn., 1978.

[98] Kolcchin, A. and Demido V. V; *"Design of Automotive Engines"*, Mir Publishers. Moscow.

[99] Shigley, J.E., *'Mechanical Engg. Design'*, McGraw-Hill International Edition, 1986.

[100] Benson, R.S. *'The Thermodynamics and Gas Dynamics of IC Engines'*, Oxford University Press, 1982.

[101] C.C. Chan and K.T. Chau *"Modern Electric Vehicle Technology"* Oxford University Press, Oxford, New York, 2001.

[102] E.H. Wakefield *"History of Automobile Hybrid Electric Vehicle,"* SAF, 1998.

[102a] Mehrdad Ehsani, Yimin Gao, Sebastien E. Gay, and Ali Emadi, "*Modern Electric, Hybrid, Electric, and Fuel Cell Vehicles,*" CRC Press, N.Y., 2005.

### Additional Research Papers References

[103] Chapman, M., Novak J.M., and Stein R.A., "Numerical Modeling of Inlet and Exhaust Flows in Multi-Cylinder Internal Combustion Engines", in T. Uzkan, Flows in Internal Combustion Engines, ASME, 1982.

[104] Poulus, S.G., and Heywood, J.B., "The Effect of Chamber Geometry on Spark-Ignition Engine Combustion," SAE paper 830334, SAE Trans., Vol. 92, 1983.

[105] Watts, P.A., and Heywood, J.B.: "Simulation Studies of the Effects of Turbotechnology and Reduced Heat Transfer on Spark Engine Operation," SAE paper 800289, 1980.

[106] Lovoie, G.A., and Blumberg, P.N., "A Fundamental Model for Predicting Fuel Consumption, $NO_x$ and H.C. emissions of the Conventional Spark-Ignited Engines", Combustion Science and Technology, Vol. 21, pp. 225-258, 1980.

[107] Novak, J.M., and Blumberg, P.N., "Parametric Simulation of Significant Design and Operationg Alternatives Affecting the Fuel Economy of Spark-Ignited Engine," SAE paper 800133, SAE Trans, Vol. 89, 1980.

[108] Hiroyasu, H., Kodfa, T., and Arai, M.,: "Development and use of Spray Combustion Modeling to Predict Diesel Engine Efficiency and Pollutant Emission (Part 2 Computational Procedure and Parametric Study), Paper 214-13, Bull. JSME, Vol. 26 no. 214, pp. 576-583, 1983.

[109] Schapertons, H., and Thiele, F., "Three-Dimensional Computations for Flowfields in DI Piston Bowls," SAE paper 860463, 1986.

[110] Finlay, IC, Marris, D., Boam, D.J. and Parks, B.J., "Factors Influencing Combustion Chamber Wall Temperature in a Liquid-cooled, Automotive Spark-Ignition Engine", Proc. Inst. of Mech. Engrs, Vol. 199, no. D3, pp. 207-214, 1985.

[111] Greeves, G., "Response of Diesel Combustion System to Increase Fuel Injection Rates" SAE paper 790037, SAE Trans., Vol. 88, 1979.

[112] Heywood et al., "Development and use of Cycle Simulation to Predict SI Engine Efficiency and No Emissions," SAE paper 790291, 1979.

# Index

## A



## B

## F

# J



# K



# L



# M

## N



## O



## P



## Q



## R